Pharmaceutical Biotechnology

Daan J.A. Crommelin • Robert D. Sindelar
Bernd Meibohm
Editors

Pharmaceutical Biotechnology

Fundamentals and Applications

Fourth Edition

Editors
Daan J.A. Crommelin, Ph.D.
Department of Pharmaceutical Sciences
Utrecht Institute for Pharmaceutical Sciences
Utrecht University
Utrecht
The Netherlands

Bernd Meibohm, Ph.D., FCP
Department of Pharmaceutical Sciences
University of Tennessee
Health Science Center
College of Pharmacy
Memphis, Tennessee
USA

Robert D. Sindelar, Ph.D., FCAHS
Department of Pharmaceutical Sciences
and Department of Medicine
Providence Health Care and
The University of British Columbia
Vancouver, BC
Canada

ISBN 978-1-4939-4339-5 ISBN 978-1-4614-6486-0 (eBook)
DOI 10.1007/978-1-4614-6486-0
Springer New York Heidelberg Dordrecht London

Softcover reprint of the hardcover 4th edition 2013

Printed on acid-free paper

Springer is part of Springer Science+Business Media (www.springer.com)

Contents

Preface

Over the past 25 years, biotechnologically derived drug products have become a major share of the therapeutically used pharmaceuticals. These drug products include proteins, including monoclonal antibodies and antibody fragments, as well as antisense oligonucleotides and DNA preparations for gene therapy. In 2001 already, biotech products accounted for more than 35 % of the New Active Substances that were launched in the USA. Twelve out of the twenty-nine approved marketing authorization applications at the European Medicines Agency (EMA) in 2009 were biotech products. Drug products such as epoetin-α (Epogen®, Eprex®, Procrit®), abciximab (ReoPro®), interferons-α (Intron®A, Roferon®A) and interferons-β (Avonex®, Rebif®, Betaseron®), anti-TNF-α agents (Enbrel®, Remicade®, Humira®), bevacizumab (Avastin®), and trastuzumab (Herceptin®) are all examples of highly successful biotech drugs that have revolutionized the pharmacotherapy of previously unmet medical needs. And last but not least, biotech drugs also have a major socioeconomic impact. In 2010, five of the ten top selling drugs in the world were biotechnologically derived drug products, with sales varying between five and eight billion US dollars.

The techniques of biotechnology are a driving force of modern drug discovery as well. Due to the rapid growth in the importance of biopharmaceuticals and the techniques of biotechnologies to modern medicine and the life sciences, the field of pharmaceutical biotechnology has become an increasingly important component in the education of today's and tomorrow's pharmacists and pharmaceutical scientists. We believe that there is a critical need for an introductory textbook on Pharmaceutical Biotechnology that provides well-integrated, detailed coverage of both the relevant science and clinical application of pharmaceuticals derived by biotechnology.

Previous editions of the textbook *Pharmaceutical Biotechnology: Fundamentals and Applications* have provided a well-balanced framework for education in various aspects of pharmaceutical biotechnology, including production, dosage forms, administration, economic and regulatory aspects, and therapeutic applications. Rapid growth and advances in the field of pharmaceutical biotechnology, however, made it necessary to revise this textbook in order to provide up-to-date information and introduce readers to the cutting-edge knowledge and technology of this field.

This fourth edition of the textbook *Pharmaceutical Biotechnology: Fundamentals and Applications* builds on the successful concept used in the preceding editions and further expands its availability as electronic versions of the full book as well as individual chapters are now readily available and downloadable though online platforms.

The textbook is structured into two sections. An initial basic science and general features section comprises chapters introducing the reader to key concepts at the foundation of the technology relevant for protein therapeutics including molecular biology, production and analytical procedures, formulation development, pharmacokinetics and pharmacodynamics, and immunogenicity and chapters dealing with regulatory, economic and pharmacy practice considerations, and with evolving new technologies and applications. The second section discusses the various therapeutic classes of protein biologics and nucleotide-based therapeutics.

All chapters of the previous edition were revised and regrouped according to therapeutic application. The section on Monoclonal Antibodies was differentiated into a section on general considerations for this important class of biologics as well as sections focused on their application in oncology, inflammation, and transplantation in order to allow for a comprehensive discussion of the substantial number of approved antibody drugs. A chapter on stem cell technologies was newly added to give greater depth to the area of cell-based technologies.

In accordance with previous editions, the new edition of *Pharmaceutical Biotechnology: Fundamentals and Applications* will have as a primary target students in undergraduate and professional pharmacy programs as well as graduate students in the pharmaceutical sciences. An additional important audience is pharmaceutical scientists in industry and academia, particularly those that have not received formal training in pharmaceutical biotechnology and are inexperienced in this field.

We are convinced that this fourth edition of *Pharmaceutical Biotechnology: Fundamentals and Applications* makes an important contribution to the education of pharmaceutical scientists, pharmacists, and other healthcare professionals as well as serving as a ready resource on biotechnology. By increasing the knowledge and expertise in the development, application, and therapeutic use of "biotech" drugs, we hope to help facilitate a widespread, rational, and safe application of this important and rapidly evolving class of therapeutics.

Utrecht, The Netherlands — *Daan J.A. Crommelin*
Vancouver, BC, Canada — *Robert D. Sindelar*
Memphis, TN, USA — *Bernd Meibohm*

Abbreviations

5-FU	Fluorouracil
6-MP	6-mercaptopurine
A	Adenine
AA	Amino acid
AAV	Adeno-associated virus
Ab	Antibody
ABVD	Cytostatic protocol
ACE	Angiotensin converting enzyme
ACER	Average cost-effectiveness ratio
ACR	American College of Rheumatology
ADA	Adenosine deaminase
ADA	Anti-drug antibody(ies)
ADCC	Antibody dependent cellular cytotoxicity
ADME	Absorption, distribution, metabolism, and elimination
ADR	Adverse drug reaction
ADR	Antibody-drug conjugate
AE(s)	Adverse event(s) or Adverse effect(s)
Ag	Antigen
AGT	Angiotensinogen
AHFS	American Hospital Formulary Service
AIDS	Acquired immunodeficiency syndrome
ALCL	Anaplastic large-cell lymphoma
ALL	Acute lymphoblastic leukemia
AMD	Age-related macular degeneration
AMI	Acute myocardial infarction
AML	Acute myeloid leukemia cells
AMR	Antibody-mediated rejection
ANC	Absolute neutrophil count
AP	Alkaline phosphatase
APC	Antigen-presenting cell
ART	Assisted reproductive technologies
AS	Ankylosing spondylitis
ASCT	Autologous stem cell transplant
ASHP	American Society of Health-System Pharmacists
ASSENT	Assessment of the safety and efficacy of a new thrombolytic
ATAs	Anti-therapeutic antibodies
ATF	Alternating tangential flow
ATMP	Advanced therapy medicinal products
AUC	Area under the concentration-time curve
AZA	Azathioprine

BCG	Bacille Calmette-Guérin
BCGF	B cell growth factor
BFU	Burst-forming unit erythroid
BHK	Baby hamster kidney cells
BILAG	British Isles Lupus Assessment Group
BLA	Biologics license application
BLys	B lymphocyte stimulator
BMD	Becker's muscular dystrophy
BMP	Bone morphogenetic protein
Bp	Base pair
B_p	Peripheral blood band cell population
BPCI	Biologics Price Competition and Innovation Act
BSE/TSE	Bovine or transmissable spongiform encephalopathy
BSI	British Standards Institute
C	Cytosine
C	Drug concentration in plasma
CAPS	Cryopyrin-associated periodic syndrome
CAR	Coxsackievirus and adenovirus receptor.
CAT	Committee for Advanced Therapies
CBA	Cost benefit analysis
CBER	Centre for Biologics Evaluation and Research
CCK	Cholecystokinin
CD	Circular dichroism
CD	Cluster designation/cluster of differentiation (term to label surface molecules of lymphocytes)
CD	Crohn's disease
CDAI	Crohn's disease activity index
CDC	Centers for Disease Control and Prevention
CDC	Complement-dependent cytotoxicity
CDI	Chronic kidney insufficiency
cDNA	Copy DNA
CDR	Complementarity-determining region
CEA	Cost effectiveness analysis
CF	Cystic fibrosis
CFR	Code of Federal Regulations.
CFTR	Cystic fibrosis transmembrane conductance regulator
CFU	Colony-forming unit
CFUe	Colony-forming unit erythroid
CG	Chorionic gonadotropin
CGM	Continuous glucose monitoring
cGMP	Good manufacturing practice
C_H	Constant heavy chain region in MAB
CHMP	Committee for Medicinal Products for Human Use
CHO	Chinese hamster ovary
CIP	Clean in place
CK	Chemokines
CKD	Chronic kidney disease
CL	Clearance
C_L	Constant light chain region in MAB
CLL	Chronic lymphocytic leukemia
C_{max}	Peak plasma/serum concentration
CMC	Chemistry, manufacturing, and controls
CMI	Cell-mediated immunity

CMV	Cytomegalovirus
CNS	Central nervous system
COBALT	Continuous infusion versus double-bolus administration of alteplase
COS	Controlled ovarian stimulation
CPP	Cell-penetrating peptide
cQT	Corrected QT
CR	Complete response rate
CRC	Colorectal cancer
CsA	Ciclosporin A
CSII	Continuous subcutaneous insulin infusion
cSNPs	SNPs occurring in gene coding regions
CT	Cholera toxin
CT	Computerized tomography
CTA	Clinical trial application
CTL	Cytotoxic T lymphocytes
CTLA	Cytotoxic T lymphocyte antigen
CTP	Carboxy terminal peptide
CUA	Cost utility analysis
CyNA	Cyclohexene nucleic acid
CYP	Cytochrome P450
CZE	Capillary zone electrophoresis
dATP	Deoxyadenosine 5′-triphosphate
DC	Dendritic cell
dCTP	Deoxycytidine 5′-triphosphate
DDA	Dioctadecyldimethylammonium bromide
DDBJ	DNA Data Bank of Japan
ddNTPs	Dideoxynucleotidetriphosphates
DF	Diafiltration
dGTP	Deoxyguanosine 5′-triphosphate
DHHS	Department of Health and Human Services.
DLS	Dynamic light scattering
DMARDs	Disease-modifying anti-rheumatic drugs
DMD	Duchenne muscular dystrophy
DNA	Deoxyribonucleic acid
DNaqse I	Human deoxyribonuclease I
dNTPs	Deoxynucleotide triphosphates
DOE	US Department of Education
DS	Degree of cross-linking
DSC	Differential scanning calorimetry
DSP	Downstream process(ing)
dsRNA	Double-stranded RNA
dTTP	Deoxythymidine-5′-triphosphate
E	Effect measure
EBV	Epstein-Barr virus
EC_{50}	Concentration of the drug that produces half of the maximum effect
EDF	Eosinophil differentiation factor
EDSS	Expanded disability status scale
EDTA	Ethylenediaminetetraacetic acid
EGF	Epidermal growth factor
EGFR	Endothelial growth factor receptor
EGS	External guide sequences

EHR	Electronic health records
EI	Electrospray ionization
ELISA	Enzyme-linked immunosorbent assay
EMA	European Medicines Agency
E_{max}	Maximum achievable effect
EMBL	European Molecular Biology Laboratory
eNOS	Endothelial nitric oxide synthase
EPAR	European Public Assessment Report
EPO	Erythropoietin (alfa)
EPOR	Erythropoietin receptor
EPR	Enhanced permeability and retention
ER	Endoplasmic reticulum
ES	Embryonic stem cell
ESA	Erythropoietin stimulating agents
ESCF	Epidemiologic study of cystic fibrosis
EUCOMM	European conditional mouse mutagenesis program
Eur. Pharm.	European Pharmacopeia
F	Systemic bioavailability compared to IV administration
FANA	2-phosphoroarabino nucleic acid
Fc	Constant region of MAB
FcRn	Neonatal Fc-receptor
FDA	US Food and Drug Administration
FDC	(US) Food, Drug, and Cosmetic Act
FEV_1	Mean forced expiratory volume in 1 second
FGF	Fibroblast growth factor
FSGS	Focal segmental glomerulosclerosis
FSH	Follicle stimulating hormone
FTET	Frozen-thawn embryo transfer
FTIR	Fourier transform infrared spectroscopy
FVC	Forced vital capacity
G	Guanine
GAD	Glutamic acid decarboxylase
GBM	Glioblastoma multiforme
GCP	Good clinical practice
G-CSF	Granulocyte colony stimulating factor
GF	Growth factors
GFP	Green fluorescent protein
GFR	Glomerular filtration rate
GHBP	Growth hormone binding protein
GHD	Growth hormone deficient
GHR	hGH receptor
GHRH	Growth hormone releasing hormone
GI	Gastrointestinal
GLA	γ-carboxyglutamic acid
GlcNAc	N-acetylglucosamine
GLP	Good laboratory practice
GLP-1	Glucagon-1-like peptide
GM-CSF	Granulocyte-macrophage colony-stimulating factor
GMO	Genetically modified organism
GMP	Good manufacturing practice
GnRH	Gonadotropin-releasing hormone
GO	Gemtuzumab ozogamicin

GON	Guanidinium-containing oligonucleotide
GRF	Growth hormone releasing factor
GSD II	Glycogen storage disease II
GSK	Glycogen synthase kinase
GST	Glutathione-S-transferase
GUSTO	Global utilization of streptokinase and tissue plasminogen activator for occluded coronary arteries
GVHD	Graft-versus-host disease
GWAS	Genome-wide Association Studies
HACA	Human anti-chimeric antibodies
HAHA	Human anti-human antibody
HAMA	Human antibodies to murine antibodies
HSA	Human serum albumin
Hb	Hemoglobin
HbA1c	Glycated hemoglobin
HBsAg	Hepatitis B surface antigen
HBV	Hepatitis B virus
HCT/Ps	Human cell, tissue, and cellular and tissue-based products
HDAC	Histone deacetylase
HEMA	Hydroxyethyl methacrylate
HEPA	High-efficiency particulate air
Her or HER	Human epidermal growth factor receptor
HGF	Hematopoietic growth factors (Chap. 18)
HGF	Hepatocyte growth factor (Chap. 24)
hGH	Human chorionic gonadotropin (Chap. 13)
hGH	Human growth hormone (Chap. 14)
HGI	Human Genome Initiative
HGNC	Human Genome Nomenclature Committee
HGP	Human Genome Project
HGPRT	Hypoxanthine-guanine-phosphoribosyl transferase
HGVbase	Human genome variation database
Hib	*Haemophilus influenzae* type b
HIC	Hydrophobic interaction chromatography
HIV	Human immunodeficiency virus
HLA	Human leukocyte antigen
HMWP	High molecular weight protein
HO-1	Heme oxygenase-1
HPLC	High-performance liquid chromatography
HPV	Human papilloma virus
HRP	Horseradish peroxidase
HSA	Human serum albumin
HSC	Hematopoietic stem cell
HSV	Herpes simplex virus
HTS	High-throughput screening
HUPO	Human Proteome Organization
HV	Hypervariable sequences in MAB
IBC	Institutional Biosafety Committees
IBD	Inflammatory bowel disease
ICAM-1	Intercellular cell adhesion molecule
ICER	Incremental cost effectiveness ratio
ICH	International conference on harmonization
ICSI	Intracytoplasmic sperm injection
IEF	Isoelectric focusing

IFN	Interferon
IGF	Insulin-like growth factor
IL	Interleukin
ILPS	Insulin lispro protamine suspension
IM	Intramuscular
IND	Investigational new drug application
INJECT	International joint efficacy comparison of thrombolytics
INN	International Non-proprietary Names
INR	International normalized ratio
InTIME	Intravenous nPA for treatment of infarcting myocardium early
IP	Intraperitoneal
IPN	Intravenous parenteral nutrition
iPS(C)	Induced pluripotent stem
IR	Infra red
IRB	Institutional review board
ISCOM	Immune stimulating complex
ISS	Idiopathic short stature
ITP	Immune thrombocytopenia purpura
ITR	Inverted terminal repeats
IU	International units
IV	Intravenous
IVF	In vitro fertilization
IVT	Intravitreal
JAK-STAT	Janus Kinase/Signal Transducers and Activators of Transcription
JIA	Juvenile idiopathic arthritis
Ka	First-order absorption rate constant
k_{app}	Apparent absorption rate constant
KOMP	Knockout mouse project (NIH)
LABA	Long-acting β-adrenoceptor agonist
LAtPA	Long acting tissue plasminogen activator
LC/MS/MS	Liquid chromatography-tandem mass spectrometry
LDL	Low-density lipoprotein
LFA	Leukocyte function antigen
LH	Luteinizing hormone
LHRH	Luteinizing hormone-releasing hormone
LIF	Leukemia inhibitory factor
LNA	Locked nucleic acids
LRP	(low density) Lipoprotein receptor-related protein
LTR	Long terminal repeats
LVEF	Left ventricular ejection fraction
M	Microfold
MAA	Marketing authorization application
MAB	Monoclonal antibodies
MAC	Membrane attack complex
MALDI	Matrix-assisted laser desorption
MBP	Maltose-binding protein
MCS	Multiple cloning site
M-CSF	Macrophage-colony stimulating factor
MDR	Multidrug resistance
MDS	Myelodysplastic syndrome
MEO	Methoxy ethyl

met-hGH	Methionine recombinant human growth hormone
MF	Morpholino phosphoroamidate
MHC	Major histocompatibility complex
miRNA	MicroRNA
MMAD	Mass median aerodynamic diameter
MMAE	Monomethyl auristatin E
MOA	Mechanism of action
MPS	Mononuclear phagocyte system
MRM	Measles-rubella-mumps
mRNA	Messenger ribonucleic acid/messenger RNA
MRT	Mean residence time
MS	Multiple sclerosis
MSC	Mesenchymal stem cell.
MTD	Maximum tolerated dose
MTX	Methotrexate
mVar	Murine variable
NBCI	National Center for Biotechnology Information
NBP	Nonionic block copolymers
ncRNA	Noncoding RNA
NDA	New drug application
NDV	Newcastle disease virus
Neo^r	Neomycin resistance gene
NF-κB	Nuclear factor kappa B
NGS	Next-generation genome sequencing
NHGRI	National Human Genome Research Institute
NHL	Non-Hodgkin lymphoma
NICE	National Institute for Health and Clinical Excellence
NIH	National Institutes of Health.
NK	Natural killer
NMR	Nuclear magnetic resonance
NOAEL	No observable adverse effect level
NorCOMM	North American conditional mouse mutagenesis project
NPH	Neutral protamine Hagedorn
NPL	Neutral protamine lispro
NSAIDs	Nonsteroidal anti-inflammatory drugs
NSCLC	Non-small cell lung cancer
OAS	2′,5′-oligoadenylate synthetase
OATP	Organic anion transporting polypeptide
OBA	Office of Biotechnology Activities
OHRP	Office for Human Research Protections
OMe	2′-O Methyl
ON	Oligonucleotide
ONJ	Osteonecrosis of the jaw
OR	Overall response rate
ORF	Open reading frames.
Ori	Origin of replication
PAGE	Polyacrylamide gel electrophoresis
PAMP	Pathogen associated molecular patterns
PASI	Psoriasis activity and severity index
PBMC	Peripheral blood mononuclear cell
PBMs	Pharmacy benefits management companies
PBPC	Peripheral blood progenitor cells
PBPK	Physiologically based pharmacokinetics

PCR	Polymerase chain reaction
PEG	Polyethylene glycol
Peg-MGDF	Pegylated megakaryocyte growth and development factor
PEI	Polyethyleneimine
PEPT1, PEPT2	Proton driven peptide transporters
PGA	Physician's global assessment
PHS	Public Health Service
PI3K	Phosphatidylinositol 3-kinase
PIC	Polyion complex
pit-hGH	Pituitary-derived growth hormone
PJIA	Polyarticular juvenile idiopathic arthritis
PK/PD	Pharmacokinetics/pharmacodynamics (modeling)
PLGA	Polylactic-coglycolic acid
PLL	Poly (L-lysine) (Chap. 24)
PLL	Prolymphocytic leukemia (Chap. 17)
PML	Progressive multifocal leukoencephalopathy
PNA	Peptide nucleic acids
PRCA	Pure red cell aplasia
PRR	Pattern recognition receptors
PS nucleotides	Phosphothioate nucleotides
PsA	Psoriatic arthritis
PsO	Plaque psoriasis
PT	Prothrombin time
PTLD	Post-transplant lymphoproliferative disorder
PTM	Posttranslational modification
PVC	Pneumococcal conjugate vaccine (Chap. 22)
PVC	Polyvinyl chloride (Chap. 9)
PVDF	Polyvinylidene difluoride
PWS	Prader-Willi syndrome
Q2W	Every two weeks
Q3W	Every three weeks
QALY	Quality-adjusted life-year
QCM	Quartz crystal microbalance
QW	Weekly
RA	Rheumatoid arthritis
rAAT	Recombinant α_1-antitrypsin
RAC	Recombinant DNA advisory committee
rAHF-PFM	Recombinant antihemophilic factor-plasma/albumin free method
RANK(L)	Receptor activator of nuclear factor *kappa* B (ligand)
RAPID	Reteplase angiographic phase II international dose-finding study/reteplase versus alteplase patency investigation during myocardial infarction
RBC	Red blood cell
RCA	Replication competent adenovirus
RCL	Replication competent lentivirus
rDNA	Recombinant DNA
RES	Reticuloendothelial system
rFVII or rFVIII or IX	Recombinant factor VII or VIII or IX
RGD	Arginine-glycine-aspartic acid
rhAT	Recombinant human anti-trypsin
rhGH	Recombinant human growth hormone
rhIFN α-2a	Recombinant human interferon α-2a

rHuEPO	Recombinant human erythropoietin
rhVEGF	Recombinant human vascular endothelial growth factor
RIA	Radioimmunoassay
rIL-2	Recombinant interleukin-2
RISC	RNA-induced silencing complex
RIT	Radioimmunotherapeutic
RME	Receptor-mediated endocytosis
RNA	Ribonucleic acid
RNAi	RNA interference
RP	Reverse primer
RP-HPLC	Reversed-phase high performance liquid chromatography
RSV	Respiratory syncytial virus
RT	Reverse transcriptase
RT-PCR	Reverse-transcriptase polymerase chain reaction
SABA	Short-acting β-adrenoceptor agonist
SARS	Severe acute respiratory syndrome
SBGN	Systems biology graphical notation
SBS	Short bowel syndrome
SC	Subcutaneous
SCCHN	Squamous cell cancers of the head and neck
SCF	Stem cell factor
scFv	Single chain Fv
SCID	Severe combined immune-deficiency
SCNT	Somatic cell nuclear transfer
SCPF	Stem cell proliferation factor
SCT	Stem cell transplant
SDR	Specificity determining residues
SDS	Sodium dodecyl sulfate
SDS-PAGE	Sodium dodecyl sulfate polyacrylamide gel electrophoresis
SEC	Size-exclusion chromatography
SGA	Small for gestational age
SHOX	Short stature homeobox containing gene deficiency on the X chromosome
SIP	Steam in place
siRNA	Small interfering RNA
SIV	Simian immunodeficiency virus
SLE	Systemic lupus erythematosus
SNP	Single nucleotide polymorphism
SOCS	Suppressors of cytokine signaling
SOS	Sinusoidal obstructive syndrome
S_p	Segmented neutrophil population
SPR	Surface plasmon resonance
SREs	Skeletal reverse events
SRI	SLE Responder Index
SRIF	Somatotropin release-inhibitory factor
SRP	Signal recognition particle
SUPAC	Scale-up and post approval changes
T	Thymine
T1 (or 2) DM	Type 1 (or 2) diabetes mellitus
$t_{1/2}$	Half-life
tcDNA	Tricyclo-DNA
TF	Tissue factor

TFF	Tangential flow filtration
TFPI	Tissue factor pathway inhibitor
T_g	Glass transition temperature
TGF	Tissue growth factor (Chap. 5)
TGF	Transforming growth factor (Chap. 21)
TGF-β	Tissue growth factor-beta
TIMI	Thrombolysis in myocardial infarction
TIW	Three-times in a week
TK	Thymidine kinase
TLR	Toll-like receptor
TLS	Tumor lysis syndrome
TNF	Tissue necrosis factor
TNF-α	Tumor necrosis factor alpha
TO	Total relative uptake
t-PA	Tissue plasminogen activator
TPMT	Thiopurine methyltransferase
TPO	Thrombopoietin
TRE	Tet response element (see context Chap. 24)
TRE	Transcription regulatory element (see context Chap. 24)
TRF	T-cell replacement factor
tRNA	Transfer ribonucleic acid/transfer RNA
TS	Turner syndrome
TSC	The SNP Consortium
TSH	Thyroid-stimulating hormone
TSP	Thrombospondin
UC	Ulcerative colitis
UF	Ultrafiltration
UFH	Unfractionated heparin
USP	United States Pharmacopeia
UTR	Untranslated region
Vd	Volume of distribution
Ve	Exclusion volume
V_H	Variable region of the heave chain of a MAB
V_L	Variable region of the light chain of a MAB
VLDL	Very low-density lipoprotein
VLP	Virus-like particles
V_{max}	Maximum catabolic/enzymatic capacity
V_{ss}	Volume of distribution at steady-state
VSV	Vesicular stomatitis virus
WBC	White blood cell
WCB	Working cell bank
WHO	World Health Organization
ΔG_U	Difference in free energy

Contributors

Val Adams Department of Pharmacy Practice and Science, University of Kentucky, College of Pharmacy, Lexington, KY, USA

Rita R. Alloway Division of Nephrology, Department of Internal Medicine, University of Cincinnati, Cincinnati, OH, USA

Tsutomu Arakawa Department of Protein Chemistry, Alliance Protein Laboratories, Thousand Oaks, CA, USA

John M. Beals Lilly Research Laboratories, Biotechnology Discovery Research, Lilly Corporate Center, Eli Lilly and Company, Indianapolis, IN, USA

C. Andrew Boswell Preclinical and Translational Pharmacokinetics, Genentech Inc., South San Francisco, CA, USA

Andrew T. Chow Department of Quantitative Pharmacology, Pharmacokinetics and Drug Metabolism, Amgen Inc., Thousand Oaks, CA, USA

Emile van Corven Department of Process Development, Crucell, Leiden, The Netherlands

Daan J.A. Crommelin Department of Pharmaceutical Sciences, Utrecht Institute for Pharmaceutical Sciences, Utrecht University, Utrecht, The Netherlands

Le N. Dao Clinical Pharmacology, Genentech Inc., South San Francisco, CA, USA

Hugh M. Davis Biologics Clinical Pharmacology, Janssen Research & Development LLC, Radnor, PA, USA

John D. Davis Clinical Pharmacology, Genentech Inc., South San Francisco, CA, USA

Michael R. DeFelippis Lilly Research Laboratories, Bioproduct Research and Development, Eli Lilly and Company, Indianapolis, IN, USA

Rong Deng Clinical Pharmacology, Genentech Inc., South San Francisco, CA, USA

Paul Fielder Development Sciences, Genentech Inc., South San Francisco, CA, USA

Amy Grimsley Department of Pharmacy Practice, Mercer University College of Pharmacy and Health Sciences, Atlanta, GA, USA

Paul Ives Manufacturing Department, SynCo Bio Partners BV, Amsterdam, The Netherlands

Department of Process Development, Crucell, Leiden, The Netherlands

Jeffrey A. Jackson Lilly Research Laboratories, Medical Affairs, Eli Lilly and Company, Indianapolis, IN, USA

Wim Jiskoot Division of Drug Delivery Technology, Leiden/Amsterdam Center for Drug Research, Leiden University, Leiden, The Netherlands

Amita Joshi Clinical Pharmacology, Genentech Inc., South San Francisco, CA, USA

Farida Kadir Postacamedic Education Pharmacists (POA), Bunnik, The Netherlands

Saraswati Kenkare-Mitra Development Sciences, Genentech Inc., South San Francisco, CA, USA

Gideon F.A. Kersten Institue for Translational Vaccinology, Bilthoven, The Netherlands

Eugene M. Kolassa Department of Pharmacy, University of Mississippi, Oxford, MS, USA

Department of Pharmacy, Medical Marketing Economics, LLC, Oxford, MS, USA

Anneke Koole Global CMC Regulatory Affairs – Biologics, MSD, Oss, The Netherlands

Paul M. Kovach Lilly Research Laboratories, Technical Services and Manufacturing Sciences, Eli Lilly and Company, Indianapolis, IN, USA

Michael Laird Department of Endocrine Care, Genentech Inc., South San Francisco, CA, USA

Robert A. Lazarus Department of Early Discovery Biochemistry, Genentech Inc., South San Francisco, CA, USA

Renato de Leeuw Regulatory Affairs Department, MSD, Oss, The Netherlands

Jing Li Clinical Pharmacology, Genentech Inc., South San Francisco, CA, USA

Barbara Lippe Endocrine Care, Genentech Inc., South San Francisco, CA, USA

Alfred Luitjens Department of Process Development, Crucell, Leiden, The Netherlands

Ram I. Mahato Department of Pharmaceutical Sciences, University of Nebraska Medical Center, Omaha, NE, USA

Mary Ann Mascelli Department of Clinical Pharmacology and Pharmacokinetics, Shire, Lexington, MA, USA

Enrico Mastrobattista Department of Pharmaceutics, Utrecht Institute for Pharmaceutical Sciences, Utrecht University, Utrecht, The Netherlands

Trevor McKibbin Department of Pharmaceutical Services, Emory University Hospital/Winship Cancer Institute, Atlanta, GA, USA

Holly B. Meadows Department of Pharmacy and Clinical Sciences, South Carolina College of Pharmacy, Medical University of South Carolina, Charleston, SC, USA

Bernd Meibohm Department of Pharmaceutical Sciences, University of Tennessee Health Science Center, College of Pharmacy, Memphis, TN, USA

Nishit B. Modi Departments of Nonclinical R&D and Clinical Pharmacology, Impax Pharmaceuticals, Hayward, CA, USA

Ronald S. Oosting Division of Pharmacology, Utrecht Institute for Pharmaceutical Sciences, Utrecht University, Utrecht, The Netherlands

Tushar B. Padwal Department of Pharmacy, University of Mississippi, Oxford, MS, USA

Department of Pharmacy, Medical Marketing Economics, LLC, Oxford, MS, USA

Juan Jose Perez-Ruixo Department of Quantitative Pharmacology, Pharmacokinetics and Drug Metabolism, Amgen Inc., Thousand Oaks, CA, USA

Sidney Pestka PBL Interferon Source, Piscataway, NJ, USA

John S. Philo Department of Biophysical Chemistry, Alliance Protein Laboratories, Thousand Oaks, CA, USA

Peggy Piascik Department of Pharmacy Practice and Science, University of Kentucky, College of Pharmacy, Lexington, KY, USA

Nicole A. Pilch Department of Pharmacy and Clinical Sciences, South Carolina College of Pharmacy, Medical University of South Carolina, Charleston, SC, USA

Colin Pouton Drug Delivery, Disposition and Dynamics, Monash Institute of Pharmaceutical Sciences, Monash University (Parkville Campus), Parkville, VIC, Australia

Jean-Charles Ryff Biotechnology Research and Innovation Network, Basel, Switzerland

Tom Sam Regulatory Affairs Department, MSD, Oss, CC, The Netherlands

Huub Schellekens Department of Pharmaceutics Utrecht, Utrecht Institute for Pharmaceutical Sciences, Utrecht University, Utrecht, The Netherlands

Raymond M. Schiffelers Laboratory Clinical Chemistry & Hematology, University Medical Center Utrecht, Utrecht, The Netherlands

Katherine S. Shah Department of Pharmaceutical Services, Emory University Hospital/Winship Cancer Institute, Atlanta, GA, USA

Vinod P. Shah VPS Consulting, LLC, North Potomac, MD, USA

Robert D. Sindelar Department of Pharmaceutical Sciences and Department of Medicine, Providence Health Care and The University of British Columbia, Vancouver, BC, Canada

Gijs Verheijden Synthon Biopharmaceuticals B.V., Nijmegen, The Netherlands

Jeffrey S. Wagener Department of Pediatrics, University of Colorado School of Medicine, Aurora, CO, USA

Hao Wu Department of Pharmaceutical Sciences, University of Tennessee Health Science Center, Memphis, TN, USA

Zhenhua Xu Biologics Clinical Pharmacology, Janssen Research & Development LLC, Spring House, PA, USA

Yi Zhang Clinical Pharmacology, Genentech Inc., South San Francisco, CA, USA

Honghui Zhou Biologics Clinical Pharmacology, Janssen Research & Development LLC, Spring House, PA, USA

1

Molecular Biotechnology: From DNA Sequence to Therapeutic Protein

Ronald S. Oosting

INTRODUCTION

Proteins are already used for more than 100 years to treat or prevent diseases in humans. It started in the early 1890s with "serum therapy" for the treatment of diphtheria and tetanus by Emile von Behring and others. The antiserum was obtained from immunized rabbits and horses. Behring received the Nobel Prize for Medicine in 1901 for this pioneering work on passive immunization. A next big step in the development of therapeutic proteins was the use of purified insulin isolated from pig or cow pancreas for the treatment of diabetes type I in the early 1920s by Banting and Best (in 1923 Banting received the Nobel Prize for this work). Soon after the discovery of insulin, the pharmaceutical company Eli Lilly started large-scale production of the pancreatic extracts for the treatment of diabetes. Within 3 years after the start of the experiments by Banting, already enough animal-derived insulin was produced to supply the entire North American continent. Compare this to the present average time-to-market of a new drug (from discovery to approval) of 13.5 years (Paul et al. 2010).

Thanks to advances in biotechnology (e.g., recombinant DNA technology, hybridoma technology), we have moved almost entirely away from animal-derived proteins to proteins with the complete human amino acid sequence.

Such therapeutic human proteins are less likely to cause side effects and to elicit immune responses. Banting and Best were very lucky. They had no idea about possible sequence or structural differences between human and porcine/bovine insulin. Nowadays, we know that porcine insulin differs only with one amino acid from the human sequence and bovine insulin differs by three amino acids (see Fig. 1.1).

R.S. Oosting, Ph.D.
Division of Pharmacology, Utrecht Institute for Pharmaceutical Sciences, Utrecht University, Universiteitsweg 99, 3584 CG Utrecht, The Netherlands
e-mail: r.s.oosting@uu.nl

Thanks to this high degree of sequence conservation, porcine/bovine insulin can be used to treat human patients. In 1982, human insulin became the first recombinant human protein approved for sale in the USA (also produced by Eli Lilly) (cf. Chap. 12). Since then a large number of biopharmaceuticals have been developed. There are now almost 200 human proteins marketed for a wide range of therapeutic areas.

PHARMACEUTICAL BIOTECHNOLOGY, WHY THIS BOOK, WHY THIS CHAPTER?

In this book we define pharmaceutical biotechnology as all technologies needed to produce biopharmaceuticals (other than (nongenetically modified) animal- or human blood-derived medicines). Attention is paid both to these technologies and the products thereof. Biotechnology makes use of findings from various research areas, such as molecular biology, biochemistry, cell biology, genetics, bioinformatics, microbiology, bioprocess engineering, and separation technologies. Progress in these fields has been and will remain a major driver for the development of new biopharmaceuticals. Biopharmaceuticals form a fast-growing segment in the world of medicines opening new therapeutic options for patients with severe diseases. This success is also reflected by the fast growth in global sales. Double-digit growth numbers were reported over the last 25 years, reaching $80 billion in 2012. Five drugs in the top ten of drugs with the highest sales are biopharmaceuticals (2010), clearly showing the therapeutic and economic importance of this class of drugs.

Until now biopharmaceuticals are primarily proteins, but therapeutic DNA or RNA molecules (think about gene therapy products, DNA vaccines, and RNA interference-based products; Chaps. 22, 23, and 24, respectively) may soon become part of our therapeutic arsenal.

Therapeutic proteins differ in many aspects from classical, small molecule drugs. They differ in size, composition, production, purification, contaminations, side

D.J.A. Crommelin, R.D. Sindelar, and B. Meibohm (eds.), *Pharmaceutical Biotechnology*,
DOI 10.1007/978-1-4614-6486-0_1, © Springer Science+Business Media New York 2013

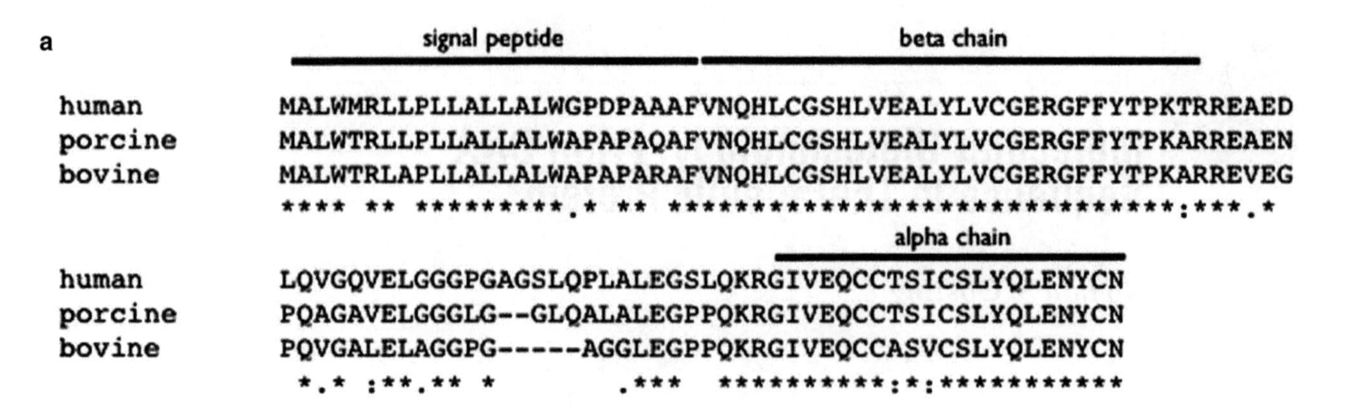

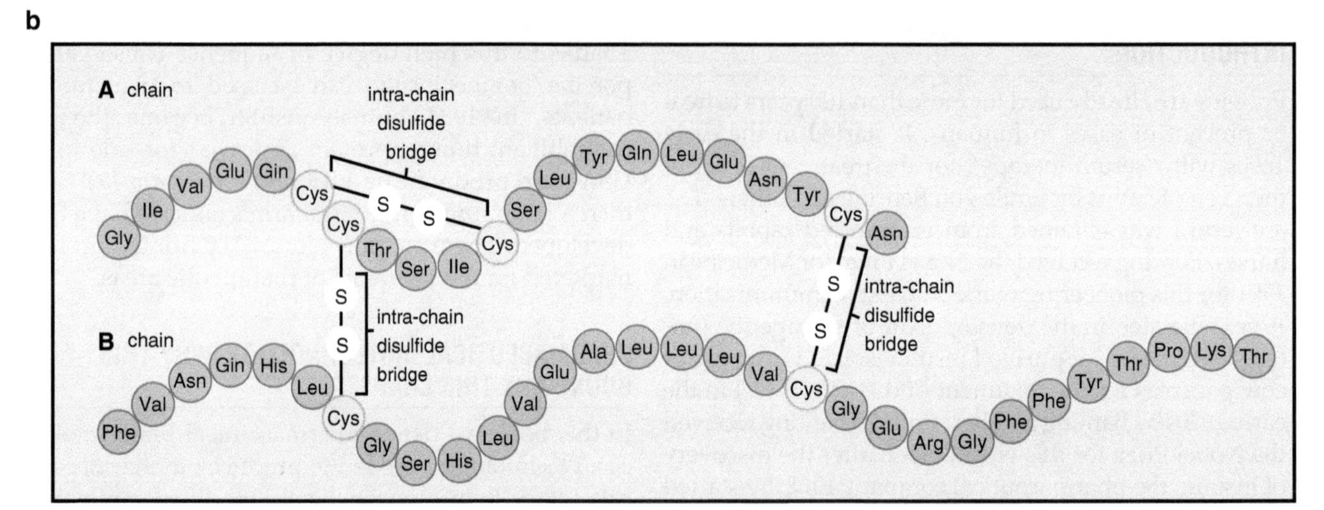

Figure 1.1 ■ (**a**) Multiple alignment (http://www.ebi.ac.uk/Tools/msa/clustalw2) of the amino acid sequences of human, porcine, and bovine preproinsulin. (*): identical residue. (**b**) Schematic drawing of the structure of insulin. The alpha and beta chain are linked by two disulphide bridges. Both the one-letter and three-letter codes for the amino acids are used in this figure: alanine (*ala*, A), arginine (*arg*, R), asparagine (*asn*, N), aspartic acid (*asp*, D), cysteine (*cys*, C), glutamic acid (*glu*, E), glutamine (*gln*, Q), glycine (*gly*, H), histidine (*his*, H), isoleucine (*ile*, I), leucine (*leu*, L), lysine (*Lys*, K), methionine (*met*, M), phenylalanine (*phe*, F), proline (*pro*, P), serine (*ser*, S), threonine (*thr*, T), tryptophan (*trp*,W), tyrosine (*tyr*, Y), and valine (*val*, V) (Figure **b** is taken from Wikipedia).

effects, stability, formulation, regulatory aspects, etc. These fundamental differences justify paying attention to therapeutic proteins as a family of medicines, with many general properties different from small molecules. These general aspects are discussed in the first set of chapters of this book ("General Topics"). After those general topics, the different families of biopharmaceuticals are dealt with in detail. This first chapter should be seen as a chapter where many of the basic elements of the selection, design, and production of biopharmaceuticals are touched upon. For in detail information the reader is referred to relevant literature and other chapters in this book.

ECONOMICS AND USE

Newly introduced biopharmaceuticals are very expensive. This is partly due to the high development cost (~$1.5 billion), but this is not different from the development costs of small molecule drugs (Paul et al. 2010), combined with high production costs and, for many therapeutic proteins, a relatively low number of patients. In addition, the relatively high price of (bio) pharmaceuticals is also due to too many failures in the drug discovery and development process. The few products that actually reach the market have to compensate for all the expenses made for failed products. For a monoclonal antibody, the probability to proceed from the preclinical discovery stage into the market is around 17 % (for small molecule drugs the probability of success is even lower, ~7 %). Economic aspects of biopharmaceuticals are discussed in Chap. 10.

As mentioned above, the number of patients for many marketed therapeutic proteins is relatively small. This has several reasons. The high price of therapeutic proteins makes that they are used primarily for the treatment of the relative severe cases. The specificity of many therapeutic proteins makes that they are only effective in subgroups of patients (personalized medicine). This is in particular true for the monoclonal antibodies used to treat cancer patients. For instance, the antibody trastuzumab (Herceptin) is only approved for breast cancer patients with high expression levels of the HER2 receptor on the tumor cells (±20 % of breast cancer cases). Other examples from the cancer field are the monoclonal antibodies cetuximab and panitumumab for the treatment of metastatic colorectal cancer. Both antibodies target the EGF receptor. Successful treatment of a patient with one of these monoclonal antibodies depends on (1) the presence of the EGF receptor on the tumor and (2) the absence of mutations in signaling proteins downstream of the EGF receptor (KRAS and BRAF). Mutations in downstream signaling proteins cause the tumor to grow independently from the EGF receptor and make the tumor nonresponsive to the antagonistic monoclonal antibodies.

Some diseases are very rare and thus the number of patients is very small. Most of these rare diseases are due to a genetic defect. Examples are cystic fibrosis (CF) and glycogen storage disease II (GSD II or Pompe disease). CF is most common in Caucasians. In Europe 1:2,000–3,000 babies are affected annually. GSD II is even rarer. It affects 1:140,000 newborns. The effects of GSD II can be reduced by giving the patients recombinant myozyme. It is clear that developing a drug for such a small patient population is commercially not very interesting. To booster drug development for the rare diseases (known as orphan drugs and orphan diseases), in the USA, Europe, and Japan, specific legislation exists.

FROM AN IN SILICO DNA SEQUENCE TO A THERAPEUTIC PROTEIN

We will discuss now the steps and methods needed to select, design, and produce a recombinant therapeutic protein (see also Fig. 1.2). We will not discuss in detail the underlying biological mechanisms. We will limit ourselves, in Box 1.1, to a short description of the central dogma of molecular biology, which describes the flow of information from DNA via RNA into a protein. For detailed information, the reader is referred to more specialized molecular biology and cell biology books (see "Recommended Reading" at the end of this chapter).

■ Selection of a Therapeutic Protein

The selection of what protein should be developed for a treatment of a particular disease is often challenging, with lots of uncertainties. This is why most big phar-

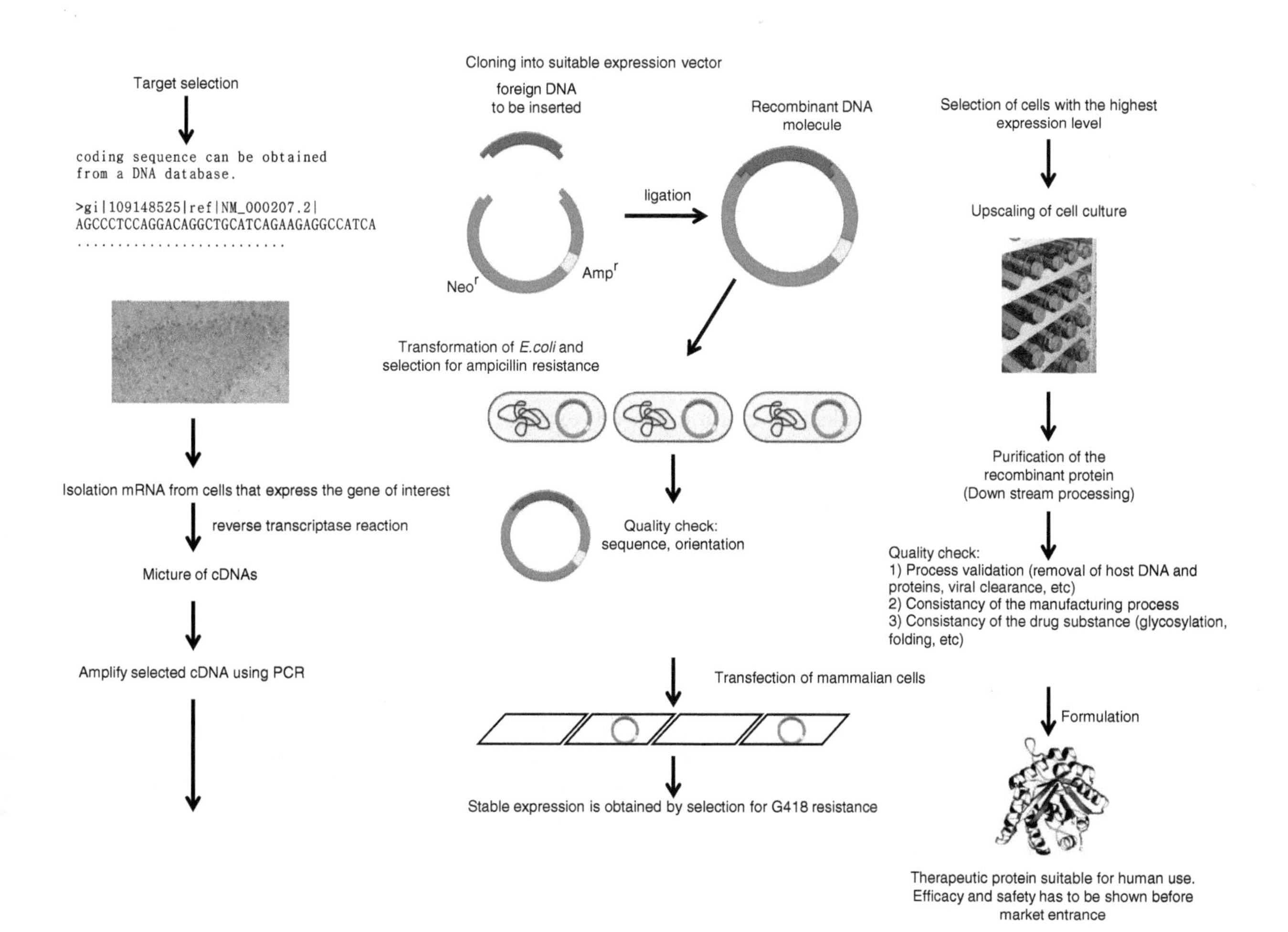

Figure 1.2 ■ Schematic representation of all the steps required to produce a therapeutic protein.

a

>gi|109148525|ref|NM_000207.2| Homo sapiens insulin (INS), transcript variant 1, mRNA

5'AGCCCTCCAGGACAGGCTGCATCAGAAGAGGCCATCAAGCAGATCACTGTCCTTCTGCCATGGCCCTGT
GGATGCGCCTCCTGCCCCTGCTGGCGCTGCTGGCCCTCTGGGGACCTGACCCAGCCGCAGCCTTTGTGAAC
CAACACCTGTGCGGCTCACACCTGGTGGAAGCTCTCTACCTAGTGTGCGGGGAACGAGGCTTCTTCTACAC
ACCCAAGACCCGCCGGGAGGCAGAGGACCTGCAGGTGGGGCAGGTGGAGCTGGGCGGGGGCCCTGGTGCAG
GCAGCCTGCAGCCCTTGGCCCTGGAGGGGTCCCTGCAGAAGCGTGGCATTGTGGAACAATGCTGTACCAGC
ATCTGCTCCCTCTACCAGCTGGAGAACTACTGCAACTAGACGCAGCCCGCAGGCAGCCCCACACCCGCCGC
CTCCTGCACCGAGAGAGATGGAATAAAGCCCTTGAACCAGCAAAA 3'

b

>gi|4557671|ref|NP_000198.1| insulin preproprotein [Homo sapiens]

(NH_2)MALWMRLLPLLALLALWGPDPAAAFVNQHLCGSHLVEALYLVCGERGFFYTPKTRREAEDLQVGQV
ELGGGPGAGSLQPLALEGSLQKRGIVEQCCTSICSLYQLENYCN-(COOH)

Figure 1.3 ■ DNA sequences are always written from the 5′ → 3′ direction and proteins sequences from the amino-terminal to the carboxy-terminal.

maceutical companies only become interested in a certain product when there is some clinical evidence that the new product actually works and that it is safe. This business model gives opportunities for startup biotech companies and venture capitalists to engage in this important early development process.

Sometimes the choice for a certain protein as a therapeutic drug is very simple. Think, for instance, about replacement of endogenous proteins such as insulin and erythropoietin for the treatment of diabetes type I and anemia, respectively. For many other diseases it is much more difficult to identify an effective therapeutic protein or target. For instance, an antibody directed against a growth factor receptor on a tumor cell may look promising based on in vitro and animal research but may be largely ineffective in (most) human cancer patients.

It is beyond the scope of this chapter to go further into the topic of therapeutic protein and target discovery. For further information the reader is referred to the large number of scientific papers on this topic, as can be searched using PubMed (http://www.ncbi.nlm.nih.gov/pubmed).

In the rest of this chapter, we will mainly focus on a typical example of the steps in the molecular cloning process and production of a therapeutic protein. At the end of this chapter, we will shortly discuss the cloning and large-scale production of monoclonal antibodies (see also Chap. 7).

Molecular cloning is defined as the assembly of recombinant DNA molecules (most often from two different organisms) and their replication within host cells.

■ DNA Sequence

The DNA, mRNA, and amino acid sequence of every protein in the human genome can be obtained from publicly available gene and protein databases, like those present at the National Center for Biotechnology Information (NCBI) in the USA and the European Molecular Biology Laboratory (EMBL). Their websites are http://www.ncbi.nlm.nih.gov/ and http://www.ebi.ac.uk/Databases/, respectively.

DNA sequences in these databases are always given from the 5′ end to the 3′ end and protein sequences from the amino- to the carboxy-terminal end (see Fig. 1.3). These databases also contain information about the gene (e.g., exons, introns, and regulatory sequences. See Box 1.1 for explanations of these terms) and protein structure (domains, specific sites, posttranslation modifications, etc.). The presence or absence of certain posttranslation modifications determines what expression hosts (e.g., *Escherichia coli* (*E. coli*), yeast, or a mammalian cell line) can be used (see below).

■ Selection of Expression Host

Recombinant proteins can be produced in *E. coli*, yeast, plants (e.g., rice and tomato), mammalian cells, and even by transgenic animals. All these expression hosts have different pros and cons.

Most marketed therapeutic proteins are produced in cultured mammalian cells. In particular Chinese hamster ovary (CHO) cells are used. On first sight, mammalian cells are not a logical choice. They are much more difficult to culture than, for instance, bacteria or yeast. On average, mammalian cells divide only once every 24 h, while cell division in *E. coli* takes ~ 30 min and in yeast ~ 1 h. In addition, mammalian cells need expensive growth media and in many cases bovine (fetal) serum as a source of growth factors (see Table 1.1 for a comparison of the various expression systems). Since the outbreak of the bovine or transmissible spongiform encephalopathy epidemic (BSE/TSE, better known as mad cow disease) under cattle in the United Kingdom, the use of bovine serum for the production of therapeutic proteins is considered a safety risk by the regulatory authorities (like the EMA in Europe and the FDA in the USA). To minimize the risk of transmitting TSE via a medicinal product, bovine serum has to be obtained from animals in countries with the lowest possible TSE risk, e.g., the USA, Australia, and New Zealand.

The main reason why mammalian cells are used as production platform for therapeutic proteins is that in these cells posttranslational modification (PTM) of the synthesized proteins resembles most closely the human situation. An important PTM is the formation of disulfide bonds between two cysteine moieties.

	Prokaryotes *E. coli*	Yeast *Pichia pastoris Saccharomyces cerevisiae*	Mammalian cells (e.g., CHO or HEK293 cells)
+	Easy manipulation Rapid growth Large-scale fermentation Simple media High yield	Grows relatively rapidly Large-scale fermentation Performs some posttranslational modifications	May grow in suspension, perform all required posttranslational modifications
–	Proteins may not fold correctly or may even aggregate (inclusion bodies) Almost no posttranslational modifications	Posttranslational modifications may differ from humans (especially glycosylation)	Slow growth Expensive media Difficult to scale up Dependence of serum (BSE)

Table 1.1 ■ Pros and cons of different expression hosts

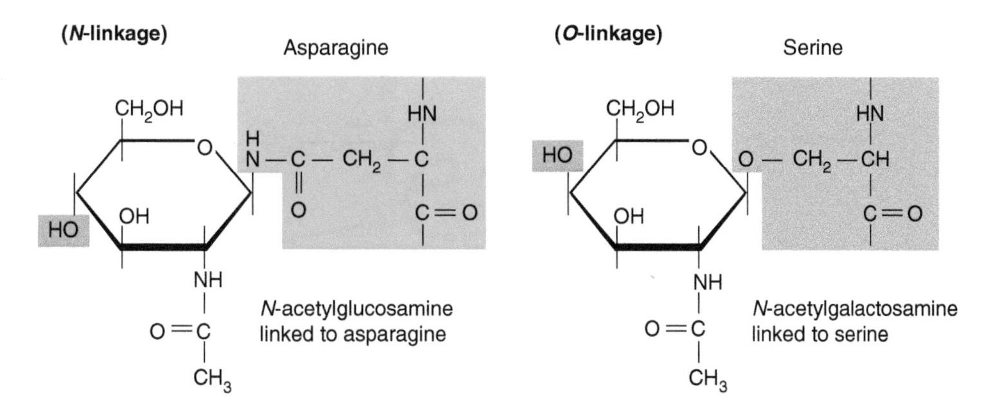

Figure 1.4 ■ Glycosylation takes place either at the nitrogen atom in the side chain of asparagine (N-linked) or at the oxygen atom in the side chain of serine or threonine. Glycosylation of asparagine takes place only when this residue is part of an Asn-X-Ser or Ans-X-Thr (X can be any residue except proline). Not all potential sites are glycosylated. Which sites become glycosylated depend also on the protein structure and on the cell type in which the protein is expressed.

Disulfide bonds are crucial for stabilizing the tertiary structure of a protein. *E. coli* is not able to make disulfide bonds in a protein, and already for this reason, *E. coli* is not very suitable for producing most of the marketed therapeutic proteins.

Another important PTM of therapeutic proteins is glycosylation. Around 70 % of all marketed therapeutic proteins, including monoclonal antibodies, are glycosylated. Glycosylation is the covalent attachment of oligosaccharides to either asparagine (N-linked) or serine/threonine (O-linked) (see Fig. 1.4). The oligosaccharide moiety of a therapeutic protein affects many of its pharmacological properties, including stability, solubility, bioavailability, in vivo activity, pharmacokinetics, and immunogenicity. Glycosylation differs between species, between different cell types within a species, and even between batches of in cell culture-produced therapeutic proteins. N-linked glycosylation is found in all eukaryotes (and also in some bacteria, but not in *E. coli*; see Nothaft and Szymanski 2010) and takes place in the lumen of the endoplasmatic reticulum and the Golgi system (see Fig. 1.5). All N-linked oligosaccharides have a common pentasaccharide core containing three mannose and two *N*-acetylglucosamine (GlcNAc) residues. Additional sugars are attached to this core. These maturation reactions take place in the Golgi system and differ between expression hosts. In yeast, the mature glycoproteins are rich in mannose, while in mammalian cells much more complex oligosaccharide structures are possible. O-linked glycosylation takes place solely in the Golgi system.

In Chap. 3 more details can be found regarding the selection of the expression system.

■ CopyDNA

The next step is to obtain the actual DNA that codes for the protein. This DNA is obtained by reverse-transcribing the mRNA sequence into copyDNA (cDNA). To explain this process, it is important to discuss first the structure of a mammalian gene and mRNA.

Most mammalian genes contain fragments of coding DNA (exons) interspersed by stretches of DNA that do not contain protein-coding information (introns). Messenger RNA synthesis starts with the making of a large primary transcript. Then, the introns are removed via a regulated process, called splicing. The mature mRNA contains only the exon sequences. Most mammalian mRNAs contain also a so-called poly-A "tail," a string of 100–300 adenosine

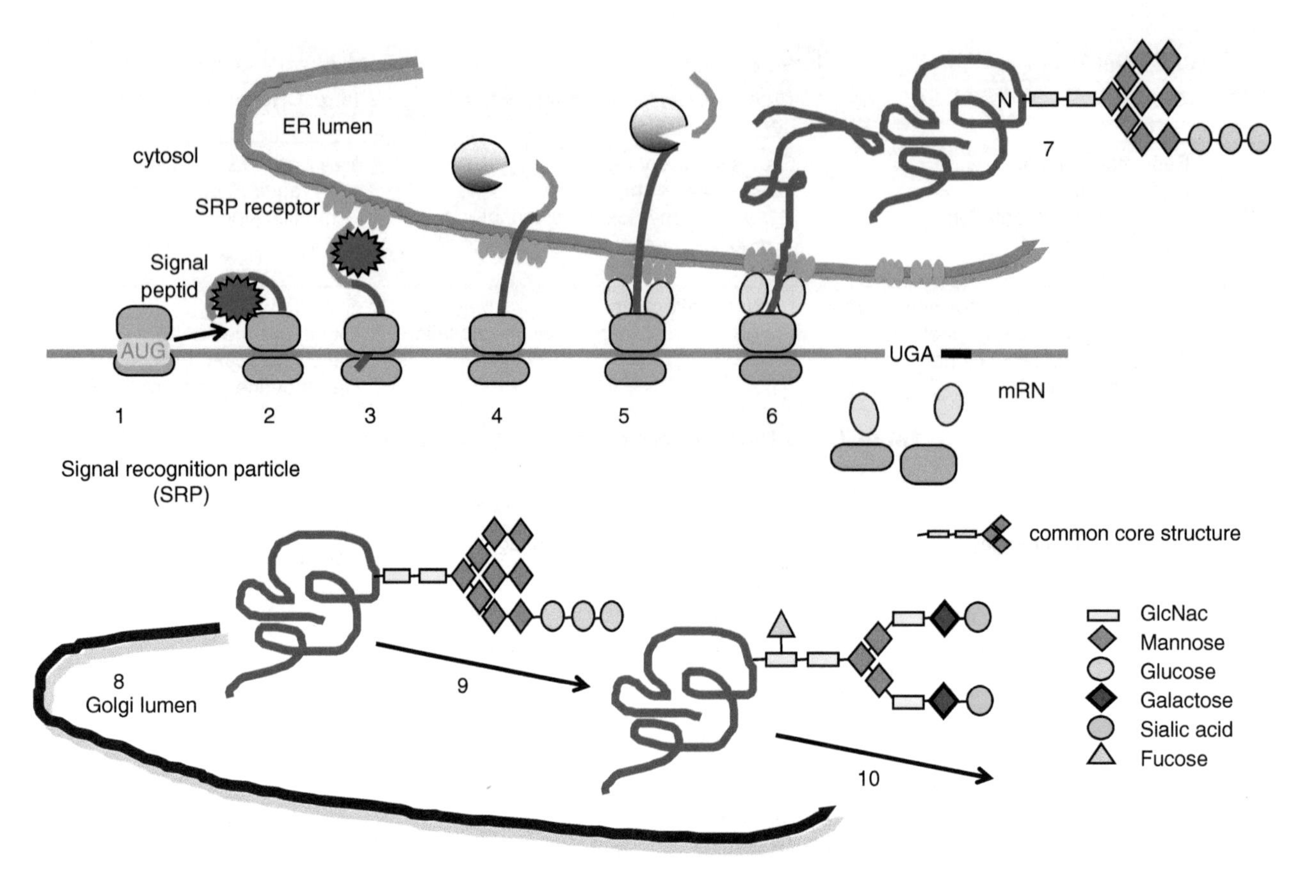

Figure 1.5 ■ Schematic drawing of the N-linked glycosylation process as occurs in the endoplasmic reticulum (ER) and Golgi system of a eukaryotic cell. (*1*) The ribozyme binds to the mRNA and translation starts at the AUG. The first ~ 20 amino acids form the signal peptide. (*2*) The signal recognition particle (SRP) binds the signal peptide. (*3*) Next, the SRP docks with the SRP receptor to the cytosolic side of the ER membrane. (*4*) The SRP is released and (*5*) the ribosomes dock onto the ER membrane. (*6*) Translation continues until the protein is complete. (*7*) A large oligosaccharide (activated by coupling to dolichol phosphate) is transferred to the specific asparagine (N) residue of the growing polypeptide chain. (*8*) Proteins in the lumen of the ER are transported to the Golgi system. (*9*) The outer carbohydrate residues are removed by glycosidases. Next, glycosyltransferases add different carbohydrates to the core structure. The complex type carbohydrate structure shown is just an example out of many possible varieties. The exact structure of the oligosaccharide attached to the peptide chain differs between cell types and even between different batches of in cell culture-produced therapeutic proteins. (*10*) Finally, secretory vesicles containing the glycoproteins are budded from the Golgi. After fusion of these vesicles with the plasma membrane, their content is released into the extracellular space.

nucleotides. These adenines are coupled to the mRNA molecule in a process called polyadenylation. Polyadenylation is initiated by binding of a specific set of proteins at the polyadenylation site at the end of the mRNA. The poly-A tail is important for transport of the mRNA from the nucleus into the cytosol, for translation, and it protects the mRNA from degradation.

An essential tool in cDNA formation is reverse transcriptase (RT). This enzyme was originally found in retroviruses. These viruses contain an RNA genome. After infecting a host cell, their RNA genome is reverse-transcribed first into DNA. The finding that RNA can be reverse-transcribed into DNA by RT is an important exception of the central dogma of molecular biology (as discussed in Box 1.1).

To obtain the coding DNA of the protein, one starts by isolating (m)RNA from cells/tissue that expresses the protein. Next, the mRNA is reverse-transcribed into copyDNA (cDNA) (see Fig. 1.6). The RT reaction is performed in the presence of an oligo-dT (a single-stranded oligonucleotide containing ~20 thymidines). The oligo-dT binds to the poly-A tail and reverse transcriptase couples deoxyribonucleotides complementary to the mRNA template, to the 3′end of the growing cDNA. In this way a so-called library of cDNAs is obtained, representing all the mRNAs expressed in the starting cells or tissue.

The next step is to amplify specifically the cDNA for the protein of interest using the polymerase chain reaction (PCR, see Fig. 1.7). A PCR reaction uses a (c) DNA template, a forward primer, a reverse primer, deoxyribonucleotides (dATP, dCTP, dGTP, and dTTP), Mg^{2+}, and a thermostable DNA polymerase. DNA polymerase adds free nucleotides only to the 3′ end of the newly forming strand. This results in elongation of the new strand in a 5′→ 3′ direction. DNA polymerase

Box 1.1 ■ The Central Dogma of Molecular Biology

The central dogma of molecular biology was first stated by Francis Crick in 1958 and deals with the information flow in biological systems and can best be summarized as "DNA makes RNA makes protein" (this quote is from Marshall Nirenberg who received the Nobel Prize in 1968 for deciphering the genetic code). The basis of the information flow from DNA via RNA into a protein is pairing of complementary bases; thus, adenine (A) forms a base pair with thymidine (*T*) in DNA or uracil in RNA and guanine (*G*) forms a base pair with cytosine (*C*).

To make a protein, the information contained in a gene is first transferred into a RNA molecule. RNA polymerases and transcription factors (these proteins bind to regulatory sequences on the DNA, like promoters and enhancers) are needed for this process. In eukaryotic cells, genes are built of exons and introns. Intron sequences (intron is derived from intragenic region) are removed from the primary transcript by a highly regulated process which is called splicing. The remaining mRNA is built solely of exon sequences and contains the coding sequence or sense sequence. In eukaryotic cells, transcription and splicing take place in the nucleus.

The next step is translation of the mRNA molecule into a protein. This process starts by binding of the mRNA to a ribosome. The mRNA is read by the ribosome as a string of adjacent 3-nucleotide-long sequences, called codons. Complexes of specific proteins (initiation and elongation factors) bring aminoacylated transfer RNAs (tRNAs) into the ribosome-mRNA complex. Each tRNAs (via its anticodon sequence) base pairs with its specific codon in the mRNA, thereby adding the correct amino acid in the sequence encoded by the gene. There are 64 possible codon sequences. Sixty-one of those encode for the 20 possible amino acids. This means that the genetic code is redundant (see Table 1.2). Translation starts at the start codon AUG, which codes for methionine and ends at one of the three possible stop codons: UAA, UGA, or UAG. The nascent polypeptide chain is then released from the ribosome as a mature protein. In some cases the new polypeptide chain requires additional processing to make a mature protein.

		2nd		**Base**			
		U	**C**	**A**	**G**		
	U	Phe	Ser	Tyr	Cys	**U**	
		Phe	Ser	Tyr	Cys	**C**	
		Leu	Ser	Stop	Stop	**A**	
1		Leu	Ser	Stop	Trp	**G**	**3**
s	**C**	Leu	Pro	His	Arg	**U**	**r**
t		Leu	Pro	His	Arg	**C**	**d**
		Leu	Pro	Gln	Arg	**A**	
b		Leu	Pro	Gln	Arg	**G**	**b**
a	**A**	Ile	Thr	Asn	Ser	**U**	**a**
s		Ile	Thr	Asn	Ser	**C**	**s**
e		Ile	Thr	Lys	Arg	**A**	**e**
		Met	Thr	Lys	Arg	**G**	
	G	Val	Ala	Asp	Gly	**U**	
		Val	Ala	Asp	Gly	**C**	
		Val	Ala	Glu	Gly	**A**	
		Val	Ala	Glu	Gly	**G**	

Table 1.2 ■ The genetic code.

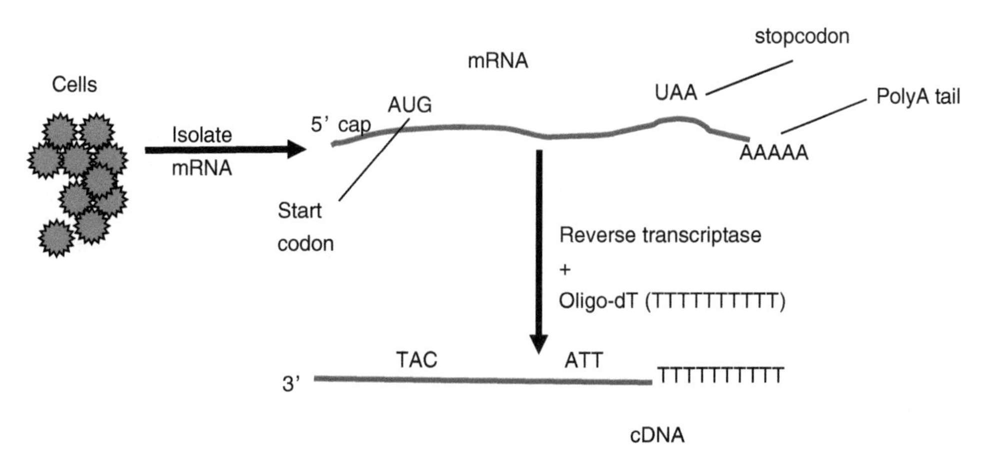

Figure 1.6 ■ Reverse transcriptase reaction.

can add a nucleotide only to a preexisting 3′-OH end, and therefore it needs a primer at which it can add the first nucleotide. PCR primers are single-stranded oligonucleotides around 20 to 30 nucleotides long, flanking opposite ends of the target DNA (see Fig. 1.8). The PCR is usually carried out for 30 cycles. Each cycle consists of three stages: a denaturing stage at ~94 °C (the double-stranded DNA is converted into single-stranded DNA), a primer annealing stage at ~60 °C (the optimal anneal temperature depends on sequences of the primers and template), and an extension stage at 72 °C. Theoretically, the amount of DNA should double during each cycle. A 30-cycle-long PCR should therefore result in a 2^{30} fold (~10^9) increase in the amount of DNA. In practice this is never reached. In particular at later cycles, the efficiency of the PCR reaction reduces.

PCR makes use of a thermostable DNA polymerase. These polymerases were obtained from Archaea living in hot springs such as those occurring in Yellowstone National Park (see Fig. 1.9) and at the ocean bottom. DNA polymerases make mistakes. When the aim is to clone and express a PCR product, a thermostable DNA polymerase should be used with 3′→5′ exonuclease "proofreading activity." One such enzyme is Pfu polymerase. This enzyme makes 1 mistake per every 10^6 base pairs, while the well-known Taq polymerase, an enzyme without proofreading activity, makes on average ten times more mistakes. As a trade-off, Pfu is much slower than Taq polymerase (Pfu adds ± 1,000 nucleotides per minute to the growing DNA chain and Taq 6,000 nucleotides/min).

■ Cloning PCR Products into an Expression Vector

There are several ways to clone a PCR product. One of the easiest ways is known as TA cloning (see Fig. 1.10). TA cloning makes use of the property of Taq polymerase to add a single adenosine to the 3′end of a PCR product. Such a PCR product can subsequently be ligated (using DNA ligase, see Molecular Biology toolbox) into a plasmid with a 5′ thymidine overhang (see Box 1.2 for a general description of expression plasmids). PCR products obtained with a DNA polymerase with proofreading activity have a blunt end, and thus they do not contain the 3′ A overhang. However, such PCR fragments can easily be A-tailed by incubating for a short period with Taq polymerase and dATP. Blunt PCR products can also directly be cloned into a linearized plasmid with 2 blunt ends. However, the efficiency of blunt-end PCR cloning is much lower than that of TA cloning. A disadvantage of TA and blunt-end cloning is that directional cloning is not possible, so the PCR fragment can be cloned either in the sense or antisense direction (see Fig. 1.10). PCR products can also be cloned by adding unique recognition sites of restriction enzymes to both ends of the PCR product. This can be done by incorporating these sites at the 5′end of the PCR primers. Although this strategy looks very straightforward, it is also not very efficient.

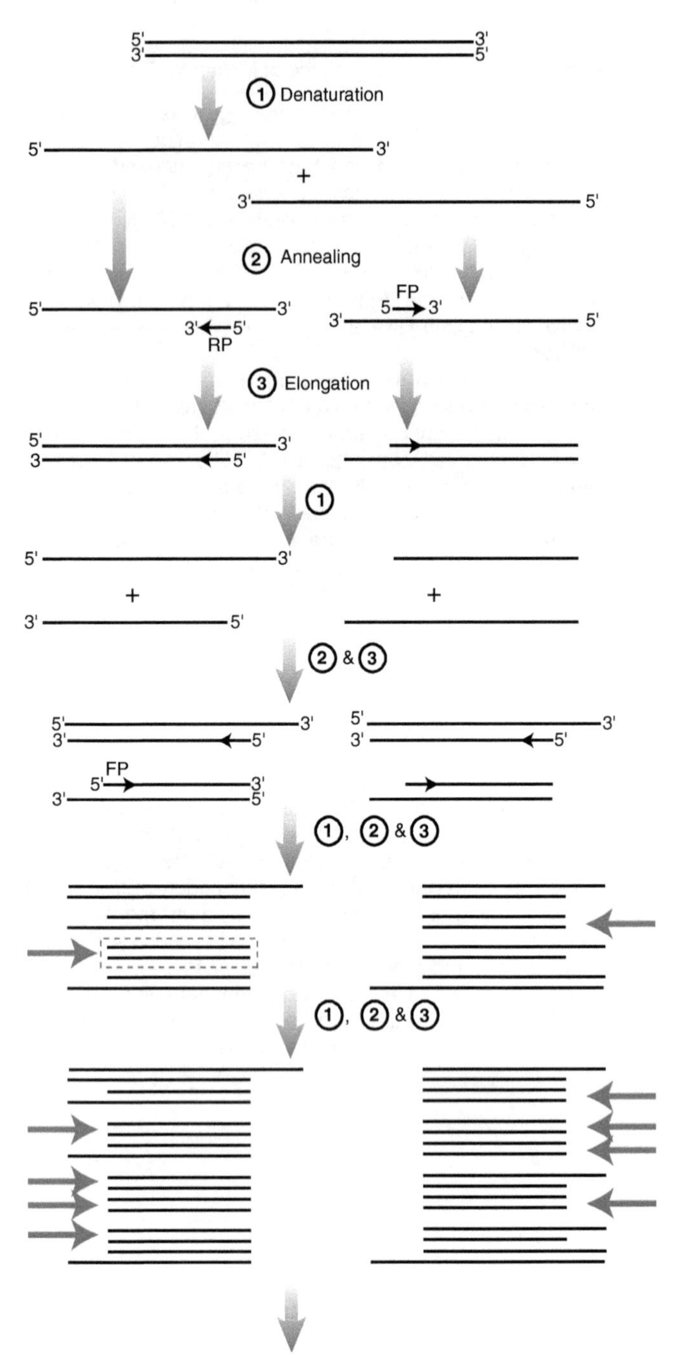

Figure 1.7 ■ The PCR process. (*1*) DNA is denatured at 94–96 °C. (*2*) The temperature is lowered to ± 60 °C. At this temperature the primers bind (anneal) to their target sequence in the DNA. (*3*) Next, the temperature is raised to 72 °C, the optimal temperature for Taq polymerase. Four cycles are shown here. A typical PCR reaction runs for 30 cycles. The *arrows* point to the desired PCR product.

After ligation, the plasmid is introduced into *E. coli* by a process called transformation. There are several ways to transform *E. coli*. Most used are the calcium

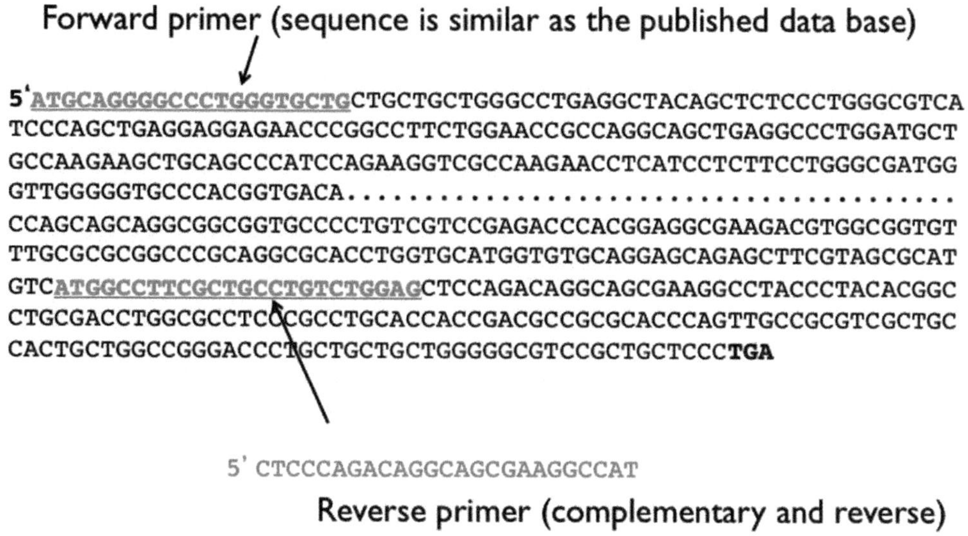

Forward primer (sequence is similar as the published data base)

```
5'ATGCAGGGGCCCTGGGTGCTGCTGCTGCTGGGCCTGAGGCTACAGCTCTCCCTGGGCGTCA
TCCCAGCTGAGGAGGAGAACCCGGCCTTCTGGAACCGCCAGGCAGCTGAGGCCCTGGATGCT
GCCAAGAAGCTGCAGCCCATCCAGAAGGTCGCCAAGAACCTCATCCTCTTCCTGGGCGATGG
GTTGGGGGTGCCCACGGTGACA....................................
CCAGCAGCAGGCGGCGGTGCCCCTGTCGTCCGAGACCCACGGAGGCGAAGACGTGGCGGTGT
TTGCGCGCGGCCCGCAGGCGCACCTGGTGCATGGTGTGCAGGAGCAGAGCTTCGTAGCGCAT
GTCATGGCCTTCGCTGCCTGTCTGGAGCTCCAGACAGGCAGCGAAGGCCTACCCTACACGGC
CTGCGACCTGGCGCCTCCCGCCTGCACCACCGACGCCGCGCACCCAGTTGCCGCGTCGCTGC
CACTGCTGGCCGGGACCCTGCTGCTGCTGGGGGCGTCCGCTGCTCCCTGA
```

```
5' CTCCCAGACAGGCAGCGAAGGCCAT
```

Reverse primer (complementary and reverse)

Figure 1.8 ■ PCR primer design.

Figure 1.9 ■ A hot spring in Yellowstone National Park. In hot spring like this one, Archaea, the bacterial source of thermostable polymerases, live.

chloride method (better known as heat shock) and electroporation (the bacteria are exposed to a very high electric pulse). Whatever the transformation method, channels in the membrane are opened through which the plasmid can enter the cell. Next, the bacteria are plated onto an agar plate with an antibiotic. Only bacteria that have taken up the plasmid with an antibiotic-resistant gene and thus produce a protein that degrades the antibiotic will survive. After an overnight incubation at 37 °C, the agar plate will contain a number of clones. The bacteria in each colony are the descendants of one bacterium. Subsequently, aliquots of a number of these colonies are grown overnight in liquid medium at 37 °C. From these cultures, plasmids can be isolated (this is known as a miniprep). The next steps will be to determine whether the obtained plasmid preparations contain an insert, and if so, to determine what the orientation is of the insert relative to the promoter that will drive the recombinant protein expression. The orientation can, for instance, be determined by cutting the obtained plasmids with a restriction enzyme that cuts only once somewhere in the plasmid and with another enzyme that cuts once somewhere in the insert. On the basis of the obtained fragment sizes (determined via agarose gel electrophoresis using appropriate molecular weight standards), the orientation of the insert in the plasmid can be determined (see Fig. 1.10).

As already discussed above, DNA polymerases make mistakes, and therefore, it is crucial to determine the nucleotide sequence of the cloned PCR fragment. DNA sequencing is a very important method in biotechnology (the developments in high-throughput sequencing have enabled the sequencing of many different genomes, including that of humans) and is therefore further explained in Box 1.3.

■ Transfection of Host Cells and Recombinant Protein Production

Introducing DNA into a mammalian cell is called transfection (and as already mentioned above, transformation in *E. coli*). There are several methods to introduce DNA into a mammalian cell line. Most often, the plasmid DNA is complexed to cationic lipids (like Lipofectamine) or polymers (like polyethyleneimines or PEI) and then pipetted to the cells. Next, the positively charged aggregates bind to the negatively charged cell membrane and are subsequently endocytosized (see Fig. 1.11). Then, the plasmid DNA has to escape from the endosome and has to find its way into the nucleus where mRNA synthesis can take place. This is actually achieved during cell division when the nuclear membrane is absent. Another way to introduce DNA into the cytosol is through electroporation.

Box 1.2. ■ Plasmids.

Schematic drawing of an expression plasmid for a mammalian cell line

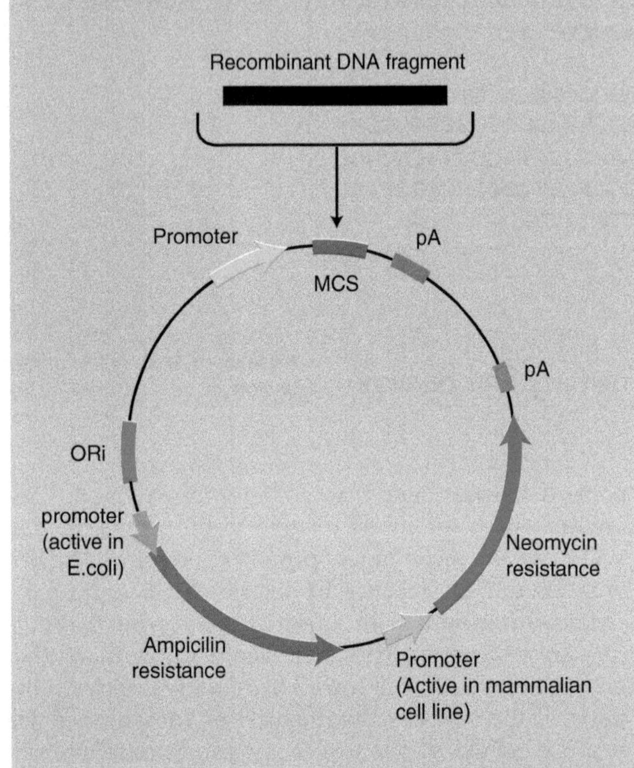

Plasmids are self-replicating circular extrachromosomal DNA molecules. The plasmids used nowadays in biotechnology are constructed partly from naturally occurring plasmids and partly from synthetic DNA. The figure above shows a schematic representation of a plasmid suitable for driving protein expression in a mammalian cell. The most important features of this plasmid are:

1. An origin of replication. The ori allows plasmids to replicate separately from the host cell's chromosome.
2. A multiple cloning site. The MCS contains recognition sites for a number of restriction enzymes. The presence of the MCS in plasmids makes it relatively easy to transfer a DNA fragment from one plasmid into another.
3. Antibiotic-resistant genes. All plasmids contain a gene that makes the recipient *E. coli* resistant to an antibiotic, in this case resistant to ampicillin. Other antibiotic-resistant genes that are often used confer resistance to tetracycline and Zeocin. The expression plasmid contains also the neomycin resistance gene. This selection marker enables selection of those mammalian cells that have integrated the plasmid DNA in their chromosome. The protein product of the neomycin resistance gene inactivates the toxin Geneticin.
4. Promoter to drive gene expression. Many expression vectors for mammalian cells contain the CMV promoter, which is taken from the cytomegaloma virus and is constitutively active. To drive recombinant protein expression in other expression hosts, other plasmids with other promoter sequences have to be used.
5. Poly (A) recognition site. This site becomes part of the newly produced mRNA and binds a protein complex that adds subsequently a poly (A) tail to the 3′ end of the mRNA. Expression vectors that are used to drive protein expression in *E. coli* do not contain a poly(A) recognition site.

Molecular biology enzyme toolbox

DNA polymerase produces a polynucleotide sequence against a nucleotide template strand using base pairing interactions (G against C and A against T). It adds nucleotides to a free 3′OH, and thus it acts in a 5′ → 3′ direction. Some polymerases have also 3′ → 5′ exonuclease activity (see below), which mediates proofreading.

Reverse transcriptase (RT) is a special kind of DNA polymerase, since it requires an RNA template instead of a DNA template.

Restriction enzymes are endonucleases that bind specific recognition sites on DNA and cut both strands. Restriction enzymes can either cut both DNA strands at the same location (blunt end) or they can cut at different sites on each strand, generating a single-stranded end (better known as a sticky end).

Examples:

HindIII	5′A[a]AGCTT	XhoI	5′C[a]TCGAG
	3′TTCGA[a]A		3′GAGCT[a]C
KpnI	5′GGTAC[a]C	EcoRV	5′GAT[a]ATC
	3′C^CATGG		3′CTA[a]TAG
NotI	5′GC[a]GGCCGC	PacI	5′TTAAT[a]TAA
	3′CGCCGG[a]CG		3′AAT[a]TAATT

[a]Location where the enzyme cuts

DNA ligase joints two DNA fragments. It links covalently the 3′-OH of one strand with the 5′-PO4 of the other DNA strand. The linkage of two DNA molecules with complementary sticky ends by ligase is much more efficient than blunt-end ligation.

Alkaline phosphatase. A ligation reaction of a blunt-end DNA fragment into a plasmid also with blunt ends will result primarily in empty plasmids, being the result of self-ligation. Treatment of a plasmid with blunt ends with alkaline phosphatase, which removes the 5′PO4 groups, prevents self-ligation.

Exonucleases remove nucleotides one at a time from the end (exo) of a DNA molecule. They act, depending on the type of enzyme, either in a 5′→3′ or 3→5′ direction and on single- or double-stranded DNA. Some polymerases have also exonuclease activity (required for proofreading). Exonucleases are used, for instance, to generate blunt ends on a DNA molecule with either a 3′ or 5′ extension.

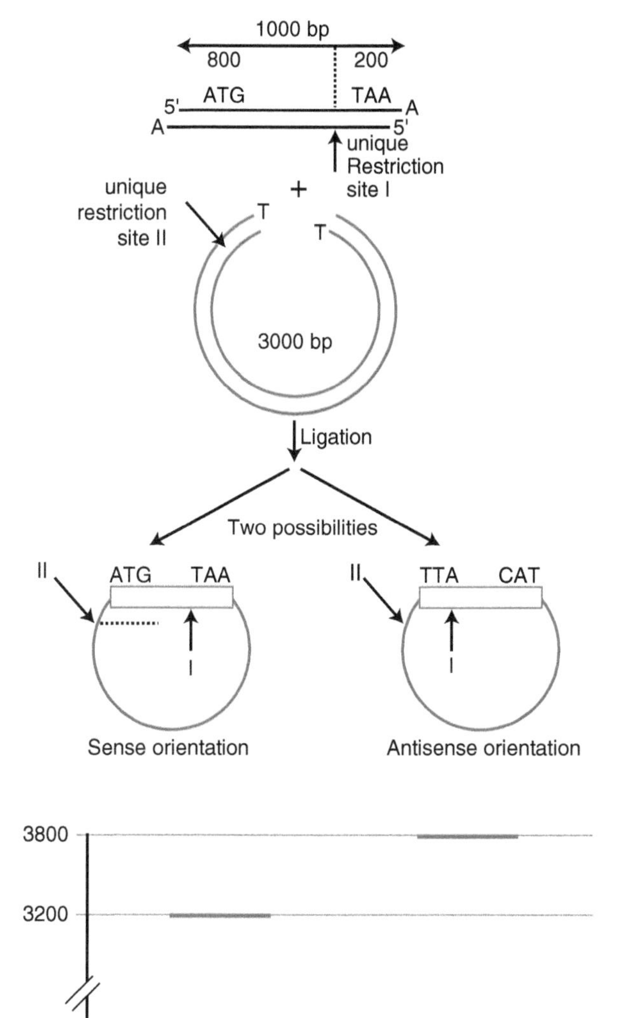

Figure 1.10 ■ Cloning of a PCR product via TA cloning (**a**). This cloning strategy makes use of the property of Taq polymerase to add an extra A to the 3′ end of the PCR product. To determine the orientation of the insert, the plasmid is cut by enzymes I and 2 (enzyme 1 cuts in the insert and enzyme 2 cuts in the plasmid). On the basis of the obtained fragment size (as determined by agarose electrophoresis), the orientation of the insert can be deduced (**b**).

During electroporation, an electric pulse is applied to the cells, which results in the formation of small pores in the plasma membrane. Through these pores the plasmid DNA can enter the cells.

Transfection leads to transient expression of the introduced gene. The introduced plasmids are rapidly diluted as a consequence of cell division or even degraded. However, it is possible to stably transfect cells leading to long expression periods. Then, the plasmid DNA has to integrate into the chromosomal DNA of the host cell. To accomplish this, a selection gene is normally included into the expression vector, which gives the transfected cells a selectable growth advantage. Only those cells that have integrated the selection

Box 1.3 ■ DNA Sequencing.

Technical breakthroughs in DNA sequencing, the determination of the nucleotide sequence, permit the sequencing of entire genomes, including the human genome. It all started with the sequencing in 1977 of the 5,386-nucleotide-long single-stranded genome of the bacteriophage φX174.

Chain-termination method and high-throughput sequencing

The most used method for DNA sequencing is the chain-termination method, also known as the dideoxynucleotide method, as developed by Frederick Sanger in the 1970s.

The method starts by creating millions of copies of the DNA to be sequenced. This can be done by isolating plasmids with the DNA inserted from bacterial cultures or by PCR. Next, the obtained double-stranded DNA molecules are denatured, and the reverse strand of one of the two original DNA strands is synthesized using DNA polymerase, a DNA primer complementary to a sequence upstream of the sequence to be determined, normal deoxynucleotidetriphosphates (dNTPs), and dideoxyNTPs (ddNTPs) that terminate DNA strand elongation. The four different ddNTPs (ddATP, ddGTP, ddCTP, or ddTTP) miss the 3'OH group required for the formation of a phosphodiester bond between two nucleotides and are each labeled with a different fluorescent dye, each emits light at different wavelengths. This reaction results in different reverse strand DNA molecules extended to different lengths. Following denaturation and removal of the free nucleotides, primers, and the enzyme, the resulting DNA molecules are separated on the basis of their molecular weight with a resolution of just one nucleotide (corresponding to the point of termination). The presence of the fluorescent label attached to the terminating ddNTPs makes a sequentially read out in the order created by the separation process possible. See also the figures below. The separation of the DNA molecules is nowadays carried out by capillary electrophoresis. The available capillary sequencing systems are able to run in parallel 96 or 384 samples with a length of 600 to 1,000 nucleotides. With the more common 96 capillary systems, it is possible to obtain around 6 million bases (Mb) of sequence per day.

Next-generation sequencing

The capillary sequencing systems are still used a lot, but they will be replaced in the future by alternative systems with a much higher output (100–1,000 times more) and at the same time a strong reduction in the costs.

The description of these really high-throughput systems is beyond the purpose of this book. An excellent review about this topic is written by Kirchner and Kelso (2010).

(continued)

Box 1.3 ■ (continued)

Schematic representation of the DNA sequencing process

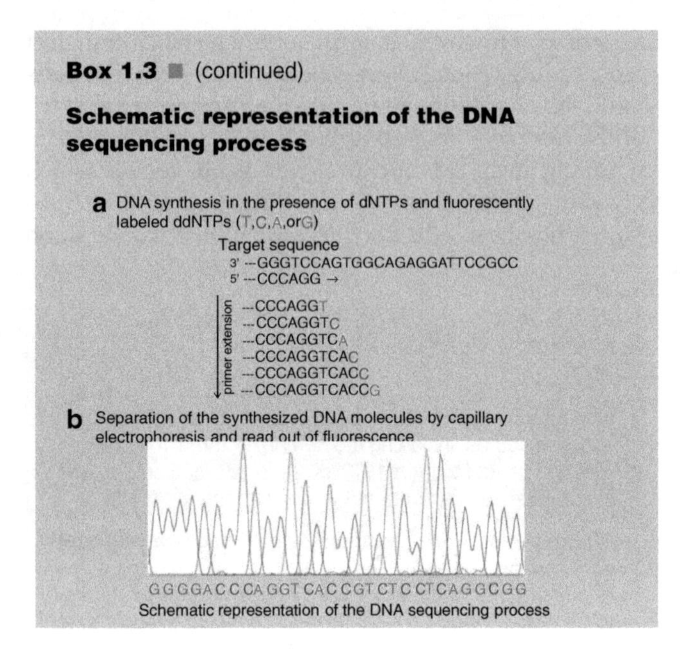

Schematic representation of the DNA sequencing process

marker (and most likely, but not necessary, also the gene of interest) into their genome will survive. Most expression plasmids for mammalian cells contain as selection marker the neomycin resistance gene (Neo^r). This gene codes for a protein that neutralizes the toxic drug Geneticin, also known as G418. The entire selection process takes around 2 weeks and results in a tissue culture dish with several colonies. Each colony contains the descendants of 1 stably transfected cell. Then, the cells from individual colonies have to be isolated and further expanded. The next step will be to quantify the recombinant protein production of the obtained cell cultures and to select those with the highest yields.

Transfection of mammalian cells is a very inefficient process (compared to transformation of *E. coli*) and needs relative large amounts of plasmid DNA. Integration of the transfected plasmid DNA into the genome is a very rare event. As a typical example, starting with 10^7 mammalian cells, one obtains usually not more than 10^2 stably expressing clones.

■ Cell Culture

A big challenge is to scale up cell cultures from lab scale (e.g., a 75 cm^2 tissue culture bottle) to a large-scale production platform (like a bioreactor). Mammalian cells are relatively weak and may easily become damaged by stirring or pumping liquid in or out a fermenter (shear stress). In this respect, *E. coli* is much sturdy, and thus this bacterium can therefore be grown in much larger fermenters.

A particular problem is the large-scale culturing of adherent (versus suspended) mammalian cells. One way to grow adherent cells in large amounts is on the surface of small beads. After a while the surface of the beads will be completely covered (confluent) with cells, and then, it is necessary to detach the cells from the beads and to redivide the cells over more (empty) beads and to transfer them to a bioreactor compatible with higher working volumes. To loosen the cells from the beads, usually the protease trypsin is used. It is very important that the trypsinization process is well timed: if it is too short, many cells are still on the beads, and if it is too long, the cells will lose their integrity and will not survive this.

Some companies have tackled the scale-up problem by "simply" culturing and expanding their adherent cells in increasing amounts of roller bottles. These bottles revolve slowly (between 5 and 60 revolutions per hour), which bathes the cells that are attached to the inner surface with medium (see Fig. 1.12). See Chap. 3 for more in-depth information.

■ Purification; Downstream Processing

Recombinant proteins are usually purified from cell culture supernatants or cell extracts by filtration and conventional column chromatography, including affinity chromatography (see Chap. 3).

The aim of the downstream processing (DSP) is to purify the therapeutic protein from (potential) endogenous and extraneous contaminants, like host cell proteins, DNA, and viruses.

It is important to mention here that slight changes in the purification process of a therapeutic protein may affect its activity and the amount and nature of the co-purified impurities. This is one of the main reasons (in addition to differences in expression host and culture conditions) why follow-on products (after expiration of the patent) made by a different company will never be identical to the original preparation and that is why they are not considered a true generic product (see also Chap. 11). A generic drug must contain the same active ingredient as the original drug, and in the case of a therapeutic protein, this is almost impossible and that is why the term "biosimilar" was invented.

Although not often used for the production of therapeutic proteins, recombinant protein purification may be simplified by linking it with an affinity tag, such as the his-tag (6 histidines). His-tagged proteins have a high affinity for Ni^{2+}-containing resins. There are two ways to add the 6 histidine residues. The DNA encoding the protein may be inserted into a plasmid encoding already a his-tag. Another possibility is to perform a PCR reaction with a regular primer and a primer with at its 5′end 6 histidine codons (CAT or CAC) (see Fig. 1.13). To enable easy removal of the his-tag from the recombinant protein, the tag may be followed by a suitable amino acid sequence that is recognized by an endopeptidase.

In *E. coli*, recombinant proteins are often produced as a fusion protein with another protein such as thioredoxin, beta-galactosidase, and glutathione

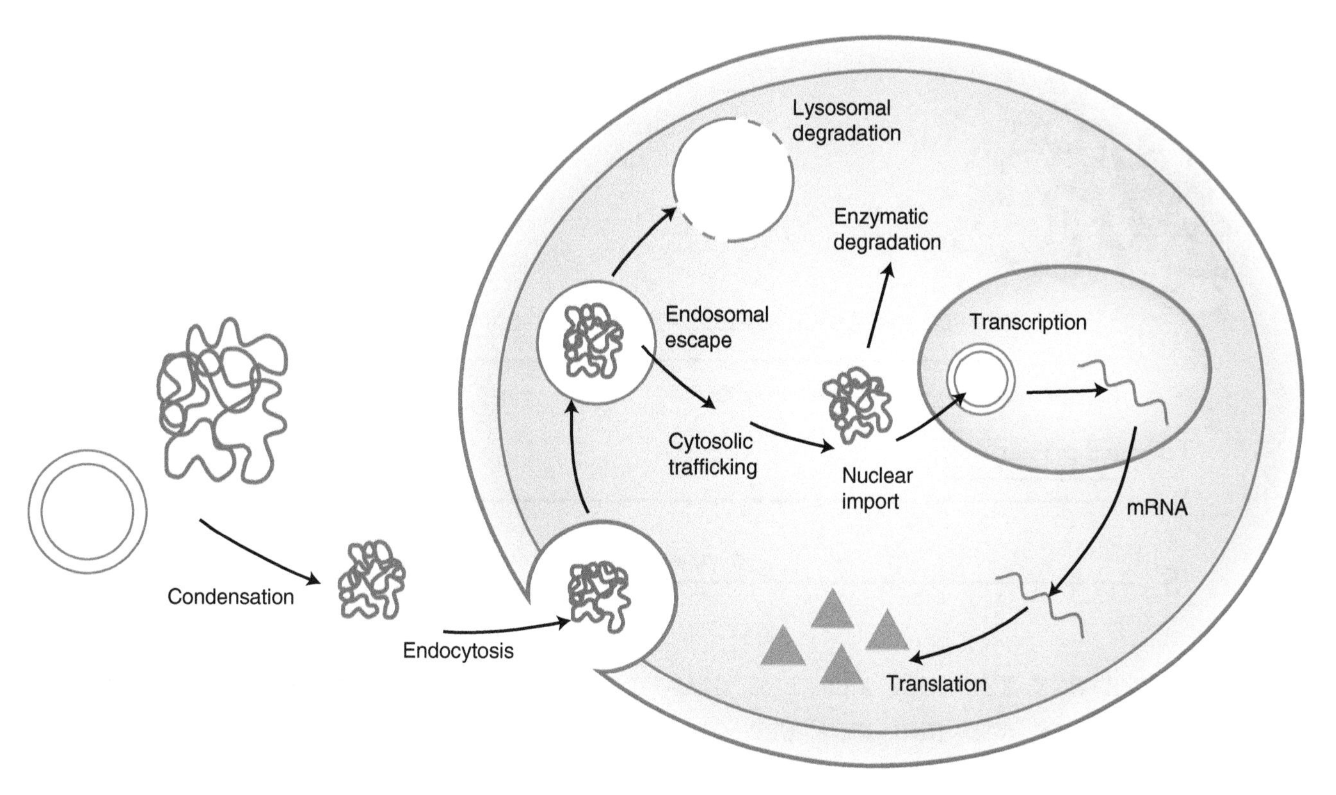

Figure 1.11 ■ Carrier-mediated transfection of mammalian cells.

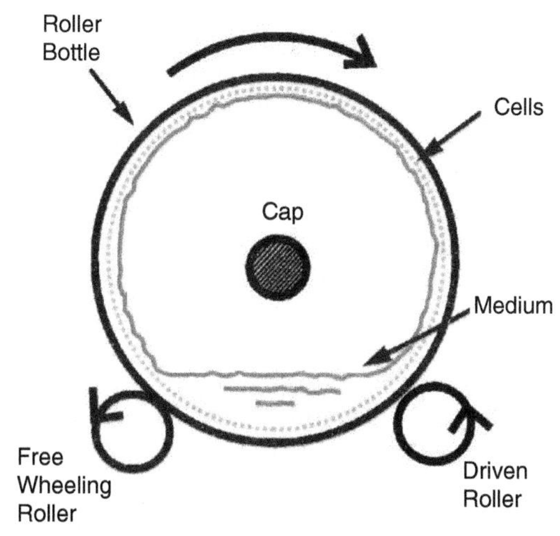

Figure 1.12 ■ Cell culturing in roller bottles.

S-transferase (GST). These fusion partners may improve the proper folding of the recombinant protein and may be used as affinity tag for purification.

MONOCLONAL ANTIBODIES

So far, we discussed the selection, design, and production of a protein starting from a DNA sequence in a genomic database. There is no database available of the entire repertoire of human antibodies. Potentially there are millions of different antibodies possible, and our knowledge about antibody-antigen interactions is not large enough to design a specific antibody from scratch.

Many marketed therapeutic proteins are monoclonal antibodies (cf. Chaps. 7, 17, 19, and 20). We will focus here on the molecular biological aspects of the design and production of (humanized) monoclonal antibodies in cell culture (primarily CHO cells are used). For a description of the structural elements of monoclonal antibodies, we refer to Chapter 7, Figs. 1.1 and 1.2.

The classic way to make a monoclonal antibody starts by immunizing a laboratory animal with a puria

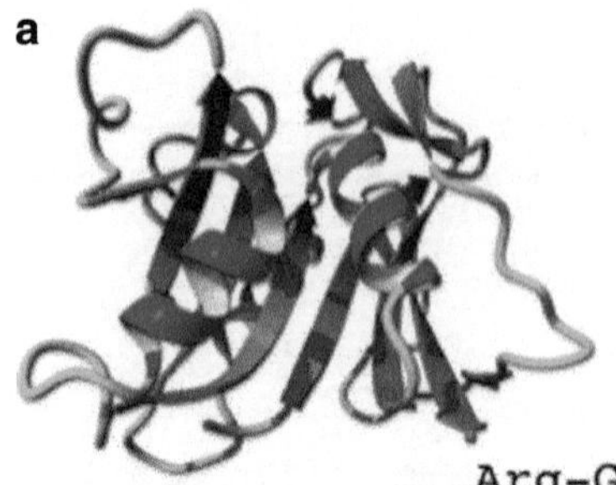

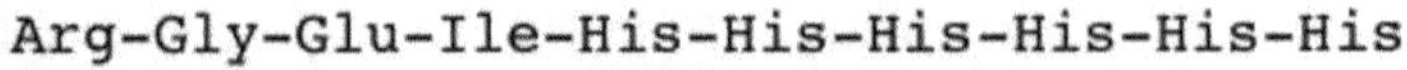

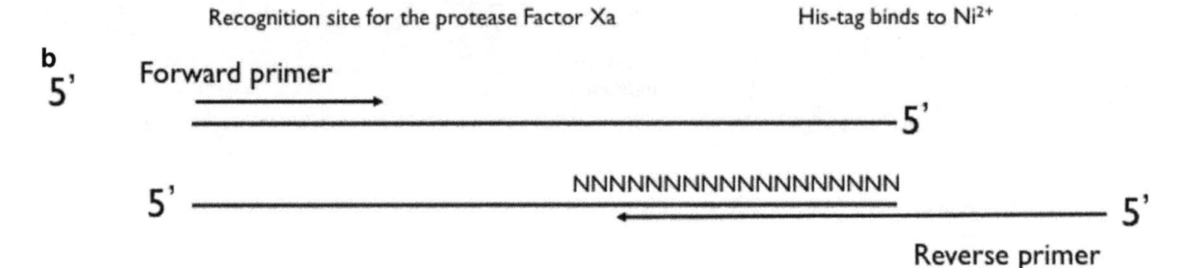

XXX XXX XXX XXX XXX XXX ILE GLU GLY ARG HIS HIS HIS HIS HIS HIS **stop**

5'NNN NNN NNN NNN NNN NNN ATT GAA GGA CGT CAT CAT CAT CAT CAT CAT TAA

REVERSE PRIMER:

5" TTA ATG ATG ATG ATG ATG ATG ACG TCC TTC AAT NNN NNN NNN NNN NNN NNN

Figure 1.13 ■ (**a**) Schematic drawing of a his-tagged fusion protein. (**b**) Design of the primers needed to generate a his-tag at the carboxy-terminal end of a protein.

fied human protein against which the antibody should be directed (see Fig. 1.14). In most cases, mice are used. The immunization process (a number of injections with the antigens and an adjuvant) will take several weeks. Then the spleens of these mice are removed and lymphocytes are isolated. Subsequently, the lymphocytes are fused using polyethylene glycol (PEG) with a myeloma cell. The resulting hybridoma cell inherited from the lymphocytes the ability to produce antibodies and from the myeloma cell line the ability to divide indefinitely. To select hybridoma cells from the excess of non-fused lymphocytes and myeloma cells, the cells are grown in HAT selection medium. This culture medium contains hypoxanthine, aminopterin, and thymidine. The myeloma cell lines used for the production of monoclonal antibodies contain an inactive hypoxanthine-guanine phosphoribosyltransferase (HGPRT), an enzyme necessary for the salvage synthesis of nucleic acids. The lack of HGPRT activity is not a problem for the myeloma cells because they can still synthesize purines de novo. By exposing the myeloma cells to the drug aminopterin also de novo synthesis of purines is blocked and these cells will not survive anymore. Selection against the unfused lymphocytes is not necessary, since these cells, like most primary cells, do not survive for a long time in cell culture. After PEG treatment, the cells are diluted and divided over several dishes. After approximately 2 weeks, individual clones are visible. Each clone contains the descendants of one hybridoma cell and will produce one particular type of antibody (that is why they are called monoclonal antibodies). The next step is to isolate hybridoma cells from individual clones and grow them in separate wells of a 96-well plate. The hybridomas secrete antibodies into the culture medium. Using a suitable test (e.g., an ELISA), the obtained culture media can be screened for antibody binding to the antigen. The obtained antibodies can then be further characterized using other tests. In this way a mouse monoclonal antibody is generated.

These mouse monoclonal antibodies cannot be used directly for the treatment of human patients. The amino acid sequence of a mouse antibody is too different from the sequence of an antibody in humans and thus will elicit an immune response. To make a mouse antibody less immunogenic, the main part of its sequence must be replaced by the corresponding human sequence. Initially, human-mouse chimeric antibodies were made. These antibodies consisted of the constant regions of the human heavy and light chain and the variable regions of the mouse antibody.

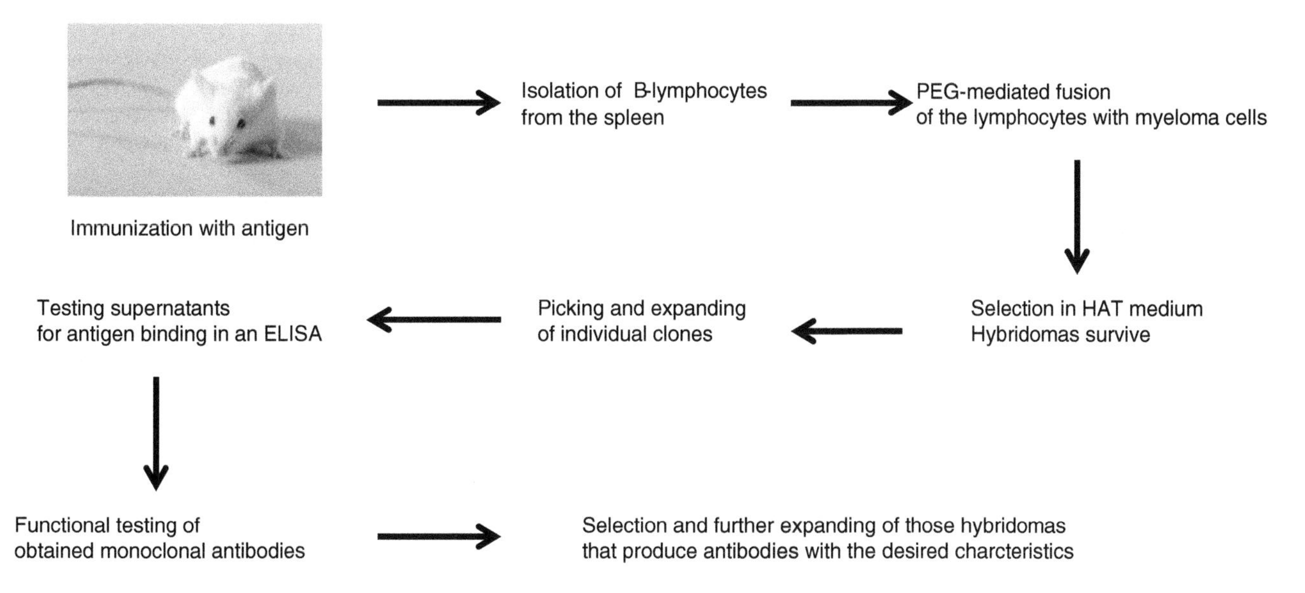

Figure 1.14 ■ The making of a mouse monoclonal antibody.

Later, so-called humanized antibodies were generated by grafting only the complementarity-determining regions (CDRs), which are responsible for the antigen-binding properties, of the selected mouse antibody onto a human framework of the variable light (V_L) and heavy (V_H) domains. The humanized antibodies are much less immunogenic than the previously used chimeric antibodies. To even further reduce immunogenicity, SDR grafting is used nowadays (Kashmiri et al. 2005). SDR stands for 'specificity determining residues'. From the analysis of the 3-D structure of antibodies, it appeared that only ~ 30 % of the amino acid residues present in the CDRs are critical for antigen binding. These residues, which form the SDR, are thought to be unique for a given antibody.

Humanization of a mouse antibody is a difficult and tricky process. It results usually in a reduction of the affinity of the antibody for its antigen. One of the challenges is the selection of the most appropriate human antibody framework. This framework determines basically the structure of the antibody and thus the orientation of the antigen recognition domains in space. Sometimes it is necessary to change some of the residues in the human antibody framework to restore antigen binding. To further enhance the affinity of the humanized antibody for its antigen, mutations within the CDR/SDR sequences are introduced. How this in vitro affinity maturation is done is beyond the scope of this chapter.

So far the generation of a humanized antibody has been described in a rather abstract way. How is it done in practice? First, the nucleotide sequence of each of the V_L and V_H regions is deduced (contains either the murine CDRs or SDRs). Next, the entire sequence is divided over four or more alternating oligonucleotides with overlapping flanks (see Fig. 1.15). These relatively long oligonucleotides are made synthetically. The reason why the entire sequence is divided over four nucleotides instead of over two or even one is that there is a limitation to the length of an oligonucleotide that can be synthesized reliably (a less than 100 % yield of each coupling step (nucleotides are added one at the time) and the occurrence of side reactions make that oligonucleotides hardly exceed 200 nucleotide residues).

To the four oligonucleotides, a heat-stable DNA polymerase and the 4 deoxyribonucleotides (dATP, dCTP, dGTP, and dTTP) are added, and the mixture is incubated at an appropriate temperature. Finally 2 primers, complementary to both ends of the fragment, are added, which enable the amplification of the entire sequence. The strategy to fuse overlapping oligonucleotides by PCR is called PCR sewing.

Finally, the PCR product encoding the humanized V_L and V_H region is cloned into an expression vectors carrying the respective constant regions and a signal peptide. The signal peptide is required for glycosylation. Subsequently, the expression constructs will be used to stably transfect CHO cells. The obtained clones will be tested for antibody production, and clones with the highest antibody production capacity will be selected for further use.

YIELDS

To give an idea about the production capacity needed to produce a monoclonal antibody, we will do now some calculations. The annual production of the most successful therapeutic monoclonal antibodies is around 1,000 kg (in 2009). In cell culture, titers of 2–6 g/L are routinely reached and the yield of the DSP is around 80 %. Thus, to produce 1,000 kg monoclonal antibody,

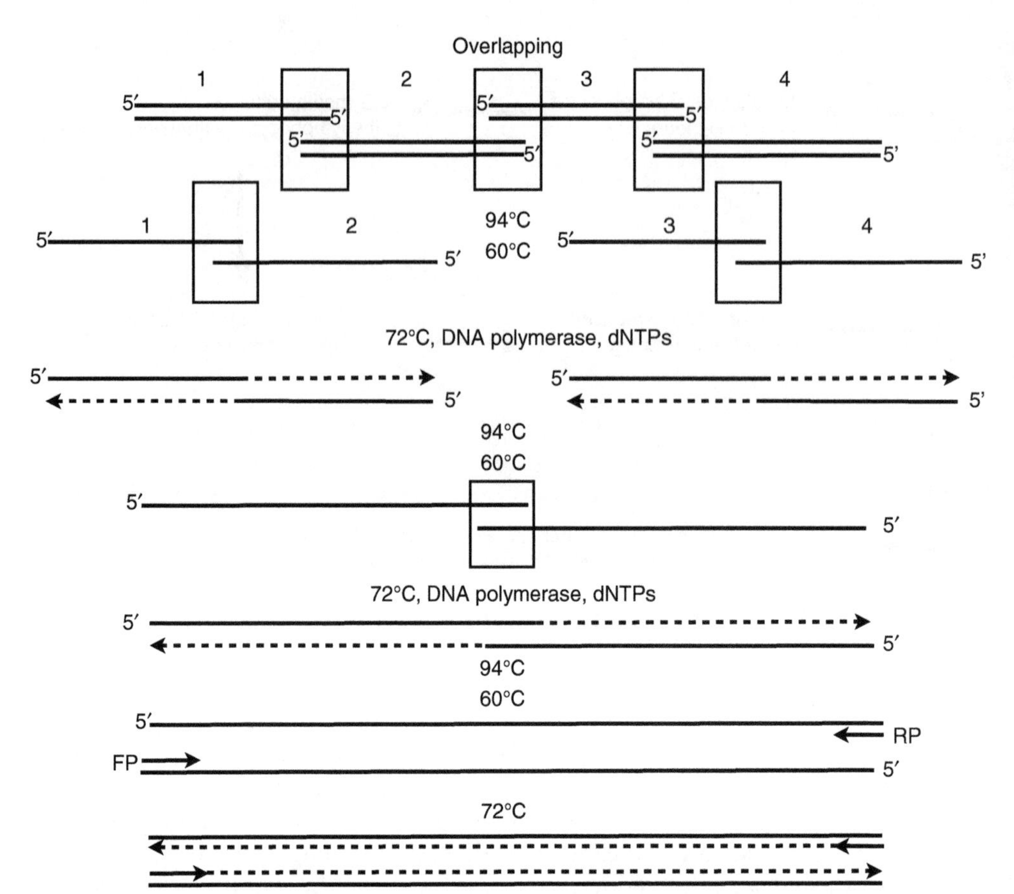

Fig. 1.15 ■ The making of a sequence containing the humanized V_L or V_H region of an antibody by PCR sewing. Both the V_L and V_H regions contain three highly variable loops (known as complementarity-determining regions 1, 2, and 3). The V_L and V_H regions are approximately 110 amino acids in size. Four alternating oligonucleotides with overlapping flanks are incubated together with a DNA polymerase and deoxynucleotides. DNA polymerase fills in the gaps. The sequences of these oligonucleotides are based upon the original mouse CDR/SDR sequences inserted into a human V_L or V_H framework. Next, the entire sequence is PCR amplified using end primers (*FP* forward primer, *RP* reverse primer). The resulting PCR fragments will be around 330 base pair in size

one needs 200,000–600,000 L of cell culture supernatant. In Chap. 3 more details can be found.

CONCLUSION

Thanks to advances in many different areas, including molecular biology, bioinformatics, and bioprocess engineering, we have moved from an animal-/human-derived therapeutic protein product towards in vitro-produced therapeutic proteins with the fully human sequence and structure. Importantly, we have now access to potentially unlimited amounts of high-quality therapeutic proteins. Of course, there will always be a risk for (viral) contaminations in the *in vitro*-produced therapeutic protein preparation, but this risk is much smaller than when the protein has to be isolated from a human source (examples from the past include the transmission of hepatitis B and C and HIV via blood-derived products and the transmission of Creutzfeldt-Jakob disease from human growth hormone preparations from human pituitaries).

As basic knowledge in molecular biology and engineering keeps on growing, the efficiency of the cloning and production process will increase in parallel.

SELF-ASSESSMENT QUESTIONS

■ Questions

1. A researcher wanted to clone and subsequently express the human histone H4 protein in *E. coli*.

   She obtained the sequence below from the NCBI, as shown below. The start and stop codons are underlined.

   >gi|29553982|ref|NM_003548.2| Homo sapiens histone cluster 2, H4a (HIST2H4A), mRNA

AGAAGCTGTCTATCGGGCTCCAGCGGTCATGTCCG
   GCAGAGGAAAGGGCGGAAAAGGCTTAGGCAA
   AGGG
GGCGCTAAGCGCCACCGCAAGGTCTTGAGAGAC
   AACATTCAGGGCATCACCAAGCCTGCCATTCG
   GCGTC

TAGCTCGGCGTGGCGGCGTTAAGCGGATCTCTGG
CCTCATTTACGAGGAGACCCGCGGTGTGCTGA
AGGT
GTTCCTGGAGAATGTGATTCGGGACGCAGTCACC
TACACCGAGCACGCCAAGCGCAAGACCGTCAC
AGCC
ATGGATGTGGTGTACGCGCTCAAGCGCCAGGGGC
GCACCCTGTACGGCTTCGGAGGCTAGGCCGCC
GCTC
CAGCTTTGCACGTTTCGATCCCAAAGGCCCTTTT
TAGGGCCGACCA.

(i) Is *E. coli* a suitable expression host for the H4 protein?

(ii) Design primers for the amplification of the coding sequence of the H4 protein by PCR.

(iii) To ease purification the researcher decided to add an affinity tag (Trp-Ser-His-Pro-Gln-Phe-Glu-Lys) to the carboxy-terminal end of the H4 protein. PCR was used to clone this tag in frame with the H4 protein. What was the sequence of the primers she probably used?

To answer this question, make use of the table below.

		2nd		Base			
		U	C	A	G		
	U	Phe	Ser	Tyr	Cys	U	
		Phe	Ser	Tyr	Cys	C	
		Leu	Ser	Stop	Stop	A	
1		Leu	Ser	Stop	Trp	G	3
s	C	Leu	Pro	His	Arg	U	r
t		Leu	Pro	His	Arg	C	d
		Leu	Pro	Gln	Arg	A	
b		Leu	Pro	Gln	Arg	G	b
a	A	Ile	Thr	Asn	Ser	U	a
s		Ile	Thr	Asn	Ser	C	s
e		Ile	Thr	Lys	Arg	A	e
		Met	Thr	Lys	Arg	G	
	G	Val	Ala	Asp	Gly	U	
		Val	Ala	Asp	Gly	C	
		Val	Ala	Glu	Gly	A	
		Val	Ala	Glu	Gly	G	

(iv) And finally, she decided to optimize the codon usage for expression in *E. coli*. What is coding optimalization? What is its purpose?

(v) Design a strategy/method to optimize the codon usage of the H4 protein.

(vi) The human H4 mRNA differs from most other human mRNAs by lacking a poly-A tail (instead the H4 mRNA is protected by a palindromic termination element), and thus the cDNA encoding this protein cannot be obtained by a reverse transcriptase reaction using an oligo-dT as primer. Describe a method to obtain the H4 cDNA.

2. Ampicillin, G418, and HAT medium are used to select for transformed *E. coli*, transfected mammalian cells, and hybridomas, respectively. Describe shortly the mechanism underlying the three mentioned selection strategies.

3. *E. coli* does not take up plasmid DNA spontaneously. However, the so-called chemical competent *E. coli* is able to take up plasmids following a heat shock (30 s 42 °C, followed by an immediate transfer to 0 °C). These competent bacteria can be obtained by extensive washing with a 100 mM $CaCl_2$ solution.

Transformation of competent *E. coli* of good quality results in $\pm\ 10^8$ colonies/μg of supercoiled plasmid DNA. The bacteria in each colony are the descendants of one bacterium that had initially taken up one plasmid molecule.

Calculate the transformation efficiency defined as the number of plasmids taken up by the competent bacteria divided by the total number of plasmids added. Make the calculation for a plasmid of 3,333 base pairs (the MW of a nucleotide is 300 g/mol and the Avogadro constant is 6×10^{23} molecules/mol).

Answers

1. (i) Information about the protein structure can be obtained from http://www.expasy.org/. The H4 protein does not contain disulfide bridges and is unglycosylated. It is therefore likely that *E. coli* is able to produce a correctly folded H4 protein.

(ii) PCR primers are usually around 18–20 nucleotides long. The sequences of the forward and reverse primer are *ATG* TCC GGC AGA GGA AAG (identical to the published sequence) and *CTA* GCC TCC GAA GCC GTA (complementary and reverse), respectively.

(iii) The forward primer will be as above. At the 5′ end of the reverse primer, additional sequences must be added. First, the DNA sequence encoding the affinity tag Trp-Ser-His-Pro-Gln-Phe-Glu-Lys must be determined using the codon usage table: TCG CAC CCA CAG TTC GAA AAG. It is important to place the tag in front of the stop codon (TAG). The sequence of the reverse primer will then be 5′- CTA CTT TTC GAA CTG TGG GTG CGA CCA GCC TCC GAA GCC GTA CAG- 3′.

(iv) For most amino acids more than one codon exist (see the codon usage table).
Differences in preferences for one of the several codons that encode the same amino acid exist between organisms. In particular in fast-growing organisms, like *E. coli*, the optimal codons reflect the composition of their transfer RNA (tRNA) pool. By changing the native codons into those codons preferred by *E. coli*, the level of heterologous protein expression may increase. Alternatively, and much easier, one could use as expression host an *E. coli* with plasmids encoding extra copies of rare tRNAs.

(v) The H4 protein is 103 amino acids long. The easiest way to change the sequence at many places along the entire length of the coding sequence/mRNA is by designing four overlapping oligonucleotides.
Next, the four overlapping oligonucleotides must be "sewed" together by a DNA polymerase in the presence of dNTPs. Finally, by the addition of two flanking primers, the entire, now optimized sequence can be amplified.

(vi) An oligo-dT will not bind to the mRNA of H4, and therefore one has to use a H4-specific primer. One could use for instance the reverse primer as designed by question 1.ii

2. (a) *Selection of transformed bacteria using ampicillin.* The antibiotic ampicillin is an inhibitor of transpeptidase. This enzyme is required for the making of the bacterial cell wall. The ampicillin resistance gene encodes for the enzyme beta-lactamase, which degrades ampicillin.
   (b) *Selection of stably transfected mammalian cells using G418.* Most expression plasmids for mammalian cells contain as selection marker the neomycin resistance gene (Neo^r). This gene codes for a protein that neutralizes the toxic drug Geneticin, also known as G418. G418 blocks protein synthesis both in prokaryotic and eukaryotic cells. Only cells that have incorporated the plasmid with the Neo^r gene into their chromosomal DNA will survive.
   (c) *Selection of hybridomas using HAT medium.* HAT medium contains hypoxanthine, aminopterin, and thymidine. The myeloma cell lines used for the production of monoclonal antibodies contain an inactive hypoxanthine-guanine phosphoribosyltransferase (HGPRT), an enzyme necessary for the salvage synthesis of nucleic acids. The lack of HGPRT activity is not a problem for the myeloma cells because they can still synthesize purines de novo. By exposing the myeloma cells to the drug aminopterin also de novo synthesis of purines is blocked and these cells will not survive anymore. Selection against the unfused lymphocytes is not necessary, since these cells, like most primary cells, do not survive for a long time in cell culture.

3. First, calculate the molecular weight of the plasmid: $333 \times 2 \times 300 = 2 \times 10^6$ g/mol. → 2×10^6 g plasmid $= 6 \times 10^{23}$ molecules. → 1 g plasmid $= 3 \times 10^{17}$ molecules. →1 μg (1×10^{-6} g) $= 3 \times 10^{11}$ molecules.
   1 μg gram plasmid results in 10^8 colonies. Thus, only one in 3,000 plasmids is taken up by the bacteria.

RECOMMENDED READING AND REFERENCES

Alberts B, Johnson A, Lewis J, Raff M, Roberts K, Walter P (2007) Molecular biology of the cell. Garland Science, New York

Berg JM, Tymoczko JL, Stryer L (2011) Biochemistry, 7th edn. WH. Freeman & CO., New York

Brekke OH, Sandlie I (2003) Therapeutic antibodies for human diseases at the dawn of the twenty-first century. Nat Rev Drug Discov 2(1):52–62

Wikepedia. Available at: http://en.wikipedia.org

Kashmiri SV, De Pascalis R, Gonzales NR, Schlom J (2005) SDR grafting-a new approach to antibody humanization. Methods 36(1):25–34

Kircher M, Kelso J (2010) High-throughput DNA sequencing–concepts and limitations. Bioessays 32(6):524–536

Leader B, Baca QJ, Golan DE (2008) Protein therapeutics: a summary and pharmacological classification. Nat Rev Drug Discov 7(1):21–39

Lodish H, Berk A, Kaiser CA, Krieger M, Scott MP (2007) Molecular cell biology, 6th edn. WH. Freeman & CO., New York

Nothaft H, Szymanski CM (2010) Protein glycosylation in bacteria: sweeter than ever. Nat Rev Microbiol 8(11):765–778

Paul SM, Mytelka DS, Dunwiddie CT, Persinger CC, Munos BH, Lindborg SR, Schacht AL (2010) How to improve R&D productivity: the pharmaceutical industry's grand challenge. Nat Rev Drug Discov 9(3):203–214

Strohl WR, Knight DM (2009) Discovery and development of biopharmaceuticals: current issues. Curr Opin Biotechnol 20(6):668–672

2

Biophysical and Biochemical Analysis of Recombinant Proteins

Tsutomu Arakawa and John S. Philo

INTRODUCTION

For a recombinant protein to become a human therapeutic, its biophysical and biochemical characteristics must be well understood. These properties serve as a basis for comparison of lot-to-lot reproducibility; for establishing the range of conditions to stabilize the protein during production, storage, and shipping; and for identifying characteristics useful for monitoring stability during long-term storage.

A number of techniques can be used to determine the biophysical properties of proteins and to examine their biochemical and biological integrity. Where possible, the results of these experiments are compared with those obtained using naturally occurring proteins in order to be confident that the recombinant protein has the desired characteristics of the naturally occurring one.

PROTEIN STRUCTURE

■ Primary Structure

Most proteins which are developed for therapy perform specific functions by interacting with other small and large molecules, e.g., cell-surface receptors, binding proteins, nucleic acids, carbohydrates, and lipids. The functional properties of proteins are derived from their folding into distinct three-dimensional structures. Each protein fold is based on its specific polypeptide sequence in which different amino acids are connected through peptide bonds in a specific way. This alignment of the 20 amino acids, called a primary sequence, has in general all the information necessary for folding into a distinct tertiary structure comprising different secondary structures such as α-helices and β-sheets (see below). Because the 20 amino acids possess different side chains, polypeptides with widely diverse properties are obtained.

All of the 20 amino acids consist of a C_α carbon to which an amino group, a carboxyl group, a hydrogen, and a side chain bind in L configuration (Fig. 2.1). These amino acids are joined by condensation to yield a peptide bond consisting of a carboxyl group of an amino acid joined with the amino group of the next amino acid (Fig. 2.2).

The condensation gives an amide group, NH, at the N-terminal side of C_α and a carbonyl group, C=O, at the C-terminal side. These groups, as well as the amino acyl side chains, play important roles in protein folding. Due to their ability to form hydrogen bonds, they make major energetic contributions to the formation of two important secondary structures, α-helix and β-sheet. The peptide bonds between various amino acids are very much equivalent, however, so that they do not determine which part of a sequence should form an α-helix or β-sheet. Sequence-dependent secondary structure formation is determined by the side chains.

The 20 amino acids commonly found in proteins are shown in Fig. 2.3. They are described by their full names and three- and one-letter codes. Their side chains are structurally different in such a way that at neutral pH, aspartic and glutamic acid are negatively charged and lysine and arginine are positively charged. Histidine is positively charged to an extent that depends on the pH. At pH 7.0, on average, about half of the histidine side chains are positively charged. Tyrosine and cysteine are protonated and uncharged at neutral pH, but become negatively charged above pH 10 and 8, respectively.

Polar amino acids consist of serine, threonine, asparagine, and glutamine, as well as cysteine, while nonpolar amino acids consist of alanine, valine, phenylalanine, proline, methionine, leucine, and isoleucine. Glycine behaves neutrally while cystine, the

T. Arakawa, Ph.D. (✉)
Department of Protein Chemistry, Alliance Protein Laboratories, 3957 Corte Cancion, Thousand Oaks, CA 91360, USA
e-mail: tarakawa2@aol.com

J.S. Philo, Ph.D.
Department of Biophysical Chemistry, Alliance Protein Laboratories, 3957 Corte Cancion, Thousand Oaks, CA 91360, USA

D.J.A. Crommelin, R.D. Sindelar, and B. Meibohm (eds.), *Pharmaceutical Biotechnology*,
DOI 10.1007/978-1-4614-6486-0_2,

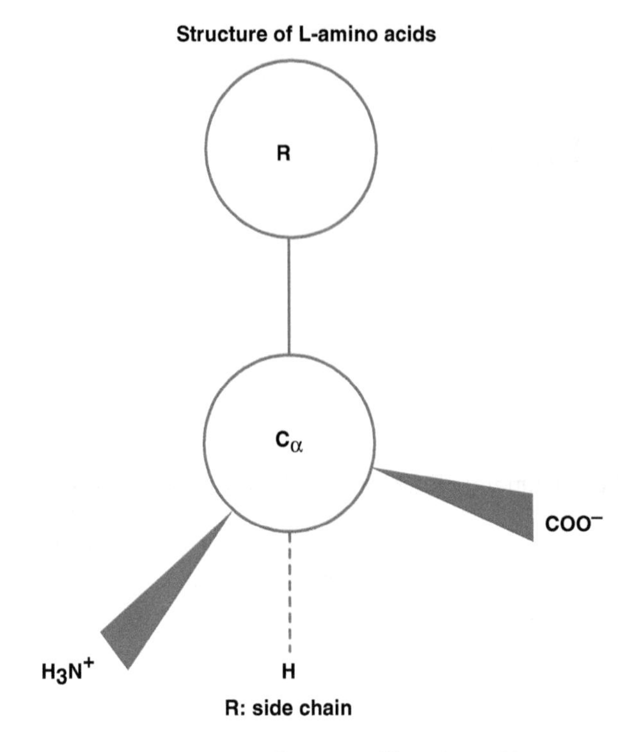

Figure 2.1 ■ Structure of L-amino acids.

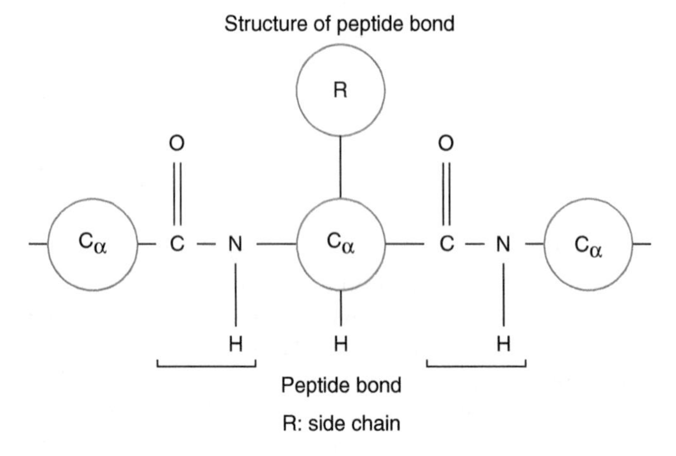

Figure 2.2 ■ Structure of peptide bond.

oxidized form of cysteine, is characterized as hydrophobic. Although tyrosine and tryptophan often enter into polar interactions, they are better characterized as nonpolar, or hydrophobic, as described later.

These 20 amino acids are incorporated into a unique sequence based on the genetic code, as the example in Fig. 2.4 shows. This is an amino acid sequence of granulocyte-colony-stimulating factor (G-CSF), which selectively regulates proliferation and maturation of neutrophils. Although the exact properties of this protein depend on the location of each amino acid and hence the location of each side chain in the three-dimensional structure, the average properties can be estimated simply from the amino acid composition, as shown in Table 2.1, i.e., a list of the total number of each type of amino acid contained in this protein molecule.

Using the pK_a values of these side chains and one amino and carboxyl terminus, one can calculate total charges (positive plus negative charges) and net charges (positive minus negative charges) of a protein as a function of pH, i.e., a titration curve. Since cysteine can be oxidized to form a disulfide bond or can be in a free form, accurate calculation above pH 8 requires knowledge of the status of cysteinyl residues in the protein. The titration curve thus obtained is only an approximation, since some charged residues may be buried and the effective pKa values depend on the local environment of each residue. Nevertheless, the calculated titration curve gives a first approximation of the overall charged state of a protein at a given pH and hence its solution property. Other molecular parameters, such as isoelectric point (pI, where the net charge of a protein becomes zero), molecular weight, extinction coefficient, partial specific volume, and hydrophobicity, can also be estimated from the amino acid composition, as shown in Table 2.1.

The primary structure of a protein, i.e., the sequence of the 20 amino acids, can lead to the three-dimensional structure because the amino acids have diverse physical properties. First, each type of amino acid has the tendency to be more preferentially incorporated into certain secondary structures. The frequencies with which each amino acid is found in α-helix, β-sheet, and β-turn, secondary structures that are discussed later in this chapter, can be calculated as an average over a number of proteins whose three-dimensional structures have been solved. These frequencies are listed in Table 2.2. The β-turn has a distinct configuration consisting of four sequential amino acids and there is a strong preference for specific amino acids in these four positions. For example, asparagine has an overall high frequency of occurrence in a β-turn and is most frequently observed in the first and third position of a β-turn. This characteristic of asparagine is consistent with its side chain being a potential site of N-linked glycosylation. Effects of glycosylation on the biological and physicochemical properties of proteins are extremely important. However, their contribution to structure is not readily predictable based on the amino acid composition.

Based on these frequencies, one can predict for particular polypeptide segments which type of secondary structure they are likely to form. As shown in

Fig. 2.5a, there are a number of methods developed to predict the secondary structure from the primary sequence of the proteins. Using G-CSF (Fig. 2.5b) as an example, regions of α-helix, β-sheets, turns, hydrophilicity, and antigen sites can be suggested.

Another property of amino acids, which impacts on protein folding, is the hydrophobicity of their side chains. Although nonpolar amino acids are basically hydrophobic, it is important to know how hydrophobic they are. This property has been determined by measuring the partition coefficient or solubility of amino acids in water and organic solvents and normalizing such parameters relative to glycine. Relative to the side chain of glycine, a single hydrogen, such

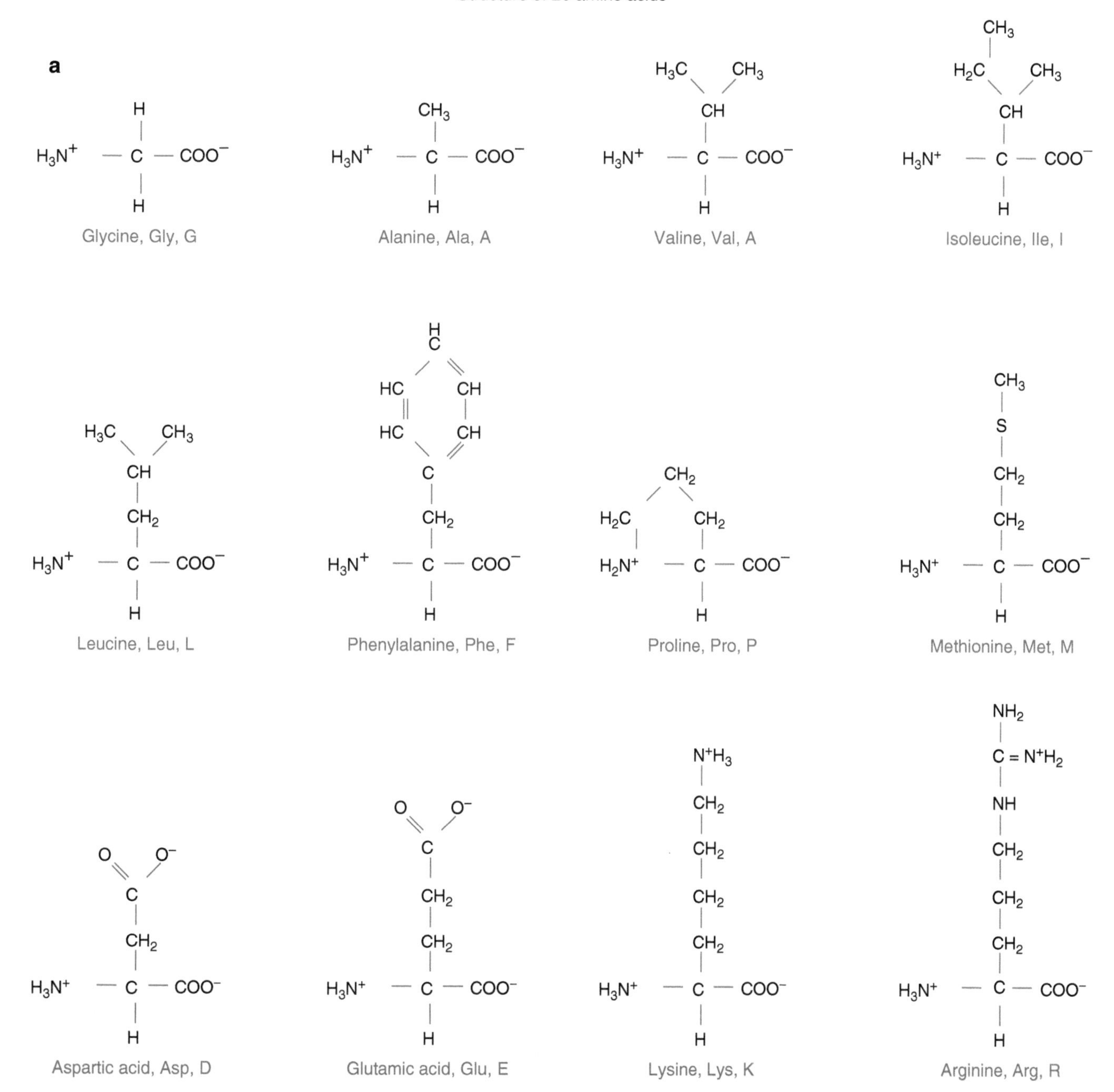

Figure 2.3 ■ NB (**a** and **b**) Structure of 20 amino acids.

Structure of 20 amino acids

b

Serine, Ser, S

Threonine, Thr, T

Tyrosine, Tyr, Y

Histidine, His, H

Cysteine, Cys, C

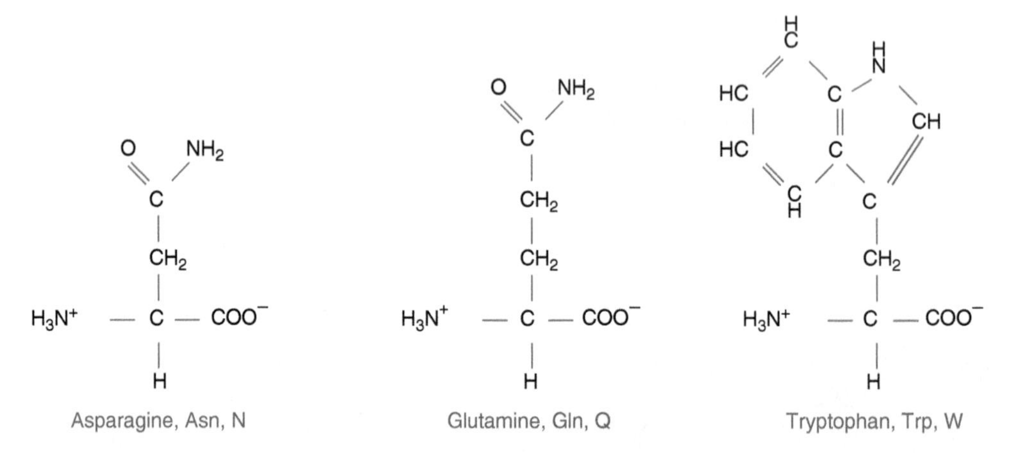

Figure 2.3 ■ (continued)

```
TPLGPASSLPQSFLLKCLEQVRKIQGDGAALQEKLCATYK   40
LCHPEELVLLGHSLGIPWAPLSSCPSQALQLAGCLSQLHS   80
GLFLYQGLLQALEGISPELGPTLDTLQLDVADFATTIWQQ  120
MEELGMAPALQPTQGAMPAFASAFQRRAGGVLVASHLQSF  160
LEVSYRVLRHLAQP
```

Figure 2.4 ■ Amino acid sequence of granulocyte-colony-stimulating factor.

Parameter		Value	
Molecular weight		18,673	
Total number of amino acids		174	
1 µg		53.5 picomoles	
Molar extinction coefficient		15,820	
1 A (280)		1.18 mg/ml	
Isoelectric point		5.86	
Charge at pH 7		–3.39	
Amino acid	**Number**	**% By weight**	**% By frequency**
A Ala	19	7.23	10.92
C Cys	5	2.76	2.87
D Asp	4	2.47	2.30
E Glu	9	6.22	5.17
F Phe	6	4.73	3.45
G Gly	14	4.28	8.05
H His	5	3.67	2.87
1 Me	4	2.42	2.30
K Lys	4	2.75	2.30
L Leu	33	20.00	18.97
M Met	3	2.11	1.72
N Asn	0	0.00	0.00
P Pro	13	6.76	7.47
Q Gln	17	11.66	9.77
R Arg	5	4.18	2.87
S Ser	14	6.53	8.05
T Thr	7	3.79	4.02
V Val	7	3.71	4.02
W Trp	2	1.99	1.15
Y Tyr	3	2.62	1.72

Table 2.1 ■ Amino acid composition and structural parameters of granulocyte-colony-stimulating factor.

normalization shows how strongly the side chains of nonpolar amino acids prefer the organic phase to the aqueous phase. A representation of such measurements is shown in Table 2.3. The values indicate that the free energy increases as the side chain of tryptophan and tyrosine are transferred from an organic solvent to water and that such transfer is thermodynamically unfavorable. Although it is unclear how comparable the hydrophobic property is between an organic solvent and the interior of protein molecules, the hydrophobic side chains favor clustering together, resulting in a core structure with properties similar to an organic solvent. These hydrophobic characteristics of nonpolar amino acids and hydrophilic characteristics of polar amino acids generate a partition of amino acyl residues into a hydrophobic core and hydrophilic surface, resulting in overall folding.

α-Helix		β-Sheet		β-Turn		β-Turn position 1		β-Turn position 2		β-Turn position 3		β-Turn position 4	
Glu	1.51	Val	1.70	Asn	1.56	Asn	0.161	Pro	0.301	Asn	0.191	Trp	0.167
Met	1.45	Lie	1.60	Gly	1.56	Cys	0.149	Ser	0.139	Gly	0.190	Gly	0.152
Ala	1.42	Tyr	1.47	Pro	1.52	Asp	0.147	Lys	0.115	Asp	0.179	Cys	0.128
Leu	1.21	Phe	1.38	Asp	1.46	His	0.140	Asp	0.110	Ser	0.125	Tyr	0.125
Lys	1.16	Trp	1.37	Ser	1.43	Ser	0.120	Thr	0.108	Cys	0.117	Ser	0.106
Phe	1.13	Leu	1.30	Cys	1.19	Pro	0.102	Arg	0.106	Tyr	0.114	Gin	0.098
Gin	1.11	Cys	1.19	Tyr	1.14	Gly	0.102	Gin	0.098	Arg	0.099	Lys	0.095
Trp	1.08	Thr	1.19	Lys	1.01	Thr	0.086	Gly	0.085	His	0.093	Asn	0.091
Ile	1.08	Gin	1.10	Gin	0.98	Tyr	0.082	Asn	0.083	Glu	0.077	Arg	0.085
Val	1.06	Met	1.05	Thr	0.96	Trp	0.077	Met	0.082	Lys	0.072	Asp	0.081
Asp	1.01	Arg	0.93	Trp	0.96	Gin	0.074	Ala	0.076	Tyr	0.065	Thr	0.079
His	1.00	Asn	0.89	Arg	0.95	Arg	0.070	Tyr	0.065	Phe	0.065	Leu	0.070
Arg	0.98	His	0.87	His	0.95	Met	0.068	Glu	0.060	Trp	0.064	Pro	0.068
Thr	0.83	Ala	0.83	Glu	0.74	Val	0.062	Cys	0.053	Gin	0.037	Phe	0.065
Ser	0.77	Ser	0.75	Ala	0.66	Leu	0.061	Val	0.048	Leu	0.036	Glu	0.064
Cys	0.70	Gly	0.75	Met	0.60	Ala	0.060	His	0.047	Ala	0.035	Ala	0.058
Tyr	0.69	Lys	0.74	Phe	0.60	Phe	0.059	Phe	0.041	Pro	0.034	Ile	0.056
Asn	0.67	Pro	0.55	Leu	0.59	Glu	0.056	Ile	0.034	Val	0.028	Met	0.055
Pro	0.57	Asp	0.54	Val	0.50	Lys	0.055	Leu	0.025	Met	0.014	His	0.054
Gly	0.57	Glu	0.37	Ile	0.47	Ile	0.043	Trp	0.013	Ile	0.013	Val	0.053

Taken and edited from Chou PY, Fasman GD (1978) Empirical predictions of protein conformation. Ann Rev Biochem 47: 251–276 with permission from Annual Reviews, Inc.

Table 2.2 ■ Frequency of occurrence of 20 amino acids in α-helix, β-sheet, and β-turn.

■ Secondary Structure

α-Helix

Immediately evident in the primary structure of a protein is that each amino acid is linked by a peptide bond. The amide, NH, is a hydrogen donor and the carbonyl, C=O, is a hydrogen acceptor, and they can form a stable hydrogen bond when they are positioned in an appropriate configuration of the polypeptide chain. Such structures of the polypeptide chain are called secondary structure. Two main structures, α-helix and β-sheet, accommodate such stable hydrogen bonds. The main chain forms a right-handed helix, because only the L-form of amino acids is in proteins and makes one turn per 3.6 residues. The overall length of α-helices can vary widely. Figure 2.6 shows an example of a short α-helix. In this case, the C=O group of residue 1 forms a hydrogen bond to the NH group of residue 5 and C=O group of residue 2 forms a hydrogen bond with the NH group of residue 6. Thus, at the start of an α-helix, four amide groups are always free, and at the end of an α-helix, four carboxyl groups are also free. As a result, both ends of an α-helix are highly polar.

Moreover, all the hydrogen bonds are aligned along the helical axis. Since both peptide NH and C=O groups have electric dipole moments pointing in the same direction, they will add to a substantial dipole moment throughout the entire α-helix, with the negative partial charge at the C-terminal side and the positive partial charge at the N-terminal side.

The side chains project outward from the α-helix. This projection means that all the side chains surround the outer surface of an α-helix and interact both with each other and with side chains of other regions which come in contact with these side chains. These interactions, so-called long-range interactions, can stabilize the α-helical structure and help it to act as a folding unit. Often an α-helix serves as a building block for the three-dimensional structure of globular proteins by bringing hydrophobic side chains to one side of a helix and hydrophilic side chains to the opposite side of the same helix. Distribution of side chains along the α-helical axis can be viewed using the helical wheel. Since one turn in an α-helix is 3.6 residues long, each residue can be plotted every 360°/3.6 = 100° around a circle (viewed from the top of α-helix), as shown in Fig. 2.7. Such a plot shows the projection of the position of the residues onto a plane perpendicular to the helical axis. One of the predicted helices in erythropoietin is shown in Fig. 2.7, using an open circle for hydrophobic side chains and an open rectangle for hydrophilic side chains. It becomes immediately obvious that one side of the α-helix is highly hydrophobic, suggesting that this side forms an internal core, while the other side is relatively hydrophilic and is hence most likely exposed to the surface. Since many biologically important proteins function by interacting with other macromolecules, the information obtained from the helical wheel is extremely useful. For example, mutations of amino acids in the solvent-exposed side may lead to identification of regions responsible for biological activity, while mutations in the internal core may lead to altered protein stability.

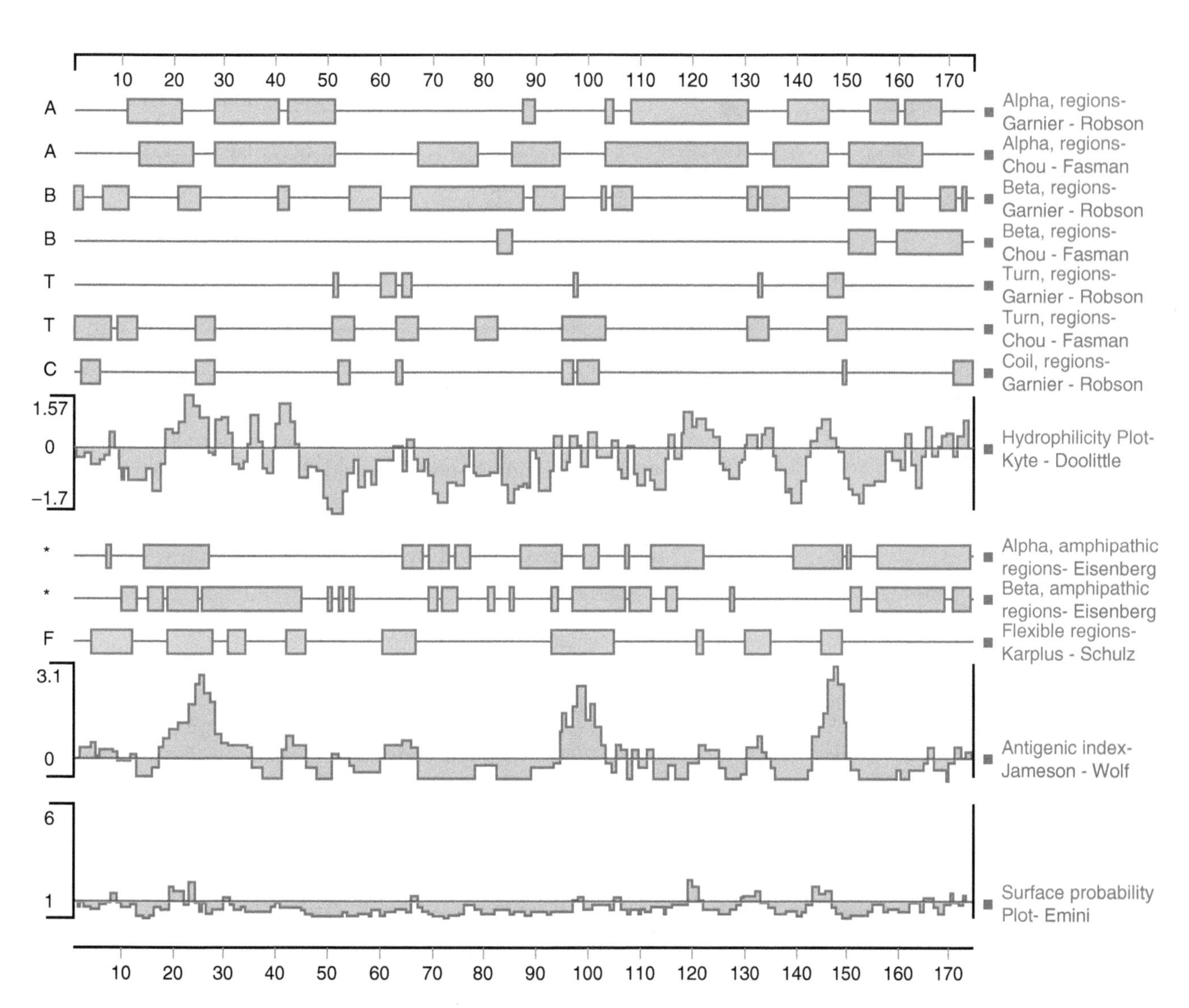

Figure 2.5 (**a**) Predicted secondary structure of granulocyte-colony-stimulating factor. Obtained using a program "DNASTAR" (DNASTAR Inc., Madison, WI). (**b**) Secondary structure of Filgrastim (recombinant G-CSF). Filgrastim is a 175-amino acid polypeptide. Its four antiparallel alpha helices (*A*, *B*, *C*, and *D*) and short 3-to-10 type helix (3_{10}) form a helical bundle. The two biologically active sites (α and α_L) are remote from modifications at the N-terminus of the α-helix and the sugar chain attached to loops *C–D*. Note: Filgrastim is not glycosylated; the sugar chain is included to illustrate its location in endogenous G-CSF.

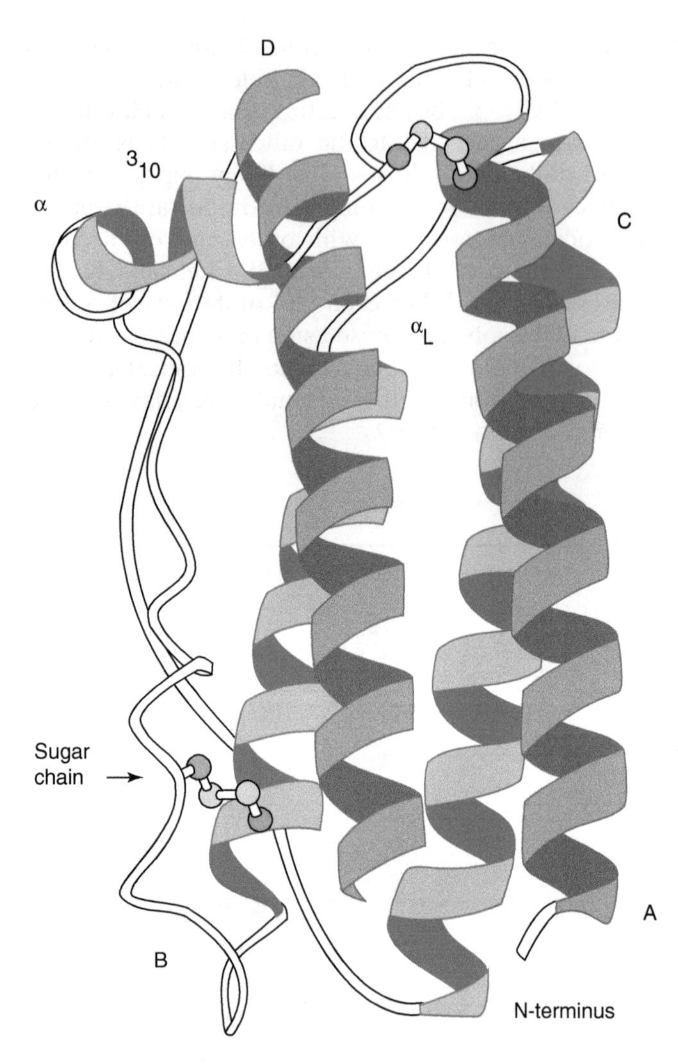

Figure 2.5 ■ (continued)

Amino acid side chain	Cal/mol
Tryptophan	3,400
Norleucine	2,600
Phenylalanine	2,500
Tyrosine	2,300
Dihydroxyphenylalanine	1,800
Leucine	1,800
Valine	1,500
Methionine	1,300
Histidine	500
Alanine	500
Threonine	400
Serine	–300

Taken from Nozaki Y, Tanford C (1971) The solubility of amino acids and two glycine peptides in aqueous ethanol and dioxane solutions. Establishment of a hydrophobicity scale. J Biol Chem 246:2211–2217 with permission from American Society of Biological Chemists

Table 2.3 ■ Hydrophobicity scale: transfer free energies of amino acid side chains from organic solvent to water.

β-Sheet

The second major secondary structural element found in proteins is the β-sheet. In contrast to the α-helix, which is built up from a continuous region with a peptide hydrogen bond linking every fourth amino acid, the β-sheet is comprised of peptide hydrogen bonds between different regions of the polypeptide that may be far apart in sequence. β-strands can interact with each other in one of the two ways shown in Fig. 2.8, i.e., either parallel or antiparallel. In a parallel β-sheet, each strand is oriented in the same direction with peptide hydrogen bonds formed between the strands, while in antiparallel β-sheets, the polypeptide sequences are oriented in the opposite direction. In both structures, the C=O and NH groups project into opposite sides of the polypeptide chain, and hence, a β-strand can interact from either side of that particular chain to form peptide hydrogen bonds with adjacent strands. Thus, more than two β-strands can contact each other either in a parallel or in an antiparallel manner, or even in combination. Such clustering can result in all the β-strands lying in a plane as a sheet. The β-strands which are at the edges of the sheet have unpaired alternating C=O and NH groups.

Side chains project perpendicularly to this plane in opposite directions and can interact with other side chains within the same β-sheet or with other regions of the molecule, or may be exposed to the solvent.

In almost all known protein structures, β-strands are right-handed twisted. This way, the β-strands adapt into widely different conformations. Depending on how they are twisted, all the side chains in the same strand or in different strands do not necessarily project in the same direction.

Loops and Turns

Loops and turns form more or less linear structures and interact with each other to form a folded three-dimensional structure. They are comprised of an amino acid sequence which is usually hydrophilic and exposed to the solvent. These regions consist of β-turns (reverse turns), short hairpin loops, and long loops. Many hairpin loops are formed to connect two antiparallel β-strands.

As shown in Fig. 2.5a, the amino acid sequences which form β-turns are relatively easy to predict, since turns must be present periodically to fold a linear sequence into a globular structure. Amino acids found most frequently in the β-turn are usually not found in α-helical or β-sheet structures. Thus, proline and glycine represent the least-observed amino acids in these typical secondary structures. However, proline has an extremely high frequency of occurrence at the second position in the β-turn, while glycine has a high preference at the third and fourth position of a β-turn.

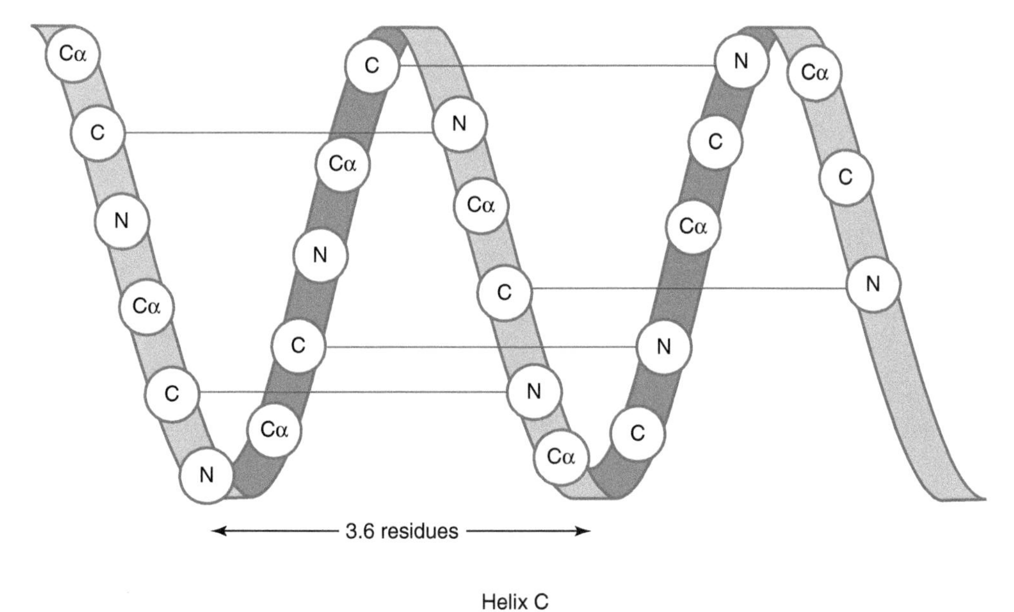

Figure 2.6 ■ Schematic illustration of the structure of α-helix.

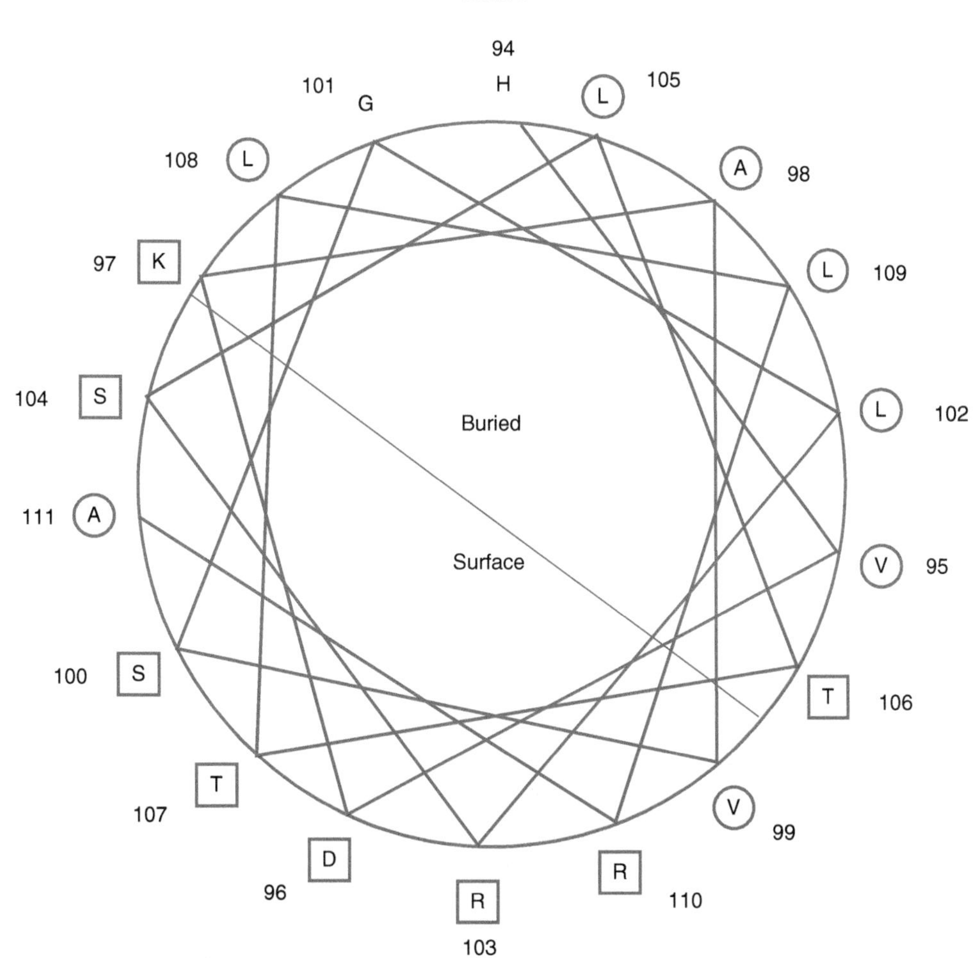

Figure 2.7 ■ Helical wheel analysis of erythropoietin sequence, from His94 to Ala111 (Elliott S, personal communication, 1990).

Although loops are not as predictable as β-turns, amino acids with high frequency for β-turns also can form a long loop. Even though difficult to predict, loops are an important secondary structure, since they form a highly solvent-exposed region of the protein molecules and allow the protein to fold onto itself.

■ Tertiary Structure

Combination of the various secondary structures in a protein results in its three-dimensional structure. Many proteins fold into a fairly compact, globular structure.

The folding of a protein molecule into a distinct three-dimensional structure determines its function.

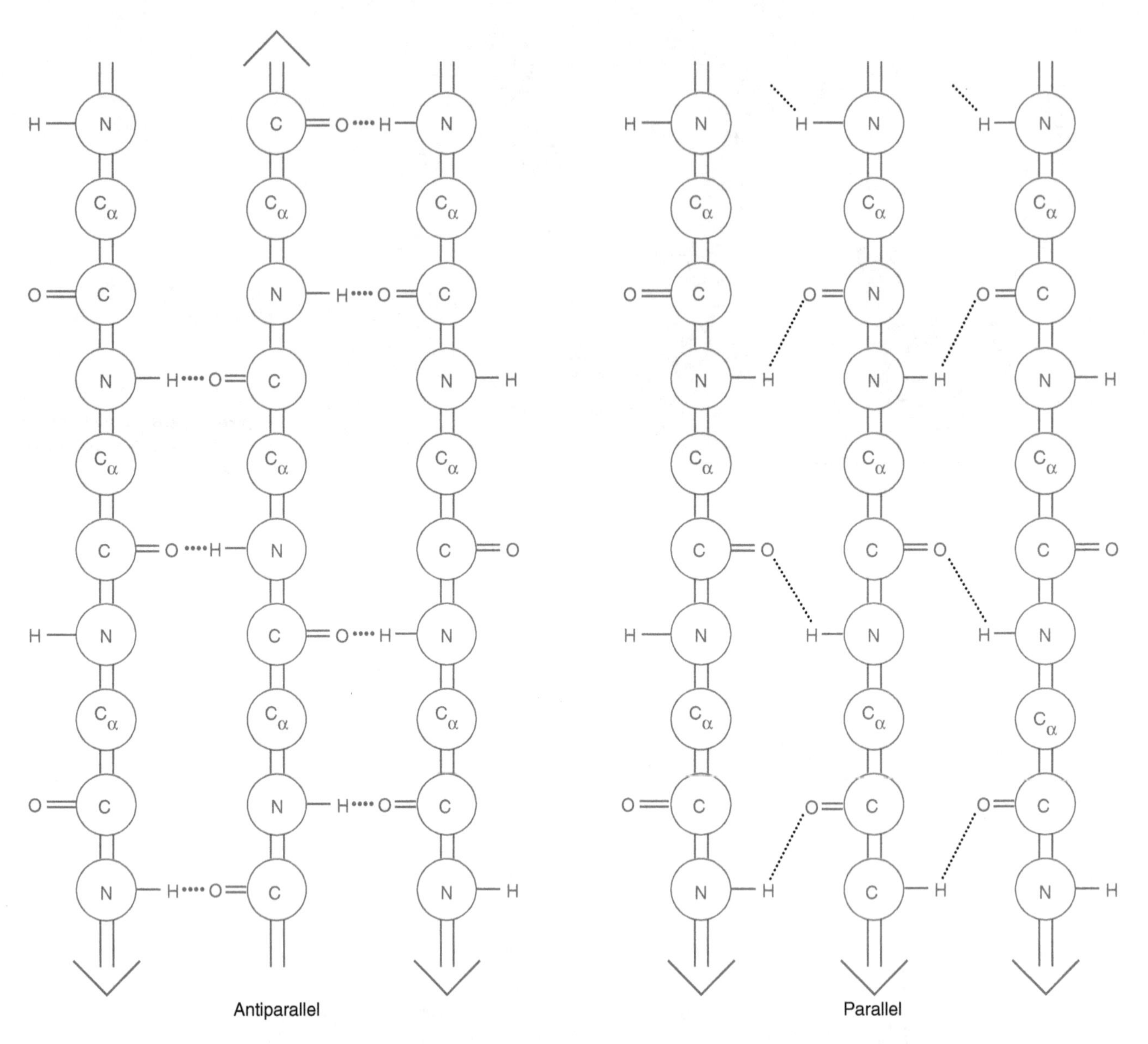

Figure 2.8 ■ Schematic illustration of the structure of antiparallel (*left side*) and parallel (*right side*) β-sheet. *Arrow* indicates the direction of amino acid sequence from the N-terminus to C-terminus.

Enzyme activity requires the exact coordination of catalytically important residues in the three-dimensional space. Binding of antibody to antigen and binding of growth factors and cytokines to their receptors all require a distinct, specific surface for high-affinity binding. These interactions do not occur if the tertiary structures of antibodies, growth factors, and cytokines are altered.

A unique tertiary structure of a protein can often result in the assembly of the protein into a distinct quaternary structure consisting of a fixed stoichiometry of protein chains within the complex. Assembly can occur between the same proteins or between different polypeptide chains. Each molecule in the complex is called a subunit. Actin and tubulin self-associate into F-actin and microtubule, while hemoglobin is a tetramer consisting of two α- and two β-subunits. Among the cytokines and growth factors, interferon-γ is a homodimer, while platelet-derived growth factor is a homodimer of either A or B chains or a heterodimer of the A and B chain. The formation of a quaternary structure occurs via non-covalent interactions or through disulfide bonds between the subunits.

■ Forces

Interactions occurring between chemical groups in proteins are responsible for formation of their specific secondary, tertiary, and quaternary structures. Either repulsive or attractive interactions can occur between different groups. Repulsive interactions consist of steric hindrance and electrostatic effects. Like charges repel each other and bulky side chains, although they do not repel each other, cannot occupy the same space. Folding is also against the natural tendency to move toward randomness, i.e., increasing entropy. Folding leads to a fixed position of each atom and hence a decrease in entropy. For folding to occur, this decrease in entropy, as well as the repulsive interactions, must be overcome by attractive interactions, i.e., hydrophobic interactions, hydrogen bonds, electrostatic attraction, and van der Waals interactions. Hydration of proteins, discussed in the next section, also plays an important role in protein folding.

These interactions are all relatively weak and can be easily broken and formed. Hence, each folded protein structure arises from a fine balance between these repulsive and attractive interactions. The stability of the folded structure is a fundamental concern in developing protein therapeutics.

Hydrophobic Interactions

The hydrophobic interaction reflects a summation of the van der Waals attractive forces among nonpolar groups in the protein interior, which change the surrounding water structure necessary to accommodate these groups if they become exposed. The transfer of nonpolar groups from the interior to the surface requires a large decrease in entropy so that hydrophobic interactions are essentially entropically driven. The resulting large positive free energy change prevents the transfer of nonpolar groups from the largely sheltered interior to the more solvent-exposed exterior of the protein molecule. Thus, nonpolar groups preferentially reside in the protein interior, while the more polar groups are exposed to the surface and surrounding environment. The partitioning of different amino acyl residues between the inside and outside of a protein correlates well with the hydration energy of their side chains, that is, their relative affinity for water.

Hydrogen Bonds

The hydrogen bond is ionic in character since it depends strongly on the sharing of a proton between two electronegative atoms (generally oxygen and nitrogen atoms). Hydrogen bonds may form either between a protein atom and a water molecule or exclusively as protein intramolecular hydrogen bonds. Intramolecular interactions can have significantly more favorable free energies (because of entropic considerations) than intermolecular hydrogen bonds, so the contribution of all hydrogen bonds in the protein molecule to the stability of protein structures can be substantial. In addition, when the hydrogen bonds occur in the interior of protein molecules, the bonds become stronger due to the hydrophobic environment.

Electrostatic Interactions

Electrostatic interactions occur between any two charged groups. According to Coulomb's law, if the charges are of the same sign, the interaction is repulsive with an increase in energy, but if they are opposite in sign, it is attractive, with a lowering of energy. Electrostatic interactions are strongly dependent upon distance, according to Coulomb's law, and inversely related to the dielectric constant of the medium. Electrostatic interactions are much stronger in the interior of the protein molecule because of a lower dielectric constant. The numerous charged groups present on protein molecules can provide overall stability by the electrostatic attraction of opposite charges, for example, between negatively charged carboxyl groups and positively charged amino groups. However, the net effects of all possible pairs of charged groups must be considered. Thus, the free energy derived from electrostatic interactions is actually a property of the whole structure, not just of any single amino acid residue or cluster.

Van der Waals Interactions

Weak van der Waals interactions exist between atoms (except the bare proton), whether they are polar or nonpolar. They arise from net attractive interactions between permanent dipoles and/or induced (temporary and fluctuating) dipoles. However, when two atoms approach each other too closely, the repulsion between their electron clouds becomes strong and counterbalances the attractive forces.

■ Hydration

Water molecules are bound to proteins internally and externally. Some water molecules occasionally occupy small internal cavities in the protein structure and are hydrogen bonded to peptide bonds and side chains of the protein and often to a prosthetic group, or cofactor, within the protein. The protein surface is large and consists of a mosaic of polar and nonpolar amino acids, and it binds a large number of water molecules, i.e., it is hydrated, from the surrounding environment. As described in the previous section, water molecules trapped in the interior of protein molecules are bound more tightly to hydrogen-bonding donors and acceptors because of a lower dielectric constant.

Solvent around the protein surface clearly has a general role in hydrating peptide and side chains but might be expected to be rather mobile and nonspecific in its interactions. Well-ordered water molecules can make significant contributions to protein stability. One water molecule can hydrogen bond to two groups distant in the primary structure on a protein molecule, acting as a bridge between these groups. Such a water molecule may be highly restricted in motion and can contribute to the stability, at least locally, of the protein, since such tight binding may exist only when these groups assume the proper configuration to accommodate a water molecule that is present only in the native state of the protein. Such hydration can also decrease the flexibility of the groups involved.

There is also evidence for solvation over hydrophobic groups on the protein surface. So-called hydrophobic hydration occurs because of the unfavorable nature of the interaction between water molecules and hydrophobic surfaces, resulting in the clustering of water molecules. Since this clustering is energetically unfavorable, such hydrophobic hydration does not contribute to the protein stability. However, this hydrophobic hydration facilitates hydrophobic interaction. This unfavorable hydration is diminished as the various hydrophobic groups come in contact either intramolecularly or intermolecularly, leading to the folding of intrachain structures or to protein-protein interactions.

Both the loosely and strongly bound water molecules can have an important impact, not only on protein stability but also on protein function. For example, certain enzymes function in nonaqueous solvent provided that a small amount of water, just enough to cover the protein surface, is present. Bound water can modulate the dynamics of surface groups, and such dynamics may be critical for enzyme function. Dried enzymes are, in general, inactive and become active after they absorb 0.2 g water per g protein. This amount of water is only sufficient to cover surface polar groups, yet may give sufficient flexibility for function.

Evidence that water bound to protein molecules has a different property from bulk water can be demonstrated by the presence of non-freezable water. Thus, when a protein solution is cooled below −40 °C, a fraction of water, ~0.3 g water/g protein, does not freeze and can be detected by high-resolution NMR. Several other techniques also detect a similar amount of bound water. This unfreezable water reflects the unique property of bound water that prevents it from adopting an ice structure.

Proteins are immersed under physiological conditions or in test tubes in aqueous solutions containing not only water but also other solution components, e.g., salts, metals, amino acids, sugars, and many other minor components. These components also interact with the protein surface and affect protein folding and stability. For examples, sugars and amino acids are known to enhance folding and stability of the proteins, as described below.

PROTEIN FOLDING

Proteins become functional only when they assume a distinct tertiary structure. Many physiologically and therapeutically important proteins present their surface for recognition by interacting with molecules such as substrates, receptors, signaling proteins, and cell-surface adhesion macromolecules. When recombinant proteins are produced in *Escherichia coli*, they often form inclusion bodies into which they are deposited as insoluble proteins. Formation of such insoluble states does not naturally occur in cells where they are normally synthesized and transported. Therefore, an in vitro process is required to refold insoluble recombinant proteins into their native, physiologically active state. This is usually accomplished by solubilizing the insoluble proteins with detergents or denaturants, followed by the purification and removal of these reagents concurrent with refolding the proteins (see Chap. 3).

Unfolded states of proteins are usually highly stable and soluble in the presence of denaturing agents. Once the proteins are folded correctly, they are also relatively stable. During the transition from the unfolded form to the native state, the protein must go through a multitude of other transition states in which it is not fully folded, and denaturants or solubilizing agents are at low concentrations or even absent.

The refolding of proteins can be achieved in various ways. The dilution of proteins at high denaturant concentration into aqueous buffer will decrease both denaturant and protein concentration simultaneously. The addition of an aqueous buffer to a protein-denaturant solution also causes a decrease in concentrations of both denaturant and protein. The difference in these procedures is that, in the first case, both denaturant and protein concentrations are the lowest at the beginning of dilution and gradually increase as the process continues. In the second case, both denaturant and protein concentrations are highest at the beginning of dilution and gradually decrease as the dilution proceeds. Dialysis or the diafiltration of protein in the denaturant against an aqueous buffer resembles the second case, since the denaturant concentration decreases as the procedure continues. In this case, however, the protein concentration remains unchanged. Refolding can also be achieved by first binding the protein in denaturants to a solid phase, i.e., to a column matrix, and then equilibrating it with an aqueous buffer. In this case, protein concentrations are not well defined. Each procedure has advantages and disadvantages and may be applicable for one protein, but not to another.

If proteins in the native state have disulfide bonds, cysteines must be correctly oxidized. Such oxidation may be done in various ways, e.g., air oxidation, glutathione-catalyzed disulfide exchange, or mixed-disulfide formation followed by reduction and oxidation or by disulfide reshuffling.

Protein folding has been a topic of intensive research since Anfinsen's demonstration that ribonuclease can be refolded from the fully reduced and denatured state in in vitro experiments. This folding can be achieved only if the amino acid sequence itself contains all information necessary for folding into the native structure. This is the case, at least partially, for many proteins. However, a lot of other proteins do not refold in a simple one-step process. Rather, they refold via various intermediates which are relatively compact and possess varying degrees of secondary structures, but which lack a rigid tertiary structure. Intrachain interactions of these preformed secondary structures eventually lead to the native state. However, the absence of a rigid structure in these preformed secondary structures can also expose a cluster of hydrophobic groups to those of other polypeptide chains, rather than to their own polypeptide segments, resulting in intermolecular aggregation. High efficiency in the recovery of native protein depends to a large extent on how this aggregation of intermediate forms is minimized. The use of chaperones or polyethylene glycol has been found quite effective for this purpose. The former are proteins, which aid in the proper folding of other proteins by stabilizing intermediates in the folding process and the latter serves to solvate the protein during folding and diminishes interchain aggregation events.

Protein folding is often facilitated by cosolvents, such as polyethylene glycol. As described above, proteins are functional and highly hydrated in aqueous solutions. True physiological solutions, however, contain not only water but also various ions and low- and high-molecular-weight solutes, often at very high concentrations. These ions and other solutes play a critical role in maintaining the functional structure of the proteins. When isolated from their natural environment, the protein molecules may lose these stabilizing factors and hence must be stabilized by certain compounds, often at high concentrations. These solutes are also used in vitro to assist in protein folding and to help stabilize proteins during large-scale purification and production as well as for long-term storage. Such solutes are often called cosolvents when used at high concentrations, since at such high concentrations they also serve as a solvent along with water molecules. These solutes encompass sugars, amino acids, inorganic and organic salts, and polyols. They may not strongly bind to proteins, but instead typically interact weakly with the protein surface to provide significant stabilizing energy without interfering with their functional structure.

When recombinant proteins are expressed in eukaryotic cells and secreted into media, the proteins are generally folded into the native conformation. If the proteins have sites for N-linked or O-linked glycosylation, they undergo varying degrees of glycosylation depending on the host cells used and level of expression. For many glycoproteins, glycosylation is not essential for folding, since they can be refolded into the native conformation without carbohydrates, nor is glycosylation often necessary for receptor binding and hence biological activity. However, glycosylation can alter important biological and physicochemical properties of proteins, such as pharmacokinetics, solubility, and stability.

■ Techniques Specifically Suitable for Characterizing Protein Folding

Conventional spectroscopic techniques used to obtain information on the folded structure of proteins are circular dichroism (CD), fluorescence, and Fourier transform infrared spectroscopies (FTIR). CD and FTIR are widely used to estimate the secondary structure of proteins. The α-helical content of a protein can be readily estimated by CD in the far-UV region (180–260 nm) and by FTIR. FTIR signals from loop structures, however, occasionally overlap with those arising from an α-helix. The β-sheet gives weak CD signals, which are variable in peak positions and intensities due to twists of interacting β-strands, making far-UV CD unreliable for evaluation of these structures. On the other hand, FTIR can reliably estimate the β-structure content as well as distinguish between parallel and antiparallel forms.

CD in the near-UV region (250–340 nm) reflects the environment of aromatic amino acids, i.e., tryptophan, tyrosine, and phenylalanine, as well as that of disulfide structures. Fluorescence spectroscopy yields information on the environment of tyrosine and tryptophan residues. CD and fluorescence signals in many cases are drastically altered upon refolding and hence can be used to follow the formation of the tertiary structure of a protein.

None of these techniques can give the folded structure at the atomic level, i.e., they give no information on the exact location of each amino acyl residue in the three-dimensional structure of the protein. This information can only be determined by X-ray crystallography or NMR. However, CD, FTIR, and fluorescence spectroscopic methods are fast and require lower protein concentrations than either NMR or X-ray crystallography and are amenable for the examination of the protein under widely different conditions. When a naturally occurring form of the protein is available, these techniques, in particular near-UV CD and fluorescence spectroscopies, can quickly address whether the refolded protein assumes the native folded structure.

Temperature dependence of these spectroscopic properties also provides information about protein folding. Since the folded structures of proteins are built upon cooperative interactions of many side chains and peptide bonds in a protein molecule, elimination of one interaction by heat can cause cooperative elimination of other interactions, leading to the unfolding of protein molecules. Thus, many proteins undergo a cooperative thermal transition over a narrow temperature range. Conversely, if the proteins are not fully folded, they may undergo noncooperative thermal transitions as observed by a gradual signal change over a wider range of temperature.

Such a cooperative structure transition can also be examined by differential scanning calorimetry. When the structure unfolds, it requires heat. Such heat absorption can be determined using this highly sensitive calorimetry technique.

Hydrodynamic properties of proteins change greatly upon folding, going from elongated and expanded structures to compact globular ones. Sedimentation velocity and size-exclusion chromatography (see section "Analytical Techniques") are two frequently used techniques for the evaluation of hydrodynamic properties, although the latter is much more accessible. The sedimentation coefficient (how fast a molecule migrates in a centrifugal field) is a function of both the molecular weight and hydrodynamic size of the proteins, while elution position in size-exclusion chromatography (how fast it migrates through pores) depends only on the hydrodynamic size (see Chap. 3). In both methods, comparison of the sedimentation coefficient or elution position with that of a globular protein with an identical molecular weight (or upon appropriate molecular-weight normalization) gives information on how compactly the protein is folded.

For oligomeric proteins, the determination of molecular weight of the associated states and acquisition of the quaternary structure can be used to assess the folded structure. For strong interactions, specific protein association requires that intersubunit contact surfaces perfectly match each other. Such an associated structure, if obtained by covalent bonding, may be determined simply by sodium dodecyl sulfate-polyacrylamide gel electrophoresis. If protein association involves non-covalent interactions, sedimentation equilibrium or light scattering experiments can assess this phenomenon. Although these techniques have been used for many decades with some difficulty, emerging technologies in analytical ultracentrifugation and laser light scattering, and appropriate software for analyzing the results, have greatly facilitated their general use, as described in detail below.

Two fundamentally different light scattering techniques can be used in characterizing recombinant proteins. "Static" light scattering measures the intensity of the scattered light. "Dynamic" light scattering measures the fluctuations in the scattered light intensity as molecules diffuse in and out of a very small scattering region (Brownian motion).

Static light scattering is often used online in conjunction with size-exclusion chromatography (SEC). The scattering signal is proportional to the product of molecular mass times weight concentration. Dividing this signal by one proportional to the concentration, such as obtained from an UV absorbance or refractive index detector, then gives a direct and absolute measure of the mass of each peak eluting from the column, independent of molecular conformation and elution position. This SEC-static scattering combination allows rapid identification of whether the native state of a protein is a monomer or an oligomer and the stoichiometry of multi-protein complexes. It is also very useful in identifying the mass of aggregates which may be present and thus is useful for evaluating protein stability.

Dynamic light scattering (DLS) measures the diffusion rate of the molecules, which can be translated into the Stokes radius, a measure of hydrodynamic size. Although the Stokes radius is strongly correlated with molecular mass, it is also strongly influenced by molecular shape (conformation), and thus, DLS is far less accurate than static scattering for measuring molecular mass. The great strength of DLS is its ability to cover a very wide size range in one measurement and to detect very small amounts of large aggregates (<0.01 % by weight). Other important advantages over static scattering with SEC are a wide choice of buffer conditions and no potential loss of species through sticking to a column.

An analytical ultracentrifuge incorporates an optical system and special rotors and cells in a high-speed centrifuge to permit measurement of the concentration of a sample versus position within a spinning centrifuge cell. There are two primary strategies: analyzing either the sedimentation velocity or the sedimentation equilibrium. When analyzing the sedimentation velocity, the rotor is spun at very high speed, so the protein sample will completely sediment and form a pellet. The rate at which the protein pellets is measured by the optical system to derive the sedimentation coefficient, which depends on both mass and molecular conformation. When more than one species is present (e.g., a monomer plus a covalent dimer degradation product), a separation is achieved based on the relative sedimentation coefficient of each species.

Proteins form not only small oligomers that can be measured by the above techniques but also much larger aggregates, called subvisible and visible particles, which are present in minute quantities. As their size approaches the size of virus, they become highly immunogenic (cf. Chap. 6) and hence determination of their size and amount becomes critical for developing pharmaceutical protein products. Such determination requires imaging of the particles, as the hydrodynamic techniques such as dynamic light scattering and sedimentation velocity have neither sensitivity nor resolution for such large aggregates heterogeneous in size distribution. Normally such particles are present in minute quantity, but yet cause serious immunogenic responses due to their large size.

Because the sedimentation coefficient is sensitive to molecular conformation and can be measured with high precision (~0.5 %), sedimentation velocity can detect even fairly subtle differences in conformation. This ability can be used, for example, to confirm that a recombinant protein has the same conformation as the natural wild-type protein or to detect small changes in structure with changes in the pH or salt concentration that may be too subtle to detect by other techniques, such as CD or differential scanning calorimetry.

In sedimentation equilibrium, a much lower rotor speed and milder centrifugal force is used than for sedimentation velocity. The protein still accumulates toward the outside of the rotor, but no pellet is formed. This concentration gradient across the cell is continuously opposed by diffusion, which tries to restore a uniform concentration. After spinning for a long time (usually 12–36 h), an equilibrium is reached where sedimentation and diffusion are balanced and the distribution of protein no longer changes with time. At sedimentation equilibrium, the concentration distribution depends only on the molecular mass and is independent of molecular shape. Thus, self-association for the formation of dimers or higher oligomers (whether reversible or irreversible) is readily detected, as are binding interactions between different proteins. For reversible association, it is possible to determine the strength of the binding interaction by measuring samples over a wide range of protein concentrations.

In biotechnology applications, sedimentation equilibrium is often used as the "gold standard" for confirming that a recombinant protein has the expected molecular mass and biologically active state of oligomerization in solution. It can also be used to determine the average amount of glycosylation or conjugation of moieties such as polyethylene glycol. The measurement of binding affinities for receptor-cytokine, antigen-antibody, or other interaction can also sometimes serve as a functional characterization of recombinant proteins (although some of these interactions are too strong to be measured by this method).

Site-specific chemical modification and proteolytic digestion are also powerful techniques for studying the folding of proteins. The extent of chemical modification or proteolytic digestion depends on whether the specific sites are exposed to the solvent or are buried in the interior of the protein molecules and are thus inaccessible to these modifications. For example, trypsin cleaves peptide bonds on the C-terminal side of basic residues. Although most proteins contain several basic residues, brief exposure of the native protein to trypsin usually generates only a few peptides, as cleavage occurs only at the accessible basic residues, whereas the same treatment can generate many more peptides when done on the denatured (unfolded) protein, since all the basic residues are now accessible (see also peptide mapping in section "Mass Spectrometry").

PROTEIN STABILITY

Although freshly isolated proteins may be folded into a distinct three-dimensional structure, this folded structure is not necessarily retained indefinitely in aqueous solution. The reason is that proteins are neither chemically nor physically stable. The protein surface is chemically highly heterogeneous and contains reactive groups. Long-term exposure of these groups to environmental stresses causes various chemical alterations. Many proteins, including growth factors and cytokines, have cysteine residues. If some of them are in a free or sulfhydryl form, they may undergo oxidation and disulfide exchange. Oxidation can also occur on methionyl residues. Hydrolysis can occur on peptide bonds and on amides of asparagine and glutamine residues. Other chemical modifications can occur on peptide bonds, tryptophan, tyrosine, and amino and carboxyl groups. Table 2.4 lists both a number of reactions that can occur during purification and storage of proteins and methods that can be used to detect such changes.

Physical stability of a protein is expressed as the difference in free energy, ΔG_U, between the native and denatured states. Thus, protein molecules are in equilibrium between the above two states. As long as this unfolding is reversible and ΔG_U is positive, it does not matter how small the ΔG_U is. In many cases, this reversibility does not hold. This is often seen when ΔG_U is decreased by heating. Most proteins denature upon heating and subsequent aggregation of the denatured molecules results in irreversible denaturation. Thus, unfolding is made irreversible by aggregation:

$$\text{Native state} \overset{\Delta G_U}{\Leftrightarrow} \text{Denatured state} \overset{k}{\Rightarrow} \text{Aggregated state}$$

	Physical property effected	Method of analysis
Oxidation	Hydrophobicity size	RP-HPLC, SDS-PAGE, size-exclusion chromatography, and mass spectrometry
Cys	Hydrophobicity	
Disulfide		
Intrachain		
Interchain		
Met, Trp, Tyr		
Peptide bond hydrolysis	Size	Size-exclusion chromatography SDS-PAGE
N to O migration	Hydrophobicity	RP-HPLC inactive in Edman reaction
Ser, Thr	Chemistry	
α-carboxy to β-carboxy migration	Hydrophobicity	RP-HPLC inactive in Edman reaction
Asp, Asn	Chemistry	
Deamidation	Charge	Ion-exchange chromatography
Asn, Gln		
Acylation	Charge	Ion-exchange chromatography Mass spectrometry
α-amino group, ε-amino group		
Esterification/carboxylation	Charge	Ion-exchange chromatography Mass spectrometry
Glu, Asp, C-terminal		
Secondary structure changes	Hydrophobicity	RP-HPLC
	Size	Size-exclusion chromatography
	Sec/tert structure	CD
	Sec/tert structure	FTIR
	Aggregation	Light scattering
	Sec/tert structure, Aggregation	Analytical ultracentrifugation

Table 2.4 ■ Common reactions affecting stability of proteins.

Therefore, any stress that decreases ΔG_U and increases k will cause the accumulation of irreversibly inactivated forms of the protein. Such stresses may include chemical modifications as described above and physical parameters, such as pH, ionic strength, protein concentration, and temperature. Development of a suitable formulation that prolongs the shelf life of a recombinant protein is essential when it is to be used as a human therapeutic.

The use of protein stabilizing agents to enhance storage stability of proteins has become customary. These compounds affect protein stability by increasing ΔG_U. These compounds, however, may also increase k and hence their net effect on long-term storage of proteins may vary among proteins, as well as on the storage conditions.

When unfolding is irreversible due to aggregation, minimizing the irreversible step should increase the stability, and often, this may be attained by the addition of mild detergents. Prior to selecting the proper detergent concentration and type, however, their effects on ΔG_U must be carefully evaluated.

Another approach for enhancing storage stability of proteins is to lyophilize, or freeze-dry, the proteins (see Chap. 4). Lyophilization can minimize the aggregation step during storage, since both chemical modification and aggregation are reduced in the absence of water. The effects of a lyophilization process itself on ΔG_U and k are not fully understood and hence such a process must be optimized for each protein therapeutic.

ANALYTICAL TECHNIQUES

In one of the previous sections on "Techniques Specifically Suitable for Characterizing Folding," a number of (spectroscopic) techniques were mentioned that can be specifically used to monitor protein folding. These were CD, FTIR, fluorescence spectroscopy, and DSC. Moreover, analytical ultracentrifugation and light scattering techniques were discussed in more detail. In this section, other techniques will be discussed.

■ Blotting Techniques

Blotting methods form an important niche in biotechnology. They are used to detect very low levels of unique molecules in a milieu of proteins, nucleic acids, and other cellular components. They can detect aggregates or breakdown products occurring during long-term storage and they can be used to detect components from the host cells used in producing recombinant proteins.

Biomolecules are transferred to a membrane ("blotting"), and this membrane is then probed with specific reagents to identify the molecule of interest. Membranes used in protein blots are made of a variety of materials including nitrocellulose, nylon, and polyvinylidene difluoride (PVDF), all of which avidly bind protein.

Liquid samples can be analyzed by methods called dot blots or slot blots. A solution containing the biomolecule of interest is filtered through a membrane which captures the biomolecule. The difference between a dot blot and a slot blot is that the former uses a circular or disk format, while the latter is a rectangular configuration. The latter method allows for a more precise quantification of the desired biomolecule by scanning methods and relating the integrated results to that obtained with known amounts of material.

Often, the sample is subjected to some type of fractionation, such as polyacrylamide gel electrophoresis, prior to the blotting step. An early technique, Southern blotting, named after the discoverer, E.M. Southern, is used to detect DNA fragments. When this procedure was adapted to RNA fragments and to proteins, other compass coordinates were chosen as labels for these procedures, i.e., northern blots for RNA and western blots for proteins. Western blots involve the use of labeled antibodies to detect specific proteins.

Transfer of Proteins

Following polyacrylamide gel electrophoresis, the transfer of proteins from the gel to the membrane can be accomplished in a number of ways. Originally, blotting was achieved by capillary action. In this commonly used method, the membrane is placed between the gel and absorbent paper. Fluid from the gel is drawn toward the absorbent paper and the protein is captured by the intervening membrane. A blot, or impression, of the protein within the gel is thus made.

The transfer of proteins to the membrane can occur under the influence of an electric field, as well. The electric field is applied perpendicularly to the original field used in separation so that the maximum distance the protein needs to migrate is only the thickness of the gel, and hence, the transfer of proteins can occur very rapidly. This latter method is called electroblotting.

Detection Systems

Once the transfer has occurred, the next step is to identify the presence of the desired protein. In addition to various colorimetric staining methods, the blots can be probed with reagents specific for certain proteins, as for example, antibodies to a protein of interest. This technique is called immunoblotting. In the biotechnology field, immunoblotting is used as an identity test for the product of interest. An antibody that recognizes the desired protein is used in this instance. Secondly, immunoblotting is sometimes used to show the absence of host proteins. In this instance, the antibodies are raised against proteins of the organism in which the recombinant protein has been expressed. This latter method can attest to the purity of the desired protein.

1. Transfer protein to membrane, e.g., by electroblotting
2. Block residual protein binding sites on membrane with extraneous proteins such as milk proteins
3. Treat membrane with antibody which recognizes the protein of interest. If this antibody is labeled with a detecting group, then go to step 5
4. Incubate membrane with secondary antibody which recognizes primary antibody used in step 3. This antibody is labeled with a detecting group
5. Treat the membrane with suitable reagents to locate the site of membrane attachment of the labeled antibody in step 4 or step 5

Table 2.5 ■ Major steps in blotting proteins to membranes.

1. Antibodies are labeled with radioactive markers such as ^{125}I
2. Antibodies are linked to an enzyme such as horseradish peroxidase (HRP) or alkaline phosphatase (AP). On incubation with substrate, an insoluble colored product is formed at the location of the antibody. Alternatively, the location of the antibody can be detected using a substrate which yields a chemiluminescent product, an image of which is made on photographic film
3. Antibody is labeled with biotin. Streptavidin or avidin is added to strongly bind to the biotin. Each streptavidin molecule has four binding sites. The remaining binding sites can combine with other biotin molecules which are covalently linked to HRP or to AP

Table 2.6 ■ Detection methods used in blotting techniques.

Table 2.5 lists major steps needed for the blotting procedure to be successful. Once the transfer of proteins is completed, residual protein binding sites on the membrane need to be blocked so that antibodies used for detection react only at the location of the target molecule, or antigen, and not at some nonspecific location. After blocking, the specific antibody is incubated with the membrane.

The antibody reacts with a specific protein on the membrane only at the location of that protein because of its specific interaction with its antigen. When immunoblotting techniques are used, methods are still needed to recognize the location of the interaction of the antibody with its specific protein. A number of procedures can be used to detect this complex (see Table 2.6).

The antibody itself can be labeled with a radioactive marker such as ^{125}I and placed in direct contact with X-ray film. After exposure of the membrane to

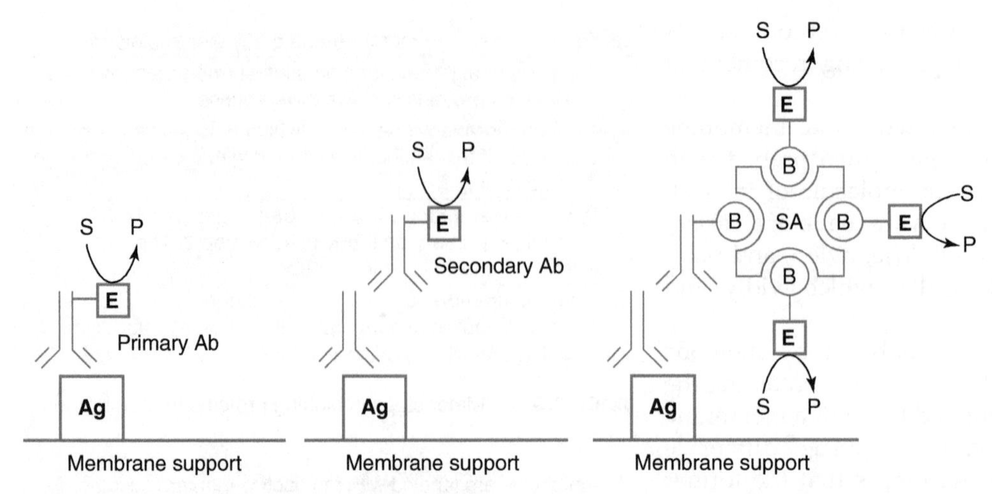

Figure 2.9 ■ Common immunoblotting detection systems used to detect antigens, Ag, on membranes. Abbreviations used: *Ab* antibody, *E* enzyme, such as horseradish peroxidase or alkaline phosphatase, *S* substrate, *P* product, either colored and insoluble or chemiluminescent, *B* biotin, *Sa* streptavidin.

the film for a suitable period, the film is developed and a photographic negative is made of the location of radioactivity on the membrane. Alternatively, the antibody can be linked to an enzyme which, upon the addition of appropriate reagents, catalyzes a color or light reaction at the site of the antibody. These procedures entail purification of the antibody and specifically label it. More often, "secondary" antibodies are used. The primary antibody is the one which recognizes the protein of interest. The secondary antibody is then an antibody that specifically recognizes the primary antibody. Quite commonly, the primary antibody is raised in rabbits. The secondary antibody may then be an antibody raised in another animal, such as goat, which recognizes rabbit antibodies. Since this secondary antibody recognizes rabbit antibodies in general, it can be used as a generic reagent to detect rabbit antibodies in a number of different proteins of interest that have been raised in rabbits. Thus, the primary antibody specifically recognizes and complexes a unique protein, and the secondary antibody, suitably labeled, is used for detection (see also section "ELISA" and Fig. 2.10).

The secondary antibody can be labeled with a radioactive or enzymatic marker group and used to detect several different primary antibodies. Thus, rather than purifying a number of different primary antibodies, only one secondary antibody needs to be purified and labeled for recognition of all the primary antibodies. Because of their wide use, many common secondary antibodies are commercially available in kits containing the detection system and follow routine, straightforward procedures.

In addition to antibodies raised against the amino acyl constituents of proteins, specific antibodies can be used which recognize unique posttranslational components in proteins, such as phosphotyrosyl residues, which are important during signal transduction, and carbohydrate moieties of glycoproteins.

Figure 2.9 illustrates a number of detection methods that can be used on immunoblots. The primary antibody, or if convenient, the secondary antibody, can have an appropriate label for detection. They may be labeled with a radioactive tag as mentioned previously. Secondly, these antibodies can be coupled with an enzyme such as horseradish peroxidase (HRP) or alkaline phosphatase (AP). Substrate is added and is converted to an insoluble, colored product at the site of the protein-primary antibody-secondary antibody-HRP product. An alternative substrate can be used which yields a chemiluminescent product. A chemical reaction leads to the production of light which can expose photographic or X-ray film. The chromogenic and chemiluminescent detection systems have comparable sensitivities to radioactive methods. The former detection methods are displacing the latter method, since problems associated with handling radioactive material and radioactive waste solutions are eliminated.

As illustrated in Fig. 2.9, streptavidin, or alternatively avidin, and biotin can play an important role in detecting proteins on immunoblots. This is because biotin forms very tight complexes with streptavidin and avidin. Secondly, these proteins are multimeric and contain four binding sites for biotin. When biotin is covalently linked to proteins such as antibodies and enzymes, streptavidin binds to the covalently bound biotin, thus recognizing the site on the membrane where the protein of interest is located.

■ Immunoassays

ELISA

Enzyme-linked immunosorbent assay (ELISA) provides a means to quantitatively measure extremely small amounts of proteins in biological fluids and serves as a tool for analyzing specific proteins during purification. This procedure takes advantage of the observation that plastic surfaces are able to adsorb low but detectable amounts of proteins. This is a solid-

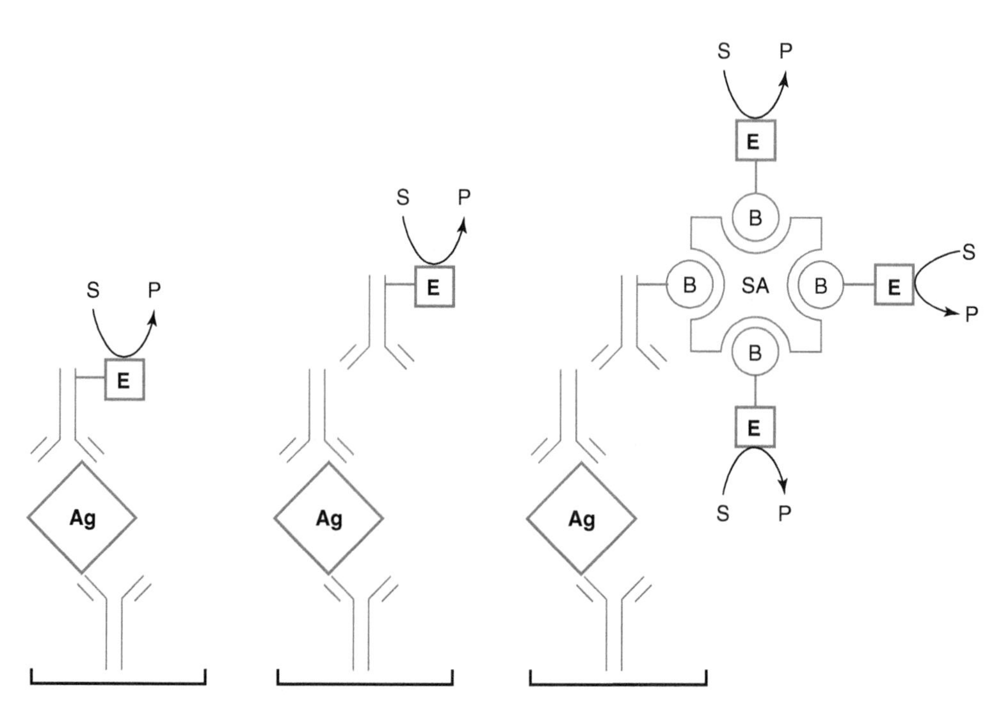

Figure 2.10 ■ Examples of several formats for ELISA in which the specific antibody is adsorbed to the surface of a microtitration plate. See Fig. 2.9 for abbreviations used. The antibody is represented by the Y-type structure. The product *P* is colored and the amount generated is measured with a spectrometer or plate reader.

phase assay. Therefore, antibodies against a certain desired protein are allowed to adsorb to the surface of microtitration plates. Each plate may contain up to 96 wells so that multiple samples can be assayed. After incubating the antibodies in the wells of the plate for a specific period of time, excess antibody is removed and residual protein binding sites on the plastic are blocked by incubation with an inert protein. Several microtitration plates can be prepared at one time since the antibodies coating the plates retain their binding capacity for an extended period. During the ELISA, sample solution containing the protein of interest is incubated in the wells and the protein (Ag) is captured by the antibodies coating the well surface. Excess sample is removed and other antibodies which now have an enzyme (E) linked to them are added to react with the bound antigen.

The format described above is called a sandwich assay since the antigen of interest is located between the antibody on the titer well surface and the antibody containing the linked enzyme. Figure 2.10 illustrates a number of formats that can be used in an ELISA. A suitable substrate is added and the enzyme linked to the antibody-antigen-antibody well complex converts this compound to a colored product. The amount of product obtained is proportional to the enzyme adsorbed in the well of the plate. A standard curve can be prepared if known concentrations of antigen are tested in this system and the amount of antigen in unknown samples can be estimated from this standard curve. A number of enzymes can be used in ELISAs. However, the most common ones are horseradish peroxidase and alkaline phosphatase. A variety of substrates for each enzyme are available which yield colored products when catalyzed by the linked enzyme. Absorbance of the colored product solutions is measured on plate readers, instruments which rapidly measure the absorbance in all 96 wells of the microtitration plate, and data processing can be automated for rapid throughput of information. Note that detection approaches partly parallel those discussed in the section on "Blotting." The above ELISA format is only one of many different methods. For example, the microtitration wells may be coated directly with the antigen rather than having a specific antibody attached to the surface. Quantitation is made by comparison with known quantities of antigen used to coat individual wells.

Another approach, this time subsequent to the binding of antigen either directly to the surface or to an antibody on the surface, is to use an antibody specific to the antibody binding the protein antigen, that is, a secondary antibody. This latter, secondary, antibody contains the linked enzyme used for detection. As already discussed in the section on "Blotting," the advantage to this approach is that such antibodies can be obtained in high purity and with the desired enzyme linked to them from commercial sources. Thus, a single source of enzyme-linked antibody can be used in assays for different protein antigens. Should a sandwich assay be used, then antibodies from different species need to be used for each side of the sandwich. A possible scenario is that rabbit antibodies are used to coat the microtitration wells; mouse antibodies, possibly a monoclonal antibody, are used to complex with the antigen; and then, a goat anti-mouse immunoglobulin containing linked HRP or AP is used for detection purposes.

As with immunoblots discussed above, streptavidin or avidin can be used in these assays if biotin is covalently linked to the antibodies and enzymes (Fig. 2.10).

If a radioactive label is used in place of the enzyme in the above procedure, then the assay is a solid-phase radioimmunoassay (RIA). Assays are moving away from the use of radioisotopes, because of problems with safety and disposal of radioactive waste and since nonradioactive assays have comparable sensitivities.

■ Electrophoresis

Analytical methodologies for measuring protein properties stem from those used in their purification. The major difference is that systems used for analysis have a higher resolving power and lower detection limit than those used in purification. The two major methods for analysis have their bases in chromatographic or electrophoretic techniques.

Polyacrylamide Gel Electrophoresis

One of the earliest methods for analysis of proteins is polyacrylamide gel electrophoresis (PAGE). In this assay, proteins, being amphoteric molecules with both positive and negative charge groups in their primary structure, are separated according to their net electrical charge. A second factor which is responsible for the separation is the mass of the protein. Thus, one can consider more precisely that the charge to mass ratio of proteins determines how they are separated in an electrical field. The charge of the protein can be controlled by the pH of the solution in which the protein is separated. The farther away the protein is from its pI value, that is, the pH at which it has a net charge of zero, the greater is the net charge and hence the greater is its charge to mass ratio. Therefore, the direction and speed of migration of the protein depend on the pH of the gel. If the pH of the gel is above its pI value, then the protein is negatively charged and hence migrates toward the anode. The higher the pH of the gel, the faster the migration. This type of electrophoresis is called native gel electrophoresis.

The major component of polyacrylamide gels is water. However, they provide a flexible support so that after a protein has been subjected to an electrical field for an appropriate period of time, it provides a matrix to hold the proteins in place until they can be detected with suitable reagents. By adjusting the amount of acrylamide that is used in these gels, one can control the migration of material within the gel. The more acrylamide, the more hindrance for the protein to migrate in an electrical field.

The addition of a detergent, sodium dodecyl sulfate (SDS), to the electrophoretic separation system allows for the separation to take place primarily as a function of the size of the protein. Dodecyl sulfate ions form complexes with proteins, resulting in an unfolding of the proteins, and the amount of detergent that is complexed is proportional to the mass of the protein. The larger the protein, the more detergent that is complexed. Dodecyl sulfate is a negatively charged ion. When proteins are in a solution of SDS, the net effect is that the own charge of the protein is overwhelmed by that of the dodecyl sulfate complexed with it, so that the proteins take on a net negative charge proportional to their mass.

Polyacrylamide gel electrophoresis in the presence of sodium dodecyl sulfates is commonly known as SDS-PAGE. All the proteins take on a net negative charge, with larger proteins binding more SDS but with the charge to mass ratio being fairly constant among the proteins. An example of SDS-PAGE is shown in Fig. 2.11. Here, SDS-PAGE is used to monitor expression of G-CSF receptor and of G-CSF (panel B) in different culture media.

Since all proteins have essentially the same charge to mass ratio, how can separation occur? This is done by controlling the concentration of acrylamide in the path of proteins migrating in an electrical field. The greater the acrylamide concentration, the more difficult it is for large protein molecules to migrate relative to smaller protein molecules. This is sometimes thought of as a sieving effect, since the greater the acrylamide concentration, the smaller the pore size within the polyacrylamide gel. Indeed, if the acrylamide concentration is sufficiently high, some high-molecular-weight proteins may not migrate at all within the gel. Since in SDS-PAGE the proteins are denatured, their hydrodynamic size, and hence the degree of retardation by the sieving effects, is directly related to their mass. Proteins containing disulfide bonds will have a much more compact structure and higher mobility for their mass unless the disulfides are reduced prior to electrophoresis.

As described above, native gel electrophoresis and SDS-PAGE are quite different in terms of the mechanism of protein separation. In native gel electrophoresis, the proteins are in the native state and migrate on their own charges. Thus, this electrophoresis can be used to characterize proteins in the native state. In SDS-PAGE, proteins are unfolded and migrate based on their molecular mass. As an intermediate case, Blue native electrophoresis is developed, in which proteins are bound by a dye, Coomassie blue, used to stain protein bands. This dye is believed to bind to the hydrophobic surface of the proteins and to add negative charges to the proteins. The dye-bound

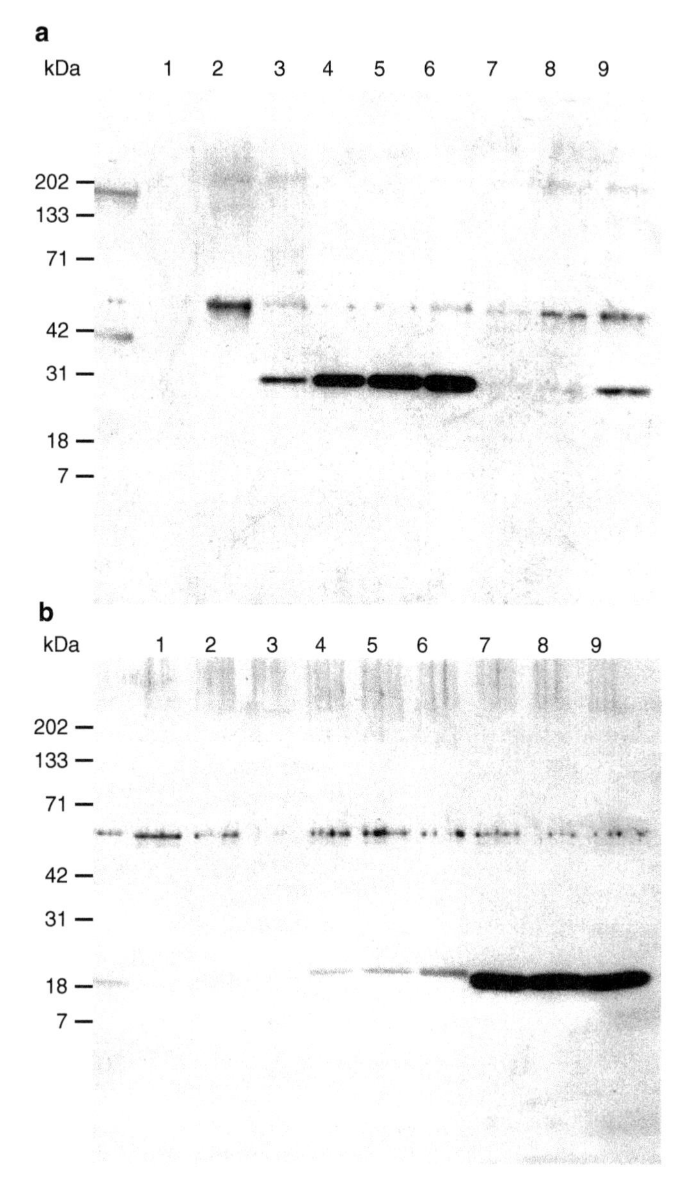

Figure 2.11 ■ SDS-PAGE of G-CSF receptor, about 35 kDa (panel **a**), and G-CSF, about 20 kDa (panel **b**). These proteins are expressed in different culture media (lanes 1–9). Positions of molecular weight standards are given on the left side. The bands are developed with antibody against G-CSF receptor (panel **a**) or G-CSF (panel **b**) after blotting.

proteins are still in the native state and migrate based on the net charges, which depend on the intrinsic charges of the proteins and the amounts of the negatively charged dye. This is particularly useful for analyzing membrane proteins, which tend to aggregate in the absence of detergents. The dye prevents the proteins from aggregation by binding to their hydrophobic surface.

Isoelectric Focusing (IEF)

Another method to separate proteins based on their electrophoretic properties is to take advantage of their isoelectric point. In a first run, a pH gradient is established within the gel using a mixture of small-molecular-weight ampholytes with varying pI values. The high pH conditions are established at the site of the cathode. Then, the protein is brought on the gel, e.g., at the site where the pH is 7. In the electrical field, the protein will migrate until it reaches the pH on the gel where its net charge is zero. If the protein were to migrate away from this pH value, it could gain a charge and migrate toward its pI value again, leading to a focusing effect.

2-Dimensional Gel Electrophoresis

The above methods can be combined into a procedure called 2-D gel electrophoresis. Proteins are first fractionated by isoelectric focusing based upon their pI values. They are then subjected to SDS-PAGE perpendicular to the first dimension and fractionated based on the molecular weights of proteins. SDS-PAGE cannot be performed before isoelectric focusing, since once SDS binds to and denatures the proteins, they no longer migrate based on their pI values.

Detection of Proteins Within Polyacrylamide Gels

Although the polyacrylamide gels provide a flexible support for the proteins, with time, the proteins will diffuse and spread within the gel. Consequently, the usual practice is to fix the proteins or trap them at the location where they migrated to. This is accomplished by placing the gels in a fixing solution in which the proteins become insoluble.

There are many methods for staining proteins in gels, but the two most common and well-studied methods are either staining with Coomassie blue or by a method using silver. The latter method is used if increased sensitivity is required. The principle of developing the Coomassie blue stain is the hydrophobic interaction of a dye with the protein. Thus, the gel takes on a color wherever a protein is located. Using standard amounts of proteins, the amount of protein or contaminant may be estimated. Quantification using the silver staining method is less precise. However, due to the increased sensitivity of this method, very low levels of contaminants can be detected. These fixing and staining procedures denature the proteins. Hence, proteins separated under native conditions, as in native or Blue native gel electrophoresis, will be denatured. To maintain the native state, the gels can be stained with copper or other metal ions.

Capillary Electrophoresis

With recent advances in instrumentation and technology, capillary electrophoresis has gained an increased presence in the analysis of recombinant proteins. Rather than having a matrix, as in polyacrylamide gel

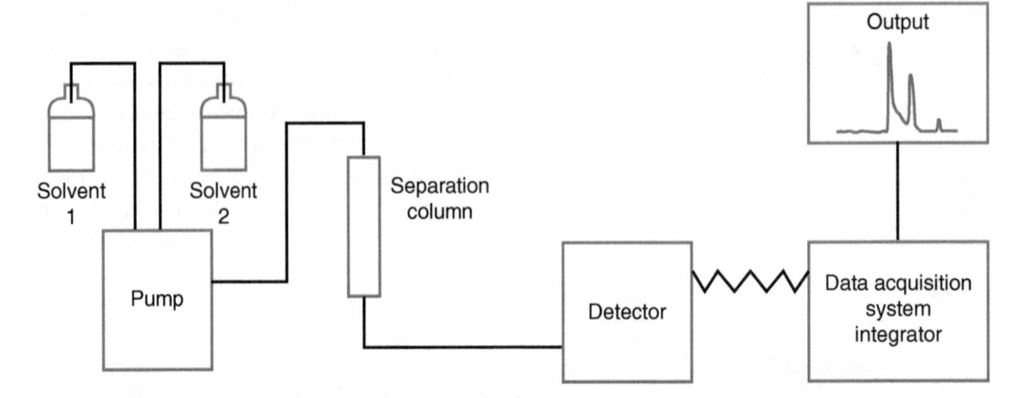

Figure 2.12 ■ Components of a typical chromatography station. The pump combines solvents one and two in appropriate ratios to generate a pH, salt concentration, or hydrophobic gradient. Proteins that are fractioned on the column pass through a detector which measures their occurrence. Information from the detector is used to generate chromatograms and the relative amount of each component.

electrophoresis through which the proteins migrate, they are free in solution in an electric field within the confines of a capillary tube with a diameter of 25–50 μm. The capillary tube passes through an ultraviolet light or fluorescence detector that measures the presence of proteins migrating in the electric field. The movement of one protein relative to another is a function of the molecular mass and the net charge on the protein. The latter can be influenced by pH and analytes in the solution. This technique has only partially gained acceptance for routine analysis, because of difficulties in reproducibility of the capillaries and in validating this system. Nevertheless, it is a powerful analytical tool for the characterization of recombinant proteins during process development and in stability studies.

■ Chromatography

Chromatography techniques are used extensively in biotechnology not only in protein purification procedures (see Chap. 3) but also in assessing the integrity of the product. Routine procedures are highly automated so that comparisons of similar samples can be made. An analytical system consists of an autosampler which will take a known amount (usually a known volume) of material for analysis and automatically places it in the solution stream headed toward a separation column used to fractionate the sample. Another part of this system is a pump module which provides a reproducible flow rate. In addition, the pumping system can provide a gradient which changes properties of the solution such as pH, ionic strength, and hydrophobicity. A detection system (or possibly multiple detectors in series) is located at the outlet of the column. This measures the relative amount of protein exiting the column. Coupled to the detector is a data acquisition system which takes the signal from the detector and integrates it into a value related to the amount of material (see Fig. 2.12). When the protein appears, the signal begins to increase, and as the protein passes through the detector, the signal subsequently decreases. The area under the peak of the signal is proportional to the amount of material which has passed through the detector. By analyzing known amounts of protein, an area versus amount of protein plot can be generated and this may be used to estimate the amount of this protein in the sample under other circumstances. Another benefit of this integrated chromatography system is that low levels of components which appear over time can be estimated relative to the major desired protein being analyzed. This is a particularly useful function when the long-term stability of the product is under evaluation.

Chromatographic systems offer a multitude of different strategies for successfully separating protein mixtures and for quantifying individual protein components (see Chap. 3). The following describes some of these strategies.

Size-Exclusion Chromatography

As the name implies, this procedure separates proteins based on their size or molecular weight or shape. The matrix consists of very fine beads containing cavities and pores accessible to molecules of a certain size or smaller, but inaccessible to larger molecules. The principle of this technique is the distribution of molecules between the volume of solution within the beads and the volume of solution surrounding the beads. Small molecules have access to a larger volume than do large molecules. As solution flows through the column, molecules can diffuse back and forth, depending upon their size, in and out of the pores of the beads. Smaller molecules can reside within the pores for a finite period

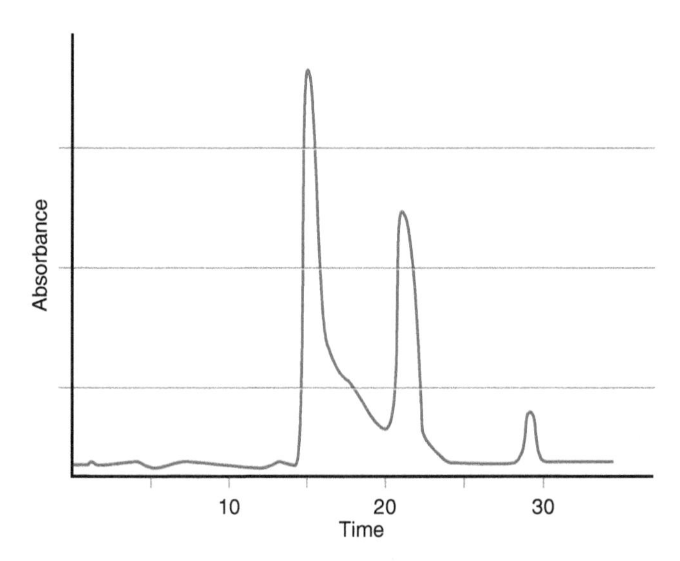

Figure 2.13 ■ Size-exclusion chromatography of a recombinant protein which on storage yields aggregates and smaller peptides.

of time whereas larger molecules, unable to enter these spaces, continue along in the fluid stream. Intermediate-sized molecules spend an intermediate amount of time within the pores. They can be fractionated from large molecules that cannot access the matrix space at all and from small molecules that have free access to this volume and spend most of the time within the beads. Protein molecules can distribute between the volume within these beads and the excluded volume based on the mass and shape of the molecule. This distribution is based on the relative concentration of the protein in the beads versus the excluded volume.

Size-exclusion chromatography can be used to estimate the mass of proteins by calibrating the column with a series of globular proteins of known mass. However, the separation depends on molecular shape (conformation) as well as mass and highly elongated proteins—proteins containing flexible, disordered regions— and glycoproteins will often appear to have masses as much as two to three times the true value. Other proteins may interact weakly with the column matrix and be retarded, thereby appearing to have a smaller mass. Thus, sedimentation or light scattering methods are preferred for accurate mass measurement (see section "Techniques Specifically Suitable for Characterizing Protein Folding"). Over time, proteins can undergo a number of changes that affect their mass. A peptide bond within the protein can hydrolyze, yielding two smaller polypeptide chains. More commonly, size-exclusion chromatography is used to assess aggregated forms of the protein. Figure 2.13 shows an example of this. The peak at 22 min represents the native protein. The peak at 15 min is aggregated protein and that at 28 min depicts degraded protein, yielding smaller polypeptide chains. Aggregation can occur when a protein molecule unfolds to a slight extent and exposes surfaces that are attracted to complementary surfaces on adjacent molecules. This interaction can lead to dimerization or doubling of molecular weight or to higher-molecular-weight oligomers. From the chromatographic profile, the mechanism of aggregation can often be implicated. If dimers, trimers, tetramers, etc., are observed, then aggregation occurs by stepwise interaction of a monomer with a dimer, trimer, etc. If dimers, tetramers, octamers, etc., are observed, then aggregates can interact with each other. Sometimes, only monomers and high-molecular-weight aggregates are observed, suggesting that intermediate species are kinetically of short duration and protein molecules susceptible to aggregation combine into very large-molecular-weight complexes.

Reversed-Phase High-Performance Liquid Chromatography

Reversed-phase high-performance liquid chromatography (RP-HPLC) takes advantage of the hydrophobic properties of proteins. The functional groups on the column matrix contain from one to up to 18 carbon atoms in a hydrocarbon chain. The longer this chain, the more hydrophobic is the matrix. The hydrophobic patches of proteins interact with the hydrophobic chromatographic matrix. Proteins are then eluted from the matrix by increasing the hydrophobic nature of the solvent passing through the column. Acetonitrile is a common solvent used, although other organic solvents such as ethanol also may be employed. The solvent is made acidic by the addition of trifluoroacetic acid, since proteins have increased solubility at pH values further removed from their pI. A gradient with increasing concentration of hydrophobic solvent is passed through the column. Different proteins have different hydrophobicities and are eluted from the column depending on the "hydrophobic potential" of the solvent.

This technique can be very powerful. It may detect the addition of a single oxygen atom to the protein, as when a methionyl residue is oxidized or when the hydrolysis of an amide moiety on a glutamyl or asparaginyl residue occurs. Disulfide bond formation or shuffling also changes the hydrophobic characteristic of the protein. Hence, RP-HPLC can be used not only to assess the homogeneity of the protein but also to follow degradation pathways occurring during long-term storage.

Reversed-phase chromatography of proteolytic digests of recombinant proteins may serve to identify this protein. Enzymatic digestion yields unique peptides that elute at different retention times or at different organic solvent concentrations. Moreover, the map, or chromatogram, of peptides arising from enzymatic digestion of one protein is quite different from the map obtained from another protein. Several different proteases, such as trypsin, chymotrypsin, and other endoproteinases, are used for these identity tests (see below under "Mass Spectrometry").

Hydrophobic Interaction Chromatography

A companion to RP-HPLC is hydrophobic interaction chromatography (HIC), although in principle, this latter method is normal-phase chromatography, i.e., here an aqueous solvent system rather than an organic one is used to fractionate proteins. The hydrophobic characteristics of the solution are modulated by inorganic salt concentrations. Ammonium sulfate and sodium chloride are often used since these compounds are highly soluble in water. In the presence of high salt concentrations (up to several molar), proteins are attracted to hydrophobic surfaces on the matrix of resins used in this technique. As the salt concentration decreases, proteins have less affinity for the matrix and eventually elute from the column. This method lacks the resolving power of RP-HPLC, but is a more gentle method, since low pH values or organic solvents as used in RP-HPLC can be detrimental to some proteins.

Ion-Exchange Chromatography

This technique takes advantage of the electronic charge properties of proteins. Some of the amino acyl residues are negatively charged and others are positively charged. The net charge of the protein can be modulated by the pH of its environment relative to the pI value of the protein. At a pH value lower than the pI, the protein has a net positive charge, whereas at a pH value greater than the pI, the protein has a net negative charge. Opposites attract in ion-exchange chromatography. The resins in this procedure can contain functional groups with positive or negative charges. Thus, positively charged proteins bind to negatively charged matrices and negatively charged proteins bind to positively charged matrices. Proteins are displaced from the resin by increasing salt, e.g., sodium chloride, concentrations. Proteins with different net charges can be separated from one another during elution with an increasing salt gradient. The choice of charged resin and elution conditions are dependent upon the protein of interest.

In lieu of changing the ionic strength of the solution, proteins can be eluted by changing the pH of the medium, i.e., with the use of a pH gradient. This method is called chromatofocusing and proteins are separated based on their pI values. When the solvent pH reaches the pI value of a specific protein, the protein has a zero net charge and is no longer attracted to the charged matrix and hence is eluted.

Other Chromatographic Techniques

Other functional groups may be attached to chromatographic matrices to take advantage of unique properties of certain proteins. These affinity methodologies, however, are more often used in the manufacturing process than in analytical techniques (see Chap. 3). For example, conventional affinity purification schemes of antibodies use protein A or G columns. Protein A or G specifically binds antibodies. Antibodies consist of variable regions and constant regions (see Chap. 7). The variable regions are antigen specific and hence vary in sequence from one antibody to another, while the constant regions are common to each subgroup of antibodies. The constant region binds to protein A or G.

Mixed-mode chromatography uses columns having both hydrophobic and charged groups, i.e., combination of ion-exchange and hydrophobic interaction chromatography. Mixed-mode columns confer protein binding under conditions at which protein binding normally does not occur. For example, protein binding to an ion-exchange column requires low ionic strength. Under identical conditions, mixed-mode columns can bind proteins through both ionic and hydrophobic interactions.

■ Bioassays

Paramount to the development of a protein therapeutic is to have an assay that identifies its biological function. Chromatographic and electrophoretic methodologies can address the homogeneity of a biotherapeutic and be useful in investigating stability parameters. However, it is also necessary to ascertain whether the protein has acceptable bioactivity. Bioactivity can be determined either in vivo, i.e., by administering the protein to an animal and ascertaining some change within its body (function), or in vitro. Bioassays in vitro monitor the response of a specific receptor or microbiological or tissue cell line when the therapeutic protein is added to the system. An example of an in vitro bioassay is the increase in DNA synthesis in the presence of the therapeutic protein as measured by the incorporation of radioactively labeled thymidine. The protein factor binds to receptors on the cell surface that triggers secondary messengers to send signals to the cell nucleus to synthesize DNA. The binding of the protein factor to the cell surface is dependent upon the amount of factor present. Figure 2.14 presents a dose–response curve of thymidine incorporation as a function of concentration of the factor. At low concentrations, the factor is too low to trigger a response. As the concentration increases, the incorporation of thymidine occurs, and at higher concentrations, the amount of thymidine incorporation ceases to increase as DNA synthesis is occurring at the maximum rate. A standard curve can be obtained using known quantities of the protein factor. Comparison of other solutions containing unknown amounts of the factor with this standard curve will then yield quantitative estimates of the factor concentration. Through experience during the development of the protein therapeutic, a value is obtained for a fully functional protein. Subsequent comparisons to this value can be used to ascertain any loss in activity during stability studies or changes in activity when amino acyl residues of the protein are modified.

Other in vitro bioassays can measure changes in cell number or production of another protein factor in response to the stimulation of cells by the protein therapeutic. The amount of the secondary protein produced can be estimated by using an ELISA.

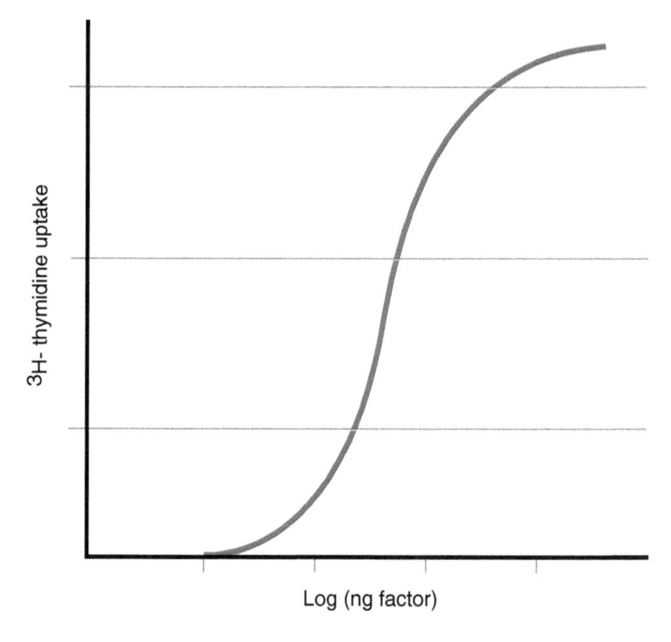

Figure 2.14 ■ An in vitro bioassay showing a mitogenic response in which radioactive thymidine is incorporated into DNA in the presence of an increasing amount of a protein factor.

■ Mass Spectrometry

Recent advances in the measurement of the molecular masses of proteins have made this technique an important analytical tool. While this method was used in the past to analyze small volatile molecules, the molecular weights of highly charged proteins with masses of over 100 kilodaltons (kDa) can now be accurately determined.

Because of the precision of this method, post-translational modifications such as acetylation or glycosylation can be predicted. The masses of new protein forms that arise during stability studies provide information on the nature of this form. For example, an increase in mass of 16 Da suggests that an oxygen atom has been added to the protein as happens when a methionyl residue is oxidized to a methionyl sulfoxide residue. The molecular mass of peptides obtained after proteolytic digestion and separation by HPLC indicates from which region of the primary structure they are derived. Such HPLC chromatogram is called a "peptide map." An example is shown in Fig. 2.15. This

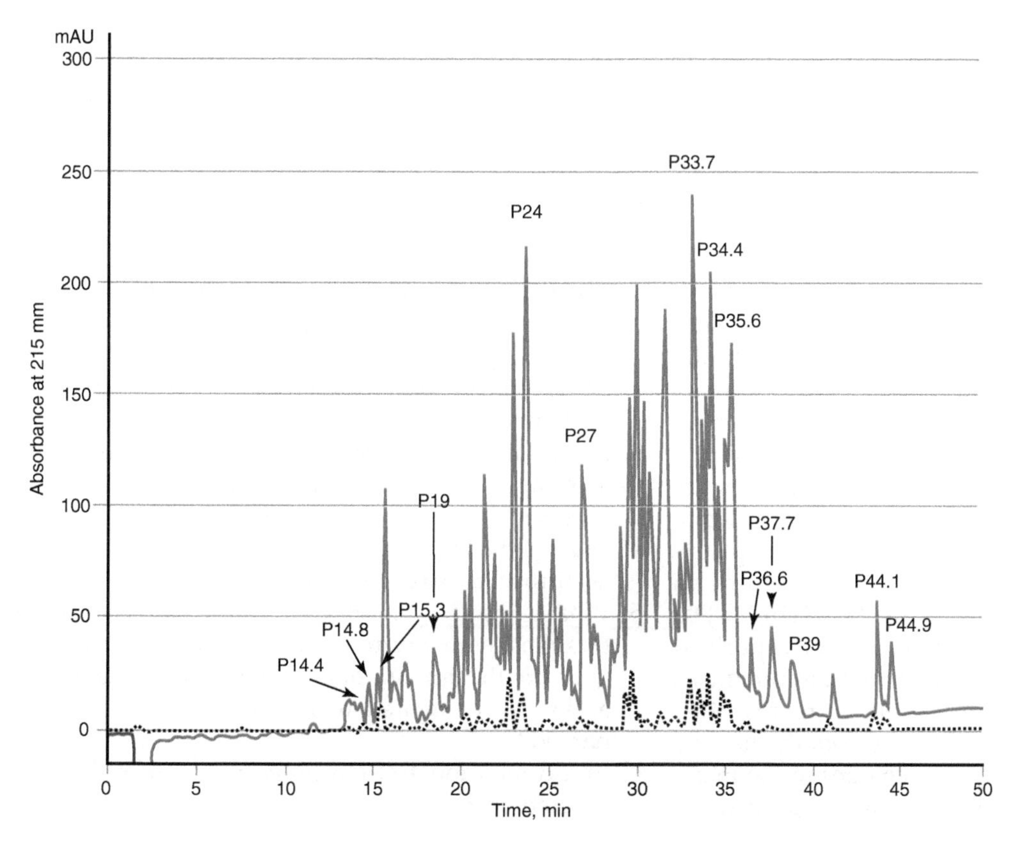

Figure 2.15 ■ Peptide map of pepsin digest of recombinant human β-secretase. Each peptide is labeled by elution time in HPLC.

is obtained by digesting a protein with pepsin and by subsequently separating the digested peptides by reverse HPLC. This highly characteristic pattern for a protein is called a "protein fingerprint." Peaks are identified by elution times on HPLC. If peptides have molecular masses differing from those expected from the primary sequence, the nature of the modification to that peptide can be implicated. Moreover, molecular mass estimates can be made for peptides obtained from unfractionated proteolytic digests. Molecular masses that differ from expected values indicate that a part of the protein molecule has been altered, that glycosylation or another modification has been altered, or that the protein under investigation still contains contaminants.

Another way that mass spectrometry can be used as an analytical tool is in the sequencing of peptides. A recurring structure, the peptide bond, in peptides tends to yield fragments of the mature peptide which differ stepwise by an amino acyl residue. The difference in mass between two fragments indicates the amino acid removed from one fragment to generate the other. Except for leucine and isoleucine, each amino acid has a different mass and hence a sequence can be read from the mass spectrograph. Stepwise removal can occur from either the amino terminus or carboxy terminus.

By changing three basic components of the mass spectrometer, the ion source, the analyzer, and the detector, different types of measurement may be undertaken. Typical ion sources which volatilize the proteins are electrospray ionization, fast atom bombardment, and liquid secondary ion. Common analyzers include quadrupole, magnetic sector, and time-of-flight instruments. The function of the analyzer is to separate the ionized biomolecules based on their mass-to-charge ratio. The detector measures a current whenever impinged upon by charged particles. Electrospray ionization (El) and matrix-assisted laser desorption (MALDI) are two sources that can generate high-molecular-weight volatile proteins. In the former method, droplets are generated by spraying or nebulizing the protein solution into the source of the mass spectrometer. As the solvent evaporates, the protein remains behind in the gas phase and passes through the analyzer to the detector. In MALDI, proteins are mixed with a matrix which vaporizes when exposed to laser light, thus carrying the protein into the gas phase. An example of MALDI-mass analysis is shown in Fig. 2.16, indicating the singly charged ion (116,118 Da) and the doubly charged ion (5,8036.2) for a purified protein. Since proteins are multi-charge compounds, a number of components are observed representing mass-to-charge forms, each differing from the next by one charge. By imputing various charges to the mass-to-charge values, a molecular mass of the protein can be estimated. The latter step is empirical since only the mass-to-charge ratio is detected and not the net charge for that particular particle.

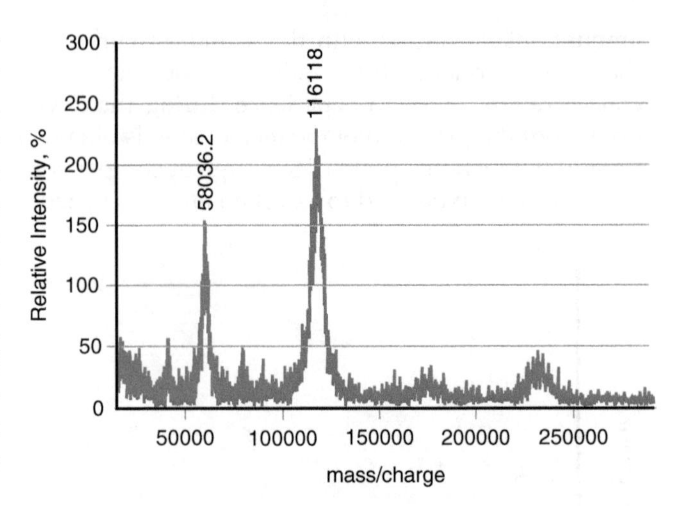

Figure 2.16 ■ MALDI-mass analysis of a purified recombinant human β-secretase. Numbers correspond to the singly charged and doubly charged ions.

CONCLUDING REMARKS

With the advent of recombinant proteins as human therapeutics, the need for methods to evaluate their structure, function, and homogeneity has become paramount. Various analytical techniques are used to characterize the primary, secondary, and tertiary structure of the protein and to determine the quality, purity, and stability of the recombinant product. Bioassays establish its activity.

SELF-ASSESSMENT QUESTIONS

■ Questions

1. What is the net charge of granulocyte-colony-stimulating factor at pH 2.0, assuming that all the carboxyl groups are protonated?
2. Based on the above calculation, do you expect the protein to unfold at pH 2.0?
3. Design an experiment using blotting techniques to ascertain the presence of a ligand to a particular receptor.
4. What is the transfer of proteins to a membrane such as nitrocellulose or PDVF called?
5. What is the assay in which the antibody is adsorbed to a plastic microtitration plate and then is used to quantify the amount of a protein using a secondary antibody conjugated with horseradish peroxidase named?
6. In 2-dimensional electrophoresis, what is the first method of separation?

7. What is the method for separating proteins in solution based on molecular size called?
8. Why are large protein particles more immunogenic?

■ Answers

1. Based on the assumption that glutamyl and aspartyl residues are uncharged at this pH, all the charges come from protonated histidyl, lysyl, arginyl residues, and the amino terminus, i.e., 5 His + 4 Lys + 5 Arg + N-terminal = 15.
2. Whether a protein unfolds or remains folded depends on the balance between the stabilizing and destabilizing forces. At pH 2.0, extensive positive charges destabilize the protein, but whether such destabilization is sufficient or insufficient to unfold the protein depends on how stable the protein is in the native state. The charged state alone cannot predict whether a protein will unfold.
3. A solution containing the putative ligand is subjected to SDS-PAGE. After blotting the proteins in the gel to a membrane, it is probed with a solution containing the receptor. The receptor, which binds the ligand, may be labeled with agents suitable for detection or, alternatively, the complex can subsequently be probed with an antibody to the receptor and developed as for an immunoblot. Note that the reciprocal of this can be done as well, in which the receptor is subjected to SDS-PAGE and the blot is probed with the ligand.
4. This method is called blotting. If an electric current is used, then the method is called electroblotting.
5. This assay is called an ELISA, enzyme-linked immunosorbent assay.
6. Either isoelectric focusing or native polyacrylamide electrophoresis. The second dimension is performed in the presence of the detergent sodium dodecyl sulfate.
7. Size-exclusion chromatography.
8. The immune systems are designed to fight against virus infections and hence generate antibodies against foreign particles with the size of the virus. When pharmaceutical proteins aggregate into the particle size, the immune system recognizes them as viruslike (cf. Chap. 6).

FURTHER READING

Butler JE (ed) (1991) Immunochemistry of solid-phase immunoassay. CRC Press, Boca Raton

Coligan J, Dunn B, Ploegh H, Speicher D, Wingfield P (eds) (1995) Current protocols in protein science. Wiley, New York

Crabb JW (ed) (1995) Techniques in protein chemistry VI. Academic, San Diego

Creighton TE (ed) (1989) Protein structure: a practical approach. IRL Press, Oxford

Crowther JR (1995) ELISA, theory and practice. Humana Press, Totowa

Dunbar BS (1994) Protein blotting: a practical approach. Oxford University Press, New York

Gregory RB (ed) (1994) Protein-solvent interactions. Marcel Dekker, New York

Hames BD, Rickwood D (eds) (1990) Gel electrophoresis of proteins: a practical approach, 2nd edn. IRL Press, New York

Jiskoot W, Crommelin DJA (eds) (2005) Methods for structural analysis of protein pharmaceuticals. AAPS Press, Arlington

Landus JP (ed) (1994) Handbook of capillary electrophoresis. CRC Press, Boca Raton

McEwen CN, Larsen BS (eds) (1990) Mass spectrometry of biological materials. Dekker, New York

Price CP, Newman DJ (eds) (1991) Principles and practice of immunoassay. Stockton, New York

Schulz GE, Schirmer RH (eds) (1979) Principles of protein structure. Springer, New York

Shirley BA (ed) (1995) Protein stability and folding. Humana Press, Totowa

3 Production and Purification of Recombinant Proteins

Farida Kadir, Paul Ives, Alfred Luitjens, and Emile van Corven

INTRODUCTION

The growing therapeutic use of proteins has created an increasing need for practical and economical processing techniques. As a result, biotechnological production methods have advanced significantly over the last decade. Also, single-use production technology which has the potential to mitigate many of the economic and quality issues arising from manufacturing these products has evolved rapidly (Hodge 2004).

When producing proteins for therapeutic use, a number of issues must be considered related to the manufacturing, purification, and characterization of the products. Biotechnological products for therapeutic use have to meet strict specifications especially when used via the parenteral route (Walter and Werner 1993).

In this chapter several aspects of production (upstream processing) and purification (downstream processing) will be dealt with briefly. For further details, the reader is referred to the literature mentioned.

UPSTREAM PROCESSING

■ Expression Systems

General Considerations

Expression systems for proteins of therapeutic interest include both pro- and eukaryotic cells (bacteria, yeast, fungi, plants, insect cells, mammalian cells) and transgenic animals. The choice of a particular system will be determined to a large extent by the nature and origin of the desired protein, the intended use of the product, the amount needed, and the cost.

In principle, any protein can be produced using genetically engineered organisms, but not every type of protein can be produced by every type of cell. In the majority of cases, the protein is foreign to the host cells that have to produce it, and although the translation of the genetic code can be performed by the cells, the posttranslation modifications of the protein might be different as compared to the original product.

About 5 % of the proteome are thought to comprise enzymes performing over 200 types of posttranslation modifications of proteins (Walsh 2006). These modifications are species and/or cell-type specific. The metabolic pathways that lead to these modifications are genetically determined by the host cell. Thus, even if the cells are capable of producing the desired posttranslation modification, like glycosylation, still the resulting glycosylation pattern might be different from that of the native protein. Correct N-linked glycosylation of therapeutically relevant proteins is important for full biological activity, immunogenicity, stability, targeting, and pharmacokinetics. Prokaryotic cells, like bacteria, are sometimes capable of producing N-linked glycoproteins. However, the N-linked structures found differ from the structures found in eukaryotes (Dell et al. 2011). Yeast cells are able to produce recombinant proteins like albumin, and yeast has been engineered to produce glycoproteins with humanlike glycan structures including terminal sialylation (reviewed by Celik and Calik 2011). Still most products on the market and currently in development use cell types that are as closely related to the original protein-producing cell type as possible. Therefore, human-derived proteins, especially mammalian cells, are chosen for production. Further developments in the field of engineering (e.g., glycosylation) may allow bacteria to reproduce some of the post-translation modification steps common to eukaryotic

F. Kadir, Ph.D.
Postacamedic Education Pharmacists (POA),
Bunnik, The Netherlands

P. Ives, Ph.D.
Manufacturing Department,
SynCo Bio Partners BV, Amsterdam, The Netherlands

Department of Process Development,
Crucell, Leiden, The Netherlands

A. Luitjens • E. van Corven, Ph.D. (✉)
Department of Process Development,
Crucell, Leiden, The Netherlands
e-mail: emile.vancorven@crucell.com

D.J.A. Crommelin, R.D. Sindelar, and B. Meibohm (eds.), *Pharmaceutical Biotechnology*,
DOI 10.1007/978-1-4614-6486-0_3,

cells (Borman 2006). Although still to be further developed, bacteria and yeast may play a role as future production systems given their ease and low cost of large-scale manufacturing.

Generalized features of proteins expressed in different biological systems are listed in Table 3.1 (see also Walter et al. 1992). However, it should be kept in mind that there are exceptions to this table for specific product/expression systems.

Transgenic Animals

Foreign genes can be introduced into animals like mice, rabbits, pigs, sheep, goats, and cows through nuclear transfer and cloning techniques. Using milk-specific promoters, the desired protein can be expressed in the milk of the female offspring. During lactation the milk is collected, the milk fats are removed, and the skimmed milk is used as the starting material for the purification of the protein.

The advantage of this technology is the relatively cheap method to produce the desired proteins in vast quantities when using larger animals like cows. Disadvantages are the long lead time to generate a herd of transgenic animals and concerns about the health of the animal. Some proteins expressed in the mammary gland leak back into the circulation and cause serious negative health effects. An example is the expression of erythropoietin in cows. Although the protein was well expressed in the milk, it caused severe health effects and these experiments were stopped.

The purification strategies and purity requirements for proteins from milk can be different from those derived from bacterial or mammalian cell systems. Often the transgenic milk containing the recombinant protein also contains significant amounts of the nonrecombinant counterpart. To separate these closely related proteins poses a purification challenge. The "contaminants" in proteins for oral use expressed in milk that is otherwise consumed by humans are known to be safe for consumption.

The transgenic animal technology for the production of pharmaceutical proteins has progressed within the last few years. The US and EU authorities approved recombinant antithrombin III (ATryn®, GTC Biotherapeutics) produced in the milk of transgenic goats. More details about this technology are presented in Chap. 8.

Plants

Therapeutic proteins can also be expressed in plants and plant cell cultures (see also Chap. 1). For instance, human albumin has been expressed in potatoes and tobacco. Whether these production vehicles are economically feasible has yet to be established. The lack of genetic stability of plants was sometimes a drawback. Stable expression of proteins in edible seeds has been obtained. For instance, rice and barley can be harvested and easily kept for a prolonged period of time as raw material sources. Especially for oral therapeutics or vaccines, this might be the ideal solution to produce large amounts of cheap therapeutics, because the "contaminants" are known to be safe for consumption (see also Chap. 21). A better understanding of the plant molecular biology together with more sophisticated genetic engineering techniques and strategies to increase yields and optimize glycan structures resulted in an increase in the number of products in development including late-stage clinical trials (reviewed by Orzaez et al. 2009, and Peters and Stoger 2011). Removal of most early bottlenecks together with the regulatory acceptance of plants as platforms to produce therapeutic proteins has resulted in a renewed interest by the industry.

Protein feature	Prokaryotic bacteria	Eukaryotic yeast	Eukaryotic mammalian cells
Concentration	High	High	High
Molecular weight	Low	High	High
S-S bridges	Limitation	No limitation	No limitation
Secretion	No	Yes/no	Yes
Aggregation state	Inclusion body	Singular, native	Singular, native
Folding	Risk of misfolding	Correct folding	Correct folding
Glycosylation	Limited	Possible	Possible
Impurities: retrovirus	No	No	Possible
Impurities: pyrogen	Possible	No	No
Cost to manufacture	Low	Low	High

Table 3.1 ■ Generalized features of proteins of different biological origin.

More details about the use of plant systems for the production of pharmaceutical proteins are presented in Chap. 7.

■ Cultivation Systems

General

In general, cells can be cultivated in vessels containing an appropriate liquid growth medium in which the cells are either immobilized and grow as a monolayer, attached to microcarriers, free in suspension, or entrapped in matrices (usually solidified with agar). The culture method will determine the scale of the separation and purification methods. Production-scale cultivation is commonly performed in fermentors, used for bacterial and fungal cells, or bioreactors, used for mammalian and insect cells. Bioreactor systems can be classified into four different types:

- Stirred tank (Fig. 3.1a)
- Airlift (Fig. 3.1b)
- Fixed bed (Fig. 3.1c)
- Membrane bioreactors (Fig. 3.1d)

Because of its reliability and experience with the design and scaling up potential, the stirred tank is still the most commonly used bioreactor. This type of bioreactor is not only used for suspension cells like CHO, HEK293, and PER.C6® cells, it is also used for production with adherent cells like Vero and MDCK cells. In the latter case the production is performed on microcarriers (Van Wezel et al. 1985).

Single-Use Systems

In the last decade the development of single-use production systems was boosted. Wave Biotech AG in Switzerland (now Sartorius Stedim Biotech) is regarded as a visionary player in the nurture of single-use technologies for mammalian cell culture. With the development of 3D single-use bags by companies like Hyclone, Xcellerex, and Sartorius, the number of single-use systems used has increased. Single-use bioreactors are nowadays used for the manufacturing of products in development and on the market. Shire (Dublin, Ireland) was the first company that used single-use bioreactors up to 2,000 L for the manufacturing of one of their products. The advantages of the single-use technology are:

- Cost-effective manufacturing technology
  By introducing single-use systems, the design is such that all items not directly related to the process can be removed from the culture system, like clean-in-place (CIP) and steam-in-place (SIP) systems. Furthermore, a reduction in capital costs is achieved by introducing single-use systems.
- Increasing number of batches
  By introduction of single-use systems, it is possible to increase the number of batches that can be produced in 1 year, due to the fact that cleaning and sterilization is not needed anymore. The turnover time needed from batch to batch is shortened.
- Provides flexibility in facility design
  When stainless steel systems are used, changes to the process equipment might impact the design of the stainless steel tanks, piping, etc. These changes will directly influence the CIP and SIP validation status of the facility. By using single-use systems, process changes can easily be incorporated as the setup of the single-use process is flexible, and CIP and SIP validation are not needed. However, in case a change will influence the process, the validated status of the process must be reconsidered and a revalidation might be needed.
- Speedup implementation and time to market
  Due to the great flexibility of the single-use systems, the speed of product to market is not influenced by process changes that might be introduced during the development process. However, the process needs to be validated during Phase 3 development. When changes are introduced after the process is validated, a revalidation might be needed. There is no difference in this respect to the traditional stainless steel setup.
- Reduction in water and wastewater costs
  Due to the fact that the systems are single use, there will be a great reduction in the total costs for cleaning, not only a water reduction but also a reduction in the number of hours needed to clean systems and setup for the next batch of product.
- Reduction in validation costs
  No yearly validation costs for cleaning and sterilization are needed anymore when single-use systems are used.

A disadvantage of the single-use system is that the operational expenses will increase, storage location for single-use bags and tubing will increase, and the dependence of the company to one supplier of single-use systems will increase. Furthermore, there is not yet a standard set in single-use systems like it is with stainless steel bioreactor systems. Suppliers are still developing their own systems.

The advantages of the stainless steel bioreactors are obvious as this traditional technology is well understood and controlled, although the stainless steel pathway had major disadvantages such as expensive design, installation, and maintenance costs combined with significant expenditures of time in facilities and equipment qualification and validation efforts.

Fermentation Protocols

The kinetics of cell growth and product formation will not only dictate the type of bioreactor used but also how the growth process is performed. Three types of fermentation protocols are commonly employed and discussed below:

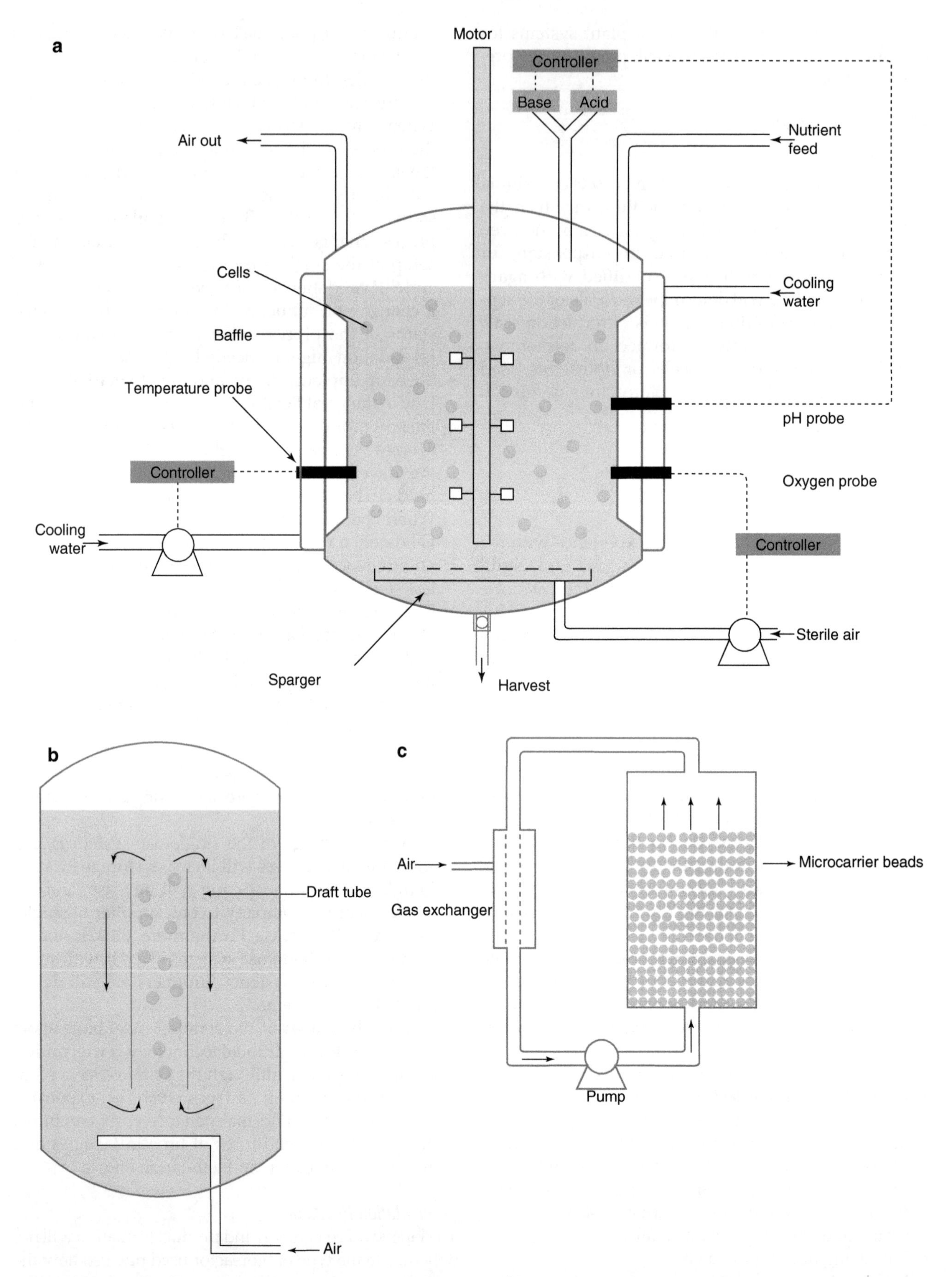

Figure 3.1 ■ (**a**) Schematic representation of stirred-tank bioreactor (Adapted from Klegerman and Groves 1992). (**b**) Schematic representation of airlift bioreactor (Adapted from Klegerman and Groves 1992). (**c**) Schematic representation of fixed-bed stirred-tank bioreactor (Adapted from Klegerman and Groves 1992). (**d**) Schematic representation of hollow fiber perfusion bioreactor (Adapted from Klegerman and Groves 1992)

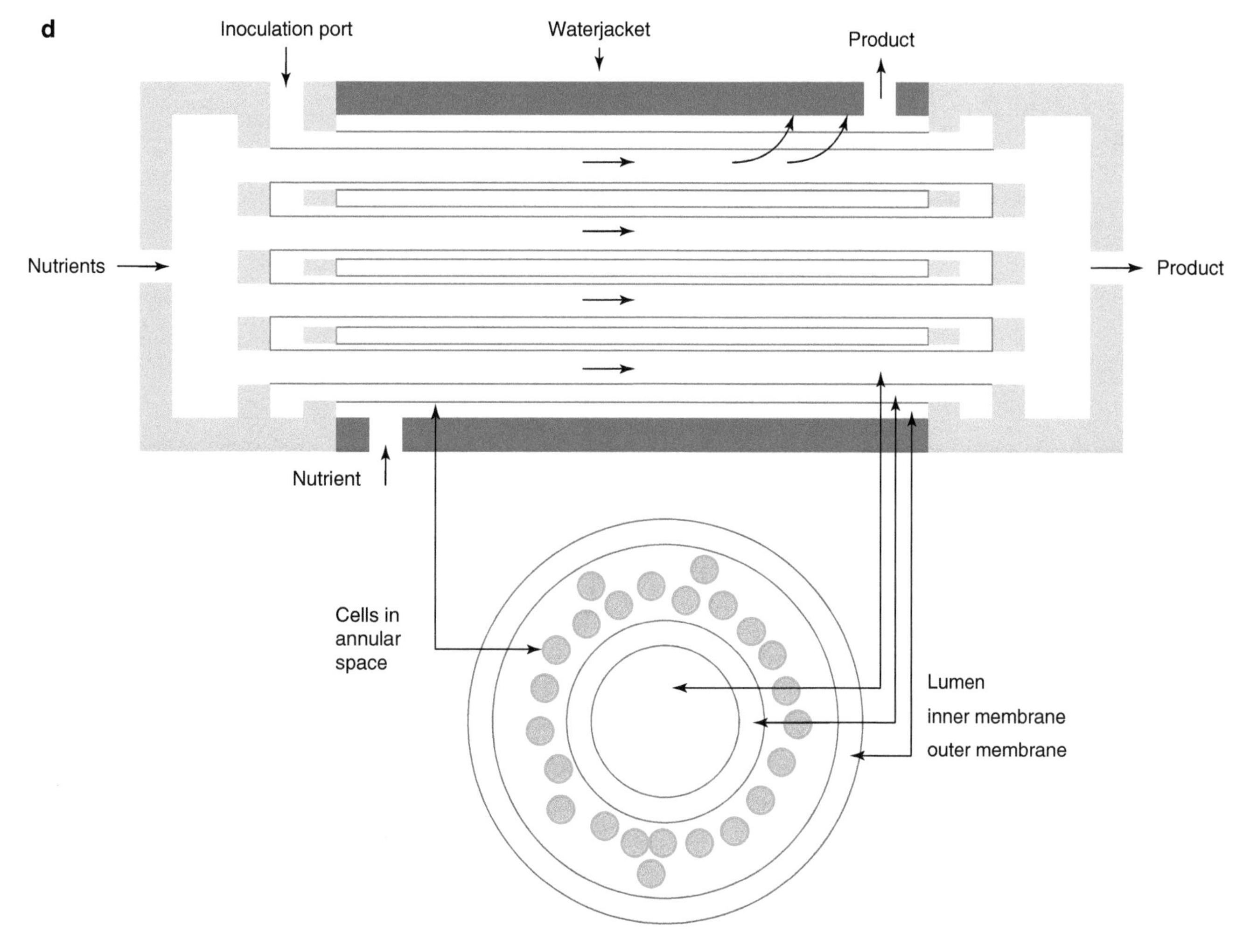

Figure 3.1 ■ (continued)

- Batch
  In a batch process, the bioreactor is filled with the entire volume of medium needed during the cell growth and/or production phase. No additional supplements are added to increase the cell growth or production during the process. Waste products, such as lactate and ammonium, and the product itself accumulate in the bioreactor. The product is harvested at the end of the process. Maximum cell density and product yields will be lower compared to a fed-batch process.
- Fed-batch
  In a fed-batch process, a substrate is supplemented to the bioreactor. The substrate consists of the growth-limiting nutrients that are needed during the cell growth phase and/or during the production phase of the process. Like the batch process, waste products accumulate in the bioreactor. The product is harvested at the end of the process. With the fed-batch process, higher cell densities and product yields can be reached compared to the batch process due to the extension of production time that can be achieved compared to a batch process. The substrate used is highly concentrated and can be added to the bioreactor on a daily basis or as a continuous feed. The fed-batch mode is currently widely used for the production of proteins. The process is well understood and characterized.
- Perfusion
  In a perfusion process, the media and waste products are continuously exchanged and the product is harvested throughout the culture period. A membrane device is used to retain the cells in the bioreactor, and waste medium is removed from the bioreactor by this device (Fig. 3.2). To keep the level medium constant in the bioreactor, fresh medium is supplemented to the bioreactor. By operating in perfusion mode, the level of waste products will be kept constant and generates a stable environment for the cells to grow or to produce. With the perfusion process, much higher cell densities can be reached and therefore higher productivity (Compton and Jensen 2007).

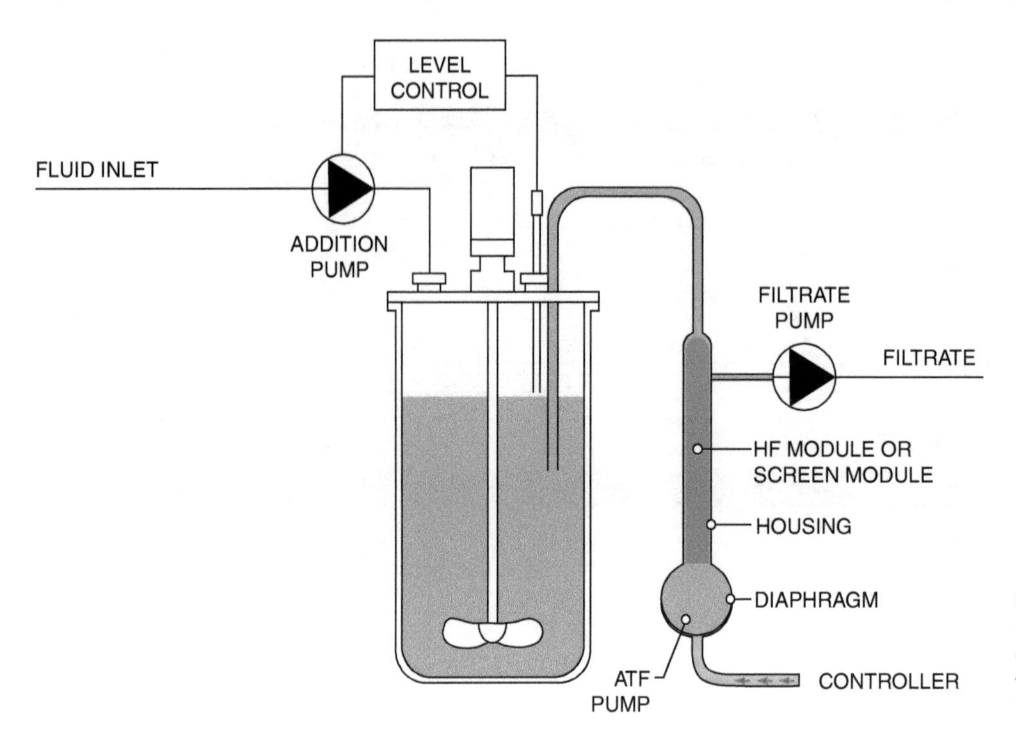

Figure 3.2 ■ Schematic representation of perfusion device coupled to a stirred-tank bioreactor. *ATF* alternating tangential flow.

In all these three protocols, the cells go through four distinctive phases (see also Chap. 1):

1. Lag phase
   In this phase the cells are adapting to the conditions in the bioreactor and do not yet grow.
2. Exponential growth phase
   During this phase, cells grow in a more or less constant doubling time for a fixed period. The mammalian cell doubling time is cell-type dependent and usually varies between 20 and 40 h. Plotting the natural logarithm of cell number against time produces a straight line. Therefore, the exponential growth phase is also called the log phase. The growth phase will be affected by growth conditions like temperature, pH, oxygen pressure, and external forces like stirring and baffles that are inserted into the bioreactor. Furthermore, the growth rate is affected by the supply of sufficient nutrients, buildup of waste nutrients, etc.
3. Stationary phase
   In the stationary phase, the growth rate of the cells slows down due to the fact that nutrients are depleted and/or build up of toxic waste products like lactate and ammonium. In this phase, constant cell numbers are found due to equal cell growth and cell death.
4. Death phase
   Cells die due to depletion of nutrients and/or presence of high concentrations of toxic products like lactate and ammonium.

Examples of animal cells that are commonly used to produce proteins of clinical interest are Chinese hamster ovary cells (CHO), immortalized human embryonic retinal cells (PER.C6® cells), baby hamster kidney cells (BHK), lymphoblastoid tumor cells (interferon production), melanoma cells (plasminogen activator), and hybridized tumor cells (monoclonal antibodies).

The cell culture has to be free from undesired microorganisms that may destroy the cell culture or present hazards to the patient by producing endotoxins. Therefore, strict measures are required for both the production procedures and materials used (WHO 2010; Berthold and Walter 1994) to prevent a possible contamination with extraneous agents like viruses, bacteria, and mycoplasma. Furthermore, strict measures are needed, especially with regard to the raw materials used, to prevent contaminations with transmissible spongiform encephalopathies (TSEs).

■ Cultivation Medium

In order to achieve optimal growth of cells and optimal production of recombinant proteins, it is of great importance not only that conditions such as stirring, pH, oxygen pressure, and temperature are chosen and controlled appropriately but also that a cell growth and protein production medium with the proper nutrients are provided for each stage of the production process.

The media used for mammalian cell culture are complex and consist of a mixture of diverse components, such as sugars, amino acids, electrolytes, vitamins, fetal calf serum, and/or a mixture of peptones,

Type of nutrient	Example(s)
Sugars	Glucose, lactose, sucrose, maltose, dextrins
Fat	Fatty acids, triglycerides
Water (high quality, sterilized)	Water for injection
Amino acids	Glutamine
Electrolytes	Calcium, sodium, potassium, phosphate
Vitamins	Ascorbic acid, -tocopherol, thiamine, riboflavine, folic acid, pyridoxin
Serum (fetal calf serum, synthetic serum)	Albumin, transferrin
Trace minerals	Iron, manganese, copper, cobalt, zinc
Hormones	Growth factors

Table 3.2 ■ Major components of growth media for mammalian cell structures.

growth factors, hormones, and other proteins (see Table 3.2). Many of these ingredients are pre-blended either as concentrate or as homogeneous mixtures of powders. To prepare the final medium, components are dissolved in purified water before filtration. The final medium is filtrated through 0.2 μm filters or through 0.1 μm filters to prevent possible mycoplasma contamination. Some supplements, especially fetal calf serum, contribute considerably to the presence of contaminating proteins and may seriously complicate purification procedures. Moreover, the composition of serum is variable. It depends on the individual animal, season of the year, suppliers' treatment, etc. The use of serum may introduce adventitious material such as viruses, mycoplasma, bacteria, and fungi into the culture system (Berthold and Walter 1994). Furthermore, the possible presence of prions that can cause transmissible spongiform encephalitis almost precludes the use of materials from animal origin. However, if use of this material is inevitable, one must follow the relevant guidelines in which selective sourcing of the material is the key measure to safety (EMA 2011). Many of these potential problems when using serum in cell culture media led to the development of fully defined, free from animal-derived material. These medium formulations were not only developed by the suppliers, there is the trend that the key players in the biotech industry develop their own fully defined medium for their specific production platforms. The advantage of this is that the industry is less dependent on medium suppliers. The fully defined media have been shown to give satisfactory results in large-scale production settings for monoclonal antibody processes. However, hydrolysates from nonanimal origin, like yeast and plant sources, are more and more used for optimal cell growth and product secretion (reviewed by Shukla and Thömmes 2010).

DOWNSTREAM PROCESSING

■ Introduction

Recovering a biological reagent from a cell culture supernatant is one of the critical parts of the manufacturing procedure for biotech products, and purification costs typically outweigh those of the upstream part of the production process. For the production of monoclonal antibodies, protein A resin accounts for some 10 % of the cost, while virus removal by filtration can account for 40 % of the cost (Gottschalk 2006).

More than a decade ago, the protein product was available in a very dilute form, e.g., 10–200 mg/L. At the most concentrations, up to 500–800 mg/L could be reached (Berthold and Walter 1994). Developments in cell culture technology through application of genetics and proteomics resulted in product titers well above 1 g/L. Product titers above 20 g/L are also reported (Monteclaro 2010). These high product titers pose a challenge to the downstream processing unit operations (Shukla and Thömmes 2010). With the low-yield processes, a concentration step is often required to reduce handling volumes for further purification. Usually, the product subsequently undergoes a series of purification steps. The first step in a purification process is to remove cells and cell debris from the process fluids. This process step is normally performed using centrifugation and/or depth filters. Depth filters are often used in combination with filter aid or diatomaceous earth. Often the clarification step is regarded as a part of the upstream process. Therefore, the first actual step in the purification process is a capture step. Subsequent steps remove the residual bulk contaminants, and a final step removes trace contaminants and sometimes variant forms of the molecule. Alternatively, the reverse strategy, where the main contaminants are captured and the product is purified in subsequent steps, might result in a more economic process, especially if the product is not excreted from the cells. In the case where the product is excreted into the cell culture medium, the product will not represent more than 1–5 % of total cellular protein, and a specific binding of the cellular proteins in a product-specific capture step will have a high impact on the efficiency of that step. If the bulk of the contaminants can be removed first, the specific capture step will be more efficient and smaller in size and therefore more economic. Furthermore, smaller subsequent unit operation steps (e.g., chromatography columns) could be used.

After purification, additional steps are performed to bring the desired product into a formulation buffer in which the product is stabilized and can be stored for the desired time until further process steps are performed. Before storage of the final bulk drug substance, the product will be sterilized. Normally this will be performed by a 0.2 μm filtration step. Formulation aspects will be dealt with in Chap. 4.

When designing an upstream and purification protocol, the possibility for scaling up should be considered carefully. A process that has been designed for small quantities is most often not suitable for large quantities for technical, economic, and safety reasons. Developing a purification process, i.e., the isolation and purification of the desired product also called the downstream process (DSP), to recover a recombinant protein in large quantities occurs in two stages: *design* and *scale-up*.

Separating the impurities from the product protein requires a series of purification steps (*process design*), each removing some of the impurities and bringing the product closer to its final specification. In general, the starting feedstock contains cell debris and/or whole-cell particulate material that must be removed. Defining the major contaminants in the starting material is helpful in the downstream process design. This includes detailed information on the source of the material (e.g., bacterial or mammalian cell culture) and major contaminants that are used or produced in the upstream process (e.g., albumin, serum, or product analogs). Moreover, the physical characteristics of the product versus the known contaminants (thermal stability, isoelectric point, molecular weight, hydrophobicity, density, specific binding properties) largely determine the process design. Processes used for production of therapeutics in humans should be safe, reproducible, robust, and produced at the desired cost of goods. The DSP steps may expose the protein molecules to high physical stress (e.g., high temperatures and extreme pH) which can alter the protein properties possibly leading to loss in efficacy. Any substance that is used by injection must be sterile. Furthermore, the endotoxin concentration must be below a certain level depending on the product. Limits are stated in the individual monographs which are to be consulted (e.g., European Pharmacopoeia: less than 0.2 endotoxin units per kg body mass for intrathecal application). Aseptic techniques have to be used wherever possible and necessitate procedures throughout with clean air and microbial control of all materials and equipment used. During validation of the purification process, one must also demonstrate that potential viral contaminants are inactivated and removed (Walter et al. 1992). The purification matrices should be at least sanitizable or, if possible, steam-sterilizable. For depyrogenation, the purification material must withstand either extended dry heat at ≥180 °C or treatment with 1–2 M sodium hydroxide (for further information, see Chap. 4). If any material in contact with the product inadvertently releases compounds, these leachables must be analyzed and their removal by subsequent purification steps must be demonstrated during process validation, or it must be demonstrated that the leachables are below a toxic level. The increased use of plastic film-based single-use production technology (e.g., sterile single-use bioreactor bags, bags to store liquids and filter housings) has made these aspects more significant in the last decade. Suppliers have reacted by providing a significant body of information regarding leachables and biocompatibility for typical solutions used during processing. The problem of leachables is especially hampering the use of affinity chromatography (see below) in the production of pharmaceuticals for human use. On small-scale affinity, chromatography is an important tool for purification and the resulting product might be used for (animal) toxicity studies, but for human use the removal of any leached ligands below a toxic level has to be demonstrated. Because free affinity ligands will bind to the product, the removal might be cumbersome.

Scale-up is the term used to describe a number of processes employed in converting a laboratory procedure into an economical, industrial process. During the scale-up phase, the process moves from the laboratory scale to the pilot plant and finally to the production plant. The objective of scale-up is to produce a product of high quality at a competitive price. Since the costs of downstream processing can be as high as 50–80 % of the total cost of the bulk product, practical and economical ways of purifying the product should be used. Superior protein purification methods hold the key to a strong market position (Wheelwright 1993).

Basic operations required for a downstream purification process used for macromolecules from biological sources are shown in Fig. 3.3.

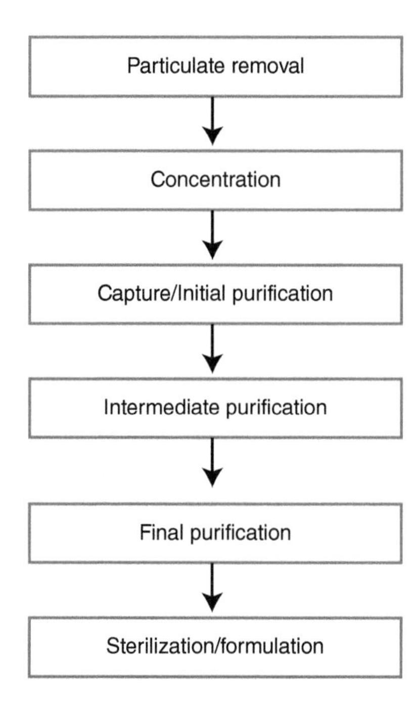

Figure 3.3 ■ Basic operations required for the purification of a biopharmaceutical macromolecule.

As mentioned before, the design of downstream processing is highly product dependent. Therefore, each product requires a specific multistage purification procedure (Sadana 1989). The basic scheme as represented in Fig. 3.3 becomes complex. A typical example of a process flow for the downstream processing is shown in Fig. 3.4. This scheme represents the processing of a glycosylated recombinant interferon (about 28 kDa) produced in mammalian cells. The aims of the individual unit operations are described.

Once the harvest volume and product concentration can be managed, the main purification phase can start. A number of purification methods are available to separate proteins on the basis of a wide variety of different physicochemical criteria such as size, charge, hydrophobicity, and solubility. Detailed information about some separation and purification methods commonly used in purification schemes is provided below.

■ Filtration/Centrifugation

Products from biotechnological industry must be separated from biological systems that contain suspended particulate material, including whole cells, lysed cell material, and fragments of broken cells generated when cell breakage has been necessary to release intracellular products. Most downstream processing flow sheets will, therefore, include at least one unit operation

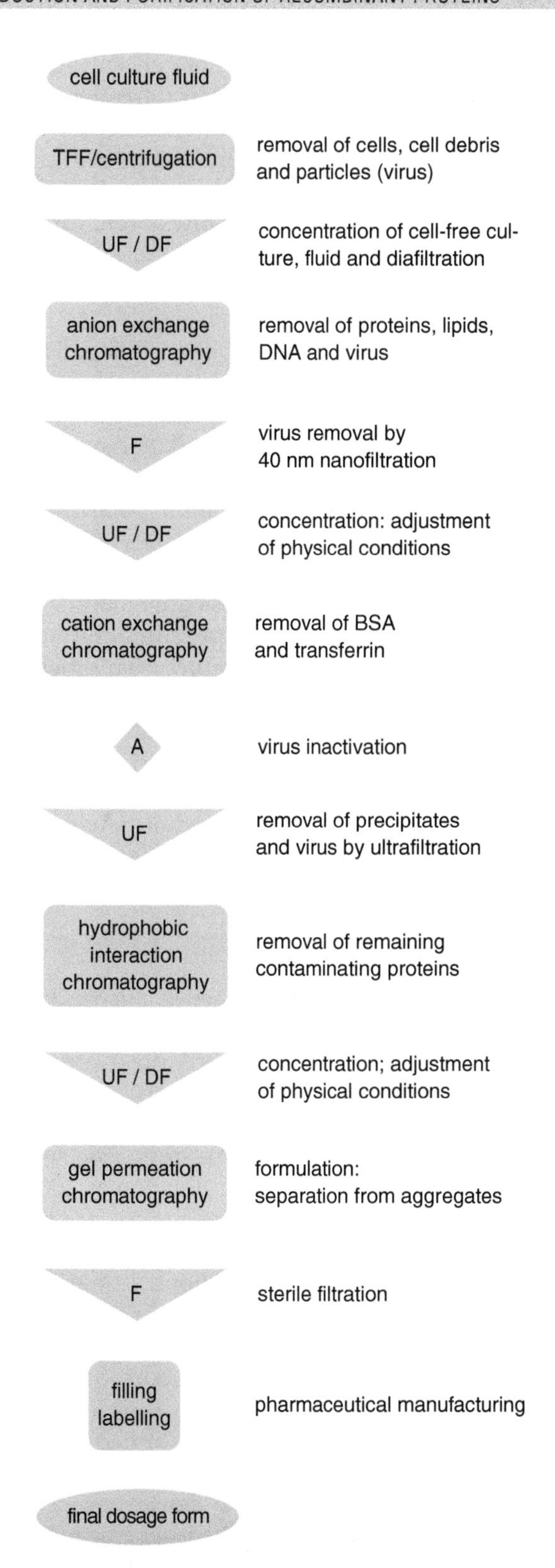

Figure 3.4 ■ Downstream processing of a glycosylated recombinant interferon, describing the purpose of the inclusion of the individual unit operations. *F* filtration, *TFF* tangential flow filtration, *UF* ultrafiltration, *DF* diafiltration, *A* adsorption (Adapted from Berthold and Walter 1994).

for the removal ("clarification") or concentration, just the opposite, of particulates. Most frequently used methods are centrifugation and filtration techniques (e.g., ultrafiltration, diafiltration, and microfiltration). However, the expense and effectiveness of such methods is highly dependent on the physical nature of the particulate material and of the product.

Filtration

Several filtration systems have been developed for separation of cells from media, the most successful being depth filtration and tangential flow systems (also referred to as "cross flow"). In the latter system, high shear across the membrane surface limits fouling, gel layer formation, and concentration polarization. In ultrafiltration, mixtures of molecules of different molecular dimensions are separated by passage of a dispersion under pressure across a membrane with a defined pore size. In general, ultrafiltration achieves little purification of protein product from other molecules with a comparable size, because of the relatively large pore-size distribution of the membranes. However, this technique is widely used to concentrate macromolecules and also to change the aqueous phase in which the particles are dispersed or in which molecules are dissolved (diafiltration) to one required for the subsequent purification steps.

Centrifugation

Subcellular particles and organelles, suspended in a viscous liquid (e.g., the particles produced when cells are disrupted by mechanical procedures), are difficult to separate either by using one fixed centrifugation step or by filtration. But, they can be isolated efficiently by centrifugation at different speeds. For instance, nuclei can be obtained by centrifugation at 400× g for 20 min, while plasma membrane vesicles are pelleted at higher centrifugation rates and longer centrifugation times (fractional centrifugation). In many cases, however, total biomass can easily be separated from the medium by centrifugation (e.g., continuous disc-stack centrifuge). Buoyant density centrifugation can be useful for separation of particles as well. This technique uses a viscous fluid with a continuous gradient of density in a centrifuge tube. Particles and molecules of various densities within the density range in the tube will cease to move when the isopycnic region has been reached. Both techniques of continuous (fluid densities within a range) and discontinuous (blocks of fluid with different density) density gradient centrifugation are used in buoyant density centrifugation on a laboratory scale. However, for application on an industrial scale, continuous centrifuges (e.g., tubular bowl centrifuges) are only used for discontinuous buoyant density centrifugation of protein products. This type of industrial centrifuge is mainly applied to recover precipitated proteins or contaminants. For influenza vaccines, continuous centrifugation is already for decades the workhorse to purify influenza viruses on an industrial scale.

■ Precipitation

The solubility of a particular protein depends on the physicochemical environment, for example, pH, ionic species, and ionic strength of the solution (see also Chap. 4). A slow continuous increase of the ionic strength (of a protein mixture) will selectively drive proteins out of solution. This phenomenon is known as "salting out." A wide variety of agents, with different "salting-out" potencies are available. Chaotropic series with increasing "salting-out" effects of negatively (I) and positively (II) charged molecules are given below:

I. SCN–, I–, CLO4–, NO3–, Br–, Cl–, CH3COO–, PO43–, SO42–

II. Ba2+, Ca2+, Mg2+, Li+, Cs+, Na+, K+, Rb+, NH4+

Ammonium sulfate is highly soluble in cold aqueous solutions and is frequently used in "salting-out" purification.

Another method to precipitate proteins is to use water-miscible organic solvents (change in the dielectric constant). Examples of precipitating agents are polyethylene glycol and trichloroacetic acid. Under certain conditions, chitosan and nonionic polyoxyethylene detergents also induce precipitation (Cartwright 1987; Homma et al. 1993; Terstappen et al. 1993). Cationic detergents have been used to selectively precipitate DNA.

Precipitation is a scalable, simple, and relatively economical procedure for the recovery of a product from a dilute feedstock. It has been widely used for the isolation of proteins from culture supernatants. Unfortunately, with most bulk precipitation methods, the gain in purity is generally limited and product recovery can be low. Moreover, extraneous components are introduced which must be eliminated later. Finally, large quantities of precipitates may be difficult to handle. Despite these limitations, recovery by precipitation has been used with considerable success for some products.

■ Chromatography

Introduction

In preparative chromatography systems, molecular species are primarily separated based on differences in distribution between two phases, one which is the stationary phase (mostly a solid phase) and the other which moves. This mobile phase may be liquid or gaseous (see also Chap. 2). Nowadays, almost all stationary phases (fine particles providing a large surface area) are packed into a column. The mobile phase is passed through by pumps. Downstream protein purification protocols usually have at least two to three chromatography steps.

Chromatographic methods used in purification procedures of biotech products are listed in Table 3.3 and are briefly discussed in the following sections.

Chromatographic Stationary Phases

Chromatographic procedures often represent the rate-limiting step in the overall downstream processing. An important primary factor governing the rate of operation is the mass transport into the pores of conventional packing materials. Adsorbents employed include inorganic materials such as silica gels, glass beads, hydroxyapatite, various metal oxides (alumina), and organic polymers (cross-linked dextrans, cellulose, agarose). Separation occurs by differential interaction of sample components with the chromatographic medium. Ionic groups such as amines and carboxylic acids, dipolar groups such as carbonyl functional groups, and hydrogen bond-donating and bond-accepting groups control the interaction of the sample components with the stationary phase, and these functional groups slow down the elution rate if interaction occurs.

Chromatographic stationary phases for use on a large scale have improved considerably over the last decades. Hjerten et al. (1993) reported on the use of compressed acrylamide-based polymer structures. These materials allow relatively fast separations with good chromatographic performance. Another approach to the problems associated with mass transport in conventional systems is to use chromatographic particles that contain some large "through pores" in addition to conventional pores (see Fig. 3.5). These flow-through or "perfusion chromatography" media enable faster convective mass transport into particles and allow operation at much higher speeds without loss in resolution or binding capacity (Afeyan et al. 1989; Fulton 1994). Another development is the design of spirally wrapped columns containing the adsorption medium. This configuration permits high throughput, high capacity, and good capture efficiency (Cartwright 1987).

The ideal stationary phase for protein separation should possess a number of characteristics, among which are high mechanical strength, high porosity, no nonspecific interaction between protein and the support phase, high capacity, biocompatibility, and high stability of the matrix in a variety of solvents. The latter is especially true for columns used for the production of clinical materials that need to be cleaned, depyrogenized, disinfected, and sterilized at regular intervals. High-performance liquid chromatography (HPLC) systems fulfill many of these criteria. Liquid phases should be carefully chosen to minimize loss of biological activity resulting from the use of some organic solvents. In HPLC small pore-size stationary phases that are incompressible are used. These particles are small, rigid, and regularly sized (to provide a high surface area). The mobile liquid phase is forced under high pressure through the column material. Reversed-phase HPLC systems, using less polar stationary phases than the mobile phases, can in a very few cases be effectively integrated into preparative scale purification schemes of proteins and can serve both as a means of concentration and purification (Benedek and Swadesh 1991).

Separation technique	Mode/principle	Separation based on
Membranes	Microfiltration	Size
	Ultrafiltration	Size
	Nanofiltration	Size
	Dialysis	Size
	Charged membranes	Charge
Centrifugation	Isopycnic banding	Density
	Non-equilibrium setting	Density
Extraction	Fluid extraction	Solubility
	Liquid/liquid extraction	Partition, change in solubility
Precipitation	Fractional precipitation	Change in solubility
Chromatography	Ion exchange	Charge
	Gel filtration	Size
	Affinity	Specific ligand-substrate interaction
	Hydrophobic interaction	Hydrophobicity
	Adsorption	Covalent/non-covalent binding

Table 3.3 ■ Frequently used separation processes and their physical basis.

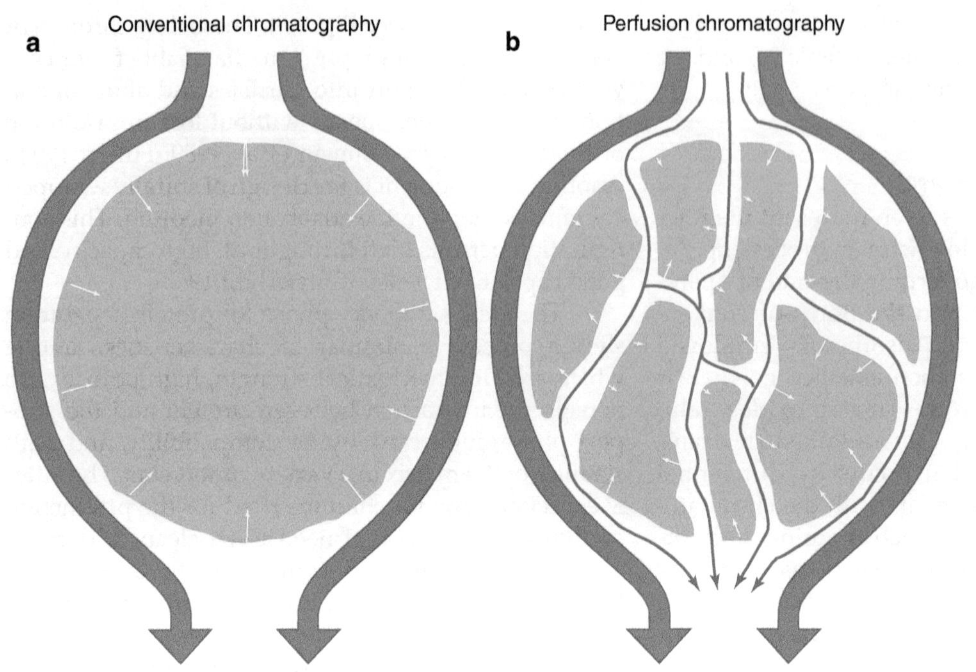

Figure 3.5 ■ The structure of conventional chromatographic particles (**a**) and the perfusion of flow through chromatographic particles (**b**) (Adapted from Fulton 1994).

Unfortunately, HPLC equipment and resin costs are high. Moreover, HPLC is poorly scalable and thus this technology is basically not applied in large-scale purification schemes.

In production environments, columns which operate at relatively low back pressure are often used. They have the advantage that they can be used in equipment constructed from plastics which, unlike conventional stainless steel equipment, resists all buffers likely to be employed in the separation of biomolecules (consider the effect of leachables from plastics). These columns are commercially available and permit the efficient separation of proteins in a single run, making this an attractive unit operation in a manufacturing process. Results can be obtained rapidly and with high resolution. A new development is the use of chromatography equipment with fully disposable flow paths that resists almost all chemicals used in protein purification including disinfection and sterilization media.

Adsorption Chromatography

In adsorption chromatography (also called "normal phase" chromatography), the stationary phase is more polar than the mobile phase. The protein of interest selectively binds to a static matrix under one condition and is released under a different condition. Adsorption chromatography methods enable high ratios of product load to stationary phase volume. Therefore, this principle is economically scalable.

Ion-Exchange Chromatography

Ion-exchange chromatography can be a powerful step early in a purification scheme. It can be easily scaled up. Ion-exchange chromatography can be used in a negative mode, i.e., the product flows through the column under conditions that favor the adsorption of contaminants to the matrix, while the protein of interest does not bind (Tennikova and Svec 1993). The type of the column needed is determined by the properties of the proteins to be purified (e.g., isoelectric point and charge density). Anion exchangers bind negatively charged molecules and cation exchangers bind positively charged molecules. In salt-gradient ion-exchange chromatography, the salt concentration in the perfusing elution buffer is increased continuously or in steps. The stronger the binding of an individual protein to the ion exchanger, the later it will appear in the elution buffer. Likewise, in pH-gradient chromatography, the pH is changed continuously or in steps. Here, the protein binds at 1 pH and is released at a different pH. As a result of the heterogeneity in glycosylation (e.g., a varying number of sialic acid moieties), glycosylated proteins may elute in a relatively broad pH range (up to 2 pH units).

In order to simplify purification, a specific amino acid tail can be added to the protein at the gene level to create a "purification handle." For example, a short tail consisting of arginine residues allows a protein to bind to a cation exchanger under conditions where almost no other cell proteins bind. However, this technique is only useful for laboratory-scale isolation of the product and cannot be used for production scale due to regulatory problems related to the removal of the arginine or other specific tags from the protein.

(Immuno)Affinity Chromatography

Affinity Chromatography

Affinity chromatography is based on highly specific interactions between an immobilized ligand and the protein of interest. Affinity chromatography is a very powerful method for the purification of proteins. Under physiological conditions, the protein binds to the ligand. Extensive washing of this matrix will remove contaminants, and the purified protein can be recovered by the addition of ligands competing for the stationary phase binding sites or by changes in physical conditions (such as low or high pH of the eluent) which greatly reduce the affinity. Examples of affinity chromatography include the purification of glycoproteins, which bind to immobilized lectins, and the purification of serine proteases with lysine binding sites, which bind to immobilized lysine. In these cases, a soluble ligand (sugar or lysine, respectively) can be used to elute the required product under relatively mild conditions. Another example is the use of the affinity of protein A and protein G for antibodies. Protein A and protein G have a high affinity for the Fc portions of many immunoglobulins from various animals. Protein A and G matrices can be commercially obtained with a high degree of purity. For the purification of, e.g., hormones or growth factors, the receptors or short peptide sequence that mimic the binding site of the receptor molecule can be used as affinity ligands. Some proteins show highly selective affinity for certain dyes commercially available as immobilized ligands on purification matrices. When considering the selection of these ligands for pharmaceutical production, one must realize that some of these dyes are carcinogenic and that a fraction may leach out during the process.

An interesting approach to optimize purification is the use of a gene that codes not only for the desired protein but also for an additional sequence that facilitates recovery by affinity chromatography. At a later stage the additional sequence is removed by a specific cleavage reaction. As mentioned before, this is a complex process that needs additional purification steps.

In general, use of affinity chromatography in the production process for therapeutics may lead to complications during validation of the removal of free ligands or protein extensions. Consequently, except for monoclonal antibodies where affinity chromatography is part of the purification platform at large scale, this technology is rarely used in the industry.

Immunoaffinity Chromatography

The specific binding of antibodies to their epitopes is used in immunoaffinity chromatography (Chase and Draeger 1993). This technique can be applied for purification of either the antigen or the antibody. The antibody can be covalently coupled to the stationary phase and act as the "receptor" for the antigen to be purified. Alternatively, the antigen, or parts thereof, can be attached to the stationary phase for the purification of the antibody. Advantages of immunoaffinity chromatography are its high specificity and the combination of concentration and purification in one step.

A disadvantage associated with immunoaffinity methods is the sometimes very strong antibody-antigen binding. This requires harsh conditions during elution of the ligand. Under such conditions, sensitive ligands could be harmed (e.g., by denaturation of the protein to be purified). This can be alleviated by the selection of antibodies and environmental conditions with high specificity and sufficient affinity to induce an antibody-ligand interaction, while the antigen can be released under mild conditions (Jones 1990). Another concern is disruption of the covalent bond linking the "receptor" to the matrix. This would result in elution of the entire complex. Therefore, in practice, a further purification step after affinity chromatography as well as an appropriate detection assay (e.g., ELISA) is almost always necessary. On the other hand, improved coupling chemistry that is less susceptible to hydrolysis has been developed to prevent leaching.

Scale-up of immunoaffinity chromatography is often hampered by the relatively large quantity of the specific "receptor" (either the antigen or the antibody) that is required and the lack of commercially available, ready-to-use matrices. The use of immunoaffinity in pharmaceutical processes will have major regulatory consequences since the immunoaffinity ligand used will be considered by the regulatory bodies as a "second product," thus will be subjected to the nearly the same regulatory scrutiny as the drug substance. Moreover, immunoaffinity ligands can have a significant effect of the final costs of goods.

Examples of proteins of potential therapeutic value that have been purified using immunoaffinity chromatography are interferons, urokinase, erythropoietin, interleukin-2, human factor VIII and X, and recombinant tissue plasminogen activator.

Hydrophobic Interaction Chromatography

Under physiological conditions, most hydrophobic amino acid residues are located inside the protein core, and only a small fraction of hydrophobic amino acids is exposed on the "surface" of a protein. Their exposure is suppressed because of the presence of hydrophilic amino acids that attract large clusters of water molecules and form a "shield." High salt concentrations reduce the hydration of a protein, and the surface-exposed hydrophobic amino acid residues become more accessible. Hydrophobic interaction chromatography (HIC) is based on non-covalent and non-electrostatic interactions between proteins and the stationary phase.

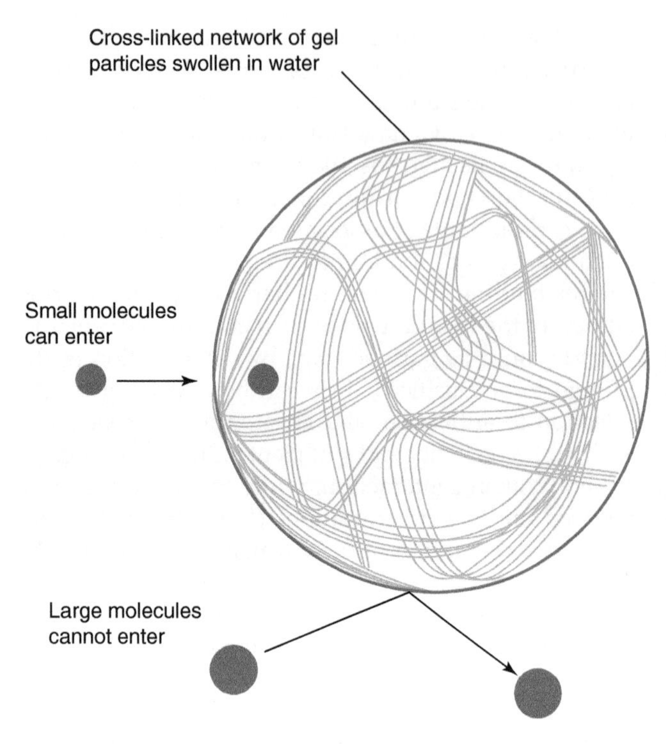

Figure 3.6 ■ Schematic representation of gel filtration (Adapted from James 1992).

HIC is a mild technique, usually yielding high recoveries of proteins that are not damaged, are folded correctly, and are separated from contaminants that are structurally related. HIC is ideally placed in the purification scheme after ion-exchange chromatography, where the protein usually is released in high ionic strength elution media (Heng and Glatz 1993).

Gel-Permeation Chromatography

Gel-permeation or size-exclusion chromatography, also known as gel filtration, separates molecules according to their shape and size (see Fig. 3.6). Inert gels with narrow pore-size distributions in the size range of proteins are available. These gels are packed into a column and the protein mixture is then loaded on top of the column and the proteins diffuse into the gel. The smaller the protein, the more volume it will have available in which to disperse. Molecules that are larger than the largest pores are not able to penetrate the gel beads and will therefore stay in the void volume of the column. When a continuous flow of buffer passes through the column, the larger proteins will elute first and the smallest molecules last. Gel-permeation chromatography is a good alternative to membrane diafiltration for buffer exchange at almost any purification stage, and it is often used in laboratory design. At production scale, the use of this technique is usually limited, because only relatively small sample volumes can be loaded on a large column (up to one-third of the column volume in the case of "buffer exchange"). It is therefore best avoided or used late in the purification process when the protein is available in a highly concentrated form. Gel filtration is commonly used as the final step in the purification to bring proteins in the appropriate buffer used in the final formulation. In this application, its use has little if no effect on the product purity characteristics.

Expanded Beds

As mentioned before, purification schemes are based on multistep protocols. This not only adds greatly to the overall production costs but also can result in significant loss of product. Therefore, there still is an interest in the development of new methods for simplifying the purification process. Adsorption techniques are popular methods for the recovery of proteins, and the conventional operating format for preparative separations is a packed column (or fixed bed) of adsorbent. Particulate material, however, can be trapped near the bed, which results in an increase in the pressure drop across the bed and eventually in clogging of the column. This can be avoided by the use of pre-column filters (0.2 μm) to save the column integrity. Another solution to this problem may be the use of expanded beds (Chase and Draeger 1993; Fulton 1994), also called fluidized beds (see Fig. 3.7). In principle, the use of expanded beds enables clarification, concentration, and purification to be achieved in a single step. The concept is to employ a particulate solid-phase adsorbent in an open bed with upward liquid flow. The hydrodynamic drag around the particles tends to lift them upwards, which is counteracted by gravity because of a density difference between the particles and the liquid phase. The particles remain suspended if particle diameter, particle density, liquid viscosity, and liquid density are properly balanced by choosing the correct flow rate. The expanded bed allows particles (cells) to pass through, whereas molecules in solution are selectively retained (e.g., by the use of ion-exchange or affinity adsorbents) on the adsorbent particles. Feedstocks can be applied to the bed without prior removal of particulate material by centrifugation or filtration, thus reducing process time and costs. Fluidized beds have been used previously for the industrial-scale recovery of antibiotics such as streptomycin and novobiocin (Fulton 1994; Chase 1994). Stable, expanded beds can be obtained using simple equipment adapted from that used for conventional, packed bed adsorption and chromatography processes. Ion-exchange adsorbents are likely to be chosen for such separations.

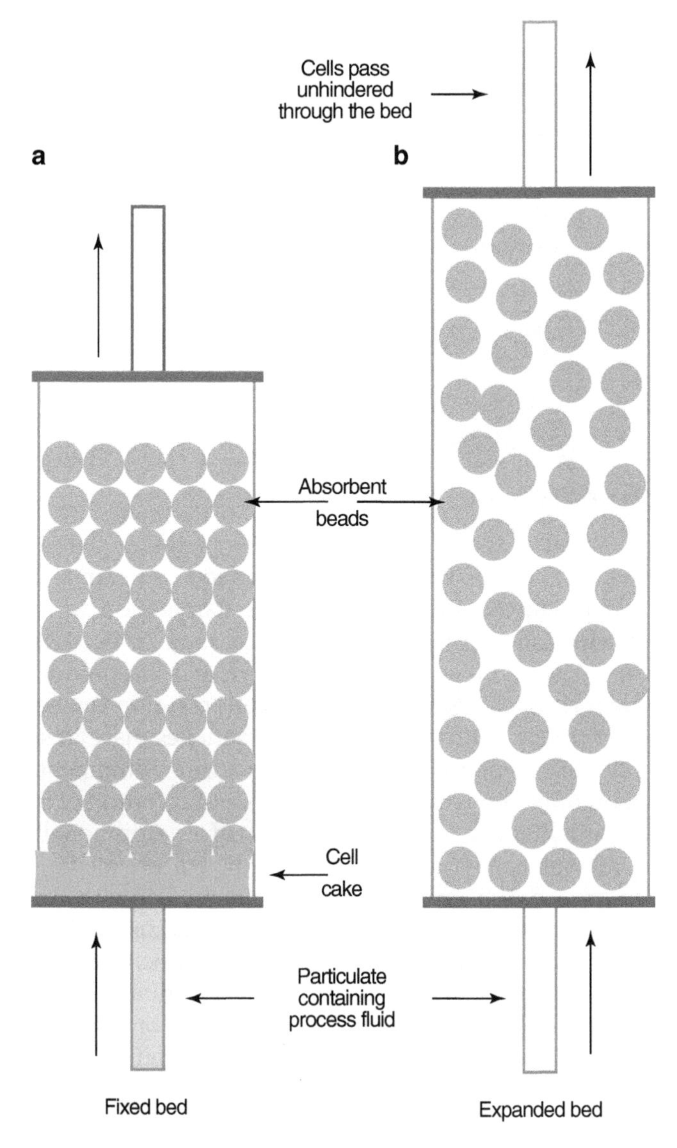

Figure 3.7 ■ Comparison between (**a**) a packed bed and (**b**) an expanded bed (Adapted from Chase and Draeger 1993).

CONTAMINANTS

For pharmaceutical applications, product purity mostly is ≥99 % when used as a parenteral (Berthold and Walter 1994; ICH 1999a). Purification processes should yield potent proteins with well-defined characteristics for human use from which "all" contaminants have been removed to a major extent. The purity of the drug protein in the final product will therefore largely depend upon the purification technology applied.

Table 3.4 lists potential contaminants that may be present in recombinant protein products from bacterial and nonbacterial sources. These contaminants can be host-related, process-related and product-related.

Origin	Contaminant
Host-related	Viruses
	Host-derived proteins acid DNA
	Glycosylation variants
	N- and C-terminal variants
	Endotoxins (from Gram-negative bacterial hosts)
Product-related	Amino acids substitution and deletion
	Denatured protein
	Conformational isomers
	Dimers and aggregates
	Disulfide pairing variants
	Deamidated species
	Protein fragments
Process-related	Growth medium components
	Purification reagents
	Metals
	Column materials

Table 3.4 ■ Potential contaminants in recombinant protein products derived from bacterial and nonbacterial hosts.

In the following sections, special attention is paid to the detection and elimination of contamination by viruses, bacteria, cellular DNA, and undesired proteins.

■ Viruses

Endogenous and adventitious viruses, which require the presence of living cells to propagate, are potential contaminants of animal cell cultures and, therefore, of the final drug product. If present, their concentration in the purified product will be very low and it will be difficult to detect them. Viruses such as retrovirus can be visualized by (nonsensitive) electron microscopy. For retroviruses, a highly sensitive RT-PCR (reverse-transcriptase polymerase chain reaction) assay is available, but for other viruses, a sensitive in vitro assay might be lacking. The risks of some viruses (e.g., hepatitis virus) are known (Walter et al. 1991; Marcus-Sekura 1991), but there are other viruses whose risks cannot be properly judged because of lack of solid experimental data. Some virus infections, such as parvovirus, can have long latent periods before their clinical effects show up. Long-term effects of introducing viruses into a patient treated with a recombinant protein should not be overlooked. Therefore, it is required that products used parenterally are free from viruses. The specific virus testing regime required will depend on the cell type used for production (Löwer 1990; Minor 1994).

Category	Types	Example
Inactivation	Heat treatment	Pasteurization
	Radiation	UV-light
	Dehydration	Lyophilization
	Cross linking agents, denaturating or disrupting agents	β-propiolactone, formaldehyde, NaOH, organic solvents (e.g., chloroform), detergents (e.g., Na-cholate)
	Neutralization	Specific, neutralizing antibodies
Removal	Chromatography	Ion-exchange, immuno-affinity, chromatography
	Filtration	Nanofiltration
	Precipitation	Cyroprecipitation

Table 3.5 ■ Methods for reducing or inactivating viral contaminants.

Viruses can be introduced by nutrients, by an infected production cell line, or they are introduced (by human handling) during the production process. The most frequent source of virus introduction is animal serum. In addition, animal serum can introduce other unwanted agents such as bacteria, mycoplasmas, prions, fungi, and endotoxins. It should be clear that appropriate screening of cell banks and growth medium constituents for viruses and other adventitious agents should be strictly regulated and supervised (Walter et al. 1991; FDA 1993; ICH 1999b; WHO 2010). Validated, orthogonal methods to inactivate and remove possible viral contaminants are mandatory for licensing of therapeutics derived from mammalian cells or transgenic animals (Minor 1994). Viruses can be inactivated by physical and chemical treatment of the product. Heat, irradiation, sonication, extreme pH, detergents, solvents, and certain disinfectants can inactivate viruses. These procedures can be harmful to the product as well and should therefore be carefully evaluated and validated (Walter et al. 1992; Minor 1994; ICH 1999b). Removal of viruses by nanofiltration is an elegant and effective technique and the validation aspects of this technology are well described (PDA 2005). Filtration through 15 nm membranes can remove even the smallest non-enveloped viruses like bovine parvovirus (Maerz et al. 1996). Another common, although less robust, method to remove viruses in antibody processes is by ion-exchange chromatography. A number of methods for reducing or inactivating viral contaminants are mentioned in Table 3.5 (Horowitz et al. 1991).

■ Bacteria

Unwanted bacterial contamination may be a problem for cells in culture or during pharmaceutical purification. Usually the size of bacteria allows simple filtration over 0.2 μm (or smaller) filters for adequate removal. In order to further prevent bacterial contamination during production, the raw materials used have to be sterilized, preferably at 121 °C or higher, and the products are manufactured under strict aseptic conditions wherever possible. Production most often takes place in so-called clean rooms in which the chances of environmental contamination is reduced through careful control of the environment, for example, filtration of air. Additionally, antibiotic agents can be added to the culture media in some cases but have to be removed further downstream in the purification process. However, the use of beta-lactam antibiotics such as penicillin is strictly prohibited due to oversensitivity of some individuals to these compounds. Because of the persistence of antibiotic residues, which are difficult to eliminate from the product, appropriately designed manufacturing plants and extensive quality control systems for added reagents (medium, serum, enzymes, etc.) permitting antibiotic-free operation are preferable.

Pyrogens (usually endotoxins of gram-negative bacteria) are potentially hazardous substances (see Chap. 4). Humans are sensitive to pyrogen contamination at very low concentrations (picograms per mL). Pyrogens may elicit a strong fever response and can even be fatal. Simple 0.2 μm filtration does not remove pyrogens. Removal is complicated further because pyrogens vary in size and chemical composition. However, sensitive tests to detect and quantify pyrogens are commercially available. Purification schemes usually contain at least one step of ion-exchange chromatography (anionic-exchange material) to remove the negatively charged endotoxins (Berthold and Walter 1994). Furthermore, materials used in process such as glassware are typically subjected to a depyrogenation step prior to use often by the use of elevated temperatures (≥180 °C) in a dry heat oven.

■ Cellular DNA

The application of continuous mammalian cell lines for the production of recombinant proteins might result in the presence of oncogene-bearing DNA fragments in the final protein product (Walter and Werner 1993; Löwer 1990). A stringent purification protocol that is capable of reducing the DNA content and fragment size to a safe level is therefore necessary (Berthold and Walter 1994; WHO 2010). There are a number of approaches available to validate that the purification method removes cellular DNA and RNA. One such approach involves incubating the cell line with radiolabeled nucleotides and determining radioactivity in the purified product obtained through the purification protocol. Other methods are dye-binding fluorescence-enhancement assays for nucleotides and PCR-based methods. If the presence of nucleic acids persists in a final preparation, then additional steps must be introduced in the purification process. The question about a safe level of nucleic acids in biotech products is difficult to answer because of minimal relevant know-how. Transfection with so-called naked DNA is very difficult and a high concentration of DNA is needed. Nevertheless, it is agreed for safety reasons that final product contamination by nucleic acids should not exceed 100 pg or 10 ng per dose depending on the type of cells used to produce the pharmaceutical (WHO 2010; European Pharmacopoeia 2011).

■ Protein Contaminants and Product Variants

As mentioned before, minor amounts of host-, process-, and product-related proteins will likely appear in biotech products. These types of contaminants are a potential health hazard because, if present, they may be recognized as antigens by the patient receiving the recombinant protein product. On repeated use the patient may show an immune reaction caused by the contaminant, while the protein of interest is performing its beneficial function. In such cases the immunogenicity may be misinterpreted as being due to the recombinant protein itself. Therefore, one must be very cautious in interpreting safety data of a given recombinant therapeutic protein.

Generally, the sources of host- and process-related protein contaminants are the growth medium used and the host proteins of the cells. Among the host-derived contaminants, the host species' version of the recombinant protein could be present (WHO 2010). As these proteins are similar in structure, it is possible that undesired proteins are co-purified with the desired product. For example, urokinase is known to be present in many continuous cell lines. The synthesis of highly active biological molecules such as cytokines by hybridoma cells might be another concern (FDA 1990). Depending upon their nature and concentration, these cytokines might enhance the antigenicity of the product.

"Known" or expected contaminants should be monitored at the successive stages in a purification process by suitable in-process controls, e.g., sensitive immunoassay(s). Tracing of the many "unknown" cell-derived proteins is more difficult. When developing a purification process, other less-specific analyses such as SDS-PAGE are usually used in combination with various staining techniques.

Product-related contaminants/variants may pose a potential safety issue for patients. These contaminants can be aggregated forms of the product, a product with heterogeneity in the disulfide bridges, N- and C-terminal variants, glycosylation variants, deamidated product, etc. Some of these contaminants/variants are described in the following paragraphs.

N- and C-Terminal Heterogeneity

A major problem connected with the production of biotech products is the problem associated with the amino (NH2)-terminus of the protein, e.g., in E. coli systems, where protein synthesis always starts with methylmethionine. Obviously, it has been of great interest to develop methods that generate proteins with an NH2-terminus as found in the authentic protein. When the proteins are not produced in the correct way, the final product may contain several methionyl variants of the protein in question or even contain proteins lacking one or more residues from the amino terminus. This is called the amino-terminal heterogeneity. This heterogeneity can also occur with recombinant proteins (e.g., α-interferon) susceptible to proteases that are either secreted by the host or introduced by serum-containing media. These proteases can clip off amino acids from the C-terminal and/or N-terminal of the desired product (amino- and/or carboxy-terminal heterogeneity). Amino- and/or carboxy-terminal heterogeneity is not desirable since it may cause difficulties in purification and characterization of the proteins. In case of the presence of an additional methionine at the N-terminal end of the protein, its secondary and tertiary structure can be altered. This could affect the biological activity and stability and may make it immunogenic. Moreover, N-terminal methionine and/or internal methionine is sensitive to oxidation (Sharma 1990).

Conformational Changes/Chemical Modifications

Although mammalian cells are able to produce proteins structurally equal to endogenous proteins, some caution is advisable. Transcripts containing the full-length coding sequence could result in conformational isomers of the protein because of unexpected secondary structures that affect translational fidelity (Sharma 1990). Another factor to be taken into account is the possible existence of equilibria between the desired form and other forms such as dimers. The correct folding of proteins after biosynthesis is important because it determines the specific activity of the protein (Berthold and Walter 1994). Therefore, it is important to determine if all molecules of a given recombinant protein secreted by a mammalian expression system are folded in their native conformation. In some cases, it may be relatively easy to detect misfolded structures, but in other cases, it may be extremely difficult. Apart from conformational changes, proteins can undergo chemical alterations, such as proteolysis, deamidation, and hydroxyl and sulfhydryl oxidations during the purification process. These alterations can result in (partial) denaturation of the protein. Vice versa, denaturation of the protein may cause chemical modifications as well (e.g., as a result of exposure of sensitive groups).

Glycosylation

Many therapeutic proteins produced by recombinant DNA technology are glycoproteins (Sharma 1990). The presence and nature of oligosaccharide side chains in proteins affect a number of important characteristics, like the proteins' serum half-life, solubility, and stability, and sometimes even the pharmacological function (Cumming 1991). Darbepoetin, a second-generation, genetically modified erythropoietin, has a carbohydrate content of 80 % compared to 40 % for the native molecule, which increases the in vivo half-life after intravenous administration from 8 h for erythropoietin to 25 h for darbepoetin (Sinclair and Elliott 2005). As a result, the therapeutic profile may be "glycosylation" dependent. As mentioned previously, protein glycosylation is not determined by the DNA sequence. It is an enzymatic modification of the protein after translation and can depend on the environment in the cell. Although mammalian cells are very well able to glycosylate proteins, it is hard to fully control glycosylation. Carbohydrate heterogeneity can occur through the size of the chain, type of oligosaccharide, and sequence of the carbohydrates. This has been demonstrated for a number of recombinant products including interleukin-4, chorionic gonadotropin, erythropoietin, and tissue plasminogen activator. Carbohydrate structure and composition in recombinant proteins may differ from their native counterparts, because the enzymes required for synthesis and processing vary among different expression systems (e.g., glycoproteins in insect cells are frequently smaller than the same glycoproteins expressed in mammalian cells) or even from one mammalian system to another.

Proteolytic Processing

Proteases play an important role in processing, maturation, modification, or isolation of recombinant proteins. Proteases from mammalian cells are involved in secreting proteins into the cultivation medium. If secretion of the recombinant protein occurs co-translationally, then the intracellular proteolytic system of the mammalian cell should not be harmful to the recombinant protein. Proteases are released if cells die or break (e.g., during cell break at cell harvest) and undergo lysis. It is therefore important to control growth and harvest conditions in order to minimize this effect. Another source of proteolytic attack is found in the components of the medium in which the cells are grown. For example, serum contains a number of proteases and protease zymogens that may affect the secreted recombinant protein. If present in small amounts and if the nature of the proteolytic attack on the desired protein is identified, appropriate protease inhibitors to control proteolysis could be used. It is best to document the integrity of the recombinant protein after each purification step.

Proteins become much more susceptible to proteases at elevated temperatures. Purification strategies should be designed to carry out all the steps at 2–8 °C (Sharma 1990) if proteolytic degradation occurs. Alternatively, Ca2+ complexing agents (e.g., citrate) can be added as many proteases depend on Ca2+ for their activity. From a manufacturing perspective, however, providing cooling to large-scale chromatographic processes, although not impossible, is a complicating factor in the manufacturing process.

BACTERIA: PROTEIN INCLUSION BODY FORMATION

In bacteria soluble proteins can form dense, finely granular inclusions within the cytoplasm. These "inclusion bodies" often occur in bacterial cells that overproduce proteins by plasmid expression. The protein inclusions appear in electron micrographs as large, dense bodies often spanning the entire diameter of the cell. Protein inclusions are probably formed by a buildup of amorphous protein aggregates held together by covalent and non-covalent bonds. The inability to measure inclusion body proteins directly may lead to the inaccurate assessment of recovery and yield and may cause problems if protein solubility is essential for efficient, large-scale purification (Berthold and Walter 1994). Several schemes for recovery of proteins from inclusion bodies have been described. The recovery of proteins from inclusion bodies requires cell breakage and inclusion body recovery. Dissolution of inclusion proteins is the next step in the purification scheme and typically takes place in extremely dilute solutions which tend to have the effect of increasing the volumes of the unit operations during the manufacturing phases. This can make process control more difficult if, for example, low temperatures are required during these steps. Generally, inclusion proteins dissolve in denaturing agents such as sodium dodecyl sulfate (SDS), urea, or guanidine hydrochloride. Because bacterial systems generally are incapable of forming disulfide bonds, a protein containing these bonds has to be refolded under oxidizing conditions to restore these bonds and to generate the biologically active protein. This so-called renaturation step is increasingly difficult if more S-S bridges are present in the molecule and the yield of renatured product could be as low as only a few percent. Once the protein is solubilized, conventional chromatographic separations can be used for further purification of the protein.

Aggregate formation at first sight may seem undesirable, but there may also be advantages as long as the protein of interest will unfold and refold properly. Inclusion body proteins can easily be recovered to yield proteins with >50 % purity, a substantial improvement over the purity of soluble proteins (sometimes below 1 % of the total cell protein). Furthermore, the aggregated forms of the proteins are more resistant to proteolysis, because most molecules of an aggregated form are not accessible to proteolytic enzymes. Thus the high yield and relatively cheap production using a bacterial system can offset a low-yield renaturation process. For a non-glycosylated, simple molecule, this still is the production system of choice.

COMMERCIAL-SCALE MANUFACTURING AND INNOVATION

A major part of the recombinant proteins on the market consist of monoclonal antibodies produced in mammalian cells. Pharmaceutical production processes have been set up since the early 1980s of the twentieth century. These processes essentially consist of production in stirred tanks bioreactors, clarification using centrifugation, and membrane technology, followed by protein A capture, low-pH virus inactivation, cation-exchange and anion-exchange chromatography, virus filtration, and UF/DF for product formulation (Shukla and Thömmes 2010). Such platform processes run consistently at very large scale (e.g., multiple 10,000 L bioreactors and higher). Product recovery is generally very high (>70 %). Since product titers in the bioreactors have increased to a level where further increases have no or a minimal effect on the cost of goods, and the cost of goods are a fraction of the sales price, the focus of these large-scale processes is changing from process development to process understanding (Kelley 2009). However, it is also anticipated that the monoclonal antibody demands may decrease due to more efficacious products like antibody-drug conjugates and the competition with generic products. A lower demand together with the increase in recombinant protein yields will lead to a decrease in bioreactor size, an increase in the need for flexible facilities, and faster turnaround times leading to a growth in the use of disposables and other innovative technologies (Shukla and Thömmes 2010).

SELF-ASSESSMENT QUESTIONS

■ Questions

1. Name four types of different bioreactors.
2. Chromatography is an essential step in the purification of biotech products. Name at least five different chromatographic purification methods.
3. What are the major safety concerns in the purification of cell-expressed proteins?
4. What are the critical issues in production and purification that must be addressed in process validation?
5. Mention at least five issues to consider in the cultivation and purification of proteins.
6. True or false: Glycosylation is an important posttranslational change of pharmaceutical proteins. Glycosylation is only possible in mammalian cells.
7. Glycosylation may affect several properties of the protein. Mention at least three possible changes in case of glycosylation.
8. Pharmacologically active biotech protein products have complex three-dimensional structures. Mention two or more important factors that affect these structures.

■ Answers

1. Stirred-tank, airlift, microcarrier, and membrane bioreactors
2. Adsorption chromatography, ion-exchange chromatography, affinity chromatography, hydrophobic interaction chromatography, gel-permeation or size-exclusion chromatography
3. Removal of viruses, bacteria, protein contaminants, and cellular DNA
4. Production: scaling up and design. Purification: procedures should be reliable in potency and quality of the product and in the removal of viral, bacterial, DNA, and protein contaminants
5. Grade of purity, pyrogenicity, N- and C-terminal heterogeneity, chemical modification/conformational changes, glycosylation, and proteolytic processing
6. False; glycosylation is also possible in yeast (and to some degree in bacteria)
7. Solubility, pKa, charge, stability, and biological activity
8. Amino acid structure, hydrogen and sulfide bridging, posttranslational changes, and chemical modification of the amino acid rest groups

REFERENCES

Afeyan N, Gordon N, Mazsaroff I, Varady L, Fulton S, Yang Y, Regnier F (1989) Flow-through particles of the high-performance liquid chromatographic separation of biomolecules, perfusion chromatography. J Chromatogr 519:1–29

Benedek K, Swadesh JK (1991) HPLC of proteins and peptides in the pharmaceutical industry. In: Fong GW, Lam SK (eds) HPLC in the pharmaceutical industry. Dekker, New York, pp 241–302

Berthold W, Walter J (1994) Protein purification: aspects of processes for pharmaceutical products. Biologicals 22:135–150

Borman S (2006) Glycosylation engineering. Chem Eng News 84:13–22

Cartwright T (1987) Isolation and purification of products from animal cells. Trends Biotechnol 5:25–30

Celik E, Calik P (2011) Production of recombinant proteins by yeast cells. Biotechnol Adv. doi:10.1016/j.biotechadv.2011.09.011

Chase HA (1994) Purification of proteins by adsorption chromatography in expanded beds. Trends Biotechnol 12:296–303

Chase H, Draeger N (1993) Affinity purification of proteins using expanded beds. J Chromatogr 597:129–145

Compton B, Jensen J (2007) Use of perfusion technology on the Rise – New modes are beginning to gain ground on Fed-Batch strategy. Genetic Engineering & Biotechnology News 27(17):48

Cumming DA (1991) Glycosylation of recombinant protein therapeutics: Control and functional implications. Glycobiology 1(2):115–130

Dell A, Galadari A, Sastre F, Hitchen P (2011) Similarities and differences in the glycosylation mechanisms in prokaryotes and eukaryotes. Int J Microbiol 2010(2010):148178

EMA (2011) Note for guidance on minimising the risk of transmitting animal spongiform encephalopathy agents via human and veterinary medicinal products (EMA/410/01 rev.3)

European Pharmacopoeia Seventh Edition. Strasbourg: Council of Europe (2011) 5.2.3. Cell substrates for the production of vaccines for human use

FD3A, Center for Biologics Evaluation and Research (1990) Cytokine and growth factor pre-pivotal trial information package with special emphasis on products identified for consideration under 21 CFR 312 Subpart E. Bethesda

FDA, Office of Biologicals Research and Review (1993) Points to consider in the characterization of cell lines used to produce biologicals. Rockville Pike/Bethesda

Fulton SP (1994) Large scale processing of macromolecules. Curr Opin Biotechnol 5:201–205

Gottschalk U (2006) The renaissance of protein purification. BioPharm Int 19(6):S8–S9

Heng M, Glatz C (1993) Charged fusions for selective recovery of ß-galactosidase from cell extract using hollow fiber ion-exchange membrane adsorption. Biotechnol Bioeng 42:333–338

Hjerten S, Mohammed J, Nakazato K (1993) Improvement in flow properties and pH stability of compressed, continuous polymer beds for high-performance liquid chromatography. J Chromatogr 646:121–128

Hodge G (2004) Disposable components enable a new approach to biopharmaceutical manufacturing. BioPharm Int 2004(15):38–49

Homma T, Fuji M, Mori J, Kawakami T, Kuroda K, Taniguchi M (1993) Production of cellobiose by enzymatic hydrolysis: removal of ß-glucosidase from cellulase by affinity precipitation using chitosan. Biotechnol Bioeng 41:405–410

Horowitz MS, Bolmer SD, Horowitz B (1991) Elimination of disease-transmitting enveloped viruses from human blood plasma and mammalian cell culture products. Bioseparation 1:409–417

ICH (International Conference on Harmonization) Topic Q6B (1999a) Specifications: test procedures and acceptance criteria for biotechnology/biological products

ICH (International Conference on Harmonization) Topic Q5A (1999b) Viral safety evaluation of biotechnology products derived from cell lines of human or animal origin

International Conference on Harmonization guideline M7 on assessment and control of DNA reactive (mutagenic) impurities in pharmaceuticals to limit potential carcinogenic risk

James AM (1992) Introduction fundamental techniques. In: James AM (ed) Analysis of amino acids and nucleic acids. Butterworth-Heinemann, Oxford, pp 1–28

Jones K (1990) Affinity chromatography, a technology update. Am Biotechnol Lab 8:26–30

Kelley B (2009) Industrialization of mAb production technology. MAbs 1(5):443–452

Klegerman ME, Groves MJ (1992) Pharmaceutical biotechnology. Interpharm Press, Inc., Buffalo Grove

Löwer J (1990) Risk of tumor induction in vivo by residual cellular DNA: quantitative considerations. J Med Virol 31:50–53

Maerz H, Hahn SO, Maassen A, Meisel H, Roggenbuck D, Sato T, Tanzmann H, Emmrich F, Marx U (1996) Improved removal of viruslike particles from purified monoclonal antibody IgM preparation via virus filtration. Nat Biotechnol 14:651–652

Marcus-Sekura CJ (1991) Validation and removal of human retroviruses. Center for Biologics Evaluation and Research, FDA, Bethesda

Minor PD (1994) Ensuring safety and consistency in cell culture production processes: viral screening and inactivation. Trends Biotechnol 12:257–261

Monteclaro F (2010) Protein expression systems, ringing in the new. Innov Pharm Technol 12:45–49

Note for Guidance (1991) *Validation* of virus removal and inactivation procedure, Ad Hoc Working Party on Biotechnology/Pharmacy, European Community, DG III/8115/89-EN

Orzaez D, Granell A, Blazquez MA (2009) Manufacturing antibodies in the plant cell. Biotechnol J 4:1712–1724

PDA Journal of Pharmaceutical Science and Technology (2005) Technical report No. 41, Virus filtration, 59, No. S-2

Peters J, Stoger E (2011) Transgenic crops for the production of recombinant vaccines and anti-microbial antibodies. Hum Vaccin 7(3):367–374

Sadana A (1989) Protein inactivation during downstream separation, part I: the processes. Biopharm 2:14–25

Sharma SK (1990) Key issues in the purification and characterization of recombinant proteins for therapeutic use. Adv Drug Deliv Rev 4:87–111

Shukla AA, Thömmes J (2010) Recent advances in large-scale production of monoclonal antibodies and related proteins. Trends Biotechnol 28:253–261

Sinclair AM, Elliott S (2005) Glycoengineering: the effect of glycosylation on the properties of therapeutic proteins. J Pharm Sci 94:1626–1635

Tennikova T, Svec F (1993) High performance membrane chromatography: highly efficient separation method for proteins in ion-exchange, hydrophobic interaction and reversed phase modes. J Chromatogr 646:279–288

Terstappen G, Ramelmeier R, Kula M (1993) Protein partitioning in detergent-based aqueous two-phase systems. J Biotechnol 28:263–275

Van Wezel AL, van der Velden-de Groot CA, de Haan HH, van den Heuvel N, Schasfoort R (1985) Large scale animal cell cultivation for production of cellular biologicals. Dev Biol Stand 60:229–236

Walsh C (2006) Posttranslational modification of proteins: expanding nature's inventory, vol xxi. Roberts and Co. Publishers, Englewood, p 490

Walter J, Werner RG (1993) Regulatory requirements and economic aspects in downstream processing of biotechnically engineered proteins for parenteral application as pharmaceuticals. In: Kroner KH, Papamichael N, Schütte H (eds) Downstream processing, recovery and purification of proteins, a handbook of principles and practice. Carl Hauser Verlag, Muenchen

Walter J, Werz W, McGoff P, Werner RG, Berthold W (1991) Virus removal/inactivation in downstream processing. In: Spier RE, Griffiths JB, MacDonald C (eds) Animal cell technology: development, processes and products. Butterworth-Heinemann Ltd. Linacre House, Oxford, pp 624–634

Walter K, Werz W, Berthold W (1992) Virus removal and inactivation, concept and data for process validation of downstream processing. Biotech Forum Europe 9:560–564

Wheelwright SM (1993) Designing downstream processing for large scale protein purification. Biotechnology 5:789–793

WHO (World Health Organization) (2010) Recommendations for the evaluation of animal cell cultures as substrates for the manufacture of biological medicinal products and for the characterization of cell banks. Technical report series, proposed replacement of 878, annex 1 (not yet published)

4 Formulation of Biotech Products, Including Biopharmaceutical Considerations

Daan J.A. Crommelin

INTRODUCTION

This chapter deals with formulation aspects of pharmaceutical proteins. Both technological questions and biopharmaceutical issues such as the choice of the delivery systems, the route of administration, and possibilities for target site-specific delivery of proteins are considered.

MICROBIOLOGICAL CONSIDERATIONS

■ Sterility

Most proteins are administered parenterally and have to be sterile. In general, proteins are sensitive to heat and other regularly used sterilization treatments; they cannot withstand autoclaving, gas sterilization, or sterilization by ionizing radiation. Consequently, sterilization of the end product is not possible. Therefore, protein pharmaceuticals have to be assembled under aseptic conditions, following the established and evolving rules in the pharmaceutical industry for aseptic manufacture. The reader is referred to standard textbooks for details (Halls 1994; Groves 1988; Klegerman and Groves 1992; Roy 2011).

Equipment and excipients are treated separately and autoclaved or sterilized by dry heat (>160 °C), chemical treatment, or gamma radiation to minimize the bioburden. Filtration techniques are used for removal of microbacterial contaminants. Prefilters remove the bulk of the bioburden and other particulate materials. The final "sterilizing" step before filling the vials is filtration through 0.2 or 0.22 μm membrane filters. Assembly of the product is done in class 100 (maximum 100 particles > 0.5 μm per cubic foot) rooms with laminar airflow that is filtered through HEPA (high-efficiency particulate air) filters. Last but not least, the "human factor" is a major source of contamination. Well-trained operators wearing protective cloths (face masks, hats, gowns, gloves, or head-to-toe overall garments) should operate the facility. Regular exchange of filters, regular validation of HEPA equipment, and thorough cleaning of the room plus equipment are critical factors for success.

■ Viral Decontamination

As recombinant DNA products are grown in microorganisms, these organisms should be tested for viral contaminants, and appropriate measures should be taken if viral contamination occurs. In the rest of the manufacturing process, no (unwanted) viral material should be introduced. Excipients with a certain risk factor such as blood-derived human serum albumin should be carefully tested before use, and their presence in the formulation process should be minimized (see Chap. 3).

■ Pyrogen Removal

Pyrogens are compounds that induce fever. Exogenous pyrogens (pyrogens introduced into the body, not generated by the body itself) can be derived from bacterial, viral, or fungal sources. Bacterial pyrogens are mainly endotoxins shed from gram-negative bacteria. They are lipopolysaccharides. Figure 4.1 shows the basic structure. This conserved structure in the full array of thousands of different endotoxins is the lipid-A moiety. Another general property shared by endotoxins is their high, negative electrical charge. Their tendency to aggregate and to form large units with M_W of over 10^6 in water and their tendency to adsorb to surfaces indicate that these compounds are amphipathic in nature. They are stable under standard autoclaving conditions but break down when heated in the dry state. For this reason equipment and container are treated at temperatures above 160 °C for prolonged periods (e.g., 30 min dry heat at 250 °C).

D.J.A. Crommelin, Ph.D.
Department of Pharmaceutical Sciences,
Utrecht Institute for Pharmaceutical Sciences,
Utrecht University,
Arthur van Schendelstraat 98,
Utrecht 3511 ME, The Netherlands
e-mail: d.j.a.crommelin@uu.nl

D.J.A. Crommelin, R.D. Sindelar, and B. Meibohm (eds.), *Pharmaceutical Biotechnology*,
DOI 10.1007/978-1-4614-6486-0_4,

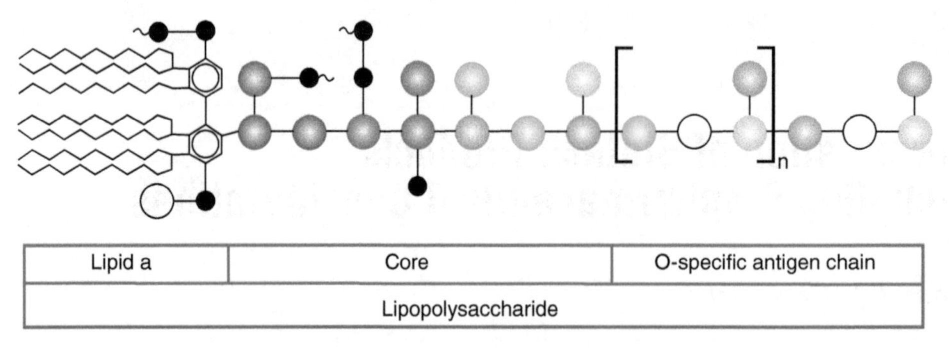

Figure 4.1 ■ Generalized structure of endotoxins. Most properties of endotoxins are accounted for by the active, insoluble "lipid A" fraction being solubilized by the various sugar moieties (*circles* with different colors). Although the general structure is similar, individual endotoxins vary according to their source and are characterized by the O-specific antigenic chain (Adapted from Groves (1988)).

Pyrogen removal of recombinant products derived from bacterial sources should be an integral part of the preparation process. Ion exchange chromatographic procedures (utilizing its negative charge) can effectively reduce endotoxin levels in solution (see also Chap. 3).

Excipients used in the protein formulation should be essentially endotoxin-free. For solutions "water for injection" (compendial standards) is (freshly) distilled or produced by reverse osmosis. The aggregated endotoxins cannot pass through the reverse osmosis membrane. Removal of endotoxins immediately before filling the final container can be accomplished by using activated charcoal or other materials with large surfaces offering hydrophobic interactions. Endotoxins can also be inactivated on utensil surfaces by oxidation (e.g., peroxide) or dry heating (e.g., 30 min dry heat at 250 °C).

Active ingredient
Solubility enhancers
Anti-adsorption and anti-aggregation agents
Buffer components
Preservatives and antioxidants
Lyoprotectants/cake formers
Osmotic agents
Carrier system (seen later on in this section)
Not necessarily all of the above are present in one particular protein formulation

Table 4.1 ■ Components found in parenteral formulations of biotech products.

EXCIPIENTS USED IN PARENTERAL FORMULATIONS OF BIOTECH PRODUCTS

In a protein formulation one finds, apart from the active substance, a number of excipients selected to serve different purposes. This process of formulation design should be carried out with great care to ensure therapeutically effective and safe products. The nature of the protein (e.g., lability) and its therapeutic use (e.g., multiple injection systems) can make these formulations quite complex in terms of excipient profile and technology (freeze-drying, aseptic preparation). Table 4.1 lists components that can be found in the presently marketed formulations. In the following sections this list is discussed in more detail. A classical review on peptide and protein excipients was published by Wang and Watson (1988).

■ Solubility Enhancers

Proteins, in particular those that are non-glycosylated, may have a tendency to aggregate and precipitate. Approaches that can be used to enhance solubility include selection of the proper pH and ionic strength conditions. Addition of amino acids such as lysine or arginine (used to solubilize tissue plasminogen activator, t-PA) or surfactants such as sodium dodecyl sulfate to solubilize non-glycosylated IL-2 can also help to increase the solubility. The mechanism of action of these solubility enhancers depends on the type of enhancer and the protein involved and is not always fully understood.

Figure 4.2 shows the effect of arginine concentration on the solubility of t-PA (alteplase) at pH 7.2 and 25 °C. This figure clearly indicates the dramatic effect of this basic amino acid on the apparent solubility of t-PA.

In the above examples, aggregation is physical in nature, i.e., based on hydrophobic and/or electrostatic interactions between molecules. However, aggregation based on the formation of covalent bridges between molecules through disulfide bonds and ester or amide linkages has been described as well (see also Table 2.4). In those cases, proper conditions should be found to avoid these chemical reactions.

■ Anti-adsorption and Anti-aggregation Agents

Anti-adsorption agents are added to reduce adsorption of the active protein to interfaces. Some proteins tend to

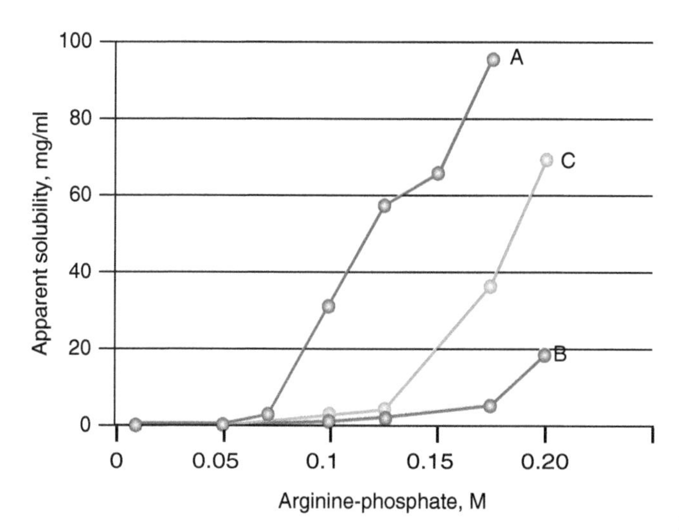

Figure 4.2 ■ Effect of arginine on type I and type II alteplase at pH 7.2 and 25 °C. *A* type I alteplase, *B* type II alteplase, *C* 50:50 mixture of type I and type II alteplase (From Nguyen and Ward (1993)).

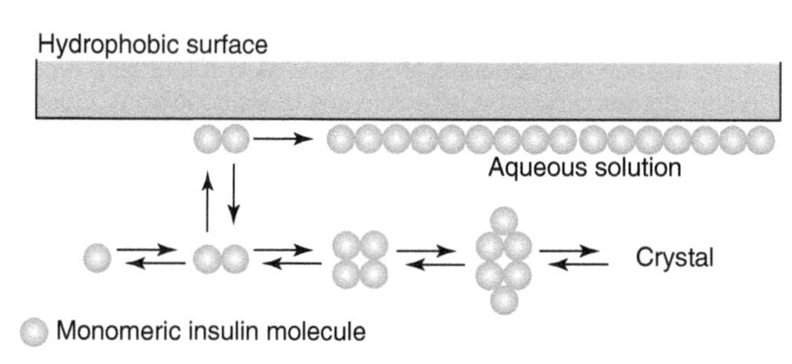

Figure 4.3 ■ Reversible self-association of insulin, its adsorption to the hydrophobic interface, and irreversible aggregation in the adsorbed protein film: ◯ represents a monomeric insulin molecule (Adapted from Thurow and Geisen (1984)).

expose hydrophobic sites, normally present in the core of the native protein structure when an interface is present. These interfaces can be water–air, water–container wall, or interfaces formed between the aqueous phase and utensils used to administer the drug (e.g., catheter, needle). These adsorbed, partially unfolded protein molecules form aggregates, leave the surface, return to the aqueous phase, form larger aggregates, and precipitate. As an example, the proposed mechanism for aggregation of insulin in aqueous media through contact with a hydrophobic surface (or water–air interface) is presented in Fig. 4.3 (Thurow and Geisen 1984).

Native insulin in solution is in an equilibrium state between monomeric, dimeric, tetrameric, and hexameric forms (see Chap. 12). The relative abundance of the different aggregation states depends on the pH, insulin concentration, ionic strength, and specific excipients (e.g., Zn^{2+} and phenol). It has been suggested that the dimeric form of insulin adsorbs to hydrophobic interfaces and subsequently forms larger aggregates at the interface. This explains why anti-adhesion agents can also act as anti-aggregation agents. Albumin has a strong tendency to adsorb to surfaces and is therefore added in relatively high concentrations (e.g., 1 %) to protein formulations as an anti-adhesion agent. Albumin competes with the therapeutic protein for binding sites and supposedly prevents adhesion of the therapeutically active agent by a combination of its binding tendency and abundant presence.

Insulin is one of the many proteins that can form fibrillar precipitates (long rod-shaped structures with diameters in the 0.1 μm range). Low concentrations of phospholipids and surfactants have been shown to exert a fibrillation-inhibitory effect. The selection of the proper pH can also help to prevent this unwanted phenomenon (Brange and Langkjaer 1993).

Apart from albumin, surfactants can also prevent adhesion to interfaces and precipitation. These molecules readily adsorb to hydrophobic interfaces with their own hydrophobic groups and render this interface hydrophilic by exposing their hydrophilic groups to the aqueous phase.

■ Buffer Components

Buffer selection is an important part of the formulation process, because of the pH dependence of protein solubility and physical and chemical stability. Buffer systems regularly encountered in biotech formulations are phosphate, citrate, and acetate. A good example of the importance of the isoelectric point (its negative logarithm = pI) is the solubility profile of human growth hormone (hGH, pI around 5) as presented in Fig. 4.4.

Even short, temporary pH changes can cause aggregation. These conditions can occur, for example, during the freezing step in a freeze-drying process, when one of the buffer components is crystallizing and the other is not. In a phosphate buffer, Na_2HPO_4 crystallizes faster than NaH_2PO_4. This causes a pronounced drop in pH during the freezing step. Other buffer components do not crystallize but form amorphous systems, and then pH changes are minimized.

■ Preservatives and Antioxidants

Methionine, cysteine, tryptophan, tyrosine, and histidine are amino acids that are readily oxidized (see Table 2.4). Proteins rich in these amino acids are liable to oxidative degradation. Replacement of oxygen by inert gases in the vials helps to reduce oxidative stress. Moreover, the addition of antioxidants such as ascorbic acid or acetylcysteine can be considered. Interestingly, destabilizing effects on proteins have been described for antioxidants as well (Vemuri et al. 1993). Ascorbic acid, for example, can act as an oxidant in the presence of a number of heavy metals.

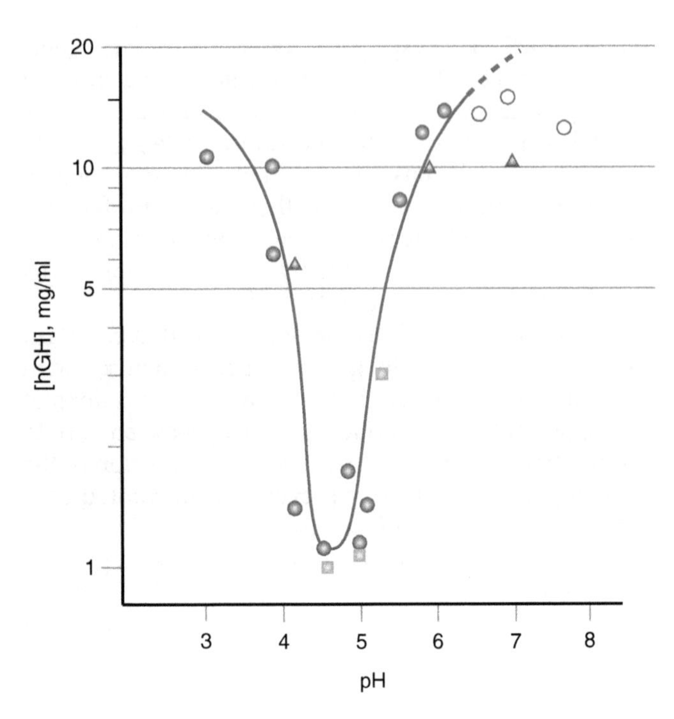

Figure 4.4 ■ A plot of the solubility of various forms of hGH as a function of pH. Samples of hGH were either recombinant hGH (*circles*), Met-hGH (*triangles*), or pituitary hGH (*squares*). Solubility was determined by dialyzing an approximately 11 mg/ml solution of each protein into an appropriate buffer for each pH. Buffers were citrate, pH 3–7, and borate, pH 8–9, all at 10 mM buffer concentrations. Concentrations of hGH were measured by UV absorbance as well as by RP-HPLC, relative to an external standard. The closed symbols indicate that precipitate was present in the dialysis tube after equilibration, whereas open symbols mean that no solid material was present, and thus the solubility is at least this amount (From Pearlman and Bewley (1993)).

Certain proteins are formulated in containers designed for multiple injection schemes. After administering the first dose, contamination with microorganisms may occur, and preservatives are needed to minimize growth. Usually, these preservatives are present in concentrations that are bacteriostatic rather than bactericidal in nature. Antimicrobial agents mentioned in the USP 29 are the mercury-containing phenylmercuric nitrate and thimerosal and p-hydroxybenzoic acids, phenol, benzyl alcohol, and chlorobutanol (USP 29 2006; Groves 1988; Pearlman and Bewley 1993). The use of mercury-containing preservatives is under discussion (FDA 2010).

■ Osmotic Agents

For proteins the regular rules apply for adjusting the tonicity of parenteral products. Saline and mono- or disaccharide solutions are commonly used. These excipients may not be inert; they may influence protein structural stability. For example, sugars and polyhydric alcohols can stabilize the protein structure through the principle of "preferential exclusion" (Arakawa et al. 1991). These additives (water structure promoters) enhance the interaction of the solvent with the protein and are themselves excluded from the protein surface layer; the protein is preferentially hydrated. This phenomenon can be monitored through an increased thermal stability of the protein. Unfortunately, a strong "preferential exclusion" effect enhances the tendency of proteins to self-associate.

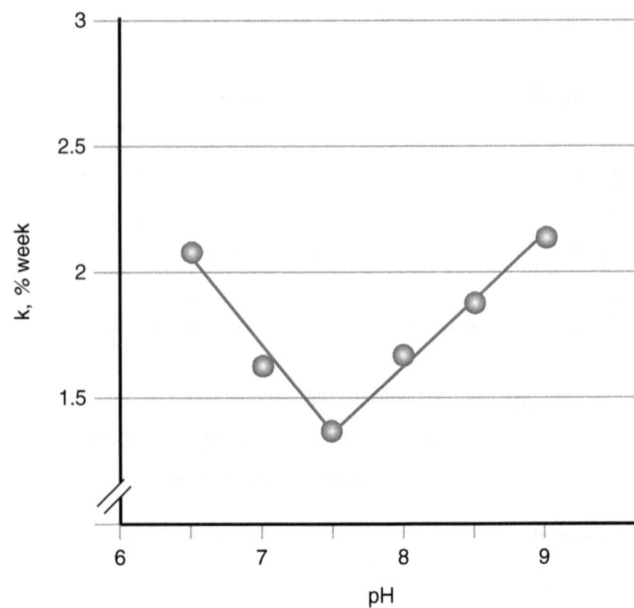

Figure 4.5 ■ pH stability profile (at 25 °C) of monomeric recombinant α_1-antitrypsin (rAAT) by size exclusion-HPLC assay, *k* degradation rate constant. Monomeric rAAT decreased rapidly in concentration both under acidic and basic conditions. Optimal stability occurred at pH 7.5 (Adjusted from Vemuri et al. (1993)).

SHELF LIFE OF PROTEIN-BASED PHARMACEUTICALS

Proteins can be stored (1) as an aqueous solution, (2) in freeze-dried form, and (3) in dried form in a compacted state (tablet). Some mechanisms behind chemical and physical degradation processes have been briefly discussed in Chap. 2.

Stability of protein solutions strongly depends on factors such as pH, ionic strength, temperature, and the presence of stabilizers. For example, Fig. 4.5 shows the pH dependence of α_1-antitrypsin and clearly demonstrates the critical importance of pH for the shelf life of proteins.

■ Freeze-Drying of Proteins

Proteins in solution often do not meet the preferred stability requirements for industrially produced pharmaceutical products (>2 years), even when kept permanently under refrigerator conditions (cold chain).

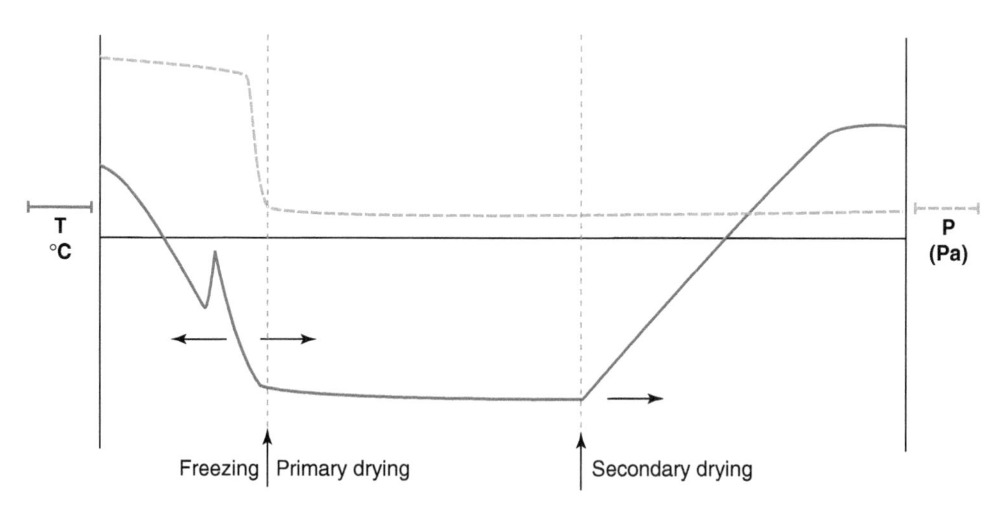

Figure 4.6 ■ Example of freeze-drying protocol for systems with crystallizing water. *T* temperature, *P* pressure.

Freezing	The temperature of the product is reduced from ambient temperature to a temperature below the eutectic temperature (T_e) or below the glass transition temperature (T_g) of the system. A T_g is encountered if amorphous phases are present
Primary drying	Crystallized and water not bound to protein/excipient is removed by sublimation. The temperature is below the T_e or T_g; the temperature is, for example, −40 °C and reduced pressures are used
Secondary drying	Removal of water interacting with the protein and excipients. The temperature in the chamber is kept below T_g and rises gradually, e.g., from −40 to 20 °C

Table 4.2 ■ Three stages in the freeze-drying process of protein formulations.

Bulking agents: mannitol/ glycine	Reason: elegance/blowout prevention[a]
Collapse temperature modifier: dextran, albumin/ gelatine	Reason: increase collapse temperature
Lyoprotectant: sugars, albumin	Reason: protection of the physical structure of the protein[b]

[a]Blowout is the loss of material taken away by the water vapor that leaves the vial. It occurs when little solid material is present in the vial

[b]Mechanism of action of lyoprotectants is not fully understood. Factors that might play a role are: (1) Lyoprotectants replace water as stabilizing agent (water replacement theory). (2) Lyoprotectants increase the Tg of the cake/frozen system. (3) Lyoprotectants will absorb moisture from the stoppers. Lyoprotectants slow down the secondary drying process and minimize the chances for overdrying of the protein. Overdrying might occur when residual water levels after secondary drying become too low. The chance for overdrying "in real life" is small

Table 4.3 ■ Typical excipients in a freeze-dried protein formulation.

The abundant presence of water promotes chemical and physical degradation processes.

Freeze-drying may provide the requested stability (Constantino and Pikal 2004). During freeze-drying water is removed through sublimation and not by evaporation. Three stages can be discerned in the freeze-drying process: (1) a freezing step, (2) the primary drying step, and (3) the secondary drying step (Fig. 4.6). Table 4.2 explains what happens during these stages.

Freeze-drying of a protein solution without the proper excipients causes, as a rule, irreversible damage to the protein. Table 4.3 lists excipients typically encountered in successfully freeze-dried protein products.

■ Freezing

In the freezing step (see Fig. 4.6) the temperature of the aqueous system in the vials is lowered. Ice crystal formation does not start right at the thermodynamic or equilibrium freezing point, but supercooling occurs. That means that crystallization often only occurs when temperatures of −15°C or lower have been reached. During the crystallization step the temperature may temporarily rise in the vial, because of the generation of crystallization heat. During the cooling stage, concentration of the protein and excipients occurs because of the growing ice crystal mass at the expense of the aqueous water phase. This can cause precipitation of one or more of the excipients, which may consequently result in pH shifts (see above and Fig. 4.7) or ionic strength changes. It may also induce protein denaturation. Cooling of the vials is done through lowering the temperature of the shelf. Selecting the proper cooling scheme for the shelf – and consequently vial – is important as it dictates the degree of supercooling and ice crystal size. Small crystals are formed during fast cooling; large crystals form at lower cooling rates. Small ice crystals are required for porous solids and fast sublimation rates (Pikal 1990).

If the system does not (fully) crystallize but forms an amorphous mass upon cooling, the temperature in the "freezing stage" should drop below Tg, the glass

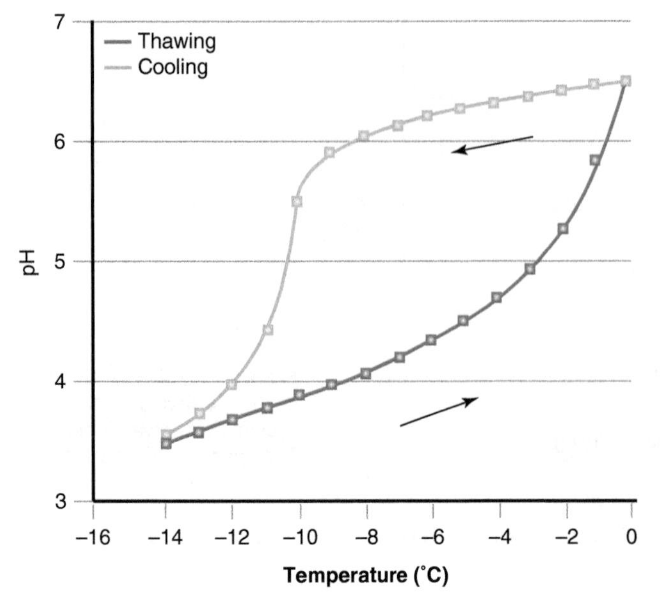

Figure 4.7 ■ Thawing/cooling; ■ thawing, ■ cooling. The effect of freezing on the pH of a citric acid–disodium phosphate buffer system (Cited in Pikal 1990).

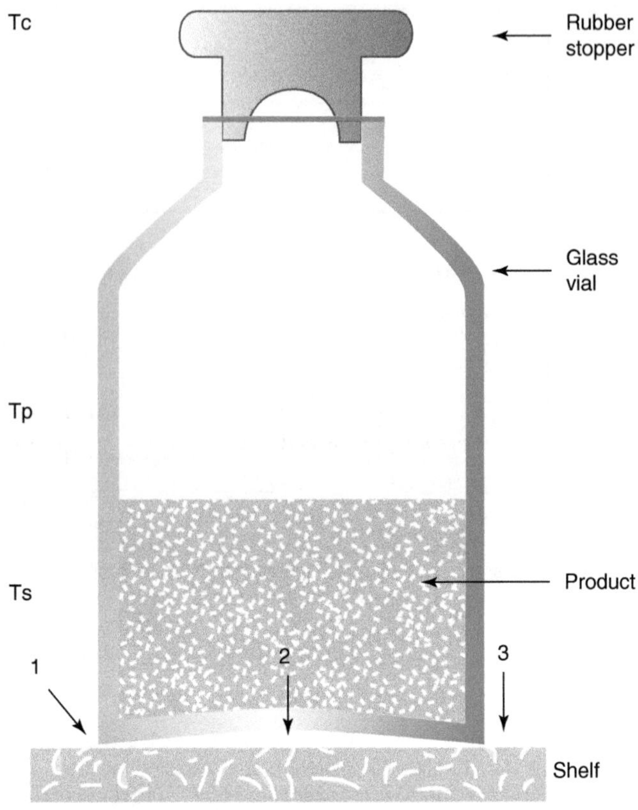

Figure 4.8 ■ Heat transfer mechanisms during the freeze-drying process: (*1*) Direct conduction via shelf and glass at points of actual contact. (*2*) Gas conduction: contribution heat transfer via conduction through gas between shelf and vial bottom. (*3*) Radiation heat transfer. *Ts* shelf temperature, *Tp* temperature sublimating product, *Tc* temperature condensor. Ts>Tp>Tc.

transition temperature. In amorphous systems the viscosity changes dramatically in the temperature range around the Tg: A "rubbery" state exists above and a glass state below the Tg.

At the start of the primary drying stage, no "free and fluid" water should be present in the vials. Minus forty degrees Celsius is a typical freezing temperature before sublimation is initiated through pressure reduction.

■ Primary Drying

In the primary drying stage (see Fig. 4.6.), sublimation of the water mass in the vial is initiated by lowering the pressure. The water vapor is collected on a condenser, with a (substantially) lower temperature than the shelf with the vials. Sublimation costs energy (about 2,500 kJ/g ice). Temperature drops are avoided by the supply of heat from the shelf to the vial, so the shelf is heated during this stage.

Heat is transferred to the vial through (1) direct shelf–vial contact (conductance), (2) radiation, and (3) gas conduction (Fig. 4.8). Gas conduction depends on the pressure: If one selects relatively high gas pressures, heat transport is promoted because of a high conductivity. But it reduces mass transfer, because of a low driving force: the pressure between equilibrium vapor pressures at the interface between the frozen mass/dried cake and the chamber pressure (Pikal 1990).

During the primary drying stage, one transfers heat from the shelf through the vial bottom and the frozen mass to the interface frozen mass/dry powder, to keep the sublimation process going. During this drying stage the vial content should never reach or exceed the eutectic temperature or glass transition temperature range. Typically a safety margin of 2–5°C is used; otherwise, the cake will collapse. Collapse causes a strong reduction in sublimation rate and poor cake formation. Heat transfer resistance decreases during the drying process as the transport distance is reduced by the retreating interface. With the mass transfer resistance (transport of water vapor), however, the opposite occurs. Mass transfer resistance increases during the drying process as the dry cake becomes thicker.

This situation makes it clear that parameters such as chamber pressure and shelf heating are not necessarily constant during the primary drying process. They should be carefully chosen and adjusted as the drying process proceeds.

The eutectic temperature and glass transition temperature are parameters of great importance to develop a rationally designed freeze-drying protocol. Information about these parameters can be obtained by microscopic observation of the freeze-drying process, differential scanning calorimetry (DSC), or electrical resistance measurements.

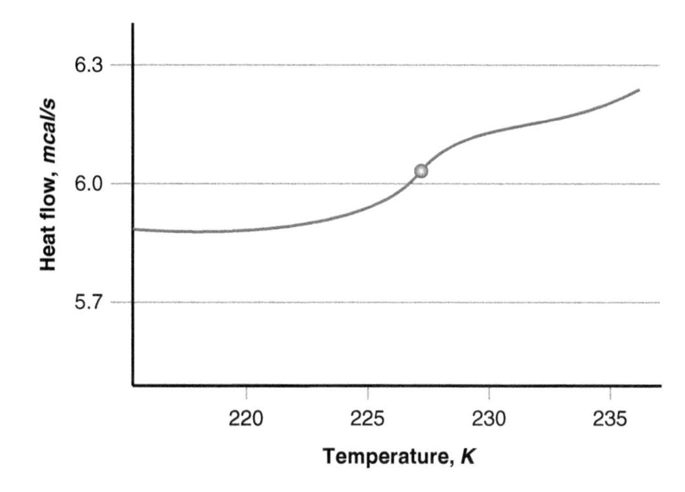

Figure 4.9 ■ Differential scanning calorimetry heating trace for a frozen solution of sucrose and sodium chloride, showing the glass transition temperature of the freeze concentrate at 227 K. For pure freeze-concentrated sucrose, T_g=241 K (1 cal=4.2 J) (From Franks et al. (1991)).

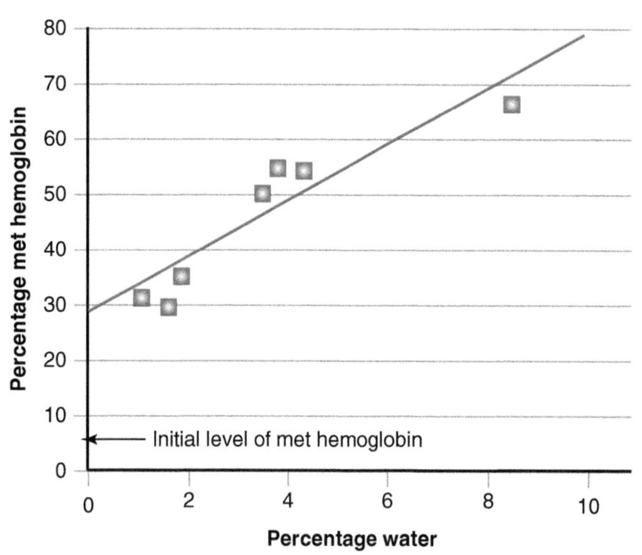

Figure 4.10 ■ The effect of residual moisture on the stability of freeze-dried hemoglobin (–6 %) formulated with 0.2 M sucrose; decomposition to methemoglobin during storage at 23 °C for 4 years (From Pikal (1990). Data reported by Pritoupil et al. (1985)).

An example of a DSC scan providing information on the Tg is presented in Fig. 4.9 (Franks et al. 1991). The Tg heavily depends on the composition of the system: excipients and water content. Lowering the water content of an amorphous system causes the Tg to shift to higher temperatures.

■ Secondary Drying

When all frozen or amorphous water that is nonprotein and non-excipient bound is removed, the secondary drying step starts (Fig. 4.6). The end of the primary drying stage is reached when product temperature and shelf temperature become equal or when the partial water pressure drops (Pikal 1990). As long as the "non-bound" water is being removed, the partial water pressure almost equals the total pressure. In the secondary drying stage, the temperature is slowly increased to remove "bound" water; the chamber pressure is still reduced. The temperature should stay all the time below the collapse/eutectic temperature, which continues to rise when residual water contents drop. Typically, the secondary drying step ends when the product has been kept at 20°C for some time. The residual water content is a critical, end point-indicating parameter. Values as low as 1 % residual water in the cake have been recommended. Figure 4.10 (Pristoupil 1985; Pikal 1990) exemplifies the decreasing stability of freeze-dried hemoglobin with increasing residual water content.

When stored in the presence of reducing lyoprotectants such as glucose and lactose, the Maillard reaction may occur: Amino groups of the proteins react with the lyoprotectant in the dry state, and the cake color turns yellow brown. The use of nonreducing sugars such as sucrose or trehalose may avoid this problem.

■ Other Approaches to Stabilize Proteins

Compacted forms of proteins are being used for certain veterinary applications, such as sustained-release formulations of growth hormones. The pellets should contain as few additives as possible. They can be applied subdermally or intramuscularly when the compact pellets are introduced by compressed air-powered rifles into the animals (Klegerman and Groves 1992).

DELIVERY OF PROTEINS: ROUTES OF ADMINISTRATION AND ABSORPTION ENHANCEMENT

■ The Parenteral Route of Administration

Parenteral administration is here defined as administration via those routes where a needle is used, including intravenous (IV), intramuscular (IM), subcutaneous (SC), and intraperitoneal (IP) injections. More information on the pharmacokinetic behavior of recombinant proteins is provided in Chap. 5. It suffices here to state that the blood half-life of biotech products can vary over a wide range. For example, the circulation half-life of tissue plasminogen activator (t-PA) is a few minutes, while monoclonal antibodies reportedly have half-lives of a few days. Obviously, one reason to develop modified proteins through site-directed mutagenesis is to enhance circulation half-life. A simple way to expand the mean residence time for short half-life proteins is to switch from IV to IM or SC administration. One should realize that by doing that, changes in disposition may occur, with a significant impact on the

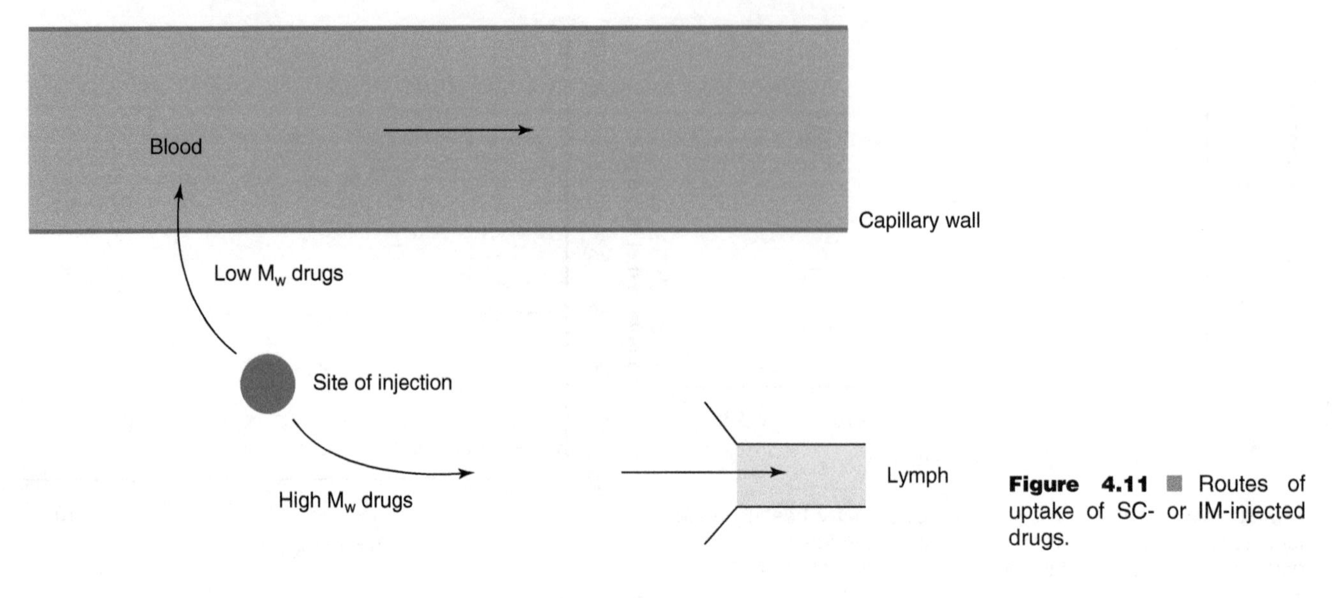

Figure 4.11 ■ Routes of uptake of SC- or IM-injected drugs.

therapeutic performance of the drug. These changes are related to (1) the prolonged residence time at the IM or SC site of injection compared to IV administration and the enhanced exposure to degradation reactions (peptidases) and (2) differences in disposition.

Regarding point 1: Prolonged residence time at the IM or SC site of injection and the enhanced exposure to degradation reactions. For instance, diabetics can become "insulin resistant" through high tissue peptidase activity (Maberly et al. 1982). Other factors that can contribute to absorption variation are related to differences in exercise level of the muscle at the injection site and also massage and heat at the injection site. The state of the tissue, for instance, the occurrence of pathological conditions, may be important as well.

Regarding point 2: Differences in disposition. Upon administration, the protein may be transported to the blood through the lymphatics or may enter the blood circulation through the capillary wall at the site of injection (Figs. 4.11 and 4.12). The fraction of the administered dose taking this lymphatic route is molecular weight dependent (Supersaxo et al. 1990). Lymphatic transport takes time (hours), and uptake in the blood circulation is highly dependent on the injection site. On its way to the blood, the lymph passes through draining lymph nodes, and contact is possible between lymph contents and cells of the immune system such as macrophages and B and T lymphocytes residing in the lymph nodes.

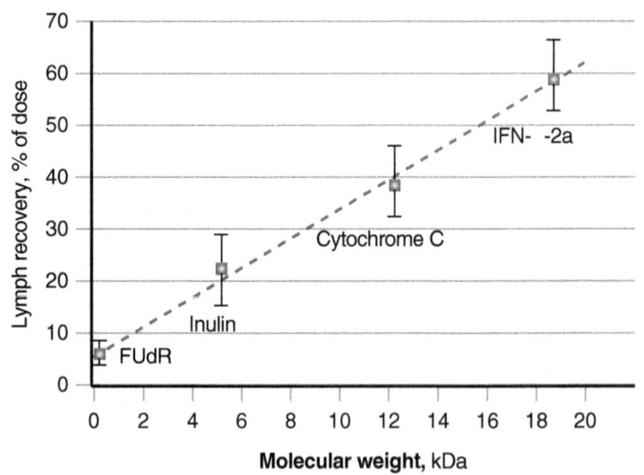

Figure 4.12 ■ Correlation between the molecular weight and the cumulative recovery of rIFN alpha-2a (M_w 19 kDa), cytochrome c (M_W 12.3 kDa), insulin (M_W 5.2 kDa), and FUDR (M_W 256.2 Da) in the efferent lymph from the right popliteal lymph node following SC administration into the lower part of the right hind leg of sheep. Each point and bar shows the mean and standard deviation of three experiments performed in separate sheep. The line drawn is the best fit by linear regression analysis calculated with the four mean values. The points have a correlation coefficient r of 0.998 ($p<0.01$) (From Supersaxo et al. (1990)).

■ The Oral Route of Administration

Oral delivery of protein drugs would be preferable, because it is patient friendly and no intervention by a healthcare professional is necessary to administer the drug. Oral bioavailability, however, is usually very low. The two main reasons for this failure of uptake are (1) protein degradation in the gastrointestinal (GI) tract and (2) poor permeability of the wall of the GI tract in case of a passive transport process.

Regarding point 1: Protein degradation in the GI tract. The human body has developed a very efficient system to break down proteins in our food to amino acids or di- or tripeptides. These building stones for body proteins are actively absorbed for use wherever necessary in the body. In the stomach, pepsins, a family of aspartic proteases, are secreted. They are particularly active between pH 3 and 5 and lose activity at higher pH values. Pepsins are endopeptidases capable of cleaving

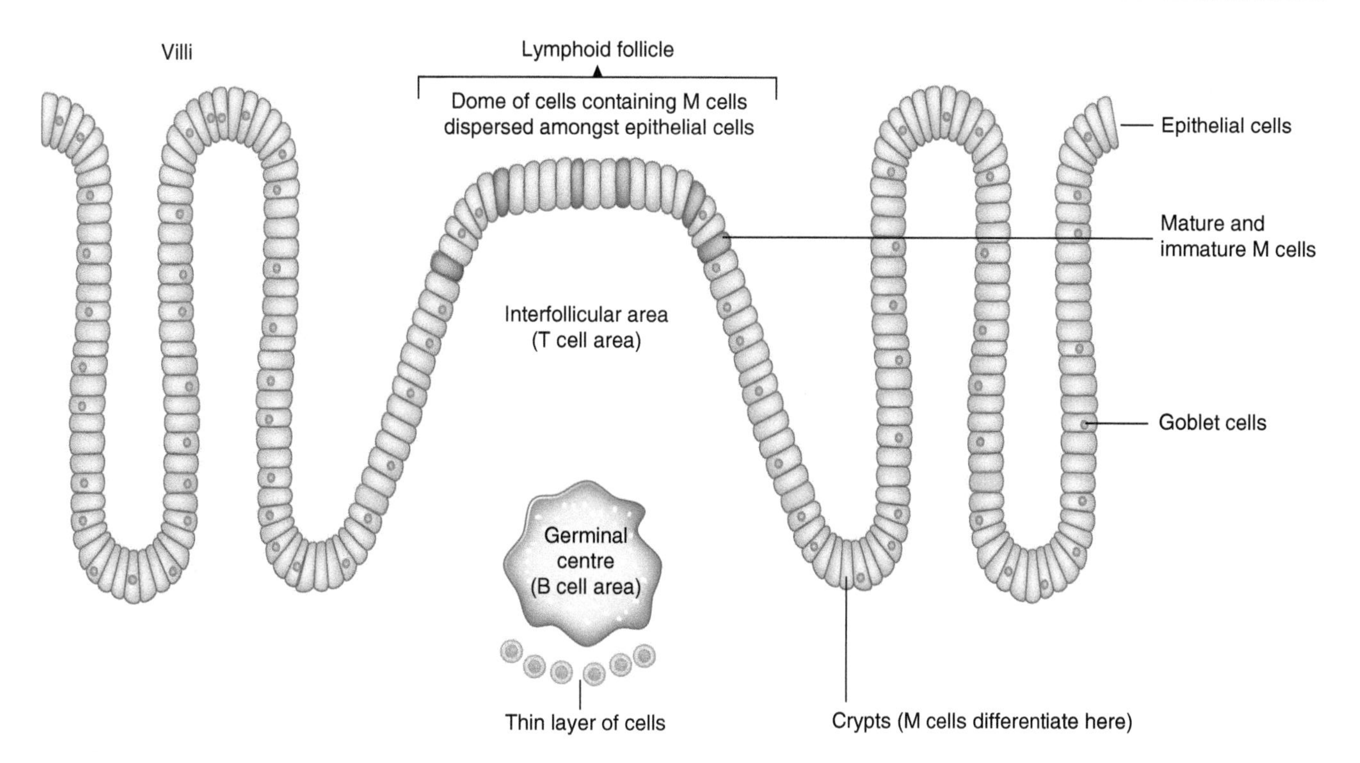

Figure 4.13 ■ Schematic diagram of the structure of intestinal Peyer's patches. M cells within the follicle-associated epithelium are enlarged for emphasis (From O'Hagan (1990)).

peptide bonds distant from the ends of the peptide chain. They preferentially cleave peptide bonds between two hydrophobic amino acids. Other endopeptidases are active in the gastrointestinal tract at neutral pH values, e.g., trypsin, chymotrypsin, and elastase. They have different peptide bond cleavage characteristics that more or less complement each other. Exopeptidases, proteases degrading peptide chains from their ends, are present as well. Examples are carboxypeptidase A and B. In the GI lumen the proteins are cut into fragments that effectively further break down to amino acids, di- and tripeptides by brush border, and cytoplasmic proteases of the enterocytes.

Regarding point 2: Permeability. High molecular weight molecules do not readily penetrate the intact and mature epithelial barrier if diffusion is the sole driving force for mass transfer. Their diffusion coefficient decreases with increasing molecule size. Proteins are no exception to this rule. Active transport of intact therapeutic recombinant proteins over the GI-epithelium has not been described yet.

The above analysis leads to the conclusion that nature, unfortunately, does not allow us to use the oral route of administration for therapeutic proteins if high (or at least constant) bioavailability is required.

However, for the category of oral vaccines, the above-mentioned hurdles of degradation and permeation are not necessarily prohibitive. For oral immunization, only a (small) fraction of the antigen (protein) has to reach its target site to elicit an immune response. The target cells are lymphocytes and antigen-presenting accessory cells located in Peyer's patches (Fig. 4.13). The B-lymphocyte population includes cells that produce secretory IgA antibodies.

These Peyer's patches are macroscopically identifiable follicular structures located in the wall of the gastrointestinal tract. Peyer's patches are overlaid with microfold (M) cells that separate the luminal contents from the lymphocytes. These M cells have little lysosomal degradation capacity and allow for antigen sampling by the underlying lymphocytes. Moreover, mucus-producing goblet cell density is reduced over Peyer's patches. This reduces mucus production and facilitates access to the M cell surface for luminal contents (Delves et al. 2011). Attempts to improve antigen delivery via the Peyer's patches and to enhance the immune response are made by using microspheres, liposomes, or modified live vectors, such as attenuated bacteria and viruses (Kersten and Hirschberg 2004, see also Chap. 22).

■ Alternative Routes of Administration

Parenteral administration has disadvantages (needles, sterility, injection skills) compared to other possible routes. Therefore, systemic delivery of recombinant proteins by alternative routes of administration (apart from the GI tract, discussed above) has been studied extensively. The nose, lungs, rectum, oral cavity, and

Route
+=relative advantage, −=relative disadvantage
Nasal
+ easily accessible, fast uptake, proven track record with a number of "conventional" drugs, probably lower proteolytic activity than in the GI tract, avoidance of first pass effect, spatial containment of absorption enhancers is possible
– reproducibility (in particular under pathological conditions), safety (e.g., ciliary movement), low bioavailability for proteins
Pulmonary
+ relatively easy to access, fast uptake, proven track record with "conventional" drugs, substantial fractions of insulin are absorbed, lower proteolytic activity than in the GI tract, avoidance of hepatic first pass effect, spatial containment of absorption enhancers (?)
– reproducibility (in particular under pathological conditions, smokers/nonsmokers), safety (e.g., immunogenicity), presence of macrophages in the lung with high affinity for particulates
Rectal
+ easily accessible, partial avoidance of hepatic first pass, probably lower proteolytic activity than in the upper parts of the GI tract, spatial containment of absorption enhancers is possible, proven track record with a number of "conventional" drugs
– low bioavailability for proteins
Buccal
+ easily accessible, avoidance of hepatic first pass, probably lower proteolytic activity than in the lower parts of the GI tract, spatial containment of absorption enhancers is possible, option to remove formulation if necessary
– low bioavailability of proteins, no proven track record yet (?)
Transdermal
+ easily accessible, avoidance of hepatic first pass effect, removal of formulation if necessary is possible, spatial containment of absorption enhancers, proven track record with "conventional" drugs, sustained/controlled release possible
– low bioavailability of proteins

Table 4.4 ■ Alternative routes of administration to the oral route for biopharmaceuticals.

skin have been selected as potential sites of application. The potential pros and cons for the different relevant routes are listed in Table 4.4. Moeller and Jorgensen (2009) and Jorgenson and Nielsen (2009) describe "the state of the art" in more detail.

The nasal, buccal, rectal, and transdermal routes all have been shown to be of little clinical relevance if systemic action is required and if simple protein formulations without an absorption-enhancing technology are used. In general, bioavailability is too low and varies too much! The pulmonary route may be the exception to this rule. Table 4.5 (from Patton et al. 1994) presents the bioavailability in rats of intratracheally administered protein solutions with a wide range of molecular weights. Absorption was strongly protein dependent, with no clear relationship with its molecular weight.

	Mw		Absolute
Molecule	**kDa**	**#AA**	**Bioavailability (%)**
α-Interferon	20	165	>56
PTH-84	9	84	>20
PTH-34	4.2	34	40
Calcitonin (human)	3.4	32	17
Calcitonin (salmon)	3.4	32	17
Glucagons	3.4	29	<1
Somatostatin	3.1	28	<1

Adapted from Patton et al. (1994)

PTH recombinant human parathyroid hormone, *#AA* number of amino acids

Table 4.5 ■ Absolute bioavailability of a number of proteins (intratracheal vs. intravenous) in rats.

In humans the drug should be inhaled instead of intratracheally administered. The first pulmonary insulin formulation was approved by FDA in January 2006 (Exubera®) but taken off the market in 2008 because of poor market penetration. Pulmonary inhalation of insulin is specifically indicated for mealtime glucose control. Uptake of insulin is faster than after a regular SC insulin injection (peak 5–60 min vs. 60–180 min). The reproducibility of the blood glucose response to inhaled insulin was equivalent to SC-injected insulin. Inhalation technology plays a critical role when considering the prospects of the pulmonary route for the systemic delivery of therapeutic proteins. Dry powder inhalers and nebulizers are being tested. The fraction of insulin that is ultimately absorbed depends on (1) the fraction of the inhaled/nebulized dose that is actually leaving the device, (2) the fraction that is actually deposited in the lung, and

Classified according to proposed mechanism of action
Increase the permeability of the absorption barrier: Addition of fatty acids/phospholipids, bile salts, enamine derivatives of phenylglycine, ester and ether type (non)-ionic detergents, saponins, salicylate derivatives, derivatives of fusidic acid or glycyrrhizinic acid, or methylated β-cyclodextrins Through iontophoresis By using liposomes
Decrease peptidase activity at the site of absorption and along the "absorption route": aprotinin, bacitracin, soybean tyrosine inhibitor, boroleucin, borovaline
Enhance resistance against degradation by modification of the molecular structure
Prolongation of exposure time (e.g., bio-adhesion technologies)

Adapted from Zhou and Li Wan Po (1991a)

Table 4.6 Approaches to enhance bioavailability of proteins.

(3) the fraction that is being absorbed, i.e., total relative uptake (TO %) = % uptake from device × % deposited in the lungs × % actually absorbed from the lungs. TO % for insulin is estimated to be about 10 % (Patton et al. 2004). The fraction of insulin that is absorbed from the lung is around 20 %. These figures demonstrate that insulin absorption via the lung may be a promising route, but the fraction absorbed is small and with the Exubera technology, the patient/medical community preferred parenteral administration.

Therefore, different approaches have been evaluated to increase bioavailability of the pulmonary and other non-parenteral routes of administration. The goal is to develop a system that temporarily decreases the absorption barrier resistance with minimum and acceptable safety concerns. The mechanistic background of these approaches is given in Table 4.6. Until now, no products utilizing one of these approaches have successfully passed clinical test programs. Safety concerns are an important hurdle. Questions center on the specificity and reversibility of the protein permeation enhancing effect and the toxicity.

■ Examples of Absorption-Enhancing Effects

The following section deals with absorption enhancement and non-parenteral administration of recombinant proteins. A number of typical examples are provided.

Table 4.7 presents an example of the (apparently complex) relationship between nasal bioavailability of some peptide and protein drugs, their molecular weight, and the presence of the absorption enhancer glycocholate (Zhou and Li Wan Po 1991b).

Molecule	#AA	Bioavailability (%) Without	Bioavailability (%) With glycocholate
Glucagon	29	<1	**70–90**
Calcitonin	32	<1	**15–20**
Insulin	51	<1	**10–30**
Met-hGH[a]	191	<1	**7–8**

Adapted from Zhou and Li Wan Po (1991b)

[a]See also Chap. 14

Table 4.7 Effect of glycocholate (absorption enhancer) and molecular weight of some proteins and peptides on nasal bioavailability.

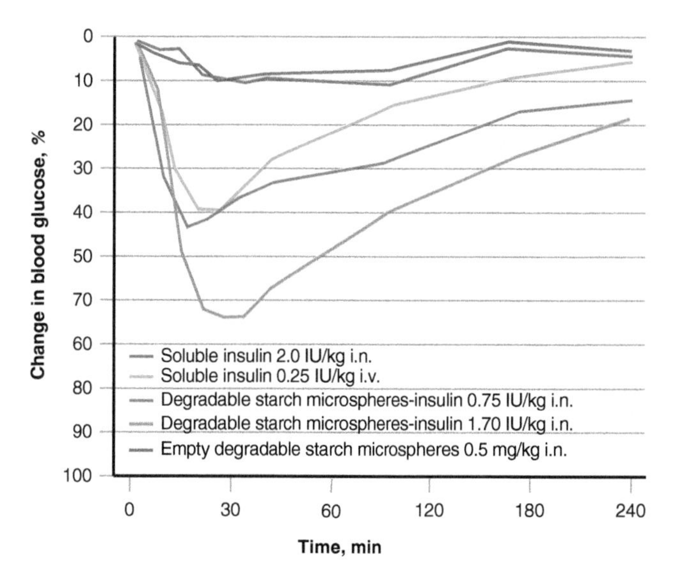

Figure 4.14 Change in blood glucose in rats after intranasal (i.n.) administration of insulin. — Soluble insulin 2.0 IU/kg i.n. — Soluble insulin 0.25 IU/kg IV. — Degradable starch microspheres – insulin 0.75 IU/kg i.n. — Degradable starch microspheres – insulin 1.70 IU/kg i.n. — Empty degradable starch microspheres – insulin 0.5 mg/kg i.n (Discussed by Edman and Björk (1992)).

Figure 4.14 (Björk and Edman 1988) illustrates another case where degradable starch microspheres loaded with insulin were used and where changes in glucose levels were monitored after nasal administration to rats.

In these examples, the effect of the presence of the absorption enhancers is clear. Major issues to be addressed are reproducibility, effect of pathological conditions (e.g., rhinitis) on absorption, and safety aspects of chronic use. Interestingly, absorption-enhancing effects were shown to be species dependent. Pronounced differences in effect were observed between rats, rabbits, and humans.

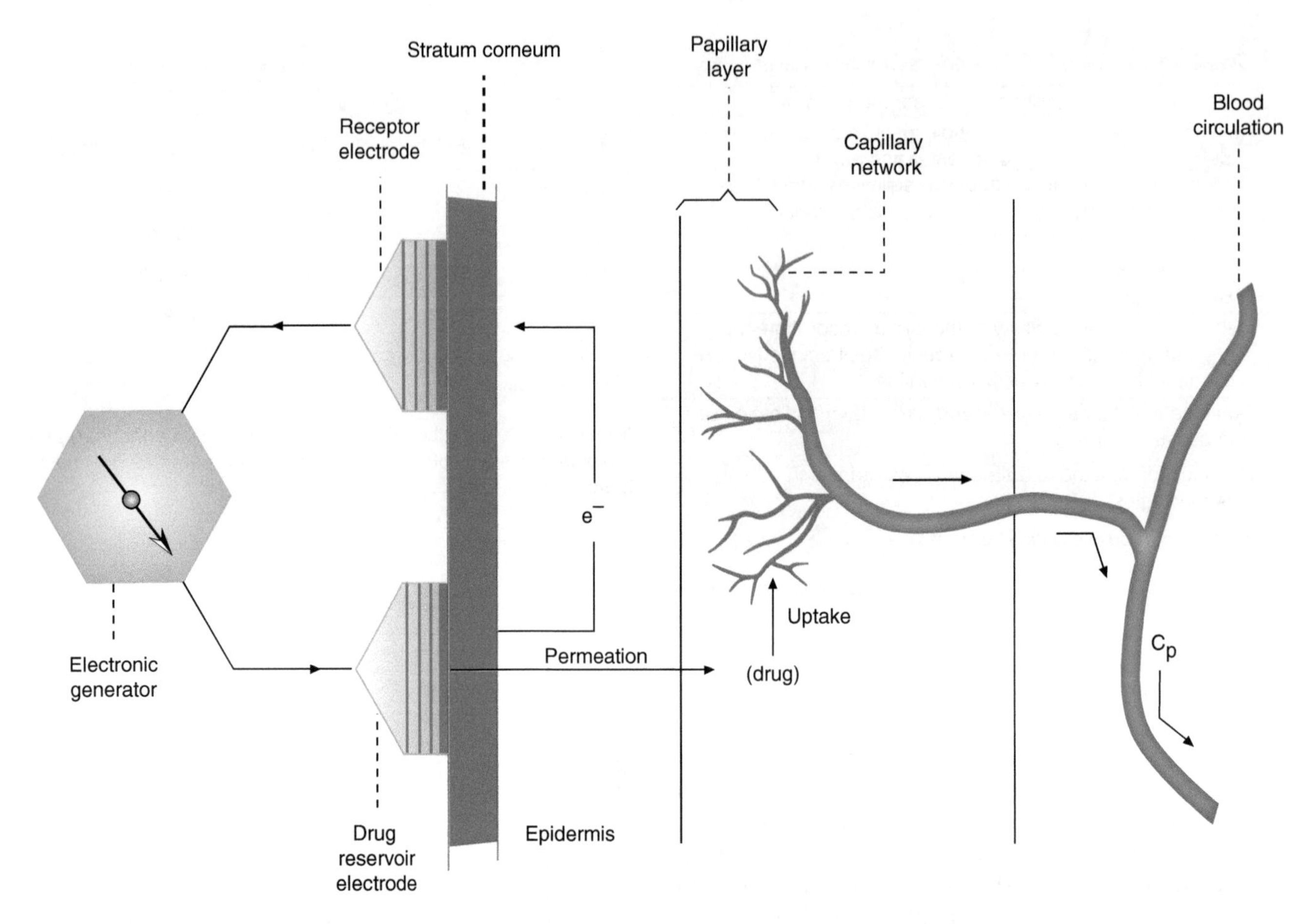

Figure 4.15 ■ Schematic illustration of the transdermal iontophoretic delivery of peptide and protein drugs across the skin (Adapted from Chien (1991)).

With iontophoresis a transdermal electrical current is induced by positioning two electrodes on different places on the skin (Fig. 4.15). This current induces a migration of (ionized) molecules through the skin. Delivery depends on the current (on/off, pulsed/direct, wave shape), pH, ionic strength, molecular weight, charge on the protein, and temperature. The protein should be charged over the full thickness of the skin (pH of hydrated skin depends on the depth and varies between pH 4 (surface) and pH 7.3), which makes proteins with pI values outside this range prime candidates for iontophoretic transport. It is not clear whether there are size restrictions (protein M_W) for iontophoretic transport. However, only potent proteins will be successful candidates. With the present technology the protein flux through the skin is in the 10 μg/cm²/h range (Sage et al. 1995).

Figure 4.16 presents the plasma profile of growth hormone-releasing factor, GRF (44 amino acids, M_W 5 kDa after SC, IV, and iontophoretic transdermal delivery to hairless guinea pigs). A prolonged appearance of GRF in the plasma can be observed. Iontophoretic delivery offers interesting opportunities if pulsed delivery of the protein is required. The device can be worn permanently and

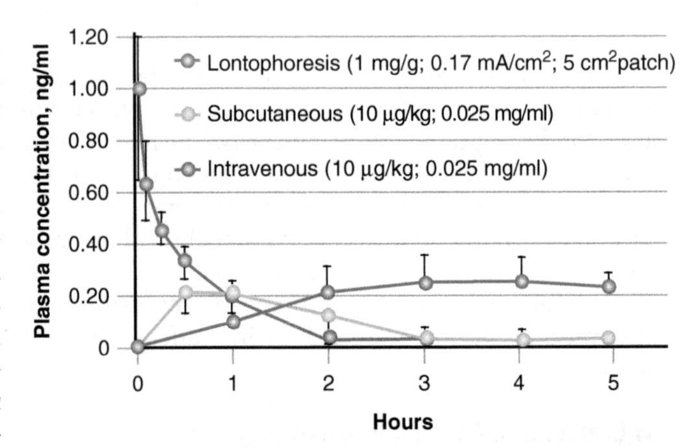

Figure 4.16 ■ Plasma concentration versus time profiles after subcutaneous, intravenous, and iontophoretic transdermal administration of GRF (1–44) to hairless guinea pigs. iontophoresis (1 mg/g; 0.17 mA/cm²; 5 cm² patch). subcutaneous (10 μg/kg; 0.025 mg/ml). intravenous (10 μg/kg; 0.025 mg/ml) (From Kumar et al. (1992)).

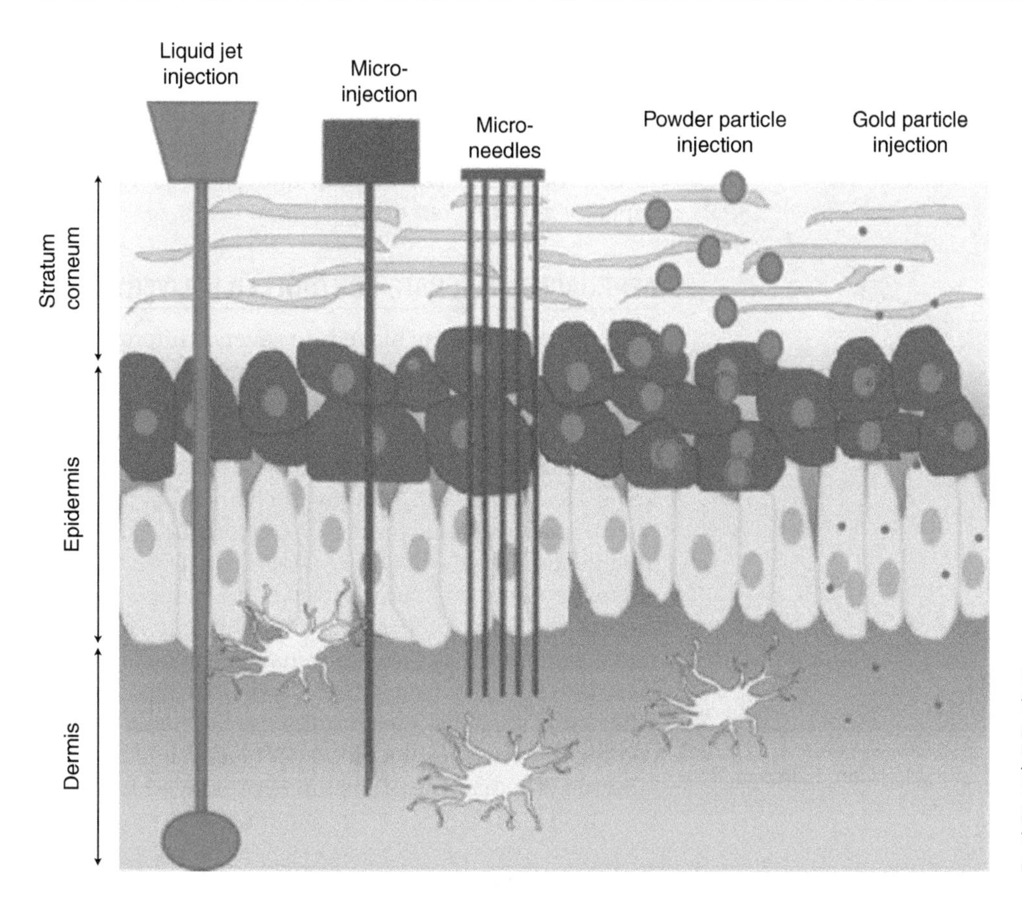

Figure 4.17 ■ Delivery systems enabling the intradermal application of vaccines: liquid jet injection, microinjection, microneedles (*arrays*), powder particle injection, and gold particle injection (Adapted from Kis et al. (2012)).

only switched on for the desired periods of time, simulating pulsatile secretion of endogenous hormones such as growth hormone and insulin. At present, no iontophoretic devices for protein delivery have been approved.

And, last but not least, a number of intracutaneous delivery systems have been developed. Figure 4.17 shows different options. In all approaches shown, the administered volumes were small, microliter level, and basically these approaches are restricted to antigen/adjuvant delivery. The classical liquid jet injectors deliver small amounts of vaccine fluid in the skin with a high velocity. The modern versions use prefilled disposable delivery units for single use to avoid contamination. Ballistic injectors shoot powder particles or gold particles covered with antigen under high pressure into the skin. Finally, microinjectors with microneedles in the (sub) mm range are being used or microneedle arrays with small individual needles in the 100 μm range. The newer versions of these microneedle arrays are self-dissolving. They are made of, e.g., sugar (derivatives) which dissolve rapidly after application (Mitragotri 2005; Kis et al. 2012). Examples are shown in Fig. 4.18a, b.

DELIVERY OF PROTEINS: APPROACHES FOR RATE-CONTROLLED AND TARGET SITE-SPECIFIC DELIVERY BY THE PARENTERAL ROUTE

Presently used therapeutic proteins widely differ in their pharmacokinetic characteristics (see Chap. 5). If they are endogenous agents such as insulin, tissue plasminogen activator, growth hormone, erythropoietin, interleukins, or factor VIII, it is important to realize why, when, and where they are secreted. There are three different ways in which cells can communicate with each other: the endocrine, paracrine, and autocrine pathway (Table 4.8).

The dose–response relationship of these mediators is often not S shaped, but, for instance, bell shaped: At high doses the therapeutic effect disappears (see Chap. 5). Moreover, the presence of these mediators may activate a complex cascade of events that needs to be carefully controlled. Therefore, key issues for their therapeutic success are (1) access to target cells, (2) retention at the target site, and (3) proper timing of delivery.

In particular, for paracrine- and autocrine-acting proteins, site-specific delivery can be highly desirable, because otherwise side effects will occur outside the

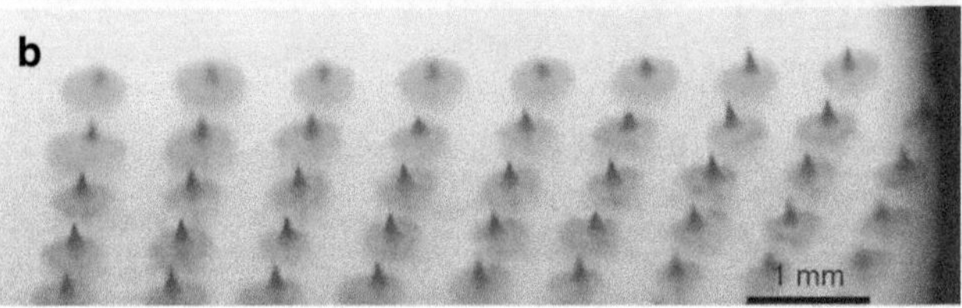

Figure 4.18 ■ (**a**) Hollow 300 μm tall silicon microneedles, (and 26-gauge syringe needle) fabricated using a combination of wet and dry etch micromachining technologies. These microneedles have wide-ranging applications in painless transdermal delivery and physiological sensing (Courtesy: Joe O'Brien & Conor O'Mahony, Tyndall National Institute) (**b**) Example of dissolvable microneedle patches. Dissolvable microneedles, composed of sugars and polymers, were fabricated in PDMS molds of master silicon microneedle arrays. The dimensions of microneedles on the array were 280 μm in height at a density of 144 needles per 1 cm². These biodegradable dissolving microneedles were fabricated using a proprietary method (UK Patent Application Number 1107642.9) (Courtesy: Anne Moore, Anto Vrdoljak, School of Pharmacy, University College Cork).

Endocrine hormones:
A hormone secreted by a distant cell to regulate cell functions distributed widely through the body. The bloodstream plays an important role in the transport process
Paracrine-acting mediators:
The mediator is secreted by a cell to influence surrounding cells, short-range influence
Autocrine-acting mediators:
The agent is secreted by a cell and affects the cell by which it is generated, (very) short-range influence

Table 4.8 ■ Communication between cells: chemical messengers.

target area. Severe side effects were reported with cytokines, such as tumor necrosis factor and interleukin-2 upon parenteral (IV or SC) administration (see Chap. 21). The occurrence of these side effects limits the therapeutic potential of these compounds. Therefore, the delivery of these proteins at the proper site, rate, and dose is a crucial part in the process of the design and development of these compounds as pharmaceutical entities. The following sections discuss first concepts developed to control the release kinetics and subsequently concepts for site-directed drug delivery.

APPROACHES FOR RATE-CONTROLLED DELIVERY

Rate control can be achieved by several different technologies similar to those used for "conventional" drugs. Insulin is an excellent example. A spectrum of options is available and accepted. Different types of suspensions and continuous/"smart" infusion systems are marketed (see Chap. 12). Moreover, chemical approaches can be used to change protein characteristics. For example, insulin half-life can be prolonged by using the long circulation time of serum albumin and its high binding affinity for fatty acids such as myristic acid. In insulin detemir (Levemir®) the C-terminal threonine of insulin is replaced by a lysine to which myristic acid is coupled. After subcutaneous injection the myristic acid–insulin combination reaches the blood circulation and binds to albumin. The half-life of insulin is prolonged from less than 10 min to over 5 h. A similar approach is used with glucagon-1-like peptide (GLP-1 (7–37)) for the treatment of diabetes. Attaching myristic acid to GLP-1 (7–37) (liraglutide marketed as Victoza®) increases the plasma half-life from 2 min to over 10 h.

Another approach that has been very successful in prolonging plasma circulation times and dosing intervals is the covalent attachment of polyoxyethylene glycol (PEG) to proteins. Figure 4.19 shows an example of this approach. Commercially highly successful examples that were developed later are pegylated interferon beta formulations (see Chap. 21).

In general, proteins are parenterally administered as an aqueous solution. Only recombinant vaccines and a number of insulin formulations are delivered as (colloidal) dispersions. At present, portable and patient-controlled pump systems are regularly used in practice. As experience with biotech drugs grows, more advanced technologies will definitely be introduced to optimize the therapeutic benefit of the drug. Table 4.9 lists some of the technologically feasible options. They are briefly touched upon below.

■ Open-Loop Systems: Mechanical Pumps

Mechanically driven pumps are common tools to administer drugs intravenously in hospitals (continuous infusion, open-loop type). They are available in different kinds of sizes/prices, portable or not, inside/outside the body, etc. Table 4.10 presents a checklist of issues to be considered when selecting the proper pump. The pump system may fail because of energy

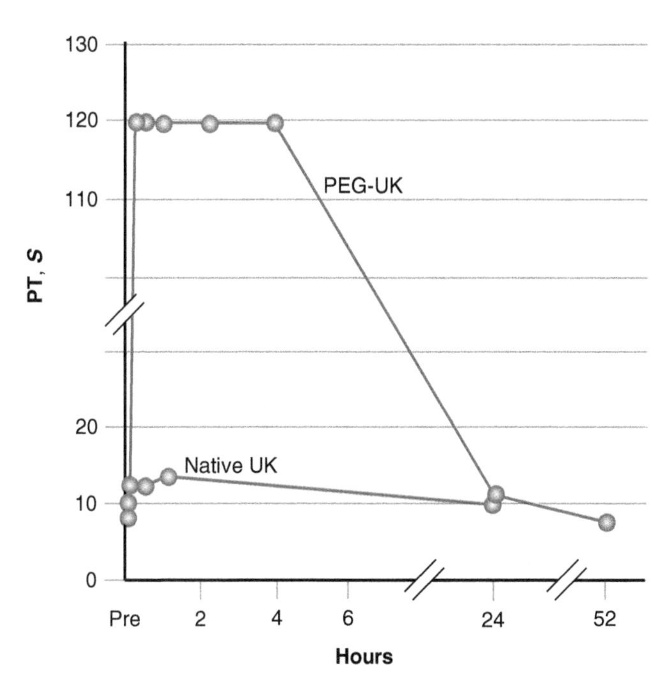

Figure 4.19 ■ Influence of chemical grafting of polyethylene glycol (*PEG*) on the ability of urokinase (*UK*) to affect the prothrombin time (*PT*) in vivo in beagles with time (Through Tomlinson (1987)).

Rate control through open-loop-type approach
Continuous infusion with pumps: mechanically or osmotically driven. Input, constant/pulsatile/wave form
Implants, biodegradable polymers; lipids. Input, limited control
Rate control through closed-loop approach/feedback system
Biosensor–pump combination
Self-regulating system
Encapsulated secretory cells

Table 4.9 ■ Controlled release systems for parenteral delivery.

failure, problems with the syringe, accidental needle withdrawal, leakage of the catheter, and problems at the injection or implantation site (Banerjee et al. 1991). Moreover, long-term protein drug stability may become a problem: The protein should be stable at 37°C or ambient temperature (internal and external device, respectively) between two refills.

Controlled administration of a drug does not necessarily imply a constant input rate. Pulsatile or variable-rate delivery is the desired mode of input for a number of protein drugs, and for these drugs pumps should provide flexible input rate characteristics. Insulin is a prime example of a protein drug, where there is a need to adjust the input rate to the needs of the body. Today by far most experiences with pump systems in an ambulatory setting have been gained with this drug. Even with high-tech pump systems, the patient still has to collect data to adjust the pump rate. This implies invasive sampling from body fluids on a regular basis, followed by calculation/setting of the required input rate. The concept of closed-loop systems integrates these three actions and creates a "natural" biofeedback system (see below).

The pump must deliver the drug at the prescribed rate(s) for extended periods of time. It should
Have a wide range of delivery rates
Ensure accurate, precise, and stable delivery
Contain reliable pump and electrical components
Contain drugs compatible with pump internals
Provide simple means to monitor the status and performance of the pump
The pump must be safe. It should
Have a biocompatible exterior if implanted
Have overdose protection
Show no leakage
Have a fail-safe mechanism
Have sterilizable interiors and exteriors (if implantable)
The pump must be convenient. It should
Be reasonably small in size and inconspicuous
Have a long reservoir life
Be easy to program

Table 4.10 ■ Listing the characteristics of the ideal pump (Banerjee et al. 1991).

■ Open-Loop Systems: Osmotically Driven Systems

The subcutaneously implantable, osmotic mini-pump developed by ALZA (Alzet mini-pump, Fig. 4.20, Alzet product information 2012) has proven to be useful in animal experiments where continuous, constant infusion is required over prolonged periods of time. The rate-determining process is the influx of water through the rigid, semipermeable external membrane dissolving the salt (osmotic agent), creating a constant osmotic influx over the semipermeable membrane. The incoming water empties the drug-containing reservoir (solution or dispersion) surrounded by a flexible impermeable membrane. The release rate depends on the characteristics of this semipermeable membrane and on osmotic pressure differences over this membrane (osmotic agents/salt inside the pump). Zero-order release kinetics exist as long as the osmotic pressure difference over the semipermeable membrane stays constant.

The protein solution (or dispersion) must be physically and chemically stable at body temperature over the full term of the experiment. Moreover, the protein solution must be compatible with the pump parts to which it is exposed. A limitation of the system is the fixed release rate, which is not always desired (see above). These devices have currently not been used on a regular basis in the clinic.

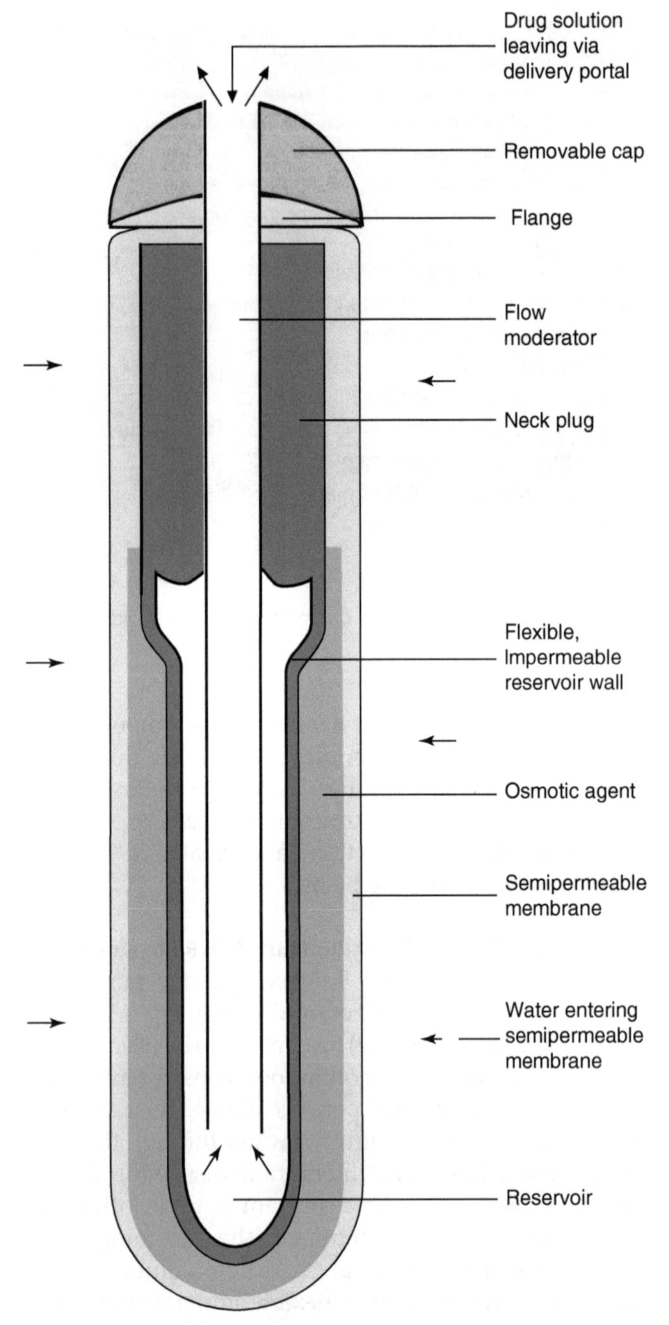

Figure 4.20 ■ Cross section of functioning ALZA Alzet osmotic mini-pump (Through Banerjee et al. (1991)).

■ Open-Loop Systems: Biodegradable Microspheres

Polylactic acid–polyglycolic acid (PLGA)-based delivery systems are being used extensively for the delivery of therapeutic peptides, in particular luteinizing hormone-releasing hormone (LHRH) agonists such as leuprolide in the therapy of prostate cancer. The first LHRH agonist-controlled release formulations were implants containing leuprolide with dose ranges of 1–3 months. Later, microspheres loaded with leuprolide were introduced, and dosing intervals were prolonged to up to 6 months. Critical success factors for the design of these controlled release systems are (1) the drug has to be highly potent (only a small dose is required over the dosing interval), (2) a sustained presence in the body is required, and (3) no adverse reactions at the injection site should occur. A glucagon-like protein-1 (GLP-1, 39 amino acids) slow release formulation (Bydureon™) based on PLGA microspheres for once a week administration to type II diabetics was released in 2012.

New strategies for controlled release of therapeutic proteins are presently under development. For example, Figs. 4.21 and 4.22 describe a dextran-based microsphere technology for SC or IM administration that often has an almost 100 % protein encapsulation efficiency. No organic solvents are being used in the preparation protocol. Thus, a direct interaction of the dissolved protein with an organic phase (as seen in many (e.g., polylactic-co-glycolic acid, PLGA) polymeric microsphere preparation schemes) is avoided. This minimizes denaturation of the protein. Figure 4.22 shows that by selecting the proper cross-linking conditions, one has a degree of control over the protein release kinetics. Release kinetics mainly depend on degradation kinetics of the dextran matrix and size of the protein molecule (Stenekes 2000). Another approach for prolonged and controlled release of therapeutic proteins is to use microspheres based on another biodegradable hydrogel material. PolyActive™ is a block copolymer consisting of polyethylene glycol blocks and polybutylene terephthalate blocks. Results of a dose finding study in humans with PolyActive™ microspheres loaded with interferon-α are shown in Fig. 4.23 (de Leede et al. 2008).

■ Closed-Loop Systems: Biosensor-Pump Combinations

If input rate control is desired to stabilize a certain body function, then this function should be monitored. Via an algorithm and connected pump settings, this data should be converted into a drug-input rate. These systems are called closed-loop systems as compared to the open-loop systems discussed above. If there is a known relationship between plasma level and pharmacological effect, these systems contain (Fig. 4.24):

1. A biosensor, measuring the (plasma) level of the biomarker
2. An algorithm, to calculate the required input rate for the delivery system
3. A pump system, able to administer the drug at the required rate over prolonged periods of time

The concept of a closed-loop delivery of proteins still has to overcome a number of conceptual and practical hurdles. A simple relationship between plasma level and therapeutic effect does not always

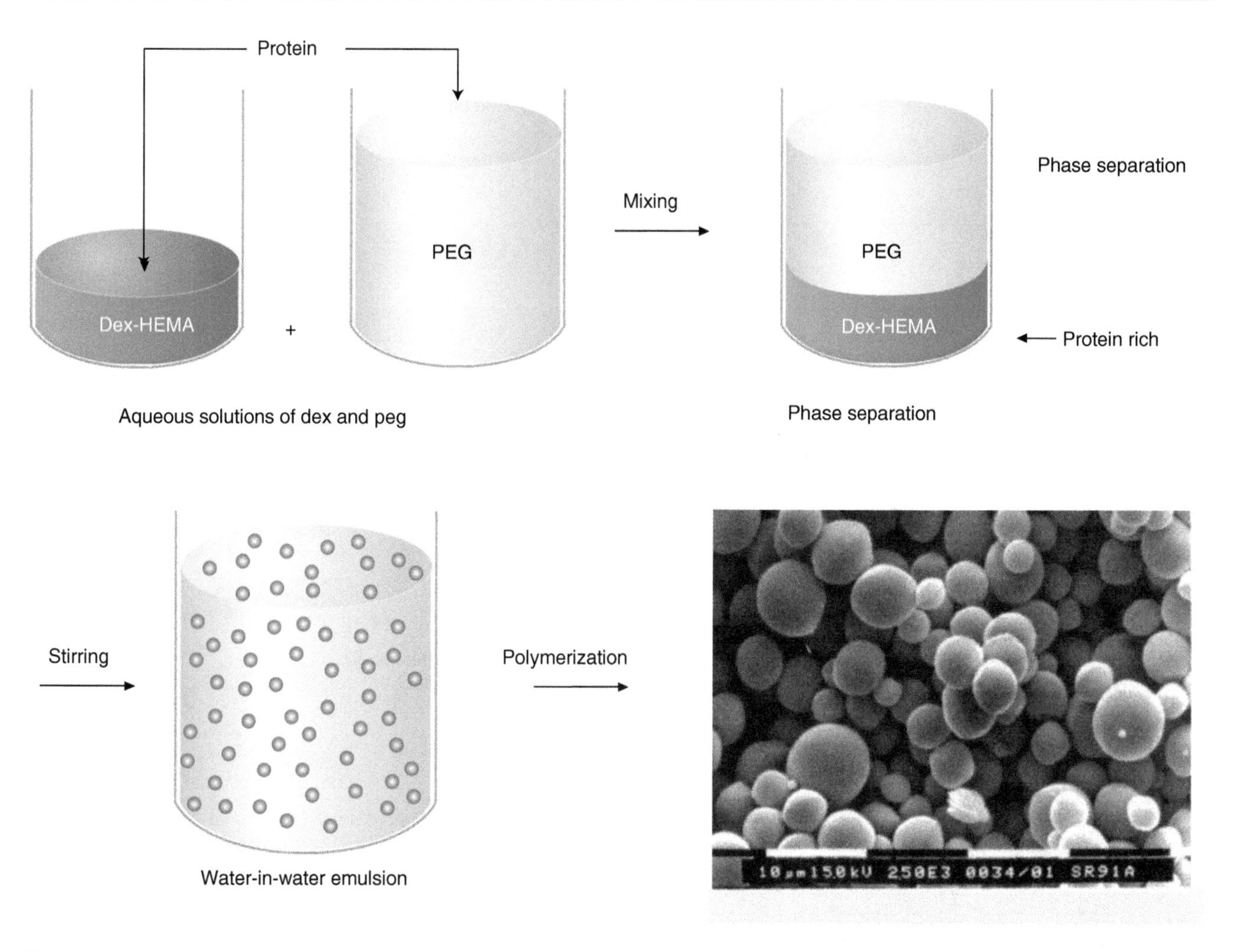

Figure 4.21 ■ Schematic representation of the microsphere preparation process for the controlled release of therapeutic proteins from dextran (DexHEMA = modified dextran = dextran hydroxyethylmethacrylate) microspheres. No organic solvents are involved, and encapsulation efficiency (percentage of therapeutic protein ending up in the microspheres) is routinely > 90 %. Polymerization: cross-linking of dextran chains through the HEMA units (Stenekes 2000).

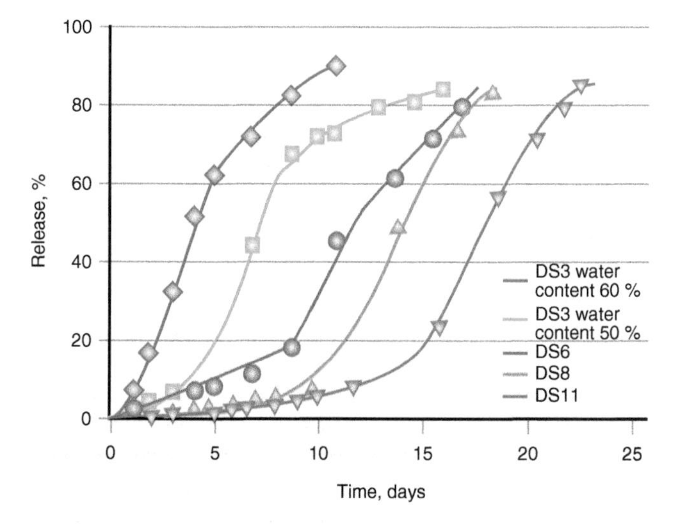

Figure 4.22 ■ Cumulative release of IgG from degrading dex-HEMA microspheres in time in vitro at pH 7, 37 °C. Water content of the dextran microspheres upon swelling: about 60 %, DS 3 (◇), and water content of about 50 %, DS 3 (□), DS 6 (○), DS 8 (△), and DS 11 (▽). The values are the mean of two independent measurements that deviated typically less than 5 % from each other. *DS* degree of cross-linking (Stenekes 2000).

exist (see Chap. 5). There are many exceptions known to this rule; for instance, "hit and run" drugs can have long-lasting pharmacological effects after only a short exposure time. Also, drug effect–blood level relationships may be time dependent, as in the case of down-regulation of relevant receptors on prolonged stimulation. Finally, if circadian rhythms exist, these will be responsible for variable PK/PD relationships as well.

If PK/PD relationships can be established as with insulin in selected groups of diabetics, then integrated biosensor–pump combinations can be used that almost act as closed-loop biofeedback systems (Schaepelynck et al. 2011; Hovorka 2011). In 2010, FDA approved an integrated diabetes management system (insulin pump, continuous glucose monitoring (CGM), and diabetes therapy management software). The CGM measures interstitial fluid levels every 5 min and sends the outcome (wireless) to a therapy management algorithm. This software program advises the patient to program the insulin pump to deliver an appropriate dose of insulin. In spite of the impressive progress made, it has

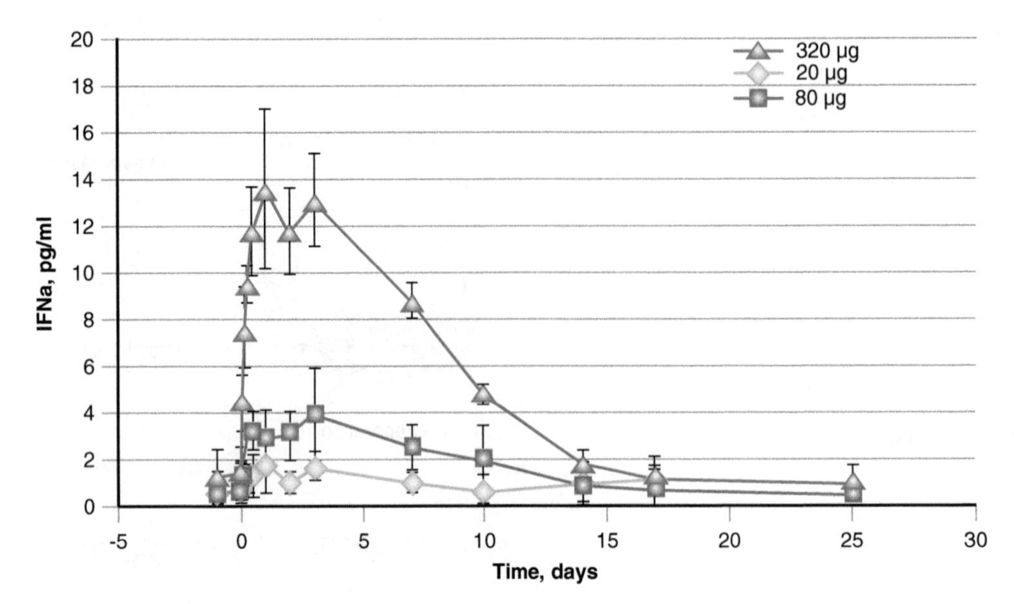

Figure 4.23 ■ Plasma profiles of interferon-α after SC injection of PolyActive™ microspheres loaded with interferon-α in volunteers in a dose finding study. The elimination half-life of "free" interferon-α is about 4–16 h. *Red* alfa interferon 20 μg. *Blue* alfa interferon 80 μg. *Dark blue* (triangles) alfa interferon 320 μg.

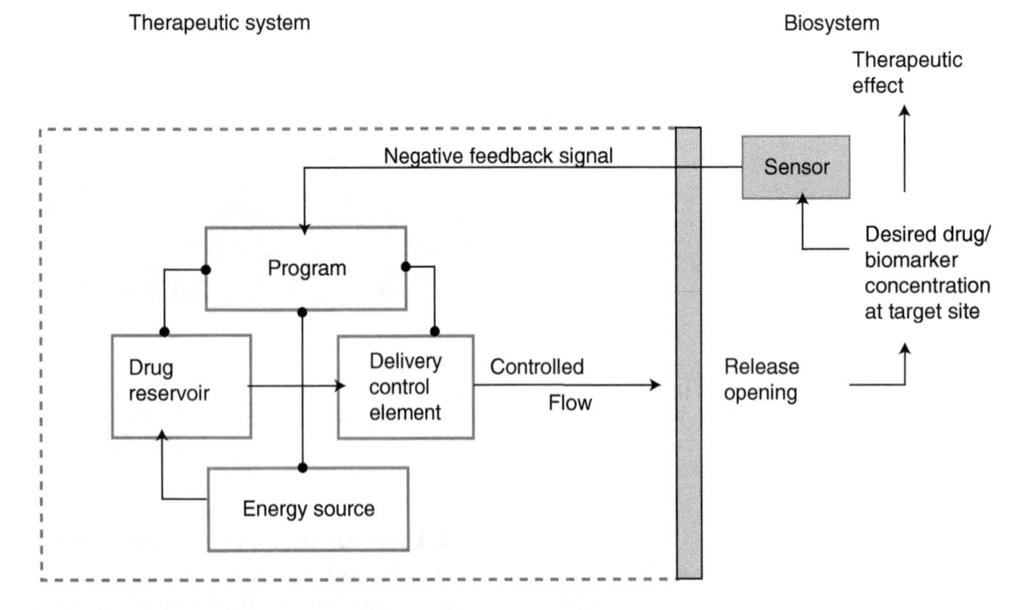

Figure 4.24 ■ Therapeutic system with closed control loop (From Heilman (1984)). (1) A biosensor, measuring the plasma level of the protein. (2) An algorithm, to calculate the required input rate for the delivery system. (3) A pump system, able to administer the drug at the required rate over prolonged periods of time.

not been possible yet to design fully closed-loop biosensors that work reliably in vivo over prolonged periods of time. Biosensor stability, robustness, and absence of histological reactions still pose problems.

■ Protein Delivery by Self-Regulating Systems

Apart from the design of biosensor–pump combinations, two other developments should be mentioned when discussing closed-loop approaches: self-regulating systems and encapsulated secretory cells. Both concepts are still under development (Heller 1993; Traitel et al. 2008).

In self-regulating systems, drug release is controlled by stimuli in the body. By far most of the research is focused on insulin release as a function of local glucose concentrations in order to stabilize blood glucose levels in diabetics. Two approaches for controlled drug release are being followed: (1) competitive desorption and (2) enzyme–substrate reactions. The competitive desorption approach is schematically depicted in Fig. 4.25.

It is based on the competition between glycosylated insulin and glucose for concanavalin (Con A) binding sites. Con A is a plant lectin with a high affinity for certain sugars. Con A attached to sepharose beads and loaded with glycosylated insulin (a bioactive form of insulin) is implanted in a pouch with a semipermeable membrane: permeable for insulin and glucose but impermeable for the sepharose beads carrying the toxic Con A. An example of the performance of a Con A-glycosylated insulin complex in pancreatectomized dogs is given in Fig. 4.26.

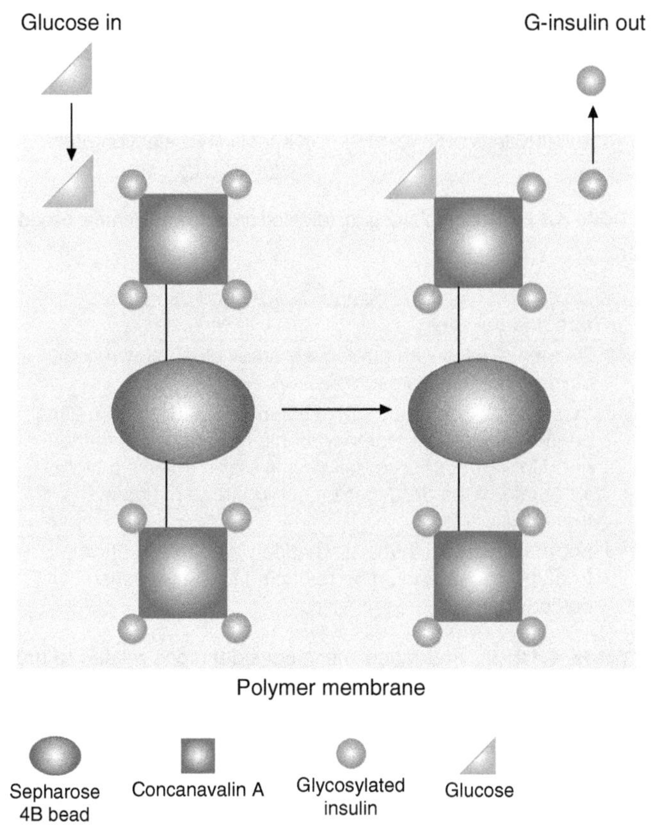

Figure 4.25 ■ Schematic design of the Con A immobilized bead/G(glycosylated)-insulin/membrane self-regulating insulin delivery system (From Kim et al. (1990)).

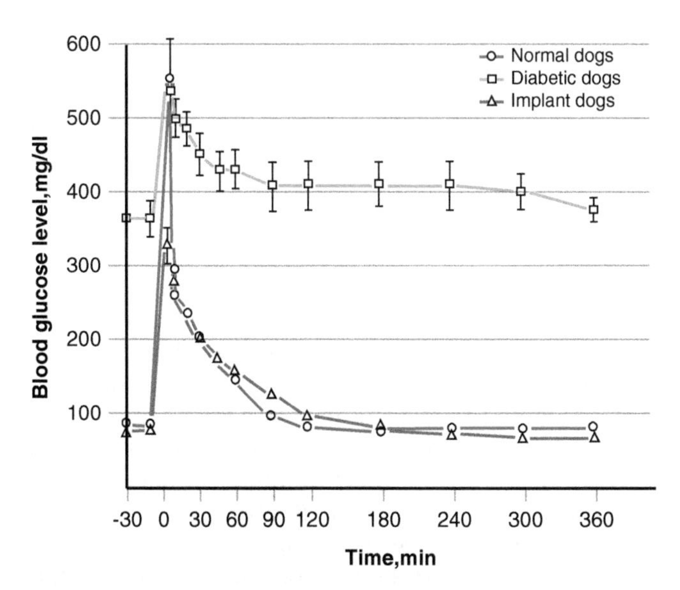

Figure 4.26 ■ Peripheral blood glucose profiles of dogs administered with bolus dextrose (500 mg/kg) during an intravenous glucose tolerance test. Normal dogs (-o-) had an intact pancreas, diabetic dogs (-□-) had undergone total pancreatectomy, and implant dogs (-△-) had been intraperitoneally implanted with a cellulose pouch containing a Con A–G–insulin complex. Blood glucose at $t=-30$ min shows the overnight fasting level 30 min prior to bolus injection of dextrose (Through Heller (1993)).

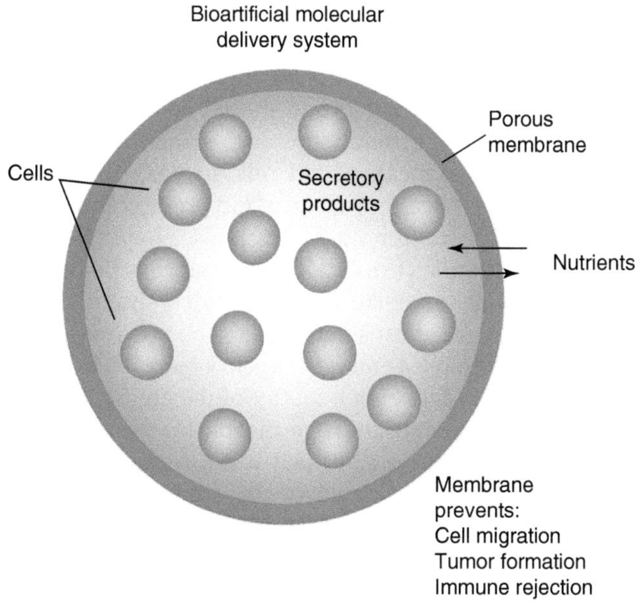

Figure 4.27 ■ Schematic illustration of a "bioartificial molecular delivery system." Secretory cells are surrounded by a semipermeable membrane prior to implantation in host tissue. Nutrients and secretory products passively diffuse through pores in the encapsulating membrane powered by concentration gradients. The use of a membrane that excludes the humoral and the cellular components of the host immune system allows immunologically incompatible cells to survive implantation without the need to administer immunosuppressive agents. Extracellular matrix material may be included depending upon the requirements of the encapsulated cells (From Tresco (1994)).

Enzyme–substrate reactions to regulate insulin release from an implanted reservoir are all based on pH drops occurring when glucose is converted to gluconic acid in the presence of the enzyme glucose oxidase. This pH drop then induces changes in the structure of acid-sensitive delivery devices such as acid-sensitive polymers, which start releasing insulin, lowering the glucose concentration, and consequently increasing the local pH and "closing the reservoir."

■ Protein Delivery by Microencapsulated Secretory Cells

The idea to use implanted, secretory cells to administer therapeutic proteins was launched long ago. A major goal has been the implantation of Langerhans cells in diabetics to restore their insulin production through biofeedback. These implanted secretory cells should be protected from the body environment, since rejection processes would immediately start, if imperfectly matched cell material is used. Besides, it is desirable to keep the cells from migrating in all different directions. When genetically modified cells are used, safety issues would be even stricter. Therefore, (micro) encapsulation of the secretory cells has been proposed (Fig. 4.27).

Thin (wall thickness in micrometer range), robust, biocompatible, and permselective polymeric membranes have been designed for these (micro) capsules (Tresco 1994). The membrane should ensure transport of nutrients (in general low M_W) from the outside medium to the encapsulated cells to keep them in a physiological, "healthy" state and to prohibit induction of undesirable immunological responses (rejection processes). Antibodies (M_W > 150 kDa) and cells belonging to the immune system (e.g., lymphocytes) should not be able to reach the encapsulated cells. The polymer membrane should have a cutoff between 50 and 150 kDa, the exact number still being a matter of debate. In the case of insulin, the membrane is permeable for this relatively small-sized hormone (5.4 kDa) and for glucose ("indicator" molecule), which is essential for proper biofeedback processes. Successful studies in diabetic animals were performed, and clinical trials have been initiated (Hernández et al. 2010).

SITE-SPECIFIC DELIVERY (TARGETING) OF PROTEIN DRUGS

Why are we still not able to beat life-threatening diseases such as cancer with our current arsenal of drugs? Causes of failure can be summarized as follows (Crommelin et al. 1992):

1. Only a small fraction of the drug reaches the target site. By far the largest fraction of the drug is distributed over nontarget organs, where it exerts side effects, and is rapidly eliminated intact from the body through the kidneys or through metabolic action (e.g., in the liver). In other words, accumulation of the drug at the target site is rather the exception than the rule.
2. Many drug molecules (in particular high M_W and hydrophilic molecules, i.e., many therapeutic proteins) do not enter cells easily. This poses a problem if intracellular delivery is required for their therapeutic activity.

Attempts are made to increase the therapeutic index of drugs through drug targeting:

1. By specific delivery of the active compound to its site of action
2. To keep it there until it has been inactivated and detoxified

Targeted drug delivery should maximize the therapeutic effect and avoid toxic effects elsewhere. Paul Ehrlich defined the basics of the concept of drug targeting already in the early days of the twentieth century. But only in the last two decades has substantial progress been made to implement this site-specific delivery concept. Recent progress can be ascribed to (1) the rapidly growing number of technological options (e.g., safe carriers and homing devices) for drug delivery; (2) many new insights gained into the pathophysiology of diseases at the cellular and molecular level, including the presence of cell-specific receptors; and, finally, (3) a better understanding of the nature of the anatomical and physiological barriers that hinder easy access to target sites. The site-specific delivery systems presently in different stages of development generally consist of three functionally separate units (Table 4.11).

An active moiety	For: therapeutic effect
A carrier	For: (metabolic) protection, changing the disposition of the drug
A homing device	For: specificity, selection of the assigned target site

Table 4.11 ■ Components for targeted drug delivery (carrier based).

1. Drugs with high total clearance are good candidates for targeted delivery
2. Response sites with a relatively small blood flow require carrier-mediated transport
3. Increases in the rate of elimination of free drug from either central or response compartments tend to increase the need for targeted drug delivery; this also implies a higher input rate of the drug–carrier conjugate to maintain the therapeutic effect
4. For maximizing the targeting effect, the release of drug from the carrier should be restricted to the response compartment

Table 4.12 ■ Pharmacokinetic considerations related to protein targeting.

Nature has provided us with antibodies, which exemplify a class of natural drug-targeting devices. In an antibody molecule one can recognize a homing device part (antigen-binding site) and "active" parts. These active parts in the molecule are responsible for activating the complement cascade or inducing interactions with monocytes when antigen is bound. The rest of the molecule can be considered as carrier.

Most of the drug (protein)-targeting work is performed with delivery systems that are designed for parenteral and, more specifically, intravenous delivery. Only a limited number of papers have dealt with the pharmacokinetics of the drug-targeting process (Hunt et al. 1986). From these kinetic models a number of conclusions could be drawn for situations where targeted delivery is, in principle, advantageous (Table 4.12).

The potential and limitations of carrier-based, targeted drug delivery systems for proteins are briefly discussed in the following sections. The focus is on concepts where monoclonal antibodies are used. They can be used as the antibody itself (also in Chap. 7), in modified form when antibodies are conjugated with an active moiety, or attached to drug-laden colloidal carriers such as liposomes.

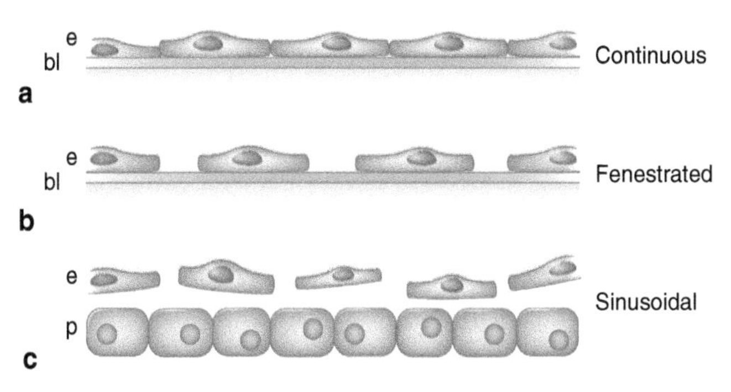

Figure 4.28 ■ Schematic illustration of the structure of different classes of blood capillaries. (**a**) Continuous capillary. The endothelium is continuous with tight junctions between adjacent endothelial cells. The subendothelial basement membrane is also continuous. (**b**) Fenestrated capillary. The endothelium exhibits a series of fenestrae which are sealed by a membranous diaphragm. The subendothelial basement membrane is continuous. (**c**) Discontinuous (sinusoidal) capillary. The overlying endothelium contains numerous gaps of varying size enabling materials in the circulation to gain access to the underlying parenchymal cells. The subendothelial basement is either absent (liver) or present as a fragmented interrupted structure (spleen, bone marrow). The fenestrae in the liver are about 0.1–0.2 μm; the pores in/between the endothelial cells and those in the basement membrane outside liver, spleen, and bone marrow are much smaller (From Poste (1985)).

Two terms are regularly used in the context of targeting: passive and active targeting. With passive targeting the "natural" disposition pattern of the carrier system is utilized for site-specific delivery. For instance, particulate carriers circulating in the blood (see below) are often rapidly taken up by macrophages in contact with the blood circulation and accumulate in the liver (Kupffer cells) and spleen. Active targeting is the concept where attempts are made to change the natural disposition of the carrier by a homing device or homing principle to select one particular tissue or cell type.

■ Anatomical, Physiological, and Pathological Considerations Relevant for Protein Targeting

Carrier-mediated transport in the body depends on the physicochemical properties of the carrier: its charge, molecular weight/size, surface hydrophobicity, and the presence of ligands for interaction with surface receptors (Crommelin and Storm 1990). If a drug enters the circulation and the target site is outside the blood circulation, the drug has to pass through the endothelial barrier. Figure 4.28 gives a schematic picture of the capillary wall structures (under physiological conditions) present at different locations in the body.

Figure 4.28 shows a diagram of intact endothelium under normal conditions. Under pathological conditions, such as those encountered in tumors and inflammation sites, endothelium can differ considerably in appearance, and endothelial permeability may be widely different from that in "healthy" tissue. Particles with sizes up to about 0.1 μm can enter tumor tissue as was demonstrated with long-circulating, nanoparticulate carrier systems (e.g., long-circulating liposomes) (Lammers et al. 2008 and below, Fig. 4.35). On the other hand, necrotic tissue can also hamper access to tumor tissue (Jain 1987). In conclusion, the body is highly compartmentalized. It should not be considered as one big pool without internal barriers for transport.

■ Soluble Carrier Systems for Targeted Delivery of Proteins

Monoclonal Antibodies (MAB) as Targeted Therapeutic Agents: Human and Humanized Antibodies (See Also with More Details: Chap. 7)

Antibodies are "natural targeting devices." Their homing ability is combined with functional activity (Crommelin et al. 1992; Crommelin and Storm 1990). MAB can affect the target cell function upon attachment. Complement can be bound via the Fc receptor and subsequently cause lysis of the target cell. Alternatively, certain Fc receptor-bearing killer cells can induce "antibody-dependent, cell-mediated cytotoxicity" (ADCC), or contact with macrophages can be established. Moreover, metabolic deficiencies can be induced in the target cells through a blockade of certain essential cell surface receptors by MAB. Structural aspects and therapeutic potential of MAB are dealt with in detail in Chap. 7.

A problem that occurs when using murine antibodies for therapy is the production of human antimouse antibodies (HAMA) after administration. HAMA induction may prohibit further use of these therapeutic MAB by neutralizing the antigen-binding site; anaphylactic reactions are relatively rare. Concurrent administration of immunosuppressive agents is a strategy to minimize side effects. More in-depth information regarding immunogenicity of therapeutic proteins is provided in Chap. 6.

There are several other ways to cope with this MAB-induced immunogenicity problem. Chapters 1 and 6 discuss immunogenicity issues as well. Here, a brief summary of the options relevant for protein targeting suffices. First of all, the use of $F(ab')_2$ or F(ab') fragments (Fig. 4.29) avoids raising an immune response against the Fc part, but the development of humanized or human MAB minimizes the induction of HAMA even further. For humanization of MAB several options can be considered. One can build chimeric (partly human, partly murine) molecules consisting of a human Fc part and a murine Fab part, with the antigen-binding sites. Or, alternatively, only the six complementarity-determining regions (CDR) of the murine antibody can be grafted in a human

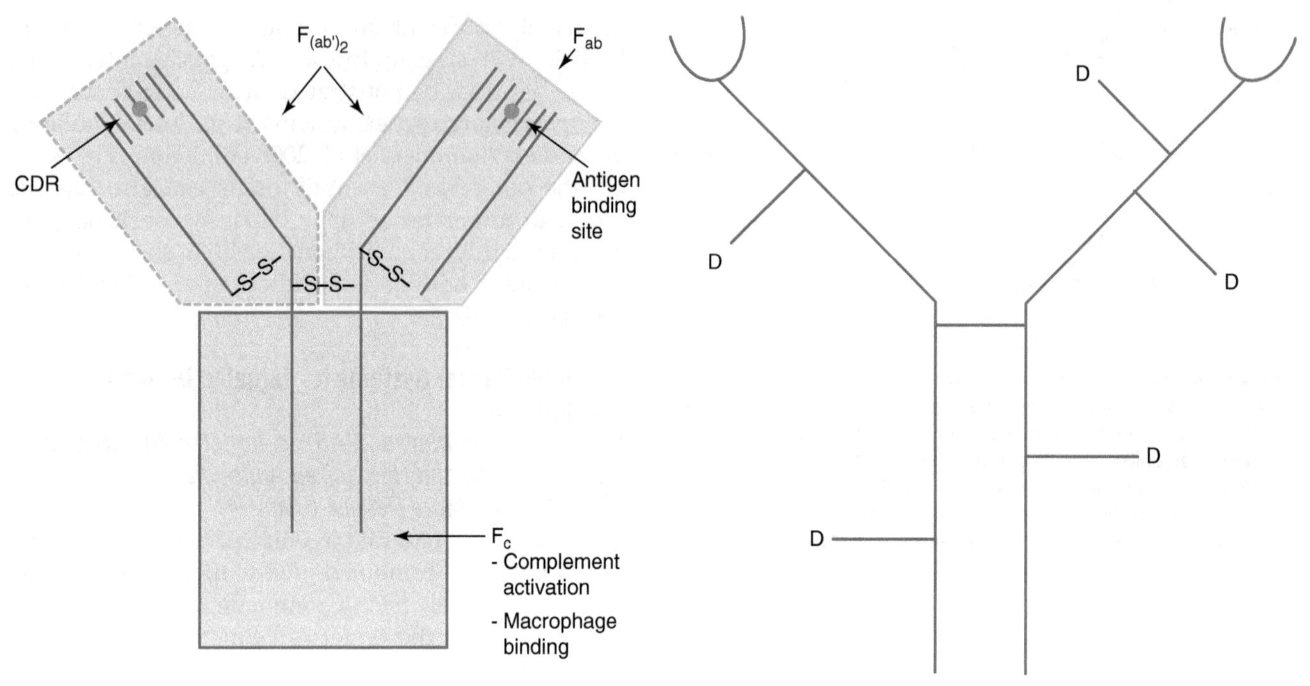

Figure 4.29 ■ Highly simplified IgG1 structure; *CDR* complementarity-determining region (see Figure 7.1 and 7.2).

Figure 4.30 ■ A schematic view of an immunoconjugate (*D* drug molecules covalently attached to antibody (fragments)); see Figure 7.3.

antibody structure. CDR grafting minimizes the exposure of the patient to murine-derived material.

Completely human MAB can be produced by transfecting human antibody genes into mouse cells, which subsequently produce the human MAB. Alternatively, transgenic mice can be used (see Chap. 6 and 8). These approaches reduce the immunogenicity compared to the existing generation of murine MAB. But even with all these human or humanized MAB, anti-idiotypic immune responses against the binding site structure of the MAB cannot be excluded.

Bispecific Antibodies (See Also Chap. 7)

To enhance the therapeutic potential of antibodies, bispecific antibodies have been designed. Bispecific antibodies are combinations of two separate antibodies to create a molecule with two different binding sites. Bispecific MABs bring target cells or tissue (one antigen-binding site) in contact with other structures (second antigen-binding site). This second antigen-binding site can bind to effector cells via cytotoxicity-triggering molecules on T cells, NK (natural killer) cells, or macrophages and thus trigger cytotoxicity.

Bispecific antibodies have reached the clinic, e.g., a bispecific antibody that targets CD3 on T cells and CD20 on lymphoma cells (Crommelin and Storm 1990; Holmes 2011). In 2009 catumaxomab (anti-CD3 and antiepithelial cell adhesion molecule) was registered in Europe for the treatment of malignant ascites.

Immunoconjugates: Combinations Between an Antibody and an Active Compound

In many cases antibodies alone or bispecific antibodies showed lack of sufficient therapeutic activity. To enhance their activity, conjugates of MAB and drugs have been designed (Fig. 4.30). These efforts mainly focus on the treatment of cancer (Crommelin and Storm 1990). To test the concept of immunoconjugates, a wide range of drugs has been covalently bound to antibodies and has been evaluated in animal tumor models. As only a limited number of antibody molecules can bind to the target cells and as the payload per MAB molecule is restricted as well, only conjugation of highly potent drugs will lead to sufficient therapeutic activity. So far gemtuzumab ozogamicin (Mylotarg®) was the first immunoconjugate on the market. It is a conjugate of a monoclonal antibody and calicheamicin (see also Chap. 17). The MAB part targets the CD33 surface antigen in CD33-positive acute myeloid leukemia cells (AML). After internalization into the cell, the highly cytotoxic calicheamicin is released. Mylotarg® was launched on the market in 2000. After extensive follow-up evaluations, it was taken off the market in 2010 as serious side effects were reported with no added antitumor benefit compared to standard therapy.

Cytostatics with a high intrinsic cytotoxicity are needed (see above). Because the kinetic behavior of active compounds is strongly affected by the conjugating antibody, not only existing and approved cytostatics

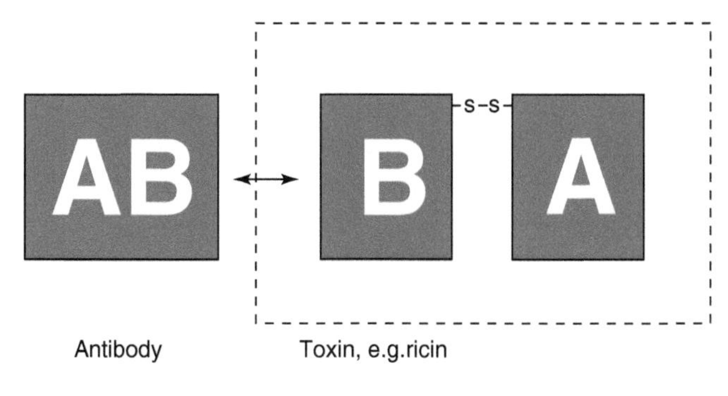

Figure 4.31 ■ Immunotoxins are composed of antibody molecules connected to a toxin, e.g., ricin. Both the integral ricin molecule and the A chain alone have been used. *AB* antibody, *A* and *B* stand for the *A* and *B* chain of the ricin toxin, respectively (not in the list of abbreviations).

1. Covalent binding of the protein/drug to the antibody may change the cytotoxic potential of the drug and decrease the affinity of the MAB for the antigen
2. The stability of the conjugate in vivo may be insufficient; fragmentation will lead to loss of targeting potential
3. The immunogenicity of the MAB and toxicity of the protein/drug involved may change dramatically

Table 4.13 ■ Potential problems encountered with immunoconjugates (Crommelin et al. 1992).

but also active compounds that were never used before as drugs, because of their high toxicity, should now be reconsidered.

Immunoconjugated toxins are now tested as chemotherapeutic agents to treat cancer (immunotoxins). Examples of the toxin family originating from plants are ricin, saporin, gelonin, and abrin. Toxins from bacterial origin are diphtheria toxin and pseudomonas endotoxin. These proteins are extremely toxic; they block enzymatically intracellular protein synthesis at the ribosomal level. For instance, ricin (M_W 66 kDa) consists of an A and a B chain that are linked through a cystine bridge (Fig. 4.31). The A chain is responsible for blocking protein synthesis at the ribosomes. The B chain is important for cellular uptake of the molecule (endocytosis) and intracellular trafficking and is deleted as it is considered redundant.

Table 4.13 lists a number of potential problems encountered with (toxin-based) immunoconjugates (Crommelin et al. 1992). In animal studies with immunoconjugated ricin, only a small fraction of these immunotoxins accumulates in solid tumor tissue (1 %). A major fraction still ends up in the liver, the main target organ for "natural" ricin. Because of the poor tissue penetration, primary targets for clinical use are target cells circulating in blood or endothelial cells. Moreover, in early clinical phase I studies (to assess the safety of the conjugates), the first generation of immunoconjugates turned out to be immunogenic, and these murine MAB have been replaced by human or humanized MAB (see above). Attempts were made to adapt the ricin molecule by genetic engineering so that liver targeting is being minimized. This can be done by blocking (removing or masking) on the ricin molecule ligands for galactose receptors on hepatocytes.

Not only MAB or fragments thereof were used as homing device/cell wall translocation moiety. Interestingly, the first and only immunotoxin approved by the FDA (denileukin diftitox) is a fusion protein of IL2 and a truncated diphtheria toxin (DAB389). Its target is the high-affinity IL2 receptor, and it is used in the treatment of cutaneous T-cell lymphoma (Choudhary et al. 2011).

1. Tumor heterogeneity
2. Antigen shedding
3. Antigen modulation

Table 4.14 ■ Factors that interfere with successful targeting of proteins to tumor cells.

■ Potential Pitfalls in Tumor Targeting

Upon IV injection, only a small fraction of the homing device-carrier-drug complex is sequestered at the target site. Apart from the compartmentalization of the body (see above: anatomical and physiological hurdles) and consequently the carrier-dependent barriers that result, several other factors account for this lack of target site accumulation (Table 4.14).

How successful are MAB in discriminating target cells (tumor cells) from nontarget cells? Do all tumor cells expose the tumor-associated antigen? These questions are still difficult to answer (Hellström et al. 1987). Tumor cell surface-specific molecules used for homing purposes are often differentiation antigens on the tumor cell wall. These structures are not unique since they occur in a lower-density level on non-target cells as well. Therefore, the target site specificity of MAB raised against these structures is more quantitative than qualitative in nature.

Another category of tumor-associated antigens are the clone-specific antigens. They are unique for the clone forming the tumor. However, the practical problem when focusing on clone-specific antibodies for drug targeting is that each patient probably needs a tailor-made MAB.

The surface "makeup" of tumor cells in a tumor or a metastasis is not constant, neither in time nor between cells in the same tumor. There are many subpopulations of tumor cells, and they express different surface molecules. This heterogeneity means that not all cells in the tumor will interact with one, single targeted conjugate.

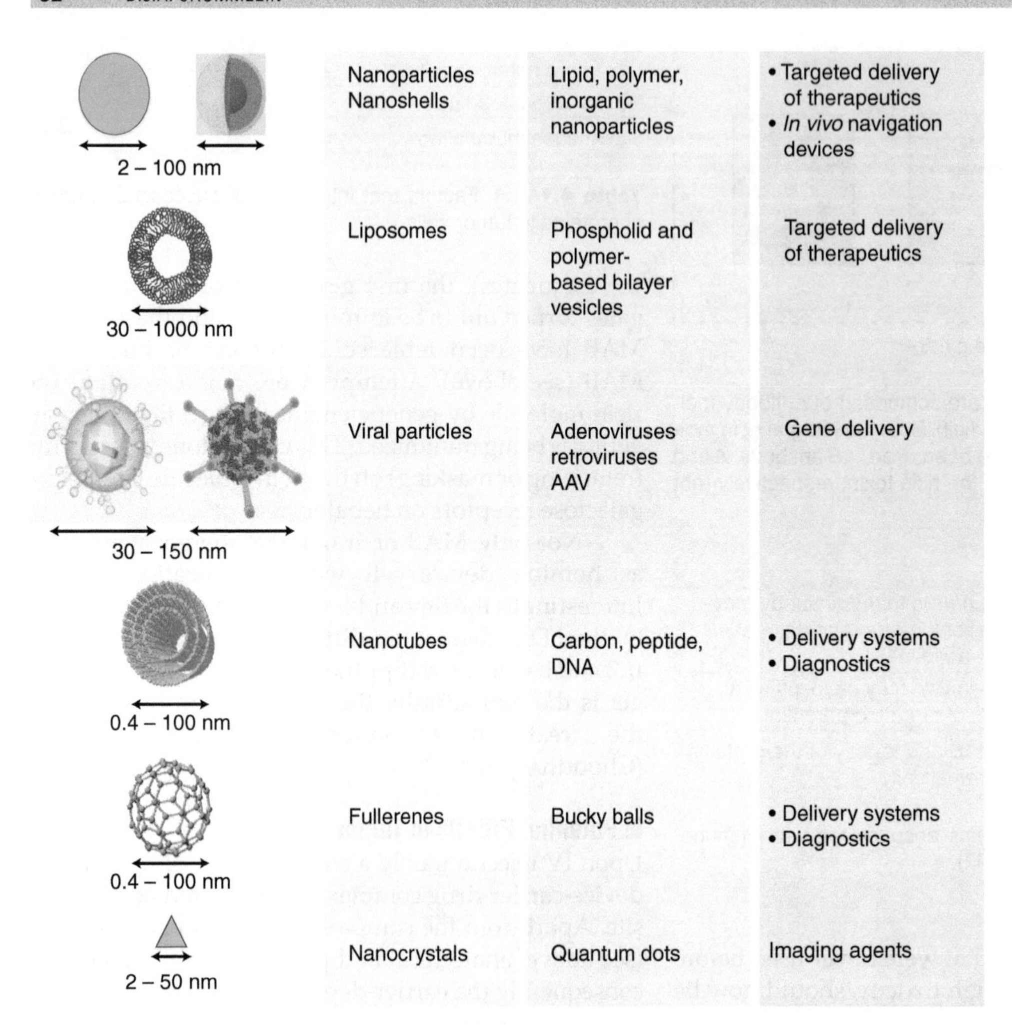

2 – 100 nm
Nanoparticles
Nanoshells
Lipid, polymer, inorganic nanoparticles
• Targeted delivery of therapeutics
• In vivo navigation devices
30 – 1000 nm
Liposomes
Phospholid and polymer-based bilayer vesicles
Targeted delivery of therapeutics
30 – 150 nm
Viral particles
Adenoviruses
retroviruses
AAV
Gene delivery
0.4 – 100 nm
Nanotubes
Carbon, peptide, DNA
• Delivery systems
• Diagnostics
0.4 – 100 nm
Fullerenes
Bucky balls
• Delivery systems
• Diagnostics
2 – 50 nm
Nanocrystals
Quantum dots
Imaging agents

Figure 4.32 ■ Carrier systems in the nanometer/colloidal size range.

Antigen shedding and antigen modulation are two other ways tumor cells can avoid recognition. Shedding of antigens means that antigens are released from the surface. They can then interact with circulating conjugates outside the target area, form an antigen–antibody complex, and neutralize the homing potential of the conjugates before the target area has been reached. Finally, antigen modulation can occur upon binding of MAB to the cell surface antigen. Modulation is the phenomenon that upon endocytosis of the (originally exposed) surface antigen–immunoconjugate complex, some of these antigens are not exposed anymore on the surface; there is no replenishment of endocytosed surface antigens.

Four strategies can be implemented to solve problems related to tumor cell heterogeneity, shedding, and modulation. (1) Cocktails of different MAB attached to the toxin can be used. (2) Another approach is to give up striving for complete target cell specificity and to induce so-called "bystander" effects. Then, the targeted system is designed in such a way that the active part is released from the conjugate after reaching a target cell but before the antigen–conjugate complex has been taken up (is endocytosed) by the target cell. (3) Not all surface antigens show shedding or modulation. If these phenomena occur, other antigen/MAB combinations should be selected that do not demonstrate these effects. (4) Injection of free MAB prior to injection of the immunoconjugate is a strategy to neutralize "free" circulating antigen. The subsequently injected conjugate should not encounter shedded, free antigen.

In conclusion, targeted (modified) MAB and MAB conjugates are assessed for their value in fighting life-threatening diseases such as cancer. During the last decade, technology has evolved quickly; many different new options became available, and time will tell how successful the concept of immunotoxins and immunoconjugates will be for the patient.

■ Nanotechnology at Work: Nanoparticles for Targeted Delivery of Proteins

A wide range of carrier systems in the colloidal/nm size range (diameters up to a few micrometers) has been proposed for protein targeting. Examples are shown in Fig. 4.32. Upon entering the bloodstream after IV injection, it is difficult for many of these nano-sized particulate systems to pass through epithelial and endothelial membranes in healthy tissue, as the size cutoff for permeation through these multilayered barriers is around 20 nm (excluding the liver; see above Fig. 4.28). Parameters that control the fate of particulate carriers in vivo are listed in Table 4.15.

As a rule, cells of the mononuclear phagocyte system (MPS), such as macrophages, recognize stable, colloidal particulate systems (< 5 μm) as "foreign body-like structures" and phagocytose them. Thus, the liver and spleen, organs rich in blood circulation-exposed macrophages, take up the majority of these particulates (Crommelin and Storm 1990). Larger (> 5 μm) intravenously injected particles tend to form emboli in lung capillaries on their first encounter with this organ.

Liposomes have gained considerable attention among the colloidal particulate systems proposed for site-specific delivery of (or by) proteins (Gregoriadis 2006). Liposomes are vesicular structures based on (phospho) lipid bilayers surrounding an aqueous core. The main component of the bilayer usually is phosphatidylcholine (Fig. 4.33).

By selecting their bilayer constituents and one of the many preparation procedures described, liposomes can be made varying in size between 30 nm (e.g., by extrusion or ultrasonication) and 10 μm, charge (by incorporation negatively or positively charged lipid molecules), and bilayer rigidity (by selecting special phospholipids or adding lipids such as cholesterol). Liposomes can carry their payload (proteins) either in the lipid core of the bilayer through partitioning, attached to the bilayer, or physically entrapped in the aqueous phase. To make liposomes target site specific, except for passive targeting to liver (Kupffer cells) and spleen macrophages, homing devices are covalently coupled to the outside bilayer leaflet. In Table 4.16 three relative advantages of liposomes over other nanoparticulate systems are given.

After injection, "standard" liposomes stay in the blood circulation only for a short time. They are taken up by macrophages in the liver and spleen, or they degrade by exchange of bilayer constituents with blood constituents. Liposome residence time in the blood circulation

Parameters
1. Size
2. Charge
3. Surface hydrophilicity
4. Presence of homing devices on their surface
5. Exchange of constitutive parts with blood components

Table 4.15 ■ Parameters controlling the fate of particulate carriers in vivo.

Advantages
1. Their relatively low toxicity, existing safety record, and experience with marketed, intravenously administered liposome products (e.g., amphotericin B, doxorubicin, daunorubicin) (Storm et al. 1993)
2. The presence of a relatively large aqueous core, which is essential to stabilize the structural features of many proteins
3. The possibility to manipulate release characteristics of liposome-associated proteins and to control disposition in vivo by changing preparation techniques and bilayer constituents (Crommelin and Schreier 1994)

Table 4.16 ■ Liposomes stand out among other particulate carrier systems, because of:

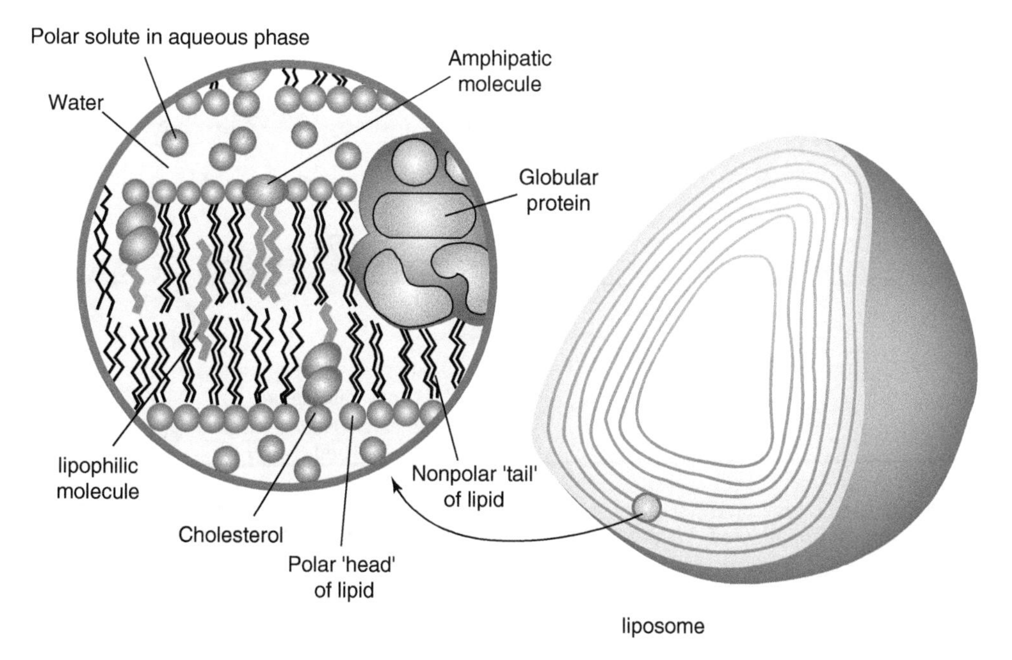

Figure 4.33 ■ An artist's view of what a multilamellar liposome looks like. The lamellae are bilayers of (phospho) lipid molecules with their hydrophobic tails oriented inwards and their polar heads directed to, and in contact with, the aqueous medium. The bilayer may accommodate lipophilic drugs inside. Hydrophilic drugs will be found in the aqueous core and in between the bilayers. Depending on their hydrophilic/hydrophobic balance and tertiary structure, proteins and peptides will be found in the aqueous phase, at the bilayer–water interface, or inside the lipid bilayer (Adapted from Fendler (1980)).

can be extended to many hours and even days upon grafting polyethylene glycol (PEG) chains on the surface and choosing for stable bilayer structures (Fig. 4.34, cf. Fig. 4.19). These long-circulating liposomes escape macrophage uptake for prolonged periods and may then be sequestered in other organs than liver and spleen alone, e.g., tumors and inflamed tissues. The reason for this sequestration is the so-called enhanced permeability and retention (EPR) effect, first described by Maeda et al. (Lammers et al. 2008) (Fig. 4.35). The term refers to the observation that the capillary bed in inflamed or tumor tissue tends to be more permeable than in not-affected, "healthy" beds and that effluent lymphatic transport from the affected beds is hampered. In Fig. 4.36 an example is shown of the use of 99mTc-labelled liposomes in the detection of inflammation sites in a patient. As a caveat, the full therapeutic benefit of the EPR effect is under discussion. For example, tumor heterogeneity with respect to capillary leakage and retarded but still considerable uptake in non-target organs (MPS) question the potential advantage over standard therapies (Bae and Park 2011).

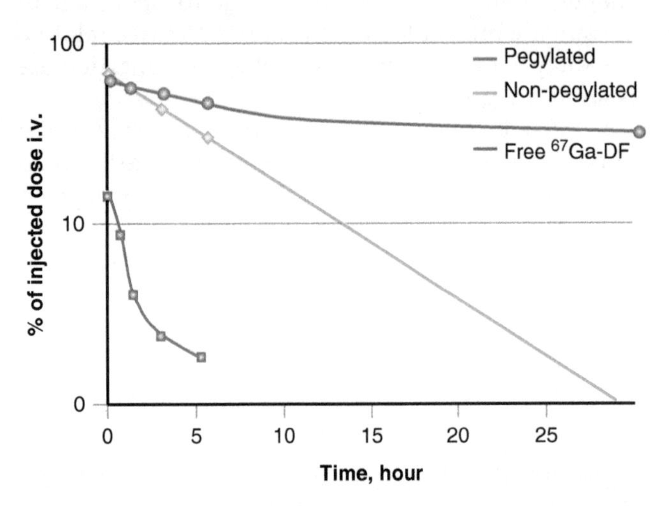

Figure 4.34 ■ Comparison of the blood levels of free label ^{67}Ga-DF, gallium-Desferal with ^{67}Ga-DF-laden pegylated (PEG) and non-pegylated liposomes upon IV administration in rats (From Woodle et al. (1990)).

On the other hand, accumulation of protein-laden liposomes in macrophages (passive targeting) offers interesting therapeutic opportunities. Reaching macrophages may help us to more effectively fight macrophage located in microbial, viral, or bacterial diseases than with our present approaches (Crommelin and Schreier 1994).

Several attempts have been made to sequester immunoliposomes (i.e., antibody (fragment)–liposome combinations) at predetermined sites in the body. Here the aim is active targeting to the desired target site instead of passive targeting to macrophages. The concept is schematically presented in Fig. 4.37.

When designing immunoliposomes, antibodies or antibody-fragments are covalently bound to the surface of liposomes through lipid anchor molecules. Non-pegylated immunoliposomes may have poor access to target sites outside the blood circulation after intravenous injection (see EPR discussion above). Therefore, target sites should be sought in the blood circulation (red blood cells, thrombi, lymphocytes, or

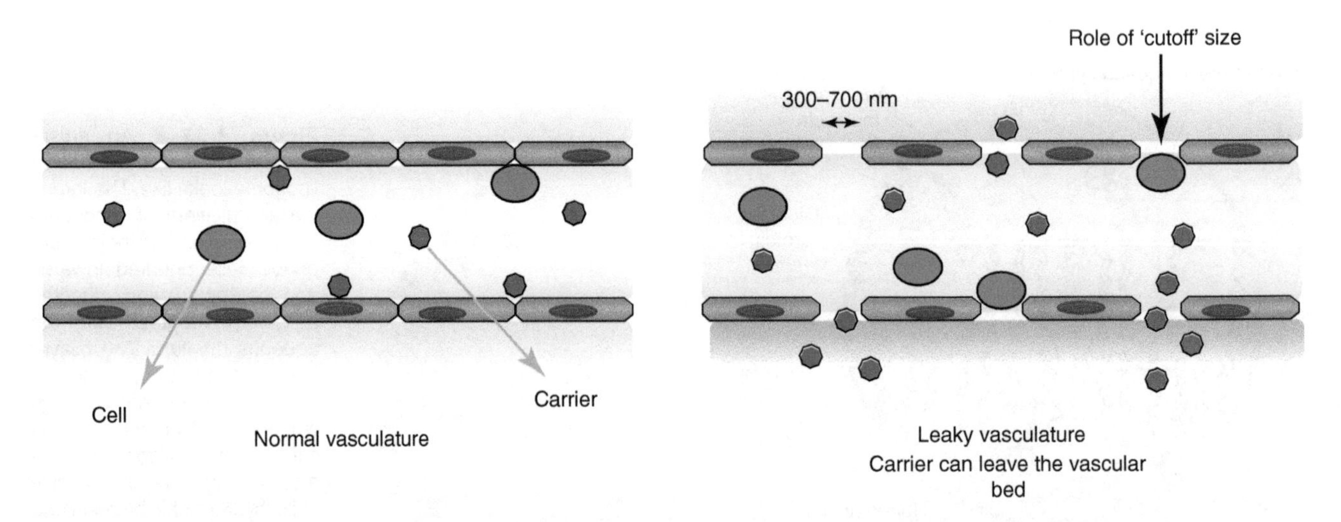

Figure 4.35 ■ The enhanced permeability and retention (EPR) effect refers to the observation that the capillary bed in inflamed or tumor tissue tends to be more permeable than in not-affected, "healthy" beds and that effluent lymphatic transport from the affected beds is hampered.

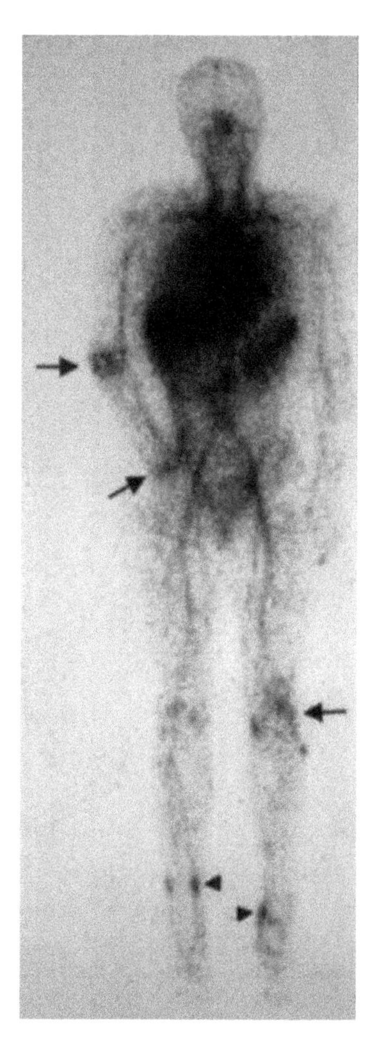

Figure 4.36 ■ ^{99m}Tc-PEG-liposomes scintigraphy of a female patient. Anterior whole body image, 24 h postinjection, shows physiological uptake in the cardiac blood, greater veins, liver, and spleen. Liposome uptake at pathological sites can be noted along synovial lining of the left elbow, left wrist, and right knee (*arrows*) and the medial site of both ankles (*arrow heads*) (Storm and Crommelin 1998).

endothelial cells exposing certain adhesion molecules when under stress, e.g., ICAM-1, intercellular cell adhesion molecule) (Crommelin et al. 1995).

Other interesting target sites are those located in cavities, where one can locally administer the drug–carrier combination. The bladder and the peritoneal cavity are such cavities. These cavities can be the sites where the diseased tissue is concentrated. For instance, with ovarian carcinomas the tumors are confined to the peritoneal cavity for most of their lifetime. After IP injection of immunoliposomes directed against human ovarian carcinomas in athymic, nude mice, a specific interaction between immunoliposomes and the human ovarian carcinoma was observed (Storm et al. 1994) (Fig. 4.38).

Attaching an immunoliposome to target cells usually does not induce a therapeutic effect per se. After establishing an immunoliposome–cell interaction, the protein drug has to exert its action on the cell. To do that, the protein has to be released in its active form. There are several pathways proposed to reach this goal (Fig. 4.39) (Peeters et al. 1987).

When the immunoliposome–cell complex encounters a macrophage, the cells plus adhering liposome are probably phagocytosed and enter the macrophage (option Fig. 4.39a). Subsequently, the liposome-associated protein drug can be released. As this will most likely happen in the "hostile" lysosomal environment, little intact protein will become available. In the situation depicted in Fig. 4.39, option b, the drug is released from the adhering immunoliposomes in the close proximity of the target cell. In principle, release rate control is achieved by selecting the proper liposomal bilayers with delayed or sustained drug release characteristics. A third approach is depicted as Fig. 4.39, option c: drug release is induced from liposomal bilayers by external stimuli (local pH change or temperature change). Finally, one can envision that

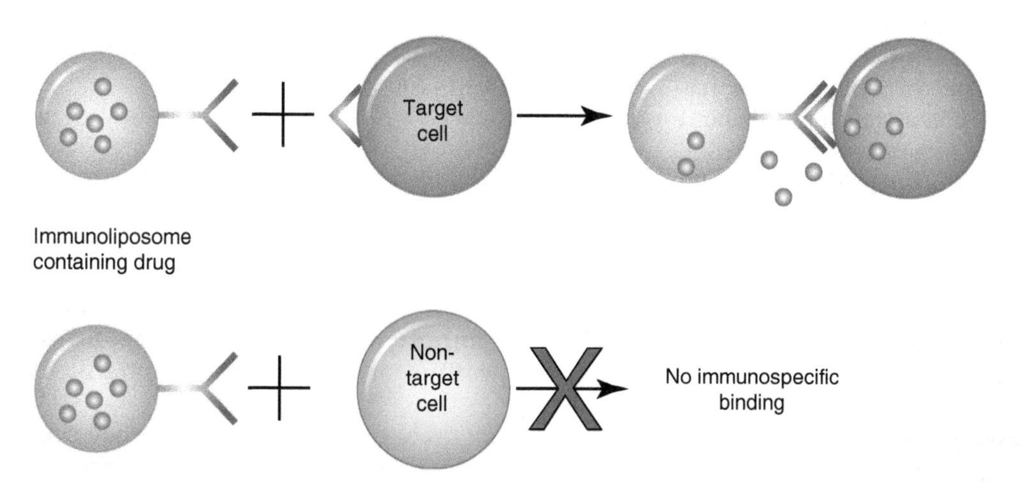

Figure 4.37 ■ Schematic representation of the concept of drug targeting with immunoliposomes (From Nässander et al. (1990)).

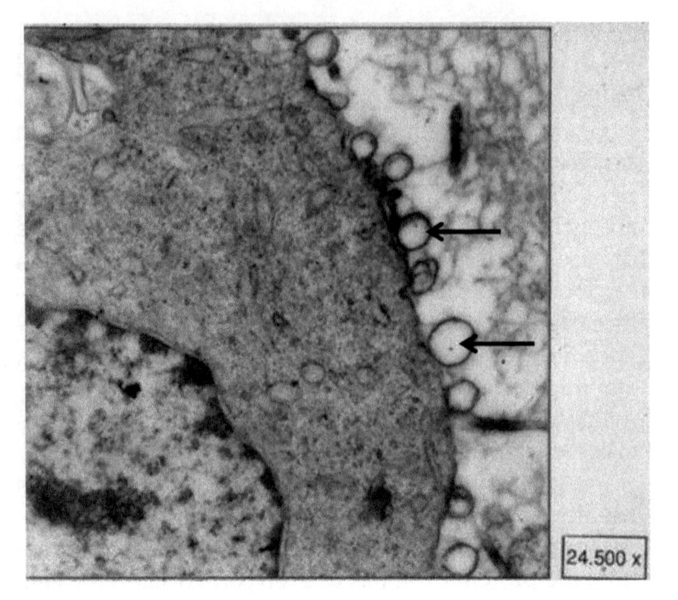

Figure 4.38 ■ Electron micrograph showing immunoliposomes (vesicular structures) attached to human ovarian carcinoma cells (see text).

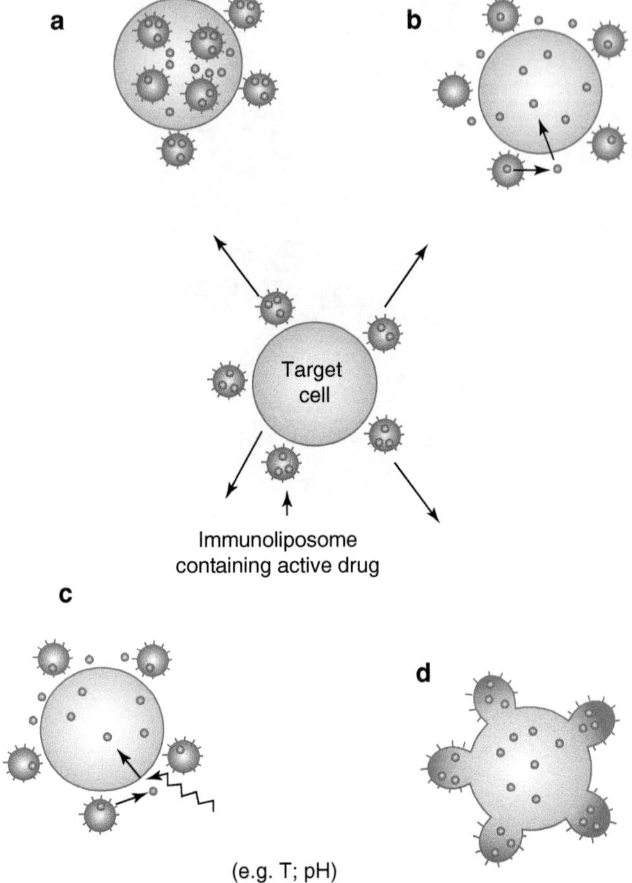

Figure 4.39 ■ Several pathways of drug internalization after immunospecific binding of the immunoliposomes to the appropriate target cell. (**a**) Uptake in liver and spleen macrophages; subsequent drug release.(**b**) Release of drug close to target cell. (**c**) Release of drug close to target cell; external triggering of release. (**d**) Fusion with target cell; subsequent drug release (From Peeters et al. (1987)).

immunoliposomes are built with intrinsic fusogenic potential, which is only activated upon attachment of the carrier to the target cell. This exciting option, Fig. 4.39d, resembles the behavior of certain viruses. Viruses offer interesting insights in pathways as to how to enter target cells and how to deliver their payload successfully in a target organelle (i.e., for viruses, the nucleus). This virus-mimicking approach led to the design of artificial viruses for targeted delivery of genetic material, but this can be extended to therapeutic proteins (Mastrobattista et al. 2006).

■ Perspectives for Targeted Protein Delivery

Protein-targeting strategies have been developing at a rapid pace. A new generation of homing devices (target cell-specific monoclonal antibodies) and a better insight into the anatomy and physiology of the human body under pathological conditions have been critical factors to achieve this success. A much better picture has emerged, not only about the potentials, but also about the limitations of the different targeting approaches.

Very little attention has been paid to typically pharmaceutical aspects of advanced drug delivery systems such as immunotoxins and immunoliposomes. These systems are now produced on a lab scale, and their therapeutic potential is currently under investigation. If therapeutic benefits have been clearly proven in preclinical and early clinical trials, then scaling up, shelf life, and quality assurance issues (e.g., reproducibility of the manufacturing process, purity of the ingredients) will still require considerable attention.

SELF-ASSESSMENT QUESTIONS

■ Questions

1. How does one sterilize biotech products for parenteral administration?
2. A pharmaceutical protein, which is poorly water soluble around its pI, has to be formulated as an injection. What conditions would one select to produce a water-soluble, injectable solution?
3. Why are many biotech proteins to be used in the clinic formulated in freeze-dried form? Why is, as a rule, the presence of lyoprotectants required? Why is it important to know the glass transition temperature or eutectic temperature of the system?
4. Why is it not necessarily wise to work at the lowest possible chamber pressures during freeze-drying?

5. Why are (with the exception of oral vaccines) no oral delivery systems for proteins available?
6. What alternative route of administration to the parenteral route would be the first to look into if a systemic therapeutic effect is pursued and if one does not wish to exploit absorption-enhancing technologies?
7. If one considers using the iontophoretic transport route for protein delivery, what are the variables to be considered?
8. What are the differences between the endocrine, paracrine, and autocrine way of cell communication? Why is information on the way cells communicate important in the drug formulation process?
9. A company decides to explore the possibility to develop a feedback system for a therapeutic protein. What information should be available for estimating the chances for success?
10. Why is the selection of the dimensions of a colloidal particulate carrier system for targeted delivery of a protein of utmost importance?
11. Design a targeted, colloidal carrier system and a protocol for its use to circumvent the three hurdles to achieve successful treatment of solid tumors (mentioned in Table 4.14).
12. What are the options for inducing therapeutic actions upon attachment of immunoliposomes to (tumor) target cells?

■ Answers

1. Through aseptic manufacturing protocols. Final filtration through 0.2 or 0.22 μm pore filters into the vials/syringes further reduces the chances of contamination of the protein solutions.
2. One has to go through the items listed in Table 4.1. As the aqueous solubility is probably pH dependent, information on the preferred pH ranges should be collected. If necessary, solubility enhancers (e.g., lysine, arginine, and/or surfactants) and stabilizers against adsorption/aggregation should be added. "As a last resort," one might consider carriers such as liposomes.
3. Chemical and physical instability of proteins in aqueous media is usually the reason to dry the protein solution.
   Freeze-drying is then the preferred technology, as other drying techniques do not give rapidly reconstitutable dry forms for the formulation and/or because elevated temperatures necessary for drying jeopardize the integrity of the protein.
   The glass transition/eutectic temperature should not be exceeded as otherwise collapse of the cake can be observed. Collapse slows down the drying process rate, and collapsed material does not rapidly dissolve upon adding water for reconstitution.
4. Because gas conduction (one of the three heat transfer routes) depends on pressure and is reduced at low pressure.
5. Because of the hostile environment in the GI tract regarding protein stability and the poor absorption characteristics of proteins (high molecular weight/often hydrophilic).
6. The pulmonary route.
7. Physical characteristics of the protein and medium, such as molecular weight, pI, ionic strength, pH, and, in addition, electrical current options (pulsed, permanent, wave shape) and desired dose level/pattern (pulsed/constant/variable).
8. This information is important because, in particular with paracrine- and autocrine-acting proteins, targeted delivery should be considered to minimize unwanted side effects.
9. Answers:
   - The desired pharmacokinetic profile (e.g., information on the PK/PD relationship/circadian rhythm)
   - Chemical and physical stability of the protein on long-term storage at body/ambient temperature
   - Availability of a biosensor system (stability in vivo, precision/accuracy)
   - Availability of a reliable pump system (see Table 4.10)
10. The body is highly compartmentalized, and access to target sites inside and outside the blood circulation is highly dependent on the size of the carrier system involved (and other factors such as the presence of diseased tissue and surface characteristics such as charge, hydrophobicity/hydrophilicity, ligands).
11. The selection should be based on the induction of bystander effects, "cocktails" of homing devices (e.g., monoclonal antibodies), and selection of non-modulating receptors and non-shedding receptors. Neutralization of free, shed tumor antigens with free, non-conjugated monoclonal antibodies by injection of these free antibodies before the administration of ligand–carrier–drug combinations would be an approach for avoiding neutralization of the carrier–homing device combination by shed antigen.
12. Figure 4.39 gives an overview of these options.

REFERENCES

Alzet product information (2012) http://www.alzet.com/products/ALZET_Pumps/index.html

Arakawa T, Kita Y, Carpenter JF (1991) Protein-solvent interactions in pharmaceutical formulation. Pharm Res 8:285–291

Bae YH, Park K (2011) Targeted drug delivery to tumors: myths, reality and possibility. J Control Release 153:198–205
Banerjee PS, Hosny EA, Robinson JR (1991) Parenteral delivery of peptide and protein drugs. In: Lee VHL (ed) Peptide and protein drug delivery. Marcel Dekker, Inc., New York, pp 487–543
Björk E, Edman P (1988) Characterization of degradable starch microspheres as a nasal delivery system for drugs. Int J Pharm 62:187–192
Brange J, Langkjaer L (1993) Insulin structure and stability. In: Wang YJ, Pearlman R (eds) Stability and characterization of protein and peptide drugs. Case histories. Plenum Press, Inc, New York, pp 315–350
Chien YW (1991) Transdermal route of peptide and protein drug delivery. In: Lee VHL (ed) Peptide and protein drug delivery. Marcel Dekker, Inc., New York, pp 667–689
Choudhari S, Mathew M, Verma RS (2011) Therapeutic potential of anticancer immunotoxins. Drug Discov Today 16:495–503
Constantino HR, Pikal MJ (2004) Lyophilization of biopharmaceuticals. AAPS Press, Arlington
Crommelin DJA, Schreier H (1994) Liposomes. In: Kreuter J (ed) Colloidal drug delivery systems. Marcel Dekker, Inc., New York, pp 73–190
Crommelin DJA, Storm G (1990) Drug targeting. In: Sammes PG, Taylor JD (eds) Comprehensive medicinal chemistry. Pergamon Press, Oxford, pp 661–701
Crommelin DJA, Bergers J, Zuidema J (1992) Antibody-based drug targeting approaches: perspectives and challenges. In: Wermuth CG, Koga N, König H, Metcalf BW (eds) Medicinal chemistry for the 21st century. Blackwell Scientific Publications, Oxford, pp 351–365
Crommelin DJA, Scherphof G, Storm G (1995) Active targeting with particulate carrier systems in the blood compartment. Adv Drug Deliv Rev 17:49–60
De Leede LGJ, Humphries JE, Bechet AC, Van Hoogdalem EJ, Verrijk R, Spencer DJ (2008) Novel controlled-release *Lemna*-derived IFN-α2b (Locteron): pharmacokinetics, pharmacodynamics, and tolerability in a phase I clinical trial. J Interferon Cytokine Res 28:113–122
Delves PJ, Martin MS, Burton DR, Roitt IM (2011) Roitt's essential immunology, 12th edn. Wiley-Blackwell, Oxford
Edman P, Björk E (1992) Nasal delivery of peptide drugs. Adv Drug Deliv Rev 8:165–177
FDA (2010) http://www.fda.gov/BiologicsBloodVaccines/SafetyAvailability/VaccineSafety/UCM096228
Fendler JH (1980) Optimizing drug entrapment in liposomes. Chemical and biophysical considerations. In: Gregoriadis G, Allison AC (eds) Liposomes in biological systems. Wiley, Chichester, p 87
Franks F, Hatley RHM, Mathias SF (1991) Materials science and the production of shelf-stable biologicals. Pharm Technol Int 3:24–34
Gregoriadis G (2006) Liposome technology, 3rd edn. Informa Healthcare, New York
Groves M (1988) Parenteral technology manual. Interpharm Press, Inc., Buffalo Grove
Halls NA (1994) Achieving sterility in medical and pharmaceutical products. Marcel Dekker, Inc., New York
Heilmann K (1984) Therapeutic systems. Rate controlled delivery: concept and development. G. Thieme Verlag, Stuttgart
Heller J (1993) Polymers for controlled parenteral delivery of peptides and proteins. Adv Drug Deliv Rev 10:163–204
Hellström KE, Hellström I, Goodman GE (1987) Antibodies for drug delivery. In: Robinson JR, Lee VHL (eds) Controlled drug delivery. Marcel Dekker, Inc., New York, pp 623–653
Hernández RM, Orive G, Murua A, Pedraz JL (2010) Microcapsules and microcarriers for in situ cell delivery. Adv Drug Deliv Rev 62:711–730
Holmes D (2011) Buy buy bispecific antibodies. Nat Rev Drug Discov 10:798–800
Hovorka R (2011) Closed-loop insulin delivery: from bench to clinical practice. Nat Rev Endocrinol 7:385–395
Hunt CA, MacGregor RD, Siegel RA (1986) Engineering targeted in vivo drug delivery. I. The physiological and physicochemical principles governing opportunities and limitations. Pharm Res 3:333–344
Jain RK (1987) Transport of molecules in the tumor interstitium: a review. Cancer Res 47:3039–3051
Jorgensen J, Nielsen HM (2009) Delivery technologies for biopharmaceuticals: peptides, proteins, nucleic acids and vaccines. Wiley, Chichester
Kersten G, Hirschberg H (2004) Antigen delivery systems. Expert Rev Vaccines 3:89–99
Kim SW, Pai CM, Makino K, Seminoff LA, Holmberg DL, Gleeson JM, Wilson DA, Mack EJ (1990) Self-regulated glycosylated insulin delivery. J Control Release 11:193–201
Kis EE, Winter G, Myschik J (2012) Devices for intradermal vaccination. Vaccine 30:523–538
Klegerman ME, Groves MJ (1992) Pharmaceutical biotechnology: fundamentals and essentials. Interpharm Press, Inc., Buffalo Grove
Kumar S, Char H, Patel S, Piemontese D, Malick AW, Iqbal K, Neugroschel E, Behl CR (1992) In vivo transdermal iontophoretic delivery of growth hormone releasing factor GRF (1–44) in hairless guinea pigs. J Control Release 18:213–220
Lammers T, Hennink WE, Storm G (2008) Tumour-targeted nanomedicines: principles and practice. Br J Cancer 99:392–397
Maberly GF, Wait GA, Kilpatrick JA, Loten EG, Gain KR, Stewart RDH, Eastman CJ (1982) Evidence for insulin degradation by muscle and fat tissue in an insulin resistant diabetic patient. Diabetol 23:333–336
Mastrobattista E, van der Aa MAEM, Hennink WE, Crommelin DJA (2006) Artificial viruses: a nanotechnological approach to gene delivery. Nat Rev Drug Discov 5:115–121
Mitragotri S (2005) Immunization without needles. Nat Rev Immunol 5:905–916
Moeller EH, Jorgensen L (2009) Alternative routes of administration for systemic delivery of protein pharmaceuticals. Drug Discovery Today: Technologies 5:89–94
Nässander UK, Storm G, Peeters PAM, Crommelin DJA (1990) Liposomes. In: Chasin M, Langer R (eds) Biodegradable polymers as drug delivery systems. Marcel Dekker, New York, pp 261–338
Nguyen TH, Ward C (1993) Stability characterization and formulation development of alteplase, a recombinant tissue plasminogen activator. In: Wang YJ, Pearlman R

(eds) Stability and characterization of protein and peptide drugs. Case histories. Plenum Press, New York, pp 91–134

O'Hagan DT (1990) Intestinal translocation of particulates - implications for drug and antigen delivery. Adv Drug Deliv Rev 5:265–285

Patton JS, Trinchero P, Platz RM (1994) Bioavailability of pulmonary delivered peptides and proteins: alpha-interferon, calcitonin and parathyroid hormones. J Control Release 28:79–85

Patton JS, Bukar JG, Eldon MA (2004) Clinical pharmacokinetics and pharmacodynamics of inhaled insulin. Clin Pharmacokinet 43:781–801

Pearlman R, Bewley TA (1993) Stability and characterization of human growth hormone. In: Wang YJ, Pearlman R (eds) Stability and characterization of protein and peptide drugs. Case histories. Plenum Press, New York, pp 1–58

Peeters PAM, Storm G, Crommelin DJA (1987) Immunoliposomes in vivo: state of the art. Adv Drug Deliv Rev 1:249–266

Pikal MJ (1990) Freeze-drying of proteins. Part I: process design. BioPharm 3(8):18–27

Poste G (1985) Drug targeting in cancer therapy. In: Gregoriadis G, Poste G, Senior J, Trouet A (eds) Receptor-mediated targeting of drugs. Plenum Press, New York, pp 427–474

Pristoupil TI (1985) Haemoglobin lyophilized with sucrose: effect of residual moisture on storage. Haematologia 18:45–52

Roy MJ (2011) Biotechnology operations: principles and practices. CRC Press, Boca Raton

Sage BH, Bock CR, Denuzzio JD, Hoke RA (1995) Technological and developmental issues of iontophoretic transport of peptide and protein drugs. In: Lee VHL, Hashida M, Mizushima Y (eds) Trends and future perspectives in peptide and protein drug delivery. Harwood Academic Publishers GmbH, Chur, pp 111–134

Schaepelynck P, Darmon P, Molines L, Jannot-Lamotte MF, Treglia C, Raccah D (2011) Advances in pump technology: insulin patch pumps, combined pumps and glucose sensors, and implanted pumps. Diabetes Metab 37:S85–S93

Stenekes R (2000) Nanoporous dextran microspheres for drug delivery. Thesis, Utrecht University

Storm G, Crommelin DJA (1998) Liposomes: quo vadis? Pharm Sci Tech Today 1:19–31

Storm G, Oussoren C, Peeters PAM, Barenholz YB (1993) Tolerability of liposomes in vivo. In: Gregoriadis G (ed) Liposome technology. CRC Press, Inc, Boca Raton, pp 345–383

Storm G, Nässander U, Vingerhoeds MH, Steerenberg PA, Crommelin DJA (1994) Antibody-targeted liposomes to deliver doxorubicin to ovarian cancer cells. J Liposome Res 4:641–666

Supersaxo A, Hein WR, Steffen H (1990) Effect of molecular weight on the lymphatic absorption of water-soluble compounds following subcutaneous administration. Pharm Res 7:167–169

Thurow H, Geisen K (1984) Stabilization of dissolved proteins against denaturation at hydrophobic interfaces. Diabetologia 27:212–218

Tomlinson E (1987) Theory and practice of site-specific drug delivery. Adv Drug Deliv Rev 1:87–198

Traitel T, Goldbart R, Kost J (2008) Smart polymers for responsive drug-delivery systems. J Biomater Sci Polym Ed 19:755–767

Tresco PA (1994) Encapsulated cells for sustained neurotransmitter delivery to the central nervous system. J Control Release 28:253–258

USP 29/NF 24 through second supplement (2006) United States Pharmacopeial Convention, Rockville

Vemuri S, Yu CT, Roosdorp N (1993) Formulation and stability of recombinant alpha1-antitrypsin. In: Wang YJ, Pearlman R (eds) Stability and characterization of protein and peptide drugs. Plenum Press, New York, pp 263–286

Wang YJ, Watson MA (1988) Parenteral formulations of proteins and peptides: stability and stabilizers. J Parenter Sci Technol 42(Suppl 4S):1–26

Woodle M, Newman M, Collins L, Redemann C, Martin F (1990) Improved long-circulating (Stealth®) liposomes using synthetic lipids. Proc Int Symp Contr Rel Bioact Mater 17:77–78

Zhou XH, Li Wan Po A (1991a) Peptide and protein drugs: I. Therapeutic applications, absorption and parenteral administration. Int J Pharm 75:97–115

Zhou XH, Li Wan Po A (1991b) Peptide and protein drugs: II. Non-parenteral routes of delivery. Int J Pharm 75: 117–130

5 Pharmacokinetics and Pharmacodynamics of Peptide and Protein Therapeutics

Bernd Meibohm

INTRODUCTION

The rational use of drugs and the design of effective dosage regimens are facilitated by the appreciation of the central paradigm of clinical pharmacology that there is a defined relationship between the administered dose of a drug, the resulting drug concentrations in various body fluids and tissues, and the intensity of pharmacologic effects caused by these concentrations (Meibohm and Derendorf 1997). This dose-exposure-response relationship and thus the dose of a drug required to achieve a certain effect are determined by the drug's pharmacokinetic and pharmacodynamic properties (Fig. 5.1).

Pharmacokinetics describes the time course of the concentration of a drug in a body fluid, preferably plasma or blood, that results from the administration of a certain dosage regimen. It comprises all processes affecting drug absorption, distribution, metabolism, and excretion. Simplified, pharmacokinetics characterizes "*what the body does to the drug*." In contrast, pharmacodynamics characterizes the intensity of a drug effect or toxicity resulting from certain drug concentrations in a body fluid, usually at the assumed site of drug action. It can be simplified to *what the drug does to the body* (Fig. 5.2) (Holford and Sheiner 1982; Derendorf and Meibohm 1999).

B. Meibohm, Ph.D., FCP (✉)
Department of Pharmaceutical Sciences,
University of Tennessee Health Science Center,
College of Pharmacy, 881 Madison Avenue, Rm. 444,
Memphis, TN 38163, USA
e-mail: bmeibohm@uthsc.edu

The understanding of the dose-concentration-effect relationship is crucial to any drug – including peptides and proteins – as it lays the foundation for dosing regimen design and rational clinical application. General pharmacokinetic and pharmacodynamic principles are to a large extent equally applicable to protein and peptide drugs as they are to traditional small molecule-based therapeutics. Deviations from some of these principles and additional challenges with regard to the characterization of the pharmacokinetics and pharmacodynamics of peptide and protein therapeutics, however, arise from some of their specific properties:

(a) Their definition by the production process in a living organism rather than a chemically exactly defined structure and purity as it is the case for small-molecule drugs
(b) Their structural similarity to endogenous structural or functional proteins and nutrients
(c) Their intimate involvement in physiologic processes on the molecular level, often including regulatory feedback mechanisms
(d) The analytical challenges to identify and quantify them in the presence of a myriad of similar molecules
(e) Their large molecular weight and macromolecule character (for proteins)

This chapter will highlight some of the major pharmacokinetic properties and processes relevant for the majority of peptide and protein therapeutics and will provide examples of well-characterized pharmacodynamic relationships for peptide and protein drugs. The clinical pharmacology of monoclonal antibodies, including special aspects in their pharmacokinetics and pharmacodynamics, will be discussed in further detail in Chap. 7. For a more general discussion on pharmacokinetic and pharmacodynamic principles, the reader is referred to several textbooks and articles that review the topic in extensive detail (see Further Reading).

D.J.A. Crommelin, R.D. Sindelar, and B. Meibohm (eds.), *Pharmaceutical Biotechnology*,
DOI 10.1007/978-1-4614-6486-0_5,

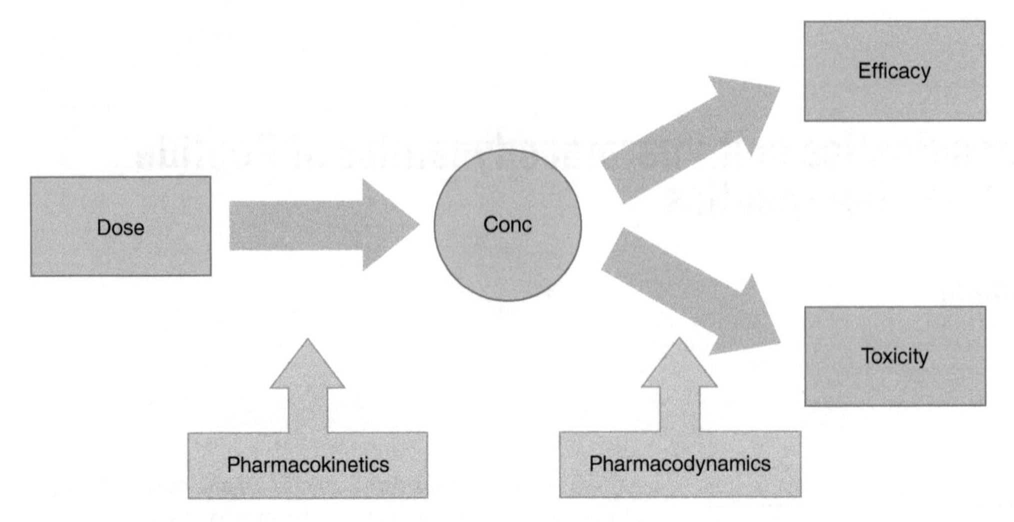

Figure 5.1 ■ The central paradigm of clinical pharmacology: the dose-concentration-effect relationship.

PHARMACOKINETICS OF PROTEIN THERAPEUTICS

The in vivo disposition of peptide and protein drugs may often be predicted to a large degree from their physiological function (Tang and Meibohm 2006). Peptides, for example, which frequently have hormone activity, usually have short elimination half-lives, which is desirable for a close regulation of their endogenous levels and thus function. Insulin, for example, shows dose-dependent elimination with a relatively short half-life of 26 and 52 min at 0.1 and 0.2 U/kg, respectively. Contrary to that, proteins that have transport tasks such as albumin or long-term immunity functions such as immunoglobulins have elimination half-lives of several days, which enables and ensures the continuous maintenance of physiologically necessary concentrations in the bloodstream (Meibohm and Derendorf 1994). This is, for example, reflected by the elimination half-life of antibody drugs such as the anti-epidermal growth factor receptor antibody cetuximab, an IgG1 chimeric antibody for which a half-life of approximately 7 days has been reported (Herbst and Langer 2002).

■ Absorption of Protein Therapeutics

Enteral Administration

Peptides and proteins, unlike conventional small-molecule drugs, are generally not therapeutically active upon oral administration (Fasano 1998; Mahato et al. 2003; Tang et al. 2004). The lack of systemic bioavailability is mainly caused by two factors: (1) high gastrointestinal enzyme activity and (2) low permeability through the gastrointestinal mucosa. In fact, the substantial peptidase and protease activity in the gastrointestinal tract makes it the most efficient body compartment for peptide and protein metabolism. Furthermore, the gastrointestinal mucosa presents a major absorption barrier for water-soluble macromolecules such as peptides and proteins (Tang et al. 2004). Thus, although various factors such as permeability, stability, and gastrointestinal transit time can affect the rate and extent of orally administered proteins, molecular size is generally considered the ultimate obstacle (Shen 2003).

Since oral administration is still a highly desirable route of delivery for protein drugs due to its convenience, cost-effectiveness, and painlessness, numerous strategies to overcome the obstacles associated with oral delivery of proteins have recently been an area of intensive research. Suggested approaches to increase the oral bioavailability of protein drugs include encapsulation into micro- or nanoparticles thereby protecting proteins from intestinal degradation (Lee 2002; Mahato et al. 2003; Shen 2003). Other strategies are chemical modifications such as amino acid backbone modifications and chemical conjugations to improve the resistance to degradation and permeability of the protein drug. Coadministration of protease inhibitors has also been suggested for the inhibition of enzymatic degradation (Pauletti et al. 1997; Mahato et al. 2003). More details on approaches for oral delivery of peptide and protein therapeutics are discussed in Chap. 4.

Parenteral Administration

Most peptide and protein drugs are currently formulated as parenteral formulations because of their poor oral bioavailability. Major routes of administration include intravenous (IV), subcutaneous (SC), and intramuscular (IM) administration. In addition, other

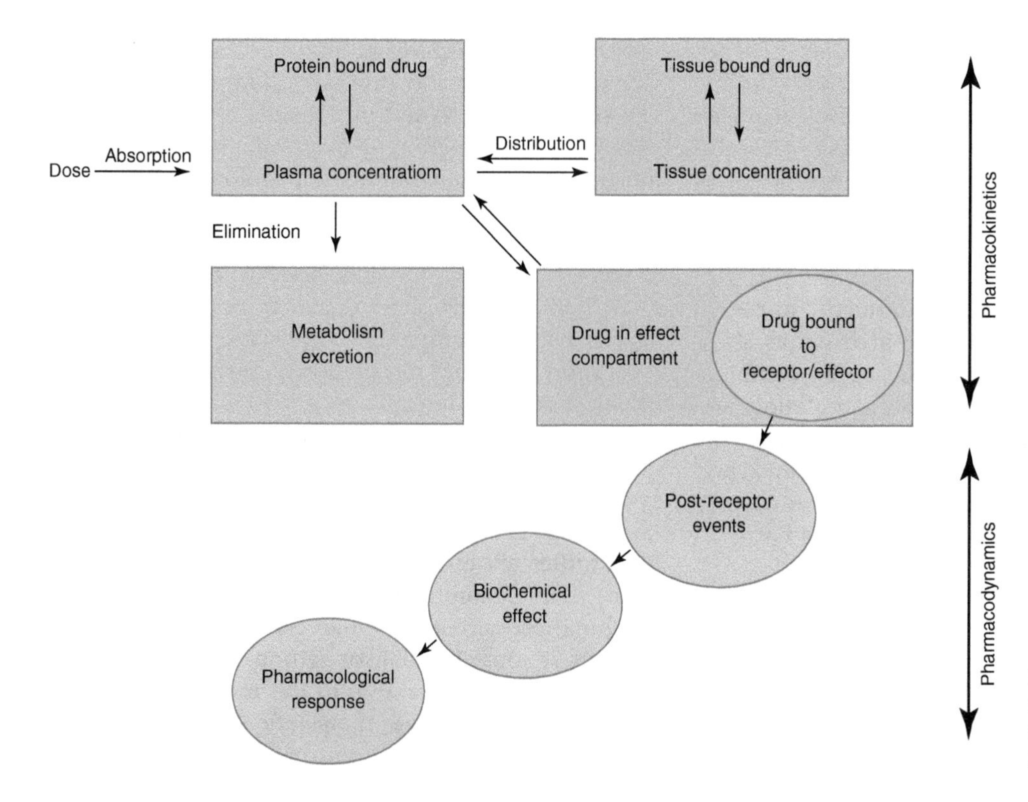

Figure 5.2 ■ Physiological scheme of pharmacokinetic and pharmacodynamic processes.

non-oral administration pathways are utilized, including nasal, buccal, rectal, vaginal, transdermal, ocular, and pulmonary drug delivery (see Chap. 4).

IV administration of peptides and proteins offers the advantage of circumventing presystemic degradation, thereby achieving the highest concentration in the biological system. Protein therapeutics given by the IV route include, among many others, the tissue plasminogen activator (t-PA) analogues alteplase and tenecteplase, the recombinant human erythropoietin epoetin-α, and the granulocyte colony-stimulating factor filgrastim (Tang and Meibohm 2006).

IV administration as either a bolus dose or constant rate infusion, however, may not always provide the desired concentration-time profile depending on the biological activity of the product. In these cases, IM or SC injections may be more appropriate alternatives. For example, luteinizing hormone-releasing hormone (LH-RH) in bursts stimulates the release of follicle-stimulating hormone (FSH) and luteinizing hormone (LH), whereas a continuous baseline level will suppress the release of these hormones (Handelsman and Swerdloff 1986). To avoid the high peaks from an IV administration of leuprorelin, an LH-RH agonist, a long-acting monthly depot injection of the drug is approved for the treatment of prostate cancer and endometriosis (Periti et al. 2002). A recent study comparing SC versus IV administration of epoetin-α in patients receiving hemodialysis reports that the SC route can maintain the hematocrit in a desired target range with a lower average weekly dose of epoetin-α compared to IV (Kaufman et al. 1998).

One of the potential limitations of SC and IM administration, however, are the presystemic degradation processes frequently associated with these administration routes, resulting in a reduced systemic bioavailability compared to IV administration. No correlation between the molecular weight of a protein therapeutic and its systemic bioavailability has so far been described in any species (Richter et al. 2012), and clinically observed bioavailability seems to be product-specific based on physicochemical properties and structure.

Bioavailability assessments for therapeutic proteins may be challenging if the protein exhibits the frequently encountered nonlinear pharmacokinetic behavior. Classic bioavailability assessments comparing systemic exposures quantified as area-under-the-concentration-time curve (AUC) resulting from extravascular versus IV administration assume linear pharmacokinetics, i.e., a drug clearance independent of concentration and the administration pathway. As this is not the case for many therapeutic proteins, especially those that undergo target-mediated drug disposition (see respective section in this chapter), bioavailability assessments using the classic approach can result in substantial bias

(Limothai and Meibohm 2011). Potential approaches suggested to minimize or overcome these effects include bioavailability assessments at doses at which the target- or receptor-mediated processes are saturated or to compare concentration-time profiles with similar shape and magnitude for extravascular and IV administration by modulating the input rate in the IV experiment.

The pharmacokinetically derived apparent absorption rate constant k_{app} for protein drugs administered via these administration routes is the combination of absorption into the systemic circulation and presystemic degradation at the absorption site, i.e., the sum of a true first-order absorption rate constant k_a and a first-order degradation rate constant. The true absorption rate constant k_a can then be calculated as

$$k_a = F \cdot k_{app}$$

where F is the systemic bioavailability compared to IV administration. A rapid apparent absorption, i.e., large k_{app}, can thus be the result of a slow true absorption and a fast presystemic degradation, i.e., a low systemic bioavailability (Colburn 1991).

For monoclonal antibodies and fusion proteins with antibody Fc fragment, interaction with the neonatal Fc receptor (FcRn) has also been identified as important to absorption processes (Roopenian and Akilesh 2007). In this context, FcRn prevents the monoclonal antibody or fusion protein from undergoing lysosomal degradation (see Chap. 7 for details) and thereby increases systemic bioavailability, but may also facilitate transcellular transport from the absorption site into the vascular space.

Other potential factors that may limit the rate and/or extent of uptake of proteins after SC or IM administration include variable local blood flow, injection trauma, and limitations of uptake into the systemic circulation related to effective capillary pore size, diffusion, and convective transport.

Several peptide and protein therapeutics including anakinra, etanercept, insulin, and pegfilgrastim are administered as SC injections. Following a SC injection, peptide and protein therapeutics may enter the systemic circulation either via blood capillaries or through lymphatic vessels (Porter and Charman 2000). In general, peptides and proteins larger than 16 kDa are predominantly absorbed into the lymphatics, whereas those under 1 kDa are mostly absorbed into the blood circulation. While diffusion is the driving force for the uptake into blood capillaries, transport of larger proteins through the interstitial space into lymphatic vessels is mediated by convective transport with the interstitial fluid following the hydrostatic and osmotic pressure differences. Since lymph flow and interstitial convective transport are substantially slower than diffusion processes, larger proteins usually show a delayed and prolonged absorption process after SC administration that can even become the rate-limiting step in their overall disposition. There appears to be a defined relationship between the molecular weight of the protein and the proportion of the dose absorbed by the lymphatics (see Fig. 4.12) (Supersaxo et al. 1990). This is of particular importance for those agents whose therapeutic targets are lymphoid cells (i.e., interferons and interleukins). Studies with recombinant human interferon α-2a (rhIFN α-2a) indicate that following SC administration, high concentrations of the recombinant protein are found in the lymphatic system, which drains into regional lymph nodes (Supersaxo et al. 1988). Due to this targeting effect, clinical studies show that palliative low-to-intermediate-dose SC recombinant interleukin-2 (rIL-2) in combination with rhIFN α-2a can be administered to patients in the ambulatory setting with efficacy and safety profiles comparable to the most aggressive IV rIL-2 protocol against metastatic renal cell cancer (Schomburg et al. 1993).

More recently, charge has also been described as an important factor in the SC absorption of proteins: While the positive and negative charges from collagen and hyaluronan in the extracellular matrix seem to be of similar magnitude, additional negative charges of proteoglycans may lead to a negative interstitial charge (Richter et al. 2012). This negative net charge and the associated ionic interactions with SC-administered proteins result in a slower transport for more positively rather than negatively charged proteins, as could be shown for several monoclonal antibodies (Mach et al. 2011).

■ Distribution of Protein Therapeutics

Distribution Mechanisms and Volumes

The rate and extent of protein distribution is largely determined by the molecule size and molecular weight, physiochemical properties (e.g., charge, lipophilicity), binding to structural or transport proteins, and their dependency on active transport processes to cross biomembranes. Since most therapeutic proteins have high molecular weights and are thus large in size, their apparent volume of distribution is usually small and limited to the volume of the extracellular space due to their limited mobility secondary to impaired passage through biomembranes (Zito 1997). In addition, there is a mutual exclusion between protein therapeutics and the structural molecules of the extracellular matrix. This fraction of the interstitial space that is not available for distribution is expressed as the exclusion volume (Ve). It is dependent on the molecular weight and charge of the macromolecule and further limits extravascular distribution. For albumin (MW 66 kDa), the Ve is ~40 % in dog muscle tissue. Active tissue uptake and binding to intra- and extravascular proteins, however, can substantially increase the apparent volume of distribution of protein drugs, as reflected by the relatively large volume of distribution of up to 2.8 L/kg for interferon β-1b (Chiang et al. 1993).

In contrast to small-molecule drugs, protein transport from the vascular space into the interstitial space of tissues is largely mediated by convection rather than diffusion, following the unidirectional fluid flux from the vascular space through paracellular pores into the interstitial tissue space (Fig. 5.3). The subsequent removal from the interstitial space is accomplished by lymph drainage back into the systemic circulation (Flessner et al. 1997). This underlines the unique role the lymphatic system plays in the disposition of protein therapeutics as already discussed in the section on absorption. The fact that the transfer clearance from the vascular to the interstitial space is smaller than the transfer clearance from the interstitial space to the lymphatic system results in lower protein concentrations in the interstitial space compared to the vascular space, thereby further limiting the apparent volume of distribution for protein therapeutics. Another, but much less prominent pathway for the movement of protein molecules from the vascular to the interstitial space is transcellular migration via endocytosis (Baxter et al. 1994; Reddy et al. 2006).

Besides the size-dependent sieving of macromolecules through the capillary walls, charge may also play an important role in the biodistribution of proteins. It has been suggested that the electrostatic attraction between positively charged proteins and negatively charged cell membranes might increase the rate and extent of tissue distribution. Most cell surfaces are negatively charged because of their abundance of glycosaminoglycans in the extracellular matrix.

After IV administration, peptides and proteins usually follow a biexponential plasma concentration-time profile that can best be described by a two-compartment pharmacokinetic model (Meibohm 2004). A biexponential concentration-time profile has, for example, been described for clenoliximab, a macaque-human chimeric monoclonal antibody specific to the CD4 molecule on the surface of T lymphocytes (Mould et al. 1999). Similarly, AJW200, a humanized monoclonal antibody to the von Willebrand factor, exhibited biphasic pharmacokinetics after IV administration (Kageyama et al. 2002). The central compartment in this two-compartment model represents primarily the vascular space and the interstitial space of well-perfused organs with permeable capillary walls, including the liver and the kidneys. The peripheral compartment is more reflective of concentration-time profiles in the interstitial space of slowly equilibrating tissues.

The central compartment in which proteins initially distribute after IV administration has thus typically a volume of distribution equal or slightly larger than the plasma volume, i.e., 3–8 L. The total volume of distribution frequently comprises with 14–20 L not more than two to three times the initial volume of distribution (Colburn 1991; Kageyama et al. 2002). An example for such a distribution pattern is the t-PA analogue tenecteplase. Radiolabeled ^{125}I-tenecteplase was described to have an initial volume of distribution of 4.2–6.3 L and a total volume of distribution of 6.1–9.9 L with liver as the only organ that had a significant uptake of radioactivity. The authors concluded that the small volume of distribution suggests primarily intravascular distribution for tenecteplase, consistent with

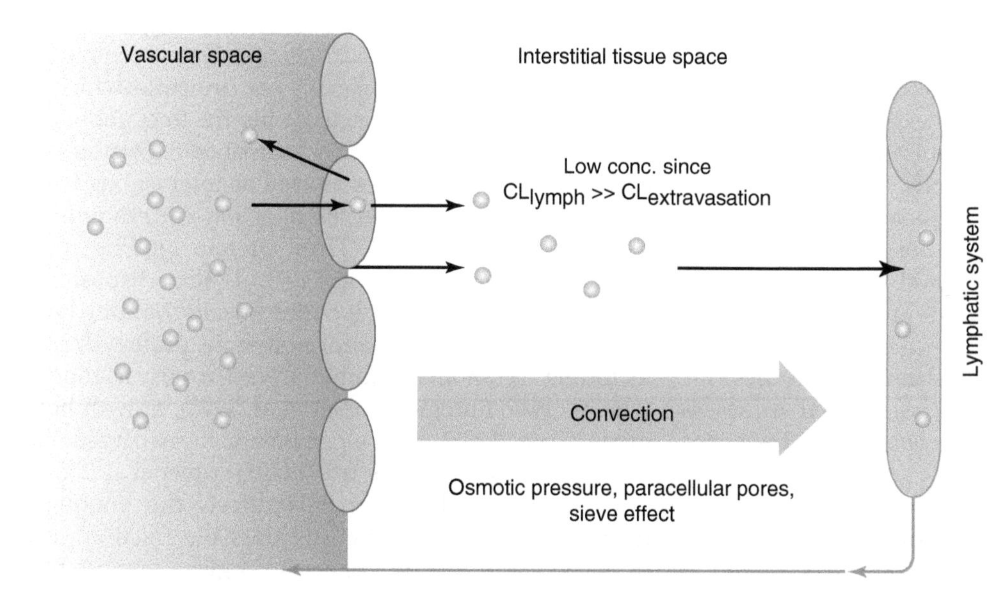

Figure 5.3 ■ Distribution mechanisms of therapeutic proteins: convective extravasation rather than diffusion as major distribution process. $CL_{extravasation}$ transfer clearance from the vascular to the interstitial, CL_{lymph} transfer clearance from the interstitial space to the lymphatic system.

the drug's large molecular weight of 65 kDa (Tanswell et al. 2002).

Epoetin-α, for example, has a volume of distribution estimated to be close to the plasma volume at 0.056 L/kg after an IV administration to healthy volunteers (Ramakrishnan et al. 2004). Similarly, volume of distribution for darbepoetin-α has been reported as 0.062 L/kg after IV administration in patients undergoing dialysis (Allon et al. 2002), and distribution of thrombopoietin has also been reported to be limited to the plasma volume (~3 L) (Jin and Krzyzanski 2004).

It should be stressed that pharmacokinetic calculations of volume of distribution may be problematic for many protein therapeutics (Tang et al. 2004; Straughn 2006). Noncompartmental determination of volume of distribution at steady state (V_{ss}) using statistical moment theory assumes first-order disposition processes with elimination occurring from the rapidly equilibrating or central compartment (Perrier and Mayersohn 1982; Straughn 1982; Veng-Pedersen and Gillespie 1984). These basic assumptions, however, are not fulfilled for numerous protein therapeutics, as proteolysis and receptor-mediated elimination in peripheral tissues may constitute a substantial fraction of the overall elimination process. If protein therapeutics are eliminated from slowly equilibrating tissues at a rate greater than their distribution process, substantial error in the volume of distribution assessment may occur. A recent simulation study could show that if substantial tissue elimination exists, a V_{ss} determined by noncompartmental methods will underestimate the "true" V_{ss} and that the magnitude of error tends to be larger the more extensively the protein is eliminated by tissue routes (Meibohm 2004; Straughn 2006; Tang and Meibohm 2006).

These challenges in characterizing the distribution of protein therapeutics can only be overcome by determining actual protein concentrations in the tissue by biopsy or necropsy or via biodistribution studies with radiolabeled compound and/or imaging techniques.

Biodistribution studies are imperative for small organic synthetic drugs, since long residence times of the radioactive label in certain tissues may be an indication of tissue accumulation of potentially toxic metabolites. Because of the possible reutilization of amino acids from protein drugs in endogenous proteins, such a safety concern does not exist for protein therapeutics. Therefore, biodistribution studies for protein drugs are usually only performed to assess drug targeting to specific tissues or to detect the major organs of elimination.

If a biodistribution study with radiolabeled protein is performed, either, an external label such as ^{125}I can be chemically coupled to the protein if it contains a suitable amino acid such as tyrosine or lysine, or internal labeling can be used by growing the production cell line in the presence of amino acids labeled with ^{3}H, ^{14}C, ^{35}S, etc. The latter method, however, is not routinely used because of the prohibition of radioactive contamination of fermentation equipment (Meibohm and Derendorf 2003). Moreover, internally labeled proteins may be less desirable than iodinated proteins because of the potential reutilization of the radiolabeled amino acid fragments in the synthesis of endogenous proteins and cell structures. Irrespective of the labeling method, but more so for external labeling, the labeled product should have demonstrated physicochemical and biological properties identical to the unlabeled molecule (Bennett and McMartin 1978).

In addition, as for all types of radiolabeled studies, it needs to be established whether the measured radioactivity represents intact labeled protein, or radiolabeled metabolites, or the liberated label. Trichloroacetic acid-precipitable radioactivity is often used to distinguish intact protein from free label or low-molecular-weight metabolites, which appear in the supernatant after centrifugation (Meibohm and Derendorf 2003). Proteins with reutilized labeled amino acids and large protein metabolites can only be distinguished from the original protein by techniques such as polyacrylamide gel electrophoresis (PAGE), high pressure liquid chromatography (HPLC), specific immunoassays, or bioassays (see Chap. 2).

Protein Binding of Protein Therapeutics

Another factor that can influence the distribution of therapeutic peptides and proteins is binding to endogenous protein structures. Physiologically active endogenous peptides and proteins frequently interact with specific binding proteins involved in their transport and regulation. Furthermore, interaction with binding proteins may enable or facilitate cellular uptake processes and thus affect the drug's pharmacodynamics.

It is a general pharmacokinetic principle, which is also applicable to proteins, that only the free, unbound fraction of a drug substance is accessible to distribution and elimination processes as well as interactions with its target structures at the site of action, for example, a receptor or ion channel. Thus, protein binding may affect the pharmacodynamics, but also disposition properties of protein therapeutics. Specific binding proteins have been identified for numerous protein drugs, including recombinant human DNase for use as mucolytic in cystic fibrosis (Mohler et al. 1993), growth hormone (Toon 1996), and recombinant human vascular endothelial growth factor (rhVEGF) (Eppler et al. 2002).

Protein binding not only affects the unbound fraction of a protein drug and thus the fraction of a drug available to exert pharmacological activity, but many times it also either prolongs protein circulation time by acting as a storage depot or it enhances protein clearance. Recombinant cytokines, for example, may after IV administration encounter various cytokine-binding proteins including soluble cytokine receptors and anti-cytokine antibodies (Piscitelli et al. 1997). In either case, the binding protein may either prolong the cytokine circulation time by acting as a storage depot or it may enhance the cytokine clearance.

Growth hormone, as another example, has at least two binding proteins in plasma (Wills and Ferraiolo 1992). This protein binding substantially reduces growth hormone elimination with a tenfold smaller clearance of total compared to free growth hormone, but also decreases its activity via reduction of receptor interactions.

Apart from these specific bindings, peptides and proteins may also be nonspecifically bound to plasma proteins. For example, metkephamid, a met-enkephalin analogue, was described to be 44–49 % bound to albumin (Taki et al. 1998), and octreotide, a somatostatin analogue, is up to 65 % bound to lipoproteins (Chanson et al. 1993).

Distribution via Receptor-Mediated Uptake

Aside from physicochemical properties and protein binding of protein therapeutics, site-specific receptor-mediated uptake can also substantially influence and contribute to the distribution of protein therapeutics, as well as to elimination and pharmacodynamics (see section on "Target-Mediated Drug Disposition").

The generally low volume of distribution should not necessarily be interpreted as low tissue penetration. Receptor-mediated specific uptake into the target organ, as one mechanism, can result in therapeutically effective tissue concentrations despite a relatively small volume of distribution. Nartograstim, a recombinant derivative of the granulocyte colony-stimulating factor (G-CSF), for example, is characterized by a specific, dose-dependent, and saturable tissue uptake into the target organ bone marrow, presumably via receptor-mediated endocytosis (Kuwabara et al. 1995).

■ Elimination of Protein Therapeutics

Protein therapeutics are generally subject to the same catabolic pathways as endogenous or dietetic proteins. The end products of protein metabolism are thus amino acids that are reutilized in the endogenous amino acid pool for the de novo biosynthesis of structural or functional proteins in the human body (Meibohm 2004). Detailed investigations on the metabolism of proteins are relatively difficult because of the myriad of potential molecule fragments that may be formed, and are therefore generally not conducted. Non-metabolic elimination pathways such as renal or biliary excretion are negligible for most proteins. If biliary excretion occurs, however, it is generally followed by subsequent metabolic degradation of the compound in the gastrointestinal tract.

Proteolysis

In contrast to small-molecule drugs, metabolic degradation of peptides and protein therapeutics by proteolysis can occur unspecifically nearly everywhere in the body. Due to this unspecific proteolysis of some proteins already in blood as well as potential active cellular uptake, the clearance of protein drugs can exceed cardiac output, i.e., >5 L/min for blood clearance and >3 L/min for plasma clearance (Meibohm 2004). The clearance of peptides or proteins in this context describes the irreversible removal of active substance from the vascular space, which includes besides metabolism also cellular uptake. The metabolic rate for protein degradation generally increases with decreasing molecular weight from large to small proteins to peptides (Table 5.1), but is also dependent on other factors such as size, charge, lipophilicity, functional groups, and glycosylation pattern as well as secondary and tertiary structure.

Proteolytic enzymes such as proteases and peptidases are ubiquitous throughout the body. Sites capable of extensive peptide and protein metabolism are not only limited to the liver, kidneys, and gastrointestinal tissue, but also include blood and vascular endothelium as well as other organs and tissues. As proteases and peptidases are also located within cells, intracellular uptake is per se more an elimination rather than a distribution process (Tang and Meibohm 2006). While peptidases and proteases in the gastrointestinal tract and in lysosomes are relatively unspecific, soluble peptidases in the interstitial space and exopeptidases on the cell surface have a higher selectivity and determine the specific metabolism pattern of an organ. The proteolytic activity of subcutaneous and lymphatic tissue, for example, results in a partial loss of activity of SC compared to IV administrated interferon-γ.

Gastrointestinal Protein Metabolism

As pointed out earlier, the gastrointestinal tract is a major site of protein metabolism with high proteolytic enzyme activity due to its primary function to digest dietary proteins. Thus, gastrointestinal metabolism is one of the major factors limiting systemic bioavailability of orally administered protein drugs. The metabolic activity of the gastrointestinal tract, however, is not limited to orally administered proteins. Parenterally administered peptides and proteins may also be metabolized in the intestinal mucosa following intestinal secretion. At least 20 % of the degradation of endogenous albumin, for example, has been reported to take place in the gastrointestinal tract (Colburn 1991).

Molecular weight	Elimination site	Predominant elimination mechanisms	Major determinant
<500	Blood, liver	Extracellular hydrolysis Passive lipoid diffusion	Structure, lipophilicity
500–1,000	Liver	Carrier-mediated uptake Passive lipoid diffusion	Structure, lipophilicity
1,000–50,000	Kidney	Glomerular filtration and subsequent degradation processes (see Fig. 5.4)	Molecular weight
50,000–200,000	Kidney, liver	Receptor-mediated endocytosis	Sugar, charge
200,000–400,000		Opsonization	α_2-macroglobulin, IgG
>400,000		Phagocytosis	Particle aggregation

After Meijer and Ziegler (1993)

Other determining factors are size, charge, lipophilicity, functional groups, sugar recognition, vulnerability for proteases, aggregation to particles, formation of complexes with opsonization factors, etc. As indicated, mechanisms may overlap. Endocytosis may occur at any molecular weight range

Table 5.1 ■ Molecular weight as major determinant of the elimination mechanisms of peptides and proteins.

Renal Protein Metabolism

The kidneys are a major site of protein metabolism for smaller-sized proteins that undergo glomerular filtration. The size-selective cutoff for glomerular filtration is approximately 60 kDa, although the effective molecule radius based on molecular weight and conformation is probably the limiting factor (Edwards et al. 1999). Glomerular filtration is most efficient, however, for proteins smaller than 30 kDa (Kompella and Lee 1991). Peptides and small proteins (<5 kDa) are filtered very efficiently, and their glomerular filtration clearance approaches the glomerular filtration rate (GFR, ~120 mL/min in humans). For molecular weights exceeding 30 kDa, the filtration rate falls off sharply. In addition to size selectivity, charge selectivity has also been observed for glomerular filtration where anionic macromolecules pass through the capillary wall less readily than neutral macromolecules, which in turn pass through less readily than cationic macromolecules (Deen et al. 2001).

The importance of the kidneys as elimination organ could, for example, be shown for interleukin-2, macrophage colony-stimulating factor (M-CSF), and interferon-α (McMartin 1992; Wills and Ferraiolo 1992).

Renal metabolism of peptides and small proteins is mediated through three highly effective processes (Fig. 5.4). As a result, only minuscule amounts of intact protein are detectable in urine.

The first mechanism involves glomerular filtration of larger, complex peptides and proteins followed by reabsorption into endocytic vesicles in the proximal tubule and subsequent hydrolysis into small peptide fragments and amino acids (Maack et al. 1985). This mechanism of elimination has been described for IL-2 (Anderson and Sorenson 1994), IL-11 (Takagi et al. 1995), growth hormone (Johnson and Maack 1977), and insulin (Rabkin et al. 1984).

The second mechanism entails glomerular filtration followed by intraluminal metabolism, predominantly by exopeptidases in the luminal brush border membrane of the proximal tubule. The resulting peptide fragments and amino acids are reabsorbed into the systemic circulation. This route of disposition applies to small linear peptides such as glucagon and LH-RH (Carone and Peterson 1980; Carone et al. 1982). Recent studies implicate the proton-driven peptide transporters PEPT1 and especially PEPT2 as the main route of cellular uptake of small peptides and peptide-like drugs from the glomerular filtrates (Inui et al. 2000). These high-affinity transport proteins seem to exhibit selective uptake of di- and tripeptides, which implicates their role in renal amino acid homeostasis (Daniel and Herget 1997).

For both mechanisms, glomerular filtration is the dominant, rate-limiting step as subsequent degradation processes are not saturable under physiologic conditions (Maack et al. 1985; Colburn 1991). Due to this limitation of renal elimination, the renal contribution to the overall elimination of proteins is dependent on the proteolytic activity for these proteins in other body regions. If metabolic activity for these proteins is high in other body regions, there is only minor renal contribution to total clearance, and it becomes negligible in the presence of unspecific degradation throughout the body. If the metabolic activity is low in other tissues or if distribution to the extravascular space is limited, however, the renal contribution to total clearance may approach 100 %.

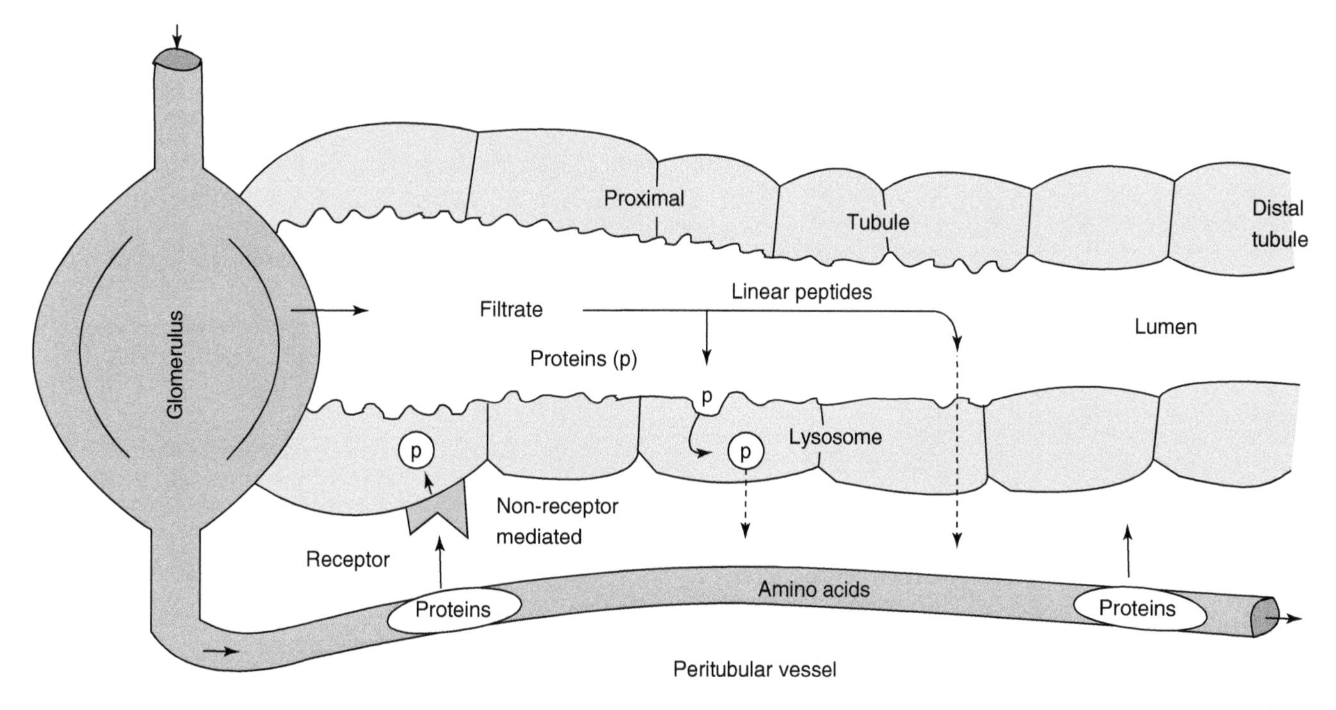

Figure 5.4 ■ Pathways of renal metabolism of peptides and proteins: glomerular filtration followed by either (**I**) intraluminal metabolism or (**II**) tubular reabsorption with intracellular lysosomal metabolism and (**III**) peritubular extraction with intracellular lysosomal metabolism (Modified from Maack et al. (1985)).

The involvement of glomerular filtration in the renal metabolism of therapeutic proteins implies that the pharmacokinetics of therapeutic proteins below the molecular weight or hydrodynamic volume cutoff size for filtration will be affected by renal impairment. Indeed, it has been reported that the systemic exposure and elimination half-life increases with decreasing glomerular filtration rate for recombinant human interleukin-10 (18 kDa), recombinant human growth hormone (22 kDa), and the recombinant human IL-1 receptor antagonist anakinra (17.3 kDa). Consistent with these theoretical considerations is also the observation that for monoclonal antibodies (150 kDa) such as rituximab, cetuximab, bevacizumab, and trastuzumab, no effect of renal impairment on their disposition has been reported (Meibohm and Zhou 2012).

The third mechanism of renal metabolism is peritubular extraction of peptides and proteins from postglomerular capillaries with subsequent intracellular metabolism. Experiments using radioiodinated growth hormone (^{125}I-rGH) have demonstrated that while reabsorption into endocytic vesicles at the proximal tubule is still the dominant route of disposition, a small percentage of the hormone may be extracted from the peritubular capillaries (Johnson and Maack 1977; Krogsgaard Thomsen et al. 1994). Peritubular transport of proteins and peptides from the basolateral membrane has also been shown for insulin (Nielsen et al. 1987).

Hepatic Protein Metabolism

Aside from renal and gastrointestinal metabolism, the liver may also play a major role in the metabolism of protein therapeutics, especially for larger proteins. Exogenous as well as endogenous proteins undergo proteolytic degradation to dipeptides and amino acids that are reused for endogenous protein synthesis. Proteolysis usually starts with endopeptidases that attack in the middle part of the protein, and the resulting oligopeptides are then further degraded by exopeptidases. The rate of hepatic metabolism is largely dependent on the specific amino acid sequence of the protein (Meibohm 2004).

The major prerequisite for hepatic protein metabolism is the uptake of proteins in the different liver cell types. An overview of the different mechanisms of hepatic uptake of proteins is listed in Table 5.2.

Small peptides may cross the hepatocyte membrane via simple passive diffusion if they have sufficient hydrophobicity. Peptides of this nature include cyclosporin (cyclic peptide) (Ziegler et al. 1988). Other cyclic and linear peptides of small size (<1.4 kDa) and hydrophobic nature (containing aromatic amino acids), such as cholecystokinin-8 (CCK-8; 8 amino acids), are taken up by the hepatocytes by carrier-mediated transport (Ziegler et al. 1988), in the case of CCK-8 by the organic anion-transporting polypeptide OATP-8 (SLCO1B3) (Ismair et al. 2001). After internalization into the cytosol, these peptides are usually metabolized by microsomal enzymes or cytosolic peptidases.

Cell type	Uptake mechanism	Proteins/peptides transported
Hepatocytes	Anionic passive diffusion, carrier-mediated transport	Cyclic and linear hydrophobic peptides (<1.4 kDa; e.g., cyclosporins, CCK-8)
	RME: Gal/GalNAc receptor (asialoglycoprotein receptor)	N-acetylgalactosamine-terminated glycoproteins, galactose-terminated glycoproteins (e.g., desialylated EPO)
	RME: low-density lipoprotein receptor (LDLR)	LDL, apoE-, and apoB-containing lipoproteins
	RME: LDLR-related protein (LRP receptor)	α_2-macroglobulin, apoE-enriched lipoproteins, lipoprotein lipase (LpL), lactoferrin, t-PA, thrombospondin (TSP), TGF-β, and IL-1β bound to α_2-macroglobulin
	RME: other receptors	IgA, glycoproteins, lipoproteins, immunoglobulin intestinal and pancreatic peptides, metallo- and hemoproteins, transferrin, insulin, glucagon, GH, EGF
	Nonselective pinocytosis (non-receptor-mediated)	Albumin, antigen-antibody complexes, some pancreatic proteins, some glycoproteins
Kupffer cells	Endocytosis	Particulates with galactose groups
Kupffer and endothelial cells	RME	IgG, N-acetylgalactosamine-terminated glycoproteins
	RME: mannose receptor	Mannose-terminated glycoproteins (e.g., t-PA, renin)
	RME: fucose receptor	Fucose-terminated glycoproteins
Endothelial cells	RME: scavenger receptor	Negatively charged proteins
	RME: other receptors	VEGF
Fat-storing cells	RME: mannose-6-phosphate receptor	Mannose-6-phosphate-terminated proteins (e.g., IGF-II)

Compiled from several sources, including reviews by Cumming (1991), Kompella and Lee (1991), and Marks et al. (1995)

RME receptor-mediated endocytosis, *IGF* insulin-like growth factor, *EGF* epidermal growth factor, *TGF* tissue growth factor, *GH* growth hormone

Table 5.2 ■ Hepatic uptake mechanisms for proteins and protein complexes.

Uptake of larger peptides and proteins can either be facilitated through pinocytosis or by receptor-mediated endocytosis. Pinocytosis is an unspecific fluid-phase endocytosis, in which molecules are taken up into cells by forming invaginations of cell membrane around extracellular fluid, that are subsequently taken up as membrane vesicles.

Receptor-mediated endocytosis is a clathrin-mediated endocytosis process via relatively unspecific, promiscuous membrane receptors (McMahon and Boucrot 2011). In receptor-mediated endocytosis, circulating proteins are recognized by specific receptor proteins. The receptors are usually integral membrane glycoproteins with an exposed binding domain on the extracellular side of the cell membrane. After the binding of the circulating protein to the receptor, the complex is already present or moves in clathrin-coated pit regions, and the membrane invaginates and pinches off to form an endocytotic coated vesicle that contains the receptor and ligand (internalization). The vesicle coat consists of proteins (clathrin, adaptin, and others), which are then removed by an uncoating adenosine triphosphatase (ATPase). The vesicle parts, the receptor, and the ligands dissociate and are targeted to various intracellular locations. Some receptors, such as the low-density lipoprotein (LDL), asialoglycoprotein, and transferrin receptors, are known to undergo recycling. Since sometimes several hundred cycles are part of a single receptor's lifetime, the associated receptor-mediated endocytosis is of high capacity. Other receptors, such as the interferon receptor, undergo degradation. This degradation leads to a decrease in the concentration of receptors on the cell surface (receptor downregulation). Others, such as insulin receptors, for example, undergo both recycling and degradation (Kompella and Lee 1991).

For glycoproteins, receptor-mediated endocytosis through sugar-recognizing C-type lectin receptors is an efficient hepatic uptake mechanism if a critical number of exposed sugar groups (mannose, galactose, fucose, N-acetylglucosamine, N-acetylgalactosamine, or glucose) is exceeded (Meijer and Ziegler 1993). Important C-type lectin receptors in the liver are the asialoglycoprotein receptor on hepatocytes and the mannose and fucose receptors on Kupffer and liver endothelial cells (Smedsrod and Einarsson 1990; Bu et al. 1992). The high-mannose glycans in the first kringle domain of t-PA, for example, have been implicated in its hepatic clearance via these receptors (Cumming 1991).

The low-density lipoprotein receptor-related protein (LRP) is a member of the LDL receptor family responsible for endocytosis of several important lipoproteins, proteases, and protease-inhibitor complexes in the liver and other tissues (Strickland et al. 1995). Examples of proteins and protein complexes for which hepatic uptake is mediated by LRP are listed in Table 5.2.

Uptake of proteins by liver cells is followed by transport to an intracellular compartment for metabolism. Proteins internalized into vesicles via an endocytotic mechanism undergo intracellular transport towards the lysosomal compartment near the center of the cell. There, the endocytotic vehicles fuse with or mature into lysosomes, which are specialized acidic vesicles that contain a wide variety of hydrolases capable of degrading all biological macromolecules. Proteolysis is started by endopeptidases (mainly cathepsin D) that act on the middle part of the proteins. Oligopeptides – as the result of the first step – are further degraded by exopeptidases. The resulting amino acids and dipeptides reenter the metabolic pool of the cell. The hepatic metabolism of glycoproteins may occur more slowly than the naked protein because protecting oligosaccharide chains need to be removed first. Metabolized proteins and peptides in lysosomes from hepatocytes, hepatic sinusoidal cells, and Kupffer cells may be released into the blood. Degraded proteins in hepatocyte lysosomes can also be delivered to the bile canaliculus and excreted by exocytosis.

Besides intracellular degradation, a second intracellular pathway for proteins is the direct shuttle or transcytotic pathway (Kompella and Lee 1991). In this case, the endocytotic vesicle formed at the cell surface traverses the cell to the peribiliary space, where it fuses with the bile canalicular membrane, releasing its contents by exocytosis into bile. This pathway bypasses the lysosomal compartment completely. It has been described for polymeric immunoglobulin A but is not assumed to be a major elimination pathway for most protein drugs.

Target-Mediated Protein Metabolism

Therapeutic proteins frequently bind with high affinity to membrane-associated receptors on the cell surface if the receptors are the target structure to which the protein therapeutic is directed. This binding can lead to receptor-mediated uptake by endocytosis and subsequent intracellular lysosomal metabolism. The associated drug disposition behavior in which the binding to the pharmacodynamic target structure affects the pharmacokinetics of a drug compound is termed "target-mediated drug disposition" (Levy 1994).

For conventional small-molecule drugs, receptor binding is usually negligible compared to the total amount of drug in the body and rarely affects their pharmacokinetic profile. In contrast, a substantial fraction of a protein therapeutic can be bound to its pharmacologic target structure, for example, a receptor. Target-mediated drug disposition can affect distribution as well as elimination processes. Most notably, receptor-mediated protein metabolism is a frequently encountered elimination pathway for many protein therapeutics (Meibohm 2004).

Receptor-mediated uptake and metabolism via interaction with these generally high-affinity, low-capacity binding sites is not limited to a specific organ or tissue type. Thus, any tissue including the therapeutically targeted cells that express receptors for the drug can contribute to the elimination of the protein therapeutic (Fig. 5.5) (Zhang and Meibohm 2012).

Since the number of protein drug receptors is limited, receptor-mediated protein metabolism can usually be saturated within therapeutic concentrations, or more specifically at relatively low molar ratios between the protein drug and the receptor (Mager 2006). As a consequence, the elimination clearance of these protein drugs is not constant but dose dependent and decreases with increasing dose. Thus, receptor-mediated elimination constitutes a major source for nonlinear pharmacokinetic behavior of numerous protein drugs, i.e., systemic exposure to the drug increases more than proportional with increasing dose (Tang et al. 2004).

Recombinant human macrophage colony-stimulating factor (M-CSF), for example, undergoes besides linear renal elimination a nonlinear elimination pathway that follows Michaelis-Menten kinetics and is linked to a receptor-mediated uptake into macrophages. At low concentrations, M-CSF follows linear pharmacokinetics, while at high concentrations nonrenal elimination pathways are saturated resulting in nonlinear pharmacokinetic behavior (Fig. 5.6) (Bauer et al. 1994).

Nonlinearity in pharmacokinetics resulting from target-mediated drug disposition has also been observed for several monoclonal antibody therapeutics, for instance for the anti-HER2 humanized monoclonal antibody trastuzumab. Trastuzumab is approved for the combination treatment of HER2 protein overexpressing metastatic breast cancer. With increasing dose level, the mean terminal half-life of trastuzumab increases and the clearance decreases, leading to overproportional increases in systemic exposure with

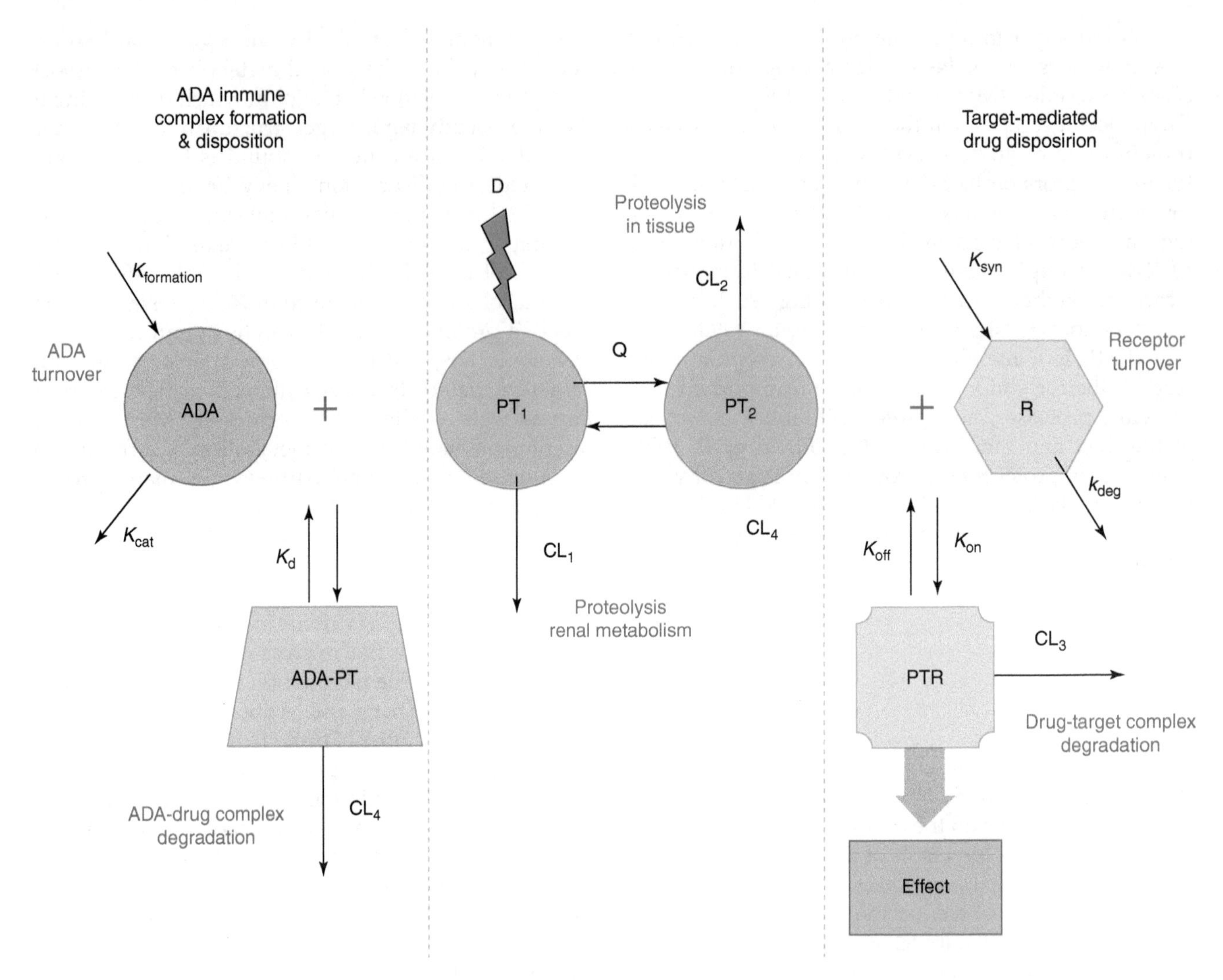

Figure 5.5 ■ Example of multiple clearance pathways affecting the pharmacokinetics of a typical protein therapeutic. Depicted is a two-compartment pharmacokinetic model with intravenous administration of a dose (*D*), concentrations of the protein therapeutic in the central (PT_1) and peripheral (PT_2) compartment, and interdepartmental clearance Q. The pharmacokinetic model includes two clearance pathways, one from the central compartment (CL_1) representative of, for example, renal metabolism or proteolytic degradation through the reticuloendothelial system and a second proteolytic degradation pathway from the peripheral compartment (CL_2) representative of, for example, proteolytic degradation through a receptor-mediated endocytosis pathway. Added to these two clearance pathways is on the right side a target-mediated disposition pathway that constitutes interaction of the protein therapeutic with its pharmacologic target receptor, which is in a homeostatic equilibrium of synthesis and degradation (synthesis rate k_{syn} and degradation rate constant k_{deg}). The dynamic equilibrium for the formation of the resulting protein therapeutic-receptor complex (*PT-R*) is determined through the association rate constant k_{on} and the dissociation rate constant k_{off}. The formation of PT-R does not only elicit the pharmacologic effect, but also triggers degradation of the complex. Thus, target binding and subsequent PT-R degradation constitute an additional clearance pathway for the protein therapeutic (CL_3). The left side of the graphic depicts the effect of an immune response to the protein therapeutic resulting in anti-drug antibody (*ADA*) formation. Again, the circulating concentration of the ADA is determined by a homeostatic equilibrium between its formation rate ($k_{formation}$) and a catabolic turnover process (rate constant k_{cat}). The ADA response results in the formation of immune complexes with the drug (*ADA-PT*). Dependent on the size and structure of the immune complexes, endogenous elimination pathways though the reticuloendothelial system may be triggered, most likely via Fcγ-mediated endocytosis. Thus, immune complex formation and subsequent degradation may constitute an additional clearance pathway (CL_4) for protein therapeutics (From Chirmule et al. (2012)).

increasing dose (Tokuda et al. 1999). Since trastuzumab is rapidly internalized via receptor-mediated endocytosis after binding to HER2, its target structure on the cell surface, saturation of this elimination pathway is the likely cause for the observed dose-dependent pharmacokinetics (Kobayashi et al. 2002).

Modulation of Protein Disposition by the FcRn Receptor

Immunoglobulin G (IgG)-based monoclonal antibodies and their derivatives as well as albumin conjugates constitute important classes of protein therapeutics with many members currently being under development or in

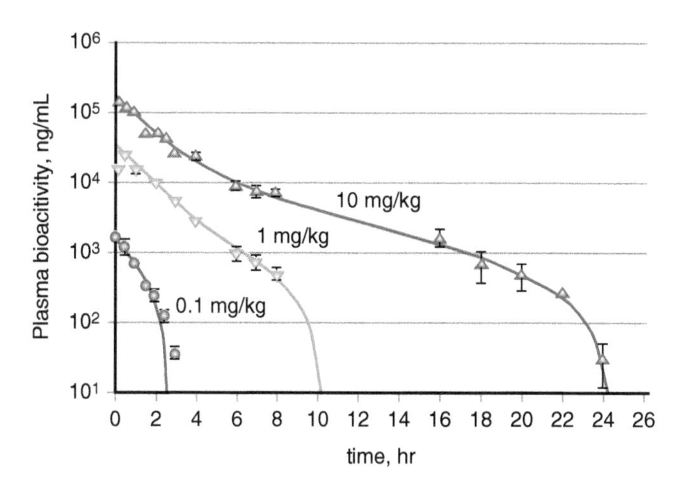

Figure 5.6 ■ Nonlinear pharmacokinetics of macrophage colony-stimulating factor (M-CSF), presented as measured (*triangles* and *circles*; mean ± SE) and modeled plasma bioactivity-time curves (*lines*) after intravenous injection of 0.1 mg/kg ($n=5$), 1.0 mg/kg ($n=3$), and 10 mg/kg ($n=8$) in rats. Bioactivity is used as a substitute for concentration (From Bauer et al. (1994), with permission from American Society for Pharmacology and Experimental Therapeutics).

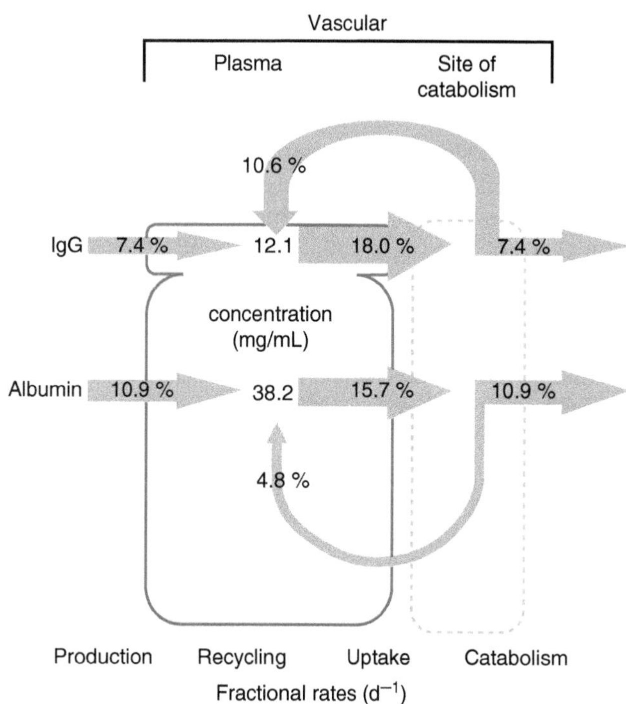

Figure 5.7 ■ Effect of FcRn-mediated recycling on IgG and albumin turnover in humans expressed as fractional rates. Shown are homeostatic plasma concentrations (12.1 and 38.2 mg/mL), fractional catabolic rates (7.4 and 10.9 %/day), the FcRn-mediated fractional recycling rates (10.6 and 4.8 %/day), and the fractional production rates (7.4 and 10.9 %/day). The figure is to scale: areas for plasma amounts and arrow widths for rates (From Kim et al. (2007), with permission from Elsevier).

therapeutic use. Interaction with the neonatal Fc receptor (FcRn) constitutes a major component in the drug disposition of IgG molecules (Roopenian and Akilesh 2007). FcRn has been well-described in the transfer of passive humoral immunity from a mother to her fetus by transferring IgG across the placenta and the proximal small intestine via transcytosis. More importantly, interaction with FcRn in a variety of cells, including endothelial cells and monocytes, macrophages, and other dendritic cells, protects IgG from lysosomal catabolism and thus constitutes a salvage pathway for IgG molecules that have been internalized in these cells types. This is facilitated by intercepting IgG in the endosomes and recycling it to the systemic circulation (Wang et al. 2008). The interaction with the FcRn receptor thereby prolongs the elimination half-life of IgG, with a more pronounced effect the stronger the binding of the Fc fragment of the antibody is to the receptor: Based on the affinity of this binding interaction, human IgG1, IgG2, and IgG4 have a half-life in humans of 18–21 days, whereas the less strongly bound IgG3 has a half-life of only 7 days, and murine IgG in humans has a half-life of 1–2 days (Dirks and Meibohm 2010).

Similar to IgG, FcRn is also involved in the disposition of albumin molecules. The kinetics of IgG and albumin recycling are illustrated in Fig. 5.7. For IgG1, approximately 60 % of the molecules taken up into lysosomes are recycled, for albumin 30 %. As FcRn is responsible for the extended presence of IgG, albumin, and other Fc- or albumin-conjugated proteins in the systemic circulation, modulation of the interaction with FcRn allows to deliberately control the half-life of these molecules (Kim et al. 2007).

■ Immunogenicity and Protein Pharmacokinetics

The antigenic potential of protein therapeutics may lead to antibody formation against the protein therapeutic during chronic therapy. This is especially of concern if animal-derived proteins are applied in human clinical studies, but also if human proteins are used in animal studies during preclinical drug development. Chapter 6 discusses in detail the phenomenon of immunogenicity and its consequences for the pharmacotherapy with protein therapeutics.

The formation of anti-drug antibodies (ADA) against a therapeutic protein may not only modulate or even obliterate the biological activity of a protein drug, but may also modify its pharmacokinetic profile. In addition, ADA-drug complex formation may lead to immune complex-mediated toxicity, particularly if the complexes get deposited in a specific organ or tissue. Glomerulonephritis has, for example, been observed after deposition of ADA-protein drug complexes in the renal glomeruli of Cynomolgus monkeys after intramuscular administration of rhINF-γ. Similar to other circulating immune complexes, ADA-protein drug complexes may trigger the regular endogenous elimination pathways for these complexes, which consist of uptake and lysosomal degradation by the reticuloendothelial system. This process has been primarily described for the liver and the spleen and seems to be mediated by Fcγ receptors.

The ADA formation may either lead to the formation of clearing ADA, that increase the clearance of the protein therapeutic, or sustaining ADA that decrease the clearance of the protein therapeutic (Fig. 5.8). For clearing ADA, the immune complex formation triggers elimination via the reticuloendothelial system, which constitutes an additional elimination pathway for the protein therapeutic (Fig. 5.5). This increase in clearance for the protein therapeutic results in a decreased systemic exposure and reduced elimination half-life. A clearing effect of ADA is often observed for large protein therapeutics such as monoclonal antibodies (Richter et al. 1999).

For sustaining ADA, the immune complex formation does not trigger the regular endogenous elimination processes, but serves as a storage depot for the protein, thereby reducing its clearance, increasing its systemic exposure, and prolonging its half-life. This behavior has often been described for small protein therapeutics where the immune complex formation, for example, prevents glomerular filtration and subsequent tubular metabolism. The elimination half-life of the protein therapeutic then is often increased to approach that of IgG (Chirmule et al. 2012).

Whether ADA-protein drug complex formation results in clearing or sustaining effects seems to be a function of its physicochemical and structural properties, including size, antibody class, ADA-antigen ratio, characteristics of the antigen, and location of the binding epitopes. For example, both an increased and decreased clearance is possible for the same protein, dependent on the dose level administered. At low doses, protein-antibody complexes delay clearance because their elimination is slower than the unbound protein. In contrast, at high doses, higher levels of protein-antibody complex result in the formation of aggregates, which are cleared more rapidly than the unbound protein.

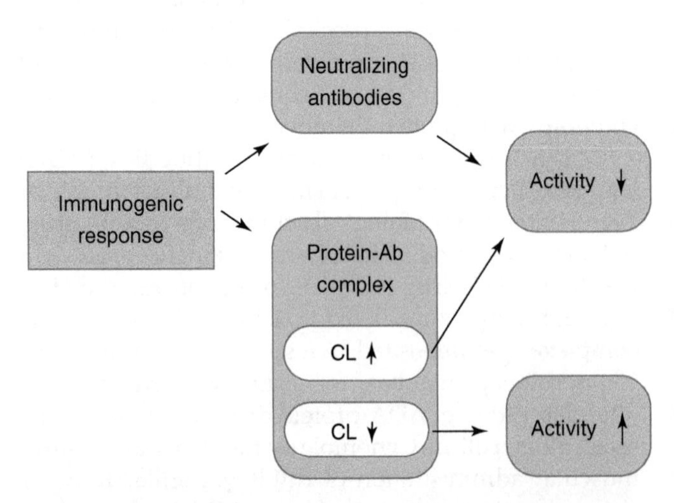

Figure 5.8 ■ Effect of anti-drug antibody formation on the pharmacokinetics and pharmacodynamics of therapeutic protein.

The enhancement of the clearance of cytokine interleukin-6 (IL-6) via administration of cocktails of three anti-IL-6 monoclonal antibodies was suggested as a therapeutic approach in cytokine-dependent diseases like multiple myeloma, B-cell lymphoma, and rheumatoid arthritis (Montero-Julian et al. 1995). The authors could show that, while the binding of one or two antibodies to the cytokine led to stabilization of the cytokine, simultaneous binding of three anti-IL-6 antibodies to three distinct epitopes induced rapid uptake of the complex by the liver and thus mediated a rapid elimination of IL-6 from the central compartment.

It should be emphasized that ADA formation is a polyclonal and usually relatively unspecific immune response to the protein therapeutic, with formation of different antibodies with variable binding affinities and epitope specificities, and that this ADA formation with its multiple-involved antibody species is different in different patients. Thus, reliable prediction of ADA formation and effects remains elusive at the current time (Chirmule et al. 2012).

The immunogenicity of protein therapeutics is also dependent on the route of administration. Extravascular injection is known to stimulate antibody formation more than IV application, which is most likely caused by the increased immunogenicity of protein aggregates and precipitates formed at the injection site. Further details on this aspect of immunogenicity are discussed in Chap. 6.

■ Species Specificity and Allometric Scaling

Peptides and proteins often exhibit distinct species specificity with regard to structure and activity. Peptides and proteins with identical physiological function may have different amino acid sequences in different species and may have no activity or be even immunogenic if used in a different species. The extent of glycosylation and/or sialylation of a protein molecule is another factor of species differences, e.g., for interferon-α or erythropoietin, which may not only alter its efficacy and immunogenicity (see Chap. 6) but also the drug's clearance.

Projecting human pharmacokinetic behavior for therapeutic proteins based on data in preclinical species is often performed using allometric approaches. Allometry is a methodology used to relate morphology and body function to the size of an organism. Allometric scaling is an empirical technique to predict body functions based on body size. Allometric scaling has found wide application in drug development, especially to predict pharmacokinetic parameters in humans based on the corresponding parameters in several animal species and the body size differences among these species and humans. Multiple allometric scaling approaches have been described with variable success rates, predominantly during the transition from preclinical to clinical drug development (Dedrick 1973; Boxenbaum 1982; Mahmood and Balian 1999; Mahmood 2002). In the most frequently used approach, pharmacokinetic parameters between different species are related via body weight using a power function:

$$P = a \cdot W^b$$

where P is the pharmacokinetic parameter scaled, W is the body weight in kg, a is the allometric coefficient, and b is the allometric exponent. a and b are specific constants for each parameter of a compound. General tendencies for the allometric exponent are 0.75 for rate constants (i.e., clearance, elimination rate constant), 1 for volumes of distribution, and 0.25 for half-lives. More recently, allometric approaches are being complemented by physiologically based pharmacokinetic modeling.

For most traditional small-molecule drugs, allometric scaling is often imprecise, especially if hepatic metabolism is a major elimination pathway and/or if there are interspecies differences in metabolism. For peptides and proteins, however, allometric scaling has frequently proven to be much more precise and reliable if their disposition is governed by relatively unspecific proteolytic degradation pathways. The reason is probably the similarity in handling peptides and proteins among different mammalian species (Wills and Ferraiolo 1992). Clearance and volume of distribution of numerous therapeutically used proteins like growth hormone or t-PA follow a well-defined, weight-dependent physiologic relationship between lab animals and humans. This allows relatively precise quantitative predictions for their pharmacokinetic behavior in humans based on preclinical findings (Mordenti et al. 1991).

Figure 5.9, for example, shows allometric plots for the clearance and volume of distribution of a P-selectin antagonist, P-selectin glycoprotein ligand-1, for the treatment of P-selectin-mediated diseases such as thrombosis, reperfusion injury, and deep vein thrombosis. The protein's human pharmacokinetic parameters could accurately be predicted using allometric power functions based on data from four species: mouse, rat, monkey, and pig (Khor et al. 2000).

Recent work on scaling the pharmacokinetics of monoclonal antibodies has suggested that allometric scaling from one nonhuman primate species, in this case the Cynomolgus monkey, using an allometric exponent of 0.85 might be superior to traditional allometric scaling approaches (Deng et al. 2011).

In any case, successful allometric scaling seems so far largely limited to unspecific protein elimination pathways. Once interactions with specific receptors are involved in drug disposition, for example, in receptor-mediated processes or target-mediated drug disposition, then allometric approaches oftentimes fail to scale drug disposition of therapeutic proteins across species due to differences in binding affinity and specificity, as well as expression and turnover kinetics of the involved receptors and targets in different species. In this situation, it becomes especially important to only consider for scaling preclinical pharmacokinetic data from "relevant" animal species for which the therapeutic protein

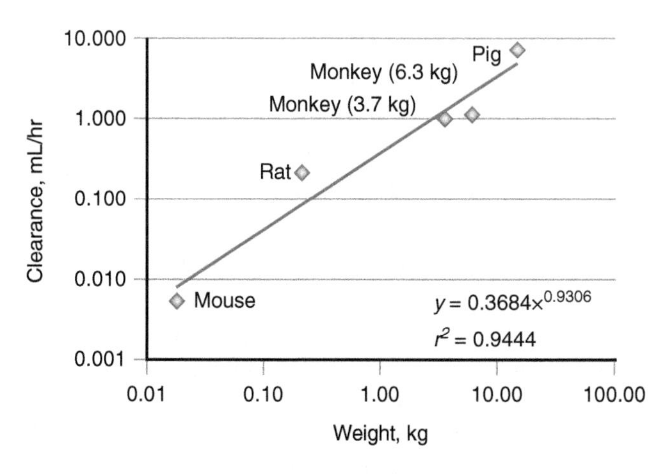

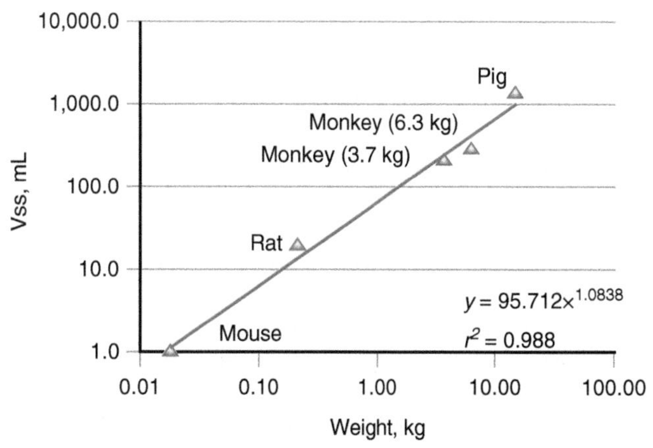

Figure 5.9 ■ Allometric plots of the pharmacokinetic parameter clearance and volume of distribution at steady state (V_{ss}) for the P-selectin antagonist rPSGL-Ig. Each data point within the plot represents an averaged value of the respective pharmacokinetic parameter in one of five species: mouse, rat, monkey (3.7 kg), monkey, (6.3 kg) and pig, respectively. The *solid line* is the best fit with a power function to relate pharmacokinetic parameters to body weight (Khor et al. 2000; with permission from American Society for Pharmacology and Experimental Therapeutics).

shows cross-reactivity between animal and human receptors or targets.

It needs to be emphasized that allometric scaling techniques are useful tools for predicting a dose that will assist in the planning of dose-ranging studies, including first-in-man studies, but are not a replacement for such studies. The advantage of including such dose prediction in the protocol design of dose-ranging studies is that a smaller number of doses need to be tested before finding the final dose level. Interspecies dose predictions simply narrow the range of doses in the initial pharmacological efficacy studies, the animal toxicology studies, and the human safety and efficacy studies.

■ Chemical Modifications for Optimizing the Pharmacokinetics of Protein Therapeutics

In recent years, approaches modifying the molecular structure of protein therapeutics have repeatedly been applied to affect the immunogenicity, pharmacokinetics, and/or pharmacodynamics of protein drugs (Kontermann 2012). These approaches include the addition, deletion, or exchange of selected amino acids within the protein's sequence, synthesis of truncated proteins with a reduced amino acid sequence, glycosylation or deglycosylation, and covalent linkage to polymers (Veronese and Caliceti 2006). The latter approach has been used for several protein therapeutics by linking them to monomethoxy polyethylene glycol (PEG) molecules of various chain lengths in a process called PEGylation (Caliceti and Veronese 2003).

The conjugation of high polymeric mass to protein drugs is generally aimed at preventing the protein being recognized by the immune system as well as reducing its elimination via glomerular filtration or proteolytic enzymes, thereby prolonging the oftentimes relatively short elimination half-life of endogenous proteins. Conjugation of protein drugs with PEG chains increases their molecular weight, but because of the attraction of water molecules by PEG even more their hydrodynamic volume, this in turn results in a reduced renal clearance and restricted volume of distribution. PEGylation can also shield antigenic determinants on the protein drug from detection by the immune system through steric hindrance (Walsh et al. 2003). Similarly, amino acid sequences sensitive towards proteolytic degradation may be shielded against protease attack. By adding a large, hydrophilic molecule to the protein, PEGylation can also increase drug solubility (Molineux 2003).

PEGylation has been used to improve the therapeutic properties of numerous protein therapeutics including interferon-α, asparaginase, and filgrastim. More details on the general concept of PEGylation and its specific application for protein therapeutics can be found in Chap. 21.

The therapeutic application of L-asparaginase in the treatment of acute lymphoblastic leukemia has been hampered by its strong immunogenicity with allergic reactions occurring in 33–75 % of treated patients in various studies. The development of pegaspargase, a PEGylated form of L-asparaginase, is a successful example for overcoming this high rate of allergic reactions towards L-asparaginase using PEG conjugation techniques (Graham 2003). Pegaspargase is well-tolerated compared to L-asparaginase, with 3–10 % of the treated patients experiencing clinical allergic reactions.

Pegfilgrastim is the PEGylated version of the granulocyte colony-stimulating factor filgrastim, which is administered for the management of chemotherapy-induced neutropenia. PEGylation minimizes filgrastim's renal clearance by glomerular filtration, thereby making neutrophil-mediated clearance the predominant route of elimination. Thus, PEGylation of filgrastim results in so-called self-regulating pharmacokinetics since pegfilgrastim has a reduced clearance and thus prolonged half-life and more sustained duration of action in a neutropenic compared to a normal patient because only few mature neutrophils are available to mediate its elimination (Zamboni 2003).

The hematopoietic growth factor darbepoetin-α is an example of a chemically modified endogenous protein with altered glycosylation pattern. It is a glycosylation analogue of human erythropoietin, with two additional N-linked oligosaccharide chains (five in total) (Mould et al. 1999). The additional N-glycosylation sites were made available through substitution of five amino acid residues in the peptide backbone of erythropoietin, thereby increasing the molecular weight from 30 to 37 kDa. Darbepoetin-α has a substantially modified pharmacokinetic profile compared to erythropoietin, resulting in a threefold longer serum half-life that allows for reduced dosing frequency (Macdougall et al. 1999). More details on hematopoietic growth factors, including erythropoietin and darbepoetin-α, are provided in Chap. 18.

PHARMACODYNAMICS OF PROTEIN THERAPEUTICS

Protein therapeutics are usually highly potent compounds with steep dose-effect curves as they are targeted therapies towards a specific, well-described pharmacologic structure or mechanism. Thus, a careful characterization of the concentration-effect relationship, i.e., the pharmacodynamics, is especially desirable for protein therapeutics (Tabrizi and Roskos 2006). Combination of pharmacodynamics with pharmacokinetics by integrated pharmacokinetic-pharmacodynamic modeling (PK/PD modeling) adds an additional level of complexity that

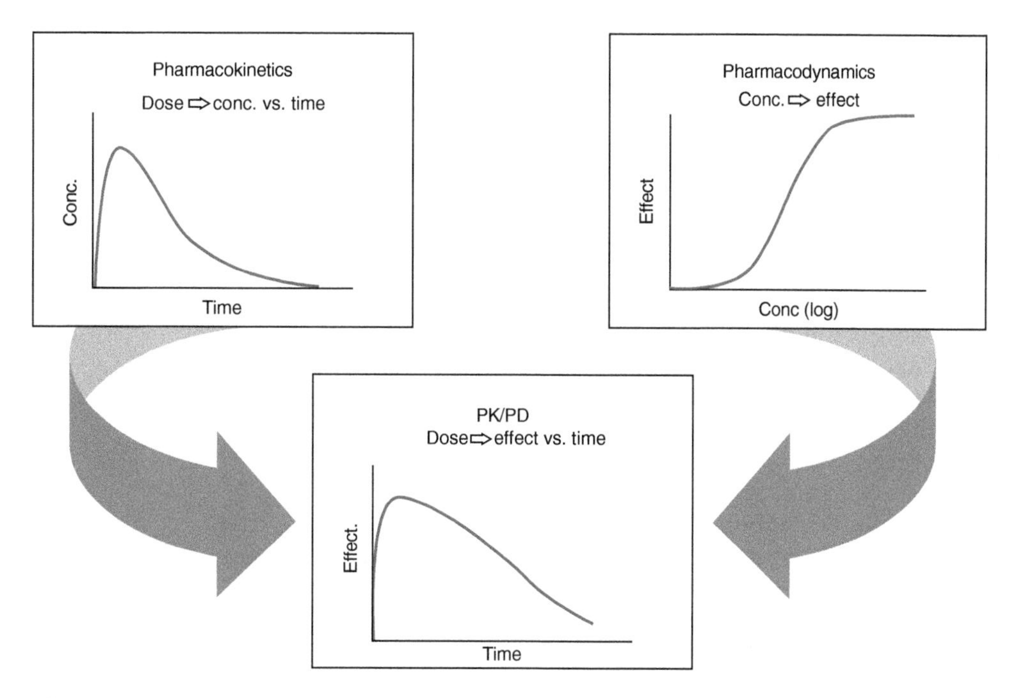

Figure 5.10 ■ General concept of PK/PD modeling. Pharmacokinetic-pharmacodynamic (PK/PD) modeling combines a pharmacokinetic model component that describes the time course of drug in plasma and a pharmacodynamic model component that relates the plasma concentration to the drug effect in order to describe the time course of the effect intensity resulting from the administration of a certain dosage regimen (From Derendorf and Meibohm (1999)).

allows furthermore characterization of the dose-exposure-response relationship of a drug and a continuous description of the time course of effect intensity directly resulting from the administration of a certain dosage regimen (Fig. 5.10) (Meibohm and Derendorf 1997; Derendorf and Meibohm 1999).

PK/PD modeling is a technique that combines the two classical pharmacologic disciplines of pharmacokinetics and pharmacodynamics. It integrates a pharmacokinetic and a pharmacodynamic model component into one set of mathematical expressions that allows the description of the time course of effect intensity in response to administration of a drug dose. This so-called integrated PK/PD model allows deriving pharmacokinetic and pharmacodynamic model parameters that characterize the dose-concentration-effect relationship for a specific drug based on measured concentration and effect data. In addition, it allows simulation of the time course of effect intensity for dosage regimens of a drug beyond actually measured data, within the constraints of the validity of the model assumptions for the simulated condition. Addition of a statistical model component describing inter- and intraindividual variation in model parameters allows expanding PK/PD models to describe time courses of effect intensity not only for individual subjects, but also for whole populations of subjects.

Integrated pharmacokinetic-pharmacodynamic (PK/PD) modeling approaches have widely been applied for the characterization of protein therapeutics (Tabrizi and Roskos 2006). Embedded in a model-based drug development approach, modeling and simulation based on integrated PK/PD does not only provide a comprehensive summary of the available data, but also enables to test competing hypotheses regarding processes altered by the drug, allows making predictions of drug effects under new conditions, and facilitates to estimate inaccessible system variables (Meibohm and Derendorf 1997; Mager et al. 2003).

Mechanism-based PK/PD modeling appreciating the physiological events involved in the elaboration of the observed effect has been promoted as superior modeling approach as compared to empirical modeling, especially because it does not only describe observations but also offers some insight into the underlying biological processes involved and thus provides flexibility in extrapolating the model to other clinical situations (Levy 1994; Derendorf and Meibohm 1999; Suryawanshi et al. 2010). Since the molecular mechanism of action of a protein therapeutic is generally well-understood, it is often straightforward to transform this available knowledge into a mechanism-based PK/PD modeling approach that appropriately characterizes the real physiological process leading to the drug's therapeutic effect.

The relationship between exposure and response may be either simple or complex, and thus obvious or hidden. However, if no simple relationship is obvious, it would be misleading to conclude a priori that no relationship exists at all rather than that it is not readily apparent (Levy 1986).

The application of PK/PD modeling is beneficial in all phases of preclinical and clinical drug development and has been endorsed by the pharmaceutical industry, academia, and regulatory agencies (Peck et al. 1994; Lesko et al. 2000; Sheiner and Steimer 2000; Meibohm and Derendorf 2002), most recently by the Critical Path Initiative of the US Food and Drug Administration (Lesko 2007). Thus, PK/PD concepts and model-based drug development play a pivotal role especially in the drug development process for biologics, and their widespread application supports a scientifically driven, evidence-based, and focused product development for protein therapeutics (Zhang et al. 2008).

While a variety of PK/PD modeling approaches has been employed for biologics, we will in the following focus on five classes of approaches to illustrate the challenges and complexities, but also opportunities to characterize the pharmacodynamics of protein therapeutics:

- Direct link PK/PD models
- Indirect link PK/PD models
- Indirect response PK/PD models
- Cell life span models
- Complex response models

It should not be unmentioned, however, that PK/PD models for protein therapeutics are not only limited to continuous responses as shown in the following, but are also used for binary or graded responses. Binary responses are responses with only two outcome levels where a condition is either present or absent, e.g., dead versus alive. Graded or categorical responses have a set of predefined outcome levels, which may or may not be ordered, for example, the categories "mild," "moderate," and "severe" for a disease state. Lee et al. (2003), for example, used a logistic PK/PD modeling approach to link cumulative AUC of the anti-TNF-α protein etanercept with a binary response, the American College of Rheumatology response criterion of 20 % improvement (ARC20) in patients with rheumatoid arthritis.

Direct Link PK/PD Models

The concentration of a protein therapeutic is usually only quantified in plasma, serum, or blood, while the magnitude of the observed response is determined by the concentration of the protein drug at its effect site, the site of action in the target tissue (Meibohm and Derendorf 1997). Effect site concentrations, however, are usually not accessible for measurement, and plasma, serum, or blood concentrations are usually used as their substitute. The relationship between the drug concentration in plasma and at the effect site may either be constant or undergo time-dependent changes. If equilibrium between both concentrations is rapidly achieved or the site of action is within plasma, serum, or blood, there is practically a constant relationship between both concentrations with no temporal delay between plasma and effect site. In this case, measured plasma concentrations can directly serve as input for a pharmacodynamic model (Fig. 5.11). The most frequently used direct link pharmacodynamic model is a sigmoid E_{max} model:

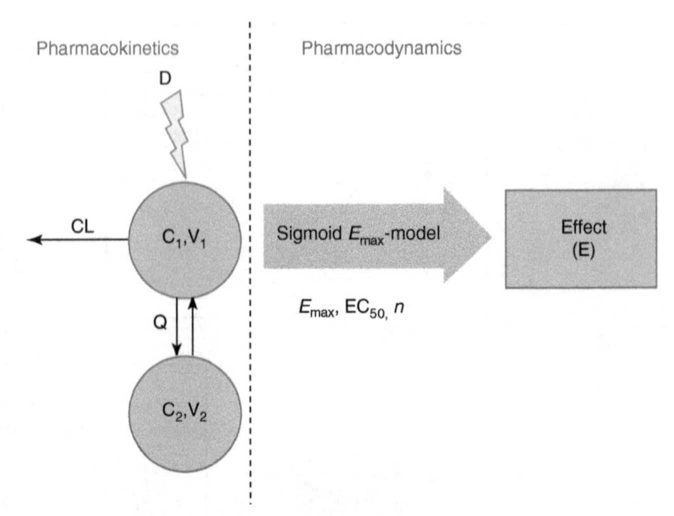

Figure 5.11 Schematic of a typical direct link PK/PD model. The PK model is a typical two-compartment model with a linear elimination clearance from the central compartment (*CL*) and a distributional clearance (*Q*). C_1 and C_2 are the concentrations in the central and peripheral compartments, and V_1 and V_2 are their respective volumes of distribution. The effect (*E*) is directly linked to the concentration in the central compartment C_1 via a sigmoid E_{max} model. The sigmoid E_{max} relationship is characterized by the pharmacodynamic parameters E_{max}, the maximum achievable effect, EC_{50}, the concentration of the drug that produced half of the maximum effect, and the Hill coefficient *n* as via the sigmoid E_{max} equation.

$$E = \frac{E_{max} \cdot C^n}{EC_{50}^n + C^n}$$

with E_{max} as maximum achievable effect, C as drug concentration in plasma, and EC_{50} the concentration of the drug that produces half of the maximum effect. The Hill coefficient n is a shape factor that allows for an improved fit of the relationship to the observed data. As represented by the equation for the sigmoid E_{max} model, a direct link model directly connects measured concentration to the observed effect without any temporal delay (Derendorf and Meibohm 1999).

A direct link model was, for example, used to relate the serum concentration of the antihuman immunoglobulin E (IgE) antibody CGP 51901 for the treat-

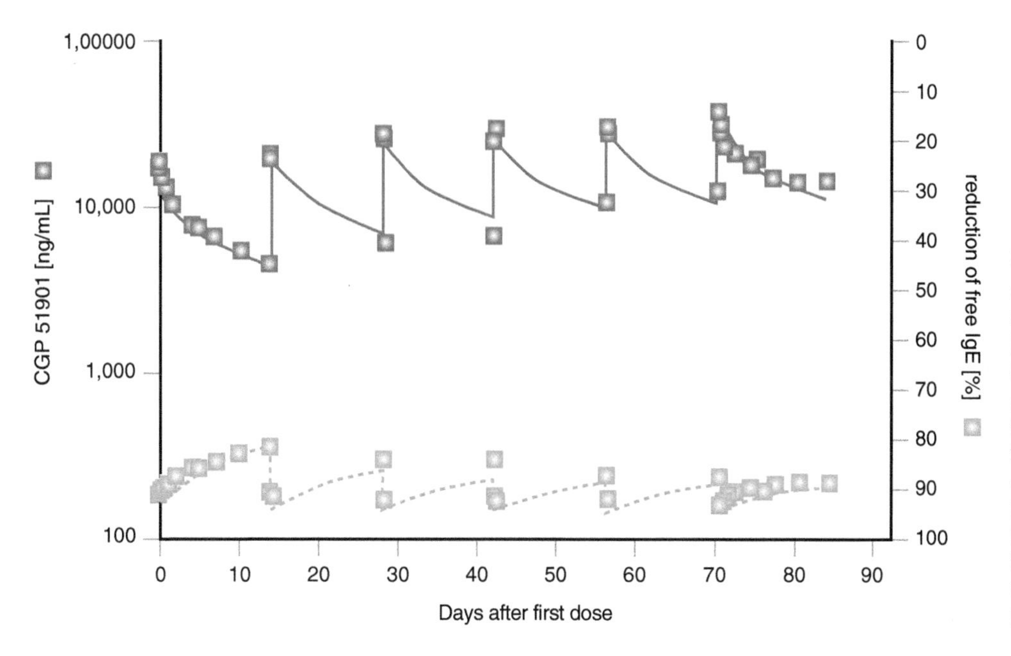

Figure 5.12 ■ Observed (■) and model-predicted (—) serum concentration of the anti-human IgE antibody CGP 51901 and observed (■) and model-predicted (- - - -) reduction of free IgE in one representative patient, given six IV doses of 60 mg biweekly. The predictions were modeled with a direct link PK/PD model (Modified from Racine-Poon et al. (1997); with permission from Macmillan Publishers Ltd.).

ment of seasonal allergic rhinitis to the reduction of free IgE via an inhibitory E_{max} model (Fig. 5.12) (Racine-Poon et al. 1997). It should be noted that the peak and trough concentrations and effects are directly related and thus occur at the same times, respectively, without time delay. Similarly, a direct link model was used to relate the effect of recombinant interleukin-10 (IL-10) on the ex vivo release of the pro-inflammatory cytokines TNF-α and interleukin-1β in LPS-stimulated leukocytes (Radwanski et al. 1998). In the first case, the site of action and the sampling site for concentration measurements of the protein therapeutic were identical, i.e., in plasma, and so the direct link model was mechanistically well justified. In the second case, the effect was dependent on the IL-10 concentration on the cell surface of leukocytes where IL-10 interacts with its target receptor. Again sampling fluid and effect site were in instant equilibrium.

■ Indirect Link PK/PD Models

The concentration-effect relationship of many protein drugs, however, cannot be described by direct link PK/PD models, but is characterized by a temporal dissociation between the time courses of plasma concentration and effect. In this case, plasma concentration maxima occur before effect maxima; effect intensity may increase despite decreasing plasma concentrations and may persist beyond the time when drug concentrations in plasma are no longer detectable. The relationship between measured concentration and observed effect follows a counterclockwise hysteresis loop. This phenomenon can either be caused by an indirect response mechanism (see next section) or by a distributional delay

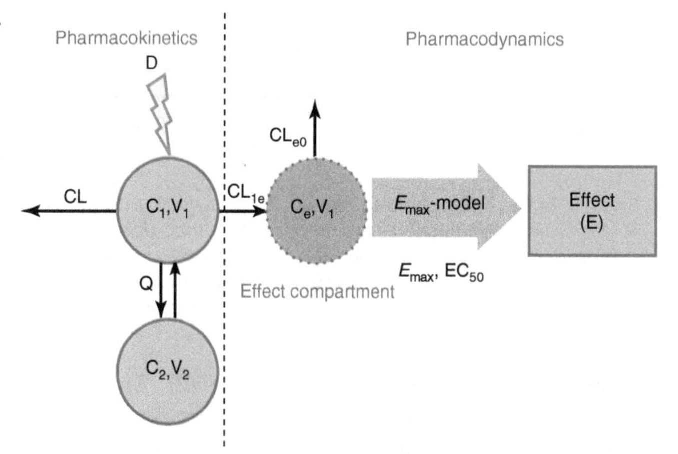

Figure 5.13 ■ Schematic of a typical indirect link PK/PD model. A hypothetical effect compartment is linked to the central compartment of a two-compartment pharmacokinetic model. The concentration in the effect compartment (C_e) drives the intensity of the pharmacodynamic effect (E) via an E_{max} relationship. CL_{1e} is the transfer clearance from the central to the effect compartment, CL_{e0} the equilibrium clearance for the effect compartment. All other PK and PD parameters are identical to those used in Fig. 5.10.

between the drug concentrations in plasma and at the effect site.

The latter one can conceptually be described by an indirect link model, which attaches a hypothetical effect compartment to a pharmacokinetic compartment model (Fig. 5.13). The effect compartment addition to the pharmacokinetic model does not account for mass balance, i.e., no actual mass transfer is implemented in the pharmacokinetic part of the PK/PD model. Instead, drug transfer with respect to the effect compartment is defined by the time course of the effect itself (Sheiner

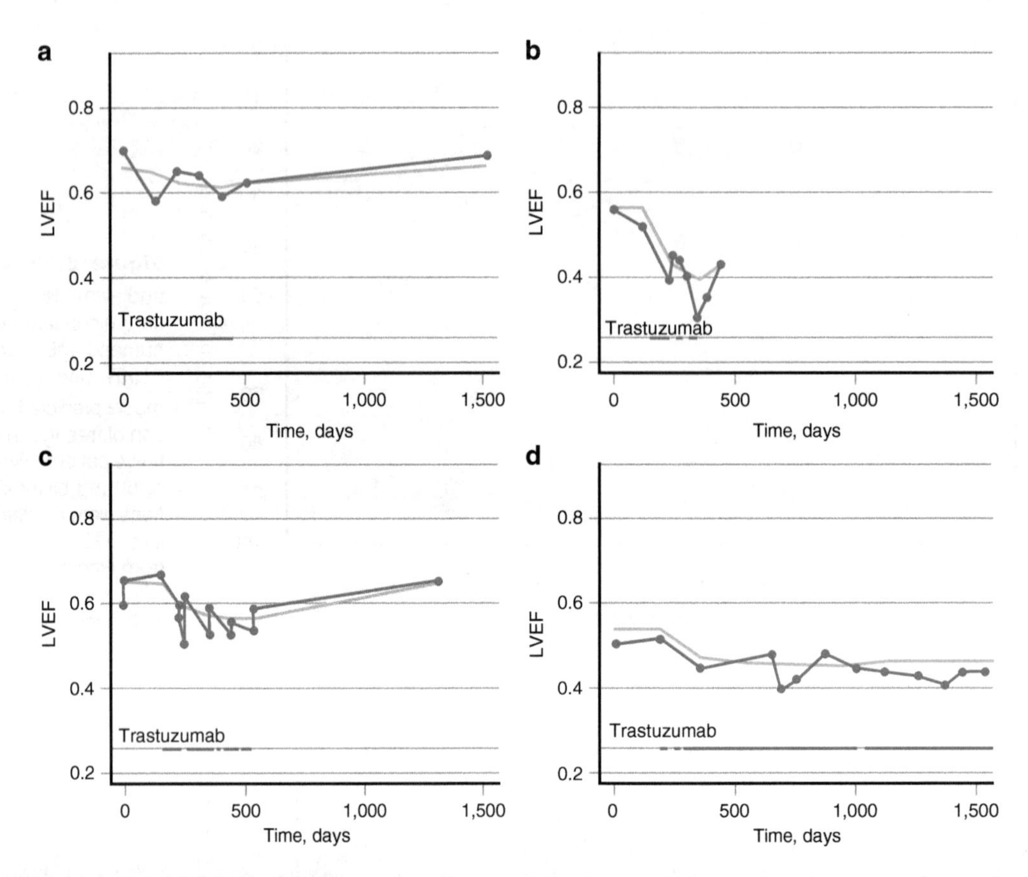

Figure 5.14 ■ Observed (*blue circles*) individual and predicted (*yellow line*) left ventricular ejection fraction (*LVEF*) vs. time (days) in four typical patients. The timing of trastuzumab dosing is indicated by horizontal red bars. Data relating to (**a**) a patient for whom long-term recovery is demonstrated, (**b**) a patient for whom no long-term recovery data were available, (**c**) a patient with LVEF values measured by concurrent use of two different clinical methodologies, and (**d**) a patient under prolonged trastuzumab treatment (From van Hasselt et al. (2011), with permission from Macmillan Publishers Ltd.).

et al. 1979; Holford and Sheiner 1982). The effect-compartment approach, however, is necessary, as the effect site can be viewed as a small part of a pharmacokinetic compartment that from a pharmacokinetic point of view cannot be distinguished from other tissues within that compartment. The concentration in the effect compartment represents the active drug concentration at the effect site that is slowly equilibrating with the plasma and is usually linked to the effect via an E_{max} model.

Although this PK/PD model is constructed with tissue distribution as the reason for the delay of the effect, the distribution clearance to the effect compartment can be interpreted differently, including other reasons of delay, such as transduction processes and secondary post-receptor events.

Pharmacotherapy with trastuzumab is associated with the occurrence of asymptomatic declines in left ventricular ejection fraction (LVEF) and the development of congestive heart failure in a small proportion of patients (van Hasselt et al. 2011). An indirect link model was applied to describe the cardiac dysfunction as an effect of trastuzumab exposure. A simplified inhibitory E_{max} model with E_{max} fixed to 1 was used to relate trastuzumab concentration in the hypothetical effect compartment of Fig. 5.13 to LVEF decline. Figure 5.14 illustrates representative time courses for LVEF under trastuzumab therapy modeled with the indirect link PKPD model. The model-derived LVEF recovery half-life after trastuzumab treatment was estimated at 49.7 days, which is consistent with a reported mean time to normalization of 1.5 months. The developed PK/PD model may be useful in developing optimal treatment and cardiac monitoring strategies for patients under trastuzumab therapy.

■ Indirect Response PK/PD Models

The effect of most protein therapeutics, however, is not mediated via a direct interaction between drug concentration at the effect site and response systems but frequently involves several transduction processes that include at their rate-limiting step the stimulation or inhibition of a physiologic process, for example the synthesis or degradation of a molecular response mediator like a hormone or cytokine. In these cases, the time courses of plasma concentration and effect are also dissociated resulting in counterclockwise hysteresis for the concentration-effect relationship, but the underlying cause is not a distributional delay as for the indirect link models, but a time-consuming indirect response mechanism (Meibohm and Derendorf 1997).

Indirect response models generally describe the effect on a representative response parameter via the dynamic equilibrium between increase or synthesis and decrease or degradation of the response, with the former being a zero-order and the latter a first-order process (Fig. 5.15). The response itself can be modulated in one of four basic variants of the models. In each variant, the synthesis or degradation process of the response is either stimulated or inhibited as a function of the effect site concentration. A stimulatory or inhibitory E_{max} model is used to describe the drug effect on the synthesis or degradation of the response (Dayneka et al. 1993; Sharma and Jusko 1998; Sun and Jusko 1999).

As indirect response models appreciate the underlying physiological events involved in the elaboration of the observed drug effect, their application is often preferred in PK/PD modeling as they have a mechanistic basis on the molecular and/or cellular level that often allows for extrapolating the model to other clinical situations.

An indirect response model was, for example, used in the evaluation of SB-240563, a humanized monoclonal antibody directed towards IL-5 in monkeys (Zia-Amirhosseini et al. 1999). IL-5 appears to play a significant role in the production, activation, and maturation of eosinophils. The delayed effect of SB-240563 on eosinophils is consistent with its mechanism of action via binding to and thus inactivation of IL-5. It was modeled using an indirect response model with inhibition of the production of response (eosinophils count) (Fig. 5.16). The obtained low EC_{50} value for reduction of circulating eosinophils combined with a long terminal half-life of the protein therapeutic of 13 days suggests the possibility of an infrequent dosing regimen for SB-240563 in the pharmacotherapy of disorders with increased eosinophil function, such as asthma.

Indirect response models were also used for the effect of growth hormone on endogenous IGF-1 concentration (Sun et al. 1999), as well as the effect of epoetin-α on two response parameters, free ferritin concentration and soluble transferrin receptor concentration (Bressolle et al. 1997). Similarly, a modified indirect response model was used to relate the concentration of the humanized anti-factor IX antibody SB-249417 to factor IX activity in Cynomolgus monkeys as well as humans (Benincosa et al. 2000; Chow et al. 2002). The drug effect in this model was introduced by interrupting the natural degradation of factor IX by sequestration of factor IX by the antibody.

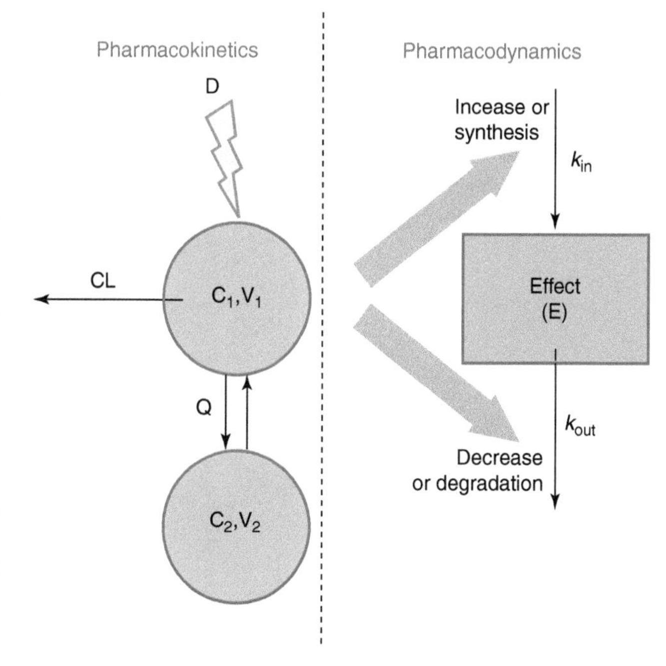

Indirect response model subtypes

Subtype I: inhibition of synthesis (k_{in})

$$\frac{dE}{dt} = k_{in} \cdot \left(1 - \frac{C_1}{EC_{50} + C_1}\right) - k_{out} \cdot E$$

Subtype II: inhibition of degradation (k_{out})

$$\frac{dE}{dt} = k_{in} - k_{out} \cdot \left(1 - \frac{C_1}{EC_{50} + C_1}\right) \cdot E$$

Subtype III: stimulation of synthesis (k_{in})

$$\frac{dE}{dt} = k_{in} \cdot \left(1 + \frac{E_{max} \cdot C_1}{EC_{50} + C_1}\right) - k_{out} \cdot E$$

Subtype IV: stimulation of degradation (k_{out})

$$\frac{dE}{dt} = k_{in} - k_{out} \cdot \left(1 + \frac{E_{max} \cdot C_1}{EC_{50} + C_1}\right) \cdot E$$

Figure 5.15 ■ Schematic of a typical indirect response PK/PD model. The effect measure (E) is maintained by a dynamic equilibrium between an increase or synthesis and a decrease or degradation process. The former is modeled by a zero-order process with rate constant k_{in}, the latter by a first-order process with rate constant k_{out}. Thus, the rate of change in effect (dE/dt) is expressed as the difference between synthesis rate (k_{in}) and degradation rate (k_{out} times E). Drug concentration (C_1) can stimulate or inhibit the synthesis or the degradation process for the effect (E) via an E_{max} relationship using one of four subtypes (model I, II, III or IV) of the indirect response model. The pharmacokinetic model and all other PK and PD parameters are identical to those used in Fig. 5.10.

■ Cell Life Span Models

A sizable number of protein therapeutics exert their pharmacologic effect through direct or indirect modulation of blood and/or immune cell types. For these kinds of therapeutics, cell life span models have been proven useful to capture their exposure-response relationship and describe and predict drug effects (Perez-Ruixo et al. 2005). Cell life span models are mechanism-based, physiologic PK/PD models that are established based on the sequential maturation and life span-driven cell turnover of their affected cell types and progenitor cell populations. Cell life span model is especially widely used for characterizing the dose-concentration-effect relationships of hematopoietic growth factors aimed at modifying erythropoiesis, granulopoiesis, or thrombopoiesis (Perez-Ruixo et al. 2005; Agoram et al. 2006). The fixed physiologic time span for the maturation of precursor cells is the major reason for the prolonged delay between drug administration and the observed response, i.e., change in the cell count in peripheral blood. Cell life span models accommodate this sequential maturation of several precursor cell populations at fixed physiologic time intervals by a series of transit compartments linked via first- or zero-order processes with a common transfer rate constant.

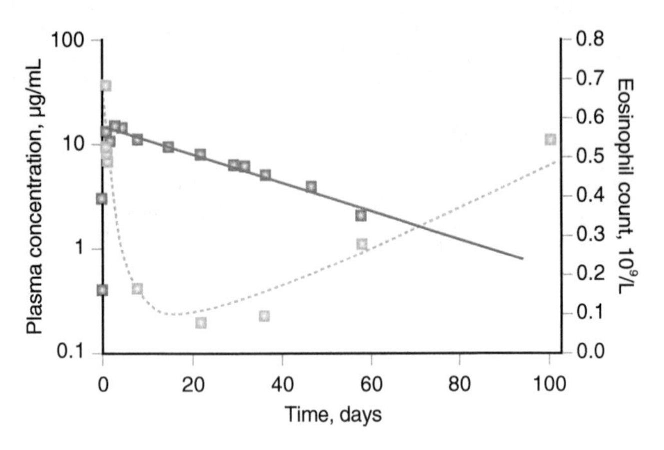

Figure 5.16 ■ Model-predicted and observed plasma concentration (observed, *blue circles*; predicted, *solid blue line*) and eosinophil count (observed, *orange squares*; predicted, *dashed orange line*) following SC administration of 1 mg/kg of the anti-IL-5 humanized monoclonal antibody SB-240563 in a Cynomolgus monkey. A mechanism-based indirect response PK/PD model was used to describe eosinophil count as a function of SB-240563 plasma concentration. The reduction in eosinophil count in peripheral blood (as effect *E*) was modeled as a reduction of the recruitment of eosinophils from the bone, i.e., an inhibition of the production rate k_{in} using the indirect response model of subtype I (see Fig. 5.15) (Zia-Amirhosseini et al. 1999; with permission from American Society for Pharmacology and Experimental Therapeutics).

A cell life span model was, for example, used to describe the effect of a multiple dose regimen of erythropoietin (EPO) 600 IU/kg given once weekly by SC injection (Ramakrishnan et al. 2004). The process of erythropoiesis and the applied PK/PD approach including a cell life span model are depicted in Figs. 5.17 and 5.18, respectively. EPO is known to stimulate the production and release of reticulocytes from the bone marrow. The EPO effect was modeled as stimulation of the maturation of two progenitor cell populations (P1 and P2 in Fig. 5.17), including also a feedback inhibition between erythrocyte count and progenitor proliferation. Development and turnover of the subsequent populations of reticulocytes and erythrocytes was modeled, taking into account their life spans as listed in Fig. 5.17. The hemoglobin concentration as pharmacodynamic target parameter was calculated from erythrocyte and reticulocyte counts and hemoglobin content per cell. Figure 5.19 shows the resulting time courses in reticulocyte count, erythrocyte count, and hemoglobin concentration.

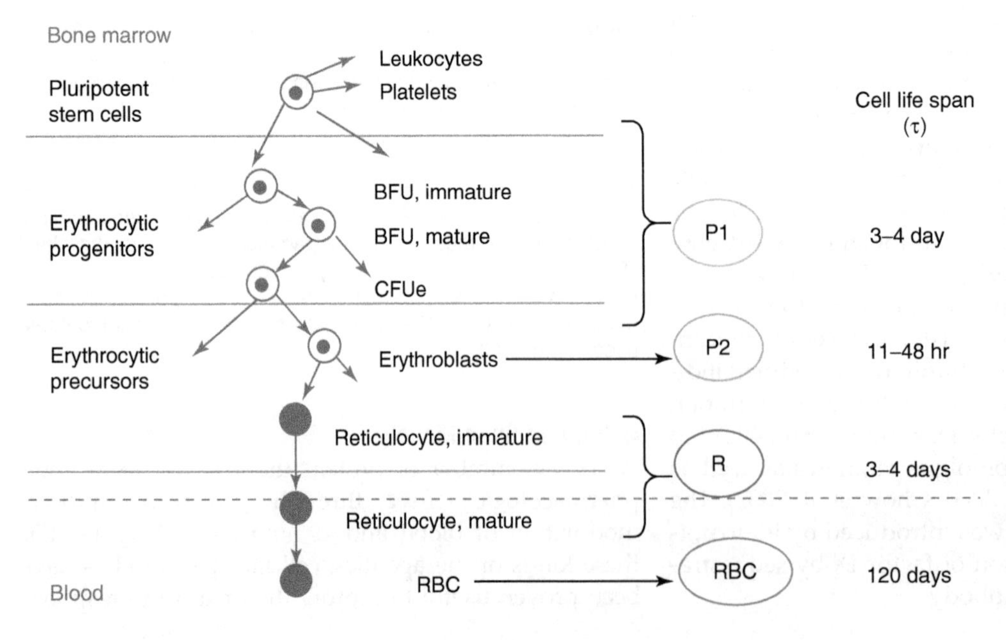

Figure 5.17 ■ Process of erythropoiesis. Erythropoietin stimulates the proliferation and differentiation of the erythrocyte progenitors (*BFU* burst-forming unit erythroid, *CFUe* colony-forming unit erythroid) as well as the erythroblasts in the bone marrow. The life spans (τ) of the various cell populations are indicated at the right (From Ramakrishnan et al. (2004), with permission from John Wiley & Sons, Inc. Copyright American College of Clinical Pharmacology 2004.).

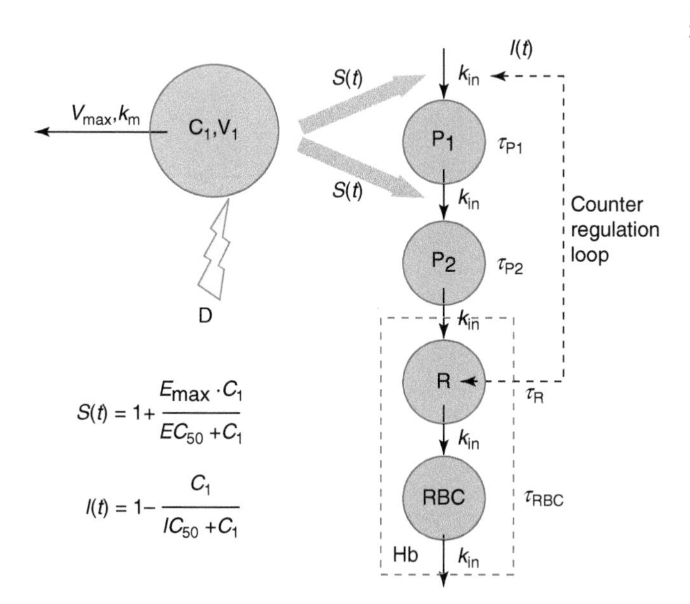

Figure 5.18 ■ A PK/PD model describing the disposition of recombinant human erythropoietin and effects on reticulocyte count, red blood cell count, and hemoglobin concentration. The PK model is a one-compartment model with Michaelis-Menten type elimination (K_m, V_{max}) from the central compartment. The PD model is a cell life span model with four sequential cell compartments, representing erythroid progenitor cells (P_1), erythroblasts (P_2), reticulocytes (R), and red blood cells (RBC). τ_{P1}, τ_{P2}, τ_R, and τ_{RBC} are the corresponding cell life spans, k_{in} the common zero-order transfer rate between cell compartments. The target parameter hemoglobin in the blood (Hb) is calculated from the reticulocyte and red blood cell count and the hemoglobin content per cell. The effect of erythropoietin is modeled as a stimulation of the production of both precursor cell populations (P_1 and P_2) in the bone marrow with the stimulation function $S(t)$. E_{max} is the maximum possible stimulation of reticulocyte production by erythropoietin, EC_{50} the plasma concentration of erythropoietin that produced half-maximum stimulation. A counter-regulatory feedback loop represents the feedback inhibition of reticulocytes on their own production by reducing the production rate of cells in the P_1 compartment via the inhibitory function $I(t)$. IC_{50} is the reticulocyte count that produced half of complete inhibition (Modified from Ramakrishnan et al. (2004)).

fertilization protocols. LH-RH antagonists suppress

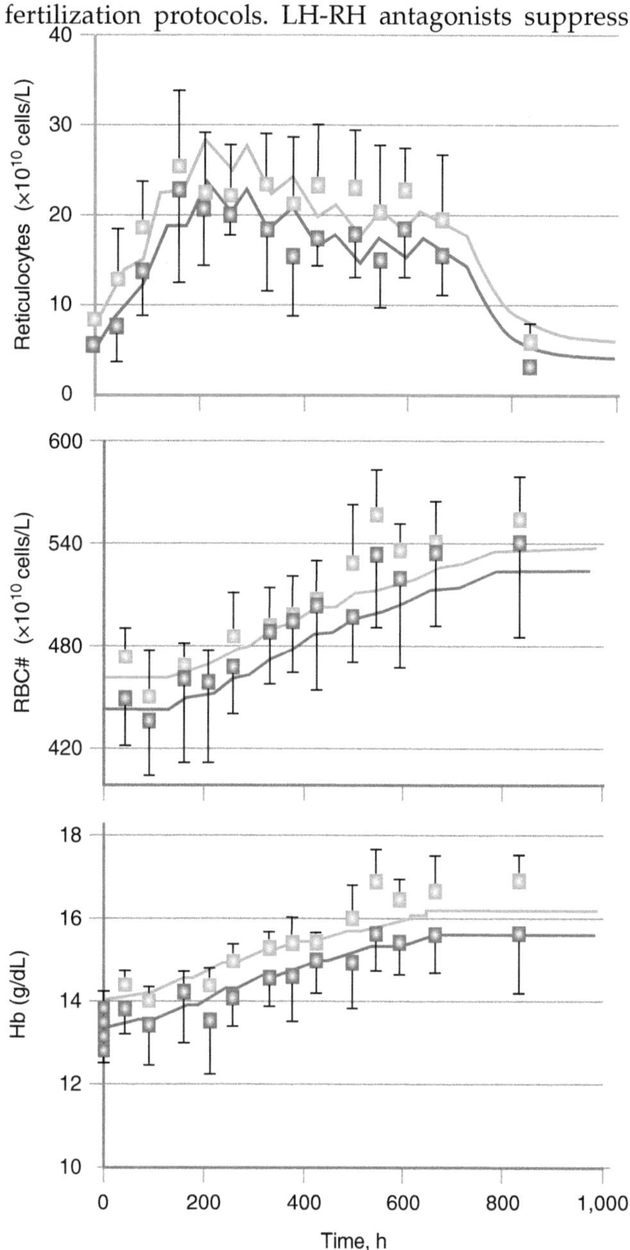

Figure 5.19 ■ Reticulocyte, red blood cell (*RBC*), and hemoglobin (*Hb*) time courses after multiple SC dosing of 600 IU/kg/weeks recombinant human erythropoietin. *Orange* and *blue squares* represent data for males and females, whereas the *orange* and *blue lines* for the reticulocytes are model fittings. The *lines* in the RBC and Hb panels are the predictions using the model-fitted curves for the reticulocytes and the life span parameters (From Ramakrishnan et al. (2004), with permission from John Wiley & Sons, Inc. Copyright American College of Clinical Pharmacology 2004.).

■ Complex Response Models

Since the effect of most protein therapeutics is mediated via complex regulatory physiologic processes including feedback mechanisms and/or tolerance phenomena, some PK/PD models that have been described for protein drugs are much more sophisticated than the four classes of models previously discussed.

One example of such a complex modeling approach has been developed for the therapeutic effects of the luteinizing hormone-releasing hormone (LH-RH) antagonist cetrorelix (Nagaraja et al. 2000, 2003; Pechstein et al. 2000). Cetrorelix is used for the prevention of premature ovulation in women undergoing controlled ovarian stimulation in in vitro fertilization protocols. LH-RH antagonists suppress the LH levels and delay the occurrence of the preovulatory LH surge, and this delay is thought to be responsible for postponing ovulation. The suppression of LH was modeled in the PK/PD approach with an indirect-response model approach directly linked

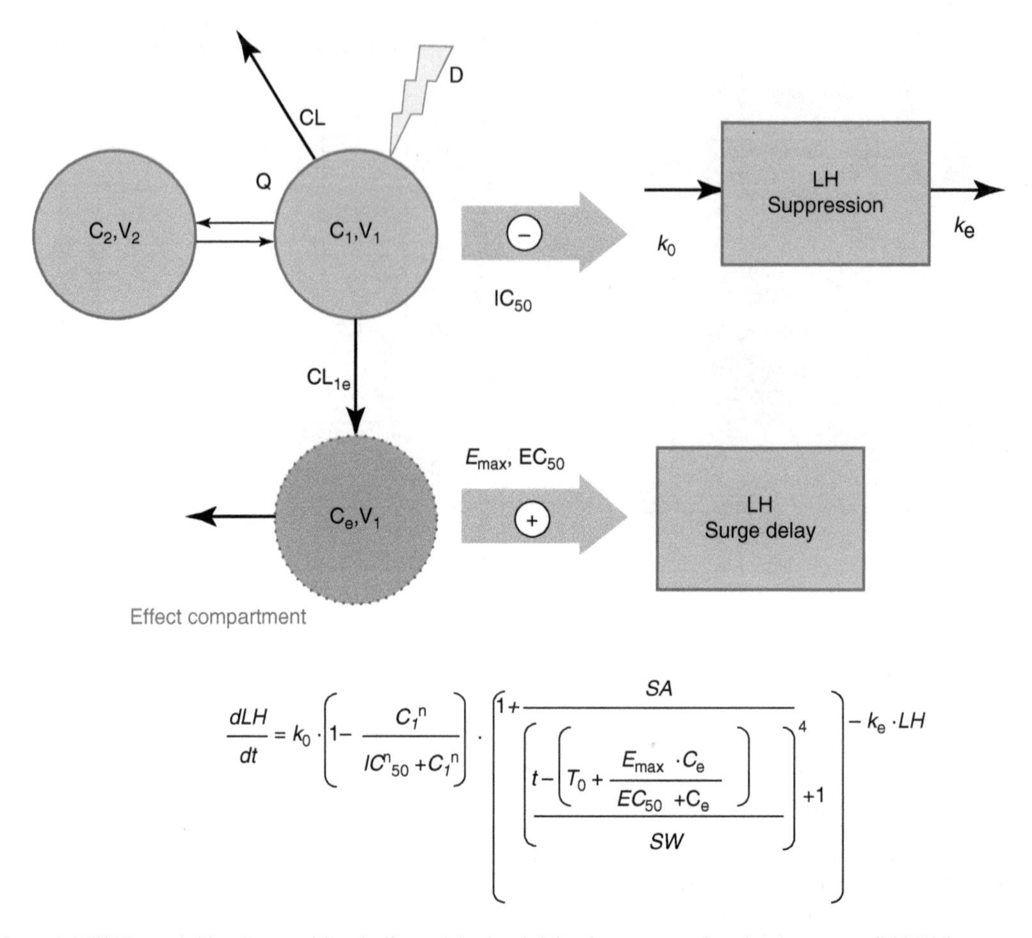

Figure 5.20 ■ A PK/PD model for the combined effect of the luteinizing hormone-releasing hormone (LH-RH) antagonist cetrorelix on luteinizing hormone (*LH*) suppression and delay of the LH surge. The PK model is a two-compartment model identical to the model described in Fig. 5.10. The PD model consists of two components: an indirect response model of subtype I to model the suppression of LH by cetrorelix and an indirect link model that models the delay in LH surge as a function of the cetrorelix concentration in a hypothetical effect compartment. Both PD model components are combined in the provided mathematical expression that described the rate of change in LH concentration (dLH/d*t*) as a function of both processes. LH is the LH concentration, k_0 and k_e are the zero-order production rate and first-order elimination rate constants for LH at baseline, C_1 and C_e are the cetrorelix concentrations in plasma and a hypothetical effect compartment, respectively, SA is the LH surge amplitude, *t* is the study time in terms of cycle day, T_0 is the study time in terms of cycle day at which the peak occurs under baseline conditions, SW is the width of the peak in time units, IC_{50} is the cetrorelix concentration that suppresses LH levels by 50 %, n is the Hill factor fixed at 2, E_{max} is the maximum delay in LH surge, and EC_{50} is the cetrorelix concentrations that produces half of E_{max}. Baseline data analysis indicated that the slope of the surge peak and SW were best fixed at values of 4 and 24 h, respectively. Other PK and PD parameters are identical to those used in Fig. 5.12 (Modified from Nagaraja et al. (2000)).

to cetrorelix plasma concentrations (Fig. 5.20) (Nagaraja et al. 2003). The shift in LH surge was linked to cetrorelix concentration with a simple E_{max} function via a hypothetical effect compartment to account for a delay in response via complex signal transduction steps of unknown mechanism of action. Figure 5.21 shows the application of this PK/PD model to characterize the LH suppression and LH surge delay after subcutaneous administration of cetrorelix to groups of 12 women at different dose levels. The analysis revealed a marked dose–response relationship for the LH surge and thus predictability of drug response to cetrorelix (Nagaraja et al. 2000).

Another example for a complex PK/PD model is the cytokinetic model used to describe the effect of pegfilgrastim on the granulocyte count in peripheral blood (Roskos et al. 2006; Yang 2006). Pegfilgrastim is a PEGylated form of the human granulocyte colony-stimulating factor (G-CSF) analogue filgrastim. Pegfilgrastim, like filgrastim and G-CSF, stimulates the activation, proliferation, and differentiation of neutrophil progenitor cells and enhances the functions of mature neutrophils (Roskos et al. 2006). Pegfilgrastim is mainly used as supportive care to ameliorate and enhance recovery from neutropenia secondary to cancer chemotherapy regimens. As already discussed in

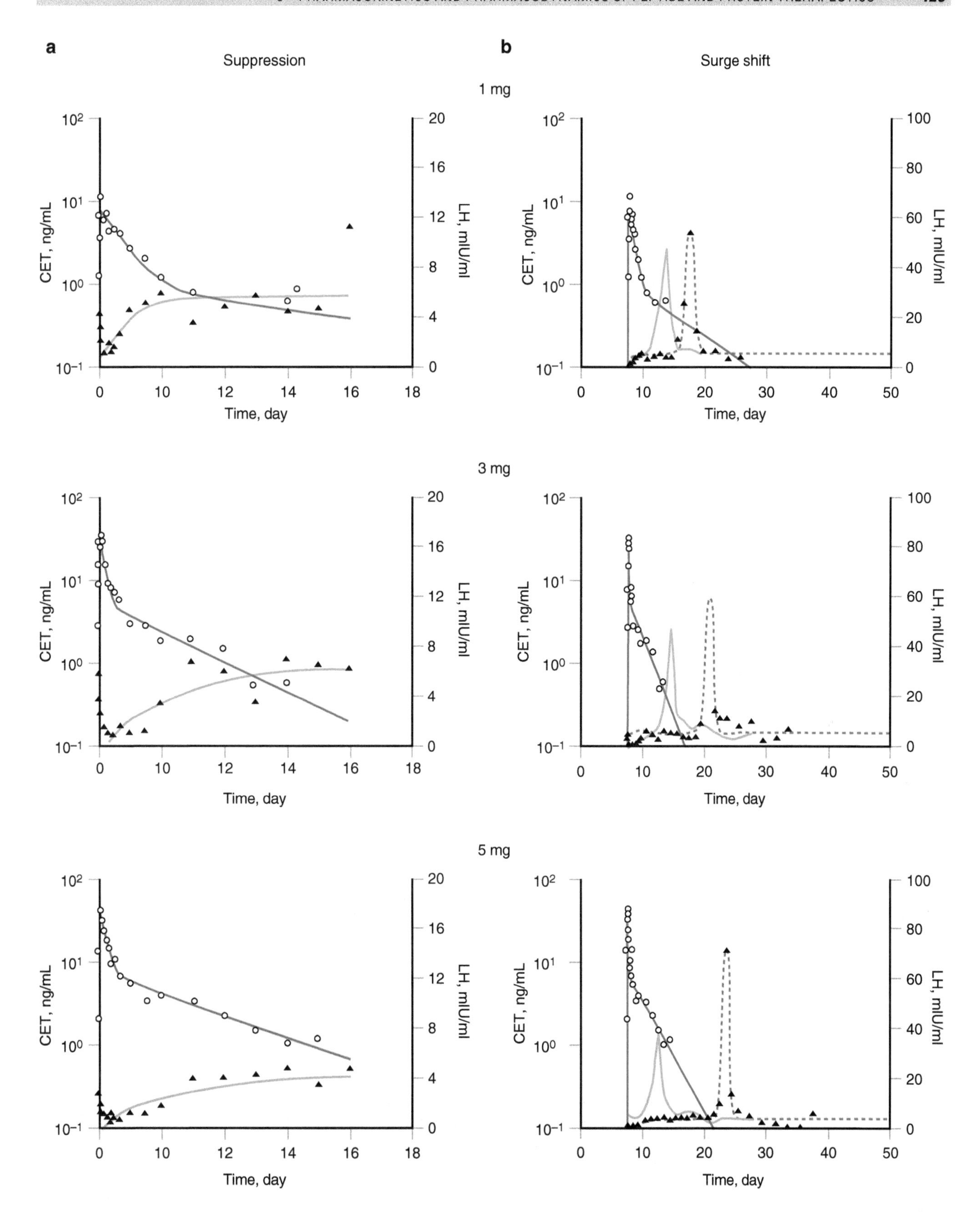

Figure 5.21 ■ Pharmacokinetic and pharmacodynamic relationship between cetrorelix (CET; ○) and LH concentrations (▲) after single doses of 1, 3, and 5 mg cetrorelix in representative subjects. Cetrorelix and LH concentrations were modeled using the PK/PD model presented in Fig. 5.20. *Left panel*: LH suppression. *Right panel*: LH suppression and LH surge profiles. The *solid blue line* represents the model-fitted cetrorelix concentration, the *dashed purple line* the model-fitted LH concentration, and the *solid orange line* in the *right panels* the pretreatment LH profile (not fitted). The cetrorelix-dependent delay in LH surge is visible as the rightward shift of the LH surge profile under cetrorelix therapy compared to the respective pretreatment LH profile (From Nagaraja et al. (2000), with permission from Macmillan Publishers Ltd.).

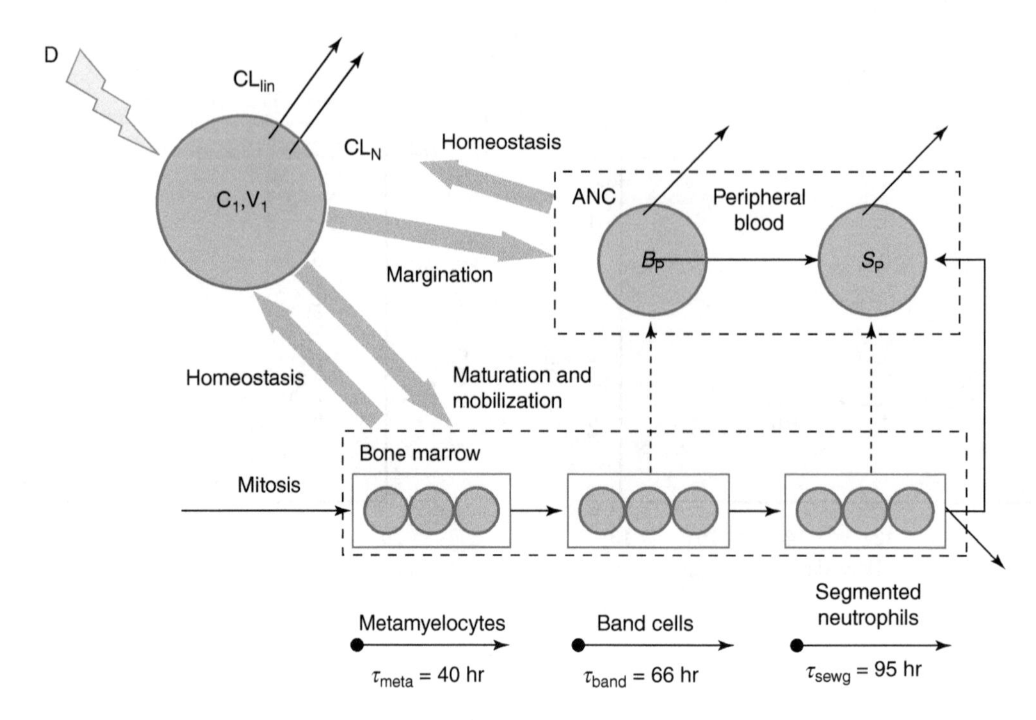

Figure 5.22 ■ A PK/PD model describing the granulopoietic effects of pegfilgrastim. The PK model is a one-compartment model with two parallel elimination pathways, a first-order elimination process (CL_{lin}) and a neutrophil-mediated elimination process (CL_N). C_1 and V_1 are the concentrations in the PK compartment and the corresponding volume of distribution. The PD model is a cytokinetic model similar to the cell life span model in Fig. 5.18. Three maturation stages of neutrophils and their respective life spans (t_{meta}, t_{band}, t_{seg}) are included in the model, metamyelocytes, band cells, and segmented neutrophils. Each maturation stage is modeled by three sequential transit compartments. Serum concentrations of pegfilgrastim stimulate mitosis and mobilization of band cells and segmented neutrophils in bone marrow, decrease maturation times for postmitotic cells in marrow, and affect margination of the peripheral blood band cell (B_P) and segmented neutrophil (S_P) populations, the sum of which is the total absolute neutrophil count (*ANC*). Changes in neutrophil counts in peripheral blood provide feedback regulation of pegfilgrastim clearance (Modified from Roskos et al. (2006)).

the section on PEGylation, pegfilgrastim follows target-mediated drug disposition with saturable receptor-mediated endocytosis by neutrophils as major elimination pathway (CL_N) and a parallel first-order process as minor elimination pathway (CL_{lin}; Fig. 5.22). The clearance for the receptor-mediated pathway is determined by the absolute neutrophil count (ANC), the sum of the peripheral blood band cell (B_p), and segmented neutrophil (S_p) populations.

A maturation-structured cytokinetic model of granulopoiesis was established to describe the relationship between pegfilgrastim serum concentration and neutrophil count (Fig. 5.22). The starting point is the production of metamyelocytes from mitotic precursors. Subsequent maturation stages are captured as band cells and segmented neutrophils in the bone marrow. Each maturation stage is modeled by three sequential transit compartments. Pegfilgrastim concentrations are assumed to increase ANC by stimulating mitosis and mobilization of band cells and segmented neutrophils from the bone marrow into the systemic circulation. Pegfilgrastim also promotes rapid margination of peripheral blood neutrophils, i.e., adhesion to blood vessels; this effect is modeled as an expansion of neutrophil dilution volume.

Figure 5.23 shows observed and modeled pegfilgrastim concentration time and ANC time profiles after escalating single SC dose administration of pegfilgrastim. The presented PK/PD model for pegfilgrastim allowed determining its EC_{50} for the effect on ANC. Based on this EC_{50} value and the obtained pegfilgrastim plasma concentrations, it was concluded that a 100 µg/kg dose was sufficient to reach the maximum therapeutic effect of pegfilgrastim on ANC (Roskos et al. 2006; Yang 2006).

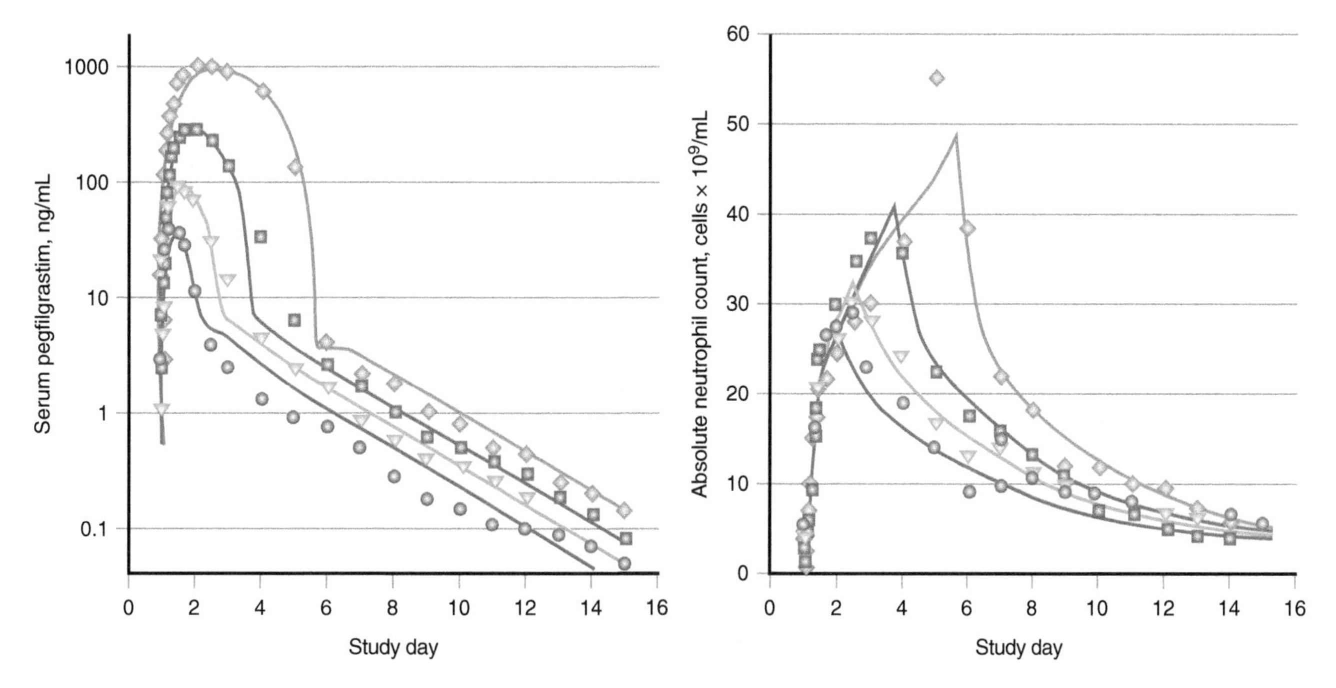

Figure 5.23 ■ Pegfilgrastim concentration time course and absolute neutrophil count (ANC) time profiles in healthy subjects after a single SC administration of 30, 60, 100, and 300 μg/kg pegfilgrastim (n=8/dose group). Measured data are presented by symbols as mean ± SEM. Lines represent modeled time courses based on the cytokinetic PK/PD model presented in Fig. 5.22 (From Roskos et al. (2006), with permission from John Wiley & Sons, Inc. Copyright American College of Clinical Pharmacology 2006.).

CONCLUSION

The pharmacokinetic and pharmacodynamic characteristics of peptides and proteins form the basis for their therapeutic application. Appreciation of the pharmacokinetic and pharmacodynamic differences between therapeutic biologics and traditional small-molecule drugs will empower the drug development scientist as well as the healthcare provider to handle, evaluate, and apply these compounds in an optimal fashion during the drug development process as well as during applied pharmacotherapy. Rationale, scientifically based drug development and pharmacotherapy based on the use of pharmacokinetic and pharmacodynamic concepts will undoubtedly propel the success and future of protein therapeutics and might ultimately contribute to provide the novel medications that may serve as the key for the aspired "personalized medicine" in the healthcare systems of the future (Nagle et al. 2003).

SELF-ASSESSMENT QUESTIONS

■ Questions

1. What are the major elimination pathways for protein drugs after administration?
2. Which pathway of absorption is rather unique for proteins after SC injection?
3. What is the role of plasma binding proteins for natural proteins?
4. How do the sugar groups on glycoproteins influence hepatic elimination of these glycoproteins?
5. In which direction might elimination clearance of a protein drug change when antibodies against the protein are produced after chronic dosing with the protein drug? Why?
6. What is the major driving force for the transport of proteins from the vascular to the extravascular space?
7. Why are protein therapeutics generally not active upon oral administration?
8. Many protein therapeutics exhibit Michaelis-Menten type, saturable elimination kinetics. What are the underlying mechanisms for this pharmacokinetic behavior?
9. Explain counterclockwise hysteresis in plasma concentration-effect plots.
10. Why is mechanism-based PK/PD modeling a preferred modeling approach for protein therapeutics?

■ Answers

1. Proteolysis, glomerular filtration followed by intraluminal metabolism or tubular reabsorption with intracellular lysosomal degradation, renal peritubular absorption followed by catabolism,

receptor-mediated endocytosis followed by metabolism in the liver and possibly other organs and tissues.
2. Biodistribution from the injection site into the lymphatic system.
3. Plasma proteins may act as circulating reservoirs for the proteins that are their ligands. Consequently, the protein ligands may be protected from elimination and distribution. In some cases, protein binding may protect the organism from undesirable, acute effects; in other cases, receptor binding may be facilitated by the binding protein.
4. In some cases, the sugar groups are recognized by hepatic receptors (e.g., galactose by the galactose receptor), facilitating receptor-mediated uptake and metabolism. In other cases, sugar chains and terminal sugar groups (e.g., terminal sialic acid residues) may shield the protein from binding to receptors and hepatic uptake.
5. Clearance may increase or decrease by forming antibody-protein complexes. A decrease of clearance occurs when the antibody-protein complex is eliminated slower than free protein. An increase of clearance occurs when the protein-antibody complex is eliminated more rapidly than the unbound protein, such as when reticuloendothelial uptake is stimulated by the complex.
6. Protein extravasation, i.e., transport from the blood or vascular space to the interstitial tissue space, is predominantly mediated by fluid convection. Protein molecules follow the fluid flux from the vascular space through pores between adjacent cells into the interstitial space. Drainage of the interstitial space through the lymphatic system allows protein therapeutics to distribute back into the vascular space.
7. The gastrointestinal mucosa is a major absorption barrier for hydrophilic macromolecule such as proteins. In addition, peptide and protein therapeutics are degraded by the extensive peptidase and protease activity in the gastrointestinal tract. Both processes minimize the oral bioavailability of protein therapeutics.
8. Receptor-mediated endocytosis is the most frequent cause of nonlinear pharmacokinetics in protein therapeutics. Its occurrence becomes even more prominent if the protein therapeutic undergoes target-mediated drug disposition, i.e., if the receptor-mediated endocytosis is mediated via the pharmacologic target of the protein therapeutic. As the binding to the target is usually of high affinity, and the protein therapeutic is often dosed to saturate the majority of the available target receptors for maximum pharmacologic efficacy, saturation of the associated receptor-mediated endocytosis as elimination pathway is frequently encountered.
9. Counterclockwise hysteresis is an indication of the indirect nature of the effects seen for many protein drugs. It can be explained by delays between the appearance of drug in plasma and the appearance of the pharmacodynamic response. The underlying cause may either be a distributional delay between the drug concentrations in plasma and at the effect site (modeled with an indirect link PK/PD model) or by time-consuming post-receptor events that cause a delay between the drug-receptor interaction and the observed drug effect, for example, the effect on a physiologic measure or endogenous substance.
10. Protein therapeutics are often classified as "targeted therapies" where the drug compound acts on one specific, well-defined response pathway. This well-documented knowledge on the mechanism of action can relatively easily be translated into a mechanism-based PK/PD modeling approach that incorporates the major physiological processes relevant for the pharmacologic effect. The advantage of mechanism-based as compared to empirical PK/PD modeling is that mechanism-based models are usually more robust and allow more reliable simulations beyond the actually measured data.

REFERENCES

Agoram B, Heatherington AC, Gastonguay MR (2006) Development and evaluation of a population pharmacokinetic-pharmacodynamic model of darbepoetin alfa in patients with nonmyeloid malignancies undergoing multicycle chemotherapy. AAPS J 8(3):E552–E563

Allon M, Kleinman K, Walczyk M et al (2002) Pharmacokinetics and pharmacodynamics of darbepoetin alfa and epoetin in patients undergoing dialysis. Clin Pharmacol Ther 72(5):546–555

Anderson PM, Sorenson MA (1994) Effects of route and formulation on clinical pharmacokinetics of interleukin-2. Clin Pharmacokinet 27(1):19–31

Bauer RJ, Gibbons JA, Bell DP, Luo ZP, Young JD (1994) Nonlinear pharmacokinetics of recombinant human macrophage colony- stimulating factor (M-CSF) in rats. J Pharmacol Exp Ther 268(1):152–158

Baxter LT, Zhu H, Mackensen DG, Jain RK (1994) Physiologically based pharmacokinetic model for specific and nonspecific monoclonal antibodies and fragments in normal tissues and human tumor xenografts in nude mice. Cancer Res 54(6):1517–1528

Benincosa LJ, Chow FS, Tobia LP et al (2000) Pharmacokinetics and pharmacodynamics of a humanized monoclonal antibody to factor IX in cynomolgus monkeys. J Pharmacol Exp Ther 292(2):810–816

Bennett HP, McMartin C (1978) Peptide hormones and their analogues: distribution, clearance from the circulation, and inactivation in vivo. Pharmacol Rev 30(3):247–292

Boxenbaum H (1982) Interspecies scaling, allometry, physiological time, and the ground plan of pharmacokinetics. J Pharmacokinet Biopharm 10(2):201–227
Bressolle F, Audran M, Gareau R, Pham TN, Gomeni R (1997) Comparison of a direct and indirect population pharmacodynamic model: application to recombinant human erythropoietin in athletes. J Pharmacokinet Biopharm 25(3):263–275
Bu G, Williams S, Strickland DK, Schwartz AL (1992) Low density lipoprotein receptor-related protein/alpha 2-macroglobulin receptor is an hepatic receptor for tissue-type plasminogen activator. Proc Natl Acad Sci U S A 89(16):7427–7431
Caliceti P, Veronese FM (2003) Pharmacokinetic and biodistribution properties of poly(ethylene glycol)-protein conjugates. Adv Drug Deliv Rev 55(10):1261–1277
Carone FA, Peterson DR (1980) Hydrolysis and transport of small peptides by the proximal tubule. Am J Physiol 238(3):F151–F158
Carone FA, Peterson DR, Flouret G (1982) Renal tubular processing of small peptide hormones. J Lab Clin Med 100(1):1–14
Chanson P, Timsit J, Harris AG (1993) Clinical pharmacokinetics of octreotide. Therapeutic applications in patients with pituitary tumours. Clin Pharmacokinet 25(5):375–391
Chiang J, Gloff CA, Yoshizawa CN, Williams GJ (1993) Pharmacokinetics of recombinant human interferon-beta ser in healthy volunteers and its effect on serum neopterin. Pharm Res 10(4):567–572
Chirmule N, Jawa V, Meibohm B (2012) Immunogenicity to therapeutic proteins: impact on PK/PD and efficacy. AAPS J 14(2):296–302
Chow FS, Benincosa LJ, Sheth SB et al (2002) Pharmacokinetic and pharmacodynamic modeling of humanized anti-factor IX antibody (SB 249417) in humans. Clin Pharmacol Ther 71(4):235–245
Colburn W (1991) Peptide, peptoid, and protein pharmacokinetics/pharmacodynamics. In: Garzone P, Colburn W, Mokotoff M (eds) Petides, peptoids, and proteins, 3rd edn. Harvey Whitney Books, Cincinnati, pp 94–115
Cumming DA (1991) Glycosylation of recombinant protein therapeutics: control and functional implications. Glycobiology 1(2):115–130
Daniel H, Herget M (1997) Cellular and molecular mechanisms of renal peptide transport. Am J Physiol 273(1 Pt 2):F1–F8
Dayneka NL, Garg V, Jusko WJ (1993) Comparison of four basic models of indirect pharmacodynamic responses. J Pharmocokinet Biopharm 21(4):457–478
Dedrick RL (1973) Animal scale-up. J Pharmacokinet Biopharm 1(5):435–461
Deen WM, Lazzara MJ, Myers BD (2001) Structural determinants of glomerular permeability. Am J Physiol Renal Physiol 281(4):F579–F596
Deng R, Iyer S, Theil FP et al (2011) Projecting human pharmacokinetics of therapeutic antibodies from nonclinical data: what have we learned? MAbs 3(1):61–66
Derendorf H, Meibohm B (1999) Modeling of pharmacokinetic/pharmacodynamic (PK/PD) relationships: concepts and perspectives. Pharm Res 16(2):176–185
Dirks NL, Meibohm B (2010) Population pharmacokinetics of therapeutic monoclonal antibodies. Clin Pharmacokinet 49(10):633–659
Edwards A, Daniels BS, Deen WM (1999) Ultrastructural model for size selectivity in glomerular filtration. Am J Physiol 276(6 Pt 2):F892–F902
Eppler SM, Combs DL, Henry TD et al (2002) A target-mediated model to describe the pharmacokinetics and hemodynamic effects of recombinant human vascular endothelial growth factor in humans. Clin Pharmacol Ther 72(1):20–32
Fasano A (1998) Novel approaches for oral delivery of macromolecules. J Pharm Sci 87(11):1351–1356
Flessner MF, Lofthouse J, el Zakaria R (1997) In vivo diffusion of immunoglobulin G in muscle: effects of binding, solute exclusion, and lymphatic removal. Am J Physiol 273(6 Pt 2):H2783–H2793
Graham ML (2003) Pegaspargase: a review of clinical studies. Adv Drug Deliv Rev 55(10):1293–1302
Handelsman DJ, Swerdloff RS (1986) Pharmacokinetics of gonadotropin-releasing hormone and its analogs. Endocr Rev 7(1):95–105
Herbst RS, Langer CJ (2002) Epidermal growth factor receptors as a target for cancer treatment: the emerging role of IMC-C225 in the treatment of lung and head and neck cancers. Semin Oncol 29(1 Suppl 4):27–36
Holford NH, Sheiner LB (1982) Kinetics of pharmacologic response. Pharmacol Ther 16(2):143–166
Inui K, Terada T, Masuda S, Saito H (2000) Physiological and pharmacological implications of peptide transporters, PEPT1 and PEPT2. Nephrol Dial Transplant 15(Suppl 6):11–13
Ismair MG, Stieger B, Cattori V et al (2001) Hepatic uptake of cholecystokinin octapeptide by organic anion-transporting polypeptides OATP4 and OATP8 of rat and human liver. Gastroenterology 121(5):1185–1190
Jin F, Krzyzanski W (2004) Pharmacokinetic model of target-mediated disposition of thrombopoietin. AAPS PharmSci 6(1):E9
Johnson V, Maack T (1977) Renal extraction, filtration, absorption, and catabolism of growth hormone. Am J Physiol 233(3):F185–F196
Kageyama S, Yamamoto H, Nakazawa H et al (2002) Pharmacokinetics and pharmacodynamics of AJW200, a humanized monoclonal antibody to von Willebrand factor, in monkeys. Arterioscler Thromb Vasc Biol 22(1):187–192
Kaufman JS, Reda DJ, Fye CL et al (1998) Subcutaneous compared with intravenous epoetin in patients receiving hemodialysis. Department of Veterans Affairs Cooperative Study Group on Erythropoietin in Hemodialysis Patients. N Engl J Med 339(9):578–583
Khor SP, McCarthy K, DuPont M, Murray K, Timony G (2000) Pharmacokinetics, pharmacodynamics, allometry, and dose selection of rPSGL-Ig for phase I trial. J Pharmacol Exp Ther 293(2):618–624
Kim J, Hayton WL, Robinson JM, Anderson CL (2007) Kinetics of FcRn-mediated recycling of IgG and albumin in human: pathophysiology and therapeutic implications using a simplified mechanism-based model. Clin Immunol 122(2):146–155

Kobayashi H, Shirakawa K, Kawamoto S et al (2002) Rapid accumulation and internalization of radiolabeled herceptin in an inflammatory breast cancer xenograft with vasculogenic mimicry predicted by the contrast-enhanced dynamic MRI with the macromolecular contrast agent G6-(1B4M-Gd)(256). Cancer Res 62(3): 860–866

Kompella U, Lee V (1991) Pharmacokinetics of peptide and protein drugs. In: Lee V (ed) Peptide and protein drug delivery. Marcel Dekker, New York, pp 391–484

Kontermann R (2012) Therapeutic proteins: strategies to modulate their plasma half-lives. Wiley, Weinheim

Krogsgaard Thomsen M, Friis C, Sehested Hansen B et al (1994) Studies on the renal kinetics of growth hormone (GH) and on the GH receptor and related effects in animals. J Pediatr Endocrinol 7(2):93–105

Kuwabara T, Uchimura T, Kobayashi H, Kobayashi S, Sugiyama Y (1995) Receptor-mediated clearance of G-CSF derivative nartograstim in bone marrow of rats. Am J Physiol 269(1 Pt 1):E1–E9

Lee HJ (2002) Protein drug oral delivery: the recent progress. Arch Pharm Res 25(5):572–584

Lee H, Kimko HC, Rogge M et al (2003) Population pharmacokinetic and pharmacodynamic modeling of etanercept using logistic regression analysis. Clin Pharmacol Ther 73(4):348–365

Lesko LJ (2007) Paving the critical path: how can clinical pharmacology help achieve the vision? Clin Pharmacol Ther 81(2):170–177

Lesko LJ, Rowland M, Peck CC, Blaschke TF (2000) Optimizing the science of drug development: opportunities for better candidate selection and accelerated evaluation in humans. J Clin Pharmacol 40(8): 803–814

Levy G (1986) Kinetics of drug action: an overview. J Allergy Clin Immunol 78(4 Pt 2):754–761

Levy G (1994) Mechanism-based pharmacodynamic modeling. Clin Pharmacol Ther 56(4):356–358

Limothai W, Meibohm B (2011) Effect of dose on the apparent bioavailability of therapeutic proteins that undergo target-mediated drug disposition. AAPS J 13(S2)

Maack T, Park C, Camargo M (1985) Renal filtration, transport and metabolism of proteins. In: Seldin D, Giebisch G (eds) The kidney. Raven Press, New York, pp 1773–1803

Macdougall IC, Gray SJ, Elston O et al (1999) Pharmacokinetics of novel erythropoiesis stimulating protein compared with epoetin alfa in dialysis patients. J Am Soc Nephrol 10(11):2392–2395

Mach H, Gregory SM, Mackiewicz A et al (2011) Electrostatic interactions of monoclonal antibodies with subcutaneous tissue. Ther Deliv 2(6):727–736

Mager DE (2006) Target-mediated drug disposition and dynamics. Biochem Pharmacol 72(1):1–10

Mager DE, Wyska E, Jusko WJ (2003) Diversity of mechanism-based pharmacodynamic models. Drug Metab Dispos 31(5):510–518

Mahato RI, Narang AS, Thoma L, Miller DD (2003) Emerging trends in oral delivery of peptide and protein drugs. Crit Rev Ther Drug Carrier Syst 20(2–3):153–214

Mahmood I (2002) Interspecies scaling: predicting oral clearance in humans. Am J Ther 9(1):35–42

Mahmood I, Balian JD (1999) The pharmacokinetic principles behind scaling from preclinical results to phase I protocols. Clin Pharmacokinet 36(1):1–11

Marks DL, Gores GJ, LaRusso NF (1995) Hepatic processing of peptides. In: Taylor MD, Amidon GL (eds) Peptide-based drug design: controlling transport and metabolism. American Chemical Society, Washington, DC, pp 221–248

McMahon HT, Boucrot E (2011) Molecular mechanism and physiological functions of clathrin-mediated endocytosis. Nat Rev Mol Cell Biol 12(8):517–533

McMartin C (1992) Pharmacokinetics of peptides and proteins: opportunities and challenges. Adv Drug Res 22:39–106

Meibohm B (2004) Pharmacokinetics of protein- and nucleotide-based drugs. In: Mahato RI (ed) Biomaterials for delivery and targeting of proteins and nucleic acids. CRC Press, Boca Raton, pp 275–294

Meibohm B, Derendorf H (1994) Pharmacokinetics and pharmacodynamics of biotech drugs. In: Kayser O, Muller R (eds) Pharmaceutical biotechnology: drug discovery and clinical applications. Wiley, Weinheim, pp 141–166

Meibohm B, Derendorf H (1997) Basic concepts of pharmacokinetic/pharmacodynamic (PK/PD) modelling. Int J Clin Pharmacol Ther 35(10):401–413

Meibohm B, Derendorf H (2002) Pharmacokinetic/pharmacodynamic studies in drug product development. J Pharm Sci 91(1):18–31

Meibohm B, Derendorf H (2003) Pharmacokinetics and pharmacodynamics of biotech drugs. In: Muller R, Kayser O (eds) Applications of pharmaceutical biotechnology. Wiley-VCH, Weinheim

Meibohm B, Zhou H (2012) Characterizing the impact of renal impairment on the clinical pharmacology of biologics. J Clin Pharmacol 52(1 Suppl):54S–62S

Meijer D, Ziegler K (1993) Biological barriers to protein delivery. Plenum Press, New York

Mohler M, Cook J, Lewis D et al (1993) Altered pharmacokinetics of recombinant human deoxyribonuclease in rats due to the presence of a binding protein. Drug Metab Dispos 21(1):71–75

Molineux G (2003) Pegylation: engineering improved biopharmaceuticals for oncology. Pharmacotherapy 23(8 Pt 2):3S–8S

Montero-Julian FA, Klein B, Gautherot E, Brailly H (1995) Pharmacokinetic study of anti-interleukin-6 (IL-6) therapy with monoclonal antibodies: enhancement of IL-6 clearance by cocktails of anti-IL-6 antibodies. Blood 85(4):917–924

Mordenti J, Chen SA, Moore JA, Ferraiolo BL, Green JD (1991) Interspecies scaling of clearance and volume of distribution data for five therapeutic proteins. Pharm Res 8(11):1351–1359

Mould DR, Davis CB, Minthorn EA et al (1999) A population pharmacokinetic-pharmacodynamic analysis of single doses of clenoliximab in patients with rheumatoid arthritis. Clin Pharmacol Ther 66(3):246–257

Nagaraja NV, Pechstein B, Erb K et al (2000) Pharmacokinetic and pharmacodynamic modeling of cetrorelix, an

LH-RH antagonist, after subcutaneous administration in healthy premenopausal women. Clin Pharmacol Ther 68(6):617–625
Nagaraja NV, Pechstein B, Erb K et al (2003) Pharmacokinetic/pharmacodynamic modeling of luteinizing hormone (LH) suppression and LH surge delay by cetrorelix after single and multiple doses in healthy premenopausal women. J Clin Pharmacol 43(3):243–251
Nagle T, Berg C, Nassr R, Pang K (2003) The further evolution of biotech. Nat Rev Drug Discov 2(1):75–79
Nielsen S, Nielsen JT, Christensen EI (1987) Luminal and basolateral uptake of insulin in isolated, perfused, proximal tubules. Am J Physiol 253(5 Pt 2):F857–F867
Pauletti GM, Gangwar S, Siahaan TJ, Jeffrey A, Borchardt RT (1997) Improvement of oral peptide bioavailability: peptidomimetics and prodrug strategies. Adv Drug Deliv Rev 27(2–3):235–256
Pechstein B, Nagaraja NV, Hermann R et al (2000) Pharmacokinetic-pharmacodynamic modeling of testosterone and luteinizing hormone suppression by cetrorelix in healthy volunteers. J Clin Pharmacol 40(3):266–274
Peck CC, Barr WH, Benet LZ et al (1994) Opportunities for integration of pharmacokinetics, pharmacodynamics, and toxicokinetics in rational drug development. J Clin Pharmacol 34(2):111–119
Perez-Ruixo JJ, Kimko HC, Chow AT et al (2005) Population cell life span models for effects of drugs following indirect mechanisms of action. J Pharmacokinet Pharmacodyn 32(5–6):767–793
Periti P, Mazzei T, Mini E (2002) Clinical pharmacokinetics of depot leuprorelin. Clin Pharmacokinet 41(7):485–504
Perrier D, Mayersohn M (1982) Noncompartmental determination of the steady-state volume of distribution for any mode of administration. J Pharm Sci 71(3):372–373
Piscitelli SC, Reiss WG, Figg WD, Petros WP (1997) Pharmacokinetic studies with recombinant cytokines. Scientific issues and practical considerations. Clin Pharmacokinet 32(5):368–381
Porter CJ, Charman SA (2000) Lymphatic transport of proteins after subcutaneous administration. J Pharm Sci 89(3):297–310
Rabkin R, Ryan MP, Duckworth WC (1984) The renal metabolism of insulin. Diabetologia 27(3):351–357
Racine-Poon A, Botta L, Chang TW et al (1997) Efficacy, pharmacodynamics, and pharmacokinetics of CGP 51901, an anti- immunoglobulin E chimeric monoclonal antibody, in patients with seasonal allergic rhinitis. Clin Pharmacol Ther 62(6):675–690
Radwanski E, Chakraborty A, Van Wart S et al (1998) Pharmacokinetics and leukocyte responses of recombinant human interleukin-10. Pharm Res 15(12):1895–1901
Ramakrishnan R, Cheung WK, Wacholtz MC, Minton N, Jusko WJ (2004) Pharmacokinetic and pharmacodynamic modeling of recombinant human erythropoietin after single and multiple doses in healthy volunteers. J Clin Pharmacol 44(9):991–1002
Reddy ST, Berk DA, Jain RK, Swartz MA (2006) A sensitive in vivo model for quantifying interstitial convective transport of injected macromolecules and nanoparticles. J Appl Physiol 101(4):1162–1169
Richter WF, Gallati H, Schiller CD (1999) Animal pharmacokinetics of the tumor necrosis factor receptor-immunoglobulin fusion protein lenercept and their extrapolation to humans. Drug Metab Dispos 27(1):21–25
Richter WF, Bhansali SG, Morris ME (2012) Mechanistic determinants of biotherapeutics absorption following SC administration. AAPS J 14(3):559–570
Roopenian DC, Akilesh S (2007) FcRn: the neonatal Fc receptor comes of age. Nat Rev Immunol 7(9):715–725
Roskos LK, Lum P, Lockbaum P, Schwab G, Yang BB (2006) Pharmacokinetic/pharmacodynamic modeling of pegfilgrastim in healthy subjects. J Clin Pharmacol 46(7):747–757
Schomburg A, Kirchner H, Atzpodien J (1993) Renal, metabolic, and hemodynamic side-effects of interleukin-2 and/or interferon alpha: evidence of a risk/benefit advantage of subcutaneous therapy. J Cancer Res Clin Oncol 119(12):745–755
Sharma A, Jusko W (1998) Characteristics of indirect pharmacodynamic models and applications to clinical drug responses. Br J Clin Pharmacol 45:229–239
Sheiner LB, Steimer JL (2000) Pharmacokinetic/pharmacodynamic modeling in drug development. Annu Rev Pharmacol Toxicol 40:67–95
Sheiner LB, Stanski DR, Vozeh S, Miller RD, Ham J (1979) Simultaneous modeling of pharmacokinetics and pharmacodynamics: application to d-tubocurarine. Clin Pharmacol Ther 25(3):358–371
Shen WC (2003) Oral peptide and protein delivery: unfulfilled promises? Drug Discov Today 8(14):607–608
Smedsrod B, Einarsson M (1990) Clearance of tissue plasminogen activator by mannose and galactose receptors in the liver. Thromb Haemost 63(1):60–66
Straughn AB (1982) Model-independent steady-state volume of distribution. J Pharm Sci 71(5):597–598
Straughn AB (2006) Limitations of noncompartmental pharmacokinetic analysis of biotech drugs. In: Meibohm B (ed) Pharmacokinetics and pharmacodynamics of biotech drugs. Wiley, Weinheim, pp 181–188
Strickland DK, Kounnas MZ, Argraves WS (1995) LDL receptor-related protein: a multiligand receptor for lipoprotein and proteinase catabolism. FASEB J 9(10):890–898
Sun YN, Jusko WJ (1999) Role of baseline parameters in determining indirect pharmacodynamic responses. J Pharm Sci 88(10):987–990
Sun YN, Lee HJ, Almon RR, Jusko WJ (1999) A pharmacokinetic/pharmacodynamic model for recombinant human growth hormone effects on induction of insulin-like growth factor I in monkeys. J Pharmacol Exp Ther 289(3):1523–1532
Supersaxo A, Hein W, Gallati H, Steffen H (1988) Recombinant human interferon alpha-2a: delivery to lymphoid tissue by selected modes of application. Pharm Res 5(8):472–476
Supersaxo A, Hein WR, Steffen H (1990) Effect of molecular weight on the lymphatic absorption of water-soluble compounds following subcutaneous administration. Pharm Res 7(2):167–169

Suryawanshi S, Zhang L, Pfister M, Meibohm B (2010) The current role of model-based drug development. Expert Opin Drug Discov 5(4):311–321
Tabrizi M, Roskos LK (2006) Exposure-response relationships for therapeutic biologics. In: Meibohm B (ed) Pharmacokinetics and pharmacodynamics of biotech drugs. Wiley, Weinheim, pp 295–330
Takagi A, Masuda H, Takakura Y, Hashida M (1995) Disposition characteristics of recombinant human interleukin-11 after a bolus intravenous administration in mice. J Pharmacol Exp Ther 275(2):537–543
Taki Y, Sakane T, Nadai T et al (1998) First-pass metabolism of peptide drugs in rat perfused liver. J Pharm Pharmacol 50(9):1013–1018
Tang L, Meibohm B (2006) Pharmacokinetics of peptides and proteins. In: Meibohm B (ed) Pharmacokinetics and pharmacodynamics of biotech drugs. Wiley, Weinheim, pp 17–44
Tang L, Persky AM, Hochhaus G, Meibohm B (2004) Pharmacokinetic aspects of biotechnology products. J Pharm Sci 93(9):2184–2204
Tanswell P, Modi N, Combs D, Danays T (2002) Pharmacokinetics and pharmacodynamics of tenecteplase in fibrinolytic therapy of acute myocardial infarction. Clin Pharmacokinet 41(15):1229–1245
Tokuda Y, Watanabe T, Omuro Y et al (1999) Dose escalation and pharmacokinetic study of a humanized anti-HER2 monoclonal antibody in patients with HER2/neu-overexpressing metastatic breast cancer. Br J Cancer 81(8):1419–1425
Toon S (1996) The relevance of pharmacokinetics in the development of biotechnology products. Eur J Drug Metab Pharmacokinet 21(2):93–103
van Hasselt JG, Boekhout AH, Beijnen JH, Schellens JH, Huitema AD (2011) Population pharmacokinetic-pharmacodynamic analysis of trastuzumab-associated cardiotoxicity. Clin Pharmacol Ther 90(1):126–132
Veng-Pedersen P, Gillespie W (1984) Mean residence time in peripheral tissue: a linear disposition parameter useful for evaluating a drug's tissue distribution. J Pharmacokinet Biopharm 12(5):535–543
Veronese FM, Caliceti P (2006) Custom-tailored pharmacokinetics and pharmacodynamics via chemical modifications of biotech drugs. In: Meibohm B (ed) Pharmacokinetics and pharmacodynamics of boptech drugs. Wiley, Weinheim, pp 271–294
Walsh S, Shah A, Mond J (2003) Improved pharmacokinetics and reduced antibody reactivity of lysostaphin conjugated to polyethylene glycol. Antimicrob Agents Chemother 47(2):554–558
Wang W, Wang EQ, Balthasar JP (2008) Monoclonal antibody pharmacokinetics and pharmacodynamics. Clin Pharmacol Ther 84(5):548–558
Wills RJ, Ferraiolo BL (1992) The role of pharmacokinetics in the development of biotechnologically derived agents. Clin Pharmacokinet 23(6):406–414
Yang BB (2006) Integration of pharmacokinetics and pharmacodynamics into the drug development of pegfilgrastim, a pegylated protein. In: Meibohm B (ed) Pharmacokinetics and pharmacodynamics of biotech drugs. Wiley, Weinheim, pp 373–394
Zamboni WC (2003) Pharmacokinetics of pegfilgrastim. Pharmacotherapy 23(8 Pt 2):9S–14S
Zhang Y, Meibohm B (2012) Pharmacokinetics and pharmacodynamics and therapeutic peptides and proteins. In: Kayzer O, Warzecha H (eds) Pharmaceutical biotechnology: drug discovery and clinical applications. Wiley-VCH, Weinheim, pp 337–368
Zhang L, Pfister M, Meibohm B (2008) Concepts and challenges in quantitative pharmacology and model-based drug development. AAPS J 10(4):552–559
Zia-Amirhosseini P, Minthorn E, Benincosa LJ et al (1999) Pharmacokinetics and pharmacodynamics of SB-240563, a humanized monoclonal antibody directed to human interleukin-5, in monkeys. J Pharmacol Exp Ther 291(3):1060–1067
Ziegler K, Polzin G, Frimmer M (1988) Hepatocellular uptake of cyclosporin A by simple diffusion. Biochim Biophys Acta 938(1):44–50
Zito SW (1997) Pharmaceutical biotechnology: a programmed text. Technomic Pub. Co, Lancaster

FURTHER READING

General Pharmacokinetics and Pharmacodynamics

Atkinson A, Abernethy D, Daniels C, Dedrick R, Markey S (2006) Principles of clinical pharmacology. Academic, San Diego
Bonate PL (2011) Pharmacokinetic-pharmacodynamic modeling and simulation. Springer, New York
Derendorf H, Meibohm B (1999) Modeling of pharmacokinetic/pharmacodynamic (PK/PD) relationships: concepts and perspectives. Pharm Res 16(2):176–185
Gabrielsson J, Hjorth S (2012) Quantitative pharmacology. Swedish Academy of Pharmaceutical Sciences, Stockholm
Gibaldi M, Perrier D (1982) Pharmacokinetics. Marcel Dekker Inc., New York
Holford NH, Sheiner LB (1982) Kinetics of pharmacologic response. Pharmacol Ther 16(2):143–166
Rowland M, Tozer TN (2011) Clinical pharmacokinetics and pharmacodynamics: concepts and applications. Lippincott Williams & Wilkins, Baltimore

Pharmacokinetics and Pharmacodynamics of Peptides and Proteins

Baumann A (2006) Early development of therapeutic biologics – pharmacokinetics. Curr Drug Metab 7:15–21
Ferraiolo BL, Mohler MA, Gloff CA (1992) Protein pharmacokinetics and metabolism. Plenum Press, New York
Kontermann R (2012) Therapeutic proteins: strategies to modulate their plasma half-lives. Wiley, Weinheim
Meibohm B (2006) Pharmacokinetics and pharmacodynamics of biotech drugs. Wiley, Weinheim
Tang L, Persky AM, Hochhaus G, Meibohm B (2004) Pharmacokinetic aspects of biotechnology products. J Pharm Sci 93(9):2184–2204

6 Immunogenicity of Therapeutic Proteins

Huub Schellekens and Wim Jiskoot

INTRODUCTION

The era of the medical application of proteins started at the end of the nineteenth century when animal sera were introduced for the treatment of serious complications of infections such as diphtheria and tetanus. The high doses used, the general lack of quality controls and a regulatory system, and the impurity of the preparations led to many serious and sometimes even fatal side effects. Many of the problems were caused by the strong immune response these foreign proteins induced, especially when readministered. People who had been treated in general had a warning in their passports or identification cards to alert physicians for a possible anaphylactic reaction after rechallenge with an antiserum. Also serum sickness caused by deposits of antigen-antibody complexes was a common complication of the serum therapy.

Also the porcine and bovine insulins introduced after 1922 induced antibodies in many patients. This was again explained by the animal origin of the products, although over the years the immunogenicity became less because of improvements in the production methods and increasing purity.

In the second half of the twentieth century, a number of human proteins from natural sources were introduced such as plasma-derived clotting factors and growth hormone produced from pituary glands of cadavers. These products were given mainly to children with an innate deficiency who therefore lacked the natural immune tolerance. Therefore, their immune response was also interpreted as a response to foreign proteins. The correlation between the factor VIII gene defect and level of deficiency with the immune response in hemophilia patients confirmed this explanation.

Thus, until the advent of recombinant DNA technology, the immunological response to therapeutic proteins could be explained as a classical immune response comparable to that of a vaccine.

THE NEW PARADIGM

In 1982 human insulin was marketed as the first recombinant DNA-derived protein for human use. Since then dozens of recombinant proteins have been introduced and some of these products such as the interferons and the epoetins are among the most widely used drugs in the world. And, although these proteins were developed as close copies of human endogenous proteins, nearly all these proteins induce antibodies, sometimes even in a majority of patients (Table 6.1). In addition, most of these products are used in patients who do not have an innate deficiency and can be assumed to have immune tolerance to the protein.

The initial assumption was that the production by recombinant technology in nonhuman host cells and the downstream processing modified the proteins and the immunological response was the classical response to a foreign protein. However, according to the current opinion, the antibody response to human homologues is based on circumventing B-cell tolerance. This phenomenon is not yet completely understood but is clearly different from the vaccine type of reaction seen with foreign proteins (Sauerborn et al. 2010).

The clinical manifestations of both types of reaction are very different. The vaccine-type response occurs within weeks and sometimes a single injection is sufficient to induce a substantial antibody response. In general high levels of neutralizing antibodies are

H. Schellekens, Ph.D. (✉)
Department of Pharmaceutics Utrecht,
Utrecht Institute for Pharmaceutical Sciences,
Utrecht University, David de Wiedgebouw, room 3.68,
Universiteitsweg 99, Utrecht 3584 CG, The Netherlands
e-mail: h.schellekens@uu.nl

W. Jiskoot, Ph.D.
Division of Drug Delivery Technology,
Leiden/Amsterdam Center for Drug Research,
Leiden University, Einsteinweg 55,
Leiden 2333 CC, The Netherlands
e-mail: w.jiskoot@lacdr.leidenuniv.nl

D.J.A. Crommelin, R.D. Sindelar, and B. Meibohm (eds.), *Pharmaceutical Biotechnology*,
DOI 10.1007/978-1-4614-6486-0_6, © Springer Science+Business Media New York 2013

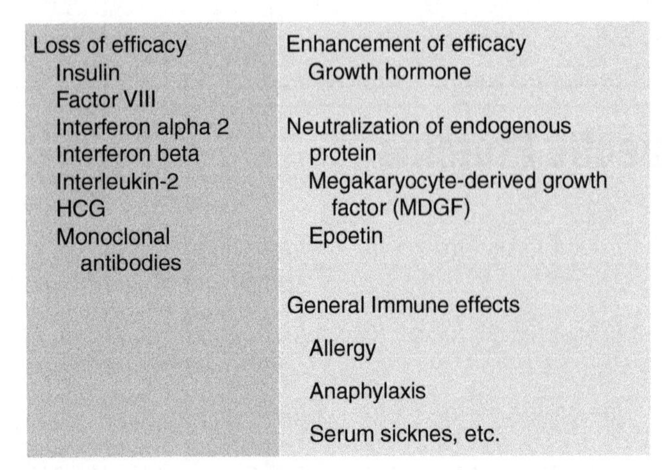

Loss of efficacy	Enhancement of efficacy
Insulin	Growth hormone
Factor VIII	
Interferon alpha 2	Neutralization of endogenous protein
Interferon beta	Megakaryocyte-derived growth factor (MDGF)
Interleukin-2	Epoetin
HCG	
Monoclonal antibodies	General Immune effects
	Allergy
	Anaphylaxis
	Serum sicknes, etc.

Table 6.1 ■ Non-exhaustive list of recombinant proteins showing immune reactions upon administration.

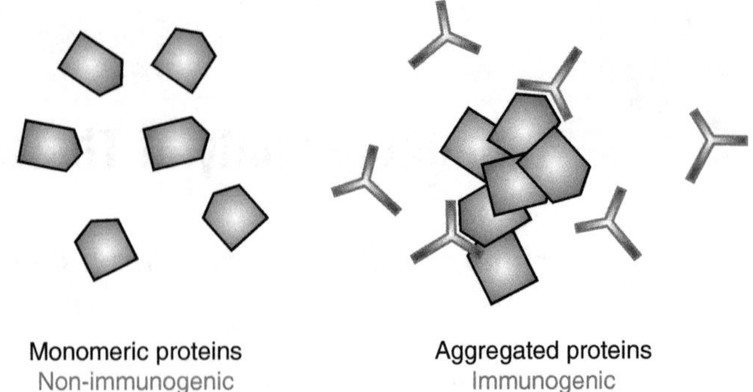

Figure 6.1 ■ Dogma: protein aggregates are immunogenic.

induced and a rechallenge leads to a booster reaction, indicating a memory response.

However, circumventing B-cell tolerance takes in general 6–12 months of chronic treatment and often only leads to the production of binding antibodies with no biological effect. The antibodies often disappear shortly after treatment has been stopped and sometimes even during treatment. This response also appears to have no memory, because rechallenging patients in whom the antibody levels have declined does not induce a response. The lack of a memory response has also been shown in the relevant immune-tolerant transgenic mouse models.

THE IMMUNOLOGICAL RESPONSE

The therapeutic proteins currently available cover the whole spectrum, from completely foreign like bacterial-derived asparaginase to completely human-like interferon-α-2b and everything in between. The foreign protein elicits antibodies by the classical pathway which includes ingestion and cleaving of the proteins into peptides by macrophages and dendritic cells, presentation of peptides by the MHC-II system and activation of B cells and boosting, and affinity maturation and isotype switching of the B cells by helper T cells (Stevanovic 2005). Furthermore, memory B cells are induced (see Chap. 22 for details).

It is much less clear how B-cell tolerance is circumvented. There are always autoreactive B cells present. When the receptor on these B cells meets its epitope on the protein in solution, this interaction does not lead to activation. When these B cells meet their epitope in a regularly repeated form, then the B-cell receptor oligomerizes and the cell is activated and starts to divide and produce antibodies. So B cells can recognize three-dimensional repeated protein structures as has been shown experimentally in a number of studies. The explanation why is based on evolution. The only naturally occurring, narrowly spaced repeated protein structures are found on the surface of viruses and some bacterial structures (Bachmann et al. 1993). Apparently, the B-cell system was also selected for its potential to respond to microbial structures independent of the system which discriminates self from nonself.

This explanation of nonself-independent response by repeated protein structures fits nicely with aggregates being recognized as the main driver of an autoreactive response by human therapeutic proteins, because in protein aggregates, certain structures are also presented in a repeated form (Fig. 6.1) (Moore and Leppert 1980).

Thus, the initial activation of B cells by aggregates can be explained and also how these cells start with producing IgM. It is not known how the isotype switching from IgM to IgG occurs. Some studies suggest that aggregates after reacting with the B-cell receptor are internalized. And by internalizing, the B cells become helper cells and start to produce cytokines which will activate other B cells. Others claim that helper T cells are involved. However, studies to show the presence of specific T-cell activity in patients producing antibodies to human therapeutic proteins have failed. In addition, a T-cell-independent mechanism is suggested by the lack of any association with HLA type in some studies and the absence of memory. However, in studies in immune-tolerant transgenic mice, depletion of T cell inhibited the response to aggregated proteins capable of circumventing B-cell tolerance.

FACTORS INFLUENCING ANTIBODY FORMATION TO THERAPEUTIC PROTEINS

Figure 6.2 depicts the different factors that influence immunogenicity (Schellekens 2002; Hermeling et al. 2004).

■ Structural Factors

The degree of nonself and presence of aggregates are the initial triggers of an antibody response to a therapeutic protein. The degree of nonself necessary to induce a vaccine-type response is highly dependent on the protein involved and the site of the divergence from the natural sequence of the endogenous protein. For insulin there are single mutations which lead to a new epitope and an antibody response, while other mutations have no influence at all. Consensus interferon-α in which more than 10 % of the amino acids diverge from the nearest naturally occurring interferon-α subtype shows not more immunogenicity than the interferon-α-2 homologue.

Glycosylation is another important structural factor for the immunogenicity of therapeutic proteins. There is little evidence that modified glycosylation, e.g., by expressing human glycoproteins in plant cells or other nonhuman eukaryotic hosts, may induce an immunogenic response. However, the level of glycosylation has a clear effect. Interferon-β produced in *E. coli* (non-glycosylated) is much more immunogenic than the same product produced in mammalian cells. The explanation is the lower solubility causing aggregation in the non-glycosylated E. coli product.

Antigenicity does not equal immunogenicity (Fig. 6.3). A protein, peptide, or glycan structure with a high affinity for antibodies may not be able to induce antibodies at all. Glycosylation is reported to influence antigenicity by antibody binding. Non-glycosylated epoetin was reported to have a higher affinity to antibodies than epoetin with a normal level of glycosylation; the opposite is true for epoetin engineered to have extra glycosylation. In most subjects who had a hypersensitivity reaction to cetuximab, IgE antibodies against cetuximab were present in serum before therapy (Arnold and Misbah 2008). These antibodies were specific for galactose-α-1,3-galactose and were apparently induced by a microbial organism.

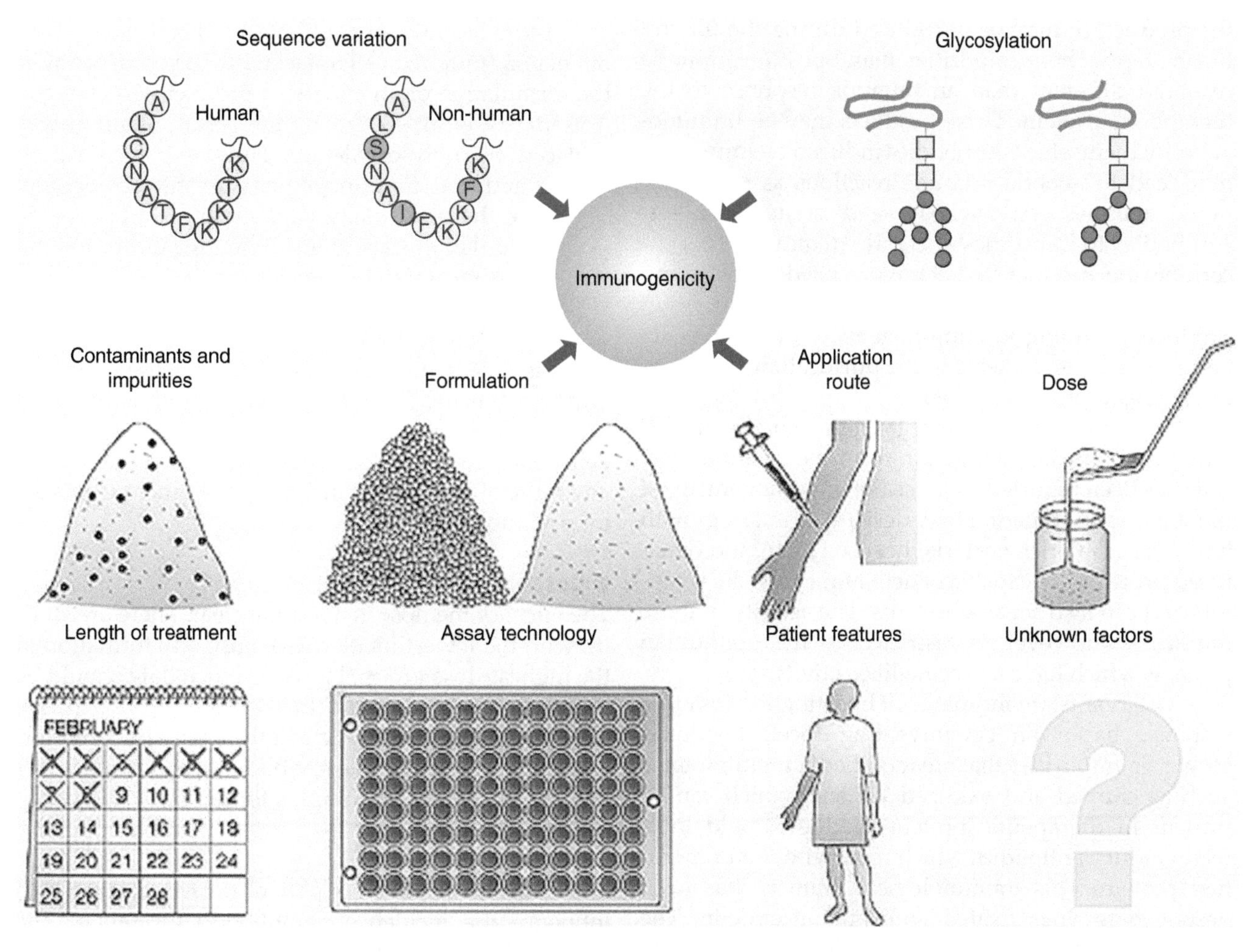

Figure 6.2 ■ Factors relevant in immunogenicity.

The difference between antigenicity and immunogenicity

- **antigenicity:** the capability of a substance to interact with (pre-existing) components of the immune system
- **immunogenicity:** the capability of a substance to elicit an immune response

Figure 6.3 ■ Immunogenicity versus antigenicity.

■ Impurities

Impurities are considered to be important factors in the immunogenicity of therapeutic protein products. Substances like host cell components, resins from chromatographic columns, or enzymes used to activate the product and monoclonal antibodies used for affinity purification may end up in the final product. Impurities may also be introduced by the components of the formulation, may leak from the container and sealing of the product, or may be introduced during the fill and finish steps. These impurities may boost an immune response different than an immune response to the therapeutic protein. These products may be immunogenic by themselves. Antibodies induced by impurities may lead to general immune reactions as skin reactions, allergies, anaphylaxis, and serum sickness. Antibodies to impurities also raise quality issues concerning the product and therefore need to be monitored. Interestingly, there are a number of examples of products declining in immunogenicity over the years because of improvements in the purification and other downstream processing steps.

Impurities may enhance immunogenicity via different mechanisms. Endotoxin from the bacterial host cells has been reported to cause the immunogenicity of the first recombinant DNA-derived human growth hormones. G-C-rich bacterial host cell DNA and denatured proteins are capable of activating Toll-like receptors and can also act as adjuvants. The activity of these impurities, however, is restricted to the nonhuman proteins which have a vaccine-like activity.

Adjuvants are incapable of boosting the immune response based on circumventing B-cell tolerance. However, impurities that are modified human proteins such as clipped and oxidized variants which can be present in therapeutic protein products could indirectly induce antibodies which react with the unmodified protein. This immunological mimicry has been described in dogs treated with human epoetin. The degree of nonself of this protein for dogs is sufficient to induce an immune response. But there is still enough homology between the human and canine erythropoietin for the antibodies to neutralize the endogenous, canine erythropoietin and causing severe anemia.

■ Formulation

Human therapeutic proteins are often highly biologically active and the doses may be at the μg level, making it a technological challenge to formulate the product to keep it stable with a reasonable shelf-life and to avoid the formation of aggregates and other product modifications. The importance of formulation in avoiding immunogenicity is highlighted in two historical cases (Fig. 6.4). In the case of interferon-α-2a, a large difference was noted among different formulations. A freeze-dried formulation containing human serum albumin as a stabilizer that according to its instructions could be kept at room temperature was particularly immunogenic. It appeared that at room temperature interferon-α-2a became partly oxidized. And the oxidized molecules formed aggregates also with the unmodified interferon and human serum albumin, and these aggregates were responsible for the immune response.

More recently an antibody-mediated severe form of anemia (pure red cell aplasia; PRCA) occurred after the formulation of an epoetin-α product was changed (Casadevall et al. 2002). Human serum albumin was replaced by polysorbate 80. How this formulation change induced immunogenicity is still not certain. However, the most likely explanation is a less stable new formulation resulting in aggregate formation when not appropriately handled.

■ Route of Administration

There is a clear effect of route of administration on the immunogenicity. The subcutaneous route is the most immunogenic and the intravenous the least immunogenic one. However, immunogenicity can be seen after any route of application including the mucosal and the intrapulmonary route.

■ Dose

The effect of the dose is not quite clear. There are studies with the lowest incidence of antibody formation in the highest-dose group. However, such data should be interpreted with caution. In the highest-dose group, there may be more products in the circulation interfering with the assay or the antibody level may be lower by increased immune complex formation.

■ Patient Features

There are also a number of patient-related factors which influence the incidence of antibody formation. The biological effect of the product can either enhance or inhibit the antibody formation as can the concomitant

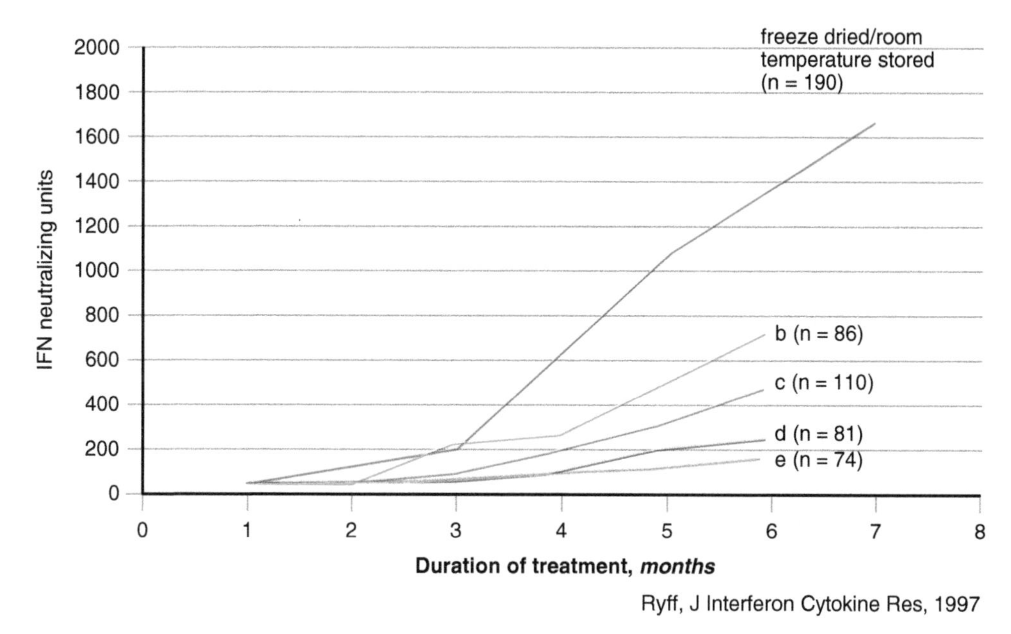

Figure 6.4 ■ Immunogenicity differences between interferon (IFN)-α formulations in patients (Ryff 1997).

treatment of patients. Sometimes the concomitant treatment is administered to inhibit antibody formation, e.g., methotrexate treatment to inhibit antibody formation to TNF inhibitors.

The underlying disease for which the patients are being treated is also important. Patients treated with interferon-α-2 for chronic hepatitis are more likely to produce antibodies than patients with solid tumors. The upsurge of epoetin-α-induced PRCA after the formulation change was only seen in patients with chronic kidney disease and not in patients with cancer.

As discussed before there are many indications that circumventing B-cell tolerance is independent of HLA type. Indeed a number of clinical studies have failed to show an association between antibody response and HLA type.

■ Assays for Antibodies

Assays are probably an important factor influencing the reported incidence of antibody induction by therapeutic proteins. In the published studies with interferon-α-2 in patients with viral infections, the incidence of antibody induction varied from 0 % to more than 60 % positive patients. This variation must be assay related. Evaluations of the performance of different test laboratories with blind panel testing showed a more than 50-fold difference in titers found in the same sera. Thus, any reliable comparison between different groups of patients when looking for a clinical effect of antibodies or studying factors influencing immunogenicity can only be done if the antibody quantification is done with in a well-validated assay in the same laboratory.

There is obviously a lack of standardization of assay methodology. There are also only a few reference and/or standard antibody preparations available. A number of white papers have appeared mainly authored by representatives of the biotechnology industry in the USA (Mire-Sluis et al. 2004). Although the area of biotechnology-derived therapeutics is still too much in development to formulate a definite assay methodology, there is a growing consensus on the general principles.

There is an agreement that a single assay is not sufficient to evaluate the immunogenicity of a new protein drug, but a number of assays need to be used in conjunction. Most antibody assay strategies are based on a two-tier approach: a screening assay to identify the antibody-positive sera followed by further characterization such as whether the antibodies are binding only or neutralizing and what is the titer, affinity, and isotype.

In general, the screening assay is a binding assay, mostly an ELISA type of assay (see Chap. 2) with the radio-immune-precipitation methodology as an alternative. Binding antibodies have mostly no biological consequences. However, assays for the more biologically important neutralizing antibodies are in general cumbersome and expensive. Thus, screening with a binding assay to select the positive sera for the neutralizing assay saves time and money.

Screening assays are designed for optimal sensitivity to avoid false negatives. For new proteins, defining an absolute sensitivity is impossible because of the lack of positive sera. An alternative approach is to set the cut-point for the assay at a 5 % false-positive level using a panel of normal human sera and/or untreated patient sera representative of the groups to be treated.

The assay for neutralizing antibodies is in general a modification of the potency assay for the therapeutic protein product. The potency assay is in most cases an in vitro cell-based assay. A predefined amount of product is added to the serum and a reduction of activity evaluated in the bioassay.

An important caveat in interpreting the neutralization assay results is the possible presence of inhibitors of the products other than antibodies (e.g., soluble receptors) in human serum or factors stimulating the bioassay which may compensate for the neutralizing activity. To overcome these problems, patient serum should be tested as control. IgG-depleted serum should also be tested for neutralizing activity to identify neutralizing factors other than antibodies. Further characterization of the antibodies may include evaluation of Ig isotype and affinity.

ISSUES SPECIFICALLY RELATED TO MONOCLONAL ANTIBODIES

The thinking about the immunogenicity of monoclonal antibodies went through the same paradigm shift as occurred with therapeutic proteins in general. The first generation of monoclonal antibodies was of murine origin. They induced an immune response in the majority of patients as foreign proteins should trigger a classical vaccine-type immune response. This so-called HAMA response (human antibodies to murine antibodies) was a major restriction in the clinical success of these murine antibodies. Over the years, however, methods were introduced in different stages to humanize monoclonal antibodies (cf. Chap. 1). Recombinant DNA technology was used to exchange the murine constant parts of the immune globulin chains with their human counterparts resulting in chimeric monoclonal antibodies. The next step was to graft murine complementarity-determining regions (CDR"s), which determine the specificity, into a human immune globulin backbone creating humanized monoclonal antibodies. And the final step was the development of transgenic animals, phage display technologies, and other developments allowing the production of human monoclonal antibodies. The assumption that human monoclonal antibodies would have no immunogenicity proved to be wrong. Although humanization has reduced the immunogenicity, even completely human monoclonal antibodies have been shown to induce antibodies. The introduction of chimeric antibodies by the exchange of the murine constant regions with their human counterparts has resulted in a substantial reduction of the induction of antibodies. Whether further humanization has resulted in an additional decrease is less clear. As discussed, the presence of aggregates has been identified as a major cause of immunogenicity of human therapeutic proteins. It is likely that with human monoclonal antibodies, aggregates are also responsible for antibody induction. In fact in the classical studies of B-cell tolerance done more than 40 years ago, aggregated immunoglobulin preparations were used to break tolerance (Weigle 1971).

Monoclonal antibodies have properties which may contribute to their immunogenicity. They can activate T cells by themselves and may boost the immune response by their Fc functions such as macrophage activation and complement activation. Indeed removal of N-linked glycosyl chains from the Fc part of the immunoglobulin may reduce Fc function and lead to a diminished immunogenicity.

What the antibodies are binding is also influencing their immunogenicity. Monoclonal antibodies targeting cell-bound antigens induce a higher level of antibody formation than those with circulating targets. Monoclonal antibodies directed to antigens on immune cells with the purpose of inducing immune suppression also suppress an immunogenic response.

Although more injections and higher doses are associated with a higher immune response, in some cases chronic treatment and higher doses were reported to be less immunogenic than episodic treatment and lower doses. The interpretation of these data is difficult because under these treatment conditions, the level of circulating product is higher and more persistent and the presence of circulating monoclonal antibodies during the time of blood sampling may mask the detection of induced antibodies. Only a few studies were performed in which the subcutaneous and intravenous route of administration of monoclonal antibodies were compared showing little difference in immunogenicity.

The immune status of the patients influences the antibody response as with other protein therapeutics. Many of the patients receiving monoclonal antibodies are immune compromised by diseases such as cancer or by immune suppressive treatment and are less likely to produce antibodies than patients with a normal immune status. Sometimes immune suppressive agents such as methotrexate are given to patients with the purpose of inhibiting an antibody response.

Another important aspect when studying the immunogenicity of monoclonal antibodies is timing of the blood sampling of patients. These products may have a relative long half-life (several weeks) and the circulating product may interfere with the detection of induced antibodies and may lead to false-negative results. Sampling sera up to 20 weeks after the patient has received the last injection may be necessary to avoid the interference of circulating monoclonal antibodies. Also natural antibodies, soluble receptors, and immune complexes may interfere with assays and lead to either false-positive or false-negative results (as explained above).

CLINICAL EFFECTS OF INDUCED ANTIBODIES

Despite the methodological drawbacks, the list of protein products with clinically relevant immunogenic side effects is growing. The most common consequence is loss of efficacy. Sometimes this loss can be overcome by increasing the dose or changing to another product.

The most dramatic and undisputed complication occurs when the antibodies to the product cross neutralize an endogenous factor with an important biological function. This has been described for a megakaryocyte growth and differentiation factor which induced antibodies cross-reacting with endogenous thrombopoietin (see Table 6.1). Volunteers and patients in a clinical trial developed severe thrombocytopenia and needed platelet transfusions. Because of this complication, the product was withdrawn from further development.

More recently the upsurge of PRCA (see above) associated with a formulation change of epoetin-α marketed outside the USA occurred. The antibodies induced by the product neutralized the residual endogenous erythropoietin in these patients resulting in a severe anemia which could only be treated with blood transfusions.

Antibodies can also influence the side effects of therapeutic proteins. The consequences are dependent on the cause of the side effects. If the adverse effects are the results of the intrinsic activity of the products, antibodies may reduce the side effects, as it is the case with interferon-α-2. Sometimes the mitigation of the side effects is even the first clinical sign of the induction of antibodies.

With some products the side effects are caused by the antibody formation. This is in general the case when the product is administered in relatively high doses, like with some monoclonal antibodies. Symptoms caused by immune complexes like delayed-type hypersensitivity and serum sickness are related to the level of antibodies induced.

The general effects caused by an immune reaction to a therapeutic protein such as acute anaphylaxis, hypersensitivity, skin reaction, and serum sickness are relatively common when large amounts of nonhuman proteins are administered. These effects are relatively rare for modern biotechnology-derived products which are highly purified human proteins administered in relatively low amounts. However, these side effects caused by an immune response are currently still relatively common during treatment with high doses of monoclonal antibodies.

PREDICTING AND REDUCING IMMUNOGENICITY

As discussed the mechanisms leading to antibody induction by therapeutic proteins are still not completely understood. As a consequence it is impossible based on our current knowledge to fully predict the immunogenicity of a new product in patients. For nonhuman proteins which induce the classical immune response, the level of nonself is a relative predictor of an immune response. However, it is not an absolute predictor. Sometimes a single amino acid change is sufficient to make a self-protein highly immunogenic. With other proteins, substantial divergence from the natural sequence has no effect. For foreign proteins a number of in vitro stimulation and binding tests and computational models are advertised as predictors of immunogenicity. However, all these tests have their limitations. T-cell proliferation assays, for example, have the drawback that many antibodies are capable of inducing some level of T-cell activation or inhibit cell proliferation. The computational algorithms which predict binding of antigens to HLA class II only give limited information on the interaction of the proteins with the immune system and also under-detect epitopes (Stevanovic 2002). These limitations are also evident when these assays or algorithms are used to reduce immunogenicity: there is hardly any convincing evidence of a clinically relevant reduction of antibody induction.

For human homologues the best predictor of immunogenicity is the presence of aggregates and to a minor degree the presence of impurities. Thus, the quality of the therapeutic protein and its formulation are important factors. There is also evidence about immunogenicity introduced by a change in formulation and a reduction in immunogenicity by avoiding aggregation and improving purification and formulation (see above).

Although animal studies are helpful in obtaining control sera and may provide insight in the possible clinical effects of immunogenicity, they are not very good predictors of immunogenicity in patients. All proteins, including the human homologues, will be in principle immunogenic in animals. Sometimes animal studies may help to study the relative immunogenicity of different products or formulations, although their predictive value for the clinic is questionable. Even monkey studies do not completely predict immunogenicity in patients. Some products which are immunogenic in nonhuman primates do not induce antibodies in humans and vice versa.

The animal model to study the factors important for circumventing tolerance is transgenic mice carrying the gene for the human protein (Hermeling et al. 2005). These animals have an immune tolerance comparable to the immune tolerance in patients. These animals have also been successfully used to identify new epitopes in modified products. Although transgenic animals have been proven to be important scientific tools to identify the factors important to circumvent tolerance and study the immunological processes involved, also this animal model will never be able to serve to predict the human response.

■ Reducing Immunogenicity

Several strategies are being applied to reduce immunogenicity besides changing the amino acid sequence of the product. Linking proteins to polymers such as polyethylene glycol (see Chap. 21) and low molecular weight dextran reduces immunogenicity. However, these modifications also make the molecules less active, necessitating higher doses. This and their increased half-life extend their exposure to the immune system, which may increase the immunogenic potential. Another approach is to reduce the immunogenic response by immunosuppressive treatments. In addition, tolerance induction is being applied, e.g., in hemophilia patients with antibodies to factor VIII.

CONCLUSIONS

The most important points of this chapter are summarized in the following bullet points:

- The immunogenicity of therapeutic proteins is a commonly occurring phenomenon.
- The clinical consequences can vary.
- Validated detection systems are essential to study the immunogenicity of therapeutic proteins.
- The prediction of immunogenicity in patients based on physicochemical characterization and animal studies is not easy.
- There is still a lot to be learned about why and how patients produce antibodies to therapeutic proteins.

The growing awareness of the importance of immunogenicity of therapeutic proteins is illustrated by the adoption of a standard requirement in regulatory dossiers for new proteins and biosimilars to evaluate their immunogenicity in clinical trials (cf. Chap. 11)

SELF-ASSESSMENT QUESTIONS

■ Questions

1. Which factors contribute to unwanted immunogenicity of therapeutic proteins?
2. What are possible clinical consequences of antibody formation against biopharmaceuticals in patients?
3. Why do aggregates of recombinant human proteins induce antibodies that cross-react with the (non-aggregated) drug?
4. Explain the fundamental difference between (a) antibody formation in children with growth hormone deficiency treated against recombinant human growth hormone and (b) antibody formation against rh-erythropoietin in patients with chronic renal failure.
5. Give an example of a case that demonstrates that the formulation of a biopharmaceutical can affect the immune response.
6. Give at least 3 approaches that can be followed to reduce the immunogenicity of a biopharmaceutical.
7. Why is standardization of assays for detection of antidrug antibodies important?
8. Why are antidrug-antibody titers against a monoclonal antibody more difficult to determine accurately than antibodies against interferon?

■ Answers

1. See Fig. 6.2.
2. Reduction of therapeutic efficacy, (seldom) enhancement of efficacy, anaphylactic reactions, cross-reactivity with endogenous protein.
3. Aggregates can circumvent B-cell tolerance against native (–like) epitopes (repetitive epitopes); the more 'native-like" the aggregate, the more likely cross-reactivity with the monomer will occur.
4. (a) is the classical immune response versus (b) circumventing B-cell tolerance.
5. The examples given in the text re erythropoietin and interferon-α.
6. Design another formulation, remove aggregates, pegylate or change the glycosylation pattern of the protein or use amino acid mutants, use human(ized) versions of the proteins, or select another route of administration. NB some of these approaches will lead to a new bioactive drug molecule and that has implications for the way authorities will judge the procedure to be followed for obtaining marketing approval ((see Chap. 11) on Regulatory Issues).
7. Different assay formats and blood sampling schedules give different answers and thus hamper direct comparison between studies. Therefore, it is difficult to compare the results obtained with different products that are tested for immunogenicity in different labs.
8. Monoclonal antibodies are often administered in high doses and have a long circulation time (days/weeks). This will likely cause interference with the assay by the circulating drug (resulting in false negatives or underestimation of antibody titers). Another possibility for interference is the occurrence of cross-reactivity of the reagents in the test for the induced antibodies and the original drug-antibody. With interferon a different situation is encountered – interferons are rapidly cleared and administered in low doses (microgram range); therefore, interferons will less likely interfere with the measurement of anti-IFN antibodies.

REFERENCES

Arnold DF, Misbah SA (2008) Cetuximab-induced anaphylaxis and IgE specific for galactose-alpha-1,3-galactose. N Engl J Med 19:358

Bachmann MF, Rohrer UH, Kundig TM, Burki K, Hengartner H, Zinkernagel RM (1993) The influence of antigen organization on B-cell responsiveness. Science 262:1448–1451

Casadevall N, Nataf J, Viron B et al (2002) Pure red-cell aplasia and antierythropoietin antibodies in patients treated with recombinant erythropoietin. N Engl J Med 346:469–475

Hermeling S, Crommelin DJA, Schellekens H, Jiskoot W (2004) Structure-immunogenicity relationships of therapeutic proteins. Pharm Res 21:897–903

Hermeling S, Jiskoot W, Crommelin DJA, Bornaes C, Schellekens H (2005) Development of a transgenic mouse model immune tolerant for human interferon β. Pharm Res 22:847–851

Mire-Sluis AR, Barrett YC, Devanarayan V et al (2004) Recommendations for the design and optimization of immunoassays used in the detection of host antibodies against biotechnology products. J Immunol Methods 289:1–16

Moore WV, Leppert P (1980) Role of aggregated human growth hormone (hGH) in development of antibodies to hGH. J Clin Endocrinol Metab 51:691–697

Ryff JC (1997) Clinical Investigation of the immunogenicity of interferon alpha 2a. J Interferon Cytokine Res 17:S29–S33

Sauerborn M, Brinks V, Jiskoot W, Schellekens H (2010) Immunological mechanism underlying the immune response to recombinant human protein therapeutics. Trends Pharmacol Sci 31(2):53–59

Schellekens H (2002) Bioequivalence and the immunogenicity of biopharmaceuticals. Nat Rev Drug Discov 1:1–7

Stevanovic S (2002) Structural basis of immunogenicity. Transpl Immunolog 10:133–136

Stevanovic S (2005) Antigen processing is predictable: from genes to T cell epitopes. Transpl Immunol 14: 171–174

Weigle WO (1971) Recent observations and concepts in immunological unresponsiveness and autoimmunity. Clin Exp Immunol 9:437–447

FURTHER READING

Brinks V, Jiskoot W, Schellekens H (2011) Immunogenicity of therapeutic proteins: the use of animal models. Pharm Res 28(10):2379–2385

Hermeling S, Crommelin DJA, Schellekens H, Jiskoot W (2007) Immunogenicity of therapeutic proteins. In: Gad SC (ed) Handbook of pharmaceutical biotechnology. Wiley, Hoboken, pp 911–931

Schellekens H (2010) The immunogenicity of therapeutic proteins. Discov Med 9:560–564

Schellekens H, Crommelin D, Jiskoot W (2007) Immunogenicity of antibody therapeutics. In: Dübel S (ed) Handbook of therapeutic antibodies. Wiley-VCH, Weinheim, pp 267–276

7

Monoclonal Antibodies: From Structure to Therapeutic Application

John D. Davis, Rong Deng, C. Andrew Boswell, Yi Zhang, Jing Li, Paul Fielder, Amita Joshi, and Saraswati Kenkare-Mitra

INTRODUCTION

The exciting field of therapeutic monoclonal antibodies (MABs) had its origins as Milstein and Koehler presented their murine hybridoma technology in 1975 (Kohler and Milstein 1975). This technology provides a reproducible method for producing monoclonal antibodies with unique target selectivity in almost unlimited quantities. In 1984, both scientists received the Nobel Prize for their scientific breakthrough, and their work was viewed as a key milestone in the history of MABs as therapeutic modalities and their other applications. Although it took some time until the first therapeutic MAB got market authorization from the FDA in 1986 (Orthoclone OKT3, Chap. 19), monoclonal antibodies are now the standard of care in several disease areas. In particular, in the areas of oncology (Chap. 17), transplantation (Chap. 19), and inflammatory diseases (Chap. 20), patients now have novel life-changing treatment alternatives for diseases which had very limited or nonexistent medical treatment options before the emergence of MABs. To date more than 30 MABs and MAB derivatives including fusion proteins and MAB fragments are available for different therapies (Table 7.1). Eight MABs and three immunoconjugates in oncology; 11 MABs, one Fab conjugate, and four Fc fusion proteins in inflammation; and three MABs and one Fc fusion protein in transplantation comprise the majority of the approved therapies.

J.D. Davis, Ph.D. (✉) • R. Deng, Ph.D. • Y. Zhang, Ph.D.
J. Li, PhD. • A. Joshi, PhD.
Clinical Pharmacology, Genentech Inc., 1 DNA Way, MS463a, South San Francisco, CA 94080, USA
e-mail: davisj29@gene.com

C.A. Boswell, Ph.D.
Preclinical and Translational Pharmacokinetics, Genentech Inc., South San Francisco, CA, USA

P. Fielder, Ph.D. • S. Kenkare-Mitra, Ph.D.
Development Sciences, Genentech Inc., South San Francisco, CA, USA

Technological evolutions have subsequently allowed much wider application of MABs via the ability to generate mouse/human chimeric, humanized, and fully humanized MABs from the pure murine origin. In particular, the reduction of the xenogenic portion of the MAB structure decreased the immunogenic potential of the murine MABs thus allowing their wider application. MABs are generally very safe drugs because of their target selectivity, thus avoiding unnecessary exposure to and consequently activity in nontarget organs. This is particularly apparent in the field of oncology, where MABs like rituximab, trastuzumab, and bevacizumab can offer a more favorable level of efficacy/safety ratios compared to common chemotherapeutic treatment regimens for some hematologic and solid tumors.

The dynamic utilization of these biotechnological methods resulted not only in new drugs, but it also triggered the development of an entirely new business model for drug research and development with hundreds of newly formed and rapidly growing biotech companies. Furthermore, the ability to selectively target disease-related molecules resulted in a new scientific area of molecular-targeted medicine, where the development of novel MABs probably contributed substantially to setting new standards for a successful drug research and development process. The term translational medicine was developed to cover the biochemical, biological, (patho) physiological understanding and using this knowledge to find intervening options to treat diseases. During this process, biomarkers (e.g., genetic expression levels of marker genes, protein expression of target proteins, molecular imaging) are used to get the best possible understanding of the biological activities of drugs in a qualitative and most importantly quantitative sense, which encompasses essentially also in the entire field of pharmacokinetics/pharmacodynamics (PK/PD). The application of those scientific methods together with the principle of molecular-targeted medicine

D.J.A. Crommelin, R.D. Sindelar, and B. Meibohm (eds.), *Pharmaceutical Biotechnology*,
DOI 10.1007/978-1-4614-6486-0_7, © Springer Science+Business Media New York 2013

Name	Therapeutic area	Type	Antibody isotype	Target		PK				
				Receptor	Type	Behavior[a]	Half-life	Clearance	Vss	Regimen
Abciximab	Cardiovascular	Fragment	Chimeric Fab: mVar-hIgG1	CD41		Linear	0.29 days (Kleiman et al. 1995)	0.068 L/h/kg (Mager et al. 2003)	0.12 L/kg	0.25 mg/kg IV bolus
Abatacept	Inflammation	Fusion protein	Extracellular domain of hCTLA-4 + hinge of hFc	CD80/CD86		Linear	13.1 days (Hervey and Keam 2006)	0.22 mL/h/kg (Hervey and Keam 2006)	0.07 L/kg	IV infusions 2 or 10 mg/kg Q2W for first three cycles and monthly thereafter
Adalimumab	Inflammation	MAB	hIgG1	TNFα	Soluble and cell bound	Linear	14.7–19.3 days (Weisman et al. 2003)	9–12 mL/h (Weisman et al. 2003)	5.1–5.75 L SC: *F*=64 %	40 mg SC every week
Alefacept	Inflammation	Fusion protein	LFA-3/hIgG1(Fc)	CD2		Nonlinear	11.3 days (Amevive 2003)	0.25 mL/h/kg (Amevive 2003)	IV:94 mL/kg IM: *F*=63 %	7.5 mg IV infusion Q1W/12 weeks or 10 mg and 15 mg IM injection
Alemtuzumab	Oncology	MAB	rCDR-hIgG1	CD52	Soluble	Nonlinear	11 h after single dose, 12 days following multiple doses (Morris et al. 2003)	NA	0.18 L/kg (Campath 2009)	IV infusion at 3 mg initially, escalated to 10 mg until tolerated, and maintained at 30 mg, 3 times a week for 12w
Basiliximab	Transplantation	MAB	Chimeric: mVar-hIgG1	CD25		NR	4.1 days (Kovarik et al. 2001)	75 mL/h (Kovarik et al. 2001)	5.5–13.9 L	IV bolus or infusion 20 mg on days 0 and 4 (40 mg total)

Belatacept	Transplantation	Fusion protein	Extracellular domain of hCTLA-4+ hinge of hFc	CD80/CD86		Linear	9.8 days (Nulojix 2011)	0.49 mL/h/kg (Nulojix 2011)	0.11 L/kg	10 mg/kg IV infusion over 30 min
Belimumab	Inflammation	MAB	Fully humanized IgG1	BLyS (B lymphocyte stimulator)	Soluble and cell bound	Linear	19.4 days (Benlysta 2011)	215 mL/day (Benlysta 2011)	5.29 L (Benlysta 2011)	10 mg/kg IV infusion Q2W for first 3 doses, then Q4W
Bevacizumab	Oncology	MAB	hIgG1	VEGF	Soluble	Linear	20 days (Avastin 2004)	0.207–0.262 L/day (Avastin 2004)	2.66–3.25 L	IV infusion 5 or 10 mg/kg Q2W or 15 mg/kg Q3W depending on indication
Brentuximab vedotin	Oncology	MAB-ADC	Chimeric IgG1	CD30	Soluble	Linear	4.43 days	1.76 L/day (Younes et al. 2010)	8.21 L	IV infusion of 1.8 mg/kg over 30 min every 3 weeks
Canakinumab	Inflammation	MAB	hIgG1	IL-1beta	Soluble	Linear	26 days (Ilaris 2011)	0.174 L/day	6.01 L F=70 %	150 mg SC
Certolizumab pegol	Inflammation	Fragment	Fab conjugated with PEG2MAL40K	TNFα		Linear	14 days (Cimzia 2011)	9.21–14.38 mL/h (Cimzia 2011)	6–8 L F=76-88 % (Cimzia 2011)	400 mg SC initially, and at Weeks 2 and 4; maintenance regimen 400 mg every 4 weeks
Cetuximab	Oncology	MAB	Chimeric: mVar-hIgG1	EGFR	Soluble	Nonlinear	4.8 days (Erbitux 2004) MRT: 12.6 days	0.02–0.08 L/h/m² (Erbitux 2004)	2–3 L/m²	IV infusion 400 mg/m² as first dose followed by weekly doses of 250 mg/m²
Daclizumab	Transplantation	MAB	Hyperchimeric: mCDRhIgG1	CD25		Linear	20 days (Zenapax 2005)	15 mL/h (Zenapax 2005)	5.9 L (Zenapax 2005)	IV infusion 1 mg/kg Q2W for 5 doses

Table 7.1 ■ The pharmacological properties of the approved therapeutic antibodies, antibody isotype, antigen target, clinical indication, and pharmacokinetic parameters.

Name	Therapeutic area	Type	Antibody isotype	Target		PK				
				Receptor	Type	Behavior[a]	Half-life	Clearance	Vss	Regimen
Denosumab	Oncology	MAB	Human IgG2	RANKL		Linear	28 days (Prolia 2010)	NA	F=62 % (Prolia 2010)	120 mg SC injection Q4W
Eculizumab	Hemolysis	MAB	Humanized IgG2/4 k	Complement protein C5	Soluble	Linear	8–14.8 days	22 mL/h (Soliris 2007)	7.7 L (Soliris 2007)	600 mg Q1W for 4 weeks, 900 mg 7 days later, then 900 mg Q2W thereafter
Efalizumab	Inflammation	MAB	mCDR-hIgG1	CD11a	Cell-bound internalized	Nonlinear	NR[a]	6.6 mL/kg/day (Joshi et al. 2006)	58 mL/kg SC: F=50 %	single 0.7 mg/kg SC conditioning dose, then 1 mg/kg weekly
Etanercept	Inflammation	Fusion protein	TNF receptor/ hIgG1(Fc)	TNFα	Soluble and cell bound	Linear	4 days (Lee et al. 2003)	120 mL/h (Lee et al. 2003)	F: 58 % Vd: 6–11 L	Biweekly 25 mg SC injection
Golimumab	Inflammation	MAB	Fully human IgG1	TNFα	Soluble and cell bound	Linear	2 weeks	4.9–6.7 mL/ day/kg (Simponi 2009)	F=53 % Vd= 58–126 mL/ kg	50 or 100 mg SC injection Q2W or Q4W
Ibritumomab tiuxetan	Oncology	MAB	Murine IgG1	CD20	Cell-bound stable	NR	47 h (Wiseman et al. 2001)	NA	NA	14.8 MBq/kg
Infliximab	Inflammation	MAB	Chimeric: mVar-hIgG1	TNFα	Soluble and cell bound	Linear	7.7–9.5 days (Cornillie et al. 2001; Remicade 2006)	9.8 mL/h	NA	Single IV infusion 1, 5, 10, or 20 mg/ kg at 1, 2, 4, 8, and 12 weeks
Ipilimumab	Oncology	MAB	Fully humanized IgG1	CTLA-4		Linear	14.7 days (Yervoy 2011)	15.3 mL/h (Yervoy 2011)	7.21 L (Yervoy 2011)	3 mg/kg IV infusion Q3W
Muromonab-CD3	Transplantation	MAB	Murine IgG2α	CD3		NR	0.75 days (Hooks et al. 1991)	NA	NA	IV bolus of 5 mg every day for 10–14 days

Natalizumab	Inflammation	MAB	Humanized IgG_4k	a_4b_1 and a_4b_7 integrins		NR	11 days (Tysabri 2006)	16 mL/h (Tysabri 2006)	NA	300 mg IV infusion Q4W
Ofatumumab	Oncology	MAB	Human IgG1κ	CD-20	Cell bound	Nonlinear	14 days (Arzerra 2009)	0.01 L/h (Arzerra 2009)	1.7–5.1 L (Arzerra 2009)	IV infusion 300 mg initial dose, followed 1 week later by 2000 mg Q1W for 7 doses, followed 4 weeks later by 2,000 mg Q4W for 4 doses
Omalizumab	Inflammation	MAB	mCDR-hIgG1	IgE	Soluble	Linear	26 days (Xolair 2003)	2.4 mL/day/kg	48 mL/kg SC: $F=62$ %	150–375 mg by subcutaneous (SC) injection every 2 or 4 weeks
Palivizumab	Antiviral	MAB	mCDR-hIgG1	RSV		NR	20 days (Synagis 2004)	NR	NA	IM 15 mg/kg Q4W
Panitumumab	Oncology	MAB	hIgG2	EGFR	Soluble	Nonlinear	7.5 days (Vectibix 2006)	4.9 mL/day/kg	82 mL/kg (Ma et al. 2009)	6 mg/kg given IV over 60 min, once every 2 weeks
Ranibizumab	Macular degeneration	Fragment	hIgG1κ	VEGF	Soluble	NA	9^b days (Lucentis 2006)	NA	NA	0.5 mg intravitreal injection Q4W
Rilonacept	Inflammation	Fusion protein	hIgG1	IL-1beta	Soluble	NA	7.6 days (Radin et al. 2010)	CL/F: 0.866 L/day	Vz/F: 9.73 L	SC loading dose of 320 mg, then 160 mg once weekly
Rituximab	Inflammation	MAB	Chimeric: mVar-hIgG1	CD20	Cell-bound stable	Linear	19 days (Rituxan 2006)	10 mL/h (Rituxan 2006)	3.1 L	375 mg/m²

Table 7.1 ■ (continued)

Name	Therapeutic area	Type	Antibody isotype	Target		PK					
				Receptor	Type	Behavior[a]	Half-life	Clearance	Vss	Regimen
Tocilizumab	Inflammation	MAB	Recombinant humanized IgG1	IL-6	Soluble and cell bound	Nonlinear	up to 13 days (Actemra® Tocilizumab 2010)	12.5 mL/h (Actemra® Tocilizumab 2010)	6.4 L (Actemra® Tocilizumab 2010)	8 mg/kg IV infusion either Q2W or Q4W depending on indication
Tositumomab	Oncology	MAB radiolabeled	Murine IgG2α	CD20	Cell-bound stable	Nonlinear	NA	68.2 mL/h (Bexxar 2003)	NA	485 mg IV infusion
Trastuzumab	Oncology	MAB	mCDR-hIgG1	Her2	Cell-bound shed	Nonlinear	1.7–12 days[c] (Herceptin 2006)	16–41 mL/h (Tokuda et al. 1999)	3.6–5.2 L	IV infusion 10–500 mg Q1W
Ustekinumab	Inflammation	MAB	Human IgG1	IL-12, IL-23		Linear	14.9–45.6 days (Stelara, 2009)	1.90–2.22 mL/kg/day (Stelara, 2009)	*V*=0.1 L/kg F ~57 % (Zhu et al. 2009)	45 mg SC initially followed by a further 45 mg dose after 4 weeks, then 45 mg q12w

CD cluster of differentiation, *CDR* complementarity determining region, *CTLA* cytotoxic T lymphocyte-associated antigen, *EGFR* epidermal growth factor receptor, *Fab* antigen-binding fragment, *Fc* constant fragment, *Ig* immunoglobulin, *LFA-1* lymphocyte function-associated antigen, *MAB* monoclonal antibody, *MRT* mean residence time, *mVar* murine variable, *RSV* respiratory syncytial virus, *TNF* tumor necrosis factor, *VEGF* vascular endothelial growth factor, *NA* information not found, *NR* not reported

[a]Where PK are nonlinear, parameters are reported at usual clinical dose

[b]Vitreous elimination half-life

[c]Dose-dependent pharmacokinetics

combined with the favorable pharmacokinetics and safety of MABs might at least partly explain why biotechnologically derived products have substantially higher success rates to become marketed therapy compared to chemically derived small molecule drugs.

This chapter tries to address the following questions: What are the structural elements of MABs? How do MABs turn functional differences into different functional activities? And how is a MAB protein turned from a potential clinical drug candidate into a therapeutic drug by using a translational medicine framework? In this sense, this chapter provides a general introduction to Chaps. 17, 19, and 20, where the currently marketed MABs and MAB derivatives are discussed in the context of their therapeutic applications. Efalizumab (anti-CD11a, Raptiva®), a MAB marketed as anti-psoriasis drug in the US and EU, was chosen to illustrate the application of pharmacokinetic/pharmacodynamic principles in the drug development process.

ANTIBODY STRUCTURE AND CLASSES

Antibodies (Abs) (immunoglobulin (Ig)) are roughly Y-shaped molecules or combinations of such molecules. There are five major classes of Ig: IgG, IgA, IgD, IgE, and IgM. Table 7.2 summarizes the characteristics of these molecules, particularly their structure (monomer, dimer, hexamer, or pentamer), molecular weight (ranging from ~150 to ~1,150 kDa), and functions (e.g., activate complement, FcγR binding). Among these classes, IgGs and their derivatives form the framework for the development of therapeutic antibodies. Figure 7.1 depicts the general structure of an IgG with its structural components as well as a conformational structure of efalizumab (anti-CD11a, Raptiva®). An IgG molecule has four peptide chains, including two identical heavy (H) chains (50 ~ 55 kDa) and two identical light (L) chains (25 kDa), which are linked via disulfide (S–S) bonds at the hinge region. The first ~110 amino acids of both chains form the variable regions (V_H and V_L) and are also the antigen-binding regions. Each V domain contains three short stretches of peptide with hypervariable sequences (HV1, HV2, and HV3), known as complementarity determining regions (CDRs), i.e., the region that binds antigen. The remaining sequences of each light chain consist of a single constant domain (C_L). The remainder of each heavy chain contains three constant regions (C_{H1}, C_{H2}, and C_{H3}). Constant regions are responsible for effector recognition and binding. IgG can be further divided into four subclasses (IgG1, IgG2, IgG3, and IgG4). The differences among these subclasses are also summarized in Table 7.2.

■ Murine, Chimeric, Humanized, and Fully Humanized MABs

With the advancement of technology, early murine MABs have been engineered further to chimeric (mouse CDR human Fc), humanized, and fully humanized MABs (Fig. 7.2). Murine MABs, chimeric MABs, humanized MABs, and fully humanized MABs have 0 %, ~60–70 %, ~90–95 %, and ~100 % sequences that are similar to human MABs, respectively. Decreasing the xenogenic portion of the MAB potentially reduces the immunogenic risks of generating anti-therapeutic antibodies (ATAs). The first therapeutic MABs were murine MABs produced via hybridomas; however, these murine antibodies easily elicited formation of neutralizing human anti-mouse antibodies (HAMA) (Kuus-Reichel et al. 1994). Muromonab-CD3 (Orthoclone OKT3), a first-generation MAB of murine origin, has shown efficacy in the treatment of acute transplant rejection and was the first MAB licensed for use in humans. It is reported that 50 % of the patients who received OKT3 produced HAMA after the first dose. HAMA interfered with OKT3's binding to T cells, thus decreasing the therapeutic efficacy of the MAB (Norman et al. 1993). Later, molecular cloning and the expression of the variable region genes of IgGs have facilitated the generation of engineered antibodies. A second generation of MABs, chimeric MABs, consists of human constant regions and mouse variable regions. The antigen specificity of chimeric MAB is the same as the parental mouse antibodies; however, the human Fc region renders a longer in vivo half-life than the parent murine MAB, and similar effector functions as the human Ab. Currently, there are 5 chimeric antibodies and fragments on the market (abciximab, basiliximab, cetuximab, infliximab, and rituximab). These antibodies can still induce human anti-chimeric antibodies (HACA). For example, about 61 % of patients who received infliximab had HACA response associated with shorter duration of therapeutic efficacy and increased risk of infusion reactions (Baert et al. 2003). The development of ATA is currently not predictable, as 6 of 17 patients with systemic lupus erythematosus receiving rituximab developed high-titer HACA (Looney et al. 2004), whereas only 1 of 166 lymphoma patients developed HACA (McLaughlin et al. 1998). Humanized MABs contain significant portions of human sequence except the CDR which is still of murine origin. There are eleven humanized antibodies on the market (alemtuzumab, bevacizumab, daclizumab, eculizumab, efalizumab, natalizumab, omalizumab, palivizumab, ranibizumab, tocilizumab, and trastuzumab). The incidence rate of antidrug antibody (i.e., human antihuman antibody (HAHA)) was greatly decreased for these humanized MABs. Trastuzumab has a reported HAHA incidence rate of

Property		IgA		IgG						
		IgA1	IgA2	IgG1	IgG2	IgG3	IgG4	IgM	IgD	IgE
Serum concentration in adult (mg/mL)		1.4–4.2	0.2–0.5	5–12	2–6	0.5–1	0.2–1	0.25–3.1	0.03–0.4	0.0001–0.0002
Molecular form		Monomer, dimer		Monomer				Pentamer, hexamer	Monomer	Monomer
Functional valency		2 or 4		2				5 or 10	2	2
Molecular weight (kDa)		160 (m), 300 (d)	160 (m), 350 (d)	150	150	160	150	950(p)	175	190
Serum half-life (days)		5–7	4–6	21–24	21–24	7–8	21–24	5–10	2–8	1–5
% Total IgG in adult serum		11–14	1–4	45–53	11–15	3–6	1–4	10	0.2	50
Function	Activate classical complement pathway	–		+	+/–	++	–	+++	–	–
	Activate alternative complement pathway	+	–	–	–	–	–	–	–	–
	Cross placenta	–		+	+/–	+	+	–	–	–
	Present on membrane of mature B cell	–		–	–	–	–	+	–	+
	Bind to Fc Receptors of phagocytes	–		++	+/–	++	+	?	–	–
	Mucosal transport	++		–	–	–	–	+	–	–
	Induces mast cell degranulation	–		–	–	–	–	–	+	–
Biological properties		Secretory Ig, binds to polymeric Ig receptor		Placental transfer, secondary antibody for most response to pathogen , binds macrophage and other phagocytic cells by Fcγ receptor				Primary antibody response, some binding to polymeric Ig receptor, some binding to phagocytes	Mature B cell marker	Allergy and parasite reactivity, binds FcεR on mast cells and basophiles

Table 7.2 ■ Important properties of endogenous immunoglobulin subclass (Goldsby et al. 1999; Kolar and Capra 2003).

only 0.1 % (1 of 903 cases) (Herceptin 2006), but daclizumab had a HAHA rate as high as 34 % (Zenapax 2005). Another way to achieve full biocompatibility of MABs is to develop fully humanized antibodies, which can be produced by two approaches: through phage-display library and by using transgenic XenoMouse® with human heavy and light chain gene fragments (Weiner 2006). Adalimumab is the first licensed fully humanized MAB generated by the phage-display library. Adalimumab was approved in 2002 and 2007 for the treatment of rheumatoid arthritis and Crohn's diseases, respectively (Humira 2007). However, despite its fully humanized Ab structure, the incidence of HAHA was about 5 % (58 of 1,062 patients) in three randomized clinical trials with adalimumab (Cohenuram and Saif 2007; Humira 2007). Panitumumab is the first approved fully humanized monoclonal antibody generated by using transgenic mouse technology. No HAHA responses have been reported yet in clinical trial after chronic dosing with panitumumab to date (Vectibix 2006; Cohenuram and Saif 2007). Of note, typically ATAs are measured using ELISA, and the

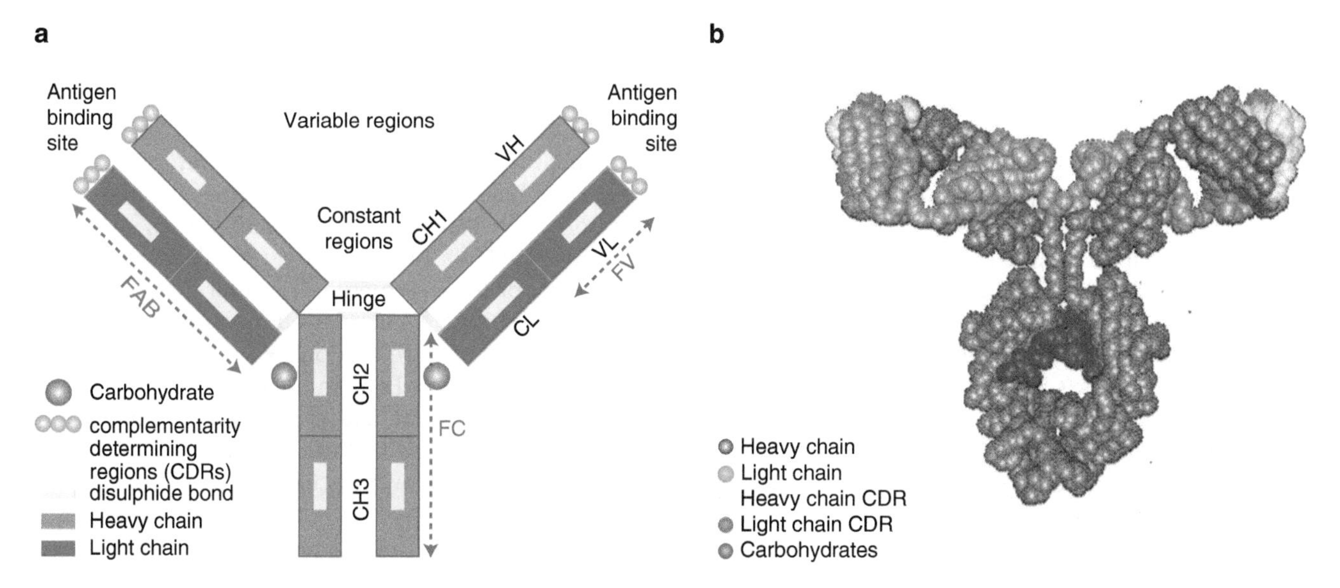

Figure 7.1 ■ (**a**) IgG1 antibody structure. Antigen is bound via the variable range of the antibody, whereas the Fc part of the IgG determines the mode of action (also called effector function). (**b**) Example efalizumab (anti-CD11a), Raptiva®. *H chain* heavy chain consisting of VH, CH1, CH2, CH3, *L chain* light chain consisting of VL, CL, *VH, VL* variable light and heavy chain, *CHn, CL* constant light and heavy chain, *Fv* variable fraction, *Fc* crystallizable fraction, *Fab* antigen-binding fraction (http://people.cryst.bbk.ac.uk/~ubcg07s/gifs/IgG.gif).

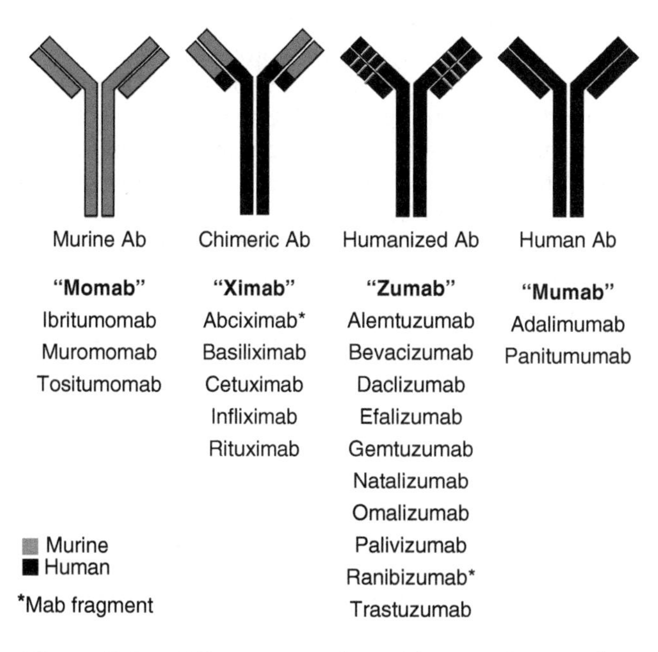

Figure 7.2 ■ Different generations of therapeutic antibodies.

reported incidence rates of ATAs for a given MAB can be influenced by the sensitivity and specificity of the assay. Additionally, the observed incidence of antibody positivity in an assay may be also influenced by several other factors including sample handling, timing of sample collection, concomitant medications, and underlying disease. For these reasons, comparison of the incidence of a specific MAB with the incidence of antibodies to other products may be misleading.

■ Key Structural Components of MABs

Proteolytic digestion of antibodies releases different fragments termed Fv (fragment variable), Fab (fragment antigen binding), and Fc (fragment crystallization). These different forms have been reviewed by others (Wang et al. 2007). These fragments can also be generated by recombinant engineering. Treatment with papain generates two identical Fab's and one Fc. Pepsin treatment generates a F(ab')2 and several smaller fragments. Reduction of F(ab')2 will produce two Fab's. The Fv consists of the heavy chain variable domain (V_H) and the light chain variable domain (V_L) held together by strong noncovalent interaction. Stabilization of the Fv by a peptide linker generates a single chain Fv (scFv).

■ Modifying Fc Structures

The Fc regions of MABs play a critical role not only in their function but also in their disposition in the body. Monoclonal antibodies elicit effector functions (antibody-dependent cellular cytotoxicity (ADCC) and complement-dependent cytotoxicity (CDC)) following interaction between their Fc regions and different Fcγ receptors and complement fixation (C1q, C3b). The CH2 domain or the hinge region joining CH1 and CH2 has been identified as the crucial regions for binding to FcγR (Presta et al. 2002). Engineered MABs with enhanced or decreased ADCC and CDC activity have been produced by manipulation of the critical Fc regions. Umana et al. (1999) engineered an anti-neuroblastomal IgG1 with enhanced ADCC activity compared with wild type (WT). Shields et al. (2001) demonstrated that selected IgG1 variants with improved binding to FcγRIIIA showed an enhancement in ADCC for peripheral blood monocyte cells or natural killer cells. These findings indicate that Fc-engineered antibodies may have important applications for improving therapeutic efficacy. It was found that the FCGR3A gene dimorphism generates two allotypes, FcγRIIIa-158V and FcγRIIIa-158F, and the polymorphism in FcγRIIIA is associated with favorable clinical response following rituximab administration in non-Hodgkin's lymphoma patients (Cartron et al. 2004; Dall'Ozzo et al. 2004). Currently, several anti-CD20 MABs with increased binding affinity to FcγRIIIA are in clinical trials. The efficacy of antibody-interleukin 2 fusion protein (Ab-IL-2) was improved by reducing its interaction with Fc receptors (Gillies et al. 1999). In addition, the Fc portion of MABs also binds to the Fc receptor (FcRn named based on discovery in neonatal rats as neonatal FcRn), an Fc receptor belonging to the major histocompatibility complex structure, which is involved in IgG transport and clearance (Junghans 1997). Engineered MABs with a decreased or increased FcRn binding affinity have been investigated for the potential of modifying the pharmacokinetic behavior of MAB (see the section on Antibody Clearance for detail).

■ Antibody Derivatives (F(ab')2, Fab, Antibody Drug Conjugates) and Fusion Proteins

The fragments of antibodies (Fab, F(ab')2, and scFv) have a shorter half-life compared with the full-sized corresponding antibodies. scFv can be further engineered into a bivalent dimer (diabody) (~60 kDa, or trimer: triabody ~90 kDa). Two diabodies can be further linked together to generate bispecific tandem diabody (tandab). Figure 7.3 illustrates the structure of different antibody fragments. Of note, abciximab and ranibizumab are two Fab approved by FDA. Abciximab is a chimeric Fab used for keeping blood from clotting with 20–30 min half-life in serum and 4 h half-life in platelets (Schror and Weber 2003). Ranibizumab, which is administrated via an intravitreal (IVT) injection, was approved for the treatment of macular degeneration in 2006 and exhibits a vitreous elimination half-life of 9 days (Albrecht and DeNardo 2006).

The half-life of Fc is more similar to that of full-sized IgGs (Lobo et al. 2004). Therefore, Fc portions of IgGs have been used to form fusions with molecules such as cytokines, growth factor enzymes, or the ligand-binding region of receptor or adhesion molecules to improve their half-life and stability. Alefacept, abatacept, and etanercept are three Fc fusion proteins on the market. Etanercept, a dimeric fusion molecule consisting of the TNF-α receptor fused to the Fc region of human IgG1, has a half-life of approximately 70–100 h (Zhou 2005), which is much longer than the TNF-α receptor itself (30 min ~2 h) (Watanabe et al. 1988).

Antibodies and antibody fragments can also be linked covalently with cytotoxic radionuclides or drugs to form radioimmunotherapeutic (RIT) agents or antibody drug conjugates (ADCs), respectively. In each case, the Ab is used as a delivery mechanism to selectively target the cytotoxic moiety to tumors (Prabhu et al. 2011). For both ADCs and RIT agents, the therapeutic strategy involves selective delivery of a cytotoxin (drug or radionuclide) to tumors via the antibody. As targeted approaches, both technologies exploit the overexpression of target on the surface of the cancer

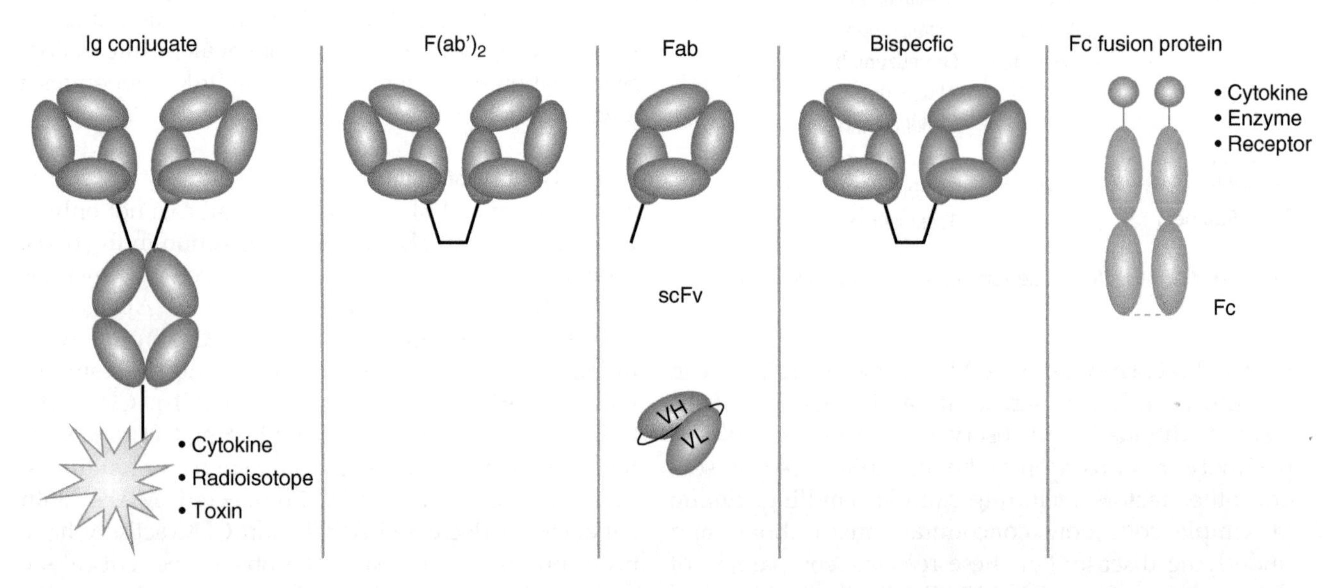

Figure 7.3 ■ Schematic representation of antibody derivatives (F(ab')2, Fab, scFv, and antibody conjugates) and Fc fusion proteins.

cells and thereby minimize damage to normal tissues. Such approaches are anticipated to minimize the significant side effects encountered when cytotoxic small molecule drugs or radionuclides are administered as single agents, thus leading to enhanced therapeutic windows. However, important distinctions exist between these two therapeutic modalities. For example, ADCs often require internalization into the endosomes and/or lysosomes for efficacy, while RIT agents are able to emit radiation, even from the cell surface, to achieve cell killing following direct binding to membrane antigens. Furthermore, RIT can deliver high levels of radiation even with very low doses of radioimmunoconjugate compared to ADCs. Importantly, most clinically successful ADC and RIT agents to date have been against hematologic tumors (Boswell and Brechbiel 2007). Various impediments to the delivery of antibodies and other macromolecules to solid tumors have been widely discussed and studied, especially in the context of microspatial distribution (Thurber et al. 2008).

Gemtuzumab ozogamicin (Mylotarg®, Dowell et al 2001), an anti-CD33 MAB linked to the cytotoxic antitumor antibiotic drug calicheamicin, became the first approved ADC in 2000 when it was granted accelerated approval for the treatment of acute myelogenous leukemia (AML). Calicheamicin binds to the minor groove of DNA, causing double-strand DNA breaks and resulting in inhibition of DNA synthesis. However, gemtuzumab ozogamicin was removed from the US market in June 2010, based on negative results in a follow-up confirmatory trial. In August 2011, the FDA approved a second ADC, brentuximab vedotin (Adcetris® 2011), for treatment of Hodgkin's lymphoma and systemic anaplastic large-cell lymphoma. Like gemtuzumab ozogamicin, brentuximab vedotin is directed against a soluble target (CD30). Most recently, in February 2013, FDA approved ado-trastuzumab emtansine Kadcyla®, a human epidermal growth factor receptor (HER2)-targeted ADC for treatment of HER2-positive breast cancer (LoRusso et al. 2011).

The only current radioimmunotherapeutic agents licensed by the FDA are ibritumomab tiuxetan (Zevalin® 2002) and tositumomab plus ^{131}I tositumomab (Bexxar® 2003), both for non-Hodgkin's lymphoma. Both of the above intact murine MABs bind CD20 and carry a potent beta particle-emitting radioisotope (^{90}Y for ibritumomab/tiuxetan and ^{131}I for tositumomab). In the case of ibritumomab, the bifunctional chelating agent, tiuxetan, is used to covalently link the radionuclide to the antibody, ibritumomab. However, another approved anti-CD20 antibody, rituximab, is included in the dosing regimen as a non-radioactive predose to improve the biodistribution of the radiolabeled antibody. Despite impressive clinical results, radioimmunotherapeutic antibodies have not generated considerable commercial success; various financial, regulatory, and commercial barriers have been cited as contributing factors to this trend (Boswell and Brechbiel 2007).

HOW DO ANTIBODIES FUNCTION AS THERAPEUTICS?

The pharmacological effects of antibodies are first initiated by the specific interaction between antibody and antigen. Monoclonal antibodies generally exhibit exquisite specificity for the target antigen. The binding site on the antigen called the epitope can be linear or conformational and may comprise continuous or discontinuous amino acid sequences. The epitope is the primary determinant of the antibody's modulatory functions, and depending on the epitope, the antibody may exert antagonist or agonist effects, or it may be nonmodulatory. The epitope may also influence the antibody's ability to induce ADCC and CDC. Monoclonal antibodies exert their pharmacological effects via multiple mechanisms that include direct modulation of the target antigen, CDC and ADCC, and delivery of a radionuclide or immunotoxin to target cells.

■ Direct Modulation of Target Antigen

Examples of direct modulation of the target antigen include anti-TNFα, anti-IgE, and anti-CD11a therapies that are involved in blocking and removal of the target antigen. Most monoclonal antibodies act through multiple mechanisms and may exhibit cooperativity with concurrent therapies.

■ Complement-Dependent Cytotoxicity (CDC)

The complement system is an important part of the innate (i.e., nonadaptive) immune system. It consists of many enzymes that form a cascade with each enzyme acting as a catalyst for the next. CDC results from interaction of cell-bound monoclonal antibodies with proteins of the complement system. CDC is initiated by binding of the complement protein, C1q, to the Fc domain. The IgG1 and IgG3 isotypes have the highest CDC activity, while the IgG4 isotype lacks C1q binding and complement activation (Presta 2002). Upon binding to immune complexes, C1q undergoes a conformational change, and the resulting activated complex initiates an enzymatic cascade involving complement proteins C2 to C9 and several other factors. This cascade spreads rapidly and ends in the formation of the membrane attack complex (MAC), which inserts into the membrane of the target and causes osmotic disruption and lysis of the target. Figure 7.4 illustrates the mechanism for CDC with rituximab (a chimeric antibody, which targets the CD20 antigen) as an example.

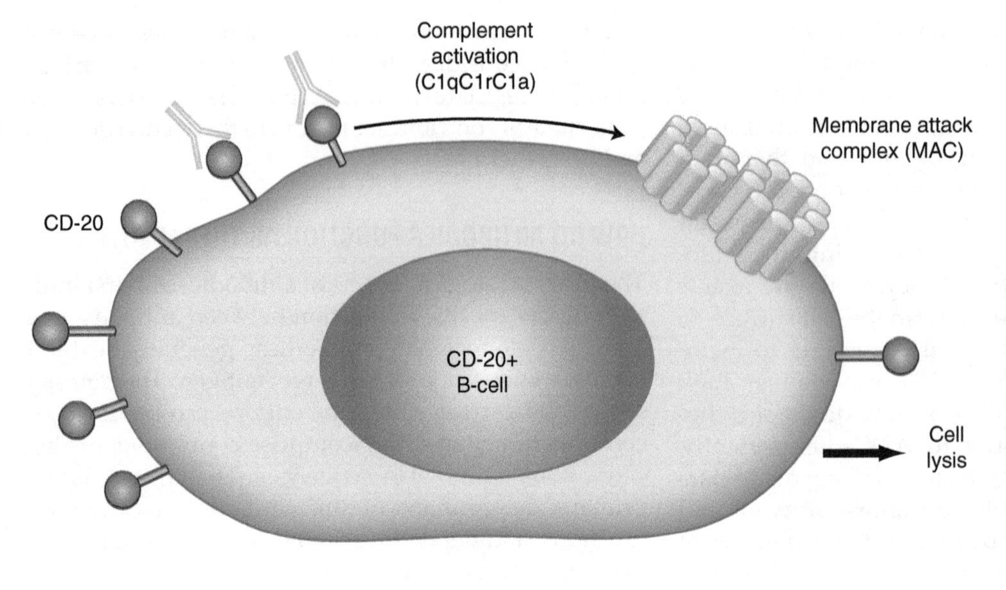

Figure 7.4 ■ An example of CDC, using a B cell lymphoma model, where the monoclonal antibody rituximab binds to the receptor and initiates the complement system, also known as the "complement cascade." The end result is a membrane attack complex (*MAC*), which leads to cell lysis and death.

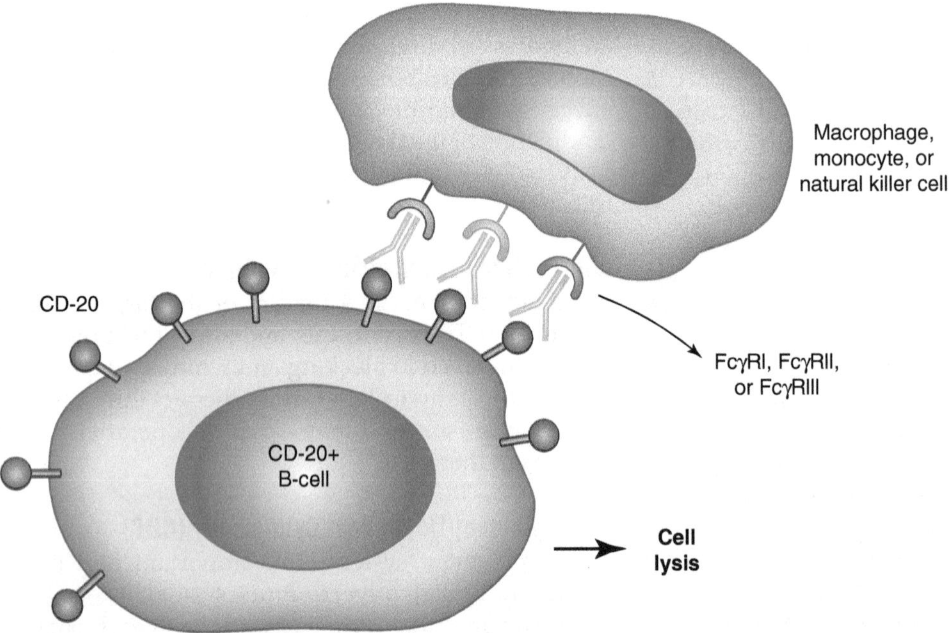

Figure 7.5 ■ An example of ADCC. In this situation rituximab targets the CD20 antigen. This antigen is expressed on a significant number of B cell malignancies. The Fc fragment of the monoclonal antibody binds the Fc receptors found on monocytes, macrophages, and NK cells. These cells in turn engulf the bound tumor cell and destroy it. NK cells secrete cytokines that lead to cell death, and they also recruit B cells.

■ Antibody-Dependent Cellular Cytotoxicity (ADCC)

ADCC is a mechanism of cell-mediated immunity whereby an effector cell of the immune system actively lyses a target cell that has been bound by specific antibodies. It is one of the mechanisms through which antibodies, as part of the humoral immune response, can act to limit and contain infection. Classical ADCC is mediated by natural killer (NK) cells, monocytes, or macrophages, but an alternate ADCC is used by eosinophils to kill certain parasitic worms known as helminths. ADCC is part of the adaptive immune response due to its dependence on a prior antibody response. The typical ADCC involves activation of NK cells, monocytes, or macrophages and is dependent on the recognition of antibody-coated infected cells by Fc receptors on the surface of these cells. The Fc receptors recognize the Fc portion of antibodies such as IgG, which bind to the surface of a pathogen-infected target cell. The Fc receptor that exists on the surface of NK cell is called CD16 or FcγRIII. Once bound to the Fc receptor of IgG, the NK cell releases cytokines such as IFN-γ and cytotoxic granules like perforin and granzyme that enter the target cell and promote cell death by triggering apoptosis. This is similar to, but independent of, responses by cytotoxic T cells. Figure 7.5 illustrates the mechanism for ADCC with rituximab as an example.

■ Apoptosis

Monoclonal antibodies achieve their therapeutic effect through various mechanisms. In addition to the abovementioned effector functions, they can have direct effects in producing apoptosis or programmed cell death. It is characterized by nuclear DNA degradation, nuclear degeneration and condensation, and the phagocytosis of cell remains.

TRANSLATIONAL MEDICINE/DEVELOPMENT PROCESS

The tight connection of basic to clinical sciences is an essential part of translational medicine trying to *translate* the knowledge of basic science into practical therapeutic applications for patients. This knowledge transfer is also often entitled as *from-bench-to-bedside* process emphasizing the transition of scientific advancements into clinical applications. This framework of translational medicine is applied during the discovery and drug development process of a specific antibody against a certain disease. It includes major steps such as identifying an important and viable pathophysiological target antigen to modify the disease in a beneficial way, producing MABs with structural elements providing optimal pharmacokinetics, and safety and efficacy by testing the MAB in nonclinical safety and efficacy models and finally in patients. An overview of the development phases of the molecules comprising the nonclinical activities is outlined in Fig. 7.6. Furthermore, the critical components of the entire development process of MABs from a PK/PD perspective is explained in detail within the following sections.

■ Preclinical Safety Assessment of MABs

Preclinical safety assessment of MABs offers unique challenges, as many of the classical evaluations employed for small molecules are not appropriate for protein therapeutics in general and MABs in particular. For example, in vitro genotoxicology tests such as the Ames and chromosome aberration assays are generally not conducted for MABs given their limited interaction with nuclear material and the lack of appropriate receptor/target expression in these systems. As MAB binding tends to be highly species specific, suitable animal models are often limited to nonhuman primates, and for this reason, many common in vivo models such as rodent carcinogenesis bioassays and some safety pharmacology bioassays are not viable for MAB therapeutic candidates. For general toxicology studies, cynomolgus and rhesus monkeys are most commonly employed and offer many advantages given their close phylogenetic relationship with humans; however, due to logistics, animal availability, and costs, group sizes tend to be much smaller than typically used for lower species thus limiting statistical power. In some cases, alternative models are employed to enable studies in rodents. Rather than directly testing the therapeutic candidate, analogous monoclonal antibodies that can bind to target epitopes in lower species (e.g., mice) can be engineered and used as a surrogate MAB for safety evaluation (Clarke et al. 2004). Often the antibody framework amino acid sequence is modified to reduce antigenicity thus enabling longer-term studies (Albrecht and DeNardo 2006; Weiner 2006; Cohenuram and Saif 2007). Another approach is to use transgenic models that express the human receptor/target of interest (Bugelski et al. 2000); although, results must be interpreted with caution as transgenic models often have altered physiology and typically lack historical background data for the model.

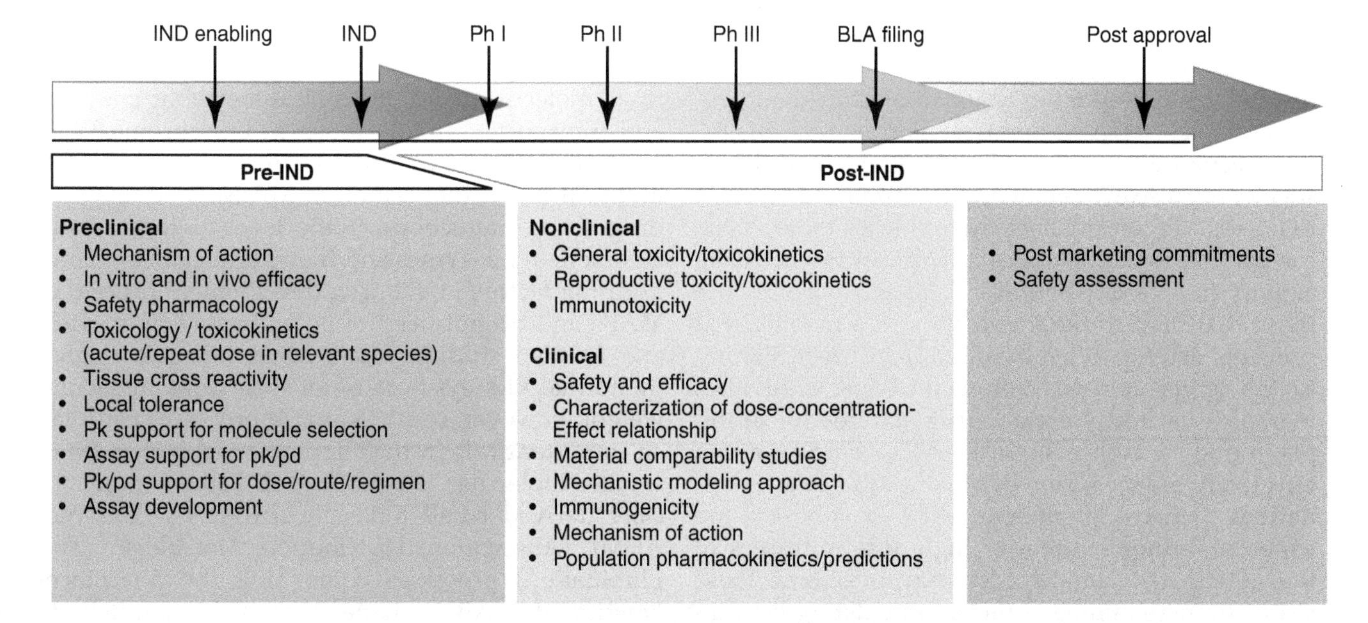

Figure 7.6 ■ Flowchart depicting PK/PD/toxicology study requirements during preclinical and clinical drug product development.

To address development issues that are specific to monoclonal antibodies and other protein therapeutics, the International Conference of Harmonization (ICH) has developed guidelines specific to the preclinical evaluation of biotechnology-derived pharmaceuticals (ICH 1997a, b).

For general safety studies, species selection is an important consideration given the exquisite species specificity often encountered with MABs. Model selection needs to be justified based on appropriate expression of the target epitope, appropriate binding affinity with the therapeutic candidate, and appropriate biologic activity in the test system. To aid in the interpretation of results, tissue cross-reactivity studies offer the ability to compare drug localization in both animal and human tissues. For MAB therapeutic candidates, a range of three or more dose levels are typically selected to attain pharmacologically relevant serum concentrations, to approximate levels anticipated in the clinic, and to provide information at doses higher than anticipated in the clinic. For most indications, it is important to include dose levels that allow identification of a no observable adverse effect level (NOAEL). If feasible, the highest dose should fall within the range where toxicity is anticipated; although, in practice, many monoclonal antibodies do not exhibit toxicity, and other factors limit the maximum dose. To best reflect human exposures, doses are often normalized and selected to match and exceed anticipated human therapeutic exposure in plasma, serum, or blood based upon the exposure parameters, area under the concentration-time curve (AUC), maximum concentrations (C_{max}), or concentration prior to next treatment (C_{trough}). The route of administration, dosing regimen, and dosing duration should be selected to best model the anticipated use in clinical trials (ICH 1997a, b).

To adequately interpret nonclinical study results, it is important to characterize anti-therapeutic antibody (ATA) responses. For human MABs, ATA responses are particularly prominent in lower species but also evident in nonhuman primates albeit to a lesser degree, making these species more viable for chronic toxicity studies. ATAs can impact drug activity in a variety of ways. Neutralizing ATAs are those that bind to the therapeutic in a manner that prevents activity, often by inhibiting direct binding to the target epitope. Non-neutralizing antibodies may also indirectly impact drug activity, for example, rapid clearance of drug–ATA complexes can effectively reduce serum drug concentrations. In situations where prominent ATA responses are expected, administration of high-dose multiples of the anticipated clinical dose may overcome these issues by maintaining sufficient circulating concentrations of active drug. To properly interpret study results, it is important to characterize ATA incidence and magnitude as the occurrence of ATA responses could mask toxicities. Alternatively, robust ATA responses may induce significant signs of toxicity such as infusion-related anaphylaxis that may not be predictive of human outcome where ATA formation is likely to be less of an issue. If ATA formation is clearly impacting circulating drug levels, ATA-positive individuals are often removed from consideration when evaluating pharmacokinetic parameters to better reflect the anticipated pharmacokinetics in human populations.

Pharmacokinetics

A thorough and rigorous PK program in the early learning phase of preclinical drug development can provide a linkage between drug discovery and preclinical development. PK information can be linked to PD by mathematical modeling, which allows characterizing the time course of the effect intensity resulting from a certain dosing regimen. Antibodies often exhibit pharmacokinetic properties that are much more complex than those typically associated with small molecule drugs (Meibohm and Derendorf 2002). In the following sections, the basic characteristics of antibody pharmacokinetics are summarized.

The pharmacokinetics of antibodies is very different from small molecules. Table 7.3 summarizes the PK differences between small molecule drugs and therapeutic antibodies regarding pharmacokinetics. Precise, sensitive, and accurate bioanalytical methods are essential for PK interpretation. However, for MABs, the immunoassays and bioassay methodologies are often less specific as compared to assays used for small molecule drugs (e.g., LC/MS/MS). Monoclonal antibodies are handled by the body very differently than small molecules. In contrast to small molecule drugs, the typical metabolic enzymes and transporter proteins such as cytochrome P450, multidrug resistance (MDR) efflux pumps are not involved in the disposition of MABs. Consequently, drug–drug interactions at the level of these drug-metabolizing enzymes and transporters are not complicating factors in the drug development process of MABs and do not need to be addressed by in vitro and in vivo studies. Intact MABs are not cleared by normal kidneys because of their large molecular weight; however, renal clearance processes can play an important role in the elimination of molecules of smaller molecular weight such as Fab's and chemically derived small molecule drugs. The different *ADME* (*A*bsorption, *D*istribution, *M*etabolism, and *E*limination) processes comprising the pharmacokinetics of MABs will be discussed separately to address their individual specifics.

Small molecule drugs	Monoclonal antibodies			
	Target is soluble antigen	Target is cell-bound antigen	Target is cell-bound antigen that is internalized and downregulated	Target is cell-/tissue-bound antigen that can be shed
PK usually independent of PD	PK often independent of PD	PK often dependent of PD		
Binding generally nonspecific (can affect multiple enzymes)	Binding very specific for target protein or antigen			
Usually linear PK Nonlinear PK problematic	Linear PK	Nonlinear PK		
Relatively short $t_{1/2}$	Long $t_{1/2}$	Low dose: short $t_{1/2}$ High dose: long $t_{1/2}$		
Not always orally available	Need parenteral dosing. SC or IM is possible			
Metabolism by P450s or other enzymes	Metabolism by nonspecific clearance mechanisms. No P450s involved	Metabolism by specific and nonspecific clearance mechanisms. No P450s involved		
Renal clearance often important	MABs: No renal clearance of intact antibody. May be cleared by damaged kidneys Antibody fragment might be eliminated by renal clearance			
Binding to tissues, high Vd	Distribution usually limited to blood and extracellular space			

Table 7.3 ■ Comparing the pharmacokinetics of small molecule drugs and monoclonal antibodies (Lobo et al. 2004; Roskos et al. 2004; Mould and Sweeney 2007).

ABSORPTION

Most of the MABs are not administrated orally because of their limited gastrointestinal stability, lipophilicity, and size all of which result in insufficient resistance against the hostile proteolytic gastrointestinal milieu and very limited permeation through the lipophilic intestinal wall. Therefore, intravenous administration is still the most frequently used route, which allows for immediate systemic delivery of large volume of drug product and provides complete systemic availability. Of note, 13 of over 30 FDA-approved antibody therapies listed in Table 7.1 are administered by an extravascular route (adalimumab (SC), alefacept (IM), canakinumab (SC), certolizumab pegol (SC), denosumab (SC), efalizumab (SC), etanercept (SC), golimumab (SC), omalizumab (SC), palivizumab (IM), ranibizumab (intravitreal), rilonacept (SC), and ustekinumab (SC)). The absorption mechanisms of SC or IM administration are poorly understood. However, it is believed that the absorption of MABs after IM or SC is likely via lymphatic drainage due to its large molecular weight, leading to a slow absorption rate (see Chap. 5). The bioavailability for antibodies after SC or IM administration has been reported to be around 50–100 % with maximal plasma concentrations observed 1–8 days following administration (Lobo et al. 2004). For example, following an IM injection, the bioavailability of alefacept was ~60 % in healthy male volunteers; its C_{max} was threefold (0.96 versus 3.1 μg/mL) lower, and its T_{max} was 30 times longer (86 versus 2.8 h) than a 30-min IV infusion (Vaishnaw and TenHoor 2002). Interestingly differences in PK have also been observed between different sites of IM dosing. The pharmacokinetics of MAB after IM injection is also dependent on the injection site. PAMAB, a fully humanized MAB against *Bacillus anthracis* protective antigen, has a significantly different pharmacokinetics between IM-GM (gluteus maximus site) and IM-VL (vastus lateralis site) injection in healthy volunteers (Subramanian et al. 2005). The bioavailability of PAMAB is 50–54 % for IM-GM injection and 71–85 % for IM-VL injection (Subramanian et al. 2005). Of note, MABs appear to have greater bioavailability after SC administration in monkeys than in humans (Oitate et al. 2011). The mean bioavailability of adalimumab is 52–82 % after a single 40 mg SC administration in healthy adult subjects, whereas it was observed to be 94–100 % in monkeys. Similarly the mean bioavailability of omalizumab is 66–71 % after a single SC dose in patients with asthma versus 88–104 % in monkeys.

DISTRIBUTION

After reaching the bloodstream, MABs undergo biphasic elimination from serum, beginning with a rapid distribution phase. The distribution volume of the rapid distribution compartment is relatively small, approximating plasma volume. It is reported that the volume of the central compartment (Vc) is about 2–3 L, and the steady-state volume of distribution (Vss) is around 3.5–7 L for MABs in humans (Lobo et al. 2004; Roskos et al. 2004). The small Vc and Vss for MABs indicate that the distribution of MABs is restricted to the blood and extracellular spaces, which is in agreement with their hydrophilic nature and their large molecular weight, limiting access to the lipophilic tissue compartments. Small volumes of distributions are consistent with relatively small tissue: blood ratios for most antibodies typically ranging from 0.1 to 0.5 (Baxter et al. 1994; Baxter et al. 1995, Berger et al. 2005). For example, the tissue to blood concentration ratios for a murine IgG1 MAB against the human ovarian cancer antigen CA125 in mice at 24 h after injection are 0.44, 0.39, 0.48, 0.34, 0.10, and 0.13 for the spleen, liver, lung, kidney, stomach, and muscle, respectively (Berger et al. 2005). Brain and cerebrospinal fluid are anatomically protected by blood–tissue barriers. Therefore, both compartments are very limited distribution compartments for Abs hindering the access for therapeutic MABs. For example, endogenous IgG levels in CSF were shown to be in the range of only 0.1–1 % of their respective serum levels (Wurster and Haas 1994). However, it has been repeatedly noted that the reported Vss obtained by traditional non-compartmental or compartmental analysis may be not correct for some MABs with high extent of catabolism within tissue (Tang et al. 2004; Lobo et al. 2004; Straughn 2006). The rate and extent of antibody distribution will be dependent on the kinetics of antibody extravasation within tissue, distribution within tissue, and elimination from tissue. Convection, diffusion, transcytosis, binding, and catabolism are important determining factors for antibody distribution (Lobo et al. 2004). Therefore, Vss might be substantially greater than the plasma volume in particular for those MABs demonstrating high binding affinity in the tissue. Effects of the presence of specific receptors (i.e., antigen sink) on the distribution for MAB have been reported by different research groups (Danilov et al. 2001; Kairemo et al. 2001). Danilov et al. (2001) found that anti-PECAM-1 (CD31) MABs show tissue to blood concentration ratios of 13.1, 10.9, and 5.96 for the lung, liver, and spleen, respectively, in rats at 2 h after injection. Therefore, the true Vss of the anti-PECAM-1 is likely to be 15-fold greater than plasma volume.

Another complexity which needs to be considered is that tissue distribution via interaction with target proteins (e.g., cell surface proteins) and subsequent internalization of the antigen-MAB complex might be dose dependent. For the murine analog MAB of efalizumab (M17), a pronounced dose-dependent distribution was demonstrated by comparing tissue to blood concentration ratios for liver, spleen, bone marrow, and lymph node after a tracer dose of radiolabeled M17 and a high-dose treatment (Coffey et al. 2005). The tracer dose of M17 resulted into substantially higher tissue to blood concentration ratios of 6.4, 2.8, 1.6, and 1.3 for the lung, spleen, bone marrow, and lymph node, respectively, in mice at 72 h after injection. Whereas, the saturation of the target antigen at the high-dose level reduced the tissue distribution to the target independent distribution and resulted consequently into substantially lower tissue to blood concentration ratios (less than 1).

FcRn may play an important role in the transport of IgGs from plasma to the interstitial fluid of tissue. However, the effects of FcRn on the MABs' tissue distribution have not been fully understood. Ferl et al. (2005) reported that a physiologically based pharmacokinetic (PBPK) model, including the kinetic interaction between the MAB and the FcRn receptor within intracellular compartments, could describe the biodistribution of an anti-CEA MAB in a variety of tissue compartments such as plasma, lung, spleen, tumor, skin, muscle, kidney, heart, bone, and liver. FcRn was also reported to mediate the IgG across the placental barriers (Junghans 1997) and the vectorial transport of IgG into the lumen of intestine (Dickinson et al. 1999) and lung (Spiekermann et al. 2002).

ANTIBODY CLEARANCE

Antibodies are mainly cleared by catabolism and broken down into peptide fragments and amino acids, which can be recycled – to be used as energy supply or for new protein synthesis. Due to the small molecular weight of antibodies fragments (e.g., Fab and Fv), elimination of these fragments is faster than intact IgGs, and they can be filtered through the glomerus and reabsorbed and/or metabolized by proximal tubular cells of the nephron (Lobo et al. 2004). Murine monoclonal anti-digoxin Fab, F(ab')2, and IgG1 have half-lives of 0.41, 0.70, and 8.10 h in rats, respectively (Bazin-Redureau et al. 1997). Several studies reported that the kidney is the major route for the catabolism of Fab and elimination of unchanged Fab (Druet et al. 1978; McClurkan et al. 1993).

Typically, IgGs have serum half-life of approximately 21 days, resulting from clearance values of about 3–5 mL/day/kg, and Vss of 50–100 mL/kg. The exception is IgG3, which has only a half-life of 7 days. The half-life of IgG is much longer than other Igs (IgA 6 days, IgE 2.5 days, IgM 5 days, IgD 3 days). The FcRn receptor has been demonstrated to be a prime determinant of the disposition of IgG antibodies (Ghetie et al. 1996; Junghans and Anderson 1996; Junghans 1997). FcRn, which protects IgG from catabolism and contributes to the long plasma half-life of IgG, was first postulated by Brambell in 1964 (Brambell et al. 1964) and cloned in the late 1980s (Simister and Mostov 1989a, b).

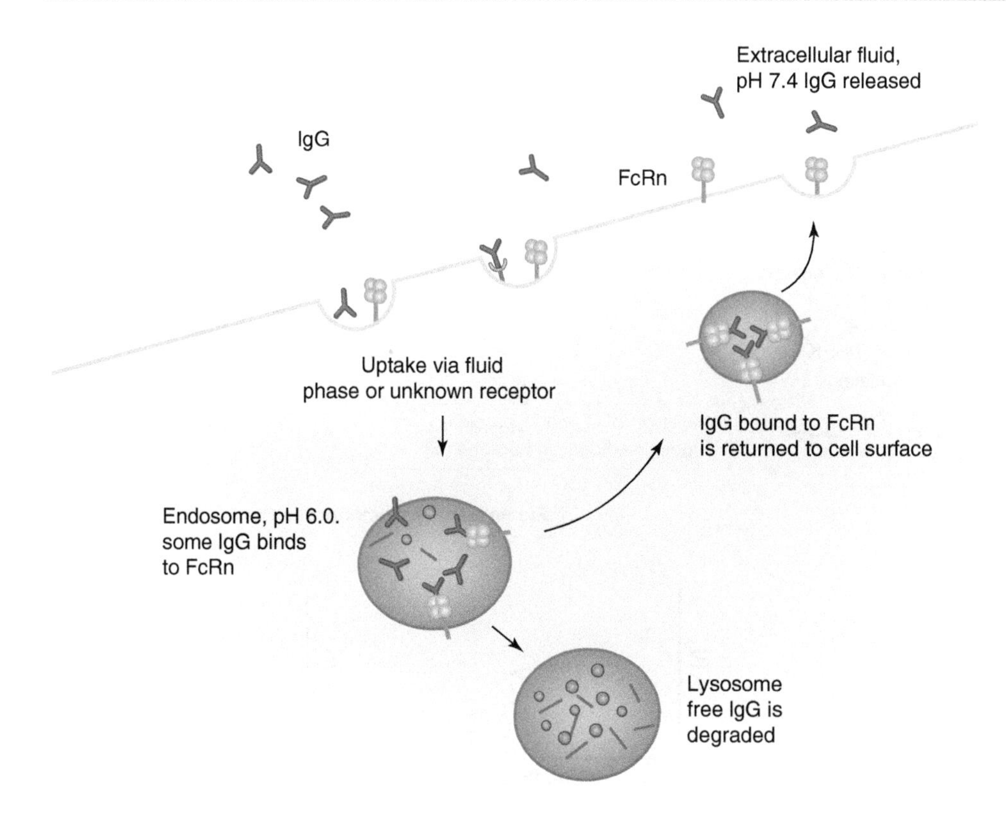

Figure 7.7 ■ Schematic disposition pathway of IgG antibodies via interaction with FcRn in endosomes. (1) IgGs enter cells by receptor-mediated endocytosis by binding of the Fc part to FcRn. (2) The intracellular vesicles (endosomes) can fuse with lysosome containing proteases. (3) Proteases can degrade non-bound IgG molecules, whereas IgG bound to FcRn is protected. (4) The intact IgG bound to FcRn is transported back to the cell surface and released back to the extracellular fluid.

FcRn is a heterodimer comprising of a β_2m light chain and a MHC class I-like heavy chain. The receptor is ubiquitously expressed in cells and tissues. Several studies have shown that IgG clearance in β_2m knockout mice (Ghetie et al. 1996; Junghans and Anderson 1996) and FcRn heavy chain knockout mice (Roopenian et al. 2003) is increased 10–15-fold, with no changes in the elimination of other Igs. Figure 7.7 illustrates how the FcRn receptor protects IgG from catabolism and contributes to its long half-life. The FcRn receptor binds to IgG in a pH-dependent manner: binding to IgG at acidic pH (6.0) at endosome and releasing IgG at physiological pH (7.4). The unbound IgG proceeds to the lysosome and undergoes proteolysis.

It has been demonstrated that IgG half-life is dependent on its affinity to FcRn receptors. The shorter half-life of IgG3 was attributed to its low binding affinity to the FcRn receptor (Junghans 1997; Medesan et al. 1997). Murine MABs have serum half-lives of 1–2 days in human. The shorter half-life of murine antibodies in human is due to their low binding affinity to the human FcRn receptor. It is reported that human FcRn binds to human, rabbit, and guinea pig IgG, but not to rat, mouse, sheep, and bovine IgG; however, mouse FcRn binds to IgG from all of these species (Ober et al. 2001). Interestingly, human IgG1 has greater affinity to murine FcRn (Petkova et al. 2006), which indicates potential limitations of using mice as preclinical models for human IgG1 pharmacokinetic evaluations. Ward's group confirmed that an engineered human IgG1 had disparate properties in murine and human systems (Vaccaro et al. 2006). Engineered IgGs with higher affinity to FcRn receptor have a two to three-fold higher half-life compared with wild type in mice and monkeys (Hinton et al. 2006; Petkova et al. 2006). Two engineered human IgG1 mutants with enhanced binding affinity to human FcRn show a considerably extended half-life compared with wild type in hFcRn transgenic mice (4.35 ± 0.53, 3.85 ± 0.55 days versus 1.72 ± 0.08 days) (Petkova et al. 2006). Hinton et al. (2006) found that the half-life of IgG1 FcRn mutants with increasing binding affinity to human FcRn at pH 6.0 is about 2.5-fold longer that the wild-type Ab in monkey (838 ± 187 h versus 336 ± 34 h).

Dose-proportional, linear clearance has been observed for MAB against soluble antigens with low endogenous levels (such as TNF-α, IFN-α, VEGF, and IL-5). For example, linear PK has been observed for a humanized MAB directed to human interleukin-5 following intravenous administration over a 6,000-fold dose range (0.05–300 mg/kg) in monkeys (Zia-Amirhosseini et al. 1999). The clearance of rhuMAB against vascular endothelial growth factor after IV dosing (2–50 mg/kg) ranged from 4.81 to 5.59 mL/day/kg and did not depend on dose (Lin et al. 1999). The mean total serum clearance and the estimated mean terminal half-life of adalimumab were reported to range from 0.012 to 0.017 L/h and 10.0 to 13.6 days,

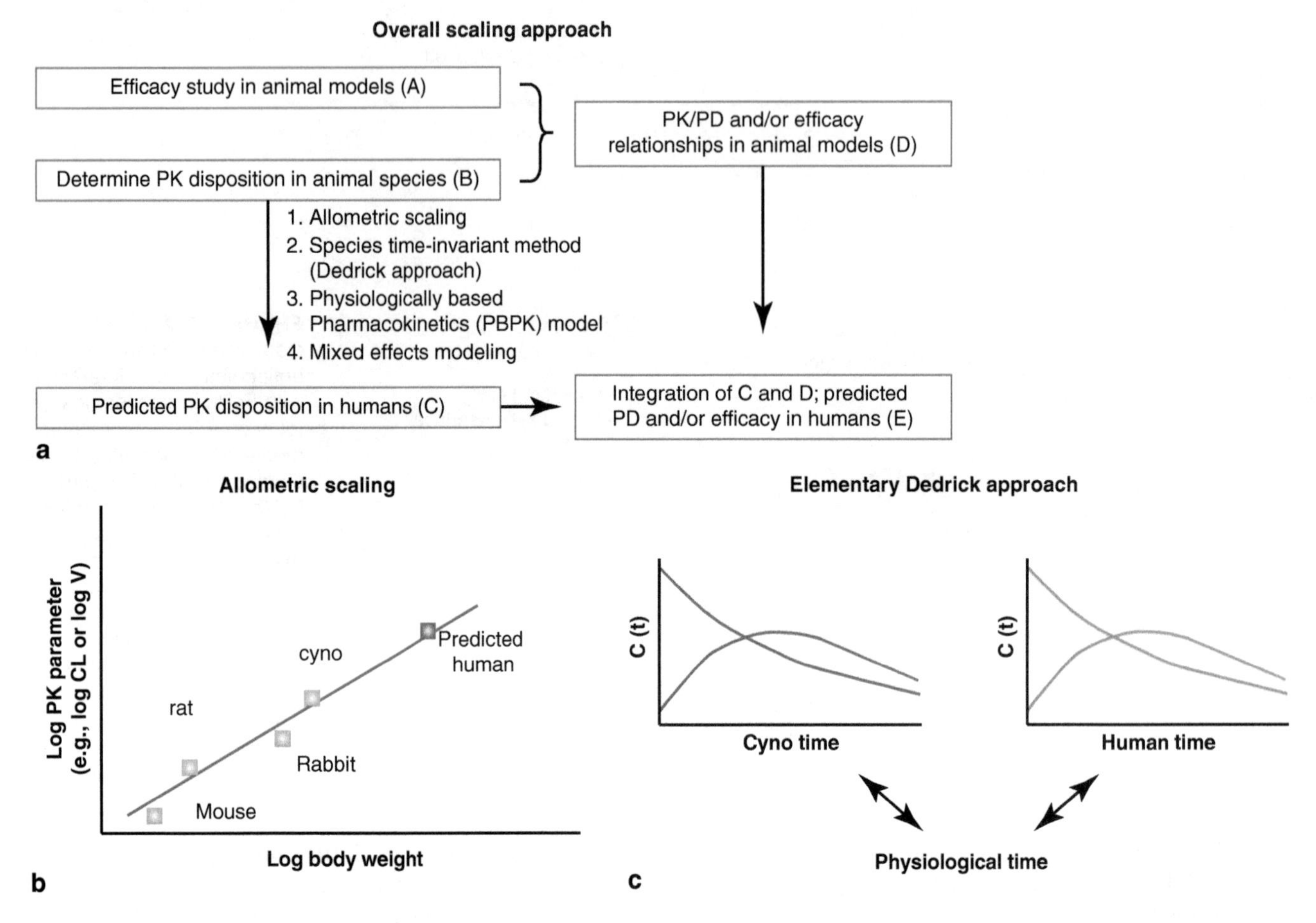

Figure 7.8 ■ PK/PD scaling approach from preclinical studies to humans. (**a**) Overall scaling approach. (**b**) Allometric scaling. (**c**) Elementary dedrick approach.

respectively, for a 5-cohort clinical trial (0.5–10 mg/kg), with an overall mean half-life of 12 days (den Broeder et al. 2002). However, MABs against soluble antigens with high endogenous levels (such as IgE) exhibit nonlinear pharmacokinetics. The pharmacokinetics of omalizumab, an antibody against IgE, is linear only at doses greater than 0.5 mg/kg (Petkova et al. 2006; Xolair 2003).

Elimination of MABs may also be impacted by interaction with the targeted cell-bound antigen, and this phenomenon was demonstrated by dose-dependent clearance and half-life. At low dose, MABs show a shorter half-life and a faster clearance due to receptor-mediated elimination. With increasing doses, receptors become saturated; the half-life gradually increases to a constant; and the clearance gradually decreases to a constant. The binding affinity (K_d), antigen density, and antigen turnover rate may influence the receptor-mediated elimination. Koon et al. (2006) found a strong inverse correlation between CD25+ cell expression and apparent daclizumab (a MAB specifically binding to CD25) half-life. It has been shown that the pharmacokinetics of murine antihuman CD3 antibodies may be determined by the disappearance of target antigen (Meijer et al. 2002). In monkeys and mice, clearance of SGN-40, a humanized monoclonal anti-CD40 antibody, was much faster at low dose, suggesting nonlinear pharmacokinetics (Kelley et al. 2006). In addition, Ng et al. (2006) demonstrated that an anti-CD4 monoclonal antibody (TRX-1) had ~5-fold faster CL at 1 mg/kg dose compared with 10 mg/kg dose (37.4 ± 2.4 versus 7.8 ± 0.6 mL/day/kg) in healthy volunteers. They also found that receptor-mediated CL via endocytosis became saturated at higher doses; nonspecific clearance of TRX-1 contributed 8.6, 27.1, and 41.7 % of total CL when dose was 1, 5, and 10 mg/kg, respectively.

In addition to FcRn and antigen–antibody interaction, other factors may also contribute to MAB elimination (Lobo et al. 2004; Roskos et al. 2004; Tabrizi et al. 2006):

1. *Immunogenicity of antibody*: The elimination of MABs in humans often increases with increasing level of immunogenicity (Ternant and Paintaud 2005; Tabrizi et al. 2006).
2. *The degree and the nature of antibody glycosylation*: The study conducted by Newkirk et al. (1996) shows that

the state of glycosylation of IgG affects the half-life in mice and that by removing the terminal sugars (sialic acid and galactose), the antibody (IgG2a) will remain in circulation significantly longer. However, Huang et al. (2006) demonstrated that a humanized anti-Aβ MAB with different glycans in the Fc region had the same clearance in mice.

3. *Susceptibility of antibody to proteolysis*: Gillies and his coworkers (2002) improved the circulating half-life of antibody-interleukin 2 immunocytokine twofold compared with wild type (1.0 h versus 0.54 h) by increasing the resistance to intracellular degradation.
4. *Effector function*, such as interactions with FcγR, could also regulate elimination and PK of MABs (Mahmood and Green 2005). Mutation of the binding site of FcγR has dramatic effects on the clearance of the Ab-IL-2 fusion protein (Gillies et al. 1999).
5. *Concomitant medications*: Methotrexate reduced adalimumab apparent clearance after single dose and multiple dosing by 29 and 44 %, respectively, in patients with rheumatoid arthritis (Humira 2007). In addition, azathioprine and mycophenolate mofetil were reported to reduce clearance of basiliximab by approximately 22 and 51 %, respectively (Simulect 2005). These interactions could be explained by the effects of small molecule drugs on the expression of Fcγ receptors. It has been found that methotrexate has the impact on the expression profiles of FcγRI on monocytes significantly in rheumatoid arthritis patients (Bunescu et al. 2004).
6. *Body weight, age, disease state, and other demography factors* can also change MAB pharmacokinetics (Mould and Sweeney 2007) (see discussion on Population Pharmacokinetics).

THERAPEUTIC MAB–DRUG INTERACTIONS

MABs and other therapeutic proteins are increasingly combined with small molecule drugs to treat various diseases. Assessment of the potential for PK- and/or PD-based MAB–drug interactions is more frequently incorporated into the drug development process (Girish et al. 2011). The exposure and/or response of concomitantly administered drugs can be altered by MABs (MAB as perpetrator). Alternately, PK and/or PD of therapeutic MABs can be affected by other drugs (MAB as victim).

Several different mechanisms have been proposed for MAB–drug interactions. Various cytokines and cytokine modulators can influence the expression and activity of cytochrome P450 (CYP) enzymes and drug transporters (Lee et al. 2010). Therefore, if a therapeutic MAB is a cytokine or cytokine modulator, it can potentially alter the systemic exposure and/or clinical response of concomitantly administered drugs that are substrates of CYPs or transporters (Huang et al. 2010) particularly those with narrow therapeutic windows. For example, an increase in cyclosporin A (CsA) trough level was observed when given in combination with muromomab (Vasquez and Pollak 1997). Similarly, basiliximab has been shown to increase CsA and tacrolimus level when used in combination (Sifontis et al. 2002). In diseases states, such as infection or inflammation, cytokines or cytokine modulators can also normalize previously changed activity of CYPs or transporters, thereby alter the exposure of coadministered drugs. Examples include tocilizumab coadministered with omeprazole and tocilizumab coadministered with simvastatin (Actemra® 2010).

MAB–drug interactions can also occur through changing the formation of anti-therapeutic antibody (ATA), which may enhance MAB clearance from the body. For example, methotrexate (MTX) reduced the apparent clearance of adalimumab by 29 and 44 % after single and repeated dosing (Humira® 2010). MTX also had similar effect on infliximab (Marni et al. 1998). PD-based interactions can result from alteration of target biology, such as information on the site of expression, relative abundance of expression, and the pharmacology of the target (Girish et al. 2011). Examples include efalizumab in combination with triple immune-suppressant therapy (Vincenti et al. 2007) and anakinra in combination with etanercept (Genovese et al. 2004).

To date, evidence of MAB–drug interactions via nonspecific clearance appears to be limited, although downregulation of Fcγ receptors by MTX is observed in patients with rheumatoid arthritis. It is possible that changes in Fcγ receptors can affect MAB clearance in the presence of MTX (Girish et al. 2011).

ADCs can also interact with drugs or MABs vis mechanisms described above. However, evidence of ADC–drug or ADC–MAB interaction appears to be absent. Lu et al. reported lack of interaction between ado-trastuzumab emtansine (T-DM1) and pertuzumab in patients with HER2-positive metastatic breast cancer (Lu et al. 2011). Similarly no interaction was observed between T-DM1 and paclitaxel or T-DM1 and docetaxel (Lu et al. 2012). With the theoretical potential for and current experiences with MAB–drug interactions, a question and risk-based integrated approach depending on the mechanism of the MABs and patient population have been progressively adopted during drug development to address important questions regarding the safety and efficacy of MAB and drug combinations (Girish et al. 2011). Various in vitro test systems have been used to provide some insight into the MAB–drug interactions, such as isolated hepatocytes and liver microsomes. However, the interpretation of these in vitro data is difficult. More importantly, prospective predictions of drug interactions based on in vitro findings have not been feasible for MABs. Therefore, clinical methods are primarily used to assess MAB–drug interactions. Three common methods are dedicated drug interaction studies, although rare, population pharmacokinetics, and clinical cocktail studies. Details of various strategies used in pharmaceutical industry were reviewed in a 2011 AAPS white paper (Girish et al. 2011).

■ Prediction of Human PK/PD Based on Preclinical Information

Prior to the first-in-human (FIH) clinical study, a number of preclinical in vivo and in vitro experiments are conducted to evaluate the PK/PD, safety, and efficacy of a new drug candidate. However, the ultimate goal is at all times to predict how these preclinical results on pharmacokinetics, safety, and efficacy translate into a given patient population. Therefore, the objective of translational research is to predict PK/PD/safety outcomes in a target patient population, acknowledging the similarities and differences between preclinical and clinical settings.

Over the years, many theories and approaches have been proposed and used for scaling preclinical PK data to humans. Allometric scaling, based on a power–law relationship between size of the body and physiological and anatomical parameters, is the simplest and most widely used approach (Dedrick 1973; Mahmood 2005, 2009). Physiologically based PK modeling (Shah and Betts 2012), species-invariant time method (Dedrick approach) (Oitate et al. 2012), and nonlinear mixed effect modeling based on allometry (Jolling et al. 2005; Martin-Jimenez and Riviere 2002) have also been used for interspecies scaling of PK. While no single scaling method has been shown to definitively predict human PK in all cases, especially for small molecule drugs (Tang and Mayersohn 2005), the PK for MABs can be predicted reasonably well, especially for MAB at doses where the dominant clearance route is likely to be independent of concentration. Most therapeutic MABs bind to nonhuman primate antigens more often than to rodent antigens, due to the greater sequence homology observed between nonhuman primates and humans. The binding epitope, in vitro binding affinity to antigen, binding affinity to FcRn, tissue cross-reactivity profiles, and disposition and elimination pathways of MABs are often comparable in nonhuman primates and humans. It has recently been demonstrated that clearance and distribution volume of MABs with linear PK in humans can be reasonably projected based on data from nonhuman primates alone, with a fixed scaling exponent ranging from 0.75 to 0.9 for clearance and a fixed scaling exponent 1 for volume (Ling et al. 2009; Wang and Prueksaritanont 2010; Deng et al. 2011; Dong et al. 2011; Oitate et al. 2011). For MABs that exhibited nonlinear pharmacokinetics, the best predictive performance was obtained above doses that saturated the target of the MAB (Dong et al. 2011). Pharmacokinetic prediction for low doses of a MAB with nonlinear elimination remains challenging and will likely require further exploration of species difference in target expression level, target antibody binding and target kinetics, as well as strategic animal in vivo PK studies, designed with relevant dose ranges. Immunogenicity is an additional challenge for prediction of MAB PK. Alterations in the PK profile due to immune-mediated clearance mechanisms in preclinical species cannot be scaled up to humans, since animal models are not predictive of human immune response to human MABs. Thus, either excluding antidrug antibody (ADA)-positive animals from PK scaling analysis or using only the early time points prior to their observation in ADA positive animals has been a standard practice in the industry.

Due to its complexity, any extrapolation of PD to humans requires more thorough consideration than for PK. Little is known about allometric relationships in PD parameters. It is expected that the physiological turnover rate constants of most general structures and functions among species should obey allometric principles, whereas capacity and sensitivity tend to be similar across species (Mager et al. 2009). Through integration of PK/PD modeling and interspecies scaling, PD effects in humans may be predicted if the PK/PD relationship is assumed to be similar between animal models and humans (Duconge et al. 2004; Kagan et al. 2010). For example, a PK/PD model was first developed to optimize the dosing regimen of a MAB against EGF/r3 using tumor-bearing nude mice as an animal model of human disease (Duconge et al. 2004). This PK/PD model was subsequently integrated with allometric scaling to calculate the dosing schedule required in a potential clinical trial to achieve a specific effect (Duconge et al. 2004).

In summary, species differences in antigen expression level, antigen–antibody binding and antigen kinetics, differences in FcRn binding between species, the immunogenicity, and other factors must be considered during PK/PD scaling of a MAB from animals to humans.

■ PK/PD in Clinical Development of Antibody Therapeutics

Several new developments have taken place in the antibody therapeutics in the last years. The emphasis in the field has grown and is obvious by the fact that many of the companies are now involved in building antibody product-based collaborations. Drug development has traditionally been performed in sequential phases, divided into preclinical as well as clinical phases I–IV. During the development phases of the molecules, the safety and PK/PD characteristics are established in order to narrow down on the compound selected for development and its dosing regimen. This information-gathering process has been characterized as two successive learning–confirming cycles (Sheiner 1997; Sheiner and Wakefield 1999).

The first cycle (phases I and IIa) comprises learning about the dose that is tolerated in healthy subjects and confirming that this dose has some measurable benefits in the targeted patients. An affirmative answer at this first cycle provides the justification for a larger and more costly second learn–confirm cycle (phases IIb and III), where the learning step is focused on how to use the drug benefit/risk ratio, whereas the confirm step is aimed at demonstrating acceptable benefit/risk in a large patient population (Meibohm and Derendorf 2002). In the following sections, the approved therapeutic antibody efalizumab is provided as a case study to understand the various steps during the development of antibodies for various indications.

A summary of the overall PK/PD data from multiple studies within the efalizumab (Raptiva®) clinical development program and an integrated overview of how these data were used for development and the selection of the approved dosage of efalizumab for psoriasis will be discussed in detail. Psoriasis is a chronic skin disease characterized by abnormal keratinocyte differentiation and hyperproliferation and by an aberrant inflammatory process in the dermis and epidermis. T cell infiltration and activation in the skin and subsequent T cell-mediated processes have been implicated in the pathogenesis of psoriasis (Krueger 2002).

Efalizumab is a subcutaneously (SC) administered recombinant humanized monoclonal IgG1 antibody that has received approval for the treatment of patients with psoriasis in more than 30 countries, including the United States and the European Union (Raptiva 2004). Efalizumab is a targeted inhibitor of T cell interactions (Werther et al. 1996). An extensive preclinical research program was conducted to study the safety and mechanism of action (MOA) of efalizumab. Multiple clinical studies have also been conducted to investigate the efficacy, safety, pharmacokinetics (PK), pharmacodynamics (PD), and MOA of efalizumab in patients with psoriasis.

PRE-PHASE I STUDIES

In the process of developing therapeutic antibodies, integrated understanding of the pharmacokinetic/pharmacodynamic (PK/PD) concepts provides a highly promising tool. A thorough and rigorous preclinical program in the early learning phase of preclinical drug development can provide a linkage between drug discovery and preclinical development. As it sets the stage for any further development activities, the obtained information at this point is key to subsequent steps (Meibohm and Derendorf 2002). At the preclinical stage, potential applications might comprise the evaluation of in vivo potency and intrinsic activity, the identification of bio-/surrogate markers, understanding the MOA, as well as dosage form/regimen selection and optimization. A few of these specific aims are described below with information on efalizumab as an example.

IDENTIFICATION OF MOA AND PD BIOMARKERS

The identification of appropriate PD endpoints is crucial to the process of drug development. Thus, biomarkers are usually tested early during exploratory preclinical development for their potential use as pharmacodynamic or surrogate endpoints.

Through an extensive preclinical research program, the MOA and PD biomarkers for efalizumab have been established. Efalizumab binds to CD11a, the α-subunit of leukocyte function antigen-1 (LFA-1), which is expressed on all leukocytes, and decreases cell surface expression of CD11a. Efalizumab inhibits the binding of LFA-1 to intercellular adhesion molecule-1 (ICAM-1), thereby inhibiting the adhesion of leukocytes to other cell types. Interaction between LFA-1 and ICAM-1 contributes to the initiation and maintenance of multiple processes, including activation of T lymphocytes, adhesion of T lymphocytes to endothelial cells, and migration of T lymphocytes to sites of inflammation, including skin. Consistent with the proposed MOA for efalizumab, in vitro experiments have demonstrated that efalizumab binds strongly to human lymphocytes with a K_d of approximately 110 ng/mL (Werther et al. 1996; Dedrick et al. 2002) and blocks the interaction of human T lymphocytes with tissue-specific cells such as keratinocytes in a concentration-dependent manner.

Upon understanding the MOA, PD effects relevant to the MOA of efalizumab are usually measured in order to identify the efficacious dosage of antibody therapeutics. As saturation of CD11a binding sites by efalizumab has been shown to increase while T cell activation is increasingly inhibited, maximum saturation of CD11a binding sites occurs at efalizumab concentrations >10 μg/mL, resulting in maximum T cell inhibition (Werther et al. 1996; Dedrick et al. 2002). Therefore, CD11a expression and saturation have been chosen as relevant PD markers for this molecule.

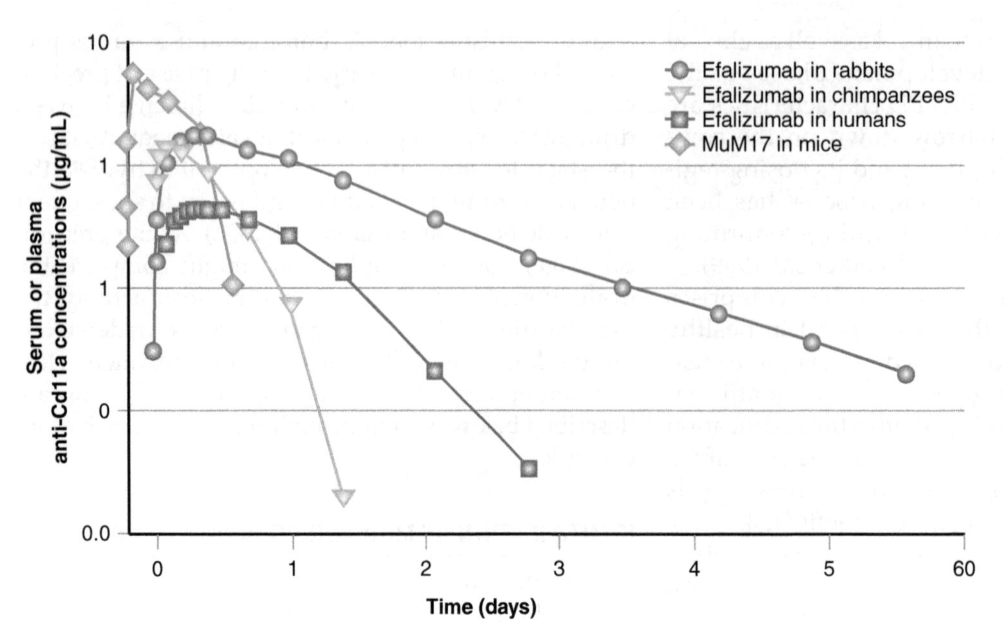

Figure 7.9 ■ Anti-CD11a molecules comparative PK profiles in humans, chimpanzees, rabbits, and mice following SC dose. Due to the species differences in binding, the pharmacokinetics of efalizumab are nonlinear (i.e., dose dependent) in humans and chimpanzees, while being linear in rabbits (nonbinding species). muM17, on the other hand binds to the mouse anti-CD11a and exhibits dose-dependent pharmacokinetics in mice.

ROLE OF SURROGATE MOLECULES

The role of surrogate molecules in assessing ADME of therapeutic antibodies is important as the antigen specificity limits ADME studies of humanized monoclonal antibodies in rodents. In the development of therapeutic antibodies, various molecules may be used to provide a comprehensive view of their PK/PD properties. Studies with surrogates might lead to important information regarding safety, mechanism of action, disposition of the drug, tissue distribution, and receptor pharmacology, which might be too cumbersome and expensive to conduct in nonhuman primates. Surrogates (mouse/rat) provide a means to gaining knowledge of PK and PD in a preclinical rodent model thus allowing rational dose optimization in the clinic. Therefore, in the case of efalizumab to complete a more comprehensive safety assessment, a chimeric rat anti-mouse CD11a antibody, muM17, was developed and evaluated as a species-specific surrogate molecule for efalizumab. muM17 binds mouse CD11a with specificity and affinity similar to those of efalizumab to human. In addition, muM17 in mice was demonstrated to have similar pharmacological activities as that of efalizumab in human (Nakakura et al. 1993; Clarke et al. 2004). Representative PK profiles of efalizumab and muM17 in various species are depicted in Fig. 7.9 to help understand the species differences in the PK behavior of molecules.

PHARMACOKINETICS OF EFALIZUMAB

A brief overview of efalizumab nonclinical PK/PD results is provided in the following sections to summarize the key observations that led to decisions in designing the subsequent clinical programs. The ADME program consisted of pharmacokinetic, pharmacodynamic (CD11a down-modulation and saturation), and toxicokinetic data from pharmacokinetic, pharmacodynamic, and toxicology studies with efalizumab in chimpanzees and with muM17 in mice. The use of efalizumab in the chimpanzee and muM17 in mice for pharmacokinetic and pharmacodynamic and safety studies was supported by in vitro activity assessments. The nonclinical data were used for pharmacokinetic and pharmacodynamic characterization, pharmacodynamic-based dose selection, and toxicokinetic support for confirming exposure in toxicology studies. Together, these data have supported both the design of the nonclinical program and its relevance to the clinical program.

The observed pharmacodynamics as well as the mechanism of action of efalizumab and muM17 is attributed to bind CD11a present on cells and tissues. The binding affinities of efalizumab to human and chimpanzee CD11a on CD3 lymphocytes are comparable confirming the use of chimpanzees as a valid nonclinical model for humans. CD11a expression has been observed to be greatly reduced on T lymphocytes in chimpanzees and mice treated with efalizumab and muM17, respectively. Expression of CD11a is restored as efalizumab and muM17 are eliminated from the plasma. The bioavailability of efalizumab in chimpanzees and muM17 in mice after an SC dose was dose-dependent and ranged from 35 to 48 % and 63 to 89 % in chimpanzees and mice, respectively. Binding to CD11a serves as a major pathway for clearance of these molecules, which leads to nonlinear pharmacokinetics depending on the relative amounts of CD11a and efalizumab or muM17 (Coffey et al. 2005).

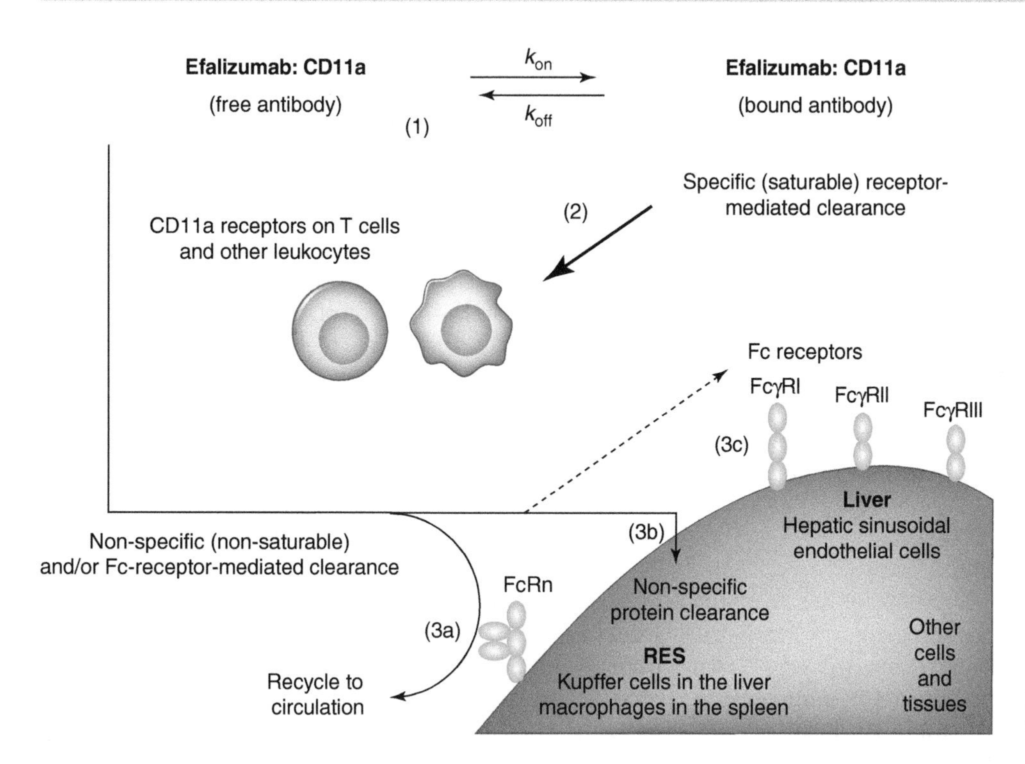

Figure 7.10 ■ Clearance pathways for efalizumab.

The disposition of efalizumab and the mouse surrogate muM17 is mainly determined by the combination of both specific interactions with the ligand CD11a and by their IgG1 framework and is discussed in detail as follows. The factors controlling the disposition of these antibodies are shown in Fig. 7.10 and include the following:

1. The binding of the free antibody with its ligand CD11a present on both circulating lymphocytes and tissues leads to its removal from circulation. Data suggests that anti-CD11a antibodies are internalized by purified T cells, and upon internalization, the antibodies appeared to be targeted to lysosomes and cleared from within the cells in a time-dependent manner. CD11a-mediated internalization and lysosomal targeting of efalizumab may constitute one pathway by which this antibody is cleared in vivo (Coffey et al. 2005).
2. Binding to CD11a is both specific and saturable as demonstrated by the dose-dependent clearance of efalizumab in chimpanzees and humans or muM17 in mice.
3. Because of its IgG1 framework, free or unbound efalizumab or muM17 levels are also likely to be influenced by:
   (a) Recycling and circulation following binding to and internalization by the neonatal Fc receptor (FcRn)
   (b) Nonspecific uptake and clearance by tissues
   (c) Binding via its Fc framework to Fcγ receptors present on hepatic sinusoidal endothelial cells

The disposition of efalizumab is governed by the species specificity and affinity of the antibody for its ligand CD11a, the amount of CD11a in the system, and the administered dose.

Based on the safety studies, efalizumab was considered to be generally well tolerated in chimpanzees at doses up to 40 mg/kg/week IV for 6 months, providing an exposure ratio of 339-fold based on cumulative dose and 174-fold based on the cumulative AUC, compared with a clinical dose of 1 mg/kg/week. The surrogate antibody muM17 was also well tolerated in mice at doses up to 30 mg/kg/week SC. In summary efalizumab was considered to have an excellent nonclinical safety profile thereby supporting the use in adult patients.

CLINICAL PROGRAM OF EFALIZUMAB: PK/PD STUDIES, ASSESSMENT OF DOSE, ROUTE, AND REGIMEN

The drug development process at the clinical stage provides several opportunities for integration of PK/PD concepts. Clinical phase I dose escalation studies provide, from a PK/PD standpoint, the unique chance to evaluate the dose–concentration–effect relationship for therapeutic and toxic effects over a wide range of doses up to or even beyond the maximum tolerated dose under controlled conditions (Meredith et al. 1991). PK/PD evaluations at this stage of drug development can provide crucial information regarding the potency and tolerability of the drug in vivo and the verification and suitability of the PK/PD concept established during preclinical studies.

Efalizumab PK and PD data are available from ten studies in which more than 1,700 patients with psoriasis received IV or SC efalizumab. In the phase I studies, PK and PD parameters were characterized by extensive sampling during treatment; in the phase III trials, steady-state trough levels were measured once or twice during the first 12-week treatment period for all the studies and during extended treatment periods for some studies. Several early phase I and II trials have examined IV injection of efalizumab, and dose-ranging findings from these trials have served as the basis for SC dosing levels used in several subsequent phase I and all phase III trials.

■ IV Administration of Efalizumab

The PK of monoclonal antibodies varies greatly, depending primarily on their affinity for and the distribution of their target antigen (Lobo et al. 2004). Efalizumab exhibits concentration-dependent nonlinear PK after administration of single IV doses of 0.03, 0.1, 0.3, 0.6, 1.0, 2.0, 3.0, and 10.0 mg/kg in a phase I study. This nonlinearity is directly related to specific and saturable binding of efalizumab to its cell surface receptor, CD11a, and has been described by a PK/PD model developed by Bauer et al. (Bauer et al. 1999) which is discussed in the following sections. The PK profiles of efalizumab following single IV doses with observed data and model predicted fit are presented in Fig. 7.11. Mean clearance (CL) decreased from 380 to 6.6 mL/kg/day for doses of 0.03 mg/kg to 10 mg/kg, respectively. The volume of distribution of the central compartment (Vc) of efalizumab was 110 mL/kg at 0.03 mg/kg (approximately twice the plasma volume) and decreased to 58 mL/kg at 10 mg/kg (approximately equal to plasma volume), consistent with saturable binding of efalizumab to CD11a in the vascular compartment. Because of efalizumab's nonlinear PK, its half-life ($t_{1/2}$) is dose dependent.

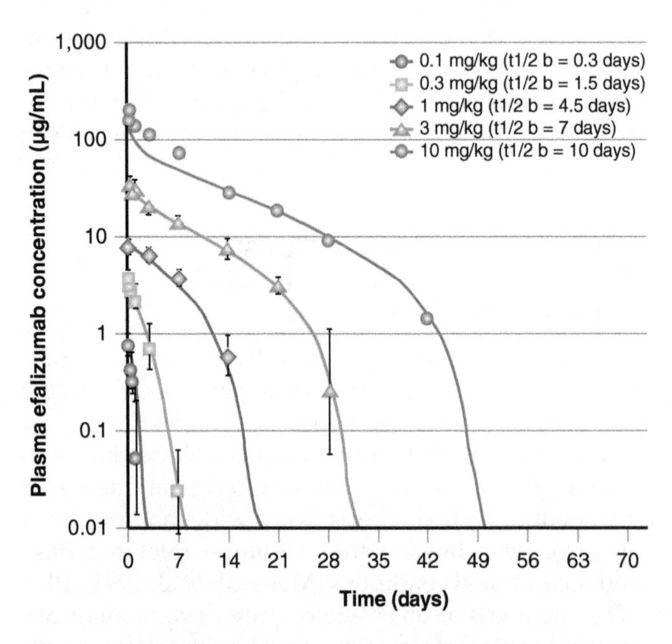

Figure 7.11 ■ Plasma concentration versus time profile for efalizumab following single IV doses in psoriasis patients.

In a phase II study of efalizumab, it was shown that at a weekly dosage of 0.1 mg/kg IV, patients did not maintain maximal down-modulation of CD11a expression and did not maintain maximal saturation. Also at the end of 8 weeks of efalizumab treatment, 0.1 mg/kg/week IV, patients did not have statistically significant histological improvement and did not achieve a full clinical response. The minimum weekly IV dosage of efalizumab tested that produced histological improvements in skin biopsies was 0.3 mg/kg/week, and this dosage resulted in submaximal saturation of CD11a binding sites but maximal down-modulation of CD11a expression. Improvements in patients' psoriasis were also observed, as determined by histology and by the Psoriasis Area and Severity Index (PASI) (Papp et al. 2001).

■ Determination of SC Doses

Although efficacy was observed in phase I and II studies with 0.3 mg/kg/week IV efalizumab, dosages of 0.6 mg/kg/week and greater (given for 7–12 weeks) provided more consistent T lymphocyte CD11a saturation and maximal PD effect. At dosages ≤0.3 mg/kg/week, large between-subject variability was observed, whereas at dosages of 0.6 or 1.0 mg/kg/week, patients experienced better improvement in PASI scores, with lower between-patient variability in CD11a saturation and down-modulation. Therefore, this dosage was used to estimate an appropriate minimum SC dose of 1 mg/kg/week (based on a 50 % bioavailability) that would induce similar changes in PASI, PD measures, and histology. The safety, PK, and PD of a range of SC efalizumab doses (0.5–4.0 mg/kg/week administered for 8–12 weeks) were evaluated initially in 2 phase I studies (Gottlieb et al. 2003). To establish whether a higher SC dosage might produce better results, several phase III clinical trials assessed a 2.0 mg/kg/week SC dosage in addition to the 1.0 mg/kg/week dosage. A dose of 1.0 mg/kg/week SC efalizumab was selected as it produced sufficient trough levels in patients to maintain the maximal down-modulation of CD11a expression and binding-site saturation between weekly doses (Joshi et al. 2006). Figure 7.12 depicts the serum efalizumab levels, CD11a expression, and available CD11a binding sites on T lymphocytes (mean ± SD) after subcutaneous administration of 1 mg/kg efalizumab.

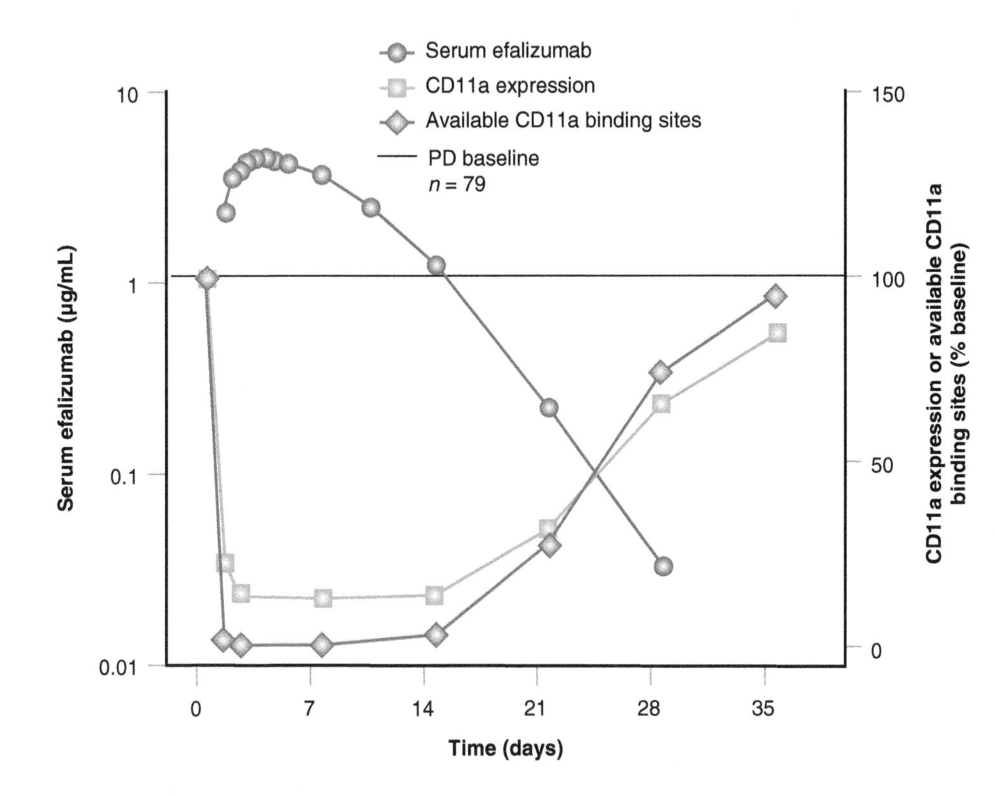

Figure 7.12 ■ PK/PD profile following efalizumab in humans (1 mg/kg SC).

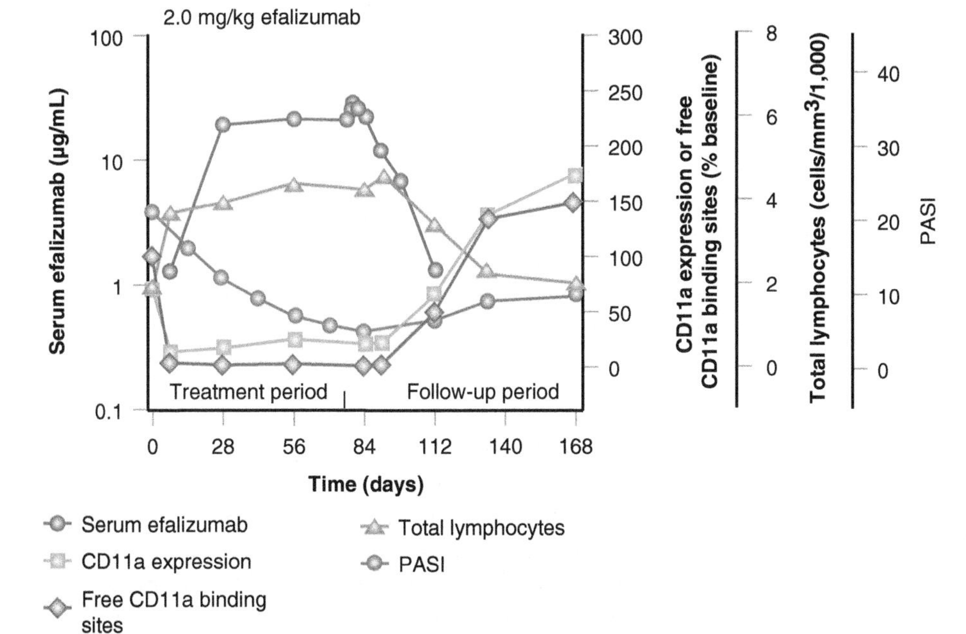

Figure 7.13 ■ Serum efalizumab, CD11a expression, and free CD11a binding sites on T lymphocytes, absolute lymphocyte counts, and Psoriasis Area and Severity Index (PASI) score (mean) following 1.0 mg/kg/week SC efalizumab for 12 weeks and 12 weeks posttreatment.

■ SC Administration of Efalizumab

The PK of SC efalizumab has been well characterized following multiple SC doses of 1.0 and 2.0 mg/kg/week (Mortensen et al. 2005; Joshi et al. 2006). A phase I study that collected steady-state PK and PD data for 12 weekly SC doses of 1.0 and 2.0 mg/kg in psoriasis patients provided most of the pharmacologic data relevant to the marketed product. Although peak serum concentration after the last dose (C_{max}) was observed to be higher for the 2.0 mg/kg/week (30.9 μg/mL) than for the 1.0 mg/kg/week dosage (12.4 μg/mL), no additional changes in PD effects were observed at the higher dosages (Mortensen et al. 2005). Following a dose of 1.0 mg/kg/week, serum efalizumab concentrations were adequate to induce maximal down-modulation of CD11a expression and a reduction in free CD11a binding sites on T lymphocytes (Fig. 7.13). Steady-state serum efalizumab levels were reached more quickly with the 1.0 mg/kg/week dosage at 4 weeks compared with the 2.0 mg/kg/week dosage at 8 weeks (Mortensen et al. 2005), which is in agreement with the average effective $t_{1/2}$ for SC efalizumab 1.0 mg/kg/week of 5.5 days (Boxenbaum and Battle 1995). The bioavailability was estimated at approximately 50 %. Population PK analyses indicated that body weight was the most significant covariate affecting efalizumab SC clearance, thus supporting body weight-based dosing for efalizumab (Sun et al. 2005).

Mechanistic Modeling Approaches

In clinical drug development, PK/PD modeling approaches can be applied as analytical tools for identifying and characterizing the dose–response relationships of drugs and the mechanisms and modulating factors involved. Additionally, they may be used as predictive tools for exploring various dosage regimens as well as for optimizing further clinical trial designs, which might allow one to perform fewer, more focused studies with improved efficiency and cost-effectiveness. The PK/PD database established during the preclinical and clinical learning phases in the development process and supplemented by population data analysis provides the backbone for these assessments.

PK/PD modeling has been used to characterize efalizumab plasma concentrations and CD11a expression on CD3-positive lymphocytes in chimpanzees and in subjects with psoriasis (Bauer et al. 1999). As the PK data revealed that CL of efalizumab was not constant across dose levels, one of the models described by Bauer et al. (1999) incorporated a Michaelis–Menten clearance term into the pharmacokinetic equations and utilized an indirect response relationship to describe CD11a turnover. However, in the above model, the exposure–response relationship of efalizumab was not addressed, and another report expanded on the developed receptor-mediated pharmacokinetic and pharmacodynamic model by incorporating data from five phase I and II studies to develop a pharmacokinetic (PK) pharmacodynamic (PD) efficacy (E) model to further increase the understanding of efalizumab interaction with CD11a on T cells and consequent reduction in severity of psoriasis (Ng et al. 2005). A general outline of the mechanistic modeling approach for various molecules is presented alongside the model for efalizumab in Fig. 7.14a. The description of the pharmacokinetic–pharmacodynamic–efficacy model of efalizumab in psoriasis patients is described below and is schematically represented in Fig. 7.14b. Details on parameters utilized in the model can be found in the paper by Ng et al. (2005).

PHARMACOKINETIC ANALYSIS

A first-order absorption, two-compartment model with both linear and Michaelis–Menten elimination was used to describe the plasma efalizumab concentration data. This model is schematically represented in Fig. 7.14b (iv).

PHARMACODYNAMIC ANALYSIS

A receptor-mediated pharmacodynamic model previously developed was used to describe the dynamic interaction of efalizumab to CD11a, resulting in the removal of efalizumab from the circulation and reduction of cell surface CD11a (Bauer et al. 1999). This model is schematically represented in Fig. 7.14b (v).

EFFICACY ANALYSIS

The severity of the disease has been assessed by the PASI score that is assumed to be directly related to the psoriasis skin production. The rate of psoriasis skin production was then modeled to be directly proportional to the amount of free surface CD11a on T cells, which is offset by the rate of skin healing (Fig. 7.14b (vi)).

MODEL RESULTS

Upon evaluation and development, the model was used to fit the PK/PD/efficacy data simultaneously. The plasma concentration-time profile of efalizumab was reasonably described by use of the first-order absorption, two-compartment model with Michaelis–Menten elimination from the central compartment. In addition, the pharmacodynamic model described the observed CD11a-time data from all the studies reasonably well. In the efficacy model, an additional CD11a-independent component to psoriasis skin production accounted for incomplete response to efalizumab therapy and the model described the observed data well. Figure 7.15 depicts the fit of the model to the PK/PD/efficacy data.

The pharmacokinetic-pharmacodynamic-efficacy model developed for efalizumab has a broad application to antibodies that target cell-bound receptors, subjected to receptor-mediated clearance, and for which coating and modulation of the receptors are expected to be related to clinical response (Mould et al. 1999). Despite the nonlinear pharmacokinetics of these agents, the model can be used to describe the time course of the pharmacodynamic effect and efficacy after different dosing regimens.

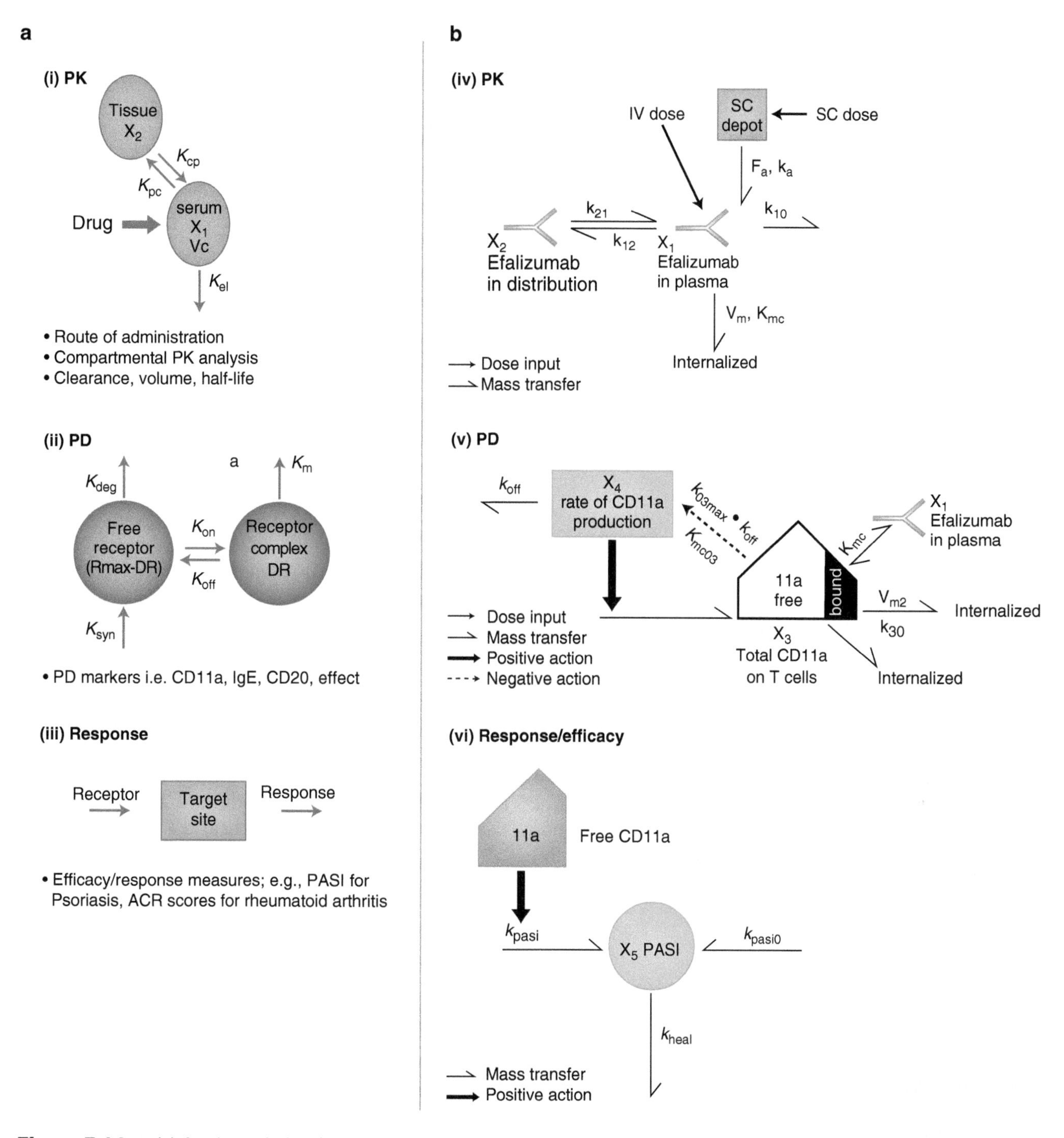

Figure 7.14 ■ (**a**) A schematic for pharmacokinetic–pharmacodynamic–response model for antibody therapeutics (*i*) PK, (*ii*) PD, and (*iii*) response. (**b**) Schematic representation of pharmacokinetic–pharmacodynamic–efficacy model of efalizumab in psoriasis patients. (*iv*) First-order absorption, two-compartment pharmacokinetic model with linear and nonlinear elimination from the central compartment. (*v*) Pharmacodynamic model with negative feedback mechanism. (*vi*) Efficacy model with CD11a-dependent and -independent pathway.

Population Pharmacokinetics of Monoclonal Antibodies

Compared to many small molecule drugs, monoclonal antibodies typically exhibit less inter- and intra-subject variability of the standard pharmacokinetic parameters such as volume of distribution and clearance. However, it is possible that certain pathophysiological conditions may result into substantially increased intra- and inter-patient variability. In addition, patients are usually not very homogeneous; patients vary in sex, age, body weight; they may have concomitant disease and may be receiving multiple drug treatments.

Even the diet, lifestyle, ethnicity, and geographic location can differ from a selected group of "normal" subjects. These covariates can have substantial influence on pharmacokinetic parameters. Therefore, good therapeutic practice should always be based on an understanding of both the influence of covariates on pharmacokinetic parameters as well as the pharmacokinetic variability in a given patient population. With this knowledge, dosage adjustments can be made to accommodate differences in pharmacokinetics due to genetic, environmental, physiological, or pathological factors, for instance, in case of compounds with a relatively small therapeutic index. The framework of application of population pharmacokinetics during drug development is summarized in the FDA guidance document entitled "*Guidance for Industry – Population Pharmacokinetics*" (www.fda.gov).

For population pharmacokinetic data analysis, there are generally two reliable and practical approaches. One approach is the standard two-stage (STS) method, which estimates parameters from the plasma drug concentration data for an individual subject during the first stage. The estimates from all subjects are then combined to obtain a population mean and variability estimates for the parameters of interest. The method works well when sufficient drug concentration-time data are available for each individual patient; typically these data are gathered in phase 1 clinical trials. A second approach, nonlinear mixed effect modeling (NonMEM), attempts to fit the data and partition the differences between theoretical and observed values into random error terms. The influence of fixed effect (i.e., age, sex, body weight) can be identified through a regression model building process.

The original scope for the NonMEM approach was its applicability even when the amount of time-concentration data obtained from each individual is sparse and conventional compartmental PK analyses are not feasible. This is usually the case during the routine visits in phase III or IV clinical studies. Nowadays the NonMEM approach is applied far beyond its original scope due to its flexibility and robustness. It has been used to describe data-rich phase I and phase IIa studies or even preclinical data to guide and expedite drug development from early preclinical to clinical studies (Aarons et al. 2001; Chien et al. 2005).

There has been increasing interest in the use of population PK and pharmacodynamic (PD) analyses for different antibody products (i.e., antibodies, antibody fragments, or antibody fusion proteins) over the past 10 years (Lee et al. 2003; Nestorov et al. 2004; Zhou et al. 2004; Yim et al. 2005; Hayashi et al. 2007; Agoram et al. 2007; Gibiansky and Gibiansky 2009; Dirks and Meibohm 2010; Zheng et al. 2011; Gibiansky and Frey 2012). One example involving analysis of population plasma concentration data involved a dimeric fusion protein, etanercept (Enbrel®). A one-compartment first-order absorption and elimination population PK model with interindividual and inter-occasion variability on clearance, volume of distribution, and absorption rate constant, with covariates of sex and race on apparent clearance and body weight on clearance and volume of distribution, was developed for etanercept in rheumatoid arthritis adult patients (Lee et al. 2003). The population PK model for etanercept was further applied to pediatric patients with juvenile rheumatoid arthritis and established the basis of the 0.8 mg/kg once weekly regimen in pediatric patients with juvenile rheumatoid arthritis (Yim et al. 2005). Unaltered etanercept PK with concurrent methotrexate in patients with rheumatoid arthritis has been demonstrated in a phase IIIb study using population PK modeling approach (Zhou et al. 2004). Thus, no etanercept dose adjustment is needed for patients taking concurrent methotrexate. A simulation exercise of using the final population PK model of subcutaneously administered etanercept in patients with psoriasis indicated that the two different dosing regimens (50 mg QWk versus 25 mg BIWk) provide a similar steady-state exposure (Nestorov et al. 2004). Therefore, their respective efficacy and safety profiles are likely to be similar as well.

An added feature is the development of a population model involving both pharmacokinetics and pharmacodynamics. Population PK/PD modeling has been used to characterize drug PK and PD with models ranging from simple empirical PK/PD models to advanced mechanistic models by using drug–receptor binding principles or other physiologically based principles. A mechanism-based population PK and PD binding model was developed for a recombinant DNA-derived humanized IgG1 monoclonal antibody, omalizumab (Xolair®) (Hayashi et al. 2007). Clearance and volume of distribution for omalizumab varied with body weight, whereas clearance and rate of production of IgE were predicted accurately by baseline IgE, and overall, these covariates explained much of the interindividual variability. Furthermore, this mechanism-based population PK/PD model enabled the estimation of not only omalizumab disposition but also the binding with its target, IgE, and the rate of production, distribution, and elimination of IgE.

Population PK/PD analysis can capture uncertainty and the expected variability in PK/PD data generated in preclinical studies or early phases of clinical development. Understanding the associated PK or PD variability and performing clinical trial simulation by incorporating the uncertainty from the existing PK/PD data allows projecting a plausible range of doses for future clinical studies and final practical uses.

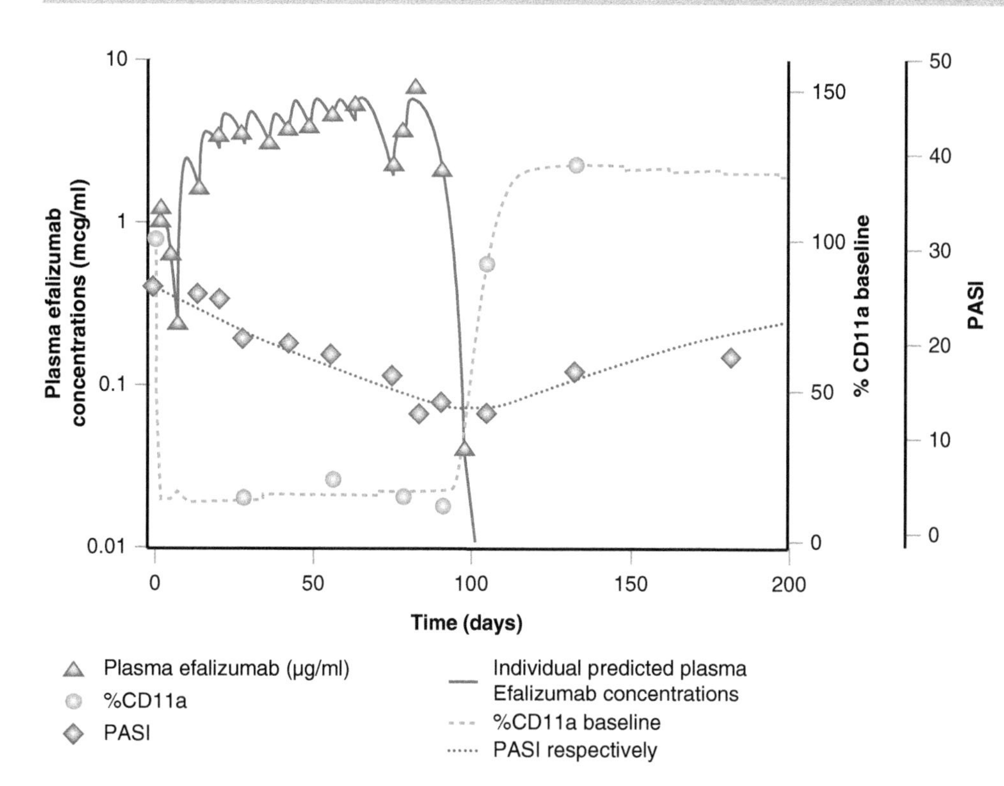

Figure 7.15 ■ Representative pharmacokinetic–pharmacodynamic–efficacy profiles from a patient receiving a 1 mg/kg weekly dose of efalizumab subcutaneously for 12 weeks. *Solid triangle*, plasma efalizumab (μg/mL); *open circles*, %CD11a; and *solid circles*, PASI. *Solid*, *dashed*, and *dotted lines* represent individual predicted plasma efalizumab concentrations, %CD11a baseline, and PASI, respectively.

FUTURE PERSPECTIVE

The success of monoclonal antibodies as new therapeutic agents in several disease areas such as oncology, inflammatory diseases, autoimmune diseases, and transplantation has triggered growing scientific, therapeutic, and business interest in the MAB technology. The market for therapeutic MABs is one of the most dynamic sectors within the pharmaceutical industry. Further growth is expected by developing MABs towards other surface protein targets, which are not covered yet by marketed MABs. Particularly, the technological advancement in the area of immunoconjugates and MAB fragments may overcome some of the limitations of MABs by providing highly potent drugs selectively to effect compartments and to extend the distribution of the active moiety, which are typically not reached by MABs. Immunoconjugates hold great promise for selective drug delivery of potent drugs with unfavorable own selectivity to target cells (e.g., highly potent cytotoxic drugs). Several of such immunoconjugates are under development to target different tumor types and are expected to reach the market in the next years. Modification of the MAB structure allows adjusting the properties according to therapeutic needs (e.g., adjusting half-life, increasing volume of distribution, changing clearance pathways). By using modified MAB derivatives, optimized therapeutic agents might become available. So far this technology has been successfully used for two antibody fragments marketed in inflammatory disease and antiangiogenesis (abciximab, ranibizumab).

Bispecific antibodies represent another promising new methodological approach to antibody therapy. Technological refinements in antibody engineering have allowed the production of bispecific antibodies that are simultaneously directed towards two distinct target antigens (Holmes 2011). For instance, the CDR consisting of the variable domains (V_L and V_H) at the tip of one arm of an IgG may be asymmetrically designed to bind to a different target than that of the other arm (Fig. 7.1). Symmetrical formats in which each arm can bind two targets are also possible.

MABs have become a key part of the pharmaceutical armamentarium, especially in the oncology and immunology settings and will continue to be a focus area for drug discovery and development.

SELF-ASSESSMENT QUESTIONS

■ Questions

1. What are the structural differences between the five immunoglobulin classes?
2. (a) What are key differences in PK/PD between MABs and small molecule drugs?
   (b) Why do IgGs typically show nonlinear PK in the lower plasma (serum) concentration range?
3. What is a surrogate MAB and how can it potentially be used in the drug development process of MABs?
4. Which other modes of actions apart from ADCC – antibody dependent cellular cytotoxicity are known for MABs? What are the key steps of ADCC?
5. Why do IgGs have a longer in vivo half-life compared with other Igs?

6. What are the development phases for antibody therapeutics? What major activities are involved in the each phase?

■ Answers

1. The following structural properties distinguish MABs:
   - The molecular form can be different for the 5 immunoglobulin classes: IgG, IgD, and IgE are monomer; IgM appears as pentamer or hexamer, and IgA are either monomer or dimer.
   - Consequently, the molecular weight of the different Igs is different (IgG 150–169 kD, IgA 160–300 kD, IgD 175 kD, IgE 190, IgM 950 kD).
2. (a) I. Metabolism of MABs appears to be simpler than for small molecules. In contrast to small molecule drugs, the typical metabolic enzymes and transporter proteins such as cytochrome P450, multidrug resistance (MDR) efflux pumps are not involved in the disposition of MABs. Therefore, drug–drug interaction studies for those disposition processes are only part of the standard safety assessment for small molecules and not for MABs. Monoclonal antibodies, which have a protein structure, are metabolized by proteases. These enzymes are ubiquitously available in mammalian organisms. In contrast, small molecule drugs are primarily metabolized in the liver.

   II. Because of the large molecular weight, intact MABs are typically not cleared by the renal elimination route in the kidneys. However, renal clearance processes can play a major role in the elimination of small molecule drugs.

   III. Pharmacokinetics of MABs usually is dependent on the binding to the pharmacological target protein and shows nonlinear behavior as consequence of its saturation kinetics.

   IV. In general, MABs have a longer half-life (in the order of days and weeks) than small molecule drugs (typically in the order of hours).

   V. The distribution of MABs is very restricted (volume of distribution in the range of 0.1 L/kg). As a consequence, MABs do have limited access to tissue compartments as potential target sites via passive, energy-independent distribution processes only (e.g., brain).

   (b) At lower concentrations, MABs generally show nonlinear pharmacokinetics due to receptor-mediated clearance processes, which are characterized by small capacity of the clearance pathway and high affinity to the target protein. Consequently at these low concentrations, MABs exhibit typically shorter half-life. With increasing doses, these receptors become saturated, and the clearance as well as elimination half-life decreases until it becomes constant. The clearance in the higher concentration range, which is dominated by linear, nontarget-related clearance processes, is therefore also called nonspecific clearance in contrast to the target-related, specific clearance.
3. A surrogate MAB has similar antigen specificity and affinity in experimental animals (e.g., mice and rats) compared to those of the corresponding human antibody in humans. It is quite common that the antigen specificity limits ADME studies of humanized monoclonal antibodies in rodents. Studies using surrogate antibodies might lead to important information regarding safety, mechanism of action, disposition of the drug, tissue distribution, and receptor pharmacology in the respective animal species, which might be too cumbersome and expensive to be conducted in nonhuman primates. Surrogate MABs (from mouse or rat) provide a means to gain knowledge of ADME and PD in preclinical rodent models and might facilitate the dose selection for clinical studies.
4. Apart from ADCC, monoclonal antibodies can exert pharmacological effects by multiple mechanisms that include direct modulation of the target antigen, complement-dependent cytotoxicity (CDC) and apoptosis.

   The key steps of ADCC are (1) opsonization of the targeted cells, (2) recognition of antibody-coated targeted cells by Fc receptors on the surface of monocytes, macrophages, natural killer cells, and other cells, and (3) destruction of the opsonized targets by phagocytosis of the opsonized targets and/or by toxic substances released after activation of monocytes, macrophages, natural killer cells, and other cells.
5. IgG can bind to neonatal Fc receptor (FcRn) in the endosome, which protects IgG from catabolism via proteolytic degradation. This protection results into a slower clearance and thus longer plasma half-life of IgGs. Consequently, changing the FcRn affinity allows to adjust the clearance of MABs (higher affinity – lower clearance), which can be employed to tailor the pharmacokinetics of these molecules.
6. Pre-IND, phase I, II, III, and IV are the major development phases for antibody therapies. Safety pharmacology, toxicokinetics, toxicology, tissue cross reactivity, local tolerance, PK support for molecules selection, assay support for PK/PD, and PK/PD support for dose/route/regimen are major activities in the pre-IND phase. General toxicity, reproductive toxicity, carcinogenicity, immunogenicity, characterization of dose–concentration–effect relationship, material comparability studies, mechanistic modeling approach, and population pharmacokinetics/

predictions are major activities from phase I to phase III. Further studies might be performed as needed after the MAB got market authorization. These studies are called phase IV studies.

REFERENCES

Aarons L, Karlsson MO et al (2001) Role of modelling and simulation in Phase I drug development. Eur J Pharm Sci 13(2):115–122

Actemra® (Tocilizumab) (2010) Prescribing information. Genentech Inc., South San Francisco

Adcetris® (Brentuximab vedotin) (2011) Prescribing information. Seattle Genetics Inc., Bothell

Agoram BM, Martin SW, van der Graaf PH (2007) The role of mechanism-based pharmacokinetic–pharmacodynamic (PK–PD) modelling in translational research of biologics. Drug Discov Today 12(23–24):1018–1024

Albrecht H, DeNardo SJ (2006) Recombinant antibodies: from the laboratory to the clinic. Cancer Biother Radiopharm 21(4):285–304

Amevive® (Alefacept) (2003) Amevive prescribing information. Biogen Inc., Cambridge

Arzerra® (ofatumumab) (2009) Arzerra prescribing information. GlaxoSmithKline, Research Triangle Park

Avastin® (2004) Avastin (Bevacizumab) prescribing information. Genentech Inc., South San Francisco

Baert F, Noman M et al (2003) Influence of immunogenicity on the long-term efficacy of infliximab in Crohn's disease. N Engl J Med 348(7):601–608

Bauer RJ, Dedrick RL et al (1999) Population pharmacokinetics and pharmacodynamics of the anti-CD11a antibody hu1124 in human subjects with psoriasis. J Pharmacokinet Biopharm 27(4):397–420

Baxter LT, Zhu H et al (1994) Physiologically based pharmacokinetic model for specific and nonspecific monoclonal antibodies and fragments in normal tissues and human tumor xenografts in nude mice. Cancer Res 54(6):1517–1528

Baxter LT, Zhu H et al (1995) Biodistribution of monoclonal antibodies: scale-up from mouse to human using a physiologically based pharmacokinetic model. Cancer Res 55(20):4611–4622

Bazin-Redureau MI, Renard CB et al (1997) Pharmacokinetics of heterologous and homologous immunoglobulin G, F(ab')2 and Fab after intravenous administration in the rat. J Pharm Pharmacol 49(3):277–281

Benlysta® (belimumab) (2011) Benlysta prescribing information. Human Genome Sciences Inc., Rockville

Berger MA, Masters GR et al (2005) Pharmacokinetics, biodistribution, and radioimmunotherapy with monoclonal antibody 776.1 in a murine model of human ovarian cancer. Cancer Biother Radiopharm 20(6):589–602

Bexxar (2003) Bexxar (Tositumomab) prescribing information. Corixa Corp/GlaxoSmithKline, Seattle/Philadelphia

Boswell CA, Brechbiel MW (2007) Development of radioimmunotherapeutic and diagnostic antibodies: an inside-out view. Nucl Med Biol 34(7):757–778

Boxenbaum H, Battle M (1995) Effective half-life in clinical pharmacology. J Clin Pharmacol 35(8):763–766

Brambell F, Hemmings W et al (1964) A theoretical model of gamma-globulin catabolism. Nature 203:1352–1355

Bugelski PJ, Herzyk DJ et al (2000) Preclinical development of keliximab, a primatized anti-CD4 monoclonal antibody, in human CD4 transgenic mice: characterization of the model and safety studies. Hum Exp Toxicol 19(4):230–243

Bunescu A, Seideman P et al (2004) Enhanced Fcgamma receptor I, alphaMbeta2 integrin receptor expression by monocytes and neutrophils in rheumatoid arthritis: interaction with platelets. J Rheumatol 31(12):2347–2355

Campath® (Alemtuzumab) (2009) Campath prescribing information. Genzyme Inc., Cambridge

Cartron G, Watier H et al (2004) From the bench to the bedside: ways to improve rituximab efficacy. Blood 104(9):2635–2642

Chien, JY, Friedrich S et al (2005) Pharmacokinetics/Pharmacodynamics and the stages of drug development: role of modeling and simulation. AAPS J. 7(3):E544–559

Cimzia® (2011) Cimzia (Certolizumab pegol) prescribing information. UCB Inc., Smyrna

Clarke J, Leach W et al (2004) Evaluation of a surrogate antibody for preclinical safety testing of an anti-CD11a monoclonal antibody. Regul Toxicol Pharmacol 40(3): 219–226

Coffey GP, Fox JA et al (2005) Tissue distribution and receptor-mediated clearance of anti-CD11a antibody in mice. Drug Metab Dispos 33(5):623–629

Cohenuram M, Saif MW (2007) Panitumumab the first fully human monoclonal antibody: from the bench to the clinic. Anticancer Drugs 18(1):7–15

Cornillie F, Shealy D et al (2001) Infliximab induces potent anti-inflammatory and local immunomodulatory activity but no systemic immune suppression in patients with Crohn's disease. Aliment Pharmacol Ther 15(4):463–473

Dall'Ozzo S, Tartas S et al (2004) Rituximab-dependent cytotoxicity by natural killer cells: influence of FCGR3A polymorphism on the concentration-effect relationship. Cancer Res 64(13):4664–4669

Danilov SM, Gavrilyuk VD et al (2001) Lung uptake of antibodies to endothelial antigens: key determinants of vascular immunotargeting. Am J Physiol Lung Cell Mol Physiol 280(6):L1335–L1347

Dedrick RL (1973) Animal scale-up. J Pharmacokinet Biopharm 1(5):435–461

Dedrick RL, Walicke P et al (2002) Anti-adhesion antibodies efalizumab, a humanized anti-CD11a monoclonal antibody. Transpl Immunol 9(2–4):181–186

den Broeder A, van de Putte L et al (2002) A single dose, placebo controlled study of the fully human anti-tumor necrosis factor-alpha antibody adalimumab (D2E7) in patients with rheumatoid arthritis. J Rheumatol 29(11):2288–2298

Deng R, Iyer S, Theil FP et al (2011) Projecting human pharmacokinetics of therapeutic antibodies from nonclinical data. What have we learned? MAbs 3(1):61–66

Dong JQ, Salinger DH, Endres CJ et al (2011) Quantitative prediction of human pharmacokinetics for monoclonal antibodies: retrospective analysis of monkey as a single species for first-in-human prediction. Clin Pharmacokinet 50:131–142

Dickinson BL, Badizadegan K et al (1999) Bidirectional FcRn-dependent IgG transport in a polarized human intestinal epithelial cell line. J Clin Invest 104(7): 903–911
Dirks NL, Meibohm B (2010) Population pharmacokinetics of therapeutic monoclonal antibodies. Clin Pharmacokinet 49(10):633–659
Dowell JA, Korth-Bradley J et al (2001) Pharmacokinetics of gemtuzumab ozogamicin, an antibody-targeted chemotherapy agent for the treatment of patients with acute myeloid leukemia in first relapse. J Clin Pharmacol 41(11):1206–1214
Druet P, Bariety J et al (1978) Distribution of heterologous antiperoxidase antibodies and their fragments in the superficial renal cortex of normal Wistar-Munich rat: an ultrastructural study. Lab Invest 39(6):623–631
Ducongé J, Castillo R et al (2004) Integrated pharmacokinetic-pharmacodynamic modeling and allometric scaling for optimizing the dosage regimen of the monoclonal ior EGF/r3 antibody. Eur J Pharm Sci 21(2–3):261–270
Erbitux® (2004) Erbitux (Cetuximab) prescribing information. Imclone Systems Inc., Bristol-Myers Squibb Company, Branchburg/Princeton
Ferl GZ, Wu AM et al (2005) A predictive model of therapeutic monoclonal antibody dynamics and regulation by the neonatal Fc receptor (FcRn). Ann Biomed Eng 33(11):1640–1652
Genovese MC, Cohen S, Moreland L, Lium D, Robbins S, Newmark R et al (2004) Combination therapy with etanercept and anakinra in the treatment of patients with rheumatoid arthritis who have been treated unsuccessfully with methotrexate. Arthritis Rheum 50(5):1412–1419
Ghetie V, Hubbard JG et al (1996) Abnormally short serum half-lives of IgG in beta 2-microglobulin-deficient mice. Eur J Immunol 26(3):690–696
Gibiansky L, Gibiansky E (2009) Target-mediated drug disposition model: relationships with indirect response models and application to population PK–PD analysis. J Pharmacokinet Pharmacodyn 36(4):341–351
Gibiansky L, Frey N (2012) Linking interleukin-6 receptor blockade with tocilizumab and its hematological effects using a modeling approach. J Pharmacokinet Pharmacodyn 39(1):5–16
Gillies SD, Lo KM et al (2002) Improved circulating half-life and efficacy of an antibody-interleukin 2 immunocytokine based on reduced intracellular proteolysis. Clin Cancer Res 8(1):210–216
Gillies SD, Lan Y et al (1999) Improving the efficacy of antibody-interleukin 2 fusion proteins by reducing their interaction with Fc receptors. Cancer Res 59(9):2159–2166
Girish S, Martin SW, Peterson MC et al (2011) AAPS workshop report: strategies to address therapeutic protein–drug interactions during clinical development. AAPS J 13(3):405–416
Goldsby RA, Kindt TJ et al (1999) Immunoglobulins: structure and function. In: Kuby immunology, 4th edn. W.H. Freeman and Company, New York
Gottlieb AB, Miller B et al (2003) Subcutaneously administered efalizumab (anti-CD11a) improves signs and symptoms of moderate to severe plaque psoriasis. J Cutan Med Surg 7(3):198–207
Hayashi N, Tsukamoto Y et al (2007) A mechanism-based binding model for the population pharmacokinetics and pharmacodynamics of omalizumab. Br J Clin Pharmacol 63(5):548–561
Herceptin® (2006) Herceptin (Trastuzumab) prescribing information. Genentech Inc., South San Francisco
Hervey PS, Keam SJ (2006) Abatacept. BioDrugs 20(1):53–61, discussion 62
Hinton PR, Xiong JM et al (2006) An engineered human IgG1 antibody with longer serum half-life. J Immunol 176(1): 346–356
Holmes D (2011) Buy buy bispecific antibodies. Nat Rev Drug Discov 10:798–800
Hooks MA, Wade CS et al (1991) Muromonab CD-3: a review of its pharmacology, pharmacokinetics, and clinical use in transplantation. Pharmacotherapy 11(1):26–37
Huang L, Biolsi S et al (2006) Impact of variable domain glycosylation on antibody clearance: an LC/MS characterization. Anal Biochem 349(2):197–207
Huang S-M, Zhao H, Lee J-I et al (2010) Therapeutic protein–drug interactions and implications for drug development. Clin Pharmacol Ther 87(4):497–503
Humira® (2007) Humira (Adalimumab) prescribing information. Abbott Laboratories, Chicago
ICH (1997a) ICH harmonized tripartite guideline M3: nonclinical safety studies for the conduct of human clinical trials for pharmaceuticals
ICH (1997b) ICH harmonized tripartite guideline S6: preclinical safety evaluation of biotechnology-derived pharmaceuticals
Ilaris® (canakimumab) (2011) Ilaris prescribing information. Novartis Corp, East Hanover
Jolling K, Perez Ruixo JJ et al (2005) Mixed-effects modelling of the interspecies pharmacokinetic scaling of pegylated human erythropoietin. Eur J Pharm Sci 24(5):465–475
Joshi A, Bauer R et al (2006) An overview of the pharmacokinetics and pharmacodynamics of efalizumab: a monoclonal antibody approved for use in psoriasis. J Clin Pharmacol 46(1):10–20
Junghans RP (1997) Finally! The Brambell receptor (FcRB). Mediator of transmission of immunity and protection from catabolism for IgG. Immunol Res 16(1):29–57
Junghans RP, Anderson CL (1996) The protection receptor for IgG catabolism is the beta2-microglobulin-containing neonatal intestinal transport receptor. Proc Natl Acad Sci U S A 93(11):5512–5516
Kadcyla® (ado-trastuzumab emtansine) (2013) Kadcyla prescribing information. Genentech Inc., South San Francisco
Kagan L, Abraham AK, Harrold JM et al (2010) Interspecies scaling of receptor-mediated pharmacokinetics and pharmacodynamics of type I interferons. Pharm Res 27:920–932
Kairemo KJ, Lappalainen AK et al (2001) In vivo detection of intervertebral disk injury using a radiolabeled monoclonal antibody against keratan sulfate. J Nucl Med 42(3):476–482

Kelley SK, Gelzleichter T et al (2006) Preclinical pharmacokinetics, pharmacodynamics, and activity of a humanized anti-CD40 antibody (SGN-40) in rodents and non-human primates. Br J Pharmacol 148(8): 1116–1123

Kleiman NS, Raizner AE et al (1995) Differential inhibition of platelet aggregation induced by adenosine diphosphate or a thrombin receptor-activating peptide in patients treated with bolus chimeric 7E3 Fab: implications for inhibition of the internal pool of GPIIb/IIIa receptors. J Am Coll Cardiol 26(7):1665–1671

Kohler G, Milstein C (1975) Continuous cultures of fused cells secreting antibody of predefined specificity. Nature 256(5517):495–497

Kolar GR, Capra JD (2003) Immunoglobulins: structure and function. In: Paul WE (ed) Fundamental immunology, 5th edn. Lippincott Williams & Wilkins, Philadelphia

Koon HB, Severy P et al (2006) Antileukemic effect of daclizumab in CD25 high-expressing leukemias and impact of tumor burden on antibody dosing. Leuk Res 30(2):190–203

Kovarik JM, Nashan B et al (2001) A population pharmacokinetic screen to identify demographic-clinical covariates of basiliximab in liver transplantation. Clin Pharmacol Ther 69(4):201–209

Krueger JG (2002) The immunologic basis for the treatment of psoriasis with new biologic agents. J Am Acad Dermatol 46(1):1–23, quiz 23–6

Kuus-Reichel K, Grauer LS et al (1994) Will immunogenicity limit the use, efficacy, and future development of therapeutic monoclonal antibodies? Clin Diagn Lab Immunol 1(4):365–372

Lee H, Kimko HC et al (2003) Population pharmacokinetic and pharmacodynamic modeling of etanercept using logistic regression analysis. Clin Pharmacol Ther 73(4):348–365

Lee JI, Zhang L, Men A et al (2010) CYP-mediated therapeutic protein-drug interactions: clinical findings, proposed mechanisms and regulatory implications. Clin Pharmacokinet 49(5):295–310

Lin YS, Nguyen C et al (1999) Preclinical pharmacokinetics, interspecies scaling, and tissue distribution of a humanized monoclonal antibody against vascular endothelial growth factor. J Pharmacol Exp Ther 288(1): 371–378

Ling J, Zhou H, Jiao Q et al (2009) Interspecies scaling of therapeutic monoclonal antibodies: initial look. J Clin Pharmacol 49(12):1382–1402

Lobo ED, Hansen RJ et al (2004) Antibody pharmacokinetics and pharmacodynamics. J Pharm Sci 93(11):2645–2668

Looney RJ, Anolik JH et al (2004) B cell depletion as a novel treatment for systemic lupus erythematosus: a phase I/II dose-escalation trial of rituximab. Arthritis Rheum 50(8):2580–2589

LoRusso PM, Weiss D, Guardino E et al (2011) Trastuzumab emtansine: a unique antibody-drug conjugate in development for human epidermal growth factor receptor 2-positive cancer. Clin Cancer Res 17(20):6437–6447

Lu D, Modi S, Elias A et al (2011) Pharmacokinetics (PK) of Trastuzumab emtansine and paclitaxel or docetaxel in patients with HER2-positive MBS previously treated with trastuzumab-containing regimen. In: 34th annual San Antonio breast cancer symposium. San Antonio

Lu D, Burris H, Wang B et al (2012) Drug interaction potential of trastuzumab emtansine in combination with pertuzumab in patients with HER2-positive metastatic breast cancer. Curr Drug Metab 13:911–922

Lucentis® (2006) Lucentis (Ranibizumab) prescribing information. Genentech Inc., South San Francisco

Ma P, Yang BB, Wang YM et al (2009) Population pharmacokinetic analysis of panitumumab in patients with advanced solid tumors. J Clin Pharmacol 49(10):1142–1156

Mager DE, Mascelli MA, Kleiman NS, Fitzgerald DJ, Abernethy DR (2003) Simultaneous modeling of abciximab plasma concentrations and ex vivo pharmacodynamics in patients undergoing coronary angioplasty. J Pharmacol Exp Ther 307(3):969–976

Mager DE, Woo S, Jusko WJ (2009) Scaling pharmacodynamics from in vitro and preclinical animal studies to humans. Drug Metab Pharmacokinet 24:16–24

Mahmood I (2005) Prediction of concentration-time profiles in humans. In: Interspecies pharmacokinetics scaling. Pine House Publishers, Rockville, pp 219–241

Mahmood I, Green MD (2005) Pharmacokinetic and pharmacodynamic considerations in the development of therapeutic proteins. Clin Pharmacokinet 44(4):331–347

Mahmood I (2009) Pharmacokinetic allometric scaling of antibodies: application to the first-in-human dose estimation. J Pharm Sci 98:3850–3861

Maini RN, Breedveld FC, Kalden JR, Smolen JS, Davis D, Macrarlane JD et al (1998) Therapeutic efficacy of multiple intravenous infusions of anti-tumor necrosis factor α monoclonal antibody combined with low-dose weekly methotrexate in rheumatoid arthritis. Arthritis Rheum 41(9):1552–1563

Martin-Jimenez T, Riviere JE (2002) Mixed-effects modeling of the interspecies pharmacokinetic scaling of oxytetracycline. J Pharm Sci 91(2):331–341

McClurkan MB, Valentine JL et al (1993) Disposition of a monoclonal anti-phencyclidine Fab fragment of immunoglobulin G in rats. J Pharmacol Exp Ther 266(3): 1439–1445

McLaughlin P, Grillo-Lopez AJ et al (1998) Rituximab chimeric anti-CD20 monoclonal antibody therapy for relapsed indolent lymphoma: half of patients respond to a four-dose treatment program. J Clin Oncol 16(8):2825–2833

Medesan C, Matesoi D et al (1997) Delineation of the amino acid residues involved in transcytosis and catabolism of mouse IgG1. J Immunol 158(5):2211–2217

Meibohm B, Derendorf H (2002) Pharmacokinetic/pharmacodynamic studies in drug product development. J Pharm Sci 91(1):18–31

Meijer RT, Koopmans RP et al (2002) Pharmacokinetics of murine anti-human CD3 antibodies in man are determined by the disappearance of target antigen. J Pharmacol Exp Ther 300(1):346–353

Meredith PA, Elliott HL et al (1991) Dose–response clarification in early drug development. J Hypertens Suppl 9(6):S356–S357

Morris EC, Rebello P et al (2003) Pharmacokinetics of alemtuzumab used for in vivo and in vitro T-cell depletion in allogeneic transplantations: relevance for early adoptive immunotherapy and infectious complications. Blood 102(1):404–406

Mortensen DL, Walicke PA et al (2005) Pharmacokinetics and pharmacodynamics of multiple weekly subcutaneous efalizumab doses in patients with plaque psoriasis. J Clin Pharmacol 45(3):286–298

Mould DR, Sweeney KR (2007) The pharmacokinetics and pharmacodynamics of monoclonal antibodies–mechanistic modeling applied to drug development. Curr Opin Drug Discov Devel 10(1):84–96

Mould DR, Davis CB et al (1999) A population pharmacokinetic-pharmacodynamic analysis of single doses of clenoliximab in patients with rheumatoid arthritis. Clin Pharmacol Ther 66(3):246–257

Nakakura EK, McCabe SM et al (1993) Potent and effective prolongation by anti-LFA-1 monoclonal antibody monotherapy of non-primarily vascularized heart allograft survival in mice without T cell depletion. Transplantation 55(2):412–417

Nestorov I, Zitnik R et al (2004) Population pharmacokinetic modeling of subcutaneously administered etanercept in patients with psoriasis. J Pharmacokinet Pharmacodyn 31(6):463–490

Newkirk MM, Novick J et al (1996) Differential clearance of glycoforms of IgG in normal and autoimmune-prone mice. Clin Exp Immunol 106(2):259–264

Ng CM, Joshi A et al (2005) Pharmacokinetic-pharmacodynamic-efficacy analysis of efalizumab in patients with moderate to severe psoriasis. Pharm Res 22(7):1088–1100

Ng CM, Stefanich E et al (2006) Pharmacokinetics/pharmacodynamics of nondepleting anti-CD4 monoclonal antibody (TRX1) in healthy human volunteers. Pharm Res 23(1):95–103

Norman DJ, Chatenoud L et al (1993) Consensus statement regarding OKT3-induced cytokine-release syndrome and human antimouse antibodies. Transplant Proc 25 (2 Suppl 1):89–92

Nulojix® (belatacept) (2011) Nulojix prescribing information. Bristol Myers Squibb Inc, Princeton

Ober RJ, Radu CG et al (2001) Differences in promiscuity for antibody-FcRn interactions across species: implications for therapeutic antibodies. Int Immunol 13(12): 1551–1559

Oitate M, Masubuchi N, Ito T et al (2011) Prediction of human pharmacokinetics of therapeutic monoclonal antibodies from simple allometry of monkey data. Drug Metab Pharmacokinet 26:423–430

Oitate M, Nakayama S, Ito T et al (2012) Prediction of human plasma concentration-time profiles of monoclonal antibodies from monkey data by species-invariant time method. Drug Metab Pharmacokinet 27:354–359. Online advance publication at http://www.jstage.jst.go.jp/article/dmpk/advpub/0/advpub_1111290286/_article

Papp K, Bissonnette R et al (2001) The treatment of moderate to severe psoriasis with a new anti-CD11a monoclonal antibody. J Am Acad Dermatol 45(5): 665–674

Petkova SB, Akilesh S et al (2006) Enhanced half-life of genetically engineered human IgG1 antibodies in a humanized FcRn mouse model: potential application in humorally mediated autoimmune disease. Int Immunol 18(12):1759–1769

Prabhu S, Boswell SC, Leipold D et al (2011) Antibody delivery of drugs and radionuclides: factors influencing clinical pharmacology. Ther Deliv 2(6):769–791

Presta LG (2002) Engineering antibodies for therapy. Curr Pharm Biotechnol 3(3):237–256

Presta LG, Shields RL et al (2002) Engineering therapeutic antibodies for improved function. Biochem Soc Trans 30(4):487–490

Prolia® (denosumab) (2010) Prolia prescribing information. Amgen Inc., Thousand Oaks

Radin A, Marbury T, Osgood G, Belomestnov P (2010) Safety and pharmacokinetics of subcutaneously administered rilonacept in patients with well- controlled end-stage renal disease. J Clin Pharmacol 50:835–841

Raptiva® (2004) Raptiva (Efalizumab) prescribing information. Genentech Inc., South San Francisco

Remicade® (2006) Remicade (Infliximab) prescribing information. Centocor Inc., Malvern

Rituxan® (2006) Rituxan (Rituximab) prescribing information. Genentech Inc./Biogen Inc., South San Francisco/Cambridge

Roopenian DC, Christianson GJ et al (2003) The MHC class I-like IgG receptor controls perinatal IgG transport, IgG homeostasis, and fate of IgG-Fc-coupled drugs. J Immunol 170(7):3528–3533

Roskos LK, Davis CG et al (2004) The clinical pharmacology of therapeutic monoclonal antibodies. Drug Dev Res 61(3):108–120

Schror K, Weber AA (2003) Comparative pharmacology of GP IIb/IIIa antagonists. J Thromb Thrombolysis 15(2): 71–80

Shah DK, Betts AM (2012) Towards a platform PBPK model to characterize the plasma and tissue disposition of monoclonal antibodies in preclinical species and human. J Pharmacokinet Pharmacodyn 39:67–86

Sheiner L, Wakefield J (1999) Population modelling in drug development. Stat Methods Med Res 8(3):183–193

Sheiner LB (1997) Learning versus confirming in clinical drug development. Clin Pharmacol Ther 61(3):275–291

Shields RL, Namenuk AK et al (2001) High resolution mapping of the binding site on human IgG1 for Fc gamma RI, Fc gamma RII, Fc gamma RIII, and FcRn and design of IgG1 variants with improved binding to the Fc gamma R. J Biol Chem 276(9):6591–6604

Sifontis NM, Benedetti E, Vasquez EM (2002) Clinically significant drug interaction between basiliximab and tacrolimus in renal transplant recipients. Transplant Proc 34:1730–1732

Simister NE, Mostov KE (1989a) An Fc receptor structurally related to MHC class I antigens. Nature 337(6203): 184–187

Simister NE, Mostov KE (1989b) Cloning and expression of the neonatal rat intestinal Fc receptor, a major histocompatibility complex class I antigen homolog. Cold Spring Harb Symp Quant Biol 54(Pt 1):571–580

Simulect® (2005) Simulect (Basiliximab) prescribing information. Novartis Pharmaceuticals, East Hanover

Simponi® (2009) Simponi (golimumab) prescribing information. Janssen Biotech Inc., Horsham

Soliris® (eculizumab) (2007) Soliris prescribing Information. Alexion Pharmaceuticals Inc, Cheshire

Spiekermann GM, Finn PW et al (2002) Receptor-mediated immunoglobulin G transport across mucosal barriers in adult life: functional expression of FcRn in the mammalian lung. J Exp Med 196(3):303–310

Straughn AB (2006) Limitations of noncompartmental pharmacokinetic analysis of biotech drugs. In: Meibohm B (ed) Pharmacokinetics and pharmacodynamics of biotech drugs. Weinheim, Wiley, pp 181–188

Subramanian GM, Cronin PW et al (2005) A phase 1 study of PAmab, a fully human monoclonal antibody against Bacillus anthracis protective antigen, in healthy volunteers. Clin Infect Dis 41(1):12–20

Sun YN, Lu JF et al (2005) Population pharmacokinetics of efalizumab (humanized monoclonal anti-CD11a antibody) following long-term subcutaneous weekly dosing in psoriasis subjects. J Clin Pharmacol 45(4): 468–476

Synagis (2004) Synagis (Palivizumab) prescribing information. MedImmune Inc./Abbott Laboratories Inc., Gaithersburg/Columbus

Tabrizi MA, Tseng CM et al (2006) Elimination mechanisms of therapeutic monoclonal antibodies. Drug Discov Today 11(1–2):81–88

Tang H, Mayersohn M (2005) Accuracy of allometrically predicted pharmacokinetic parameters in humans: role of species selection. Drug Metab Dispos 33(9): 1288–1293

Tang L, Persky AM, Hochhaus G, Meibohm B (2004) Pharmacokinetic aspects of biotechnology products. J Pharm Sci 93(9):2184–2204

Ternant D, Paintaud G (2005) Pharmacokinetics and concentration-effect relationships of therapeutic monoclonal antibodies and fusion proteins. Expert Opin Biol Ther 5(Suppl 1):S37–S47

Thurber GM, Schmidt MM, Wittrup KD (2008) Antibody tumor penetration: transport opposed by systemic and antigen-mediated clearance. Adv Drug Deliv Rev 60(12):1421–1434

Tokuda Y, Watanabe T et al (1999) Dose escalation and pharmacokinetic study of a humanized anti-HER2 monoclonal antibody in patients with HER2/neu-overexpressing metastatic breast cancer. Br J Cancer 81(8):1419–1425

Tysabri® (2006) Tysabri (Natalizumab) prescribing information. Elan Pharmaceuticals Inc./Biogen Idec Inc., San Diego/Cambridge

Umana P, Jean-Mairet J et al (1999) Engineered glycoforms of an antineuroblastoma IgG1 with optimized antibody-dependent cellular cytotoxic activity. Nat Biotechnol 17(2):176–180

Vaccaro C, Bawdon R et al (2006) Divergent activities of an engineered antibody in murine and human systems have implications for therapeutic antibodies. Proc Natl Acad Sci U S A 103(49):18709–18714

Vaishnaw AK, TenHoor CN (2002) Pharmacokinetics, biologic activity, and tolerability of alefacept by intravenous and intramuscular administration. J Pharmacokinet Pharmacodyn 29(5–6):415–426

Vasquez EM, Pollak R (1997) OKT3 therapy increases cyclosporine blood levels. Clin Transplant 11(1):38–41

Vectibix® (2006) Vectibix (Panitumumab) prescribing information. Amgen Inc, Thousand Oaks

Vincenti F, Mendez R, Pescovitz M, Rajagopalan PR, Wilkinson AH, Butt K et al (2007) A phase I/II randomized open-label multi-center trial of efalizumab, a humanized ani-CD11a, anti-LFA-1 in renal transplantation. Am J Transplant 7:1770–1777

Wang W, Prueksaritanont T (2010) Prediction of human clearance of therapeutic proteins: simple allometric scaling method revisited. Biopharm Drug Dispos 31:253–263

Wang W, Singh S et al (2007) Antibody structure, instability, and formulation. J Pharm Sci 96(1):1–26

Watanabe N, Kuriyama H et al (1988) Continuous internalization of tumor necrosis factor receptors in a human myosarcoma cell line. J Biol Chem 263(21): 10262–10266

Weiner LM (2006) Fully human therapeutic monoclonal antibodies. J Immunother 29(1):1–9

Weisman MH, Moreland LW et al (2003) Efficacy, pharmacokinetic, and safety assessment of adalimumab, a fully human anti-tumor necrosis factor-alpha monoclonal antibody, in adults with rheumatoid arthritis receiving concomitant methotrexate: a pilot study. Clin Ther 25(6):1700–1721

Werther WA, Gonzalez TN et al (1996) Humanization of an anti-lymphocyte function-associated antigen (LFA)-1 monoclonal antibody and reengineering of the humanized antibody for binding to rhesus LFA-1. J Immunol 157(11):4986–4995

Wiseman GA, White CA, Sparks RB et al (2001) Biodistribution and dosimetry results from a phase III prospectively randomized controlled trial of Zevalin radioimmunotherapy for low-grade, follicular, or transformed B-cell non-Hodgkin's lymphoma. Crit Rev Oncol Hematol 39(1–2):181–194

Wurster U, Haas J (1994) Passage of intravenous immunoglobulin and interaction with the CNS. J Neurol Neurosurg Psychiatry 57(Suppl):21–25

Xolair® (2003) Xolair (Omalizumab) [prescribing information]. Genentech Inc/Novartis Pharmaceuticals Corp., South San Francisco/East Hanover

Yervoy® (2011) Yervoy (ipilimumab) prescribing information. Bristol Myers Squibb Inc, Princeton

Yim DS, Zhou H et al (2005) Population pharmacokinetic analysis and simulation of the time-concentration profile of etanercept in pediatric patients with juvenile rheumatoid arthritis. J Clin Pharmacol 45(3):246–256

Younes A, Bartlett NL, Leonard JP et al (2010) Brentuximab vedotin (SGN-35) for relapsed CD30-positive lymphomas. N Engl J Med 363:1812–1821

Zenapax® (2005) Zenapax (Daclizumab) prescribing information. Hoffman-La Roche Inc, Nutley

Zevalin® (2002) Zevalin (ibritumomab tiuxetan) prescribing information. Spectrum Pharmaceuticals, Irvine

Zheng Y, Scheerens H, Davis JC et al (2011) Translational pharmacokinetics and pharmacodynamics of an FcRn-variant anti-CD4 monoclonal antibody from preclinical model to phase I study. Clin Pharmacol Ther 89(2):283–290

Zhou H (2005) Clinical pharmacokinetics of etanercept: a fully humanized soluble recombinant tumor necrosis factor receptor fusion protein. J Clin Pharmacol 45(5): 490–497

Zhou H, Mayer PR et al (2004) Unaltered etanercept pharmacokinetics with concurrent methotrexate in patients with rheumatoid arthritis. J Clin Pharmacol 44(11): 1235–1243

Zhu Y, Hu C, Lu M et al (2009) Population pharmacokinetic modeling of ustekinumab, a human monoclonal antibody targeting IL-12/23p40, in patients with moderate to severe plaque psoriasis. J Clin Pharmacol 49(2): 162–175

Zia-Amirhosseini P, Minthorn E et al (1999) Pharmacokinetics and pharmacodynamics of SB-240563, a humanized monoclonal antibody directed to human interleukin-5, in monkeys. J Pharmacol Exp Ther 291(3):1060–1067

8 Genomics, Other "Omic" Technologies, Personalized Medicine, and Additional Biotechnology-Related Techniques

Robert D. Sindelar

INTRODUCTION

The products resulting for biotechnologies continue to grow at an exponential rate, and the expectations are that an even greater percentage of drug development will be in the area of the biologics. In 2011, worldwide there were over 800 new biotech drugs and treatments in development including 23 antisense, 64 cell therapy, 50 gene therapy, 300 monoclonal antibodies, 78 recombinant proteins, and 298 vaccines (PhRMA 2012). Pharmaceutical biotechnology techniques are at the core of most methodologies used today for drug discovery and development of both biologics and small molecules. While recombinant DNA technology and hybridoma techniques were the major methods utilized in pharmaceutical biotechnology through most of its historical timeline, our ever-widening understanding of human cellular function and disease processes and a wealth of additional and innovative biotechnologies have been, and will continue to be, developed in order to harvest the information found in the human genome. These technological advances will provide a better understanding of the relationship between genetics and biological function, unravel the underlying causes of disease, explore the association of genomic variation and drug response, enhance pharmaceutical research, and fuel the discovery and development of new and novel biopharmaceuticals. These revolutionary technologies and additional biotechnology-related techniques are improving the very competitive and costly process of drug development of new medicinal agents, diagnostics, and medical devices. Some of the technologies and techniques described in this chapter are both well established and commonly used applications of biotechnology producing potential therapeutic products now in development including clinical trials. New techniques are emerging at a rapid and unprecedented pace and their full impact on the future of molecular medicine has yet to be imagined.

Central to any meaningful discussion of pharmaceutical biotechnology and twenty-first century healthcare are the "omic" technologies. The completion of the Human Genome Project has provided a wealth of new knowledge. Researchers are turning increasingly to the task of converting the DNA sequence data into information that will improve, and even revolutionize, drug discovery (see Fig. 8.1) and patient-centered pharmaceutical care. Pharmaceutical scientists are poised to take advantage of this scientific breakthrough by incorporating state-of-the-art genomic and proteomic techniques along with the associated technologies utilized in bioinformatics, metabonomics/metabolomics, epigenomics, systems biology, pharmacogenomics, toxicogenomics, glycomics, and chemical genomics into a new drug discovery, development, and clinical translation paradigm. These additional techniques in biotechnology and molecular biology are being rapidly exploited to bring new drugs to market and each topic will be introduced in this chapter.

It is not the intention of this author to detail each and every biotechnology technique exhaustively, since numerous specialized resources already meet that need. Rather, this chapter will illustrate and enumerate various biotechnologies that should be of key interest to pharmacy students, practicing pharmacists, and pharmaceutical scientists because of their effect on many aspects of pharmacy, drug discovery, and drug development.

R.D. Sindelar, Ph.D., FCAHS
Department of Pharmaceutical Sciences
and Department of Medicine,
Providence Health Care and
The University of British Columbia,
1081 Burrard Street, Vancouver, BC V6Z 1Y6, Canada
e-mail: rsindelar@providencehealth.bc.ca,
robert.sindelar@ubc.ca

D.J.A. Crommelin, R.D. Sindelar, and B. Meibohm (eds.), *Pharmaceutical Biotechnology*,
DOI 10.1007/978-1-4614-6486-0_8,

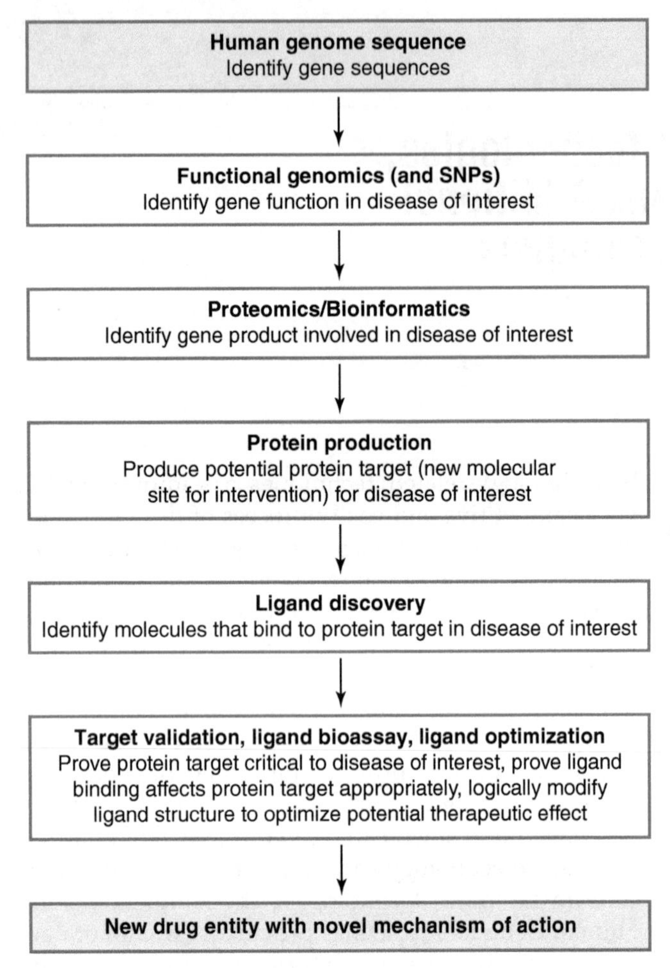

Figure 8.1 ■ The genomic strategy for new drug discovery.

AN INTRODUCTION TO "OMIC" TECHNOLOGIES

Since the discovery of DNA's overall structure in 1953, the world's scientific community has rapidly gained a detailed knowledge of the genetic information encoded by the DNA of a cell or organism so that today we are beginning to "personalize" this information. In the 1980s and 1990s, biotechnology techniques produced novel therapeutics and a wealth of information about the mechanisms of various diseases such as cancer at the genetic and molecular level, yet the etiologies of other complex diseases such as obesity and heart disease remained poorly understood. Recently, however, researchers utilizing exciting and groundbreaking "omic" technologies and working closely with clinicians have begun to make serious progress not only toward a molecular-level understanding of the etiology of complex diseases but to clearly identify that there are actually many genetically different diseases called by the single name of cancer, diabetes, depression, etc. Later in this chapter, we will explore the concepts of phenotype. Important here is that most human diseases are manifested through very complex phenotypes that result from genetic, environmental, and other factors. In a large part, the answers were hidden in what was unknown about the human genome. Despite the increasing knowledge of DNA structure and function in the 1990s, the genome, the entire collection of genes and all other functional and nonfunctional DNA sequences in the nucleus of an organism, had yet to be sequenced. DNA may well be the largest, naturally occurring molecule known. Successfully meeting the challenge of sequencing the entire human genome is one of history's great scientific achievements and heralds enormous potential (Venter et al. 2001; International Human Genome Sequencing Consortium 2001). While the genetic code for transcription and translation has been known for years, sequencing the human genome provides a blueprint for all human proteins and the sequences of all regulatory elements that govern the developmental interpretation of the genome. The potential significance includes identifying genetic determinants of common and rare diseases, providing a methodology for their diagnosis, suggesting interesting new molecular sites for intervention (see Fig. 8.1), and the development of new biotechnologies to bring about their eradication. Unlocking the secrets of the human genome may lead to a paradigm shift in clinical practice toward true targeted molecular medicine, better disease taxonomy, and patient-personalized therapy.

■ Genomics

The term genomics is the comprehensive analysis and understanding of DNA structure and function and broadly refers to the analysis of all genes within the genome of an organism. Sequencing the human genome and the genomes of other organisms has led to an enhanced understanding of not only DNA structure and function but also a fundamental understanding of human biology and disease. While it is a complex and complicated journey from DNA sample to DNA sequence stored in a database, a multitude of technologies and approaches along with impressive enhancements in instrumentation and computation have been employed to sequence genomic DNA faster and less expensively. While many industry analysts predicted a tripling of pharmaceutical R&D productivity due to the sequencing of the human genome, it is the "next-generation" genome sequencing technology and the quest for the "$1000 genome" that will move genomic technology effectively into the clinic (Davies 2010).

Likewise, the field of genomics is having a fundamental impact on modern drug discovery and development. While validation of viable drug targets identified by genomics has been challenging, great progress has occurred (Yang et al. 2009). No matter

Human genome project goals	Base pair resolution
Detailed genetic linkage map Comments: poorest resolution; depicts relative chromosomal locations of DNA markers, genes, or other markers and the spacing between them on each chromosome	2 Mb
Complete physical map Comments: instead of relative distances between markers, maps actual physical distance in base pairs between markers; lower resolution = actual observance of chromosomal banding under microscope; higher resolution is "restriction map" generated in presence of restriction enzymes	0.1 Mb
Complete DNA sequence Comments: the ultimate goal; determine the base sequence of the genes and markers found in mapping techniques along with the other segments of the entire genome; techniques commonly used include DNA amplification methods such as cloning, PCR and other techniques described in Chap. 1 along with novel sequencing and bioinformatics techniques	1 bp

Mb megabase = 1 million base pairs, *bp* base pair

Table 8.1 ■ The increasing levels of genetic resolution obtained from structural genomic studies of the HGP.

whether it is a better understanding of disease or improved drug discovery, the genomic revolution has been the foundation for an explosion in "omic" technologies that find applications in research to address poorly treated and neglected diseases.

Structural Genomics and the Human Genome Project

Genetic analysis initially focused on the area of structural genomics, essentially, the characterization of the macromolecular structure of a genome utilizing computational tools and theoretical frameworks. Structural genomics intersects the techniques of DNA sequencing, cloning, PCR, protein expression, crystallography, and big data analysis. It focuses on the physical aspects of the genome through the construction and analysis of gene sequences and gene maps. Proposed in the late 1980s, the publicly funded Human Genome Project (HGP) or Human Genome Initiative (HGI) was officially sanctioned in October 1990 to map the structure and to sequence human DNA (US DOE 2012a). As described in Table 8.1, HGP structural genomics was envisioned to proceed through increasing levels of genetic resolution: detailed human genetic linkage maps [approximately 2 megabase pairs (Mb = million base pairs) resolution], complete physical maps (0.1 Mb resolution), and ultimately complete DNA sequencing of the approximately three billion base pairs (23 pairs of chromosomes) in a human cell nucleus [1 base pair (bp) resolution]. Projected for completion in 2003, the goal of the project was to learn not only what was contained in the genetic code but also how to "mine" the genomic information to cure or help prevent the estimated 4,000 genetic diseases afflicting humankind. The project would identify all the approximately 20,000–25,000 genes in human DNA, determine the base pair sequence and store the information in databases, create new tools and improve existing tools for data analysis, and address the ethical, legal, and societal issues (ELSI) that may arise from the project. Earlier than projected, a milestone in genomic science was reached on June 26, 2000, when researchers at the privately funded Celera Genomics and the publicly funded International Human Genome Sequencing Consortium (the international collaboration associated with the HGP) jointly announced that they had completed sequencing 97–99 % of the human genome. The journal *Science* rates the mapping of the human genome as its "breakthrough of the year" in its December 22, 2000, issue. The two groups published their results in 2001 (Venter et al. 2001; The Genome International Sequencing Consortium 2001).

While both research groups employed the original cloning-based Sanger technique for DNA sequencing (now approximately 30 years old), the genomic DNA sequencing approaches of the HGP and Celera Genomics differed. HGP utilized a "nested shotgun" approach. The human DNA sequence was "chopped" into segments of ever decreasing size and the segments put into rough order. Each DNA segment was further divided or blasted into smaller fragments. Each small fragment was individually sequenced and the sequenced fragments assembled according to their known relative order. The Celera researchers employed a "whole shotgun" approach where they broke the whole genome into small fragments. Each fragment was sequenced and assembled in order by identifying where they overlapped. Each of the two sequencing approaches required unprecedented computer resources (the field of bioinformatics is described later in this chapter).

Regardless of genome sequencing strategies, the collective results are impressive. More than 27 million high-quality sequence reads provided fivefold coverage of the entire human genome. Genomic studies have identified over one million single-nucleotide polymorphisms (SNPs), binary elements of genetic variability (SNPs are described later in this chapter). While original estimates of the number of human genes in the genome varied consistently between 80,000–120,000, the genome researchers unveiled a number far short of biologist's predictions; 32,000 (Venter et al. 2001; The International Human Genome Sequencing Consortium 2001). Within months, others suggested that the human genome possesses between 65,000 and 75,000 genes (Wright et al. 2001). Approximately 20,000–25,000 genes is now most often cited number (Lee et al. 2006).

Next-Generation Genome Sequencing (NGS) and the $1,000 Genome

The full spectrum of human genetic variation ranges from large chromosomal changes down to the single base pair alterations. The challenge for genomic scientists is to discover the full extent of genomic structural variation, referred to as genotyping, so that the variations and genetic coding may be associated with the encoded trait or traits displayed by the organism (the phenotype). And they wish to do this using as little DNA material as possible, in as short time and for the least cost, all important characteristics of a useful point-of-care clinical technology. The discovery and genotyping of structural variation has been at the core of understanding disease associations as well as identifying possible new drug targets (Alkan et al. 2011). DNA sequencing efficiency has facilitated these studies. In the decade since the completion of the HGP, sequencing efficiency has increased by approximately 100,000-fold and the cost of a single genome sequence has decreased from nearly $1 million in 2007 to $1,000 just recently (Treangen and Salzberg 2011). The move toward low-cost, high-throughput sequencing is essential for the implementation of genomics into personalized medicine and will likely alter the future clinical landscape. Next-generation genome sequencing methodologies, which differ from the original cloning-based Sanger technique, are high-throughput, imaging-based systems with vastly increased speeds and data output. There is no clear definition for *next-generation genome sequencing, also known generally as NGS*, but most are characterized by the direct and parallel sequencing of large numbers of amplified and fragmented DNA without vector-based cloning. The fragmented DNA tends to have sequence reads of 30–400 base pairs. There are now numerous examples of single-molecule techniques utilizing commercially available DNA sequencers (Cherf et al. 2012; Woollard et al. 2011). Early in 2012, the DNA sequencing companies Illumina and Life Technologies each announced new products that can sequence an entire human genome in 1 day for approximately $1,000 (BusinessWeek 2012).

Functional Genomics and Comparative Genomics

Functional genomics is the subfield of genomics that attempts to answer questions about the function of specific DNA sequences at the levels of transcription and translation, i.e., genes, RNA transcripts, and protein products (Raghavachari 2012). Research to relate genomic sequence data determined by structural genomics with observed biological function is predicted to fuel new drug discoveries thorough a better understanding of what genes do, how they are regulated, and the direct relationship between genes and their activity. The DNA sequence information itself rarely provides definitive information about the function and regulation of that particular gene. After genome sequencing, a functional genomic approach is the next step in the knowledge chain to identify functional gene products that are potential biotech drug leads and new drug discovery targets (see Fig. 8.1).

To relate functional genomics to therapeutic clinical outcomes, the human genome sequence must reveal the thousands of genetic variations among individuals that will become associated with diseases or symptoms in the patient's lifetime. Sequencing alone is not the solution, simply the end of the beginning of the genomic medicine era. Determining gene functionality in any organism opens the door for linking a disease to specific genes or proteins, which become targets for new drugs, methods to detect organisms (i.e., new diagnostic agents), and/or biomarkers (the presence or change in gene expression profile that correlates with the risk, progression, or susceptibility of a disease). Success with functional genomics will facilitate the ability to observe a clinical problem, take it to the benchtop for structural and functional genomic analysis, and return personalized solutions to the bedside in the form of new therapeutic interventions and medicines.

The face of biology has changed forever with the sequencing of the genomes of numerous organisms. Biotechnologies applied to the sequencing of the human genome are also being utilized to sequence the genomes of comparatively simple organisms as well as other mammals. Often, the proteins encoded by the genomes of more simple organisms and the regulation of those genes closely resemble the proteins and gene regulation in humans. Now that the sequencing of the entire genome is a reality, the chore of sorting through human, pathogen, and other organism diversity factors and correlating them with genomic data to provide real pharmaceutical benefits is an active area of research. Comparative genomics is the field of genomics that studies the relationship of genome structure and function across different biological species or strains and thus, provides information about the evolutionary processes that act upon a genome (Raghavachari 2012). Comparative genomics exploits both similarities and differences in the regulatory regions of genes, as well as RNA and proteins of different organisms to infer how selection has acted upon these elements.

Since model organisms are much easier to maintain in a laboratory setting, researchers are actively pursuing "comparative" genomic studies between multiple organisms. Unlocking genomic data for each of these organisms provides valuable insight into the molecular basis of inherited human disease. *S. cerevisiae*, a yeast, is a good model for studying cancer and is a common organism used in rDNA methodology. For example, it has become well known that women who inherit a gene mutation of the *BRCA1* gene have a high risk, perhaps as high as 85 %, of developing breast cancer before the age of 50 (Petrucelli et al. 2011). The first diagnostic product generated from genomic data was the *BRCA1* test for breast cancer predisposition. The gene product of *BRCA1* is a well-characterized protein implicated in both breast and ovarian cancer. Evidence has accumulated suggesting that the Rad9 protein of *S. cerevisiae* is distantly, but significantly, related to the *BRCA1* protein. The fruit fly possesses a gene similar to *p53*, the human tumor suppressor gene. Studying *C. elegans*, an unsegmented vermiform, has provided much of our early knowledge of apoptosis, the normal biological process of programmed cell death. Greater than 90 % of the proteins identified thus far from a common laboratory animal, the mouse, have structural similarities to known human proteins.

Similarly, mapping the whole of a human cancer cell genome will pinpoint the genes involved in cancer and aid in the understanding of cell changes and treatment of human malignancies utilizing the techniques of both functional and comparative genomics (Collins and Barker 2007). In cancer cells, small changes in the DNA sequence can cause the cell to make a protein that doesn't allow the cell to function as it should. These proteins can make cells grow quickly and cause damage to neighboring cells, becoming cancerous. The genome of a cancer cell can also be used to stratify cancer cells identifying one type of cancer from another or identifying a subtype of cancer within that type, such as HER2+ breast cancer. Understanding the cancer genome is a step toward personalized oncology. Numerous projects are underway around the world. Two such projects include the US NIH Cancer Genome Atlas Project (U.S. NIH 2012) and the Sanger Institute Cancer Genome Project (Sanger Institute 2012).

Comparative genomics is being used to provide a compilation of genes that code for proteins that are essential to the growth or viability of a pathogenic organism, yet differ from any human protein (cf. Chap. 22). For example, the worldwide effort to rapidly sequence the severe acute respiratory syndrome (SARS)-associated coronavirus genome to speed up diagnosis, prevent a pandemic, and guide vaccine creation was a great use of genomics in infectious disease. NGS will likely provide an opportunity to place genomics directly into the clinic to enable infectious disease point-of-care applications and thus, selective and superior patient outcomes. Also, genomic mining of new targets for drug design using genomic techniques may aid the quest for new antibiotics in a clinical environment of increasing incidence of antibiotic resistance.

A valuable resource for performing functional and comparative genomics is the "biobank," a collection of biological samples for reference purposes. Repositories of this type also might be referred to as biorepositories or named after the type of tissue depending on the exact type of specimens (i.e., tissue banks). Genomic techniques are fostering the creation of DNA banks, the collection, storage, and analysis of hundreds of thousands of specimens containing analyzable DNA. All nucleated cells, including cells from blood, hair follicles, buccal swabs, cancer biopsies, and urine specimens, are suitable DNA samples for analysis in the present or at a later date. DNA banks are proving to be valuable tools for genetics research (Thornton et al. 2005). While in its broadest sense such repositories could incorporate any collection of plant or animal samples, some of the most developed biobanks in the world are devoted to research on various types of cancer. While DNA banks devoted to cancer research have grown the fastest, there also has been an almost explosive growth in biobanks specializing in research on autism, schizophrenia, heart disease, diabetes, and many other diseases.

■ "Omic"-Enabling Technology: Bioinformatics

Structural genomics, functional genomics, proteomics, pharmacogenomics, and other "omic" technologies have generated an enormous volume of genetic and biochemical data to store and analyze. Living in an era of faster computers, bigger and better data storage, and improved methods of data analysis have led to the bioinformation superhighway that has facilitated the "omic" revolution. Scientists have applied advances in information technology, innovative software algorithms, and massive parallel computing to the ongoing research in biotechnology areas such as genomics to give birth to the fast growing field of bioinformatics (Lengauer and Hartmann 2007; Singh and Somvanshi 2012). The integration of new technologies and computing approaches in the domain of bioinformatics is essential to accelerating the rate of discovery of new breakthroughs that will improve health, well-being, and patient care. Bioinformatics is the application of computer technologies to the biological sciences with the object of discovering knowledge. With bioinformatics, a researcher can now better exploit the tremendous flood of genomic and proteomic data, and more cost-effectively data mine for a drug discovery "needle" in that massive data "haystack." In this case, data mining refers to the bioinformatics approach of "sifting" through volumes of raw data, identifying and

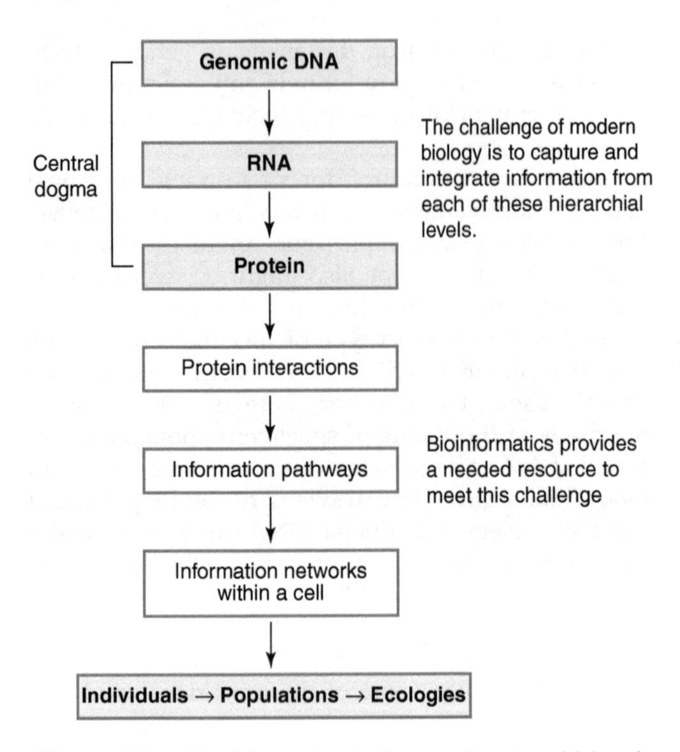

Figure 8.2 ■ The information challenges of systems biology in the genomic era.

extracting relevant information, and developing useful relationships among them.

Modern drug discovery and the commensurate need to better understand and define disease is utilizing bioinformatics techniques to gather information from multiple sources (such as the HGP, functional genomic studies, proteomics, phenotyping, patient medical records, and bioassay results including toxicology studies), integrate the data, apply life science developed algorithms, and generate useful target identification and drug lead identification data. As seen in Fig. 8.2, the hierarchy of information collection goes well beyond the biodata contained in the genetic code that is transcribed and translated. A recent National Research Council report for the US National Academies entitled "Toward Precision Medicine: Building a Knowledge Network for Biomedical Research and a New Taxonomy of Disease" calls for a new data network that integrates emerging research on the molecular basis of diseases with the clinical data from individual patients to drive the development of a more accurate taxonomy of disease that ultimately improves disease diagnosis and patient outcomes (U.S. National Academies 2011). The report notes that challenges include both scientific (technical advances needed to correlate genetic and environmental findings with incidence of disease) and legal and ethical challenges (privacy issues, electronic health records or EHR, etc.).

The entire encoded human DNA sequence alone requires computer storage of approximately 10^9 bits of information: the equivalent of a thousand 500-page books! GenBank (managed by the National Center for Biotechnology Information, NCBI, of the National Institutes of Health), the European Molecular Biology Laboratory (EMBL), and the DNA Data Bank of Japan (DDBJ) are three of the many centers worldwide that collaborate on collecting nucleic acid sequences. These databanks (both public and private) store tens of millions of sequences (Wu et al. 2011a). Once stored, analyzing the volumes of data (i.e., comparing and relating information from various sources) to identify useful and/or predictive characteristics or trends, such as selecting a group of drug targets from all proteins in the human body, presents a Herculean task. This approach has the potential of changing the fundamental way in which basic science is conducted and valid biological conclusions are reached.

Bioinformatics in its multifaceted implementations may be thought of as a technique of "electronic biology" (eBiology), conceptual biology, in silico biology or computational biology. A data-driven tool, the integration of bioinformatics with functional knowledge of the complex biological system under study, remains the critical foundation of any of the omic technologies described above and to follow.

The profession of pharmacy has readily recognized that optimal patient-centered care requires an effective integration of drug information and patient information into a system now known as "pharmacy informatics" (Anderson et al. 2010). Patient information includes data from genomics, proteomics, individual patient characteristics, patient safety, evidence-based medicine, and electronic health records. Drug information includes that found in the primary literature, drug information databases, internet resources, hospital information systems, pharmacy information systems, drug discovery literature, and pharmacogenomic studies. While it is beyond the scope of this chapter to explore pharmacy informatics further, this is becoming an important area for pharmacists to be knowledgeable (Fox 2010).

■ Transcriptomics

Remember that the central dogma of molecular biology is DNA to RNA via the process of transcription and RNA to protein via the process of translation. The transcriptome is the collection of all RNA transcribed elements for a given genome, not only the collection of transcripts that are subsequently translated into proteins (mRNAs). Noncoding transcripts such as noncoding microRNAs (miRNAs) are part of the transcriptome (cf. Chap. 23). The transcriptome represents just a small part of a genome, for instance, only

5 % of the human genome (Lu et al. 2005). The term transcript*omics* refers to the omic technology that examines the complexity of RNA transcripts of an organism under a variety of internal and external conditions reflecting the genes that are being actively expressed at any given time (with the exception of mRNA degradation phenomena such as transcriptional attenuation) (Subramanian et al. 2005). Therefore, the transcriptome can vary with external environmental conditions, while the genome is roughly fixed for a given cell line (excluding mutations). The transcriptomes of stem cells and cancer cells are of particular interest to better understand the processes of cellular differentiation and carcinogenesis. High-throughput techniques based on microarray technology are used to examine the expression level of mRNAs in a given cell population.

■ Proteomics, Structural Proteomics, and Functional Proteomics

Proteomics is the study of an organism's complete complement of proteins. Proteomics seeks to define the function and correlate that with expression profiles of all proteins encoded within an organism's genome or "proteome" (Veenstra 2010). While functional genomic research will provide an unprecedented information resource for the study of biochemical pathways at the molecular level, certainly a vast array of the approximately 20,000 genes identified in sequencing the human genome will be shown to be functionally important in various disease states (see druggable genome discussion above). These key identified proteins will serve as potential new sites for therapeutic intervention (see Fig. 8.1). The application of functional proteomics in the process of drug discovery has created a field of research referred to as pharmacoproteomics that tries to compare whole protein profiles of healthy persons versus patients with disease. This analysis may point to new and novel targets for drug discovery and personalized medicine (D'Alessandro and Zolla 2010). The transcription and translation of approximately 20,000 human genes can produce hundreds of thousands of proteins due to posttranscriptional regulation and posttranslational modification of the protein products. The number, type, and concentration may vary depending on cell or tissue type, disease state, and other factors. The proteins' function(s) is dependent on the primary, secondary, and tertiary structure of the protein and the molecules they interact with. Less than 30 years old, the concept of proteomics requires determination of the structural, biochemical, and physiological repertoire of all proteins. Proteomics is a greater scientific challenge than genomics due to the intricacy of protein expression and the complexity of 3D protein structure (structural proteomics) as it relates to biological activity (functional proteomics). Protein expression, isolation, purification, identification, and characterization are among the key procedures utilized in proteomic research.

To perform these procedures, technology platforms such as 2D gel electrophoresis, mass spectrometry, chip-based microarrays (discussed later in this chapter), X-ray crystallography, protein nuclear magnetic resonance (nmr), and phage displays are employed. Initiated in 2002, the Human Proteome Organization (HUPO) completed the first large-scale study to characterize the human serum and plasma proteins, i.e., the human serum and plasma proteome (States et al. 2006). They have spent the past 3 years developing a strategy for the first phase of the Human Proteome Project (Paik et al. 2012). The international consortium Chromosome-Centric Human Proteome Project is attempting to define the entire set of encoded proteins in each human chromosome (Paik et al. 2012). Pharmaceutical scientists anticipate that many of the proteins identified by proteomic research will be entirely novel, possessing unknown functions. This scenario offers not only a unique opportunity to identify previously unknown molecular targets, but also to develop new biomarkers and ultrasensitive diagnostics to address unmet clinical needs (Veenstra 2010). Today's methodology does not allow us to identify valid drug targets and new diagnostic methodologies simply by examining gene sequence information. However, "in silico proteomics," the computer-based prediction of 3D protein structure, intermolecular interactions, and functionality is currently a very active area of research.

Often, multiple genes and their protein products are involved in a single disease process. Since few proteins act alone, studying protein interactions will be paramount to a full understanding of functionality. Also, many abnormalities in cell function may result from overexpression of a gene and/or protein, under-expression of a gene and/ or protein, a gene mutation causing a malformed protein, and posttranslational modification changes that alter a protein's function. Therefore, the real value of human genome sequence data will only be realized after every protein coded by the approximately 20,000 genes has a function assigned to it.

■ "Omic"-Enabling Technology: Microarrays

The biochips known as DNA microarrays and oligonucleotide microarrays are a surface collection of hundreds to thousands of immobilized nucleic acid sequences or oligonucleotides in a grid created with specialized equipment that can be simultaneously examined to conduct expression analysis (Amaratunga et al. 2007; Semizarov 2009a). Biochips

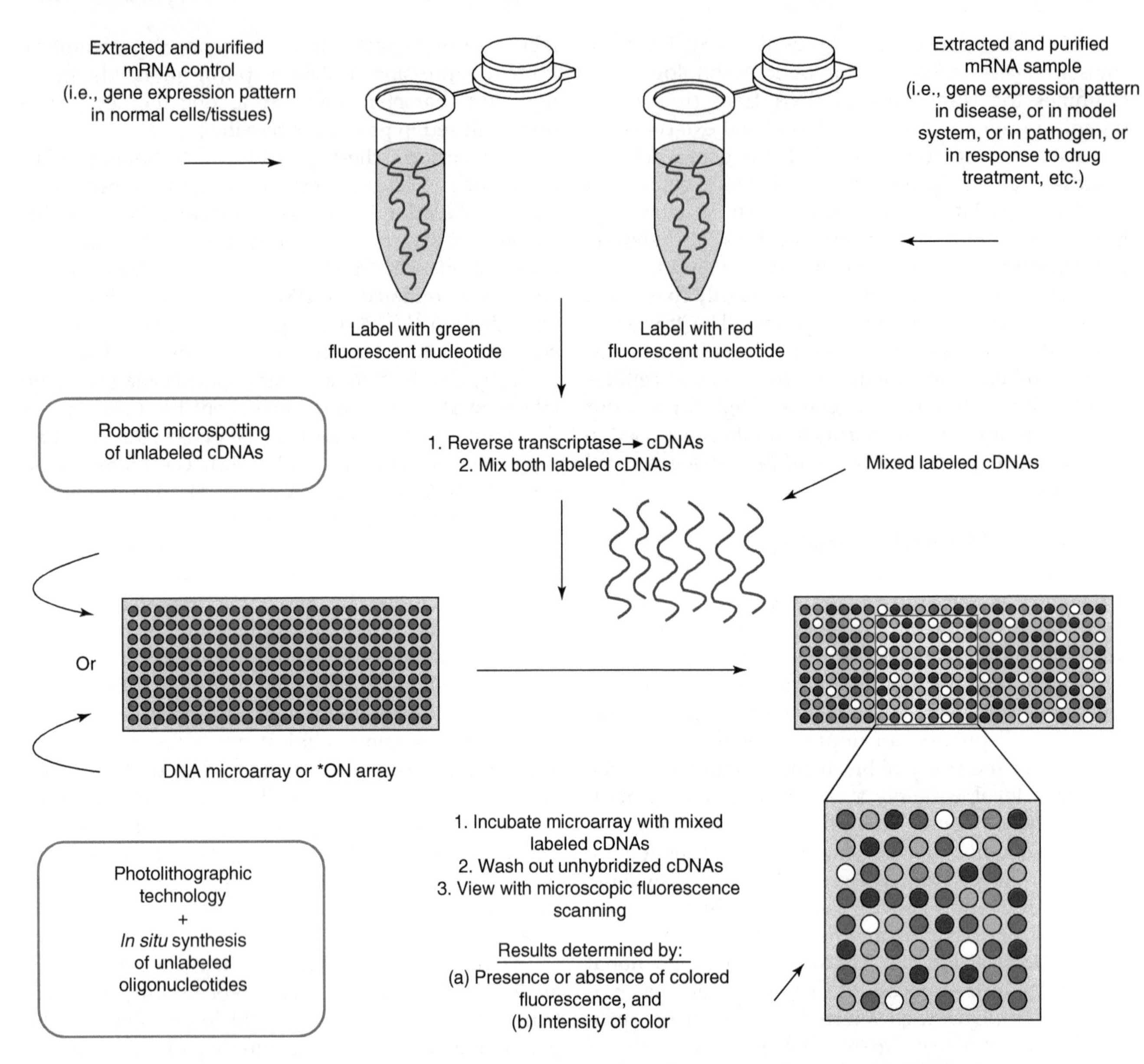

Figure 8.3 ■ Principle of operation of a representative DNA microarray or oligonucleotide (*ON) microarray.

may contain representatives of a particular set of gene sequences (i.e., sequences coding for all human cytochrome P450 isozymes) or may contain sequences representing all genes of an organism. They can produce massive amounts of genetic information (Semizarov 2009a). While the in vitro diagnostics market has been difficult to enter, Roche Diagnostics AmpliChip CYP 450 is a FDA-approved diagnostic tool able to determine a patient's genotype with respect to two genes that govern drug metabolism. This information obtained may be useful by a physician to select the appropriate drug and/or dosage for a given patient in the areas of cardiovascular disease, high blood pressure, depression, and others (according to the company).

Commonly, arrays are prepared on nonporous supports such as glass microscope slides. DNA microarrays generally contain high-density microspotted cDNA sequences approximately 1 kb in length representing thousands of genes. The field was advanced significantly when technology was developed to synthesize closely spaced oligonucleotides on glass wafers using semiconductory industry photolithographic masking techniques (see Fig. 8.3). Oligonucleotide microarrays (often called oligonucleotide arrays or DNA chips) contain closely spaced synthetic gene-specific oligonucleotides representing thousands of gene sequences. Microarrays can provide expression analysis for mRNAs. Screening of DNA variation is also possible. Thus, biochips can provide polymorphism detection and genotyping as well as hybridization-based expression monitoring (Semizarov 2009a).

Microarray analysis has gained increasing significance as a direct result of the genome sequencing studies. Array technology is a logical tool for studying functional genomics since the results obtained may link function to expression. Microarray technology's potential to study key areas of molecular medicine and drug discovery is unlimited at this stage of development. For example, gene expression levels of thousands of mRNA species may be studied simultaneously in normal versus cancer cells, each incubated with potential anticancer drug candidates. Related microarray technologies include protein microarrays, tissue microarrays, cell microarrays (also called transfection microarrays), chemical compound microarrays, and antibody microarrays. The principles are the same, while the immobilized collections differ accordingly.

■ "Omic"-Enabled Technology: Brief Introduction to Biomarkers

Biomarkers are clinically relevant biological features used as indicators of a biologic state, a disease, predisposition to a disease, disease progression, or disease regression (DePrimo 2007). Detection of or concentration change of a biomarker may indicate a particular disease state (e.g., the presence of a certain antibody may indicate an infection), physiology, or toxicity. A change in expression or state of a protein biomarker may correlate with the risk or progression of a disease, with the susceptibility of the disease to a given treatment or the drug's safety profile. Implemented in the form of a medical device, a measured biomarker becomes an in vitro diagnostic tool (Williams et al. 2006). While it is well beyond this chapter to provide a detailed discussion of biomarkers, it is important to note that omic technologies including omic-enabled technologies such as microarrays are being developed as clinical measuring devices for biomarkers. Biomarkers enable characterization of patient populations undergoing clinical trials or drug therapy and may accelerate drug development. Modern drug discovery often simultaneously involves biomarker discovery and diagnostic development (Frank and Hargreaves 2003). Drug development scientists are hopeful that the development of appropriate biomarkers will facilitate "go" and "no go" decisions during a potential therapeutic agents development process (Pritchard and Jurima-Romet 2010). Biomarker discovery is closely tied to the other applications of genomics previously described in this chapter. As an indicator of normal biological processes, pathogenic processes, or pharmacological responses to therapeutic intervention, biomarkers may serve as a substitute for a clinical end point and thus be a surrogate end point (Semizarov 2009b). Biomarkers are now available for a wide range of diseases and conditions including Alzheimer's and Parkinson's disease (Maetzler and Berg 2010), cardiac injury (McLean and Huang 2010), lung injury (Kodavanti 2010), drug-induced liver injury (Ozer et al. 2010), acute kidney injury (Dieterle and Sistare 2010), immunotoxicity (Dietert 2010), various cancers (Kelloff and Sigman 2012), pediatric care (Goldman et al. 2011), and a host of other diseases and biological conditions.

A "theranostic" is a rapid diagnostic, possibly a microarray, measuring a clinically significant biomarker, which may identify patients most likely to benefit or be harmed by a new medication (Warner 2004). Bundled with a new drug (and likely developed in parallel with that drug), the theranostic's diagnosis of the requisite biomarker (e.g., the overexpression of the HER2 gene product in certain breast cancer patients) influences the physician's therapeutic decisions [i.e., prescribing the drug trastuzumab (Herceptin) for HER2 receptor-positive breast cancer patients]. Thus, the diagnostic and the therapy are distinctly coupled = theranostic. The theranostic predicts clinical success of the drug. This example used to introduce the concept of a theranostic is possibly the best example of personalized medicine (see later in this chapter), achieving the best medical outcomes by choosing treatments that work well with a person's genomic profile or with certain characteristics.

■ Metabonomics and Metabolomics

The metabolome consists of the complete set of small molecules that are involved in the energy transmission in the cells by interacting with other biological molecules following metabolic pathways. These metabolites may be metabolic intermediates, hormones and other signaling molecules, and secondary metabolites (Nicholson and Wilson 2003; Patti et al. 2012). The techniques and processes for identifying clinically significant biomarkers of human disease and drug safety have fostered the systematic study of the unique chemical fingerprints that specific cellular processes leave behind, specifically their small molecule metabolite profiles. In January 2007 scientists at the University of Alberta and the University of Calgary finished a draft of the human metabolome (Wishart et al. 2007). They have catalogued and characterized 2,500 metabolites, 1,200 drugs, and 3,500 food components that can be found in the human body. Thus, while genomics and proteomics do not tell the whole story of what might be happening within a cell, metabolic profiling can give an instantaneous snapshot of the physiology of that cell.

High-performance liquid chromatography coupled with sophisticated nuclear magnetic resonance

(NMR) and mass spectrometry (MS) techniques is used to separate and quantify complex metabolite mixtures found in biological fluids to get a picture of the metabolic continuum of an organism influenced by an internal and external environment. The field of metabonomics is the holistic study of the metabolic continuum at the equivalent level to the study of genomics and proteomics. However, unlike genomics and proteomics, microarray technology is little used since the molecules assayed in metabonomics are small molecule end products of gene expression and resulting protein function. The term metabolomics has arisen as the metabolic composition of a cell at a specified time, whereas metabonomics includes both the static metabolite composition and concentrations and the full-time course fluctuations. Coupling the information being collected in biobanks, large collections of patient's biological samples and medical records, with metabonomic and metabolomic studies, will not only detect why a given metabolite level is increasing or decreasing but may reliably predict the onset of disease. Recent research and discoveries in oncology have led to reconsiderations regarding metabolic dysfunctions in cancer cell proliferation and differentiation. Metabolomic studies may be able to interrogate cancer cells for oxidative stress, a leading cause of genetic instability underpinning carcinogenesis, therefore indicative windows during the life of a cancerous cell for optimal therapeutic intervention (D'Alessandro and Zolla 2012). Also, the techniques are finding use in drug safety screening, identification of clinical biomarkers, and systems biology studies (see below).

■ Pharmacogenetics and Pharmacogenomics

It has been noted for decades that patient response to the administration of a drug was highly variable within a diverse patient population. Efficacy as determined in clinical trials is based upon a standard dose range derived from the large population studies. Better understanding of the molecular interactions occurring within the pharmacokinetics phase of a drug's action, coupled with new genetics knowledge and then genomic knowledge of the human have advanced us closer to a rational means to optimize drug therapy. Optimization with respect to the patients' genotype, to ensure maximum efficacy with minimal adverse effects, is the goal. Environment, diet, age, lifestyle, and state of health all can influence a person's response to medicines, but understanding an individual's genetic makeup is thought to be the key to creating personalized drugs with greater efficacy and safety. Approaches such as the related pharmacogenetics and pharmacogenomics promise the advent of "personalized medicine," in which drugs and drug combinations are optimized for each individual's unique genetic makeup. This chapter will only serve as an introduction, as entire classes are now offered and many books and review articles have been written about pharmacogenetics and pharmacogenomics (Lindpainter 2007; Knoell and Sadee 2009; Grossman and Goldstein 2010; Zdanowicz 2010; Pirmohamed 2011; Brazeau and Brazeau 2011a).

Single-Nucleotide Polymorphisms (SNPs)

While comparing the base sequences in the DNA of two individuals reveals them to be approximately 99.5 % identical, base differences, or polymorphisms, are scattered throughout the genome. The best-characterized human polymorphisms are single-nucleotide polymorphisms (SNPs) occurring approximately once every 1,000 bases in the three billion base pair human genome (Kassam et al. 2005). The DNA sequence variation is a single nucleotide – A, T, C, or G – in the genome difference between members of a species (or between paired chromosomes in an individual). For example, two sequenced DNA fragments from different individuals, AAGTTCCTA to AAGTTC<u>T</u>TA, contain a difference in a single nucleotide. Commonly referred to as "snips," these subtle sequence variations account for most of the genetic differences observed among humans. Thus, they can be utilized to determine inheritance of genes in successive generations. Technologies available from several companies allow for genotyping hundreds of thousands of SNPs for typically under $1,000 in a couple of days.

Research suggests that, in general, humans tolerate SNPs as a probable survival mechanism. This tolerance may result because most SNPs occur in noncoding regions of the genome. Identifying SNPs occurring in gene coding regions (cSNPs) and/or regulatory sequences may hold the key for elucidating complex, polygenic diseases such as cancer, heart disease, and diabetes and understanding the differences in response to drug therapy observed in individual patients (Grossman and Goldstein 2010; Pirmohamed 2011; US DOE 2012b). Some cSNPs do not result in amino acid substitutions in their gene's protein product(s) due to the degeneracy of the genetic code. These cSNPs are referred to as synonymous cSNPs. Other cSNPs, known as non-synonymous, can produce conservative amino acid changes, such as similarity in side chain charge or size or more significant amino acid substitutions.

While SNPs themselves do not cause disease, their presence can help determine the likelihood that an individual may develop a particular disease or malady. SNPs, when associated with epidemiological and pathological data, can be used to track susceptibilities to common diseases such as cancer, heart disease, and diabetes (Davidson and McInerney 2009). Biomedical researchers have recognized that discovering SNPs linked to diseases will lead potentially to the identification of new drug targets and diagnostic tests. The identification and mapping of hundreds of thousands of SNPs for use in large-scale association studies may turn the SNPs into biomarkers of disease and/or drug response. Genetic factors such as SNPs are believed to likely influence the etiology of diseases such as hypertension, diabetes, and lipidemias directly and via effects on known risk factors (Davidson and McInerney 2009). For example, in the chronic metabolic disease type 2 diabetes, a strong association with obesity and its pathogenesis includes defects of both secretion and peripheral actions of insulin. The association between type 2 diabetes and SNPs in three genes was detected in addition to a cluster of new variants on chromosome 10q. However, heritability values range only from 30 to 70 % as type 2 diabetes is obviously a heterogeneous disease etiologically and clinically. Thus, SNPs, in the overwhelming majority of cases, will likely not be indicators of disease development by themselves.

The projected impact of SNPs on our understanding of human disease led to the formation of the SNP Consortium in 1999, an international research collaboration involving pharmaceutical companies, academic laboratories, and private support. In the USA, the DOE and the NIH Human Genome programs helped establish goals to identify and map SNPs. The goals included the development of rapid large-scale technologies for SNP identification, the identification of common variants in the coding regions of most identified genes, the creation of an SNP map of at least 100,000 elements that may serve as future biomarkers, the development of knowledge that will aid future studies of sequence variation, and the creation of public resources of DNA samples, cell lines, and databases (US DOE 2012b). SNP databases include a database of the SNP Consortium (TSC), the dbSNP database from the National Center for Biotechnology Information (NCBI), and the Human Genome Variation Database (HGVbase).

Pharmacogenetics Versus Pharmacogenomics

In simplest terms, pharmacogenomics is the whole genome application of pharmacogenetics, which examines the single gene interactions with drugs. Tremendous advances in biotechnology are causing a dramatic shift in the way new pharmaceuticals are discovered, developed, and monitored during patient use. Pharmacists will utilize the knowledge gained from genomics and proteomics to tailor drug therapy to meet the needs of their individual patients employing the fields of pharmacogenetics and pharmacogenomics (Kalow 2009; Knoell and Sadee 2009; Grossman and Goldstein 2010; Zdanowicz 2010; Pirmohamed 2011; Brazeau and Brazeau 2011a).

Pharmacogenetics is the study of how an individual's genetic differences influence drug action, usage, and dosing. A detailed knowledge of a patient's pharmacogenetics in relation to a particular drug therapy may lead to enhanced efficacy and greater safety. Pharmacogenetic analysis may identify the responsive patient population prior to administration, i.e., personalized medicine. The field of pharmacogenetics is over 50 years old, but is undergoing renewed, exponential growth at this time. Of particular interest in the field of pharmacogenetics is our understanding of the genetic influences on drug pharmacokinetic profiles such as genetic variations affecting liver enzymes (i.e., cytochrome P450 group) and drug transporter proteins and the genetic influences on drug pharmacodynamic profiles such as the variation in receptor protein expression (Abla and Kroetz 2009; Frye 2009; Johnson 2009; Kalow 2009; Wang 2009).

In contrast, pharmacogenomics is linked to the whole genome, not an SNP in a single gene. It is the study of the entire genome of an organism (i.e., human patient), both the expressed and the non-expressed genes in any given physiologic state. Pharmacogenomics combines traditional pharmaceutical sciences with annotated knowledge of genes, proteins, and single-nucleotide polymorphisms. It might be viewed as a logical convergence of the stepwise advances in genomics with the growing field of pharmacogenetics. Incorrectly, the definitions of pharmacogenetics and pharmacogenomics are often used interchangeably. Whatever the definitions, they share the challenge of clinical translation, moving from bench top research to bedside application for patient care.

Genome-Wide Association Studies (GWAS)

The methods of genome-wide association studies (GWAS), also known as whole genome association studies, are powerful tools to identify genetic loci that affect, for instance, drug response or susceptibility to adverse drug reactions (Davidson and McInerney 2009; Wu et al. 2011a). These studies are an examination of the many genetic variations found in different individuals to determine any association between a variant (genotype) and a biological trait (phenotype). The majority of GWAS typically study associations between SNPs and drug response or SNPs and major disease. While the first GWAS was published only in 2005, they have emerged as important tools with, as per data from the NHGRI GWAS Catalog, hundreds of thousands of individuals now tested in over 1,200 human GWAS examining over 200 diseases and traits and 6,229 SNPs as of early 2012 (Hindroff et al. 2012). While believed to be a core driver in the vision for personalized medicine, GWAS when coupled to the HapMap Project (an international effort to identify and map regions of DNA sequence nearly identical within the broad population) to date have been plagued by inconsistencies in genotypes, difficulties in assigning phenotypes, and overall quality of the data (Hong et al. 2010; Miclaus et al. 2010; Wu et al. 2011a, b, c). Challenges have included difficulties identifying the key genetic loci due to two or more genes with small and additive effects on the trait (epistasis), the trait caused by gene mutations at several different chromosomal loci (locus heterogeneity), environmental causes modifying expression of the trait or responsible for the trait, and undetected population structure in the study such as those arising when some study members share a common ancestral heritage (Brazeau and Brazeau 2011b). The practical use of this approach and its introduction into the everyday clinical setting remain a challenge, but will undoubtedly be aided by new next-generation sequencing techniques, enhanced bioinformatics capabilities, and better genomic understanding.

■ On the Path to Personalized Medicine: A Brief Introduction

Much of modern medical care decision-making is based upon observations of successful diagnosis and treatment at the larger population level. There is an expectation, however, that healthcare is starting to undergo a revolutionary change as new genomic and other "omic" technologies become available to the clinic that will better predict, diagnose, monitor, and treat disease at the level of the specific patient. A goal is match individual patients with the most effective and safest drugs and doses. Direct-to-consumer genomic tests became more readily available (such as 23andMe, Navigenics, and deCODE Genetics) (McGuire et al. 2010). Academic medical centers have begun to demonstrate the feasibility of routine clinical genotyping as a means of informing pharmacotherapeutic treatment selection in oncology (Tursz et al. 2011). Likewise, demonstration projects in pharmacogenomics entered pharmacy practice in several settings (Koomer and Ansong 2010; Crews et al. 2011; Padgett et al. 2011). Pharmacy education curricula are evolving to prepare graduates practice in a personalized medicine environment (Lee et al. 2009; Krynetskiy and Calligaro 2009; Koomer 2010; Murphy et al. 2010; Zembles 2010). This approach is entirely consistent with the concept of patient-centered care to improve patient outcomes (Clancy and Collins 2010; Waldman and Terzic 2011; Kaye et al. 2012).

Modern genomics, proteomics, metabolomics, pharmacogenomics, epigenomics (to be discussed later in this chapter), and other technologies, implemented in the clinic in faster and less expensive instrumentation and methodologies, are now being introduced to identify genetic variants, better inform healthcare providers about their individual patient, tailor evidence-based medical treatment, and suggest rational approaches toward preventative care. The hopes and realities of personalized medicine (sometimes referred to as part of "molecular medicine"), pharmacotherapy informed by a patient's individual genomic and proteomic information, are global priorities (Knoell and Sadee 2009; Grossman and Goldstein 2010; Rahbar et al. 2011). As a pharmaceutical biotechnology text, our limited discussion here will focus on personalized medicine in a primarily pharmacogenomic and pharmacogenetic context. However, other genomic-type technologies including GWAS, next-generation sequencing, proteomics, and metabolomics will be crucial for the successful implementation of personalized medicine. The hope is that "omic" science will bring predictability to the optimization of drug selection and drug dosage to assure safe and effective pharmacotherapy (Fig. 8.4).

For our discussion, it is again important to recognize that pharmacogenetics and pharmacogenomics are subtly different (Brazeau and Brazeau 2011a). Pharmacogenomics introduces the additional element of our present technical ability to pinpoint patient-specific DNA variation using genomic techniques. The area looks at the genetic composition or genetic variations of an organism and their connection to drug response. Variations in target pathways are studied to understand how the variations are manifested and how they influence response. While overlapping fields of study, pharmacogenomics is a much newer term that correlates an individual patient's DNA variation (SNP level of variation knowledge rather than gene level of variation knowledge) with his or her response to pharmacotherapy. Personalized medicine will employ both technologies.

Optimized personalized medicine utilizing pharmacogenomic knowledge would not only spot disease before it occurs in a patient or detect a critical variant that will influence treatment but should increase drug efficacy upon pharmacotherapy and reduce drug toxicity. Also, it would facilitate the drug development process (see Fig. 8.1) including improving clinical development outcomes, reducing overall cost of drug development, and leading to development of new diagnostic tests that impact on therapeutic decisions (Grossman and Goldstein 2010; Zineh and Huang 2011). Individualized optimized pharmacotherapy would first require a detailed genetic analysis of a patient, assembling a comprehensive list of SNPs. Pharmacogenomic tests most likely in the form of microarray technology and based upon clinically validated biomarkers would be administered to pre-identify responsive patients before dosing with a specific agent. Examples of such microarray-based

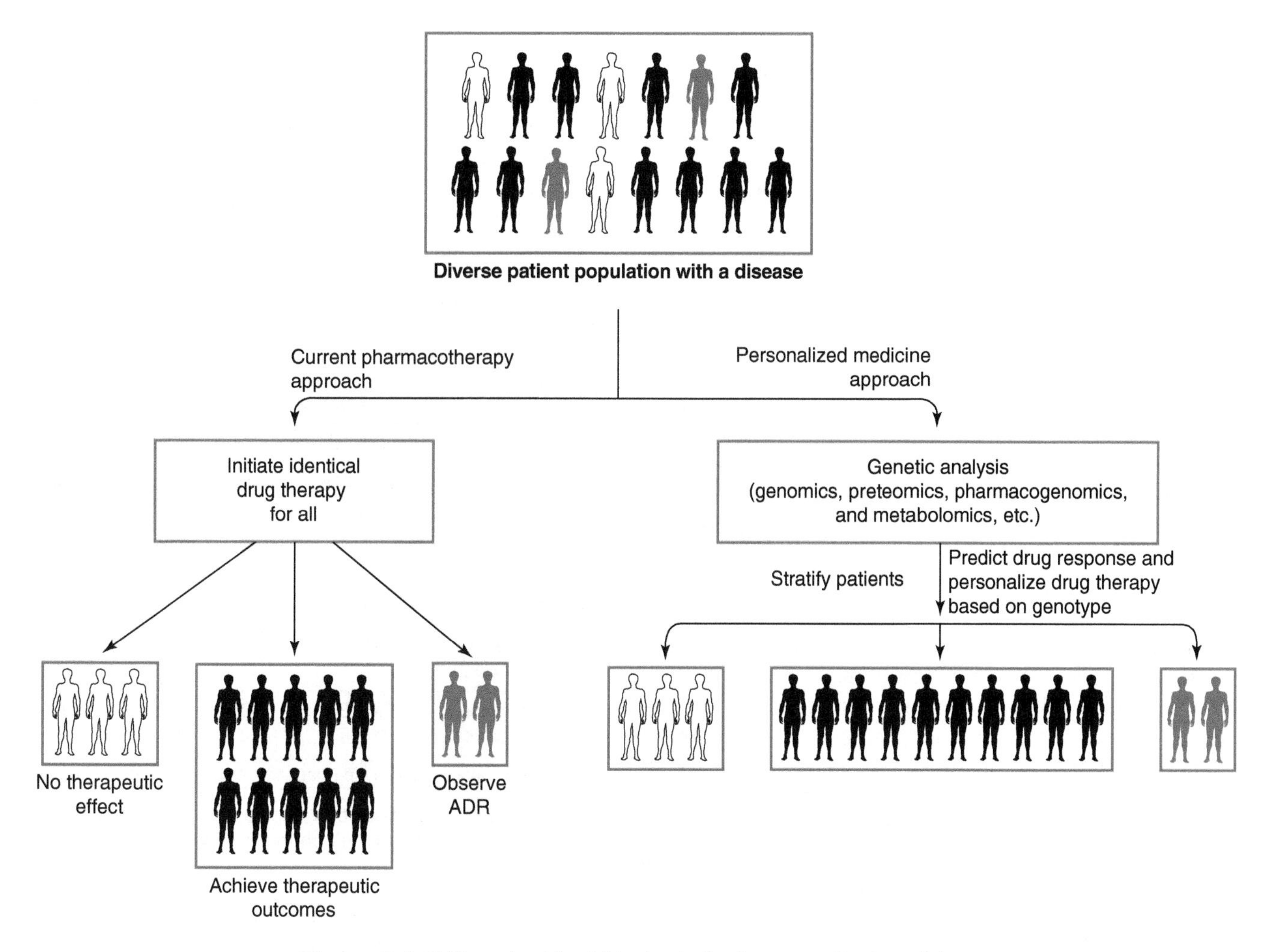

Figure 8.4 ■ The role of "omic" technologies in personalized medicine.

diagnostics are the FDA-approved AmpliChip P450 from Roche to screen a patient for the presence of any of 27 SNPs in CYP2D6 and CYP2C19, the Infiniti 2C9 & VKORC1 Multiplex Assay for warfarin therapy from AutoGenomics, and the Pathwork Tissue of Origin test of 15 common malignant tumor types to better focus treatment options. The impact of the patient's SNPs on the use of new or existing drugs would thus be predicted and individualized drug therapy would be identified that assures maximal efficacy and minimal toxicity (Topol 2010).

Personalized medicine would also require knowledge of an individual patient's genomic profile to help identify potential drug responders and nonresponders. This might be accomplished by testing for the presence or absence of critical biomarkers that may be associated with prediction of response rates. The US FDA provides an online list of all FDA-approved drugs with pharmacogenomic information in their labels (black boxes). Some, but not all, of the labels include specific actions to be taken based on genetic information. The drug labels contain information on genomic biomarkers that may be predictive of drug exposure and clinical response rate, risk of adverse reactions, genotype-specific dosing, susceptibility to a specific mechanism of drug action, or polymorphic drug target and disposition genes. Rather than reproducing this table in whole or part in this text, the reader may access it in its constantly updated form at www.fda.gov/Drugs/ScienceResearch/ResearchAreas/Pharmacogenetics/ucm083378.htm.

It is well understood that beyond genomics and proteomics, a patient's behavioral and environmental factors influence clinical outcomes and susceptibility to disease. Emerging fields of nutrigenomics and envirogenomics are studying these additional layers of complexity. Personalized medicine will become especially important in cases where the cost of testing is less than either the cost of the drug or the cost of correcting adverse drug reactions caused by the drug. Pharmaceutical care would begin by identifying a patient's susceptibility to a disease, then administering the right drug to the right patient at the right time. For example, the monoclonal antibody trastuzumab

(Herceptin) is a personalized breast cancer therapy specifically targeted to the HER2 gene product (25–30 % of human breast cancers overexpress the human epidermal growth factor receptor, HER2 protein) (Kolesar 2009). Exhibiting reduced side effects as compared to standard chemotherapy due to this protein target specificity, trastuzumab is not prescribed to treat a breast cancer patient unless the patient has first tested positive for HER2 overexpression. While currently an immunohistochemical assay, not a sophisticated DNA microarray assay, the example shows the power of such future tests.

The success of targeted therapy for personalized medicine has fostered the concept that the era of the blockbuster drug may be over and will be replaced by the "niche buster" drug, a highly effective medicine individualized for a small group of responding patients identified by genomic and proteomic techniques. Also, while numerous articles predicted that pharmacogenomics would revolutionize medicine, the initial predictions have not been lived up to the hype due to statistical, scientific, and commercial hurdles. With more than 11 million SNP positions believed to be present in the human population, large-scale detection of genetic variation holds the key to successful personalized medicine (Pennisi 2010; Reardon 2011; Baker 2012). Correlation of environmental factors, behavioral factors, genomic and proteomic factors (including pharmacogenomic and metabolomic factors), and phenotypical observables across large populations remains a daunting data-intensive challenge. Yet, pharmacogenetics and pharmacogenomics are having an impact on modern medicine.

Enzyme	Common variant	Potential consequence
CYP1A2	CYP1A2*1F	Increased inducibility
CYP1A2	CYP1A2*1K	Decreased metabolism
CYP2A6	CYP2A6*2	Decreased metabolism
CYP2B6	CYP2B6*5	No effect
CYP2B6	CYP2B6*6	Increased metabolism
CYP2B6	CYP2B6*7	Increased metabolism
CYP2C8	CYP2C8*2	Decreased metabolism
CYP2C9	CYP2C9*2	Altered affinity
CYP2C9	CYP2C9*3	Decreased metabolism
CYP2D6	CYP2D6*10	Decreased metabolism
CYP2D6	CYP2D6*17	Decreased metabolism
CYP2E1	CYP2E1*2	Decreased metabolism
CYP3A7	CYP3A7*1C	Increased metabolism
Flavin-containing monooxygenase 3	FMO3*2	Decreased metabolism
Flavin-containing monooxygenase 3	FMO3*4	Decreased metabolism

Data from references noted in the Pharmacogenomics section of Chap. 8

Table 8.2 ■ Some selected examples of common drug metabolism polymorphisms and their pharmacokinetic consequences.

Human Genomic Variation Affecting Drug Pharmacokinetics

Genetic variation associated with drug metabolism and drug transport, processes resulting from products of gene expression (metabolic enzymes and transport proteins, respectively) play a critical role in determining the concentration of a drug in its active form at the site of its action and also at the site of its possible toxic action(s). Thus, pharmacogenetic and pharmacogenomic analysis of drug metabolism and drug transport is important to a better clinical understanding of and prediction of the effect of genetic variation on drug effectiveness and safety (Abla and Kroetz 2009; Frye 2009; Wang 2009; Cox 2010; Weston 2010).

It is well recognized that specific drug metabolizer phenotypes may cause adverse drug reactions. For instance, some patients lack an enzymatically active form, have a diminished level, or possess a modified version of CYP2D6 (a cytochrome P450 allele) and will metabolize certain classes of pharmaceutical agents differently to other patients expressing the native active enzyme. All pharmacogenetic polymorphisms examined to date differ in frequency among racial and ethnic groups. For example, CYP2D6 enzyme deficiencies may occur in ≤2 % Asian patients, ≤5 % black patients, and ≤11 % white patients (Frye 2009). A diagnostic test to detect CYP2D6 deficiency could be used to identify patients that should not be administered drugs metabolized predominantly by CYP2D6. Table 8.2 provides some selected examples of common drug metabolism polymorphisms and their pharmacokinetic consequences.

With the burgeoning understanding of the genetics of warfarin metabolism, warfarin anticoagulation therapy is becoming a leader in pharmacogenetic analysis for pharmacokinetic prediction (Limdi and Rettie 2009; Momary and Crouch 2010; Bungard et al. 2011; McDonagh et al. 2011). Adverse drug reactions (ADRs) for warfarin account for 15 % of all ADRs in the USA, second only to digoxin. Warfarin dose is adjusted with the goal of achieving an INR (International Normalized Ratio = ratio of patient's prothrombin time as compared to that of a normal control) of 2.0–3.0. The clinical challenge is to limit hemorrhage, the primary ADR, while achieving the optimal degree of protection against thromboembolism. Deviation in the INR has been shown to be the strongest risk factor for bleeding complications. The major routes of metabolism of warfarin are by CYP2C9 and CYP3A4. Some of the compounds, which have been identified to influence positively or negatively warfarin's INR, include cimetidine, clofibrate, propranolol, celecoxib (a competitive inhibition of CYP2C9), fluvoxamine (an inhibitor of several CYP enzymes), various antifungals and antibiotics (e.g., miconazole, fluconazole, erythromycin), omeprazole, alcohol, ginseng, and garlic. Researchers have determined that the majority of individual patient variation observed clinically in response to warfarin therapy is genetic in nature, influenced by the genetic variability of metabolizing enzymes, vitamin K cycle enzymes, and possibly transporter proteins. The CYP2C9 genotype polymorphisms alone explain about 10 % of the variability observed in the warfarin maintenance dose. Figure 8.5 shows the proteins involved in warfarin action and indicates the pharmacogenomic variants that more significantly influence warfarin therapy optimal outcome.

Studies at both the basic research and clinical level involve the effect of drug transport proteins on the pharmacokinetic profile of a drug (Abla and Kroetz 2009). Some areas of active study of the effect of genetic variation on clinical effectiveness include efflux transporter proteins (for bioavailability, CNS exposure, and tumor resistance) and neurotransmitter uptake transporters (as valid drug targets). Novel transporter proteins are still being identified as a result of the Human Genome Project and subsequent proteomic research. More study is needed on the characterization of expression, regulation, and functional properties of known and new transporter proteins to better assess the potential for prediction of altered drug response based on transporter genotypes.

Human Genomic Variation Affecting Drug Pharmacodynamics

Genomic variation affects not only the pharmacokinetic profile of drugs, it also strongly influences the pharmacodynamic profile of drugs via the drug target. To understand the complexity of most drug responses, factors influencing the expression of the protein target directly with which the drug interacts must be studied. Targets include the drug receptor involved in the response as well as the proteins associated with disease risk, pathogenesis, and/or toxicity including infectious disease (Johnson 2009; Rogers 2009; Webster 2010). There are increasing numbers of prominent examples of inherited polymorphisms influencing drug pharmacodynamics. To follow on the warfarin example above (see Fig. 8.5), the majority of individual patient variation observed clinically in response to anticoagulant therapy is genetic in nature. However, the CYP2C9 genotype polymorphisms alone only explain about 10 % of the variability observed in the warfarin

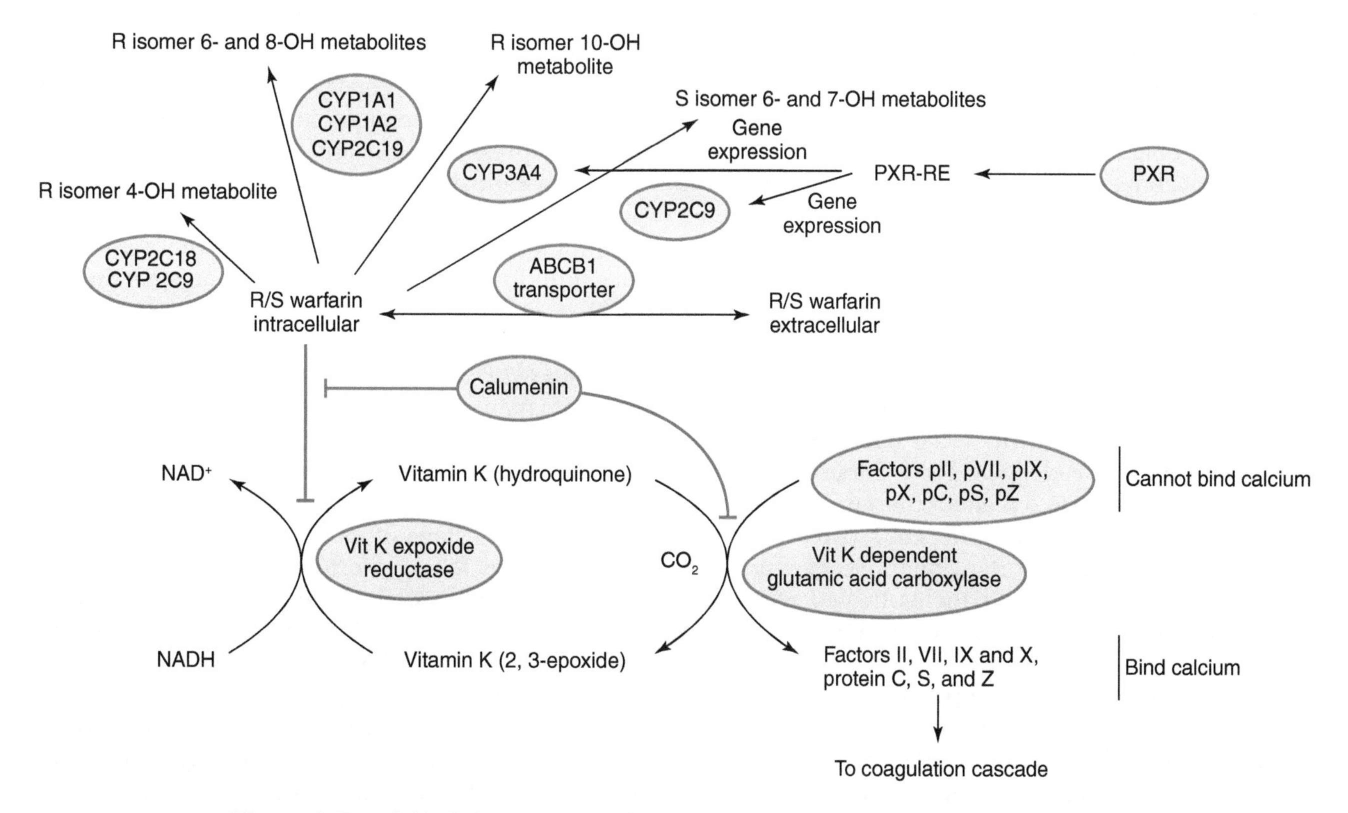

Figure 8.5 ■ Critical pharmacogenomic variants affecting warfarin drug action and ADR.

maintenance dose. Warfarin effectiveness is also influenced by the genetic variability of vitamin K cycle enzymes. The drug receptor for warfarin is generally recognized as vitamin K epoxide reductase, the enzyme that recycles vitamin K in the coagulation cascade. Vitamin K epoxide reductase complex1 (VKORC1) has been determined to be highly variant with as much as 50 % of the clinical variability observed for warfarin resulting from polymorphisms of this enzyme.

Associations have been implicated between drug response and genetic variations in targets for a variety of drugs including antidepressants (G-protein β3), antipsychotics (dopamine D2, D3, D4; serotonin 5HT2A, 5HT2C), sulfonylureas (sulfonylurea receptor protein), and anesthetics (ryanodine receptor) (Johnson 2009). In addition, similar associations have been studied for drug toxicity and disease polymorphisms including abacavir (major histocompatability proteins; risk of hypersensitivity), cisapride and terfenadine (HERG, KvLQT1, Mink, MiRP1; increased risk of drug-induced torsade de pointes), and oral contraceptives (prothrombin and factor V; increased deep vein thrombosis) (Johnson 2009). Likewise, similar associations for efficacy are known such as statins (apolipoprotein E; enhanced survival prolongation with simvastatin) and tacrine (apolipoprotein E; clinical improvement of Alzheimer's symptoms) (Johnson 2009).

Value of Personalized Medicine in Disease

Due to the intimate role of genetics in carcinogenesis, personalized medicine is rapidly becoming a success story in oncology based on genetic profiling using proteomic analyses of tumor biopsies (Garnett 2012; Kelloff and Sigman 2012; Shaw and Johnson 2012). As described above, targeted cancer therapies such as trastuzumab (Herceptin) are successful and are viewed as the way of the future. Also, clinically important polymorphisms predict increased toxicity in patients with cancer being treated with the chemotherapeutic drugs, for example, 6-mercaptopurine (thiopurine methyltransferase *2, *3A, and *3C variants), 5-fluorouracil (5-FU) (dihydropyrimidine dehydrogenase *2A variant), and irinotecan (UGT1A1*28 allele; FDA-approved Invader UGT1A1 Molecular Assay diagnostic available to screen for presence of this allele associated with irinotecan toxicity) (Kolesar 2009). Likewise, clinically important pharmacogenetics predicts efficacy in oncology patients treated with 5-FU (thymidylate synthase *2 and *3C variants) (Petros and Sharma 2009).

A classic application of pharmacogenetics is our present understanding of the potentially fatal hematopoietic toxicity that occurs in some patients administered standard doses of the antileukemic agents azathioprine, mercaptopurine, and thioguanine (Zhou 2006; Petros and Sharma 2009). These drugs are metabolized by the enzyme thiopurine methyltransferase (TPMT) to the inactive S-methylated products. Gene mutations (polymorphisms) may occur in as many as 11 % of patients resulting in decreased TPMT-mediated metabolism of the thiopurine drugs. A diagnostic test for TPMT is now available and used clinically. Identified patients with poor TPMT metabolism may need their drug dose lowered 10–15-fold. Mechanisms of multidrug resistance to cancer drugs are influenced by genetic differences. A number of polymorphisms in the MDR-1 gene coding for P-glycoprotein, the transmembrane protein drug efflux pump responsible for multidrug resistance, have been identified. One, known as the T/T genotype and correlated with decreased intestinal expression of P-glycoprotein and increased drug bioavailability, has an allele frequency of 88 % in African-American populations, yet only approximately 50 % in Caucasian-American populations (Kolesar 2009).

Pharmacogenetic and pharmacogenomic analysis of patients is being actively studied in many disease states. However, a detailed discussion goes beyond what this introduction may provide. The reader is encouraged to read further in the pharmacogenetic/pharmacogenomic-related references at the end of this chapter. Some examples include infectious disease (genetic predisposition to infection in the host; Rogers 2009), cardiovascular disease (genes linked to heart failure and treating hypertension, warfarin anticoagulant therapy, lipid lowering drugs; Zineh and Pacanowski 2009), psychiatry (the roles of drug metabolism and receptor expression in drug response rates for antidepressants and antipsychotic drugs, weight gain from antipsychotics; Ellingrod et al. 2009), asthma (leukotriene inhibitors and beta-agonists; Blake et al. 2009), and transplantation (cyclosporine metabolism and multidrug resistance efflux mechanisms; Burckert 2009).

Challenges in Personalized Medicine

There are many keys to success for personalized medicine that hinge on continued scientific advancement. While it is great for the advancement of the genomic sciences, some have questioned how good it is for patients at this stage of its development due to exaggerated claims falling short of the predictive and preventative healthcare paradigm promised (Browman et al. 2011; Nature Biotechnology Editorial Staff 2012). The pace of advancement has been slower than promised (Zuckerman and Milne 2012). There are also economic, societal, and ethical issues that must be addressed to successfully implement genetic testing-based individualized pharmacotherapy (Huston 2010). It is fair to state that most drugs will not be effective in all patients all of the time. Thus, the pressure of payers to move from a "payment for product" to a "payment for clinically significant health outcomes" model is reasonable. The use of omic health technologies and health informatics approaches to stratify patient populations for drug effectiveness and drug safety is a laudable goal. However, the technologies are currently quite expensive and the resulting drug response predictability is now just being validated clinically. Cost-effectiveness and cost-benefit analyses are limited at this date (Chalkidou and Rawlins 2011). Also, the resulting environment created by these technologies in the context of outcomes expectations and new drug access/reimbursement models will give rise to a new pharmaceutical business paradigm that is still evolving and not well understood.

In 2005, the FDA approved what some referred to as the "first racially targeted drug," BiDil (isosorbide/hydralazine; from NitroMed) (Branca 2005). Omic technologies were not generally involved in the development and approval process. Based on the analysis of health statistics suggesting that the rate of mortality in blacks with heart disease is twice as high in whites in the 45–64 age group, a clinical trial of this older drug combination in 1,050 African-Americans was conducted and the 43 % improvement in survival in the treatment arm resulted in FDA approval of the drug exclusively in African-Americans. Yet, modern anthropology and genetics have shown that while race does exist as a real social construct, there are no genetically distinguishable human racial groups (Ossorio 2004). Thus, attributing observed differences in biomedical outcomes and phenotypical observations to genetic differences among races is problematic and ethically challenging. Race is likely just a surrogate marker for the environmental and genetic causes of disease and response to pharmacotherapy. Now, factor in the introduction of omic technologies broadly into healthcare in a manner to segregate patient populations based on genomic and proteomic characteristics. It is obvious that these modern technologies pose provocative consequences for public policy (including data protection, insurability, and access to care), and these challenges must be addressed by decision makers, scientists, healthcare providers, and the public for personalized medicine to be successful (for further insight into this complex area, please read Brazeau and Brazeau 2011c). In conclusion, even with challenges and questioned progress, personalized medicine is a global concern and an unprecedented opportunity if the science and the clinic can both succeed.

Epigenetics and Epigenomics

DNA is the heritable biomolecule that contains the genetic information resulting in phenotype from parent to offspring. Modern genomics, GWAS and SNP analyses, confirm this and identify genetic variants that may be associated with a different phenotype. However, genome-level information alone does not generally predict phenotype at an individual level (Daxinger and Whitelaw 2012). For instance, researchers and clinicians have known for some time that an individual's response to a drug is affected by their genetic makeup (DNA sequence, genotype) and a set of disease and environmental characteristics working alone or in concert to determine that response. Research in animal models has suggested that in addition to DNA sequence, there are a number of other "levels" of information that influence transcription of genomic information. As you are aware, every person's body contains trillions of cells, all of which have essentially the same genome and, therefore, the same genes. Yet some cells are optimized for development into one or more of the 200+ specialized cells that make up our bodies: muscles, bones, brain, etc. For this to transpire from within the same genome, some genes must be turned on or off at different points of cell development in different cell types to affect gene expression, protein production, and cell differentiation, growth, and function. There is a rapidly evolving field of research known as epigenetics (or epigenomics) that can be viewed as a conduit between genotype and phenotype. Epigenetics literally means "above genetics or over the genetic sequence." It is the factor or factors that influence cell behavior by means other than via a direct effect on the genetic machinery. Epigenetic regulation includes DNA methylation and covalent histone modifications (Fig. 8.6) and is mitotically and/or meiotically heritable changes in gene expression that result without a change in DNA sequence (Berger et al. 2009). Epigenomics is the merged science of genomics and epigenetics (Raghavachari 2012). Functionally, epigenetics acts to regulate gene expression, gene silencing during genomic imprinting, apoptosis, X-chromosome inactivation, and tissue-specific gene activation (such as maintenance of stem cell pluripotency) (Garske and Denu 2009).

The more we understand epigenetics and epigenomics, the more we are likely to understand those phenotypic traits that are not a result of genetic information alone. Epigenetics/epigenomics may also explain low association predictors found in some pharmacogenetic/pharmacogenomic studies. Etiology of disease, such as cancer, likely involves both genetic variants and epigenetic modifications that could result from environmental effects (Bjornsson et al. 2004; Jirtle and Skinner 2007). Age also likely influences epigenetic modifications as studies of identical twins show greater differences in global DNA methylation in older rather than younger sets of twins (Feinberg et al. 2010).

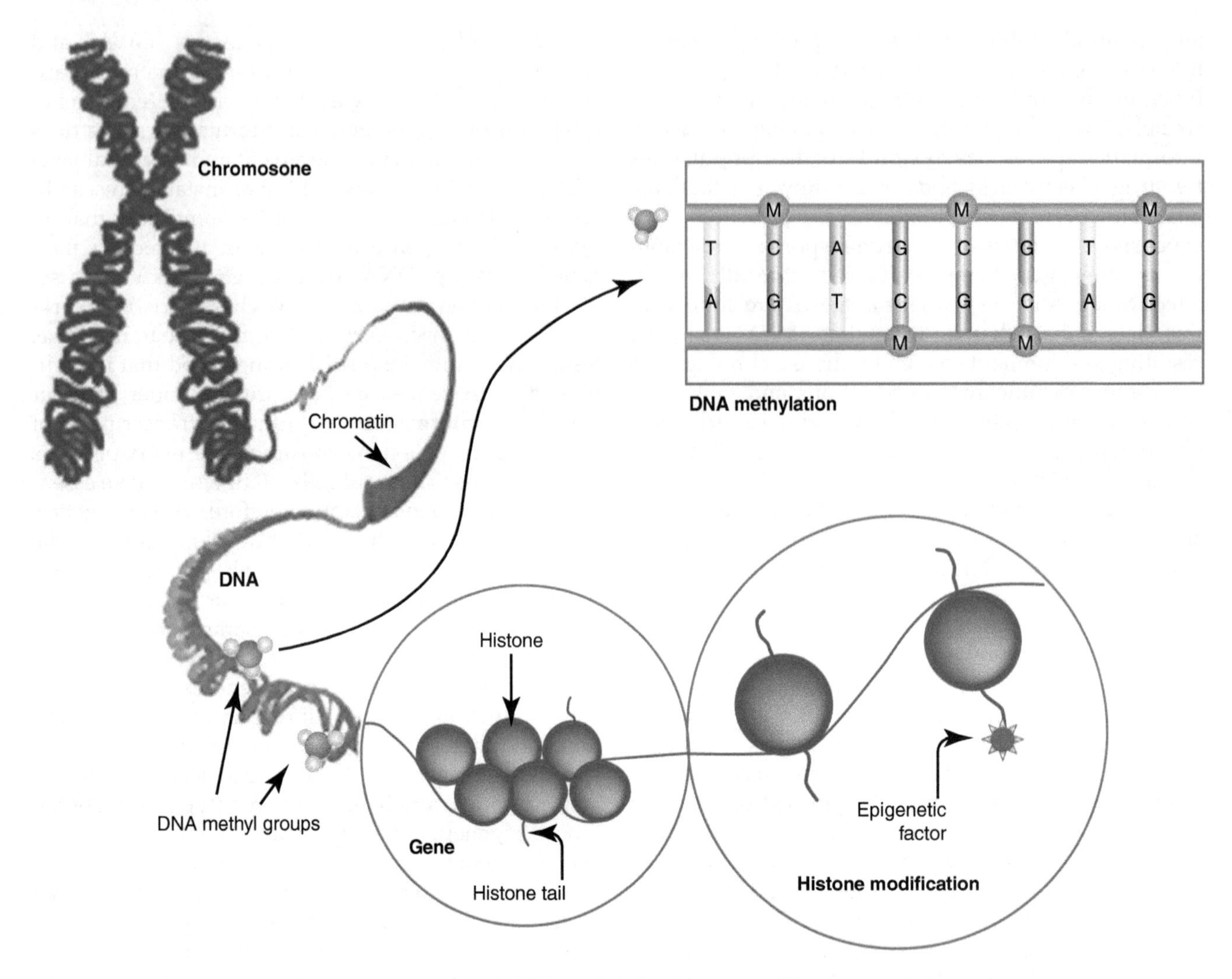

Figure 8.6 ■ Epigenetic regulation via DNA methylation, histone modifications, and chromatin structure.

Abnormal epigenetic regulation is likely a feature of complex diseases such as diabetes, cancer, and heart disease (Chen and Zhang 2011; Hamm and Costa 2011; Rakyan et al. 2011). Therefore, epigenetics targets are being explored for drug design, especially those observed in cancer (Woster 2010). The first generation of FDA-approved epigenetics-based drugs is available with two DNA demethylating agents (5-azacytidine and decitabine) and two histone deacetylase (HDAC) inhibitors (vorinostat and romidepsin). These have been approved mainly for the treatment of blood cancers, in particularly myelodysplastic syndromes (MDS).

One of the most studied and best understood molecular mechanisms of epigenetic regulation is methylation of cytosine residues at specific positions in the DNA molecule (Fig. 8.6) (Portela and Esteller 2010). Another mechanism of epigenomic control appears to occur at the level of chromatin. In the cell, DNA is wrapped around 8 different histone proteins to form chromatin. Packaging of DNA into chromatin can render large regions of the DNA inaccessible and prevent processes such as DNA transcription from occurring. Epigenetic regulation of histone proteins can be by chemical modification including acetylation, methylation, sumoylation and ubiquitylation (Herceg and Murr 2011). Each can cause structural changes in chromatin affecting DNA accessibility. Non-protein-coding RNAs, known as ncRNAs, have also been shown to contribute to epigenetic regulation as have mRNAs which can be processed and participate in various interference pathways (Collins and Schonfeld 2011).

■ Toxicogenomics

Toxicogenomics, related to pharmacogenomics, combines toxicology, genetics, molecular biology, and environmental health to elucidate the response of living organisms to stressful environments or xenobiotic agents based upon their genetic makeup (Rockett 2003). While toxicogenomic studies how the genome responds to toxic exposures, pharmacogenetics studies how an individual's genetic makeup affects his/her response to environmental stresses and toxins such as carcinogens, neurotoxins, and reproductive toxins (Smith 2010). Toxicogenomics can be very useful in drug discovery and development as new drug candidates can be screened through a combination of gene expression profiling and toxicology to understand gene response, identify general mechanisms of toxicity, and possibly predict drug safety (Furness 2002; Blomme 2009). There have been suggestions that toxicogenomics may decrease the time needed for toxicological investigations of new drug candidates and reduce both cost and animal usage versus conventional toxicity studies.

Genomic techniques utilized in toxicogenomic studies include gene expression level profiling, SNP analysis of the genetic variation, proteomics, and/or metabolomic methods so that gene expression, protein production, and metabolite production may be studied (Raghavachari 2012). The rapid growth in next-generation DNA sequencing capability may drive a conversion from microarrays now most commonly used for SNP analysis) to NGS technology.

Toxicogenomic studies attempt to discover associations between the development of drug toxicities and genotype. Clinicians and researchers are attempting to correlate genetic variation in one population to the manifestations of toxicity in other populations to identify and then to predict adverse toxicological effects in clinical trials so that suitable biomarkers for these adverse effects can be developed. Using such methods, it would then theoretically possible to test an individual patient for his or her susceptibility to these adverse effects before prescribing a medication. Patients that would show the marker for an adverse effect would be switched to a different drug. Therefore, toxicogenomics will become increasingly more powerful in predicting toxicity as new biomarkers are identified and validated. First described in 1999, the field is in its infancy, yet emerging rapidly. Much of the new toxicogenomic technology is developing in the pharmaceutical industry and other corporate laboratories.

■ Glycomics and Glycobiology

The novel scientific field of glycomics, or glycobiology, may be defined most simply as the study of the structure, synthesis, and biological function of all glycans (may be referred to as oligosaccharides or polysaccharides, depending on size) and glycoconjugates in simple and complex systems (Varkin et al. 1999; Fukuda and Hindsgaul 2000; Raghavachari 2012). The application of glycomics or glycobiology is sometimes called glycotechnology to distinguish it from biotechnology (referring to glycans rather than proteins and nucleic acids). However, many in the biotech arena consider glycobiology one of the research fields encompassed by the term biotechnology. In the postgenomic era, the intricacies of protein glycosylation, the mechanisms of genetic control, and the internal and external factors influencing the extent and patterns of glycosylation are important to understanding protein function and proteomics. Like proteins and nucleic acids, glycans are biopolymers. While once referred to as the last frontier of pharmaceutical discovery, recent advances in the biotechnology of discovering, cloning, and harnessing sugar cleaving and synthesizing enzymes have enabled glycobiologists to analyze and manipulate complex carbohydrates more easily (Walsh and Jefferis 2006).

Many of the proteins produced by animal cells contain attached sugar moieties, making them glycoproteins. The majority of protein-based medicinal agents contain some form of posttranslational modification that can profoundly affect the biological activity of that protein. Bacterial hosts for recombinant DNA could produce the animal proteins with identical or nearly identical amino acid sequences. However, early work in bacteria lacked the ability to attach sugar moieties to proteins (a process called glycosylation). New methodologies may help overcome this issue (cf. Chap. 3). Many of the non-glycosylated proteins differ in their biological activity as compared to the native glycoprotein. The production of animal proteins that lacked glycosylation provided an unexpected opportunity to study the functional role of sugar molecules on glycoproteins. There has been extensive progress in glycoengineering of yeast to humanize N-glycosylation pathways resulting in therapeutic glycoprotein expression in yeasts (Wildt and Gerngross 2005).

The complexity of the field can best be illustrated by reviewing the building blocks of glycans, the simple carbohydrates called saccharides or sugars and their derivatives (i.e., amino sugars). Simple carbohydrates can be attached to other types of biological molecules to form glycoconjugates including glycoproteins (predominantly protein), glycolipids and proteoglycans (about 95 % polysaccharide and 5 % protein). While carbohydrate chemistry and biology have been active areas of research for centuries, advances in biotechnology have provided techniques and added energy to the study of glycans. Oligosaccharides found conjugated to proteins (glycoproteins) and lipids (glycolipids) display a tremendous structural diversity. The linkages of the monomeric units in proteins and in nucleic acids are generally consistent in all such molecules. Glycans, however, exhibit far greater variability in the linkage between monomeric units than that found in the other

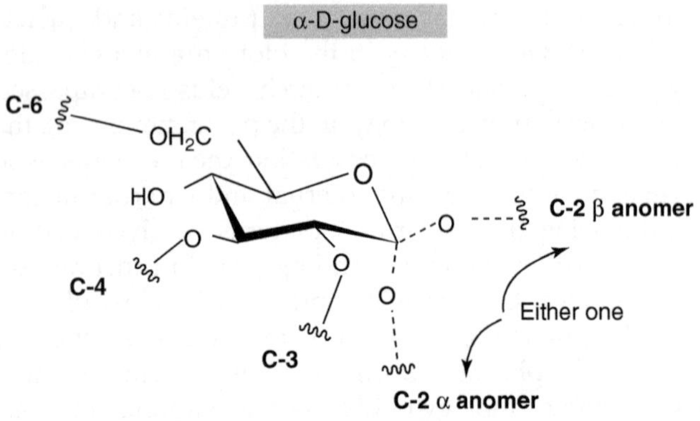

Figure 8.7 ■ Illustration of the common linkage sites to create biopolymers of glucose. Linkages at four positions: C-2, C-3, C-4, and C-6 and also can take one of two possible anomeric configurations at C-2 (α and β).

biopolymers. As an example, Fig. 8.7 illustrates the common linkage sites to create polymers of glucose. Glucose can be linked at four positions: C-2, C-3, C-4, and C-6 and also can take one of two possible anomeric configurations at C-2 (α and β). The effect of multiple linkage arrangements is seen in the estimate of (Kobata 1996). He has estimated that for a 10-mer (oligomer of length 10), the number of structurally distinct linear oligomers for each of the biopolymers is DNA (with 4 possible bases), 1.04×10^6; protein (with 20 possible amino acids), 1.28×10^{13}; and oligosaccharide (with eight monosaccharide types), 1.34×10^{18}.

Glycosylation and Medicine

Patterns of glycosylation significantly affect the biological activity of proteins (Wildt and Gerngross 2005; Walsh and Jefferis 2006). Many of the therapeutically used recombinant DNA-produced proteins are glycosylated including erythropoietin, glucocerebrosidase, and tissue plasminogen activator. Without the appropriate carbohydrates attached, none of these proteins will function therapeutically as does the parent glycoprotein. Glycoforms (variations of the glycosylation pattern of a glycoprotein) of the same protein may differ in physicochemical and biochemical properties. For example, erythropoietin has one O-linked and three N-linked glycosylation sites. The removal of the terminal sugars at each site destroys in vivo activity and removing all sugars results in a more rapid clearance of the molecule and a shorter circulatory half-life (Takeuchi et al. 1990). Yet, the opposite effect is observed for the deglycosylation of the hematopoietic cytokine granulocyte-macrophage colony-stimulating factor (GM-CSF) (Cebon et al. 1990). In that case, removing the carbohydrate residues increases the specific activity sixfold. The sugars of glycoproteins are known to play a role in the recognition and binding of biomolecules to other molecules in disease states such as asthma, rheumatoid arthritis, cancer, HIV infection, the flu, and other infectious diseases.

■ Lipidomics

Lipids, the fundamental components of membranes, play multifaceted roles in cell, tissue, and organ physiology. The relatively new research area of lipidomics may be defined as the large-scale study of pathways and networks of cellular lipids in biological systems (Wenk 2005; Raghavachari 2012). The metabolome would include the major classes of biological molecules: proteins (and amino acids), nucleic acids, and carbohydrates. The "lipidome" would be a subset of the metabolome that describes the complete lipid profile within a cell, tissue, or whole organism. In lipidomic research, a vast amount of information (structures, functions, interactions, and dynamics) quantitatively describing alterations in the content and composition of different lipid molecular species is accrued after perturbation of a cell, tissue, or organism through changes in its physiological or pathological state. The study of lipidomics is important to a better understanding of many metabolic diseases, as lipids are believed to play a role in obesity, atherosclerosis, stroke, hypertension, and diabetes.

Lipid profiling is a targeted metabolomic platform that provides a comprehensive analysis of lipid species within a cell or possibly a tissue. The progress of modern lipidomics has been greatly accelerated by the development of sensitive analytical techniques such as electrospray ionization (ESI) and matrix-assisted laser desorption/ionization (MALDI). Currently, the isolation and subsequent analysis of lipid mixtures is hampered by extraction and analytical limitations due to characteristics of lipid chemistry.

■ Nutrigenomics

The well-developed tools and techniques of genomics and bioinformatics have been applied to the examination of the intricate interplay of mammalian diet and genetic makeup. Nutrigenomics or nutritional genomics has been defined as the influence of genetic variation on nutrition. This appears to result from gene expression and/or gene variation (e.g., SNP analysis) on a nutrient's absorption, distribution, metabolism, elimination, or biological effects (Laursen 2010). This includes how nutrients impact on the production and action of specific gene products and how the expressed proteins in turn affect the response to nutrients. Nutrigenomic studies aim to develop predictive means to optimize nutrition, with respect to an individual's genotype. Areas of study include dietary supplements, common foods and beverages, mother's milk, as well as diseases such as cardiovascular, obesity, and diabetes.

While still in its infancy, nutrigenomics is thought to be a critical science for personalized health and public health over the next decade (Kaput et al. 2007).

■ Other "Omic" Technologies

Pharmaceutical scientists and pharmacists may hear about other "omic" technologies in which the "omic" terms derive from the application of modern genomic techniques to the study of various biological properties and processes. For example, interactomics is the data-intensive broad system study of the interactome, which is the interaction among proteins and other molecules within a cell. Proteogenomics has been used as a broadly encompassing term to describe the merging of genomics, proteomics, small molecules, and informatics. Cellomics has been defined as the study of gene function and the proteins they encode in living cells utilizing light microscopy and especially digital imaging fluorescence microscopy.

■ "Omics" Integrating Technology: Systems Biology

The Human Genome Project and the development of bioinformatics technologies have catalyzed fundamental changes in the practice of modern biology and helped unveil a remarkable amount of information about many organisms (Aderem and Hood 2001; Price et al. 2010). Biology has become an information science defining all the elements in a complex biological system and placing them in a database for comparative interpretation. As seen in Fig. 8.2, the hierarchy of information collection goes well beyond the biodata contained in the genetic code that is transcribed and translated. Systems biology involves a generally complex interactive system. This research area is often described as a noncompetitive or precompetitive technology by the pharmaceutical industry because it is believed to be a foundational technology that must be better developed to be successful at the competitive technology of drug discovery and development. It is the study of the interactions between the components of a biological system and how these interactions give rise to the function and behavior of that system. Systems biology is essential for our understanding of how all the individual parts of intricate biological networks are organized and function in living cells. The biological system may involve enzymes and metabolites in a metabolic pathway or other interacting biological molecules affecting a biological process. Molecular biologists have spent the past 50+ years teasing apart cellular pathways down to the molecular level. Characterized by a cycle of theory, computational modeling, and experiment to quantitatively describe cells or cell processes, systems biology is a data-intensive endeavor that results in a conceptual framework for the analysis and understanding of complex biological systems in varying contexts (Rothberg et al. 2005; Klipp et al. 2005; Meyer et al. 2011). Statistical mining, data alignment, probabilistic and mathematical modeling, and data visualization into networks are among the mathematical models employed to integrate the data and assemble the systems network (Gehlenborg et al. 2010). New measurements are stored with existing data, including extensive functional annotations, in molecular databases, and model assembly provides libraries of network models (Schrattenholz et al. 2010).

As the biological interaction networks are extremely complex, so are graphical representations of these networks. After years of research, a set of guidelines known as the Systems Biology Graphical Notation (SBGN) has been generally accepted to be the standard for graphical representation by all researchers. These standards are designed to facilitate the interchange of systems biology information and storage. Due to the complexity of these diagrams depending on the interactions examined and the level of understanding, a figure related to systems biology has not been included in this chapter. However, the reader is referred to the following website authored by the SBGN organization for several excellent examples of complex systems biology-derived protein interaction networks: http://www.sbgn.org/Documents/Examples

The inability to visualize the complexity of biological systems has in the past impeded the identification and validation of new and novel drug targets. The accepted SBGN standards should facilitate the efforts of pharmaceutical scientists to validate new and novel targets for drug design (Hood and Perlmutter 2004).

Since the objective is a model of all the interactions in a system, the experimental techniques that most suit systems biology are those that are system-wide and attempt to be as complete as possible. High-throughput "omic" technologies such as genomics, epigenomics, proteomics, pharmacogenomics, transcriptomics, metabolomics, and toxicogenomics are used to collect quantitative data for the construction and validation of systems models. Pharmaceutical and clinical end points include systems level biomarkers, genetic risk factors, aspects of personalized medicine, and drug target identification (Price et al. 2010; Yuryev 2012). In the future, application of systems biology approaches to drug discovery promises to have a profound impact on patient-centered medical practice, permitting a comprehensive evaluation of underlying predisposition to disease, disease diagnosis, and disease progression. Also, realization of personalized medicine and systems medicine will require new analytical approaches such as systems biology to decipher extraordinarily large, and extraordinarily noisy, data sets (Price et al. 2010).

TRANSGENIC ANIMALS AND PLANTS IN DRUG DISCOVERY, DEVELOPMENT, AND PRODUCTION

For thousands of years, man has selectively bred animals and plants either to enhance or to create desirable traits in numerous species. The explosive development of recombinant DNA technology and other molecular biology techniques have made it possible to engineer species possessing particular unique and distinguishing genetic characteristics. As described in Chap. 3, the genetic material of an animal or plant can be manipulated so that extra genes may be inserted (transgenes), replaced (i.e., human gene homologs coding for related human proteins), or deleted (knockout). Theoretically, these approaches enable the introduction of virtually any gene into any organism. A greater understanding of specific gene regulation and expression will contribute to important new discoveries made in relevant animal models. Such genetically altered species have found utility in a myriad of research and potential commercial applications including the generation of models of human disease, protein drug production, creation of organs and tissues for xenotransplantation, a host of agricultural uses, and drug discovery (Dunn et al. 2005; Clark and Pazdernik 2012a, b).

■ Transgenic Animals

As describe in Chap. 3, the term transgenic animal describes an animal in which a foreign DNA segment (a transgene) is incorporated into their genome. Later, the term was extended to also include animals in which their endogenous genomic DNA has had its molecular structure manipulated. While there are some similarities between transgenic technology and gene therapy, it is important to distinguish clearly between them. Technically speaking, the introduction of foreign DNA sequences into a living cell is called gene transfer. Thus, one method to create a transgenic animal involves gene transfer (transgene incorporated into the genome). Gene therapy (cf. Chap. 24) is also a gene transfer procedure and, in a sense, produces a transgenic human. In transgenic animals, however, the foreign gene is transferred indiscriminately into all cells, including germ line cells. The process of gene therapy differs generally from transgenesis since it involves a transfer of the desired gene in such a way that involves only specific somatic and hematopoietic cells, and not germ cells. Thus unlike in gene therapy, the genetic changes in transgenic organisms are conserved in any offspring according to the general rules of Mendelian inheritance.

The production of transgenic animals is not a new technology (Dunn et al. 2005; Clark and Pazdernik 2012b; Khan 2012a). They have been produced since the 1970s. However, modern biotechnology has greatly improved the methods of inducing the genetic transformation. While the mouse has been the most studied animal species, transgenic technology has been applied to cattle, fish (especially zebra fish), goats, poultry, rabbits, rats, sheep, swine, cats, dogs, horses, mules, deer, and various lower animal forms such as mosquitoes (Table 8.3). Transgenic animals have already made valuable research contributions to studies involving regulation of gene expression, the function of the immune system, genetic diseases, viral diseases, cardiovascular disease, and the genes responsible for the development of cancer. Transgenic animals have proven to be indispensable in drug lead identification, lead optimization, preclinical drug development, and disease modeling.

Year	Cloned animals
1996	Sheep (Dolly)
1997	Mouse
1998	Cattle
1999	Goat
2000	Pig
2001	Cat
2002	Rabbit
2003	Rat
2003	Mule
2003	Horse
2003	Deer
2005	Dog

Table 8.3 ■ Some cloned animals.

Production of Transgenic Animals by DNA Microinjection and Random Gene Addition

The production of transgenic animals has most commonly involved the microinjection (also called gene transfer) of 100–200 copies of exogenous transgene DNA into the larger, more visible male pronucleus (as compared to the female pronucleus) of a recipient fertilized embryo (see Fig. 8.8) (Clark and Pazdernik 2012b; Khan 2012a). The transgene contains both the DNA encoding the desired target amino acid sequence along with regulatory sequences that will mediate the expression of the added gene. The microinjected eggs are then implanted into the reproductive tract of a female and allowed to develop into embryos. The foreign DNA generally becomes randomly inserted at a single site on just one of the host chromosomes (i.e., the

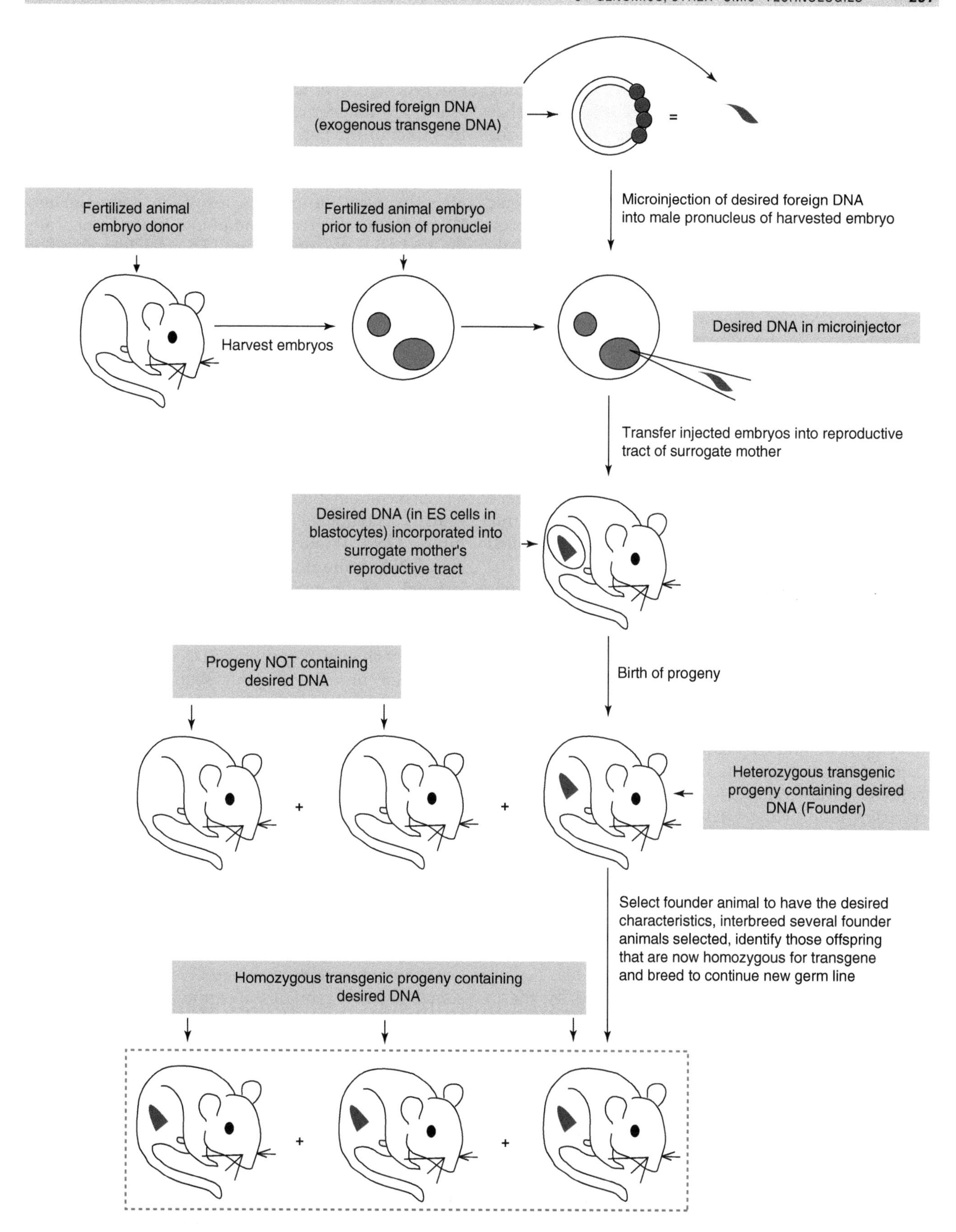

Figure 8.8 ■ Schematic representation of the production of transgenic animals by DNA microinjection.

founder transgenic animal is heterozygous). Thus, each transgenic founder animal (positive transgene incorporated animals) is a unique species. Interbreeding of founder transgenic animals where the transgene has been incorporated into germ cells may result in the birth of a homozygous progeny provided the transgene incorporation did not induce a mutation of an essential endogenous gene. All cells of the transgenic animal will contain the transgene if DNA insertion occurs prior to the first cell division. However, usually only 20–25 % of the offspring contain detectable levels of the transgene. Selection of neonatal animals possessing an incorporated transgene can readily be accomplished either by the direct identification of specific DNA or mRNA sequences or by the observation of gross phenotypic characteristics.

Production of Transgenic Animals by Retroviral Infection

The production of the first genetically altered laboratory mouse embryos was by insertion of a transgene via a modified retroviral vector (Clark and Pazdernik 2012b; Khan 2012a) (see Chap. 24 for more detailed description of retroviral vectors in gene therapy). The non-replicating viral vector binds to the embryonic host cells, allowing subsequent transfer and insertion of the transgene into the host genome. Many of the experimental human gene therapy trials employ the same viral vectors. Advantages of this method of transgene production are the ease with which genes can be introduced into embryos at various stages of development, and the characteristic that only a single copy of the transgene is usually integrated into the genome. Disadvantages include possible genetic recombination of the viral vector with other viruses present, the size limitation of the introduced DNA (up to 7 kb of DNA, less than the size of some genes), and the difficulty in preparing certain viral vectors.

Production of Transgenic Animals by Homologous Recombination in Embryonic Stem Cells Following Microinjection of DNA

Transgenic animals can also be produced by the in vitro genetic alteration of pluripotent embryonic stem cells (ES cells) (see Fig. 8.9; cf. Chap. 25) (Clark and Pazdernik 2012b; Khan 2012a). ES cell technology is more efficient at creating transgenics than microinjection protocols. ES cells, a cultured cell line derived from the inner cell mass (blastocyst) of a blastocyte (early preimplantation embryo), are capable of having their genomic DNA modified while retaining their ability to contribute to both somatic and germ cell lineages. The desired gene is incorporated into ES cells by one of several methods such as microinjection. This is followed by introduction of the genetically modified ES cells into the blastocyst of an early preimplantation embryo, selection, and culturing of targeted ES cells which are transferred subsequently to the reproductive tract of the surrogate host animal. The resulting progeny is screened for evidence that the desired genetic modification is present and selected appropriately. In mice, the process results in approximately 30 % of the progeny containing tissue genetically derived from the incorporated ES cells. Interbreeding of selected founder animals can produce species homozygous for the mutation.

While transforming embryonic stem cells is more efficient than the microinjection technique described first, the desired gene must still be inserted into the cultured stem cell's genome to ultimately produce the transgenic animal. The gene insertion could occur in a random or in a targeted process. Nonhomologous recombination, a random process, readily occurs if the desired DNA is introduced into the ES cell genome by a gene recombination process that does not require any sequence homology between genomic DNA and the foreign DNA. While most ES cells fail to insert the foreign DNA, some do. Those that do are selected and injected into the inner cell mass of the animal blastocyst and thus eventually lead to a transgenic species. In still far fewer ES cells, homologous recombination occurs by chance. Segments of DNA base sequence in the vector find homologous sequences in the host genome, and the region between these homologous sequences replaces the matching region in the host DNA. A significant advance in the production of transgenic animals in ES cells is the advent of targeted homologous recombination techniques.

Homologous recombination, while much more rare to this point in transgenic research than nonhomologous recombination, can be favored when the researcher carefully designs (engineers) the transferred DNA to have specific sequence homology to the endogenous DNA at the desired integration site and also carefully selects the transfer vector conditions. This targeted homologous recombination at a precise chromosomal position provides an approach to very subtle genetic modification of an animal or can be used to produce knockout mice (to be discussed later).

A modification of the procedure involves the use of hematopoietic bone marrow stem cells rather than pluripotent embryonic stem cells. The use of ES cells results in changes to the whole germ line, while hematopoietic stem cells modified appropriately are expected to repopulate a specific somatic cell line or lines (more similar to gene therapy).

The science of cloning and the resulting ethical debate surrounding it is well beyond the scope of this

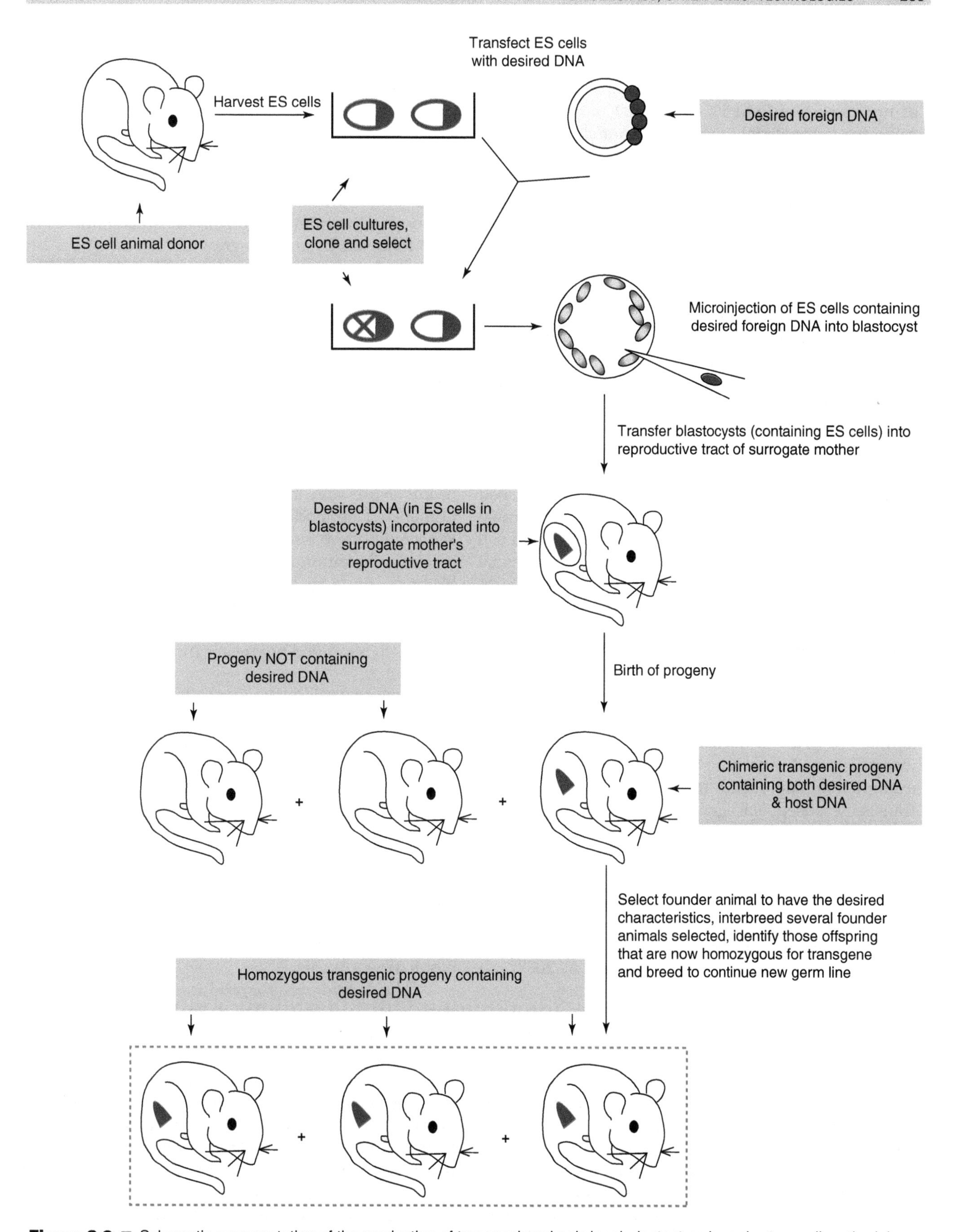

Figure 8.9 ■ Schematic representation of the production of transgenic animals by pluripotent embryonic stem cell methodology.

chapter. Yet it is important to place the concept of animal cloning within the pharmaceutically important context of transgenic animal production. The technique of microinjection (and its variations) has formed the basis for commercial transgenic animal production. While successful, the microinjection process is limited to the creation of only a small number of transgenic animals in a given birth. The slow process of conventional breeding of the resulting transgenic progeny must follow to produce a larger number of transgenic animals with the same transgene as the original organism. To generate a herd (or a flock, etc.), an alternative approach would be advantageous. The technique of nuclear transfer, the replacement of the nuclear genomic DNA of an oocyte (immature egg) or a single-cell fertilized embryo with that from a donor cell, is such an alternative breeding methodology. Animal "cloning" can result from this nuclear transfer technology. Judged the journal *Science's* most important breakthrough of 1997 creating the sheep Dolly, the first cloned mammal, from a single cell of a 6-year old ewe was a feat many had thought impossible. Dolly was born after nuclear transfer of the genome from an adult mammary gland cell (Khan 2012a). Since this announcement, commercial and exploratory development of nuclear transfer technology has progressed rapidly with various species cloned. It is important to note that the cloned sheep Dolly was NOT a transgenic animal. While Dolly was a clone of an adult ewe, she did not possess a transgene. However, cloning could be used to breed clones of transgenic animals or to directly produce transgenic animals (if prior to nuclear transfer, a transgene was inserted into the genome of the cloning donor). For example, human factor IX (a blood factor protein) transgenic sheep was generated by nuclear transfer from transfected fetal fibroblasts (Schniecke et al. 1997). Several of the resulting progeny were shown to be transgenic (i.e., possessing the human factor IX gene), and one was named Polly. Thus, animal cloning can be utilized not only for breeding but also for the production of potential human therapeutic proteins and other useful pharmaceutical products (Shultz et al. 2007).

■ Transgenic Plants

A variety of biotechnology genetic engineering techniques have been employed to create a wealth of transgenic plant species as mentioned in Chap. 3: cotton, maize, soybean, potato, petunia, tobacco, papaya, rose, and others (Clark and Pazdernik 2012a; Khan 2012b). Agricultural enhancements have resulted by engineering plants to be more herbicide tolerant, insect resistant, fungus resistant, virus resistant, and stress tolerant (Baez 2005). Of importance for human health and pharmaceutical biotechnology, gene transfer technology is routinely used to manipulate bulk human protein production in a wide variety of transgenic plant species. It is significant to note that transgenic plants are attractive bulk bioreactors because their posttranslational modification processes often result in plant-derived recombinant human proteins with greater glycosylation pattern similarity to that found in the corresponding native human proteins than would be observed in a corresponding mammalian production system. The transgenic seeds would result in seedlings capable of expressing the human protein. Transplantation to the field followed by normal growth, harvest of the biomass, and downstream isolation and protein purification results in a valuable alternative crop for farming. Tobacco fields producing pharmaceutical grade human antibodies (sometimes referred to as "plantibodies") and edible vaccines contained in transgenic potatoes and tomatoes are not futuristic visions, but current research projects in academic and corporate laboratories. Farmers' fields are not the sole sites for human protein production from flora. For example, cultured eukaryotic microalgae is also being developed into a useful expression system, especially for human and humanized antibodies (Mayfield and Franklin 2005). With antibody-based targeted therapeutics becoming increasingly important, the use of transgenic plants will likely continue to expand once research helps solve problems related to the isolation of the active protein drug and issues concerning cross fertilization with non-genetically modified organisms (non-GMOs).

Biopharmaceutical Protein Production in Transgenic Animals and Plants: "Biopharming"

The use of transgenic animals and plants as bioreactors for the production of pharmaceutically important proteins may become one of the most important uses of engineered species once numerous practical challenges are addressed (Baez 2005; Klimyuk et al. 2005). Table 8.4 provides a list of some selected examples of biopharmaceuticals from transgenic animals and plants. Utilizing conventional agronomic and farming techniques, transgenic animals and plants offer the opportunity to produce practically unlimited quantities of biopharmaceuticals.

The techniques to produce transgenic animals have been used to develop animal strains that secrete high levels of important proteins in various end organs such as milk, blood, urine, and other tissues. During such large animal "gene farming," the transgenic animals serve as bioreactors to synthesize recoverable quantities of therapeutically useful proteins. Among the advantages of expressing protein in animal milk is that the protein is generally produced in sizable quantities and can be harvested manually or mechanically by simply milking the animal. Protein purification from the milk requires the usual separation techniques for proteins. In general, recombinant genes coding for the desired protein product are fused to the regulatory

Species	Protein product	Potential indication(s)
Cow	Collagen	Burns, bone fracture
Cow	Human fertility hormones	Infertility
Cow	Human serum albumin	Surgery, burns, shock, trauma
Cow	Lactoferrin	Bacterial GI infection
Goat	α-1-antiprotease inhibitor	Inherited deficiency
Goat	α-1-antitrypsin	Anti-inflammatory
Goat	Antithrombin III (ATryn)	Associated complications from genetic or acquired deficiency
Goat	Growth hormone	Pituitary dwarfism
Goat	Human fertility hormones	Infertility
Goat	Human serum albumin	Surgery, burns, shock, trauma
Goat	LAtPA2	Venous status ulcers
Goat	Monoclonal antibodies	Colon cancer
Goat	tPA2	Myocardial infarct, pulmonary embolism
Pig	Factor IX	Hemophilia
Pig	Factor VIII	Hemophilia
Pig	Fibrinogen	Burns, surgery
Pig	Human hemoglobin	Blood replacement for transfusion
Pig	Protein C	Deficiency, adjunct to tPA
Rabbit	Insulin-like growth factor	Wound healing
Rabbit	Interleukin-2	Renal cell carcinoma
Rabbit	Protein C	Deficiency, adjunct to tPA
Sheep	α–1-antitrypsin	Anti-inflammatory
Sheep	Factor VIII	Hemophilia
Sheep	Factor IX	Hemophilia
Sheep	Fibrinogen	Burns, surgery
Sheep	Protein C	Deficiency, adjunct to tPA
Tobacco	IgG	Systemic therapy (rabies virus, hepatitis B virus)
Tobacco	TGF-β2	Ovarian cancer
Tobacco	Vitronectin	Protease
Tobacco	RhinoR	Fusion of human adhesion protein and human IgA for common cold
Safflower	Insulin	Diabetes
Corn	Meripase	Cystic fibrosis
Duckweed	Lacteron	Controlled release of α-interferon for hepatitis B and C
Potato	Poultry vaccine	Avian influenza (H5N1)

Data from references noted in this section of Chap. 8

Abbreviations: *tPA* tissue plasminogen activator, *LAtPA* long acting tissue plasminogen activator, *TGF-β3* tissue growth factor-beta

Table 8.4 ■ Some examples of human proteins under development in transgenic animals and plants.

sequences of the animal's milk-producing genes. The animals are not endangered by the insertion of the recombinant gene. The logical fusion of the protein product gene to the milk-producing gene targets the transcription and translation of the protein product exclusively in mammary tissues normally involved in milk production and does not permit gene activation in other, non-milk-producing tissues in the animal.

Transgenic strains are established and perpetuated by breeding the animals since the progeny of the original transgenic animal (founder animal) usually also produce the desired recombinant protein.

Yields of protein pharmaceuticals produced transgenically are expected to be 10–100 times greater than those achieved in recombinant cell culture. Protein yields from transgenic animals are generally good [conservative estimates of 1 g/l (g/L) with a 30 % purification efficiency] with milk yield from various species per annum estimated at: cow=10,000 L; sheep=500 L; goat=400 L; and pig=250 L (Rudolph 1995). PPL Therapeutics has estimated that the cost to produce human therapeutic proteins in large animal bioreactors could be as much as 75 % less expensive than cell culture. In addition, should the desired target protein require posttranslational modification, the large mammals used in milk production of pharmaceuticals would be a bioreactor capable of adding those groups (unlike a recombinant bacterial culture).

Some examples of human long peptides and proteins under development in the milk of transgenic animals include growth hormone, interleukin-2, calcitonin, insulin-like growth factor, alpha 1 antitrypsin (AT), clotting factor VIII, clotting factor IX, tissue plasminogen activator (tPA), lactoferrin, gastric lipase, vaccine derived from Escherichia coli LtB toxin subunit, protein C, and various human monoclonal antibodies (such as those from the Xenomouse) (Garner and Colman 1998; Rudolph 2000). The first approved biopharmaceutical from transgenic animals is recombinant human antithrombin which remains on the market today (trade name ATryn). Produced in a herd of transgenic dairy goats, rhAT is expressed in high level in the milk. The human AT transgene was assembled by linking the AT cDNA to a normal milk protein sequence (XhoI site of the goat beta casein vector). See Table 8.4 for additional examples.

Using genetic engineering techniques to create transgenic plants, "pharming" for pharmaceuticals is producing an ever-expanding list of drugs and diagnostic agents derived from human genes (Baez 2005; Opar 2011). Some examples of human peptides and proteins under development in transgenic plants include TGF-beta, vitronectin, thyroid-stimulating hormone receptor, insulin, glucocerebrosidase, apolipoprotein A-1, and taliglucerase alfa. See Table 8.4 for additional examples.

■ Xenotransplantation: Transplantable Transgenic Animal Organs

An innovative use of transgenics for the production of useful proteins is the generation of clinically transplantable transgenic animal organs, the controversial cross-species transplant (Khan 2012c). The success of human-to-human transplantation of heart, kidney, liver, and other vascularized organs (allotransplantation) created the significant expectation and need for donor organs. Primate-to-human transplantation (xenotransplantation) was successful, but ethical issues and limited number of donor animals were significant barriers. Transplant surgeons recognized early on that organs from the pig were a rational choice for xenotransplantation (due to physiological, anatomical, ethical, and supply reasons) if the serious hyperacute rejection could be overcome. Several research groups in academia and industry have pioneered the transgenic engineering of pigs expressing both human complement inhibitory proteins and key human blood group proteins (antigens) (McCurry et al. 1995; Dunn et al. 2005; Van Eyck et al. 2010). Cloning has now produced transgenic pigs for xenotransplantation. Cells, tissues, and organs from these double transgenic animals appear to be very resistant to the humoral immune system-mediated reactions of both primates and likely humans. These findings begin to pave the way for potential xenograft transplantation of animal components into humans with a lessened chance of acute rejection. A continuing concern is that many animals, such as pigs, have shorter life spans than humans, meaning that their tissues age at a quicker rate.

■ Knockout Mice

While many species including mice, zebra fish, and nematodes have been transformed to lose genetic function for the study of drug discovery and disease modeling, mice have proven to be the most useful. Mice are the laboratory animal species most closely related to humans in which the knockout technique can be easily performed, so they are a favorite subject for knockout experiments. While a mouse carrying an introduced transgene is called a transgenic mouse, transgenic technologies can also produce a knockout animal. A knockout mouse, also called a gene knockout mouse or a gene-targeted knockout mouse, is an animal in which an endogenous gene (genomic wild-type allele) has been specifically inactivated by replacing it with a null allele (Sharpless and DePinho 2006; Wu et al. 2011b; Clark and Pazdernik 2012b). A null allele is a nonfunctional allele of a gene generated by either deletion of the entire gene or mutation of the gene resulting in the synthesis of an inactive protein. Recent advances in intranuclear gene targeting and embryonic stem cell technologies as described above are expanding the capabilities to produce knockout mice routinely for studying certain human genetic diseases or elucidating the function of a specific gene product.

The procedure for producing knockout mice basically involves a four-step process. A null allele (i.e., knockout allele) is incorporated into one allele of murine ES cells. Incorporation is generally quite low;

approximately one cell in a million has the required gene replacement. However, the process is designed to impart neomycin and ganciclovir resistance only to those ES cells in which homologous gene integration has resulted. This facilitates the selection and propagation of the correctly engineered ES cells. The resulting ES cells are then injected into early mouse embryos creating chimeric mice (heterozygous for the knockout allele) containing tissues derived from both host cells and ES cells. The chimeric mice are mated to confirm that the null allele is incorporated into the germ line. The confirmed heterozygous chimeric mice are bred to homogeneity producing progeny that are homozygous knockout mice. Worldwide, three major mouse knockout programs are proceeding in collaboration to create a mutation in each of the approximately 20,000 protein-coding genes in the mouse genome using a combination of gene trapping and gene targeting in mouse embryonic stem (ES) cells (Staff 2007). These include (1) KOMP (KnockOut Mouse Project, http://www.knockoutmouse.org), funded by the NIH; (2) EUCOMM (EUropean Conditional Mouse Mutagenesis Program, http://www.eucomm.org), funded by the FP6 program of the EC; and (3) NorCOMM (North American Conditional Mouse Mutagenesis Project, http://norcomm.phenogenomics.ca/index.htm), a Canadian project funded by Genome Canada and partners. To date, over 4,000 targeted knockouts of genes have been accomplished. This comprehensive and publicly available resource will aid researchers examining the role of each gene in normal physiology and development and shed light on the pathogenesis of abnormal physiology and disease.

The continuing discoveries of the three worldwide mouse knockout consortia and independent research laboratories around the world will further create better models of human monogenic and polygenic diseases such as cancer, diabetes, obesity, cardiovascular disease, and psychiatric and neurodegenerative diseases. For example, knockout mice have been engineered that have extremely elevated cholesterol levels while being maintained on normal chow diets due to their inability to produce apolipoprotein E (apoprotein E). Apoprotein E is the major lipoprotein component of very low-density lipoprotein (VLDL) responsible for liver clearance of VLDL. These engineered mice are being examined as animal models of atherosclerosis useful in cardiovascular drug discovery and development. Table 8.5 provides a list of some additional selected examples of knockout mouse disease models.

The knockout mouse is becoming the basic tool for researchers to determine gene function in vivo in numerous biological systems. For example, knockout mouse technology has helped transform our understanding of the immune response. The study of single and multiple gene knockout animals has provided new perspectives on T-cell development, costimulation, and activation. "Humanized mice," transgenic severe combined immunodeficiency (SCID) mice grafted with human cells and tissues, enable research in regenerative medicine, infectious disease, cancer, and human hematopoiesis. In addition, high-throughput DNA sequencing efforts, positional cloning programs, and novel embryonic stem cell-based gene discovery research areas all exploit the knockout mouse as their laboratory.

Genetic engineering	Gene	Disease model
Knockout	*BRCA1*, *BRACA2*	Breast cancer
Knockout	Apolipoprotein E	Atherosclerosis
Knockout	Glucocerebrosidase	Gaucher's disease
Knockout	HPRT	Lesch-Nyhan syndrome
Knockout	Hexokinase A	Tay-Sachs disease
Knockout	Human CFTR	Cystic fibrosis
Knockout	p53	Cancer suppressor gene deletion
Knockout	P-glycoprotein	Multidrug resistance (MDR)
Knockout	α-globin and β-globin	Sickle cell anemia
Knockout	Urate oxidase	Gout
Knockout	Retinoblastoma-1	Familial retinoblastoma
Transgene	c-*neu* oncogene	Cancer
Transgene	c-*myc* oncogene	Cancer
Transgene	growth hormone	Dwarfism
Transgene	H-*ras* oncogene	Cancer
Transgene	Histocompatibility antigens	Autoimmunity
Transgene	HIV *tat*	Kaposi sarcoma
Transgene	Human APP	Alzheimer's disease
Transgene	Human β-globin	Thalassemia
Transgene	Human CD4 expression	HIV infection
Transgene	Human β-globin mutant	Sickle cell anemia
Transgene	Human CETP	Atherosclerosis
Transgene	LDL receptor	Hypercholesterolemia

Abbreviations: *BRCA1*, *BRCA2* suspected breast cancer genes, *CETP* cholesterol (cholesteryl) ester transfer protein, *CFTR* cyctic fibrosis transport regulator, *HIV* human immunodeficiency virus, *HPRT* hypoxanthine phosphoribosyl transferase, *LDL* low-density lipoprotein

Table 8.5 ■ Some selected examples of genetically engineered animal disease models.

Engineered animal models are proving invaluable to pharmaceutical research since small animal models of disease may be created and validated to mimic a disease in human patients. Mouse, rat, and zebra fish are the most common models explored and used. Genetic engineering can predispose an animal to a particular disease under scrutiny, and the insertion of human genes into the animal can initiate the development of a more clinically relevant disease condition. In human clinical studies, assessments of efficacy and safety often rely on measured effects for surrogate biomarkers and adverse event reporting. Validated transgenic animal models of human disease allow for parallel study and possible predictability prior to entering clinical trials. Also, it is possible to screen potential drug candidates in vivo against a human receptor target inserted into an animal model. The number of examples of transgenic animal models of human disease useful in drug discovery and development efforts is growing rapidly (Sharpless and DePinho 2006; Schultz et al. 2007; Wu et al. 2011b; Clark and Pazdernik 2012b). Such models have the potential to increase the efficiency and decrease the cost of drug discovery and development by reducing the time it takes to move a candidate medicinal agent from discovery into clinical trials. Table 8.5 provides a list of some selected examples of genetically engineered animal models of human disease.

SITE-DIRECTED MUTAGENESIS

Site-directed mutagenesis, also called site-specific mutagenesis or oligonucleotide-directed mutagenesis, is a protein engineering technique allowing specific amino acid residue (site-directed) alteration (mutation) to create new protein entities (Johnson and Reitz 1998). Mutagenesis at a single amino acid position in an engineered protein is called a point mutation. Therefore, site-directed mutagenesis techniques can aid in the examination at the molecular level of the relationship between 3D structure and function of interesting proteins. The technique is commonly used in protein engineering. This technique resulted in a Nobel Prize for one of the early researchers in this field, Dr. Michael Smith (Hutchison et al. 1978).

Figure 8.10 suggests an excellent example of possible theoretical mutations of the active site of a model serine protease enzyme that could be engineered to

a. Catalytic triad (the catalytic machinery) at active site of a wild type, parent serine protease

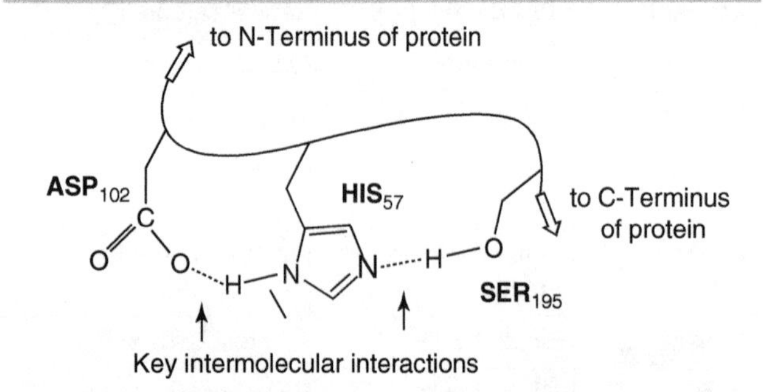

b. Theoretical site-directed mutagenesis studies HIS_{57} to PHE_{57} mutant

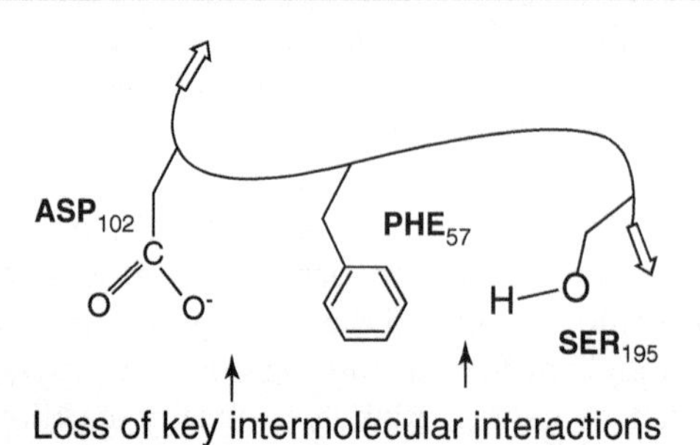

c. Theoretical site-directed mutagenesis studies ASP_{102} to ASN_{102} and SER_{195} to ALA_{195} mutant

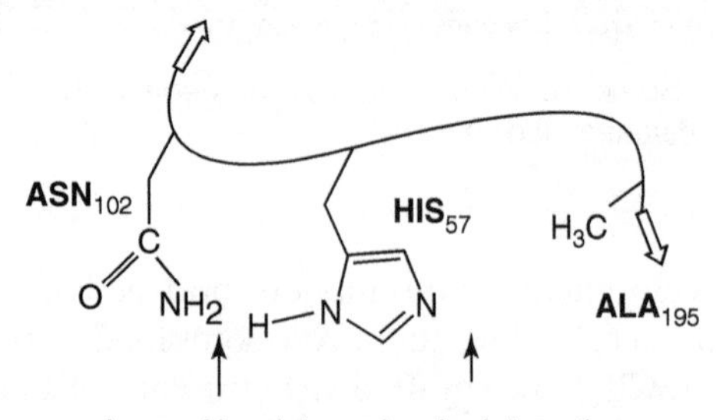

d. ASP_{102} to ASN_{102} mutant from site-directed mutagenesis (from Craik et al. 1987)

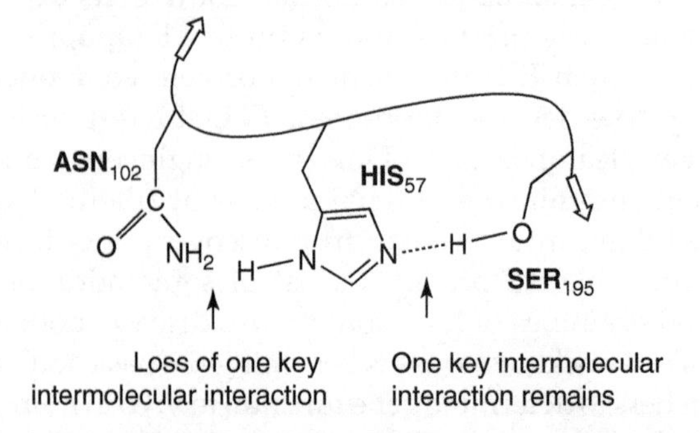

Figure 8.10 ■ Some possible site-directed mutations of the amino acids composing the catalytic triad of a serine protease: influence on key hydrogen bonding. (**a**) Catalytic triad (the catalytic machinery) at active site of a wild type, parent serine protease. (**b**) Theoretical site-directed mutagenesis studies HIS_{57} to PHE_{57} mutant. (**c**) Theoretical site-directed mutagenesis studies ASP_{102} to ASN_{102} and SER_{195} to ALA_{195} mutant. (**d**) ASP_{102} to ASN_{102} mutant from site-directed mutagenesis (From Craik et al. 1987).

probe the mechanism of action of the enzyme. Structures B and C of Fig. 8.10 represent a theoretical mutation to illustrate the technique. Craik and coworkers have actually tested the role of the aspartic acid residue in the serine protease catalytic triad Asp, His, and Ser. They replaced Asp^{102} (carboxylate anion side chain) of trypsin with Asn (neutral amide side chain) by site-directed mutagenesis and observed a pH-dependent change in the catalytic activity compared to the wild-type parent serine protease (see Fig. 8.10, structure D) (Craik et al. 1987). Site-directed mutagenesis studies also provide invaluable insight into the nature of intermolecular interactions of ligands with their receptors. For example, studies of the effect of the site-directed mutagenesis of various key amino acid residues on the binding of neurotransmitters to G-protein-coupled receptors have helped in defining more accurate models for alpha-adrenergic, D2-dopaminergic, 5HT2a-serotonergic, and both M1 and M3 muscarinic receptors (Bikker et al. 1998).

SYNTHETIC BIOLOGY

Modern biotechnology tools have allowed for a number of ways to study very complex biological systems. For example, as described above systems biology examines complex biological systems as interacting and integrated complex networks. The new and developing field of study known as synthetic biology explores how to build artificial complex biological systems employing many of the same tools and experimental techniques favored by system biologists. Synthetic biology looks at both the strategic redesign and/or the fabrication of existing biological systems and the design and construction of biological components and systems that do not already exist in nature (Khalil and Collins 2010). Synthetic genomics is a subset of synthetic biology that focuses on the redesign and fabrication of new genetic material constructed from raw chemicals. The focus of synthetic biology is often on ways of taking parts of natural biological systems, characterizing and simplifying them, and using them as a component of a highly unnatural, engineered, biological system. Synthetic biology studies may provide a more detailed understanding of complex biological systems down to the molecular level. Being able to design and construct a complex system is also one very practical approach to understanding that system under various conditions. The levels a synthetic biologist may work at include the organism, tissue and organ, intercellular, intracellular, biological pathway, and down to the molecular level.

There are many exciting applications for synthetic biology that have been explored or hypothesized across various fields of scientific study including designed and optimized biological pathways, natural product manufacturing, new drug molecule synthesis, and biosensing (Ruder et al. 2011). From an engineering perspective, synthetic biology could lead to the design and building of engineered biological systems that process information, modify existing chemicals, fabricate new molecules and materials, and maintain and enhance human health and our environment. Because of the obvious societal concerns that synthetic biology experiments raise, the broader science community has engaged in considerable efforts at developing guidelines and regulations and addressing the issues of intellectual property and governance and the ethical, societal, and legal implications. Several bioethics research institutes published reports on ethical concerns and the public perception of synthetic biology. A report from the United States Presidential Commission for the Study of Bioethical Issues called for enhanced federal oversight on this emerging technology.

In 2006, a research team, at the J. Craig Venter Institute constructed and filed a patent application for a synthetic genome of a novel synthetic minimal bacterium named *Mycoplasma laboratorium* (Smith et al. 2003; Glass et al. 2007). The team was able to construct an artificial chromosome of 381 genes, and the DNA sequence they have pieced together is based upon the bacterium Mycoplasma genitalium. The original bacterium had a fifth of its DNA removed and was able to live successfully with the synthetic chromosome in place. Venter's goal is to make cells that might take carbon dioxide out of the atmosphere and produce methane, used as a feedstock for other fuels.

Chang and coworkers published pioneering work utilizing a synthetic biology approach to assemble two heterologous pathways for the biosynthesis of plant-derived terpenoid natural products (Chang 2007). Terpenoids are a highly diverse class of lipophilic natural products that have historically provided a rich source for discovery of pharmacologically active small molecules, such as the anticancer agent paclitaxel (Taxol) and the antimalarial artemisinin. Unfortunately, these secondary metabolites are typically produced in low abundance in their host organism, and their isolation consequently suffers from low yields and high consumption of natural resources. A key step is developing methods to carry out cytochrome P450 (P450)-based oxidation chemistry in vivo. Their work suggests that potentially, entire metabolic pathways can be designed in silico and constructed in bacterial hosts.

BIOTECHNOLOGY AND DRUG DISCOVERY

Pharmaceutical scientists have taken advantage of every opportunity or technique available to aid in the long, costly, and unpredictable drug discovery process.

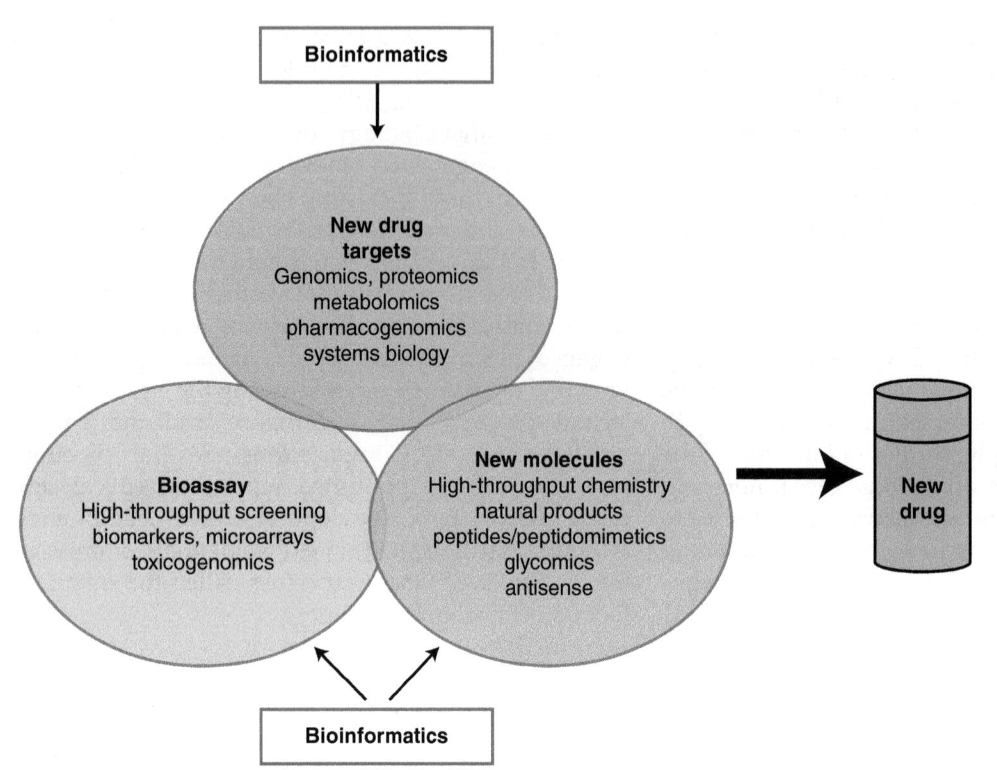

Figure 8.11 ■ Elements of modern drug discovery: impact of biotechnology.

In essence, Chap. 8 is an overview of some of the many applications of biotechnology and related techniques useful in drug discovery or design, lead optimization, and development. In addition to recombinant DNA and hybridoma technology, the techniques described throughout Chap. 8 have changed the way drug research is conducted, refining the process that optimizes the useful pharmacological properties of an identified novel chemical lead and minimizing the unwanted properties. The promise of genomics, proteomics, metabolomics, pharmacogenomics/pharmacogenetics, epigenomics, toxicogenomics, systems biology, and bioinformatics to radically change the drug discovery paradigm is eagerly anticipated (Beeley et al. 2000; Basu and Oyelere 2003; Loging et al. 2007). Figure 8.11 shows schematically the interaction of three key elements that are essential for modern drug discovery: new targets identified by genomics, proteomics, and related technologies; validation of the identified targets; rapid, sensitive bioassays utilizing high-throughput screening methods; and new molecule creation and optimization employing a host of approaches. The key elements are underpinned at each point by bioinformatics. Several of the technologies, methods, and approaches listed in Fig. 8.11 have been described previously in this chapter. Others will be described below.

■ Screening and Synthesis

Traditionally, drug discovery programs relied heavily upon random screening followed by analog synthesis and lead optimization via structure-activity relationship studies. Discovery of novel, efficacious, and safer small molecule medicinal agents with appropriate "drug-like characteristics is an increasingly costly and complex process" (Williams 2007). Therefore, any method allowing for a reduction in time and money is extremely valuable. Advances in biotechnology have contributed to a greater understanding of the cause and progression of disease and have identified new therapeutic targets forming the basis of novel drug screens. New technical discoveries in the fields of proteomics for target discovery and validation and systems biology are expected to facilitate the discovery of new agents with novel mechanisms of action for diseases that were previously difficult or impossible to treat. In an effort to decrease the cost of identifying and optimizing useful, quality drug leads against a pharmaceutically important target, researchers have developed newer approaches including high-throughput screening and high-throughput synthesis methods.

Advances in Screening: High-Throughput Screening (HTS)

Recombinant DNA technology has provided the ability to clone, express, isolate, and purify receptor enzymes, membrane-bound proteins, and other binding proteins in larger quantities than ever before. Instead of using receptors present in animal tissues or partially purified enzymes for screening, in vitro bioassays now utilize the exact human protein target. Applications of biotechnology to in vitro screening include the improved preparation of (1) cloned membrane-bound receptors expressed in cell lines carrying few endogenous receptors; (2) immobilized preparations of receptors, antibodies, and other ligand-binding proteins; and (3) soluble enzymes and extracellular cell-surface expressed protein receptors. In most cases today, biotechnology contributes directly to the understanding, identification, and/or the generation of the drug target being screened (e.g., radioligand binding displacement from a cloned protein receptor).

Previously, libraries of synthetic compounds along with natural products from microbial fermentation, plant extracts, marine organisms, and invertebrates provide a diversity of molecular structures that were screened randomly. Screening can be made more directed if the compounds to be investigated are selected on the basis of structural information about the receptor or natural ligand. The development of sensitive radioligand binding assays and the access to fully automated, robotic screening techniques have accelerated the screening process.

High-throughput screening (HTS) provides for the bioassay of thousands of compounds in multiple assays at the same time (Cik and Jurzak 2007; Rankovic et al. 2010). The process is automated with robots and utilizes multi-well microtiter plates. While 96-well microtiter plates are a versatile standard in HTS, the development of 1,536- and 3,456-well nanoplate formats and enhanced robotics brings greater miniaturization and speed to cell-based and biochemical assays. Now, companies can conduct 100,000 bioassays a day. In addition, modern drug discovery and lead optimization with DNA microarrays allows researchers to track hundreds to thousands of genes.

Enzyme inhibition assays and radioligand binding assays are the most common biochemical tests employed. The technology has become so sophisticated and the interactive nature of biochemical events so much better understood (through approaches such as systems biology) that HT whole cell assays have become commonplace. Reporter gene assays are routinely utilized in HTS (Ullmann 2007). Typically, a reporter gene, that is a reporter that indicates the presence or absence of a particular gene product that in turn reflects the changes in a biological process or pathway, is transfected into a desired cell. When the gene product is expressed in the living cell, the reporter gene is transcribed and the reporter is translated to yield a protein that is measured biochemically. A common reporter gene codes for the enzyme luciferase, and the intensity of the resulting green fluorescent protein (i.e., a quantitative measure of concentration) is a direct function of the assayed molecule's ability to stimulate or inhibit the biologic process or signaling pathway under study. A further advance in HT screening technologies for lead optimization is rapid, high-content pharmacology. This HT screening approach can be used to evaluate solubility, adsorption, toxicity, metabolism, and other important drug characteristics.

High-Throughput Chemistry: Combinatorial Chemistry and Multiple Parallel Synthesis

Traditionally, small drug molecules were synthesized by joining together structural pieces in a set sequence to prepare one product. One of the most powerful tools to optimize drug discovery is automated high-throughput synthesis. When conducted in a combinatorial approach, high-throughput synthesis provides for the simultaneous preparation of hundreds or thousands of related drug candidates (Fenniri 2000; Sucholeiki 2001; Mason and Pickett 2003; Pirrung 2004). The molecular libraries generated are screened in high-throughput screening assays for the desired activity, and the most active molecules are identified and isolated for further development.

There are two overall approaches to high-throughput synthesis: combinatorial chemistry that randomly mixes various reagents (such as many variations of reagent A with many variations of reagent B to give random mixtures of all products in a reaction vessel) and parallel synthesis that selectively conducts many reactions parallel to each other (such as many variations of reagent A in separate multiple reaction vessels with many variations of reagent B to give many single products in separate vessels) (Mitscher and Dutta 2003; Seneci 2007; Ashton and Maloney 2007). True combinatorial chemistry or sometimes referred to as combichem, applies methods to substantially reduce the number of synthetic operations or steps needed to synthesize large numbers of compounds. Combichem is conducted on solid supports (resins) to facilitate the manipulations required to reduce labor during purification of multiple products in the same vessel. Differing from combinatorial chemistry, multiple parallel synthesis procedures apply automation to the synthetic process to address the many separate reaction vessels needed, but the number of operations needed to carry out a synthesis is practically the same as the conventional approach. Thus, the potential productivity of multiple parallel methods is not as high as combinatorial chemistries. Parallel chemistries can be conducted on solid-phase supports or in solution to facilitate purification. Figure 8.12 provides an illustration of a combinatorial mix-and-match process in which a simple building block (a starting material such as an amino acid, peptide, heterocycle, other small molecule, etc.) is joined to one or more other simple building blocks in every possible combination. Assigning the task to automated synthesizing equipment results in the rapid creation of large collections or libraries (as large as 10,000 compounds) of diverse molecules. Ingenious methods have been devised to direct the molecules to be

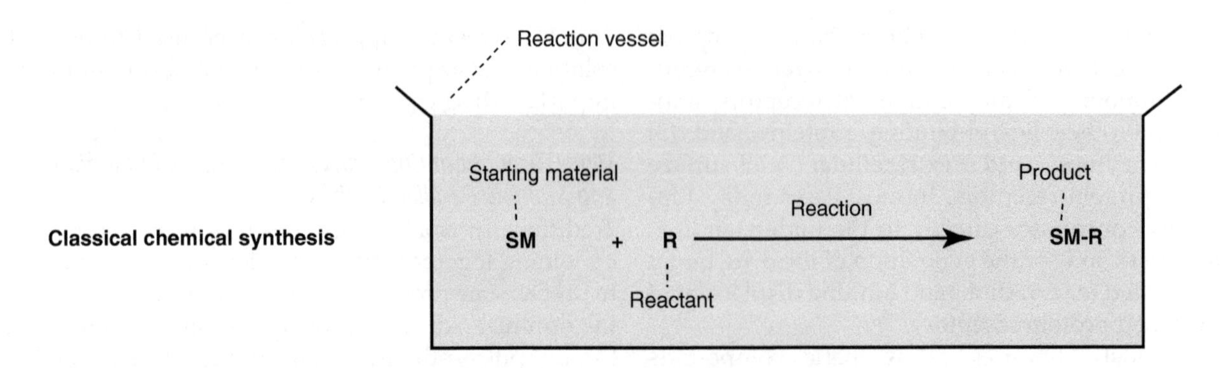

In classical chemical synthesis, a coupling reaction of one starting material (SM) with one reactant (R) wouldyield just one product, SM-R. One or several reactions may be run simultaneously in separate reaction vessels

Combinatorial chemical synthesis: Example = Coupling reaction

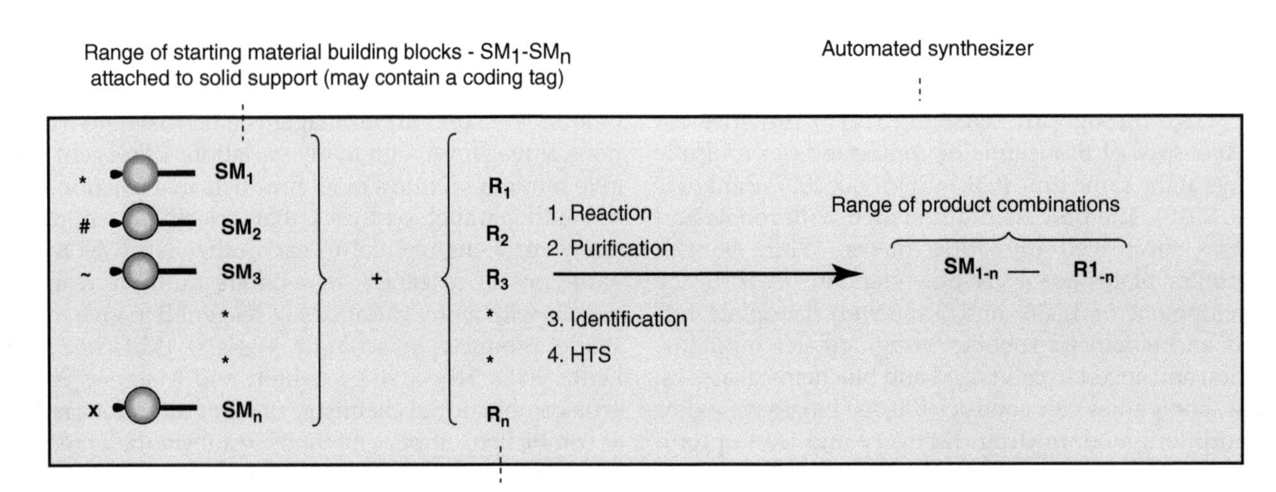

In a combinatorial chemical synthesis such as a coupling reaction, a range of starting material building blocks are reacted with a range of reactant building blocks yielding any or all possible product combinations, $SM_{1\text{-}n}$ $R_{1\text{-}n}$. The automated reactions may occur in the same reaction vessel (and coding tag used to separate/identify) or may each occur in small, separate reaction vessels (parallel synthesis)

Figure 8.12 ■ A schematic representation of a coupling reaction: difference between classical chemical synthesis and combinatorial chemistry.

synthesized, to identify the structure of the products, to purify the products via automation, and to isolate compounds. When coupled with high-throughput screening, thousands of compounds can be generated, screened, and evaluated for further development in a matter of weeks.

Building blocks include amino acids, peptides, nucleotides, carbohydrates, lipids, and a diversity of small molecule scaffolds or templates (Mason and Pickett 2003). A selection of reaction types used in combinatorial chemistry to produce compound libraries is found in Table 8.6.

■ Chemical Genomics

Chemical genomics is an emerging "omic" technology not discussed above. The development of high-throughput screening and high-throughput chemistry coupled with "omic" technologies has changed the drug discovery paradigm and the approach for the investigation of target pharmacology (Kubinyi and Müller 2004; Kubinyi 2007; Flaumenhaft 2007; Rankovic et al. 2010) (see Fig. 8.11). In modern drug discovery, chemical genomics (sometimes called chemogenomics or more generally included as a subset of chemical biology) involves the screening of large chemical

Compound types
α,β-Unsaturated ketones
α-Hydroxy acids
Acyl piperidines
Azoles
β-Mercaptoketones
β-Turn mimetics
Benzisothiazolones
Benzodiazepines
Biaryls
Cyclopentenones
Dihydropyridines
γ-Butyropyridines
Glycosylamines
Hydantoins
Isoxazoles
Modified oligonucleotides
Peptoids
Piperazinediones
Porphyrins
Porphyrins
1,3-Propanediols
Protease inhibitors
Pyrrolidines
Sulfamoylbenzamides
Tetrahydrofurans
Thiazole
Thiazolidinones

Table 8.6 ■ A sample of the diversity of compounds capable of being synthesized by combinatorial chemistry methods.

libraries (typically combinatorially derived "druggable" small molecule libraries covering a broad expanse of "diversity space") against all genes or gene products, such as proteins or other targets (i.e., chemical universe screened against target universe). In Fig. 8.11, basically chemical genomics would occur when the "new molecules" to be tested in the "bioassay" developed from the "new drug targets" "omic" approaches came from large chemical libraries typically of small molecules synthesized by high-throughput chemistry. As part of the National Institutes of Health (NIH) Roadmap for Biomedical Research, the National Human Genome Research Institute (NHGRI) will lead an effort to offer public sector biomedical researchers access to libraries of small organic molecules that can be used as chemical probes to study cellular pathways in greater depth. It remains difficult to predict which small molecule compounds will be most effective in a given situation. Researchers can maximize the likelihood of a successful match between a chemical compound, its usefulness as a research tool, and its desired therapeutic effect by systematically screening libraries containing thousands of small molecules. Drug candidates are expected from the correlations observed during functional analysis of the molecule – gene product interactions. Genomic profiling by the chemical library may also yield relevant new targets and mechanisms. Chemical genomics is expected to be a critical component of drug lead identification and proof of principle determination for selective modulators of complex enzyme systems including proteases, kinases, G-protein-coupled receptors, and nuclear receptors.

CONCLUSION

Tremendous advances have occurred in biotechnology since Watson and Crick determined the structure of DNA. Improved pharmaceuticals, novel therapeutic agents, unique diagnostic products, and new drug design tools have resulted from the escalating achievements of pharmaceutical biotechnology. While recombinant DNA technology and hybridoma techniques received most of the press in the late 1980s and early1990s, a wealth of additional and innovative biotechnologies and approaches have been, and will continue to be, developed in order to enhance pharmaceutical research. Genomics, proteomics, transcriptomics, microarrays, pharmacogenomics/genetics, epigenomics, personalized medicine, metabonomics/metabolomics, toxicogenomics, glycomics, systems biology, genetically engineered animals, high-throughput screening, and high-speed combinatorial synthesis are directly influencing the pharmaceutical sciences and are well positioned to significantly impact modern pharmaceutical care. Application of these and yet to be discovered biotechnologies will continue to reshape effective drug therapy as well as improve the competitive, challenging process of drug discovery and development of new medicinal agents and diagnostics. Pharmacists, pharmaceutical scientists, and pharmacy students should be poised to take advantage of the products and techniques made available by the unprecedented scope and pace of discovery in biotechnology in the twenty-first century.

SELF-ASSESSMENT QUESTIONS

■ Questions

1. What were the increasing levels of genetic resolution of the human genome planned for study as part of the HGP?
2. What is functional genomics?
3. What is proteomics?
4. What are SNPs?
5. What is the difference between pharmacogenetics and pharmacogenomics?
6. Define metabonomics.
7. What is a DNA microarray?

8. What phase(s) of drug action is affected by genetic variation?
9. Define personalized medicine.
10. What is a biomarker?
11. Define systems biology.
12. Why are engineered animal models valuable to pharmaceutical research?
13. What two techniques are commonly used to produce transgenic animals?
14. What is a knockout mouse?
15. What are two approaches to high-throughput synthesis of drug leads and how do they differ?

Answers

1. HGP structural genomics was envisioned to proceed through increasing levels of genetic resolution: detailed human genetic linkage maps [approx. 2 megabase pairs (Mb = million base pairs) resolution], complete physical maps (0.1 Mb resolution), and ultimately complete DNA sequencing of the approximately 3.5 billion base pairs (23 pairs of chromosomes) in a human cell nucleus [1 base pair (bp) resolution].
2. Functional genomics is a new approach to genetic analysis that focuses on genome-wide patterns of gene expression, the mechanisms by which gene expression is coordinated, and the interrelationships of gene expression when a cellular environmental change occurs.
3. A new research area called proteomics seeks to define the function and correlate that with expression profiles of all proteins encoded within a genome.
4. While comparing the base sequences in the DNA of two individuals reveals them to be approximately 99.9 % identical, base differences, or polymorphisms, are scattered throughout the genome. The best-characterized human polymorphisms are single-nucleotide polymorphisms (SNPs) occurring approximately once every 1,000 bases in the 3.5 billion base pair human genome.
5. Pharmacogenetics is the study of how an individual's genetic differences influence drug action, usage, and dosing. A detailed knowledge of a patient's pharmacogenetics in relation to a particular drug therapy may lead to enhanced efficacy and greater safety. While sometimes used interchangeably (especially in pharmacy practice literature), pharmacogenetics and pharmacogenomics are subtly different. Pharmacogenomics introduces the additional element of our present technical ability to pinpoint patient-specific DNA variation using genomic techniques. While overlapping fields of study, pharmacogenomics is a much newer term that correlates an individual patient's DNA variation (SNP level of variation knowledge rather than gene level of variation knowledge) with his or her response to pharmacotherapy.
6. The field of metabonomics is the holistic study of the metabolic continuum at the equivalent level to the study of genomics and proteomics.
7. The biochips known as DNA microarrays and oligonucleotide microarrays are a surface collection of hundreds to thousands of immobilized DNA sequences or oligonucleotides in a grid created with specialized equipment that can be simultaneously examined to conduct expression analysis.
8. Genomic variation affects not only the pharmacokinetic profile of drugs (via drug metabolizing enzymes and drug transporter proteins), it also strongly influences the pharmacodynamic profile of drugs via the drug target.
9. Pharmacotherapy informed by a patient's individual genomic and proteomic information. Sometimes referred to as giving the right drug to the right patient in the right dose at the right time.
10. Biomarkers are clinically relevant substances used as indicators of a biologic state. Detection or concentration change of a biomarker may indicate a particular disease state physiology or toxicity. A change in expression or state of a protein biomarker may correlate with the risk or progression of a disease, with the susceptibility of the disease to a given treatment or the drug's safety profile.
11. Systems biology is the study of the interactions between the components of a biological system, and how these interactions give rise to the function and behavior of that system.
12. Engineered animal models are proving invaluable since small animal models of disease are often poor mimics of that disease in human patients. Genetic engineering can predispose an animal to a particular disease under scrutiny, and the insertion of human genes into the animal can initiate the development of a more clinically relevant disease condition.
13. (a) DNA microinjection and random gene addition and (b) homologous recombination in embryonic stem cells.
14. A knockout mouse, also called a gene knockout mouse or a gene-targeted knockout mouse, is an animal in which an endogenous gene (genomic wild-type allele) has been specifically inactivated by replacing it with a null allele.
15. There are two overall approaches to high-throughput synthesis. True combinatorial chemistry applies methods to substantially reduce the number of synthetic operations or steps needed to synthesize large numbers of compounds. Combichem, as it is sometimes referred to, is conducted on solid supports (resins) to facilitate the needed manipulations that reduce labor. Differing from combinatorial chemistry, parallel procedures

apply automation to the synthetic process, but the number of operations needed to carry out a synthesis is practically the same as the conventional approach. Thus, the potential productivity of parallel methods is not as high as combinatorial chemistries. Parallel chemistries can be conducted on solid-phase supports or in solution.

Acknowledgment I wish to acknowledge the tremendous contribution of Dr. Arlene Marie Sindelar, my wife, to some of the graphics found in figures in all four editions of this textbook.

REFERENCES

Abla N, Kroetz DL (2009) Drug transporter pharmacogenetics. In: American College of Clinical Pharmacy Board (ed) Pharmacogenomics: applications to patient care, 2nd edn. American College of Clinical Pharmacy, Kansas City, pp 70–89

Aderem A, Hood L (2001) Immunology in the post-genomic era. Nat Immunol 2:373–375

Alkan C, Coe BP, Eichler EE (2011) Genome structural variation discovery and genotyping. Nat Rev Genet 12:363–375

Amaratunga D, Gohlmann H, Peeters PJ (2007) Microarrays. In: Kubinyi H (ed) Comprehensive medicinal chemistry II, vol 3. Elsevier, Amsterdam, pp 87–106

Anderson PO, McGuinness SM, Bourne PE (2010) What is pharmacy informatics? In: Anderson PO, McGuinness SM, Bourne PE (eds) Pharmacy informatics. CRC Press, Boca Raton, pp 3–5

Ashton M, Maloney B (2007) Solution phase parallel chemistry. In: Kubinyi H (ed) Comprehensive medicinal chemistry II, vol 3. Elsevier, Amsterdam, pp 761–790

Baez J (2005) Biopharmaceuticals derived from transgenic plants and animals. In: Knablein J (ed) Modern biopharmaceuticals, vol 3. Wiley-VCH, Weinheim, pp 833–892

Baker M (2012) Structural variation: the genome's hidden architecture. Nat Methods 9:133–137

Basu S, Oyelere AK (2003) Application of recombinant DNA technology in medicinal chemistry and drug discovery. In: Abraham DJ (ed) Burger's medicinal chemistry and drug discovery, vol 2, 6th edn. John Wiley & Sons, Inc, New York, pp 81–114

Beeley LJ, Duckworth DM, Southan C (2000) The impact of genomics on drug discovery. Prog Med Chem 37:1–43

Berger SL, Kouzarides T, Shiekhattar R, Shilatifard A (2009) An operational definition of epigenetics. Genes Dev 23:781–783

Bikker JA, Trumpp-Kallmeyer S, Humblet C (1998) G-protein coupled receptors: models, mutagenesis, and drug design. J Med Chem 41:2911–2927

Bjornsson HT, Fallin MD, Feinberg AP (2004) An integrated epigenetic and genetic approach to common human disease. Trends Genet 20:350–358

Blake K, Wang J, Lima J (2009) Respiratory diseases. In: American College of Clinical Pharmacy Board (ed) Pharmacogenomics: applications to patient care, 2nd edn. American College of Clinical Pharmacy, Kansas City, pp 230–255

Blomme E (2009) Fundamental principles of toxicogenomics. In: Semizarov D, Blomme E (eds) Genomics in drug discovery and development. John Wiley & Sons, Inc, Hoboken, pp 167–217

Branca MA (2005) BiDil raises questions about race as a marker. Nat Rev Drug Discov 4:615–616

Brazeau DA, Brazeau GA (2011a) Introduction: pharmacogenomics and pharmacogenetics – a historical look. In: Principles of the human genome and pharmacogenomics. American Pharmacists Association, Washington, D.C., pp 1–10

Brazeau DA, Brazeau GA (2011b) Information flow in biological systems. In: Principles of the human genome and pharmacogenomics. American Pharmacists Association, Washington, D.C., pp 11–34

Brazeau DA, Brazeau GA (2011c) Ethical challenges and opportunities in pharmacogenetics and pharmacogenomics. In: Principles of the human genome and pharmacogenomics. American Pharmacists Association, Washington, D.C., pp 81–95

Browman G, Hebert PC, Coutts J, Stanbrook MB, Flegel K, MacDonald NE (2011) Personalized medicine: a windfall for science, but what about patients? Can Med Assoc J 183:E1277

Bungard TJ, Yakiwchuk E, Foisy M, Brocklebank C (2011) Drug interactions involving warfarin: practice tool and practical management tips. Can Pharm J 144:21–25

Burckert GJ (2009) Transplantation. In: American College of Clinical Pharmacy Board (ed) Pharmacogenomics: applications to patient care, 2nd edn. American College of Clinical Pharmacy, Kansas City, pp 256–270

BusinessWeek (2012) Illumina and life technologies add genome-in-a-day machines. Available at: http://www.businessweek.com/news/2012-01-18/illumina-life-technologies-add-genome-in-a-day-machines.html. Accessed 19 Jan 2012

Cebon J, Nicola N, Ward M, Gardner I, Dempsey P, Layton J, Duhrsen U, Burgess AW, Nice E, Morstyn G (1990) Granulocyte-macrophage colony stimulating factor from human lymphocytes. J Biol Chem 265:4483–4491

Chalkidou K, Rawlins M (2011) Pharmacogenetics and the cost-effectiveness analysis: a two-way street. Drug Discov Today 16:873–883

Chang MCY, Eachus RA, Trieu W, Ro DK, Keasling JD (2007) Engineering *Escherichia coli* for production of functionalized terpenoids using plant P450s. Nat Chem Biol 3:274–277

Chen M, Zhang L (2011) Epigenetic mechanisms in developmental programming of adult disease. Drug Discov Today 16:1007–1018

Cherf GM, Lieberman KR, Rashid H, Lam CE, Karplus K, Akeson M (2012) Nat Biotechnol 30:344–348

Cik M, Jurzak M (2007) High-throughput and high-content screening. In: Kubinyi H (ed) Comprehensive medicinal chemistry II, vol 3. Elsevier, Amsterdam, pp 679–696

Clancy C, Collins FS (2010) Patient-centered outcomes research institute: the intersection of science and health care. Sci Transl Med 2:11–13, Collect

Clark DP, Pazdernik NJ (2012a) Transgenic plants and plant biotechnology. In: Clark DP, Pazdernik NJ (eds) Biotechnology: academic cell update. Elsevier, Burlington, pp 397–423

Clark DP, Pazdernik NJ (2012b) Transgenic animals. In: Clark DP, Pazdernik NJ (eds) Biotechnology: academic cell update. Elsevier, Burlington, pp 425–455

Collins LJ, Barker AD (2007) Mapping the cancer genome. Sci Am 296:50–57

Collins LJ, Schonfeld B (2011) The epigenetics of non-coding RNA. In: Tollefsbol T (ed) Handbook of epigenetics: the new molecular and medical genetics. Elsevier, London, pp 49–61

Cox AG (2010) Pharmacogenomics and drug transport/efflux. In: Zdanowicz MM (ed) Concepts in pharmacogenomics. American Society of Health-systems Pharmacists, Bethesda, pp 129–153

Craik CS, Roczniak S, Largman C, Rutter WJ (1987) The catalytic role of the active site aspartic acid in serine proteases. Science 237:909–913

Crews KR, Cross SJ, McCormick JN, Baker DK, Molinelli AR, Mullins R, Relling MV, Hoffman JM (2011) Development and implementation of a pharmacist-managed clinical pharmacogenetics service. Am J Health Syst Pharm 68:143–150

D'Alessandro A, Zolla L (2010) Pharmacoproteomics: a chess game on a protein field. Drug Discov Today 15:1015–1023

D'Alessandro A, Zolla L (2012) Metabolomics and cancer discovery: let the cells do the talking. Drug Discov Today 17:3–9

Davidson RG, McInerney JD (2009) Principles of genetic medicine. In: American College of Clinical Pharmacy Board (ed) Pharmacogenomics: applications to patient care, 2nd edn. American College of Clinical Pharmacy, Kansas City, pp 341–379

Davies K (2010) The $1,000 genome. Free Press, a division of Simon and Schuster, Inc., New York

Daxinger L, Whitelaw E (2012) Understanding transgenerational epigenetic inheritance via the gametes in mammals. Nat Rev Genet 13:153–162

DePrimo SE (2007) Biomarkers. In: Kubinyi H (ed) Comprehensive medicinal chemistry II, vol 3. Elsevier, Amsterdam, pp 69–85

Dieterle F, Sistare FD (2010) Biomarkers of acute kidney injury. In: Vaidya VS, Bonventre JV (eds) Biomarkers in medicine, drug discovery, and environmental health. John Wiley & Sons, Inc, Hoboken, pp 237–279

Dietert RR (2010) Biomarkers of immunotoxicity. In: Vaidya VS, Bonventre JV (eds) Biomarkers in medicine, drug discovery, and environmental health. John Wiley & Sons, Inc, Hoboken, pp 307–322

Dunn DA, Kooyman DL, Pinkert CA (2005) Transgenic animals and their impact on the drug discovery industry. Drug Discov Today 10:757–767

Ellingrod VL, Bishop JR, Thomas K (2009) Psychiatry. In: American College of Clinical Pharmacy Board (ed) Pharmacogenomics: applications to patient care, 2nd edn. American College of Clinical Pharmacy, Kansas City, pp 204–229

Feinberg AP, Irizarry RA, Fradin D, Aryee MJ, Murakami P, Aspelund T, Eirksdottir G, Harris TB, Launer L, Gudnason V, Fallin MD (2010) Personalized epigenomic signatures that are stable over time and covary with body mass index. Sci Transl Med 2:45–51, Collect

Fenniri H (2000) Combinatorial chemistry: a practical approach. Oxford University Press, Oxford

Flaumenhaft R (2007) Chemical biology. In: Kubinyi H (ed) Comprehensive medicinal chemistry II, vol 3. Elsevier, Amsterdam, pp 129–149

Fox BI (2010) Informatics and the medication use process. In: Fox BI, Thrower MR, Felkey BG (eds) Building core competencies in pharmacy informatics. American Pharmacists Association, Washington, D.C., pp 11–31

Frank R, Hargreaves R (2003) Clinical biomarkers in drug discovery and development. Nat Rev Drug Discov 2:564–565

Frye RF (2009) Pharmacogenetics of oxidative drug metabolism and its clinical applications. In: American College of Clinical Pharmacy Board (ed) Pharmacogenomics: applications to patient care, 2nd edn. American College of Clinical Pharmacy, Kansas City, pp 32–53

Fukuda M, Hindsgaul O (2000) Molecular and cellular glycobiology. Oxford University Press, Oxford

Furness LM (2002) Genomics applications that facilitate the understanding of drug action and toxicity. In: Licinio J, Wong M-L (eds) Pharmacogenomics, the search for individualized therapies. Wiley-BCH Verlag GmbH & Co. KgaA, Weinheim, pp 83–125

Garner I, Colman A (1998) Therapeutic proteins form livestock. In: Clark AJ (ed) Animal breeding-technology for the 21st century. Harwood Academic Publishers, Amsterdam, pp 215–227

Garnett MJ (2012) Exploiting genetic complexity in cancer to improve therapeutic strategies. Drug Discov Today 17:188–193

Garske AL, Denu JM (2009) New frontiers in epigenetic modifications. In: Sippl W, Jung M (eds) Epigenetic targets in drug discovery. Wiley-VCH Verlag GmbH & Co. KGaA, Weinheim, pp 3–22

Gehlenborg N, O'Donoghue SI, Baliga NS, Goesmann A, Hibbs MA, Kitano H, Kohlbacher O, Neuweger H, Schneider R, Tenenbaum D, Gavin A-C (2010) Visualization of omics data for systems biology. Nat Methods 7:S56–S58

Glass JI, Smith HO, Hutchison III CA, Alperovich NY, Assad-Garcia N (2007) Minimal bacterial genome. United States Patent Application 20070122826, May 31, 2007

Goldman J, Becker ML, Jones B, Clements M, Leeder JS (2011) Development of biomarkers to optimize pediatric patient management: what makes children different? Biomark Med 5:781–794

Grossman I, Goldstein DB (2010) Pharmacogenetics and pharmacogenomics. In: Ginsburg GS, Willard HF (eds) Essentials of genomic and personalized medicine. Academic Press, San Diego, pp 175–190

Hamm CA, Costa FF (2011) The impact of epigenomics on future drug design and new therapies. Drug Discov Today 16:626–635

Herceg Z, Murr R (2011) Mechanisms of histone modifications. In: Tollefsbol T (ed) Handbook of epigenetics: the new molecular and medical genetics. Elsevier, London, pp 25–45

Hindroff LA, MacArthur J (European Bioinformatics Institute), Wise A, Junkins HA, Hall PN, Klemm AK, Manolio TA (2012) A catalog of published genome-wide association studies. Available at: http://www.genome.gov/gwastudies. Accessed 14 April 2012

Hong H, Shi L, Su Z, Ge W, Jones WD, Czika W, Miclaus K, Lambert CG, Vega SC, Zhang J, Ning B, Liu J, Green B, Xu L, Fang H, Perkins R, Lin SM, Jafari N, Park K, Ahn T, Chierici M, Furlanello C, Zhang L, Wolfinger RD, Goodsaid F, Tong W (2010) Pharmacogenom J 10: 364–374

Hood L, Perlmutter RM (2004) The impact of systems approaches on biological problems in drug discovery. Nat Biotechnol 22:1215–1217

Huston SA (2010) Ethical consideration in pharmacogenomics. In: Zdanowicz MM (ed) Concepts in pharmacogenomics. American Society of Health-systems Pharmacists, Bethesda, pp 375–397

Hutchison CA, Phillips S, Edgell MH, Gillham S, Jahnke P, Smith M (1978) Mutagenesis at a specific position in a DNA sequence. J Biol Chem 253:651–656

Jirtle RL, Skinner MK (2007) Environmental epigenomics and disease susceptibility. Nat Rev Genet 8: 253–262

Johnson JA (2009) Drug target pharmacogenetics. In: American College of Clinical Pharmacy Board (ed) Pharmacogenomics: applications to patient care, 2nd edn. American College of Clinical Pharmacy, Kansas City, pp 90–106

Johnson AC, Reitz M (1998) Site-directed mutagenesis. In: Greene JJ, Rao VB (eds) Recombinant DNA principles and methodologies. Marcel Dekker, Inc, New York, pp 699–719

Kalow W (2009) Pharmacogenetics: a historical perspective. In: American College of Clinical Pharmacy Board (ed) Pharmacogenomics: applications to patient care, 2nd edn. American College of Clinical Pharmacy, Kansas City, pp 22–31

Kaput J, Perlina A, Hatipoglu B, Bartholomew A, Nikolsky Y (2007) Nutrigenomics: concepts and applications to pharmacogenomics and clinical medicine. Pharmacogenomics 8:369–390

Kassam S, Meyer P, Corfield A, Mikuz G, Sergi C (2005) Single nucleotide polymorphisms (SNPs): history, biotechnological outlook and practical applications. Curr Pharmacogenomics 3:237–245

Kaye J, Curren L, Anderson N, Edwards K, Fullerton SM, Kanellopoulou N, Lund D, MacArthur DG, Mascalzoni D, Shepherd J, Taylor PL, Terry SF, Winter SF (2012) From patients to partners: participant-centric initiatives in biomedical research. Nat Rev Genet 13:371–376

Kelloff GJ, Sigman CC (2012) Cancer biomarkers: selecting the right drug for the right patient. Nat Rev Drug Discov 11:201–214

Khalil AS, Collins JJ (2010) Synthetic biology: applications come of age. Nat Rev Genet 11:367–379

Khan FA (2012a) Animal biotechnology. In: Khan FA (ed) Biotechnology fundamentals. CRC Press, Boca Raton, pp 201–235

Khan FA (2012b) Agricultural biotechnology. In: Khan FA (ed) Biotechnology fundamentals. CRC Press, Boca Raton, pp 163–199

Khan FA (2012c) Ethics in biotechnology. In: Khan FA (ed) Biotechnology fundamentals. CRC Press, Boca Raton, pp 397–415

Klimyuk V, Marillonnet S, Knablein J, McCaman M, Gleba Y (2005) Biopharmaceuticals derived from transgenic plants and animals. In: Knablein J (ed) Modern biopharmaceuticals, vol 3. Wiley-VCH, Weinheim, pp 893–917

Klipp E, Herwig R, Kowald A, Wierling C, Lehrach H (2005) Basic principles. In: Klipp E, Herwig R, Kowald A, Wierling C, Lehrach H (eds) Systems biology in practice. Wiley-BCH Verlag GmbH & Co. KgaA, Weinheim, pp 3–17

Knoell DL, Sadee W (2009) Applications of genomics in human health and complex disease. In: AACP Editorial Board (ed) Pharmacogenomics: applications to patient care, 2nd edn. American College of Clinical Pharmacy, Kansas City, pp 1–19

Kobata A (1996) Function and pathology of the sugar chains of human immunoglobulin G. Glycobiology 1:5–8

Kodavanti UP (2010) Lung injury biomarkers. In: Vaidya VS, Bonventre JV (eds) Biomarkers in medicine, drug discovery, and environmental health. John Wiley & Sons, Inc, Hoboken, pp 157–201

Kolesar JM (2009) Drug target pharmacogenomics of solid tumors. In: American College of Clinical Pharmacy Board (ed) Pharmacogenomics: applications to patient care, 2nd edn. American College of Clinical Pharmacy, Kansas City, pp 123–134

Koomer A (2010) Pharmacogenomics in pharmacy education. In: Zdanowicz MM (ed) Concepts in pharmacogenomics. American Society of Health-systems Pharmacists, Bethesda, pp 363–374

Koomer A, Ansong MA (2010) The role of pharmacists in pharmacogenomics. In: Zdanowicz MM (ed) Concepts in pharmacogenomics. American Society of Health-systems Pharmacists, Bethesda, pp 341–350

Krynetskiy E, Calligaro IL (2009) Introducing pharmacy students to pharmacogenomic analysis. Am J Pharm Educ 73:1–7

Kubinyi H (2007) Chemogenomics. In: Kubinyi H (ed) Comprehensive medicinal chemistry II, vol 3. Elsevier, Amsterdam, pp 921–937

Kubinyi H, Müller H (2004) Chemogenomics in drug discovery a medicinal chemistry perspective. In: Kubinyi H, Müller G, Mannhold R, Folkers G. Weinheim: Wiley-VCH Verlag GmbH & Co. KGaA

Laursen L (2010) Big science at the table. Nature 468:S2–S4

Lee LJ, Hughes TR, Frey BJ (2006) How many new genes are there? Science 311:1709

Lee KC, Ma JD, Kuo GM (2009) Pharmacogenomics: bridging the gap between science and practice. Pharm Today 15(12):36–48

Lengauer T, Hartmann C (2007) Bioinformatics. In: Kubinyi H (ed) Comprehensive medicinal chemistry II, vol 3. Elsevier, Amsterdam, pp 315–348

Limdi NA, Rettie AE (2009) Warfarin pharmacogenetics. In: American College of Clinical Pharmacy Board (ed) Pharmacogenomics: applications to patient care, 2nd edn. American College of Clinical Pharmacy, Kansas City, pp 143–164

Lindpainter K (2007) Pharmacogenomics. In: Kubinyi H (ed) Comprehensive medicinal chemistry II, vol 3. Elsevier, Amsterdam, pp 51–68

Loging W, Harland L, Williams-Jones B (2007) High-throughput electronic biology: mining information for drug discovery. Nat Rev Drug Discov 6:220–230

Lu J, Getz G, Miska EA, Alvarez-Saavedra E, Lamb J, Peck D, Sweet-Cordaro A, Elbert BL, Mak RH, Ferando AA (2005) MicroRNA expression profiles classify human cancers. Nature 435:834–838

Maetzler W, Berg D (2010) Biomarkers of Alzheimer's and Parkinson's disease. In: Vaidya VS, Bonventre JV (eds) Biomarkers in medicine, drug discovery, and environmental health. John Wiley & Sons, Inc, Hoboken, pp 91–117

Mason JS, Pickett SD (2003) Combinatorial library design, molecular similarity, and diversity applications. In: Abraham DJ (ed) Burger's Medicinal chemistry and drug discovery, vol 1, 6th edn. John Wiley & Sons, Inc, New York, pp 187–242

Mayfield SP, Franklin SE (2005) Expression of human antibodies in eukaryotic micro-algae. Vaccine 23(15):1828–1832

McCurry KR, Kooyman DL, Alvarado CG, Cotterell AH, Martin MJ, Logan JS, Platt JL (1995) Human complement regulatory proteins protect swine-to-primate cardiac xenografts from humoral injury. Nat Med 1:423–427

McDonagh EM, Whirl-carillo M, Garten Y, Altman RB, Klein TE (2011) From pharmacogenomics knowledge acquisition to clinical applications: the PharmGKB as a clinical pharmacogenomics biomarker resource. Biomark Med 5:795–806

McGuire AL, Evans BJ, Caulfield T, Burke W (2010) Regulating direct-to-consumer personal genome testing. Science 330:181–182

McLean AS, Huang SJ (2010) Biomarkers of cardiac injury. In: Vaidya VS, Bonventre JV (eds) Biomarkers in medicine, drug discovery, and environmental health. John Wiley & Sons, Inc, Hoboken, pp 119–155

Meyer P, Alexopoulos LG, Bonk T, Califano A, Cho CR, de la Fuente A, de Graaf D, Hartemink AJ, Hoeng J, Ivanov NV, Koeppl H, Linding R, Marbach D, Norel R, Peitsch MC, Rice JM, Royyuru A, Schacherer F, Sprengel J, Stolle K, Vitkup D, Stolovitzky G (2011) Verification of systems biology research in the age of collaborative competition. Nat Biotechnol 29:811–813

Miclaus K, Chierici M, Lambert C, Zhang L, Vega S, Hong H, Yin S, Furlanello C, Wolfinger R, Goodsaid F (2010) Variability in GWAS analysis: the impact of genotype calling algorithm inconsistencies. Pharmacogenomics J 10:324–335

Mitscher LA, Dutta A (2003) Combinatorial chemistry and multiple parallel synthesis. In: Abraham DJ (ed) Burger's medicinal chemistry and drug discovery, vol 2, 6th edn. John Wiley & Sons, Inc, New York, pp 1–35

Momary KM, Crouch MA (2010) Cardiovascular disease. In: Zdanowicz MM (ed) Concepts in pharmacogenomics. American Society of Health-systems Pharmacists, Bethesda, pp 183–211

Murphy JE, Green JS, Adams LA, Squire RB, Kuo GM, McKay A (2010) Pharmacogenomics in the curricula of colleges and schools of pharmacy in the United States. Am J Pharm Educ 74:1–10

Nature Biotechnology Editorial Staff (2012) What happened to personalized medicine? Nat Biotechnol 30:1

Nicholson JK, Wilson ID (2003) Understanding 'global' systems biology: metabonomics and the continuum of metabolism. Nat Rev Drug Discov 2:668–676

Opar A (2011) 'Pharmers' hope for first plant drug harvest. Nat Rev Drug Discov 10:81–82

Ossorio P (2004) Societal and ethical issues in pharmacogenomics. In: Allen WL, Johnson JA, Knoell DL, Kolesar JM, McInerney JD, McLeod HL, Spencer HT, Tami JA (eds) Pharmacogenomics: applications to patient care. American College of Clinical Pharmacy, Kansas City, pp 399–442

Ozer JS, Reagan WJ, Schomaker S, Palandra J, Baratta M, Ramaiah S (2010) Translational biomarkers of acute drug-induced liver injury: the current state, gaps, and future opportunities. In: Vaidya VS, Bonventre JV (eds) Biomarkers in medicine, drug discovery, and environmental health. John Wiley & Sons, Inc, Hoboken, pp 203–236

Padgett L, O'Connor S, Roederer M, McLeod H, Ferreri S (2011) Pharmacogenomics in a community pharmacy: ACT now. J Am Pharm Assoc 51:189–193

Paik Y-K, Jeong S-K, Omenn GS, Uhlen M, Hanash S, Cho SY, Lee H-J, Na K, Choi E-Y, Yan F, Zhang F, Zhang Y, Snyder M, Cheng Y, Chen R, Marko-Varga G, Deutsch EW, Kim H, Kwon J-Y, Aebersold R, Bairoch A, Taylor AD, Kim KY, Lee E-Y, Hochstrasser D, Legrain P, Hancock WS (2012) A chromosome-centric human proteome project to characterize the sets of proteins encodedin the genome. Nature Biotechnol 30: 221–223

Patti GJ, Yanes O, Siuzdak G (2012) Metabolomics: the apogee of the omics trilogy. Nat Rev Mol Cell Biol 13:263–269

Pennisi E (2010) 1000 Genomes Project gives new map of genetic diversity. Science 330:574–575

Petros WP, Sharma M (2009) Pharmacogenomics of hematologic malignancies. In: American College of Clinical Pharmacy Board (ed) Pharmacogenomics: applications to patient care, 2nd edn. American College of Clinical Pharmacy, Kansas City, pp 135–142

Petrucelli N, Daly MB, Feldman GL (2011) *BRCA1* and *BRCA2* hereditary breast and ovarian cancer. Gene Reviews [Internet]. Available at: http://www.ncbi.nlm.nih.gov/books/NBK1247/. Accessed 12 Feb 2012

PhRMA (2012) 2011 report: medicines in development – biotechnology. Available at: http://www.phrma.org/research/new-medicines. Accessed 5 Feb 2012

Pirmohamed M (2011) Pharmacogenetics: past, present and future. Drug Discov Today 16:852–861

Pirrung MC (2004) Molecular diversity and combinatorial chemistry. Elsevier, Amsterdam

Portela A, Esteller M (2010) Epigenetic modifications and human disease. Nat Biotechnol 28:1057–1068

Price ND, Edelman LB, Lee I, Yoo H, Hwang D, Carlson G, Galas DJ, Heath JR, Hood L (2010) Systems biology and systems medicine. In: Ginsburg GS, Willard HF (eds) Essentials of genomic and personalized medicine. Academic Press, San Diego, pp 131–141

Pritchard JF, Jurima-Romet M (2010) Enabling go/no go decisions. In: Bleavins MR, Carini C, Jurima-Romet M, Rahbari R (eds) Biomarkers in drug development – a handbook of practice, application, and strategy. John Wiley & Sons, Inc, Hoboken, pp 31–39

Raghavachari N (2012) Overview of omics. In: Barh D, Blum K, Madigan MA (eds) OMICS-biomedical perspectives and applications. CRC Press, Boca Raton, pp 1–19

Rahbar A, Rivers R, Boja E, Kinsinger C, Mesri M, Hiltke T, Rodriguez H (2011) Realizing individualized medicine: the road to translating proteomics from the laboratory to the clinic. Personal Med 8:45–57

Rakyan VK, Down TA, Balding DJ, Beck S (2011) Epigenome-wide association studies for common human diseases. Nat Rev Genet 12:529–541

Rankovic Z, Jamieson C, Morphy R (2010) High throughput screening approach to lead discovery. In: Rankovic Z, Morphy R (eds) Lead generation approaches in drug discovery. John Wiley & Sons, Inc, Hoboken, pp 21–71

Reardon J (2011) The 'persons' and 'genomics' of personal genomics. Personal Med 8:95–107

Rockett JC (2003) The future of toxicogenomics. In: Burczynski ME (ed) An introduction to toxicogenomics. CRC Press LLC, Boca Raton, pp 299–317

Rogers PD (2009) Infectious diseases. In: American College of Clinical Pharmacy Board (ed) Pharmacogenomics: applications to patient care, 2nd edn. American College of Clinical Pharmacy, Kansas City, pp 165–175

Rothberg BEG, Pena CEA, Rothberg JM (2005) A systems biology approach to target identification and validation for human chronic disease drug discovery. In: Knablein J (ed) Modern biopharmaceuticals, design, development and optimization, vol 1. Wiley-BCH Verlag GmbH & Co. KgaA, Weinheim, pp 99–125

Rudolph NS (1995) Advances continue in production of proteins in transgenic animal milk. Genet Eng News, October 15, 8–9

Rudolph NS (2000) Biopharmaceutical production in transgenic livestock. Trends Biotechnol 17:367–374

Sanger Institute (2012) The cancer genome project. Available at: http://www.sanger.ac.uk/genetics/CGP/. Accessed 6 Mar 2012

Schniecke AE, Kind AJ, Ritchie WA, Mycock K, Scott AR, Ritchie M, Wilmut I, Colman A, Campbell KHS (1997) Human factor IX transgenic sheep produced by transfer of nuclei from transfected fetal fibroblasts. Science 278:2130–2133

Schrattenholz A, Groebe K, Soskic V (2010) Syetems biology approaches and tools for analysis of interactomes and multi-target drugs. In: Yan Q (ed) Systems biology in drug discovery and development: methods and protocols. Springer, New York, pp 29–58

Semizarov D (2009a) Genomics technologies as tools in drug discovery. In: Semizarov D, Blomme E (eds) Genomics in drug discovery and development. John Wiley & Sons, Inc, Hoboken, pp 25–103

Semizarov D (2009b) Genomic biomarkers. In: Semizarov D, Blomme E (eds) Genomics in drug discovery and development. John Wiley & Sons, Inc, Hoboken, pp 105–166

Seneci P (2007) Combinatorial chemistry. In: Kubinyi H (ed) Comprehensive medicinal chemistry II, vol 3. Elsevier, Amsterdam, pp 315–348

Sharpless NE, DePinho RA (2006) The mighty mouse: genetically engineered mouse models in cancer drug development. Nat Rev Drug Discov 5:741–754

Shaw EC, Johnson PWM (2012) Stratified medicine for cancer therapy. Drug Discov Today 17:261–268

Shultz LD, Ishikawa F, Greiner DL (2007) Humanized mice in translational biomedical research. Nat Rev Immunol 7:118–130

Singh AA, Somvanshi P (2012) Bioinformatics – a brief introduction to changing trends in modern biology. In: Barh D, Blum K, Madigan MA (eds) OMICS-biomedical perspectives and applications. CRC Press, Boca Raton, pp 23–39

Smith HE (2010) Toxicogenomics. In: Zdanowicz MM (ed) Concepts in pharmacogenomics. American Society of Health-systems Pharmacists, Bethesda, pp 321–337

Smith HO, Hutchison CA III, Pfannkoch C, Venter JC (2003) Generating a synthetic genome by whole genome assembly: φX174 bacteriophage from synthetic oligonucleotides. Proc Natl Acad Sci U S A 100:15440–15445

Staff (2007) A mouse for all reasons – The International Mouse Knockout Consortium. Cell 128:9–13

States DJ, Omenn GS, Blackwell TW, Fermin D, Eng J, Speicher DW, Hanash SM (2006) Challenges in deriving high-confidence protein identifications from data gathered by HUPO plasma proteome collaborative study. Nat Biotechnol 24:333–338

Subramanian A, Tamayo P, Mootha VK, Mukherjee S, Ebert BL, Gillette MA, Paulovich A, Pomeroy SL, Golub TR, Lander ES, Mesirov JP (2005) Gene set enrichment analysis: a knowledge-based approach for interpreting genome-wide expression profiles. Proc Natl Acad Sci U S A 102:15545–15550

Sucholeiki I (2001) High-throughput synthesis: principles and practices. Marcel Dekker, Inc, New York

Takeuchi M, Takasaki S, Shimada M, Kobata A (1990) Role of sugar chains in the in vitro biological activity in human erythropoietin produced in recombinant Chinese hamster ovary cells. J Biol Chem 265:12127–12130

The International Human Genome Sequencing Consortium (2001) Initial sequencing and analysis of the human genome. Nature 409:860–921

Thornton M, Gladwin A, Payne R, Moore R, Cresswell C, McKechnie D, Kelly S, March R (2005) Automation and validation of DNA-banking systems. Drug Discov Today 10:1369–1375

Topol EJ (2010) Pharmacy benefit managers, pharmacies, and pharmacogenomic testing: prescription for progress? Sci Transl Med 2:14–16, collect

Treangen TJ, Salzberg SL (2011) Repetitive DNA and next-generation sequencing: computational challenges and solutions. Nat Rev Genet 13:36–46

Tursz T, Andre F, Lazar V, Lacroix L, Soria J-C (2011) Implications of personalized medicine – perspective from a cancer center. Nat Clin Coll, Personalized cancer medicine 8(3):S46–S52

U.S. DOE (2012a) Human genome project information. Available at: http://www.ornl.gov/sci/techresources/Human_Genome/home.shtml. Accessed 3 Jan 2012

U.S. DOE (2012b) Human genome project information. Available at: http://www.ornl.gov/sci/techresources/Human_Genome/faq/snps.shtml. Accessed 27 Mar 2012

U.S. National Academies (2011) Toward precision medicine: building a knowledge network for biomedical research and a new taxonomy of disease. The National Academies Press, Washington, D.C., pp 1–4, http://www.nap.edu/catalog.php?record_id=1328

U.S. NIH (2012) The cancer genome atlas. Available at: http://cancergenome.nih.gov/. Accessed 7 Mar 2012

Ullmann D (2007) Fluorescence screening assays. In: Kubinyi H (ed) Comprehensive medicinal chemistry II, vol 3. Elsevier, Amsterdam, pp 599–615

Van Eyck AS, Bouzin C, Feron O (2010) Both host and graft vessels contribute to revascularization of xenografted human ovarian tissue in a murine model. Fertil Steril 93:1676–1685

Varkin A, Cummings R, Esko J, Freeze H, Hart G, Marth J (1999) Essentials of glycobiology. Cold Spring Harbor Laboratory Press, LaJolla

Veenstra TD (2010) Proteomics for biomarker discovery. In: Vaidya VS, Bonventre JV (eds) Biomarkers in medicine, drug discovery, and environmental health. John Wiley & Sons, Inc, Hoboken, pp 25–46

Venter JC et al (2001) The sequence of the human genome. Science 291:1304–1351

Waldman SA, Terzic A (2011) Patient-centric clinical pharmacology advances the path to personalized medicine. Biomark Med 5:697–700

Walsh G, Jefferis R (2006) Post-translational modifications in the context of therapeutic proteins. Nat Biotechnol 24:1241–1252

Wang L (2009) Phase II drug-metabolizing enzymes. In: American College of Clinical Pharmacy Board (ed) Pharmacogenomics: applications to patient care, 2nd edn. American College of Clinical Pharmacy, Kansas City, pp 54–69

Warner S (2004) Diagnostics + therapy = theranostics. Scientist 18:38–39

Webster KD (2010) Pharmacogenomics: pharmacodynamics and pharmacogenomics. In: Zdanowicz MM (ed) Concepts in pharmacogenomics. American Society of Health-systems Pharmacists, Bethesda, pp 155–180

Wenk MR (2005) The emerging field of lipidomics. Nat Rev Drug Discov 4:594–610

Weston GS (2010) The pharmacogenetics of drug metabolism. In: Zdanowicz MM (ed) Concepts in pharmacogenomics. American Society of Health-systems Pharmacists, Bethesda, pp 85–127

Wildt S, Gerngross TU (2005) The humanization of N-glycosylation pathways in yeast. Nat Rev Microbiol 3:119–126

Williams SA, Slavin DE, Wagner JA, Webster CJ (2006) A cost effectiveness approach to the qualification and acceptance of biomarkers. Nat Rev Drug Discov 5:897–902

Wishart DS, Tzur D, Knox C, Eisner R, Guo AC, Young N, Cheng D, Jewell K, Arndt D, Sawhney S, Fung C, Nikolai L, Lewis M, Coutouly M-A, Forsythe I, Tang P, Shrivastava S, Jeroncic K, Stothard P, Amegbey G, Block D, Hau DD, Wagner J, Miniaci J, Clements M, Gebremedhin M, Guo N, Zhang Y, Duggan GE, MacInnis GD, Weljie AM, Dowlatabadi R, Bamforth F, Clive D, Greiner R, Li L, Marrie T, Sykes BD, Vogel HJ, Querengesser L (2007) HMDB: the human metabolome database. Nucleic Acids Res 35(Database Issue):D521–D526

Woollard PM, Mehta NAL, Vamathevan JJ, Van Horn S, Bonde BK, Dow DJ (2011) The application of next-generation sequencing technologies to drug discovery and development. Drug Discov Today 16:512–519

Woster PM (2010) Epigenetic targets and cancer drug discovery. Annu Rep Med Chem 45:245–260

Wright FA et al (2001) A draft annotation and overview of the human genome. Genome Biol 2:1–18

Wu R, Tong C, Wang Z, Mauger D, Tantisira K, Szefler SJ, Cinchilli VM, Isreal E (2011a) A conceptual framework for pharmacodynamics genome-wide association studies in pharmacogenomics. Drug Discov Today 16:884–890

Wu W, Zhang HH, Welsh MJ, Kaufman PB (2011b) Bioinformation superhighway and computer databases of nucleic acids and proteins. In: Gene technology, 3rd edn. CRC Press, Boca Raton, pp 133–153

Wu W, Zhang HH, Welsh MJ, Kaufman PB (2011c) Strategies for gene double knockout. In: Gene technology, 3rd edn. CRC Press, Boca Raton, pp 351–380

Yang Y, Adelstein SJ, Kassis AI (2009) Target discovery from data mining approaches. Drug Discov Today 14:147–154

Yuryev A (2012) Pathways analysis in drug discovery. In: Young DL, Michelson S (eds) Systems biology in drug discovery and development. John Wiley & Sons, Inc, Hoboken, pp 289–302

Zdanowicz MM (2010) Pharmacogenomics: past, present, and future. In: Zdanowicz MM (ed) Concepts in pharmacogenomics. American Society of Health-systems Pharmacists, Bethesda, pp 3–17

Zembles T (2010) An inservice program on pharmacogenetics to individualize drug therapy. Am J Pharm Educ 74:1–5

Zhou S (2006) Clinical pharmacogenomics of thiopurine S-methyltransferase. Curr Clin Pharmacol 1:119–128

Zineh I, Huang S-M (2011) Biomarkers in drug development and regulation: a paradigm for clinical implementation of personalized medicine. Biomark Med 5:705–713

Zineh I, Pacanowski M (2009) Cardiovascular diseases. In: American College of Clinical Pharmacy Board (ed) Pharmacogenomics: applications to patient care, 2nd edn. American College of Clinical Pharmacy, Kansas City, pp 176–203

Zuckerman R, Milne C-P (2012) Market watch: industry perspectives on personalized medicine. Nat Rev Drug Discov 11:178–179

9

Dispensing Biotechnology Products: Handling, Professional Education, and Product Information

Peggy Piascik and Val Adams

INTRODUCTION

Preparation, dispensing, and patient education regarding appropriate use of pharmaceuticals are primarily the responsibility of the pharmacist. Traditionally, parenteral products have been available in ready-to-use containers or required dilution with water or saline prior to use with no other special handling requirements. Hospital pharmacists, in particular, have prepared and dispensed parenteral products for individual patients for many years. While many pharmacists are skilled in handling parenteral products, biotechnology products present additional challenges since they are proteins subject to denaturation and thus require special handling techniques. These challenges will be explained in greater detail in this chapter. Practice issues with biotechnology products may be handled in slightly different ways depending on laws and pharmacy practice standards in each country. This chapter is written primarily from the view of practice in the United States since that is the primary experience of the authors.

PHARMACIST READINESS

Pharmacists may be unprepared to provide pharmaceutical care services to patients who require therapy with biotechnology drugs for a variety of reasons including (1) lack of knowledge about the tools of biotechnology; (2) lack of understanding of the therapeutic aspects of recombinant protein products; (3) lack of familiarity with the side effects and patient education considerations; (4) lack of familiarity with the storage, handling, and reconstitution of proteins; and (5) difficulty of handling reimbursement issues.

Pharmacists may view biotechnology drugs as quite different from traditional parenteral products and familiar oral dosage forms. However, in most respects, the services offered by pharmacists when preparing and dispensing biotechnology products are the same as those provided for traditional tablets or injectable products. To determine the knowledge and skills a pharmacist requires to work with biotechnology drugs, one must first consider who will be storing, preparing, dispensing, and administering the agent. Many agents will be prepared by a pharmacist or other health-care provider and the drug administered by a nurse, while others will be prepared by the patient and self-administered. Pharmacists who work in clinics or with home health-care providers need to understand how to store, prepare, and dispense the product to a nurse with instructions to maintain potency and sterility until it is administered to the patient. The knowledge and skill set is similar but has some significant differences from the skills required by a community pharmacist who must be able to teach the patient how to store, reconstitute, and administer the biotechnology agent.

The decision as to who will store, prepare, and administer the drug is typically determined from a business perspective. In order for a clinic to administer and be paid for a drug, the drug must not be "usually self-administered by the patients who take them" (Department of Health and Human Services 2010). The Medicare Benefit Policy manual outlines a process to make this determination. From a practical standpoint, drugs administered in the clinic are billed using a "J" code. Based on logistics, it is relatively safe to say that a drug that does not have a "J code" will be prepared and administered by the patient (see Table 9.3 for examples of biotechnology drugs with and without a J code). It is important, regardless of the product being dispensed, to ensure that the pharmacist and patient understand the use, dosage regimen, and potential adverse effects of the product. Patients who will be preparing and self-administering the drug must know the proper storage and handling instructions as well as receive specific training on the administration of the drug and proper disposal of

P. Piascik, Ph.D. (✉) • V. Adams, PharmD
Department of Pharmacy Practice and Science,
University of Kentucky, College of Pharmacy,
789 South Limestone Street, Lexington, KY 40536-0596, USA
e-mail: piascik@email.uky.edu

D.J.A. Crommelin, R.D. Sindelar, and B. Meibohm (eds.), *Pharmaceutical Biotechnology*,
DOI 10.1007/978-1-4614-6486-0_9, © Springer Science+Business Media New York 2013

unused medication. When patients do not understand the administration and monitoring requirements of biotechnology products, training sessions for patients and caregivers should be considered to ensure appropriate patient care.

As more novel protein products have come to market and the indications for existing agents have expanded, pharmacists are increasingly required to deal with these protein pharmaceuticals. While the first protein/peptide recombinant products were used primarily in hospital settings, many of these agents are now commonplace in ambulatory settings. The traditional community pharmacy may now dispense products like colony-stimulating factors, growth hormone, and interferons to name a few.

Traditional routes of delivery for pharmaceuticals have been challenged by the unique characteristics of biotech product delivery. Community pharmacies may struggle to maintain sufficient inventory of high-cost products, with in-depth knowledge of the products and its characteristics and with product administration. Assisting patients with reimbursement issues is a time-consuming, complicated process. Physicians also have difficulty with inventory and with slow reimbursement. Managed care organizations may have difficulty tracking claims for these products.

As a result, the majority of patients receiving biotech drugs are now managed by home health, home infusion, or specialty pharmacy services. Specialty pharmacies have evolved to manage outpatient biotechnology therapies for patients Suchanek (2005). The services offered by these pharmacies go far beyond dispensing biotech products. These pharmacies have expertise in the following areas (Caremark 2006):

- Insurance coverage and drug costs
- Pipeline monitoring and management
- Utilization management
- Promoting adherence to drug regimen
- Disease state management
- Risk Evaluation and Mitigation Strategies (REMS) requirements

Payers, particularly managed care organizations, now contract with specialty pharmacies to provide biotech and other expensive agents to solve many of the problems these products pose for the payer. The specialty pharmacy market grew 15.3 % from 2009 to 2010. In 2010, $39.2 billion of specialty pharmaceuticals was dispensed by specialty, retail, and mail order pharmacies (Fein 2010).

■ Types of Information Needed by Pharmacists

What types of information do pharmacists require to be confident providers of biotech drugs and services? For pharmacists who have been out of school for many years, a contemporary understanding of the immune system, autoimmune diseases, and mechanisms by which drugs modify the immune system is essential. Several appropriate books that can provide a basic background in immunology are listed in Table 9.1. Additionally, practitioners may enroll in organized courses or continuing education programs that can provide up-to-date information in the discipline of immunology. Current pharmacy students and recent graduates should be sufficiently trained in basic immunology as part of their professional curriculum.

Cellular and Molecular Immunology. 7th ed.
Abbas AK, Lichtman AH, Pillai S. Philadelphia: W.B. Saunders Company, 2012: 545 pp.
Softbound book providing basic immunology concepts and clinical issues. Includes access to online edition
Immunology: A Short Course. 6th ed.
Coico R, Sunshine G, Benjamini E. New Jersey: John Wiley and Sons, Inc., 2009: 391 pp
Softbound elementary text with review questions for each chapter
Concepts in Immunology and Immunotherapeutics. 4th ed.
Smith, BT. Bethesda, MD: American Society of Hospital Pharmacists, 2008: 304 pp.
Review of basic immunology including therapeutic applications
Medical Immunology Made Memorable
Playfair JHL, Lydyard PM: New York: Churchill Livingstone, 2000: 108 pp.
Softbound, simple overview of basic immunology, immunopathology, and clinical immunology
Roitt's Essential immunology. 12th ed.
Delves, PJ, Martin, S, Burton, D, Roitt, I. Oxford; Boston: Wiley-Blackwell Publishing, 2011: 560 pp.
Softbound basic immunology textbook
Janeway's Immunobiology. 8th ed.
Murphy K. New York: Garland Science, 2011: 888 pp.
Softback text that presents immunology at the introductory level. Also available in e-book format

Table 9.1 ■ Selected texts to enhance immunology knowledge.

Many pharmacists, upon hearing the word biotechnology, imagine a discipline too technical or complicated to be understood by the typical practitioner. Pharmacists must recognize that biotechnology primarily refers to a set of tools that has allowed great strides to be made in basic research, the understanding of disease and development of new therapeutic agents. It is essential for pharmacists to have a basic understanding of recombinant DNA technology and monoclonal antibody technology. However, it is not necessary that pharmacy practitioners know how to use these tools in the laboratory but rather how the use of these tools provides new therapeutic agents and a greater understanding of disease processes.

Pharmacists may need to review or learn anew about protein chemistry and those characteristics that affect therapeutic activity, product storage, and routes of administration of these drugs. Apart from this textbook, several publications, videotapes, and continuing professional education programs from industry and academic institutions are available to pharmacists for learning about the technical aspects of product storage and handling. Pharmacists also need to become familiar with the drug delivery systems currently in use for biotech drugs as well as those that are in development (see Chap. 4).

■ Sources of Information for Pharmacists

Many pharmacists do not know where to obtain the information that will allow them to be good providers of products of biotechnology. This textbook provides much of the essential background information in one source.

An excellent source of information on biotechnology in general, and specific products in particular, is the biotech drug industry. Many manufacturer-sponsored programs describe approved biotech products and those likely to come to market in the near future. Manufacturer programs provide extensive information about the disease states for which their products are indicated as well as product-specific information. Manufacturers are prepared to help pharmacists in the most effective provision of products and services to hospital-based and ambulatory patients. However, many pharmacists are unaware of these services and how to obtain them. A web search of specific products will lead to the product and manufacturer's websites where this information can be accessed.

The information provided by manufacturers can help pharmacists to confidently provide biotechnology products to their patients. The services provided generally fall into three categories: customer/medical services and support, educational materials, and reimbursement information. Manufacturers may have a separate number for reimbursement questions. Table 9.2 lists the manufacturer's toll-free assistance numbers and web addresses for obtaining product and reimbursement information in North America. Vaccines and insulin products are not included in this table since these products were previously available in a nonrecombinant form and pharmacists are generally well familiar with these products. Moreover, the recombinant forms of these products are generally not as costly as other types of biologic agents.

■ The Pharmacist and Handling of Biotech Drugs

The pharmacist is responsible for the storage, preparation, and dispensing of biotechnology drugs as well as patient education regarding the use of these products. In many cases, pharmacists must have additional training in order to be prepared for this role. This is especially true for pharmacists who practice in the ambulatory setting since these products are increasingly available for self-administration in the home. Pharmacies of the future may stock pumps, patches, timed-release tablets, liposomes, implants, and vials of tailored monoclonal antibodies. With advances in gene therapy and pharmacogenomics, it is possible that the pharmacist may eventually prepare and dispense products tailored for specific patients.

This chapter discusses the general principles that pharmacists need to understand about storage, handling, preparation, administration of biotech products, and issues related to outpatient/home care. Specific examples will be discussed for illustrative purposes. Table 9.3 lists selected products along with specific handling requirements for each. For specific products or recent updates to these requirements, contact the manufacturer. For additional information regarding drug handling and preparation, the pharmacist may consult publications such as the American Hospital Formulary Service (AHFS) Drug Information published annually by the Association of Health-Systems Pharmacists or the *King Guide to Parenteral Admixture* Catania (2010), a quarterly updated guide to IV drug compatibility and stability. In addition to hardcover publications with frequent updates, both of these references are available online at www.ashp.org/ahfs/ and www.kingguide.com. Pharmacy benefits management companies (PBMs) usually own specialty pharmacy companies and provide valuable information via their websites. The three largest specialty pharmacy companies in 2010 were Accredo Health owned by Medco Health Solutions, CuraScript Pharmacy owned by Express Scripts, and CVS Caremark (Fein 2010).

Manufacturer	Professional services	Reimbursement hotline/indigent patient programs	Manufacturer website
Amgen	1-800-772-6436	1-800-272-9376	www.amgen.com
			www.imminex.com
Astellas	1-800-727-7003	1-800-477-6472	www.astellas.us
Baxter Healthcare	1-800-422-9837	1-800-548-4448	www.baxter.com
Bayer Healthcare	1-888-842-	1-800-288-8374	www.bayerhealthcare.com
Bedford Laboratories	1-800-521-5169		www.bedfordlabs.com
Biogen Idec	1-800-456-2255	1-800-456-2255	www.biogenidec.com
BioMarin	1-866-906-6100	1-866-906-6100	www.bmrn.com
Bristol-Myers Squibb	1-800-332-2056	1-800-736-0003	www.bms.com
ChiRhoClin	1-877-272-4888		www.chirhoclin.com
CSL Behring	1-610-878-4000		www.cslbehring.com
Genentech	1-800-821-8590	1-800-530-3083	www.gene.com
		1-866-422-2377	www.genentechaccesssolutions.com
Genzyme	1-800-745-4447	1-800-745-4447	www.genzyme.com
GlaxoSmithKline	1-888-825-5249	1-888-825-5249	www.gsk.com
Insmed	1-866-464-7539	1-866-464-7539	www.insmed.com
InterMune	1-415-466-2200		www.intermune.com
Janssen	1-800-526-7736	1-800-652-6227	www.janssenbiotech.com
Ligand	1-858-550-7896		www.ligand.com
Lilly	1-877-237-8197	1-800-545-5979	www.lilly.com
Merck	1-800-444-2080	1-800-727-5400	www.merck.com
Novo Nordisk	1-800-727-6500	1-877-668-6777	www.novomedlink.com
		1-866-310-7549	www.novonordisk-us.com
OSI	1-631-962-0600		www.osip.com
Pfizer	1-800-505-4426	1-866-706-2400	www.pfizer.com
			www.pfizerhelpfulanswers.com
Roche	1-800-821-8590		www.roche.com
Sanofi	1-800-981-2491	1-800-221-4025	www.sanofi.us
Tercica	1-650-624-4900		www.tercica.com
UniGene	1-973-265-1100		www.unigene.com

Table 9.2 ■ Toll-free assistance numbers and websites for selected biopharmaceutical manufacturers in the USA and Canada.

STORAGE

Biotech products have unique storage requirements when compared to the majority of products that pharmacists dispense. The shelf-life of these products is often considerably shorter than for traditional compounds. For example, interferon-α2a (Roche Laboratories 2005) is only stable in a refrigerator in the ready-to-use solution for 2 years. After the first dose, cartridges may be stored at less than 25 °C for *up to* 28 days although refrigeration is recommended. Since most biologic products need to be kept at refrigerated temperatures (as discussed below), some pharmacies may need to increase cold storage space in order to accommodate the storage needs.

■ Temperature Requirements

Since biotech products are primarily proteins, they are subject to denaturation when exposed to extreme temperatures. In general, most biotech products are shipped by the manufacturer in gel ice containers and need to be

Generic name	Brand name	Storage temperature	Reconstitution solution	Stability after reconstitution		Dilution/stability	J code[a]
				RT	Ref		
Abatacept	Orencia®	2–8 °C	SWFI	24 h	24 h	24 h (further diluted in NS)	Yes
Adalimumab	Humira®	2–8 °C	RTU	NA	NA	NA	Yes
Testosterone	AndroGel 1 %	25 °C	RTU	NA	NA	NA	
Alteplase	Activase®	2–25 °C	Dil	8 h	8 h	NA	Yes
Alteplase	Cathflo® Activase®	2–8 °C	SWFI	8 h	8 h	NA	Yes
Bevacizumab	Avastin®	2–8 °C	NS	NA	8 h	NA	Yes
Cetuximab	Erbitux®	2–8 °C	RTU	NA	NA	NA	Yes
Darbepoetin alfa	Aranesp®	2–8 °C	RTU	NA	NA	NA	Yes
Denosumab	Prolia® Xgeva®	2–8 °C	RTU	NA	NA	NA	Yes
Epoetin alfa SDV	Epogen® Procrit®	2–8 °C	SBWFI containing benzyl alcohol 0.9 % in a 1:1 ratio	14 d (except for 40,000 units/mL vials which are stable for 7 d)	NA	Dilutions of 1:10 and 1:20 (1 part epoetin:19 parts sodium chloride): 18 h	Yes
Epoetin alfa MDV	Epogen® Procrit®	2–8 °C aie and between doses	RTU	NA	NA	Dilutions of 1:10 in $D_{10}W$ with human albumin 0.05 or 0.1 %: 24 h	Yes
Etanercept	Enbrel®	2–8 °C	SBWFI	NA	14 d	NA	Yes
Factor VIIa recombinant	NovoSeven® RT	2–25 °C	Histidine Diluent	3 h	3 h	NA	No
Filgrastim	Neupogen®	2–8 °C	D_5W	24 h	14 d	24 h	Yes
Infliximab	Remicade®	2–8 °C	SWFI	NA	NA	3 h	Yes
Interferon alfa- 2b	Intron®A	2–8 °C	SWFI		24 h	24 h (further diluted in NS)	Yes
Interferon-β1a prefilled syringe	Avonex®, Rebif®	2–8 °C	RTU	NA	NA	NA	Yes
Interferon-β1a reconstitutable vial	Avonex®, Rebif	2–8 °C	SWFI	NA	6 h	NA	Yes
Interferon-β1b	Betaseron®	25 °C	NaCl 0.54 %	NA	3 h	NA	Yes
Ipilimumab	Yervoy®	2–8 °C	NS or D_5W	24 h	24 h	NA	Yes
Paclitaxel (protein bound)	Abraxane®	25 °C	NS	8 h	8 h	NA	Yes
Palivizumab	Synagis®	2–8 °C	RTU	NA	NA	NA	No
Peginterferon alfa-2a	Pegasys® Convenience Pack	2–8 °C	RTU	NA	NA	NA	No
Peginterferon alfa-2b	PegIntron®	25 °C	SWFI	NA	24 h	NA	No

Table 9.3 ■ Storage, stability, and reconstitution of selected biotechnology products.

Generic name	Brand name	Storage temperature	Reconstitution solution	Stability after reconstitution		Dilution/stability	J code[a]
				RT	Ref		
Peginterferon alfa-2b	Sylatron™	25 °C	SWFI	NA	24 h	NA	No
Peginterferon alfa-2b	Redipen®	2–8 °C	RTU	NA	24 h	NA	No
Pegfilgrastim	Neulasta®	2–8 °C	RTU	NA	NA	NA	Yes
Dornase alfa	Pulmozyme®	2–8 °C	RTU	NA	NA	NA	Yes
Ranibizumab	Lucentis®	2–8 °C	RTU	NA	NA	NA	Yes
Rituximab	Rituxan®	2–8 °C	NS or D_5W	24 h	24 h	NA	No
Teriparatide	Forteo®	2–8 °C	RTU	NA	NA	NA	No
Trastuzumab	Herceptin®	2–8 °C	SBWFI	NA	28 d	24 h (further diluted in NS)	Yes

Biologic products listed in the top 200 drugs in the US market by sales, 2010

Table key: *aie* after initial entry into vial, *d* days; *dil* supplied diluent, *h* hours, *mdv* applies only to multidose vials, *NA* not applicable/not available, *NS* normal saline, *Ref* under refrigeration, *RT* room temperature, *RTU* ready to use, *SBWFI* sterile bacterial water for injection, *SDV* applies only to single-dose vials, *SWFI* sterile water for injection

[a]Products have a J code for the first quarter of 2012 according to the document, 2012 ASP Drug Pricing Files Medicare Part B Drug Average Sales Price, listed on cms.gov

Table 9.3 ■ (continued)

stored at 2–8 °C (Banga and Reddy 1994). Once reconstituted, they should be stored under refrigeration until just prior to use. There are a few exceptions to this rule. For example, alteplase (tissue plasminogen activator) lyophilized powder is stable at room temperature for several years at temperatures not to exceed 30 °C (86 °F). However, after reconstitution, the product should be used within 8 h (Genentech 2011). For individual product temperature requirements, the product insert, product website, or the manufacturer should be contacted. Table 9.3 lists temperature requirements for selected frequently prescribed products.

The variability between products with respect to temperature is exemplified by granulocyte colony-stimulating factor (G-CSF, filgrastim; Amgen Inc 2007) and erythropoietin (Amgen 2011) which are stable in ready-to-use form at room temperature for 24 h and 14 days, respectively. Granulocyte macrophage colony-stimulating factor (GM-CSF, sargramostim; Genzyme Corporation 2009) is packaged as a lyophilized powder but still requires refrigeration and once reconstituted is stable at room temperature for 30 days or in the refrigerator for 2 years. Aldesleukin (interleukin-2) is stable for 48 h at room temperature or under refrigeration (Prometheus Therapeutics 2011). Betaseron (interferon-β1b) must be stored in a refrigerator and should be used within 3 h after reconstitution (Bayer HealthCare 2010). While most products require refrigeration to maintain stability due to denaturation by elevated temperatures, extreme cold such as freezing may be just as harmful to most products. The key is to avoid extremes in temperature whether it is heat or cold (Banga and Reddy 1994).

■ Storage in Dosing and Administration Devices

Many biotech products can adhere to either plastic or glass containers such as syringes, polyvinyl chloride (PVC) intravenous bags, infusion equipment, and glass intravenous bottles. The effectiveness of the product may be reduced by three- or fourfold due to adherence. In order to decrease the amount of adherence, human serum albumin (HSA) is usually added to the solutions (see Chap. 4). The relative loss through adherence is concentration dependent, i.e., the more concentrated the final solution, the less significant the adherence becomes. The amount of HSA added varies with the product (Banga and Reddy 1994; Koeller and Fields 1991). Some products that require the addition of HSA include filgrastim, sargramostim, aldesleukin, erythropoietin, and interferon-α. In the case of filgrastim, the addition of 2 mg/mL of HSA to the final solution is required for concentrations of 5–15 μg/mL (Amgen Inc 2007). One miligram of HSA per 1 mL 0.9 % sodium chloride injection is added to achieve a final concentration of 0.1 % HSA for sargramostim concentrations of <10 μg/mL (Genzyme Corporation 2009). For aldesleukin 0.1 %, HSA is required for all concentrations (Prometheus Therapeutics 2011). For erythropoietin, 2.5 mg HSA is present per mL in each single-dose and multidose vial (Amgen 2011). One milligram per milliliter of HSA is added to interferon-α in single-dose and multidose vials and pens (Schering Laboratories 2004).

For additional information or to find information for other products, check the current product information or contact the manufacturer.

■ Storage in IV Solutions

Biotech product stability may vary when stored in different types of containers and syringes. Some products are only stable in plastic syringes, e.g., somatropin and erythropoietin, while others are stable in glass, polyvinyl chloride, and polypropylene, e.g., aldesleukin. Batch prefilling of syringes is possible. However, it is important to make sure that the product you wish to provide in prefilled syringes is stable in the type of syringe you wish to use. This may present a challenge to specialty pharmacy programs. Determining how far in advance doses may be prepared is also an important consideration. G-CSF is stable in Becton Dickinson (B-D) disposable plastic syringes for up to 7 days (Amgen Inc 2007), while erythropoietin is stable for up to 14 days (Amgen 2011). Aldesleukin is recommended to be administered in PVC although glass has been used in clinical trials with comparable results (Prometheus Therapeutics 2011). Solutions are stable for 48 h when refrigerated. GM-CSF and G-CSF can be administered in either PVC or polypropylene (Genzyme Corporation 2009).

■ Light Protection

Many biotech products are sensitive to light. Manufacturer's information usually suggests that products be protected from strong light until the product is used. Dornase-α is packaged in protective foil pouches by the manufacturer to protect it from light degradation and should be stored in these original light protective containers until use. For patients who travel, the manufacturer will provide special travel pouches on request (Genentech, Inc 2011). Alteplase in the lyophilized form also needs to be protected from light but is not light sensitive when in solution (Genentech 2011). Pharmacists must be aware of the specific storage requirements with respect to light for each of the products stocked in the pharmacy.

HANDLING

■ Mixing and Shaking

Improper handling of protein products can lead to denaturation. Shaking and severe agitation of most of these products will result in degradation. Therefore, special techniques must be observed in preparing biotech products for use. Biotech products should not be shaken when adding any diluent as this may cause the product to breakdown. Once the diluent is added to the container, the vial should be swirled rather than vigorously shaken. Some shaking during transport may be unavoidable and proper inspection of products should occur to make sure the products have not been damaged during transit. When a product is affected by excessive shaking, physical separation or frothing within the vial of liquid products can usually be observed. For lyophilized products, agitation is not harmful until the product has been or is reconstituted. In distributing individual products to patient or ward areas, pneumatic tubes should be avoided.

■ Travel Requirements

When patients travel with these products, certain precautions should be observed. The drugs should be stored in insulated, cool containers. This can be accomplished by using ice packs to keep the biotech drug at the proper temperature in warmer climates, whereas the insulated container in colder climates may be all that is required. When traveling in subfreezing weather, the products should be protected from freezing (temperatures below 2 °C). Keeping biotech drugs at proper temperature during automobile travel may present a problem with temperatures inside a parked car often exceeding 37 °C (100 °F) on a warm day. Patients and delivery personnel must take care not to leave products that are not in insulated containers inside the car, trunk, or glove compartment while shopping or making deliveries. When ice is used, care should be taken not to place the product directly on the ice. Dry ice should be avoided since it has the potential for freezing the product. When traveling by air, biotech products should be taken onto the plane in insulated packages and not placed in a cargo container. Airplane cargo containers may be cold enough to cause freezing (Banga and Reddy 1994; Koeller and Fields 1991).

PREPARATION

When preparing biotech products, aseptic technique must be employed as it is with traditional parenteral products. The product should be prepared in a clean room designed for this purpose with laminar airflow hoods, and other practices consistent with USP 797. Most of the products require reconstitution with sterile water or bacteriostatic water for injection depending on stability data. The compatibility of individual products varies and limited data is available. As mentioned previously, when adding diluent to these products, care should be taken not to shake them, but to swirl the container or roll it between the palms of the hands. In the case of lyophilized products, introduction of the diluent should be directed down the side of the vial and not directly on the powder to avoid denaturing the protein. It is important to mention that stability does not mean sterility. Biotech products require the same precautions as any other parenteral product. Sterility is particularly important when prefilling and premixing various doses for administration at home. Once the manufacturer's sterile packaging is entered, sterility can no longer be assured nor will the manufacturer be responsible for any subsequent related problems. Many biotech drugs are not compatible with preservative agents, and single-use vials do not contain a preservative. Individual manufacturers have not addressed the issue of sterility and each institution or organization must determine its own policy on this issue. Many of the currently available biotechnology-produced products are provided as single-dose vials and should not be reused. This does not, however, prevent preparing batches ("batching") of unit-of-use doses in order to be efficient. Many of the patients receiving these agents are likely to have suppressed immune systems and are vulnerable to infection. Therefore, a policy involving the maintenance of sterility of biotech products should be developed by each health-care organization, especially hospitals and specialty pharmacies. When products are made in a sterile environment under aseptic procedures, they should remain sterile until used and thus could be stored for as long as physical compatibility data dictates. However, most institutions have shorter expiration dates, which are generally 72 h or less, on reconstituted products. These expiration dates have been conservatively set due to lack of good sterility data to the contrary. Sterility studies should be performed in order to determine if reconstituted products could be stored for a longer period of time and still maintain sterility. For products reconstituted for home use, in the pharmacy sterile products area, a 7 day expiration date is used provided the product is stable and can be stored in the refrigerator. The American Society of Health-System Pharmacists has published a technical assistance bulletin on sterile products which should be consulted for developing policies on storage of reconstituted parenteral products (American Society of Health-System Pharmacists 2000). Patients need to be informed about specific storage requirements and expiration dates to assure sterility and stability.

ADMINISTRATION

Prior to administering these products, pharmacists will need to use caution in reviewing dosing regimens. A potential source of medication error is the variation in units of measure for the various products. Some products are dosed in micrograms/kilogram (μg/kg) rather than milligrams/kilogram (mg/kg). Dosage calculations need to be carefully checked to avoid potential errors. Biotech products frequently receive approval for new indications after they have been on the market for a few years. The dosing regimen for these indications may be different than the original indication. Therefore, it is important to confirm the diagnosis and indication for products with multiple indications and dosing regimens. For example, adalimumab is dosed at 40 mg subcutaneously every other week to treat rheumatoid arthritis. The initial dose for plaque psoriasis is 80 mg subcutaneously, followed by a weekly dose of 40 mg. The initial dose for Crohn's disease is 160 mg subcutaneously, given as 4 injections on day 1 or 2 injections/day over 2 consecutive days, followed by an 80 mg dose 2 weeks later and a weekly maintenance dose of 40 mg every other week beginning on day 29 (Abbott Laboratories 2011).

Another example of variations in dosing regimen is denosumab. For treatment of osteoporosis in postmenopausal females, the dose of Prolia® is 60 mg every month. For prevention of skeletal-related events in bone metastases from solid tumors, Xgeva™ is administered 120 mg every 4 weeks. The manufacturer recognizes the risk of errors in dosing and has given the product different names to help prevent mistakes in dosing regimens.

■ Routes of Administration

Biotech products are primarily administered parenterally although routes of administration may be used. For example, Pulmozyme (dornase alfa) is administered by inhalation (Genentech 1994). Some products may be given by either the intravenous or subcutaneous route, while others are restricted to the subcutaneous or intramuscular routes. In some cases, manufacturers have information on unapproved routes of administration or other unpublished information that may be available by contacting the individual manufacturer. In any case, the manufacturer should always be consulted in order to obtain supporting evidence for a particular route that is not approved, but may be more convenient for the patient. For example, G-CSF should be administered by the subcutaneous or intravenous route only, while GM-CSF is given by intravenous infusion, over a 2 h period (McEvoy 1993). Aldesleukin is approved for intravenous administration only. However, subcutaneous administration has been used by some as an unlabeled route of administration (McDermott et al. 2005). Erythropoietin should only be administered by the intravenous or subcutaneous routes (Amgen Inc 2007), while alteplase is only approved for the intravenous route (Genentech Inc 2011; McEvoy 2011). Alteplase has also been administered by the intracoronary, intra-arterial, and intraorbital routes as well (McEvoy 2011).

■ Filtration

Filtering biotech products is not generally recommended since most of these proteins will adhere to the filter. Some hospitals and home infusion companies routinely use in-line filters for all intravenous solutions to minimize the introduction of particulate matter into the patient. In the case of biotech products, they should be infused below the filter to avoid a potential decrease in the amount of drug delivered to the patient (Banga and Reddy 1994; Koeller et al. 1991). Some manufacturers recommend infusing products using an in-line low protein-binding filter (≤1.2 μm).

■ Flushing Solutions

Biotechnology products are usually flushed with either saline or dextrose 5 % in water. The product literature should be consulted and care should be taken to assure that the proper solution is used with each agent. In general, biotech drugs should not be administered with other drugs since, in most cases, data does not exist that demonstrates whether biotech products are compatible with other drugs or fluids.

■ Prophylaxis to Prevent Infusion Reactions

Some products have protocols to treat and/or prevent infusion reactions for repeat infusions. For example, the infliximab protocol to treat an infusion reaction includes reducing the infusion rate, initiating a normal saline infusion, use of symptomatic treatment (normally consisting of acetaminophen and diphenhydramine), and vital sign monitoring every 10 min until resolution of the reaction. For subsequent infusions, pretreatment with acetaminophen and diphenhydramine 90 min prior to the infusion is standard procedure. Patients who had severe reactions may receive corticosteroids (Abbott Laboratories 2011).

BIOSIMILARS: ALMOST TO MARKET

A huge potential exists for the development of biosimilar products or "generic" versions of existing biotech drugs. Factors driving the development of biosimilars include the growing number of products, the size of the biotech drug market, and the high cost of existing patent-protected biologic products. Biosimilar legislation and some biosimilar products have been available in Europe for several years. The present state of the regulatory aspects of biosimilars (through FDA and EMA) is dealt with in Chap. 11.

Making choices for health-care professionals is not new in the biotech market as it already contains several types of insulins, growth hormones, and second-generation products such as Aranesp and Neulasta. Pharmacists and formulary committees need to choose between a variety of biotech drugs produced in different cell lines with differences in physical properties but intended to produce the same therapeutic effect. The ability to achieve a similar therapeutic effect for patients with a particular chronic disease using a biosimilar product is only one important consideration of comparing biosimilar products to the innovator drug. Biosimilars will also differ from the innovator drug in the manufacturing process. For example, a different cell line may be used to produce the recombinant protein. It is possible that the innovator and biosimilar drug may therefore differ in the immunogenicity of the product. Patients may be more or less likely to develop an immune response to the biosimilar agent. Health professionals will need to be involved in the clinical trials, patient monitoring, and postmarketing surveillance of biosimilars to determine the interchangeability of products and the patient care considerations that may be involved in using biosimilar agents.

OUTPATIENT/HOME CARE ISSUES

As mentioned previously, the management of patients in the outpatient and home settings is now an accepted aspect of health-care delivery. Home infusion and specialty pharmacy services dispense all forms of parenteral and enteral products including biotech drugs. These pharmacies have grown exponentially in the last 20 years due to cost savings for third-party payers, technological advances that allow these services to occur in the home, and patient preference to be treated at home rather than an in-patient setting.

■ Patient Assessment and Education

Before a patient can be a candidate for home therapy, an assessment of the patient's capabilities must occur. The patient, family member, or caregiver will need to be able to administer the medication and comply with all of the storage, handling, and preparation requirements. If the patient is incapable, then a caregiver (usually a relative, spouse, or friend) needs to be recruited to assist the patient. The pharmacy staff or other health professional may also make home visits to assist the patient in these tasks. The use of aseptic technique is usually new to the patients and in some cases may be overwhelming. The health-care provider must be sure that the patient or caregiver is competent and willing to follow these procedures. Self-instructional guides on specific products may be available from the manufacturer and, if so, should be provided to the patient providing they have the proper equipment for viewing.

Proper storage facilities will need to be available in the patient's home as well as a clean area for preparation and administration. Ideally, the patient will be able to prepare each dose immediately prior to the time of administration. If this is not possible, the pharmacy will have to prepare prefilled syringes and provide appropriate storage and handling requirements to the patient. The patient will also need to be educated regarding the proper handling of the syringes as well as other required supplies and materials such as needles, syringes, and alcohol wipes. Proper disposal of these hazardous wastes must also be reviewed. Specific issues related to patient teaching include rotating injection sites, product handling, drug storage including transporting and traveling with biotech drugs, expiration dates, refrigeration, cleansing the injection site with alcohol, disposal of needles and syringes, potential adverse effects, and expected therapeutic outcomes.

■ Monitoring

For patients who receive biotech drug therapy in the home, it is particularly important that close patient monitoring occurs. This will require frequent phone calls to the patient and periodic home visits. Monitoring parameters should include adverse events, progress to expected outcomes, assessment of administration technique, review of storage and handling procedures, and adherence to aseptic technique.

REIMBURSEMENT

Reimbursement issues include third-party billing information and availability of forms, cost-sharing programs that limit the annual cost of therapy, financial assistance programs for patients who would otherwise have difficulty paying for therapy, and reimbursement assurance programs that are designed to remove reimbursement barriers when reimbursement has been denied. Any detailed discussion of reimbursement issues is beyond the scope of this book and is subject to practice location. This discussion will deal only with the availability of information to pharmacists to appropriately handle reimbursement for products and services in the United States.

Pharmacists need to know current third-party payment policies including those conditions under which insurance companies will disallow claims. Some examples include off-label prescribing or administration of the product in the home rather than administration in a hospital or physician's office. Prior authorization is usually required particularly with managed care or prepaid plans. Manufacturers will often assist the patient by contacting the carrier to verify coverage, providing sample prior-approval letters, and following up on claims to determine the claim's status, and continuing to follow the case until it is resolved.

Manufacturers can also provide information that may convince the third-party payer to reconsider a denied claim. Some companies will intervene with the third-party payer to evaluate the case for denial and provide additional clinical documentation or coding information and will follow the appeal to conclusion. Pharmacists can act as facilitators to get qualified patients enrolled in programs to provide free medication to those who have insufficient insurance coverage or are otherwise unable to purchase the therapy. Manufacturers' websites and toll-free numbers for reimbursement issues are provided in Table 9.2. Websites and toll-free numbers for some of the patient assistance programs are provided in Table 9.4. The Partnership for Patient Assistance website provides information on a variety of patient assistance programs as well as the requirements to qualify for various programs.

EDUCATIONAL MATERIALS

Therapy with biotech drugs is a rapidly growing, ever-changing area of therapeutics. Pharmacists need to keep abreast of current information about existing agents such as new indications, management of adverse effects, results of studies describing drug interactions, or changes in information regarding product stability and reconstitution. Pharmacists will also be interested in the status of new agents as they move through the FDA approval process. Some good periodical sources of practical information about products of biotechnology are listed in Table 9.5.

■ Educational Materials for Health Professionals

Manufacturer and specific product websites provide a variety of educational materials including continuing education programs for physicians, pharmacists, and nurses. These programs often focus on specific disease states as well as drug therapy. The programs sometimes include slides, videos, and brochures. Since most biotechnology products are parenteral products, several manufacturers have produced videotapes that show the proper procedure for product administration, stor-

Partnership for patient assistance	1-888-477-2669	www.pparx.org
Rx assist	1-401-729-3284	www.rxassist.org
Together Rx access	1-800-444-4106	www.togetherrxaccess.org
Needy meds	No phone number; large database of patient assistance programs	www.needymeds.org

Table 9.4 ■ Toll-free assistance numbers and websites for patient assistance programs.

Biotechnology Medicines in Development
Communications Division, Pharmaceutical Research and Manufacturers of America, Washington, D.C., 202-835-3400, updated approximately every 18 months
http://www.phrma.org/research/new-medicines
"*The Pink Sheet*"
Bridewater, NJ, published weekly
http://www.elsevierbi.com/publications/the-pink-sheet
BioWorld Today, Atlanta
Bioworld Publishing Group, newspaper, 5 issues per week; 800-688-2421
www.bioworld.com
Bio/Technology, New York
Nature Publishing Co., a monthly journal dealing with all aspects of biotechnology
www.nature.com/nbt
Genetic Engineering News, New York
GEN Publishing, bimonthly publication Tel. (914)-740-2200; Fax (914)-740-2201
www.genengnews.com

Table 9.5 ■ Information sources for current trends in biotechnology.

age, and handling. These instructional tapes are beneficial not just for patients but also for health professionals who may not be skilled in injection techniques.

■ Educational Materials for Patients

Detailed patient information booklets exist for many of the products both in print and by downloading from the Internet. Patient education materials can assist the patient and family members in learning more about his or her disease and how it will be treated. Education allows the patient to participate more actively in the therapy and to feel a greater level of control over the process. By contacting the manufacturer and acquiring patient educational materials, pharmacists can offer support to the patient in learning to use a new product. Many patients are already overwhelmed by dealing with a diagnosis of serious or chronic disease. Learning about a new therapy, especially if it involves the necessity of self-injection, can cause additional stress for the patient and family.

Most commercially available biotech drugs now have individual websites to provide updated information to patients. These sites usually contain the following types of information: disease background, reimbursement information, dosing information, references, frequently asked questions, administration and storage information, and information specifically for health professionals. These websites also offer tools such as journals for patients to record administration of doses and monitoring information to assist health professionals in following the patient's progress. The websites also refer patients to disease-related associations and organizations whose services include a link to local chapters, meetings, and support groups. These groups may provide support to the patient while he or she adjusts to the diagnosis and treatment of a potentially serious disease.

■ The Internet and Biotech Information

The Internet is a valuable site rich in up-to-date information concerning all aspects of pharmaceutical biotechnology. Sites include virtual libraries/catalogs, online journals (usually requiring a subscription), biomedical newsletters, and biotechnology-specific home pages. Since the number of biotech-related sites is constantly increasing, only a small sampling of sites of interest could be provided in Table 9.6.

CONCLUDING REMARKS

The handling of biotechnology products requires similar skills and techniques as required for the preparation of other parenteral drugs, but there are often different nuances to the handling, preparation, and administration of biotechnology-produced pharmaceuticals. The pharmacist can become an educator regarding the pharmaceutical aspects of biotechnology products and can serve as a valuable resource to other health-care professionals. In addition, biotech products give the pharmacist the opportunity to provide enhanced patient care services since patient education and monitoring is required. To carry out this role successfully, the pharmacist will need to keep abreast of new developments as new literature and products become available.

SELF-ASSESSMENT QUESTIONS

■ Questions

1. What are some of the causes of pharmacist reluctance to handling biotech products?
2. In what areas of study do pharmacists and pharmacy students need to engage to be best prepared to provide pharmaceutical care services to patients receiving biotechnology therapeutic agents?

Internet site	Type of site	Web address
A Doctor's Guide to the Internet	Biomedical news	www.docguide.com
BioCentury	Biotechnology industry new	www.biocentury.com
BioPharma	Database	www.biopharma.com
Genetic Engineering News	Online journal	www.genengnews.com
Nature Biotechnology	Online journal	www.nature.com/nbt
Pharmaceutical Research and Manufacturers of America	Professional organization	www.phrma.org
Reuter's Health Information Services, Inc.	Biomedical news	www.reutershealth.com

Table 9.6 ■ Examples of biotech-related Internet sites.

3. What resources are available to pharmacy practitioners to learn more about biotechnology and the drug products of biotechnology?
4. How do the storage requirements of biotech products differ from the majority of products pharmacists normally dispense?
5. What is the most common temperature for the storage of biotech pharmaceuticals?
6. Why is human serum albumin added to the solution of many biotech drugs?
7. Why should biotech products not be shaken when adding any diluent?
8. During travel, what precautions should also be observed with biotech products?
9. Should biotech products be filtered prior to administration?
10. What assessments must be done by the pharmacist before a patient can be considered a candidate for home therapy with a biotech product?
11. What types of professional services information are provided by manufacturers of biotech drugs?
12. What issues will the pharmacist need to consider when comparing innovator drugs to biosimilars?

■ Answers

1. Lack of understanding of the basics of biotechnology; lack of understanding of the therapeutics of recombinant protein products; unfamiliarity with the side effects and patient counseling information; lack of familiarity with the storage, handling, and reconstitution of proteins; and the difficulty of handling reimbursement issues.
2. Basic biotechnology/immunological methods; protein chemistry; therapeutics of biotechnology agents; and storage, handling, reconstitution, and administration of biotechnology products.
3. Biotechnology/immunology texts, continuing education programs, manufacturers' information and toll-free assistance, biotechnology-oriented journals, and the Internet.
4. The shelf-life of these products is often considerably shorter than has been the case with more traditional compounds. These products need to be kept at refrigerated temperatures. There are, of course, exceptions to this rule.
5. In general, most biotech products are shipped by the manufacturer in gel ice containers and need to be stored at 2–8 °C. Once reconstituted, they should be kept under refrigeration until just prior to use.
6. Most biotech products may adhere to either plastic or glass containers such as syringes and polyvinyl chloride (PVC) intravenous bags reducing effectiveness of the product. Human serum albumin is usually added to the solutions to prevent adherence.
7. Shaking may cause the product to breakdown (aggregation). Usually when this happens one can observe physical separation or frothing within the vial of liquid products.
8. They should be stored in insulated, cool containers. This can be accomplished by using ice packs to keep the biotech drug at the proper temperature in warmer climates, whereas the insulated container in colder climates may be all that is required. In fact, when traveling in subfreezing weather, the products should be protected from freezing.
9. Filtering biotech products is not generally recommended since most of the proteins will adhere to the filter.
10. Before a patient can be a candidate for home therapy, an assessment of the patient's capabilities must occur. The patient, family member, or caregiver will need to be able to inject the medication and comply with all of the storage, handling, and preparation requirements.
11. Medical information services provided by manufacturers of biotech drugs are similar to the product, medical and patient management services provided by drug companies for traditional drug products. Information provided via this service generally includes appropriate indications, side effects, contraindications to use, results of clinical trials, and investigational uses. Upon request, manufacturers can supply a product monograph

and selected research articles that provide valuable information about each product.

12. In addition to ensuring that the biosimilar drug produces the same therapeutic effect, differences in manufacturing that may affect the patient will need to be considered. The most significant of these is potential immunogenicity of the product.

REFERENCES

Abbott Laboratories (2011) Humira® full prescribing information. North Chicago

American Society of Health-System Pharmacists (2000) ASHP guidelines on quality assurance for pharmacy-prepared sterile products. Am J Health Syst Pharm 57(12):1150–1169

Amgen Inc. (2007) Neupogen® full prescribing information. Thousand Oaks

Amgen Inc. (2011) Epogen® full prescribing information. Thousand Oaks

Banga AK, Reddy IK (1994) Biotechnology drugs: pharmaceutical issues. Pharm Times 60:68–76

Bayer HealthCare (2010) Betaseron® full prescribing information. Montville

Caremark (2006) Trends Rx Report, Focus on specialty pharmacy. Northbrook

Catania P (2010) King guide to parenteral admixture. King Guide Publications, Inc. St Louis

Department of Health and Human Services (2010) CMS manual system, Pub 100–02 Medicare Benefit Policy, Transmittal 123

Fein AJ (2010) 2010–11 Economic report on retail and specialty pharmacies. Pembroke Consulting, Philadelphia

Genentech, Inc. (1994) Pulmozyme® full prescribing information. South San Francisco

Genentech, Inc. (2011) Activase® full prescribing information. South San Francisco

Genzyme Corporation (2009) Leukine® full prescribing information. Cambridge

Koeller J, Fields S (1991) The pharmacist's role with biotechnology products. The Upjohn Company, Kalamazoo

McDermott DF, Regan MM, Clark JI et al (2005) Randomized phase III trial of high-dose interleukin-2 versus subcutaneous interleukin-2 and interferon in patients with metastatic renal cell carcinoma. J Clin Oncol 23(1):133–141

McEvoy GK (2011) American hospital formulary service drug information. American Society of Health-System Pharmacists, Bethesda

Prometheus Therapeutics (2011) Proleukin® full prescribing information. Emeryville

Roche Laboratories (2005) Summary of product characteristics. Accessed at: http://www.rocheuk.com/ProductDB/Documents/rx/spc/Roferon-A_Cartridge_SPC.pdf. Accessed 31 Oct 2006

Schering Laboratories (2004) Intron® full prescribing information. Kenilworth, Schering Laboratories

Suchanek D (2005) The rise and role of specialty pharmacy. Biotechnol Healthc 2:31–35

FURTHER READING

See Tables 9.1, 9.5, and 9.6 for suggested readings

10 Economic Considerations in Medical Biotechnology

Eugene M. Kolassa and Tushar B. Padwal

INTRODUCTION

The biotechnology revolution has coincided with another revolution in health care: the emergence of finance and economics as major issues in the use and success of new medical technologies. Health care finance has become a major social issue in nearly every nation, and the evaluation and scrutiny of the pricing and value of new treatments has become an industry unto itself. The most tangible effect of this change is the establishment of the so-called third hurdle for approval of new agents in many nations, after proving safety and efficacy. Beyond the traditional requirements for demonstrating the efficacy and safety of new agents, some nations, and many private health-care systems, now demand data on the economic costs and benefits of new medicines. Although currently required only in a few countries, methods to extend similar prerequisites are being examined by the governments of most developed nations. Many managed care organizations in the USA now prefer that an economic dossier be submitted along with the clinical dossier to make coverage decisions.

The licensing of new agents in most non-US nations has traditionally been accompanied by a parallel process of price and reimbursement approval, and the development of an economic dossier has emerged as a means of securing the highest possible rates of reimbursement. In recent years, sets of economic guidelines have been developed and adopted by the regulatory authorities of several nations to assist them in their decisions to reimburse new products. As many of the products of biotechnology are used to treat costly disorders and the products themselves are often costly to discover and produce, these new agents have presented new problems to those charged with the financing of medical care delivery. The movement to require an economic rationale for the pricing of new agents brings new challenges to those developing such agents. These requirements also provide firms with new tools to help determine which new technologies will provide the most value to society as well as contribute the greatest financial returns to those developing and marketing the products.

THE VALUE OF A NEW MEDICAL TECHNOLOGY

The task of determining the value of a new agent should fall somewhere within the purview of the marketing function of a firm. Although some companies have established health-care economic capabilities within the clinical research structure of their organizations, it is essential that the group that addresses the value of a new product does so from the perspective of the market and not of the company or the research team. This is important for two reasons. First, evaluating the product candidate from the perspective of the user, and not from the team that is developing it, can minimize the bias that is inherent in evaluating one's own creations. Second, and most importantly, a market focus will move the evaluation away from the technical and scientifically interesting aspects of the product under evaluation and toward the real utility the product might bring to the medical care marketplace. Although the scientific, or purely clinical, aspects of a new product should never be ignored, when the time comes to measure the economic contribution of a new agent, those developing the new agent must move past these considerations. It is the tangible effects that a new treatment will have on the patient and the health-care system that determine its value, not the technology supporting it. The phrase to keep in mind is "value in use."

The importance of a marketing focus when evaluating the economic effects of a new agent, or product candidate, cannot be overstated. Failing to consider the

E.M. Kolassa, Ph.D. (✉) • T.B. Padwal
Department of Pharmacy, University of Mississippi, Oxford, MS, USA

Department of Pharmacy,
Medical Marketing Economics, LLC,
1200 Jefferson Ave, Suite 200, Oxford, MS 38655, USA
e-mail: mkolassa@m2econ.com

D.J.A. Crommelin, R.D. Sindelar, and B. Meibohm (eds.), *Pharmaceutical Biotechnology*,
DOI 10.1007/978-1-4614-6486-0_10,

product's value in use can result in overly optimistic expectations of sales performance and market acceptance. Marketing is often defined as the process of identifying and filling the needs of the market. If this is the case, then the developers of new pharmaceutical technologies must ask two questions: "What does the market need?" and "What does the market want?" Analysis of the pharmaceutical market in the first decade of the twenty-first century will show that the market needs and wants:

- Lower costs
- Controllable costs
- Predictable cost
- Improved outcomes

Note that this list does not include new therapeutic agents. From the perspective of many payers, authorities, clinicians, and buyers, a new agent, in and of itself, is a problem. The effort required to evaluate a new agent and prepare recommendations to adopt or reject it takes time away from other efforts. For many in the health-care delivery system, a new drug means more work—not that they are opposed to innovation, but newness in and of itself, regardless of the technology behind it, has no intrinsic value. The value of new technologies is in their efficiency and their ability to render results that are not available through other methods or at costs significantly lower than other interventions. Documenting and understanding the economic effects of new technologies on the various health-care systems help the firm to allocate its resources more appropriately, accelerate the adoption of new technologies into the health-care system, and reap the financial rewards of its innovation.

There are many different aspects of the term "value," depending upon the perspective of the individual or group evaluating a new product and the needs that are met by the product itself. When developing new medical technologies, it is useful to look to the market to determine the aspects of a product that could create and capture the greatest amount of value. Two products that have entered the market in recent years provide good examples of the different ways in which value is assessed.

Activase® (tPA) from Genentech, one of the first biotechnology entrants in health care; entered the market priced at nearly ten times the price level of streptokinase, its nearest competitor. This product, which is used solely in the hospital setting, significantly increased the cost of medical treatment of patients suffering myocardial infarctions. But the problems associated with streptokinase and the great urgency of need for treatments for acute infarctions were such that many cardiologists believed that any product that proved useful in this area would be worth the added cost. The hospitals, which in the USA are reimbursed on a capitated basis for the bulk of such procedures, were essentially forced to subsidize the use of the agent, as they were unable to pass the added cost of tPA to many of their patients' insurers. The pricing of the product created a significant controversy, but the sales of Activase and its successors have been growing consistently since its launch. The key driver of value for tPA has been, and continues to be, the urgency of the underlying condition. The ability of the product to reduce the rate of immediate mortality is what drives its value. Once the product became a standard of care, incidentally, reimbursement rates were increased to accommodate it, making its economic value positive to hospitals.

An early biotechnology product that delivered a different type of value is the colony-stimulating factor from Amgen (G-CSF, Neupogen®). which was priced well below its economic value. The product's primary benefit is in the reduction of serious infections in cancer patients, who often suffer significant decreases in white blood cells due to chemotherapy. By bolstering the white blood cell count, Neupogen allows oncologists to use more efficacious doses of cytotoxic oncology agents while decreasing the rate of infection and subsequent hospitalization for cancer patients. It has been estimated that the use of Neupogen reduces the expected cost of treating infections by roughly $6,000 per cancer patient per course of therapy. At a price of roughly $1,400 per course of therapy, Neupogen not only provides better clinical care but also offers savings of approximately $4,600 per patient. The economic benefits of the product have helped it to gain use rapidly with significantly fewer restrictions than products such as tPA, whose economic value is not as readily apparent.

These two very successful products both provide clear clinical benefits, but their sources of value are quite different. The value of a new product may come from several sources, depending on the needs of clinicians and their perceptions of the situations in which they treat patients. Some current treatments bring risk, either because of the uncertainty of their effects on the patient (positive or negative) or because of the effort or cost required to use or understand the treatments. A new product that reduces this risk will be perceived as bringing new value to the market. In such cases, the new product removes or reduces some negative aspects of treatment. Neupogen, by reducing the chance of infection and reducing the average cost of treatment, brought new value to the marketplace in this manner.

Value can come from the enhancement of the positive aspects of treatment as well. A product that has a higher rate of efficacy than current therapies is the most obvious example of such a case. But any product that provides benefits in an area of critical need, where

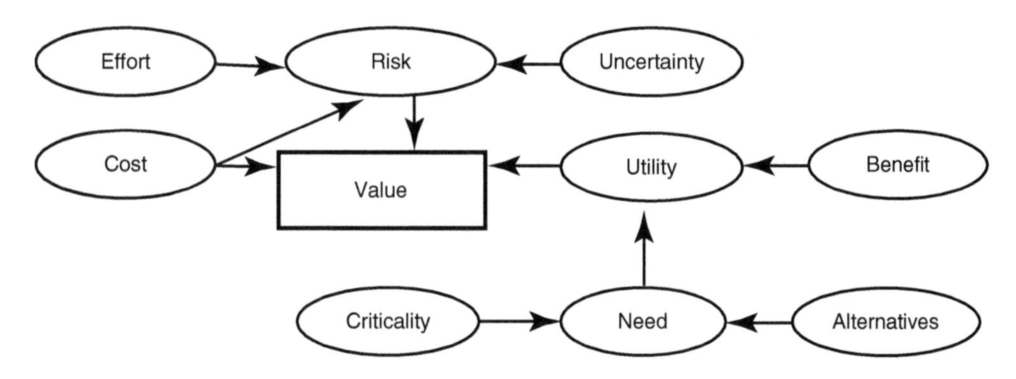

Figure 10.1 ■ Generalized model of value (Copyright © 2003, Medical Marketing Economics, LLC, Oxford, MS).

few or no current treatments are available, will be seen as providing immediate value. This was, and remains, the case for tPA.

Any new product under development should be evaluated with these aspects of value in mind. A generalized model of value, presented in Fig. 10.1 below, can be used to determine the areas of greatest need in the marketplace for a new agent and to provide guidance in product development. By talking with clinicians, patients, and others involved in current treatments and keeping this model in mind, the shortcomings of those current approaches can be evaluated and the sources of new incremental value can be determined.

Understanding the source of the value brought to the market by a new product is crucial to the development of the eventual marketing strategy. Using Fig. 10.1 as a guide, the potential sources of value can be determined for a product candidate and appropriate studies, both clinical and economic, can be designed to measure and demonstrate that value.

AN OVERVIEW OF ECONOMIC ANALYSIS FOR NEW TECHNOLOGIES

A thorough economic analysis should be used to guide the clinical research protocol to ensure that the end points measured are commercially relevant and useful. The analysis should describe important elements of the market to the firm, helping decision makers to understand the way decisions are made and providing guidance in affecting those decisions. Later, the results of economic analyses should inform and guide marketing and pricing decisions as the product is prepared for launch as well as help customers to use the product efficiently and effectively.

To prepare a thorough economic analysis, the researchers must first have a comprehensive understanding of the flow of patients, services, goods, and money through the various health-care systems. This process should begin as soon as the likely indications for a new product have been identified and continue throughout the product's development. The first step is to create basic economic models of the current treatment for the disorder(s) for which the product is likely to be indicated. This step will be used to provide better information to fine-tune financial assumptions and to provide critical input into the clinical development process to assure that the clinical protocols are designed to extract the greatest commercial potential from the product. If the product is likely to be used for more than one indication and/or if there is the potential that several different levels of the same indication (e.g., mild, moderate, and severe) would be treated by the same product, separate models should be prepared for each indication and level.

The purpose of the basic model is to provide a greater understanding of the costs associated with the disorder and to identify areas and types of cost that provide the greatest potential for the product to generate cost savings. For example, the cost of a disorder that currently requires a significant amount of laboratory testing offers the potential for savings, and thus better pricing, if the new product can reduce or eliminate the need for tests. Similarly, some indications are well treated, but the incidence of side effects is sufficiently high to warrant special attention. When developing a new agent, it is as important to understand the source of the value to be provided as it is to understand the clinical effects of the agent.

PHARMACOECONOMICS

The field of economic evaluation of medical technologies goes by several names, depending on the discipline of the researchers undertaking the study and the type of technology being measured. For pharmaceutical and biotechnology products, the field has settled on the name of pharmacoeconomics, and an entire discipline has emerged to fill the needs of the area. Contributions to the development of the field have come from several disciplines, including economics, pharmacy administration, and many of the behavioral sciences.

Pharmacoeconomics has been defined as "the description and analysis of the costs of drug therapy to the health care systems and society" (Townsend 1987). Clinical studies assess the efficacy of a biotechnology product; likewise, pharmacoeconomic studies help to evaluate the efficiency of biotechnologically derived drug. In a complete pharmacoeconomic assessment, both the costs and consequences are identified, measured, and compared with other available medical interventions. The increase in the health-care expenditure in the United States has resulted in excessive demand for cost-containment measures. Managed care organizations are striving hard to control drug spending and other health-care-related costs. Payers are moving from an open formulary system to a more closed formulary system, leading to additional emphasis on pharmacoeconomic assessment.

Importance of Pharmacoeconomics

To understand the importance of pharmacoeconomics in the biotechnology industry, it is necessary to understand the differences between the biotechnology products and traditional pharmaceutical products. Szcus and Schneeweiss (2003) have highlighted these differences. They observed that biotechnologically derived products are more expensive than traditional pharmaceutical products and that many biotechnology products are termed "orphan drugs" as they are used in small- or moderate-size patient populations. At times these products could be the only option to treat underlying disease condition. Given the high production costs and selling prices of biotechnology products, it is critical for these products to demonstrate adequate cost-effectiveness to justify their high cost. Therefore, pharmacoeconomics analysis is one of the major tools for payers to differentiate between a high-priced traditional pharmaceutical products and costly biotechnology products in certain instances.

Pharmacoeconomic analysis plays a crucial role in disease management. Chang and Nash (1998) outlined the role of pharmacoeconomics in disease management, which includes evaluation and identification of cost-effective medications for the treatment of particular disease conditions. This information can be and is used by payers and hospital personnel to make potential formulary decisions. In such instances, drugs with unfavorable pharmacoeconomics evaluations are unlikely to remain on formularies or will be moved to a restricted status. In addition to formulary decisions, disease management programs often include clinical guidelines that are designed primarily on cost-effectiveness of medications Joshnson and Nash (1996). When communicated properly, economic analysis can lead physicians to change their prescribing behavior thus decreasing unexplained variation in the treatment of the same disease. Walkom and colleagues (2006) studied the role of pharmacoeconomics in formulary decision making and found growing importance of pharmacoeconomic evaluations in formulary decision making.

When used appropriately, pharmacoeconomics analysis should help us to answer questions such as:

- What drugs should be included in the outpatient formulary?
- Should these same drugs be included on a hospital formulary?
- What is the best drug for particular disease in terms of efficacy and cost?
- What is the best drug for a pharmaceutical manufacturer to invest time and money?
- What are the relative cost and benefits of comparable treatment options?

To address the above questions, it becomes necessary for us to understand different costs considered in pharmacoeconomics analysis and the underlying techniques used to perform these pharmacoeconomic evaluations.

Understanding Costs

A comprehensive evaluation of relevant cost and consequences differentiates pharmacoeconomics from traditional cost-containment strategies and drug use evaluations. Costs are defined as the value of the resource consumed by a program or treatment alternative. Health economists use different costs in pharmacoeconomic evaluations, which can be grouped under direct costs, indirect costs, intangible costs, and opportunity costs.

In pharmacoeconomic evaluations, a comparison of two or more treatments extends beyond a simplistic comparison of drug acquisition cost. Including different costs, when appropriate, provides a more accurate estimate of the total economic impact of treatment alternatives and disease management programs in distinguished patients or populations

Direct Costs

Direct costs are the resources consumed in the prevention or treatment of a disease. The direct costs are further divided into direct medical costs and direct nonmedical costs.

The direct medical costs include expenditures on drugs, medical equipments, laboratory testing, hospital supplies, physician visits, and hospitalization costs. Direct medical costs could be further divided into fixed costs and variable costs. Fixed costs generally represent the overhead costs and are relatively constant. Fixed costs include expenditures on rent, utilities, insurance, accounting, and other administrative activities. These costs are often not included in the pharmacoeconomic evaluations because their use or total cost is unlikely to change as a direct result of a specific intervention. On the other hand, variable costs are an integral part of pharmacoeconomic analysis. Variable costs include drugs, fees for professional services, and supplies. These variable costs increase or decrease depending on the volume.

Direct nonmedical costs are out-of-pocket costs paid by patients (or their caregivers) for nonmedical services which are generally outside healthcare sector. Direct nonmedical costs included expenditure on transportation to and from the hospital, clinic or physician office, additional trips to emergency rooms, expenses on special diet, family care expenses, and other various forms of out-of-pocket expenses.

Indirect Costs

Indirect costs are those costs that result from morbidity or mortality. Indirect costs assess the overall economic impact of an illness on a patient's life. Typical indirect costs include the loss of earnings due to temporary or permanent disability, loss of income to family member who gave up their job temporarily or permanently to take care of patient, and loss in productivity due to illness. Indirect medical costs are more related to patients and often unknown to or unappreciated by providers and payers.

Intangible Costs

Intangible costs are the most difficult to quantify in monetary terms. These costs represent the nonfinancial outcome of disease and medical care. The examples of intangible costs include pain, suffering, and emotional disturbance due to underlying conditions. Though these costs are identified in an economic analysis, they are not formally calculated. At times intangible costs are converted into a common unit of outcome measurement such as a quality-adjusted life-year (QALY).

Opportunity Costs

Opportunity costs are often discussed in the economic literature. Opportunity cost is defined as the value of the alternative that was forgone. In simple terms, suppose a person spends $100 to buy a drug to treat a particular disease condition, then the opportunity to use the same $100 to obtain a different medical intervention or treatment for the same disease condition, or for some nonmedical purpose, is lost. This is referred to as an opportunity cost. Although not often included in traditional pharmacoeconomic analysis, opportunity costs are often considered implicitly by patients when cost sharing (e.g., co-pays and coinsurance) is increased in a health benefit plan.

UNDERSTANDING PHARMACOECONOMIC METHODS

The purpose of this section is to provide an overview of pharmacoeconomic techniques currently used to evaluate drugs or treatment options. Table 10.1 represents the list of pharmacoeconomic methods. Selection of a particular technique depends on the objective of the study and outcome units which are compared. Grauer and colleagues (2003) stated that "the fundamental task of economic evaluation is to identify, measure, value and compare the costs and consequences of the alternatives being considered."

■ Cost of Illness

Cost of illness analysis is an important pharmacoeconomic tool to examine the economic burden of a particular disease. This technique takes into consideration the direct and indirect costs of a particular disease. A cost of illness analysis thus identifies the overall cost of a particular disease in a defined population. Bootman, Townsend, and McGhan (1991) argue that cost of illness analysis helps to evaluate the humanistic impact of disease and quantify the resources used in the treatment of disease prior to the discovery of new intervention. This information could be effectively used by pharmacoeconomic researchers to establish a baseline for comparison of new treatment or intervention. Cost of illness analysis is not used to compare two alternative treatment options, but to estimate the financial burden of the disease under consideration. Thus, the monetary benefits of prevention and treatment strategies could be measured against the baseline value estimated by cost of illness. In essence, a cost of illness analysis provides the foundation for the measurement of the economic consequences of any treatment for the disorder in question. For example, Segel points out (2006) that a

Table 10.1 ■ Economic evaluation methodologies.

Method	Cost unit	Outcome unit
Cost of illness	Currency	Not assessed
Cost-minimization	Currency	Assumed to be equivalent in comparative groups
Cost-benefit	Currency	Currency
Cost-effectiveness	Currency	Natural units (life-years gained, mg/dl, blood glucose, mm Hg blood pressure)
Cost-utility	Currency	Quality-adjusted life-years or other utility

study on the cost-effectiveness of donepezil, published in 1999, relied on a cost of illness study of Alzheimer's disease published a few years earlier. Without the initial COI study the cost-effectiveness work would have been exponentially more difficult and costly.

■ Cost-Minimization Analysis

Cost-minimization analysis is the simplest pharmacoeconomic evaluation technique. The primary objective of the cost-minimization analysis is to determine the least costly alternative. Cost-minimization analysis is used to compare two or more treatment alternatives that are equal in efficacy. An example of cost-minimization analysis would be a comparison of branded product to a generic equivalent. It is assumed that the outcomes associated with the two drugs are equivalent; therefore, costs alone could be compared directly. The cost included in this economic evaluation must extend beyond drugs acquisition cost and should include all relevant costs incurred for preparing and administering drugs.

Argenta et al. (2011) performed a cost-minimization analysis to evaluate the direct costs of venous thromboembolism treatment with unfractionated heparin (UFH) and enoxaparin from the institutional perspective. The drug acquisition costs, laboratory tests, hospitalization costs, and drug administration costs were included to estimate the medical cost. Statistically nonsignificant differences were observed between unfractionated heparin and enoxaparin groups in the number of bleeding events, blood transfusion, and death. The daily cost per patient for UHF was $12.63 and for enoxaparin was $9.87. Depending on the mean time of use, the total cost for UHF was $88.39 as compared to $69.11 for enoxaparin. Therefore, it was concluded that enoxaparin provided higher cost saving as compared to unfractionated heparin for the treatment of hospitalized patients with venous thromboembolism.

■ Cost-Benefit Analysis

In cost-benefit analysis (CBA), costs and benefits are both measured in currency. In a CBA all the benefits obtained from the program or intervention are converted into some currency value (e.g., US dollars or euros). Likewise all program costs are identified and assigned a specific currency value. At times the costs are discounted to their present value. To determine the cost-benefit of a program, the costs are subtracted from the benefit. If the net benefit value is positive, then it can be concluded that the program is worth undertaking from an economic perspective.

The results of the cost-benefit analysis could be expressed either as cost-benefit ratio or a net benefit. For example, a cost associated with a medical program is $1,000, and the outcome/benefit resulting from the program is $9,000. Therefore, subtracting the cost $1,000 from the benefits $9,000 will yield net benefit of $8,000. When comparing many treatment alternatives, an alternative with the greatest net benefit could be considered as most efficient in terms of use of resources. In CBA, all costs and benefits resulting from the program should be included.

A typical use of CBA is in the decision of whether a national health benefit should include the administration of a specific vaccine. In this case the cost of vaccinating the population and treating a smaller number of cases of the disease would be compared with the costs that would be incurred if the disease were not to be prevented. At times, however, it is much more difficult to assign a monetary value to benefits. For example, the benefit of a patient's satisfaction with the treatment or improvement in patient's quality of life is very difficult to convert to a monetary sum. This presents a considerable problem. At times these variables are considered as "intangible benefits," and the decision is left to the researcher to include in final analysis. Because of this CBA is seldom used as a pharmacoeconomic method to evaluate a specific treatment, although many who perform different types of analyses often mistakenly refer to their work as "cost-benefit."

■ Cost-Effectiveness Analysis (CEA)

Cost-effectiveness analysis is a method used to compare treatment alternatives or programs where cost is measured in currency and outcomes/consequences are measured in units of effectiveness or natural units. Therefore, cost-effectiveness analysis helps to establish and promote the most efficient drug therapy for the treatment of particular disease condition. The results of cost-effectiveness analysis are expressed as average cost-effectiveness ratios or as the incremental cost of one alternative over another. CEA is useful in comparing different therapies that have the same outcome units, such as an increase in life expectancy or decrease in blood pressure for hypertension drugs. CEA is a frequently used tool for evaluating different drug therapies to treat a particular disease condition. This type of analysis helps in determining the optimal alternative, which is not always the least costly alternative. CEA has an advantage that it does not require the conversion of health outcomes to monetary units.

CEA is often used to guide formulary management decisions. For example, consider a biotechnology product "X" that provides a 90 % efficacy or cure rate for a specific disorder. The total treatment cost for 100 patients with product X is $750,000. Likewise assume that another biotechnology product "Y" prescribed for the same disorder shows 95 % efficacy; however, the treatment cost of 100 patients with product Y is $1,000,000. The average cost-effectiveness ratio (ACER) of product X is calculated by dividing the cost $750,000 by the outcome, 90 cures, to yield an ACER of $8,333 per cure. Similarly the ACER for product Y is $10,526 per cure. From this analysis, it is evident that using product Y would cost an additional $2,192 per cure, which is the difference between ACERs of product Y and product X.

At times the incremental cost-effectiveness ratio (ICER) is important in drug selection decisions. From the above example, to calculate the ICER, total cost of product Y ($1,000,000) is subtracted from total cost of product X ($750,000). This is then divided by the cures from product X, 90, subtracted from the cures resulting for product Y, 95. Therefore, the incremental cost for each additional cure with product Y is $250,000 divided by 5 cures or $50,000 per cure. The incremental cost-effectiveness ratio poses the question of whether one additional cure is worth spending $50,000. The additional cost of cure might be justified by the severity of the disease or condition; this is a decision that is best made with the full knowledge of the economic implications. This provides an example of a situation in which the economic analysis is used to help guide the decision, but not to make the decision. Table 10.2 represents the ICER for product X and Y.

Raftery and colleagues (2007) conducted an interesting cost-effectiveness analysis to compare ranibizumab (Lucentis) and bevacizumab (Avastin) for the treatment of age-related macular degeneration (AMD). It is interesting to note that both drugs are produced and marketed by a single manufacturer. Lucentis was licensed for AMD in the USA in 2006 and in EU in 2007. The cost of Lucentis is $2,000 per injection. Avastin is licensed for cancer treatment but is not approved for AMD. The cost of Avastin is $17–$50 per injection. Avastin is used off-label for the treatment of AMD. Roche/Genentech owns both drugs and has stated no plan to license Avastin for AMD. Researchers developed a model to assess the cost-effectiveness of these drugs for the treatment of AMD. The price ratio of Lucentis to Avastin was 39:1, based on the US price of US$1,950 per injection for Lucentis and US$50 per injection for Avastin. In the model the efficacy of Avastin relative to Lucentis was varied from 0.1 to 0.9, to test for the sensitivity of the price differences to efficacy differences. The results showed that the efficacy of Avastin would need to be very low for Lucentis to demonstrate cost-effectiveness for AMD. It was also observed that only when the relative efficacy was reduced to 0.4 the cost per QALY (quality-adjusted life-years) fell to £31,092, which is a cost-utility analysis threshold in the UK. At an efficacy rate of 0.8 the cost per QAYL was well over £100,000, which is significantly higher than the NICE (National Institute for Health and Clinical Excellence) threshold. Therefore, it was concluded that Lucentis is not cost-effective compared with Avastin at current prices unless it is at least 2.5 times more efficacious. The observational studies of Avastin have showed comparable efficacy between the two products (Avery et al. 2006).

■ Cost-Utility Analysis (CUA)

Cost-utility analysis was developed to factor quality of life into economic analysis by comparing the cost of the therapy/intervention with the outcomes measured in quality-adjusted life-years (QALY). The quality-adjusted life-years are calculated by multiplying the length of time in a specific health state by the utility of that health state—the utility of a specific health state is, in essence, the desirability of life in a specific health state compared with life in perfect health. A utility rating of 0.9 would mean that the health state in question is 90 % as desirable as perfect health, while a utility rating of 0.5 would mean that health state is only half a desirable. Death is given a utility score of 0.0. The results of cost-utility analysis are expressed in terms of cost per quality-adjusted life-years gained as a result of given treatment/intervention. CUA is beneficial when comparing therapies that produce improvements in different or multiple health outcomes. Cost per QALY can be measured and evaluated across several different treatment scenarios, allowing for comparisons of disparate therapies.

Goulart and Ramsey (2011) evaluated cost utility of bevacizumab (Avastin) and chemotherapy versus chemotherapy alone for the treatment of advanced non-small-cell lung cancer (NSCLC). Bevacizumab is currently approved for the treatment of NSCLC in combination with chemotherapy based on 2 months median survival proved in clinical trials. Researchers developed a model to determine quality-adjusted life-years and direct medical cost incurred due to treatment with bevacizumab in combination with chemotherapy. The utilities used in calculating QALY were obtained

Product	Efficacy (%)	No. of patients treated	Total costs ($)	ACER ($)	ICER ($)
Product X	90	100	750,000	8,333	50,000
Product Y	95	100	1,000,000	10,526	

Table 10.2 ■ Incremental Cost Effectiveness Ratio (ICER) for two products.

from the literature and costs were obtained from Medicare. The results of the study showed that bevacizumab is not cost-effective when added to chemotherapy. It was found that bevacizumab with chemotherapy increased the mean QALYs by only 0.13 (roughly the equivalent of 1.5 months of perfect health), at an incremental lifetime cost of US$72,000 per patient. The incremental cost-utility ratio (ICUR) was found to be US$560,000/QAYL. The results of these analyses could be potentially used by payers while allocating resources for the treatment of NSCLC care. Table 10.3 represents the base case results of cost-utility analysis.

The economic value of the product may have elements besides the basic economic efficiency implied by the break-even level just discussed. Quality differences, in terms of reduced side effects, greater efficacy, or other substantive factors, can result in increases in value beyond the break-even point calculated in a simple cost comparison. Should these factors be present, it is crucial to capture their value in the price of the product, but how much value should be captured?

It is important to recognize that a product can provide a significant economic benefit in one indication but none in another. Therefore, it is prudent to perform these studies on all indications considered for a new product. A case in point is that of epoetin alfa (EPO). EPO was initially developed and approved for use in dialysis patients, where its principle benefit is to reduce, or even eliminate, the need for transfusion. Studies have shown that EPO doses that drive hematocrit levels to between 33 and 36 % result in significantly lower total patient care costs than lower doses of EPO or none at all (Collins et al. 2000). The same product, when used to reduce the need for transfusion in elective surgery, however, has been shown not to be cost-effective (Coyle et al. 1999). Although EPO was shown to reduce the need for transfusion in this study, the cost of the drug far outweighed the savings from reduced transfusions as well as reductions in the transmission and treatment of blood-borne pathogens. Economic efficiency is not automatically transferred from one indication to another.

The lack of economic savings in the surgical indication does not necessarily mean that the product should not be used, only that users must recognize that in this indication use results in substantially higher costs while in dialysis it actually reduces the total cost of care.

Payers within the US healthcare system have begun to use similar methods of evaluation. Although it cannot be stated with certainty that the US system will adopt this approach to coverage wholeheartedly, the consistent news reports of new drugs costing tens, and hundreds, of thousands of dollars would indicate that the importance of delivering demonstrable value will increase in that market as well.

In the pharmaceutical marketing environment of the foreseeable future, it is wise to first consider determining the true medical need for the intervention. Then, if the need is real, to consider surrendering some

Table 10.3 ■ Base case results of Cost-Utility analysis.

Outcomes	CPB	CP	Differences
Effectiveness			
Life expectancy (years)	1.24	1.01	0.23
Progression-free survival (years)	0.72	0.47	0.25
QALYs	0.66	0.53	0.13
*Lifetime costs per patients (US$)**			
Drug utilization	70,284.75	646.96	69,637.79
Drug administration	4,239.87	1,495.24	2,744.63
Fever and neutropenia	25.32	4.37	20.95
Severe bleeding	19.65	1.33	18.32
Other adverse events	39.06	32.09	6.97
Outpatient visits	1,017.90	609.41	408.49
Progressive disease	40,283.71	41,500.96	−1,217.25
Total	115,910.26	44,290.36	71,619.90
ICER (US$/life-years gained)			308,981.58
ICUR (US$/QALY gained)			**559,609.48**

*Cost in 2010 US dollars

CP carboplatin and paclitaxel, *CPB* carboplatin, paclitaxel, and bevacizumab, *ICER* incremental cost-effectiveness ratio, *ICUR* incremental cost-utility ratio, *QALYs* quality-adjusted life-years value to the market—pricing of the product at some point below its full economic value. This is appealing for several reasons:

- The measurement of economics is imprecise and the margin for error can be large.
- If the market is looking for lower costs, filling that need enhances the market potential of the product.
- From a public relations and public policy perspective, launching a new product with the message that it provides savings to the system can also provide positive press and greater awareness.

CONCLUSIONS

As societies continue to focus on the cost of health-care interventions, we must all be concerned about the economic and clinical implications of the products we bring into the system. Delivering value, in the form of improved outcomes, economic savings, or both, is an important part of the pharmaceutical science. Understanding the value that is delivered and the different ways in which it can be measured should be the responsibility of everyone involved with new product development.

REFERENCES

Argenta C, Ferreira MA, Sander GB, Moreira LB (2011) Short-term therapy with enoxaparin or unfractionated heparin for venous thromboembolism in hospitalized patients: utilization study and cost-minimization analysis. Value Health 14:S89–S92

Avery RL, Pieramici DJ, Rabena MD et al (2006) Intravitreal bevacizumab (Avastin) for neovascular age-related macular degeneration. Ophthalmology 113(3):363–372

Bootman JL, Townsend JT, McGhan WF (1991) Principles of Pharmacoeconomics, Harvey Whitney Books

Chang K, Nash D (1998) The role of pharmacoeconomic evaluations in disease management. Pharmacoeconomics 14(1):11–17

Collins AJ, Li S, Ebben J, Ma JZ, Manning W (2000) Hematocrit levels and associated medicare expenditures. Am J Kidney Dis 36(2):282–293

Coyle D, Lee K, Laupacis A, Fergusson D (1999) Economic analysis of erythropoietin in surgery. Canadian Coordinating Office for Health Technology Assessment, Ottawa

Goulart B, Ramsey S (2011) A trial-based assessment of the utility of Bevacizumab and chemotherapy versus chemotherapy alone for advanced non-small cell lung cancer. Value Health 14:836–845

Grauer DW, Lee J, Odom TD, Osterhaus JT, Sanchez LA (2003) Pharmacoeconomics and outcomes: applications for patient care, 2nd edn. American college of clinical pharmacy, Kansas City

Joshnson N, Nash D (eds) (1996) The role of pharmacoeconomics in outcomes management. American Hospital Publishing, Chicago

Raftery J, Clegg A, Jones J, Chuen S et al (2007) Ranibizumab (Lucentis) versus bevacizumab (Avastin): modeling cost effectiveness. Br J Ophthalmol 91:1244–1246

Segel JE. Cost-of-Illness Studies—A Primer, RTI-UNC Center of Excellence in Health Promotion Economics, Jan 2006. Available at: http://www.rti.org/pubs/coi_primer.pdf

Szcus TD, Schneeweiss S (2003) Pharmacoeconomics and its role in the growth of the biotechnology industry. J Commercial Biotechnol 10(2):111–122

Townsend R (1987) Postmarketing drug research and development. Drug Intell Clin Pharm 21(1pt 2):134–136

Walkom E, Robertson J, Newby D, Pillay T (2006) The role of pharmacoeconomics in formulary decision-making: considerations for hospital and managed care pharmacy and therapeutics committees. Formulary 41:374–386

FURTHER READING

Further Reading on Pharmacoeconomic Methods and Pricing Issues

Bonk RJ (1999) Pharmacoeconomics in perspective. Pharmaceutical Products Press, New York

Drummond MF, O'Brien BJ, Stoddart GL, Torrance GW (1997) Methods for the economic evaluation of health care programmes, 2nd edn. Oxford University Press, Oxford

Kolassa EM (1997) Elements of pharmaceutical pricing. Pharmaceutical Products Press, New York

Further Reading on Pharmacoeconomics of Biotechnology Drugs

Dana WJ, Farthing K (1998) The pharmacoeconomics of high-cost biotechnology products. Pharm Pract Manag Q 18(2):23–31

Hui JW, Yee GC (1998a) Pharmacoeconomics of biotechnology drugs (part 1 of 2). J Am Pharm Assoc 38(1):93–97

Hui JW, Yee GC (1998b) Pharmacoeconomics of biotechnology drugs (part 2 of 2). J Am Pharm Assoc 38(2):231–233

Reeder CE (1995) Pharmacoeconomics and health care outcomes management: focus on biotechnology (special supplement). Am J Health Syst Pharm 52(19,S4):S1–S28

11 Regulatory Framework for Biosimilars

Vinod P. Shah and Daan J.A. Crommelin

INTRODUCTION

The term "biopharmaceuticals" is used to describe biotechnologically derived drug products. Biopharmaceuticals are protein-based macromolecules and include insulin, human growth hormone, the families of the cytokines and of the monoclonal antibodies antibody fragments, and nucleotide-based systems such as antisense oligonucleotides, siRNA, and DNA preparations for gene delivery. A "generic" version of biopharmaceuticals may be introduced after patent expiration of the innovator's product. However, the generic paradigm as it has developed for low molecular weight actives over the years cannot be used for biopharmaceuticals. In the European Union and the US regulatory systems, the term "biosimilar" was coined for copies of brand name, new biopharmaceuticals. Different than for small molecule generic versions, in most cases, comparative clinical testing of the biosimilar product must include a robust evaluation of safety and confirmation of efficacy in appropriate patient populations.

The aim of this chapter is to provide a comprehensive view on current regulatory policies related to the approval of a biosimilar product. Table 11.1 provides definitions of terms relevant to this chapter.

BACKGROUND

The mission of a regulatory authority is to "Assure that safe, effective, and high-quality drugs are marketed in the country and are available to the people." Safety of an innovator's drug product, be it a small molecule or biopharmaceutical, is established through preclinical studies in animals and controlled clinical studies in humans. Efficacy is established through clinical studies in patients. In order to have a better understanding of the regulatory process involved, it is essential to appreciate the basic difference between small molecule drugs and macromolecular biopharmaceuticals and their approval process (cf. Tables 11.2 and 11.3). Small molecules are chemically synthesized and can be fully characterized. On the other hand, most biopharmaceuticals are produced in a living system such as a microorganism or plant or animal cells and are difficult to fully characterize. The proteins are typically complex molecules and are unlikely to be shown to be structurally identical to a reference product. Differences in a manufacturing process may lead to alteration in the protein structure (Crommelin 2003). Protein structures can differ in at least three ways: in their primary amino acid sequence, in (posttranslational) modification to those amino acids sequences (e.g., glycosylation), and in higher-order structure (folding patterns). Advances in analytical sciences enable protein products to be extensively characterized with respect to their physicochemical and biological properties. However, current methodology may not detect all relevant structural and functional differences between two proteins (Chap. 2).

For the approval of a (small molecule) generic product, it must be pharmaceutically equivalent (same dosage form, strength of active, route of administration, and labeling as brand-name drug) and bioequivalent. Depending on the active, it should have the same in vitro dissolution, pharmacokinetic, pharmacodynamic, and clinical outcome profile as the brand name, innovator's drug. For biological products, such a simple assessment of pharmaceutical equivalence and bioequivalence alone is not an option.

Both for a generic and for a biosimilar product, a complete set of information on Chemistry, Manufacturing, and Controls (CMC section) is needed in the dossier submitted for approval to ensure that the drug substance and the drug product are pure, potent,

V.P. Shah, Ph.D. (✉)
VPS Consulting, LLC, 11309 Dunleith Place,
North Potomac, MD 20878, USA
e-mail: dr.vpshah@comcast.net; dr.vpshah@gmail.com

D.J.A. Crommelin, Ph.D.
Department of Pharmaceutical Sciences,
Utrecht Institute for Pharmaceutical Sciences,
Utrecht University, Arthur van Schendelstraat 98,
Utrecht 3511 ME, The Netherlands
e-mail: d.j.a.crommelin@uu.nl

D.J.A. Crommelin, R.D. Sindelar, and B. Meibohm (eds.), *Pharmaceutical Biotechnology*,
DOI 10.1007/978-1-4614-6486-0_11,

FDA	US Food Drug Administration
EMA	European Medicines Agency
Low molecular weight drug	Classical medicinal product prepared by chemical synthesis
Generic product	Non-patented medicinal product of low molecular weight and therapeutically equivalent
Biopharmaceutical drug	The term "biopharmaceuticals" is used to describe biotechnologically derived drug products. Biopharmaceuticals are protein-based macromolecules and include insulin, human growth hormone, the families of the cytokines and of the monoclonal antibodies, antibody fragments, and nucleotide-based systems such as antisense oligonucleotides, siRNA, and DNA preparations for gene delivery[a]
Biosimilar product	A biosimilar is a biological product that is highly similar to an already approved biological product, notwithstanding minor differences in clinically inactive components, and for which there are no clinically meaningful differences between the biosimilar and the approved biological product in terms of safety, purity, and potency
Second-generation biopharmaceuticals	The second-generation biopharmaceutical product is derived from an approved biopharmaceutical product which has been deliberately modified to change one or more of the product's characteristics

[a]For the FDA, "biopharmaceuticals" are part of the "biological product" group, which also included viruses, sera, vaccines, and blood products

Table 11.1 ■ Definitions.

Table 11.2 ■ Difference between small molecular weight drugs and biopharmaceuticals.

Small molecular weight drugs	Biopharmaceuticals
Low molecular weight	High molecular weight
Simple chemical structure	Complex three-dimensional structure
Chemically synthesized	Produced by living organism
Easy characterization	Difficult to impossible to fully characterize
Synthetically pure	Often heterogeneous
Rarely produce immune response	Prone to eliciting an immune response

Table 11.3 ■ Drug approval for generic product and biosimilar product (FDA).

Generic (small molecule) product	Biosimilar product
Drug Price Competition and Patent Restoration Act of 1984	Biologics Price Competition and Innovation Act of 2009
FDC Act 505(j)	PHS Act 351(k)
Pharmaceutical equivalent	Pharmaceutical equivalent
Bioequivalent	Non-clinical comparison (animal studies)
Pharmacokinetics	Physicochemical analysis
Pharmacodynamics	Clinical comparison (in humans)
Clinical comparison (not standard)	
In vitro analysis	

and of high quality. The CMC section should include full analytical characterization, a description of the manufacturing process and test methods, and stability data. In addition to the establishment of safety and efficacy of the drug product, the approval process requires manufacturing of the drug product under controlled current good manufacturing practice (cGMP) conditions. The cGMP requirement ensures identity, potency, purity, and quality of the final product.

For biosimilars, a full comparison with the innovator's product characteristics (e.g., the primary, secondary, tertiary and quaternary structure, post-translational modifications) and functional activity(ies) should be considered. A comprehensive understanding of all steps in the manufacturing process, process controls, and the use of a Quality-by-Design approach will facilitate consistent manufacturing of a high-quality product.

As stated before, for complex proteins, the full characterization and assessment of equality of the biosimilar and innovator's product is not possible with our present arsenal of analytical techniques (see Chap. 2). This means that for establishing biosimilarity, as a rule, clinical studies are required in the regulatory protocols described in the next section.

REGULATORY FRAMEWORK IN THE USA

FDA defines a generic drug as a copy that is the same as a brand-name drug in dosage form, strength, route of administration, quality, purity, safety, performance, and intended use. A generic product is a copy of the brand name, innovator's product, except for the inactive ingredients. The required dossier for market authorization focuses on only two aspects. The generic (small molecule) drug product should be pharmaceutically equivalent and bioequivalent to the brand-name product and is therefore therapeutically equivalent and interchangeable with the brand-name drug product. A generic product has the same active ingredient, and therefore, the safety and efficacy of the product is already established. The only question is the efficacy of the generic formulation, and this is assured by the bioequivalence study in healthy volunteers or patients.

In the case of small molecules, the identity of the active substance is established through a validated chemical synthesis route, full analysis of the active agent, impurity profiling, etc. In the case of biopharmaceuticals, this is, generally speaking, not possible. This biopharmaceutical product contains the active ingredient which is similar (but not necessarily equal) in characteristics to the reference product. For this reason, the generic biopharmaceutical products are referred to as biosimilar products. A biosimilar product has a pharmacological and therapeutic activity that is similar to the reference (comparator) product (FDA Guidance).

■ Scale-Up and Postapproval Changes (SUPAC) Concept and Comparability

Very often, a biopharmaceutical manufacturer has to make changes in the manufacturing process, e.g., changes in the formulation or a change in the manufacturing process. The regulatory regime is then similar to scale-up and postapproval changes (SUPAC) in small molecules. This requires assurance of pharmaceutical equivalence. At times, depending upon the nature of the change, a comparability study (see below) may be required. The FDA Guidance on Comparability is summarized in the insert Box 11.1. The "FDA Guidance on Comparability" protocol is used for assuring product quality of the approved product after certain changes have been made in the manufacturing process (comparability assessment, see below). The quality of these products is assured by chemical analysis and/or by using a comparability clinical study protocol (ICH 2005).

Box 11.1 ■ Key Points of the FDA Guidance on Comparability.

The concept of comparability studies is described in the FDA Guidance on Comparability.

- The FDA comparability study protocol is generally followed for changes *related to existing products within a given manufacturer after the drug product is approved.* The demonstration of comparability does not necessarily mean that the quality attributes of the pre-change and post-change products are identical but that they are similar enough and that the existing knowledge is sufficiently predictive to ensure that any differences in quality attributes have no adverse impact upon safety or efficacy of the drug product.
- Process changes among biotech products frequently include changing master cell banks. The changes in manufacturing process are judged by characterization studies, possible preclinical, pharmacokinetic, and pharmacodynamic and/or clinical studies. The principles of comparability are used to permit pioneer manufacturers to change the production process, cell line, formulations, and manufacturing site.
- This concept takes into consideration the unique features of protein products including complexity of the structures to allow flexibility in comparability testing.
- The comparison relies heavily on analytical characterization. A determination of comparability can be based on a combination of analytical testing, biological assays, and in some cases non-clinical and clinical data.
- The comparability is based on the expectation that the safety and efficacy expected for the product from a modified process/formulation or site change is similar to that from the original process.
- The principal objective of the comparability protocol is to reduce or eliminate the need for human clinical studies, especially after certain scale-up and postapproval changes in composition, manufacturing process, and change in manufacturing site. The comparability exercise is to demonstrate similarity.

A question often raised and debated by a generic manufacturer is whether the approach of using a comparability protocol can be extrapolated and adopted for approval of biosimilar products manufactured by a different manufacturer. The regulatory answer today is no. It cannot be used for the approval of a biosimilar product. This is based on the following way of reasoning. Demonstrating that a proposed product is biosimilar to a reference product is more complex than assessing the comparability of a product before and after manufacturing changes made by the same manufacturer. This is because a manufacturer who modifies its own manufacturing process has extensive knowledge and information about the product, process, controls, and acceptance parameters. In contrast, the manufacturer of a proposed biosimilar product has no direct knowledge of the manufacturing process of the reference product. The proposed product manufacturer may have a different cell line, different raw materials, equipment, process, controls, and acceptance criteria than the reference product manufacturer.

Biosimilars and Drug Product Approval

A biosimilar is a biological product that is highly similar to an already approved biological product, notwithstanding minor differences in clinically inactive components, and for which there are no clinically meaningful differences between the biosimilar and the approved biological product in terms of safety, purity, and potency. Approval of the biosimilar product will be based on scientific considerations in demonstrating biosimilarity to a reference product. The scientific considerations will be on the basis of a risk-based "totality-of-the evidence" approach in comparing the proposed (test) and the reference product. The requirements for approval of biosimilar products should be based on the structural complexity and clinical knowledge of and experience with the reference biopharmaceutical product. For example, products such as growth hormone have known and relatively simple chemical structures. In addition, extensive manufacturing and clinical experience are available for these products. On the other hand, recFactor VIII is a large, highly complex molecule with several isoforms. Because of the varying complexity of the biotech-derived products, the requirements for the approval process should be structured on a case-by-case basis. The following information is required for product approval:

- Structural information – primary, secondary, tertiary, and, if relevant, quaternary structure information, including information regarding the glycosylation pattern, if relevant.
- Manufacturing process
- Quality attributes and clinical activities
- Pharmacokinetic-pharmacodynamic information, mechanism of drug action
- Clinical experience, efficacy and toxicity information

Box 11.2 ■ FDA Routes for Biosimilars.

From a regulatory perspective, a copy of a biopharmaceutical product can be identified as a generic product. But in practice, it is unlikely, due to additional complexity, that complete equality can be demonstrated. The copy of a biopharmaceutical product is required to have a similar safety and efficacy profile as the brand-name/innovator product, and therefore, it is referred to as "biosimilar."

There are two pathways that can be followed for biosimilars: (1) following the Food, Drug, and Cosmetic Act (FDC) Act and (2) following the Public Health Service (PHS) Act.

According to the FDC Act:

- 505(b)(1) – Full reports of investigations of safety and efficacy are needed. This results in a full NDA, with right of reference. This means that information about safety and efficacy can be used by others to document safety and efficacy of the product.
- The PHS Act – "351(k)" application – For demonstrating that the therapeutic protein product is biosimilar to a reference product, using the "totality-of-the-evidence" (see below) approach. It describes a stepwise approach in the development of biosimilar products (e.g., rec. proteins).

Biosimilar products are different from second-generation biopharmaceuticals, e.g., pegylated G-CSF and interferon-alpha and darbepoetin. The second-generation biopharmaceuticals have improved pharmacological properties/biological activity compared to an already approved biopharmaceutical product which has been deliberately modified. The second-generation products are marketed with the claim of improved clinical superiority. The second-generation biopharmaceuticals require a full New Drug Application and are not interchangeable with the brand-name product.

The Biologics Price Competition and Innovation Act of 2009 (BPCI Act) was enacted as part of the Affordable Care Act on March 23, 2010. The BPCI Act creates an abbreviated licensure pathway for biological products shown to be biosimilar to, or interchangeable with, an FDA-licensed biological referenced product. The objectives of the BPCI Act are conceptually similar to those of the Drug Price Competition and Patent Term Restoration Act of 1984, commonly referred to as the Hatch-Waxman Act which established abbreviated pathways for the approval of drug products under the Federal Food, Drug, and Cosmetic Act. Section 351(k) of the PHS Act, added by the BPCI Act, sets forth the requirements for an application for a proposed biosimilar product and an application or a supplement for a proposed interchangeable product. A 351(k) application must contain information demonstrating that the biological product is biosimilar to a reference product based upon the data derived from analytical studies, animal studies, and clinical studies. To meet the *higher* standard of interchangeability, sufficient information must be provided to demonstrate biosimilarity. Interchangeable products may be substituted for the reference product without intervention of the prescribing health-care provider (note: this is the US interpretation of the term "interchangeable." It is interpreted differently in other parts of the world, e.g., in European countries). The BPCI Act also includes several exclusivity terms.

▪ Guidance Documents on Biosimilars

The regulatory process for biosimilars is an evolutionary process. February 2012, the FDA released three draft guidance documents on biosimilar product development: (1) scientific considerations in demonstrating biosimilarity to a reference product, (2) quality considerations in demonstrating biosimilarity to a reference protein product and biosimilars, and (3) questions and answers regarding drug implementation of the Biologics Price Competition and Innovation Act of 2009 (February 2012). The guidance is intended to assist sponsors in demonstrating that the proposed therapeutic protein product is biosimilar to a reference product under section 351(k) of the PHS Act.

▪ Scientific Considerations in Demonstrating Biosimilarity to a Reference Product

The draft guidance (February 2012) suggests the "totality-of-the-evidence" approach to determining biosimilarity. The guidance recommends a stepwise approach to demonstrate biosimilarity, which includes a comparison of the proposed product (test product) and the reference product with respect to structure, function, animal toxicity, human pharmacokinetics and pharmacodynamics, clinical immunogenicity, and clinical safety and effectiveness. Demonstrating that a proposed product is biosimilar to a reference product is more complex than assessing the comparability of a product before and after manufacturing changes made by the same manufacturer. As earlier stated, this is because a manufacturer who modifies his own manufacturing process has extensive knowledge and information about the product, process, controls, and acceptance parameters. In contrast, the manufacturer of a proposed biosimilar product has no direct knowledge of the manufacturing process of the reference product. The proposed product manufacturer may have a different cell line, use different raw materials, equipment, process, controls, and acceptance criteria than the reference product manufacturer.

The sponsor of a proposed product must include information demonstrating that there are no clinically meaningful differences between the biological product and the reference product in terms of safety, purity, and potency of the product. Human pharmacokinetic and pharmacodynamic (if applicable) studies comparing the proposed biological product to the reference product are fundamental components in documenting biosimilarity. Assessment of clinical immunogenicity is important to evaluate potential differences between the proposed product and the reference product. Comparative safety and effectiveness data are necessary to support biosimilarity.

Complete characterization of the protein product (e.g., the primary, secondary, tertiary, and quaternary structure (if relevant)), posttranslational modifications, and functional activity(ies) should be considered. A comprehensive understanding of all steps in the manufacturing process, process controls, and use of Quality-by-Design approach will facilitate consistent manufacturing of a high-quality product.

▪ Characterization

Characterization of the active moiety and impurity/contaminant profiling plays a significant role in the development process of a biosimilar drug product. The technological advances in instrumentation have played a major role in improved identification and characterization of biotech products (Table 11.4 and Chap. 2). It is acknowledged that no one analytical method can fully characterize the biotech product. A collection of orthogonal analytical methods is needed to piece together a complete picture of a biotech product. Determining how a small, homogeneous protein is folded in absolute terms can be accomplished using crystal X-ray diffraction. This can be difficult, expensive, and nonrealistic to perform in the formulated drug product and on a routine basis. Moreover, X-ray diffraction analysis does not pick up low levels of conformational contaminants. The most basic aspect of assessing its identity is to determine its covalent, primary structure using LC/MS/MS, peptide mapping/amino acid sequencing (e.g., via the Edman-degradation protocol), and disulfide bond-locating methods. Through circular dichroism, Fourier transform infrared and fluorescence spectroscopy, immunological methods, chromatographic techniques, etc., differences in secondary and higher-order structures can be monitored. Selective analytical methods are used to determine the purity as well as impurities of the biotech product. Methods here include again chromatographic techniques and gel electrophoresis, capillary electrophoresis, isoelectric focusing, static and dynamic light scattering, and ultracentrifugation. Inadequate characterization can result in failure to detect product changes that can impact the safety and efficacy of a product. The characterization factors that impact safety and efficacy of the product should be identified in the product development process.

▪ Clinical Studies

As a rule, clinical studies are required to assure safety and efficacy of the product.

The range of variation in the reference product used in clinical studies can be used for the definition of the boundaries within which the reference product has been shown to be safe and efficacious. These boundaries can be related to composition or to processing/formulation parameters.

The comparative clinical trial exercise is a stepwise procedure that should begin with pharmacokinetic

UV absorption
Circular dichroism spectroscopy
Fourier transform IR
Fluorescence spectroscopy
NMR spectroscopy
Calorimetric approaches
Bio-assays
Immunochemical assays
ELISA
Immunoprecipitation
Biosensor (SPR, QCM)
Potency testing
In cell lines
In animals
Chromatographic techniques
RP-HPLC
SEC-HPLC
Hydrophobic interaction HPLC
Ion-exchange HPLC
Peptide mapping
Electrophoretic techniques
SDS-PAGE
IEF
CZE
Field flow fractionaction
Ultracentrifugation
Static and dynamic light scattering
Electron microscopy
X-ray techniques
Mass spectrometry

Adapted from Crommelin et al. (2003)
IR infrared, *NMR* nuclear magnetic resonance, *SPR* surface plasmon resonance, *QCM* quartz crystal microbalance, *IEF* isoelectric focusing, *CZE* capillary zone electrophoresis

Table 11.4 ■ Analytical techniques for monitoring protein structure.

and pharmacodynamic studies followed by clinical efficacy and safety trials. The choice of the design of the pharmacokinetic study, i.e., single dose and/or multiple doses, should be justified. Normally comparative clinical trials are required for demonstration of clinical efficacy and safety. However, in certain cases, comparative pharmacokinetic/pharmacodynamic studies between the biosimilar product and the reference product using biomarkers may prove to be adequate.

EMA REGULATORY FRAMEWORK

The regulatory process for the approval of biopharmaceuticals and biosimilars in the EU follows a centralized (i.e., not a national competence of the member states) route through the European Medicines Agency (EMA). The EMA started issuing guidance documents on the regulatory process for biosimilars in 2004. The overarching guideline for biosimilar drug products is CHMP/437/04. From 2006 on, EMA published a series of guidance documents on biosimilar medical products containing specific biotechnology-derived proteins as active substance, e.g., on recombinant erythropoietin, somatropin, granulocyte colony-stimulating factor (G-CSF), and human insulin. These guidelines define key concepts/principles of biotechnology-derived proteins and discuss quality issues and non-clinical and clinical issues and are regularly updated (http://www.ema.europa.eu/ema/index.jsp?curl=pages/regulation/general/general_content_000408.jsp&mid=WC0b01ac058002958c).

Manufacturers of registered biotech products may propose changes to the manufacturing process. The EMA has adopted a "Guideline on comparability of biotechnology-derived medicinal products after a change in the manufacturing process" (updated version in 2007; cf. FDA legislation, see before). This guideline advises companies with a marketing authorization for a biotech product how to demonstrate similarity regarding quality, safety, and efficacy between the product before and after the change (FDA Guidance, EMEA Guidance).

■ Immunogenicity (Related to Overarching EMA Document)

Immunogenicity is a major concern for all biotech products (see Chap. 6). The immune response against therapeutic proteins differs between products since the immunogenic potential is influenced by many factors. These factors include, but are not limited to, the protein-, the product-, and process-related impurities, excipients, stability, dosing regimen, and the patient. Immunogenicity should be studied in patients. At present, neither animal studies nor physicochemical characterization protocols can (fully) predict immunogenicity. All biotech products must be tested for their immunogenic response in humans. Our immune system can detect alterations in a product missed by analytical methods. Therefore, postmarketing surveillance programs are essential to monitor immunological reactions.

■ Status of Biosimilars in the EU

Table 11.5 lists the biosimilars approved by the EMA (mid-2011). This list is growing. The market share of biosimilars is still small. Around 60 million $ in 2008, with an expected growth up to 2–5 billion $ in 2015 (Sheppard 2010). A lot will depend on the legislation, in particular in the USA, and the rate of acceptance of the biosimilar paradigm by the medical/pharmaceutical field. The advent of the biosimilars to the market led to questions on what grounds to choose, either for the originator's product or for the biosimilar. The *European Journal of Hospital Pharmacists* published a document: "*Points to consider in the evaluation of*

Product name	Active substance	Therapeutic area	Authorization date	Manufacturer/company name
Abseamed	Epoetin alfa	Kidney failure Chronic anemia Cancer	28/08/2007	Medice Arzneimittel Pütter GmbH & Co KG
Binocrit	Epoetin alfa	Kidney failure Chronic anemia	28/08/2007	Sandoz GmbH
Biograstim	Filgrastim	Hematopoietic stem cell Transplantation Neutropenia Cancer	15/09/2008	CT Arzneimittel GmbH
Epoetin alfa Hexal	Epoetin alfa	Kidney failure Chronic anemia Cancer	28/08/2007	Hexal AG
Filgrastim Hexal	Filgrastim	Neutropenia Cancer Hematopoietic stem cell Transplantation	06/02/2009	Hexal AG
Filgrastim ratiopharm	Filgrastim	Neutropenia Hematopoietic stem cell Transplantation Cancer	15/09/2008 Withdrawn on 3 May 2011	Ratiopharm GmbH
Nivestim	Filgrastim	Hematopoietic stem cell Transplantation Cancer Neutropenia	08/06/2010	Hospira UK Ltd
Omnitrope	Somatropin	Turner syndrome Dwarfism Pituitary Prader-Willi syndrome	12/04/2006	Sandoz GmbH
Ratiograstim	Filgrastim	Neutropenia Hematopoietic stem cell Transplantation Cancer	15/09/2008	Ratiopharm GmbH
Retacrit	Epoetin zeta	Hematopoietic stem cell Transplantation Cancer Neutropenia	18/12/2007	Hospira UK Limited
Silapo	Epoetin zeta	Chronic Anemia Blood Transfusion Autologous Cancer Kidney failure	18/12/2007	Stada R & D AG
Tevagrastim	Filgrastim	Neutropenia Cancer Hematopoietic stem cell Transplantation	15/09/2008	Teva Generics GmbH
Valtropin	Somatropin	Dwarfism Pituitary Turner syndrome	24/04/2006	BioPartners GmbH
Zarzio	Filgrastim	Cancer Hematopoietic stem cell Transplantation Neutropenia	06/02/2009	Sandoz GmbH

Table 11.5 ■ EMA-approved biosimilars (status 2011).

biopharmaceuticals." This is a document that the healthcare provider can use to make a documented choice (Kraemer et al. 2008).

THE CHALLENGE AND THE FUTURE

A major problem today is the inadequate definition of the relationship between the complex structure and function of protein pharmaceuticals. Analytical tools are becoming increasingly sensitive and provide more detailed information regarding the molecular structure. This may allow for certain biopharmaceuticals to be shown to be pharmaceutically equivalent, therapeutically equivalent, and interchangeable on the basis of a validated analytical definition alone. For instance, for relatively small protein molecules like insulin, a well-characterized biopharmaceutical product, equivalence may be established using presently available analytical techniques (Table 11.4).

But for larger and complex proteins, it is not possible to fully characterize the molecule and establish equivalence. In such a scenario, clinical safety and efficacy studies will be needed to establish equivalence.

Science-based regulatory policies are being developed. They are dynamic. They have evolved over the years and will continue to do so based on the development of superior analytical techniques to characterize the products, on introducing improved manufacturing practices and controls and on growing clinical and regulatory experience. Rigorous standards of ensuring product safety and efficacy must be maintained, and at the same time, unnecessary and/or unethical duplication trials must be avoided.

The approval of a biosimilar product should depend on the complexity of the molecule. A gradation scheme should be designed for the drug approval process rather than using a "one size fits all" model. From simple chemically synthesized molecules to highly complex molecules, e.g., from a chemically synthesized simple molecule like acetaminophen to cyclosporine, to insulin, to human growth hormone, to interleukins, to erythropoietin type of growth factors, to albumin, to monoclonal antibodies, and to factor VIII, different regulatory regimens are required.

The regulatory processes for biopharmaceuticals are following an evolutionary route. We learn from new information coming in every day; we evaluate the data, adjust the rules, and develop new protocols to make sure that the patient keeps on receiving high-quality, safe, and effective biopharmaceuticals.

SELF-ASSESSMENT QUESTIONS

Question 1: Human growth hormone has a molecular weight of around 22 kDa (see Chapter 14) and erythropoietin of 34 kDa (see Chapter 18). Why does the EMA request different clinical protocols for approval of a biosimilar product for these protein drugs?

Question 2: (a) When an US company builds a second manufacturing plant to increase the supply of its biopharmaceutical product, does it have to follow the biosimilar-approval route to obtain market authorization for the product produced in its new plant? (b) Does it have to perform clinical studies to show equivalence?

Answer 1: Human growth hormone is a non-glycosylated protein with a well-established primary sequence; erythropoietin is heavily glycosylated with a number of isoforms with more analytical challenges. The EMA guidance documents giving more details can be found on the EMA website.

Answer 2: (a) No, the FDA Guidance on Comparability will apply. (b) That depends on the complexity of the molecule. For relatively simple, non-glycosylated biopharmaceuticals, the clinical study requirements will be different than for large, highly glycosylated biopharmaceuticals.

REFERENCES

Crommelin DJA, Storm G, Verrijk R, de Leede L, Jiskoot W, Hennink WE (2003) Shifting paradigms: biopharmaceuticals versus low molecular weight drugs. Int J Pharm 266:3–16

EMEA Guidances: http://www.emea.eu.net

FDA Guidance concerning demonstration of comparability human biological products, including therapeutic biotechnology-derived products. Apr 1996

FDA Guidances: http://www.fda.gov/cder/guidance/index.htm

ICH: Q5E comparability biotechnological/biological products subject to changes in their manufacturing process. June 2005

Kraemer I, Tredree R, Vulto A (2008) Points to consider in the evaluation of biopharmaceuticals. EJHP Practice 14:73–76

Sheppard A (2010) The biosimilars market today and tomorrow. Pharm Tech Europe 22:72–80

12 Insulin

John M. Beals, Michael R. DeFelippis, Paul M. Kovach, and Jeffrey A. Jackson

INTRODUCTION

Insulin was discovered by Banting and Best in 1921 (Bliss 1982). Soon afterwards manufacturing processes were developed to extract the insulin from porcine and bovine pancreas. From 1921 to 1980, efforts were directed at increasing the purity of the insulin and providing different formulations for altering time action for improved glucose control (Brange 1987a, b; Galloway 1988). Purification was improved by optimizing extraction and processing conditions and by implementing chromatographic processes (size exclusion, ion exchange, and reversed-phase (Kroeff et al. 1989)) to reduce the levels of both general protein impurities and insulin-related proteins such as proinsulin and insulin polymers. Formulation development focused on improving chemical stability by moving from acidic to neutral formulations and by modifying the time-action profile through the use of various levels of zinc and protamine. The evolution of recombinant DNA technology led to the widespread availability of human insulin, which has eliminated issues with sourcing constraints while providing the patient with a natural exogenous source of insulin. Combining the improved purification methodologies and recombinant DNA (rDNA) technology, manufacturers of insulin are now able to provide the purest human insulin ever made available, >98 %. Further advances in rDNA technology, coupled with a detailed understanding of the molecular properties of insulin and knowledge of its endogenous secretion profile, enabled the development of insulin analogs with improved pharmacology relative to existing human insulin products.

CHEMICAL DESCRIPTION

Insulin, a 51-amino acid protein, is a hormone that is synthesized as a proinsulin precursor in the β-cells of the pancreas and is converted to insulin by enzymatic cleavage. The resulting insulin molecule is composed of two polypeptide chains that are connected by two interchain disulfide bonds (Fig. 12.1) (Baker et al. 1988). The A-chain is composed of 21 amino acids, and the B-chain is composed of 30 amino acids. The interchain disulfide linkages occur between A^7–B^7 and A^{20}–B^{19}, respectively. A third intra-chain disulfide bond is located in the A-chain, between residues A^6 and A^{11}.

In addition to human insulin and insulin analog products, which are predominately used today as first-line therapies for the treatment of diabetes, bovine and porcine insulin preparations have also been made commercially available (Table 12.1). However, all major manufacturers of insulin have discontinued production of these products, marking an end to future supply of animal-sourced insulin products. Difficulties obtaining sufficient supplies of bovine or porcine pancreata and recent concerns over transmissible spongiform encephalopathies associated with the use of animal-derived materials are major reasons for the product deletions.

The net charge on the insulin molecule is produced from the ionization potential of four glutamic acid residues, four tyrosine residues, two histidine residues, a lysine residue, and an arginine residue, in conjunction with two α-carboxyl and two α-amino groups. Insulin has an isoelectric point (pI) of 5.3 in the

J.M. Beals, Ph.D. (✉)
Lilly Research Laboratories,
Biotechnology Discovery Research,
Lilly Corporate Center, Eli Lilly and Company,
Indianapolis, IN 46285, USA
e-mail: beals_john_m@lilly.com

M.R. DeFelippis, Ph.D.
Lilly Research Laboratories,
Bioproduct Research and Development,
Eli Lilly and Company, Indianapolis, IN, USA

P.M. Kovach, Ph.D.
Lilly Research Laboratories, Technical Services
and Manufacturing Sciences, Eli Lilly and Company,
Indianapolis, IN, USA

J.A. Jackson, M.D.
Lilly Research Laboratories, Medical Affairs,
Eli Lilly and Company, Indianapolis, IN, USA

D.J.A. Crommelin, R.D. Sindelar, and B. Meibohm (eds.), *Pharmaceutical Biotechnology*,
DOI 10.1007/978-1-4614-6486-0_12,

denatured state; thus, the insulin molecule is negatively charged at neutral pH (Kaarsholm et al. 1990). This net negative charge state of insulin has been used in formulation development, as will be discussed later.

In addition to the net charge on insulin, another important intrinsic property of the molecule is its ability to readily associate into dimers and higher-order associated states (Figs. 12.2 and 12.3) (Pekar and Frank 1972). The driving force for dimerization appears to be the formation of favorable hydrophobic interactions at the C-terminus of the B-chain (Ciszak et al. 1995). Insulin can associate into discrete hexameric complexes in the presence of various divalent metal ions, such as zinc at 0.33 g-atom/monomer (Goldman and Carpenter 1974), where each zinc ion (a total of two) is coordinated by a His^{B10} residue from three adjacent monomers. Physiologically, insulin is stored as a zinc-containing hexamer in the β-cells of the pancreas. As will be discussed later, the ability to form discrete hexamers in the presence of zinc has been used to develop therapeutically useful formulations of insulin.

Commercial insulin preparations also contain phenolic excipients (e.g., phenol, m-cresol, or methylparaben) as antimicrobial agents. As represented in Figs. 12.2 and 12.3d, these phenolic species also bind to specific sites on insulin hexamers, causing a conformational change that increases the chemical stability of insulin in commercial preparations (Brange

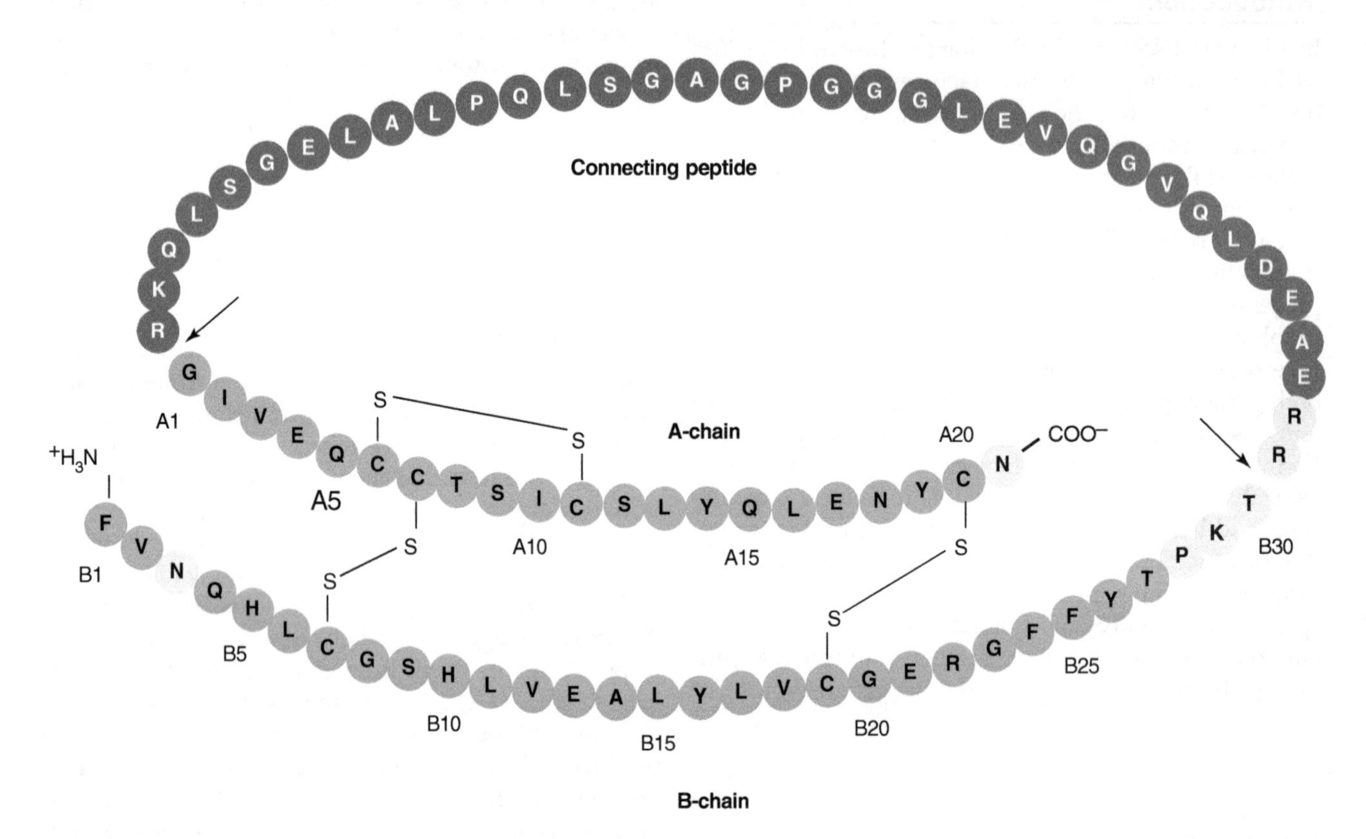

Figure 12.1 ■ Primary sequence of insulin. The shaded amino acids represent sites of sequence alterations denoted in Table 12.1.

Species	A^{21}	B^{3}	B^{28}	B^{29}	B^{30}	B^{31}	B^{32}
Human (Humulin®, Novolin®)	Asn	Asn	Pro	Lys	Thr	–	–
Insulin lispro (Humalog®)	Asn	Asn	Lys	Pro	Thr	–	–
Insulin aspart (NovoRapid®, NovoLog®)	Asn	Asn	Asp	Lys	Thr	–	–
Insulin glulisine (Apidra®)	Asn	Lys	Pro	Glu	Thr		
Insulin glargine (Lantus®)	Gly	Asn	Pro	Lys	Thr	Arg	Arg
Insulin detemir (Levemir®)	Asn	Asn	Pro	Lys-(*N*-tetradecanoyl)			

Table 12.1 ■ Amino acid substitutions in insulin analogs compared to human insulin.

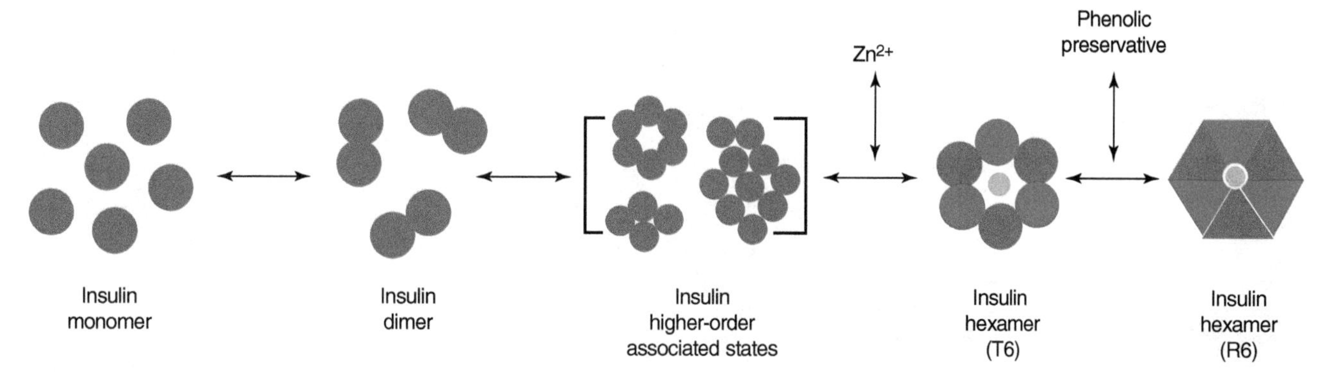

Figure 12.2 ■ Schematic representation of insulin association in the presence and absence of zinc and phenolic, antimicrobial preservatives.

and Langkjaer 1992). X-ray crystallographic studies have identified the location of six phenolic ligand binding sites on the insulin hexamer and the nature of the conformational change induced by the binding of these ligands (Derewenda et al. 1989). The phenolic ligands are stabilized in a binding pocket between monomers of adjacent dimers by hydrogen bonds with the carbonyl oxygen of Cys^{A6} and the amide proton of Cys^{A11} as well as numerous van der Waals contacts. The binding of these ligands stabilizes a conformational change that occurs at the N-terminus of the B-chain in each insulin monomer, shifting the conformational equilibrium of residues B1 to B8 from an extended structure (T state) to an α-helical structure (R state). This conformational change is referred to as the T<->R transition (Brader and Dunn 1991) and is illustrated in Fig. 12.3c, d.

In addition to the presence of zinc and phenolic preservatives, modern insulin formulations may contain an isotonicity agent (glycerol or NaCl) and/or a buffer (e.g., sodium phosphate). The former is used to minimize subcutaneous tissue damage and pain on injection. The latter is present to minimize pH drift in some pH-sensitive formulations.

PHARMACOLOGY AND FORMULATIONS

Normal insulin secretion in the nondiabetic person falls into two categories: (1) insulin that is secreted in response to a meal and (2) the background or *basal* insulin that is continually secreted between meals and during the nighttime hours (Fig. 12.4). The pancreatic response to a meal typically results in peak serum insulin levels of 60–80 μU/mL, whereas basal serum insulin levels fall within the 5–15 μU/mL range (Galloway and Chance 1994). Because of these vastly different insulin demands, considerable effort has been expended to develop insulin formulations that meet the pharmacokinetic (PK) and pharmacodynamic (PD) requirements of each condition. More recently, insulin analogs and insulin analog formulations have been developed to improve PK and PD properties.

■ Regular and Rapid-Acting Soluble Preparations

Initial soluble insulin formulations were prepared under acidic conditions and were chemically unstable. In these early formulations, considerable deamidation was identified at Asn^{A21}, and significant potency loss was observed during prolonged storage under acidic conditions. Efforts to improve the chemical stability of these soluble formulations led to the development of neutral, zinc-stabilized solutions.

The insulin in these neutral, regular formulations is chemically stabilized by the addition of zinc (~0.4 % relative to the insulin concentration) and phenolic preservatives. As mentioned above, the addition of zinc leads to the formation of discrete hexameric structures (containing 2Zn atoms per hexamer) that can bind six molecules of phenolic preservatives, e.g., m-cresol (Figs. 12.2 and 12.3c). The binding of these excipients increases the stability of insulin by inducing the formation of a specific hexameric conformation (R_6), in which the B1 to B8 region of each monomer is in an α-helical conformation (Fig. 12.3d). This in turn decreases the availability of residues involved in deamidation and high molecular weight polymer formation (Brange et al. 1992a, b).

The pharmacodynamic profile of this soluble formulation (Type R) is listed in Table 12.2. The neutral, regular formulations show peak insulin activity between 2 and 3 h with a maximum duration of 5–8 h. As with other formulations, the variations in time action can be attributed to factors such as dose, site of injection, temperature, and the patient's physical activity. Despite the soluble state of insulin in these formulations, a delay in activity is still observed. This delay has been attributed to the time required for the hexamer to dissociate into the dimeric and/or monomeric substituents prior to absorption from the

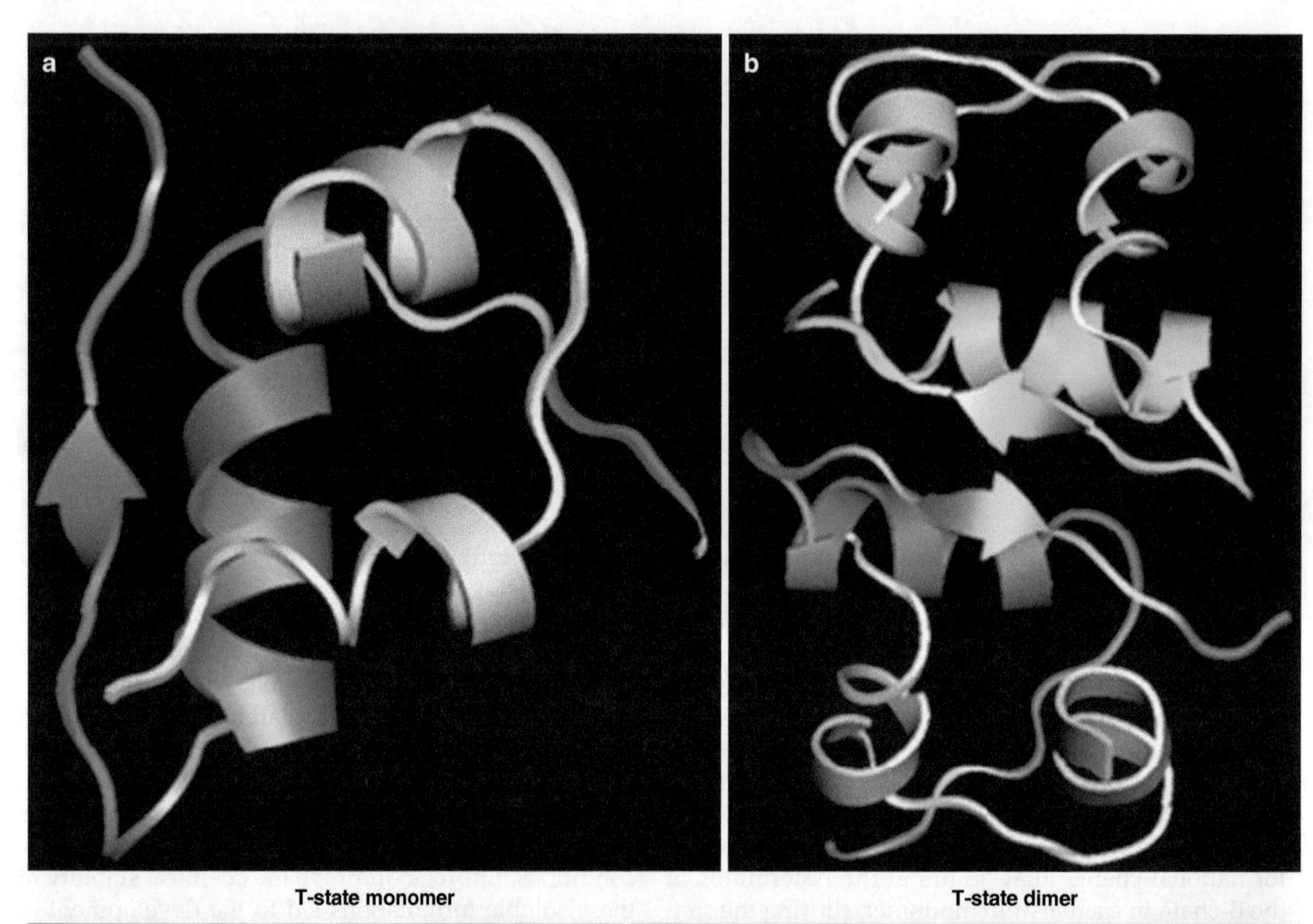

c

d

T-state hexamer

R-state hexamer

Figure 12.3 ■ (**a**) A cartoon representation of the secondary and tertiary structures of a T-state monomer of insulin, with the B1–B8 region in an extended conformation. The A-chain is colored white and the B-chain is colored blue. (**b**) A cartoon representation of the secondary and tertiary structures of a T-state dimer of insulin. The A-chains are colored white and the B-chains are colored blue and cyan. (**c**) A cartoon representation of the secondary and tertiary structures of a T-state hexamer of insulin. The A-chains are colored white, the B-chains are colored blue and cyan, and zinc is colored green. (**d**) A cartoon representation of the secondary and tertiary structures of R-state hexamer of insulin in the presence of preservative. The A-chains are colored white, the B-chains are colored blue and cyan, zinc is colored green, and preservative is colored magenta.

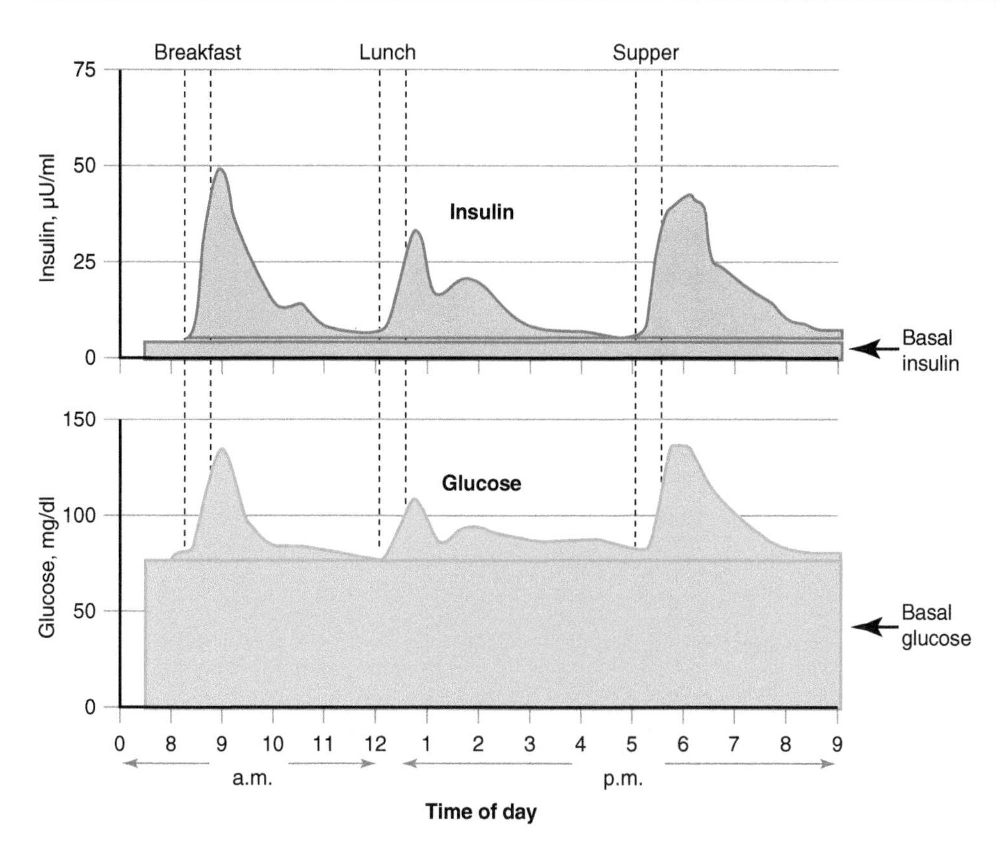

Figure 12.4 ■ A schematic representation of glucose and insulin profiles during the day in nondiabetic individuals (Adapted and reprinted from Schade et al. 1983).

interstitium. This dissociation requires the diffusion of the preservative and insulin from the site of injection, effectively diluting the protein and shifting the equilibrium from hexamers to dimers and monomers (Fig. 12.5) (Brange et al. 1990). Recent studies exploring the relationship of molecular weight and cumulative dose recovery of various compounds in the popliteal lymph following subcutaneous injection suggest that lymphatic transport may account for approximately 20 % of the absorption of insulin from the interstitium (Supersaxo et al. 1990; Porter and Charman 2000; Charman et al. 2001). The remaining balance of insulin is predominately absorbed through capillary diffusion.

Monomeric insulin analogs were designed to achieve a more natural response to prandial glucose level increases while providing dosing convenience to the patient. The pharmacological properties of these soluble formulations are listed in Table 12.3. The development of monomeric analogs of insulin for the treatment of insulin-dependent diabetes mellitus has focused on shifting the self-association properties of insulin to favor the monomeric species and consequently minimizing the delay in time action (Brange et al. 1988, 1990; Brems et al. 1992). One such monomeric analog, $Lys^{B28}ProB^{29}$-human insulin (insulin lispro; CAS Number 133107-64-9; Humalog® or Liprolog®; Eli Lilly & Co.) has been developed and does have a more rapid time-action profile, with a peak activity of approximately 1 h (Howey et al. 1994). The sequence inversion at positions B28 and B29 yields an analog with reduced self-association behavior compared to human insulin (Fig. 12.1; Table 12.1); however, insulin lispro can be stabilized in a preservative-dependent hexameric complex that provides the necessary chemical and physical stability required by insulin preparations. Despite the hexameric complexation of this analog, insulin lispro retains its rapid time action. Based on the crystal structure of the insulin lispro hexameric complex, Ciszak et al. (1995) have hypothesized that the reduced dimerization properties of the analog, coupled with the preservative dependence, yield a hexameric complex that readily dissociates into monomers after rapid diffusion of the phenolic preservative into the subcutaneous tissue at the site of injection (Fig. 12.6). Consequently, the substantial dilution (10^5) of the human insulin zinc hexamers is not necessary for the analog to dissociate from hexamers to monomers/dimers, which is required for absorption.

It is important to highlight that the properties engineered into insulin lispro (Humalog®) not only provide the patient with a more convenient therapy but also improve control of postprandial hyperglycemia and reduce the frequency of severe hypoglycemic events (Holleman et al. 1997; Anderson et al. 1997).

Since the introduction of insulin lispro, two additional rapid-acting insulin analogs have been introduced to the market. The amino acid modifications made to the human insulin sequence to produce these analogs are depicted in Table 12.1. Like insulin lispro,

Type[b]	Description	Appearance	Components	Action (h)[a]		
				Onset	Peak	Duration
R[c]	Regular soluble insulin injection	Clear solution	Metal ion: zinc (10–40 mcg/mL)	0.5–1	2–4	5–8
			Buffer: none			
			Preservative: m-cresol (2.5 mg/mL)			
			Isotonicity agent: glycerin (16 mg/mL)			
			pH: 7.25–7.6			
N	NPH insulin isophane suspension	Turbid or cloudy suspension	Metal ion: zinc (21–40 mcg/mL)	1–2	2–8	14–24
			Buffer: dibasic sodium phosphate (3.78 mg/mL)			
			Preservatives: m-cresol (1.6 mg/mL), phenol (0.73 mg/mL)			
			Isotonicity agent: glycerin (16 mg/mL)			
			Modifying protein: protamine (~0.35 mg/mL)			
			pH: 7.0–7.5			
70/30	70 % insulin isophane suspension, 30 % regular insulin injection	Turbid or cloudy suspension	Metal ion: zinc (21–35 mcg/mL)	0.5	2–4	14–24
			Buffer: dibasic sodium phosphate (3.78 mg/mL)			
			Preservatives: m-cresol (1.6 mg/mL), phenol (0.73 mg/mL)			
			Isotonicity agent: glycerin (16 mg/mL)			
			Modifying protein: protamine (~0.241 mg/mL)			
			pH: 7.0–7.8			
50/50	50 % insulin isophane suspension, 50 % regular insulin injection	Turbid or cloudy suspension	Metal ion: zinc (21–35 mcg/mL)	0.5	2–4	14–24
			Buffer: dibasic sodium phosphate (3.78 mg/mL)			
			Preservatives: m-cresol (1.6 mg/mL), phenol (0.73 mg/mL)			
			Isotonicity agent: glycerin (16 mg/mL)			
			Modifying protein: protamine (~0.172 mg/mL)			
			pH: 7.0–7.8			
L[d]	Lente insulin zinc suspension	Turbid or cloudy suspension	Metal ion: zinc (120–250 mcg/mL)	1–2	3–10	20–24
			Buffer: sodium acetate (1.6 mg/mL)			
			Preservative: methylparaben (1.0 mg/mL)			
			Isotonicity agent: sodium chloride (7.0 mg/mL)			
			Modifying protein: none			
			pH: 7.0–7.8			
U[d]	Ultralente extended insulin zinc suspension	Turbid or cloudy suspension	Metal ion: zinc (120–250 mcg/mL)	0.5–3	4–20	20–36
			Buffer: sodium acetate anhydrous (1.6 mg/mL)			
			Preservative: methylparaben (1.0 mg/mL)			
			Isotonicity agent: sodium chloride (7.0 mg/mL)			
			Modifying protein: none			
			pH: 7.0–7.8			

[a]The time-action profiles of Lilly insulins are the average onset, peak action, and duration of action that are taken from a composite of studies. The onset, peak, and duration of insulin action depend on numerous factors, such as dose, injection site, presence of insulin antibodies, and physical activity. The action times listed represent the generally accepted values in the medical community

[b]US designation

[c]Another notable designation is S (Britain). Other soluble formulations have been designed for pump use and include Velosulin® and HOE 21PH®

[d]Discontinued

Table 12.2 ■ A list of neutral U-100 insulin formulations.

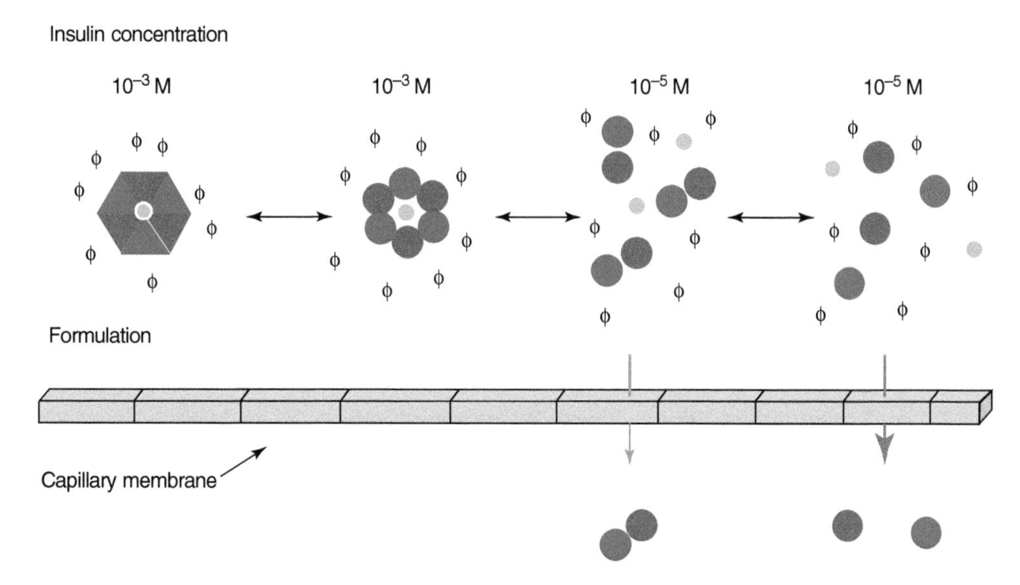

Figure 12.5 ■ A schematic representation of insulin dissociation after subcutaneous administration.

Type[b]	Description	Appearance	Components	Action (h)[a]		
				Onset	Peak	Duration
Humalog®	Rapid-acting soluble insulin analog for injection	Aqueous, clear, and colorless solution	Metal ion: zinc (19.7 mcg/mL) Buffer: dibasic sodium phosphate (1.88 mg/mL) Preservatives: m-cresol (3.15 mg/mL), phenol (trace) Isotonicity agent: glycerin (16 mg/mL) pH: 7.0–7.8	0.25–0.5	0.5–2.5	≤5
Humalog® Mix75/25™	75 % insulin lispro protamine suspension and 25 % insulin lispro for injection	Turbid or cloudy suspension	Metal ion: zinc (25 mcg/mL) Buffer: dibasic sodium phosphate (3.78 mg/mL) Preservative: m-cresol (1.76 mg/mL), phenol (0.715 mg/mL) Isotonicity agent: glycerin (16 mg/mL) Modifying protein: protamine (0.28 mg/mL) pH: 7.0–7.8	0.25–0.5	0.2–2.5	14–24
NovoLog®	Rapid-acting soluble insulin analog for injection	Aqueous, clear, and colorless solution	Metal ion: zinc (19.6 mcg/mL) Buffer: disodium hydrogen phosphate dihydrate (1.25 mg/mL) Preservative: m-cresol (1.72 mg/mL), phenol (1.50 mg/mL) Isotonicity agents: glycerin (16 mg/mL), sodium chloride (0.58 mg/mL) pH: 7.2–7.6	0.25[c]	0.75–1.5[c]	3–5[c]

Table 12.3 ■ A list of human-based U-100 insulin analog formulations.

Type[b]	Description	Appearance	Components	Action (h)[a]		
				Onset	Peak	Duration
Novolog® Mix 70/30	70 % insulin aspart protamine suspension and 30 % insulin aspart for injection	Turbid or cloudy suspension	Metal ion: zinc (19.6 mcg/mL) Buffer: dibasic sodium phosphate (1.25 mg/mL) Preservatives: m-cresol (1.72 mg/mL), phenol (1.50 mg/mL) Isotonicity agents: sodium chloride (0.58 mg/mL), mannitol (36.4 mg/mL) Modifying protein: protamine (0.33 mg/mL) pH: 7.2–7.44	<0.5[d]	1–4[d]	≤24[d]
Apidra®	Rapid-acting soluble insulin analog for injection	Aqueous, clear, and colorless solution	Metal ion: none Buffer: tromethamine (6 mg/mL) Preservative: m-cresol (3.15 mg/mL) Isotonicity agent: sodium chloride (5 mg/mL) Stabilizing agent: polysorbate 20, (0.01 mg/mL) pH: ~7.3	~0.3[c]	0.5–1.5[c]	~5.3[c]
Humalog® Basal (EU) or Humalog® NPL (Japan)	Intermediate-acting insulin lispro protamine suspension for injection	Turbid or cloudy suspension	Metal ion: zinc (0.0225 mcg/mL) Buffer: dibasic sodium phosphate (3.78 mg/mL) Preservatives: m-cresol (1.76 mg/mL), phenol (0.8 mg/mL) Isotonicity agent: glycerin (16 mg/mL) Modifying protein: protamine (0.376 mg/mL) pH: 7.0–7.5	1–2	2–8	14–24
Lantus®	Long-acting soluble insulin analog for injection	Aqueous, clear, and colorless solution	Metal ion: zinc (30 mcg/mL) Buffer: none Preservative: m-cresol (2.7 mg/mL) Isotonicity agent: glycerin (20 mg 85 %/mL) Modifying protein: none pH: ~ 4		Constant release with no pronounced peak[c]	10.8 to >24.0[d]
Levemir®	Long-acting soluble insulin analog for injection	Aqueous, clear, and colorless solution	Metal ion: zinc (65.4 mcg/mL) Buffer: dibasic sodium phosphate (0.89 mg/mL) Preservatives: m-cresol (2.06 mg/mL), phenol (1.8 mg/mL) Isotonicity agents: mannitol (30 mg/mL), sodium chloride (1.17 mg/mL) Modifying protein: none pH: 7.4		3–14[c]	5.7–23.2[c]

[a]The time-action profiles of Lilly insulins are the average onset, peak action, and duration of action taken from a composite of studies. The onset, peak, and duration of insulin action depend on numerous factors, such as dose, injection site, presence of insulin antibodies, and physical activity. The action times listed represent the generally accepted values in the medical community

[b]US designation

[c]DRUGDEX® System [Internet database]. Greenwood Village, Colo: Thomson Micromedex. Updated periodically

[d]PDR® Electronic Library™ [Internet database]. Greenwood Village, Colo: Thomson Micromedex. Updated periodically

Table 12.3 ■ (continued)

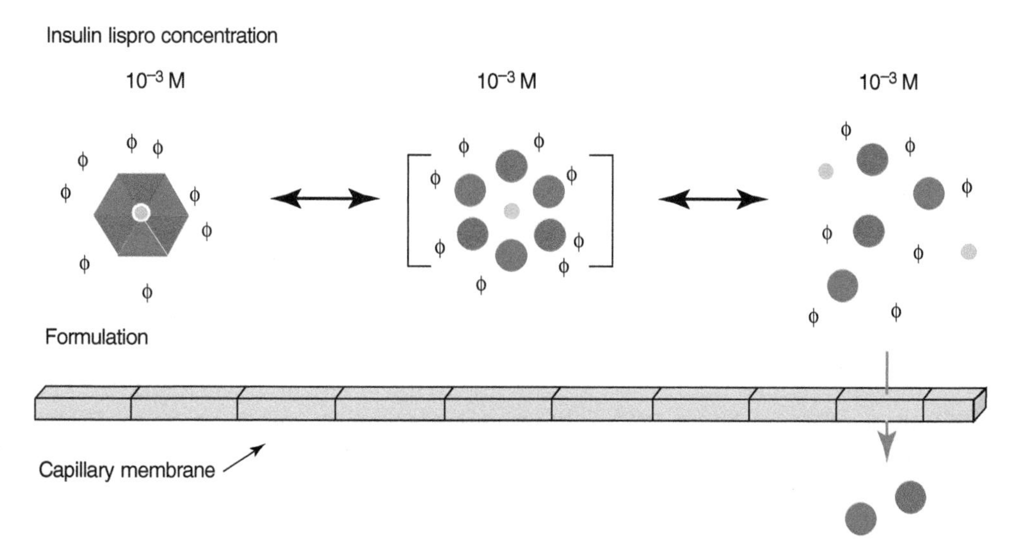

Figure 12.6 ■ A schematic representation of insulin lispro dissociation after subcutaneous administration.

both analogs are supplied as neutral pH solutions containing phenolic preservative. The design strategy for Asp^{B28}-human insulin (insulin aspart; CAS Number 116094-23-6; NovoRapid® or NovoLog®; Novo Nordisk A/S) (Brange et al. 1988, 1990) involves the replacement of Pro^{B28} with a negatively charged aspartic acid residue. Like $Lys^{B28}ProB^{29}$-human insulin, Asp^{B28}-human insulin has a more rapid time action following subcutaneous injection (Heinemann et al. 1997). This rapid action is achieved through a reduction in the self-association behavior compared to human insulin (Brange et al. 1990; Whittingham et al. 1998). The other rapid-acting analog, Lys^{B3}-Glu^{B29}-human insulin (insulin glulisine; CAS Number 160337-95-1; Apidra®, Sanofi-Aventis), involves a substitution of the lysine residue at position 29 of the B-chain with a negatively charged glutamic acid. Additionally, this analog replaces the Asn^{B3} with a positively charged lysine. Scientific reports describing the impact of these changes on the molecular properties of this analog are lacking. However, the glutamic acid substitution occurs at a position known to be involved in dimer formation (Brange et al. 1990) and may result in disruption of key interactions at the monomer-monomer interface. The Asn residue at position 3 of the B-chain plays no direct role in insulin self-association (Brange et al. 1990), but it is flanked by two amino acids involved in the assembly of the Zn^{2+} insulin hexamer. Despite the limited physicochemical information on insulin glulisine, studies conducted in persons with either type 1 (T1DM) or type 2 diabetes (T2DM) (Dreyer et al. 2005; Dailey et al. 2004) confirm that the analog displays similar pharmacological properties as insulin lispro. Interestingly, insulin glulisine is not formulated in the presence of zinc as are the other rapid-acting analogs. Instead, insulin glulisine is formulated in the presence of a stabilizing agent (polysorbate 20) (Table 12.3). The surfactant in the formulation presumably minimizes higher-order association. Since the purely monomeric formulated Apidra® demonstrates only a slightly faster PK profile and no difference in PD properties from Novolog® (insulin aspart) and Humalog® (insulin lispro), the hexameric breakdown of the two latter formulations must be rapid relative to the rate-limiting step, subcutaneous absorption (Home 2012).

In addition to the aforementioned rapid-acting formulations, manufacturers have designed soluble formulations for use in external or implanted infusion pumps. In most respects, these formulations are very similar to regular insulin (i.e., hexameric association state, preservative, and zinc); however, buffer and/or surfactants may be included in these formulations to minimize the physical aggregation of insulin that can lead to clogging of the infusion sets. In early pump systems, gas-permeable infusion tubing was used with the external pumps. Consequently, a buffer was added to the formulation in order to minimize pH changes due to dissolved carbon dioxide. Infusion tubing composed of materials having greater resistance to carbon dioxide diffusion is currently being used and the potential for pH-induced precipitation of insulin is greatly reduced. All three of the commercially available rapid-acting insulin analogs are approved for use in external infusion pumps.

■ Ultrarapid Initiatives

The efforts of developing an artificial pancreas, an external pump that can rapidly control blood glucose coupled with a continuous blood glucose monitor, are driving the need for insulins with increasingly faster time action. To this end, numerous approaches are being explored, including modification of the subcutaneous tissue using permeation enhancers to increase insulin dispersion and accelerate absorption

(Muchmore and Vaughn 2010), disruption of the hexameric state of insulin and masking surface charges to facilitate more rapid absorption of monomers (Heinemann et al. 2012), and the use of "biochaperones" to assist transport of insulin across the capillary membrane (Soula et al. 2010). None of these approaches have produced commercially available products at this time.

■ Intermediate-Acting Insulin Preparations

The only intermediate-acting insulin preparation available is NPH. This formulation achieves extended time action by necessitating the dissolution of a crystalline form of insulin. This dissolution is presumed to be the rate-limiting step in the absorption of intermediate insulin. Consequently, the time action of the formulation is prolonged by further delaying the dissociation of the hexamer into dimers and monomers.

NPH refers to neutral protamine Hagedorn, named after its inventor H. C. Hagedorn (1936), and is a neutral crystalline suspension that is prepared by the cocrystallization of insulin with protamine. Protamine consists of a closely related group of very basic peptides that are isolated from fish sperm. Protamine is heterogeneous in composition; however, four primary components have been identified and show a high degree of sequence homology (Hoffmann et al. 1990). In general, protamine is ~30 amino acids in length and has an amino acid composition that is primarily composed of arginine, 65–70 %. Using crystallization conditions identified by Krayenbuhl and Rosenberg (1946), oblong tetragonal NPH insulin crystals with volumes between 1 and 20 μm^3 can be consistently prepared from protamine and insulin (Deckert 1980). These formulations, by design, have very minimal levels of soluble insulin in solution. The condition at which no measurable protamine or insulin exists in solution after crystallization is referred to as the isophane point.

NPH has an onset of action from 1 to 2 h, peak activity from 6 to 12 h, and duration of activity from 18 to 24 h (Table 12.2). As with other formulations, the variations in time action are due to factors such as dose, site of injection, temperature, and the patient's physical activity. In T2DM patients, NPH can be used as either once-daily or twice-daily therapy; however, in T1DM patients, NPH is predominately used as a twice-daily therapy. NPH can be readily mixed with regular insulin either extemporaneously by the patient or as obtained from the manufacturer in a premixed formulation (Table 12.2). Premixed insulin, e.g., 70/30 or 50/50 NPH/regular, has been shown to provide the patient with improved dose accuracy and consequently improved glycemic control (Bell et al. 1991). In these preparations, a portion of the soluble regular insulin will reversibly adsorb to the surface of the NPH crystals through an electrostatically mediated interaction under formulation conditions (Dodd et al. 1995); however, this adsorption is reversible under physiological conditions and consequently has no clinical significance (Galloway et al. 1982; Hamaguchi et al. 1990; Davis et al. 1991). Due, in part, to the reversibility of the adsorption process, NPH/regular mixtures are uniquely stable and have a 3-year shelf life.

The rapid-acting insulin analog, insulin lispro, can be extemporaneously mixed with NPH; however, such mixtures must be injected immediately upon preparation due to the potential for exchange between the soluble and suspension components upon long-term storage. Exchange refers to the release of human insulin from the NPH crystals into the solution phase and concomitant loss of the analog into the crystalline phase. The presence of human insulin in solution could diminish the rapid time-action effect of the analog. One way to overcome the problem of exchange is to prepare mixtures containing the same insulin species in both the suspension and the solution phases, analogous to human insulin regular/NPH preparations. However, this approach requires an NPH-like preparation of the rapid-acting analog.

An NPH-like suspension of insulin lispro has been prepared, and its physicochemical properties relative to human insulin NPH have been described (DeFelippis et al. 1998). In order to prepare the appropriate crystalline form of the analog, significant modifications to the NPH crystallization procedure are required. The differences between the crystallization conditions have been proposed to result from the reduced self-association properties of insulin lispro.

Pharmacological studies reported for the insulin lispro NPH-like suspension, formerly referred to as neutral protamine lispro (NPL) (DeFelippis et al. 1998; Janssen et al. 1997), indicate that the PK and PD properties of this analog suspension are analogous to human insulin NPH (Table 12.3). Clinical trials of insulin lispro protamine suspension (ILPS) alone in T2DM and in combination with insulin lispro in T1DM have been reported (Strojek et al. 2010; Fogelfeld et al. 2010; Chacra et al. 2010). In T2DM patients, the PK/PD profile of ILPS can support a once-daily therapy regimen (Hompesch et al. 2009), In addition, studies with ILPS in T1DM patients have shown a more predictable response than insulin glargine due to reduced intrasubject variability (Ocheltree et al. 2010). Moreover, the availability of ILPS allows for the preparation of homogeneous, biphasic mixture preparations containing intermediate-acting ILPS and rapid-acting solutions of insulin lispro that are not impacted by exchange between solution and crystalline forms. ILPS is also available as a stand-alone basal analog in several EU countries and Japan.

As with insulin lispro, premixed formulations of the insulin aspart have been prepared in which rapid-acting soluble insulin aspart has been combined with a protamine-retarded crystalline preparation of insulin aspart (Balschmidt 1996). Clinical data on insulin lispro mixtures and those composed of insulin aspart have been reported in the literature (Weyer et al. 1997; Heise et al. 1998). The pharmacological properties of the rapid-acting analogs are preserved in these stable mixtures (Table 12.3). Premixed formulations of both rapid-acting analogs are now commercially available in many countries.

Immunogenicity issues with protamine have been documented in a small percentage of diabetic patients (Kurtz et al. 1983; Nell and Thomas 1988). Individuals who show sensitivity to the protamine in NPH formulations (or premixed formulations of insulins lispro and aspart) are routinely switched to other long-acting insulin formulations, e.g., Lantus® or Levemir®, to control their basal glucose levels.

Long-Acting Insulin Formulations

The normal human pancreas secretes approximately 1 unit of insulin (0.035 mg) per hour to maintain basal glycemic control (Waldhäusl et al. 1979). Adequate basal insulin levels are a critical component of diabetes therapy because they regulate hepatic glucose output, which is essential for proper maintenance of glucose homeostasis during the diurnal cycling of the body. Consequently, long-acting insulin formulation must provide a very different PK profile than "mealtime" insulin formulation.

There are two long-acting insulin analog preparations currently commercially available, Lantus® (insulin glargine) and Levemir® (insulin detemir), which were approved in the 2000s (Table 12.1; Fig. 12.1). The approval of these solution-based analog preparations made the zinc-insulin crystalline Ultralente obsolete, and it was subsequently removed from the marketplace. Lantus® derives its protracted time-action profiles from the slow and relatively constant dissolution of solid particles that form as result of a pH shift of the acidic formulation to neutral pH in the subcutaneous tissue. This slow dissolution precedes the dissociation of insulin into absorbable units, and thus the *rate of absorption* (units per hour) into the bloodstream is significantly decreased in comparison to that of prandial or bolus (mealtime) formulations. Levemir®, on the other hand, achieves its protracted effect by a combination of structural interactions and physiological binding events (Havelund et al. 2004).

Insulin glargine (Gly^{A21}, Arg^{B31}, Arg^{B32}-human insulin; CAS Number 160337-95-1; Lantus®; Sanofi-Aventis) is a long-acting insulin analog, whose amino acid sequence modifications are highlighted in Table 12.1 and Fig. 12.1. This analog differs from human insulin in that the amino acid asparagine is replaced with glycine at position Asn^{A21} and two arginine residues have been added to the C-terminus of the B-chain. The impact of the additional arginine residues is to shift the isoelectric point from a pH of 5.4–6.7, thereby producing an insulin analog that is soluble at acidic pH values, but is less soluble at the neutral pH of subcutaneous tissue. Lantus® is a solution formulation prepared under acidic conditions, pH 4.0. The introduction of glycine at position Asn^{A21} yields a protein with acceptable chemical stability under acidic formulation conditions, since the native asparagine is susceptible to acid-mediated degradation and reduced potency. Thus, the changes to the molecular sequence of insulin have been made to improve chemical stability and to modulate absorption from the subcutaneous tissue, resulting in an analog that has approximately the same potency as human insulin. The Lantus® formulation is a clear solution that incorporates zinc and m-cresol (preservative) at a pH value of 4. Consequently, Lantus® does not need to be resuspended prior to dosing. Immediately following injection into the subcutaneous tissue, the insulin glargine precipitates due to the pH change, forming a slowly dissolving precipitate. This results in a relatively constant rate of absorption over 24 h with no pronounced peak (Table 12.3). This profile allows once-daily dosing as a patient's basal insulin. As with all insulin preparations, the time course of Lantus® may vary in different individuals or at different times in the same individual, and the rate of absorption is dependent on blood supply, temperature, and the patient's physical activity. Lantus® should not be diluted or mixed with any other solution or insulin, as will be discussed below.

Insulin detemir (Lys^{B29}(*N*-tetradecanoyl)des(B30) human insulin; CAS Number 169148-63-4; Levemir®; Novo Nordisk A/S) utilizes acylation of insulin with a fatty acid moiety as a means to achieve a protracted pharmacological effect. As shown in Table 12.1 and Fig. 12.1, the B30 threonine residue of human insulin is eliminated in insulin detemir, and a 14-carbon, myristoyl fatty acid is covalently attached to the ε-amino group of Lys^{B29}. The analog forms a zinc hexamer at neutral pH in a preserved solution. Clinical studies have reported that insulin detemir displays lower PK and PD variability than NPH and/or insulin glargine (Hermansen et al. 2001; Vague et al. 2003; Heise et al. 2004; Porcellati et al. 2011). An approximate description of the PD profile of Levemir® is listed in Table 12.3. This analog appears to display a slower onset of action than NPH without a pronounced peak (Heinemann et al. 1999). However, whether the duration of the protracted effect can truly be considered sufficient to warrant classification of insulin detemir as a long-acting insulin remains a subject of debate since published clinical studies of this insulin analog are typically referenced to intermediate-acting NPH.

Binding of the tetradecanoyl-acylated insulin to albumin was originally proposed as the underlying mechanism behind the observed prolonged effect for insulin detemir analog; however, recent investigations on insulin detemir have determined that the mechanism is more complex (Havelund et al. 2004). It has been proposed that subcutaneous absorption is initially delayed as a result of hexamer stability and dihexamerization. Such interactions between hexamers are likely a consequence of the symmetrical arrangement of fatty acid moieties around the outside of the hexamers (Whittingham et al. 2004), as shown by X-ray crystallographic studies. These associated forms further bind to albumin within the injection site depot. Additional prolongation may result due to albumin binding.

Although Lantus® and Levemir® have improved basal insulin therapy, both products fail to achieve the goal of a once-daily administered basal insulin product with both full 24-h coverage and low variability. Moreover, the desire to eliminate or minimize nocturnal hypoglycemia has driven the exploration of improved basal insulin therapies. Consequently, there are five basal insulin programs of note in Phase III clinical testing, according to ClinicalTrials.gov (http://clinicaltrial.gov). As of March 2012, Sanofi-Aventis is testing a new formulation of insulin glargine (ClinicalTrials.gov Identifier: NCT01499082); although no peer-reviewed literature is available, Sanofi-Aventis disclosed at the 32nd Cowen Annual Health Care Conference in Boston that the new glargine formulation provides a unique flat PK/PD profile with lower injection volume (Zerhouni 2012). Eli Lilly and Company is testing two basal insulin candidates, LY2605541 and LY2963016. The company has yet to disclose the nature of these basal insulin candidates; however, as of March 2012, LY2605541 is slated for six Phase III trials*, and LY2963016 is slated for two Phase III studies. Novo Nordisk, as of March 2012, had filed with regulatory agencies, insulin degludec (NN1250), an ultra-long basal insulin, and insulin degludec plus (NN5401), a soluble basal insulin derivative combined with a bolus insulin. Insulin degludec is a new acylated insulin, wherein desB30 human insulin is modified at position Lys^{B29} with a derivatized fatty acid moiety defined as 29B-[N^6-[N-(15-carboxy-1-oxopentadecyl)-L-γ-glutamyl]-L-lysine] (CAS Number 844439-96-9). The protracted time action of degludec is derived from the utilization of a depot release strategy specific to the derivatized insulin wherein, after injection of the soluble insulin degludec formulation, di-hexamers agglomerate to form multi-hexamers to add an additional rate-limiting step to the release of absorbable insulin monomers and dimers from the subcutaneous tissue. The preliminary data indicate that insulin degludec had comparable glycemic control to insulin glargine with a reduced hypoglycemia profile in T1DM patients (Birkeland et al. 2011) and T2DM patients (Zinman et al. 2011). Moreover, in the latter study the protracted time action of insulin degludec was exemplified by showing the product could be administered every 2 days with efficacy and safety.

■ Concentrated Insulin Formulations

Concentrated U-500 beef regular insulin (500 U/mL) first became available in the USA in 1952 (Iletin®, Eli Lilly and Company). It was initially used to manage very high insulin requirements of diabetes patients with insulin antibody insulin resistance. Pork U-500 regular insulin (Iletin II®, Lilly) replaced the beef formulation in 1980. Over time, with progressive improvements in insulin formulations and purity, severe insulin resistance (insulin requirements of >200 U/day or ≥2 U/kg/day) became associated with T2DM and severe obesity, parallel epidemics currently in the USA and worldwide. Recombinant human U-500 regular insulin was introduced in 1997 (Humulin® R U-500, Eli Lilly and Company in the USA, Actrapid® U-500, Novo Nordisk in the UK [voluntarily withdrawn in 2008]). Providing an appropriate amount of insulin for these patients using U-100 insulins may be logistically difficult and may require eight or more separate syringes or pen injections daily, making patient adherence to therapy and attaining glycemic control difficult (Lane et al. 2009; Segal et al. 2010).

Early pharmacological studies demonstrated reduced absorption associated with increasing concentrations of insulin (Binder 1969; Binder et al. 1984). Galloway et al. (1981) showed no statistically significant differences in PK serum insulin levels with increasing concentrations of pork regular insulin (at 0.25 U/kg) from U-40 to U-500; however, time to peak glucose responses were mildly delayed, and peak effect was variably reduced as concentration increased. The first PK/PD study of human U-500 vs. U-100 regular insulin in healthy obese subjects was recently published (de la Peña et al. 2011). Overall insulin exposure and overall effect were similar at both 50- and 100-U doses (0.5 and 1.0 U/kg) with both formulations. However, the two formulations were not bioequivalent: peak insulin concentration (C_{max}) and effect (R_{max}) were significantly prolonged for U-500 vs. U-100 for both doses. Time to peak concentration (t_{max}) and time to maximal effect (tR_{max}) were significantly longer for U-500 vs. U-100 only at the 100-U dose. Duration of action (tR_{last}) was prolonged for U-500 at both doses vs. U-100 (50 U: 19.7 vs. 18.3 h; 100 U: 21.5 vs. 18.3 h; $p<0.05$ for both). The

*During the preparation of this book chapter, the structure of LY2605541 was disclosed at conference proceedings as insulin lispro PEGylated at LysB28 with a 20kDa PEG (Hansen et al. 2012).

onset of action (t_{onset}) was within 20 min for both formulations and supports the clinical use of human U-500 regular 30 min before meals to leverage the prandial effect. Basal insulin needs are expected to be covered by the long "tail" of action of the U-500 formulation (de la Peña et al. 2011).

Although no randomized controlled trials of U-500 insulin have been completed (A randomized controlled trial comparing twice-daily and thrice-daily U-500 in insulin-resistant T2DM was initiated in the USA in 2013 [CT.gov NCT 01774968]), case series (review by Lane et al. 2009; Ziesmer et al. 2012; Boldo and Comi 2012) have generally demonstrated reductions in HbA1c (glycated hemoglobin) of 1.0–1.7 % over 3–98 months of use. Paradoxically, insulin dose generally did not statistically increase after conversion to human U-500 regular insulin, although one large case series did report an increase in total daily dose by 0.44 U/kg (Boldo and Comi 2012). Weight gain with treatment was variable, up to 4.2–6.8 kg (Lane et al. 2009; Boldo and Comi 2012). Reports of severe hypoglycemia have been infrequent, although an increase in non-severe hypoglycemia was reported in one large series (Boldo and Comi 2012). Most series have used twice-daily or thrice-daily regimens (Lane et al. 2009; Ziesmer et al. 2012; Boldo and Comi 2012). A simplified dosing algorithm was published by Segal et al. (2010).

Safety concerns with concentrated insulin therapy in diabetes patients, besides hypoglycemia and weight gain, mainly relate to the risk of dose confusion due to lack of a dedicated injection device for U-500 insulin. Thus, U-100 insulin syringes or tuberculin (volumetric) syringes have to be used; careful notation of unit markings (e.g., a 100-unit dose would be drawn to the 20 unit marking on a U-100 insulin syringe) or volume markings in mL (e.g., a 100-unit dose would be drawn to 0.2 mL on a tuberculin syringe), respectively, is required. Dosing conversion tables and formulas are useful, as have been included in the revised product label (March 2011) and recent clinical reviews (Lane et al. 2009; Segal et al. 2010). Pharmacists need to ensure that patients have had appropriate education on how to measure and administer doses (Segal et al. 2010). The U-500 insulin vial and labeling of vial and box are distinctive from U-100 insulins, with black-and-white lettering, brown diagonal stripes, and larger size (20 mL containing 10,000 U).

PHARMACEUTICAL CONCERNS

■ Chemical Stability of Insulin Formulations

Insulin has two primary routes of chemical degradation upon storage and use: hydrolytic transformation of amide to acid groups and formation of covalent dimers and higher-order polymers. Primarily the pH, the storage temperature, and the components of the specific formulation influence the rate of formation of these degradation products. The purity of insulin formulations is typically assessed by high-performance liquid chromatography using reversed-phase and size exclusion separation modes (USP Monographs: Insulin 2012). In acidic solution, the main degradation reaction is the transformation of asparagine (Asn) at the terminal 21 position of the A-chain to aspartic acid. This reaction is relatively facile at low pH, but is extremely slow at neutral pH (Brange et al. 1992b). This was the primary degradation route in early soluble (acidic) insulin formulations. However, the development of neutral solutions and suspensions has diminished the importance of this degradation route. Stability studies of neutral solutions indicate that the amount of A21 desamido insulin does not change upon storage. Thus, the relatively small amounts of this bioactive material present in the formulation arise either from the source of insulin or from pharmaceutical process operations.

The deamidation of the AsnB3 of the B-chain is the primary degradation mechanism at neutral pH. The reaction proceeds through the formation of a cyclic imide that results in two products, aspartic acid (Asp) and iso-aspartic acid (iso-Asp) (Brennan and Clarke 1994). This reaction occurs relatively slowly in neutral solution (approximately 1/12 the rate of A21 desamido formation in acid solution) (Brange et al. 1992b). The relative amounts of these products are influenced by the flexibility of the B-chain, with approximate ratios of Asp:iso-Asp of 1:2 and 2:1 for solution and crystalline formulations, respectively. As noted earlier, the use of phenolic preservatives provides a stabilizing effect on the insulin hexamer that reduces the formation of the cyclic imide, as evidenced by reduced deamidation. The rate of formation also depends on temperature; typical rates of formation are approximately 2 % per year at 5 °C. Studies have shown B3 deamidated insulin to be essentially fully potent (R.E. Chance, personal communication).

High molecular weight protein (HMWP) products form at both refrigerated and room temperature storage conditions. Covalent dimers that form between two insulin molecules are the primary condensation products in marketed insulin products. There is evidence that insulin-protamine heterodimers also form in NPH suspensions (Brange et al. 1992a). At higher temperatures, the probability of forming higher-order insulin oligomers increases. The rate of formation of HMWP is less than that of hydrolytic reactions; typical rates are less than 0.5 % per year for soluble neutral regular insulin formulations at 5 °C. The rate of formation can be affected by the strength of the insulin formulation or by the addition of glycerol as an isotonicity agent. The latter increases the rate of HMWP formation presumably by introducing impurities such as glyceraldehyde. HMWP formation is believed to also occur as a result of a reaction between the N-terminal B1 phenylalanine amino group and the C-terminal A21 asparagine of a second insulin molecule via a cyclic anhydride (or succinimide, based on unpublished results of the authors) intermediate (Darrington and Anderson 1995). Reaction with the intermediate may also occur via the N-terminus of the A-chain or side-chain epsilon amine of the lysine residue located near the C-terminus of the B-chain. Disulfide exchange leading to polymer formation is also possible at basic pH; however, the rate for these reactions is very slow under neutral pH formulation conditions. The quality of excipients such as glycerol is also critical because small amounts of aldehyde and other glycerol-related chemical impurities can accelerate the formation of HMWP. The biopotency of HMWP is significantly less (1/12–1/5 of insulin) than monomeric species (Brange 1987c).

Only limited chemical stability data has been published in the scientific literature for the insulin analog formulations containing insulin lispro, insulin aspart, insulin glulisine, insulin glargine, or insulin detemir; however, it is reasonable to presume that similar chemical degradation pathways are present to varying extents in these compounds. Nevertheless, since some analogs are formulated under acidic conditions, e.g., Lantus® is formulated at pH 4.0, or have been modified with hydrophobic moieties, e.g., Levemir®, it is reasonable to presume that alternate chemical degradation pathways may be operable. It should be noted that the amino acid substitution of glycine for asparagine at position 21 of the insulin glargine A-chain is expected to effectively eliminate the potential for deamidation that would occur under the acidic pH conditions used in the Lantus® formulation.

■ Physical Stability of Insulin formulations

The physical stability of insulin formulations is mediated by noncovalent aggregation of insulin. Hydrophobic forces typically drive the aggregation, although electrostatics plays a subtle but important role. Aggregation typically leads to a loss in potency of the formulation, and therefore conditions promoting this type of physical degradation (i.e., extreme mechanical agitation or exposure to air-liquid interfaces often in combination with elevated temperatures) should be avoided for all insulin products. A particularly severe type of nonreversible aggregation results in the formation of insulin fibrils. The mechanism of insulin fibrillation is widely believed to result from destabilization of hexamers (i.e., the predominant self-associated form of most insulin solution preparations) causing an increase in the population of monomers that can partially unfold and initiate the aggregation process (Jansen et al. 2005). Physical attributes of insulin formulations are readily assessed by visual observation for macroscopic characteristics as well as by instrumental methods such as light and differential phase contrast microscopy. Insulin fibrillation can be confirmed using atomic force microscopy (Jansen et al. 2005). Various particle-sizing techniques also may be used to characterize physical degradation phenomena. Fluorescence spectroscopy using specific dyes has proven useful in monitoring the time course of insulin fibrillation process (Nielsen et al. 2001).

In general, insulin solutions have good physical stability. Physical changes in soluble formulations may be manifested as color or clarity change or, in extreme situations, increases in solution viscosity, a phenomenon referred to as gelation, or the formation of a precipitate that could be an indication of fibrillation. Insulin suspensions, such as NPH, are the most susceptible to changes in physical stability. Such physical instability typically occurs as a result of both elevated temperature and mechanical stress to the insulin preparation. The increase in temperature favors hydrophobic interactions, while mechanical agitation serves to provide mixing and stress across interfacial boundaries. Nucleation and higher-order forms of aggregation in suspensions can lead to conditions described as visible clumping of the insulin microcrystalline particles or adherence of the aggregates to the inner wall of the glass storage container. The latter phenomenon is referred to as frosting. In severe cases, resuspension may be nearly impossible because of caking of the suspension in the vial. Temperatures above ambient (>25 °C) can accelerate the aggregation process, especially those at or above body temperature (37 °C). Normal mechanical mixing of suspensions to achieve dispersion of the microcrystalline insulin particles prior to administration is not deleterious to physical stability. However, vigorous shaking or mixing should be avoided. Consequently, this latter constraint has, in part, led to the observation that patients do not place enough effort into resuspension. Thus, proper emphasis must be placed on training the patient in resuspension of crystalline, amorphous, and premixed suspension formulations of insulin and insulin analogs. The necessity of rigorous resuspension may be the first sign of aggregation and should prompt a careful examination of the formulation to verify its suitability for use.

As with the chemical stability data, published information regarding the physical stability of the newer insulin analog formulations containing insulin lispro, insulin aspart, insulin glulisine, insulin glargine, or insulin detemir is limited. However, it is reasonable to assume that similar controls are practiced for preventing exposures to extreme agitation and thermal excursions to minimize undesirable physical transformations such as precipitation, aggregation, gelation, or fibrillation.

CLINICAL AND PRACTICE ASPECTS

■ Vial Presentations

Insulin is commonly available in 10-mL vials. In the United States, a strength of U-100 (100 U/mL) is the standard, whereas outside the USA both U-100 and U-40 (40 U/mL) are commonly used. Recent introduction of U-100 insulins (Humalog®, Humulin® N, R, and 70/30) in 5-mL vials (filled to 3 mL: 300 U) has met a need for smaller volumes and less waste in hospital usage. It is essential to obtain the proper strength and formulation of insulin in order to maintain glycemic control. In addition, brand/method of manufacture is important. Any change in insulin should be made cautiously and only under medical supervision (Galloway 1988; Brackenridge 1994). Common formulations, such as regular and NPH, are listed in Table 12.2, and the newer insulin analog formulations are listed in Table 12.3. Mixtures of rapid- or fast-acting with intermediate-acting insulin formulations are popular choices for glycemic control. The ratio is defined as ratio of protamine-containing fraction/rapid- or fast-acting fraction, e.g., Humalog Mix 75/25 where 75 % of a dose is available as 75 % ILPS and 25 % insulin lispro for injection. With regard to NPH regular mixtures, caution must be used in the nomenclature because it may vary depending on the country of sale and the governing regulatory body. In the USA, for example, the predominant species is listed first as in N/R 70/30, but in Europe the same formulation is described as R/N 30/70 (Soluble/Isophane) where the base ("normal") ingredient is listed first. Currently, an effort is being made to standardize worldwide to the European nomenclature. Human insulin mixtures available in the USA include N/R 70/30 and 50/50, while Europe has R/N 15/85, 25/75, 30/70, and 50/50 available from Eli Lilly and Company, Novo Nordisk, and Sanofi-Aventis.

■ Injection Devices

Insulin syringes should be purchased to match the strength of the insulin that is to be administered (e.g., for U-100 strength use 30-, 50-, or 100-unit syringes designated for U-100). The gauge of needles available for insulin administration has been reduced to very fine gauges (30–32 G) in order to minimize pain during injection. In addition to finer gauge needles, the length of needles has shortened to a minimum of 5 mm, in part, to prevent unintended IM injection. Recently, studies have shown that skin thickness is rarely >3 mm and that needles of 4–5 mm consistently deliver insulin into the subcutaneous adipose tissue (Gibney et al. 2010). The use of a new needle for each dose maintains the sharp point of the needle and ensures a sterile needle for the injection.

In recent years, the availability of insulin pen devices has made dosing and compliance easier for the patient with diabetes. The first pen injector used a 1.5-mL cartridge of U-100 insulin (NovoPen® by Novo Nordisk in 1985). A needle was attached to the end of the pen, and the proper dose was selected and then injected by the patient. The cartridge was replaced when the contents were exhausted, typically 3–7 days. Currently, 3.0-mL cartridges in U-100 strength for regular, NPH, and the range of R/N mixtures, as well as the various rapid- and long-acting insulin analogs, have become the market standard, particularly disposable pen devices with prefilled insulin reservoirs, with regard to size and strength. The advantages of the pen devices are primarily better compliance for the patient through a variety of factors including more accurate and reproducible dose control, easier transport of the drug, more discrete dose administration, timelier dose administration, and greater convenience.

■ Continuous Subcutaneous Insulin Infusion: External Pumps

As previously mentioned, solution formulations of human insulin specifically designed for continuous subcutaneous insulin infusion (CSII) are commercially available. CSII systems were traditionally used by a small population of diabetic patients but have become more popular with the recent introduction of rapid-acting insulin analogs. Currently, all three rapid-acting insulin analog formulations have received regulatory approval for this mode of delivery. Specific in vitro data demonstrating physicochemical stability for CSII has been reported for Humalog® (DeFelippis et al. 2006; Sharrow et al. 2012), Novolog®, and Apidra® (Senstius et al. 2007a, b). Pump devices contain glass or plastic reservoirs that must be hand filled from vial presentations by the patient. Some pumps have been specifically designed to accept the same glass 3-mL cartridges used in pen injector systems. Due to concerns over the impact of elevated temperature exposure and mechanical stress on the integrity of the insulin molecule along with the potential increased risk of microbial contamination, the patient information leaflets for the rapid-acting insulin analog products specify time intervals for changing the CSII infusion set as well as the infusion site. The package information leaflets should be consulted for the maximum duration each product may remain in the CSII reservoir. This time period varies with 7, 6, or 2 days listed for Humalog®, Novolog®, and Apidra®, respectively. As always, the patient information leaflets supplied with these products should be consulted for the most current information related to in-use periods.

Noninvasive Delivery

Since the discovery of insulin, there has been a strong desire to overcome the need for injection-based therapy (cf. Chap. 4). Progress has been made in the form of needle-free injector systems (Robertson et al. 2000), but these devices have not gained widespread acceptance presumably because administration is not entirely pain-free, device costs are high, and other factors make it less desirable than traditional injection. Extensive research efforts have also focused on noninvasive routes of administration with attempts made to demonstrate the feasibility of transdermal, nasal, buccal, ocular, pulmonary, oral, and even rectal delivery of insulin (Heinemann et al. 2001). Unfortunately, most attempts failed to progress beyond the proof of concept stage because low bioavailability, dose–response variability, and other adverse factors seriously called into question commercial viability. This situation has changed to some extent for pulmonary and buccal delivery of insulin. Several pulmonary delivery systems specifically aimed at insulin administration have advanced sufficiently through development to enable more extensive studies in human clinical trials, and comprehensive reviews examining this work in detail are available (Patton et al. 1999, 2004; Cefalu 2004). One of these insulin pulmonary delivery system, referred to as Exubera®, received regulatory approval in both Europe and the United States (White et al. 2005). Exubera® consisted of a dry powder insulin formulation composed of small geometric diameter particles produced by spray drying (Eljamal et al. 2003). The powder formulation was packaged into individual blisters and combined with an active device that incorporates a mechanical energy source to achieve dispersion and aerosolization of the particles. While the pharmacological properties reported for this Exubera® were deemed appropriate to meet prandial insulin requirements, the product was ultimately withdrawn from the market shortly after being introduced. Several reasons for the limited use by providers and patients that prompted this action include (1) need for follow-up of pulmonary function tests, (2) large delivery device, (3) insulin dose in capsule marked in mg rather than the traditional units of insulin, and (4) cost and lack of payers for reimbursement or a higher tier (co-pay) reimbursement (Garg and Kelly 2009). Consequently, the pulmonary insulin development programs of other major insulin manufacturers were terminated prior to seeking evaluation by regulatory authorities for potential marketing approval. Only one pulmonary delivery technology from Mannkind Corp. currently remains as an active development program (Pfützner et al. 2002; Richardson and Boss 2007; Peyrot and Rubin 2010; Heinemann 2012).

In addition to pulmonary insulin, a buccal insulin product, referred to as Oralin™, has been developed consisting of a solution formulation of insulin containing various absorption enhancers needed to achieve mucosal absorption (Modi et al. 2002), and a metered-dose inhaler is used to administer a fine mist into the oral cavity. Clinical study results evaluating this buccal delivery system in healthy subjects as well as patients with T1DM and T2DM have been reported (Modi et al. 2002; Cernea et al. 2004). The regulatory approval status of Oralin™ is still limited to only a few countries; however, clinical investigations are continuing presumably to acquire data needed to support additional marketing authorizations in other locations.

The future of noninvasive insulin administration is presently uncertain. Withdrawal of Exubera® was clearly a major setback for pulmonary delivery, and the situation for Oralin™ suggests a rather challenging path to regulatory approval. The lack of any significant developments in other noninvasive routes of delivery may reflect a general realization of the limited practicality of such products. Indeed, a recent examination of the scientific literature suggests there is an apparent decline in research efforts focusing on noninvasive insulin delivery (Heinemann 2012).

Storage

Insulin formulations should be stored in a cool place that avoids sunlight. Vials or cartridges that are not in active use should be stored under refrigerated (2–8 °C) conditions. Vials or cartridges in active use may be stored at ambient temperature. The in-use period for insulin formulations ranges from 28 to 42 days depending upon the product and its chemical, physical, and microbiological stability during use. High temperatures, such as those found in non-air-conditioned vehicles in the summer or other non-climate-controlled conditions, should be avoided due to the potential for chemical and/or physical changes to the formulation properties. Insulin formulations should not be frozen; if this occurs, the product should be disposed of immediately, since either the formulation or the container-closure integrity may be compromised. Insulin formulations should never be purchased or used past the expiration date on the package. Further information on storage and use of specific insulin products are contained in their respective patient information leaflets.

Usage

Resuspension

Insulin suspensions (e.g., NPH, ILPS, premixtures) should be resuspended by gentle back-and-forth mixing and rolling of the vial between the palms to obtain a uniform, milky suspension. The patient should be advised of the resuspension technique for specific insoluble insulin and insulin analog formulations, which is detailed in the package insert. The homogeneity of suspensions is critical to obtaining an accurate dose. Any suspension that fails to provide a homogeneous dispersion of particles should not be used. Insulin formulations contained in cartridges in pen injectors may be suspended in a similar manner; however, the smaller size of the container and shape of the pen injector may require slight modification of the resuspension method to ensure complete resuspension. A bead (glass or metal) is typically added to cartridges to aid in the resuspension of suspension formulations.

Dosing

Dose withdrawal should immediately follow the resuspension of any insulin suspension. The patient should be instructed by their doctor, pharmacist, or nurse educator in proper procedures for dose administration. Of particular importance are procedures for disinfecting the container top and injection site. The patient is also advised to use a new needle and syringe for each injection. Reuse of these components, even after cleaning, may lead to contamination of the insulin formulation by microorganisms or by other materials, such as cleaning agents.

Extemporaneous Mixing

As discussed above in the section on "Intermediate-Acting Insulin," regular insulin can be mixed in the syringe with NPH and is stable enough to be stored for extended periods of time.

With regard to extemporaneous mixing of the newer insulin analogs, caution must be used. Lantus®, due to its acidic pH, should not be mixed with other fast- or rapid-acting insulin formulations which are formulated at neutral pH. If Lantus® is mixed with other insulin formulations, the solution may become cloudy due to isoelectric point (pI) precipitation of both the insulin glargine and the fast- or rapid-acting insulin resulting from pH changes. Consequently, the PK/PD profile, e.g., onset of action and time to peak effect, of Lantus® and/or the mixed insulin may be altered in an unpredictable manner. With regard to rapid-acting insulin analogs, extemporaneous mixing with human insulin NPH formulations is acceptable if used immediately. Under no circumstances should these formulations be stored as mixtures, as human insulin and insulin analog exchange can occur between solution and the crystalline matter, thereby potentially altering time-action profiles of the solution insulin analog. With regard to Levemir®, the human prescription drug label states that the product should not be diluted or mixed with any other insulin or solution to avoid altered and unpredictable changes in PK or PD profile (e.g., onset of action, time to peak effect).

SELF-ASSESSMENT QUESTIONS

■ Questions

1. Which insulin analog formulations cannot be mixed and stored? Why?
2. What are the primary chemical and physical stability issues with human insulin formulations?

■ Answers

1. Lantus®, a long-acting insulin formulation which is formulated at pH 4.0, should not be mixed with rapid- or fast-acting insulin, which are formulated under neutral pH. If Lantus® is mixed with other insulin formulations, the solution may become cloudy due to pI precipitation of both the insulin glargine and the fast- or rapid-acting insulin resulting from pH changes. Consequently, the PK/PD profile, e.g., onset of action and time to peak effect, of Lantus® and/or the mixed insulin may be altered in an unpredictable manner.
2. The two primary modes of chemical degradation are deamidation and HMWP formation. These routes of chemical degradation occur in all formulations. However, they are generally slower in suspension formulations. Physical instability is most often observed in insulin suspension formulations and pump formulations. In suspension formulations, particle agglomeration can occur resulting in the visible clumping of the crystalline and/or amorphous insulin. The soluble insulin in pump formulations can also precipitate or aggregate.

REFERENCES

Anderson JH Jr, Brunelle RL, Keohane P, Koivisto VA, Trautmann ME, Vignati L, DiMarchi R (1997) Mealtime treatment with insulin analog improves postprandial hyperglycemia and hypoglycemia in patients with non-insulin-dependent diabetes mellitus. Multicenter Insulin Lispro Study Group. Arch Intern Med 157:1249–1255

Baker EN, Blundell TL, Cutfield JF, Cutfield SM, Dodson EJ, Dodson GG, Hodgkin DM, Hubbard RE, Isaacs NW, Reynolds CD, Sakabe K, Sakabe N, Vijayan NM (1988) The structure of 2Zn pig insulin crystals at 1.5Å resolution. Philos Trans R Soc Lond B Biol Sci 319:369–456

Balschmidt P (1996) Asp^{B28} Insulin crystals. US Patent 5,547,930

Bell DS, Clements RS Jr, Perentesis G, Roddam R, Wagenknecht L (1991) Dosage accuracy of self-mixed vs premixed insulin. Arch Intern Med 151:2265–2269

Binder C (1969) Absorption of injected insulin. A clinical-pharmacologic study. Acta Pharmacol Toxicol (Copenh) 27(Suppl 2):1–84

Binder C, Lauritzen T, Faber O, Pramming S (1984) Insulin pharmacokinetics. Diabetes Care 7:188–199

Birkeland KI, Home PD, Wendisch U, Ratner RE, Johansen T, Endahl LA, Lyby K, Jendle JH, Roberts AP, DeVries JH, Meneghini LF (2011) Insulin degludec in type 1 diabetes: a randomized controlled trial of a new-generation ultra-long-acting insulin compared with insulin glargine. Diabetes Care 34:661–665

Bliss M (1982) Who discovered insulin. In: The discovery of insulin. McClelland and Stewart Limited, Toronto, pp 189–211

Boldo A, Comi RJ (2012) Clinical experience with U500 insulin: risks and benefits. Endocr Pract 18:56–61

Brackenridge B (1994) Diabetes medicines: insulin. In: Brackenridge B (ed) Managing your diabetes. Eli Lilly and Company, Indianapolis, pp 36–50

Brader ML, Dunn MF (1991) Insulin hexamers: new conformations and applications. Trends Biochem Sci 16:341–345

Brange J (1987a) Insulin preparations. In: Galenics of insulin. Springer, Berlin, pp 17–39

Brange J (1987b) Production of bovine and porcine insulin. In: Galenics of insulin, Springer, Berlin, pp 1–5

Brange J (1987c) Insulin preparations. In: Galenics of insulin. Springer, Berlin, pp 58–60

Brange J, Langkjaer L (1992) Chemical stability of insulin. 3. Influence of excipients, formulation, and pH. Acta Pharm Nord 4:149–158

Brange J, Ribel U, Hansen JF, Dodson G, Hansen MT, Havelund S, Melberg SG, Norris F, Norris K, Snel L et al (1988) Monomeric insulins obtained by protein engineering and their medical implications. Nature 333:679–682

Brange J, Owens DR, Kang S, Vølund A (1990) Monomeric insulins and their experimental and clinical applications. Diabetes Care 13:923–954

Brange J, Havelund S, Hougaard P (1992a) Chemical stability of insulin. 2. Formation of higher molecular weight transformation products during storage of pharmaceutical preparations. Pharm Res 9:727–734

Brange J, Langkjaer L, Havelund S, Vølund A (1992b) Chemical stability of insulin. 1. Hydrolytic degradation during storage of pharmaceutical preparations. Pharm Res 9:715–726

Brems DN, Alter LA, Beckage MJ, Chance RE, DiMarchi RD, Green LK, Long HB, Pekar AH, Shields JE, Frank BH (1992) Altering the association properties of insulin by amino acid replacement. Protein Eng 6:527–533

Brennan TV, Clarke S (1994) Deamidation and isoasparate formation in model synthetic peptides. In: Aswad DW (ed) Deamidation and isoaspartate formation in peptides and proteins. CRC Press, Boca Raton, pp 65–90

Cefalu WT (2004) Concept, strategies, and feasibility of noninvasive insulin delivery. Diabetes Care 27:239–246

Cernea S, Kidron M, Wohlgelernter J, Modi P, Raz I (2004) Comparison of pharmacokinetic and pharmacodynamic properties of single-dose oral insulin spray and subcutaneous insulin injection in healthy subjects using the euglycemic clamp technique. Clin Ther 26:2084–2091

Chacra AR, Kipnes M, Ilag LL, Sarwat S, Giaconia J, Chan J (2010) Comparison of insulin lispro protamine suspension and insulin detemir in basal-bolus therapy in patients with type 1 diabetes. Diabet Med 27:563–569

Charman SA, McLennan DN, Edwards GA, Porter CJH (2001) Lymphatic absorption is a significant contributor to the subcutaneous bioavailability of insulin in a sheep model. Pharm Res 18:1620–1626

Charvet R, Soula G, Mora G, Soula O, Soula R (2010) Fast-acting insulin formulations, US Patent Application 2010249020A

Ciszak E, Beals JM, Frank BH, Baker JC, Carter ND, Smith GD (1995) Role of the C-terminal B-chain residues in insulin assembly: the structure of hexameric LysB28ProB29-human insulin. Structure 3:615–622

Dailey G, Rosenstock J, Moses RG, Ways K (2004) Insulin glulisine provides improved glycemic control in patients with type 2 diabetes. Diabetes Care 27:2363–2368

Darrington RT, Anderson BD (1995) Effects of insulin concentration and self-association on the partitioning of its A-21 cyclic anhydride intermediate to desamido insulin and covalent dimer. Pharm Res 12: 1077–1084

Davis SN, Thompson CJ, Brown MD, Home PD, Alberti KG (1991) A comparison of the pharmacokinetics and metabolic effects of human regular and NPH mixtures. Diabetes Res Clin Pract 13:107–117

De la Peña A, Riddle M, Morrow LA, Jiang HH, Linnebjerg H, Scott A, Win KM, Hompesch M, Mace KF, Jacobson JG, Jackson JA (2011) Pharmacokinetics and pharmacodynamics of high-dose human regular U-500 insulin versus human regular U-100 insulin in healthy obese subjects. Diabetes Care 34:2496–2501

Deckert T (1980) Intermediate-acting insulin preparations: NPH and lente. Diabetes Care 3:623–626

DeFelippis MR, Bakaysa DL, Bell MA, Heady MA, Li S, Pye S, Youngman KM, Radziuk J, Frank BH (1998) Preparation and characterization of a cocrystalline suspension of [LysB28, ProB29]-human insulin analogue. J Pharm Sci 87:170–176

DeFelippis MR, Bell MA, Heyob JA, Storms SM (2006) In vitro stability of insulin lispro in continuous subcutaneous insulin infusion. Diabetes Technol Ther 8:358–368

Derewenda U, Derewenda Z, Dodson EJ, Dodson GG, Reynolds CD, Smith GD, Sparks C, Swenson D (1989) Phenol stabilizes more helix in a new symmetrical zinc insulin hexamer. Nature 338:594–596

Dodd SW, Havel HA, Kovach PM, Lakshminarayan C, Redmon MP, Sargeant CM, Sullivan GR, Beals JM (1995) Reversible adsorption of soluble hexameric insulin onto the surface of insulin crystals cocrystallized with protamine: an electrostatic interaction. Pharm Res 12:60–68

Dreyer M, Prager R, Robinson A, Busch K, Ellis G, Souhami E, Van Leendert R (2005) Efficacy and safety of insulin glulisine in patients with type 1 diabetes. Horm Metab Res 37:702–707

Eljamal M, Patton JS, Foster LC, Platz RM (2003) Powdered Pharmaceutical Formulation Having Improved Dispersibility, US Patent 6,582,729

Fogelfeld L, Dharmalingam M, Robling K, Jones C, Swanson D, Jacober SJ (2010) A randomized, treat-to-target trial comparing insulin lispro protamine suspension and insulin detemir in insulin-naive patients with type 2 diabetes. Diabet Med 27:181–188

Galloway JA (1988) Chemistry and clinical use of insulin. In: Galloway JA, Potvin JH, Shuman CR (eds) Diabetes mellitus, 9th edn. Lilly Research Laboratories, Indianapolis, pp 105–133

Galloway JA, Chance RE (1994) Improving insulin therapy: achievements and challenges. Horm Metab Res 26:591–598

Galloway JA, Spradlin CT, Nelson RL, Wentworth SM, Davidson JA, Swarner JL (1981) Factors influencing the absorption, serum insulin concentration, and blood glucose responses after injections of regular insulin and various insulin mixtures. Diabetes Care 4:366–376

Galloway JA, Spradlin CT, Jackson RL, Otto DC, Bechtel LD (1982) Mixtures of intermediate-acting insulin (NPH and Lente) with regular insulin: an update. In: Skyler JS (ed) Insulin update: 1982. Exerpta Medica, Princeton, pp 111–119

Garg SK, Kelly WC (2009) Insulin delivery via lungs-is it still possible? Diabetes Technol Ther 11 (Suppl 2):S1–S3

Gibney MA, Arce CH, Byron KJ, Hirsch LJ (2010) Skin and subcutaneous adipose layer thickness in adults with diabetes at sites used for insulin injections: implications for needle length recommendations. Curr Med Res Opin 26:1519–1530

Goldman J, Carpenter FH (1974) Zinc binding, circular dichroism, and equilibrium sedimentation studies on insulin (bovine) and several of its derivatives. Biochemistry 13:4566–4574

Hagedorn HC, Jensen BN, Krarup NB, Wodstrup I (1936) Protamine insulinate. JAMA 106:177–180

Hamaguchi T, Hashimoto Y, Miyata T, Kishikawa H, Yano T, Fukushima H, Shichiri M (1990) Effect of mixing short and intermediate NPH insulin or Zn insulin suspension acting human insulin on plasma free insulin levels and action profiles. J Jpn Diabetes Soc 33:223–229

Havelund S, Plum A, Ribel U, Jonassen I, Vølund A, Markussen J, Kurtzhals P (2004) The mechanism of protraction of insulin detemir, a long-acting, acylated analog of human insulin. Pharm Res 21:1498–1504

Heinemann L (2012) New ways of insulin delivery. Int J Clin Pract 66(Suppl 175):35–39

Heinemann L, Weyer C, Rave K, Stiefelhagen O, Rauhaus M, Heise T (1997) Comparison of the time-action profiles of U40- and U100-regular human insulin and the rapid-acting insulin analogue B28 Asp. Exp Clin Endocrinol Diabetes 105:140–144

Heinemann L, Sinha K, Weyer C, Loftager M, Hirschberger S, Heise T (1999) Time-action profile of the soluble, fatty acid acylated, long-acting insulin analogue NN304. Diabet Med 16:332–338

Heinemann L, Pfützner A, Heise T (2001) Alternative routes of administration as an approach to improve insulin therapy: update on dermal, oral, nasal and pulmonary insulin delivery. Curr Pharm Des 7:1327–1351

Heinemann L, Nosek L, Flacke F, Albus K, Krasner A, Pichotta P, Heise T, Steiner S (2012) U-100, pH-neutral formulation of VIAject®: faster onset of action than insulin lispro in patients with type 1 diabetes. Diabetes Obes Metab 14:222–227

Heise T, Weyer C, Serwas A, Heinrichs S, Osinga J, Roach P, Woodworth J, Gudat W, Heinemann L (1998) Time-action profiles of novel premixed preparations of insulin lispro and NPL insulin. Diabetes Care 21:800–803

Heise T, Nosek L, Rønn BB, Endahl L, Heinemann L, Kapitza C, Draeger E (2004) Lower within-subject variability of insulin detemir in comparison to NPH insulin and insulin glargine in people with type 1 diabetes. Diabetes 53:1614–1620

Hermansen K, Madsbad S, Perrild H, Kristensen A, Axelsen M (2001) Comparison of the soluble basal insulin analog insulin detemir with NPH insulin: a randomized open crossover trial in type 1 diabetic subjects on basal-bolus therapy. Diabetes Care 24:296–301

Hoffmann JA, Chance RE, Johnson MG (1990) Purification and analysis of the major components of chum salmon protamine contained in insulin formulations using high-performance liquid chromatography. Protein Expr Purif 1:127–133

Holleman F, Schmitt H, Rottiers R, Rees A, Symanowski S, Anderson JH (1997) Reduced frequency of severe hypoglycemia and coma in well-controlled IDDM patients treated with insulin lispro. The Benelux-UK Insulin Lispro Study Group. Diabetes Care 20: 1827–1832

Home PD (2012) The pharmacokinetics and pharmacodynamics of rapid-acting insulin analogues and their clinical consequences. Diabetes Obes Metab 14: 780–788

Hompesch M, Ocheltree SM, Wondmagegnehu ET, Morrow LA, Kollmeier AP, Campaigne BN, Jacober SJ (2009) Pharmacokinetics and pharmacodynamics of insulin lispro protamine suspension compared with insulin glargine and insulin detemir in type 2 diabetes. Curr Med Res Opin 25:2679–2687

Howey DC, Bowsher RR, Brunelle RL, Woodworth JR (1994) [Lys(B28), Pro(B29)]-human insulin: a rapidly-absorbed analogue of human insulin. Diabetes 43:396–402

Jansen R, Dzwolak W, Winter R (2005) Amyloidogenic self-assembly of insulin aggregates probed by high resolution atomic force microscopy. Biophys J 88: 1344–1353

Janssen MM, Casteleijn S, Devillé W, Popp-Snijders C, Roach P, Heine RJ (1997) Nighttime insulin kinetics and glycemic control in type 1 diabetic patients following administration of an intermediate-acting lispro preparation. Diabetes Care 20:1870–1873

Kaarsholm NC, Havelund S, Hougaard P (1990) Ionization behavior of native and mutant insulins: pK perturbation of B13-Glu in aggregated species. Arch Biochem Biophys 283:496–502

Krayenbuhl C, Rosenberg T (1946) Crystalline protamine insulin. Rep Steno Hosp (Kbh) 1:60–73

Kroeff EP, Owen RA, Campbell EL, Johnson RD, Marks HI (1989) Production scale purification of biosynthetic human insulin by reversed-phase high-performance liquid chromatography. J Chromatogr 461:45–61

Kurtz AB, Gray RS, Markanday S, Nabarro JD (1983) Circulating IgG antibody to protamine in patients treated with protamine-insulins. Diabetologia 25:322–324

Lane WS, Cochran EK, Jackson JA, Scism-Bacon JL, Corey IB, Hirsch IB, Skyler JS (2009) High-dose insulin therapy: is it time for U-500 insulin? Endocr Pract 15:71–79

Modi P, Mihic M, Lewin A (2002) The evolving role of oral insulin in the treatment of diabetes using a novel RapidMistSystem. Diabetes Metab Res Rev 18(Suppl 1):S38–S42

Muchmore DB, Vaughn DE (2010) Review of the mechanism of action and clinical efficacy of recombinant human hyaluronidase coadministration with current prandial insulin formulations. J Diabetes Sci Technol 4:419–428

Nell LJ, Thomas JW (1988) Frequency and specificity of protamine antibodies in diabetic and control subjects. Diabetes 37:172–176

Nielsen L, Khurana R, Coats A, Frokjaer S, Brange J, Vyas S, Uversky VN, Fink AL (2001) Effect of environmental factors on the kinetics of insulin fibril formation: elucidation of the molecular mechanism. Biochemistry 40:6036–6046

Ocheltree SM, Hompesch M, Wondmagegnehu ET, Morrow L, Win K, Jacober SJ (2010) Comparison of pharmacodynamic intrasubject variability of insulin lispro protamine suspension and insulin glargine in subjects with type 1 diabetes. Eur J Endocrinol 163:217–223

Patton JS, Bukar J, Nagarajan S (1999) Inhaled insulin. Adv Drug Deliv Rev 35:235–247

Patton JS, Bukar JG, Eldon MA (2004) Clinical pharmacokinetics and pharmacodynamics of inhaled insulin. Clin Pharmacokinet 43:781–801

Pekar AH, Frank BH (1972) Conformation of proinsulin. A comparison of insulin and proinsulin self-association at neutral pH. Biochemistry 11:4013–4016

Peyrot M, Rubin RR (2010) Effect of technosphere inhaled insulin on quality of life and treatment satisfaction. Diabetes Technol Ther 12:49–55

Pfutzner A, Mann AE, Steiner SS (2002) Technosphere/insulin–a new approach for effective delivery of human insulin via the pulmonary route. Diabetes Technol Ther 4:589–594

Porcellati F, Bolli GB, Fanelli CG (2011) Pharmacokinetics and pharmacodynamics of basal insulins. Diabetes Technol Ther 13(Suppl 1):S15–S24

Porter CJ, Charman SA (2000) Lymphatic transport of proteins after subcutaneous administration. J Pharm Sci 89:297–310

Richardson PC, Boss AH (2007) Technosphere insulin technology. Diabetes Technol Ther 9(Suppl 1):S65–S72

Robertson KE, Glazer NB, Campbell RK (2000) The latest developments in insulin injection devices. Diabetes Educ 26:135–152

Schade DS, Santiago JV, Skyler JS, Rizza RA (1983) Intensive insulin therapy. Medical Examination Publishing, Princeton, p 24

Segal AR, Brunner JE, Burch FT, Jackson JA (2010) Use of concentrated insulin human regular (U-500) for patients with diabetes. Am J Health Syst Pharm 67:1526–1535

Senstius J, Harboe E, Westermann H (2007a) In vitro stability of insulin aspart in simulated continuous subcutaneous insulin infusion using a MiniMed 508 pump. Diabetes Technol Ther 9:75–79

Senstius J, Poulsen C, Hvass A (2007b) Comparison of in vitro stability for insulin aspart and insulin glulisine during simulated use in infusion pumps. Diabetes Technol Ther 9:517–521

Sharrow SD, Glass LC, Dobbins MA (2012) 14-day in vitro chemical stability of insulin lispro in the MiniMed paradigm pump. Diabetes Technol Ther 14:264–270

Strojek K, Shi C, Carey MA, Jacober SJ (2010) Addition of insulin lispro protamine suspension or insulin glargine to oral type 2 diabetes regimens: a randomized trial. Diabetes Obes Metab 12:916–922

Supersaxo A, Hein WR, Steffen H (1990) Effect of molecular weight on the lymphatic absorption of water-soluble compounds following subcutaneous administration. Pharm Res 7:167–169

USP Monographs: Insulin (2013) USP36-NF31: 3911–3913

Vague P, Selam JL, Skeie S, De Leeuw I, Elte JW, Haahr H, Kristensen A, Draeger E (2003) Insulin detemir is associated with more predictable glycemic control and reduced risk of hypoglycemia than NPH insulin in patients with type 1 diabetes on a basal-bolus regimen with premeal insulin aspart. Diabetes Care 26:590–596

Waldhäusl W, Bratusch-Marrain P, Gasic S, Kom A, Nowotny P (1979) Insulin production rate following glucose ingestion estimated by splanchnic C-peptide output in normal man. Diabetologia 17:221–227

Weyer C, Heise T, Heinemann L (1997) Insulin aspart in a 30/70 premixed formulation. Pharmacodynamic properties of a rapid-acting insulin analog in stable mixture. Diabetes Care 20:1612–1614

White S, Bennett DB, Cheu S, Conley PW, Guzek DB, Gray S, Howard J, Malcolmson R, Parker JM, Roberts P, Sadrzadeh N, Schumacher JD, Seshadri S, Sluggett GW, Stevenson CL, Harper NJ (2005) EXUBERA: pharmaceutical development of a novel product for pulmonary delivery of insulin. Diabetes Technol Ther 7:896–906

Whittingham JL, Edwards DJ, Antson AA, Clarkson JM, Dodson GG (1998) Interactions of phenol and m-cresol in the insulin hexamer, and their effect on the association properties of B28 pro –> Asp insulin analogues. Biochemistry 37:11516–11523

Whittingham JL, Jonassen I, Havelund S, Roberts SM, Dodson EJ, Verma CS, Wilkinson AJ, Dodson GG (2004) Crystallographic and solution studies of N-lithocholyl insulin: a new generation of prolonged-acting human insulins. Biochemistry 43:5987–5995

Zerhouni E (2012) Sanofi. 32nd Cowen Annual Health Care Conference, Boston. http://en.sanofi.com/investors/events/other_events/2012/Presentation_2012-03-06_Cowen_Zerhouni.aspx. Accessed 2 Jul 2013

Ziesmer AE, Kelly KC, Guerra PA, George KG, Dunn KL (2012) U500 regular insulin use in insulin-resistant type 2 diabetic veteran patients. Endocr Pract 18: 34–38

Zinman B, Fulcher G, Rao PV, Thomas N, Endahl LA, Johansen T, Lindh R, Lewin A, Rosenstock J, Pinget M, Mathieu C (2011) Insulin degludec, an ultra-long-acting basal insulin, once a day or three times a week versus insulin glargine once a day in patients with type 2 diabetes: a 16-week, randomised, open-label, phase 2 trial. Lancet 377:924–931

RECOMMENDED READING

American Diabetes Association (2011) Practical insulin: a handbook for prescribing providers, 3rd edn. American Diabetes Association, New York

Bliss M (1982) The discovery of insulin. McClelland and Stewart Limited, Toronto

Brange J (1987) Galenics of insulin. Springer, Berlin

Burant C (ed) (2008) Medical management of type 2 diabetes, 6th edn. American Diabetes Association, New York

Cooper T, Ainsburg A (2010) Breakthrough: Elizabeth Hughes, the discovery of insulin, and the making of a medical miracle. St. Martin's Press, New York

Galloway JA, Potvin JH, Shuman CR (1988) Diabetes mellitus, 9th edn. Lilly Research Laboratories, Indianapolis

Wolfsdorf JI (2009) Intensive diabetes management, 4th edn. American Diabetes Association, New York

13 Follicle-Stimulating Hormone

Tom Sam, Renato de Leeuw, Gijs Verheijden, and Anneke Koole

INTRODUCTION

About 15 % of all couples experience infertility at some time during their reproductive lives. Nowadays, infertility can be treated by the use of assisted reproductive technologies (ART), such as in vitro fertilization (IVF) and intracytoplasmic sperm injection (ICSI). A common element of these programs is the treatment with follicle-stimulating hormone (FSH) to increase the number of oocytes retrievable for the IVF or ICSI procedure (multifollicular development). Patients suffering from female infertility because of chronic anovulation may also be treated with FSH, then with the aim to achieve monofollicular development.

Natural FSH is produced and secreted by the anterior lobe of the pituitary, a gland at the base of the brain. Its target is the FSH receptor at the surface of the granulosa cells that surround the oocyte. FSH acts synergistically with estrogens and luteinizing hormone (LH) to stimulate proliferation of these granulosa cells, which leads to follicular growth. As the primary function of FSH in the female is the regulation of follicle growth and development, this process explains why deficient endogenous production of FSH may cause infertility. In males, FSH plays a pivotal role in spermatogenesis.

T. Sam, Ph.D. (✉)
Regulatory Affairs Department, MSD,
Molenstraat 110, Oss 5342 CC, The Netherlands
e-mail: tom.sam@merck.com

R. de Leeuw
Regulatory Affairs Department,
MSD, Oss, The Netherlands

G. Verheijden, Ph.D.
Synthon Biopharmaceuticals B.V.,
Nijmegen, The Netherlands

A. Koole
Global CMC Regulatory Affairs – Biologics,
MSD, Oss, The Netherlands

FSH preparations for infertility treatment are traditionally derived from urine from (post) menopausal women. As over 100,000 l of urine may be required for a single batch, many thousands of donors are needed. Hence, the source of urinary FSH is heterogeneous and the sourcing cumbersome. Moreover, in addition to FSH, these urinary preparations contain impurities including pharmaceutically active proteins such as LH. Recombinant DNA technology allows the reproducible manufacturing of FSH preparations of high purity and specific activity, devoid of urinary contaminants. Recombinant FSH is produced using a Chinese hamster ovary (CHO) cell line, transfected with the genes encoding for the two human FSH subunits (Van Wezenbeek et al. 1990; Howles 1996). The isolation procedures render a product of high purity (at least 99 %), devoid of LH activity and very similar to natural FSH.

Currently, there are several clinically approved recombinant FSH-containing drug products on the various markets. The most widely approved products are Gonal-F®, manufactured by Merck Serono S.A., and Puregon®, with the brand name of Follistim® in the USA and Japan, manufactured by N.V. Organon, now part of Merck Sharp and Dohme. Regulatory authorities have issued two distinct International Nonproprietary Names (INN) for the two corresponding recombinant FSH drug substances, i.e., follitropin-α (Gonal-F®) and follitropin-β (Puregon®/Follistim®). In addition, a few other recombinant FSH preparations (claimed to be "biosimilar" to follitropin-α or follitropin-β) are under development or available in a very limited number of countries.

FSH IS A GLYCOPROTEIN HORMONE

Follicle-stimulating hormone belongs to a family of structurally related glycoproteins which also includes luteinizing hormone (LH); chorionic gonadotropin (CG), collectively called the gonadotropins; and thyroid-stimulating hormone (TSH, also named thyrotropin). Each hormone is a heterodimeric protein

D.J.A. Crommelin, R.D. Sindelar, and B. Meibohm (eds.), *Pharmaceutical Biotechnology*,
DOI 10.1007/978-1-4614-6486-0_13, © Springer Science+Business Media New York 2013

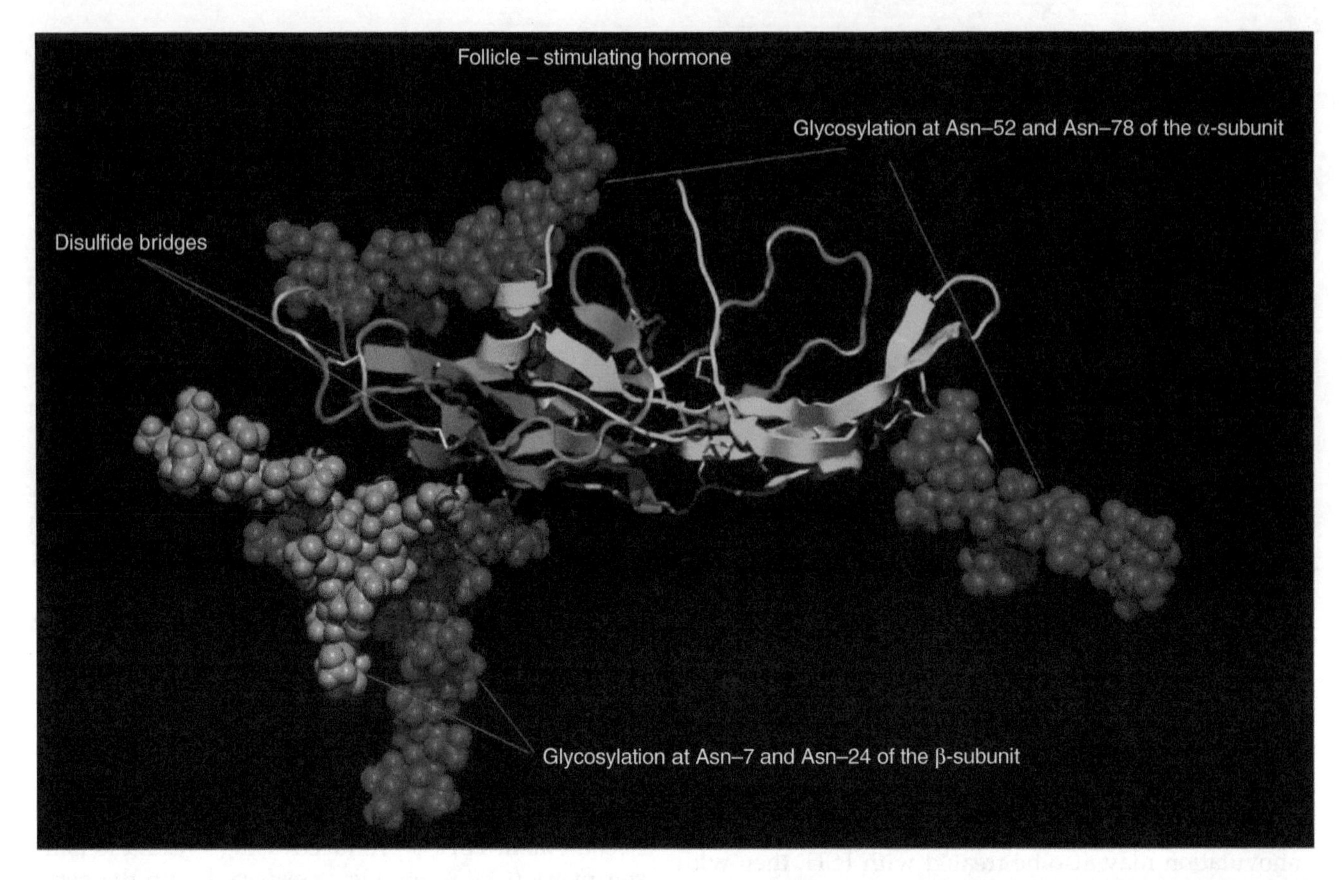

Figure 13.1 ■ A three-dimensional model of FSH. The ribbons represent the polypeptide backbones of the α-subunit (*green ribbon*) and the β-subunit (*blue ribbon*). The carbohydrate side chains (*violet and pink space-filled globules*) cover large areas of the surface of the polypeptide subunits.

consisting of two non-covalently associated glycoprotein subunits, denoted α and β. The α-subunit contains five intra-subunit disulfide bonds and is identical for all these glycoproteins, and it is the β-subunit (having six intra-subunit disulfide bonds) that provides each hormone with its specific biological function.

The glycoprotein subunits of FSH consist of two polypeptide backbones with carbohydrate side chains attached to the two asparagine (Asn) amino acid residues on each subunit. The oligosaccharides are attached to Asn-52 and Asn-78 on the α-subunit (92 amino acids) and to Asn-7 and Asn-24 on the β-subunit (111 amino acids). The glycoprotein FSH has a molecular mass of approximately 35 kDa. For the FSH preparation to be biologically active, the two subunits must be correctly assembled into their three-dimensional heterodimeric protein structure and posttranslationally modified (Fig. 13.1).

Assembly and glycosylation are intracellular processes that take place in the endoplasmic reticulum and in the Golgi apparatus. This glycosylation process leads to the formation of a population of hormone isoforms differing in their carbohydrate side-chain composition. The carbohydrate side chains of FSH are essential for its biological activity since they (1) influence FSH receptor binding, (2) play an important role in the signal transduction into the FSH target cell, and (3) affect the plasma residence time of the hormone.

Recombinant FSH contains approximately one third carbohydrate on a mass per mass basis. The carbohydrate side chains are composed of mannose, fucose, N-acetyl-glucosamine, galactose, and sialic acid. Structure analysis by ^{1}H-NMR spectroscopy on oligosaccharides enzymatically cleaved from follitropin-β reveals minor differences with natural FSH. For instance, the bisecting GlcNAc residues are lacking in the recombinant molecule, simply because the FSH-producing CHO cells do not possess the enzymes to incorporate these residues. Furthermore, the carbohydrate side chains of recombinant FSH exclusively contain α2-3-linked sialic acid, whereas in the natural hormone α1-6-linked sialic acid occurs as well. However, all carbohydrate side chains identified in recombinant FSH are moieties normally found in other natural human glycoproteins.

Whereas FSH only contains N-linked carbohydrates, human chorionic gonadotropin (hCG) also carries 4 O-linked (at serine or threonine residues) carbohydrates, all located at the carboxy-terminal peptide (CTP) of its beta subunit. This glycosylated CTP is the major difference with the beta subunit of LH and is demonstrated to be responsible for the much longer plasma residence time of hCG compared to natural LH (Matzuk et al. 1990).

PRODUCTION OF RECOMBINANT FSH

The genes coding for the human FSH α-subunit and β-subunit were inserted in cloning vectors (plasmids) to enable efficient transfer into recipient cells. These vectors also contained promoters that could direct transcription of foreign genes in recipient cells. CHO cells were selected as recipient cells since they were easily transfected with foreign DNA and are capable of synthesizing complex glycoproteins. Furthermore, they could be grown in cell cultures on a large scale. To construct an FSH-producing cell line, N.V. Organon, the manufacturer of Puregon®/Follistim®, used one single vector containing the coding sequences for both subunit genes (Olijve et al. 1996). Merck Serono S.A., the manufacturer of Gonal-F®, used two separate vectors, one for each subunit gene (Howles 1996). Following transfection, a genetically stable transformant producing biologically active recombinant FSH was isolated. For the CHO cell line used for manufacturing Puregon®/Follistim®, it was shown that approximately 150–450 gene copies were present.

To establish a master cell bank (MCB), identical homogeneous cell preparations of the selected clone are stored in individual vials and cryopreserved until needed. Subsequently, a working cell bank (WCB) is established by the expansion of cells derived from a single vial of the MCB, and aliquots are put in vials and cryopreserved as well. Each time a production run is started, cells from one or more vials of the WCB are cultured.

Both recombinant FSH products are isolated from cell culture supernatants. These supernatants are collected from a perfusion-type bioreactor containing recombinant FSH-producing CHO cells grown on microcarriers. This is because the CHO cell lines used are anchorage-dependent cells, which implies that a proper surface must be provided for cell growth. The reactor is perfused with growth-promoting medium during a period that may continue for up to three months (see also Chap. 3). The downstream purification processes for the isolation of the two recombinant FSH products are different. For Puregon®/Follistim®, a series of chromatographic steps, including anion and cation exchange chromatography, hydrophobic chromatography, and size-exclusion chromatography, is used. Recombinant FSH in Gonal-F® is obtained by a similar process of five chromatographic steps but also includes an immunoaffinity step using a murine FSH-specific monoclonal antibody (European Public Assessment Report, Gonal-F 2011). In both production processes, each purification step is rigorously controlled in order to ensure the batch-to-batch consistency of the purified product.

DESCRIPTION OF RECOMBINANT FSH

■ Structural Characteristics

Like urinary-sourced (natural) FSH, the recombinant versions exist in several distinct molecular forms (isohormones), with identical polypeptide backbones but with differences in oligosaccharide structure, in particular in the degree of terminal sialylation. These isohormones can be separated by chromatofocusing or isoelectric focusing on the basis of their different isoelectric points (pI, as has been demonstrated for follitropin-β (De Leeuw et al. 1996) (Fig. 13.2)). The typical pattern for FSH indicates an isohormone distribution between pI values of 6 and 4. To obtain structural information at the subunit level, the two subunits were separated by RP-HPLC and treated to release the N-linked carbohydrate side chains. Fractions with low pI values (acidic fractions) displayed a high content of tri- and tetrasialo oligosaccharides and a low content of neutral and monosialo oligosaccharides. For fractions with a high pI (basic fractions) value, the reverse was found. The β-subunit carbohydrate side chains appeared to be more heavily sialylated and branched than the α-subunit carbohydrate side chains. The low pI value isohormones of follitropin-β have a high sialic acid/galactose ratio and are rich in tri- and tetra-antennary N-linked carbohydrate side chains, as compared with the side chains of the high pI value isohormones.

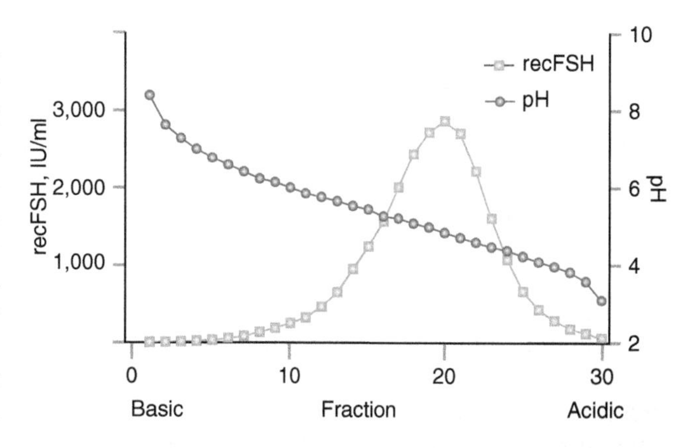

Figure 13.2 ■ Isohormone profile of recombinant follicle-stimulating hormone (follitropin-β) after preparative free-flow focusing (De Leeuw et al. 1996). The FSH concentration was determined by a two-site immunoassay that is capable of quantifying the various isohormones equally well.

One of the tools for further characterization is the immunoassay. Due to the specific recognition characteristics of the antibodies used, this assay determines FSH-specific structural features and provides a relative measure for the quantity of FSH, as it is not sensitive to the differences in glycosylation.

■ Biological Properties of Recombinant FSH Isohormones

An FSH preparation can be biologically characterized with several essentially different assays, each having its own specific merits (Mannaerts et al. 1992). The receptor-binding assay provides information on the proper conformation for interaction with the FSH receptor. Receptor-binding studies with calf testis membranes have shown that FSH isoform activity in follitropin-β decreases when going from high to low pI isoforms. The in vitro bioassay measures the capability of FSH to transduce signals into target cells (the intrinsic bioactivity). The in vitro bioactivity, assessed in the rat Sertoli cell bioassay, also decreases when going from high to low pI isoforms. The in vivo bioassay provides the overall bioactivity of an FSH preparation. It is determined by the number of molecules, the plasma residence time, the receptor-binding activity, and the signal transduction. Interestingly, in contrast to the receptor-binding and in vitro bioassays, the in vivo biological activity determined in rats shows an approximate 20-fold increase between isoforms with a pI value of 5.49, as compared to those with a pI of 4.27. These results indicate that the basic isohormones exhibit the highest receptor-binding and signal transduction activity, whereas the acidic isohormones are the more active forms under in vivo conditions. This notion also warrants pharmacokinetic studies to further characterize the biological properties of FSH preparations.

■ Pharmacokinetic Behavior of Recombinant FSH Isohormones

The pharmacokinetic behavior of follitropin-β and its isohormones was investigated in Beagle dogs that were given an intramuscular bolus injection of a number of FSH isohormone fractions, each with a specific pI value. With a decrease in pI value from 5.49 (basic) to 4.27 (acidic), the AUC increased and the clearance decreased, each more than tenfold (Fig. 13.3). A more than twofold difference in elimination half-life between the most acidic and the most basic FSH isohormone fraction was calculated. The absorption rate of the two most acidic isoforms was higher than the absorption rates of all other isoforms. The AUC and the clearance for the follitropin-β preparation, being a mixture of all isohormone fractions, corresponded with the center of the isohormone profile (Fig. 13.3). In contrast, the elimination of the follitropin-β preparation occurred at a rate similar to that of the most acidic fractions, indicating that the elimination rate is largely determined by the removal of the most acidic isoforms from the plasma.

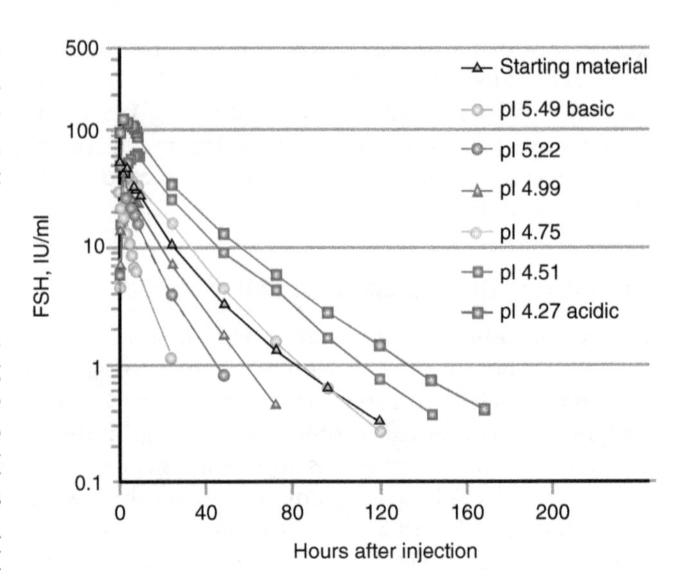

Figure 13.3 ■ Kinetic behavior of FSH isoforms after a single intramuscular injection (20 IU/kg) in Beagle dogs.

Thus, for follitropin-β isohormone fractions, a clear correlation exists between pI value and pharmacokinetic behavior. Increasing acidity leads to an increase in the extent of absorption and elimination half-life and to a decrease in clearance.

PHARMACEUTICAL FORMULATIONS

Recombinant FSH preparations distinguish themselves from the earlier urinary FSH preparations by their high purity (at least 99 %). However, pure proteins are relatively unstable and are therefore often lyophilized, unless some specific stabilizing measures can be taken. FSH preparations are available in different strengths and presentation forms, both as freeze-dried products (powder, cake) and as solution for injection (see Table 13.1). Follitropin-α was originally formulated with sucrose (bulking agent, lyoprotectant), sodium dihydrogen phosphate/disodium hydrogen phosphate, phosphoric acid, and sodium hydroxide (for pH adjustment). In 2002, L-methionine (antioxidant) and polysorbate 20 (to prevent adsorption losses) were added to the single-dose formulation. Follitropin-β is formulated with sucrose, sodium citrate (stabilizer), polysorbate 20 (lyoprotectant and agent to prevent adsorption losses), and hydrochloride/sodium hydroxide (for pH adjustment). The lyophilized preparations are to be reconstituted before use to obtain a ready-for-use solution for injection. In addition to the

Market preparation	Presentation	Container	Strength
Gonal-F® (follitropin-α)	Powder	Ampoule	75 and 150 IU
Merck Serono	Powder	Vial	37.5 IU (2.8 μg), 75 IU (5.5 μg), 150 IU (11 μg), 300 IU, and 600 IU
	Powder	Multidose vial	300 IU (22 μg), 450 IU (33 μg), 1,050 IU (77 μg)
	Solution for injection	Cartridge	300 IU (22 μg), 450 IU(33 μg), 900 IU (66 μg)
Puregon®/Follistim® (follitropin-β)	Solution for injection	Vial	50, 75, 100, 150, 200, and 225 IU
Merck/Merck Sharp and Dohme	Solution for injection	Cartridge/pen injector	150, 300, 600, and 900 IU
Elonva® (corifollitropin alfa) Merck/Merck Sharp and Dohme	Solution for injection	Prefilled syringe equipped with an automatic safety system	100 and 150 μg

Table 13.1 ■ The presentation forms of recombinant FSH products.

freeze-dried presentation form, a solution for injection with several strengths of follitropin-β could be developed. To stabilize the solutions, 0.25 mg of L-methionine had to be added. Furthermore, the solution in the cartridge contains benzyl alcohol as preservative. For follitropin-α, a multidose solution for injection in a prefilled pen became available in 2004. This solution contains poloxamer 188 instead of polysorbate 20, and m-cresol has been added as preservative.

The Puregon®/Follistim® solution for injection is available in vials and is very suitable for titration because of the large range of available strengths as expressed in IU's. Pen injectors have been developed with multidose cartridges containing solution for injection, giving the patient improved convenience.

The solutions for injection should be stored in the refrigerator for a maximum of 3 years with the container kept in the outer carton to protect the solution from light. The patient can keep the solutions at room temperature for a maximum of 3 months. The multidose solution of follitropin-α has a shelf life of 2 years and can be stored for 1 month at room temperature.

CLINICAL ASPECTS

Recombinant FSH products on the market have been approved for two female indications. The first indication is anovulation (including polycystic ovarian disease) in women who are unresponsive to clomiphene citrate (an estrogen receptor modulator). The second indication is controlled ovarian hyperstimulation to induce the development of multiple follicles in medically assisted reproduction programs, such as in vitro fertilization (IVF) and intracytoplasmic sperm injection (ICSI). In addition, recombinant FSH may be used in men with congenital or required hypogonadotropic hypogonadism to stimulate spermatogenesis.

For the treatment of anovulatory patients (aiming at monofollicular growth), it is recommended to start Puregon®/Follistim® treatment with 50 IU per day for 7–14 days and gradually increase dosing with steps of 50 IU if no sufficient response is seen. This gradual dose-increasing schedule is followed in order to prevent multifollicular development and the induction of ovarian hyperstimulation syndrome (a serious condition of unwanted hyperstimulation). In the most commonly applied treatment regimens in IVF, endogenous gonadotropin levels are suppressed by a GnRH agonist or by the more recently approved GnRH antagonists (Cetrotide® and Orgalutran®/Ganirelix acetate injection®). It is recommended to start Puregon® treatment with 100–200 IU of recombinant FSH followed by maintenance doses of 50–350 IU. The availability of a surplus of collected oocytes allows the vitrification of embryos for replacement in frozen-thawn embryo transfer (FTET) cycles. Similar treatment regimens are recommended for Gonal-F®.

After subcutaneous administration, follitropin-β has an elimination half-life of approximately 33 h (Voortman et al. 1999). Steady-state levels of follitropin-β are therefore obtained after 4–5 daily doses reaching therapeutically effective plasma concentrations of FSH. Follitropin-β is administered via the subcutaneous route with good local tolerance. Bioavailability is approximately 77 %. In a large fraction of patients treated with follitropin-β, no formation of antibodies against recombinant FSH- or CHO-cell-derived proteins was observed. Injections of the follitropin-β preparations can be given by the patient herself or her partner.

A NEWLY DEVELOPED FSH ANALOG

The need for daily injections of FSH, especially in combination with GnRH agonists, is a burden for the women treated in an ART regimen. Therefore, several

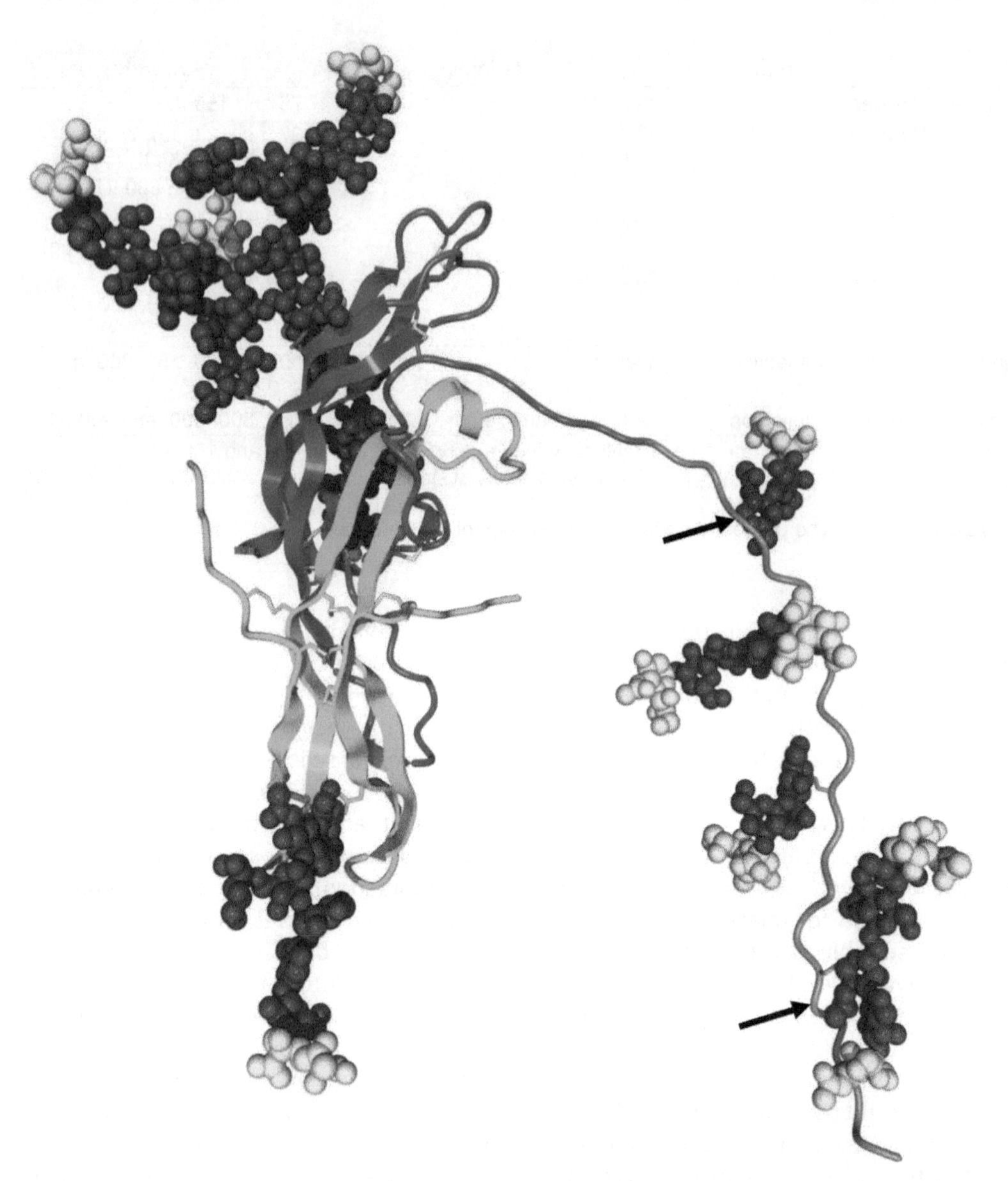

Figure 13.4 ■ A three-dimensional model of corifollitropin alfa. The ribbons represent the polypeptide backbones of the α-subunit (*green ribbon*) and the β-subunit (*blue ribbon*). The carbohydrate side chains (*pink and yellow space-filled globules*) represent N-linked and O-linked carbohydrates. The sialic acid carbohydrates are depicted in yellow. The *arrows* indicate additional O-linked carbohydrate sites (Courtesy MLCE Kouwijzer and R Bosch).

different approaches have been undertaken to arrive at FSH preparations that need fewer injections, such as slow-release formulations, addition of N-linked carbohydrates, and other chemical modifications including pegylation (Fauser et al. 2009). An elegant approach pioneered by Irving Boime and collaborators is based on the longer in vivo half-life of hCG compared to LH. Using genetic engineering, the beta subunit of FSH was extended by one or two CTPs of hCG. It was demonstrated that fewer injections with preparations containing such molecules were needed to induce similar pharmacodynamic effects in laboratory animals. Subsequently, a new cell line was generated by Organon (now part of Merck Sharp & Dohme) that produced corifollitropin alfa (the INN of this molecule), an FSH analog in which the beta subunit was extended by a single CTP (28 amino acids). Thorough biochemical analysis demonstrated the expected amino acid sequence of the alpha subunit and the extended beta subunit but revealed two additional O-linked glycosylation sites in corifollitropin alfa (Henno van den Hooven, Ton Swolfs, personal communication) compared to the 4–5 sites reported in hCG (Fig. 13.4). Nonclinical evaluation demonstrated that the receptor-binding and transactivation profile of this new molecular entity was specific and comparable to that of rec-FSH without intrinsic TSH-receptor or LH-receptor activation. However, the in vivo half-life was increased 1.5–2-fold in the species tested, and a 2–4-fold increase of bioactivity was found across all in vivo pharmacodynamic parameters tested (Verbost et al. 2011). These observations were corroborated by a very extensive data set obtained in a broad panel of clinical trials (phase I, II, and III), including the largest comparator controlled trial of its kind in fertility (the comparator being rec-FSH) (Devroey et al. 2009; Fauser et al. 2010). A single subcutaneous dose of corifollitropin alfa (Elonva®) can be used to initiate and sustain multifollicular growth for 7 days while the efficacy and safety of this novel biopharmaceutical were

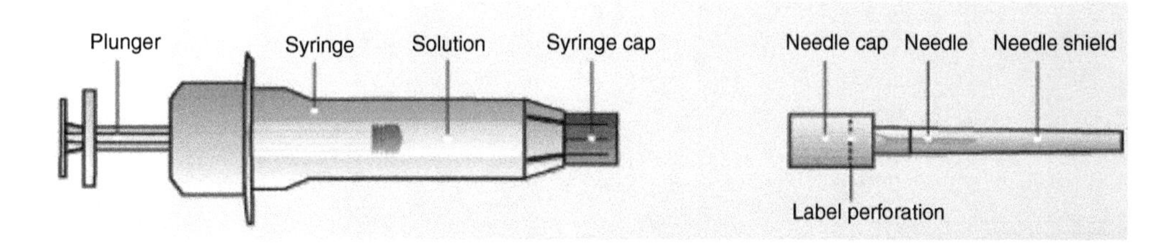

Figure 13.5 ■ Prefilled syringe with corifollitropin alfa solution to be assembled with a needle assembly. The syringe is equipped with an automatic safety system to prevent needle sticking.

similar to that of daily injections with recombinant FSH. Whereas normally more than 7 days of FSH treatment has to be given after the first injection, in about one third of the women treated with FSH-CTP, no additional FSH treatment was needed.

Dedicated clinical research revealed no clinically relevant immunogenicity against the FSH analog (Norman et al. 2011), despite being a fusion protein. Hence, by virtue of its ~2-fold increased in vivo half-life, corifollitropin alfa has demonstrated to provide a valuable alternative for FSH by acting as a sustained follicle stimulant. Elonva® is approved (EU) for controlled ovarian stimulation (COS) in combination with a GnRH (gonadotropin-releasing hormone) antagonist for the development of multiple follicles in women participating in an assisted reproductive technology (ART) program. It is supplied in prefilled syringes equipped with an automatic safety system to prevent needle stick injuries after use and is packed together with a sterile injection needle. Each prefilled syringe contains 0.5 ml solution for injection (Fig. 13.5).

FSH provides a great example of the evolution of biopharmaceuticals, starting from the natural form (urine derived), via close imitations thereof (recombinant FSH), towards further improved biopharmaceuticals (FSH analogs, corifollitropin alfa being the only CTP form that made it to the market). Such developments in pharmaceutical biotechnology are clearly to the benefit of the patients in need for effective, safe, and convenient treatment options.

REFERENCES

De Leeuw R, Mulders J, Voortman G, Rombout F, Damm J, Kloosterboer L (1996) Structure-function relationship of recombinant follicle stimulating hormone (Puregon®). Mol Hum Reprod 2:361–369

Devroey P, Boostanfar R, Koper NP, Jzerman PI, Mannaerts BMJL, Fauser BC, for the ENGAGE Investigators (2009) A double-blind, non-inferiority RCT comparing corifollitropin alfa and recombinant FSH during the first seven days of ovarian stimulation using a GnRH antagonist protocol. Hum Reprod 24:3063–3072

European Public Assessment Report Gonal-F (Follitropin alpha), CPMP/415/95, revision 18, 11-April 2011. European Agency for the Evaluation of Medicinal Products

Fauser BCJM, Mannaerts BMJL, Devroey P, Leader A, Boime I, Baird DT (2009) Advances in recombinant DNA technology: corifollitropin alfa, a hybrid molecule with sustained follicle-stimulating activity and reduced injection frequency. Hum Reprod Update 15:309–321

Fauser BCJM, Alper MM, Ledger W, Schoolcraft WB, Zandvliet A, Mannaerts BMJL (2010) Pharmacokinetics and follicular dynamics of corifollitropin alfa versus recombinant FSH during controlled ovarian stimulation for in vitro fertilisation. Reprod Biomed Online 21:593–601

Howles CM (1996) Genetic engineering of human FSH (Gonal-F®). Hum Reprod Update 2:172–191

Mannaerts BMJL, De Leeuw R, Geelen J, Van Ravenstein A, Van Wezenbeek P, Schuurs A, Kloosterboer L (1992) Comparative in vitro and in vivo studies on the biological properties of recombinant human follicle stimulating hormone. Endocrinology 129:2623–2630

Matzuk MM, Hsueh AJ, Lapolt P, Tsafriri A, Keene JL, Boime I (1990) The biological role of the carboxy-terminal extension of human chorionic gonadotropin (corrected) beta-subunit. Endocrinology 126:376–383

Norman RJ, Zegers-Hochschild F, Salle BS, Elbers J, Heijnen E, Marintcheva-Petrova M, Mannaerts B (2011) Repeated ovarian stimulation with corifollitropin alfa in patients in a GnRH antagonist protocol: no concern for immunogenicity. Hum Reprod 26(8):2200–2208

Olijve W, de Boer W, Mulders JWM, van Wezenbeek PMGF (1996) Molecular biology and biochemistry of human recombinant follicle stimulating hormone (PuregonR). Mol Hum Reprod 2:371–382

Van Wezenbeek P, Draaier J, Van Meel F, Olyve W (1990) Recombinant follicle stimulating hormone. I. Construction, selection and characterization of a cell line. In: Crommelin DJA, Schellekens H (eds) From clone to clinic developments in biotherapy, vol I. Kluwer, Dordrecht, pp 245–251

Verbost P, Sloot WN, Rose UM, de Leeuw R, Hanssen RGJM, Verheijden GFM (2011) Pharmacologic profiling of corifollitropin alfa, the first developed sustained follicle stimulant. Eur J Pharmacol 651:227–233

Voortman G, van de Post J, Schoemaker RC, van Gerwen J (1999) Bioequivalence of subcutaneous injections of recombinant human follicle stimulating hormone (Puregon®) by Pen-injector and syringe. Hum Reprod 14:1698–1702

FURTHER READING

Mannaerts BMJL (2013) Innovative drug development for infertility therapy, Ph.D. thesis, Utrecht, The Netherlands, ISBN 978-90-393-5939-6

Seyhan A, Ata B (2011) The role of corifollitropin alfa in controlled ovarian stimulation for IVF in combination with GnRH antagonist. Int J Womens Health 3:243–255

14 Human Growth Hormone

Le N. Dao, Barbara Lippe, and Michael Laird

INTRODUCTION

Human growth hormone (hGH) is a protein hormone essential for normal growth and development in humans. hGH affects many aspects of human metabolism, including lipolysis, the stimulation of protein synthesis, and the inhibition of glucose metabolism. Human growth hormone was first isolated and identified in the late 1950s from extracts of pituitary glands obtained from cadavers and from patients undergoing hypophysectomy. The first clinical use of these pituitary-extracted hGHs for stimulation of growth in hypopituitary children occurred in 1957 and 1958 (Raben 1958). From 1958 to 1985 the primary material used for clinical studies was pituitary-derived growth hormone (pit-hGH). Human growth hormone was first cloned in 1979 (Goeddel et al. 1979; Martial et al. 1979). The first use in humans of recombinant human growth hormone (rhGH) was reported in the literature in 1982 (Hintz et al. 1982). The introduction of rhGH coincided with reports of a number of cases of Creutzfeldt-Jakob disease, a fatal degenerative neurological disorder, in patients receiving pituitary-derived hGH. Concern over possible contamination of the pituitary-derived hGH preparations by the prion responsible for Creutzfeldt-Jakob disease led to the removal of pit-hGH products from the market in the US in 1985 followed by the FDA approval of rhGH later in the year. The initial rhGH preparations were produced in bacteria (*E. coli*) but, unlike endogenous hGH, contained an N-terminal methionine group (met-rhGH). Natural sequence recombinant hGH products have subsequently been produced in bacteria, yeast, and mammalian cells.

HGH STRUCTURE AND ISOHORMONES

The major, circulating form of hGH is a non-glycosylated, 22 kDa protein composed of 191 amino acid residues linked by disulfide bridges in two peptide loops (Fig. 14.1). The three dimensional structure of hGH includes four antiparallel alpha-helical regions (Fig. 14.2) and three mini-helices. Helix 4 and Helix 1 have been determined to contain the primary sites for binding to the growth hormone receptor. In addition, two of the three mini-helices located within the connecting link between Helix 1 and 2 have been shown to play an important role in the binding of growth hormone to its receptor (Root et al. 2002; Wells et al. 1993). Endogenous growth hormone contains a variety of other isoforms including a 20 kDa monomer, disulfide-linked dimers, oligomers, proteolytic fragments, and other modified forms (Boguszewski 2003; Lewis et al. 2000). The 20 kDa monomer, dimers, oligomers, and other modified forms occur as a result of different gene products, different splicing of hGH mRNA, and posttranslational modifications. These isoforms are generally expressed at lower amounts compared to the 22 kDa protein (Baumann 2009).

There are two hGH genes in humans, the "normal" hGH-N gene and the "variant" hGH-V gene. The hGH-N gene is expressed in the pituitary gland. The hGH-V gene is expressed in the placenta and is responsible for the production of several variant forms of hGH found in pregnant women. Non-glycosylated and glycosylated isoforms of hGH-V have been identified (Ray et al. 1989; Baumann 1991).

PHARMACOLOGY

■ Growth Hormone Secretion and Regulation

Growth hormone is secreted in a pulsatile manner from somatotrophs in the anterior pituitary. Multiple feedback loops are present in normal regulation of hGH

L.N. Dao (✉)
Clinical Pharmacology, Genentech Inc., 1 DNA Way, MS463a, South San Francisco, CA 94080, USA
e-mail: le1@gene.com

B. Lippe, M.D.
Endocrine Care, Genentech Inc, South San Francisco, CA, USA

M. Laird, Ph.D.
Department of Bioprocess Development, Division of Medical Affairs, Genentech Inc, South San Francisco, CA, USA

D.J.A. Crommelin, R.D. Sindelar, and B. Meibohm (eds.), *Pharmaceutical Biotechnology*,
DOI 10.1007/978-1-4614-6486-0_14,

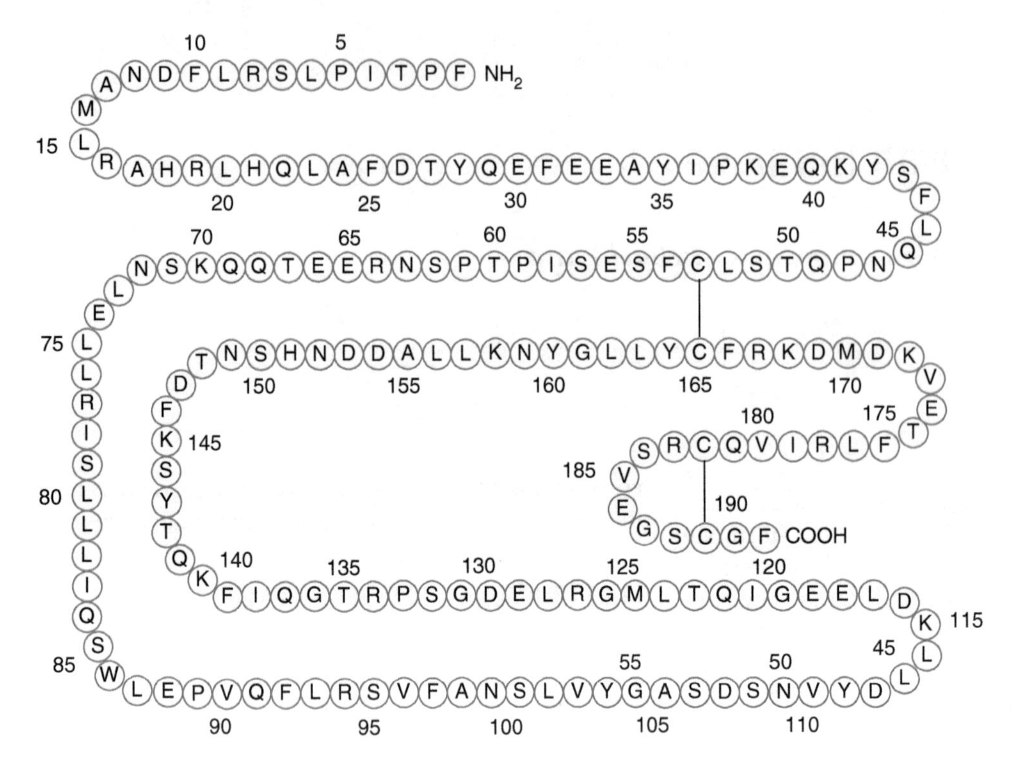

Figure 14.1 ■ Primary structure of recombinant human growth hormone.

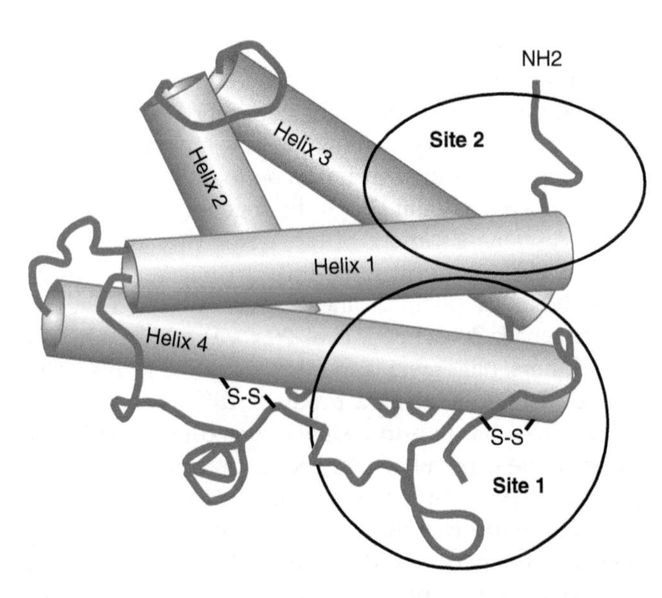

Figure 14.2 ■ Schematic 3-D structure of hGH showing four antiparallel α-helices and receptor binding sites 1 and 2. Approximate positions of the two disulfide bridges (*S-S*) are also indicated (Modified from Wells et al. (1993)).

secretion (Casanueva 1992; Giustina and Veldhuis 1998) (Fig. 14.3). Growth hormone release from the pituitary is regulated by a "short loop" of two-coupled hypothalamic peptides – a stimulatory peptide, growth hormone releasing hormone (GHRH), and an inhibitory peptide, somatostatin (also known as somatotropin release-inhibitory factor (SRIF)). GHRH and somatostatin are, in turn, regulated by neuronal input to the hypothalamus and the GH secretagogue, ghrelin (Kojima et al. 2001). There is possibly also an "ultra-short loop" in which hGH release is feedback regulated by growth hormone receptors present on the somatotrophs of the pituitary themselves. Growth hormone secretion is also regulated by a "long loop" of indirect peripheral signals including negative feedback via insulin-like growth factor (IGF-1) and positive feedback via ghrelin. Growth hormone-induced peripheral IGF-1 inhibits somatotroph release of hGH and stimulates somatostatin release.

Growth hormone secretion changes during human development, with the highest production rates observed during gestation and puberty (Giustina and Veldhuis 1998; Brook and Hindmarsh 1992). Growth hormone production declines approximately 10–15 % each decade from age 20 to 70. Endogenous hGH secretion also varies with sex, age, nutritional status, obesity, physical activity, and in a variety of disease states. Endogenous hGH is secreted in periodic bursts over a 24-h period with great variability in burst frequency, amplitude, and duration. There is little detectable hGH released from the pituitary between bursts. The highest endogenous hGH serum concentrations of 10–30 ng/mL usually occur at night when the secretory bursts are largest and most frequent.

■ Growth Hormone Biologic Actions

hGH has well-defined growth-promoting and metabolic actions. hGH stimulates the growth of cartilage

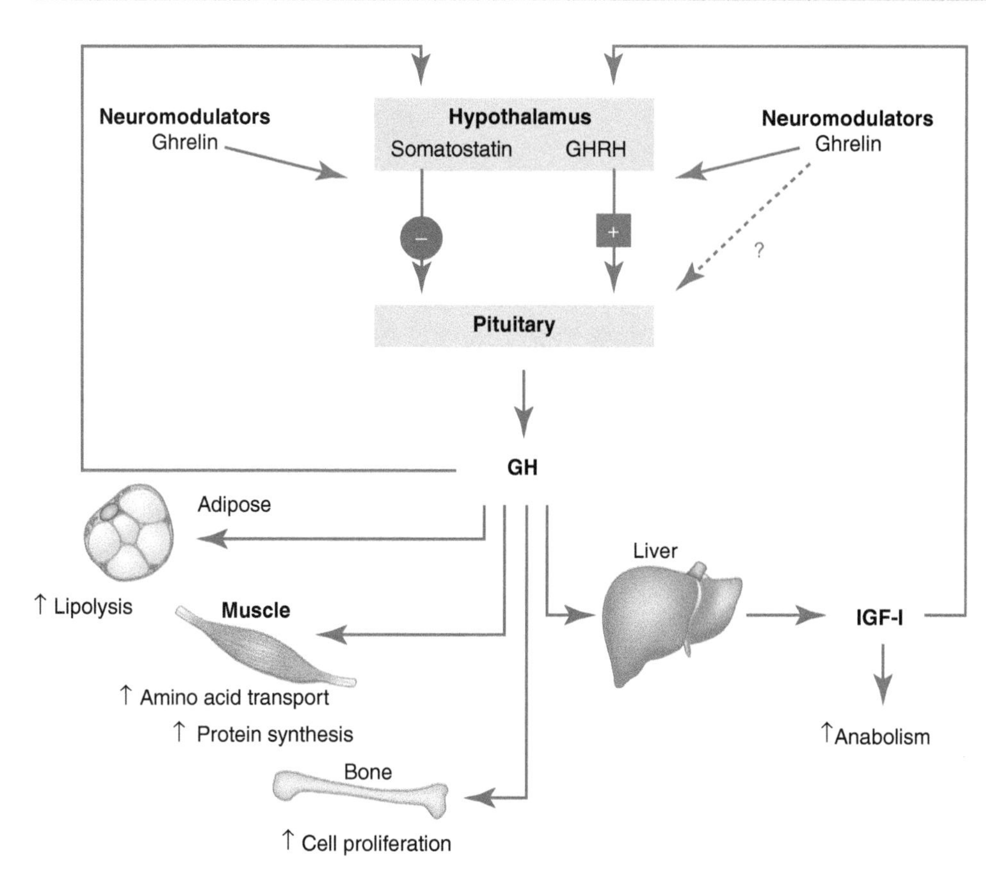

Figure 14.3 ■ Schematic representation of hGH regulation and biologic actions in man. "Short loop" regulation of hGH secretion occurs between the hypothalamus and pituitary. GHRH stimulates GH release. Somatostatin inhibits GH release. "Long loop" regulation of hGH secretion occurs through peripheral feedback signals, primarily negative feedback from insulin-like growth factor 1 (IGF-1). hGH acts directly on muscle, bone, and adipose tissue. Other anabolic actions are generally mediated through IGF-1.

and bone directly, through hGH receptors in those tissues, and indirectly, via local increases in IGF-1 (Isaksson et al. 2000; Bouillon 1991). Metabolic actions, which may be directly controlled by hGH, include the elevation of circulating glucose levels (diabetogenic effect) and acute increases in circulating concentrations of free fatty acids (lipolytic effect). Other hGH anabolic and metabolic actions believed to be mediated through increases in local or systemic IGF-1 concentrations include the following: increases in net muscle protein synthesis (anabolic effect), skeletal muscle growth, chondroblasts and osteoblasts proliferation, and linear growth; modulation of reproduction in both males and females; maintenance, control, and modulation of lymphocyte functions; increases in glomerular filtration rate and renal plasma flow rate (osmoregulation); influences on the release and metabolism of insulin, glucagon, and thyroid hormones (T3, T4); and possible direct effects on pituitary function and neural tissue development (Casanueva 1992; Strobl and Thomas 1994; Le Roith et al. 1991).

■ hGH Receptor and Binding Proteins

The hGH receptor (GHR) is a member of the hematopoietic cytokine receptor family. It has an extracellular domain consisting of 246 amino acids, a single 24-amino-acid transmembrane domain, and a 350-amino-acid intracellular domain (Fisker 2006). The extracellular domain has at least six potential N-glycosylation sites and is usually extensively glycosylated. GHRs are found in most tissues in humans. However, the greatest concentration of receptors in humans and other mammals occurs in the liver (Mertani et al. 1995).

As much as 40–45 % of monomeric hGH circulating in plasma is bound to one of two binding proteins (GHBP) (Fisker 2006). Binding proteins decrease the clearance of hGH from the circulation (Baumann 1991) and may also serve to dampen the biological effects of hGH by competing with cell receptors for circulating free hGH. The major form of GHBP in humans is a high-affinity (Ka = 10^{-9} to 10^{-8} M), low-capacity form which preferentially binds the 22 kDa form of hGH (Baumann 1991; Herington et al. 1986). Another low-affinity (Ka = 10^{-5} M), high-capacity GHBP is also present which binds the 20 kDa form with equal or slightly greater affinity than the 22 kDa form. In humans, the high-affinity GHBP is identical to the extracellular domain of the hGH receptor and arises by proteolytic cleavage of hGH receptors by a process called ectodomain shedding. Since the high-affinity binding protein is derived from hGH receptors, circulating levels of GHBP generally reflect hGH receptor status in many tissues (Fisker 2006; Hansen 2002).

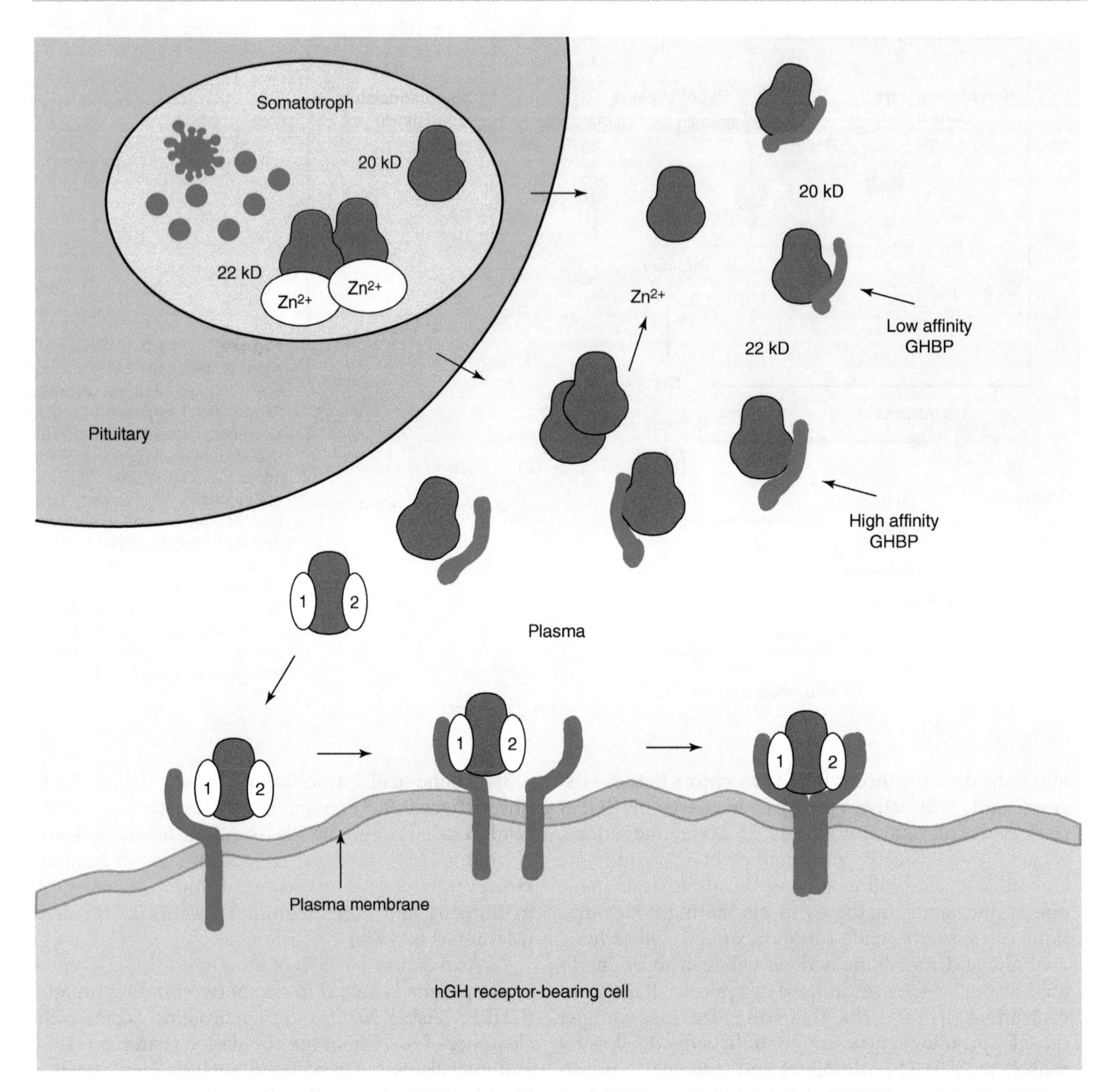

Figure 14.4 ■ Growth hormone secreted isoforms, binding proteins, and receptor interactions. Both 22 and 20 kDa forms are secreted by the pituitary. Pituitary hGH is stored bound to zinc (*Zn*$^{2+}$) which is released upon secretion from the pituitary. Secreted hGH is free or bound to either the low- or high-affinity GHBP in plasma. Receptor activation involves dimerization of two receptor molecules with 1 molecule of hGH (Modified from Wells et al. (1993)).

■ Molecular Endocrinology and Signal Transduction

X-ray crystallographic studies and functional studies of the extracellular domain of the hGH receptor suggest that two receptor molecules form a dimer with a single growth hormone molecule by sequentially binding to Site 1 on Helix 4 of hGH and then to Site 2 on Helix 3 (Fig. 14.4) (Wells et al. 1993). Signal transduction may occur by activation/phosphorylation of JAK-2 tyrosine kinase followed by activation/phosphorylation of multiple signaling cascades (Herrington and Carter-Su 2001; Piwien-Pilipuk et al. 2002; Brooks and Waters 2010).

■ Dosing Schedules and Routes

The dosing levels and routes for exogenously administered growth hormone were first established for pit-hGH in growth hormone deficient (GHD) patients (Laursen 2004; Jorgensen 1991). The initial pit-hGH regimen, three times weekly by intramuscular (IM) injection, was based on a number of factors including patient compliance and limited availability of hGH derived from cadaver pituitaries and the assumption that intramuscular injections would be less immunogenic. Subsequent clinical evaluations found a very strong patient preference for subcutaneous (SC) administration and data supporting no increased immunogenicity. Furthermore, increased growth rates were observed with daily SC injections compared to the three times weekly injection schedules (MacGillivray et al. 1996). The abdomen, deltoid area, and thigh are commonly used subcutaneous injection sites. Current dosing schedules are usually daily SC injections, often self-administered with a variety of injection devices.

■ Pharmacokinetics and Metabolism

The earliest pharmacokinetic studies were conducted with pituitary-derived hGH (pit-hGH). The pharmacokinetic profiles of pit-hGH, met-rhGH, and rhGH have been compared (Hansen 2002; Laursen 2004) and shown to be very similar. The pharmacokinetics of hGH have been studied in normal, healthy children and adults and a variety of patient populations (Hansen 2002; Laursen 2004; Jorgensen 1991).

Exogenously administered pit-hGH, met-rhGH, and rhGH are rapidly cleared following intravenous (IV) injection with terminal half-lives of approximately 15–20 min (Hansen 2002; Laursen 2004). Distribution volumes usually approximate the plasma volume. hGH clearance in normal subjects ranges from 2.2 to 3.0 mL/kg/min. hGH clearance decreases with increasing serum GH concentrations, most likely due to saturation of hGH receptors at concentrations >10–15 μg/L (Hansen 2002). Comparative analyses of total hGH clearance have not shown consistent population differences based on age, sex, or body composition. However, hGH clearance is controlled by a complex interaction between free hGH, GHBP-bound hGH, and GH receptor status (Hansen 2002). Individual subject variations in GHBP or GH receptor levels may result in substantial differences in hGH clearance.

Human growth hormone is slowly, but relatively completely, absorbed after either IM or SC injection. Time to peak concentration ranges from 2 to 4 h following IM bolus administration and 4 to 6 h following SC bolus administration (Laursen 2004; Jorgensen 1991). Subcutaneously administered rhGH is approximately 50–80 % bioavailable (Laursen 2004). The rate of absorption of hGH is slightly faster after injection in the abdomen compared with the thigh (Laursen 2004), but the extent of absorption is comparable. Elimination half-lives following extravascular administration (2–5 h) are usually longer than the IV terminal half-lives indicating absorption rate-limited kinetics.

hGH pharmacokinetics in the presence of growth hormone deficiency, diabetes, obesity, and critical illness or diseases of the thyroid, liver, and kidney have been evaluated. Results suggest disposition is not significantly altered compared with normal subjects except in severe liver or kidney dysfunction (Hansen 2002; Haffner et al. 1994; Owens et al. 1973; Cameron et al. 1972). The reduction in clearance observed in severe liver (30 %) or kidney dysfunction (40–75 %) is consistent with the role of the liver and kidney as major organs of hGH elimination.

Both the kidney and the liver have been shown to be important in the clearance of hGH in humans (Hansen 2002). The relative contribution of each organ has not been rigorously quantitated in humans, but the preponderance of studies in laboratory animals and in isolated perfused organ systems suggests a dominant role for the kidney at pharmacologic levels of hGH. Receptor-mediated uptake of hGH by the liver is the major extrarenal clearance mechanism (Harvey 1995).

PROTEIN MANUFACTURE, FORMULATION, AND STABILITY

Commercially available hGH preparations are summarized in Table 14.1. All recombinant growth hormones except Serostim®/Saizen®/Zorbtive® are produced in bacteria (*E. coli*) or yeast (*S. cerevisiae*). Serostim®/ Saizen®/Zorbtive® are produced in mammalian cells (C127 mouse cells). Growth hormone produced in the cytoplasm of *E. coli* may contain an N-terminal methionine residue. Natural sequence rhGH is produced either by enzymatic cleavage of the methionine residue during the purification process or by secreting the rhGH into the periplasmic space where the signal peptide is removed by the cell and the native N-terminus of rhGH is revealed. rhGH can be produced in the periplasm in a soluble, properly folded form (Chang et al. 1989) or as refractile/inclusion bodies which require the insoluble rhGH to be extracted, denatured, and refolded (Shin et al. 1998) (cf. Chap. 3). The rhGH is released from the cells by osmotic shock (periplasm) or mechanical lysis (cytoplasm & periplasm), and the protein is recovered and purified. rhGH synthesized in yeast and mammalian cells is transported across the endoplasmic reticulum and secreted directly into the culture medium from which it is recovered and purified (Catzel et al. 2003).

Historically, the potency of hGH products was expressed in International Units per mg (IU/mg). The initial standard, established in 1982 for pit-hGH preparations, was 2 IU/mg. The standard for rhGH products was 2.6 IU/mg until September 1994. The current WHO standard, established in September 1994, is 3.0 IU/mg. Dosages are usually expressed as IU/kg or IU/m^2 in Europe and Japan and as mg/kg in the USA. However, the use of IU dosages is no longer necessary due to the high level of purity and consistent potency of recombinant hGH products.

Source	Brand names	Product	Container	Injection device	Manufacturer
Recombinant protein produced in bacteria *(E. coli)*	Genotropin® Genotonorm	Lyophilized powder Multiple-dose cartridge: 5 and 12 mg Single-dose Miniquick cartridge: 0.2–2 mg	Two-chamber cartridge	Genotropin Pen® Genotropin Miniquick® Genotropin Mixer® GoQuick™ pen	Pfizer
	Norditropin®	Liquid FlexPro® and NordiFlex® cartridges: 5 mg/1.5 mL, 10 mg/1.5 mL, 15 mg/1.5 mL NordiFlex® pen: 30 mg/3 mL	Cartridge	Norditropin FlexPro® Norditropin NordiFlex® NordiFlex PenMate® NordiPen® with Norditropin® SimpleXx® NordiPenMate® Norditropin NordiPen® Norditropin NordiLet®	Novo Nordisk
	Nutropin®	Lyophilized powder: 10 mg	Vial	Single-use syringe	Genentech, Inc
	Nutropin AQ®	Liquid Vial: 10 mg Cartridges: 5 mg/2 mL, 10 mg/2 mL, 20 mg/2 mL	Vial and cartridge	Single-use syringe Nutropin AQ® pen Nutropin AQ® NuSpin	
	Humatrope® Umatrope	Lyophilized powder Vial: 5 mg/5 mL Cartridge: 6 mg/3 mL, 12 mg/3 mL, 24 mg/3 mL	Vial and cartridge	Single-use syringe HumatroPen™	Eli Lilly & Co.
	Zomacton® Bio-Tropin® SciTropin® Growject®	Lyophilized powder	Vial	Single-use syringe Needle-free device:Zomajet 2 Vision	Ferring
	Tev-Tropin™	Lyophilized powder Vial: 5 mg	Vial	Single-use syringe Needle-free device:Tjet®	Teva Pharmaceuticals
	Omnitrope® Omnitrop	Lyophilized powder Vial: 5.8 mg	Vial	Single use syringe	Sandoz
	Omnitrope® Omnitrop	Liquid Cartridge: 5 mg/1.5 mL 10 mg/1.5 mL	Cartridge	Omnitrope® pen	

Recombinant protein produced in yeast cells (*S. cerevisiae*)	Valtropin®	Lyophilized powder Vial: 5 mg	Vial	Single-use syringe	BioPartners GmbH LG Life Sciences
Recombinant protein produced in mammalian cells (C127 mouse cell line-derived)	Serostim® (AIDS wasting)	Lyophilized powder Vials: 4 mg, 5 mg, 6 mg, 8.8 mg	Vial with click.easy® reconstitution device	Syringe one.click® Needle-free devices: cool.click and cool.click® 2	Serono
	Saizen® (growth inadequacy)	Lyophilized powder Vial: 5 mg, 8.8 mg	Vial with click.easy® reconstitution device	Single-use syringe easypod® one.click® Needle-free devices: cool.click and cool.click® 2	
	Zorbtive® (short bowel syndrome)	Lyophilized powder Vial: 4 mg, 5 mg, 6 mg, 8.8 mg	Vial	Single-use syringe	

Note: This table represents most rhGH products available globally and is not meant to be an exhaustive list of available marketed hGH products

Table 14.1 ■ Recombinant hGH products.

All current rhGH products are available as lyophilized or liquid preparations. Lyophilized formulations usually include 5 or 10 mg of protein in a glycine and mannitol or sucrose-containing phosphate buffer excipient. The materials are usually reconstituted with sterile water for injection for single use or with bacteriostatic water or bacteriostatic saline for multiple injection use. Liquid formulations of rhGH (Nutropin AQ®, Omnitrope ®, Norditropin® SimpleXx®) contain excipients such as mannitol or sodium chloride, histidine or citrate buffer, poloxamer 188 or polysorbate 20, and phenol or benzyl alcohol. Product stability has been very good with shelf lives of approximately 2 years at 2–8 °C. Omnitrope® (US/EU) and Valtropin (EU only), lyophilized rhGH preparations, were approved for marketing as the first "biosimilar" rhGH products in 2006. Two strengths of Omnitrope® (5 and 10 mg cartridges) were approved for marketing in 2008 for use in two pen devices [Omnitrope Pen® 5 and Omnitrope Pen® 10 (US/EU)].

CLINICAL USAGE

Clinical usage of rhGH has been reviewed for pediatric and adult indications (Franklin and Geffner 2011). Investigations of clinical usage of hGH have focused, generally, on two major areas of hGH biologic action: (1) linear growth promotion and (2) modulation of metabolism. Growth-promoting indications in children which have been approved for the market include growth hormone deficiency, idiopathic short stature, growth failure associated with chronic renal insufficiency, growth failure in children born small for gestational age, and short stature in Prader-Willi syndrome, Turner's syndrome, Noonan syndrome, and short stature homeobox-containing gene deficiency on the X chromosome (SHOX). Modulation of metabolism is the primary biologic action in long-term replacement therapy in adults with GH deficiency of either childhood or adult onset or for GH supplementation in AIDS wasting or cachexia and in short bowel syndrome. Contraindications to rhGH use include use in patients with active malignancy, active proliferative or severe nonproliferative retinopathy, acute critical illness, children with Prader-Willi syndrome (PWS) who are severely obese or have severe respiratory impairment, children with closed epiphyses, and hypersensitivity to somatropin or excipients.

■ Growth Hormone Deficiency (GHD)

The major indication for therapeutic use of hGH is the long-term replacement treatment for children with classic growth hormone deficiency in whom growth failure is due to a lack of adequate endogenous hGH secretion. Children with GHD fall into a variety of etiologic categories including genetic defects in the hypothalamic pituitary axis, developmental anomalies of the brain with or without identifiable syndromes, acquired events such as CNS lesions (craniopharyngioma most commonly) or from the treatment of CNS tumors (medulloblastoma, gliobastoma, etc.), trauma, or other events requiring CNS irradiation. When diagnosed in otherwise healthy children, it is called idiopathic. In patients suspected of having GHD, the diagnosis is usually defined based on an inadequate response to two hGH provocation tests implying a functional deficiency in the production or secretion of hGH from the pituitary gland. In patients with documented organic causes, especially if panhypopituitarism is present, two stimulation tests may not be required. Usual doses range from 0.24 to 0.30 mg/kg/week administered as daily SC injections in prepubertal children. Doses up to 0.7 mg/kg/week have been approved for GHD adolescent subjects to improve final height based on a clinical trial (Mauras et al. 2000) showing hGH treatment results in increased growth velocity and enhancement in final adult height. For most GHD children the growth response is greatest in the first year of treatment and correlates positively with hGH dose, degree of short stature, and frequency of injections and negatively with chronological age at onset of treatment. hGH therapy in children is usually continued until growth has been completed, as evidenced by epiphyseal fusion. rhGH treatment of idiopathic GH-deficient children has a positive overall safety profile documented in long-term clinical registries (Bell et al. 2010; Darendeliler et al. 2007). However, in those with organic causes, the safety profile may be dependent on the underlying medical condition and its prior treatment. In childhood cancer survivors, an increased risk of a second neoplasm has been reported in patients treated with somatropin after their first neoplasm (Sklar 2002), and this information is communicated under warnings and precautions in the product labels.

■ Idiopathic Short Stature (ISS)

Idiopathic short stature (ISS) comprises a heterogeneous group of growth failure states for which the proximate cause remains unknown. Postulated defects include impaired spontaneous hGH secretion, hGH resistance due to low levels of hGH receptors, or other defects in either secreted hGH, hGH receptors or post-receptor events, or other genetic defects yet to be elucidated. Nevertheless, studies have documented that many patients in the "idiopathic" group respond to exogenous growth hormone treatment with acceleration of growth and improvement in final height. As a consequence of several long-term multicenter trials to final height, rhGH was approved for the treatment of

ISS in the United States (Hintz et al. 1999; Leschek et al. 2004). The risk benefit assessment of improved growth in what is generally considered to be a population of "healthy" children continues to be debated, and while short-term safety has been documented (Bell et al. 2010), long-term safety is still a question that is being addressed by a multi-country study in Europe looking at adult patients treated with rhGH as children. Based on preliminary data and published reports (Carel et al. 2012; Savendahl et al. 2012), the FDA and the EMA preliminary assessments are that there should be no change in prescribing information with emphasis on adherence to approved dosages. Final assessments are expected to be available in 2013.

■ Turner Syndrome (TS)

Turner syndrome is a disease of females caused by partial or total loss of one sex chromosome and is characterized by decreased intrauterine and postnatal growth, short final adult height, incomplete development of the ovaries and secondary sexual characteristics, and other physical abnormalities. Although serum levels of hGH and IGF-1 are not consistently low in this population, hGH treatment (0.375 mg/kg/week), alone or in combination with oxandrolone, significantly improves growth rate and final adult height in this patient group, and its use in Turner syndrome is now an accepted indication (Rosenfeld et al. 1998; Kappelgaard and Laursen 2011).

■ Prader-Willi Syndrome (PWS)

Prader-Willi syndrome is a genetic disease usually caused by the functional deletion of a gene on the paternal allele of chromosome 15. Clinical manifestations in childhood include lack of satiety, obesity, hypotonia, short stature, hypogonadism, and behavior abnormalities. PWS children have especially high rates of morbidity due to obesity-related illnesses. Growth hormone treatment (0.24–0.48 mg/kg/week) has been shown to improve height and, perhaps more importantly, improve body composition, physical strength, and agility in PWS children (Allen and Carrel 2004). However, growth hormone treatment is contraindicated in the face of severe obesity or severe respiratory impairment since sudden deaths in this group of patients, when treated with growth hormone, have been reported.

■ Small for Gestational Age (SGA)

Growth hormone is approved for use in long-term treatment of growth failure in children born small for gestational age who fail to manifest catch-up growth. Children born at birth weights and birth lengths more than two standard deviations below the mean are considered small for gestational age. Children who fail to catch up by age two to three are at risk for growing into adults with substantial height deficits (Rappaport 2004). Growth hormone treatment at doses of 0.24–0.48 mg/kg/week can induce catch-up growth, with the potential to normalize height at an earlier age and the potential to improve adult height.

■ Chronic Renal Insufficiency (CRI)/Chronic Kidney Disease (CKD)

Children with renal insufficiency, which is termed chronic kidney disease (CKD) grow slowly, possibly related to defects in metabolism, nutrition, metabolic bone disease, and/or defects in the IGF-1/hGH axis. Basal serum hGH concentrations may be normal or high, and IGF-1 responses to hGH stimulation is usually normal. However, there are reported abnormalities in the IGF-binding protein levels in CKD patients suggesting possible problems with GH/IGF-1 action. Growth hormone therapy (0.35 mg/kg/week) in children with chronic renal insufficiency results in significant increases in height velocity (Greenbaum et al. 2004). Increases are best during the first year of treatment for younger children with stable renal disease. Responses are less for children on dialysis. Growth hormone has not been approved for children posttransplant.

■ Noonan Syndrome

First called male Turner syndrome due to similar phenotypic characteristics, Noonan syndrome, an autosomal dominant disorder, was later recognized as a separate condition. It occurs in both males and females. Its key features are short stature (although some patients will achieve normal adult height), congenital heart disease, most commonly pulmonic stenosis or hypertrophic cardiomyopathy, a short and often webbed neck, ptosis, and chest/sternum deformities. While the GH/IGF axis is intact in Noonan syndrome, and the mechanisms through which the mutations cause short stature are unknown, clinical trials with rhGH demonstrated significant increases in growth rate and modest increases in final height (Osio et al. 2005), resulting in FDA approval of this indication in 2007. Pediatric patients with short stature associated with Noonan syndrome are given up to a rhGH dose of 0.066 mg/kg/day. Safety concerns in treatment with rhGH have been raised with respect to progression of cardiomyopathy although data to date do not support this clinical concern.

■ Short Stature Homeobox-Containing Gene (SHOX)

The SHOX gene is located on the pseudoautosomal region of the X chromosome and the homologous distal region on the Y chromosome. Healthy males and females express two active copies of the SHOX gene, one from each of the sex chromosomes. Females with

TS missing an X chromosome or part of an X chromosome have one copy of the SHOX gene. A significant percentage of the growth failure in TS females is secondary to this gene loss. The variably expressed SHOX gene in long bones tends to be in the mesomelic segments. Mutations resulting in haploinsufficiency of SHOX are also responsible for the short stature in some patients with a pseudoautosomal dominant condition of mesomelic dyschondrosteosis called Leri-Weill syndrome. In addition, sporadic mutations are also responsible for short stature in a small percentage of patients who would otherwise be characterized as idiopathic short stature (Rao et al. 1997). A multicenter study of a heterogeneous group of patients with SHOX haploinsufficiency demonstrated a clinically significant effect of rhGH on growth in children with SHOX mutations compared to the untreated control group. In addition, the efficacy of rhGH treatment was similar to that seen in a comparable group of girls with TS, leading to approval of the SHOX indication (Blum et al. 2007). The FDA-approved rhGH dose for SHOX deficiency is 0.35 mg/kg/week. To date no specific safety signals attributable to rhGH treatment have emerged in these children.

■ Growth Hormone Deficient Adults

Early limitations in hGH supply severely limited treatment of adults with GHD. With the increased supply of recombinant rhGH products, replacement therapy for adults was evaluated and, ultimately, approved as a clinical indication. Growth hormone has been approved for two growth hormone deficient adult populations: (a) adults with childhood-onset GHD and (b) adults with adult-onset GHD usually due to pituitary tumors, CNS irradiation, or head trauma. Growth hormone treatment (starting dose of 0.006–0.025 5 mg/kg/week in patients under 35 years old and 0.0125 mg/kg/week in patients over 35 years old or a starting dose of 0.2–0.4 mg/day and progressing based on IGF and clinical symptoms) reduces body fat, increases lean body mass, and increases exercise capacity. Increases in bone density have been observed in some bone types although treatment duration greater than 1 year may be necessary to see significant effects. hGH treatment consistently elevates both serum IGF-1 and insulin levels. Women have also been shown to require higher doses to normalize IGF-1 levels than men, especially women taking oral estrogens.

■ Clinical Malnutrition and Wasting Syndromes

Growth hormone is approved for treatment of short bowel syndrome (SBS) in adults, a congenital or acquired condition in which less than ~200 cm of small intestine is present. Short bowel syndrome patients have severe fluid and nutrient malabsorption and are often dependent upon intravenous parenteral nutrition (IPN). Administration of growth hormone, 0.1 mg/kg/day to a maximum of 8 mg for 4 weeks, alone or in combination with glutamine, reduces the volume and frequency of required IPN (Keating and Wellington 2004). Growth hormone is indicated for use in adult patients who are also receiving specialized nutritional support. Usage for periods >4 weeks, or in children, has not been investigated. Usage of growth hormone for SBS remains controversial due to potential risks associated with IGF-1-related fibrosis and cancer (Theiss et al. 2004).

Growth hormone is also approved for use in wasting associated with AIDS. Growth hormone treatment (~0.1 mg/kg daily, max. 6 mg/day), when used with controlled diets, increases body weight and nitrogen retention. rhGH treatment is also under investigation for HIV-associated lipodystrophy, a syndrome of fat redistribution and metabolic complications resulting from the highly active antiretroviral therapy commonly used in HIV infection (Burgess and Wanke 2005).

■ Other Conditions Under Investigation

Growth hormone levels and IGF decline with age, prompting the initiation of multiple clinical trials for use in adults over age 60 (Di Somma et al. 2011). However, clear long-term efficacy in muscle strength or improvements in activities of daily life have not been sufficiently demonstrated to gain regulatory approval for this indication. The use of hGH therapy to ameliorate the negative nitrogen balance seen in patients following surgery, injury, or infections has been investigated in a number of studies (Takala et al. 1999; Jeevanandam et al. 1995; Ponting et al. 1988; Voerman et al. 1995). However, due to the increased mortality found in a study of severe critical illness (Takala et al. 1999) and the subsequent contraindication for use in acute critical illness, very few registration trials examining the use of hGH for these conditions have been initiated. Studies of hGH effects in burns have shown significant effectiveness in acceleration of healing in skin graft donor sites and improvements in growth in burned children (Herndon and Tompkins 2004). Growth hormone has been shown to significantly reduce multiple disease symptoms and improve well-being and growth in children and adults with Crohn's disease, a chronic inflammatory disorder of the bowel (Theiss et al. 2004; Slonim et al. 2000; Denson et al. 2010). Growth hormone has also shown benefit in cardiovascular recovery and function in congestive heart failure (Colao et al. 2004). Recent studies indicate that growth hormone treatment improves growth, pulmonary function, and clinical status in children with cystic fibrosis (Stalvey et al. 2012).

■ Safety Concerns

hGH has been widely used for many years and has been proven to have a positive safety profile for most pediatric indications (Growth Hormone Research Society 2001). However, sudden death in some patients with PWS and severe obesity associated with rhGH treatment resulted in a contraindication to its use in severely obese or respiratory compromised PWS children (Eiholzer 2005). Adverse events have been reported in a small number of children and include benign intracranial hypertension, glucose intolerance, and the rare development of anti-hGH antibodies. In most cases, the formation of anti-hGH antibodies following rhGH treatment has not been positively correlated with a loss in efficacy.

Growth hormone therapy is also not associated with increased risk of primary malignancies or tumor recurrence (Growth Hormone Research Society 2001; Sklar et al. 2002). However, an increase in secondary malignancies in childhood cancer survivors, especially those treated with CNS irradiation, has been described (Sklar 2002).

Growth hormone inhibits 11βhydroxysteroid dehydrogenase type 1 (11βHSD-1) activity in adipose/hepatic tissue and may impact the metabolism of cortisol and cortisone (Gelding et al. 1998). Treatment with rhGH could potentially unmask undiagnosed central (secondary hypoadrenalism) or increase the requirement for maintenance or stress doses of replacement corticosteroid in those already diagnosed with adrenal insufficiency.

Growth hormone has caused significant, dose-limiting fluid retention in adult populations resulting in increased body weight, swollen joints and arthralgias, and carpal tunnel syndrome (Carroll and van den Berghe 2001). Symptoms were usually transient and resolved upon reduction of hGH dosage or upon discontinuation of the hGH treatment. Growth hormone administration has been associated with increased mortality in clinical trials in critically ill, intensive-care patients with acute catabolism (Takala et al. 1999) and is, therefore, contraindicated for use in critically ill patients.

Growth hormone's anabolic and lipolytic effects have made it attractive as a performance enhancement drug among athletes. Illicit hGH usage has been anecdotally reported for the last 20 years. Detection of rhGH abuse proximate to the time of testing is now possible due to the development of assays which rely on detecting changed ratios of exogenous rhGH (22 kDa only) and endogenous hGH (22 kDa, 20 kDa and other forms). Screening for proximate rhGH abuse, based on the new ratio assays, was included in the 2006 Olympic Games for the first time (McHugh et al. 2005).

CONCLUDING REMARKS

The abundant supply of rhGH, made possible by recombinant DNA technology, has allowed enormous advances to be made in understanding the basic structure, function, and physiology of hGH over the past 20 years. As a result of those advances, recombinant hGH has been developed into a safe and efficacious therapy for a variety of growth and metabolic disorders in children and adults. Continuing basic research in GH and IGF-1 biology, genomics, and GH-related diseases and continuing clinical investigation into additional uses in pediatric growth disorders or disorders of metabolism may yield as yet new indications for treatment.

SELF-ASSESSMENT QUESTIONS

■ Questions

1. One molecule of hGH is required to sequentially bind to two receptor molecules for receptor activation. What consequences might the requirement for sequential dimerization have on observed dose–response relationships?
2. Growth hormone is known or presumed to act directly upon which tissues?
3. You are investigating the use of hGH as an adjunct therapy for malnutrition/wasting in a clinical population which also has severe liver disease. What effects would you expect the liver disease to have on the observed plasma levels of hGH after dosing and on possible efficacy (improvement in nitrogen retention, prevention of hypoglycemia, etc.)?

■ Answers

1. Sequential dimerization will potentially result in a "bell-shaped" dose–response curve, i.e., response is stimulated at low concentrations and inhibited at high concentrations. The inhibition of responses at high concentrations is due to blocking of dimerization caused by the excess hGH saturating all the available receptors. Inhibition of in vitro hGH binding is observed at high hGH (mM) concentrations. Reductions in biological responses (total IGF-1 increase and weight gain) have also been seen with increasing hGH doses in animal studies. However, inhibitory effects of high concentrations of hGH are not seen in treatment of human patients since hGH dose levels are maintained within normal physiological ranges and never approach inhibitory levels.
2. Growth hormone is known to act directly on both bone and cartilage and possibly also on muscle and adipose tissue. Growth hormone effects on other tissues appear to be mediated through the IGF-1 axis or other effectors.
3. Severe liver disease may reduce the clearance of the exogenously administered hGH, and observed

plasma levels may be higher and persist longer compared to patients without liver disease. However, the increased drug exposure may not result in increased anabolic effects. The desired anabolic effects require the production/release of IGF-1 from the liver. Both IGF-1 production and the number of hGH receptors may be reduced due to the liver disease. To understand the results (or lack of results) from the treatment, it is important to monitor effect parameters (i.e., IGF-1 and possibly IGF-1 binding protein levels, liver function enzymes) in addition to hGH levels.

REFERENCES

Allen DB, Carrel AL (2004) Growth hormone therapy for Prader-Willi Syndrome: a critical appraisal. J Pediatr Endocrinol Metab 17:1297–1306

Baumann G (1991) Growth hormone heterogeneity: genes, isohormones, variants and binding proteins. Endocr Rev 12:424–449

Baumann GP (2009) Growth hormone isoforms. Growth Horm IGF Res 19:333–340

Bell J, Parker KL, Swinford RD, Hoffman AR, Maneatis T, Lippe B (2010) Long-term safety of recombinant human growth hormone in children. J Clin Endocrinol Metab 95:167–177

Blum WF, Crowe BJ, Quigley CA, Jung H, Cao D, Ross JL, Braun L, Rappold G, Shox Study Group (2007) Growth hormone is effective in treatment of short stature associated with short stature homeobox-containing gene deficiency: two-year results of a randomized, controlled, multicenter trial. J Clin Endocrinol Metab 92:219–228

Boguszewski CL (2003) Molecular heterogeneity of human GH: from basic research to clinical implications. J Endocrinol Invest 26:274–288

Bouillon R (1991) Growth hormone and bone. Horm Res 36(Suppl 1):49–55

Brook CGD, Hindmarsh PC (1992) The somatotropic axis in puberty. Endocrinol Metab Clin North Am 21:767–782

Brooks AJ, Waters MJ (2010) The growth hormone receptor: mechanism of activation and clinical implications. Nat Rev Endocrinol 6(9):515–525

Burgess E, Wanke C (2005) Use of recombinant human growth hormone in HIV-associated lipodystrophy. Curr Opin Infect Dis 18:17–24

Cameron DP, Burger HG, Catt KJ et al (1972) Metabolic clearance of human growth hormone in patients with hepatic and renal failure, and in the isolated perfused pig liver. Metabolism 21:895–904

Carel J-C, Ecosse E, Landier F, Meguellati-Hakkas D, Kaguelidou F, Rey G, Coste J (2012) Long-term mortality after recombinant growth hormone treatment for isolated growth hormone deficiency or childhood short stature: preliminary report of the French SAGhE study. J Clin Endocrinol Metab 97:416–425. doi:10.1210/jc.2011-1995

Carroll PV, van den Berghe G (2001) Safety aspects of pharmacological GH therapy in adults. Growth Horm IGF Res 11:166–172

Casanueva F (1992) Physiology of growth hormone secretion and action. Endocrinol Metab Clin North Am 21: 483–517

Catzel D, Lalevski H, Marquis CP, Gray PP, Van Dyk D, Mahler SM (2003) Purification of recombinant human growth hormone from CHO cell culture supernatant by Gradiflow preparative electrophoresis technology. Protein Expr Purif 32(1):126–234

Chang JY, Pai RC, Bennett WF, Bochner BR (1989) Periplasmic secretion of human growth hormone by Escherichia coli. Biochem Soc Trans 17(2):335–337

Colao A, Vitale G, Pivonello R et al (2004) The heart: an end-organ of GH action. Eur J Endocrinol 151:S93–S101

Darendeliler F, Karagiannis G, Wilton P (2007) Headache, idiopathic intracranial hypertension and slipped capital femoral epiphysis during growth hormone treatment: a safety update from KIGS. Horm Res 68(Suppl 5):41–47

Denson LA, Kim MO, Bezold R et al (2010) A randomized controlled trial of growth hormone in active pediatric Crohn disease. J Pediatr Gastroenterol Nutr 51: 130–139

Di Somma C, Brunelli V, Savanelli MC et al (2011) Somatopause: state of the art. Minerva Endocrinol 36: 243–255

Eiholzer U (2005) Deaths in children with Prader-Willi syndrome: a contribution to the debate about the safety of growth hormone treatment in children with PWS. Horm Res 63(1):33–39

Fisker S (2006) Physiology and pathophysiology of growth hormone binding protein: methodological and clinical aspects. Growth Horm IGF Res 16:1–28

Franklin SL, Geffner ME (2011) Growth hormone: the expansion of available products and indications. Endocrinol Metab Clin North Am 38:587–611

Gelding SV, Taylor NF, Wood PJ, Noonan K, Weaver JU, Wood DF, Monson JP (1998) The effect of growth hormone replacement therapy on cortisol-cortisone interconversion in hypopituitary adults: evidence for growth hormone modulation of extrarenal 11 beta-hydroxysteroid dehydrogenase activity. Clin Endocrinol (Oxf) 48(2):153–162

Giustina A, Veldhuis JD (1998) Pathophysiology of the neuroregulation of growth hormone secretion in experimental animals and in the human. Endocr Rev 19(6): 717–797

Goeddel DV, Heyreker HL, Hozumi T et al (1979) Direct expression in Escherichia coli of a DNA sequence coding for human growth hormone. Nature 281: 544–548

Greenbaum LA, Del Rio M, Bamgbola F et al (2004) Rationale for growth hormone therapy in children with chronic kidney disease. Adv Chronic Kidney Dis 11(4): 377–386

Consensus:. Critical evaluation of the safety of recombinant human growth hormone administration: statement from the Growth Hormone Research Society (2001) J Clin Endocrinol Metab 86(5):1868–1870

Haffner D, Schaefer F, Girard J et al (1994) Metabolic clearance of recombinant human growth hormone in health and chronic renal failure. J Clin Invest 93:1163–1171

Hansen TK (2002) Pharmacokinetics and acute lipolytic actions of growth hormone: impact of age, body composition, binding proteins and other hormones. Growth Horm IGF Res 12:342–358

Harvey S (1995) Growth hormone metabolism. In: Harvey S, Scanes CG, Daughaday WH (eds) Growth hormone. CRC Press, Inc, Boca Raton, pp 285–301

Herington AC, Ymer S, Stevenson J (1986) Identification and characterization of specific binding proteins for growth hormone in normal human sera. J Clin Invest 77: 1817–1823

Herndon DN, Tompkins RG (2004) Support of the metabolic response to burn injury. Lancet 363:1895–1902

Herrington J, Carter-Su C (2001) Signaling pathways activated by the growth hormone receptor. Trends Endocrinol Metab 12(6):252–257

Hintz RL, Rosenfeld RG, Wilson DM et al (1982) Biosynthetic methionyl human growth hormone is biologically active in adult man. Lancet 1:1276–1279

Hintz RL, Attie KM, Baptista J, Roche A (1999) Effect of growth hormone treatment on adult height of children with idiopathic short stature. N Engl J Med 340:502–507

Isaksson OG, Ohlsson C, Bengtsson B et al (2000) GH and bone-experimental and clinical studies. Endocr J 47(Suppl):S9–S16

Jeevanandam M, Ali MR, Holaday NJ et al (1995) Adjuvant recombinant human hormone normalizes plasma amino acids in parenterally fed trauma patients. J Parenter Enteral Nutr 19:137–144

Jorgensen JOL (1991) Human growth hormone replacement therapy: pharmacological and clinical aspects. Endocr Rev 12:189–207

Kappelgaard AM, Laursen T (2011) The benefits of growth hormone therapy in patients with Turner syndrome, Noonan syndrome, and children born small for gestational age. Growth Horm IGF Res 21(6):305–313

Keating GM, Wellington K (2004) Somatropin (zorbtive™) in short bowel syndrome. Drugs 64(12):1375–1381

Kojima M, Hosoda H, Matsuo H et al (2001) Ghrelin: discovery of the natural endogenous ligand for the growth hormone secretagogue receptor. Trends Endocrinol Metab 12(3):118–126

Laursen T (2004) Clinical pharmacological aspects of growth hormone administration. Growth Horm IGF Res 14:16–44

Le Roith D, Adamo M, Werner H, Roberts CT Jr (1991) Insulin-like growth factors and their receptors as growth regulators in normal physiology and pathologic states. Trends Endocrinol Metab 2:134–139

Leschek EW, Ross SR, Yanovski JA, Troendle JF, Quigley CA, Chipman JJ, Crowe BJ et al (2004) Effect of growth hormone treatment on adult height in peripubertal children with idiopathic short stature: a randomized, double blind, placebo-controlled trial. J Clin Endocrinol Metab 89:3140–3148

Lewis UJ, Sinhda YN, Lewis GP (2000) Structure and properties of members of the hGH family: a review. Endocr J 47:S1–S8

MacGillivray MH, Baptista J, Johanson A (1996) Outcome of a four-year randomized study of daily versus three times weekly somatropin treatment in prepubertal naïve growth hormone deficient children. J Clin Endocrinol Metab 81:1806–1809

Martial JA, Hallewell RA, Baxter JD (1979) Human growth hormone: complementary DNA cloning and expression in bacteria. Science 205:602–607

Mauras N, Attie KM, Reiter EO et al (2000) High dose recombinant human growth hormone (GH) treatment of GH-deficient patients in puberty increases near-final height: a randomized, multicenter trial. J Clin Endocrinol Metab 85:3653–3660

McHugh CM, Park RT, Sonksen PH et al (2005) Challenges in detecting the abuse of growth hormone in sport. Clin Chem 51(9):1587–1593

Mertani HC, Delehaye-Zervas MC, Martini JF et al (1995) Localization of growth hormone receptor messenger RNA in human tissues. Endocrine 3:135–142

Osio D, Dahlgren J, Wikland KA, Westphal O (2005) Improved final height with long-term growth hormone treatment in Noonan syndrome. Acta Paediatr 94(9):1232–1237

Owens D, Srivastava MC, Tompkins CV et al (1973) Studies on the metabolic clearance rate, apparent distribution space and plasma half-disappearance time of unlabelled human growth hormone in normal subjects and in patients with liver disease, renal disease, thyroid disease and diabetes mellitus. Eur J Clin Invest 3: 284–294

Piwien-Pilipuk G, Huo JS, Schwartz J (2002) Growth hormone signal transduction. J Pediatr Endocrinol Metab 15:771–786

Ponting GA, Halliday D, Teale JD et al (1988) Postoperative positive nitrogen balance with intravenous hyponutrition and growth hormone. Lancet 1:438–440

Raben MS (1958) Treatment of a pituitary dwarf with human growth hormone. J Clin Endocrinol Metab 18: 901–903

Rao E, Weiss B, Fukami M et al (1997) Pseudoautosomal deletions encompassing a novel homeobox gene cause growth failure in idiopathic short stature and Turner syndrome. Nat Genet 16:54–63

Rappaport R (2004) Growth and growth hormone in children born small for gestational age. Growth Horm IGF Res 14:S3–S6

Ray J, Jones BK, Liebhaber SA, Cooke NE (1989) Glycosylated human growth hormone variant. Endocrinology 125(1):566–568

Root AW, Root MJ et al (2002) Clinical pharmacology of human growth hormone and its secretagogues. Curr Drug Targets Immune Endocr Metabol Disord 2: 27–52

Rosenfeld RG, Attie KM, Frane J et al (1998) Growth hormone therapy of Turner's syndrome: beneficial effect on adult height. J Pediatr 132:319–324

Savendahl L, Maes M, Albersson K, Borgstrom B, Carel J-C, Henrad S, Speybroeck N, Thomas M, Xandwijken G, Hokken-Koelega A (2012) Long-term mortality and causes of death in isolated GHD, ISS and SGA patients treated with recombinant growth hormone during childhood in Belgium, the Netherlands, and Sweden: preliminary report of a 3 countries participating in the EU SAGhE study. J Clin Endocrinol Metab 97:E213–E217. doi:10.1210/jc.2011-2882

Shin NK, Kim DY, Shin CS, Hong MS, Lee J, Shin HC (1998) High-level production of human growth hormone in Escherichia coli by a simple recombinant process. J Biotechnol 62(2):143–151

Sklar CA, Mertens AC, Mitby P et al (2002) Risk of disease recurrence and second neoplasms in survivors of children cancer treated with growth hormone: a report from the Childhood Cancer Survivor Study. J Clin Endocrinol Metab 87(7):3136–3141

Slonim AE, Bulone L, Damore MB et al (2000) A preliminary study of growth hormone therapy for Crohn's disease. N Engl J Med 342:1633–1637

Stalvey MS, Anbar RD, Konstan MVV, Jacobs JR, Bakker B, Lippe B, Geller DE (2012) A multi-center controlled trial of growth hormone treatment in children with cystic fibrosis. Pediatr Pulmonol 47:252–263. doi:10.1002/ppul.21546

Strobl JS, Thomas MJ (1994) Human growth hormone. Pharm Rev 46:1–34

Takala J, Ruokonen E, Webster NR et al (1999) Increased mortality associated with growth hormone treatment in critically ill adults. N Engl J Med 341(11):785–792

Theiss AL, Fruchtman S, Lund PK (2004) Growth factors in inflammatory bowel disease. The actions and interactions of growth hormone and insulin-like growth factor-I. Inflamm Bowel Dis 10(6):871–880

Voerman BJ, van Schijndel RJM S, Goreneveld ABJ et al (1995) Effects of human growth hormone in critically ill non-septic patients: results from a prospective, randomized, placebo-controlled trial. Crit Care Med 23:665–673

Wells JA, Cunningham BC, Fuh G et al (1993) The molecular basis for growth hormone-receptor interactions. Recent Prog Horm Res 48:253–275

FURTHER READING

Boguszewski CL (2003) Molecular heterogeneity of human GH: from basic research to clinical implications. J Endocrinol Invest 26:274–288

Brooks AJ, Waters MJ (2010) The growth hormone receptor: mechanism of activation and clinical implications. Nat Rev Endocrinol 6(9):515–525

Fisker S (2006) Physiology and pathophysiology of growth hormone binding protein: methodological and clinical aspects. Growth Horm IGF Res 16:1–28

Giustina A, Veldhuis JD (1998) Pathophysiology of the neuroregulation of growth hormone secretion in experimental animals and in the human. Endocr Rev 19(6):717–797

Harvey S, Scanes CG, Daughaday WH (eds) (1995) Growth hormone. CRC Press, Inc, Boca Raton

Harris M, Hofman PL, Cutfield WS (2004) Growth hormone treatment in children. Pediatr Drugs 6(2):93–106

Laursen T (2004) Clinical pharmacological aspects of growth hormone administration. Growth Horm IGF Res 14:16–44

Simpson H, Savine R, Sonksen P et al (2002) Growth hormone replacement therapy for adults: into the new millennium. Growth Horm IGF Res 12:1–33

15 Recombinant Coagulation Factors and Thrombolytic Agents

Nishit B. Modi

INTRODUCTION

Coagulation and fibrinogenolysis exist in a mutually compensatory or balanced state. Endogenous regulatory mechanisms ensure that the processes of hemostasis and blood coagulation at a site of injury, and the subsequent fibrinolysis of the blood clot, are localized and well regulated. This ensures a rapid and efficient hemostatic response at a site of injury while avoiding thrombogenic events at sites distant from the site of injury or the hemostatic response from persisting beyond its physiologic need. This chapter will focus on recombinant products that are available to facilitate coagulation and for thrombolysis.

Two models of blood hemostasis, a cascade model and a cell-based model, have been proposed. A schematic of the cascade and cell-based models of coagulation is presented in Fig. 15.1.

The initial model of coagulation was proposed in the 1960s, encompassing a series of steps, or cascade, where enzymes cleaved a zymogen to generate the subsequent enzyme. In the cascade model, coagulation was divided into the intrinsic and extrinsic pathways. The extrinsic pathway was located outside the blood and consisted of tissue factor-dependent cofactors and enzymes and factor VIIa. The intrinsic system was localized within the blood and could be initiated through contact activation of factor XII, which leads to activation of subsequent components. The two pathways converged into the common pathway, leading to the generation of thrombin. While the cascade model was useful in explaining how the coagulation enzymatic steps occurred in vitro and in helping interpret laboratory tests, it did not adequately explain the hemostatic process as it occurs in vivo.

More recently, a cell-based model of coagulation has been proposed (Hoffman 2003; Hoffman and Monroe 2005). This cell-based model emphasizes the interaction of clotting factors with cell surfaces and appears to explain some of the unresolved issues with the cascade model. The cell-based model of coagulation comprises four phases: initiation, amplification, propagation, and termination. The initiation phase is localized to cells expressing tissue factor (TF), which are generally localized outside the vasculature. Upon injury, blood is exposed to cells bearing TF, and factor VIIa rapidly binds to exposed TF. The TF-VIIa complex activates additional factor VII to factor VIIa which in turn activates small amounts of factor IX and factor X. Factor Xa binds factor Va to form the prothrombinase complex, which cleaves prothrombin to thrombin. During amplification, the small amount of thrombin generated diffuses away from the TF-bearing cells and activated platelets, exposing receptors and binding sites for activated clotting factors. Once platelets are activated, the release of granule contents leads to recruitment of additional platelets to the site of injury, leading to the propagation phase on the surface of activated platelets. The propagation step culminates in a burst of thrombin generation of sufficient magnitude to clot fibrinogen that is converted to fibrin. Once the fibrin platelet clot has formed over the site of injury, the clotting process must be terminated to prevent thrombotic occlusion.

Normally hemostasis is a highly efficient and tightly regulated process to ensure that it occurs quickly and is localized. Abnormalities that result in a delay in blood coagulation are associated with a bleeding tendency termed hemophilia. Hemophilia is an X-linked recessive disorder that affects approximately 400,000 people worldwide (Shapiro et al. 2005). Hemophilia A (classical hemophilia) patients have decreased, defective, or absent production of factor

N.B. Modi, Ph.D.
Departments of Nonclinical R&D
and Clinical Pharmacology,
Impax Pharmaceuticals, Hayward, CA, USA
e-mail: n.modi@sbcglobal.net

D.J.A. Crommelin, R.D. Sindelar, and B. Meibohm (eds.), *Pharmaceutical Biotechnology*,
DOI 10.1007/978-1-4614-6486-0_15, © Springer Science+Business Media New York 2013

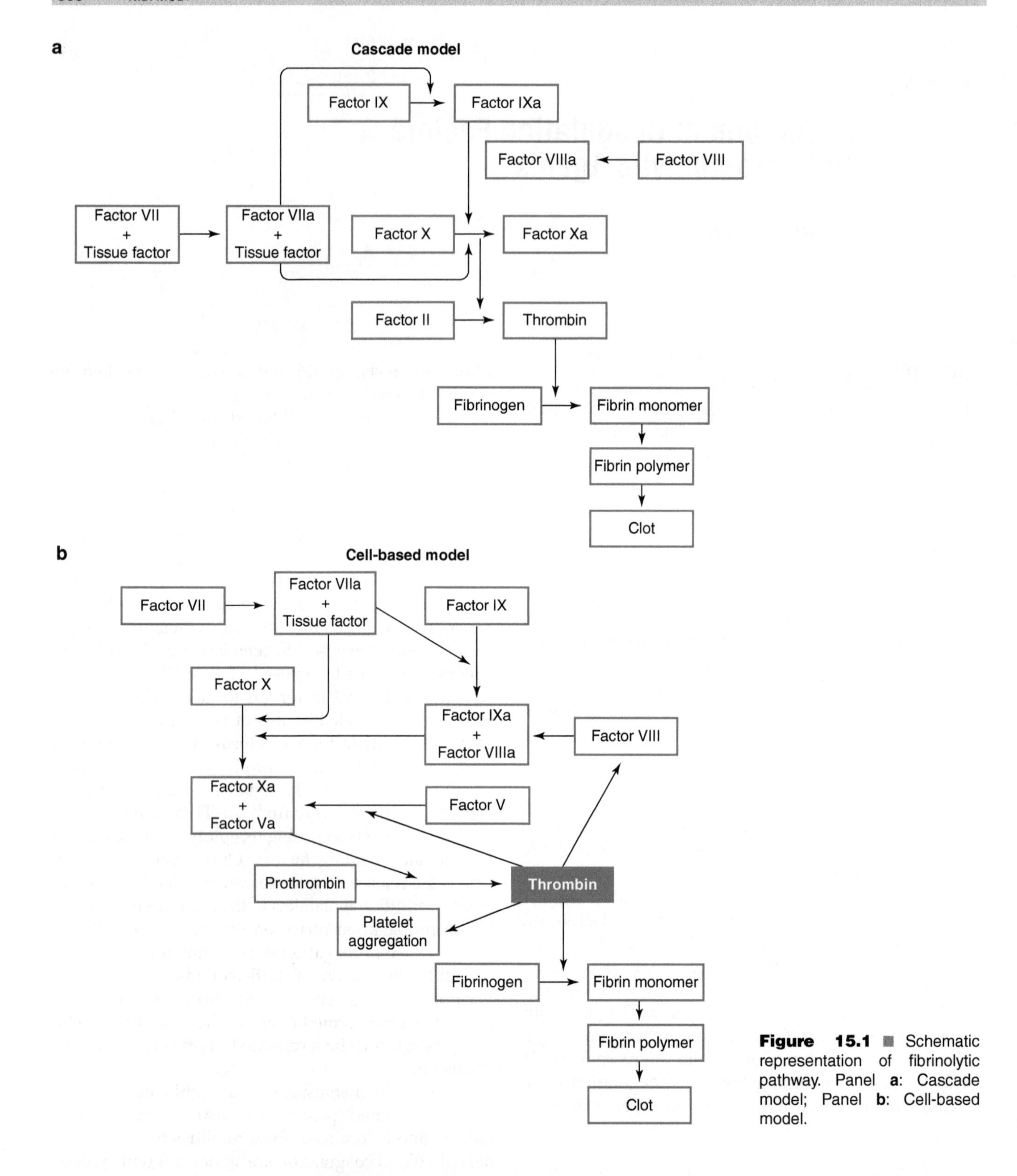

Figure 15.1 ■ Schematic representation of fibrinolytic pathway. Panel **a**: Cascade model; Panel **b**: Cell-based model.

VIII. It affects approximately 1 in 5,000–10,000 males. Patients with hemophilia B lack factor IX, affecting 1 in 20,000–30,000 males. Factor XI deficiency, originally termed hemophilia C, an autosomal recessive disorder, is less common (affecting 1 in 100,000 patients in the American population) and in most cases is a mild bleeding disorder. The availability of recombinant coagulation factors has been a major advance in the treatment of hemophilia, providing the promise of unlimited supply, ease of use, improved safety, and reducing the risk of infections transmitted by transfusion.

FACTOR VIII

Factor VIII (antihemophilia factor) is a plasma protein that functions as a cofactor by increasing the maximum catabolic capacity (V_{max}) in the activation of factor X by factor IXa in the presence of calcium ions and negatively charged phospholipid. The congenital absence of factor VIII is termed hemophilia A and afflicts approximately 1 in 5,000–10,000 males.

■ Structure

Factor VIII is synthesized as a single-chain polypeptide of 2,332 amino acids. Shortly after synthesis, cleavage occurs, and most plasma factor VIII circulates as an 80-kD light chain (A3-C1-C2 domains) associated with a series of 90–210-kD heavy chains (A1-A2-B domains) in a metal ion-dependent complex. There are 25 potential N-linked glycosylation sites and 22 cysteines (Vehar et al. 1984a).

■ Pharmacology

The concentration of factor VIII in plasma is about 200 ng/mL (Hoyer 1981). It is not known where factor VIII is synthesized although evidence suggests that several different tissues, including the spleen, liver, and kidney, may play a role. Factor VIII is normally covalently associated with a 50-fold excess of von Willebrand factor. Von Willebrand factor protects factor VIII from proteolytic cleavage and allows concentration at sites of hemostasis. Circulating von Willebrand factor is bound by exposed subendothelium and activated platelets at sites of injury, which allows localization of von Willebrand factor and factor VIII.

Factor VIII circulates in the body as a large precursor polypeptide devoid of coagulant activity. Cleavage by thrombin at Arg372-Ser373, Arg740-Ser741, and Arg1689-Ser1690 results in its procoagulant structure (Vehar et al. 1984a). While cleavage at Arg740 is not essential for coagulant activity, cleavage at the other two sites is necessary. Although factor VIII is synthesized as a single-chain polypeptide, the single-chain polypeptide is cleaved shortly after synthesis. Most of the factor VIII in plasma exists as an 80-kD light chain and a series of heavy chains. Factor VIII circulates as a heterodimer of the 80-kD light chain and a variable (90–210 kD) heavy chain in a metal ion-dependent complex. Small acidic regions in the C-terminal portion of A1 and A2 and in the N-terminal part of A3 are required for optimal coagulation activity, while the B-domain is not directly necessary for procoagulant activity.

RECOMBINANT FACTOR VIII

Recombinant factor VIII (rFVII) is available from three sources: Baxter, Bayer, and Pfizer. Recombinant factor VIII products may be divided into three classes based on the use of human or mammalian-derived raw materials (Table 15.1). Recombinant factor VIII from Baxter (Advate®, Recombinate®) and Pfizer (ReFacto®, Xyntha®) is produced using transfected Chinese hamster ovary (CHO) cells, whereas that from Bayer (Kogenate®, Kogenate FS, Helixate® FS) is produced using transfected baby hamster kidney cells. A major difference between rFVIII from Bayer and Baxter is the presence of a Galα1 → 3Gal carbohydrate moiety in the Baxter product (Hironaka et al. 1992). The recombinant product from Baxter and Bayer consists of full-length factor VIII which, like plasma-derived factor VIII, consists of a dimer of the 80-kD light chain and a heterogeneous heavy chain of 90- to 210-kD (Schwartz et al. 1990). The Pfizer products (ReFacto® and Xyntha®) are a deletion mutant in which the heavy chain lacks nearly the entire B-domain, which is not needed for clotting activity (Roddie and Ludlam 1997). After proteolytic cleavage by thrombin, the activated B-domain-depleted molecule is essentially identical to the activated full-length native rFVIII.

■ Pharmaceutical Considerations

Advate® recombinant antihemophilic factor-Plasma/Albumin Free Method (rAHF-PFM, Baxter) is formulated as a sterile, non-pyrogenic, lyophilized cake for intravenous injection and is provided in single-dose vials, containing nominally 250, 500, 1,000, or 1,500 international units (IU). Biological potency is determined using an in vitro assay that employs a factor VIII concentrate standard that is referenced to the World Health Organization (WHO) International Standard for factor VIII:C concentrates. The specific activity is 4,000–10,000 IU/mg of protein. The final product contains no preservative nor added human or animal components in the formulation. Recombinant antihemophilic factor is administered only by intravenous infusion following reconstitution with 5 mL sterile water for injection. The product contains mannitol, trehalose, sodium, histidine, Tris, calcium, polysorbate 80, and glutathione. Plastic syringes must be used since the protein can adhere to glass syringes.

Recombinant factor VIII (Kogenate®, Bayer; Recombinate®, Baxter; ReFacto®, Xyntha®, Pfizer) is supplied as sterile, single-dose vials containing 250–3,000 IU of factor VIII. The preparation is lyophilized and stabilized with human albumin (Kogenate® and Recombinate®) or polysorbate 80 (ReFacto®, Xyntha®).

Product (manufacturer)	Viral inactivation procedure(s)	Available strengths (IU)
First Generation (media enriched with human or animal plasma proteins)		
Recombinate® (Baxter Healthcare)	Immunoaffinity, ion-exchange chromatography	250, 500, 1,000
Human albumin as stabilizer	Bovine serum albumin used in culture medium for CHO cells	
Second Generation (human albumin-free final formulations)		
Kogenate FS® (Bayer Healthcare)	Immunoaffinity chromatography	250, 500, 1,000, 2,000
Helixate® FS (Bayer for CSL Behring)	Ion exchange	
Sucrose as stabilizer	Solvent detergent (TNBP/polysorbate 80)	
	Ultrafiltration	
ReFacto® (Wyeth) B-domain deleted	Ion exchange	250, 500, 1,000, 2,000
Sucrose as stabilizer	Solvent detergent (TNBP/Triton X-100)	
(Not available after 2009)	Nanofiltration	
Third Generation (no human or animal protein used in culture medium or manufacturing process; does contain trace amounts of murine monoclonal antibody)		
Advate® (Baxter Healthcare)	Immunoaffinity chromatography	250, 500, 1,000, 1,500, 2,000, 3,000
Trehalose as stabilizer	Ion exchange	
	Solvent detergent (TNBP/polysorbate 80)	
Xyntha® (Pfizer) B-domain deleted	Nanofiltration	250, 500, 1,000, 2,000
Sucrose as stabilizer	Solvent detergent	
	Polysorbate 80	

Table 15.1 ■ Summary of recombinant factor VIII concentrates.

Recently a reformulated product, Kogenate® FS (Bayer), has become available. This product is similar to its predecessor Kogenate antihemophilic factor but incorporates a revised purification and formulation process that eliminates the addition of human albumin as a stabilizer, instead of using histidine. The products contain no preservatives and should be stored at 2–8 °C. The lyophilized powder may be stored at room temperature (up to 25 °C) for up to 3 months without loss of biological activity. Freezing should be avoided. Factor VIII should be reconstituted with the diluent provided. The reconstituted product must be administered intravenously by direct syringe injection or drip infusion within 3 h of reconstitution.

■ Clinical Usage

Recombinant factor VIII (Kogenate® and Kogenate® FS, Bayer Corporation; Recombinate® and Advate®, Baxter; ReFacto®, Xyntha®, Pfizer) is indicated for the control and prevention of bleeding episodes in adults and children (0–16 years) with hemophilia A, the perioperative management in adults and children with hemophilia A, and the routine prophylaxis to prevent or reduce the frequency of bleeding episodes in adults and children with hemophilia A. Recombinant factor VIII is not indicated for the treatment of von Willebrand disease.

The pharmacokinetics of rFVIII is summarized in Table 15.2. The increase in factor VIII concentration is dose proportional and the disposition is similar following single and chronic dosing.

Dosage of rFVIII (in IU) must be individualized to the needs of the patient, the severity of the deficiency and of the hemorrhage, the presence of inhibitors, and to the desired increase in factor VIII activity (in IU/dL, or percentage of normal). The required dosage of factor VIII may be estimated as follows:

$$\text{Required dose (IU)} = \text{Body weight (kg)} \times \text{desired rise in factor VIII} \left(\frac{\text{IU}}{\text{dL}} \text{ or \% of normal} \right) \times 0.5 \left(\frac{\text{IU}}{\text{kg}} \text{ per } \frac{\text{IU}}{\text{dL}} \right)$$

	Dose (IU/kg)	Population	N	C_{max} (IU/dL)	CL (mL/h/kg)	V_{ss} (mL/kg)	MRT (h)	$t_{1/2}$ (h)	Reference
Plasma and albumin-free (rAHF-PFM, Baxter Bioscience)	50	Previously treated pediatric patients <6 years of age	47[a]	95.6±23.3	4.3±1.4	51.4±12.9	12.5±3.1	9.9±1.9	Blanchette et al. (2008)
Plasma and albumin-free (rAHF-PFM, Advate®, Baxter Bioscience)	50	Pediatric, 1–6 years age	52	92	4.3	50	NR	9.4	Björkman et al. (2010)
	50	Previously treated patients, 10–65 years	100	122	3.3	48	NR	11.2	
rFVIII (Kogenate FS®, Bayer)	50	Children (<18 years age)	20	NR	4.1	59.2	15.1	10.7	Barnes et al. (2006)
BDD-rFVIII, (ReFacto®, Pfizer)	25	Patients (>12 years age) with severe hemophilia	21	NR	3.0±1.0	62.7±18.7[b]	24.3±17	17.7±13	Morfini et al. (2003)

Abbreviations: *AUC* area under the curve, C_{max} maximum concentration, *CL* clearance, *MRT* mean residence time, V_{dss} volume of distribution, $t_{1/2}$ half-life, *NR* not reported
[a]Per protocol population
[b]V_{darea}

Table 15.2 ■ Clinical pharmacokinetic profile of recombinant factor VIII.

Local manufacturer's prescribing information should be consulted for full details of indications, dosage regimens, administration, precautions and warning, and regional differences in prescribing.

Safety

Trace amounts of mouse or hamster protein may be present in certain rFVIII as contaminants from the manufacturing process. Therefore, caution should be exercised when administering rFVIII to individuals with known hypersensitivity to plasma-derived antihemophilic factor or with hypersensitivity to biological preparations with trace amounts of murine or hamster proteins.

FACTOR VIIA

The development of recombinant factor VIIa was motivated by the fact that a small fraction of patients with hemophilia (15–20 % of patients with hemophilia A and 2–5 % of patients with hemophilia B) develop antibodies (inhibitors) to factor VIII and factor IX, respectively. High titers of inhibitors make it impossible to give sufficient coagulation factor to overcome the inhibitor, and therapy is ineffective or is associated with unacceptable side effects. Factor VIIa plays a central role in coagulation according to the cell-based concept of coagulation. In the absence of tissue factor, factor VII has very low proteolytic activity whereas factor VIIa requires either tissue factor or activated platelets (tissue factor independent) to generate thrombin. Hereditary deficiency in factor VII is rare, affecting one symptomatic individual per 500,000 population (Lapecorella and Mariani 2008).

■ Structure

Factor VII is a vitamin K-dependent glycosylated serine protease proenzyme that is synthesized in the liver. It has 406 amino acids and the molecular weight is ~50 kDa. The protein is not functionally active unless it is γ-carboxylated. There are two sites of N-linked glycosylation on factor VII. Factor VII is synthesized in the liver as a proenzyme that becomes activated and cleaved upon hydrolysis at Arg-152 and Ile-153. The two polypeptide chains comprise a light (152 residues) and heavy (254 residues) chain. The light chain contains an N-terminal γ-carboxyglutamic acid (Gla) domain, in which all 10 Gla residues are posttranslationally γ-carboxylated followed by two domains homologous to epidermal growth factor (EGF). The heavy chain comprises a serine protease domain which is homologous to trypsin.

Pharmacology

The concentration of factor VII in plasma is approximately 0.5 mcg/mL. The factor VII zymogen forms a high-affinity complex with cell-bound tissue factor. Activated factor VII (FVIIa) cleaves factor X and factor IX, eventually leading to the formation of thrombin.

RECOMBINANT FACTOR VIIA

Recombinant factor VII is expressed in baby hamster kidney cells as a single-chain form and is spontaneously autoactivated to factor VIIa during purification with almost 100 % yield (Thim et al. 1988). The purification of rFVII from the cell culture medium involves pH adjustment of the medium and loading onto a Q-Sepharose FF column to concentrate the protein, treatment with a detergent to ensure virus inactivation, loading onto an immunoaffinity column to purify rFVII using a monoclonal antibody that recognizes the FVII Gla domain in its functional conformation, final purification, and complete activation of rFVII to rFVIIa through the use of two anion-exchange chromatography steps (Jurlander et al. 2001). The resulting FVIIa is then formulated, dispensed into vials, and freeze dried. The characterization of the protein indicates that it is very similar to plasma-derived factor VIIa with regard to amino acid sequence. In plasma-derived FVIIa, all 10 possible Gla residues are fully γ-carboxylated, whereas rVIIa contains 9 fully and 1 partially (50 %) γ-carboxylated residue (Thim et al. 1988). Recombinant FVIIa is approved for use in hemophilia with inhibitors and in patients with acquired hemophilia in Europe, the United States, and Japan.

Pharmacokinetics and Pharmacodynamics

The single-dose pharmacokinetics of recombinant factor VIIa was first investigated in 15 patients with hemophilia with severe factor VIII or factor IX deficiency (Lindley et al. 1994). Following an intravenous dose of 17.5, 35, and 70 μg/kg, the plasma clearance was 30.3, 32.4, and 36.1 mL/h/kg, respectively. The pharmacokinetics was linear, and no difference in clearance was noted between nonbleeding and bleeding episodes. Median clearance was 31.0 mL/h in nonbleeding episodes and 32.6 mL/h in bleeding episodes. The median half-life was 2.9 h in nonbleeding episodes and 2.3 h in bleeding episodes. Table 15.3 summarizes the pharmacokinetics of rFVIIa in various populations. In patients with hemophilia or factor VII deficiency, the clearance, volume of distribution, elimination half-life, and mean residence time appear independent of dose. Plasma clearance appears similar in healthy adults and patients with hemophilia whereas clearance is generally faster in children than in adults. There are no significant gender differences. The pharmacokinetics of rFVIIa appears similar in healthy Caucasian and Japanese subjects (Table 15.4). In a summary of the pharmacokinetics, Klitgaard and Nielsen categorized subjects falling in two groups in terms of rVIIa pharmacokinetics: in healthy volunteers, adult patients with hemophilia, and nonbleeding patients with cirrhosis, plasma clearance is low (30–40 mL/h/kg), whereas in children with hemophilia, patients with congenital FVII deficiency, and patients with active, high levels of bleeding, plasma clearance appears higher (60–90 mL/h/kg) (Klitgaard and Nielsen 2007).

In the absence of FVIII or FIX, doses of rFVIIa greater than 25 nM are required to induce hemostasis. Although the concentrations of rFVIIa for maximum prolongation of clot lysis time varies widely, the median concentration was 73.0 U/mL (Lisman et al. 2002), roughly equivalent to a rFVIIa dose of 90–120 μg/kg (Hedner 2006; 2007).

Pharmaceutical Considerations

NovoSeven® (Novo Nordisk) is supplied as a white lyophilized powder in single-use glass vials formulated with sodium chloride, calcium chloride dihydrate, glycylglycine, polysorbate 80, and mannitol. The pH is adjusted to 5.3–6.3. The product does not contain any stabilizing protein. Before reconstitution, NovoSeven® should be stored refrigerated (2–8 °C) avoiding exposure to direct sunlight. NovoSeven® is distributed in vials of 1.2 mg (60,000 IU), 2.4 mg (120,000 IU), or 4.8 mg (240,000 IU) to be reconstituted with sterile water for injection. After reconstitution with the appropriate volume of diluent, each vial contains approximately 0.6 mg/mL. Following reconstitution, NovoSeven® may be stored refrigerated or at room temperature for up to 3 h. NovoSeven® is intended for intravenous bolus injection and should not be mixed with infusion solutions.

A formulation of rFVIIa that is stable at room temperature (rFVII-RT, NovoSeven® RT, Novo Nordisk) is formulated with sodium chloride, calcium chloride dihydrate, glycylglycine, polysorbate 80, mannitol, sucrose, and methionine. The formulation was shown to be bioequivalent to rFVIIa based on data from a crossover study comparing the two formulations at 90 μg/kg (Bysted et al. 2007). NovoSeven® RT is distributed in vials of 1 mg (50,000 IU), 2 mg (100,000 IU), 5 mg (250,000 IU), or 8 mg (400,000 IU) with a diluent comprising 10 mmol solution of L-histidine in water for injection.

Dose (mcg/kg)	Population	N	C_{max} (U/mL)	CL (mL/h/kg)	V_{ss} (mL/kg)	MRT (h)	$t_{1/2}$ (h)	Reference
<20	Healthy volunteers anticoagulated with acenocoumarol		NR	30.9±5.1	80±8	NR	2.4±0.2	Girard et al. (1998)
>20	Healthy volunteers anticoagulated with acenocoumarol		NR	34.5±7.0	94±15	NR	2.5±0.3	Girard et al. (1998)
90 rFVIIa	Healthy male volunteers	22	55.4±7.9	37.6±6.0	111±18	3.0±0.3	3.5±0.3	Bysted et al. (2007)
90 rfVIIa-RT	Healthy male volunteers	22	52.8±7.3	40.4±6.2	123±20	3.1±0.3	3.5±0.3	Bysted et al. (2007)
17.5–70	Adult hemophiliac, bleeding	5–21	9.9±6.3	36.6±8.7	104±25	2.7±0.5	2.5±0.5	Lindley et al. (1994)
17.5–70	Adult hemophiliac, nonbleeding	25–29	11.4±7.7	32.1±12.2	110±39	3.5±0.6	2.8±0.5	Lindley et al. (1994)
15	Patients with severe congenital deficiency in factor VII	5	NR	64.9±22.7	210±70	3.3±0.5	2.4±0.4	Berrettini et al. (2001)
30	Patients with severe congenital deficiency in factor VII	5	NR	67.7±17.9	230±70	3.5±0.6	2.6±0.6	Berrettini et al. (2001)
90–180	Children with hemophilia A	12	NR	78	196	2.5	2.3	Villar et al. (2004)
90	Adults with hemophilia A	5	NR	53	159	3.0	2.3	Villar et al. (2004)
17.5–70	Adults with hemophilia A or B with or without inhibitors	15	10.8±7.1	32.8±11.7	108.9±37.2	3.3±0.6	2.7±0.5	Lindley et al. (1994)
100–200	Trauma subjects 16–65 years of age	230	NR	40	120	NR	2.4	Klitgaard et al. (2006)
5	Adult nonbleeding patients with cirrhosis and prolonged PTT (>2 s above upper limit of reference range)	10	2.3±0.6	32.9±16.9	NR	3.1±1.3	2.4	Bernstein et al. (1997)
20		10	6.1±3.3	43.7±20.0	NR	4.0±2.3	3.2	Bernstein et al. (1997)
80		10	30.8±6.1	34.9±16.5	NR	3.1±0.5	2.9	Bernstein et al. (1997)

Abbreviations: *AUC* area under the curve, C_{max} maximum concentration, *CL* clearance, *MRT* mean residence time, V_{dss} volume of distribution, $t_{1/2}$ half-life, *NR* not reported

Table 15.3 ■ Clinical pharmacokinetics of recombinant factor VIIa.

Dose (mcg/kg)	Population	N	C_{max} (U/mL)	CL (mL/h/kg)	V_{ss} (mL/kg)	$t_{1/2}$ (h)	Reference
40	Caucasian	11	13.4±2.3	34.3±4.9	145±37	5.2±2.7	Fridberg et al. (2005)
	Japanese	11	12.1±1.4	33.3±5.0	165±42	6.0±2.4	
80	Caucasian	12	24.1±1.9	36.6±5.2	132±13	3.9±1.1	
	Japanese	11	25.2±2.4	33.7±3.0	130±18	4.3±0.9	
160	Caucasian	11	45.9±6.3	37.2±6.3	139±22	4.1±1.0	
	Japanese	10	48.0±5.9	34.5±4.6	135±14	4.0±0.6	

C_{max} maximum plasma concentration, *CL* plasma clearance, V_1 initial volume of distribution, V_{ss} steady-state volume of distribution, $t_{1/2}$ half-life

Table 15.4 ■ Comparison of recombinant-activated FVII pharmacokinetics in Caucasians and Japanese.

■ Clinical Usage

Recombinant factor VIIa (NovoSeven® and NovoSeven® RT) is indicated for the treatment of bleeding episodes in patients with congenital hemophilia A (deficiency of factor VIII) or B (deficiency of factor IX) with inhibitors. It is also indicated for use in patients who are congenitally deficient in factor VII and for prophylaxis of surgical bleeding in patients with hemophilia A or B and in patients who have acquired hemophilia with inhibitors to factor VIII or IX. In the European Union and in Japan, recombinant factor VIIa is also indicated for the control of Glanzmann's thrombasthenia with antibodies to platelet membrane glycoprotein IIb/IIIa and/or human leukocyte antigen and with past or present refractoriness to platelet transfusion.

The recommended dose of recombinant factor VIIa for patients with hemophilia A or B with inhibitors is 90 μg/kg every 2 h by bolus infusion until hemostasis is achieved or until treatment is judged to be inadequate. The dose for Glanzmann's thrombasthenia approved in the European Union is 90 μg/kg every 2 h for a minimum of 3 doses. In the EU a single-dose injection of 270 μg/kg was recently approved. For patients with factor VII deficiency, the initial dosage is 15–30 μg/kg every 4–6 h until hemostasis is achieved. Local manufacturer's prescribing information should be consulted for full details of indications, dosage regimens, administration, precautions and warning, and regional differences in prescribing.

Safety

The main safety concern with rFVIIa is the risk of thromboembolic adverse events because rFVIIa is administered in supraphysiological (~1,000-fold) doses. A review of safety data up to April 2003, involving more than 700,000 administrations of 90 μg/kg, indicates a low level of serious adverse events (1 %) including thrombotic events such as myocardial infarction, stroke, pulmonary embolism, deep vein thrombosis, and disseminated intravascular coagulation (Abshire and Kenet 2004). In an updated review covering the period from May 2003 to December 2006, with approximately 800,000 doses of rFVIIa (90 μg/kg), there were a total of 30 thromboembolic events and 6 thromboembolic event-associated fatal events (Abshire and Kenet 2008).

Recombinant factor VIIa should not be administered to patients with known hypersensitivity to recombinant factor VIIa or any of the components of recombinant factor VIIa. Recombinant factor VIIa is contraindicated in patients with known hypersensitivity to mouse, hamster, or bovine proteins.

■ Recent Developments

Recombinant VII is rapidly cleared with a short terminal half-life (2.4 h), requiring multiple, frequent administrations (2–3 doses given at 2- or 3-h intervals). Various approaches, including rFVII variants with site-directed amino acid substitution, conjugation with polyethylene glycol or carrier protein such as albumin or Fc component of IgG, and incorporation into pegylated liposomes, have been explored to prolong the residence time.

NN1731, a variant of rFVIIa with three amino acid substitutions (V158D, E296V, M298Q) designed to stabilize the molecule in the active conformation without tissue factor has been evaluated in healthy subjects (Møss et al. 2009).

A recent report indicated the development of a recombinant FVIIa-albumin fusion protein (Schulte 2008). In vitro characterization demonstrated that the specific molar activity of the rVIIa-albumin fusion protein was 70 % of that of wild-type rFVIIa. A pharmacokinetic study in rats showed that the rVIIa-albumin fusion protein had a half-life that was 5.8 times longer than NovoSeven®. An in vivo FVII depletion model in rats involving treatment with phenprocoumon showed that the rFVIIa-albumin fusion corrected the coagulation time comparable to NovoSeven® and showed a longer duration of effect.

Several groups have evaluated the use of liposomes (cf. Chap. 4) to deliver FVIIa. Recent studies have shown that rFVII formulated with pegylated liposomes improved the hemostatic efficacy in vitro (Yatuv et al. 2008), an animal model (Yatuv et al. 2010), and more recently in phase I/II clinical study (Spira et al. 2010).

FACTOR IX

Factor IX is activated by factor VII/tissue factor complex in the extrinsic pathway and by factor XIa in the intrinsic pathway (Fig. 15.1). Activated factor IX, in combination with activated factor VIII, activates factor X, resulting in the conversion of prothrombin to thrombin. Thrombin then converts fibrinogen to fibrin, forming a blood clot at a site of hemorrhage.

RECOMBINANT COAGULATION FACTOR IX

Recombinant factor IX is a 415 amino acid glycoprotein with a molecular weight of ~55 kD. Recombinant coagulation factor IX (BeneFIX®, Pfizer) is produced in a CHO cell line. The transfected cell line secretes recombinant factor IX in the culture medium from which the protein is purified via several steps (Monahan and Di Paola 2010). The condition medium undergoes ultrafiltration/diafiltration to concentrate the protein and establish a consistent buffer. This is followed by four sequential column chromatography steps. The first chromatographic step involves binding to Q-Sepharose Fast Flow resin and subsequent elution with calcium chloride. The second step is running over Cellufine Sulfate, a heparin analog, to achieve further affinity purification. The third step involves running over a column consisting of a synthetic form of calcium phosphate in macroporous particles to retain factor IX based on charge and removes lower activity forms of rFIX. The final column involves immobilized copper to retain factor IX and remove trace host cell and other contaminants. The purified rFIX is subsequently passed through an ultrafiltration membrane and undergoes final diafiltration/ultrafiltration to concentrate and exchange the rFIX to the final formulation buffer.

While plasma-derived factor IX carries a Thr/Ala dimorphism at position 148, the primary amino acid sequence of recombinant factor IX is identical to the Ala148 allelic form. As a result of posttranslational modifications, recombinant and plasma-derived factor IX differ in a number of respects (White et al. 1997). First, plasma-derived factor IX carries 12 gamma-carboxyglutamic acid (Gla) residues in its amino-terminal Gla-domain, whereas 40 % of recombinant factor IX is undercarboxylated, lacking gamma-carboxylation at Glu40. Other differences between recombinant and plasma-derived factor IX are in the activation peptide region (residues 146–180), which is cleaved off upon factor IX activation. These include the lack of sulfation at Tyr155 and of phosphorylation at Ser158, as well as different N-linked glycosylation patterns at Asn157 and Asn167.

The potency of recombinant factor IX is determined in international units using an in vitro clotting assay. One international unit is the amount of factor IX activity present in a milliliter of pooled normal human plasma. The specific activity of BeneFIX® is greater than or equal to 200 IU/mg protein.

■ Pharmacology

Pharmacokinetic and pharmacodynamic studies have indicated that increases in recombinant factor IX plasma concentrations are correlated with factor IX activity. Comparison of recombinant factor IX and plasma-derived factor IX in a dog model of hemophilia B indicated that while plasma-derived factor IX had a higher AUC and C_{max} compared with recombinant factor IX, the efficacy of the two products was similar (Keith et al. 1995).

■ Pharmaceutical Considerations

In 2007, a reformulation of recombinant factor IX (BeneFIX®) was implemented, replacing the original formulation. The reformulation increased the ionic strength but still retaining iso-osmolality, allowed a more concentrated preparation, and replaced sterile water for reconstitution with 0.234 % sodium chloride. Additional improvements included a needleless reconstitution device, a prefilled diluent syringe, and a similar dilution volume for all dosage strengths. BeneFIX® is supplied as a sterile, non-pyrogenic, lyophilized powder in single-use vials containing nominally 250, 500, 1,000, 2,000, or 3,000 IU per vial.

The product labeled for room temperature storage may be stored at room temperature (not to exceed 30 °C) or under refrigeration (2–8 °C). The product labeled for refrigerated storage should be stored at 2–8 °C. Prior to the expiration date, the product may be stored at room temperature, not to exceed 30 °C for up to 6 months. Freezing should be avoided to prevent damage to the diluent vial.

Recombinant factor IX is administered by intravenous infusion after reconstitution of the lyophilized powder with the supplied prefilled diluent (0.234 % sodium chloride solution) syringe. The reconstituted solution may be stored at room temperature prior to administration, but BeneFIX® should be administered within 3 h of reconstitution since it does not contain a preservative. Upon reconstitution, BeneFIX® contains polysorbate 80 which is known to increase the rate of di-(2-ethylhexyl)phthalate extraction from polyvinyl chloride. This should be considered during preparation and administration of BeneFIX®, including any storage following reconstitution.

After reconstitution, BeneFIX® should be injected intravenously over several minutes. BeneFIX® should be administered using the tubing provided in the kit. In addition, the solution should be withdrawn from the vial using the vial adapter provided. Reconstituted BeneFIX® should not be administered in the same tubing or container with other medicinal products.

■ Clinical Usage

BeneFIX® is indicated for the control and prevention of bleeding episodes in adults and pediatric patients with hemophilia B (congenital factor IX deficiency or Christmas disease) and for perioperative management in adult and pediatric patients with hemophilia B.

BeneFIX® is contraindicated in patients who have manifested life-threatening, immediate hypersensitivity reactions, including anaphylaxis, to the product or its components, including hamster protein.

The in vivo recovery using BeneFIX® was 25–30 % less than the recovery using highly purified plasma-derived factor IX, whereas there was no difference in the biological half-life (White et al. 1997). The pharmacokinetics of recombinant factor IX is summarized in Table 15.5. Several reports in previously untreated hemophilia B patients and patients previously treated with plasma-derived FIX indicate a very high rate of efficacy (>80 %) with rFIX.

The dosage and duration of substitution treatment depend on the severity of factor IX deficiency, the location and extent of bleeding, the clinical condition, patient age, and the desired recovery in factor IX. The initial estimated dose may be determined as follows:

$$\begin{aligned}\text{Required units} &= \text{Body weight}\,(\text{kg}) \\ &\times \text{desired factor IX increase}\,(\text{IU / dL or \% of normal}) \\ &\times \text{reciprocal of observed recovery}\,(\text{IU / kg per IU / dL})\end{aligned}$$

In clinical studies, one IU of BeneFIX® per kilogram of body weight (average recovery) increased circulating factor IX as follows:

Adults: 0.8 ± 0.2 IU/dL (range 0.4–1.2)
Pediatrics (<15 years): 0.7±0.3 IU/dL (range 0.2–2.1)

Higher doses of factor IX may be necessary in patients with inhibitors. If the expected levels of factor IX are not attained, or if bleeding is not controlled, biological testing may be merited to determine if factor IX inhibitors are present.

Local manufacturer's prescribing information should be consulted for full details of indications, dosage regimens, administration, precautions and warning, and regional differences in prescribing.

Safety

Since BeneFIX® is produced in a CHO cell line, it is contraindicated in patients with known history of hypersensitivity to hamster protein and other constituents in the preparation. During uncontrolled open-label clinical studies with recombinant factor IX, adverse events reported in more than 2 % of patients included nausea, taste perversion, injection site reaction, injection site pain, headache, dizziness, allergic rhinitis, rash, hives, flushing, fever, and shaking.

■ Recent Developments

When treating bleeding episodes, several infusions of factor IX may be required to maintain sufficient coagulation factor levels. Recombinant factor IX products with a longer half-life may provide less frequent dosing, potentially improving compliance.

N9-GP is a modified recombinant serum-free factor IX obtained through site-directed glycopegylation wherein a 40 kDA polyethylene glycol is attached to the FIX activation peptide. Upon activation, the activation peptide, including the attached PEG, is cleaved off releasing the active FIX. A recent pharmacokinetic study in patients with hemophilia B and a FIX activity ≤2 % showed that N9-GP had a half-life that

Age (years)	*N*	IVR (IU/dL per U/kg)	CL (mL/h/kg)	V_{ss} (mL/kg)	MRT (h)	$t_{1/2}$ (h)	Reference
4–9	11	0.61 ± 0.2	10.4 ± 2.3	270 ± 70	26 ± 4.9	20 ± 4.2	Björkman et al. (2001)
10–19	10	0.79 ± 0.3	8.3 ± 2.3	210 ± 70	25 ± 5.3	20 ± 4.1	
20–29	12	0.67 ± 0.2	8.5 ± 1.2	220 ± 60	26 ± 6.2	19 ± 4.9	
30–39	12	0.84 ± 0.2	7.2 ± 1,4	190 ± 40	27 ± 7.7	20 ± 6.5	
40–49	7	0.80 ± 0.3	7.6 ± 1,7	200 ± 50	27 ± 4.7	19 ± 4.2	
50–56	3	0.88 ± 0.2	7.5 ± 0.3	180 ± 80	24 ± 9.6	17 ± 7.1	

Table 15.5 ■ Clinical pharmacokinetic profile of recombinant factor IX.

was approximately five times longer than the patient's previous FIX product and the AUC was approximately eightfold higher compared to pdFIX and tenfold higher compared to rFIX (Negrier et al. 2011).

Recombinant factor IX-Fc fusion protein (rFIXFc) comprises a single molecule of recombinant FIX attached to the constant region of IgG. The presence of the Fc domain allows binding to the neonatal Fc receptor, protecting the Fc-containing molecular from catabolism. In mice, rats, monkeys, and FIX-deficient mice and dogs, rFIXFc was shown to have a three- to fourfold extended half-life compared to rFIX (Peters et al. 2010). In FIX-deficient mice, rFIXFc and rFIX at comparable doses resulted in a similar correction of the clotting deficiency, but rFIXFc had a prolonged effect to 96 h. These results were extended to humans in a phase I/2a open-label study in previously treated patients with hemophilia B (Shapiro et al. 2011). A dose proportional increase in C_{max} was noted, and the elimination half-life was ~54–58 h for rFIXFc compared to the reported half-life of 19 h for BeneFIX®.

FACTOR XIII

Factor XIII is a plasma transglutaminase and the terminal enzyme in the clotting cascade, increasing clot strength by cross-linking fibrin, and increases fibrinolytic resistance by incorporating α_2-plasmin into the clot matrix. Congenital factor XIII deficiency is a rare autosomal recessive disorder affecting one in five million persons (Di Paola et al. 2001).

RECOMBINANT COAGULATION FACTOR XIII

Recombinant factor XIII is an investigational product produced in yeast *Saccharomyces cerevisiae* as a nonglycosylated FXIII A_2 homodimer which is equivalent to cellular FXIII normally found in platelets and readily forms the heterotetrameric A_2B_2 complex in the presence of free FXIII B subunits. Recombinant factor XIII is captured by concentration of the fermentation broth, homogenization of cells, and purification by several chromatography steps.

■ Clinical Usage

The pharmacokinetics of recombinant factor XIII has been studied in healthy volunteers (Reynolds et al. 2005; Visich et al. 2005) and patients with congenital factor XIII deficiency (Lovejoy et al. 2006). Following a single intravenous injection of 50 U/kg rFXIII in healthy volunteers, the estimated half-life was 270–320 h, the volume of distribution ranged from 40 to 75 mL/kg, and FXIII activity increased by 1.77 % per 1 U/kg rFXIII administered (Reynolds et al. 2005). In a multiple-dose study investigating doses of 10 and 25 U/kg in healthy volunteers, the elimination half-life ranged from 228 to 346 h for the 10 U/kg dose and 167 to 197 h for the 25 U/kg dose, and a three- to four-fold accumulation of rFXIII was noted following five daily doses (Visich et al. 2005). In a phase I escalating-dose study in patients with factor XIII deficiency, rFXIII-A_2 complexed with endogenous FXIII B subunits with a half-life of 8.5 days, similar to endogenous FXIII. The median dose response was a 2.4 % increase in FXIII activity per U/kg rFXIII administered (Lovejoy et al. 2006).

RECOMBINANT THROMBOLYTIC AGENTS

■ Tissue-Type Plasminogen Activator

Deposition of fibrin and platelets in the vasculature leads to thromboembolic diseases that are responsible for considerable mortality and morbidity. Early thrombolytic therapy can decrease mortality and improve coronary-artery patency in patients with acute myocardial infarction (AMI). During fibrinolysis, the inactive zymogen plasminogen is enzymatically converted to the active moiety, plasmin, which in turn digests the insoluble fibrin matrix of a thrombus to soluble fibrin degradation products. Tissue-type plasminogen activator (t-PA) exhibits fibrin-specific plasminogen activation with minimal systemic fibrinogenolysis. The relative absence of systemic fibrinogenolysis with t-PA means that there are fewer systemic side effects compared to other plasminogen activators. Mean t-PA antigen concentrations at rest in humans are approximately 5 μg/mL and can increase 1.5- to 2-fold in venous occlusion (Holvoet et al. 1987).

■ Structure

Native t-PA is a serine protease synthesized by vascular endothelial cells as a single-chain polypeptide of 527 amino acids with a molecular mass of 64 kD (Pennica et al. 1983). Approximately 6–8 % of the molecular mass consists of carbohydrate. A schematic of the primary structure of human t-PA is shown in Fig. 15.2. There are 17 disulfide bridges and an additional free cysteine at position 83 and 4 putative N-linked glycosylation sites recognized by the consensus sequence Asn-X-Ser/Thr at residues 117, 184, 218, and 448 (Pennica et al. 1983). In addition, the presence of a fucose attached to Thr61 via an O-glycosidic linkage has been reported (Harris et al. 1991). Two forms of t-PA that differ by the absence or presence of a carbohydrate at Asp184 have been characterized: Type I t-PA is glycosylated at asparagine 117, 184, and 448; whereas Type II t-PA lacks a glycosylation at asparagine 184. The asparagine at amino acid 218 is normally not occupied in either form of t-PA (Vehar et al. 1984b). Asparagine 117 contains a high-mannose

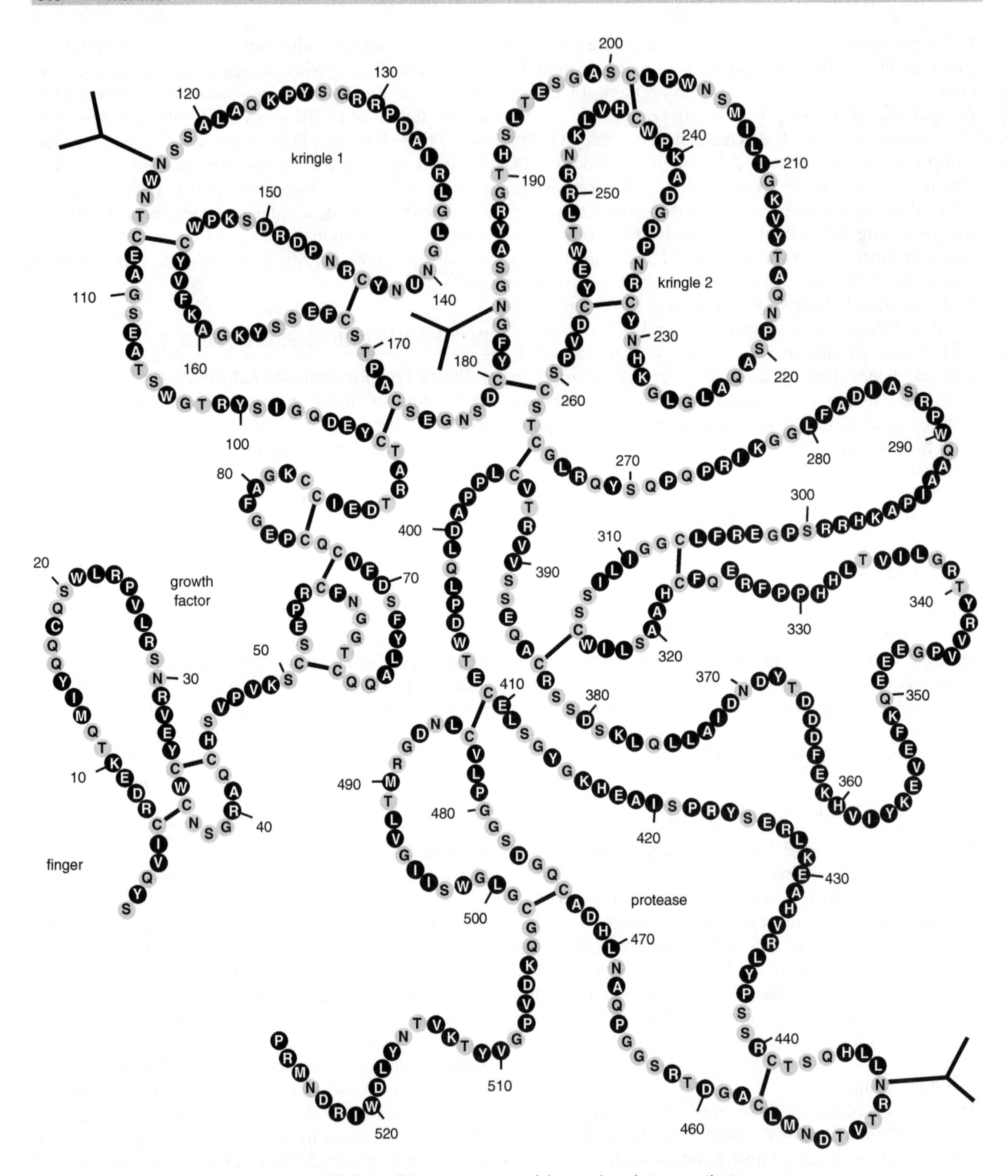

Figure 15.2 ■ Primary structure of tissue-plasminogen activator.

oligosaccharide whereas Asn184 and Asn448 are of the complex carbohydrate type (Spellman et al. 1989). Complex N-linked glycan structures contain a disaccharide Galβ(1,4)GlcNac and terminate in sialic acid residues, while an oligomannose (high mannose)-type glycan contains only mannose in the outer arms.

During fibrinolysis, the single-chain t-PA polypeptide is cleaved between Arg275 and Ile276 by

Administration regimen	Health status	C_{max} (μg/mL)	CL (L/min)	V_1 (L)	V_{ss} (L)	$t_{1/2}\alpha$ (min)	$t_{1/2}\beta$ (min)	$t_{1/2}\gamma$ (h)	Reference
0.25 mg/kg/30 min	Healthy	0.96±0.18	0.64±0.05	4.6±0.3	8.1±0.8	4.4±0.2	39±2.6	–	Tanswell et al. (1989)
0.5 mg/kg/30 min	Healthy	1.8±0.25							
100 mg/2.5 h	AMI	3.3±0.95	0.38±0.07	2.8±0.9	9.3±5.0	3.6±0.9	15±5.4	3.7±1.4	Seifried et al. (1989)
100 mg/1.5 h	AMI	4±1	0.57±0.1	3.4±1.5	8.4±5	3.4±1.4	72±68	–	Tanswell et al. (1992)
100 mg/1.5 h	AMI	–	0.45±0.17	7.2±4	28.9±22	–	144±100	–	Modi et al. (2000)
100 mg/1.5 h	AMI		0.39	6.7	17.3	7.4	22.3	228	Kostis et al. (2002)
Bolus	AMI	9.8±3.6	0.48±0.15	4.5±1.3	31±18	4.8±1.0	17±6.3	9.1±3.1	Tebbe et al. (1989)

C_{max} maximum plasma concentration, *CL* plasma clearance, V_1 initial volume of distribution, V_{ss} steady-state volume of distribution, $t_{1/2}$ half-life

Table 15.6 ■ Pharmacokinetic parameters (mean±SD) for alteplase antigen following intravenous administration in healthy volunteers and patients with AMI.

plasmin to yield 2-chain t-PA. Two-chain t-PA consists of a heavy chain (A-chain) derived from the amino terminus and a light chain (B-chain) linked by a single disulfide bridge between Cys264 and Cys395. The A-chain consists of the finger, growth factor, and 2 kringle domains. The finger domain and the second kringle are responsible for t-PA binding to fibrin and for the activation of plasminogen. The function of the first kringle is not known. The B-chain contains the serine protease domain consisting of the His-Asp-Ser triad that cleaves plasminogen (Pennica et al. 1983).

FIRST-GENERATION RECOMBINANT THROMBOLYTIC AGENTS: RECOMBINANT T-PA (RT-PA)

Recombinant t-PA (rt-PA) (alteplase) is identical to endogenous human t-PA. Like melanoma-derived t-PA, rt-PA lacks glycosylation at Asn218 and exists in two forms that differ by the absence or presence of a carbohydrate at residue Asn184. Type II t-PA has a slightly higher specific activity in vitro compared with Type I t-PA.

■ Pharmacokinetics of rt-PA

The pharmacokinetics of rt-PA has been studied in mice, rats, rabbits, primates, and humans. After intravenous administration, the plasma concentrations decline rapidly with an initial dominant half-life of less than 5 min in all species. Plasma clearance ranges from 27 mL/min in rabbits (Hotchkiss et al. 1988) to 620 mL/min in humans (Tanswell et al. 1989). Recombinant t-PA exhibits nonlinear (Michaelis-Menten) pharmacokinetics at high plasma concentrations (Tanswell et al. 1990). The estimated Michaelis-Menten constant (K_m) and V_{max} values estimated by simultaneously fitting multiple plasma concentration-time curves following several doses were 12–15 μg/mL and 3.7 μg/mL/h, respectively, with little species variation in these parameters. The pharmacokinetics is essentially linear in cases where plasma concentrations do not exceed 10–20 % of K_m (i.e., 1.5–3 μg/mL). A pharmacokinetic summary of alteplase following intravenous administration in humans is presented in Table 15.6. These data show that rt-PA has an initial volume of distribution approximating plasma volume and a rapid plasma clearance. The initial half-life was less than 5 min. There was no difference in the pharmacokinetics following the different infusion regimens. A lower plasma clearance was noted following intravenous bolus injection, suggesting saturation of clearance mechanisms.

The primary route of alteplase clearance is via receptor-mediated clearance mechanisms in the liver. Three cell types in the liver are responsible for the clearance of t-PA: parenchymal cells, endothelial cells, and Kupffer cells. Kupffer cells and endothelial cells mediate t-PA clearance via the mannose receptor. Parenchymal cells clear t-PA via a carbohydrate-independent, receptor-mediated mechanism. Data suggest that this carbohydrate-independent clearance is mediated by the low-density lipoprotein receptor-related protein (LRP) (Bu et al. 1993).

Pharmaceutical Considerations

Recombinant human t-PA (Alteplase; Activase®, Genentech, Inc; Actilyse®, Boehringer Ingelheim) is supplied as a sterile, white to off-white lyophilized powder in 50 mg vials containing vacuum and 100 mg vials without vacuum. Cathflo Activase is supplied as a sterile, lyophilized powder in 2 mg vials. Recombinant t-PA is practically insoluble in water, and arginine is included in the formulation to increase aqueous solubility. Phosphoric acid and/or sodium hydroxide may be used to adjust the pH. The sterile lyophilized powder should be stored at controlled room temperatures not to exceed 30 °C or refrigerated at 2–8 °C, and it should be protected from excessive light.

The powder is reconstituted by aseptically adding the accompanying sterile water for injection to the vial, resulting in a colorless to pale yellow transparent solution containing 1 mg/mL rt-PA, with a pH of approximately 7.3 and an osmolality of approximately 215 mOs/kg. Recombinant t-PA is stable in solution over a pH range of 5–7.5. Since the reconstituted solution does not contain any preservatives, it should be used within 8 h of preparation and should be stored at 2–30 °C. The solution is incompatible with bacteriostatic water for injection. Other solutions such as sterile water for injection or preservative-containing solutions should not be used for further dilution. The 1 mg/mL solution can be diluted further with an equal volume of 0.9 % sodium chloride for injection, USP or 5 % dextrose injection USP to yield a solution with a concentration of 0.5 mg/mL. This solution is compatible with glass bottles and polyvinyl chloride bags.

Clinical Usage

Recombinant human t-PA (alteplase t-PA) is indicated for use in the management of acute myocardial infarct (AMI) in adults for the improvement of ventricular function following AMI, reduction of the incidence of congestive heart failure, reduction of mortality associated with AMI, and for the management of acute massive pulmonary embolism in adults. It is also indicated for the management of acute ischemic stroke in adults for improving neurological recovery and reducing the incidence of disability if therapy is initiated within 3 h after the onset of stroke symptoms and after exclusion of intracranial hemorrhage by cranial computerized tomography (CT) scan or other diagnostic imaging method sensitive for the presence of hemorrhage.

Alteplase (Cathflo® Activase®) is indicated for the restoration of function to central venous access devices as assessed by the ability to withdraw blood.

Two dose regimens, the 90-min accelerated regimen and the 3-h regimen, have been studied in patients experiencing AMI; controlled studies comparing the clinical outcome of the two regimens have not been conducted. For the accelerated regimen, the recommended dose is based on patient weight, not to exceed 100 mg alteplase. For patients weighing more than 67 kg, the recommended dose regimen is 100 mg as a 15-mg intravenous bolus injection, followed by 50 mg infused over 30 min and then 35 mg infused over the next 60 min. For patients weighing no more than 67 kg, the recommended dose regimen is a 15-mg intravenous bolus injection, followed by 0.75 mg/kg infused over 30 min not to exceed 50 mg and then 0.5 mg/kg over the next 60 min not to exceed 35 mg.

For the 3-h regimen, the recommended dose is 100 mg administered as 60 mg in the first hour (6–10 mg as a bolus) and 20 mg over each of the second and third hours. For patients weighing less than 65 kg, the dose is 1.25 mg/kg over 3 h. Infarct artery-related patency rates of 70–77 % are achieved at 90 min with this 3-h regimen (Verstraete et al. 1985). Patency grades of blood flow in the infarct-related artery are defined by the Thrombolysis in Myocardial Infarction (TIMI) scale and are assessed angiographically with TIMI grade 0 representing no flow; grade 1, minimal flow; grade 2, sluggish flow; and grade 3, complete or full, brisk flow.

The efficacy of the accelerated 90-min regimen was demonstrated in an international, multicenter trial, Global Utilization of Streptokinase and Tissue Plasminogen Activator for Occluded Coronary Arteries (GUSTO), that enrolled approximately 41,000 patients (The GUSTO Investigators 1993a, b). The GUSTO trial demonstrated a higher infarct-related artery patency rate at 90 min in the group treated with rt-PA with heparin compared with streptokinase with either intravenous or subcutaneous heparin. The patency in the alteplase group was 81.3 % compared with 53.5–59.0 % in the streptokinase groups. In addition, the alteplase group had a reduced mortality (an additional 10 lives saved per 10,000 patients treated). The intracranial hemorrhage rate was approximately 1 %.

In a multicenter, open-label study in 461 patients with AMI randomized to receive 100 mg alteplase over 90 min or two 50-mg bolus doses 30 min apart, the 90-min angiographic patency rate was 74.5 % for the double-bolus group and 81.4 % in the infusion group ($p = 0.08$). The 30-day mortality rates were 4.5 % in the bolus group and 1.7 % in the infusion group (not significantly different) (Bleich et al. 1998). Similarly, the Continuous Infusion Versus Double-Bolus Administration of Alteplase (COBALT) trial in 7,169 patients with AMI showed a higher incidence of 30-day mortality in the double-bolus alteplase group (7.98 %) compared with the accelerated-infusion group (7.53 %). There was also a slightly higher incidence of intracranial hemorrhage in the double-bolus group (COBALT Investigators 1997).

For acute ischemic stroke, the recommended dose is 0.9 mg/kg not to exceed 90 mg infused over 60 min, with 10 % of the total dose administered as an initial intravenous bolus over 1 min. The safety and efficacy of this regimen with concomitant use of heparin and aspirin during the first 24 h has not been investigated.

The recommended dose for treatment of pulmonary embolism is 100 mg administered by intravenous infusion over 2 h. Heparin therapy should be instituted or reinstituted near the end of or immediately following alteplase infusion when the partial thromboplastin time or thrombin time returns to twice normal or less.

Safety Concerns

Since thrombolytic therapy increases the risk of bleeding, alteplase is contraindicated in patients with a history of cerebrovascular accidents or patients who have any kind of active internal bleeding, intracranial neoplasm, arteriovenous malformation, or aneurism or who have had recent intracranial or intraspinal surgery or trauma.

SECOND-GENERATION RECOMBINANT THROMBOLYTIC AGENTS

The rapid clearance of rt-PA from the circulation by the liver necessitates administration as an intravenous infusion. Although alteplase provides more rapid thrombolysis and superior patency compared with streptokinase and urokinase at therapeutic doses, there is some fibrinogenolysis and the administration scheme is relatively complicated. Thus, there is room for further improvements in efficacy and safety. Considerable nonclinical and clinical research has been underway to identify rt-PA variants that are fibrin specific and that have a simpler administration regimen compared with alteplase. Strategies that have been used to develop t-PA variants have included domain deletions, glycosylation changes, or site-directed amino acid substitutions. A number of these second-generation thrombolytic agents are currently in late-stage clinical trials or have been approved for marketing.

▪ Reteplase

Reteplase is a 355-amino deletion variant of t-PA, consisting of the protease and kringle 2 domains of human t-PA. It is expressed in *Escherichia coli* cells as a single-chain, nonglycosylated, 39.6-kDa peptide.

▪ Pharmacology

Like alteplase, reteplase is a fibrin-specific activator of plasminogen. In vitro, the plasminogenolytic activity of reteplase is 2- to 3.8-fold lower than alteplase on a molar basis (Kohnert et al. 1993), which may be attributed to the absence of the finger domain in reteplase. Reteplase had a similar in vitro maximal efficacy (E_{max}) compared with alteplase. However, the molar concentration required to produce 50 % clot lysis (EC_{50}) was 6.4-fold higher for reteplase than for alteplase (Martin et al. 1993). The data also suggested that in vitro, reteplase has a lower thrombolytic potency in lysing aged and platelet-rich clots compared with alteplase.

Due to the deletion of the finger region, epidermal growth factor domain, and kringle 1, as well as the carbohydrate side chains, the hepatic clearance of reteplase is reduced. A summary of the pharmacokinetics of reteplase in humans is presented in Table 15.7.

▪ Pharmaceutical Considerations

Reteplase is supplied as a sterile, white, lyophilized powder for intravenous injection after reconstitution with "sterile water for injection, USP" supplied as part of the kit. Following reconstitution, the pH of the solution is 6.0. Reteplase contains no antibacterial preservatives and should be reconstituted immediately before use. The solution should be used within 4 h when stored at 2–30 °C.

▪ Clinical Usage

Reteplase (Retavase®, Rapilysin®) is indicated for use in the management of AMI in adults for the improvement of ventricular function following AMI, the reduction of the incidence of congestive heart failure and the reduction of mortality associated with AMI.

The potency of reteplase is expressed in units using a reference standard that is specific for reteplase and is not comparable with units used for other thrombolytic agents. Reteplase is administered as a double-bolus injection regimen consisting of 10 U each. Each bolus is administered as an intravenous bolus injection over 2 min via an intravenous line in which no other medications are being administered simultaneously. The second bolus injection is given 30 min after the first. Heparin and reteplase are incompatible when combined in solution and should not be administered simultaneously through the same intravenous line. If reteplase is administered through an intravenous line containing heparin, normal saline or 5 % dextrose solution should be flushed through the intravenous line before and following reteplase.

The International Joint Efficacy Comparison of Thrombolytics (INJECT) trial evaluated the effects of reteplase (10+10 U) and streptokinase (1.6 million Units over 60 min) on 35-day mortality in 6,010 AMI patients in a double-blind randomized fashion. The 35-day mortality was 9.0 % for patients treated with reteplase and 9.5 % for those treated with streptokinase with no difference between the two groups. The incidence of stroke was also similar between the groups.

Dose	Population	N	C_{max}	CL (L/h)	$t_{1/2}\alpha$ (min)	$t_{1/2}\beta$ (h)	Reference
Reteplase[a]							
10 U	AMI patients	4	4,620	6.24	19.2	6.3	Seifried et al. (1992)
15 U	AMI patients	9	5,060	8.34	18.8	6.3	Seifried et al. (1992)
15 U	AMI patients	9	5,170	8.70	21.4	5.0	Grünewald et al. (1997)
10 U+5 U	AMI patients	7	3,610	9.12	16.3	5.4	Grünewald et al. (1997)
10 U+10 U	AMI patients	8	3,370	6.90	17.0	5.5	Grünewald et al. (1997)
Tenecteplase[b]							
30 mg	AMI patients	48	10.0	98.5	21.5	1.93	Modi et al. (2000)
40 mg	AMI patients	31	10.9	119	23.8	2.15	Modi et al. (2000)
50 mg	AMI patients	20	15.2	99.9	20.1	1.50	Modi et al. (2000)
Lanoteplase							
15 kU/kg	AMI patients	4	NR	3.42	36	12	Kostis et al. (2002)
30 kU/kg	AMI patients	2	NR	3.12	31	7.4	Kostis et al. (2002)
60 kU/kg	AMI patients	8	NR	3.42	32	8.7	Kostis et al. (2002)
120 kU/kg	AMI patients	8	NR	2.4	47	10.5	Kostis et al. (2002)
10 kU/kg	Healthy young male	9	NR	2.6	32	9.4	Vachharajani et al. (2011)
10 kU/kg	Healthy young female	10	NR	2.8	30	8.7	Vachharajani et al. (2011)
10 kU/kg	Healthy elderly male	10	NR	3.0	28	8.5	Vachharajani et al. (2011)
10 kU/kg	Healthy elderly female	10	NR	2.0	31	9.8	Vachharajani et al. (2011)

Abbreviations: *AUC* area under the curve, C_{max} maximum concentration, *CL* clearance, *MRT* mean residence time, V_{dss} volume of distribution, $t_{1/2}$ half-life, *NR* not reported
C_{max} values are reported in [a]U/mL or [b]ng/mL

Table 15.7 ■ Pharmacokinetic parameters for second-generation recombinant thrombolytic agents.

However, more patients treated with reteplase experienced hemorrhagic strokes.

Two open-label angiographic studies (Reteplase Angiographic Phase II International Dose-finding study [RAPID 1] and Reteplase versus Alteplase Patency Investigation During myocardial infarction [RAPID 2]) have compared reteplase with alteplase. In RAPID 1 patients were treated with reteplase (10+10 U, 15 U, or 10+5 U) or the standard alteplase regimen (100 mg over 3 h) within 6 h of symptoms. Ninety-minute TIMI grade 3 flow was seen in 63 % of patients in the 10+10 U reteplase group and 49 % of the patients in the standard regimen alteplase group.

RAPID 2 was an open-label, randomized trial in 320 patients comparing 10+10 U reteplase and accelerated alteplase within 12 h of symptom onset. Percentages of patients with TIMI grade 3 flow at 90 min were 59.9 % in the reteplase group and 45.2 % in the alteplase group. There was no significant difference in the 35-day mortality between the two groups. Neither trial was powered to compare the efficacy or safety with respect to mortality or incidence of stroke.

The more favorable results for reteplase compared with alteplase noted in smaller trials were not replicated in a large, randomized, double-blind trial. In the GUSTO III trial, 15,059 patients were randomized in a 2:1 fashion to receive reteplase in 2 bolus doses of 10 U 30 min apart or up to 100 mg alteplase infused over 90 min. The 30-day mortality rates were 7.47 % for reteplase and 7.24 % for alteplase (The Global Use of Strategies to Open Occluded Coronary Arteries [GUSTO III] Investigators 1997). The stroke rate was 1.64 % for reteplase and 1.79 % for alteplase ($p = 0.50$). Reteplase, while easier to administer than accelerated alteplase, did not demonstrate any survival advantage.

Safety Concerns

As with other thrombolytic agents, reteplase is contraindicated in cases of active internal bleeding, history of cerebrovascular accident, recent intracranial or intraspinal surgery or trauma, intracranial neoplasm, arteriovenous malformation, or aneurism, in cases of known bleeding diathesis, or in severe uncontrolled hypertension.

■ Tenecteplase

Tenecteplase (TNKase®, Genentech, Inc; Metalyse®, Boehringer Ingelheim) is a t-PA variant that has amino acid substitutions in three regions of t-PA. Replacement of threonine at amino acid 103 by asparagine (T103N) incorporates a complex oligosaccharide carbohydrate structure at this position. The replacement of arginine at position 117 by glutamine (N117Q) results in the removal of the high-mannose carbohydrate present at this site. A tetra-alanine substitution at positions 496–499 (KHRR496–499AAAA) contributes to increased fibrin specificity. These three design modifications result in a thrombolytic that, compared to the parent t-PA molecule, is approximately 10- to 14-fold more fibrin specific, is 80-fold more resistant to local inactivation, and has an 8-fold slower clearance in rabbits (Keyt et al. 1994).

■ Pharmacology

Like alteplase, tenecteplase has Type I and Type II glycoforms. Type I has 3 carbohydrate structures at asparagine 103, 184, and 448; and Type II lacks the carbohydrate at asparagine 184. Carbohydrate structures on tenecteplase are all of the complex oligosaccharide type with no high-mannose structures. For this reason, the rapid mannose receptor-mediated clearance observed for alteplase does not occur with tenecteplase. Rather, tenecteplase is thought to be cleared by galactose receptors present in liver sinusoidal cells.

Enzymatic removal of terminal sialic acid from tenecteplase has been shown to increase the clearance in rabbits and is likely due to increased exposure of underlying galactose sugars. This desialylation effect is more profound with tenecteplase than with alteplase and is probably due to the predominant mannose-receptor-mediated clearance for alteplase. A second possible clearance pathway for tenecteplase is a noncarbohydrate-mediated mechanism via the low-density lipoprotein receptor-related protein (LRP) that is also a clearance pathway for alteplase (Camani et al. 1998).

The thrombolytic potency of tenecteplase was five- to tenfold greater than alteplase in animal models of coronary-artery thrombosis (Benedict et al. 1995) and embolic stroke (Thomas et al. 1994). The slower clearance of tenecteplase results in a longer exposure of the clot to the thrombolytic agent, which likely offsets the slightly lower activity. The higher fibrin specificity of tenecteplase results in lower systemic activation of plasminogen and an observed conservation of fibrinogen.

■ Pharmaceutical Considerations

Tenecteplase is supplied as a sterile, white to off-white, lyophilized powder in a 40-mg (Metalyse®) or 50-mg vial (Metalyse®, TNKase®) under a partial vacuum. Each vial is packaged with sterile water for injection for reconstitution and syringe with a dual cannula device. It should be stored at controlled room temperature not to exceed 30 °C, or it should be stored refrigerated at 2–8 °C. Tenecteplase is intended for intravenous bolus injection following reconstitution with sterile water for injection. Each vial nominally contains a 5 % overfill, L-arginine, phosphoric acid, and polysorbate 20. The biological potency of tenecteplase is determined by an in vitro clot lysis assay and is expressed in tenecteplase-specific activity units. The specific activity of tenecteplase has been defined as 200 units/mg protein.

■ Clinical Usage

Tenecteplase is indicated for the reduction of mortality associated with AMI. The recommended total dose of tenecteplase should not exceed 50 mg and is based on patient weight according to the following weight-adjusted dosing table (Wang-Clow et al. 2001):

Patient weight (kg)	Tenecteplase dose (mg)
<60	30
≥60 to <70	35
≥70 to <80	40
≥80 to <90	45
≥90	50

Treatment should be initiated as soon as possible after the onset of AMI symptoms. Tenecteplase is contraindicated in patients with known bleeding diathesis or active internal bleeding, history of cerebrovascular accident, recent intracranial or intraspinal surgery or trauma, intracranial neoplasm, arteriovenous malformation, aneurysm, or severe uncontrolled hypertension due to an increased risk of bleeding.

The clinical pharmacokinetics of tenecteplase have been examined in two studies (TIMI 10A and TIMI 10B). TIMI 10A was a phase 1 pilot safety study in patients with AMI. Pharmacokinetic data were obtained in 82 patients following intravenous bolus doses of 5–50 mg. Tenecteplase plasma concentrations decreased in a biphasic manner with an initial half-life of 11–20 min and a terminal half-life of 41–138 min. Mean plasma clearance of tenecteplase ranged from 125 to 216 mL/min and decreased with increasing dose (Modi et al. 1998).

TIMI 10B was a dose-finding phase 2 efficacy study comparing 30-, 40-, and 50-mg doses of bolus tenecteplase to 100 mg alteplase administered via the accelerated-infusion regimen. The pharmacokinetic data from TIMI 10B are summarized in Table 15.7. Tenecteplase plasma clearance was approximately

100 mL/min compared to 453 mL/min for accelerated alteplase. In contrast to TIMI 10A, no dose-dependent decrease in plasma clearance was noted in TIMI 10B, likely as a result of the narrower dose range examined. Additionally, the plasma clearances noted in TIMI 10B were slightly lower than those noted in TIMI 10A at comparable doses (Modi et al. 2000). The 30-, 40-, and 50-mg doses in TIMI 10B produced TIMI grade 3 flow in 54.3, 62.8, and 65.8 % of the patients, respectively. TIMI grade 3 flow was seen in 62.7 % of patients in the accelerated alteplase group, not significantly different from that in the 40-mg tenecteplase group. An additional finding of this dose-finding efficacy trial was that dose-adjusted dosing is important in achieving optimal reperfusion (Cannon et al. 1998). In addition, tenecteplase resulted in a lower change from baseline in systemic coagulation factors compared with alteplase.

The safety and efficacy of tenecteplase were studied in a large double-blind, randomized trial (Assessment of the Safety and Efficacy of a New Thrombolytic [ASSENT-2]). This trial in 16,949 AMI patients showed that the 30-day mortality rates for single-bolus tenecteplase and accelerated alteplase were almost identical (6.18 % for tenecteplase and 6.15 % for alteplase) (Assessment of the Safety and Efficacy of a New Thrombolytic (ASSENT-2) Investigators 1999). Intracranial hemorrhage rates were similar in both groups (0.9 %), but fewer noncerebral hemorrhages and a lower need for blood transfusion were noted in the tenecteplase group. In conclusion, tenecteplase and 90-min alteplase are equivalent in terms of mortality and rates of intracranial hemorrhage. The single-bolus regimen for tenecteplase may facilitate thrombolytic therapy.

Safety Concerns

Tenecteplase is contraindicated in cases of active internal bleeding, history of cerebrovascular accident, recent intracranial or intraspinal surgery or trauma, intracranial neoplasm, arteriovenous malformation, or aneurism, in cases of known bleeding diathesis, or in severe uncontrolled hypertension.

The most common complication during tenecteplase therapy is bleeding. In clinical studies of tenecteplase, patients were treated with both aspirin and heparin. Heparin may contribute to the bleeding risk. Use of tenecteplase with other antiplatelet agents has not been adequately studied. In the ASSENT-2 study, the incidence of intracranial hemorrhage was 0.9 % and any stroke was 1.8 %

▪ Lanoteplase

Lanoteplase (also referred to as ΔFE1X PA, BMS-200980, SUN9216, and nPA) is currently not commercialized. Lanoteplase is a t-PA variant in which the fibronectin fingerlike and epidermal growth factor domains have been removed (Collen et al. 1988). In addition, an asparagine to glutamine substitution at amino acid 117 provides reduced clearance (Hansen et al. 1988). The clearance appears to be mediated by the low-density lipoprotein receptor-related protein and asialoglycoprotein receptors, and the mannose receptor plays a smaller role (Komoriya et al. 2007). The clinical pharmacokinetics of lanoteplase is summarized in Table 15.7.

The Intravenous nPA for Treatment of Infarcting Myocardium Early (InTIME) study compared lanoteplase with accelerated alteplase. Patients were randomized to receive intravenous bolus doses of 15, 30, 60, or 120 kU/kg (not to exceed 12,000 kU) of lanoteplase or accelerated alteplase (den Heijer et al. 1998). A statistically significant increase in the proportion of patients with TIMI grade 3 flow at 60 min was noted with increasing lanoteplase dose ($p<0.001$). There was no difference in the 30-day composite end point of death, heart failure, major bleeding, or nonfatal infarction (Ross 1999).

A larger randomized, multicenter equivalence trial (InTIME-II) in 15,078 patients compared 120 kU/kg lanoteplase with accelerated alteplase (The InTIME Investigators 2000). The primary end point was 30-day mortality with an incidence of 6.75 % for lanoteplase and 6.61 % for alteplase. The incidence of stroke was not statistically significantly different between treatment groups (1.87 % for lanoteplase and 1.53 % for alteplase). The incidence of hemorrhagic stroke was 0.64 % for alteplase and 1.12 % for lanoteplase ($p=0.004$).

CONCLUSIONS

Recombinant technology has brought about significant advances in the treatment of coagulation disorders and in the availability of thrombolytic agents for the treatment of thrombotic disorders.

Current efforts focus on modifying the proteins to enhance their pharmacokinetic properties and to reduce immunogenicity. For recombinant coagulation factors, this includes pegylation or incorporating into pegylated liposomes, polysialylation, and attachment to the Fc region of IgG to increase half-life or mutagenesis to enhance resistance to degradation or clearance. Notably, cost continues to be a significant limitation in making recombinant coagulation factors available to all patients. Potentially, the development of biosimilars and a clear regulatory pathway to make these available may ease this limitation in the future.

Although significant strides have been made in reducing mortality due to acute myocardial infarction, there continue to be some drawbacks with current thrombolytic agents including the need for large doses, limited fibrin specificity, and risk of bleeding and reocclusion. Future efforts will focus on developing recombinant thrombolytic agents to improve efficacy and safety such as selective antibody-targeted plasminogen activators and chimeric molecules and to develop effective agents and regimens for the treatment of stroke and pulmonary embolism.

SELF-ASSESSMENT QUESTIONS

■ Questions

Question 1: A number of second-generation thrombolytic agents have either been approved or are in late stages of development. Discuss some of the limitations that the second-generation thrombolytic agents are designed to address.

Question 2: Design a rFVIII therapeutic regimen for a 35-kg patient with a laceration. Assume that the desired plasma concentration of factor VIII is 30 IU/dL.

Question 3: What criteria should factor VIII dosage be based on?

■ Answers

Answers 1: Although alteplase demonstrated an increased patency rate in the infarct-related artery and a decrease in mortality, several areas were identified where further improvements could be made in the treatment of acute myocardial infarction. Second-generation thrombolytic agents are designed to address some of these shortcomings.

(i) Due to the rapid clearance of rt-PA from the circulation by the liver, the current administration is via intravenous infusion over 90 min or 3 h. Second-generation thrombolytic agents have a slower plasma clearance allowing administration as a single- or double-bolus regimen (see Tables 15.1, 15.2, 15.3, 15.4, 15.5, 15.6, and 15.7).

(ii) Although alteplase is more fibrin selective compared to streptokinase and urokinase, there is still a 30–50 % fall in systemic fibrinogen levels. Second-generation thrombolytic agents are more fibrin specific and could result in further reduction in systemic fibrinogenolysis.

Answers 2: Dose = 30 IU/dL × 50 mL/kg (volume of distribution) × 35 kg = 525 IU.

Answers 3: Dosage should be individualized based on the needs of the patient, severity of deficiency, presence of inhibitors, and the desired increase in factor VIII.

REFERENCES

Abshire T, Kenet G (2004) Recombinant factor VIIa: review of efficacy, dosing regimens and safety in patients with congenital and acquired factor VIII or IX inhibitors. J Thromb Haemost 2:899–909

Abshire T, Kenet G (2008) Safety update on the use of recombinant factor VIIa and the treatment of congenital and acquired deficiency of factor VIII or IX with inhibitors. Haemophilia 14:898–902

Assessment of the Safety and Efficacy of a New Thrombolytic (ASSENT-2) Investigators (1999) Single-bolus tenecteplase compared with front-loaded alteplase in acute myocardial infarction: the ASSENT-2 double-blind randomized trial. Lancet 354:716–722

Barnes C, Lillicrap D, Pazmino-Canizares J, Blanchette VS, Stain AM, Clark D, Hensmen C, Carcao M (2006) Pharmacokinetics of recombinant factor VIII (Kogenate-FS®) in children and causes of inter-patient pharmacokinetic variability. Haemophilia 12(Suppl 4):40–49

Benedict CR, Refino CJ, Keyt BA, Pakala R, Paoni NF, Thomas GR, Bennett WF (1995) New variant of human tissue plasminogen activator (tPA) with enhanced efficacy and lower incidence of bleeding compared with recombinant human tPA. Circulation 92:3032–3040

Bernstein DE, Jeffers L, Erhardtsen E, Reddy KR, Glazer S, Squiban P, Bech R, Hedner U, Schiff ER (1997) Recombinant factor VIIa corrects prothrombin time in cirrhotic patients: a preliminary study. Gastroenterology 113:1930–1937

Berrettini M, Mariani G, Schiavoni M, Rocini A, Di Paolantonio T, Longo G, Morfini M (2001) Pharmacokinetic evaluation of recombinant, activated factor VII in patients with inherited factor VII deficiency. Haematologica 86: 640–645

Björkman S, Shapiro AD, Berntorp E (2001) Pharmacokinetics of recombinant factor IX in relation to age of the patient: implications for dosing in prophylaxis. Haemophilia 7:133–139

Björkman S, Blanchette VS, Fischer K, Oh M, Spotts G, Schroth P, Fritsch S, Patrone L, Ewenstein BM, Advate Clinical Program Group, Collin PW (2010) Comparative pharmacokinetics of plasma- and albumin-free recombinant factor VIII in children and adults: the influence of blood sampling schedule on observed age-related differences and implications for dose tailoring. J Thromb Haemost 8:730–736

Blanchette VS, Shapiro AD, Liesner RJ, Hernández-Navarro F, Warrier I, Schroth PC, Spotts G, Ewenstein BM (2008) Plasma and albumin-free recombinant factor VIII: pharmacokinetics, efficacy and safety in previously treated pediatric patients. J Thromb Haemost 6:1319–1326

Bleich SD, Adgey AA, McMechan SR, Love TW (1998) An angiographic assessment of alteplase: double-bolus and front-loaded infusion regimens in myocardial infarction. Am Heart J 136:741–748

Bu G, Maksymovitch EA, Schwartz AL (1993) Receptor-mediated endocytosis of tissue-type plasminogen activator by low density lipoprotein receptor-related protein on human hepatoma HepG2 cells. J Biol Chem 268:13002–13009

Bysted BV, Scharling B, Møller T, Hansen BL (2007) A randomized, double-blind trial demonstrating bioequivalence of the current recombinant activated factor VII formulation and a new robust 25 degrees C stable formulation. Haemophilia 13:527–532

Camani C, Gavin O, Bertossa C, Samatani E, Kruithof EK (1998) Studies on the effect of fucosylated and non-fucosylated finger/growth factor constructs on the clearance of tissue-type plasminogen activator mediated by the low-density lipoprotein-receptor-related protein. Eur J Biochem 251:804–811

Cannon CP, Gibson CM, McCabe CH, Adgey AA, Schweiger MJ, Sequeira RF, Grollier G, Giugliano RP, Frey M, Mueller HS, Steingart RM, Weaver WD, Van de Werf F, Braunwald E (1998) TNK-tissue plasminogen activator compared with front-loaded alteplase in acute myocardial infarction: results of the TIMI 10B trial. Circulation 98:2805–2814

Collen D, Stassen JM, Larsen G (1988) Pharmacokinetics and thrombolytic properties of deletion mutants of human tissue-type plasminogen activator in rabbits. Blood 71:216–219

den Heijer P, Vermeer F, Ambrosioni E, Sadowski Z, López-Sendón JL, von Essen R, Beaufils P, Thadani U, Adgey J, Pierard L, Brinker J, Davies RF, Smallin RW, Wallentin L, Caspi A, Pangerl A, Trickett L, Hauck C, Henry D, Chew P (1998) Evaluation of a weight-adjusted single-bolus plasminogen activator in patients with myocardial infarction: a double-blind, randomized angiographic trial of lanoteplase versus alteplase. Circulation 98:2117–2125

Di Paola J, Nugent D, Young G (2001) Current therapy for rare factor deficiencies. Haemophilia 7(Suppl 1):16–22

Fridberg M, Heder U, Roberts HR, Erhardtsen E (2005) A study of the pharmacokinetics and safety of recombinant activated factor VII in healthy Caucasian and Japanese subjects. Blood Coagul Fibrinolysis 16:259–266

Girard P, Nony P, Erhardtsen E, Delair S, Ffrench P, Dechavanne M, Boissel J-P (1998) Population pharmacokinetics of recombinant factor VIIa in volunteers anticoagulated with acenocoumarol. Thromb Haemost 80:109–113

Grünewald M, Müller M, Ellbrück D, Osterhues H, Kochs M, Mohren M, Schirmer G, Ziesche S, Güloglu A, Bock R, Seifried E (1997) Double- versus single-bolus thrombolysis with reteplase for acute myocardial infarction: a pharmacokinetic and pharmacodynamic study. Fibrinolysis Proteolysis 11:137–145

Hansen L, Blue Y, Barone K, Collen D, Larsen GR (1988) Functional effects of asparagine-linked oligosaccharide on natural and variant human tissue-type plasminogen activator. J Biol Chem 263:15713–15719

Harris RJ, Leonard CK, Guzzetta AW, Spellman MW (1991) Tissue plasminogen activator has an O-linked fucose attached to threonine-61 in the epidermal growth factor domain. Biochemistry 30:2311–2314

Hedner U (2006) Mechanism of action of factor VIIa in the treatment of coagulopathies. Semin Thromb Hemost 32(Suppl 1):77–85

Hedner U (2007) Recombinant factor VIIa: its background, development and clinical use. Curr Opin Hematol 14:225–229

Hironaka T, Furukawa K, Esmon PC, Fournel MA, Sawada S, Kato M, Minaga T, Kobata A (1992) Comparative study of the sugar chains of factor VIII purified from human plasma and from the culture media of recombinant baby hamster kidney cells. J Biol Chem 267:8012–8020

Hoffman M (2003) A cell-based model of coagulation and the role of factor VIIa. Blood Rev 17(Suppl 1):S1–S5

Hoffman MM, Monroe DM (2005) Rethinking the coagulation cascade. Curr Hematol Rep 4:391–396

Holvoet P, Boes J, Collen D (1987) Measurement of free, one-chain tissue-type plasminogen activator in human plasma with an enzyme-linked immunosorbent assay based on an active site-specific murine monoclonal antibody. Blood 69:284–289

Hotchkiss A, Refino CJ, Leonard CK, O'Connor JV, Crowley C, McCabe J, Tate K, Nakamura G, Powers D, Levinson A, Mohler M, Spellman MW (1988) The influence of carbohydrate structure on the clearance of recombinant tissue-type plasminogen activator. Thromb Haemost 60:255–261

Hoyer LW (1981) The factor VIII complex: structure and function. Blood 58:1–13

Jurlander B, Thim L, Klausen NK, Persson E, Kjalke M, Rexen P, Jørgensen TB, Østergaard PB, Erhardsen E, Bjørn SE (2001) Recombinant activated factor VII (rFVIIa): characterization, manufacturing, and clinical development. Semin Thromb Hemost 27:373–384

Keith JC, Ferranti TJ, Misra B, Fredrick T, Rup B, McCarthy K, Faulkner R, Bush L, Schaub RG (1995) Evaluation of recombinant human factor IX: pharmacokinetic studies in the rat and the dog. Thromb Haemost 73:101–105

Keyt BA, Paoni NF, Refino CJ, Berleau L, Nguyen H, Chow A, Lai J, Pena L, Pater C, Ogez J, Etcheverry T, Botstein D, Bennett WF (1994) A faster-acting and more potent form of tissue plasminogen activator. Proc Natl Acad Sci (U S A) 91:3670–3674

Klitgaard T, Nielsen TG (2007) Overview of the human pharmacokinetics of recombinant activated factor VII. Br J Clin Pharmacol 65:3–11

Klitgaard T, Tabanera y Palacios R, Boffard KD, Iau PTC, Warren B, Rizoli S, Rossaint R, Kluger Y, Riou B (2006) Pharmacokinetics of recombinant activated factor VII in trauma patients with severe bleeding. Crit Care 10:R104

Kohnert U, Horsch B, Fischer S (1993) A variant of tissue plasminogen activator (t-PA) comprised of the kringle 2 and the protease domain shows a significant difference in the in vitro rate of plasmin formation as compared to the recombinant human t-PA from transformed Chinese hamster ovary cells. Fibrinolysis 7:365–372

Komoriya K, Kato Y, Hayashi Y, Ohsuye K, Nishigaki R, Sugiyama Y (2007) Characterization of the hepatic disposition of lanoteplase, a rationally designed variant of tissue plasminogen activator in rodents. Drug Metab Dispos 35:469–475

Kostis JB, Dockens RC, Thadani U, Bethala V, Pepine C, Leimbach W, Vachharajani N, Raymond RH, Stouffer BC, Tay LK, Shyu WC, Liao W (2002) Comparison of pharmacokinetics of lanoteplase and alteplase during acute myocardial infarction. Clin Pharmacokinet 41:445–452

Lapecorella M, Mariani G (2008) Factor VII deficiency: defining the clinical picture and optimizing therapeutic options. Haemophilia 14:1170–1175

Lindley CM, Sawyer WT, Macik BG, Lusher J, Harrison JF, Baird-Cox K, Birch K, Glazer S, Roberts HR (1994) Pharmacokinetics and pharmacodynamics of recombinant factor VIIa. Clin Pharmacol Ther 55:638–648

Lisman T, Mosnier LO, Lambert T, Mauser-Bunschoten EP, Meijers JC, Nieuwenhuis HK, de Groot PG (2002) Inhibition of fibrinolysis by recombinant factor VIIa in plasma from patients with severe hemophilia A. Blood 99:175–179

Lovejoy AE, Reynolds TC, Visich JE, Butine MD, Young G, Belvedere MA, Blain RC, Pederson SM, Ishak LM, Nugent DJ (2006) Safety and pharmacokinetics of recombinant factor XIII-A_2 administration in patients with congenital factor XIII deficiency. Blood 108:57–62

Martin U, Sponer G, Strein K (1993) Differential fibrinolytic properties of the recombinant plasminogen activator BM 06.022 in human plasma and blood clot systems in vitro. Blood Coagul Fibrinolysis 4:235–242

Modi NB, Eppler S, Breed J, Cannon CP, Braunwald E, Love TW (1998) Pharmacokinetics of a slower clearing tissue plasminogen activator variant, TNK-tPA, in patients with acute myocardial infarction. Thromb Haemost 79:134–139

Modi NB, Fox NL, Clow FW, Tanswell P, Cannon CP, Van de Werf F, Braunwald E (2000) Pharmacokinetics and pharmacodynamics of tenecteplase: results from a phase II study in patients with acute myocardial infarction. J Clin Pharmacol 40:508–515

Monahan PE, Di Paola J (2010) Recombinant factor IX for clinical and research use. Semin Thromb Hemost 36:498–509

Morfini M, Cinotti S, Bellatreccia A, Paladino E, Gringeri A, Mannucci PM (2003) A multicenter pharmacokinetic study of the B-domain deleted recombinant factor VIII concentrate using different assays and standards. J Thromb Haemost 1:2283–2289

Møss J, Scharling B, Ezban M, Møller Sørensen T (2009) Evaluation of the safety and pharmacokinetics of a fast-acting recombinant FVIIa analogue, NN1731, in healthy male subjects. J Thromb Haemost 7:299–305

Negrier C, Knobe K, Tiede A, Giangrande P, Møss J (2011) Enhanced pharmacokinetic properties of a glycoPEGylated recombinant factor IX: a first human dose trial in patients with hemophilia B. Blood 118:2695–2701

Pennica D, Holmes WE, Kohr WJ, Harkins RN, Vehar GA, Ward CA, Bennett WF, Yelverton E, Seeburg PH, Heyneker HL, Goeddel DV, Collen D (1983) Cloning and expression of human tissue-type plasminogen activator cDNA in E. coli. Nature 301:214–221

Peters RT, Low SC, Kamphaus GD, Dumont JA, Amari JV, Lu Q, Zarbis-Papastoitsis G, Reidy TJ, Merricks EP, Nichols TC, Bitonti AJ (2010) Prolonged activity of factor IX as a monomeric Fc fusion protein. Blood 115:2057–2064

Reynolds TC, Butine MD, Visich JE, Gunewardena KA, MacMahon M, Pederson S, Bishop PD, Morton KM (2005) Safety, pharmacokinetics, and immunogenicity of single-dose rFXIII administered to healthy volunteers. J Thromb Haemost 3:922–928

Roddie PH, Ludlam CA (1997) Recombinant coagulation factors. Blood Rev 11:169–177

Ross AM (1999) New plasminogen activators: a clinical review. Clin Cardiol 22:165–171

Schulte S (2008) Use of albumin fusion technology to prolong the half-life of recombinant factor VIIa. Thromb Res 122(Suppl 4):S14–S19

Schwartz RS, Abildgaard CF, Aledort LM et al (1990) Human recombinant DNA-derived antihemophilic factor (factor VIII) in the treatment of hemophilia A. N Engl J Med 323:1800–1805

Seifried E, Tanswell P, Ellbrück D, Haerer W, Schmidt A (1989) Pharmacokinetics and haemostatic status during consecutive infusions of recombinant tissue-type plasminogen activator in patients with acute myocardial infarction. Thromb Haemost 61:497–501

Seifried E, Müller MM, Martin U, König R, Hombach V (1992) Bolus application of a novel recombinant plasminogen activator in acute myocardial infarction patients: pharmacokinetics and effects on the hemostatic system. Ann N Y Acad Sci 667:417–420

Shapiro AD, Korth-Bradley J, Poon MC (2005) Use of pharmacokinetics in the coagulation factor treatment of patients with haemophilia. Haemophilia 11:571–582

Shapiro AD, Ragni MV, Valentino LA, Key NS, Josephson NC, Powell JS, Cheng G, Thompson AR, Goyal J, Tubridy KL, Peters RT, Dumont JA, Euwart D, Li L, Hallén B, Gozzi P, Bitonti AJ, Jiang H, Luk A, Pierce GF (2011) Recombinant factor IX-Fc fusion protein (rFIXFc) demonstrates safety and prolonged activity in a phase 1/2a study in hemophilia B patients. Blood 119:666–672

Spellman MW, Basa LJ, Leonard CK, Chakel JA, O'Connor JV, Wilson S, Van Halbeek H (1989) Carbohydrate structures of human tissue plasminogen activator expressed in Chinese Hamster Ovary cells. J Biol Chem 264:14100–14111

Spira J, Plyushch O, Zozulya N, Yatuv R, Dayan I, Bleicher A, Robinson M, Baru M (2010) Safety, pharmacokinetics and efficacy of factor VIIa formulated with PEGylated liposomes in haemophilia A patients with inhibitors to factor VIII – an open label, exploratory, cross-over, phase I/II study. Haemophilia 16:910–918

Tanswell P, Seifried E, Su PC, Feuerer W, Rijken DC (1989) Pharmacokinetics and systemic effects of tissue-type plasminogen activator in normal subjects. Clin Pharmacol Ther 46:155–162

Tanswell P, Heinzel G, Greischel A, Krause J (1990) Nonlinear pharmacokinetics of tissue-type plasminogen activator in three animal species and isolated perfused rat liver. J Pharmacol Exp Ther 255:318–324

Tanswell P, Tebbe U, Neuhaus KL, Gläsle-Schwarz L, Wojcik J, Seifried E (1992) Pharmacokinetics and fibrin specificity of alteplase during accelerated infusions in acute myocardial infarction. J Am Coll Cardiol 19:1071–1075

Tebbe U, Tanswell P, Seifried E, Feuerer W, Scholz KH, Herrmann KS (1989) Single-bolus injection of recombinant tissue-type plasminogen activator in acute myocardial infarction. Am J Cardiol 64:448–453

The Continuous Infusion Versus Double-Bolus Administration of Alteplase (COBALT) Investigators (1997) A comparison of continuous infusion of alteplase with double-bolus administration for acute myocardial infarction. N Engl J Med 337:1124–1130

The Global Use of Strategies to Open Occluded Coronary Arteries (GUSTO III) Investigators (1997) A comparison of reteplase with alteplase for acute myocardial infarction. N Engl J Med 337:1118–1123

The Global Use of Strategies to Open Occluded Coronary Arteries (GUSTO) Angiographic Investigators (1993b) The effects of tissue plasminogen activator, streptokinase, or both on coronary-artery patency, ventricular function, and survival after acute myocardial infarction. N Engl J Med 329:1615–1622

The Global Use of Strategies to Open Occluded Coronary Arteries (GUSTO) Investigators (1993a) An international randomized trial comparing four thrombolytic strategies for acute myocardial infarction. N Engl J Med 329:673–682

The InTIME Investigators (2000) Intravenous NPA for the treatment of infracting myocardial early. InTIME-II, a double-blind comparison of single-bolus lanoteplase vs accelerated alteplase for the treatment of patients with acute myocardial infarction. Eur Heart J 21:2005–2013

Thim L, Bjoern S, Christensen M, Nicolaisen EM, Lund-Hansen T, Pedersen AH, Hedner U (1988) Amino acid sequence and posttranslational modifications of human factor VIIa from plasma and transfected baby hamster kidney cells. Biochemistry 27:7785–7793

Thomas GR, Thibodeaux H, Errett CJ, Badillo JM, Keyt BA, Refino CJ, Zivin JA, Bennett WF (1994) A long-half-life and fibrin-specific form of tissue plasminogen activator in rabbit models of embolic stroke and peripheral bleeding. Stroke 25:2072–2078

Vachharajani NN, Raymond RH, Shyu WC, Stouffer BC, Boulton DW (2011) The effects of age and gender on the pharmacokinetics and pharmacodynamics in healthy subjects of the plasminogen activator, lanoteplase. Br J Clin Pharmacol 72:775–786

Vehar GA, Keyt B, Eaton D, Rodriguez H, O'Brien DP, Rotblat F, Oppermann H, Keck RG, Wood WI, Harkins RN, Tuddenham EG, Lawn RM, Capon DJ (1984a) Structure of human factor VIII. Nature 312:337–342

Vehar GA, Kohr WJ, Bennett WF, Pennica D, Ward CA, Harkins RN, Collen D (1984b) Characterization studies on human melanoma cell tissue plasminogen activator. Bio/Tech 2:1051–1057

Verstraete M, Bory M, Collen D, Erbel R, Lennane RJ, Mathey D, Michels HR, Schartl M, Uebis R, Bernard R, Brower RW, De Bono DP, Huhmann W, Lubsen J, Meyer J, Rutsch W, Schmidt W, Von Essen R (1985) Randomised trial of intravenous recombinant tissue-type plasminogen activator versus intravenous streptokinase in acute myocardial infarction : Report from the European Cooperative Study Group for Recombinant Tissue-type Plasminogen Activator. 325:842–847.

Villar A, Aronis S, Morfini M, Santagostino E, Auerswald G, Thomsen HF, Erhardtsen E, Giangrande PLF (2004) Pharmacokinetics of activated recombinant coagulation factor VII (Novoseven®) in children vs. adults with hemophilia A. Haemophilia 10:352–359

Visich JE, Zuckerman LA, Butine MD, Gunewardena KA, Wild R, Morton KM, Reynolds TC (2005) Safety and pharmacokinetics of recombinant factor XIII in healthy volunteers: a randomized, placebo-controlled, double-blind, multi-dose study. Thromb Haemost 94:802–807

Wang-Clow F, Fox NL, Cannon CP, Gibson CM, Berioli S, Bluhmki E, Danays T, Braunwald E, Van de Werf F, Stump DC (2001) Determination of a weight-adjusted dose of TNK-tissue plasminogen activator. Am Heart J 141:33–40

White GC, Beebe A, Nielsen B (1997) Recombinant factor IX. Thromb Haemost 78:261–265

Yatuv R, Dayan I, Carmel-Goren L, Robinson M, Aviv I, Goldenber-Furmanov M, Baru M (2008) Enhancement of factor VIIa haemostatic efficacy by formulation with PEGylated liposomes. Haemophilia 14:476–483

Yatuv R, Robinson M, Dayan I, Baru M (2010) Enhancement of the efficacy of therapeutic proteins by formulation with PEGylated liposomes; a case of FVIII, FVIIa and G-CSF. Expert Opin Drug Deliv 7:187–201

16 Recombinant Human Deoxyribonuclease I

Robert A. Lazarus and Jeffrey S. Wagener

INTRODUCTION

Human deoxyribonuclease I (DNase I) is an endonuclease that catalyzes the hydrolysis of extracellular DNA and is just one of the numerous types of nucleases found in nature (Horton 2008; Yang 2011). It is the most extensively studied member of a recently discovered family of DNase I-like nucleases (Lazarus 2002; Baranovskii et al. 2004; Shiokawa and Tanuma 2001); the homologous bovine DNase I has received even greater attention historically (Laskowski 1971; Moore 1981; Chen and Liao 2006). Mammalian DNases have been broadly divided into several families initially based upon their products, pH optima, and divalent metal ion requirements. These include the neutral DNase I family (EC 3.1.21.1), the acidic DNase II family (EC 3.1.22.1), as well as apoptotic nucleases such as DFF40/CAD and endonuclease G (Lazarus 2002; Evans and Aguilera 2003; Widlak and Garrard 2005). The human DNase I gene resides on chromosome 16p13.3 and contains 10 exons and 9 introns, which span 15 kb of genomic DNA (Kominato et al. 2006). DNase I is synthesized as a precursor and contains a 22-residue signal sequence that is cleaved upon secretion, resulting in the 260-residue mature enzyme. It is secreted by the pancreas and parotid glands, consistent with its proposed primary role of digesting nucleic acids in the gastrointestinal tract. However, it is also present in blood and urine as well as other tissues, suggesting additional functions.

Recombinant human DNase I (rhDNase I, rhDNase, Pulmozyme®, dornase alfa) has been developed clinically where it is aerosolized into the airways for treatment of pulmonary disease in patients with cystic fibrosis (CF) (Suri 2005). Cystic fibrosis is an autosomal recessive disease caused by mutations in the CF transmembrane conductance regulator (CFTR) gene (Kerem et al. 1989; Riordan et al. 1989). Mutations of this gene result in both abnormal quantity and function of an apical membrane protein responsible for chloride ion transfer. The CFTR protein is a member of the ATP-binding cassette transporter superfamily (member ABCC7) and in addition to transporting chloride has many other functions including the regulation of epithelial sodium channels, ATP-release mechanisms, anion exchangers, sodium bicarbonate transporters, and aquaporin water channels found in airways, intestine, pancreas, sweat duct, and other fluid-transporting tissues (Guggino and Stanton 2006). Clinical manifestations of the disease include chronic obstructive airway disease, increased sweat electrolyte excretion, male infertility due to obstruction of the vas deferens, and exocrine pancreatic insufficiency.

In the airways, abnormal CFTR results in altered secretions and mucociliary clearance, leading to a cycle of obstruction, chronic bacterial infection, and inflammation. This neutrophil-dominated airway inflammation begins early in the patient's life and, while initially it helps to control infection, the degree of inflammation is both excessive and poorly regulated such that airway damage develops over time from the release of neutrophil-derived oxidants and proteases. Additionally, necrosis of neutrophils leads to the accumulation of extracellular DNA and actin, increasing the viscosity of mucous and creating further obstruction, and a downward spiral of lung damage, loss of lung function, and ultimately premature death (Fig. 16.1).

The use of rhDNase I has been investigated in other diseases where exogenous DNA is implicated in the disease pathology. rhDNase I was studied in systemic lupus erythematosus (SLE) where degradation or prevention of immune complexes containing anti-DNA antigens may have therapeutic benefit (Lachmann 2003; Davis et al. 1999). rhDNase I has also

R.A. Lazarus, Ph.D. (✉)
Department of Early Discovery Biochemistry,
Genentech Inc., 1 DNA Way, MS27,
South San Francisco, CA 94080, USA
e-mail: lazarus.bob@gene.com

J.S. Wagener, M.D.
Department of Pediatrics,
University of Colorado School of Medicine,
Aurora, CO, USA

D.J.A. Crommelin, R.D. Sindelar, and B. Meibohm (eds.), *Pharmaceutical Biotechnology*,
DOI 10.1007/978-1-4614-6486-0_16, © Springer Science+Business Media New York 2013

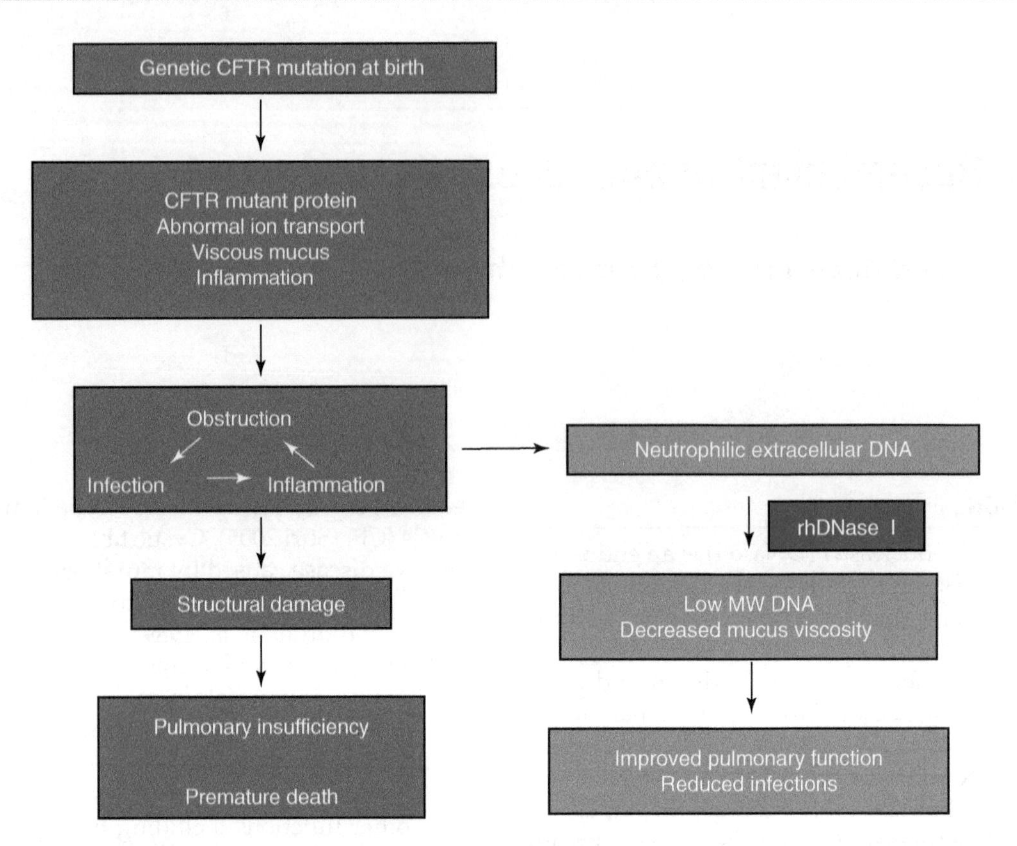

Figure 16.1 ■ Cystic fibrosis and rhDNase I. CFTR genetic mutation at birth leads to either reduced or improperly folded CFTR protein, which results in altered ion transport, viscous mucus, and inflammation in the airways. Eventually, this leads to obstruction of the airways, bacterial infection, and further inflammation. After neutrophils arrive to fight the infection, they die and release cellular contents, one of which is DNA. Persistent obstruction, infection, and inflammation leads to structural damage and eventually pulmonary insufficiency and premature death. rhDNase I is aerosolized into the airways where it degrades DNA to lower molecular weight fragments, thus reducing CF mucus viscosity and allowing expectoration, which improves lung function and reduces bacterial infections.

been studied in a variety of other diseases where extracellular DNA has been postulated to play a pathological role, including mechanical ventilation (Riethmueller et al. 2006), atelectasis (Hendriks et al. 2005), chronic sinusitis (Cimmino et al. 2005), primary ciliary dyskinesia (Desai et al. 1995; ten Berge et al. 1999; El Abiad et al. 2007), other non-CF lung diseases in children (Boogaard et al. 2007a), and empyema (Simpson et al. 2003; Rahman et al. 2011). rhDNase I has also been studied to examine its effectiveness in the presence or absence of antibiotics against biofilm producing strains of *Staphylococcus aureus* and *Staphylococcus epidermidis* (Kaplan et al. 2012). rhDNase I exhibits potent antibiofilm and antimicrobial-sensitizing activities at clinically achievable concentrations.

■ Historical Perspective and Rationale

Macromolecules that contribute to the physical properties of lung secretions include mucus glycoproteins, filamentous actin, and DNA. Experiments in the 1950s and 1960s revealed that DNA is present in very high concentrations (3–14 mg/mL) only in infected lung secretions (Matthews et al. 1963). This implied that the DNA that contributes to the high viscoelastic nature of CF sputum is derived from neutrophils responding to chronic infections (Potter et al. 1969). These DNA-rich secretions also bind aminoglycoside antibiotics commonly used for treatment of pulmonary infections and thus may reduce their efficacy (Ramphal et al. 1988; Bataillon et al. 1992).

Early in vitro studies in which lung secretions were incubated for several hours with partially purified bovine pancreatic DNase I showed a large reduction in viscosity (Armstrong and White 1950; Chernick et al. 1961). Based on these observations, bovine pancreatic DNase I (dornavac or pancreatic dornase) was approved in the United States for human use in 1958. Numerous uncontrolled clinical studies in patients with pneumonia and one study in patients with CF suggested that bovine pancreatic DNase I was effective in reducing the viscosity of lung secretions (Lieberman 1968). However, severe adverse reactions occurred occasionally, perhaps due to allergic reactions to a foreign protein or from contaminating proteases, since up to 2 % of trypsin and

Figure 16.2 ■ Primary amino acid sequence of rhDNase I. Active site residues are highlighted in *blue*, cysteine residues that form disulfide bonds are shown in *yellow*, N-linked glycosylation sites are highlighted in *pink*, and residues involved in Ca^{2+} coordination are shown in *beige*. The 22-residue signal sequence that is cleaved prior to secretion is not shown.

chymotrypsin were present in the final product (Raskin 1968; Lieberman 1962). Both bovine DNase I products were eventually withdrawn from the market.

In the late 1980s, human deoxyribonuclease I was cloned from a human pancreatic cDNA library, sequenced and expressed recombinantly using mammalian cell culture in Chinese hamster ovary (CHO) cells to reevaluate the potential of DNase I as a therapeutic for cystic fibrosis (Shak et al. 1990). In vitro incubation of purulent sputum from CF patients with catalytic concentrations of rhDNase I reduced its viscoelasticity (Shak et al. 1990). The reduction in viscoelasticity was directly related to both rhDNase I concentration and reduction in the size of the DNA in the samples. Therefore, reduction of high molecular weight DNA into smaller fragments by treatment with aerosolized rhDNase I was proposed as a mechanism to reduce the mucus viscosity and improve mucus clearability from obstructed airways in patients. It was hoped that improved clearance of the purulent mucus would enhance pulmonary function and reduce recurrent exacerbations of respiratory symptoms requiring parenteral antibiotics. This proved to be the case and rhDNase I was approved by the Food and Drug Administration in 1993. Since that time the clinical use of rhDNase I has continued to increase with over 67 % of CF patients currently receiving chronic therapy (Konstan et al. 2010).

PROTEIN CHEMISTRY, ENZYMOLOGY, AND STRUCTURE

The protein chemistry of human DNases including DNase I has been recently reviewed (Lazarus 2002; Baranovskii et al. 2004). Recombinant human DNase I is a monomeric, 260-amino acid glycoprotein (Fig. 16.2) produced by mammalian CHO cells (Shak et al. 1990). The protein has four cysteines, which are oxidized into two disulfides between Cys101–Cys104 and Cys173–Cys209 as well as two potential N-linked glycosylation sites at Asn18 and Asn106 (Fig. 16.2). rhDNase I is glycosylated at both sites and migrates as a broad band on polyacrylamide gel electrophoresis gels with an approximate molecular weight of 37 kDa, which is significantly higher than the predicted molecular mass from the amino acid sequence of 29.3 kDa. rhDNase I is an acidic protein and has a calculated pI of 4.58. The primary amino acid sequence is identical to that of the native human enzyme purified from urine.

DNase I cleaves double-stranded DNA, and to a much lesser degree single-stranded DNA, nonspecifically by nicking phosphodiester linkages in one of the strands between the 3′-oxygen atom and the phosphorus to yield 3′-hydroxyl and 5′-phosphoryl oligonucleotides with inversion of configuration at the phosphorus. rhDNase I enzymatic activity is dependent upon the presence of divalent metal ions for structure, as there are two tightly bound Ca^{2+} atoms and catalysis, which requires either Mg^{2+} or Mn^{2+} (Pan and Lazarus 1999). The active site includes two histidine residues (His134 and His252) and two acidic residues (Glu78 and Asp 212), all of which are critical for the general acid–base catalysis of phosphodiester bonds since alanine substitution of any of these results in a total loss of activity (Ulmer et al. 1996). Other residues involved in the coordination of divalent metal ions at the active site and DNA contact residues have also been identified by mutational analysis and include Asn7, Glu39, Asp168, Asn170, and Asp251 (Pan et al. 1998a; Parsiegla et al. 2012). The two Ca^{2+} binding sites require acidic or polar residues for coordination of Ca^{2+}; for site 1 these include

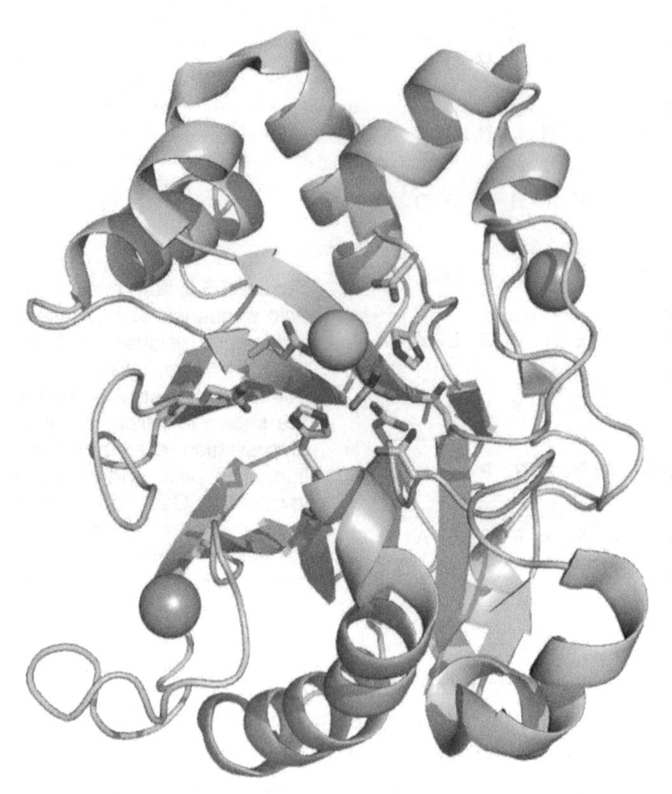

Figure 16.3 ■ Cartoon representation of rhDNase I depicting the active site of the enzyme. The active site residues are shown as *beige sticks* where oxygen and nitrogen atoms are in *red* and *blue*, respectively. The Mg^{2+} ion is shown as a *cyan sphere* and Ca^{2+} ions are shown as *grey* spheres. The phosphate ion is shown in sticks with the phosphorus atom in *orange*. The figure was made using PyMOL (www.pymol.org) using accession code 4AWN for rhDNase I (Parsiegla et al. 2012).

Asp201 and Thr203 and for site 2 these include Asp99, Asp107, and Glu112 (Pan and Lazarus 1999). DNase I is a relatively stable enzyme and shows optimal activity at pH 5.5–7.5. It is inactivated by heat and is potently inhibited by EDTA and G-actin. Surprisingly, DNase I is also inhibited by NaCl and has only ca. 30 % of the maximal activity in physiological saline.

rhDNase I belongs to the DNase I-like structural superfamily according to SCOP and is also related to the endonuclease-exonuclease-phosphatase family (Andreeva et al. 2008; Dlakic 2000; Wang et al. 2010). The X-ray crystal structure of rhDNase I was initially solved at 2.2 Å resolution and superimposes with the biochemically more widely studied bovine DNase I, which shares 78 % sequence identity, with an rms deviation for main chain atoms of 0.56 Å (Wolf et al. 1995). DNase I is a compact α/β protein having a core of two tightly packed six-stranded β-sheets surrounded by eight α-helices and several loop regions (Fig. 16.3). Bovine DNase I has also been crystallized in complex with G-actin (Kabsch et al. 1990) as well as with several short oligonucleotides, revealing key features of DNA recognition in the minor groove and catalytic hydrolysis (Suck 1994).

More recently, the structure of rhDNase I containing a divalent Mg^{2+} and a phosphate in the active site has been solved at 1.95 Å resolution (Parsiegla et al. 2012). The combined structural and mutagenesis data suggest a Mg^{2+}-assisted pentavalent phosphate transition state during catalysis of rhDNase I, where Asp168 may play a key role as a catalytic general base. Asn170 is also in close proximity to both the attacking water molecule and the phosphoryl oxygen. His134 and His252 appear to act as general acids in stabilizing the pentavalent transition state. There is also a critical catalytic role for rhDNase I Asn7, a residue that is highly conserved among mammalian DNase I enzymes and members of the DNase I-like superfamily. The Mg^{2+} cation resides at the computationally predicted site IVb (Gueroult et al. 2010) and interacts with Asn7, Glu39, and Asp251 via a complex set of water-mediated hydrogen bond interactions. A comprehensive analysis of the rhDNase I with members of the DNase I-like structural superfamily (Andreeva et al. 2008; Dlakic 2000) such as the apurinic/apyrimidinic endonucleases from human (APE1) (Mol et al. 2000) and *Neisseria meningitidis* (Nape) (Carpenter et al. 2007), sphingomyelin phosphodiesterase (SMase) from *Bacillus cereus* (Ago et al. 2006), or the C-terminal domain of human CNOT6L nuclease (Wang et al. 2010) solved in complex with cations or DNA have revealed new insights into the catalytic mechanism of DNA hydrolysis.

Several variants of rhDNase I with greatly improved enzymatic properties have been engineered by site-directed mutagenesis (Pan and Lazarus 1997; Pan et al. 1998a). The methods for production of the variants and the assays to characterize them have been reviewed recently (Pan et al 2001; Sinicropi and Lazarus 2001). The rationale for improving activity was to increase binding affinity to DNA by introducing positively charged residues (Arg or Lys) on rhDNase I loops at the DNA binding interface to form a salt bridge with phosphates on the DNA backbone. These so-called "hyperactive" rhDNase I variants are substantially more active than wild-type rhDNase I and are no longer inhibited by physiological saline. The greater catalytic activity of the hyperactive variants is due to a change in the catalytic mechanism from a "single nicking" activity in the case of wild-type rhDNase I to a "processive nicking" activity in the hyperactive rhDNase I variants (Pan and Lazarus 1997), where gaps rather than nicks result in a higher frequency of double strand cleavages.

It is interesting to note that significantly greater activity can result from just a few mutations on the surface that are not important for structural integrity. For whatever reason DNase I is not as efficient an enzyme as it could be for degrading DNA into small fragments. Furthermore the inhibition by G-actin can be eliminated by a single amino acid substitution (see below). Thus, DNase I is under some degree of regulation in vivo. One can only speculate that nature may have wanted to avoid an enzyme with too much DNA degrading activity that could result in undesired mutations in the genome.

PHARMACOLOGY

■ In Vitro Activity in CF Sputum

In vitro, rhDNase I hydrolyzes the DNA in sputum of CF patients and reduces sputum viscoelasticity (Shak et al. 1990). Effects of rhDNase I were initially examined using a relatively crude "pourability" assay. Pourability was assessed qualitatively by inverting the tubes and observing the movement of sputum after a tap on the side of the tube. Catalytic amounts of rhDNase I (50 μg/mL) greatly reduced the viscosity of the sputum, rapidly transforming it from a viscous gel to a flowing liquid. More than 50 % of the sputum moved down the tube within 15 min of incubation, and all the sputum moved freely down the tube within 30 min. The qualitative results of the pourability assay were confirmed by quantitative measurement of viscosity using a Brookfield Cone-Plate viscometer (Fig. 16.4). The reduction of viscosity by rhDNase I is rhDNase I concentration-dependent and is associated with reduction in size of sputum DNA as measured by agarose gel electrophoresis (Fig. 16.5).

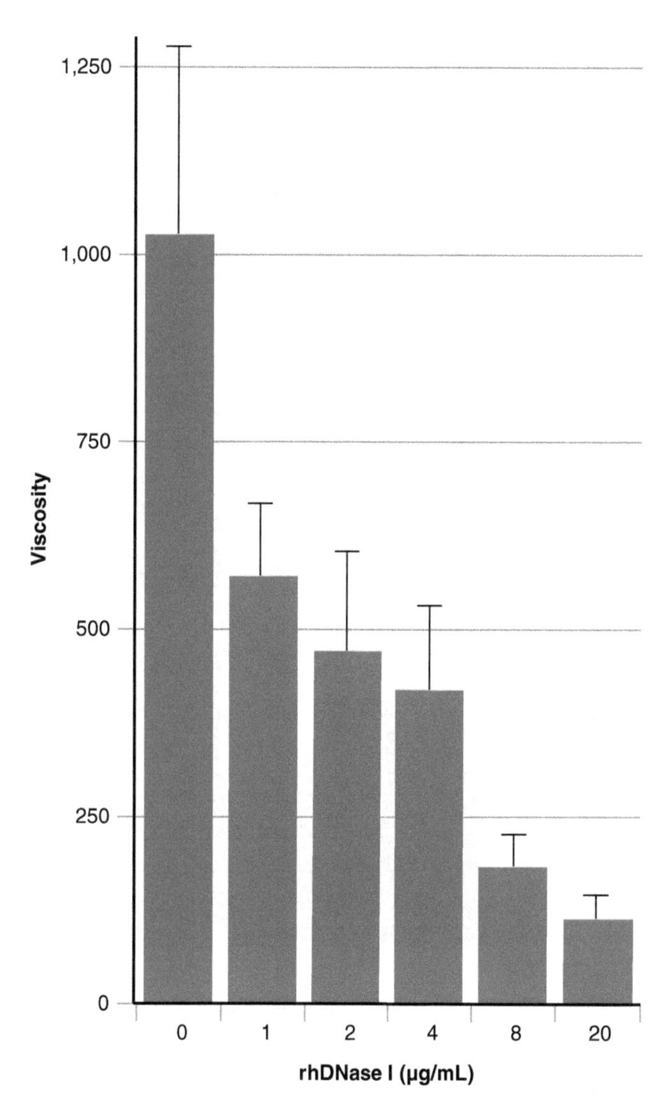

Figure 16.4 ■ In vitro reduction in viscosity (in centipoise) of cystic fibrosis sputum by cone-plate viscometry. Cystic fibrosis sputum was incubated with various concentrations of rhDNase I of 15 min at 37 °C.

Additional in vitro studies of CF mucus samples treated with rhDNase I demonstrated a dose-dependent improvement in cough transport and mucociliary transport of CF mucus using a frog palate model and a reduction in adhesiveness as measured by mucus contact angle (Zahm et al. 1995). The improvements in mucus transport properties and adhesiveness were associated with a decrease in mucus viscosity and mucus surface tension, suggesting rhDNase I treatment may improve the clearance of mucus from airways. The in vitro viscoelastic properties of rhDNase I have also been studied in combination with normal saline, 3 % hypertonic saline, or nacystelyn, the L-lysine salt of N-acetyl cysteine (King et al. 1997; Dasgupta and King 1996). The major impact of rhDNase I on CF sputum is to decrease spinnability, which is the thread forming ability of mucus under the influence of low amplitude stretching. CF sputum spinnability decreases 25 % after 30 min incubation with rhDNase I (King et al. 1997). rhDNase I in normal saline and saline alone both increased the cough clearability index. With the combination of rhDNase I and 3 % hypertonic saline, there was minimal effect on spinnability however mucus rigidity and cough clearability improved greater than with either agent alone. The predicted mucociliary clearance did not significantly increase with 3 % saline either alone or in combination with rhDNase I. Combining rhDNase I with nacystelyn has an additive benefit on spinnability, but no effect on mucous rigidity or cough clearability (Dasgupta and King 1996). These effects of rhDNase I can be variable in vivo and do not necessarily correlate with the level of DNA in sputum. For example, sputum from CF patients that clinically responded to rhDNase I contains significantly higher levels of magnesium ions compared with sputum from patients who do not have a clear response (Sanders et al. 2006). Although this response is consistent with the requirement for divalent cations and their mode of action on DNase I (Campbell and Jackson 1980), the mechanism of increased rhDNase I activity by magnesium ions has been attributed to altering the polymerization state of actin such that equilibrium favors increased F-actin and decreased G-actin (see below).

The mechanism of action of rhDNase I to reduce CF sputum viscosity has been ascribed to DNA hydrolysis (Shak et al. 1990). However, an alternative mechanism involving depolymerization of filamentous actin (F-actin)

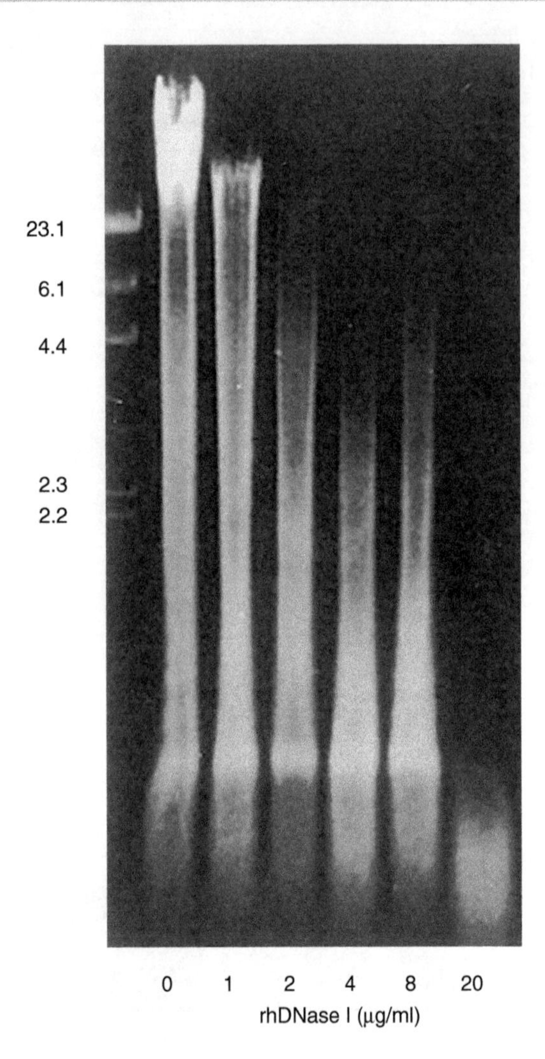

Figure 16.5 ■ In vitro reduction in sputum DNA size as measured by agarose gel electrophoresis. Cystic fibrosis sputum was incubated with increasing concentrations (0–20 μg/mL) of rhDNase I for 150 min at 37 °C. Molecular weight standards for DNA in kb are indicated.

has been suggested since F-actin contributes to the viscoelastic properties of CF sputum and the actin-depolymerizing protein gelsolin also reduces sputum viscoelasticity (Vasconcellos et al. 1994). F-actin is in equilibrium with its monomeric form (G-actin), which binds to rhDNase I with high affinity and is also a potent inhibitor of DNase I activity (Lazarides and Lindberg 1974). DNase I is known to depolymerize F-actin by binding to G-actin with high affinity, shifting the equilibrium in favor of rhDNase I/G-actin complexes (Hitchcock et al. 1976). To elucidate the mechanism of rhDNase I in CF sputum, the activity of two types of rhDNase I variants were compared in CF sputum (Ulmer et al. 1996). Active site variants were engineered that were unable to catalyze DNA hydrolysis but retained wild-type G-actin binding. Actin-resistant variants that no longer bound G-actin but retained wild-type DNA hydrolytic activity were also characterized. The active site variants did not degrade DNA in CF sputum and did not decrease sputum viscoelasticity (Fig. 16.6). Since the active site variants retained the ability to bind G-actin, these results argue against depolymerization of F-actin as the mechanism of action. In contrast, the actin-resistant variants were more potent than wild-type DNase I in their ability to degrade DNA and reduce sputum viscoelasticity (Fig. 16.6). The increased potency of the actin-resistant variants indicated that G-actin was a significant inhibitor of wild-type DNase I in CF sputum and confirmed that hydrolysis of DNA was the mechanism by which rhDNase I decreases sputum viscoelasticity. The mechanism for reduction of sputum viscosity by gelsolin was subsequently determined to result from an unexpected second binding site on actin that competes with DNase I, thus relieving the inhibition by G-actin (Davoodian et al. 1997). Additional in vitro studies characterizing the relative potency of actin-resistant and hyperactive rhDNase I variants in serum and CF sputum have been reported (Pan et al. 1998b).

■ In Vivo Activity in CF Sputum

In vivo confirmation of the proposed mechanism of action for rhDNase I has been obtained from direct characterization of apparent DNA size (Fig. 16.7) and measurements of enzymatic and immunoreactive (ELISA) activity of rhDNase I (Fig. 16.8) in sputum from cystic fibrosis patients (Sinicropi et al. 1994a). Sputum samples were obtained 1–6 h post-dose from adult cystic fibrosis patients after inhalation of 5–20 mg of rhDNase I. rhDNase I therapy produced a sustained reduction in DNA size in recovered sputum (Fig. 16.7), in good agreement with the in vitro data.

Inhalation of the therapeutic dose of rhDNase I produced sputum levels of rhDNase I which have been shown to be effective in vitro (Fig. 16.8) (Shak 1995). The recovered rhDNase I was also enzymatically active. Enzymatic activity was directly correlated with rhDNase I concentrations in the sputum. Viscoelasticity was reduced in the recovered sputum, as well. Furthermore, results from scintigraphic studies in using twice daily 2.5 mg of rhDNase I in CF patients suggested possible reductions in pulmonary obstruction and increased rates of mucociliary sputum clearance from the inner zone of the lung compared to controls (Laube et al. 1996). This finding was not confirmed in a crossover design study using once-daily dosing, suggesting that improvement of mucociliary clearance may require higher doses (Robinson et al. 2000).

■ Pharmacokinetics and Metabolism

Nonclinical pharmacokinetic data in rats and monkeys suggest minimal systemic absorption of rhDNase I following aerosol inhalation of clinically equivalent doses.

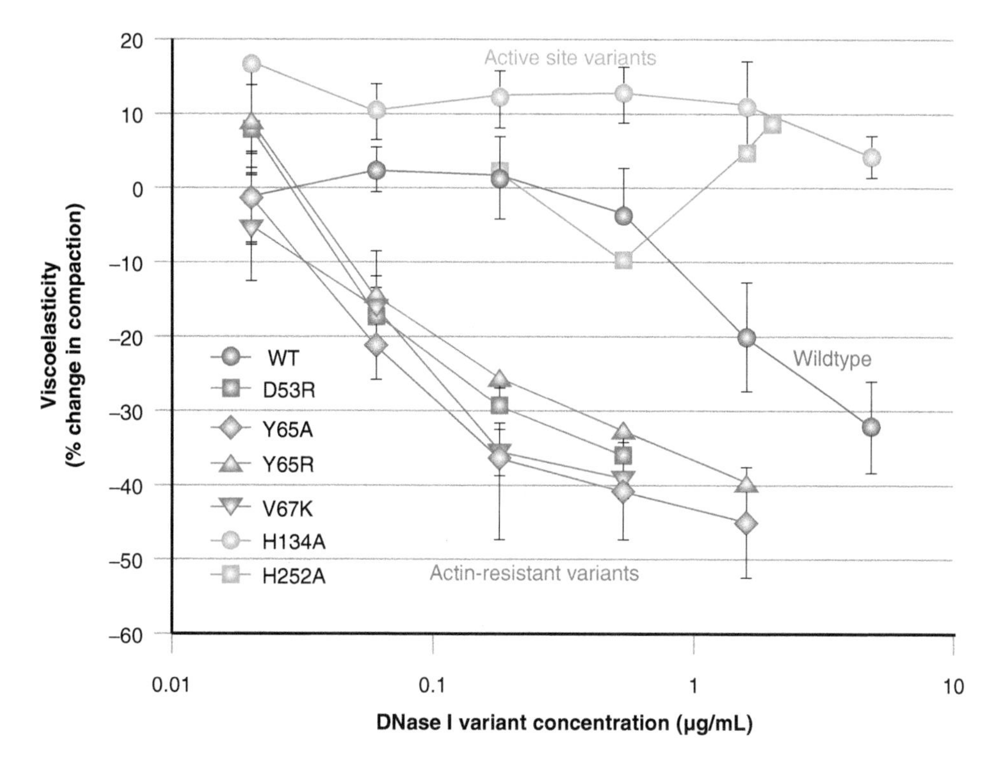

Figure 16.6 ■ Mechanism of action in CF mucus for rhDNase I. The change in viscoelasticity in CF mucus as a function of DNase concentration was determined for wild-type rhDNase I, two active site variants that no longer catalyze DNA hydrolysis and four variants that are no longer inhibited by G-actin (Ulmer et al. 1996).

rhDNase I is cleared from the systemic circulation without any accumulation in tissues following acute exposure (Green 1994). Additionally, nonclinical metabolism studies suggest that the low rhDNase I concentrations present in serum following inhalation will be bound to binding proteins (Green 1994; Mohler et al. 1993). The low concentrations of endogenous DNase I normally present in serum and the low concentrations of rhDNase I in serum following inhalation are inactive due to the ionic composition and presence of binding proteins in serum (Prince et al. 1998).

When 2.5 mg of rhDNase I was administered twice daily by inhalation to 18 CF patients, mean sputum concentrations of 2 μg/mL DNase I were measurable within 15 min after the first dose on Day 1 (Fig. 16.9). Mean sputum concentrations declined to an average of 0.6 μg/mL 2 h following inhalation. The peak rhDNase I concentration measured 2 h after inhalation on Days 8 and 15 increased to 3.0 and 3.6 μg/mL, respectively. Sputum rhDNase I concentrations measured 6 h after inhalation on Days 8 and 15 were similar to Day 1. Predose trough concentrations of 0.3–0.4 μg/mL rhDNase I measured on Day 8 and Day 15 (sample taken approximately 12 h after the previous dose) were, however, higher than Day 1, suggesting possible modest accumulation of rhDNase I with repeated dosing. Inhalation of up to 10 mg three times daily of rhDNase I by 4 CF patients for 6 consecutive days did not result in significant elevation of serum concentrations of DNase above normal endogenous levels (Aitken et al. 1992; Hubbard

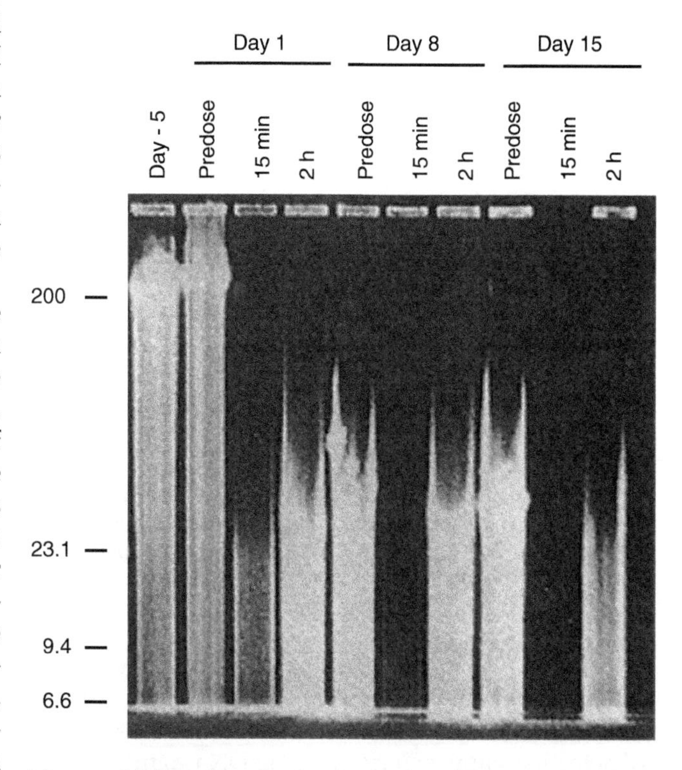

Figure 16.7 ■ Sustained reduction in DNA length in sputum recovered from a CF patient treated with 2.5 mg rhDNase I BID for up to 15 days. Samples were analyzed by pulsed field agarose field gel electrophoresis. Molecular weight standards for DNA in kb are indicated.

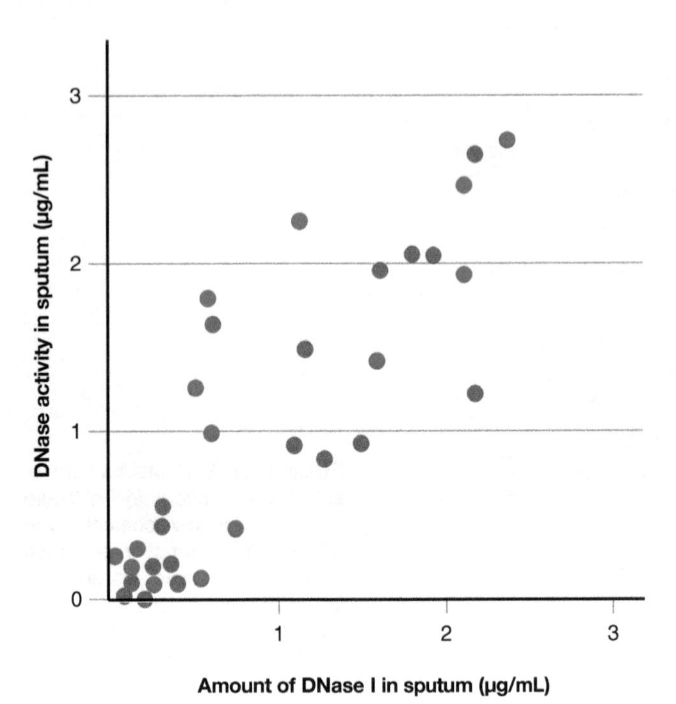

Figure 16.8 ■ Immunoreactive concentrations and enzymatic activity of rhDNase I in sputum following aerosol administration of either 10 mg (●) or 20 mg (●) rhDNase I to patients with cystic fibrosis. Each data point is a separate sample measured in duplicate.

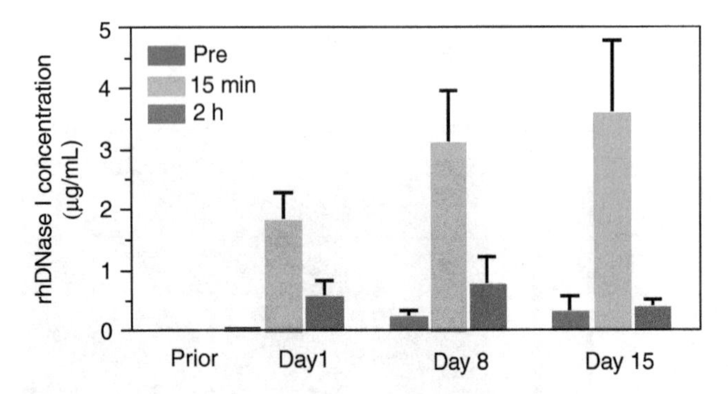

Figure 16.9 ■ rhDNase I concentration in sputum following administration of 2.5 mg of rhDNase I twice daily by inhalation to CF patients. The blue, orange, and purple bars represent concentrations at pre-dose, 15 min post-dose, and 2 h post-dose, respectively. The mean ± SD is shown with $N = 18$.

et al. 1992). After administration of up to 2.5 mg of rhDNase I twice daily for 6 months to 321 CF patients, no accumulation of serum DNase was noted (assay limit of detection = approximately 0.5 ng DNase/mL serum).

PROTEIN MANUFACTURING AND FORMULATION

rhDNase I is expressed in mammalian cell culture and purified to homogeneity using a variety of chromatographic steps. The development of the formulation of rhDNase I is especially important in that a suitable formulation is required to take into account protein stability, aerosolization properties, tonicity, and the sealed container for storage (Shire 1996). rhDNase I (Pulmozyme®, dornase alfa) is manufactured by Genentech, Inc. and formulated as a sterile, clear, and colorless aqueous solution containing 1.0 mg/mL dornase alfa, 0.15 mg/mL calcium chloride dihydrate, and 8.77 mg/mL sodium chloride. The solution contains no preservative and has a nominal pH of 6.3. Pulmozyme® is administered by the inhalation of an aerosol mist produced by a compressed air-driven nebulizer system. Pulmozyme® is supplied as single-use ampoules, which deliver 2.5 mL of solution to the nebulizer.

The choice of formulation components was determined by a need to provide 1–2 years of stability and to meet additional requirements unique to aerosol delivery (Shire 1996). A simple colorimetric assay for rhDNase I activity was used to evaluate the stability of rhDNase I in various formulations (Sinicropi et al. 1994b). In order to avoid adverse pulmonary reactions, such as cough or bronchoconstriction, aerosols for local pulmonary delivery should be formulated as isotonic solutions with minimal or no buffer components and should maintain pH >5.0. rhDNase I has an additional requirement for calcium to be present for optimal enzymatic activity. Limiting formulation components raised concerns about pH control, since protein stability and solubility can be highly pH-dependent. Fortunately, the protein itself provided sufficient buffering capacity at 1 mg/mL to maintain pH stability over the storage life of the product.

DRUG DELIVERY

The droplet or particle size of an aerosol is a critical factor in defining the site of deposition of the drug in the patient's airways (Gonda 1990). A distribution of particle or droplet size of 1–6 µm is optimal for the uniform deposition of rhDNase I in the airways (Cipolla et al. 1994). Jet nebulizers are the simplest method of producing aerosols in the desired respirable range. However, recirculation of protein solutions under high shear rates in the nebulizer bowl can present risks to the integrity of the protein molecule. rhDNase I survived recirculation and high shear rates during the nebulization process with no apparent degradation in protein quality or enzymatic activity (Cipolla et al. 1994). Ultrasonic nebulizers produce greater heat than jet nebulizers and protein breakdown prevents their use with rhDNase I. Significant advances in nebulizer technology have occurred since the original approval of rhDNase I. Newer nebulizers using a vibrating, perforated membrane do not produce protein breakdown and provide more rapid and efficient delivery of particles in the respirable range (Scherer et al. 2011).

Approved jet nebulizers produce aerosol droplets in the respirable range (1–6 μm) with a mass median aerodynamic diameter (MMAD) of 4–6 μm. The delivery of rhDNase I with a device that produces smaller droplets leads to more peripheral deposition in the smaller airways and thereby improves efficacy (Geller et al. 1998). Results obtained in 749 CF patients with mild disease confirmed that patients randomized to the Sidestream nebulizer powered by the Mobil Aire Compressor (MMAD=2.1 μm) tended to have greater improvement in pulmonary function than patients using the Hudson T nebulizer with Pulmo-Aide Compressor (MMAD=4.9 μm). These results indicate that the efficacy of rhDNase I is dependent, in part, on the physical properties of the aerosol produced by the delivery system. Nebulizers with vibrating mesh technology (Pari eFlow®) produce similarly small particles, suggesting these may result in further improved efficacy of rhDNase I. Furthermore, "smart" nebulizers are now available that coach the patient on taking a proper breath to improve delivery to the lower airways. Delivery of rhDNase I improves and lung function improves more when these nebulizers are used (Bakker et al. 2011).

CLINICAL USE

■ Indication and Clinical Dosage

rhDNase I (Pulmozyme®, dornase alfa) is currently approved for use in CF patients, in conjunction with standard therapies, to reduce the frequency of respiratory infections requiring parenteral antibiotics and to improve pulmonary function (Fig. 16.1). The recommended dose for use in most CF patients is one 2.5 mg dose inhaled daily.

■ Cystic Fibrosis

rhDNase I was evaluated in a large, randomized, and placebo-controlled trial of clinically stable CF patients, 5 years of age or older, with baseline forced vital capacity (FVC) greater than or equal to 40 % of predicted (Fuchs et al. 1994). All patients received additional standard therapies for CF. Patients were treated with placebo or 2.5 mg of rhDNase I once or twice a day for 6 months. When compared to placebo, both once daily and twice daily doses of rhDNase I resulted in a 28–37 % reduction in respiratory tract infections requiring use of parenteral antibiotics (Fig. 16.10). Within 8 days of the start of treatment with rhDNase I, mean forced expiratory volume in 1 s (FEV_1) increased 7.9 % in patients treated once a day and 9.0 % in those treated twice a day compared to the baseline values. The mean FEV_1 observed during long-term therapy increased 5.8 % from baseline at the 2.5 mg daily dose level and 5.6 % from baseline at the 2.5 mg twice daily dose level (Fig. 16.11). The risk of respiratory tract infection was reduced even in patients whose pulmonary function (FEV_1) did not improve.

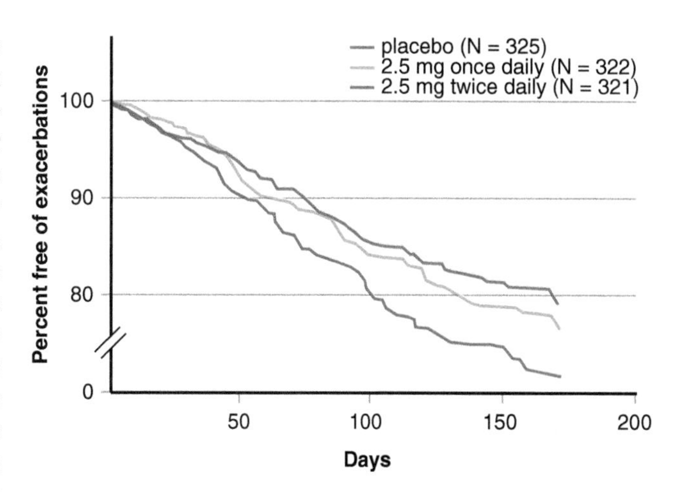

Figure 16.10 ■ Proportion of patients free of exacerbations of respiratory symptoms requiring parenteral antibiotic therapy from a 24-week study.

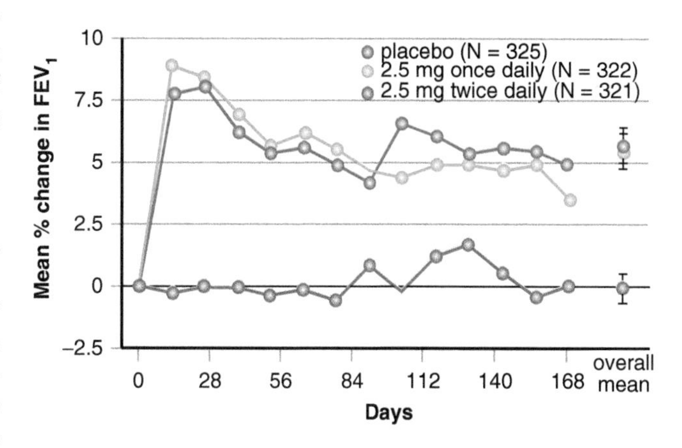

Figure 16.11 ■ Mean percent change in FEV_1 from baseline through a 24-week study.

This finding may be due to improved clearance of mucus from the small airways in the lung, which will have little effect on FEV_1 (Shak 1995). Supporting this concept is the finding that rhDNase I improves the lung clearance index in 6–18-year-old CF patients with normal lung function (Amin et al. 2011). Alternatively, use of rhDNase I may be altering the neutrophilic inflammatory response that occurs early in the course of CF lung disease, similar to the use of other anti-inflammatory therapies (Konstan and Ratjen 2012). The administration of rhDNase I also lessened shortness of breath, increased the general perception of well-being, and reduced the severity of other cystic fibrosis-related symptoms. Based on these findings, the US Cystic Fibrosis Foundation strongly recommends the use of rhDNase I in patients 6 years old and older with moderate to severe lung disease and recommends its use in patients with mild lung disease (Flume et al. 2007).

The safety and deposition, but not efficacy, of rhDNase I has been studied in CF patients <5 years old (3 months to 5 years) since therapy may provide clinical benefit for young CF patients with mild disease (Wagener et al. 1998). After 2 weeks of daily administration of 2.5 mg rhDNase I, levels of rhDNase I deposited in the lower airways were similar for children <5 years compared to a group of 5–10-year-olds. Moreover, rhDNase I was well tolerated in the younger age group with an adverse event frequency similar to that in the older age group. To further understand how rhDNase I might alter the progression of lung disease, children with mild CF related lung disease were treated for 2 years in a randomized controlled trial (Quan et al. 2001). Children with a mean age of 8.4 years and an FVC greater than 85 % of predicted were treated once daily with either placebo or 2.5 mg rhDNase I. After 96 weeks, lung function was significantly better in the treated group compared with placebo, particularly for tests measuring function of smaller airways. Respiratory exacerbations were also reduced in the treated group.

Clinical trials have indicated that rhDNase I therapy can be continued or initiated during an acute respiratory exacerbation (Wilmott et al. 1996). rhDNase I however does not produce a pulmonary function benefit when used short-term in the most severely ill CF patients (FVC less than 40 % of predicted) (Shah et al. 1995; McCoy et al. 1996). Short-term dose ranging studies demonstrated that doses in excess of 2.5 mg twice daily did not provide further significant improvement in FEV_1 (Aitken et al. 1992; Hubbard et al. 1992; Ramsey et al. 1993). Patients who have received drug on a cyclical regimen (i.e., administration of rhDNase I 10 mg twice daily for 14 days, followed by a 14-day washout period) showed rapid improvement in FEV_1 with the initiation of each cycle and a return to baseline with each rhDNase I withdrawal (Eisenberg et al. 1997). rhDNase I use improves quality of life as measured by the validated CFQ-R questionnaire (Rozov et al. 2010).

Concomitant therapy with rhDNase I and other standard CF therapies often show additive effects. The intermittent administration of aerosolized tobramycin was approved for use in CF patients with or without concomitant use of rhDNase I (Ramsey et al. 1999). Aerosolized tobramycin was well tolerated, enhanced pulmonary function, and decreased the density of *P. aeruginosa* in sputum. In combination with rhDNase I a larger treatment effect was noted but did not reach statistical significance. No differences in safety profile were observed following aerosolized tobramycin in patients that did or did not use rhDNase I. Chronic use of azithromycin has also been studied in CF patients chronically infected with *P. aeruginosa* (Saiman et al. 2005). Similar improvement in lung function and reduction in respiratory exacerbations was seen in patients receiving rhDNase I as those not, suggesting an additive, but not synergistic benefit of the two therapies used together. The combination of hypertonic saline therapy with chronic use of rhDNase I has similar additive benefits (Elkins et al. 2006), in agreement with the previously mentioned in vitro studies. Finally, combining ivacaftor, which potentiates chloride transport in CF patients with the G551D gene mutation, with chronic rhDNase I use produces an additive benefit in both lung function and respiratory exacerbations (Ramsey et al 2011). Notably, there was no evidence of a change in adverse events related to combination therapy in any of these studies.

Following FDA approval of rhDNase I in 1993, a large epidemiologic study of CF patients was initiated (the Epidemiologic Study of Cystic Fibrosis or ESCF), which continued until 2005 (Morgan et al. 1999). This study was designed to evaluate practice patterns in CF patients and has included data from over 24,000 patients. Recent analysis of ESCF data showed that chronic use of rhDNase I is associated with a decreased rate of decline in lung function overtime (Konstan et al. 2011). This reduced rate of decline in lung function is similar to findings with oral ibuprofen (Konstan et al. 1995, 2007) and inhaled corticosteroids (Ren et al. 2008), suggesting that there may be a long-term anti-inflammatory benefit with the use of rhDNase I. This potential anti-inflammatory effect is supported by a randomized trial in 105 CF patients with mild lung disease (FEV_1 >80 % predicted) (Paul et al. 2004). Based on an initial bronchoscopy and alveolar lavage, patients were divided into two groups, those without airway inflammation ($n = 20$) and those with. The patients with inflammation were then randomized to treatment with rhDNase I ($n = 46$) or not ($n = 39$). Follow-up bronchoscopy and lavage was performed at 18 and 36 months. In the patients treated with rhDNase I, there was no change in inflammation as measured by elastase and IL-8 levels and neutrophil number. Patients not treated with rhDNase I and patients who did not have inflammation at baseline all had worsening neutrophilic inflammation on follow-up. Although this study was not designed to evaluate the rate of lung function decline, treated patients dropped FEV_1 by 1.99 % predicted per year compared to a 3.26 % predicted drop per year in the untreated subjects. Finally, rhDNase I is associated with a 15 % reduction in the odds of subsequent year mortality in patients with CF (Sawicki et al. 2012).

Non-cystic Fibrosis Respiratory Disease

Although originally considered beneficial for the treatment of non-CF related bronchiectasis (Wills et al. 1996), rhDNase I had no effect on pulmonary function or the frequency of respiratory exacerbations in a randomized controlled trial (O'Donnell et al. 1998). In another randomized controlled trial of rhDNase I, young children had shorter periods of ventilatory support following cardiac surgery when rhDNase I was instilled twice daily into the endotracheal tube (Riethmueller et al.

2006). Complicating atelectasis was less frequent in the treated group, consistent with numerous case reports suggesting that rhDNase I decreases, and can be used to treat, atelectasis when directly instilled into the airway (Hendriks et al. 2005). This effectiveness in treating atelectasis seems particularly true for newborns with lung disease requiring mechanical ventilation (MacKinnon et al. 2011; Dilmen et al. 2011; Fedakar et al. 2012; Altunhan et al. 2012). Limited benefit has been seen in children with asthma (Puterman and Weinberg 1997) although not in adults (Silverman et al. 2012), consistent with the lack of neutrophil dominated inflammation in asthma. Finally, while there is increased free DNA in the secretions of infants with respiratory syncytial virus caused bronchiolitis, early suggestions of benefit (Nasr et al. 2001) have not translated into reduced hospitalization or the need for oxygen (Boogaard et al. 2007b).

Other Medical Conditions

In principle rhDNase I may be useful for treating any condition where high levels of extracellular DNA and associated viscoelastic properties are pathological. A number of other clinical diseases with high potential for this have been investigated, although only to a limited extent. Pulmonary empyema involves the collection of purulent material in the pleural space and the use of rhDNase I instilled into the pleural space has been proposed (Simpson et al. 2003). In one large, multicenter clinical trial for the treatment of empyema in adults, twice daily intrapleural administration over 3 days was evaluated in four groups: 5 mg rhDNase, 10 mg tissue plasminogen activator (t-PA), with the combination of both, and a double placebo. Patients receiving the combination therapy had improved fluid drainage and a reduced frequency of surgical referral (Rahman et al. 2011).

rhDNase I has also been instilled into the nasal sinuses after surgery for chronic infections (Cimmino et al. 2005; Raynor et al. 2000). While daily use over 28 days of nasal nebulized rhDNase I in patients with CF did not produce a significant change in the sinuses on MRI, there was a significant clinical improvement as measured by quality of life questionnaire (Mainz et al. 2011).

Safety

The administration of rhDNase I has not been associated with an increase in major adverse events. Most adverse events were not more common with rhDNase I than with placebo treatment and probably reflect complications related to the underlying lung disease. Most events associated with dosing were mild, transient in nature, and did not require alterations in dosing. Observed symptoms included hoarseness, pharyngitis, laryngitis, rash, chest pain, and conjunctivitis. Within all the studies a small percentage (average 2–4 %) of patients treated with rhDNase I developed serum antibodies to rhDNase I. None of these patients developed anaphylaxis and the clinical significance of serum antibodies to rhDNase I is unknown. rhDNase I has also been associated with a slight increased risk of allergic bronchopulmonary aspergillosis in CF patients, although this most likely represents the chronic use of a wet nebulizer and not a complication of rhDNase I (Jubin et al. 2010).

SUMMARY

DNase I, a secreted human enzyme whose normal function is thought to be for digestion of extracellular DNA, has been developed as a safe and effective adjunctive agent in the treatment of pulmonary disease in cystic fibrosis patients. rhDNase I reduces the viscoelasticity and improves the transport properties of viscous mucus both in vitro and in vivo. Inhalation of aerosolized rhDNase I reduces the risk of infections requiring antibiotics and improves pulmonary function and the well-being of CF patients with mild to moderate disease. Studies also suggest that rhDNase I has benefit in infants and young children with CF and in patients with early disease and "normal" lung function. Additional studies may assess the usefulness of rhDNase I in early-stage CF pulmonary disease and other diseases where extracellular DNA may play a pathological role.

SELF-ASSESSMENT QUESTIONS

Questions

1. Mutations of the CF gene result in abnormalities of the CF transmembrane conductance regulator protein. How do abnormalities of this protein lead to eventual lung damage?
2. How does rhDNase I result in improved pulmonary function in patients with cystic fibrosis?
3. rhDNase I is strongly recommended by the US Cystic Fibrosis Foundation for use in CF patients with moderate and severe lung disease. What other CF patients may benefit from using this therapy?
4. In addition to improving lung function in CF patients, what benefits have been demonstrated in clinical trials of rhDNase I?
5. What other medical conditions may benefit from treatment with rhDNase I?
6. Why should ultrasonic nebulizers not be used to administer rhDNase I? What other types of nebulizers might be effective for delivering rhDNase I?

Answers

1. Abnormal CFTR protein results in abnormal airway surface liquid, which leads to a vicious cycle of airway obstruction, infection, and inflammation. The

chronic, excessive neutrophilic inflammation in the airway results is release of neutrophil elastase, oxidants, and extracellular DNA. These substances result in worsening obstruction and progressive airway damage.

2. rhDNase I cleaves extracellular DNA in the airway of CF patients, resulting in improved airway clearance of secretions. Additionally, rhDNase I may have anti-inflammatory properties, resulting in a slower progression of lung damage and reduced rate of decline in lung function. This anti-inflammatory benefit is more likely present in patients with earlier, less severe lung disease.
3. rhDNase I is also recommended in CF patients over age 6 with mild lung disease. It has also been demonstrated as safe in younger patients, although efficacy has not been studied.
4. rhDNase I use decreases the frequency of respiratory exacerbations. rhDNase I also lessened shortness of breath, increased the general perception of well-being, and reduced the severity of other cystic fibrosis-related symptoms.
5. Although only approved by the FDA for treatment of lung disease in patients with CF, controlled trials have also shown efficacy for treating empyema (in combination with t-PA) and sinus disease in patients with CF.
6. Ultrasonic nebulizers generate heat, which breaks down the protein in rhDNase I. Vibrating permeable membrane nebulizers and "smart" nebulizers do not damage the drug and may improve the effectiveness of rhDNase.

REFERENCES

Ago H, Oda M, Takahashi M, Tsuge H, Ochi S, Katunuma N, Miyano M, Sakurai J (2006) Structural basis of the sphingomyelin phosphodiesterase activity in neutral sphingomyelinase from *Bacillus cereus*. J Biol Chem 281:16157–16167

Aitken ML, Burke W, McDonald G, Shak S, Montgomery AB, Smith A (1992) Recombinant human DNase inhalation in normal subjects and patients with cystic fibrosis. A phase 1 study. JAMA 267:1947–1951

Altunhan H, Annagür A, Pekcan S, Ors R, Koç H (2012) Comparing the efficacy of nebulizer recombinant human DNase and hypertonic saline as monotherapy and combined treatment in the treatment of persistent atelectasis in mechanically ventilated newborns. Pediatr Int 54:131–136

Amin R, Subbarao P, Lou W, Jabar A, Balkovec S, Jensen R, Kerrigan S, Gustafsson P, Ratjen F (2011) The effect of dornase alfa on ventilation in homogeneity in patients with cystic fibrosis. Eur Respir J 37:806–812

Andreeva A, Howorth D, Chandonia JM, Brenner SE, Hubbard TJ, Chothia C, Murzin AG (2008) Data growth and its impact on the SCOP database: new developments. Nucleic Acids Res 36:D419-25

Armstrong JB, White JC (1950) Liquefaction of viscous purulent exudates by deoxyribonuclease. Lancet 2: 739–742

Bakker EM, Volpi S, Salonini E, van der Wiel-Kooij EC, Sintnicolaas CJJCM, Hop WCJ, Assael BM, Merkus PJFM, Tiddens HAWM (2011) Improved treatment response to dornase alfa in cystic fibrosis patients using controlled inhalation. Eur Respir J 38:1328–1335

Baranovskii AG, Buneva VN, Nevinsky GA (2004) Human deoxyribonucleases. Biochemistry (Mosc) 69:587–601

Bataillon V, Lhermitte M, Lafitte JJ, Pommery J, Roussel P (1992) The binding of amikacin to macromolecules from the sputum of patients suffering from respiratory diseases. J Antimicrob Chemother 29:499–508

Boogaard R, de Jongste JC, Merkus PJ (2007a) Pharmacotherapy of impaired mucociliary clearance in non-CF pediatric lung disease. A review of the literature. Pediatr Pulmonol 42:989–1001

Boogaard R, Hulsmann AR, van Veen L, Vaessen-Verberne AA, Yap YN, Sprij AJ, Brinkhorst G, Sibbles B, Hendriks T, Feith SW, Lincke CR, Brandsma AE, Brand PL, Hop WC, de Hoog M, Merkus PJ (2007b) Recombinant human deoxyribonuclease in infants with respiratory syncytial virus bronchiolitis. Chest 131:788–795

Campbell VW, Jackson DA (1980) The effect of divalent cations on the mode of action of DNase I. The initial reaction products produced from covalently closed circular DNA. J Biol Chem 255:3726–3735

Carpenter EP, Corbett A, Thomson H, Adacha J, Jensen K, Bergeron J, Kasampalidis I, Exley R, Winterbotham M, Tang C, Baldwin GS, Freemont P (2007) AP endonuclease paralogues with distinct activities in DNA repair and bacterial pathogenesis. EMBO J 26:1363–1372

Chen WJ, Liao TH (2006) Structure and function of bovine pancreatic deoxyribonuclease I. Protein Pept Lett 13:447–453

Chernick WS, Barbero GJ, Eichel HJ (1961) In vitro evaluation of effect of enzymes on tracheobronchial secretions from patients with cystic fibrosis. Pediatrics 27: 589–596

Cimmino M, Nardone M, Cavaliere M, Plantulli A, Sepe A, Esposito V, Mazzarella G, Raia V (2005) Dornase alfa as postoperative therapy in cystic fibrosis sinonasal disease. Arch Otolaryngol Head Neck Surg 131:1097–1101

Cipolla D, Gonda I, Shire SJ (1994) Characterization of aerosols of human recombinant deoxyribonuclease I (rhDNase) generated by jet nebulizers. Pharm Res 11:491–498

Dasgupta B, King M (1996) Reduction in viscoelasticity in cystic fibrosis sputum in vitro using combined treatment with nacystelyn and rhDNase. Pediatr Pulmonol 22:161–166

Davis JC Jr, Manzi S, Yarboro C, Rairie J, McInnes I, Averthelyi D, Sinicropi D, Hale VG, Balow J, Austin H, Boumpas DT, Klippel JH (1999) Recombinant human DNase I (rhDNase) in patients with lupus nephritis. Lupus 8:68–76

Davoodian K, Ritchings BW, Ramphal R, Bubb MR (1997) Gelsolin activates DNase I in vitro and cystic fibrosis sputum. Biochemistry 36:9637–9641

Desai M, Weller PH, Spencer DA (1995) Clinical benefit from nebulized human recombinant DNase in Kartagener's syndrome. Pediatr Pulmonol 20:307–308
Dilmen U, Karagol BS, Oguz SS (2011) Nebulized hypertonic saline and recombinant human DNase in the treatment of pulmonary atelectasis in newborns. Pediatr Int 53:328–331
Dlakic M (2000) Functionally unrelated signalling proteins contain a fold similar to Mg^{2+}-dependent endonucleases. Trends Biochem Sci 25:272–273
Eisenberg JD, Aitken ML, Dorkin HL, Harwood IR, Ramsey BW, Schidlow DV, Wilmott RW, Wohl ME, Fuchs HJ, Christiansen DH, Smith AL (1997) Safety of repeated intermittent courses of aerosolized recombinant human deoxyribonuclease in patients with cystic fibrosis. J Pediatr 131:118–124
El Abiad NM, Clifton S, Nasr SZ (2007) Long-term use of nebulized human recombinant DNase I in two siblings with primary ciliary dyskinesia. Respir Med 101: 2224–2226
Elkins MR, Robinson M, Rose BR, Harbour C, Moriarty CP, Marks GB, Belousova EG, Xuan W, Bye PT (2006) A controlled trial of long-term inhaled hypertonic saline in patients with cystic fibrosis. N Engl J Med 354: 229–240
Evans CJ, Aguilera RJ (2003) DNase II: genes, enzymes and function. Gene 322:1–15
Fedakar A, Aydogdu C, Fedakar A, Ugurlucan M, Bolu S, Iskender M (2012) Safety of recombinant human deoxyribonuclease as a rescue treatment for persistent atelectasis in newborns. Ann Saudi Med 32: 131–136
Flume PA, O'Sullivan BP, Robinson KA, Goss CH, Mogayzel PJ Jr, Willey-Courand DB, Bujan J, Finder J, Lester M, Quittell L, Rosenblatt R, Vender RL, Hazle L, Sabadosa K, Marshall B (2007) Cystic fibrosis pulmonary guidelines: chronic medications for maintenance of lung health. Am J Respir Crit Care Med 176: 957–969
Fuchs HJ, Borowitz DS, Christiansen DH, Morris EM, Nash ML, Ramsey BW, Rosenstein BJ, Smith AL, Wohl ME (1994) Effect of aerosolized recombinant human DNase on exacerbations of respiratory symptoms and on pulmonary function in patients with cystic fibrosis. N Engl J Med 331:637–642
Geller DE, Eigen H, Fiel SB, Clark A, Lamarre AP, Johnson CA, Konstan MW (1998) Effect of smaller droplet size of dornase alfa on lung function in mild cystic fibrosis. Pediatr Pulmonol 25:83–87
Gonda I (1990) Aerosols for delivery of therapeutic and diagnostic agents to the respiratory tract. Crit Rev Ther Drug Carrier Syst 6:273–313
Green JD (1994) Pharmaco-toxicological expert report Pulmozyme rhDNase Genentech, Inc. Hum Exp Toxicol 13:S1–S42
Gueroult M, Picot D, Abi-Ghanem J, Hartmann B, Baaden M (2010) How cations can assist DNase I in DNA binding and hydrolysis. PLoS Comput Biol 6:e1001000
Guggino WB, Stanton BA (2006) New insights into cystic fibrosis: molecular switches that regulate CFTR. Nat Rev Mol Cell Biol 7:426–436
Hendriks T, de Hoog M, Lequin MH, Devos AS, Merkus PJ (2005) DNase and atelectasis in non-cystic fibrosis pediatric patients. Crit Care 9:R351–R356
Hitchcock SE, Carisson L, Lindberg U (1976) Depolymerization of F-actin by deoxyribonuclease I. Cell 7: 531–542
Horton NC (2008) DNA nucleases. In: Rice PA, Correll CC (eds) Protein-nucleic acid interactions: structural biology. Royal Society of Chemistry Publishing, Cambridge, pp 333–363
Hubbard RC, McElvaney NG, Birrer P, Shak S, Robinson WW, Jolley C, Wu M, Chernick MS, Crystal RG (1992) A preliminary study of aerosolized recombinant human deoxyribonuclease I in the treatment of cystic fibrosis. N Engl J Med 326:812–815
Jubin V, Ranque S, Le bel NS, Sarles J, Dubus J-C (2010) Risk factors for *Aspergillus* colonization and allergic bronchopulmonary aspergillosis in children with cystic fibrosis. Pediatr Pulmonol 45:764–771
Kabsch W, Mannherz HG, Suck D, Pai EF, Holmes KC (1990) Atomic structure of the actin: DNase I complex. Nature 347:37–44
Kaplan JB, LoVetri K, Cardona ST, Madhyastha S, Sadovskaya I, Jabbouri S, Izano EA (2012) Recombinant human DNase I decreases biofilm and increases antimicrobial susceptibility in Staphylococci. J Antibiot (Tokyo) 65:73–77
Kerem B, Rommens JM, Buchanan JA, Markiewicz D, Cox TK, Chakravarti A, Buchwald M, Tsui LC (1989) Identification of the cystic fibrosis gene: genetic analysis. Science 245:1073–1080
King M, Dasgupta B, Tomkiewicz RP, Brown NE (1997) Rheology of cystic fibrosis sputum after in vitro treatment with hypertonic saline alone and in combination with recombinant human deoxyribonuclease I. Am J Respir Crit Care Med 156:173–177
Kominato Y, Ueki M, Iida R, Kawai Y, Nakajima T, Makita C, Itoi M, Tajima Y, Kishi K, Yasuda T (2006) Characterization of human deoxyribonuclease I gene (DNASE1) promoters reveals the utilization of two transcription-starting exons and the involvement of Sp1 in its transcriptional regulation. FEBS J 273: 3094–3105
Konstan MW, Ratjen F (2012) Effect of dornase alfa on inflammation and lung function: potential role in the early treatment of cystic fibrosis. J Cyst Fibros 11:78–83
Konstan MW, Byard PJ, Hoppel CL, Davis PB (1995) Effect of high-dose ibuprofen in patients with cystic fibrosis. N Engl J Med 332:848–854
Konstan MW, Schluchter MD, Xue W, Davis PB (2007) Clinical use of Ibuprofen is associated with slower FEV_1 decline in children with cystic fibrosis. Am J Respir Crit Care Med 176:1084–1089
Konstan MW, VanDevanter DR, Rasouliyan L, Pasta DJ, Yegin A, Morgan WJ, Wagener JS (2010) Trends in the use of routine therapies in cystic fibrosis: 1995–2005. Pediatr Pulmonol 45:1167–1172
Konstan MW, Wagener JS, Pasta DJ, Millar SJ, Jacobs JR, Yegin A, Morgan WJ (2011) Clinical use of dornase alfa is associated with a slower rate of FEV_1 decline in cystic fibrosis. Pediatr Pulmonol 46:545–553

Lachmann PJ (2003) Lupus and desoxyribonuclease. Lupus 12:202–206

Laskowski M Sr (1971) Deoxyribonuclease I. In: Boyer PD (ed) The enzymes, vol 4, 3rd edn. Academic, New York, pp 289–311

Laube BL, Auci RM, Shields DE, Christiansen DH, Lucas MK, Fuchs HJ, Rosenstein BJ (1996) Effect of rhDNase on airflow obstruction and mucociliary clearance in cystic fibrosis. Am J Respir Crit Care Med 153:752–760

Lazarides E, Lindberg U (1974) Actin is the naturally occurring inhibitor of deoxyribonuclease I. Proc Natl Acad Sci USA 71:4742–4746

Lazarus RA (2002) Human deoxyribonucleases. In: Creighton TE (ed) Wiley encyclopedia of molecular medicine. Wiley, New York, pp 1025–1028

Lieberman J (1962) Enzymatic dissolution of pulmonary secretions. An in vitro study of sputum from patients with cystic fibrosis of pancreas. Am J Dis Child 104: 342–348

Lieberman J (1968) Dornase aerosol effect on sputum viscosity in cases of cystic fibrosis. JAMA 205:312–313

MacKinnon R, Wheeler KI, Sokol J (2011) Endotracheal DNase for atelectasis in ventilated neonates. J Perinatol 31:799–801

Mainz JG, Schiller I, Ritschel C, Mentzel H-J, Riethmuller J, Koitschev A, Schneider G, Beck JF, Wiedemann B (2011) Sinonasal inhalation of dornase alfa in CF: a double-blind placebo-controlled cross-over pilot trial. Auris Nasus Larynx 38:220–227

Matthews LW, Specter S, Lemm J, Potter JL (1963) The overall chemical composition of pulmonary secretions from patients with cystic fibrosis, bronchiectasis and laryngectomy. Am Rev Respir Dis 88:119–204

McCoy K, Hamilton S, Johnson C (1996) Effects of 12-week administration of dornase alfa in patients with advanced cystic fibrosis lung disease. Chest 110: 889–895

Mohler M, Cook J, Lewis D, Moore J, Sinicropi D, Championsmith A, Ferraiolo B, Mordenti J (1993) Altered pharmacokinetics of recombinant human deoxyribonuclease in rats due to the presence of a binding protein. Drug Metab Dispos 21:71–75

Mol CD, Izumi T, Mitra S, Tainer JA (2000) DNA-bound structures and mutants reveal abasic DNA binding by APE1 and DNA repair coordination. Nature 403:451–456

Moore S (1981) Pancreatic DNase. In: Boyer PD (ed) The enzymes, vol 14, 3rd edn. Academic, New York, pp 281–296

Morgan WJ, Butler SM, Johnson CA, Colin AA, FitzSimmons SC, Geller DE, Konstan MW, Light MJ, Rabin HR, Regelmann WE, Schidlow DV, Stokes DC, Wohl ME, Kaplowitz H, Wyatt MM, Stryker S (1999) Epidemiologic study of cystic fibrosis: design and implementation of a prospective, multicenter, observational study of patients with cystic fibrosis in the U.S. and Canada. Pediatr Pulmonol 28:231–241

Nasr SZ, Strouse PJ, Soskolne E, Maxvold NJ, Garver KA, Rubin BK, Moler FW (2001) Efficacy of recombinant human deoxyribonuclease I in the hospital management of respiratory syncytial virus bronchiolitis. Chest 120:203–208

O'Donnell AE, Barker AF, Ilowite JS, Fick RB (1998) Treatment of idiopathic bronchiectasis with aerosolized recombinant human DNase I. rhDNase Study Group. Chest 113:1329–1334

Pan CQ, Lazarus RA (1997) Engineering hyperactive variants of human deoxyribonuclease I by altering its functional mechanism. Biochemistry 36:6624–6632

Pan CQ, Lazarus RA (1999) Ca^{2+}–dependent activity of human DNase I and its hyperactive variants. Protein Sci 8:1780–1788

Pan CQ, Ulmer JS, Herzka A, Lazarus RA (1998a) Mutational analysis of human DNase I at the DNA binding interface: implications for DNA recognition, catalysis, and metal ion dependence. Protein Sci 7:628–636

Pan CQ, Dodge TH, Baker DL, Prince WS, Sinicropi DV, Lazarus RA (1998b) Improved potency of hyperactive and actin-resistant human DNase I variants for treatment of cystic fibrosis and systemic lupus erythematosus. J Biol Chem 273:18374–18381

Pan CQ, Sinicropi DV, Lazarus RA (2001) Engineered properties and assays for human DNase I mutants. Methods Mol Biol 160:309–321

Parsiegla G, Noguere C, Santell L, Lazarus RA, Bourne Y (2012) The structure of human DNase I bound to magnesium and phosphate ions points to a catalytic mechanism common to members of the DNase I-like superfamily. Biochemistry 51:10250–10258

Paul K, Rietschel E, Ballmann M, Griese M, Worlitzsch D, Shute J, Chen C, Schink T, Döring G, van Koningsbruggen S, Wahn U, Ratjen F (2004) Effect of treatment with dornase alpha on airway inflammation in patients with cystic fibrosis. Am J Respir Crit Care Med 169: 719–725

Potter JL, Specter S, Matthews LW, Lemm J (1969) Studies on pulmonary secretions. 3. The nucleic acids in whole pulmonary secretions from patients with cystic fibrosis bronchiectasis and laryngectomy. Am Rev Respir Dis 99:909–915

Prince WS, Baker DL, Dodge AH, Ahmed AE, Chestnut RW, Sinicropi DV (1998) Pharmacodynamics of recombinant human DNase I in serum. Clin Exp Immunol 113:289–296

Puterman AS, Weinberg EG (1997) rhDNase in acute asthma. Pediatr Pulmonol 23:316–317

Quan JM, Tiddens HA, Sy JP, McKenzie SG, Montgomery MD, Robinson PJ, Wohl ME, Konstan MW (2001) A two-year randomized, placebo-controlled trial of dornase alfa in young patients with cystic fibrosis with mild lung function abnormalities. J Pediatr 139: 813–820

Rahman NM, Maskell NA, West A, Teoh R, Arnold A, Mackinlay C, Peckham D, Davies CW, Ali N, Kinnear W, Bentley A, Kahan BC, Wrightson JM, Davies HE, Hooper CE, Lee YC, Hedley EL, Crosthwaite N, Choo L, Helm EJ, Gleeson FV, Nunn AJ, Davies RJ (2011) Intrapleural use of tissue plasminogen activator and DNase in pleural infection. N Engl J Med 365:518–526

Ramphal R, Lhermitte M, Filliat M, Roussel P (1988) The binding of anti-pseudomonal antibiotics to macromolecules from cystic fibrosis sputum. J Antimicrob Chemother 22:483–490

Ramsey BW, Astley SJ, Aitken ML, Burke W, Colin AA, Dorkin HL, Eisenberg JD, Gibson RL, Harwood IR, Schidlow DV, Wilmott RW, Wohl ME, Meyerson LJ, Shak S, Fuchs H, Smith AL (1993) Efficacy and safety of short-term administration of aerosolized recombinant human deoxyribonuclease in patients with cystic fibrosis. Am Rev Respir Dis 148:145–151

Ramsey BW, Pepe MS, Quan JM, Otto KL, Montgomery AB, Williams-Warren J, Vasiljev KM, Borowitz D, Bowman CM, Marshall BC, Marshall S, Smith AL (1999) Intermittent administration of inhaled tobramycin in patients with cystic fibrosis. N Engl J Med 340:23–30

Ramsey BW, Davies J, McElvaney NG, Tullis E, Bell SC, Dřevínek P, Griese M, McKone EF, Wainwright CE, Konstan MW, Moss R, Ratjen F, Sermet-Gaudelus I, Rowe SM, Dong Q, Rodriguez S, Yen K, Ordoñez C, Elborn JS; VX08-770-102 Study Group (2011) A CFTR Potentiator in Patients with Cystic Fibrosis and the G551D Mutation. New Engl J Med 365:1663–1672

Raskin P (1968) Bronchospasm after inhalation of pancreatic dornase. Am Rev Respir Dis 98:697–698

Raynor EM, Butler A, Guill M, Bent JP 3rd (2000) Nasally inhaled dornase alfa in the postoperative management of chronic sinusitis due to cystic fibrosis. Arch Otolaryngol Head Neck Surg 126:581–583

Ren CL, Pasta DJ, Rasouliyan L, Wagener JS, Konstan MW, Morgan WJ (2008) Relationship between inhaled corticosteroid therapy and rate of lung function decline in children with cystic fibrosis. J Pediatr 153:746–751

Riethmueller J, Borth-Bruhns T, Kumpf M, Vonthein R, Wiskirchen J, Stern M, Hofbeck M, Baden W (2006) Recombinant human deoxyribonuclease shortens ventilation time in young, mechanically ventilated children. Pediatr Pulmonol 41:61–66

Riordan JR, Rommens JM, Kerem B, Alon N, Rozmahel R, Grzelczak Z, Zielenski J, Lok S, Plavsic N, Chou JL, Drumm ML, Iannuzzi MC, Collins FS, Tsui LC (1989) Identification of the cystic fibrosis gene: cloning and characterization of complementary DNA. Science 245: 1066–1073

Robinson M, Hemming AL, Moriarty C, Eberl S, Bye PT (2000) Effect of a short course of rhDNase on cough and mucociliary clearance in patients with cystic fibrosis. Pediatr Pulmonol 30:16–24

Rozov T, de Oliveira VZ, Santana MA, Adde FV, Mendes RH, Paschoal IA, Reis FJC, Higa LYS, de Castro Toledo Jr AC, Pahl M (2010) Dornase alfa improves the health-related quality of life among Brazilian patients with cystic fibrosis – a one-year prospective study. Pediatr Pulmonol 45:874–882

Saiman L, Mayer-Hamblett N, Campbell P, Marshall BC (2005) Heterogeneity of treatment response to azithromycin in patients with cystic fibrosis. Am J Respir Crit Care Med 172:1008–1012

Sanders NN, Franckx H, De Boeck K, Haustraete J, De Smedt SC, Demeester J (2006) Role of magnesium in the failure of rhDNase therapy in patients with cystic fibrosis. Thorax 61:962–968

Sawicki GS, Signorovitch JE, Zhang J, Latremouille-Viau D, von Wartburg M, Wu EQ, Shi L (2012) Reduced mortality in cystic fibrosis patients treated with tobramycin inhalation solution. Pediatr Pulmonol 47:44–52

Scherer T, Geller DE, Owyang L, Tservistas M, Keller M, Boden N, Kesser KC, Shire SJ (2011) A technical feasibility study of dornase alfa delivery with eFlow vibrating membrane nebulizers: aerosol characteristics and physicochemical stability. J Pharm Sci 100:98–109

Shah PI, Bush A, Canny GJ, Colin AA, Fuchs HJ, Geddes DM, Johnson CA, Light MC, Scott SF, Tullis DE, De Vault A, Wohl ME, Hodson ME (1995) Recombinant human DNase I in cystic fibrosis patients with severe pulmonary disease: a short-term, double-blind study followed by six months open-label treatment. Eur Respir J 8:954–958

Shak S (1995) Aerosolized recombinant human DNase I for the treatment of cystic fibrosis. Chest 107:65S–70S

Shak S, Capon DJ, Hellmiss R, Marsters SA, Baker CL (1990) Recombinant human DNase I reduces the viscosity of cystic fibrosis sputum. Proc Natl Acad Sci USA 87:9188–9192

Shiokawa D, Tanuma S (2001) Characterization of human DNase I family endonucleases and activation of DNase gamma during apoptosis. Biochemistry 40: 143–152

Shire SJ (1996) Stability characterization and formulation development of recombinant human deoxyribonuclease I [Pulmozyme, (dornase alfa)]. In: Pearlman R, Wang YJ (eds) Pharmaceutical biotechnology: formulation, characterization and stability of protein drugs, vol 9. Plenum Press, New York, pp 393–426

Silverman RA, Foley F, Dalipi R, Kline M, Lesser M (2012) The use of rhDNAse in severely ill, non-intubated adult asthmatics refractory to bronchodilators: a pilot study. Respir Med 106:1096–1102

Simpson G, Roomes D, Reeves B (2003) Successful treatment of empyema thoracis with human recombinant deoxyribonuclease. Thorax 58:365–366

Sinicropi DV, Lazarus RA (2001) Assays for human DNase I activity in biological matrices. Methods Mol Biol 160:325–333

Sinicropi DV, Prince WS, Lofgren JA, Williams M, Lucas M, DeVault A (1994a) Sputum pharmacodynamics and pharmacokinetics of recombinant human DNase I in cystic fibrosis. Am J Respir Crit Care Med 149:A671

Sinicropi D, Baker DL, Prince WS, Shiffer K, Shak S (1994b) Colorimetric determination of DNase I activity with a DNA-methyl green substrate. Anal Biochem 222: 351–358

Suck D (1994) DNA recognition by DNase I. J Mol Recognit 7:65–70

Suri R (2005) The use of human deoxyribonuclease (rhDNase) in the management of cystic fibrosis. BioDrugs 19:135–144

ten Berge M, Brinkhorst G, Kroon AA, de Jongste JC (1999) DNase treatment in primary ciliary dyskinesia – assessment by nocturnal pulse oximetry. Pediatr Pulmonol 27:59–61

Ulmer JS, Herzka A, Toy KJ, Baker DL, Dodge AH, Sinicropi D, Shak S, Lazarus RA (1996) Engineering actin-resistant human DNase I for treatment of cystic fibrosis. Proc Natl Acad Sci USA 93:8225–8229

Vasconcellos CA, Allen PG, Wohl ME, Drazen JM, Janmey PA, Stossel TP (1994) Reduction in viscosity of cystic fibrosis sputum in vitro by gelsolin. Science 263: 969–971

Wagener JS, Rock MJ, McCubbin MM, Hamilton SD, Johnson CA, Ahrens RC (1998) Aerosol delivery and safety of recombinant human deoxyribonuclease in young children with cystic fibrosis: a bronchoscopic study. J Pediatr 133:486–491

Wang H, Morita M, Yang X, Suzuki T, Yang W, Wang J, Ito K, Wang Q, Zhao C, Bartlam M, Yamamoto T, Rao Z (2010) Crystal structure of the human CNOT6L nuclease domain reveals strict poly(A) substrate specificity. EMBO J 29:2566–2576

Widlak P, Garrard WT (2005) Discovery, regulation, and action of the major apoptotic nucleases DFF40/CAD and endonuclease G. J Cell Biochem 94: 1078–1087

Wills PJ, Wodehouse T, Corkery K, Mallon K, Wilson R, Cole PJ (1996) Short-term recombinant human DNase in bronchiectasis. Effect on clinical state and in vitro sputum transportability. Am J Respir Crit Care Med 154:413–417

Wilmott RW, Amin RS, Colin AA, DeVault A, Dozor AJ, Eigen H, Johnson C, Lester LA, McCoy K, McKean LP, Moss R, Nash ML, Jue CP, Regelmann W, Stokes DC, Fuchs HJ (1996) Aerosolized recombinant human DNase in hospitalized cystic fibrosis patients with acute pulmonary exacerbations. Am J Respir Crit Care Med 153:1914–1917

Wolf E, Frenz J, Suck D (1995) Structure of human pancreatic DNase I at 2.2 Å resolution. Protein Eng 8:79

Yang W (2011) Nucleases: diversity of structure, function and mechanism. Q Rev Biophys 44:1–93

Zahm JM, Girod de Bentzmann S, Deneuville E, Perrot-Minnot C, Dabadie A, Pennaforte F, Roussey M, Shak S, Puchelle E (1995) Dose-dependent in vitro effect of recombinant human DNase on rheological and transport properties of cystic fibrosis respiratory mucus. Eur Respir J 8:381–386

17 Monoclonal Antibodies in Cancer

Amy Grimsley, Katherine S. Shah, and Trevor McKibbin

INTRODUCTION

Cancer is the second leading cause of death in the United States with one of every four deaths attributable to cancer (Jemal et al. 2006). The first references to cancer date back to Egyptian papyrus circa 1600 BC. The introduction of nitrogen mustards in the 1940s can be considered the origin of modern, systemic antineoplastic therapies (Papac 2001). Rapid improvements in the understanding of cancer biology, medicinal chemistry, and biopharmaceutical technology have provided rationally designed drugs exploiting differences in normal and malignant cells. Monoclonal antibodies (MABs) bind to a specific epitope. This allows for a targeting approach for the development of effective anticancer compounds with relatively less and/or nonoverlapping toxicity compared to other cytotoxic drugs used to treat cancer. In cancer treatment, MABs have been developed that exert a wide array of pharmacologic effects. This chapter focuses on FDA-approved MABs for the treatment of cancer and cancer-related symptoms. Antibodies are organized based on their target. Table 17.1 summarizes the current FDA-approved MABs for cancer indications, year of approval, target, and the indications that are discussed within the chapter.

A. Grimsley, PharmD
Department of Pharmacy Practice,
Mercer University College of
Pharmacy and Health Sciences,
Atlanta, GA, USA

K.S. Shah, PharmD • T. McKibbin, PharmD, MS, BCPS (✉)
Department of Pharmaceutical Services,
Emory University Hospital/Winship Cancer Institute,
1365 Clifton RD, Building C,
Atlanta 30322, GA, USA
e-mail: trevor.mckibbin@emory.edu

CLASSES OF MONOCLONAL ANTIBODIES: CD ANTIGENS

■ Alemtuzumab

Pharmacology and Pharmacokinetics

Alemtuzumab is an unconjugated, humanized, IgG_1 kappa MAB directed against the 21–28 kDa cell surface glycoprotein CD52 (Frampton and Wagstaff 2003). Most lymphocytes (including 95 % of B and T cells at various stages of differentiation), monocytes, natural killer cells, macrophages, and eosinophils, as well as cells lining the male reproductive tract, express CD52; however, it is not found on erythrocytes, platelets, or stem cells (Liu and O'Brien 2004). While CD52 is highly expressed in some forms of chronic lymphocytic leukemia (CLL), non-Hodgkin's lymphoma (NHL), and acute lymphoblastic leukemia (ALL), it is not shed or internalized, making it an ideal therapeutic target (Liu and O'Brien 2004). Malignant CD52 expression occurs not only in CLL, low-grade lymphomas, and T-cell malignancies but also in some cases of myeloid, monocytic, and acute lymphoblastic leukemias. The compound exerts its anticancer effects by binding to CD52 antigenic sites and stimulating cross-linking by antibodies, which promotes antibody-dependent cellular cytotoxicity and direct cellular apoptosis via natural killer activity as shown in Fig. 17.1 (O'Brien et al. 2005).

Pharmacokinetic parameters of alemtuzumab were determined in a phase I dose-escalation trial. Patients with B-cell CLL and NHL were given alemtuzumab intravenously once weekly for a maximum of 12 weeks, and plasma levels were obtained (Frampton and Wagstaff 2003). A dose-proportional increase in maximum plasma concentration (Cmax) and area under the concentration-time curve (AUC) was observed. The median half-life ($t_{1/2}$) was approximately 12 days. A subsequent pharmacokinetic analysis was conducted in CLL patients who received alemtuzumab 30 mg intravenously three times weekly. A high degree of interpatient

D.J.A. Crommelin, R.D. Sindelar, and B. Meibohm (eds.), *Pharmaceutical Biotechnology*,
DOI 10.1007/978-1-4614-6486-0_17,

Generic name	Approval year	Origin	Target	Indications
Cluster of differentiation (CD) targeted				
Alemtuzumab	2001	Humanized	CD-52	B-cell CLL
Gemtuzumab	2000[a]	Humanized	CD-33	AML –compassionate use program only
Rituximab	1997	Chimeric	CD-20	Non-Hodgkin's lymphoma
Yttrium-90 (^{90}Y) ibritumomab tiuxetan	2002	Murine	CD-20	B-cell non-Hodgkin's lymphoma
Iodine-131 (^{131}I) tositumomab	2003	Murine	CD-20	Non-Hodgkin's lymphoma
Brentuximab vedotin	2011	Chimeric	CD-30	Hodgkin's lymphoma
Ofatumumab	2009	Human	CD-20	CLL
Epidermal growth factor receptor (EGFR) targeted				
Cetuximab	2004	Chimeric	EGFR	Colorectal cancer, SCCHN
Panitumumab	2006	Human	EGFR	Colorectal cancer
Trastuzumab	1998	Humanized	HER2/neu	Breast cancer, gastric cancer
Vascular endothelial growth factor (VEGF) targeted				
Bevacizumab	2004	Humanized	VEGF	NSCLC, colorectal cancer, glioblastoma multiforme, metastatic renal cell carcinoma
Receptor activator of nuclear factor kappa B ligand (RANKL)				
Denosumab	2010	Human	RANKL	Cancer metastatic to the bone
Cytotoxic T-lymphocyte antigen-4 (CTLA-4)				
Ipilimumab	2011	Human	CTLA-4	Metastatic melanoma

CLL chronic lymphocytic leukemia, *AML* acute myeloid leukemia, *VEGF* vascular endothelial growth factor, *NSCLC* non-small cell lung cancer, *EGFR* endothelial growth factor receptor, *SCCHN* squamous cell carcinoma of the head and neck
[a]Approval subsequently denied.

Table 17.1 ■ FDA-approved monoclonal antibodies in cancer.

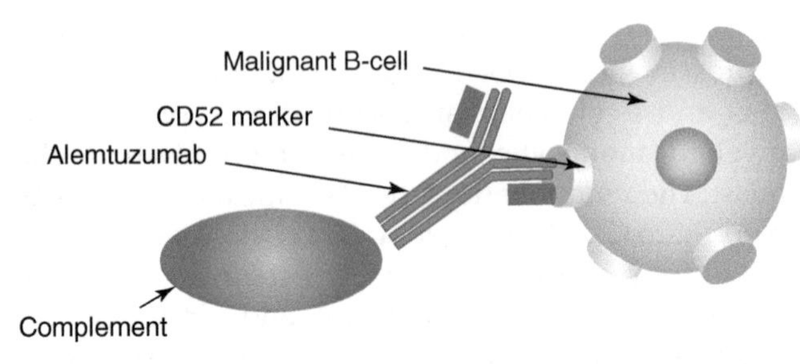

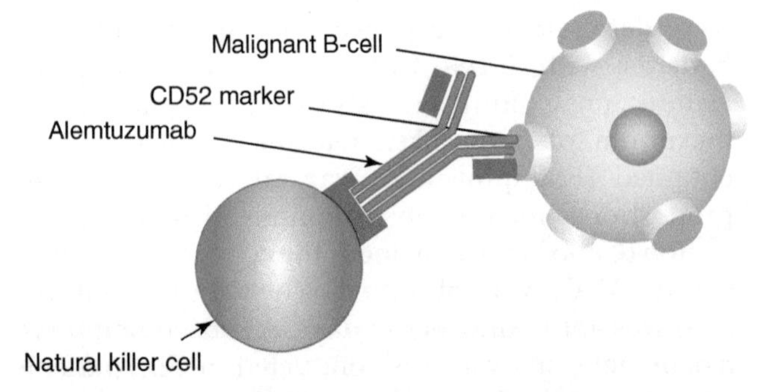

Figure 17.1 ■ Alemtuzumab mechanism of action. Diagram showing alemtuzumab bound to the CD52 surface marker on chronic lymphocytic leukemia cells where it triggers complement-dependent cytotoxicity and natural killer cell action (*Source*: The Association of the British Pharmaceutical Industry).

variability was observed and patients exhibited a trend of gradually rising plasma concentrations during initial therapy, which continued until steady state was achieved. This typically occurred after 6 weeks. Authors noted that the rise in alemtuzumab concentrations corresponded to a simultaneous decline in circulating CD52-positive malignant lymphocytes.

Alemtuzumab has also been given subcutaneously (Montagna et al. 2011). In one comparative study, 29 patients with relapsed CLL received intravenous alemtuzumab 30 mg thrice weekly, while 20 patients received the same dose subcutaneously. The authors noted that over time, maximal trough concentrations progressed to similar levels in both groups; however, accumulation of antibody in the blood was slower in the subcutaneous group, with these patients requiring slightly higher cumulative doses to achieve similar concentrations (Montagna et al. 2011). In this study, the mean steady-state volume of distribution (Vd) among both groups of patients during initial treatment was 0.185 L/kg, which expanded to 0.252 L/kg during the terminal phase. The large Vd is consistent with the notion that alemtuzumab distributes beyond the plasma compartment to an extravascular lymphocytic compartment. Mean terminal half-life was 6.1 days in this population and clearance appeared to correlate with antigenic burden (Dirks and Meibohm 2010). Patients with undetectable CLL cells exhibited a single elimination phase with a longer half-life, whereas patients with bulkier tumors cleared alemtuzumab more rapidly. The most dominant factor influencing alemtuzumab pharmacokinetic parameters appears to be CD52 concentration, which accounts for much of the interpatient variability.

Indications and Clinical Efficacy

While it is utilized in a variety of disease states including many T-cell malignancies, alemtuzumab is FDA approved for use in B-cell CLL patients who have been treated with alkylating agents and failed fludarabine therapy. Alemtuzumab was commercially available under the branded name Campath® for the treatment of leukemias until September 2012. Given recent encouraging results in phase III studies in Multiple Sclerosis the manufacturer has removed Campath® from the market to prevent unauthorized use. Though Campath is available free of charge for the treatment of malignancy, alemtuzumab is being re-branded to Lemtrada™ for the treatment of Multiple Sclerosis. Three major studies have assessed alemtuzumab in this population (Keating and Hallek 2002; Rai et al. 2002; Osterborg et al. 2002). An international collaboration of centers in the United States and Europe published the largest of these studies. Ninety-two fludarabine-resistant patients, of whom 76 % had Rai stage III or IV disease, were treated with 12 weeks of intravenous alemtuzumab (Keating and Hallek 2002). The overall response rate was 33 %, with 2 % achieving a complete response (CR). Median survival was 16 months. Patients with bulky lymphadenopathy were less likely to respond, possibly indicating poor tumor penetration of alemtuzumab (Keating and Hallek 2002; Liu and O'Brien 2004). Toxicity was moderate and consisted mainly of infections and infusion reactions. A study evaluating alemtuzumab treatment in 24 poor-prognosis, fludarabine-resistant chronic lymphocytic leukemia patients confirmed these findings (Rai et al. 2002). After approximately 16 weeks of treatment with alemtuzumab (target dose 30 mg three times weekly), the overall response rate was 33 % and the median time to progression was 19.6 months.

Because of the high incidence of infusion-related reactions encountered with alemtuzumab intravenous infusion, subcutaneous administration has been explored as a potential alternative. A pivotal trial evaluated subcutaneous alemtuzumab as first-line therapy in 41 patients with advanced, previously untreated CLL (Lundin et al. 2002). The subcutaneous alemtuzumab resulted in an overall response rate of 87 %, with 19 % CR. While injection site reactions were seen in 90 % of patients, these were grades 1–2 in severity and typically disappeared with continued treatment (often within 2 weeks). The more severe infusion reactions encountered with intravenous dosing, such as dyspnea, hypotension, and nausea, were absent; however, some patients did experience fever and rigors (Lundin et al. 2002).

Combination therapy with alemtuzumab has been investigated; one study evaluated six patients with refractory disease who were treated with fludarabine and alemtuzumab concurrently (Kennedy and Hillmen 2002). Five patients responded, with one patient achieving a complete response. Another combination that has been explored is alemtuzumab with rituximab (Faderl et al. 2003; Nabhan et al. 2004). Nabhan and colleagues administered rituximab 375 mg/m^2 weekly for four cycles, adding alemtuzumab thrice weekly on weeks 2–5 in 12 patients with relapsed CLL. One patient achieved a partial response (PR), while 90 % of patients had stable disease. Therapy was relatively well tolerated, with no treatment-related deaths; however, 75 % of patients experienced grade 2 rigors and 33 % exhibited grade 3/4 fevers. A second trial evaluating the combination of rituximab and alemtuzumab used a similar schedule in 48 relapsed CLL and PLL patients. Response rates were strong, with 65 % achieving a PR. However, infection was common, occurring in 56 % of patients (Nabhan et al. 2004). Longer follow-up and additional studies will help to fully elucidate the role of combination therapy with alemtuzumab, further addressing the issue of additive myelosuppression and infectious risk with this agent. Selected clinical studies of alemtuzumab are summarized in Table 17.2.

Safety

The most common adverse effects associated with alemtuzumab, which are also listed as black box warnings, are infusion-related reactions, infectious complications, and hematologic toxicities. Infusion reactions are common and are reported in approximately 90 % of patients (Keating and Hallek 2002; Stilgenbauer et al. 2009). Rigors, fever, nausea, vomiting, and rash are often seen with initial infusions; however, these typically decrease with subsequent drug exposure. Rarely, hypotension and dyspnea are encountered (Liu and O'Brien 2004). Premedication with antihistamines is recommended to reduce this possibility. Subcutaneous administration of alemtuzumab also significantly lessens the risk of infusion-related adverse reactions (Montillo et al. 2006; Stilgenbauer et al. 2009). Subcutaneous administration may also be associated with transient local skin reactions. Another substantial adverse effect commonly associated with alemtuzumab is infection; lymphocyte counts decrease rapidly after treatment,

Investigators	Disease(s), number of patients	Alemtuzumab dosing regimen	CR/PR (%)	Median overall survival	Significant adverse events (grade 3/4)
	Relapsed/refractory B-cell CLL, *n*=93	3 mg IV until tolerated, then 10 mg IV until tolerated, then 30 mg IV thrice weekly for up to 12 weeks	CR 2 PR 31	16 months	Infection 26.9 %
	B- or T-cell CLL after failing fludarabine, *n*=24	10 mg IV until tolerated, then 30 mg IV thrice weekly for up to 16 weeks	PR 33	35.8 months	Neutropenia 20.8 % Infection 41.7 %
	Relapsed/refractory CLL, *n*=29	3 mg IV escalated as tolerated to 30 mg IV thrice weekly for up to 12 weeks	CR 4 PR 38	*Median response duration*=12 months	Neutropenia 41 % Thrombocytopenia 27 % Hypotension 3 % Infection 17 %
Lundin et al. (2002)	Primary B-cell CLL, *n*=41	3 mg SC escalated to 10 mg SC and 30 mg SC as tolerated; then, 30 mg SC thrice weekly for 18 weeks maximum	CR 19 PR 68	Not reached yet; 8–44+ months	Neutropenia 21 % Pain at injection site 7 % Thrombocytopenia 5 % Infection 12 %
Pawson et al. (1997)	Relapsed T-cell PLL, *n*=15	10 mg IV escalated to 30 mg IV thrice weekly as tolerated[a]	CR 60 PR 13	Not reached yet	Hematologic 27 % Infection 33 %
	Relapsed lymphoproliferative disorders, including CLL and T-cell PLL; *n*=78	3 mg IV escalated to 10 mg IV and 30 mg IV as tolerated; then 30 mg IV thrice weekly for 12 weeks maximum	CR 13 PR 22	12 months	Neutropenia 27 % Thrombocytopenia 32 % Dyspnea 7 %
	Relapsed/refractory CLL or PLL, *n*=23	3 mg IV escalated to 10 mg IV and 30 mg IV as tolerated; then 30 mg IV thrice weekly for 12 weeks maximum	CR 35 PR 18	N/A	Neutropenia 9 % Thrombocytopenia 9 % Infection 9 %

CR complete response, *PR* partial response
[a]One patient received subcutaneous alemtuzumab

Table 17.2 ■ Selected clinical trials with alemtuzumab (Campath-1H).

resulting in a severe and extended durations of lymphopenia. This profound T-cell depletion leads to an increased risk of opportunistic infections, particularly CMV reactivation. Additionally, *Herpes simplex* virus infection, *Pneumocystis carinii* pneumonia, candidiasis, and septicemia have all been reported (Keating and Hallek 2002). These commonly manifest between 3 and 8 weeks from treatment, during the nadir of the T-lymphocyte count. It is recommended to assess CD4+ counts after beginning treatment and until ≥200 cells/μL. Prolonged prophylaxis with antibacterial and antiviral medications is highly encouraged. Myelosuppression, on the other hand, consisting of anemia, neutropenia, and thrombocytopenia, tends to be moderate and transient, with grade 4 neutropenia occurring in about 20 % of cases (Keating and Hallek 2002; Kennedy and Hillmen 2002; Liu and O'Brien 2004). A complete blood count should be monitored at least weekly. Dosage modifications for alemtuzumab are recommended for cases of severe myelosuppression.

In rare instances, cardiac toxicity has been reported with alemtuzumab, consisting of atrial fibrillation and left ventricular dysfunction (Lenihan et al. 2004). These case reports have occurred in patients with mycosis fungoides/Sezary syndrome, and authors suggested that patients with T-cell malignancies may be at increased risk of cardiac toxicity from alemtuzumab. However, other reports have found no link between alemtuzumab and cardiac toxicity in patients with mycosis fungoides/Sezary syndrome (Lundin et al. 2005).

Alemtuzumab treatment should be administered according to a dose-escalation schedule. If infusion-related toxicities are ≤grade 2, the patient may proceed on to the next dose. An initial dose of 3 mg should be given as a 2-h IV infusion daily; if this initial dose is tolerated, the dose should be increased to 10 mg IV over 2 h; if the 10 mg dose is tolerated, the maintenance dose of 30 mg IV over 2 h may be initiated. Patients receiving alemtuzumab subcutaneously follow this same dose-escalation schema. To help prevent infusion reactions, each dose is preceded by a dose of acetaminophen and an antihistamine. Corticosteroids and other supportive care measures may be administered if severe infusion-related events occur.

■ Gemtuzumab

Pharmacology and Pharmacokinetics

Gemtuzumab ozogamicin (GO) was one of the first commercially available bispecific monoclonal antibodies. This recombinant, humanized, IgG_4 MAB to cell surface marker CD33 is covalently bonded by a bifunctional linker to the potent cytotoxic antibiotic, calicheamicin. Immature and mature myeloid cells, as well as erythroid, megakaryocytic, and multipotent progenitor cells, express the 67-kDa glycosylated transmembrane protein CD33. In addition, this protein is expressed on the surface of most leukemic blast cells found in acute myelogenous leukemia (AML) as well as myelodysplastic syndromes (MDS) (van Der Velden et al. 2001). However, CD33 is not expressed on stem cells, nor is it expressed outside of the hematopoietic system, making it an excellent therapeutic target. The cytotoxic antibiotic calicheamicin is a natural antineoplastic compound derived from *Micromonospora echinospora*. It is made up of two molecules of the enediyne antitumor antibiotic n-acetyl-γ-calicheamicin dimethyl hydrazine (Sievers et al. 1999). This compound, along with its metabolites, has antineoplastic activity that is 1,000 times more potent than doxorubicin (Giles et al. 2003).

GO exerts its clinical effects through direct binding to the CD33 antigen. Following a 9 mg/m^2 dose, CD33 antigenic sites are maximally saturated within 3 h (van Der Velden et al. 2001). Endocytosis quickly follows, resulting in rapid internalization of the antibody-antigen complex. Additional expression of new CD33 antigenic sites occurs after internalization of the GO molecule, leading to further accumulation and increased concentration of intracellular GO. Once inside the cell, GO is directed to lysosomes which cleave the molecule via acid hydrolysis, liberating the calicheamicin compound. Calicheamicin then binds to double-stranded DNA helixes in the minor groove, causing site-specific double strand cleavage at oligopyrimidine-oligopurine tracts. Induction of apoptosis is observed after approximately 72–96 h. In addition to direct induction of apoptosis from calicheamicin, antibody-dependent cell-mediated cytotoxicity and complement-mediated cytotoxicity also stimulate leukemic cell death.

Clinical studies investigating the pharmacokinetic parameters of GO have been conducted in adults with AML in first relapse (Dowell et al. 2001, Korth-Bradley et al. 2001). Initial phase I pharmacokinetic trials found that a dose of 9 mg/m^2 fully saturated CD33 sites in all patients regardless of disease burden. Phase II studies confirmed the efficacy of GO in refractory AML patients and helped consolidate the treatment schedule of two 9 mg/m^2 infusions separated by approximately 14 days. Measurements of serial plasma concentrations have confirmed a distinct difference in pharmacokinetic parameters between the first and second doses, largely thought to be due to a decline in circulating leukemic blast cells that express CD33. A study conducted by Dowell and colleagues in 59 adult patients with relapsed AML found that maximum plasma concentrations (C_{max}) of both MABs and calicheamicin typically occurred shortly after the end of the 2-h infusion; additionally, C_{max} values were generally higher after the second dose (Dowell et al. 2001). Values for volume of distribution changed as well, averaging approximately 20.9 L after the first dose and only 9.9 L after the second. This decrease in Vd is likely also due to a decline in the number of circulating cells expressing CD33. In addition, the relatively low distribution volumes suggest that GO does not distribute beyond the plasma compartment, but rather remains bound to CD33 antigenic sites within the vascular space. This has been confirmed by radiolabeled studies that demonstrate that organs with a large blood pool, such as the spleen and liver, are primarily responsible for uptake and distribution of the antibody. Another pharmacokinetic evaluation of GO by Korth-Bradley and colleagues (2001) compared the kinetic parameters of GO in different populations. Although a great deal of interpatient variability was observed, the authors concluded that there were no significant differences in C_{max}, time to C_{max}, AUC, clearance, or Vd between males and females, nor were there any significant differences between those over 60 and those under 60 years of age. Clearance of GO from the plasma occurs mainly through uptake by CD33-positive cells and subsequent internalization and is therefore influenced by antigen concentration. Elimination half-life of the drug is fairly long and increases upon second exposure. Median half-life of the antibody component is 72.4 h after the first dose and 93.7 h after the second, while the median half-life of the calicheamicin component is 45.1 h after the first dose and 61.1 h after the second (Dowell et al. 2001a). Accumulation between doses was not found to be significant, as evidenced by concentrations equivalent to 1 % of C_{max} measured just prior to the second dose.

Clinical Considerations

Gemtuzumab ozogamicin (Mylotarg®) was approved in May 2000 under the FDA's accelerated approval program. A confirmatory, post approval clinical trial was undertaken in 2004. The trial was designed to determine whether adding GO to standard chemotherapy demonstrated an improvement in clinical benefit (overall survival) to AML patients. The trial was halted early when no improvement in clinical benefit was observed and after a greater number of deaths occurred in the group of patients who received Mylotarg® compared with those receiving chemotherapy alone. Due to the lack of survival advantage and serious safety concerns such as increased post-marketing rates of sinusoidal obstructive syndrome (SOS), the FDA withdrew GO from the market in June 2010; GO is not commercially available to new patients. Patients who are currently receiving GO may complete their therapy following consultation with their prescribing physician. Any future use of GO in the United States will require submission of an investigational new drug application to the FDA. Gemtuzumab ozogamicin was previously indicated for the treatment of CD33-positive AML in first relapse for patients greater than or equal to 60 years of age who are not candidates for other chemotherapy.

■ Rituximab, Yttrium-90 (^{90}Y) Ibritumomab Tiuxetan, Iodine-131 (^{131}I) Tositumomab

Pharmacology and Pharmacokinetics

Rituximab, the first MAB approved for the treatment of cancer, was approved for use in 1997. It is a chimeric murine/human MAB directed against the CD20 antigen found on nearly all B lymphocytes. Rituximab is approved in the United States for the following indications: treatment of relapsed or refractory, B-cell CD20-positive, low-grade or follicular NHL; first-line treatment of follicular or diffuse large B-cell (DLBCL) CD20-positive NHL in combination with chemotherapy; treatment of low-grade, CD20-positive B-cell NHL in patients achieving a response or stable disease to first-line chemotherapy; treatment of moderate to severe rheumatoid arthritis in combination with methotrexate; and treatment of Wegener's granulomatosis and microscopic polyangiitis in combination with glucocorticoids. Rituximab is widely utilized in an off-label fashion for numerous clinical conditions, and the final three indications represent the many current and future non-oncologic uses of this antibody. However, as this chapter focuses on oncologic indications, the non-oncology uses of rituximab will not be further discussed and the reader is referred to Chaps. 19 and 20 for the non-oncology use of rituximab (Fig. 17.2).

All normal B cells and greater than 90 % of malignant B cells contain the CD20 antigen (Maloney et al. 1994). CD20 is the human B-lymphocyte-restricted differentiation antigen, Bp35, and is a hydrophobic transmembrane protein. CD20 is involved with cell cycle initiation, regulation, and differentiation by activation of B cells from the G0 (resting) phase to the G1 (gap 1) phase, and CD20 has also been shown to operate as a calcium ion channel. Rituximab is thought to mediate death of CD20-positive tumor cells via activation of the complement cascade through at least three distinct mechanisms. Specifically, antibody-dependent cell-mediated cytotoxicity, direct effects via CD20 ligation, and complement-mediated apoptosis are all believed to play a role (Maloney et al. 1997). The depletion of B cells, via the CD20 antigen, can be extensive and prolonged.

The pharmacokinetics of rituximab were assessed in non-Hodgkin's lymphoma patients at a dosage of 375 mg/m^2 intravenously given weekly for 4 weeks (Maloney et al. 1994). The mean serum half-life increased throughout the study (76.3 h after the first week to 205.8 h after the fourth week). This increase is thought to occur secondary to depletion of the CD20 antigen. Without an antigen to bind to, rituximab's clearance will be reduced. After 4 weeks of treatment, rituximab may be detectable in a patient's serum for up to 6 months. After 8 weeks of weekly rituximab infusions using the same dose, the mean maximum concentration was found to increase from 243 μg/mL after the first infusion to 550 μg/mL after the final infusion.

In one study of 166 patients, B cells were depleted within the first three doses of rituximab and the depletion was maintained throughout 6–9 months in the majority of patients. B-cell levels should return to normal levels by 12 months after the last dose of rituximab.

When B cells become activated, they mature into plasma cells, the terminally differentiated B cell, and actively secrete immune globulins. With the use of rituximab, statistically significant reductions in IgG, IgM, and IgA have been observed. However, in the majority of cases, the levels remain normal.

Indications and Clinical Efficacy

Rituximab has been studied as a single agent in low-, intermediate-, and high-grade lymphomas. A review of low-grade lymphoma studies using single agent rituximab shows the ranges of overall response rate (OR), which is the complete response rate (CR) plus the partial response rate, and CR to be 27–73 % and 0–23 %, respectively. A review of intermediate- and high-grade lymphomas using single agent rituximab shows the ranges of OR and CR to be 14–73 % and 0–44 %, respectively.

Perhaps, the most studied chemotherapy used in combination with rituximab is the CHOP regimen which consists of cyclophosphamide, doxorubicin, vincristine, and prednisone. A phase III trial evaluated the addition of rituximab to CHOP versus CHOP alone in elderly patients with DLBCL (Coiffier et al. 2002). The combination provided statistically significant

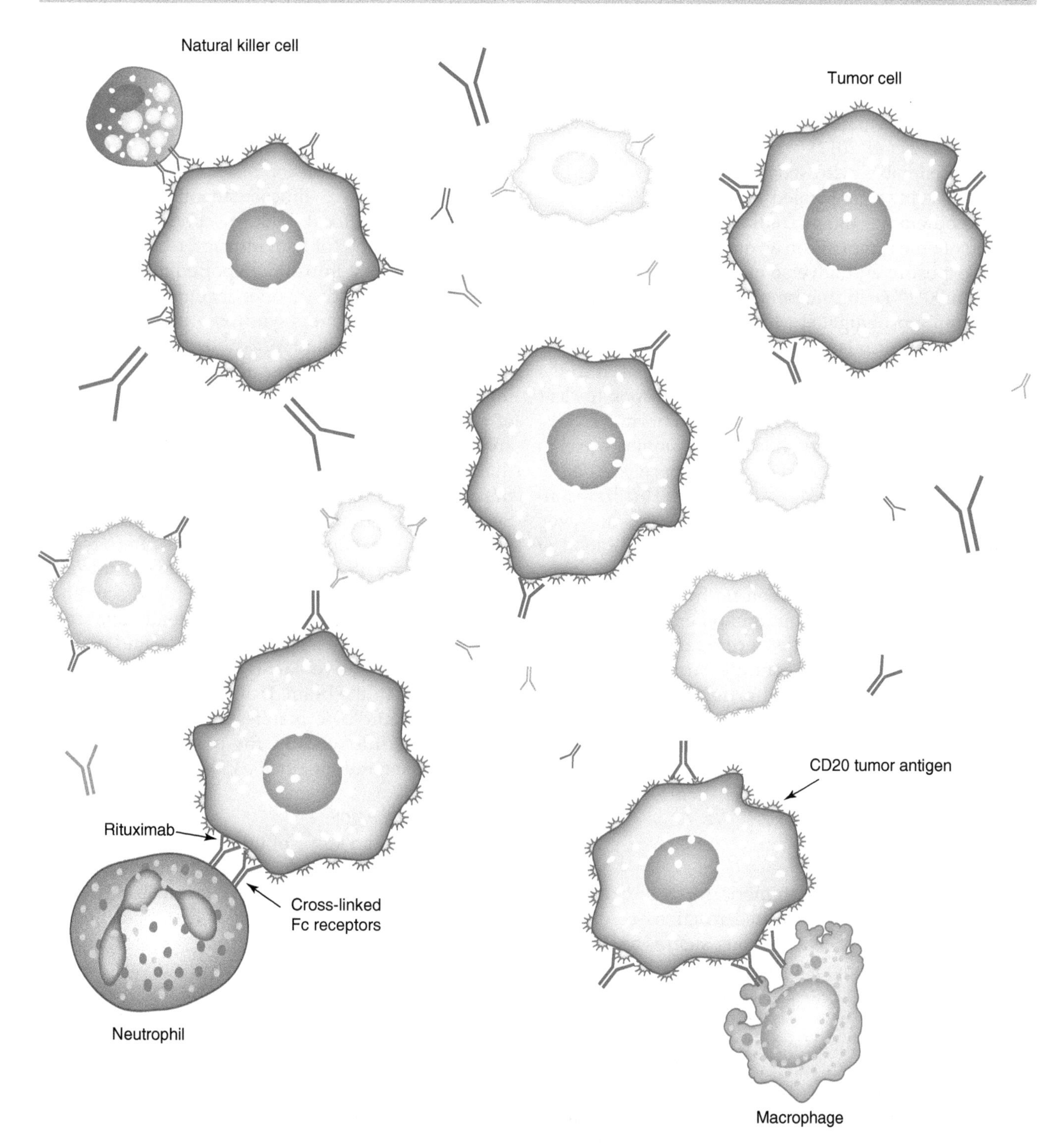

Figure 17.2 ■ Rituximab mechanism of action (*Source*: Point Therapeutics).

increases in clinical endpoints including CR, event-free survival, and overall survival, without increasing treatment-related toxicity. Rituximab in combination with chemotherapy regimens (mostly CHOP or CHOP-like regimens) achieves OR and CR in the range of 29–100 % and 11–85 % of patients, respectively (Coiffier 2002; McLaughlin et al. 1998; Plosker and Figgitt 2003).

Rituximab is increasingly used in the maintenance setting after achieving remission, particularly in low-grade lymphomas. The use of maintenance rituximab for 2 years significantly improves progression-free survival in patients with both previously untreated and relapsed follicular lymphoma who have responded to induction treatment. The safety of rituximab maintenance is consistent with the known safety profile of rituximab, and long-term follow-up in maintenance trials will determine the effect on overall survival. At this time, the optimal schedule and duration of maintenance therapy remain to be determined (Hochster et al. 2009; Huang et al. 2012).

Rituximab is also used in conjunction with a MAB designed with a conjugated radionuclide. This radioimmunotherapy allows radiation to be delivered directly to the tumor site in an attempt to limit the radiation exposure and adverse effects to healthy tissues. Yttrium-90 (^{90}Y) ibritumomab tiuxetan was the first radioimmunoconjugated MAB approved by the FDA for treatment of relapsed or refractory follicular, low-grade, or transformed NHL. Rituximab is used in the ^{90}Y ibritumomab tiuxetan therapeutic regimen to clear the peripheral blood of CD20 found on normal B cells. This facilitates binding of the ^{90}Y ibritumomab tiuxetan to the CD20 antigen located on the tumor cells. Another radioimmunoconjugated MAB, iodine-131 (^{131}I) tositumomab, is a similar therapeutic regimen; however, tositumomab is used to clear the peripheral blood of CD20 instead of rituximab.

To date, there have been no head-to-head comparisons of these two radioimmunotherapies. However, ^{90}Y-ibritumomab tiuxetan was compared to rituximab in a phase III, randomized controlled trial for rituximab-naïve patients with relapsed or refractory low-grade, follicular, or transformed B-cell NHL (Witzig et al. 2002a, b). ^{90}Y-ibritumomab tiuxetan produced a statistically significant increase in CR (30 % vs 16 %) and OR (80 % vs 56 %, $p = 0.002$). However, there were no differences in time to progression.

Safety

Because rituximab is a chimeric MAB and contains mouse protein, infusion-related reactions can be significant. Infusion-related reactions, such as fever, chills, and myalgias, along with hypersensitivity reactions, such as bronchospasm, pulmonary infiltrates, hypotension, and angioedema, have occurred during rituximab infusions. Patients are more likely to experience these reactions when receiving their first infusion of rituximab, when tumor burden is most likely at its highest. The incidence decreases with subsequent infusions. Should an infusion reaction occur, the rituximab infusion should be discontinued immediately. If the infusion reaction was not severe and symptoms have resolved, the infusion may be restarted at half the previous rate at which the reaction occurred. All infusions should be preceded by premedication with an antihistamine and acetaminophen. For patients experiencing previous infusion-related reactions, additional premedication with corticosteroids may be required for subsequent dosing. Typical rituximab infusion times range from 3 to 6 h based upon dosage and tolerability, although many clinicians are choosing to administer it via a shorter, fixed rate if no previous hypersensitivity reactions have previously occurred (Salar et al. 2006; Sehn et al. 2007).

Rituximab's package labeling contains four black box warnings: fatal infusion reactions, tumor lysis syndrome (TLS), severe mucocutaneous reactions, and progressive multifocal leukoencephalopathy (PML). The majority (80 %) of fatal infusion reactions occur in relation to the first infusion. Reported infusion-related sequelae preceding death often include bronchospasm, acute respiratory distress syndrome, myocardial infarction, ventricular fibrillation, cardiogenic shock, and/or anaphylactoid events. Those at higher risk for developing TLS are patients with NHL who have a high amount of circulating malignant cells or large tumor burden and those who have received concomitant cisplatin due to further risk of development of acute renal failure or death. Severe mucocutaneous reactions, such as Stevens-Johnson syndrome and toxic epidermal necrolysis, have occurred within 1–13 weeks after receiving rituximab. Patients who experience a severe mucocutaneous reaction should be permanently discontinued from rituximab. PML is a rare and usually fatal condition that results from a JC virus infection and is characterized by progressive damage or inflammation of the white matter in the brain. If any patient presents with new-onset neurologic manifestations while receiving rituximab, a full workup for possible PML should ensue (Plosker and Figgitt 2003; Wood 2001).

Radioimmunotherapy

Due to the limited scope of this chapter, the two conjugated MABs, yttrium-90 (^{90}Y) ibritumomab tiuxetan and iodine-131 (^{131}I) tositumomab, have only briefly been mentioned in the context of the clinical efficacy section involving rituximab therapy. These agents have a limited therapeutic scope as their primary use is in radioimmunotherapy for treating indolent NHL. Yttrium-90 (^{90}Y) ibritumomab tiuxetan is comprised of a murine IgG_1 MAB ibritumomab that is linked to the radioisotope yttrium-90 by stable chelation via the linker, tiuxetan (Hagenbeek and Lewington 2005). Yttrium-90 is a beta-emitter of high energy with a long half-life (64 h). The drug is given as a single treatment consisting two components given approximately 1 week apart (i.e., rituximab administration followed by ^{90}Y-ibritumomab tiuxetan). Rituximab acts to reduce the number of healthy B cells so that ^{90}Y-ibritumomab tiuxetan will not destroy noncancerous cells. When ^{90}Y-ibritumomab tiuxetan is administered, it attaches to the CD20 proteins on the cell surface of B cells and releases energy from the yttrium radioisotope, killing the B cell. Patients with normal platelet function (i.e., $>150 \times 10^9$/L) should receive 0.4 mCi/kg of body

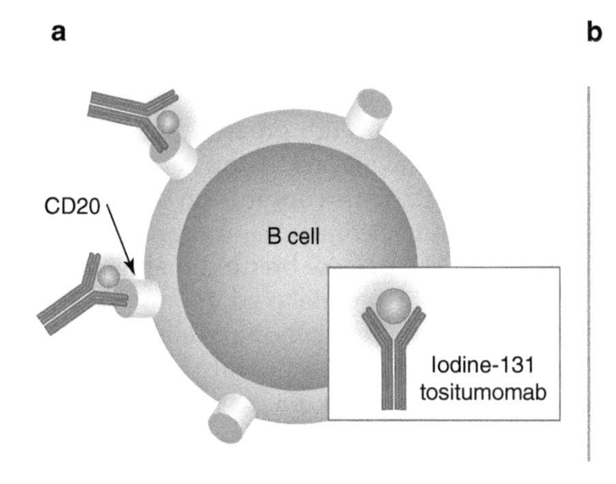

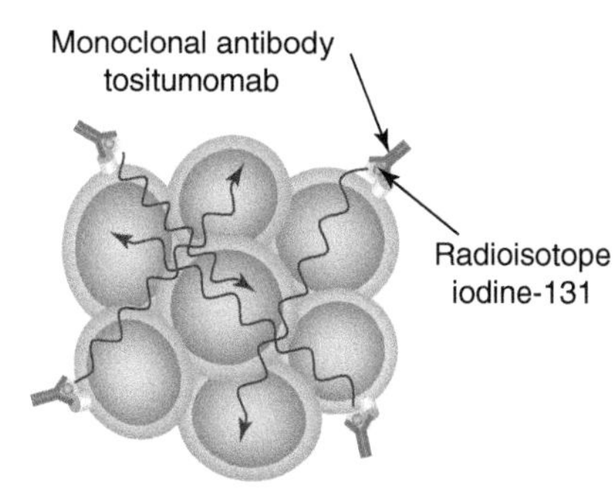

Figure 17.3 ■ Iodine-131 (^{131}I) tositumomab mechanism of Action. (**a**) Iodine-131 tositumomab binds to the CD20 antigen on normal and malignant B lymphocytes. (**b**) "Cross-fire" effect of iodine-131 in ^{131}I- tositumomab will cause damage to tumor cells as well as adjacent normal tissue (*Source*: GlaxoSmithKline).

weight up to a maximum of 32 mCi. Those with a platelet count between 100 and 149 × 10^9/L should receive a reduced dose of 0.3 mCi/kg, which has been shown to have equal efficacy to higher doses (Hagenbeek and Lewington 2005). The main safety concerns include myelosuppression with nadirs reached in 4–8 weeks after administration of therapy.

Iodine-131 tositumomab is another conjugated MAB that acts in a similar mechanism to ^{90}Y-ibritumomab tiuxetan although it emits both beta and gamma radiation. Specifically, it binds to the CD20 antigen found on B cells to kill them via two mechanisms: (1) activating an immune host response to B cells and (2) causing apoptosis in B cells to which it is bound as shown in Fig. 17.3.

■ Brentuximab Vedotin

Pharmacology and Pharmacokinetics

Brentuximab vedotin is an antibody-drug conjugate (ADC) that is directed against the CD30 antigen. CD30 is highly expressed on the surface of Hodgkin's lymphoma and anaplastic large cell lymphoma tumor cells (Francisco et al. 2003; Wahl et al. 2002; Younes et al. 2010). It is a member of the tumor necrosis superfamily. It was thought to be a promising target for MAB-based therapies since the normal expression of CD30 is limited for the most part to activated immune cells. Unfortunately, the CD30 MABs have shown little activity in clinical trials as compared to the anticancer activities demonstrated in preclinical studies (Wahl et al. 2002). In an effort to increase the potency of CD30-targeted MAB therapy, an ADC was created. It is made up of three parts which include a synthetic analog of dolastatin 10, monomethyl auristatin E (MMAE), that was conjugated to a chimeric IgG_1 antibody, cAC10, by a protease-cleavable linker that covalently attaches MMAE to cAC10, to create cAC10-vcMMAE (Francisco et al. 2003). The protease-cleavable linker is a highly stable dipeptide linker that is selectively cleaved by lysosomal enzymes after internalization. MMAE is a small molecule that is a microtubule-disrupting agent. It is suggested that the anticancer activity of brentuximab vedotin is due to the binding of the ADC to CD30-expressing cells, followed by internalization of the ADC-CD30 complex, and the release of MMAE via proteolytic cleavage by cathepsin, a lysosomal protease. The free MMAE binds to tubulin within the cell and disrupts the microtubule network which leads to G2/M cell cycle arrest and apoptotic death of the cells (Francisco et al. 2003).

Brentuximab is the first of the second-generation ADC to be approved. The principle behind ADC development is the concept that combining a cytotoxic drug to a MAB specific for an antigen that is more highly expressed on cancer cells could direct high doses of the cytotoxic drug specifically to the cancer cells and potentially avoid a large portion of the normal cells. Unfortunately, it has been difficult to obtain the desired outcome of this concept with various first-generation ADCs not demonstrating sufficient activity (Francisco et al. 2003). The second-generation ADCs are designed based on an improved understanding of several characteristics of ADCs. First, the chosen antigen should be one that would allow for substantial tumor selectivity. Second, the cytotoxic drug should be extremely potent since only a small amount of the ADC will come into contact with the tumor cell. Lastly, the linker technology should be of such that the ADC is stable while in circulation but effectively releases the active drug once it is internalized into the tumor cells (Francisco et al. 2003).

During preclinical experiments, in vitro data reflected that lysosomal enzymes effectively cleaved the dipeptide linker after brentuximab vedotin binding to CD30 and internalization into the cell (Francisco et al. 2003). The fully active drug MMAE was released into the cell cytosol and resulted in growth arrest in the G2/M phase, apoptosis, and cell death. The 50 % inhibitory concentration [IC_{50}] was<10 ng/mL and brentuximab was 300-fold less potent against antigen-negative cells.

The effect of brentuximab vedotin on the QTc interval was evaluated in an open-label, single-arm trial. Brentuximab vedotin was administered at 1.8 mg/kg to 46 evaluable patients with CD30-expressing hematologic malignancies.

The terminal half-life observed was approximately 4–6 days with steady state achieved in 21 days with every 3-week dosing. There was minimal to no accumulation of ADC with multiple doses on an every 3-week schedule. At the MTD of 1.8 mg/kg, the mean $AUC_{0\text{–}21days}$ was 76.65 day μg/mL with a maximum mean concentration of 32 μg/mL, obtained at a median time of 0.089 days (Younes et al. 2010). The maximum concentration for MMAE was achieved from approximately 2 to 3 days from administration with steady-state levels within 21 days with every 3-week dosing. Continued administration of brentuximab vedotin did result in decreased exposure to MMAE, with 50–80 % of the exposure of the first dose being observed with subsequent doses. At the dose of 1.8 mg/kg, the mean $AUC_{0\text{-}21d}$ for MMAE was 0.036 day μg/mL with a maximum mean concentration of 0.005 μg/mL, obtained at 2.09 days (Younes et al. 2010) The mean steady-state volume was 6–10 L in humans. Approximately 68–82 % of MMAE is bound to human plasma proteins. Only a small portion of the MMAE that is released from brentuximab vedotin is metabolized. MMAE is metabolized by CYP3A4/5. It does appear that brentuximab vedotin also inhibits CYP3A4/5. The half-life of the ADC is 4–6 days, close to the half-life of MMAE of 3–4 days, and this is consistent with the steady-state kinetics occurring after approximately 21 days. The elimination of MMAE is limited by the rate it is released from ADC. In patients who received a dose of 1.8 mg/kg of brentuximab vedotin, about 1/4 of the total MMAE that was administered was recovered in feces and urine. Of the MMAE that was recovered, approximately 72 % was in the feces with the majority of MMAE unchanged. Data from population pharmacokinetic analysis does not demonstrate a significant impact of age, gender, or race on the pharmacokinetics of brentuximab vedotin.[8]

Indications and Clinical Efficacy

Brentuximab vedotin received FDA approval in August 2011 to treat Hodgkin's lymphoma and systemic anaplastic large cell lymphoma (ALCL). It is indicated in patients with Hodgkin's lymphoma after failure of autologous stem cell transplant (ASCT) or after the failure of at least two prior multi-agent chemotherapy regimens in patients who are not ASCT candidates. It is also indicated to treat patients with systemic anaplastic large cell lymphoma after failure of at least one prior multi-agent chemotherapy regimen. The recommended dose of brentuximab vedotin is 1.8 mg/kg. It is administered as an intravenous infusion over 30 min every 3 weeks. For patients that weigh greater than 100 kg, the dose should be calculated using a weight of 100 kg (Ansell 2011). Patients continue on therapy up to a maximum of 16 cycles, disease progression, or intolerable toxicity. The dose of brentuximab vedotin should be reduced and/or delayed for peripheral neuropathy and neutropenia. Brentuximab vedotin received accelerated FDA approval for treating Hodgkin's lymphoma based on results from two phase II trials (Gopal et al. 2012; Younes et al. 2012). The first was an open-label, single-arm clinical trial which included 102 patients that had relapsed after receiving an autologous stem cell transplant. The overall response rate was approximately 73 %, and 32 % of the patients achieved a complete response. The median duration of response was 6.7 months. The second phase II open-label, single-arm trial was conducted in 58 patients with relapsed systemic anaplastic large cell lymphoma. All patients were relapsed after receiving prior therapy. Of the 58 patients, 72 % were anaplastic lymphoma kinase-negative. The overall response rate was 86 %. The complete response rate was 57 % and partial response rate was 29 %. The median duration of response was 12.6 months (Foyil et al. 2012).

Safety

The administration of brentuximab vedotin did not prolong the mean QTc interval>10 ms from baseline. It was stated that small increases in the mean QTc interval (<10 ms) could not be excluded due to the fact that the study did not include a placebo arm and a positive control arm. The safety of brentuximab vedotin was evaluated during two phase II trials (Furtado and Rule 2012; Skarbnik and Smith 2012; Younes et al. 2010). Brentuximab vedotin monotherapy was administered to 102 patients with relapsed or refractory Hodgkin's lymphoma and 58 patients with relapsed or refractory sACLC at a dose of 1.8 mg/kg every 3 weeks. In Hodgkin's lymphoma patients, the most common treatment-related adverse effects (AEs) of any grade that occurred in>15 % were peripheral sensory neuropathy (43 %), fatigue (40 %), nausea (35 %), neutropenia (19 %), diarrhea (18 %), and pyrexia (16 %). Grade 3 treatment-related AEs that occurred in >1 % of patients were neutropenia (14 %), peripheral sensory neuropathy (5 %), thrombocytopenia and hyperglycemia (3 %), and fatigue (2 %). The only grade 4 treatment-related AEs included neutropenia (4 %), and thrombocytopenia, abdominal pain, and pulmonary embolism that were reported at 1 % each. In the 58 patients with sACLC, the most common treatment-related AEs reported were peripheral sensory neuropathy (41 %), nausea (40 %), fatigue (38 %), pyrexia (34 %), diarrhea (29 %), rash (24 %), constipation (22 %), and neutropenia (21 %). Grade 3 peripheral sensory neuropathy was reported in 17 % of patients and there were no reports of grade 4 or greater treatment-related AEs. Infusion reactions were reported in phase I and II trials. There were two cases of anaphylaxis in phase I trials and 12 % of patients reported grade 1 and 2 infusion-related reactions during phase II trials. The most common adverse reactions that were associated with infusions were chills (4 %); nausea, dyspnea, and pruritus (each at 3 %); and pyrexia and cough (both 2 %).[8] It is currently not recommended to premedicate all patients prior to brentuximab vedotin infusion. If a patient experiences an infusion-related reaction, they should receive premedication prior to subsequent doses, which can consist of acetaminophen, an antihistamine, and a corticosteroid (Furtado and Rule 2012; Gopal et al. 2012; Pro et al. 2012).

In January 2012, the FDA issued a new boxed warning for brentuximab vedotin based on post-marking reports of additional cases of progressive multifocal leukoencephalopathy (PML). Brentuximab vedotin should be held in patients with symptoms of PML and discontinued in patients diagnosed with PML (Wagner-Johnston et al. 2012). The use of brentuximab vedotin is contraindicated due to pulmonary toxicities. The rate of noninfectious pulmonary toxicity was higher with brentuximab given concurrently with bleomycin than the historical incidence of pulmonary toxicity reported with ABVD (Haddley 2012; Minich 2012; Oki and Younes 2012).

MMAE is metabolized in the liver via CYP3A4, and thus, brentuximab vedotin should not be administered with CYP3A4 inhibitors or inducers. Administration with strong inducers or inhibitors can alter the exposure of MMAE.

CLASSES OF MONOCLONAL ANTIBODIES: VASCULAR ENDOTHELIAL GROWTH FACTOR (VEGF) INHIBITORS

Angiogenesis inhibitors have become standard therapies in multiple malignancies. As tumors enlarge, the centers become hypoxic and stimulate angiogenic growth factors, as shown in Fig. 17.4. Vascular endothelial growth factor (VEGF) is thought to be one of the most potent growth factors and has been shown to induce neovascularization for malignant cells in an autocrine fashion. High levels of VEGF have also been correlated with poor prognosis, disease recurrence, and metastases in a variety of neoplasms.

■ Bevacizumab

Pharmacology and Pharmacokinetics

Bevacizumab acts by binding and neutralizing the VEGF-A isoform. The depletion of VEGF downregulates the VEGF/VEGF receptor pathway, resulting in inhibition of new vessel formation and induction of a more normal tumor vasculature pattern leading to decreased vascular permeability (Ferrara 2004). The restoration of a more normal functioning vasculature within the tumor environment may improve delivery of concomitant chemotherapy and oxygen (Jain 2005). Bevacizumab is a humanized anti-VEGF monoclonal IgG_1 antibody. It was developed from a murine anti-human VEGF MAB and is 93 % human and 7 % murine. Similar to other IgG antibodies, a two-compartment model is used to describe the pharmacokinetics of bevacizumab. It has a relatively long elimination half-life and a limited volume of distribution (50–60 mL/kg). The initial half-life is 1.4 days with a prolonged terminal half-life of 20 days (Dirks and Meibohm 2010; Lu et al. 2008). The prolonged terminal half-life of bevacizumab permits for dosing schedules of every 2–3 weeks, allowing for bevacizumab to be dosed on the same schedule as most chemotherapy regimens.

Indications and Clinical Efficacy

Bevacizumab was first approved by the FDA in 2004 for the treatment of metastatic colorectal cancer and has since received approvals for the treatment of metastatic renal cell cancer, nonsquamous non-small cell lung cancer, and glioblastoma multiforme (Braghiroli et al. 2012). Bevacizumab was also granted accelerated approval for the treatment of breast cancer. This approval was subsequently withdrawn after trials failed to demonstrate an overall survival benefit.

In the treatment of metastatic colorectal cancer, bevacizumab is used primarily in addition to combination chemotherapy including the FOLFOX (folinic acid, 5-fluorouracil, oxaliplatin) and FOLFIRI (folinic acid, 5-fluorouracil, irinotecan) regimens (Braghiroli et al. 2012). The addition of bevacizumab at a dosage of 5 mg/kg every 2 weeks to first-line chemotherapy in metastatic setting improves response rates, progression-free survival, and overall survival. Data also support the use of bevacizumab after failure of first-line chemotherapy in patients that have not yet received bevacizumab.

Lung cancer remains the number one cause of cancer-related death in the United States with over 150,000 attributable deaths annually. Non-small-cell lung cancer (NSCLC) accounts for 85 % of lung cancer cases and approximately 70 % of cases present with advanced disease. VEGF is among the proangiogenic factors contributing to blood vessel growth in NSCLC

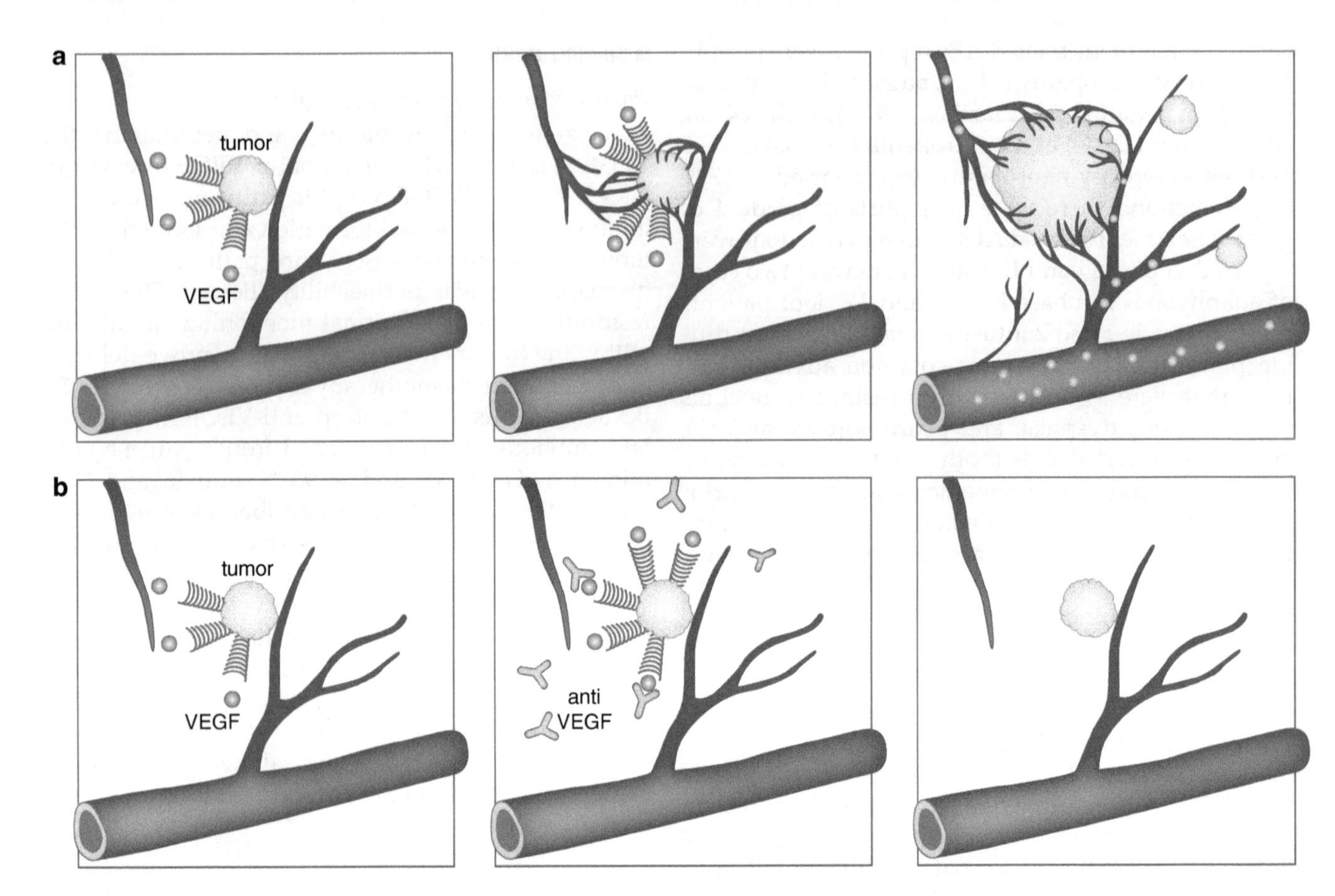

Figure 17.4 ■ Mechanism of action of vascular endothelial growth factor (VEGF) (**a**) and theorized mechanism of action of VEGF inhibitors (**b**) (*Source*: AJHP and GlaxoSmithKline).

(Planchard 2011). A phase II randomized clinical trial in patients with previously untreated advanced and recurrent NSCLC provided the first clinical evidence of benefit of antiangiogenic treatment in patients with NSCLC. Patients were randomized to one of three treatment arms: carboplatin plus paclitaxel, carboplatin plus paclitaxel with bevacizumab 7.5 mg/kg every 3 weeks, or carboplatin plus paclitaxel with bevacizumab 15 mg/kg (Johnson et al. 2004). Carboplatin and paclitaxel were continued for a maximum of four cycles, and bevacizumab was continued until disease progression or intolerable toxicity. The bevacizumab 15 mg/kg dose added to carboplatin and paclitaxel increased the response rate from 18.8 to 31.5 % and prolonged the median progression-free survival time by approximately 3 months. The patients with NSCLC of squamous cell histology tumors were noted to be at an excessive risk of bleeding complication from bevacizumab. The bleeding risk was partially attributed to the central location typical of squamous cell histology tumors. Patients with squamous cell histology were excluded from further clinical trials and it remains a contraindication for treatment with bevacizumab.

The phase III trial, E4599, was a landmark trial in the treatment of NSCLC (Sandler et al. 2006). It randomized patients with nonsquamous NSCLC to treatment with carboplatin and paclitaxel with or without bevacizumab at a dosage of 15 mg/kg every 3 weeks. Patients treated with bevacizumab had a significantly higher response rate (35 % vs 15 %), progression-free survival (6.2 months vs 4.5 months), and overall survival (12.3 months vs 10.3 months). A second phase III trial, AVAiL, evaluated bevacizumab (either 7.5 or 15 mg/kg every 3 weeks) in addition to a cisplatin plus gemcitabine chemotherapy regimen (Reck et al. 2012). The original primary endpoint was overall survival, but this was changed to progression-free survival after results from E4599 were released. The trial met its endpoint with a marginal improvement in progression-free survival of 0.6 months for the low-dose group and 0.4 months for the high-dose group. Overall survival was not significantly improved; however, the trial was underpowered for an adequate survival analysis. The results of this trial failed to confirm the findings of E4599. Further investigations continue to define the role of bevacizumab in the treatment of nonsquamous NSCLC.

Malignant gliomas account for approximately 70 % of malignant primary brain tumors, and glioblastoma multiforme (GBM) accounts for 60–70 % of malignant gliomas (Wen and Kesari 2008). In patients with newly diagnosed GBM, the 5-year survival is consistently less than 5 %. GBMs are characteristically highly vasculature tumors suggesting a role for antiangiogenic therapies. Bevacizumab received accelerated approval for the treatment of GBM on the basis of randomized phase 2 trials. The BRAIN trial randomized 167 patients to bevacizumab 10 mg/kg every 2 weeks alone or the same dosage combined with irinotecan (Friedman et al. 2009). The overall response rate was 28.2 and 37.8 % in the bevacizumab and bevacizumab plus irinotecan arms, respectively. The median duration of progression-free survival was 4.2–5.6 months, respectively. Overall survival at 12 months was 38 % in both groups. A second phase II trial randomized patients to treatment with bevacizumab 10 mg/kg every 2 weeks in combination with either temozolomide or irinotecan. Approximately 40 % of patients in both treatment arms remained progression-free at 6 months. Safety data specific to this population of patients suggest that the rate of hemorrhage is significantly increased by the concomitant use of anticoagulants. Additionally, preoperative bevacizumab may significantly impair wound healing after second and third craniotomy even if surgery is more than 4 weeks after the last dose of bevacizumab (Clark et al. 2011). While definitive efficacy data endpoints of bevacizumab in GBM compared to other standard therapies remain to be seen, the current data available was sufficient to warrant FDA approval. The European Medicines Agency did not consider the current data sufficient owing to a lack of appropriate controls in trials to date and insufficient data correlating overall response rate to longer-term benefit such as improvement in overall survival (Specenier 2012).

Safety

Common adverse events attributable to bevacizumab include hypertension, proteinuria, and increased chemotherapy-induced neutropenia (Braghiroli et al. 2012; Specenier 2012). Although occurring less commonly, bevacizumab contributes to an increased risk for both thrombotic events and bleeds. Most commonly bleeding events that do occur are minor and limited to epistaxis. Because bevacizumab may delay wound healing, it is recommended to allow a minimum of 28 days from the last dosing of bevacizumab prior to any major surgical procedure.

CLASSES OF MONOCLONAL ANTIBODIES: ENDOTHELIAL GROWTH FACTOR RECEPTOR (EGFR) INHIBITORS

The ErbB family of tyrosine kinase receptors consists of four members: EGFR (erbB-1, Her-1), Her-2/neu (erB-2/neu), erbB-3 (Her-3), and erbB-4 (Her-4). Signaling via EGFR and Her-2 in tumor cells is responsible for diverse cellular functions including proliferation, survival, adhesion, and DNA damage repair. Overexpression of EGFR family receptors has been noted in a wide variety of human cancers including breast, colon, gastric, rectal, lung, and squamous cell cancers of the head and neck. Four available monoclonal antibodies target members of this receptor family: trastuzumab and pertuzumab target the Her-2/neu receptor, while cetuximab and panitumumab are selective for EGFR.

Trastuzumab

Pharmacology and Pharmacokinetics

Trastuzumab is a recombinant DNA-derived humanized MAB (IgG_1 kappa) that selectively binds to the extracellular domain of the human Her-2 receptor with high affinity. This receptor-antibody interaction, through a series of other cellular actions, induces autophosphorylation of the tyrosine kinase internal domain resulting in decreased tumorigenic potential and possibly reversal of chemoresistance. Weekly administration of trastuzumab exhibits dose-dependent pharmacokinetics (Baselga et al. 1999; Leyland-Jones et al. 2003). When 10–500 mg doses (administered by short-duration intravenous infusions) were studied in women with metastatic breast cancer, the mean half-life increased and clearance decreased with increasing doses. The observed average half-life was 1.7 days for the 10 mg dose and 12 days for the 500 mg dose. However, in studies analyzing the commonly used regimen for trastuzumab of an initial loading dose of 4 mg/kg followed by a 2 mg/kg weekly maintenance dose, the mean half-life was 5.8 days. Studies also suggest that age and serum creatinine do not affect the disposition of trastuzumab. It is also important to note that when trastuzumab is administered in combination with paclitaxel, a 1.5-fold elevation in serum concentrations of trastuzumab is observed as compared to when trastuzumab is administered in combination with anthracycline and cyclophosphamide (Fig. 17.5).

Indications and Clinical Efficacy

Trastuzumab is indicated for the treatment of Her-2-positive breast cancer in the adjuvant and the metastatic setting and is also indicated for the treatment of Her-2-positive gastric cancer (Andersson et al. 2011; Bang et al. 2010; Guarneri et al. 2008; Seidman et al. 2008; Valero et al. 2011). In patients with breast cancer expressing Her-2, the addition of trastuzumab treatment for 1 year to adjuvant therapy improves overall survival. In the metastatic disease, adding trastuzumab to therapy will also increase disease response and improve survival. With the recent addition of pertuzumab, most metastatic breast cancer patients that would receive trastuzumab will now receive both therapies, as outlined in the following section on pertuzumab. The Her-2 receptor is also

How Herceptin® slows cancer's growth

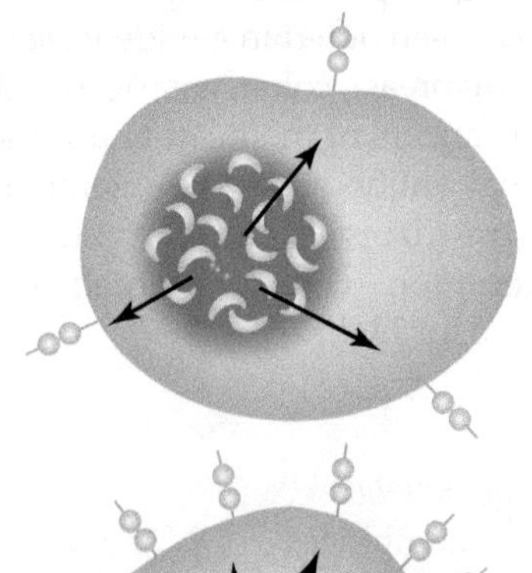

Normal cell
in a normal breast tissue cell, the HER-2 gene produces protein receptors on the cell surface. The receptors signal the cell to divide and multiply appropriately.

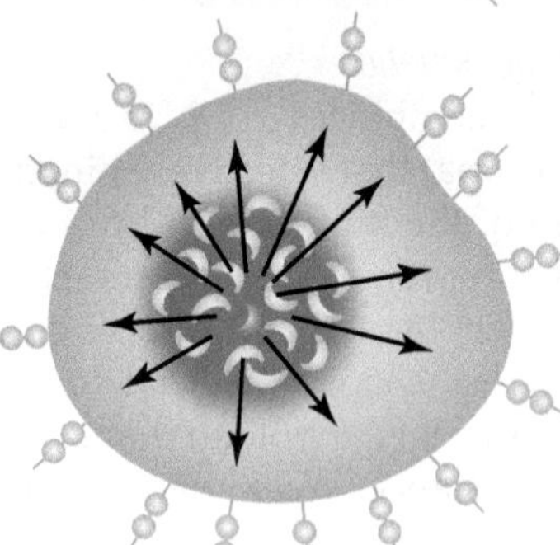

Cancer cell producing too much HER-2 protein
extra protein receptors trigger cells to grow uncontrollably, causing aggressive, hard-to-treat breast cancers. Excess HER-2 is found in 25 –30 % of advanced breast cancers.

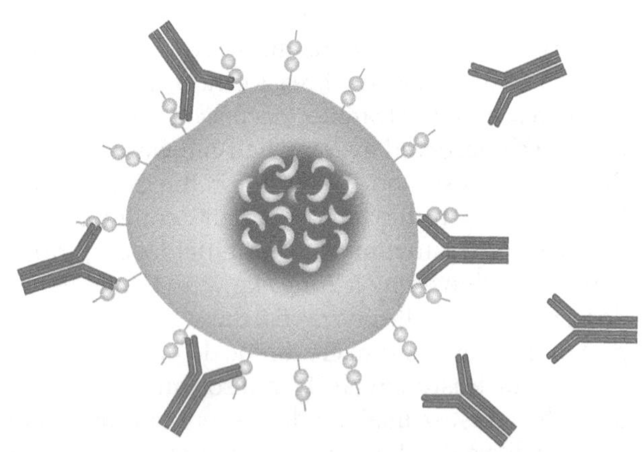

Treatment with Herceptin
Herceptin antibodies latch on to HER-2 protein receptors, blocking the sites and shutting down the excess growth signals. HER-2 treatment has extended survival of some patients with advanced breast cancer.

Figure 17.5 ■ Trastuzumab mechanism of action (*Source*: Dana-Farber Cancer Institute).

expressed on approximately 20 % of gastric cancers and the addition of trastuzumab to chemotherapy for gastric cancer has been demonstrated to improve overall survival by just under 3 months (Bang et al. 2010).

Safety

Trastuzumab has a black box warning for cardiomyopathy because of its potential to cause ventricular dysfunction and congestive heart failure. The severity and occurrence of cardiomyopathy was higher in patients who received anthracyclines and cyclophosphamide in combination with trastuzumab. Patients who require trastuzumab therapy must receive a full cardiac workup prior to the initiation of therapy and left ventricular function must be monitored during treatment. The most common adverse reaction is infusion reactions (usually mild to moderate), which rarely require discontinuation of therapy. Other adverse effects associated with trastuzumab are anemia and leukopenia, nausea/vomiting, diarrhea, and upper respiratory infections (Andersson et al. 2011; Guarneri et al. 2008; Seidman et al. 2008; Valero et al. 2011).

■ Pertuzumab

Pharmacology and Pharmacokinetics

Pertuzumab is a recombinant humanized MAB that selectively binds to the extracellular domain of the human HER-2 receptor. In contrast to trastuzumab, which binds to domain IV, pertuzumab binds to domain II and inhibits heterodimerization of HER2 with other HER family members. This inhibition prevents ligand activated HER activation and subsequent activation of intracellular pathways associated with proliferation and survival of cancer cells (Keating 2012). Pertuzumab has demonstrated activity against a number of cancer cell lines (breast, lung, ovarian, prostate, and colorectal). Tumor growth inhibition of >80 % was seen in breast cancer cell lines exposed to 5–25 µg/mL (Keating 2012) Improved anticancer activity was observed in patients treated with pertuzumab in combination with trastuzumab compared to either drug alone (Cortes et al. 2012).

A linear two-compartment model is used to describe the pharmacokinetics of pertuzumab (Ng et al. 2006). The mean volume of distribution at steady state is 4.23 L, with a clearance of 169 mL/day following a 420 mg dose. The mean elimination half-life is 19 days, allowing for every 3 week dosing. Simulated serum concentrations were similar with weight based and fixed dosing. A fixed dose regimen of an 840 mg loading dose followed by a maintenance dose of 420 mg every 3 weeks results in C_{min} serum concentration greater than the target of 20 µg/mL (Ng et al. 2006; Yamamoto et al. 2009). Steady-state is reached after the first maintenance dose (Keating 2012).

Indications and Clinical Efficacy

Pertuzumab is indicated for the treatment of HER-2 positive metastatic breast cancer in combination with trastuzumab and docetaxel for patients who have not received prior anti-HER2 therapy or chemotherapy for metastatic disease. In a randomized, controlled clinical trial of HER2 positive, recurrent or unresectable metastatic breast cancer patients, the addition of pertuzumab to trastuzumab and doxetaxel resulted in an increase of progression-free survival of 6.1 months and an overall survival benefit, HR = 0.66 (95 %, CI 0.52–0.89, $p = 0.0008$) (Baselga et al. 2012; Swain et al. 2013).

Patients discontinuing docetaxel due to adverse events could continue on treatment with the combination of pertuzumab and trastuzumab. Subset analysis revealed a benefit regardless of hormone receptor status, and regardless of prior adjuvant or neoadjuvant therapy.

Safety

Overall, pertuzumab appears well tolerated, adding minimally to the toxicity of docetaxel and trastuzumab (Baselga et al. 2012). Adverse events occurring in >5 % of patients in the pertuzumab arm compared to placebo include: diarrhea, rash, mucosal inflammation, febrile neutropenia, and dry skin. Regular cardiac ejection fraction monitoring (every 3 months) is recommended, mostly due to the concomitant administration of trastuzumab, as pertuzumab did not add to the incidence of left ventricular systolic dysfunction. While infusion-related reactions have been reported, the routine use of premedication is not needed unless an infusion-related reaction has occurred and the patient is being re-challenged with pertuzumab treatment.

Cetuximab

Pharmacology and Pharmacokinetics

Cetuximab is a chimeric IgG_1 MAB composed of the Fv region of a murine anti-EGFR antibody with human IgG heavy and kappa light chain constant regions. It binds to the extracellular domain of EGFR with an affinity 5–10-fold greater than that of endogenous ligands (epidermal growth factor, amphiregulin, transforming growth factor). Cetuximab exerts its activity by blocking endogenous ligands, with a resultant inhibition of EGFR signaling. It also induces internalization of the receptor, which can further downregulate EGFR signaling. Further, binding of cetuximab to EGFR can target cytotoxic immune effector cells toward the EGFR-expressing tumor cells (Fig. 17.6).

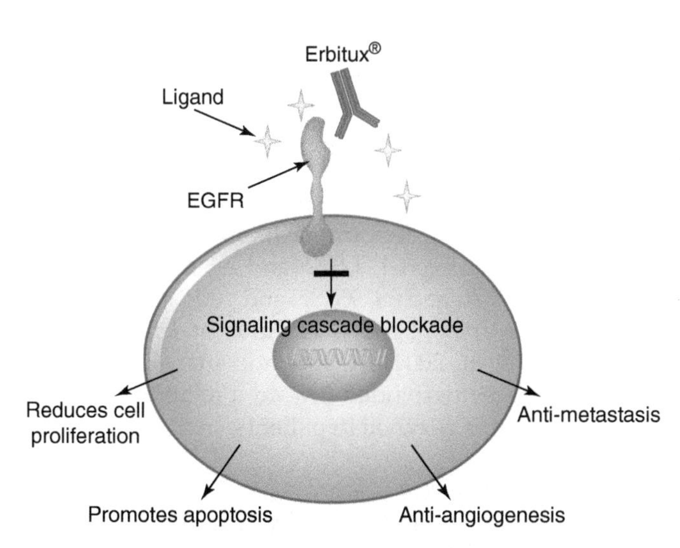

Figure 17.6 ■ Cetuximab mechanism of action. Cetuximab targets EGFR and binds to EGFR with higher affinity than its natural ligands. Binding results in the internalization of the antibody receptor complex without activation of the intrinsic tyrosine kinase. Consequently, signal transduction through this cell pathway is blocked, which inhibits tumor growth and leads to apoptosis. Other mechanisms of action include the inhibition of the production of angiogenic factors and synergistic activity with both radiotherapy and chemotherapy (*Source*: Merck KGaA).

Cetuximab pharmacokinetics were best described by a 2-compartment model with a saturable Michaelis-Menten-type elimination (Dirks et al. 2008). The AUC shows a greater than proportionate increase when the dose is increased from 20 to 400 mg/m^2. The drug clearance decreased from 0.08 to 0.02 $L/h/m^2$ with 20 and 200 mg/m^2 doses, with little change at higher doses. Steady-state levels were reached by week three of cetuximab infusions with a 114-h mean half-life when the recommended regimen of 400 mg/m^2 (loading dose) followed by weekly 250 mg/m^2 was administered (Delbaldo et al. 2005; Shirao et al. 2009). An extended dosing interval of 500 mg/m^2 every 2 weeks has been evaluated and demonstrated overall similar exposure and trough levels of cetuximab. The every 2-week dosing regimen may improve patient convenience, particularly when combined with every 2-week chemotherapy regimens. However, weekly dosing remains recommended in treatment guidelines because this dosing was used in the phase III trials.

Indications and Clinical Efficacy

Cetuximab is indicated in the treatment of metastatic colorectal cancer and squamous cell cancers of the head and neck. In the pivotal BOND trial, 329 patients with metastatic EGFR-expressing chemo-refractory CRC were randomized to receive cetuximab plus irinotecan or cetuximab alone (Cunningham et al. 2004). Both study arms received a 400 mg/m^2 loading dose followed by 250 mg/m^2 weekly until either the patient had intolerable toxicities or disease progression occurred. The overall response rate was 22.9 % in patients receiving combination therapy (irinotecan plus cetuximab) and 10.8 % in the cetuximab monotherapy group. In addition, two prespecified subpopulations were analyzed: an irinotecan-oxaliplatin failure group (irinotecan refractory patients who had previously been treated with and failed an oxaliplatin-containing regimen) and the irinotecan refractory group. The irinotecan-oxaliplatin failure subpopulation had a 23.8 % response rate and a 2.9 month median time to disease progression for the cetuximab plus irinotecan study arm, and an 11.4 % response rate and 1.5 month time to progression for the cetuximab monotherapy study arm. There was no correlation between the level of the EGFR expression and response rate. However, KRAS mutation has since been discovered to be a predictive marker of benefit to treatment with EGFR-targeting MABs in patients with colorectal cancer.

KRAS is a G protein downstream of EGFR and is a critical component of EGFR signaling, propagating EGFR signaling events. Mutations in exon 2 or KRAS occurs in approximately 40 % of patients with colorectal cancer and can cause KRAS to be constitutively activated, which can render EGFR-targeting drugs ineffective (Karapetis et al. 2008). The CRYSTAL trial was a randomized, open-label, phase III trial in the first-line treatment setting that randomized metastatic colorectal cancer patients to treatment with infusional 5-FU, folinic acid, and irinotecan (FOLFIRI) alone or FOLFIRI plus cetuximab. In the patients testing positive for KRAS mutation, the progression-free survival was 7.4 and 7.7 months in the cetuximab and control arms, respectively. In the KRAS wild-type patients, the progression-free survival was 9.9 and 8.4 months in the cetuximab and control arms, respectively. Patients with colorectal cancer considered to be potential candidates for cetuximab should have tumor tissue tested for KRAS mutation status. If a mutation is present, the patient will not benefit from treatment with cetuximab.

Cetuximab is also used in the treatment of squamous cell cancers of the head and neck (SCCHN) in both patients with locoregionally advanced disease and in patients with recurrent and/or metastatic disease. The KRAS mutation rarely occurs in SCCHN and testing for it in this population is not recommended. In a phase III trial, cetuximab plus radiation therapy was compared to radiation alone in patients with squamous cell carcinoma of the oropharynx, hypopharynx, or larynx (Bonner et al. 2006; Specenier and Vermorken 2011). The standard dose of cetuximab, 400 mg/m^2 followed by 250 mg/m^2 weekly, was administered concomitantly with radiation therapy; control group patients received radiation therapy alone. Cetuximab prolonged the duration of locoregional disease control from 14.9 to 24.4 months. Median overall survival was improved from 29.3 to 49 months with cetuximab. The reported rates of radiation dermatitis were similar among the treatment groups in this trial. However, there have been subsequent reports of radiation dermatitis, including skin necrosis. Some authors have reported a 10-fold increase in the rates of radiation dermatitis with cetuximab and radiation compared to radiation alone (Koutcher et al. 2009; Studer et al. 2011). The standard chemotherapy and radiation treatment is a combination of cisplatin and radiation; a comparison of cetuximab and radiation to a cisplatin-treated control group is needed to determine which option is preferable.

Safety

Common adverse events associated with cetuximab include acneiform rash, diarrhea, hypomagnesemia, hypocalcemia, and infusion reactions (Specenier and Vermorken 2011). Acneiform rash develops in approximately 70–80 % of patients and is most prevalent on the trunk and face. Several treatment algorithms have been proposed and generally consist of treatments used for acne (Pinto et al. 2011). Mild rashes are treated with topical antibiotics (e.g., clindamycin gel or lotion). Treatment may be escalated to a course of treatment with oral antibiotics, primarily tetracyclines, if needed. For severe rashes, oral antibiotics in combination with a pulse of corticosteroids may be needed and treatment with cetuximab may need to be delayed until the rash resolves to a lower grade. Avoiding skin irritants, sun exposure, perfumes, and harsh cleansers may help to prevent and lessen the rash. Infusion reactions can be severe and occur in approximately 2–5 % of patients. Premedication with an antihistamine is required, and European product specifications also recommend the use of a corticosteroid premedication.

■ Panitumumab

Pharmacology and Pharmacokinetics

Panitumumab is a fully human, IgG_2 kappa MAB to the extracellular domain of EGFR. Although panitumumab shares the same target as cetuximab, the two monoclonal entities have two main differences. First, panitumumab is fully humanized, reducing the risk of immunogenicity. Second, in vitro studies have shown that panitumumab has a stronger affinity and specificity for EGFR compared to cetuximab (Cohenuram and Saif 2007).

In terms of pharmacodynamics, one preclinical study demonstrated that the tumor growth inhibition of panitumumab was related to a threshold EGFR level, where xenografts that contained at least 17,000 receptors per cell were treatable with panitumumab while those with less than 11,000 were not treatable (Yang et al. 2001). Similar to other EGFR inhibitors in clinical trials, increasing dose corresponded to an increasing frequency of an acneiform rash. In a phase I trial, patients that received 1, 1.5, 2, and 2.5 mg/kg weekly of panitumumab had 68, 95, 87, and 100 % incidence of rash with its apex at 3–5 weeks after starting therapy (Rowinsky et al. 2004). A post hoc analysis revealed that rash intensity trended toward a relationship with progression-free survival in patients.

Indications and Clinical Efficacy

Panitumumab is approved by the FDA for the treatment of EGFR-expressing, metastatic CRC with disease progression on or following fluoropyridine-, oxaliplatin-, and irinotecan-containing chemotherapy regimens. A randomized phase III trial compared supportive care alone or treatment with panitumumab (6 mg/kg intravenously) every 2 weeks in 463 patients with previously treated metastatic colorectal cancer;

prior treatment with cetuximab was not allowed. The mean progression-free survival was 96 days for the panitumumab group and 60 days for the other group (Saif and Cohenuram 2006). Eight percent of patients in the treatment arm exhibited a partial response and no observable response was observed in the control arm. There was no difference in overall survival between the two groups, although this may have been confounded by a significant proportion of patients from the BSC group that later crossed over to receive panitumumab.

The Panitumumab Advanced Colorectal Cancer Evaluation Study (PACCE) evaluated first-line FOLFOX with bevacizumab and FOLFIRI with bevacizumab with or without panitumumab. The trial was discontinued after an interim analysis indicated that the addition of panitumumab shortened progression-free survival and increased toxicity (Saif and Cohenuram 2006), suggesting that the addition of EGFR inhibition to anti-vascular-targeted therapy was detrimental.

Safety

Panitumumab has a black box warning for severe dermatologic toxicities and infusion reactions. Severe skin toxicities (grade 3 or higher) were reported in 12 % of patients to include acneiform dermatosis, pruritus, erythema, rash, and skin exfoliation. Severe infusion reactions (occurring in 1 % of patients) were also observed and include anaphylactic reactions, bronchospasms, fever, chills, and hypotension. However, the infusion reactions are less frequent than with cetuximab and no premedication is required (Cohenuram and Saif 2007; Saif and Cohenuram 2006). Pulmonary fibrosis has also occurred with panitumumab therapy. The more common adverse reactions include abdominal pain, hypomagnesemia, acneiform eruption (occurring in greater than 70 % of patients) and other skin rashes, paronychia, fatigue, nausea, vomiting, and diarrhea (Cohenuram and Saif 2007; Saif and Cohenuram 2006).

CLASSES OF MONOCLONAL ANTIBODIES: ANTIHUMAN CYTOTOXIC T-LYMPHOCYTE ANTIGEN 4 (CTLA-4)

The adaptive immune response to both pathogens and tumors is complex and multifactorial. Both CD4 helper and CD8 cytotoxic T cells contribute to adaptive immunity. Full T-cell activation requires multiple cosignals including major histocompatibility complex and costimulatory molecules (B7.1 and/or B7.2) (Lipson and Drake 2011). Human cytotoxic T-lymphocyte antigen 4 (CTLA-4) (CD152) plays a critical role in T-cell activation (Thompson and Allison 1997). CTLA-4 knockout mice expire of lymphoproliferative disorders within 3–4 weeks of birth (Waterhouse et al. 1995). CTLA-4 blockade leads to increased T-cell activation and blockade of CTLA-4 with a monoclonal antibody, ipilimumab, which became the first strategy to provide improvements in overall survival in patients with metastatic melanoma.

■ Ipilimumab

Pharmacology and Pharmacokinetics

Ipilimumab is a monoclonal antibody of the IgG_1 isotype that binds to human and cynomolgus CTLA-4 (Hanaizi et al. 2012). In preclinical animal models and toxicology studies, the major adverse reactions were immune mediated and included colitis, dermatitis, and infusion reactions. The pharmacokinetics of ipilimumab evaluated 498 patients with advanced melanoma. Doses ranged from 0.3 to 10 mg/kg administered once every 3 weeks for 4 doses. In this dose range, Cmax, Cmin, and AUC were dose proportional. Steady state was reached at the third dose. While clearance increases with increasing body weight and increasing lactate dehydrogenase (LDH), dose adjustments beyond the mg/kg dosing are not recommended (Hanaizi et al. 2012). The pivotal trials that provided the bulk of the clinical safety and efficacy data used a dosage of 3 mg/kg every 3 weeks for four doses, while other trials have used doses of 10 mg/kg; current clinical usage of ipilimumab is with the 3 mg/kg dosing (Hanaizi et al. 2012).

Indications and Clinical Efficacy

The first phase III trial to demonstrate an overall survival benefit in advanced melanoma was a 3-arm randomized controlled trial evaluating an experimental cancer vaccine, gp100, ipilimumab, or the combination of gp100 and ipilimumab (Hodi et al. 2010). Median overall survival was approximately 10 months in the patients receiving ipilimumab, which was a significant improvement compared to the 6.4 months in patients receiving gp100 monotherapy. A second phase III trial randomized patients to dacarbazine (850 mg/m^2) or dacarbazine plus ipilimumab. This trial used a dose of ipilimumab of 10 mg/kg (Robert et al. 2011). Patients receiving ipilimumab had an improvement in overall survival of approximately 2.1 months. No data regarding use in ocular melanoma are available yet. The addition of ipilimumab to the available standard of care treatments expands options available to patients and is the only therapy with a proven survival benefit without requiring selection based on B-Raf mutation.

Safety

The most frequently observed adverse events include diarrhea, rash, pruritus, and fatigue. Immune-mediated adverse events primarily involve the gastrointestinal tract (colitis) and the skin (rash and pruritus) (Hodi et al. 2010). Less frequently, the liver, endocrine glands, and nervous system may also be involved. For mild and moderate symptoms, gastrointestinal and skin toxicities are generally managed symptomatically (Hanaizi et al. 2012). More severe symptoms may require a course of corticosteroids. This should be avoided if possible because of the potential to lessen the efficacy of the therapy. Hypopituitarism was reported in 4 % of patients and adrenal insufficiency, and hypothyroidism were each reported in 2 % of patients.

CLASSES OF MONOCLONAL ANTIBODIES: RECEPTOR ACTIVATOR OF NUCLEAR FACTOR KAPPA BETA LIGAND INHIBITOR

■ Denosumab

In breast and prostate cancer, nearly two-thirds to three-fourths of advanced disease will metastasize to the bone and approximately one-third of lung cancer patients will develop metastatic bone disease (Coleman 2004a, b). Bone metastases cause considerable skeletal morbidity and can reduce the patients' quality of life and decrease survival. Skeletal-related events (SREs) is a term that is used to describe a collection of adverse events associated with bone metastases and include pathologic fractures, the need for surgery or radiation to the bones and spinal cord and nerve root compression, and hypercalcemia of malignancy. Approximately half or more of patients with metastatic bone disease from breast, prostate, and lung cancer will experience an SRE within 2 years (Coleman 2004a, b).

For the last decade, bisphosphonate therapy with either pamidronate or zoledronic acid has been the cornerstone of treatment of bone metastases. Zoledronic acid is the only bisphosphonate to be evaluated for the prevention of skeletal complications in patients with bone metastases secondary to solid tumors other than breast or prostate cancer. It has been considered the standard of care for patients with breast cancer and the only bisphosphonate to show efficacy in patients with prostate cancer. Because of this, bisphosphonate therapy is the standard against which all novel treatments for bone metastases are compared. Because of poor bioavailability, bisphosphonate treatments for bone metastases are only available intravenously.

The receptor activator of nuclear factor κB ligand (RANKL) is produced by osteoblasts (and possibly osteoclasts as well) and binds to and activates RANK on osteoclasts. Once activated, RANK promotes osteoclast formation, function, and survival, which in turn promotes bone resorption. Denosumab binds to RANKL and prevents activation of RANK by RANKL thereby reducing bone resorption. In addition, the inhibition of bone resorption by osteoclasts reduces the production of growth factors that could enhance tumor growth as well as the release of osteoblast-inducing growth factors by the tumor (Romas 2009).

Denosumab is a fully humanized immunoglobulin G_2 (IgG_2) MAB that binds with high affinity and specificity to RANKL. The FDA approval of denosumab expanded the options available for the treatment of metastatic bone disease in patients with solid tumors.

Pharmacology and Pharmacokinetics

At doses below 60 mg, denosumab displays nonlinear pharmacokinetics but approximately dose-proportional increases in exposure at higher doses. Denosumab can be detected in the serum within an hour after subcutaneous administration. After a single dose, it can still be detected in the serum for up to 9 months. After multiple subcutaneous injections of denosumab at the approved dose of 120 mg every 4 weeks in patients with cancer and metastatic bone disease, up to 2.8-fold accumulation in serum denosumab concentrations was observed. Steady state was reached by 6 months and the mean serum trough concentration was 20.5 ± 13.5 μg/mL with a mean elimination half-life of 28 days (Gibiansky et al. 2012). In a population pharmacokinetic analysis of data from 14 clinical trials that included 1,076 individuals, the absolute bioavailability was 61 % following a subcutaneous injection and a mean absorption half-life of 2.7 days. Central volume of distribution was reported to be 2.61 L and the linear clearance was 3.3 mL/h. The clearance and volume of distribution were proportional to body weight. The steady-state exposures following repeat subcutaneous administration to 45 and 120 kg subjects were 48 % higher and 46 % lower, respectively, than exposure of the typical 66 kg individual. The linear fraction of the elimination of denosumab is thought to be mediated via the reticuloendothelial system.

Indications and Clinical Efficacy

Three phase III trials were conducted to review the efficacy of denosumab on SREs in patients with bone metastases (Henry et al. 2011; Stopeck et al. 2010). The three studies evaluated patients with breast cancer, prostate cancer, other solid tumors, and multiple myeloma. The same randomized, active-controlled, double-blind, double-dummy study design was utilized for all three trials. Patients were randomized in a 1:1 fashion to receive denosumab 120 mg subcutaneously plus intravenous placebo or intravenous zoledronic acid 4 mg plus subcutaneous placebo (with the dose adjusted for creatinine clearance) every 4 weeks. In the breast cancer and prostate cancer study, denosumab was noninferior with a trend to superior for the primary endpoint of time to first SRE as compared to zoledronic acid. In the third study, denosumab was noninferior to zoledronic acid for time to first SRE. There was no difference in overall survival or disease progression within the two groups; however, in the subset of patients with multiple myeloma, it appeared to have a decrease in overall survival in the denosumab group. Therefore, denosumab is not indicated for the prevention of SREs in patients with multiple myeloma.

Safety

In the phase III clinical trials comparing denosumab to zoledronic acid in patients with bone metastases, there was a similar overall incidence of adverse events, including severe and serious adverse events (Henry et al. 2011; Stopeck et al. 2010). Over 96 % of patients in both arms reported an adverse event. However, patients were receiving concomitant medications, such as chemotherapy, and many of the adverse events reflected toxicities associated with the other medications. Adverse events of interest attributed to denosumab included infections, hypocalcemia, renal toxicity, acute-phase reaction, and osteonecrosis of the jaw (ONJ). When compared to zoledronic acid, denosumab was similar in infectious adverse events, with 43.4 % in the denosumab group as compared to 42.9 % in zoledronic acid. More patients receiving denosumab developed hypocalcemia (9.6 % vs 5.0 %) and there were similar rates of ONJ (1.8 % vs 1.3 %). On the other hand, more patients who received zoledronic acid experienced acute-phase reactions (20.2 % vs 8.7 %) as well as renal toxicities (11.8 % vs 9.6 %).

There are no contraindications to the use of denosumab, but data do not currently support its use in patients with multiple myeloma. Calcium levels, if low, should be corrected prior to the initiation of denosumab and monitored throughout therapy. Patients should be informed of the symptoms of ONJ which can manifest as jaw pain, osteomyelitis, osteitis, bone erosion, tooth or periodontal infection, toothache, gingival ulceration, or gingival erosion. An oral examination should be preformed and all appropriate preventive dentistry should take place prior to initiation of denosumab. Once a patient is on denosumab, it is recommended that they have periodical oral examinations (Henry et al. 2011; Stopeck et al. 2010).

CONCLUSION

MABs have become a cornerstone in the clinical management of a variety of solid and hematologic cancers. With a variety of different antibodies or antibody-based molecules in different stage of preclinical and clinical development, this role of MABs in cancer therapy will likely be further expanded over the next decade and will provide additional new benefits for patients suffering from cancer.

SELF-ASSESSMENT QUESTIONS

■ Questions

1. From the name "tositumomab and 131I tositumomab," what can one infer about the type of drug and origin?
2. The epidermal growth factor receptor inhibitors have a unique side effect profile which may also demonstrate a pharmacodynamic effect. Describe the profile and what development of this side effect may mean in terms of treatment effectiveness.
3. Describe the theory of angiogenesis and how vascular endothelial growth factor inhibitors may counteract this important mechanism of cancer development.
4. Bevacizumab is an angiogenesis inhibitor. List what is known about bevacizumab in terms of thrombotic and bleeding concerns. Are there any guidelines on duration of time between bevacizumab use after major surgery? What are they?
5. Describe the clinical literature that supported the FDA approval of panitumumab? What indication does it currently have?
6. Keeping rituximab's mechanism of action in mind, which of the following disease states would rituximab likely not show any benefit and why?
   (a) Autoimmune hemolytic anemia
   (b) Cutaneous T-cell lymphoma
   (c) Immune thrombocytopenic purpura
   (d) Rheumatoid arthritis
7. Which of the epidermal growth factor inhibitors require that patients test positive for the KRAS wild-type?
8. Trastuzumab and pertuzumab require a positive test for the expression of which receptor?
9. List the three black box warnings associated with rituximab use.

■ Answers

1. From the name "tositumomab and 131I tositumomab," one can infer several characteristics of this drug. First, it is a monoclonal antibody of murine origin, as designated by its suffix of "omab." Second, the drug is conjugated or radiolabeled since the drug name contains a second word containing one of the periodic elements.
2. The epidermal growth factor receptors (EGFR) inhibitors all share a common side effect profile that is dermatologic in origin. Generally, patients will present with an acneiform rash that cannot be successfully treated with over-the-counter acne agents. This rash is due to the fact that EGFR is overexpressed in many cancers as well as normal skin and hair follicles. Therefore, in some cases, the development of a rash may be associated with clinical efficacy of the drug.

3. For tumors to grow larger than 2 mm^3, they must begin to grow their own blood supply, both to provide oxygen and carry away wastes, a process known as angiogenesis. Several growth factors are necessary to stimulate angiogenesis; one of the most potent is vascular endothelial growth factor (VEGF). Bevacizumab is a VEGF inhibitor that prevents VEGF from binding to receptors, which subsequently prevents angiogenesis and tumor growth. Many think that VEGF inhibitors will be very successful in early stage disease where they can prevent large tumor growth, although bevacizumab is currently used more in a metastatic and late stage setting.
4. Evidence suggests that at least 28 days must elapse between a major surgery and subsequent bevacizumab administration. This is because antiangiogenesis inhibitors are associated with vascular dysfunction by their mechanism of action. There have been wound healing concerns, excessive bleeding, and even clotting concerns with the use of bevacizumab in clinical trials. Concomitant use of warfarin was shown to be safe in one recent clinical trial.
5. Panitumumab is currently indicated for "the treatment of patients with EGFR-expressing, metastatic colorectal carcinoma with disease progression on or following fluoropyrimidine-, oxaliplatin-, and irinotecan-containing chemotherapy regimens" (i.e., third- or fourth-line use). This approval was based on a phase III trial ($n=463$) that randomized patients to receive panitumumab monotherapy or best supportive care. The mean progression-free survival was 96 days for the panitumumab group and 60 days for the BSC alone group. Eight percent of patients in the treatment arm exhibited a partial response and no observable response was detected in the control arm.
6. Cutaneous T-cell lymphoma. This is because rituximab is a chimeric monoclonal antibody that binds to the antigen CD20 (cluster of differentiation 20), which is found on B lymphocytes (B cells).
7. In the treatment of colorectal cancer, both panitumumab and cetuximab are rendered ineffective by the activating KRAS mutation. Therefore, tumor tissue should be tested for KRAS and only those patients with a wild-type KRAS are likely to benefit from EGFR inhibitor MAB therapy.
8. Use of trastuzumab and pertuzumab requires a positive test for the HER-2/neu protein (i.e., a positive result either on fluorescence in situ hybridization (FISH) or immunohistochemistry (IHC) 2+) as clinical efficacy in the pivotal approval trials was related to overexpression of this protein.
9. Fatal infusion reactions, tumor lysis syndrome (TLS), and severe mucocutaneous reactions.

REFERENCES

Andersson M, Lidbrink E, Bjerre K et al (2011) Phase III randomized study comparing docetaxel plus trastuzumab with vinorelbine plus trastuzumab as first-line therapy of metastatic or locally advanced human epidermal growth factor receptor 2-positive breast cancer: the HERNATA study. J Clin Oncol 29(3):264–271

Ansell SM (2011) Brentuximab vedotin: delivering an antimitotic drug to activated lymphoma cells. Expert Opin Investig Drugs 20(1):99–105

Bang YJ, Van Cutsem E, Feyereislova A et al (2010) Trastuzumab in combination with chemotherapy versus chemotherapy alone for treatment of HER2-positive advanced gastric or gastro-oesophageal junction cancer (ToGA): a phase 3, open-label, randomised controlled trial. Lancet 376(9742):687–697

Baselga J, Tripathy D, Mendelsohn J et al (1999) Phase II study of weekly intravenous trastuzumab (Herceptin) in patients with HER2/neu-overexpressing metastatic breast cancer. Semin Oncol 26(4 Suppl 12):78–83

Baselga J, Cortes J, Kim SB et al (2012) Pertuzumab plus trastuzumab plus docetaxel for metastatic breast cancer. N Engl J Med 366(2):109–119

Bonner JA, Harari PM, Giralt J et al (2006) Radiotherapy plus cetuximab for squamous-cell carcinoma of the head and neck. N Engl J Med 354(6):567–578

Braghiroli MI, Sabbaga J, Hoff PM (2012) Bevacizumab: overview of the literature. Expert Rev Anticancer Ther 12(5):567–580

Clark AJ, Butowski NA, Chang SM et al (2011) Impact of bevacizumab chemotherapy on craniotomy wound healing. J Neurosurg 114(6):1609–1616

Cohenuram M, Saif MW (2007) Panitumumab the first fully human monoclonal antibody: from the bench to the clinic. Anticancer Drugs 18(1):7–15

Coiffier B (2002) Rituximab in the treatment of diffuse large B-cell lymphomas. Semin Oncol 29(1 Suppl 2):30–35

Coiffier B, Lepage E, Briere J et al (2002) CHOP chemotherapy plus rituximab compared with CHOP alone in elderly patients with diffuse large-B-cell lymphoma. N Engl J Med 346(4):235–242

Coleman RE (2004a) Bisphosphonates: clinical experience. Oncologist 9(Suppl 4):14–27

Coleman RE (2004b) The role of bisphosphonates in breast cancer. Breast 13(Suppl 1):S19–S28

Cortes J, Fumoleau P, Bianchi GV et al (2012) Pertuzumab monotherapy after trastuzumab-based treatment and subsequent reintroduction of trastuzumab: activity and tolerability in patients with advanced human epidermal growth factor receptor 2-positive breast cancer. J Clin Oncol 30(14):1594–1600

Cunningham D, Humblet Y, Siena S et al (2004) Cetuximab monotherapy and cetuximab plus irinotecan in irinotecan-refractory metastatic colorectal cancer. N Engl J Med 351(4):337–345

Delbaldo C, Pierga JY, Dieras V et al (2005) Pharmacokinetic profile of cetuximab (Erbitux) alone and in combination with irinotecan in patients with advanced EGFR-positive adenocarcinoma. Eur J Cancer 41(12):1739–1745

Dirks NL, Meibohm B (2010) Population pharmacokinetics of therapeutic monoclonal antibodies. Clin Pharmacokinet 49(10):633–659

Dirks NL, Nolting A, Kovar A et al (2008) Population pharmacokinetics of cetuximab in patients with squamous cell carcinoma of the head and neck. J Clin Pharmacol 48(3):267–278

Dowell JA, Korth-Bradley J, Liu H et al (2001) Pharmacokinetics of gemtuzumab ozogamicin, an antibody-targeted chemotherapy agent for the treatment of patients with acute myeloid leukemia in first relapse. J Clin Pharmacol 41(11):1206–1214

Faderl S, Thomas DA, O'Brien S et al (2003) Experience with alemtuzumab plus rituximab in patients with relapsed and refractory lymphoid malignancies. Blood 101(9): 3413–3415

Ferrara N (2004) Vascular endothelial growth factor: basic science and clinical progress. Endocr Rev 25(4): 581–611

Foyil KV, Kennedy DA, Grove LE et al (2012) Extended retreatment with brentuximab vedotin (SGN-35) maintains complete remission in patient with recurrent systemic anaplastic large-cell lymphoma. Leuk Lymphoma 53(3):506–507

Frampton JE, Wagstaff AJ (2003) Alemtuzumab. Drugs 63(12):1229–1243; discussion 1245–1226

Francisco JA, Cerveny CG, Meyer DL et al (2003) cAC10-vcMMAE, an anti-CD30-monomethyl auristatin E conjugate with potent and selective antitumor activity. Blood 102(4):1458–1465

Friedman HS, Prados MD, Wen PY et al (2009) Bevacizumab alone and in combination with irinotecan in recurrent glioblastoma. J Clin Oncol 27(28):4733–4740

Furtado M, Rule S (2012) Emerging pharmacotherapy for relapsed or refractory Hodgkin's lymphoma: focus on brentuximab vedotin. Clin Med Insights Oncol 6: 31–39

Gibiansky L, Sutjandra L, Doshi S et al (2012) Population pharmacokinetic analysis of denosumab in patients with bone metastases from solid tumours. Clin Pharmacokinet 51(4):247–260

Giles F, Estey E, O'Brien S (2003) Gemtuzumab ozogamicin in the treatment of acute myeloid leukemia. Cancer 98(10):2095–2104

Gopal AK, Ramchandren R, O'Connor OA et al (2012) Safety and efficacy of brentuximab vedotin for Hodgkin lymphoma recurring after allogeneic stem cell transplant. Blood 120(3):150–158

Guarneri V, Frassoldati A, Bruzzi P et al (2008) Multicentric, randomized phase III trial of two different adjuvant chemotherapy regimens plus three versus twelve months of trastuzumab in patients with HER2-positive breast cancer (Short-HER Trial; NCT00629278). Clin Breast Cancer 8(5):453–456

Haddley K (2012) Brentuximab vedotin: its role in the treatment of anaplastic large cell and Hodgkin's lymphoma. Drugs Today (Barc) 48(4):259–270

Hagenbeek A, Lewington V (2005) Report of a European consensus workshop to develop recommendations for the optimal use of (90)Y-ibritumomab tiuxetan (Zevalin) in lymphoma. Ann Oncol 16(5):786–792

Hanaizi Z, van Zwieten-Boot B, Calvo G et al (2012) The European Medicines Agency review of ipilimumab (Yervoy) for the treatment of advanced (unresectable or metastatic) melanoma in adults who have received prior therapy: summary of the scientific assessment of the Committee for Medicinal Products for Human Use. Eur J Cancer 48(2):237–242

Henry DH, Costa L, Goldwasser F et al (2011) Randomized, double-blind study of denosumab versus zoledronic acid in the treatment of bone metastases in patients with advanced cancer (excluding breast and prostate cancer) or multiple myeloma. J Clin Oncol 29(9): 1125–1132

Hochster H, Weller E, Gascoyne RD et al (2009) Maintenance rituximab after cyclophosphamide, vincristine, and prednisone prolongs progression-free survival in advanced indolent lymphoma: results of the randomized phase III ECOG1496 Study. J Clin Oncol 27(10): 1607–1614

Hodi FS, O'Day SJ, McDermott DF et al (2010) Improved survival with ipilimumab in patients with metastatic melanoma. N Engl J Med 363(8):711–723

Huang BT, Zeng QC, Yu J et al (2012) How to determine post-RCHOP therapy for risk-tailored adult patients with diffuse large B-cell lymphoma, addition of maintenance rituximab or observation: multicenter experience. J Cancer Res Clin Oncol 138(1): 125–132

Jain RK (2005) Normalization of tumor vasculature: an emerging concept in antiangiogenic therapy. Science 307(5706):58–62

Jemal A, Siegel R, Ward E et al (2006) Cancer statistics, 2006. CA Cancer J Clin 56(2):106–130

Johnson DH, Fehrenbacher L, Novotny WF et al (2004) Randomized phase II trial comparing bevacizumab plus carboplatin and paclitaxel with carboplatin and paclitaxel alone in previously untreated locally advanced or metastatic non-small-cell lung cancer. J Clin Oncol 22(11):2184–2191

Karapetis CS, Khambata-Ford S, Jonker DJ et al (2008) K-ras mutations and benefit from cetuximab in advanced colorectal cancer. N Engl J Med 359(17):1757–1765

Keating GM (2012) Pertuzumab: in the first-line treatment of HER2-positive metastatic breast cancer. Drugs 72(3):353–360

Keating M, Hallek M (2002) Alemtuzumab, the first monoclonal antibody (MAB) directed against CD52. Med Oncol 19(Suppl):S1–S2

Kennedy B, Hillmen P (2002) Immunological effects and safe administration of alemtuzumab (MabCampath) in advanced B-cLL. Med Oncol 19(Suppl):S49–S55

Korth-Bradley JM, Dowell JA, King SP et al (2001) Impact of age and gender on the pharmacokinetics of gemtuzumab ozogamicin. Pharmacotherapy 21(10): 1175–1180

Koutcher LD, Wolden S, Lee N (2009) Severe radiation dermatitis in patients with locally advanced head and neck cancer treated with concurrent radiation and cetuximab. Am J Clin Oncol 32(5):472–476

Lenihan DJ, Alencar AJ, Yang D et al (2004) Cardiac toxicity of alemtuzumab in patients with mycosis fungoides/Sezary syndrome. Blood 104(3):655–658

Leyland-Jones B, Gelmon K, Ayoub JP et al (2003) Pharmacokinetics, safety, and efficacy of trastuzumab administered every three weeks in combination with paclitaxel. J Clin Oncol 21(21):3965–3971

Lipson EJ, Drake CG (2011) Ipilimumab: an anti-CTLA-4 antibody for metastatic melanoma. Clin Cancer Res 17(22):6958–6962

Liu NS, O'Brien S (2004) Monoclonal antibodies in the treatment of chronic lymphocytic leukemia. Med Oncol 21(4):297–304

Lu JF, Bruno R, Eppler S et al (2008) Clinical pharmacokinetics of bevacizumab in patients with solid tumors. Cancer Chemother Pharmacol 62(5):779–786

Lundin J, Kimby E, Bjorkholm M et al (2002) Phase II trial of subcutaneous anti-CD52 monoclonal antibody alemtuzumab (Campath-1H) as first-line treatment for patients with B-cell chronic lymphocytic leukemia (B-CLL). Blood 100(3):768–773

Lundin J, Kennedy B, Dearden C et al (2005) No cardiac toxicity associated with alemtuzumab therapy for mycosis fungoides/Sezary syndrome. Blood 105(10):4148–4149

Maloney DG, Liles TM, Czerwinski DK et al (1994) Phase I clinical trial using escalating single-dose infusion of chimeric anti-CD20 monoclonal antibody (IDEC-C2B8) in patients with recurrent B-cell lymphoma. Blood 84(8):2457–2466

Maloney DG, Grillo-Lopez AJ, White CA et al (1997) IDEC-C2B8 (Rituximab) anti-CD20 monoclonal antibody therapy in patients with relapsed low-grade non-Hodgkin's lymphoma. Blood 90(6):2188–2195

McLaughlin P, Grillo-Lopez AJ, Link BK et al (1998) Rituximab chimeric anti-CD20 monoclonal antibody therapy for relapsed indolent lymphoma: half of patients respond to a four-dose treatment program. J Clin Oncol 16(8):2825–2833

Minich SS (2012) Brentuximab vedotin: a new age in the treatment of Hodgkin lymphoma and anaplastic large cell lymphoma. Ann Pharmacother 46(3):377–383

Montagna M, Montillo M, Avanzini MA et al (2011) Relationship between pharmacokinetic profile of subcutaneously administered alemtuzumab and clinical response in patients with chronic lymphocytic leukemia. Haematologica 96(6):932–936

Montillo M, Tedeschi A, Miqueleiz S et al (2006) Alemtuzumab as consolidation after a response to fludarabine is effective in purging residual disease in patients with chronic lymphocytic leukemia. J Clin Oncol 24(15):2337–2342

Nabhan C, Patton D, Gordon LI et al (2004) A pilot trial of rituximab and alemtuzumab combination therapy in patients with relapsed and/or refractory chronic lymphocytic leukemia (CLL). Leuk Lymphoma 45(11):2269–2273

Ng CM, Lum BL, Gimenez V et al (2006) Rationale for fixed dosing of pertuzumab in cancer patients based on population pharmacokinetic analysis. Pharm Res 23(6):1275–1284

O'Brien S, Albitar M, Giles FJ (2005) Monoclonal antibodies in the treatment of leukemia. Curr Mol Med 5(7):663–675

Oki Y, Younes A (2012) Brentuximab vedotin in systemic T-cell lymphoma. Expert Opin Biol Ther 12(5):623–632

Osterborg A, Mellstedt H, Keating M (2002) Clinical effects of alemtuzumab (Campath-1H) in B-cell chronic lymphocytic leukemia. Med Oncol 19(Suppl):S21–S26

Papac RJ (2001) Origins of cancer therapy. Yale J Biol Med 74(6):391–398

Pawson R, Dyer MJ, Barge R et al (1997) Treatment of T-cell prolymphocytic leukemia with human CD52 antibody. J Clin Oncol 15(7):2667–2672

Pinto C, Barone CA, Girolomoni G et al (2011) Management of skin toxicity associated with cetuximab treatment in combination with chemotherapy or radiotherapy. Oncologist 16(2):228–238

Planchard D (2011) Bevacizumab in non-small-cell lung cancer: a review. Expert Rev Anticancer Ther 11(8):1163–1179

Plosker GL, Figgitt DP (2003) Rituximab: a review of its use in non-Hodgkin's lymphoma and chronic lymphocytic leukaemia. Drugs 63(8):803–843

Pro B, Advani R, Brice P et al (2012) Brentuximab vedotin (SGN-35) in patients with relapsed or refractory systemic anaplastic large-cell lymphoma: results of a phase II study. J Clin Oncol 30(18):2190–2196

Rai KR, Freter CE, Mercier RJ et al (2002) Alemtuzumab in previously treated chronic lymphocytic leukemia patients who also had received fludarabine. J Clin Oncol 20(18):3891–3897

Reck M, Barlesi F, Crino L et al (2012) Predicting and managing the risk of pulmonary haemorrhage in patients with NSCLC treated with bevacizumab: a consensus report from a panel of experts. Ann Oncol 23(5):1111–1120

Robert C, Thomas L, Bondarenko I et al (2011) Ipilimumab plus dacarbazine for previously untreated metastatic melanoma. N Engl J Med 364(26):2517–2526

Romas E (2009) Clinical applications of RANK-ligand inhibition. Intern Med J 39(2):110–116

Rowinsky EK, Schwartz GH, Gollob JA et al (2004) Safety, pharmacokinetics, and activity of ABX-EGF, a fully human anti-epidermal growth factor receptor monoclonal antibody in patients with metastatic renal cell cancer. J Clin Oncol 22(15):3003–3015

Saif MW, Cohenuram M (2006) Role of panitumumab in the management of metastatic colorectal cancer. Clin Colorectal Cancer 6(2):118–124

Salar A, Casao D, Cervera M et al (2006) Rapid infusion of rituximab with or without steroid-containing chemotherapy: 1-yr experience in a single institution. Eur J Haematol 77(4):338–340

Sandler A, Gray R, Perry MC et al (2006) Paclitaxel-carboplatin alone or with bevacizumab for non-small-cell lung cancer. N Engl J Med 355(24):2542–2550

Sehn LH, Donaldson J, Filewich A et al (2007) Rapid infusion rituximab in combination with corticosteroid-containing chemotherapy or as maintenance therapy is well tolerated and can safely be delivered in the community setting. Blood 109(10):4171–4173

Seidman AD, Berry D, Cirrincione C et al (2008) Randomized phase III trial of weekly compared with every-3-weeks paclitaxel for metastatic breast cancer, with trastuzumab for all HER-2 overexpressors and random assignment to trastuzumab or not in HER-2 nonoverexpressors: final results of Cancer and Leukemia Group B protocol 9840. J Clin Oncol 26(10):1642–1649

Shirao K, Yoshino T, Boku N et al (2009) A phase I escalating single-dose and weekly fixed-dose study of cetuximab pharmacokinetics in Japanese patients with solid tumors. Cancer Chemother Pharmacol 64(3): 557–564

Sievers EL, Appelbaum FR, Spielberger RT et al (1999) Selective ablation of acute myeloid leukemia using antibody-targeted chemotherapy: a phase I study of an anti-CD33 calicheamicin immunoconjugate. Blood 93(11):3678–3684

Skarbnik AP, Smith MR (2012) Brentuximab vedotin in anaplastic large cell lymphoma. Expert Opin Biol Ther 12(5):633–639

Specenier P (2012) Bevacizumab in glioblastoma multiforme. Expert Rev Anticancer Ther 12(1):9–18

Specenier P, Vermorken JB (2011) Cetuximab in the treatment of squamous cell carcinoma of the head and neck. Expert Rev Anticancer Ther 11(4):511–524

Stilgenbauer S, Zenz T, Winkler D et al (2009) Subcutaneous alemtuzumab in fludarabine-refractory chronic lymphocytic leukemia: clinical results and prognostic marker analyses from the CLL2H study of the German Chronic Lymphocytic Leukemia Study Group. J Clin Oncol 27(24):3994–4001

Stopeck AT, Lipton A, Body JJ et al (2010) Denosumab compared with zoledronic acid for the treatment of bone metastases in patients with advanced breast cancer: a randomized, double-blind study. J Clin Oncol 28(35): 5132–5139

Studer G, Brown M, Salgueiro EB et al (2011) Grade 3/4 dermatitis in head and neck cancer patients treated with concurrent cetuximab and IMRT. Int J Radiat Oncol Biol Phys 81(1):110–117

Swain SM, Kim SB, Cortés J et al (2013) Pertuzumab, trastuzumab, and docetaxel for HER2-positive metastatic breast cancer (CLEOPATRA study): overall survival results form a randomized, double-blind, placebo-controlled phase 3 study. Lancet Oncol 14:461–471

Thompson CB, Allison JP (1997) The emerging role of CTLA-4 as an immune attenuator. Immunity 7(4):445–450

Valero V, Forbes J, Pegram MD et al (2011) Multicenter phase III randomized trial comparing docetaxel and trastuzumab with docetaxel, carboplatin, and trastuzumab as first-line chemotherapy for patients with HER2-gene-amplified metastatic breast cancer (BCIRG 007 study): two highly active therapeutic regimens. J Clin Oncol 29(2):149–156

van Der Velden VH, te Marvelde JG, Hoogeveen PG et al (2001) Targeting of the CD33-calicheamicin immunoconjugate Mylotarg (CMA-676) in acute myeloid leukemia: in vivo and in vitro saturation and internalization by leukemic and normal myeloid cells. Blood 97(10):3197–3204

Wagner-Johnston ND, Bartlett NL, Cashen A et al (2012) Progressive multifocal leukoencephalopathy in a patient with Hodgkin lymphoma treated with brentuximab vedotin. Leuk Lymphoma 53(11):2283–2286

Wahl AF, Klussman K, Thompson JD et al (2002) The anti-CD30 monoclonal antibody SGN-30 promotes growth arrest and DNA fragmentation in vitro and affects anti-tumor activity in models of Hodgkin's disease. Cancer Res 62(13):3736–3742

Waterhouse P, Penninger JM, Timms E et al (1995) Lymphoproliferative disorders with early lethality in mice deficient in Ctla-4. Science 270(5238):985–988

Wen PY, Kesari S (2008) Malignant gliomas in adults. N Engl J Med 359(5):492–507

Witzig TE, Flinn IW, Gordon LI et al (2002a) Treatment with ibritumomab tiuxetan radioimmunotherapy in patients with rituximab-refractory follicular non-Hodgkin's lymphoma. J Clin Oncol 20(15):3262–3269

Witzig TE, Gordon LI, Cabanillas F et al (2002b) Randomized controlled trial of yttrium-90-labeled ibritumomab tiuxetan radioimmunotherapy versus rituximab immunotherapy for patients with relapsed or refractory low-grade, follicular, or transformed B-cell non-Hodgkin's lymphoma. J Clin Oncol 20(10): 2453–2463

Wood AM (2001) Rituximab: an innovative therapy for non-Hodgkin's lymphoma. Am J Health Syst Pharm 58(3):215–229, quiz 230–212

Yamamoto N, Yamada Y, Fujiwara Y et al (2009) Phase I and pharmacokinetic study of HER2-targeted rhuMAB 2C4 (Pertuzumab, RO4368451) in Japanese patients with solid tumors. Jpn J Clin Oncol 39(4):260–266

Yang XD, Jia XC, Corvalan JR et al (2001) Development of ABX-EGF, a fully human anti-EGF receptor monoclonal antibody, for cancer therapy. Crit Rev Oncol Hematol 38(1):17–23

Younes A, Bartlett NL, Leonard JP et al (2010) Brentuximab vedotin (SGN-35) for relapsed CD30-positive lymphomas. N Engl J Med 363(19):1812–1821

Younes A, Gopal AK, Smith SE et al (2012) Results of a pivotal phase II study of brentuximab vedotin for patients with relapsed or refractory Hodgkin's lymphoma. J Clin Oncol 30(18):2183–2189

18 Hematopoietic Growth Factors: Focus on Erythropoiesis-Stimulating Agents

Juan Jose Pérez-Ruixo and Andrew T. Chow

INTRODUCTION

Hematopoiesis is an intricate, well-regulated, and homeostatic multistep process that allows immature precursor cells in the bone marrow to proliferate, differentiate, mature, and become functional blood cells that transport oxygen and carbon dioxide; contribute to host immunity; and facilitate blood clotting. In the early 1900s, scientists recognized the presence of circulating factors that regulate hematopoiesis. It took approximately 50 years to develop in vitro cell culture systems in order to definitively prove that the growth and survival of early blood cells require the presence of specific circulating factors, called hematopoietic growth factors (HGF). The presence of many HGF with different targets at extremely small amounts in blood, bone marrow, and urine confounded the search for a single HGF with a specific activity. Scientific progress was slow until it became possible to purify sufficient quantities to evaluate the characteristics and biologic potential of the isolated materials. The introduction of recombinant DNA technology triggered a flurry of studies and an information explosion, which confirmed hematopoiesis is mediated by a series of HGF that acts individually and in various combinations involving complex feedback mechanisms. Today, many HGF have been isolated; some have been studied extensively, and a few have been manufactured for clinical use.

Different mature blood cells have been identified, each derived from primitive hematopoietic stem cells in the bone marrow. The most primitive pool of pluripotent stem cells comprises approximately 0.1 % of the nucleated cells of the bone marrow, and 5 % of these cells may be actively cycling at a given time. The stem cell pool maintains itself, seemingly without extensive depletion, by asymmetrical cell division. When a stem cell divides, one daughter cell remains in the stem cell pool and the other becomes a committed colony-forming unit (CFU). CFU proliferates at a greater rate than the other stem cells and are more limited in self-renewal than pluripotent hematopoietic stem cells. Proliferation and differentiation are regulated by different mechanisms that necessarily involve HGF, which eventually convert the dividing cells into a population of terminally differentiated functional cells committed to the myeloid or the lymphoid pathway. Functional hematopoietic-derived blood cells from the myeloid pathway are red blood cells (erythrocytes), granulocytes (neutrophils, eosinophils, and basophils), monocytes and macrophages, tissue mast cells, and platelets (thrombocytes). Cells committed to the lymphoid pathway give rise to B- or T-lymphocytes and plasma cells.

Most HGF are glycosylated single-chain polypeptides encoded by a specific gene. Production of a recombinant HGF protein is accomplished by first identifying and isolating the particular HGF gene coding region, inserting the HGF DNA into a plasmid, and then expressing the recombinant growth factor protein in a biologic system (e.g., bacteria, yeast, or mammalian cells). The carbohydrate content of HGF varies by the particular protein and production method, which affects not only the molecular weight of glycoprotein but potentially the specific biologic activity and the circulating half-life as well. A summary of the HGF and their activities is provided in Table 18.1.

This chapter will focus on reviewing the molecular structure, mechanism of action, pharmacokinetics and pharmacodynamics, clinical indications, and adverse events of erythropoietin-stimulating agents (ESA) and will also briefly discuss other HGF.

ERYTHROPOIESIS-STIMULATING AGENTS

Erythropoietin (EPO) is a 30.4 kD glycoprotein hormone secreted by the kidneys in response to tissue hypoxia, which stimulates red blood cell (RBC)

J.J. Pérez-Ruixo, Ph.D. (✉) • A.T. Chow, Ph.D.
Department of Quantitative Pharmacology, Pharmacokinetics and Drug Metabolism, Amgen Inc, Thousand Oaks, CA, USA
e-mail: juanjose@amgen.com

D.J.A. Crommelin, R.D. Sindelar, and B. Meibohm (eds.), *Pharmaceutical Biotechnology*, DOI 10.1007/978-1-4614-6486-0_18,

Factor	Molecular weight (kDa)	Target cells	Actions
Erythropoietin (EPO)	34–39	Erythroid progenitors	Increase red blood cell counts
Granulocyte colony-stimulating factor (G-CSF)	18	Granulocyte progenitors and mature neutrophils	Increase neutrophil counts
Granulocyte-macrophage colony-stimulating factor (GM-CSF)	14–35	Granulocyte-macrophage progenitors and eosinophil progenitors	Increase neutrophil, eosinophil, and monocyte counts
Stem cell factor (SCF)		Granulocyte-erythroid progenitors, lymphoid progenitors, and natural killer cells	Increase pluripotent stem cells and progenitor cells for all other cell types
Thrombopoietin (TPO)	35	Stem cells, megakaryocytes, and erythroid progenitors	Increase platelet counts

Table 18.1 ■ Hematopoietic growth factors and their activities.

production. EPO requires glycosylation to regulate erythrocyte production by activating the EPO receptor (EPOR) and stimulating the proliferation and differentiation of erythrocytic progenitors in the bone marrow, which leads to reticulocytosis and an increase in the number of erythrocytes and the concentration of hemoglobin in the blood. The gene that encodes EPO is located on chromosome 7. The cloning of the *EPO* gene in the early 1980s allowed for the development of recombinant erythropoietins and analogs (erythropoiesis-stimulating agents [ESAs]), offering an alternative to transfusion as a method of raising hemoglobin levels that has successfully been used for over 20 years to treat anemia in millions of anemic patients.

Epoetin alfa (Epogen®), the first commercial form of recombinant human erythropoietin (rHuEPO) marketed in the USA, EU, Japan, and China, and epoetin beta (Recormon®, NeoRecormon®) marketed outside of the USA are both expressed in Chinese hamster ovary cells. Both have the same 165 amino acid sequence, which is identical to human EPO, and contain two disulfide bonds and three N-linked and one O-linked sialic acid-containing carbohydrate chains (Halstenson et al. 1991) and lead to the same biological effects as endogenous EPO (Egrie et al. 1986). No important differences in clinical efficacy are apparent between epoetin alfa and beta (Jelkmann 2000). Darbepoetin alfa (Aranesp®) is a hyperglycosylated erythropoietin analog with five amino acid changes and two additional N-linked carbohydrate chains, which has the same mechanism of action as EPO (Elliott et al. 2004a). However, darbepoetin alfa has a threefold increased serum half-life (Macdougall et al. 1999; Elliott et al. 2003; Sinclair and Elliott 2005) and increased in vivo potency (Egrie et al. 2003), allowing for more convenient modes of administration, including extended dosing intervals. (Vansteenkiste et al. 2002; Nissenson et al. 2002) It is marketed in more than 50 countries and is indicated to treat the anemia of patients with chronic kidney disease and chemotherapy-induced anemia in cancer patients.

Recently, a large methoxy polyethylene glycol (PEG) polymer chain was integrated into the epoetin beta molecule via amide bonds between either the N-terminal amino group or the ε-amino group of lysine (predominantly Lys (Kaufman et al. 1998) and Lys (Yan et al. 2012)) by means of a succinimidyl butanoic acid linker (Macdougall 2005). The resulting pegylated epoetin beta molecule has been marketed in the EU as Mircera® to treat the anemia of patients with chronic kidney disease (CKD), but its clinical development as treatment for chemotherapy-induced anemia has been stopped (Gascon et al. 2010). The pegylated epoetin beta stimulates erythropoiesis by binding to EPOR; however, the EPOR binding affinity is reduced (Jarsch et al. 2008). This biologic disadvantage is counterbalanced with an extended half-life in humans, which allows for extended dosing intervals in CKD patients (Chanu et al. 2010), similar to the dosing interval of darbepoetin. Peginesatide, a synthetic ESA comprised of two identical 21 amino acid chains covalently bonded and linked to a single PEG chain, has no amino acid sequence homology to EPO but mimics the structure of EPO and stimulates erythropoiesis by binding to EPOR (Fan et al. 2006). It was recently approved in the USA as Omontys® to treat anemia due to chronic kidney disease in adult patients on dialysis, but not chemotherapy-induced anemia. Peginesatide has a similar EPOR binding affinity as rHuEPO, but a longer half-life in the presence of renal insufficiency, which also allows extended dosing intervals in CKD patients (Locatelli and Del Vecchio 2009). However, in February 2013, the manufacturer and the US FDA announced an immediate recall of peginesatide because of postmarketing reports of serious hypersensitivity reactions, including anaphylaxis, which may be life threatening or fatal.

Five rHuEPO biosimilars manufactured by two companies have been approved in the EU. Abseamed®,

Binocrit®, and Epoetin alfa HEXAL® all produced by Rentschler Biotechnologie GmbH but marketed by three different companies are epoetin alfa biosimilars of the reference product Eprex®. Comparable safety and efficacy between these three biosimilars and Eprex® was demonstrated in randomized controlled trials in hemodialysis patients with renal anemia. Although the EMA regulatory guidelines for rHuEPO biosimilars recommend that comparable efficacy and safety are demonstrated with two randomized trials in the nephrology setting, these biosimilars were approved based on a single nephrology trial. Two additional biosimilar versions of Eprex®, Retacrit® and Silapo® are manufactured by Norbitec GmbH, under the international nonproprietary name (INN) of epoetin zeta. The comparability of epoetin zeta to Eprex® was demonstrated in two randomized clinical trials, a correction phase study and a maintenance phase study, involving hemodialysis patients with renal anemia. In the correction phase study, the comparability between epoetin zeta and Eprex® over the evaluation period was demonstrated for mean hemoglobin levels, but not for mean dose. Similar results were reported in the maintenance phase study, suggesting a possible difference in the bioactivity of epoetin zeta and Eprex®. Data from studies in cancer patients receiving chemotherapy and treated with epoetin alfa biosimilars and epoetin zeta were also submitted for approval, but these studies were not adequately powered to demonstrate therapeutic equivalence to the reference product in this patient population. However, epoetin alfa biosimilars and epoetin zeta were approved for indications in renal anemia, chemotherapy-induced anemia, and for predonation of blood prior to surgery for autologous transfusion (Schellekens and Moors 2010).

■ Regulation of Erythropoietin

The primary site of EPO synthesis in adults is the peritubular cells of the kidney (Jelkmann 2000; Jelkmann 1992). The liver is a secondary site of EPO production, with synthesis occurring in both hepatocytes and fibroblastoid interstitial cells (Spivak 1998). No preformed stores of EPO exist, and serum EPO concentrations are maintained at a constant concentration by homeostatic turnover, which consists of the basal production and elimination of the hormone (Fisher 2003). Within a healthy individual, the serum EPO concentration tends to be controlled tightly; however, large interindividual variability is evident from the normal range, 5–35 IU/L (Fisher 2003). Maintenance of normal serum concentrations of endogenous EPO requires the synthesis of about 2–3 IU/kg/day, or approx. 1,000–1,500 IU/week for a 70-kg man. Sex differences and regular-to-moderate athletic training do not appear to affect endogenous EPO serum concentrations. The blood flow in the kidney has a circadian rhythm in normal individuals; therefore, the endogenous production of EPO has diurnal variations with the highest levels in the evening and at night (Wide et al. 1989).

The overexpression of EPO occurs in a number of adaptive and pathologic conditions. In response to acute hypoxic stress, such as severe blood loss or severe anemia, EPO production rate can increase 100- to 1,000-fold. Numerous studies have shown an exponential increase in serum EPO, with increasing degrees of anemia, although the maximal bone marrow response to such stimulation is only a four- to sixfold increase in RBC production rate (Jelkmann 2000). Overproduction of EPO with accompanying erythrocytosis may be an adaptive response to conditions that produce chronic tissue hypoxia, such as living at high altitude, chronic respiratory diseases, cyanotic heart disease, sleep apnea, smoking, localized renal hypoxia, radiotherapy, or hemoglobinopathies with increased oxygen affinity. Paraneoplastic production of EPO from tumors and cysts, including renal carcinomas, benign renal tumors, Wilms' tumors, hepatomas, liver carcinomas, cerebellar hemangioblastomas, adrenal gland tumors, and leiomyomas, can also result in high serum concentrations of EPO. Following bone marrow ablation, aplastic anemia, or anemia in patients with hypoplastic marrows, serum EPO levels are disproportionately increased relative to slightly decreased hemoglobin levels. Conversely, individuals with hyperactive marrow owing to hemolytic anemia had disproportionately low serum EPO levels and rapid EPO serum disappearance.

In chronic kidney disease, up to 60 % of patients have hemoglobin concentrations below 11 g/dL before beginning dialysis (Jungers et al. 2002). Multiple mechanisms contribute to the low hemoglobin levels (Fisher 2003), but the most important is the inability of the diseased kidneys to produce an appropriate EPO response for the given degree of anemia or an inability to meet the increased RBC demands of uremic patients (Adamson and Eschbach 1990). In addition, the uremic state itself appears to blunt the bone marrow response to EPO, perhaps through polyamines, inflammatory cytokines, and/or parathyroid hormone mediators (Fisher 2003). It is of interest that serum EPO concentrations in chronically anemic dialysis patients increase to some extent in response to acute hypoxic stress (from either acute bleeding or systemic hypoxemia), suggesting that kidney failure does not result in a complete inability to produce EPO (Kato et al. 1994).

In cancer patients, anemia is of multifactorial etiology (Fisher 2003), and there are three distinct types of anemia: cancer-related anemia (nontreatment related), anemia related to myelosuppressive chemotherapy, and anemia related to other causes such as bleeding, nutritional deficiency, or iron deficiency among others.

As with other anemias of chronic disease, including those associated with chronic infection and inflammatory disorders, the anemia of cancer is characterized by a decreased production of endogenous EPO (Miller et al. 1990), cytokine-induced suppression of bone marrow function, disordered iron absorption and metabolism (Bron et al. 2001), and decreased erythrocyte survival. In the anemia related to chemotherapy treatment, the amount of endogenous EPO transiently increases up to sixfold within the 48 h after the administration of chemotherapy and returns to baseline within a week (Glaspy et al. 2005). After myeloablative chemotherapy, severe thrombocytopenia and bleeding might contribute to a significant loss of RBC. Finally, the anemias associated with infant prematurity, pregnancy, allogeneic bone marrow transplantation, and HIV infection are often characterized by inappropriately low EPO concentrations (Spivak 1998).

■ Pharmacokinetics

Absorption

After a single rHuEPO s.c. dose, absorption is slow, leading to peak serum concentrations at 5–30 h and a longer terminal half-life (24–79 h) than that obtained after intravenous (i.v.) administration (McMahon et al. 1990). These results indicate the presence of flip-flop pharmacokinetics, where the rate of absorption is slower than the rate of elimination. Thus, absorption is rate limiting, and the observed terminal half-life after s.c. dosing reflects the absorption rate rather than elimination rate.

Following s.c. administration, protein therapeutics typically enter into the systemic circulation via the blood capillaries or the lymphatic system (Porter and Charman 2000; McLennan et al. 2006). The lymphatic system is considered to be the primary route of absorption from the s.c. injection site for protein therapeutics greater than 16 kD due to the restricted vascular access afforded by the continuous endothelial layer of blood capillaries (Supersaxo et al. 1990). In both healthy subjects and cancer patients, the fraction of dose absorbed via the lymphatics is about 80–90 % and increases at doses higher than 300 IU/kg (Olsson-Gisleskog et al. 2007; Ait-Oudhia et al. 2011; Krzyzanski et al. 2005; Ramakrishnan et al. 2004). The s.c. absorption rates of rHuEPO vary according to the administration site, with a more rapid and extensive absorption when rHuEPO was injected into the thigh compared with the abdomen or arm (Jensen et al. 1994). This relatively small difference is most likely reflecting regional differences in blood and lymph flow and not considered to be clinically relevant as the pharmacodynamic profile (i.e., reticulocytes time course) did not evidence any difference across the site of administration.

The s.c. absorption of darbepoetin alfa in humans is also slow, with peak concentrations reached at 34–58 h post-dose, followed by a generally monophasic decline. Similarly to rHuEPO, darbepoetin alfa also displays flip-flop pharmacokinetics, with a longer half-life after s.c. dosing than after i.v. dosing (Agoram et al. 2007). The mean terminal half-life of darbepoetin alfa, 73 h, is associated with large variability between patients, consistent with the variability observed for other ESAs (Glaspy et al. 2005). The mean absorption time of darbepoetin alpha is 56 h, substantially longer than the mean absorption time reported for rHuEPO (Olsson-Gisleskog et al. 2007).

The reported 20–30 % reduction in the darbepoetin alfa absorption rate per decade of age (Agoram et al. 2007) is consistent with the estimated effect of age on the rHuEPO absorption rate in healthy subjects (Olsson-Gisleskog et al. 2007) and cancer patients (Ait-Oudhia et al. 2011) and reflects the longer terminal half-life and the larger exposure to both drugs in older patients. It has been hypothesized (Agoram et al. 2007) that the age-dependent reduction in lymphatic flow rate could be the physiological reason behind this relationship, as it has also been reported for monoclonal antibodies administered by s.c. route (Sutjandra et al. 2011; Kakkar et al. 2011). The data available also suggest that the pharmacokinetic profile of rHuEPO and darbepoetin alfa after s.c. administration is similar in adults and children; however, s.c. absorption in children may be more rapid than in adults for both drugs (Heatherington 2003).

Bioavailability

Initial bioavailability estimates for rHuEPO after s.c. administration range from about 15 to 40 % and are similar for epoetin alfa and beta (Deicher and Horl 2004). When the pharmacokinetics of s.c. rHuEPO and darbepoetin alfa were studied over a wider dose range in healthy volunteers and after accounting for the rHuEPO nonlinear clearance, exposure was found to increase more than proportional with dose (Olsson-Gisleskog et al. 2007; Agoram et al. 2007; Cheung et al. 1998, 2001). The s.c. bioavailability of darbepoetin alfa increases from 57 to 69 % when the 200 μg dose is increased up to 400 μg, while the s.c. bioavailability of rHuEPO increases from 54 to 65 % when the 40 kIU dose is increased up to 80 kIU. The apparent increase in s.c. bioavailability with dose of ESA might indicate saturable pre-systemic processes. Nevertheless, despite the apparent low bioavailability, s.c. administration of ESA has been reported to produce equivalent or better efficacy relative to i.v. administration, although there is a wide range of inter-patient variability (Kaufman et al. 1998).

Distribution

During i.v. infusion, serum rHuEPO and darbepoetin alfa concentrations rise rapidly and then decline in a bi-exponential manner (Olsson-Gisleskog et al. 2007; Doshi et al. 2010). The peak serum rHuEPO and darbepoetin alfa concentrations correlate linearly with dose. A rHuEPO dose of 50 IU/kg produces concentrations of about 1,000 mIU/mL 15 min after the end of the infusion, while a darbepoetin alfa dose of 0.75 μg/kg generates serum concentrations of about 10–20 ng/mL after the end of the infusion (Doshi et al. 2010). As expected from its large molecular weight, the volume of distribution of rHuEPO is similar to the plasma volume (40–60 mL/kg), suggesting confinement of rHuEPO within the plasma circulation (McMahon et al. 1990; Olsson-Gisleskog et al. 2007). The data available also suggest that the volume of distribution, normalized by body weight, in adults and children is similar after i.v. administration of rHuEPO and darbepoetin alfa.

Elimination

Despite the long clinical experience with ESAs, the mechanism(s) of their clearance have not been fully elucidated, and there is a paucity of information regarding which organ(s) and tissue(s) are important in the metabolism of these drugs. Two ESA clearance pathways have been suggested to explain ESA elimination: (1) a capacity-limited clearance pathway utilizing EPO receptor-mediated endocytosis by erythroid progenitor cells and (2) a EPOR-independent linear clearance, although still not well understood, reflecting other mechanism(s). In vivo studies demonstrate that the kidney, liver, and lymph exert a negligible effect on in vivo EPOR-independent clearance. Clearly our understanding of the nature of the EPOR-independent clearance pathways is incomplete. However, it is important to recognize that renal excretion and hepatic metabolism of ESAs plays a minor role in their elimination and altered renal or hepatic function does not warrant dose adjustments.

Recently, Gross and Lodish investigated the trafficking and degradation of rHuEPO by EPOR-expressing cells (BsF3) in cell culture and found that rHuEPO was subjected to EPO receptor-mediated endocytosis followed by degradation in lysosomes (Gross and Lodish 2006). The rHuEPO receptor-binding, dissociation, and trafficking properties affected the relative rate of rHuEPO cellular uptake and intracellular degradation (Walrafen et al. 2005). About 57 % of surface-bound rHuEPO was internalized ($k_{in} = 0.06$ min^{-1}) and, after internalization, 60 % of the ligand was recycled intact to the cell surface while 40 % was degraded. In spite of the in vitro results suggesting the role of EPOR on ESA clearance, the in vivo evidence is indirect and mostly arises from chemotherapy studies in patients treated with rHuEPO and darbepoetin alfa (Chapel et al. 2001). Chemotherapy-based approaches may also affect EPOR-independent clearance mechanisms, due to destruction of macrophages or neutrophils. The reduction in the number of these cells may explain, or at least contribute to, the decrease in ESA clearance observed after chemotherapy treatment.

Studies investigating the pharmacokinetics of rHuEPO analogs with different EPOR binding activity, suggested that EPOR-independent pathway plays a major role in the ESA clearance since decreasing the number of receptors with chemotherapy or, blocking the EPOR pathway with analogs without binding activity, were unable to completely shut down the elimination of rHuEPO. In addition, since pegylation has been shown to mainly affect the EPOR-independent clearance pathway, EPOR-mediated clearance may not be the dominant route of ESA elimination (Agoram et al. 2009).

It has been shown that carbohydrate side chains of EPO are necessary for persistence and in vivo biologic activity of the molecule, but not for in vitro receptor binding or stimulation of proliferation. Indeed rHuEPO molecules with increased sialic acid content have less affinity for the EPOR (Sinclair and Elliott 2005; Elliott et al. 2004b). Darbepoetin alfa is a hyperglycosylated analog of rHuEPO, with 3- to 5-fold lower affinity for the EPOR compared to rHuEPO (Gross and Lodish 2006; Elliott et al. 2004b), but has 3- to 4-fold longer serum half-life and greater in vivo activity than rHuEPO (Egrie et al. 2003). Surface-bound darbepoetin alfa was internalized at the same rate than rHuEPO, and after internalization, 60 % of each ligand was re-secreted intact and 40 % degraded (Gross and Lodish 2006). While in vitro experiments suggested that relative to rHuEPO darbepoetin may have reduced clearance in vivo because of reduced EPOR-mediated endocytosis and degradation, darbepoetin alfa has other biophysical characteristics, such as increased molecular size and decreased isoelectric point, suggesting that the reduced clearance might be better explained by other mechanisms. In this context, studies investigating the pharmacokinetics of rHuEPO analogs with different EPOR binding activity suggest that hyperglycosylation mainly impacts the EPOR-independent clearance pathway, which also supports the hypothesis that EPOR-mediated clearance may not play a dominant role in ESA elimination (Agoram et al. 2009).

A population pharmacokinetic meta-analysis of rHuEPO in 533 healthy subjects enrolled in 16 clinical studies, where a wide range of i.v. and s.c. rHuEPO doses were administered, has helped in quantifying the two separate elimination pathways and understanding the influence of demographic characteristics and other covariates on the pharmacokinetic parameters of rHuEPO (Olsson-Gisleskog et al. 2007). At low concentrations, including the endogenous EPO concentrations observed at baseline or in ESA-untreated states, the nonlinear clearance operates at full capacity, giving a total clearance of about 0.9 L/h. As concentrations increase, the nonlinearity of pharmacokinetics becomes more important and, at the concentration of 394 IU/L, the clearance is 0.6 L/h. When the concentration are above 3,546 IU/L, the nonlinear clearance of rHuEPO was fully (>90 %) saturated and the total clearance decreased to almost one third, being mainly represented by the linear component. At concentrations higher than 3,546 IU/L, rHuEPO pharmacokinetics is approximately linear. The concentration-dependent clearance appears to be independent of the type of rHuEPO (epoetin alfa vs. epoetin beta) or population (healthy subjects or patients with chronic renal failure).

A further indication of the possible involvement of EPOR binding in the disposition of rHuEPO can be found when investigating the rHuEPO pharmacokinetics after multiple dosing. A rHuEPO time-dependent clearance, with a 10–30 % increase after several weeks of treatments with no subsequent changes (McMahon et al. 1990; Cheung et al. 1998; Yan et al. 2012) has been attributed to the limited number of EPOR located on the finite, but expandable, number of bone marrow erythroid progenitors. The pharmacodynamic action of rHuEPO increases BFU-E and CFU-E cell expansion and, consequently, the number of EPOR, which in turn results in an increase in rHuEPO clearance, a decrease in the apparent volume of distribution and a reduction in terminal half-life. The term pharmacodynamic-mediated drug disposition (PDMDD) has been coined to describe these types of TMDD models where pharmacodynamics affects the size of the target pool and influences the drug clearance, as has been described for ESAs. Consequently, the pharmacokinetics of rHuEPO is considered nonlinear because it is concentration dependent and nonstationary (time dependent) (Yan et al. 2012).

The rHuEPO pharmacokinetic models for healthy subjects can be applied to patients with anemia due to renal insufficiency; however, it may have limited predictive value when applied to patients receiving chemotherapy. The consequences of the chemotherapy effect on the pharmacokinetics of rHuEPO in oncology patients are derived from the reduced number of EPOR available to clear rHuEPO in progenitor cells and the reduction of non-EPOR-mediated clearance (Olsson-Gisleskog et al. 2007). In cancer patients treated with chemotherapy, a correlation between the decline in the absolute reticulocyte count and the decrease in the clearance of rHuEPO over time has been observed (Ait-Oudhia et al. 2011). As a consequence, the rHuEPO elimination process becomes slower than the absorption process, and the flip-flop phenomenon observed in healthy subjects disappears when rHuEPO is given s.c. to cancer patients receiving chemotherapy (Olsson-Gisleskog et al. 2007; Ait-Oudhia et al. 2011). Furthermore, this phenomenon has clinical implications with respect to the synchronicity of ESA and chemotherapy administration, suggesting asynchronous dosing might be superior (Glaspy et al. 2005).

■ Pharmacodynamics

After rHuEPO is administered, it binds to the EPOR on the surface of the BFU-E, CFU-E, and proerythroblast and activates the signal transduction pathways previously described. CFU-E cells have the highest EPOR density (1,000 receptors per cell) and are the most sensitive to EPO. Experimental data suggest that approximately only 5–10 % of EPOR must be continuously occupied with rHuEPO in order to prevent apoptosis and stimulate proliferation and differentiation of erythroid precursors. Then, CFU-Es will differentiate into normoblasts (including proerythroblast, basophilic erythroblast, polychromatophilic erythroblast, and orthochromatic erythroblast) and, upon normoblast denucleation, reticulocytes will be formed and reside in the bone marrow for 1 day before they are released into the bloodstream, where they circulate for about 1 day before maturing to erythrocytes. In healthy adults, the RBC life span is about 120 days and shows a relatively narrow distribution. The RBC life span is similar in cancer patients but markedly reduced in patients with chronic kidney disease, and it has been estimated to be about 60–65 days in dialysis patients and slightly longer in nondialysis patients (82 days), with a moderate interindividual variability (Uehlinger et al. 1992; Chanu et al. 2010).

Previous studies have demonstrated that highly glycosylated rHuEPO has increased in vivo biological activity and serum half-life, but decreased receptor binding affinity (Egrie and Browne 2001). Given these relationships, a comparison of clearance among different ESAs has to be interpreted in conjunction with EPOR binding affinity and/or in vivo activity. Darbepoetin alfa stimulates erythropoiesis by the same mechanisms as those previously discussed for endogenous EPO and rHuEPO. In vitro, the affinity of darbepoetin alfa for the EPOR is one third to one fifth of the rHuEPO affinity; (Gross and Lodish 2006) however, the increase in mean residence time of darbepoetin alfa results in a prolonged period above an erythropoietic threshold that more than compensates for the reduced receptor affinity, yielding an increased in vivo activity (Elliott et al. 2003; Egrie et al. 2003).

Besides this information, there is limited clinical data quantifying ESA potency in humans, which makes it difficult to establish the net balance of in vivo activity associated with simultaneous modifications of drug clearance and potency. Furthermore, in most of the clinical PK/PD relationships published, ESA pharmacodynamic actions have been linked to serum concentration, under the assumption that the pharmacodynamic effect is proportional to the amount of drug receptor complex. The PK/PD models developed for rHuEPO and darbepoetin alfa suggest that the reduction in darbepoetin alfa binding affinity to human EPOR relative to rHuEPO is translated into a less than two-fold reduction in vivo potency (EC_{50}=0.41 ng/mL vs. EC_{50}=58 IU/L). Additionally, darbepoetin alfa clearance is at least one third of rHuEPO clearance, while the maximum stimulatory effect of erythropoiesis (E_{max}) is similar for both ESAs (Krzyzanski et al. 2005). The darbepoetin alfa reduction in clearance is more pronounced than its reduction of in vivo potency; therefore, the longer half-life associated with darbepoetin alfa predominates over its reduction of in vivo potency, which explains the overall improvement in dose efficiency of darbepoetin alfa relative to rHuEPO.

Different mechanisms have been proposed to explain the pharmacodynamic tolerance of the rHuEPO effect. Besides the increase in rHuEPO clearance over time due to the increase in the number of EPOR, an oxygen-mediated feedback mechanism, erythroid precursor pool depletion, and iron-restricted erythropoiesis have been also proposed as tolerance mechanisms (Krzyzanski et al. 2005; Ramakrishnan et al. 2004). The oxygen feedback mechanism is regulated through an oxygen-sensing system: a high hemoglobin level leads to an increased oxygen level and eventually inhibits the production of endogenous EPO. Erythroid progenitor cells are EPO dependent; they cannot survive without EPO. On the other hand, extensive rHuEPO treatment results in anemia due to depletion of the erythroid precursor pool (Piron et al. 2001). This anemia is not due to low endogenous EPO levels but rather exhaustion of erythroid progenitors (Krzyzanski et al. 2005; Perez-Ruixo et al. 2009). Furthermore, iron-restricted erythropoiesis occurs in the presence of absolute iron deficiency, functional iron deficiency, and/or iron sequestration. Absolute iron deficiency is a common nutritional deficiency in women's health, pediatrics, and the elderly. Functional iron deficiency occurs in patients with significant EPO-mediated erythropoiesis or therapy with ESAs, even when storage iron is present. Iron sequestration, mediated by hepcidin, is an underappreciated but common cause of iron-restricted erythropoiesis in patients with chronic inflammatory disease. It has been shown that iron supplementation improves the hematopoietic response of ESAs used for chemotherapy-induced anemia. In multiple-dosing regimens, even though the endogenous EPO production might be suppressed, the total concentration of EPO is still high, and tolerance may occur due to precursor pool depletion and/or iron-restricted erythropoiesis. However, the oxygen feedback mechanism might be present, especially at the end of dosing intervals in regimens that extend longer than four rHuEPO half-lives.

■ Indications for Cancer Patients and Potential Adverse Events

Unless otherwise indicated, the information pertaining to ESA indications in cancer patients provided in this section is derived from the product prescribing information package inserts as well as the National Comprehensive Cancer Network for cancer- and chemotherapy-induced anemia. ESAs are indicated for the treatment of anemia due to the effects of concomitantly administered chemotherapy for a duration of ≥2 months in patients with metastatic, nonmyeloid malignancies. However, ESA treatment is not indicated for patients receiving hormonal agents, biologics, or radiotherapy, unless they are receiving concomitant myelosuppressive chemotherapy. Notably, ESA therapy should not be used to treat anemia associated with malignancy or anemia of cancer in patients with either solid or nonmyeloid hematological malignancies who are not receiving concurrent chemotherapy (Rizzo et al. 2008). Furthermore, ESA treatment is not indicated for patients receiving myelosuppressive therapy when the anticipated outcome is cure, due to the absence of studies that adequately characterize the impact of ESA therapy on progression-free and overall survival. ESA therapy is also not indicated for the treatment of anemia in cancer patients due to other factors such as absolute or functional iron deficiency, folate deficiencies, hemolysis, or gastrointestinal bleeding. ESA use in cancer patients has not been demonstrated in controlled clinical trials to improve symptoms of anemia, quality of life, fatigue, or patient well-being.

Depending on the clinical situation and the severity of anemia, red blood cell transfusion could be an alternative option to ESA therapy (Rizzo et al. 2008). Otherwise, a s.c. rHuEPO dose of 150 IU/kg three times in a week (TIW) or 40 kIU weekly (QW) is recommended to increase hemoglobin and decrease transfusions in patients with chemotherapy-associated anemia when the hemoglobin concentration is approaching, or has fallen below, 10 g/dL. Alternatively, s.c. rHuEPO dose of 80 kIU biweekly (Q2W) or 120 kIU every 3 weeks (Q3W) can be used as initial dosing because these two dosage schedules have not been found to have any differences in efficacy with respect to the approved TIW and QW dosing schedules. The dose of ESA therapy should be titrated for each patient to achieve and maintain the lowest hemoglobin level sufficient to avoid the need for blood transfusion. Therefore, the TIW s.c. dose of rHuEPO should be increased to 300 IU/kg if no reduction in transfusion requirements or rise in hemoglobin after 8 weeks of treatment has been observed. Similarly, the QW dose should increase to 60 kIU if no increase in hemoglobin by at least 1 g/dL after 4 weeks of treatment is observed. In addition, if hemoglobin exceeds 11 g/dL, but not 12 g/dL, the dose should be reduced by 25 %. However, if hemoglobin exceeds 12 g/dL, therapy should be held until hemoglobin falls below 11 g/dL and then restarted at a 25 % dose reduction. The pediatric dosing guidance is based on an initial i.v. dose of 600 IU/kg QW (maximum 40 kIU). If there is no increase in hemoglobin by at least 1 g/dL after 4 weeks of treatment (in the absence of RBC transfusion), the rHuEPO dose should be increased to 900 IU/kg (maximum 60 kIU) in order to maintain the lowest hemoglobin level sufficient to avoid RBC transfusion.

The recommended initial s.c. dose of darbepoetin alfa is 2.25 μg/kg QW or 500 μg once every 3 weeks (Q3W). The initial darbepoetin alfa s.c. dose of 2.25 μg/kg QW should be increased to 4.5 μg/kg QW if hemoglobin increase is less than 1 g/dL after 6 weeks of treatment. In addition, if hemoglobin increases by more than 1 g/dL in any 2-week period or when the hemoglobin reaches a level needed to avoid transfusion, the dose should be reduced by 40 %. If hemoglobin exceeds a level needed to avoid transfusion, therapy should be held until hemoglobin approaches a level where transfusions may be required then restarted at a 40 % dose reduction. A s.c. darbepoetin alfa dose of 100 μg QW, 200 μg Q2W, or 300 μg Q3W can be used as alternative initial dosing since differences in efficacy have not been found. These initial dose levels should be increased to 150–200 μg QW, 300 μg Q2W, or 500 μg Q3W, respectively. At this time the safety and efficacy of darbepoetin alfa in children receiving chemotherapy has not been established.

Although no specific serum rHuEPO level has been established which predicts which patients would be unlikely to respond to epoetin alfa therapy, treatment is not recommended for patients with grossly elevated serum rHuEPO levels (e.g., greater than 200 mUnits/mL). The hemoglobin should be monitored on a weekly basis in patients receiving ESA therapy until hemoglobin becomes stable and then at regular intervals thereafter.

Patients with multiple myeloma, especially those with renal failure, may benefit from adjunctive ESA therapy to treat anemia. Endogenous EPO levels should be monitored in order to assist in planning ESA therapy. No high-quality, published studies support the exclusive use of epoetin or darbepoetin in anemic myeloma, non-Hodgkin's lymphoma, or chronic lymphocytic leukemia in the absence of chemotherapy. Treatment with chemotherapy and/or corticosteroids should be initiated first. If a rise in hemoglobin does not result, treatment with epoetin or darbepoetin may begin in patients with particular caution exercised with chemotherapeutic agents and disease states where the risk of thromboembolism is increased. Blood transfusion is also an option (Rizzo et al. 2008). The current standard of care for symptomatic anemia in patients with myelodysplastic syndrome (MDS) is supportive care with RBC transfusion. Patients with serum EPO levels less than or equal to 500 IU/L, normal cytogenetics, and less than 15 % marrow-ringed sideroblasts may respond to relatively high doses of rHuEPO (40–60 kIU s.c.TIW) or darbepoetin alfa (150–300 μg QW s.c.). Evidence supports the use of epoetin or darbepoetin in patients with anemia associated with low-risk myelodysplasia (Rizzo et al. 2008). Supportive care with RBC transfusion is the standard of care for symptomatic anemia in patients with hematologic malignancies (non-Hodgkin's lymphoma, chronic lymphocytic leukemia). There is insufficient data to recommend ESA therapy for patients responding to treatment with good prognosis and persistent transfusion-dependent anemia.

Iron supplementation improves the hematopoietic response of ESAs used for chemotherapy-induced anemia. A recent meta-analysis of randomized, controlled trials, comparing parenteral or oral iron and no iron, when added to ESAs in anemic cancer patients, evidenced that overall parenteral iron reduces the risk of transfusions by 23 % and increases the chance of hematopoietic response by 29 % when compared with ESAs alone. On the contrary, oral iron does not increase hematopoietic response or transfusion rate. The significance of these results is that the proportion of nonresponders to ESAs treated with parenteral iron will have strongly improved quality of life and cost ameliorated (Petrelli et al. 2012).

Several studies have reported a possible decreased survival rate in cancer patients receiving ESA for correction of anemia. Analyses of eight randomized studies in patients with cancer found a decrease in overall survival and/or locoregional disease control associated with ESA therapy for correction of anemia with an off-label target hemoglobin level greater than 12 g/dL. These results were confirmed in three recent meta-analyses (Bennett et al. 2008; Bohlius et al. 2009; Tonelli et al. 2009) and refuted in other two meta-analyses (Ludwig et al. 2009; Glaspy et al. 2010). There are also observational data and data from randomized studies that show no increase in mortality with ESA use according to prescribing label specifically in patients receiving chemotherapy. In addition, an increased risk for thromboembolic events has been reported with ESA therapy in cancer patients. Besides the intrinsic risk associated with the malignancy itself, the chemotherapy, and other concomitant factors, results from several meta-analyses established a significant association between the increased risk for thrombotic events and ESA use, with relative risk point estimates ranging from 1.48 to 1.69 (Bennett et al. 2008; Tonelli et al. 2009; Ludwig et al. 2009; Glaspy et al. 2010). The increased risk for mortality and thrombotic events in cancer patients receiving ESA therapy is specified in the black box warning included in the FDA label. Seizures and antibody-associated pure red cell aplasia (PRCA) have occurred in chronic renal failure patients receiving ESA therapy. While it is unclear whether cancer patients receiving ESA therapy are at risk of seizures and/or PRCA, ESA treatments should be closely monitored.

MYELOID HEMATOPOIETIC GROWTH FACTORS

■ Granulocyte Colony-Stimulating Factor (G-CSF)

The chemical properties of the myeloid hematopoietic growth factors, G-CSF and GM-CSF, have been characterized (Table 18.2) and extensively reviewed (Armitage 1998). The gene that encodes G-CSF is located on chromosome 17; the mature G-CSF polypeptide has 174 amino acids and is produced in monocytes, fibroblasts, endothelial cells, and bone marrow stromal cells. Filgrastim, a non-glycosylated r-metHuG-CSF, is marketed by several companies under various trade names throughout the world, and several filgrastim biosimilars have also been approved. Lenograstim, a glycosylated rHuG-CSF, is not marketed in the United States but is marketed in other countries under several trade names. Pegfilgrastim, a sustained-duration form of filgrastim to which a 20 kDa polyethylene glycol molecule is covalently bound to the N-terminal methionine residue, is marketed as Neulasta® in the European Union, the United States, and other countries, and several pegfilgrastim biosimilars are in development. Although not all indications are approved in every country, filgrastim, lenograstim, and pegfilgrastim are indicated for the prevention and treatment of chemotherapy-induced febrile neutropenia in cancer patients receiving chemotherapy, mobilization of stem cells for transplantation in oncology patients, and support of induction/consolidation chemotherapy for AML and hematopoiesis after bone marrow transplantation, among others (Aapro et al. 2011).

Similar to filgrastim, pegfilgrastim increases the proliferation and differentiation of neutrophils from committed progenitor cells, induces maturation, and enhances the survival and function of mature neutrophils, resulting in dose-dependent increases in neutrophils counts. Filgrastim is primarily eliminated by glomerular filtration in the kidney and binding to the G-CSF receptor on the cell surface of neutrophils and neutrophil precursors, with subsequent internalization of the growth factor-receptor complexes via endocytosis and degradation inside the cells. Pegylation of filgrastim renders renal clearance insignificant, and neutrophil-mediated clearance becomes the predominant elimination pathway. After subcutaneous administration, pegfilgrastim exhibits

	G-CSF	GM-CSF
Nonproprietary name	Filgrastim, lenograstim, and pegfilgrastim	Molgramostim and sargramostim
Chromosome location	17	4
Amino acids	174[a]	127 or 128[b]
Glycosylation	*O*-linked (lenograstim)	*N*-linked (sargramostim)
Pegylation	Pegfilgrastim	None
Source of gene	Bladder carcinoma cell line (filgrastim, pegfilgrastim) and squamous carcinoma cell line (lenograstim)	Human monocyte cell line (molgramostim) and mouse T-lymphoma cell line (sargramostim)
Expression system	*E. coli* (bacteria): filgrastim and pegfilgrastim	*E. coli* (bacteria): molgramostim
	Chinese hamster ovary cell line (mammalian): lenograstim	*Saccharomyces cerevisiae* (yeast): sargramostim

[a]Native G-CSF has two forms, one with 177, which is less active than the other form with 174 amino acids; filgrastim has an N-terminal methionine
[b]Molgramostim has 128 amino acids; sargramostim 127

Table 18.2 ■ Characteristics of the marketed myeloid growth factors, rhG-CSF, and rhGM-CSF.

nonlinear and nonstationary pharmacokinetics due to pharmacodynamic-mediated drug disposition. During chemotherapy-induced neutropenia, the clearance of pegfilgrastim is significantly reduced, and the concentration of pegfilgrastim is sustained until the onset of neutrophil recovery. Data from a pivotal study confirmed that a once-per-chemotherapy-cycle injection of pegfilgrastim at 6 mg was as safe and effective as 11 daily injections of filgrastim at 5 μg/kg in reducing neutropenia and its complications in patients with breast cancer receiving four cycles of doxorubicin/docetaxel chemotherapy (Green et al. 2003). Because of the highly efficient regulation of pegfilgrastim clearance via neutrophils and neutrophil precursors, a single fixed dose of pegfilgrastim can be given once per chemotherapy cycle in conjunction with a variety of myelosuppressive chemotherapy regimens (Yang and Kido 2011). Extensive clinical reviews on the myeloid growth factors have been published elsewhere (Keating 2011; Crawford et al. 2009).

■ Granulocyte-Macrophage Colony-Stimulating Factor (GM-CSF) and Stem Cell Factor (SCF)

The granulocyte-macrophage colony-stimulating factor (GM-CSF) is a polypeptide of 128 amino acids encoded by a gene located on chromosome 4, secreted by macrophages, T cells, mast cells, NK cells, endothelial cells, and fibroblasts. Molgramostim (marketed in the EU) and sargramostim (marketed in the USA) are two versions of rHuGM-CSF rarely used today.

Similarly to G-CSF and GM-CSF, stem cell factor (SCF), encoded on chromosome 12, is a membrane-bound polypeptide of 248 amino acids that proteolytically release a soluble SCF containing 165 amino acids. SCF is an early-acting hematopoietic growth factor that stimulates the proliferation of primitive hematopoietic and non-hematopoietic cells. In vitro, SCF alone has minimal colony-stimulating activity on hematopoietic progenitor cells; however, it synergistically increases colony-forming or stimulatory activity of other HGF. Unlike most hematopoietic growth factors, SCF circulates in relatively high concentrations in normal human plasma. Ancestim® is a non-glycosylated version of the soluble r-metHuSCF marketed in Canada, Australia, and New Zealand and is occasionally used in combination with G-CSF to increase the mobilization of peripheral blood progenitor cells (PBPC) for harvesting and support of autologous transplantation after myeloablative chemotherapy in patients with cancer. Comprehensive reviews of r-metHuSCF have been recently published (Langley 2004).

■ Megakaryocyte Hematopoietic Growth Factors

Megakaryocytopoiesis is a continuous developmental process of platelet production regulated by a complex network of HGF. In this process, hematopoietic stem cells undergo proliferation, differentiation, and maturation, generating megakaryocytes and platelets. Platelet production is controlled by signaling through the hematopoietic c-Mpl receptor. The ligand for this receptor, thrombopoietin (TPO) is the primary regulator of megakaryocyte development and subsequent platelet formation. TPO is a HGF encoded on chromosome 3 and produced in the liver and bone marrow stroma. Depending on the source, the mature polypeptide has between 305 and 355 amino acids, which may undergo cleavage to a smaller polypeptide that retains biologic activity. Upon binding to the c-Mpl receptor, TPO triggers several cellular signal transduction processes, which involve the JAK-STAT and TYK2 tyrosine kinase pathways, the mitogen-activated protein kinase (MAPK) pathway, the phosphatidylinositol 3-kinase (PI3K) pathway, and the nuclear factor kappa B (NF-κB) pathway.

Early recombinant forms of TPO, rHu-TPO and the pegylated megakaryocyte growth and development factor (Peg-MGDF), showed promising results in clinical trials. However, later studies failed to meet their clinical endpoints, because the recombinant proteins generated antibodies that cross-reacted with c-Mpl ligands and resulted in paradoxical thrombocytopenia (Li et al. 2001). Further clinical development of these molecules was therefore suspended. An extensive compilation of the biology of rHu-TPO and Peg-MGDF has been published (Kuter et al. 1997).

Romiplostim (Nplate®), previously known as AMG 531, is a novel biological thrombopoiesis-stimulating agent that was developed to overcome the problem of cross-reacting autoantibodies by use of a peptide sequence with no homology to endogenous TPO to activate the c-Mpl receptor. Structurally, romiplostim is a 59 kD fusion protein that consists of two identical subunits, each containing a human IgG_1 Fc domain covalently linked at the C-terminus to a peptide consisting of two c-Mpl binding domains. The four copies of the TPO mimetic peptide stimulate megakaryocytopoiesis by binding the TPO receptor, yet because they bear no sequence homology with TPO, there is a reduced potential for the generation of anti-TPO antibodies. In vitro, romiplostim competes with TPO for binding to the c-Mpl receptor on normal platelets and Mpl-transfected cells (BaF3-Mpl cells). Upon binding to the c-Mpl receptor, romiplostim activates the Janus kinase/signal transducers and activators of transcription (JAK-STAT) and other pathways in the same way as endogenous TPO (Broudy and Lin 2004). When cocultured with murine bone marrow cells, romiplostim promotes the growth of CFU-megakaryocytes and promotes the proliferation as well as the maturation of megakaryocytes. During preclinical development, romiplostim led to robust platelet responses in mice, rats, rabbits, and monkeys. The pharmacokinetics and pharmacodynamics of romiplostim in animals, healthy subjects, and patients with immune thrombocytopenia purpura (ITP) have been extensively investigated during clinical development (Wang et al. 2010, 2011; Yan and Krzyzanski 2013; Perez-Ruixo et al. 2012). Currently, romiplostim has been approved in the USA and the EU for the treatment of patients with immune thrombocytopenia purpura (Bussel et al. 2006), and an extensive review of the use of romiplostim in ITP patients has been recently published (Keating 2012). At this time, romiplostim or other protein-based c-Mpl ligands are not approved for clinical use in cancer patients; however, clinical trial data in oncology patients have been recently reported (Kantarjian et al. 2010a, b).

SELF-ASSESSMENT QUESTIONS

■ Questions

1. What do hematopoietic factors do?
2. What are the major lineages or types of mature blood cells?
3. Generally, chemically describe the hematopoietic growth factors.
4. How do hematopoietic growth factors function?
5. What are the in vivo actions of rhG-CSF and rhGM-CSF in patients with advanced cancer?
6. What is the physiologic role of EPO?
7. What are the currently commercially available hematopoietic growth factors?
8. What are the indications for rhG-CSF?
9. What are the indications for rhGM-CSF?
10. What are the indications for rhEPO?
11. What are the indications for rhSCF?

■ Answers

1. Hematopoietic growth factors regulate both hematopoiesis and the functional activity of blood cells (including proliferation, differentiation, and maturation). Some hematopoietic growth factors mobilize progenitor cells to move from the bone marrow to the peripheral blood.
2. The myeloid pathway gives rise to red blood cells (erythrocytes), platelets, monocytes/macrophages, and granulocytes (neutrophils, eosinophils, and basophils). The lymphoid pathway gives rise to lymphocytes.
3. They are glycoproteins, which can be distinguished by their amino acid sequence and glycosylation (carbohydrate linkages). Hematopoietic growth factors have folding patterns that are dictated by physical interactions and covalent cysteine-cysteine disulfide bridges. Correct folding is necessary for biologic activity. Most hematopoietic growth factors are single-chain polypeptides weighing approximately 14–35 kDa. The carbohydrate content varies depending on the growth factor and production method, which in turn affects the molecular weight but not necessarily the biologic activity.
4. Hematopoietic growth factors act by binding to specific cell surface receptors. The resultant complex sends a signal to the cell to express genes, which in turn induce cellular proliferation, differentiation, or activation. A hematopoietic growth factor may also act indirectly if the cell expresses a gene that causes the production of a different hematopoietic growth factor or another cytokine, which in turn binds to and stimulates a different cell.
5. Both growth factors cause a transient leucopenia that is followed by a dose-dependent increase in the number of circulating mature and immature neutrophils. Both growth factors enhance the in vitro function of neutrophils obtained from treated patients. rhGM-CSF, but not rhG-CSF, also increases the number of circulating monocytes/macrophages and eosinophils, as well as in vitro monocyte cytotoxicity and cytokine production.
6. EPO maintains a normal red blood cell count by causing committed erythroid progenitor cells to proliferate and differentiate into normoblasts. EPO also shifts marrow reticulocytes into circulation.
7. Five hematopoietic growth factors are commercially available, rhG-CSF (filgrastim, lenograstim, pegfilgrastim), rhGM-CSF (molgramostim and sargramostim), rhEPO (epoetin alfa, epoetin beta, darbepoetin alfa), rhSCF (ancestim), and rhIL-11 (oprelvekin).
8. Approval for marketing varies by country and not all countries have all labeled uses. rhG-CSF is indicated for neutropenia associated with myelosuppressive cancer chemotherapy, bone marrow transplantation, and severe chronic neutropenia; rhG-CSF is also indicated to mobilize peripheral blood progenitor cells (PBPC) for PBPC transplantation; and rhG-CSF is indicated for the reversal of clinically significant neutropenia and subsequent maintenance or adequate neutrophil counts in patients with HIV infection during treatment with antiviral and/or other myelosuppressive medications.
9. rhGM-CSF is indicated for neutropenia associated with bone marrow transplantation and antiviral therapy for AIDS-related cytomegalovirus. rhGM-CSF is also indicated for failed bone marrow transplantation or delayed engraftment and for use in mobilization and after transplantation of autologous PBPCs.
10. rhEPO is indicated to treat anemia associated with chronic renal failure, zidovudine-induced anemia in HIV-infected patients, and chemotherapy-induced

anemia. rhEPO is also indicated to reduce allogeneic blood transfusions and hasten erythroid recovery in surgery patients.

11. rhSCF is used in combination with filgrastim to increase PBPC yield in hard-to-mobilize patients.

Acknowledgements Parts of this chapter are updated versions of previously published portions of several other chapters that include:

1. Heatherington AC (2003) Clinical pharmacokinetic properties of rHuEPO: a review. In: Molineux G, Foote MA, Elliott S (eds) Erythropoietins and erythropoiesis: molecular, cellular, preclinical, and clinical biology. Birkhauser, Basel, pp 87–112
2. Elliot S, Heatherington AC, Foote MA (2004) Erythropoietic factors: clinical pharmacology and pharmacokinetics. In: Morstyn G, Foote MA, and Lieschke GJ (eds) Hematopoietic growths factors in oncology. Humana Press, Totowa, pp 97–123
3. Foote AN (2008) Hematopoietic growth factors. In: Crommelin DJA, Sindelar RD, Meibohm B (eds) Pharmaceutical biotechnology. Fundamentals and applications, 3rd edn. Informa Healthcare USA, New York, pp 225–242
4. Doshi S, Perez-Ruixo JJ, Jang GR, Chow A, Elliot S (2008) Pharmacocinétique de les agents stimulant l'érythropoïèse. In: Rossert J, Casadevall N, Gisselbrecht C (eds) Les agents stimulant l'érythropoïèse. Paris, France
5. Doshi S, Perez-Ruixo JJ, Jang GR, Chow AT (2009) Pharmacokinetics of erythropoiesis-stimulating agents. In: Molineux G, Foote MA, Elliott S (eds) Erythropoietins and erythropoiesis: molecular, cellular, preclinical, and clinical biology. 2nd edn. Birkhäuser Verlag AG, Basel, pp 195–224

REFERENCES

Aapro MS, Bohlius J, Cameron DA et al (2011) 2010 update of EORTC guidelines for the use of granulocyte-colony stimulating factor to reduce the incidence of chemotherapy-induced febrile neutropenia in adult patients with lymphoproliferative disorders and solid tumours. Eur J Cancer 47:8–32

Adamson JW, Eschbach JW (1990) Treatment of the anemia of chronic renal failure with recombinant human erythropoietin. Annu Rev Med 41:349–360

Agoram B, Sutjandra L, Sullivan JT (2007) Population pharmacokinetics of darbepoetin alfa in healthy subjects. Br J Clin Pharmacol 63:41–52

Agoram B, Aoki K, Doshi S et al (2009) Investigation of the effects of altered receptor binding activity on the clearance of erythropoiesis-stimulating proteins: nonerythropoietin receptor-mediated pathways may play a major role. J Pharm Sci 98:2198–2211

Ait-Oudhia S, Vermeulen A, Krzyzanski W (2011) Non-linear mixed effect modeling of the time-variant disposition of erythropoietin in anemic cancer patients. Biopharm Drug Dispos 32:1–15

Armitage JO (1998) Emerging applications of recombinant human granulocyte colony-stimulating factor. Blood 92:4491–4508

Bennett CL, Silver SM, Djulbegovic B et al (2008) Venous thromboembolism and mortality associated with recombinant erythropoietin and darbepoetin administration for the treatment of cancer-associated anemia. JAMA 299:914–924

Bohlius J, Schmidlin K, Brillant C et al (2009) Recombinant human erythropoiesis-stimulating agents and mortality in patients with cancer: a meta-analysis of randomised trials. Lancet 373:1532–1542

Bron D, Meuleman N, Mascaux C (2001) Biological basis of anemia. Semin Oncol 28:1–6

Broudy VC, Lin NL (2004) AMG531 stimulates megakaryopoiesis in vitro by binding to Mpl. Cytokine 25:52–60

Bussel JB, Kuter DJ, George JN et al (2006) AMG 531, a thrombopoiesis-stimulating protein, for chronic ITP. N Engl J Med 355:1672–1681

Chanu P, Gieschke R, Charoin JE, Pannier A, Reigner B (2010) Population pharmacokinetic – pharmacodynamic model for a C.E.R.A. in both ESA-naive and ESA-treated chronic kidney disease patients with renal anemia. J Clin Pharmacol 50:507–520

Chapel S, Veng-Pedersen P, Hohl RJ, Schmidt RL, McGuire EM, Widness JA (2001) Changes in erythropoietin pharmacokinetics following busulfan-induced bone marrow ablation in sheep: evidence for bone marrow as a major erythropoietin elimination pathway. J Pharmacol Exp Ther 298:820–824

Cheung WK, Goon BL, Guilfoyle MC, Wacholtz MC (1998) Pharmacokinetics and pharmacodynamics of recombinant human erythropoietin after single and multiple subcutaneous doses to healthy subjects. Clin Pharmacol Ther 64:412–423

Cheung WK, Minton N, Gunawardena K (2001) Pharmacokinetics and pharmacodynamics of epoetin alfa once weekly and three times weekly. Eur J Clin Pharmacol 57:411–418

Crawford J, Armitage J, Balducci L et al (2009) Myeloid growth factors. J Natl Compr Canc Netw 7:64–83

Deicher R, Horl WH (2004) Differentiating factors between erythropoiesis-stimulating agents: a guide to selection for anemia of chronic kidney disease. Drugs 64:499–509

Doshi S, Chow A, Pérez Ruixo JJ (2010) Exposure-response modeling of darbepoetin alfa in anemic patients with chronic kidney disease not receiving dialysis. J Clin Pharmacol 50(9 Suppl):75S–90S

Egrie JC, Browne JK (2001) Development and characterization of novel erythropoiesis stimulating protein (NESP). Br J Cancer 84(Suppl 1):3–10

Egrie JC, Strickland TW, Lane J et al (1986) Characterization and biological effects of recombinant human erythropoietin. Immunobiology 72:213–224

Egrie JC, Dwyer E, Browne JK, Hitz A, Lykos MA (2003) Darbepoetin alfa has a longer circulating half-life and greater in vivo potency than recombinant human erythropoietin. Exp Hematol 31:290–299

Elliott S, Lorenzini T, Asher S et al (2003) Enhancement of therapeutic protein in vivo activities through glycoengineering. Nat Biotechnol 21:414–421

Elliott S, Heatherington AC, Foote MA (2004a) Erythropoietic factors. In: Morstyn G, Foote MA, Lieschke GJ (eds) Hematopoietic growth factors in oncology: basic science and clinical therapeutics. Humana Press Inc, Totowa, pp 97–123

Elliott S, Egrie J, Browne J et al (2004b) Control of rHuEPO biological activity: the role of carbohydrate. Exp Hematol 32:1146–1155

Fan Q, Leuther KK, Holmes CP, Fong KL, Zhang J, Velkovska S, Chen MJ, Mortensen RB, Leu K, Green JM, Schatz PJ, Woodburn KW (2006) Preclinical evaluation of Hematide, a novel erythropoiesis stimulating agent, for the treatment of anemia. Exp Hematol 34:1303–1311

Fisher JW (2003) Erythropoietin: physiology and pharmacology update. Exp Biol Med 228:1–14

Gascon P, Pirker R, Del Mastro L, Durrwell L (2010) Effects of CERA (continuous erythropoietin receptor activator) in patients with advanced non-small-cell lung cancer (NSCLC) receiving chemotherapy: results of a phase II study. Ann Oncol 21:2029–2039

Glaspy J, Henry D, Patel R et al (2005) The effects of chemotherapy on endogenous erythropoietin levels and the pharmacokinetics and erythropoietic response of darbepoetin alfa: a randomised clinical trial of synchronous versus asynchronous dosing of darbepoetin alfa. Eur J Cancer 41:1140–1149

Glaspy J, Crawford J, Vansteenkiste J et al (2010) Erythropoiesis stimulating agents in oncology: a study-level meta-analysis of survival and other safety outcomes. Br J Cancer 102:301–315

Green MD, Koelbl H, Baselga J et al (2003) A randomized double-blind multicenter phase III study of fixed-dose single-administration pegfilgrastim versus daily filgrastim in patients receiving myelosuppressive chemotherapy. Ann Oncol 14:29–35

Gross AW, Lodish HF (2006) Cellular trafficking and degradation of erythropoietin and novel erythropoiesis stimulating protein (NESP). J Biol Chem 281:2024–2032

Halstenson CE, Macres M, Katz SA et al (1991) Comparative pharmacokinetics and pharmacodynamics of epoetin alfa and epoetin beta. Clin Pharmacol Ther 50:702–712

Heatherington AC (2003) Clinical pharmacokinetic properties of rHuEPO: a review. In: Molineux G, Foote MA, Elliott S (eds) Erythropoietins and erythropoiesis: molecular, cellular, preclinical, and clinical biology. Birkhauser, Basel, pp 87–112

Jarsch M, Brandt M, Lanzendörfer M, Haselbeck A (2008) Comparative erythropoietin receptor binding kinetics of C.E.R.A. and epoetin-beta determined by surface plasmon resonance and competition binding assay. Pharmacology 81:63–69

Jelkmann W (1992) Erythropoietin: structure, control of production, and function. Physiol Rev 72:449–489

Jelkmann W (2000) Use of recombinant human erythropoietin as an antianemic and performance enhancing drug. Curr Pharm Biotechnol 1:11–31

Jensen JD, Jensen LW, Madsen JK (1994) The pharmacokinetics of recombinant human erythropoietin after subcutaneous injection at different sites. Eur J Clin Pharmacol 46:333–337

Jungers PY, Robino C, Choukroun G, Nguyen-Khoa T, Massy ZA, Jungers P (2002) Incidence of anaemia, and use of Epoetin therapy in pre-dialysis patients: a prospective study in 403 patients. Nephrol Dial Transplant 17:1621–1627

Kakkar T, Sung C, Gibiansky L et al (2011) Population PK and IgE pharmacodynamic analysis of a fully human monoclonal antibody against IL4 receptor. Pharm Res 28:2530–2542

Kantarjian H, Fenaux P, Sekeres MA et al (2010a) Safety and efficacy of romiplostim in patients with lower-risk myelodysplastic syndrome and thrombocytopenia. J Clin Oncol 28:437–444

Kantarjian HM, Giles FJ, Greenberg PL et al (2010b) Phase 2 study of romiplostim in patients with low- or intermediate-risk myelodysplastic syndrome receiving azacitidine therapy. Blood 116:3163–3170

Kato A, Hishida A, Kumagai H, Furuya R, Nakajima T, Honda N (1994) Erythropoietin production in patients with chronic renal failure. Ren Fail 16:645–651

Kaufman JS, Reda DJ, Fye CL et al (1998) Subcutaneous compared with intravenous epoetin in patients receiving hemodialysis. Department of Veterans Affairs Cooperative Study Group on Erythropoietin in Hemodialysis Patients. N Engl J Med 339:578–583

Keating GM (2011) Lenograstim: a review of its use in chemotherapy-induced neutropenia, for acceleration of neutrophil recovery following haematopoietic stem cell transplantation and in peripheral blood stem cell mobilization. Drugs 71:679–707

Keating GM (2012) Romiplostim. Drugs 72:415–435

Krzyzanski W, Jusko WJ, Wacholtz MC, Minton N, Cheung WK (2005) Pharmacokinetic and pharmacodynamic modeling of recombinant human erythropoietin after multiple subcutaneous doses in healthy subjects. Eur J Pharm Sci 26:295–306

Kuter DJ, Hunt P, Sheridan W, Zucker-Franklin D (eds) (1997) Thrombopoiesis and thrombopoeitin. Humana Press Inc, Totowa, p 412

Langley KE (2004) Stem cell factor and its receptor, c-Kit. In: Morstyn G, Foote MA, Lieschke GJ (eds) Hematopoietic growth factors in oncology. Humana Press Inc, Totowa, pp 153–184

Li J, Yang C, Xia Y et al (2001) Thrombocytopenia caused by the development of antibodies to thrombopoietin. Blood 98:3241–3248

Locatelli F, Del Vecchio L (2009) Hematide™ for the treatment of chronic kidney disease-related anemia. Expert Rev Hematol 2:377–383

Ludwig H, Crawford J, Osterborg A et al (2009) Pooled analysis of individual patient-level data from all randomized, double-blind, placebo controlled trials of darbepoetin alfa in the treatment of patients with chemotherapy-induced anemia. J Clin Oncol 27:2838–2847

Macdougall IC (2005) CERA (Continuous Erythropoietin Receptor Activator): a new erythropoiesis-stimulating agent for the treatment of anemia. Curr Hematol Rep 4:436–440

Macdougall IC, Gray SJ, Elston O et al (1999) Pharmacokinetics of novel erythropoiesis stimulating protein compared with epoetin alfa in dialysis patients. J Am Soc Nephrol 10:2392–2395

McLennan DN, Porter CJ, Edwards GA, Heatherington AC, Martin SW, Charman SA (2006) The absorption of darbepoetin alfa occurs predominantly via the lymphatics following subcutaneous administration to sheep. Pharm Res 23:2060–2066

McMahon FG, Vargas R, Ryan M et al (1990) Pharmacokinetics and effects of recombinant human erythropoietin after intravenous and subcutaneous injections in healthy volunteers. Blood 76:1718–1722

Miller CB, Jones RJ, Piantadosi S, Abeloff MD, Spivak JL (1990) Decreased erythropoietin response in patients with the anemia of cancer. N Engl J Med 322: 1689–1692

Nissenson AR, Swan SK, Lindberg JS et al (2002) Randomized, controlled trial of darbepoetin alfa for the treatment of anemia in hemodialysis patients. Am J Kidney Dis 40:110–118

Olsson-Gisleskog P, Jacqmin P, Perez-Ruixo JJ (2007) Population pharmacokinetics meta-analysis of recombinant human erythropoietin in healthy subjects. Clin Pharmacokinet 46:159–173

Perez-Ruixo JJ, Krzyzanski W, Bouman-Thio E et al (2009) Pharmacokinetics and pharmacodynamics of the erythropoietin mimetibody construct CNTO 528 in healthy subjects. Clin Pharmacokinet 48:601–613

Perez-Ruixo JJ, Green B, Doshi S, Wang YM, Mould DR (2012) Romiplostim dose-response in patients with immune thrombocytopenia. J Clin Pharmacol 52:1540–1551

Petrelli F, Borgonovo K, Cabiddu M, Lonati V, Barni S (2012) Addition of iron to erythropoiesis-stimulating agents in cancer patients: a meta-analysis of randomized trials. J Cancer Res Clin Oncol 138:179–187

Piron M, Loo M, Gothot A, Tassin F, Fillet G, Beguin Y (2001) Cessation of intensive treatment with recombinant human erythropoietin is followed by secondary anemia. Blood 97:442–448

Porter CJ, Charman SA (2000) Lymphatic transport of proteins after subcutaneous administration. J Pharm Sci 89:297–310

Ramakrishnan R, Cheung WK, Wacholtz MC, Minton N, Jusko WJ (2004) Pharmacokinetic and pharmacodynamic modeling of recombinant human erythropoietin after single and multiple doses in healthy volunteers. J Clin Pharmacol 44:991–1002

Rizzo JD, Somerfield MR, Hagerty KL et al (2008) Use of epoetin and darbepoetin in patients with cancer: 2007 American Society of Clinical Oncology/American Society of Hematology clinical practice guideline update. J Clin Oncol 26:132–149

Schellekens H, Moors E (2010) Clinical comparability and European biosimilar regulations. Nat Biotechnol 28:28–31

Sinclair AM, Elliott S (2005) Glycoengineering: the effect of glycosylation on the properties of therapeutic proteins. J Pharm Sci 94:1626–1635

Spivak JL (1998) The biology and clinical applications of recombinant erythropoietin. Semin Oncol 25(3 suppl 7):7–11

Supersaxo A, Hein WR, Steffen H (1990) Effect of molecular weight on the lymphatic absorption of water-soluble compounds following subcutaneous administration. Pharm Res 7:167–169

Sutjandra L, Rodriguez RD, Doshi S et al (2011) Population pharmacokinetic meta-analysis of denosumab in healthy subjects and postmenopausal women with osteopenia or osteoporosis. Clin Pharmacokinet 50:793–807

Tonelli M, Hemmelgarn B, Reiman T et al (2009) Benefits and harms of erythropoiesis-stimulating agents for anemia related to cancer: a meta-analysis. CMAJ 180:E62–E71

Uehlinger DE, Goth FA, Sheiner LB (1992) A pharmacodynamic model of erythropoietin therapy for uremic anemia. Clin Pharmacol Ther 51:76–89

Vansteenkiste J, Pirker R, Massuti B et al (2002) Double-blind, placebo-controlled, randomized phase III trial of darbepoetin alfa in lung cancer patients receiving chemotherapy. J Natl Cancer Inst 94:1211–1220

Walrafen P, Verdier F, Kadri Z, Chrétien S, Lacombe C, Mayeux P (2005) Both proteasomes and lysosomes degrade the activated erythropoietin receptor. Blood 105:600–608

Wang YM, Krzyzanski W, Doshi S, Xiao JJ, Pérez-Ruixo JJ, Chow AT (2010) Pharmacodynamics-mediated drug disposition (PDMDD) and precursor pool lifespan model for single dose of romiplostim in healthy subjects. AAPS J 12:729–740

Wang YM, Sloey B, Wong T, Khandelwal P, Melara R, Sun YN (2011) Investigation of the pharmacokinetics of romiplostim in rodents with a focus on the clearance mechanism. Pharm Res 28:1931–1938

Wide L, Bengtsson C, Birgegard G (1989) Circadian rhythm of erythropoietin in human serum. Br J Haematol 72:85–90

Yan X, Lowe PJ, Fink M, Berghout A, Balser S, Krzyzanski W (2012) Population pharmacokinetic and pharmacodynamic model-based comparability assessment of a recombinant human epoetin alfa and the biosimilar HX575. J Clin Pharmacol 52:1624–1644

Yan X, Krzyzanski W (2013) Quantitative assessment of minimal effective concentration of erythropoiesis-stimulating agents. CPT Pharmacometrics Syst Pharmacol. Aug 7;2:e62. doi:10.1038/psp.2013.39.

Yang BB, Kido A (2011) Pharmacokinetics and pharmacodynamics of pegfilgrastim. Clin Pharmacokinet 50: 295–306

19 Monoclonal Antibodies in Solid Organ Transplantation

Nicole A. Pilch, Holly B. Meadows, and Rita R. Alloway

INTRODUCTION

Administration of targeted immunosuppression, in the form of genetically engineered antibodies, is commonplace in solid organ transplantation. Polyclonal antibodies, such as rabbit antithymocyte globulin, offer global immunosuppression by targeting several cell surface antigens on B and T lymphocytes. However, secondary to their broad therapeutic targets, they are associated with infection, infusion-related reactions, inter-batch variability, and posttransplant malignancies. Nevertheless, polyclonal antibodies are still commonly administered for induction and treatment of allograft rejection and offer an important role in current solid organ transplantation, which is beyond the scope of this chapter.

In an attempt to target solid organ transplant immunosuppression, monoclonal antibodies directed against key steps in specific immunologic pathways were introduced. The first agent, muromonab-CD3 (OKT3), was initially introduced in the early 1980s for the treatment of allograft rejection (Morris 2004). The use of monoclonal antibodies has evolved and expanded over the past two decades and today monoclonal antibodies are routinely included as part of the overall immunosuppression regimen. Both the innate and adaptive immune systems have multiple components and signal transduction pathways aimed at protecting the host from a foreign body, such as transplanted tissue. The ultimate goal of posttransplant immunosuppression is tolerance, a state in which the host immune system recognizes the foreign tissue but does not react to it. This goal has yet to be achieved under modern immunosuppression secondary to immune system redundancy as well as the toxicity of currently available agents. Therefore, monoclonal antibodies are used to provide targeted, immediate immunomodulation aimed at attenuating the overall immune response. Specifically, monoclonal antibodies have been used to (1) decrease the inherent immunoreactivity of the potential transplant recipient prior to engraftment, (2) induce global immunosuppression at the time of transplantation allowing for modified introduction of other immunosuppressive agents (calcineurin inhibitors or corticosteroids), (3) spare exposure to maintenance immunosuppressive agents, and (4) treat acute allograft rejection. Monoclonal antibody selection, as well as dose, is based on patient-specific factors, such as indication for transplantation, type of organ being transplanted, and the long-term immunosuppression objective. To understand the approach that the transplant clinician uses to determine which agent to administer and when, it is necessary to briefly describe how immunoreactivity can be predicted and review the immunological basis for the use and development of monoclonal antibodies in solid organ transplantation.

IMMUNOLOGIC TARGETS: RATIONAL DEVELOPMENT/ USE OF MONOCLONAL ANTIBODIES IN ORGAN TRANSPLANT

The rational use of monoclonal antibodies in transplantation is focused on the prevention of host immune recognition of donor tissue. There are two ways in which allograft tissue can be immediately impaired secondary to the host immune response: complement-dependent antibody-mediated cell lysis (antibody-mediated rejection) and T-cell-mediated parenchymal

N.A. Pilch, PharmD (✉)
Department of Pharmacy and Clinical Sciences, South Carolina College of Pharmacy, Medical University of South Carolina, 150 Ashley Ave, 6th Fl Rutldge Tower Annex, Charleston, SC 29425, USA
e-mail: weimert@musc.edu

H.B. Meadows, PharmD
Department of Pharmacy and Clinical Sciences, South Carolina College of Pharmacy, Medical University of South Carolina, Charleston, SC, USA

R.R. Alloway, PharmD
Division of Nephrology, Department of Internal Medicine, University of Cincinnati, Cincinnati, OH, USA

D.J.A. Crommelin, R.D. Sindelar, and B. Meibohm (eds.), *Pharmaceutical Biotechnology*,
DOI 10.1007/978-1-4614-6486-0_19, © Springer Science+Business Media New York 2013

Monoclonal antibody	Molecular weight	Animal epitope	Molecular target	Target cells	Use
Alemtuzumab (Campath-1H®)	150 kDa	Murine/ human	CD52	Peripheral blood lymphocytes, natural killer cells, monocytes, macrophages, thymocytes	Induction Antibody-mediated rejection
Daclizumab (Zenapax®)	14.4 kDa	Murine/ human	CD25 alpha subunit	IL2-dependent T-lymphocyte activation	Induction
Basiliximab (Simulect®)	14.4 kDa	Murine/ human	CD25 alpha subunit	IL2-dependent T-lymphocyte activation	Induction
Muromonab-OKT3 (Orthoclone-OKT3®)	75 kDa	Murine	CD3	T lymphocytes (CD2, CD4, CD8)	Treatment of polyclonal antibody-resistant cellular-mediated rejection
Rituximab (Rituxan®)	145 kDa	Murine/ human	CD20	B lymphocytes	Desensitization Antibody-mediated rejection Focal segmental glomerulosclerosis
Belatacept (Nulojix®)	90 kDa	Humanized	CD80 and CD86	T lymphocytes	Maintenance immunosuppression
Eculizumab (Soliris®)	148 kDa	Murine/ human	C5	Block formation of membrane attack complex	Desensitization Antibody-mediated rejection Hemolytic uremic syndrome

Table 19.1 ■ Use of monoclonal antibodies in solid organ transplantation.

destruction leading to localized allograft inflammation and arteritis (cellular-mediated rejection) (Halloran 2004). Pre-transplant screening for antibodies against donor tissues has significantly reduced the incidence and severity of antibody-mediated rejection. However, as will be discussed, preferential destruction of cells that produce these antibodies using monoclonal technology, such as rituximab, prior to transplant has become an option for recipients with preformed alloantibodies. Prevention and treatment of cellular-mediated rejection, therefore, is the main focus of maintenance immunosuppression and the rationale for use of monoclonal antibodies in the posttransplant period. Cellular-mediated rejection is characterized by initial recognition of donor tissue by T cells. This leads to a complex signal transduction pathway traditionally described as three signals (Halloran 2004):

- Signal 1: Donor antigens are presented to T cells leading to activation, characterized by T-cell proliferation.
- Signal 2: CD80 and CD86 complex with CD28 on the T-cell surface activating signal transduction pathways (calcineurin, mitogen-activated protein kinase, protein kinase C, nuclear factor kappa B) which leads to further T-cell activation, cytokine release, and expression of the interleukin-2 (IL2) receptor (CD25).
- Signal 3: IL-2 and other growth factors cause the activation of the cell cycle and T-cell proliferation (Halloran 2004).

Monoclonal antibodies have been developed against various targets within this pathway to prevent propagation and lymphocyte proliferation providing profound immunosuppression (Table 19.1). Monoclonal antibodies that were originally developed for treatment of various malignancies have also been employed as immunosuppressant agents in solid organ recipients. Use of these agents must be balanced with maintenance immunosuppression to minimize the patient's risk of infection or malignancy from over-immunosuppression. Table 19.2 describes common adverse effects associated with maintenance immunosuppressant medications. Table 19.3 summarizes recent trends regarding the use of monoclonal antibodies for induction immunosuppression in solid organ transplantation.

■ Monoclonal Antibodies Administered Pre-transplant

Immunologic barriers to solid organ transplantation are common. Improved management of end-stage organ disease has increased the number of potential organ recipients and produced a significant shortage of organs available for transplant in comparison to the growing demand. Therefore, clinicians have sought to transplant across previously contraindicated immunologic barriers. In addition, more patients are surviving through their first transplant and are now waiting for a subsequent transplant. Monoclonal antibodies are now

	Hypertension	Hyperlipidemia	Hyperglycemia	Hematologic	Renal dysfunction	Dermatologic
Corticosteroids	+	++	++	–	–	++
Cyclosporine	+++	+++	++	+	+++	++
Tacrolimus	+++	+++	+++	+++	+++	++
Mycophenolate mofetil[a]	–	–	–	+++		–
Sirolimus	++	+++	–	+++	+	+++
Everolimus	++	+++	–	+++	+	+++
Belatacept	–	–	–	–	–	–

Incidence based on manufacturer package insert clinical trial approval reports, +<1 %, ++ 1–10 %, +++>10 %
[a]Adverse effects reported for mycophenolate mofetil (CellCept®) are based on clinical trials using this agent in combination with cyclosporine or tacrolimus and corticosteroids, values modified to account for concurrent agents

Table 19.2 ■ Complications of current maintenance immunosuppressants.

Organ	Who receive induction (%)	Alemtuzumab (%)	Basiliximab (%)	Daclizumab (%)	Muromonab (%)
Kidney	72	7	20	10	0
Pancreas	80	43	15	5	0
Heart	47	0	10	15	4
Lung	50	3	23	15	0
Liver	11	2	6	5	0
Intestine	50	19	0	9	0

Based on reported immunosuppression trends from 1994 to 2004, with data Adapted from Meier-Kriesche et al. (2006)

Table 19.3 ■ Current trends of monoclonal antibody induction use in solid organ transplantation.

being employed prior to transplant to desensitize the recipient's immune system. Desensitization is a strategy where immunosuppression is administered prior to transplant to prevent hyperacute or early rejection in patients who are known to have circulating antibodies against other human antigens. This strategy is generally reserved for patients who are "highly sensitized" during their evaluation for transplant. As the long-term significance of these sensitizing events is better understood, varying degrees of "desensitization" therapy are initiated based upon varying levels of sensitization. The long-term impact of this empirical therapy is yet to be defined. Specifically, as a patient develops end-stage organ disease, their medical and immunologic profiles are characterized. Blood samples from these potential recipients are screened for the presence of antibodies against the major histocompatibility complexes (MHC) on the surface of other human cells, specifically human leukocyte antigens (HLA). Potential recipients who have received blood products, previous organ transplants, or have a history of pregnancy are at higher risk for the development of antibodies against HLA. In addition, all humans have preformed IgG and IgM antibodies against the major blood group antigens (A, B, AB, and A1) (Reid and Olsson 2005). These antibodies will recognize donor tissue and quickly destroy (hyperacute rejection) the implanted organ if the tissue contains previously recognized HLA within minutes to hours following transplant. Therefore, it is necessary to evaluate the presence of preformed circulating antibodies against HLA in the potential organ recipients. Some centers will implement desensitization, which incorporates monoclonal antibodies prior to transplant to diminish the production of antibodies against a new organ, allowing for transplant across this immunologic barrier.

■ Monoclonal Antibodies Administered at the Time of Transplant

Current maintenance immunosuppression is aimed at various targets within the immune system to halt its signal transduction pathway. Available agents, although effective, are associated with significant patient and allograft adverse effects, which are correlated with long-term exposure (Table 19.2). The leading cause of death in noncardiac transplant recipients is a cardiovascular event. These cardiovascular events have been linked to long-term corticosteroid exposure. In addition, chronic administration of calcineurin inhibitors (cyclosporine and tacrolimus) is also associated with acute and chronic kidney dysfunction leading to hemodialysis or need for a kidney transplant. Monoclonal antibodies given at the time of transplant (induction) have been used to decrease the need for corticosteroids and allow for the delay or a reduction in the amount of calcineurin inhibitor used. Determination of the solid organ transplant recipient's immunologic risk at the time of transplant is necessary to determine which monoclonal antibody to use in order to minimize the risk of early acute rejection and graft loss. Recipients are stratified based on several donor, allograft, and recipient variables to determine their immunologic risk. Patients at high risk for acute rejection or those in which maintenance immunosuppression is going to be minimized should receive a polyclonal or monoclonal antibody that provides cellular apoptosis, for example, alemtuzumab or rabbit antithymocyte globulin. Recipients at low risk for acute rejection may receive a monoclonal antibody which provides immunomodulation without lymphocyte depletion, such as basiliximab.

Several important pharmacokinetic parameters must be considered when these agents are administered to the various organ transplant recipients. The volume of distribution, biological half-life, and total-body clearance can differ significantly from a kidney transplant recipient to a heart transplant recipient. Clinicians must consider when to administer monoclonal antibodies in different transplant populations to maximize efficacy and minimize toxicity. For example, heart and liver transplant recipients tend to lose large volumes of blood around the time of transplant; therefore, intraoperative administration may not be the optimal time to administer a monoclonal antibody since a large portion may be lost during surgery. Monoclonal antibodies are also removed by plasma exchange procedures, such as plasmapheresis, which may be performed during the perioperative period in solid organ transplant recipients (Nojima et al. 2005).

■ Monoclonal Antibodies Administered Following Transplant

Monoclonal antibodies given following transplantation are used to treat allograft rejection and more recently as maintenance immunosuppressants. Administration of these agents is mainly reserved for severe allograft rejection in which the immunologic insult must be controlled quickly. Under normal homeostatic conditions the humoral immune system provides immediate control of infectious pathogens through secretion of antibodies. Cell-mediated immunity, in addition to fighting infections, provides surveillance against the production of mutant cells capable of oncogenesis. Interruption of either of these immune systems through the use of monoclonal antibodies places these patients at significant risk for infection and malignancy. Careful post-administration assessment of infection and posttransplant malignancy is commonplace. While those monoclonal antibodies employed as maintenance immunosuppressants have been developed to decrease the toxicity of long-term exposure to traditional agents such as calcineurin inhibitors, which can lead to chronic kidney damage in all organ transplant recipients, the use of these monoclonal antibodies is not without their own risks.

SPECIFIC AGENTS USED IN SOLID ORGAN TRANSPLANT

■ Muromonab

Muromonab was the first monoclonal antibody used in solid organ transplantation. Muromonab is a murine monoclonal antibody directed against human CD3 receptor, which is situated on the T-cell antigen receptor of mature T cells, inducing apoptosis of the target cell (Wilde and Goa 1996). Cells which display the CD3 receptor include CD2-, CD4-, and CD8-positive lymphocytes (Ortho Biotech 2004). Other investigators suggest that muromonab may also induce CD3 complex shedding, lymphocyte adhesion molecule expression causing peripheral endothelial adhesion, and cell-mediated cytolysis (Wilde and Goa 1996; Ortho Biotech 2004; Buysmann et al. 1996; Magnussen et al. 1994; Wong et al. 1990). Muromonab is approved for the treatment of kidney allograft rejection and steroid-resistant rejection in heart transplant recipients (Ortho biotech 2004). Muromonab was initially employed as an induction agent for kidney transplant recipients, in conjunction with cyclosporine, azathioprine, and corticosteroids. When compared to patients who received no muromonab induction, the rate of acute rejection was lower and the time to first acute rejection was substantially greater (Wilde and Goa 1996). Liver recipients with renal dysfunction at the time of transplant who received muromonab induction were also able to run their posttransplant cyclosporine levels lower without an increased incidence of acute rejection (Wilde and Goa 1996). Therefore, administration of OKT3 enabled preservation of renal function in the setting of reduced calcineurin inhibitor exposure when compared to those who did not receive muromonab (Wilde and Goa 1996). The use of OKT3 as an induction agent is nearly extinct with the introduction of newer agents that have more favorable side effect profiles.

Today, muromonab is of historical value as it is no longer being manufactured. Although prior to its withdrawal from the market, it was reserved for treatment of refractory rejection. Muromonab is extremely effective at halting most corticosteroid as well as polyclonal antibody-resistant rejections. These rejections are treated with 5 mg of muromonab given daily for 7–14 days (Ortho Biotech 2004). The dose and duration of therapy is often dependent on clinical or biopsy resolution of rejection or may be correlated with circulating CD3 cell concentrations in the serum.

Most patients who are exposed to OKT3 will develop human anti-mouse antibodies (HAMA) following initial exposure. These IgG antibodies may lead to decreased efficacy of subsequent treatment courses, but premedication with corticosteroids or antiproliferative agents during initial therapy may reduce their development (Wilde and Goa 1996). Following administration, in vitro data indicates that a serum concentration of 1000 μg/L is required to inhibit cytotoxic T-cell function (Wilde and Goa 1996). In vivo concentrations near the in vivo threshold immediately (1 h) following administration but diminish significantly by 24 h (Wilde and Goa 1996). Steady-state concentrations of 900 ng/mL can be achieved after three doses, with a plasma elimination half-life of 18 h when used for treatment of rejection and 36 h when used for induction (Wilde and Goa 1996; Ortho Biotech 2004).

Muromonab administration is associated with significant acute and chronic adverse effects. Immediately following administration, patients will experience a characteristic OKT3 cytokine release syndrome. The etiology of this syndrome is characterized by the pharmacodynamic interaction the OKT3 molecule has at the CD3 receptor. Muromonab will stimulate the target cell following its interaction with the CD3 receptor prior to inducing cell death. Consequently, CD3 cell stimulation leads to cytokine production and release, which is compounded by acute cellular apoptosis leading to cell lysis and release of the intracellular contents. The cytokine release syndrome associated with muromonab manifests as high fever, chills, rigors, diarrhea, capillary leak, and in some cases aseptic meningitis (Wilde and Goa 1996). Capillary leak has been correlated with increased tumor necrosis factor release leading to an initial increase in cardiac output secondary to decreased peripheral vascular resistance, followed by a reduction in right heart filling pressures which leads to a decrease in stroke volume (Wilde and Goa 1996). Sequelae of this cytokine release syndrome can occur immediately, within 30–60 min, and last up to 48 h following administration (Wilde and Goa 1996; Ortho Biotech 2004). This syndrome appears to be the most severe following the initial dose when the highest inoculum of cells is present in the patient's serum or when preformed antibodies against the mouse epitope exist. Subsequent doses appear to be better tolerated, though cytokine release syndrome has been reported after five doses, typically when the dose has been increased or the CD3-positive cell population has rebounded from previous dose baseline (Wilde and Goa 1996). Pretreatment against the effects of this cytokine release is necessary to minimize the host response. Specifically, corticosteroids to prevent cellular response to cytokines, nonsteroidal anti-inflammatory agents to prevent sequelae of the arachidonic acid cascade, acetaminophen to halt the effects of centrally acting prostaglandins, and diphenhydramine to attenuate the recipient's response to histamine.

In addition to immediate adverse effects, the potency of muromonab has been associated with a high incidence of posttransplant lymphoproliferative disease and viral infections. For all patients, the 10-year cumulative incidence of posttransplant lymphoproliferative disease is 1.6 % (Opelz and Dohler 2004). Review of large transplant databases revealed that deceased donor kidney transplant recipients who received muromonab for induction or treatment had a cumulative incidence of posttransplant lymphoproliferative disease that was three times higher than those who did not received muromonab or other T-cell depleting induction (Opelz and Dohler 2004). This observation may be multifactorial. It is well known that posttransplant lymphoproliferative disease may be induced secondary to Epstein-Barr viral B-cell malignant transformation. Muromonab's potent inhibition of T lymphocytes over a sustained period of time diminishes the immune system's normal surveillance and destruction of malignant cell lines, consequently leading to unopposed transformed B-cell proliferation and subsequent posttransplant lymphoma (Opelz and Dohler 2004).

Early use and development of muromonab in solid organ transplantation was beneficial for the novel development and use of newer monoclonal agents. The immunodepleting potency of muromonab, combined with the significant risk for malignancy, has made its use obsolete in the setting of modern transplantation. However, this agent still serves as a template for treatment of severe allograft rejection and the use of monoclonal antibodies posttransplant.

■ Interleukin-2 Receptor Antagonists

Interleukin-2 antagonists were the next monoclonal antibodies to be used and were specifically developed for use in solid organ transplantation. As previously mentioned, monoclonal antibody use and development in solid organ transplantation is rational. The IL-2 receptor was targeted for several reasons. Interleukin-2, the ligand for the IL-2 receptor, is a highly conserved protein, with only a single gene locus on chromosome 4 (Church 2003). Animal IL-2 knockout models have decreased lymphocyte function at 2–4 weeks of age and early mortality at 6–9 weeks of age (Chen et al. 2002). These models also display significantly diminished myelopoiesis leading to severe anemia and global bone marrow failure (Chen et al. 2002). This observation confirms the significant role that IL-2 and the IL-2 receptor complex play in immunity. The function and biological effect of IL-2 binding to the IL-2 receptor was first reported by Robb and colleagues in 1981 (Robb et al. 1981). This in vitro study evaluated murine lymphocytes and found that the IL-2 receptor is only present on activated cells (CD4+ and CD8+) (Church 2003). Uchiyama and colleagues (1981) reported one of the first monoclonal antibodies developed against activated human T cells. This compound displayed in vitro preferential activity against activated T cells, including terminally mature T cells, but did not exhibit activity against B cells or monocytes (Uchiyama et al. 1981). Later it was determined that this antibody actually bound to the alpha subunit of the activated T-cell receptor, CD25 (Church 2003). The actual T-cell receptor is made up of three subunits, alpha, beta, and gamma. When the beta and gamma subunits combine, they can only be stimulated by high concentrations of IL-2; however, in conjunction with the alpha subunit, the receptor shows high affinity for IL-2 and can be stimulated at very low concentrations. The expression of IL-2 and the IL-2 receptor alpha region is highly regulated at the DNA transcription level and is induced following T-cell activation (Shibuya et al. 1990). The alpha subunit is continuously expressed during allograft rejection, T-cell-mediated autoimmune diseases, and malignancies (Church 2003). The beta and gamma subunits, however, have constitutive expression, resulting in low levels of expression in resting T lymphocytes (Vincenti et al. 1997, 1998). There is no constitutive expression of IL-2 or the alpha receptor subunit (Shibuya et al. 1990; Noguchi et al. 1993). Both, the beta and gamma subunits, have similar molecular structures and are members of the cytokine receptor superfamily, but are structurally dissimilar to the alpha subunit (Noguchi et al. 1993). Therefore, the alpha subunit (CD25) became a rational target for monoclonal development since it is only expressed on activated T cells. Blockade of the CD25 receptor was to halt the activity of IL-2, thereby decreasing proliferation and clonal expansion of T cells when activated by foreign donor antigens.

DACLIZUMAB

In 1997, daclizumab became the first anti-CD25 monoclonal antibody approved for use in the prevention of allograft rejection in kidney transplant recipients, when combined with cyclosporine and corticosteroids. Daclizumab was the first "humanized" monoclonal antibody approved in the United States for human administration (Tsurushita et al. 2005). The daclizumab molecule is a humanized IgG1 adapted from a mouse antibody against the alpha portion of the IL-2 receptor (Uchiyama et al. 1981). Daclizumab was developed as an alternative to the initial mouse antibody developed against the IL-2 receptor. The mouse antibody led to the development of human anti-mouse antibodies (HAMA) and inability to administer subsequent doses. Although daclizumab bound with one-third the affinity for the T-cell receptor site when compared to the original mouse molecule, it was still able to exhibit a high-binding capacity ($Ka = 3 \times 10^9\ M^{-1}$) (Tsurushita et al. 2005; Queen et al. 1988). A daclizumab serum concentration of 1 μg/mL is required for 50 % inhibition of antigen-induced T-cell proliferation (Junghans et al. 1990). Early, phase I clinical trials in kidney transplant recipients, who received corticosteroids in combination with cyclosporine and azathioprine, used five doses of daclizumab (Vincenti et al. 1997). Pharmacokinetic studies revealed a mean serum half-life of 11.4 days, a steady-state volume of distribution of 5 l, and displayed weight-dependent elimination. There was no change in the number of circulating CD3-positive cells following administration. Five doses of 1 mg per kg body weight given every other week were required to produce the serum concentrations needed to achieve 90 % inhibition of T-cell proliferation for 12 weeks. One patient did develop neutralizing antibodies against the daclizumab molecule after receiving weekly doses for 2 weeks. Saturation of the IL-2 receptor did not change. Intravenous doses were well tolerated with no infusion-related reactions. No infection or malignancies were reported up to 1 year following daclizumab administration. The authors concluded that daclizumab stayed within the intravascular space and doses should be based on patient weight at the time of transplant (Vincenti et al. 1997). Subsequent premarketing clinical trials confirmed these results and dosing schematic and were able to show that daclizumab administration reduced the incidence of acute rejection by 13 % in low-risk kidney transplant recipients (Vincenti et al. 1998). Following daclizumab's approval, several trials have been conducted using various dosing regimens and immunosuppression combinations within various solid organ recipients. Secondary to low utilization in solid organ transplant, however, its manufacturing has recently been halted.

BASILIXIMAB

Basiliximab was developed as a more potent anti-IL-2 receptor antagonist when compared to daclizumab and may have several logistical advantages.

Basiliximab, in combination with cyclosporine and corticosteroids, was approved for the prevention of acute allograft rejection in renal transplant recipients in May of 1998. Basiliximab is a murine/human (chimeric) monoclonal antibody directed against the alpha subunit of the IL-2 receptor on the surface of activated T lymphocytes. The antibody is produced from genetically engineered mouse myeloma cells. The variable region of the purified monoclonal antibody is comprised of murine hypervariable region, RFT5, which selectively binds to the IL-2 receptor alpha region. The constant region is made up of human IgG1 and kappa light chains (Novartis Pharmaceuticals 2005). Since the variable region is the only portion with a nonhuman epitope, there appears to be low antigenicity and increased circulating half-life associated with its administration (Amlot et al. 1995). Following administration, basiliximab rapidly binds to the alpha region of the IL-2 receptor and serves as a competitive antagonist against IL-2. The estimated receptor binding affinity (Ka) is 1×10^{10} M^{-1}, which is three times more potent than daclizumab (Novartis Pharmaceuticals 2005). Complete inhibition of the CD25 receptor occurs after the serum concentration of basiliximab exceeds 0.2 μg/mL and inhibition correlated with increasing dose (Novartis Pharmaceuticals 2005; Kovarik et al. 1996). Initial dose finding studies of basiliximab were similar to daclizumab. Basiliximab, combined with cyclosporine and corticosteroids, was administered to adult kidney transplant recipients for the prevention of acute cellular rejection.

Kovarik and colleagues (1997) performed a multicenter, open-label pharmacodynamic analysis evaluating basiliximab dose escalation in adult patients undergoing primary renal transplantation. Patients received a total of 40 or 60 mg of basiliximab in combination with cyclosporine, corticosteroids, and azathioprine. Thirty-two patients were evaluated and were primarily young (34±12 years), Caucasian (29/32) males (23/32). Basiliximab infusions were well tolerated without changes in blood pressure, temperature, or hypersensitivity reactions. Thirty patients underwent pharmacokinetic evaluation. Basiliximab blood concentrations showed biphasic elimination with an average terminal half-life of 6.5 days. Significant intra- and interpatient variability in observed volume of distribution and drug clearance was observed. This could not be corrected through body weight adjustment. Gender did not appear to influence the pharmacokinetic parameters of basiliximab; however, this cohort contained only a small number of female recipients that may have limited the detection of a difference.

Results also indicated that the use of basiliximab with a combination of cyclosporine, corticosteroids, and azathioprine may be an inadequate immunosuppression regimen to prevent acute rejection, especially if cyclosporine initiation is delayed posttransplant. A total of 22 patients had an acute rejection episode, 16 patients in the 40 mg groups and 6 in the 60 mg group. These rejections appeared within the first 2 weeks following transplantation with a mean time to rejection of 11 days. The study was designed for cyclosporine to begin on day 10 posttransplant. Also, three patients experienced graft loss, two of which were immunologically mediated. There was no difference in the basiliximab serum concentration in the patients who experienced rejection versus those who did not. The authors concluded that increased cyclosporine concentrations, which would inhibit IL-2 production, within the first few days posttransplant may increase the efficacy of basiliximab when used for induction (Kovarik et al. 1996).

The clinical efficacy of basiliximab has been confirmed in several prospective post-marketing trials. Currently, the recommended basiliximab dosing regimen is a total dose of 40 mg, with 20 mg administered 2 h prior to transplanted organ reperfusion and a subsequent 20 mg dose on postoperative day 4.

IL-2 receptor antagonists are currently used in all solid organ transplant populations for induction (Table 19.3), but are only approved for use in kidney transplant recipients. Administration does not reduce the total number of circulating lymphocytes or the number of T lymphocytes expressing other markers of activations, such as CD26, CD38, CD54, CD69, or HLA-DR (Chapman and Keating 2003). Consequently, it is necessary that additional immunosuppressive agents, such a calcineurin inhibitors and antiproliferative agents, be administered as soon as possible to decrease the risk of early acute rejection.

The advantage of IL-2 receptor antagonists is that they confer a decreased risk of infusion-related reactions, posttransplant infection, and malignancy when compared to immunodepleting agents. The use of these agents has increased since the introduction of more potent maintenance immunosuppressant agents, and they are now the agents of choice in kidney, lung, liver, and pancreas transplant recipients. Although these agents have been evaluated in organ recipients who are at high risk for acute rejection, they are mainly reserved for patients who are at low to moderate risk. Also, these agents are still being evaluated for use in immunosuppression protocols which withdraw or avoid corticosteroids or calcineurin inhibitors.

There may be an increased risk of anti-idiotypic IgE anaphylactic reaction in patients who receive repeat courses of IL-2 receptor antagonists. Two published case reports describe patients who had been previously exposed to an IL-2 receptor antagonist and upon subsequent exposure developed dyspnea, chest tightness, rash, and angioedema. However, in one case where basiliximab was the offending agent, daclizumab was successfully administered following a negative skin test. Therefore, caution may be warranted in patients who receive a dose of an IL-2 antagonist without concomitant corticosteroids following previous exposure in the past 6 months when circulating antibodies are expected to be present.

■ Alemtuzumab

Alemtuzumab is a recombinant DNA-derived, humanized, rat IgG1κ monoclonal antibody targeting the 21–28 kDa cell surface protein glycoprotein CD52, which is produced in a Chinese hamster ovary cell suspension (Genzyme Corporation 2009; Kneuchtle et al. 2004). Initially, the first anti-CD52 antibodies were developed from rat hybrid antibodies that were produced to lyse lymphocytes in the presence of complement (Morris and Russell 2006). Campath-1 M was the first agent developed. This molecule was a rat IgM antibody which produced little biological effect. In contrast, the rat IgG (Campath-1G) produced profound lymphopenia (Morris and Russell 2006). In order to prevent the formation of antibodies against the rat IgG, the molecule was humanized and called alemtuzumab or Campath-1H (Morris and Russell 2006). The biologic effects of alemtuzumab are the same as Campath-1G and include complement-mediated cell lysis, antibody-mediated cytotoxicity, and target cell apoptosis (Magliocca and Knechtle 2006). The CD52 receptors account for 5 % of lymphocyte surface antigens (Morris and Russell 2006). Cells which express the CD52 antigen include T and B lymphocytes, natural killer cells, monocytes, macrophages, and dendritic cells (Genzyme Corporation 2009; Bloom et al. 2006). However, plasma cells and memory type cells appear to be unaffected by alemtuzumab (Magliocca and Knechtle 2006). Following administration, a marked decrease in circulating lymphocytes is observed. Use in the hematology population indicates that this effect is dose dependent (see Chap. 17). However, single doses of 30 mg or two doses of 20 mg are currently used in the solid organ transplant population.

The plasma elimination half-life after single doses is reported to be around 12 days, and the molecule may be removed by posttransplant plasmapheresis (for more details, please see Chap. 17) (Magliocca and Knechtle 2006). The biological activity of alemtuzumab, however, may last up to several months. One in vivo study of kidney transplant recipients aimed to observe the recovery and function of lymphocytes following administration of 40 mg of alemtuzumab (Bloom et al. 2006). Authors reported a 2-log reduction in peripheral lymphocytes following administration. Absolute lymphocyte counts at 12 months remained markedly depleted, falling below 50 % of their original baseline. Monocytes and B lymphocytes were the first cell lines to recover at 3–12 months post-administration. T lymphocytes returned to 50 % of their baseline value by 36 months.

Currently, alemtuzumab is only FDA approved for the treatment of B-cell chronic lymphocytic leukemia. The first report of alemtuzumab use in solid organ transplantation appeared in 1991. Friend and colleagues (1991) published a case series on the use of alemtuzumab to reverse acute rejection in renal transplant recipients. Shortly thereafter, Calne and colleagues (1999) issued the first report of alemtuzumab use as an induction agent. The authors reported the results of 31 consecutive renal transplant recipients. Patients received two 20 mg doses of alemtuzumab; the first dose was given in the operating room and the second dose was given on postoperative day 1. Patients were initiated on low-dose cyclosporine monotherapy 72 h after transplant, with a target trough range of 75–125 ng/mL. Six patients experienced corticosteroid responsive rejection (20 %). Three of these were maintained on corticosteroids and azathioprine following rejection, while the other three remained on cyclosporine monotherapy. Allografts remained functional in 94 % (29/31) of patients at 15–28 months posttransplant (Calne et al. 1999).

The largest multicenter randomized controlled trial assessing alemtuzumab induction in low- and high-risk renal transplant recipients showed that biopsy-confirmed acute rejection was reduced in low-risk patients receiving alemtuzumab when compared to basiliximab after 3 years of follow-up. In high-risk renal transplant patients, alemtuzumab and Thymoglobulin® appeared to have similar efficacy. However, patients who received alemtuzumab had increased rates of late rejections (between 12 and 36 months) when compared to conventional therapies (8 % versus 3 %, $p = 0.03$). All patients were withdrawn from steroids by postoperative day 5. Adverse effects were similar with more leukopenia observed in the alemtuzumab group (54 %) compared to basiliximab (29 %), and more serious adverse effects related to malignancy were seen with alemtuzumab (5 %) when compared to a composite of all basiliximab- and Thymoglobulin®-treated patients (1 %). However, overall adverse events related to malignancy were similar between treatment groups (Hanaway et al. 2011).

Currently, the most data on the use of alemtuzumab in solid organ transplantation are with kidney transplant recipients. However, alemtuzumab is currently being used for induction and for treatment of rejection in other organs as well (Morris and Russell 2006). In the most recent review of immunosuppression trends in the United States, alemtuzumab use markedly increased from 2001 to 2004, with use primarily limited to induction of immunosuppression (see Table 19.3).

In 2004, alemtuzumab was the predominant agent used for induction in both pancreas and intestinal transplant recipients (Meier-Kriesche et al. 2006). Use in liver transplant has been limited but has appeared in a couple of published trials. Specific findings from these trials indicate that patients without hepatitis C were able to tolerate lower levels of calcineurin inhibitors which corresponded to lower serum creatinine levels at 1-year posttransplant (Tzakis et al. 2004). In contrast, administration of alemtuzumab positively correlated with early recurrence of hepatitis C viral replication (Marcos et al. 2004).

Alemtuzumab in heart transplantation has been rarely reported in the literature with only 2 % of heart transplant patients receiving alemtuzumab for induction in 2004 (Meier-Kriesche et al. 2006). Teuteberg and colleagues recently published a retrospective study on 1-year outcomes on the use of alemtuzumab for induction in cardiac transplantation at a single center. Freedom from rejection was higher in the alemtuzumab group (versus no induction); however, survival at 1 year was similar between groups with more adverse effects in the alemtuzumab group (Teuteberg et al. 2010). Despite this recent publication, there remains a paucity of data in the cardiac transplant population regarding alemtuzumab for induction immunosuppression, which has resulted in limited use in this population.

A retrospective review of 5-year outcomes on the use of alemtuzumab induction in lung transplant recipients at a single center showed an improvement in patient and graft survival with alemtuzumab compared to no induction or daclizumab induction and higher rates of freedom from cellular rejection than no induction or Thymoglobulin® or daclizumab induction (Shyu et al. 2011). The results of the previous study are consistent with another retrospective study that showed decreased rejection rates with alemtuzumab induction in comparison to Thymoglobulin® and daclizumab in lung transplant patients (McCurry et al. 2005). In 2004, 3 % of lung transplant recipients received alemtuzumab for induction (Meier-Kriesche et al. 2006); however, this number may be increasing as more data emerges regarding alemtuzumab use in the lung transplant population.

Alemtuzumab induction has allowed for early withdrawal of corticosteroids in several clinical trials, thereby decreasing long-term steroid exposure. This may lead to improved clinical outcomes since the use of steroids has been correlated with an increased incidence of cardiovascular disease, endocrine, and metabolic side effects. However, the long-term benefit of steroid withdrawal after alemtuzumab induction requires further study. Several trials have also shown success with using low-dose calcineurin inhibitors with alemtuzumab induction. However, early trials in which calcineurin inhibitor avoidance was initiated, the rate of early acute antibody-mediated rejection was 17 % compared to 10 % under traditional immunosuppression which included calcineurin inhibitors (Magliocca and Knechtle 2006).

The infusion of alemtuzumab is well tolerated. In general, induction doses are administered immediately preceding reperfusion of the transplanted allograft. Pretreatment with corticosteroids, diphenhydramine, and acetaminophen is generally advised to prevent sequelae from cellular apoptosis. However, cytokine release associated with alemtuzumab is insignificant in comparison to other agents (Morris and Russell 2006).

Until recently, there were few published experiences detailing long-term outcomes in patients who received alemtuzumab induction (Magliocca and Knechtle 2006). Initially clinicians were concerned that the profound lymphodepletion that was observed following administration would lead to a significant increase in the number of severe infections. Therefore, lymphocyte response to donor antigens following alemtuzumab administration was also evaluated in vitro (Bloom et al. 2006). Lymphocytes from patients treated with alemtuzumab were able to respond to donor antigens and cytokines. However, a small subset of patients were hyporesponsive, which is similar to the control patients observed in this study (Bloom et al. 2006). In addition, several reports detailing the use of alemtuzumab thus far suggest that both infection and malignancy rates are minimal when compared to other agents used for the same indication (Morris and Russell 2006; Magliocca and Knechtle 2006). These findings are confirmed with the prospective 3-year data published by Hanaway et al. in kidney transplant recipients as well as the retrospective 5-year data published by Shyu et al. in lung transplant recipients (Hanaway et al. 2011; Shyu et al. 2011).

At present, a concern associated with alemtuzumab administration is an increased incidence of autoimmune diseases. The exact incidence and etiology of autoimmune diseases following alemtuzumab administration in solid organ transplant is currently unknown, although the most well-designed trial with 3-year follow-up to date did not report autoimmune diseases developing in kidney transplant recipients receiving alemtuzumab for induction (Hanaway et al. 2011). Initial reports of autoimmune diseases associated with alemtuzumab administration came from the multiple sclerosis population. A single center observed the development of Grave's disease in 9 out of 27 patients who received alemtuzumab (Coles et al. 1999).

Thyroid function in all patients was normal prior to alemtuzumab and the mean time to development of autoimmune hyperthyroidism was 19 months (range 9–31 months) (Coles et al. 1999). Autoimmune hyperthyroidism was first reported in a kidney transplant recipient who received alemtuzumab induction 4 years earlier (Kirk et al. 2006). Watson and colleagues (2005) published a 5-year experience with alemtuzumab induction, in which they reported a 6 % (2/33) incidence of autoimmune disease development following administration. One patient developed hyperthyroidism in the early posttransplant period, and one patient developed hemolytic anemia, which was refractory to corticosteroids. With the increased use of alemtuzumab in solid organ transplantation, it is important to continually assess the risk of autoimmune disease development in this population.

■ Rituximab

Rituximab is a chimeric murine/human IgG1 monoclonal antibody directed at the CD20 cell surface protein (Tobinai 2003). Rituximab is currently FDA approved for the CD20-positive forms of non-Hodgkin's lymphoma and chronic lymphocytic leukemia (CLL) and Wegener granulomatosis, microscopic polyangiitis, and refractory rheumatoid arthritis (see Chaps. 17 and 20) (Genentech 2011). The CD20 antigen is a 35-kDa phosphoprotein expressed on B cells, from pre-B cells to mature B cells. This protein is not expressed on hematopoietic stem cells, plasma cells, T lymphocytes, or other tissues (Tobinai 2003). The CD20 protein is a calcium channel and is responsible for B-cell proliferation and differentiation (Tobinai 2003). Early monoclonal antibodies developed against CD20 revealed that antibody binding did not result in modulation of activity or shedding of the surface protein, making the development of a humanized anti-CD20 antibody rational (Tobinai 2003). Rituximab was originally developed to treat B-cell lymphomas, as the vast majority of malignant B cells express the CD20 receptor. Following continuously infused, high doses of engineered anti-CD20 monoclonal antibodies clearance of CD20-positive cells occurred within 4 h of administration (Press et al. 1987). Circulating B-cell clearance was immediate; however, lymph node and bone marrow B-cell clearance were dose dependent.

Rituximab was initially used in solid organ transplant recipients to treat posttransplant lymphoproliferative disorder (PTLD). PTLD is a malignancy that develops following exposure to high levels of T-cell depleting immunosuppression (see section "Immunologic Targets: Rational Development/Use of Monoclonal Antibodies in Organ Transplant"). Under normal physiologic conditions, both the humoral and cellular immune systems work in concert to fight infection. In addition, cytotoxic T lymphocytes survey the body for malignant cells. Current immunosuppression and induction therapy are focused on decreasing communication and proliferation of T lymphocytes, which may lead to unopposed B-cell proliferation. The most significant risk factors for the development of PTLD are the use of potent T-cell depleting therapies as well as an Epstein-Barr virus (EBV) negative recipient serostatus. Approximately 60–70 % of PTLD cases are associated with EBV. Certain B cells that are infected with EBV or other viruses may go into unopposed cellular differentiation leading to PTLD (Evens et al. 2010).

This disorder was first reported in five living donor renal transplant recipients in 1969 with four of the five patients dying from their disease. The fifth patient survived following radiation and reduction in immunosuppression (Penn et al. 1969). The incidence of posttransplant malignancy, specifically PTLD, increased as the number of solid organ transplants increased. Specific agents linked to the development of PTLD included OKT3 and rabbit antithymocyte globulin (Swinnen et al. 1990; Evens et al. 2010). The initial treatment for PTLD is a reduction in maintenance immunosuppression, to allow T-cell surveillance to resume and aid in the destruction of malignancy causing cells. However, pharmacotherapeutic agents have been used successfully in patients who fail to respond to decreased immunosuppression. Rituximab is the most studied medication for the treatment of PTLD and can be considered in patients with CD20-positive tumors. Rituximab was initially used in the 1990s to target B-cell-specific forms of PTLD that did not involve the central nervous system (Faye et al. 1998; Cook et al. 1999; Davis and Moss 2004). The molecular size of rituximab generally precludes its use for central nervous system tumors with <5 % of rituximab penetrating the blood brain barrier, although some recent reports have shown success with rituximab for the treatment of CNS PTLD (Patrick et al. 2011; Kordelas et al. 2008; Jagadeesh et al. 2012). Administration of rituximab in patients with peripheral lymphomas resulted in clearance of malignant B cells for up to 12 months (Davis and Moss 2004). Currently, rituximab is reserved for patients with CD20-positive PTLD who fail to respond to reduction in maintenance immunosuppression. Rituximab can be used alone or in combination with chemotherapy in patients with severe or refractory PTLD.

Rituximab has also been employed as a desensitizing agent (see section "Monoclonal Antibodies Administered Pre-transplant") prior to solid organ transplant. Doses of 375 mg per m^2 administered prior to transplant enabled transplantation across ABO incompatible blood types and transplantation of highly sensitized patients. Often rituximab is given in combination with other immunosuppressants to halt the production of new B lymphocytes and prevent the formation of new plasma cells. Desensitization protocols involve administration of pooled immunoglobulin followed by plasmapheresis to remove donor-specific antibody complexes. Rituximab is administered following the course of plasmapheresis for two reasons: (1) rituximab is removed by plasmapheresis and (2) rituximab only targets B lymphocytes, not the plasma cells currently secreting antibody. Therefore, timing of administration is crucial to the success of the desensitization protocol (Pescovitz 2006).

Following transplant, rituximab is also used for the treatment of acute, refractory antibody-mediated rejection. Antibody-mediated rejection is characterized by host recognition of donor antigens followed by T-cell proliferation and antigen presentation to B cells. B cells then undergo clonal expansion and differentiation into mature plasma cells, which secrete anti-donor antibody. This immune process may occur before or after transplantation. Often the presence of antibodies against donor tissue is discovered prior to transplant, during final crossmatch, thus preventing hyperacute rejection. In some cases, low levels of antibody or memory B cells exist which can facilitate antibody-mediated rejection within the first several weeks following transplant. Rituximab, therefore, is used to induce apoptosis of the B cells producing or capable of producing antibodies against the allograft. Unfortunately, the CD20 receptor is absent on mature plasma cells; therefore, rituximab can only stop new B cells from forming. Plasmapheresis is necessary to remove antibodies produced by secreting plasma cells. It is important to remember that rituximab may be removed by plasmapheresis and timing of administration is necessary to ensure optimal drug exposure. The optimal number of doses and length of therapy necessary to suppress antibody-mediated rejection is unknown (Pescovitz 2006; Stegall and Gloor 2010).

In 2005 and 2006, rituximab was shown to improve the clinical course of renal transplant patients with recurrent focal segmental glomerulosclerosis (FSGS) in patients who were receiving rituximab for the treatment of PTLD (Nozu et al. 2005; Pescovitz et al. 2006). A subsequent study described 7 pediatric patients who had a relapse of proteinuria after transplantation and who failed to respond to initial plasmapheresis. After failure of plasmapheresis, patients received rituximab for treatment of refractory FSGS. Three patients had complete resolution of proteinuria; urine protein decreased by 70 % in one patient and by 50 % in one patient. One patient failed to respond to therapy and one patient was unable to tolerate the rituximab infusion. This study confirmed that rituximab is a possible treatment option for recurrent FSGS (Strologo et al. 2009). Additional studies are needed to further delineate the role of rituximab in the treatment of recurrent FSGS.

Eculizumab

Eculizumab is a recombinant-humanized IgG2/4 monoclonal antibody with murine complementarity-determining regions grafted onto the framework of the human antibody on the light- and heavy-chain variable regions. Eculizumab binds with specificity and with high affinity to C5, a complement protein. By binding to C5, eculizumab prevents cleavage of C5 to C5a and C5b, which prevents the formation of the membrane attack complex. Currently, eculizumab is approved for use in the treatment of paroxysmal nocturnal hemoglobinuria and atypical hemolytic uremic syndrome (Alexion Pharmaceuticals 2011; McKeage 2011).

Because antibody-mediated rejection (AMR) is associated with complement activation evidenced by $C4d^+$ staining on biopsy, the use of eculizumab for the prevention and treatment of AMR holds promise (Stegall and Gloor 2010). The first case describing the use of eculizumab for the treatment of severe AMR was published in 2009. The patient was a highly sensitized kidney transplant recipient who received desensitization therapy before and after transplant. However, he became anuric with a biopsy that was positive for AMR approximately 8 days after transplant. After clinical failure of plasmapheresis and intravenous immunoglobulin, eculizumab was initiated. Intravenous immunoglobulin was also given in order to decrease donor-specific antibodies, and rituximab was given in order to prevent B-cell proliferation. Donor-specific antibodies did not decrease initially; however, C5d-9 staining was reduced on biopsy, and AMR was completely resolved on follow-up biopsies (Locke et al. 2009).

The use of eculizumab for the prevention of AMR has also been reported. In one study, patients with a positive crossmatch to their living kidney donor received plasmapheresis and eculizumab preoperatively and were compared to a historical control who received only plasmapheresis pre- and postoperatively. The treatment group also received eculizumab post-transplant for at least 4 weeks. Treatment continued in patients who did not have a decrease in donor-specific antibody. The incidence of AMR at 3 months was 7 % in the eculizumab group compared to 41 % in the historical control group (Stegall and Gloor 2010).

Recent evidence has proven that complement activation is involved in the development of hemolytic uremic syndrome. There have been a few case reports that show that eculizumab can improve the outcomes of patients who develop hemolytic uremic syndrome after renal transplant (Van den Hoogen and Hilbrands 2011).

There is limited data on the use of eculizumab in solid organ transplantation at this time. However, it is likely that its role in the prevention and treatment of AMR, hemolytic uremic syndrome after transplantation, and other possible indications will be more clearly defined by the next decade.

■ Belatacept

In an effort to achieve the "immunotolerant" state posttransplant, research has been focused in the area of co-stimulation blockade. Simplistically, when a T cell is exposed to an antigen particle expressed on an antigen presenting cell through the T-cell receptor, additional co-stimulation is required for full activation of the T cell (Wekerle and Grinyo 2012). If co-stimulation is blocked, then the T cell becomes unresponsive and in essence tolerant. CD28 is expressed on human T cells and is upregulated on activated T cells, while its ligands, on the surface of the antigen presenting cell, are CD80 and CD86 (Wekerle and Grinyo 2012). Cytotoxic T-lymphocyte antigen-4 (CTLA-4) was identified as a compound that would bind the same ligands as CD28 but to a much higher affinity (Wekerle and Grinyo 2012). A modification of CTLA-4, giving it higher binding affinity for CD80/86, was fused with a mutated (no longer able to fix complement) human IgG1, yielding belatacept (Wekerle and Grinyo 2012). Therefore, belatacept binds to CD80 and CD86 with high affinity, blocking their interaction with CD28 on T cells. An artifact of belatacept is that it also blocks intrinsic CTLA-4, which normally acts as an inhibitory ligand on the surface of activated T cells, responsible for limiting the proliferation of the immune response (Wekerle and Grinyo 2012). Blockade of CTLA-4 could prevent tolerance from being achieved when administered posttransplant; however, phase II trials indicate that the synthesis of CD4+ CD25+ regulatory T cells is not interrupted following belatacept exposure (Gupta and Womer 2010). Belatacept is an intravenous infusion, dosed based on actual body weight, and is unaffected by renal or hepatic function, which is administered frequently during the first 1–3 months posttransplant then monthly thereafter (Martin et al. 2011).

Belatacept has been mainly studied and demonstrated efficacy in kidney transplant recipients in combination with basiliximab induction and mycophenolate mofetil/prednisone maintenance immunosuppression. Belatacept has been touted as calcineurin inhibitor sparing and therefore more renal protective posttransplant. Recently the 3-year results of the BENEFIT study were published detailing the safety and efficacy of belatacept versus cyclosporine in combination with mycophenolate mofetil and prednisone (Vincenti et al. 2012). The BENEFIT trial evaluated 663 kidney transplant recipients who received low intensity (0–3 months; 10 mg/kg on days 1 and 5, 10 mg/kg on weeks 2, 4, 8, 12, 3–36 months 5 mg/kg every 4 weeks; $n=226$), moderate intensity (0–6 months) 10 mg/kg on days 1 and 5, 10 mg/kg on weeks 2, 4, 6, 8, 10, 12, 16, 20, and 24; 7–36 months 5 mg/kg ($n=219$) belatacept or cyclosporine ($n=221$) in combination with mycophenolate mofetil and prednisone. Graft survival at 3 years was 92 % in the low- and moderate-intensity groups and 89 % in the cyclosporine group. A total of 6 patients died, 2 in each group, and 9 patients lost their graft (4 in the low intensity, 3 in the moderate intensity, and 2 in the cyclosporine group). Calculated glomerular filtration rate was 66 ± 27 mL/min/1.73 m^2 in the low intensity, 65 ± 26 mL/min/1.73 m^2 in the moderate intensity, and 44 ± 24 mL/min/1.73 m^2 in the cyclosporine group, $p<0.0001$. Acute rejection mainly occurred in the first-year posttransplant with a cumulative rate of 17 % in the low intensity and 24 % in the moderate intensity versus 10 % in the cyclosporine group. PTLD occurred in five patients who received belatacept versus one patient in the cyclosporine group (Vincenti et al. 2012). Similar results were found at 3 years in extended criteria kidney transplant recipients (Pestana et al. 2012). When more intensive belatacept dosing was used in combination with mycophenolate mofetil ($n=33$) or sirolimus ($n=26$) versus tacrolimus with mycophenolate mofetil ($n=30$) following rabbit antithymocyte globulin and early corticosteroid withdrawal (4 days), acute rejection rates were low (12 % belatacept-mycophenolate, 4 % belatacept-sirolimus, and 3 % in the tacrolimus-mycophenolate). Graft survival was 100 % at 1 year in the tacrolimus group versus 91 % in the belatacept-mycophenolate group and 92 % in the belatacept-sirolimus group; however, graft function was roughly 8 mL/min/1.73 m^2 higher in the belatacept groups. However, less than 80 % of patients in the belatacept groups remained steroid-free at 12 months versus 93 % in the tacrolimus group (Ferguson et al. 2011). Patients 6–36 months post-kidney transplant were also enrolled in a conversion trial in which they were randomized to continue their current immunosuppression or be converted to belatacept to evaluate if an improvement in renal function could be obtained following discontinuation of a calcineurin inhibitor (Rostaing et al. 2011). An average improvement in glomerular filtration rate was noted in the belatacept group (7 mL/min versus 2.1 mL/min, $p=0.0058$) at 12 months following conversion. Six patients did develop acute rejection following their conversion to belatacept, but these rejections did not result in graft loss (Rostaing et al. 2011).

Monoclonal antibody	Dose[a]	US cost per course (AWP)[b]
Alemtuzumab	30 mg × 1	$6,354
Basiliximab	20 mg × 2	$5,605
Rituximab	375 mg/m^2 weekly × 4 doses	$20,682
Belatacept	10 mg/kg days 1 and 5	$42,090 for the first year
	10 mg/kg after 2 and 4 weeks	$28,798 subsequent years
	10 mg/kg after 8 and 12 weeks 5 mg/kg after 16 weeks and every 4 weeks thereafter	
Eculizumab	1,200 mg × 1[c]	$74,880
	600 mg × 1 then	
	600 mg weekly × 3	

[a]Based on 70 kg dosing weight, rounded to nearest vial size
[b]Actual wholesale price (AWP) Adapted from Red Book; Thomson Reuters (2012)
[c]Dosing is based on Stegall et al. (2011) study. Adequate dose for transplantation has not yet been established

Table 19.4 ■ Per dose cost comparison between monoclonal antibodies currently used in solid organ transplantation.

Evidence for the use of belatacept is currently lacking in nonrenal transplant recipients and high immunologic risk and non-Caucasian organ recipients. Additionally, patients who are EBV positive are at high risk of developing posttransplant lymphoproliferative disease in the central nervous system. This observation warranted a black box warning to be issued in the belatacept package insert detailing that the use of belatacept is contraindicated in patients who are EBV negative (Bristol Myers Squibb Company 2011).

CONCLUSION

Currently, there are several challenges remaining in solid organ transplantation. These challenges may be grouped as follows. One challenge is optimizing patient-specific immunosuppression based on risk factors for acute rejection. Monoclonal antibodies provide targeted immunosuppression that when used in conjunction with specific maintenance immunosuppressants may allow more specific therapy. Another challenge is preventing over-immunosuppression, which may lead to infection and malignancy. Although monoclonal antibodies provide targeted therapy, the toxicity and potency must be balanced with over-immunosuppression. Consideration of the mechanism of action of both the monoclonal antibody and maintenance immunosuppression must be evaluated to ensure that appropriate antimicrobial prophylaxis and malignancy screening tools are utilized to minimize the patient's risk. Finally, increasing patient and graft survival through reducing the incidence of adverse effects associated with long-term exposure to maintenance immunosuppression, such as cardiovascular events or kidney dysfunction, is necessary. Monoclonal, along with polyclonal antibodies, may allow for withdrawal or minimization of specific maintenance immunosuppressants that lead to the increased incidence of these long-term adverse effects. Oftentimes the use of specific monoclonal antibodies in institutional protocols is driven by cost (Table 19.4) with careful consideration of the goals of therapy.

SELF-ASSESSMENT QUESTIONS

■ Questions

1. Monoclonal antibodies are used for several reasons in solid organ transplantation. What benefit do they provide over polyclonal antibodies?
2. The rational development and use of monoclonal antibodies in solid organ transplantation is focused on the prevention of host recognition of donor tissue (rejection). What are the two ways in which the host immune system recognizes donor tissue and may cause tissue damage?
3. What are the molecular targets for monoclonal antibodies currently used in solid organ transplantation?
4. Monoclonal antibodies are used at various times in solid organ transplantation. Describe the reasons why a monoclonal antibody would be administered before transplant, at the time of transplant, or following transplant?
5. There are several important pharmacokinetic parameters that must be considered when administering monoclonal antibodies to solid organ transplant recipients. What are these pharmacokinetic parameters?
6. Muromonab has a characteristic infusion-related reaction. Why does this reaction occur and how can it be attenuated?

7. Daclizumab and basiliximab are two monoclonal antibodies directed against the alpha subunit of the interleukin-2 receptor. What is the difference between these two antibodies?
8. There are several benefits, as well as several risks associated with the use of monoclonal antibodies in solid organ transplantation. What are these benefits and risks?

Answers

1. Monoclonal antibodies provide targeted immunosuppression. The advantage monoclonal antibodies offer over polyclonal antibodies is that the receptor target is known. Polyclonal antibody development involves the introduction of human lymphocytes into an animal host immune system. The animal will then develop polyclonal antibodies directed against human lymphocyte cell surface targets. As a consequence, each inter-batch variability and potency may vary. Although significant outcome data exists with the use of polyclonal antibodies, monoclonal antibodies have a known target allowing for in vivo and in vitro pharmacokinetic and pharmacodynamic data to aid incorporation into novel immunosuppression regimens.
2. The two ways in which the host immune system recognizes donor tissue. Complement-dependent antibody-mediated rejection occurs when the host (recipient) develops or has preformed antibodies against the donor tissue. Preformed antibodies will aggregate to the implanted tissue and initiate the complement cascade, which facilitates cell lysis. The majority o these antibodies are usually directed against the major histocompatibility complexes (MHC) located on the surface of the donor tissue. An absolute contraindication to transplantation is the presence of preformed antibodies against MHC complex I, which is located on the surface of all nucleated cells. The second way in which the host immune system attacks donor tissue is through T-cell-mediated rejection. This occurs when the donor tissue is recognized as foreign by host antigen presenting cells. Antigen presenting cells present donor tissue antigens to the T cells which stimulates T-cell proliferation and graft infiltration leading to inflammation and arteritis.
3. Alemtuzumab (Campath-1H®) targets the CD52 receptor, located on peripheral blood lymphocytes, natural killer cells, monocytes, macrophages, and thymocytes.

   Daclizumab (Zenapax®) targets the CD25 alpha subunit of the IL-2 receptor, located on activated T lymphocytes.

   Basiliximab (Simulect®) targets the CD25 alpha subunit of the IL-2 receptor, located on activated T lymphocytes.

   Muromonab-OKT3 (Orthoclone-OKT3®) targets the CD3 receptor located on CD2-, CD4-, and CD8-positive lymphocytes.

   Rituximab (Rituxan®) targets the CD20 receptor located on B lymphocytes.

   Eculizumab (Soliris®) targets C5 in the complement pathway.
4. The administration of monoclonal antibodies prior to transplant is called desensitization. This strategy is reserved for "highly sensitized" patients, meaning they have high titers of circulating antibodies against donor-specific antigens. Monoclonal antibodies that target cells which produce these antibodies are employed, in conjunction with plasmapheresis and pooled human immune globulins. Removal of these antibodies may facilitate successful transplantation across this immunologic barrier.

   Monoclonal antibodies administered at the time of transplant are called induction. Induction is provided at the time of transplant to decrease the ability of the host immune system to respond to implantation of foreign tissue. In addition, monoclonal antibodies which provide profound T-cell depletion given at the time of transplant may facilitate the need for certain maintenance immunosuppressants.

   Following transplantation, monoclonal antibodies may be used to treat cell-mediated or antibody-mediated rejection. Cell and antibody infiltrates found in biopsy specimens in correlation with the clinical status of the patient will dictate the type, dose, and duration of the monoclonal antibody chosen.
5. The volume of distribution, biological half-life, and total-body clearance can differ significantly between solid organ transplant recipients. Careful consideration of these pharmacokinetic parameters must be employed to maximize the efficacy and minimize the toxicity associated with administration of these agents. For example, weight-based dosing in obese patients must be carefully considered, and biological markers of efficacy should be evaluated to determine the appropriate dose and dosing schedule. In addition, monoclonal antibodies are also removed by plasma exchange procedures, such as plasmapheresis, which may be performed during the perioperative period. Therefore, it would be prudent to administer the monoclonal antibody following the plasma exchange prescription to avoid removal of the drug and avoid a possible decrease in efficacy.
6. Muromonab's infusion-related reaction occurs because when the molecule binds to the CD3 receptor. It actually activates the cell prior to inducing apoptosis. T-cell activation leads to increased production of inflammatory cytokines and when the

cell undergoes apoptosis these cytokines are released causing a "cytokine release syndrome." This cytokine release syndrome is characterized by fever, chills, rigors, diarrhea, and potentially capillary leak leading to pulmonary edema. Often times this reaction is the worst when the largest number of cells are present, namely, the first dose. However, this reaction can occur after several days of dosing. This reaction can be attenuated by administration of corticosteroids, histamine blockers, and cyclooxygenase antagonists. Pharmacotherapy aimed at reducing the production or the interaction of cytokines with their receptors may decrease the severity of the cytokine release syndrome.

7. *Structure activity relationship*: Daclizumab has a binding capacity of 3×10^9 M^{-1} versus basiliximab which has a binding capacity of 1×10^{10} M^{-1}. Therefore, basiliximab is three times more potent than daclizumab.
8. *Dosing*: Daclizumab is dosed based on weight, while basiliximab is given as a 20 mg dose. The dosing schedule varies based on the type of solid organ transplanted as well as concomitant immunosuppression given. These agents, however, are only approved for prevention of acute rejection in kidney transplant recipients.
9. Benefits include targeted immunosuppression, no batch variability, and low antigenicity in humanized products. The risks associated with any type of immunosuppression include an increased risk for infection, as well as malignancy. Patients who receive monoclonal antibodies which specifically target a cell line, such as muromonab, are associated with a significantly increased risk of posttransplant lymphoproliferative disease. Appropriate antimicrobial prophylaxis and vigilant screening for posttransplant malignancy may allow for safe and effective use of these monoclonal antibodies in solid organ transplantation.

REFERENCES

Alexion Pharmaceuticals: Eculizumab (Soliris) Package Insert. Alexion Pharmaceuticals, Cheshire. Last updated: 2011

Amlot PL, Rawlings E, Fernando ON, Griffin PJ, Heinrich G, Schreier MH, Castaigne JP, Moore R, Sweny P (1995) Prolonged action of a chimeric interleukin-2 receptor (CD25) monoclonal antibody used in cadaveric renal transplantation. Transplantation 60:748–756

Bloom DD, Hu H, Fechner JH, Knechtle SJ (2006) T-lymphocyte alloresponses of Campath-1H treated kidney transplant patients. Transplantation 81:81–87

Bristol Myers Squibb Company: Belatacept (Nulojix) Package Insert. Bristol Myers Squibb, Princeton. Last updated: June 2011

Buysmann S, Bemelman FJ, Schellekens PT, van Kooyk Y, Figdor CG, ten Berge IJ (1996) Activation and increased expression of adhesion molecules on peripheral blood lymphocytes is a mechanism for the immediate lymphocytopenia after administration of OKT3. Blood 87:404–411

Calne R, Moffatt SD, Friend PJ, Jamieson NV, Bradley JA, Hale G, Firth J, Bradley J, Smith KG, Waldmann H (1999) Campath IH allows low-dose cyclosporine monotherapy in 31 cadaveric renal allograft recipients. Transplantation 68:1613–1616

Chapman TM, Keating GM (2003) Basiliximab: a review of its use as induction therapy in renal transplantation. Drugs 63:2803–2835

Chen J, Astle CM, Harrison DE (2002) Hematopoietic stem cell functional failure in interleukin-2-deficient mice. J Hematother Stem Cell Res 11:905–912

Church AC (2003) Clinical advances in therapies targeting the interleukin-2 receptor. QJM 96:91–102

Coles AJ, Wing M, Smith S, Coraddu F, Greer S, Taylor C, Weetman A, Hale G, Chatterjee VK, Waldmann H, Compston A (1999) Pulsed monoclonal antibody treatment and autoimmune thyroid disease in multiple sclerosis. Lancet 354:1691–1695

Cook RC, Connors JM, Gascoyne RD, Fradet G, Levy RD (1999) Treatment of post-transplant lymphoproliferative disease with rituximab monoclonal antibody after lung transplantation. Lancet 354:1698–1699

Davis JE, Moss DJ (2004) Treatment options for post-transplant lymphoproliferative disorder and other Epstein-Barr virus-associated malignancies. Tissue Antigens 63:285–292

Evens AM, Roy R, Sterrenberg D, Moll MZ, Chadburn A, Gordon LI (2010) Post-transplantation lymphoproliferative disorders: diagnosis, prognosis, and current approaches to therapy. Curr Oncol Rep 12:383–394

Faye A, Van Den Abeele T, Peuchmaur M, Mathieu-Boue A, Vilmer E (1998) Anti-CD20 monoclonal antibody for post-transplant lymphoproliferative disorders. Lancet 352:1285

Ferguson R, Grinyo J, Vincenti F, Kaufman DB, Woodle ES, Marder BA, Citterio F, Marks WH, Agarwal M, Wu D, Dong Y, Garg P (2011) Immunosuppression with belatacept-based, corticosteroid-avoiding regimens in de novo kidney transplant recipients. Am J Transplant 11:66–76

Friend PJ, Waldmann H, Hale G, Cobbold S, Rebello P, Thiru S, Jamieson NV, Johnston PS, Calne RY (1991) Reversal of allograft rejection using the monoclonal antibody, Campath-1G. Transplant Proc 23:2253–2254

Genentech: Rituximab (Rituxan) Package Insert. Genentech, Inc, San Francisco. Last updated: 2011

Genzyme Corporation: Alemtuzumab (Campath) Package Insert. Genzyme Corporation, Cambridge. Last updated: 2009

Gupta G, Womer KL (2010) Profile of belatacept and its potential role in prevention of graft rejection following renal transplantation. Drug Des Devel Ther 4:375–382

Halloran PF (2004) Immunosuppressive drugs for kidney transplantation. N Engl J Med 351:2715–2729

Hanaway MJ, Woodle ES, Mulgaonkar S, Peddi VR, Kaufman DB, First MR, Croy R, Holman J (2011) Alemtuzumab induction in renal transplantation. N Engl J Med 364:1909–1919

Jagadeesh D, Woda BA, Draper J, Evens AM (2012) Post transplant lymphoproliferative disorders: risk, classification, and therapeutic recommendations. Curr Treat Options Oncol 13(1):122–136

Junghans RP, Waldmann TA, Landolfi NF, Avdalovic NM, Schneider WP, Queen C (1990) Anti-Tac-H, a humanized antibody to the interleukin 2 receptor with new features for immunotherapy in malignant and immune disorders. Cancer Res 50:1495–1502

Kirk AD, Hale DA, Swanson SJ, Mannon RB (2006) Autoimmune thyroid disease after renal transplantation using depletional induction with alemtuzumab. Am J Transplant 6:1084–1085

Kneuchtle SJ, Fernandez LA, Pirsch JD et al (2004) Campath-1H in renal transplantation: the University of Wisconsin experience. Surgery 136:754–760

Kordelas L, Trenschel R, Koldehoff M, Elmaagacli A, Beelan DW (2008) Successful treatment of EBV PTLD with CNS lymphomas with the monoclonal anti-CD20 antibody rituximab. Onkologie 31:691–693

Kovarik JM, Rawlings E, Sweny P, Fernando O, Moore R, Griffin PJ, Fauchald P, Albrechtsen D, Sodal G, Nordal K, Amlot PL (1996) Pharmacokinetics and immunodynamics of chimeric IL-2 receptor monoclonal antibody SDZ CHI 621 in renal allograft recipients. Transpl Int 9:S32–S33

Kovarik J, Wolf P, Cisterne JM, Mourad G, Lebranchu Y, Lang P, Bourbigot B, Cantarovich D, Girault D, Gerbeau C, Schmidt AG, Soulillou JP (1997) Disposition of basiliximab, an interleukin-2 receptor antibody, in recipients of mismatched cadaver renal allografts. Transplantation 64:1701–1705

Locke JE, Magro CM, Singer AL, Segev DL, Haas M, Hillel AT, King KE, Kraus E, Lees LM, Melancon JK, Stewart ZA, Warren DS, Zachary AA, Montgomery RA (2009) The use of antibody to complement protein C5 for salvage treatment of severe antibody-mediated rejection. Am J Transplant 9:231–235

Magliocca JF, Knechtle SJ (2006) The evolving role of alemtuzumab (Campath-1H) for immunosuppressive therapy in organ transplantation. Transpl Int 19:705–714

Magnussen K, Klug B, Moller B (1994) CD3 antigen modulation in T-lymphocytes during OKT3 treatment. Transplant Proc 26:1731

Marcos A, Eghtesad B, Fung JJ, Fontes P, Patel K, Devera M, Marsh W, Gayowski T, Demetris AJ, Gray EA, Flynn B, Zeevi A, Murase N, Starzl TE (2004) Use of alemtuzumab and tacrolimus monotherapy for cadaveric liver transplantation: with particular reference to hepatitis C virus. Transplantation 78:966–971

Martin ST, Tichy EM, Gabardi S (2011) Belatacept: a novel biologic for maintenance immunosuppression after renal transplantation. Pharmacotherapy 31:394–407

McCurry KR, Iacono A, Zeevi A, Yousem S, Girnita A, Husain S, Zaldonis D, Johnson B, Hattler BG, Starzl TE (2005) Early outcomes in human lung transplantation with Thymoglobulin or Campath-1H for recipient pretreatment followed by posttransplant tacrolimus near-monotherapy. J Thorac Cardiovasc Surg 130:528–537

McKeage K (2011) Eculizumab: a review of its use in paroxysmal nocturnal haemoglobinuria. Drugs 71:2327–2345

Meier-Kriesche HU, Li S, Gruessner RWG, Fung JJ, Bustami RT, Barr ML, Leichtman AB (2006) Immunosuppression: evolution in practice and trends, 1994–2004. Am J Transplant 6:1111–1131

Morris PJ (2004) Transplantation–a medical miracle of the 20th century. N Engl J Med 351:2678–2680

Morris PJ, Russell NK (2006) Alemtuzumab (Campath-1H): a systematic review in organ transplantation. Transplantation 81:1361–1367

Noguchi M, Adelstein S, Cao X, Leonard WJ (1993) Characterization of the human interleukin-2 receptor gamma gene. J Biol Chem 268:13601–13608

Nojima M, Yoshimoto T, Nakao A, Itahana R, Kyo M, Hashimoto M, Shima H (2005) Sequential blood level monitoring of basiliximab during multisession plasmapheresis in a kidney transplant recipient. Transplant Proc 37:875–878

Novartis Pharmaceuticals: Basiliximab (Simulect) Package Insert. Novartis Pharmaceuticals Corporation, East Hanover. Last updated: 2005

Nozu K, Iijima K, Fujisawa M, Nakagawa A, Yoshikawa N, Matsuo M (2005) Rituximab treatment for posttransplant lymphoproliferative disorder (PTLD) induces complete remission of recurrent nephritic syndrome. Pediatr Nephrol 20:1660–1663

Opelz G, Dohler B (2004) Lymphomas after solid organ transplantation: a collaborative transplant study report. Am J Transplant 4:222–230

Ortho Biotech: Muromonab (Orthoclone) Package Insert . Ortho Biotech, Raritan. Last updated: 2004

Patrick A, Wee A, Hedderman A, Wilson D, Weiss J, Govani M (2011) High-dose intravenous rituximab for multifocal, monomorphic primary central nervous system posttransplant lymphoproliferative disorder. J Neurooncol 103:739–743

Penn I, Hammond W, Brettschneider L, Starzl TE (1969) Malignant lymphomas in transplantation patients. Transplant Proc 1:106–112

Pescovitz MD (2006) Rituximab, an anti-CD20 monoclonal antibody: history and mechanism of action. Am J Transplant 6:859–866

Pescovitz MD, Book BK, Sidner RA (2006) Resolution of recurrent focal segmental glomerulosclerosis proteinuria after rituximab treatment. N Engl J Med 354:1961–1963

Pestana JOM, Grinyo JM, Vanrenterghen Y, Becker T, Campistol JM, Florman S, Garcia VD, Kamar N, Lang P, Manfro RC, Massari P, Rial MD, Schnitzler MA, Vitko S, Duan T, Block A, Harler MB, Durrbach A (2012) Three year outcomes from BENEFIT-EXT: a phase III study of belatacept versus cyclosporine in recipients of extended criteria donor kidneys. Am J Transplant 12(3):630–639

Press OW, Appelbaum F, Ledbetter JA, Martin PJ, Zarling J, Kidd P, Thomas ED (1987) Monoclonal antibody 1F5 (anti-CD20) serotherapy of human B cell lymphomas. Blood 69:584–591

Queen C, Schneider WP, Selick HE, Payne PW, Landolfi NF, Duncan JF, Avdalovic NM, Levitt M, Junghans RP, Waldmann TA (1988) A humanized antibody that binds to the interleukin 2 receptor. Proc Natl Acad Sci U S A 86:10029–10033

Reid ME, Olsson ML (2005) Human blood group antigens and antibodies. In: Hoffman R, Benz EJ (eds) Hematology: basic principles and practice, 4th edn. Churchill Livingstone, Philadelphia, pp 2370–2374

Robb RJ, Munck A, Smith KA (1981) T cell growth factor receptors: quantitation, specific and biological relevance. J Exp Med 154:1455–1474

Rostaing L, Massari P, Garcia VD, Mancilla-Urrea E, Nainan G, del Carmen RM, Steinberg S, Vincenti F, Shi R, Di Russo G, Thomas D, Grinyo J (2011) Switching from calcineurin inhibitor based regimens to a belatacept based regimen in renal transplant recipients: a randomized phase II study. Clin J Am Soc Nephrol 6:430–439

Shibuya H, Yoneyama M, Nakamura Y, Harada H, Hatakeyama M, Minamoto S, Kno T, Doi T, White R, Taniguchi T (1990) The human interleukin-2 receptor beta-chain gene: genomic organization, promoter analysis and chromosomal assignment. Nucleic Acids Res 18:3697–3703

Shyu S, Dew MA, Pilewski JM, Dabbs AJD, Zaldonis DB, Studer SM, Crespo MM, Toyoda Y, Bermudez CA, McCurry KR (2011) Five-year outcomes with alemtuzumab induction after lung transplantation. J Heart Lung Transplant 30:743–754

Stegall MD, Gloor JM (2010) Deciphering antibody-mediated rejection: new insights into mechanisms and treatment. Curr Opin Organ Transplant 15:8–10

Stegall MD, Diwan T, Raghavaiah S, Cornell LD, Burns J, Dean PG, Cosio FG, Gandhi MJ, Kremers W, Gloor JM (2011) Terminal complement inhibition decreases antibody-mediated rejection in sensitized renal transplant recipients. Am J Transplant 11:245–2413

Strologo LD, Guzzo I, Laurenzi C, Vivarelli M, Parodi A, Barbano G, Camilla R, Scozzola F, Amore A, Ginevri F, Murer L (2009) Use of rituximab in focal glomerulosclerosis relapses after renal transplantation. Transplantation 88:417–420

Swinnen LJ, Costanzo-Nordin MR, Fisher SG, O'Sullivan EJ, Johnson MR, Heroux AL, Dizikes GJ, Pifarre R, Fisher RI (1990) Increased incidence of lymphoproliferative disorder after immunosuppression with the monoclonal antibody OKT3 in cardiac-transplant recipients. N Engl J Med 323:1723–1728

Teuteberg JJ, Shullo MA, Zomak R, Toyoda Y, McNamara DM, Bermudex C, Kormos RL, McCurry KR (2010) Alemtuzumab induction prior to cardiac transplantation with lower intensity maintenance immunosuppression: One-year outcomes. Am J Transplant 10: 382–388

Tobinai K (2003) Rituximab and other emerging antibodies as molecular target-based therapy of lymphoma. Int J Clin Oncol 8:212–223

Tsurushita N, Hinton PR, Kumar S (2005) Design of humanized antibodies: from anti-Tac to Zenapax. Methods 36:69–83

Tzakis AG, Tryphonopoulos P, Kato T, Nishida S, Levi DM, Madariaga JR, Gaynor JJ, De Faria W, Regev A, Esquenazi V, Weppler D, Ruiz P, Miller J (2004) Preliminary experience with alemtuzumab (Campath-1H) and low-dose tacrolimus immunosuppression in adult liver transplantation. Transplantation 77:1209–1214

Uchiyama T, Border S, Waldmann TA (1981) A monoclonal antibody (anti-Tac) reactive with activated and functionally mature human T cells. J Immunol 126:1393–1397

Van den Hoogen MWF, Hilbrands LB (2011) Use of monoclonal antibodies in renal transplantation. Immunotherapy 3:871–880

Vincenti F, Lantz M, Birnbaum J, Garovoy M, Mould D, Hakimi J, Nieforth K, Light S (1997) A phase I trial of humanized anti-interleukin 2 receptor antibody in renal transplantation. Transplantation 63:33–38

Vincenti F, Kirkman R, Light S, Bumgardner G, Pescovitz M, Halloran P, Neylan J, Wilkinson A, Ekberg H, Gaston R, Backman L, Burdick J (1998) Interleukin-2-receptor blockade with daclizumab to prevent acute rejection in renal transplantation. Daclizumab Triple Therapy Study Group. N Engl J Med 338:161–165

Vincenti F, Larsen CP, Alberu J, Bresnahan B, Garcia VD, Kothari J, Lang P, Urrea EM, Massari P, Mondragon-Ramirez G, Reyes-Acevedo R, Rice K, Rostaing L, Steinberg S, Xing J, Agarwal M, Harler MB, Charpentier B (2012) Three-year outcomes from BENEFIT, a randomized, active-controlled, parallel-group study in adult kidney transplant recipients. Am J Transplant 12:210–217

Watson CJ, Bradley JA, Friend PJ, Firth J, Taylor CJ, Bradley JR, Smith KG, Thiru S, Jamieson NV, Hale G, Waldmann H, Calne R (2005) Alemtuzumab (CAMPATH 1H) induction therapy in cadaveric kidney transplantation—efficacy and safety at five years. Am J Transplant 5:1347–1533

Wekerle T, Grinyo JM (2012) Belatacept: from rational design to clinical application. Transpl Int 25:139–150

Wilde MI, Goa KL (1996) Muromonab CD3: a reappraisal of its pharmacology and use of prophylaxis of solid organ transplant rejection. Drugs 51:865–894

Wong JT, Eylath AA, Ghobrial I, Colvin RB (1990) The mechanism of anti-CD3 monoclonal antibodies. Mediation of cytolysis by inter-T cell bridging. Transplantation 50:683–689

20 Monoclonal Antibodies and Antibody-Based Biotherapeutics in Inflammatory Diseases

Honghui Zhou, Zhenhua Xu, Mary Ann Mascelli, and Hugh M. Davis

INTRODUCTION

Immune-mediated inflammatory diseases encompass a broad and diverse spectrum of serious chronic disorders, many of which have significant need for safe and effective pharmacotherapies. The classes of conventional drugs used to treat immune-mediated inflammatory diseases include nonsteroidal anti-inflammatory drugs (NSAIDs), corticosteroids, sulfasalazine, 5-aminosalicylates, methotrexate, azathioprine, and 6-mercaptopurine; however, in many instances, these agents have exhibited limited efficacy or are associated with significant serious on-target and off-target side effects. The initial rationale and promise of complex biologics, such as monoclonal antibodies (MABs), as pharmacotherapy was focused on oncology and organ transplantation (Ehrlich 1891; Gura 2002); however, recent decades have witnessed the successful development of a number of complex biologics as both anti-allergic and anti-inflammatory therapies. Five of the top-selling MABs are for the treatment of chronic inflammatory conditions, and this area of research and development is rapidly expanding. Complex biologics are a subclass of protein therapeutics. They are large molecular weight glycoproteins designed and produced through recombinant DNA technology and require production in eukaryotic cells using bioreactor technology. These modalities have provided many targeted and efficacious therapeutic options for patients and are providing significant insights into the underlying complex pathological processes of these disorders, which, in turn, are identifying new targets for treatment of these diseases. A significant translational insight derived from the clinical development programs of complex biologic pharmacotherapy is the dysregulation of common pro-inflammatory mediators, such as tumor necrosis factor alpha (TNFα), across diverse rheumatologic, dermatologic, and gastroenterologic pathologies. In addition, the observation of patient subsets within a disease that are refractory to a particular therapy indicates that dysregulation of different mediators can drive the underlying pathological processes within a disease.

Complex biologics embody structural, biochemical, and pharmacologic properties distinct from other biologic or chemically synthesized molecular drugs. In general, they exhibit relatively long half-lives and high affinity and target specificity. Their pharmacokinetic and mechanistic properties translate into potent and sustained pharmacodynamic effects. Currently approved complex biologic therapies for autoimmune/inflammatory disorders include chimeric (e.g., infliximab and rituximab), humanized (e.g., omalizumab and daclizumab), fully human MABs (e.g., ustekinumab and adalimumab), and fusion proteins (e.g., etanercept and abatacept). The mechanistic properties of these agents block cytokines such as TNF-α (infliximab, golimumab, adalimumab, etanercept, certolizumab), interleukin (IL)-12/23 (ustekinumab), IL-6 receptor (tocilizumab), or IL-1 (anakinra); inhibit lymphocyte activation (belimumab); lymphocyte migration (natalizumab); or directly deplete lymphocyte subsets (rituximab). One complex biologic is currently approved for the treatment of allergic asthma, omalizumab (Xolair®), which acts by neutralizing soluble immunoglobulin E (IgE). Recent reports indicate that targeting IL-5 may also be efficacious in a subset of patients with refractory eosinophilic asthma (Haldar et al. 2009). Encouragingly, rituximab has also shown activity in Phase II clinical trials in relapsing/remitting multiple sclerosis (MS) (Hauser et al. 2008). Table 20.1 summarizes

H. Zhou, Ph.D. (✉) • Z. Xu, Ph.D.
Biologics Clinical Pharmacology,
Janssen Research & Development, LLC,
1400 McKean Road, Spring House, PA 19477, USA
e-mail: hzhou2@its.jnj.com

M.A. Mascelli, Ph.D.
Department of Clinical Pharmacology and Pharmacokinetics,
Shire, Lexington, MA, USA

H.M. Davis, Ph.D.
Biologics Clinical Pharmacology,
Janssen Research & Development,
LLC, Radnor, PA, USA

D.J.A. Crommelin, R.D. Sindelar, and B. Meibohm (eds.), *Pharmaceutical Biotechnology*,
DOI 10.1007/978-1-4614-6486-0_20,

Product	Target	Indication	Pharmacokinetics	Recommended dose regimen
Abatacept (Orencia®)	CTLA-4	RA	Abatacept exhibits linear pharmacokinetics with mean terminal half-life being 13.1–14.3 days. The mean (range) systemic CL was estimated to be 0.22 (0.13–0.47) mL/h/kg. The bioavailability of SC abatacept was 78.6 %. Following IV infusions of 10 mg/kg q4w in adult RA patients, the mean (range) steady-state peak and trough concentrations were 295 (171–398) and 24 (1–66) μg/mL, respectively. Following SC injections of 125 mg weekly, the mean (range) steady-state peak and trough concentrations were 48.1 (9.8–132.4) and 32.5 (6.6–113.8) μg/mL, respectively.	*IV infusion for RA*: For adult RA patients weighing <60 kg, the recommended dose of IV abatacept is 500 mg. For patients weighing 60–100 kg, a 750-mg dose should be administered. For patients weighing >100 kg, a 1,000-mg dose should be given. Abatacept should be administered as IV infusion at weeks 0, 2, and 4 and q4w thereafter. *SC injection for RA*: 125 mg should be administered weekly following an initial IV loading dose (according to body weight categories described above).
Alefacept (Amevive®)	CD2	PsO	The mean clearance of alefacept was 0.25 mL/h/kg with mean terminal half-life being approximately 270 h. The mean bioavailability of alefacept following an intramuscular injection was 63 %.	Alefacept should be administered as intramuscular injection. *PsO*: The recommended dose of alefacept is 15 mg once weekly for 12 weeks. The CD4+ T-lymphocyte count should be measured before initial dosing.
Adalimumab (Humira®)	TNFα	RA, PsA, AS, PsO, and CD	Adalimumab exhibits linear pharmacokinetics with mean terminal half-life being about 2 weeks. The systemic clearance of adalimumab is approximately 12 mL/h. The mean bioavailability of SC adalimumab was 64 %. Following SC administration of 40 mg q2w in adult RA patients, the mean steady-state trough concentrations were about 5 and 8–9 μg/mL without and with concomitant methotrexate, respectively. Methotrexate reduced adalimumab apparent clearance after single and multiple dosing by 29 % and 44 %, respectively, in patients with RA.	Adalimumab should be administered as SC injection. *RA, PsA, and AS*: The recommended dose of adalimumab is 40 mg SC q2w. *PsO*: The recommended dose is 80 mg initial dose followed by 40 mg SC q2w. *CD*: The recommended dose is 160 mg initial dose, 80 mg dose 2 weeks later, followed by 40 mg SC q2w.
Belimumab (Benlysta®)	BLyS	SLE	The median systemic clearance of belimumab was estimated to be 215 mL/day with a median terminal half-life being 19.4 days.	Belimumab should be administered as IV infusion. *SLE*: The recommended dose of belimumab is 10 mg/kg q2w for the first three doses, followed by 10 mg/kg q4w thereafter.
Canakinumab (Ilaris®)	IL-1	CAPS	Canakinumab exhibits linear pharmacokinetics with mean terminal half-life being 26 days. The median systemic clearance for a CAPS patient weighing 70 kg was estimated to be 0.174 L/day. The mean bioavailability of SC canakinumab was 70 %.	Canakinumab should be administered as SC injection. *CAPS*: The recommended dose of canakinumab is 150 mg SC q8w.
Certolizumab pegol (Cimzia®)	TNFα	RA and CD	Certolizumab pegol exhibits linear pharmacokinetics with mean terminal half-life being about 14 days. The systemic clearance following IV administration to healthy subjects ranged from 9.21 to 14.38 mL/h. The mean bioavailability of SC certolizumab pegol was 80 %.	Certolizumab pegol should be administered as SC injection. *RA*: The recommended dose of certolizumab pegol is 400 mg initially and 400 mg at weeks 2 and 4, followed by a lower maintenance dosing regimen of 200 mg q2w thereafter. For maintenance dosing, 400 mg q4w can also be considered. *CD*: The recommended dose of certolizumab pegol is 400 mg initially and 400 mg at weeks 2 and 4. For patients achieving clinical response, the recommended maintenance dose regimen is 400 mg q4w.

Etanercept (Enbrel®)	TNFα	RA, PsA, AS, and PsO	Etanercept exhibits linear pharmacokinetics with mean terminal half-life being 102 ± 30 h. The mean clearance was 160 ± 80 mL/h. Following SC administration of 50 mg etanercept weekly in adult RA patients, the mean ± SD steady-state peak and trough concentrations were 2.4 ± 1.5 and 1.2 ± 0.7 μg/mL.	Etanercept should be administered as SC injection. *RA, PsA, and AS*: The recommended dose of etanercept is 50 mg SC weekly. *PsO*: The recommended dose of etanercept is 50 mg twice weekly for 3 months, followed by 50 mg SC weekly.
Golimumab (Simponi®)	TNFα	RA, PsA, AS	Golimumab exhibits linear pharmacokinetics with mean terminal half-life being about 2 weeks. The mean systemic clearance was estimated to be 4.9–6.7 mL/day/kg. The mean bioavailability of SC golimumab was 53 %. Following SC administration of 50 mg q4w with concomitant methotrexate in adult RA patients, the mean steady-state trough concentrations were 0.4–0.6 μg/mL. Concomitant use of methotrexate reduced the apparent clearance of golimumab by approximately 36 %.	Golimumab should be administered as SC injection. *RA, PsA, and AS*: The recommended dose of golimumab is 50 mg SC monthly.
Infliximab (Remicade®)	TNFα	RA, PsA, AS, PsO, CD, UC	Infliximab exhibits linear pharmacokinetics with median terminal half-life being 7.7–9.5 days. Following IV infusions of a maintenance dose of 3–10 mg/kg q8w, median steady-state serum infliximab concentrations ranged from approximately 0.5–6 μg/mL.	Infliximab should be administered as IV infusion. *RA*: The recommended dose of infliximab is 3 mg/kg at 0, 2, and 6 weeks, followed by 3 mg/kg q8w thereafter. Some patients may benefit from increasing the dose up to 10 mg/kg or treating as often as q4w. *PsA and PsO*: The recommended dose of infliximab is 5 mg/kg at 0, 2, and 6 weeks, followed by 5 mg/kg q8w thereafter. *AS*: The recommended dose of infliximab is 5 mg/kg at 0, 2, and 6 weeks, followed by 5 mg/kg q6w thereafter. *CD and UC*: The recommended dose of infliximab is 5 mg/kg at 0, 2, and 6 weeks, followed by 5 mg/kg q8w thereafter. Some CD patients who initially respond to treatment may benefit from increasing the dose to 10 mg/kg if they later lose their response.
Natalizumab (Tysabri®)	Integrins	CD and MS	In patients with CD, the systemic clearance of natalizumab was 22 ± 22 mL/h with mean half-life being 10 ± 7 days. The mean steady-state trough concentration was 10 ± 9 μg/mL after q4w dosing. In patients with MS, the systemic clearance of natalizumab was 16 ± 5 mL/h with mean half-life being 11 ± 4 days. The mean steady-state trough concentrations ranged from 23 to 29 μg/mL after q4w dosing.	Natalizumab should be administered as IV infusion. *CD and MS*: The recommended dose of natalizumab is 300 mg q4w.
Omalizumab (Xolair®)	IgE	Asthma	Omalizumab exhibits target-mediated nonlinear pharmacokinetics. The mean apparent clearance of omalizumab was estimated to be 2.4 ± 1.1 mL/kg/day in asthma patients with a mean terminal half-life of 26 days. The mean bioavailability of SC omalizumab was 62 %.	Omalizumab should be administered as SC injection. *Asthma*: The recommended dose of omalizumab is 150–375 mg SC q2w or q4w, with dosage and dosing frequency determined by pretreatment serum IgE levels and body weight.

Table 20.1 ■ Monoclonal antibodies and antibody-based biotherapeutics in immune-mediated inflammatory diseases.

Product	Target	Indication	Pharmacokinetics	Recommended dose regimen
Rilonacept (Arcalyst)	IL-1	CAPS	Following weekly SC doses of 160 mg in patients with CAPS, serum rilonacept concentration appeared to reach steady state by 6 weeks with average trough levels being approximately 24 μg/mL.	Rilonacept should be administered as SC injection. *CAPS*: The recommended dose of rilonacept is 320 mg SC as a loading dose, followed by a maintenance dose regime of 160 mg SC weekly.
Rituximab (Rituxan®)	CD20	RA	The estimated clearance of rituximab was 0.335 L/day in patients with RA with mean terminal elimination half-life being 18.0 days.	Rituximab should be administered as IV infusion. Treatment courses of rituximab should be administered every 24 weeks or based on clinical evaluation, but no more frequent than every 16 weeks. *RA*: The recommended dose of each course for the treatment of RA is two 1,000 mg intravenous infusions separated by 2 weeks.
Tocilizumab (Actemra®)	IL-6 receptor	RA	Tocilizumab exhibits nonlinear pharmacokinetics with mean apparent terminal half-life being 151 ± 59 h after a single dose of 10 mg/kg in RA patients. The mean clearance was 0.29 ± 0.10 mL/h/kg at a dose level of 10 mg/kg. Following repeated dosing of 8 mg/kg IV q4w in adult RA patients, the model-predicted mean (SD) peak and trough concentrations at steady state were 183 ± 85.6 and 9.7 ± 10.5 μg/mL, respectively.	Tocilizumab should be administered as IV infusion. *RA*: The recommended starting dose of tocilizumab is 4 mg/kg IV q4w, followed by an increase to 8 mg/kg q4w based on clinical response.
Ustekinumab (Stelara®)	IL12/IL23	PsO	Ustekinumab exhibits linear pharmacokinetics with mean terminal half-life being 14.9–45.6 days. The mean systemic clearance in psoriasis patients ranged from 1.90 ± 0.28 to 2.22 ± 0.63 mL/day/kg. Following SC administration of 45 or 90 mg q12w in psoriasis patients, the mean steady-state trough concentrations were 0.31 ± 0.33 and 0.64 ± 0.64 μg/mL, respectively.	Ustekinumab should be administered as SC injection. *PsO*: The recommended dose of ustekinumab is 45 mg initially and 4 weeks later, followed by 45 mg q12w for patients weighing ≤100 kg; the recommended dose for patients weighing >100 kg is 90 mg initially and 4 weeks later, followed by 90 mg q12w.

Abbreviations: AS ankylosing spondylitis, *BLyS* B-lymphocyte stimulator, *CAPS* cryopyrin-associated periodic syndromes, *CD* Crohn's disease, *CD2* T-lymphocyte antigen CD2, *CD20* B-lymphocyte antigen CD20, *CTLA-4* cytotoxic T-lymphocyte-associated antigen 4, *IgE* immunoglobulin E, *IL* interleukin, *IV* intravenous, *MS* multiple sclerosis, *PsA* psoriatic arthritis, *PsO* plaque psoriasis, *q12w* every 12 weeks, *q2w* every 2 weeks, *q4w* every 4 weeks, *q8w* every 8 weeks, *RA* rheumatoid arthritis, *SC* subcutaneous, *SLE* systemic lupus erythematosus, *TNFα* tumor necrosis factor alpha, *UC* ulcerative colitis

Table 20.1 ■ (continued)

the 15 monoclonal antibodies and antibody-based biotherapeutics that are currently approved for the treatment of immune-mediated inflammatory diseases.

One challenge in the long-term administration of complex biologics for chronic immune-mediated disorders is to balance the potent and sustained pharmacologic action against the risk of serious side effects (mostly due to their exaggerated pharmacologic effects). The expectation for these newer therapies is that they can be used earlier in the course of disease to not only maintain control over episodic disease flares but also prevent the less reversible organ damage posed by long-term uncontrolled chronic inflammation or even reversal of joint damage caused by rheumatoid arthritis (Taylor et al. 2004). Though the list of indications continues to grow for these agents and newer agents are approved every year, this chapter will focus on describing the pharmacologic properties of a subset of approved biologic therapies for major classes of inflammatory disorders.

ARTHRITIDES

Arthritide is a class of chronic autoimmune inflammatory conditions of unknown etiology, characterized by pain and stiffness of the affected joints and tissue (Davis and Mease 2008; McInnes and Schett 2011). Arthritides consist of a variety of clinical diseases, such as rheumatoid arthritis (RA), psoriatic arthritis (PsA), and ankylosing spondylitis (AS), involving skeletal joints. Juvenile idiopathic arthritis (JIA) is a chronic inflammatory arthropathy with an age of onset of <16 years. Although these arthritic diseases may have different clinical manifestations, they are considered to have similar underlying immunologic etiology. RA is the most common of the autoimmune arthritides, affecting at least 1 % of the general population in the United States. If not properly treated, the chronic inflammation can result in progressive and irreversible joint destruction. Although early intervention with corticosteroids and disease-modifying antirheumatic drugs (DMARDs) has been proven to attenuate inflammation, these therapies may not effectively slow the progression of joint damage. In addition, these conventional therapies often have significant off-target side effects and do not provide sufficient clinical benefit for a large subset of patients (O'Dell 2004).

A number of complex biologics are approved for patients with RA, PsA, AS, or JIA (Table 20.1). These therapeutic proteins provide valuable treatment options for patients, particularly for those who experience significant side effects and/or have inadequate clinical efficacy with conventional DMARDs. These biologic agents significantly improve the signs and symptoms of the disease, effectively inhibit (and sometimes even reverse) the progression of joint damage, and greatly improve physical functions and quality of life. The mechanisms of action for these therapeutic proteins involve neutralization of tumor necrosis factor alpha [TNFα] (infliximab, etanercept, adalimumab, certolizumab, and golimumab), blockade of interleukin-1 [IL-1] (anakinra) or interleukin-6 receptor [IL-6R] (tocilizumab), inhibition of T-cell activation (abatacept), and depletion of B lymphocytes (rituximab). Anti-TNFα is considered the gold standard biologic therapy for RA; however, the availability of therapeutic proteins with different mechanisms of action provides alternative options when patients do not achieve adequate response to a certain class of biologic therapy. Compared to conventional DMARDs, the most attractive advantage for antibody-based therapeutic proteins is their more targeted mechanism of action of binding to the target with high specificity, consequently producing greater efficacy and fewer adverse effects. In addition, antibody-based therapeutic proteins usually have long half-lives (up to 2–3 weeks), which allow for infrequent dosing which is desirable for the patients in treatment of their chronic diseases.

The development of anti-TNFα therapeutic agents shows the advances of technologies and versatility in antibody generation and engineering over the past two decades. The approved anti-TNFα agents, comprised of infliximab, etanercept, adalimumab, golimumab, and certolizumab pegol, are presently one of the most successful classes of therapeutic proteins with many approved clinical indications. Etanercept is a dimeric fusion protein consisting of the p75 human TNFα receptor linked to the Fc portion of human IgG1. It is the first anti-TNFα biologic agent approved for arthritide indication. Infliximab is a chimeric IgG1 monoclonal antibody containing ~25 % mouse sequence and ~75 % human sequence. Certolizumab pegol is a Fab antibody fragment linked to polyethylene glycol that enhances solubility and prolongs elimination half-life. Adalimumab and golimumab are two fully human IgG1 monoclonal antibodies, which were created through the use of phage display libraries and the expression of human immunoglobulin genes by transgenic mice, respectively. Fully human antibodies were developed to minimize immunogenicity; however, patients treated with either adalimumab or golimumab still develop antidrug antibodies. Future efforts are still required to generate therapeutic monoclonal antibodies that not only have the human sequence as the primary structure, but also have secondary and tertiary structures like natural human immunoglobulins. Although no head-to-head comparative trials are currently available, these anti-TNFα agents appear to have similar efficacy for the treatment of adult RA patients (Salliot et al. 2011). These anti-TNFα agents have different elimination half-lives and offer a variety of dosing options. Infliximab is administered intravenously, while the other four anti-TNFα agents are administered subcutaneously. Etanercept has the shortest half-life (~4 days) and needs to be dosed once or twice a week. Infliximab is administered intravenously every 4–8 weeks, while adalimumab, certolizumab, and golimumab are administered subcutaneously every 2 weeks, every 2–4 weeks, or monthly, respectively (Tracey et al. 2008).

Although these antibody-based therapeutic proteins offer more targeted therapy, there are clearly adverse effects that need to be closely monitored (Bongartz et al. 2006; Brown et al. 2002; Ellerin et al. 2003). Certain adverse events deserve special attention. Infections are the result of inhibition of the protective functions of the targeted cytokines and related immune cells. Serious infections due to bacterial, mycobacterial, invasive fungal, viral, protozoal, or other opportunistic pathogens have been reported in patients with RA, PsA, AS, or JIA who received TNFα blockers and other immunosuppressant therapeutic proteins. Patients should be closely monitored for the development of signs and symptoms of infection during and after treatment with these therapeutic proteins, including the development of tuberculosis in patients who tested negative for latent tuberculosis infection prior to the initiation of therapy. Malignancy, albeit rare, has been another important concern when using these immunosuppressant therapeutic proteins. In controlled clinical trials of TNFα blockers, more cases of lymphoma and leukemia have been observed among patients receiving anti-TNFα treatment compared to patients in the control groups; however, there are confounders when assessing the risk of malignancy associated with the use of these therapeutic proteins in patients with chronic inflammatory diseases. Patients with chronic inflammatory diseases, particularly patients with highly active disease and/or chronic exposure to immunosuppressant therapies, may be at higher risk (up to several folds) than the general population for the development of lymphoma and leukemia, even in the absence of TNF-blocking therapy (Smedby et al. 2008). Other notable adverse events associated with these complex proteins include demyelinating disorders, liver enzyme elevation, autoimmune diseases (such as lupus), immunogenicity (formation of antibodies to the therapeutic protein), infusion/injection site reactions, and other hypersensitivity reactions. Overall, a large number of clinical trials have demonstrated that the benefits outweigh the risks in patients with various arthritides, including RA, PsA, AS, and JIA.

■ Abatacept

Abatacept (Orencia®) is a soluble fusion protein that consists of the extracellular domain of human cytotoxic T-lymphocyte-associated antigen 4 (CTLA-4) linked to the Fc portion of human IgG1. Activated T lymphocytes are implicated in the pathogenesis of RA. Abatacept inhibits T-lymphocyte activation by binding to CD80 and CD86, thereby blocking interaction with CD28 (Orencia®, US prescribing information 2011).

The efficacy and safety of abatacept have been assessed in various adult RA populations (DMARD (including methotrexate)-inadequate responders, anti-TNFα-inadequate responders, and methotrexate-naïve patients) and JIA. In a randomized, double-blind, controlled Phase III study in patients with active RA despite methotrexate therapy, 6-month treatment with intravenous abatacept in combination with methotrexate at the recommended dosing regimen resulted in 68, 40, and 20 % of RA patients achieving ACR20, ACR50, and ACR70, respectively (ACR20, ACR50, and ACR70 refer to 20, 50, and 70 % improvement from baseline, respectively, according to the American College of Rheumatology criteria), while the control group (methotrexate plus placebo) had only 40, 17, and 7 % of patients achieving ACR20, ACR50, and ACR70 responses, respectively. Recently, another Phase III study demonstrated similar efficacy and safety for treating RA patients using a subcutaneous route of administration as an alternative to the previously established intravenous route of administration.

Abatacept is indicated for the treatment of adult patients with moderately to severely active RA. It is also indicated for reducing signs and symptoms in pediatric patients 6 years of age and older with moderately to severely active polyarticular JIA.

■ Adalimumab

Adalimumab (Humira®) is a recombinant human IgG1 monoclonal antibody, which was created using phage display technology resulting in a fully human antibody. Adalimumab binds specifically to TNF-α and blocks its interaction with the p55 and p75 cell surface TNF receptors. Adalimumab does not bind or inactivate lymphotoxin (Humira®, US prescribing information 2011).

The efficacy and safety of adalimumab have been assessed in various adult RA populations (DMARD (including methotrexate)-inadequate responders and methotrexate-naïve patients), JIA, PsA, and AS. In a randomized, double-blind, controlled Phase III study in patients with active RA despite methotrexate therapy, 6-month treatment with subcutaneous adalimumab in combination with methotrexate at the recommended dosing regimen resulted in 63, 39, and 21 % of RA patients achieving ACR20, ACR50, and ACR70, respectively, while the control group (methotrexate plus placebo) had only 30, 10, and 3 % of patients with ACR20, ACR50, and ACR70 responses, respectively.

Of the rheumatic diseases, adalimumab is indicated for the treatment of adult patients with moderately to severely active RA, active PsA, or active AS. Adalimumab is also indicated for reducing signs and symptoms in pediatric patients 4 years of age and older with moderately to severely active polyarticular JIA.

■ Certolizumab Pegol

Certolizumab pegol (Cimzia®) is a recombinant, humanized antibody Fab fragment that is conjugated to an approximately 40 kDa polyethylene glycol. Certolizumab pegol binds to human TNFα with high affinity and selectively neutralizes TNFα activity, but does not neutralize lymphotoxin (Cimzia®, US prescribing information 2011).

The efficacy and safety of certolizumab pegol have been assessed in adult RA patients who had active disease despite methotrexate therapy or who had failed at least one DMARD other than methotrexate. In a randomized, double-blind, controlled Phase III study in patients with active RA despite methotrexate therapy, 6-month treatment with subcutaneous certolizumab pegol in combination with methotrexate at the recommended dosing regimen resulted in 59, 37, and 21 % of RA patients achieving ACR20, ACR50, and ACR70, respectively, while the control group (methotrexate plus placebo) had only 14, 8, and 3 % of patients with ACR20, ACR50, and ACR70 response, respectively.

Of the rheumatic diseases, certolizumab pegol is indicated for the treatment of adult patients with moderately to severely active RA.

■ Etanercept

Etanercept (Enbrel®) is a dimeric fusion protein consisting of the extracellular ligand-binding portion of the human 75 kDa (p75) TNF receptor linked to the Fc portion of human IgG1. Etanercept inhibits binding of TNF-α and TNF-β (lymphotoxin alpha) to cell surface TNFRs, rendering TNF biologically inactive (Enbrel®, US prescribing information 2011).

The efficacy and safety of etanercept have been assessed in various adult RA populations (DMARD (including methotrexate)-inadequate responders and methotrexate-naïve patients), active JIA, active PsA, and active AS. In a randomized, double-blind, controlled Phase III study in patients with active RA despite methotrexate therapy, 6-month treatment with subcutaneous etanercept in combination with methotrexate at the recommended dosing regimen resulted in 71, 39, and 15 % of RA patients achieving ACR20, ACR50, and ACR70, respectively, while the control group (methotrexate plus placebo) had only 27, 3, and 0 % of patients exhibiting ACR20, ACR50, and ACR70 responses, respectively.

Etanercept is indicated for the treatment of adult patients with moderately to severely active RA, active PsA, or active AS. Etanercept is also indicated for reducing signs and symptoms in pediatric patients 2 years of age and older with moderately to severely active polyarticular JIA.

■ Golimumab

Golimumab (Simponi®) is a fully human IgG1κ monoclonal antibody which was created using genetically engineered mice immunized with human TNF. Golimumab binds to both the soluble and transmembrane bioactive forms of human TNFα and therefore inhibits the biologic activity of TNFα. There is no evidence of golimumab binding to other TNF superfamily ligands; in particular, golimumab does not bind or neutralize human lymphotoxin. Golimumab does not lyse human monocytes expressing transmembrane TNF in the presence of complement or effector cells (Simponi®, US prescribing information 2011).

The efficacy and safety of golimumab have been assessed in various adult RA populations (methotrexate-inadequate responders, methotrexate-naïve patients, and patients with previous use of other anti-TNFα agents), PsA and AS. In a randomized, double-blind, controlled Phase III study in patients with active RA despite methotrexate therapy, 6-month treatment with subcutaneous golimumab in combination with methotrexate at the recommended dosing regimen resulted in 60, 37, and 20 % of RA patients achieving ACR20, ACR50, and ACR70, respectively, while the control group (methotrexate plus placebo) had only 28, 14, and 5 % of patients exhibiting ACR20, ACR50, and ACR70 responses, respectively.

Golimumab is indicated for the treatment of adult patients with moderately to severely active RA, active PsA, or active AS.

■ Infliximab

Infliximab (Remicade®) is a chimeric IgG1κ monoclonal antibody that is composed of human constant and murine variable regions. Infliximab neutralizes the biologic activity of TNFα by binding with high affinity to the soluble and transmembrane forms of TNFα and inhibits binding of TNFα with its receptors. Infliximab does not neutralize TNFβ (lymphotoxin-α) (Remicade®, US prescribing information 2011).

The efficacy and safety of infliximab have been assessed in various adult RA populations (methotrexate-inadequate responders and methotrexate-naïve patients), PsA, and AS. In a randomized, double-blind, controlled Phase III study in patients with active RA despite methotrexate therapy, 30-week treatment with intravenous infliximab (3 mg/kg) in combination with methotrexate resulted in 50, 27, and 8 % of RA patients achieving ACR20, ACR50, and ACR70, respectively, while the control group (methotrexate plus placebo) had only 20, 5, and 0 % of patients exhibiting ACR20, ACR50, and ACR70 responses, respectively.

Within the rheumatic disease category, infliximab is indicated for reducing signs and symptoms, inhibiting the progression of structural damage, and improving physical function in patients with moderately to severe active RA. Infliximab is also indicated for reducing signs and symptoms in patients with active AS and in patients with active PsA, inhibiting the progression of structural damage, and improving physical function in patients with PsA.

■ Rituximab

Rituximab (Rituxan®) is a genetically engineered chimeric murine/human monoclonal IgG1κ antibody that binds specifically to the antigen CD20 on pre-B and mature B lymphocytes. In the pathogenesis of RA, B cells may be involved in the autoimmune/inflammatory process through production of rheumatoid factor (RF) and other autoantibodies, antigen presentation, T-cell activation, and/or proinflammatory cytokine production. Rituximab binds to the CD20 antigen on B lymphocytes, and the Fc domain recruits immune effector functions to mediate B-cell lysis, resulting in B-cell depletion of circulating and tissue-based B cells (Rituxan®, US prescribing information 2012).

The efficacy and safety of rituximab have been assessed in adult patients with moderately to severely active RA who had a prior inadequate response to at least one anti-TNFα agent. In a randomized, double-blind, controlled Phase III study in RA patients, treatment with one course of intravenous rituximab (2 doses of 1,000 mg separated by 2 weeks) in combination with methotrexate resulted in 51, 27, and 12 % of RA patients achieving ACR20, ACR50, and ACR70, respectively, at week 24, while the control group (methotrexate plus placebo) had only 18, 5, and 1 % of patients exhibiting ACR20, ACR50, and ACR70 responses, respectively.

Rituximab is indicated as a second-line biologic therapy for the treatment of adult patients with moderately to severely active RA who have had an inadequate response to one or more TNF antagonist therapies. The estimated median terminal half-life of rituximab is 18 days in RA patients; however, the pharmacodynamic effect on B cells lasts much longer than the drug level. B-cell recovery begins at approximately 6 months, and median B-cell levels return to normal by 12 months following completion of treatment with rituximab. Consequently, treatment courses of rituximab should be administered every 24 weeks or based on clinical evaluation, but no more frequent than every 16 weeks.

■ Tocilizumab

Tocilizumab (Actemra®) is a recombinant humanized antihuman interleukin 6 (IL-6) receptor monoclonal antibody of IgG 1κ. IL-6 is a pleiotropic proinflammatory cytokine that is involved in the pathogenesis of RA. Tocilizumab binds specifically to both soluble and membrane-bound IL-6 receptors (sIL-6R and mIL-6R) and has been shown to inhibit IL-6-mediated signaling through these receptors (Actemra®, US prescribing information 2011).

The efficacy and safety of tocilizumab have been assessed in various adult RA populations (DMARD (including methotrexate)-inadequate responders, and methotrexate-naïve patients, and patients with an inadequate clinical response or intolerant to one or more TNF antagonist therapies) and systemic JIA. In a randomized, double-blind, controlled Phase III study in such RA patients, treatment with intravenous tocilizumab (4 or 8 mg/kg) in combination with methotrexate resulted in 51–56 %, 25–32 %, and 11–13 % of RA patients achieving ACR20, ACR50, and ACR70, respectively, at week 24, while the control group (methotrexate plus placebo) had only 27, 10, and 2 % of patients exhibiting ACR20, ACR50, and ACR70 responses, respectively.

Tocilizumab is indicated as a second-line biologic therapy for the treatment of adult patients with moderately to severely active RA who have had an inadequate response to one or more TNF antagonist therapies. It is also indicated for the treatment of pediatric patients 2 years of age and older with active systemic JIA.

SYSTEMIC LUPUS ERYTHEMATOSUS

Systemic lupus erythematosus (SLE) is an autoimmune disease characterized by the involvement of multiple organ systems, a clinical pattern of unpredictable exacerbations and remissions, and the presence of autoantibodies. The immunopathogenic characteristic of this disease is polyclonal B-cell activation that leads to hyperglobulinemia, autoantibody production, and immune complex formation, which in turn leads to inflammation and damage that can affect multiple organ systems. Generalized SLE symptoms may include fever, fatigue, rash, oral ulceration, hair loss, and arthralgias. US prevalence estimates for various types of lupus, including SLE, vary greatly, with estimates as high as 100 per 100,000 persons affected. Approximately 80–90 % of patients with SLE are women.

The primary causes of SLE remain unclear. Current therapies tend to be generally immunosuppressive, often providing suboptimal control over disease manifestation and long-term outcomes due to ineffectiveness or side effects. These SLE therapies have targeted nonspecific sites for inflammatory reduction (NSAIDs, antimalarials) and immune system suppression (corticosteroids, azathioprine, cyclophosphamide, methotrexate, mycophenolate).

Belimumab

Belimumab (Benlysta®), a novel human IgG1λ monoclonal antibody to B-lymphocyte stimulator (BLyS), acts by binding to the soluble form of BLyS with high affinity. Once belimumab is bound, it prevents BLyS from binding TACI, BCMA, and BAFF-R, causing an interruption of BLyS-induced B-lymphocyte proliferation and differentiation (Benlysta®, US prescribing information 2012).

The safety and effectiveness of belimumab were evaluated in three randomized, double-blind, placebo-controlled studies in patients with SLE according to criteria from the American College of Rheumatology. Patients were on a stable SLE standard of care treatment regimen comprising any of the following (alone or in combination): corticosteroids, antimalarials, NSAIDs, and immunosuppressives. The first Phase III study evaluated doses of 1, 4, and 10 mg/kg belimumab plus standard of care compared to placebo plus standard of care over 52 weeks in patients with SLE. The co-primary endpoints were percent changes in SELENA-SLEDAI score at week 24 and time to flare over 52 weeks. Exploratory analysis of this study identified a subgroup of patients (72 %), who were autoantibody positive, in whom belimumab appeared to offer benefit. The results of this study informed the design of the other two Phase III studies (BLISS-52 and BLISS-76) and led to the selection of a target population and indication that was limited to autoantibody-positive SLE patients. BLISS-52 and BLISS-76 were both randomized, double-blind, placebo-controlled trials in patients with SLE. The studies were similar in design except BLISS-52 had a 52-week duration and BLISS-76 had a 76-week duration. Both studies compared belimumab 1 and 10 mg/kg plus standard of care to placebo plus standard of care. Eligible patients had active SLE disease, defined as a SELENA-SLEDAI score ≥6, and positive autoantibody test results at screening. The primary efficacy determination was a composite endpoint (SLE Responder Index or SRI) that defined response as meeting each of the following criteria at week 52 compared to baseline:

- ≥4-point reduction in the SELENA-SLEDAI score
- No new British Isles Lupus Assessment Group (BILAG) A organ domain score or two new BILAG B organ domain scores
- No worsening (<0.30-point increase) in Physician's Global Assessment (PGA) score

In both BLISS-52 and BLISS-76, the proportion of SLE patients achieving an SRI response, as defined for the primary endpoint, was significantly higher in the belimumab 10 mg/kg group than in the placebo group in both studies. The effect on the SRI was not consistently significantly different for the belimumab 1 mg/kg group relative to placebo in both trials. In Phase III trials, reports suggest that overall adverse event rates, infections, treatment discontinuations due to adverse events, and fatalities were not significantly different between belimumab- and placebo-treated patients; however, serious and severe infusion-related reactions were reported more often in belimumab-treated patients.

Belimumab was approved for the treatment of adult patients with active autoantibody-positive SLE who are receiving standard therapy.

PSORIASIS

Psoriasis is the most common chronic, immune-mediated skin disorder, affecting approximately 2 % of the world's population (Nestle et al. 2009). Thickened epidermal layers resulting from excessive keratinocyte cell proliferation characterize psoriasis. The majority of sufferers are afflicted with psoriasis for most of their lives. Symptoms typically present between the ages of 15 and 35, with the majority of individuals diagnosed before the age of 40. Plaque psoriasis is the most common form, affecting approximately 85–90 % of individuals with the condition. The disease manifests as raised, well-demarcated, erythematous, and frequently pruritic and painful plaques with silvery scales (Christophers 2001; Griffiths and Barker 2007). Approximately 25 % of individuals with psoriasis develop moderate to severe disease with widely disseminated lesions. In clinical development and in managing patient care, the Psoriasis Activity and Severity Index (PASI) is commonly used as an instrument to measure and evaluate patient care and treatment effects of anti-psoriasis therapies (Feldman and Krueger 2005).

Prior to the availability of complex biologics in psoriasis, multiple therapeutic options existed for the treatment of the disease; however, a significant unmet need remained for a safe, highly effective, convenient systemic therapy for patients with moderate to severe forms of the disease (Papp et al. 2011). Psoralen plus ultraviolet A light therapy, while effective, is inconvenient and is associated with an increased risk of skin malignancies and photo damage. Significant safety concerns and organ toxicity are associated with chronic administration of conventional systemic agents such as methotrexate, cyclosporine, and acitretin, thus limiting their use in long-term psoriasis management.

Three anti-TNF biologic agents are approved for use in psoriasis: etanercept (Enbrel®), adalimumab (Humira®), and infliximab (Remicade®). Other approved agents for the treatment of psoriasis include ustekinumab (Stelara®), a human MAB directed against the IL-12/23 p40 receptor, and alefacept (Amevive®), which binds T and NK cells via CD2 (Table 20.1). Efalizumab (Raptiva®) is a humanized IgG_1 MAB directed against CD11 and inhibits leukocyte function. Efalizumab was approved in 2003 for the treatment of moderate to severe psoriasis but was voluntarily withdrawn in 2009 due to reports of progressive multifocal leukoencephalopathy (PML).

■ Alefacept

Alefacept (Amevive®) is a CD2-directed LFA-3/Fc fusion protein that consists of the extracellular CD2-binding portion of the human leukocyte function antigen-3 (LFA-3) linked to the Fc portion of human IgG1. Alefacept is produced by recombinant DNA technology in a Chinese Hamster Ovary (CHO) mammalian cell expression system. Activation of T lymphocytes involving the interaction between LFA-3 on antigen-presenting cells and CD2 on T lymphocytes plays a role in the pathophysiology of chronic plaque psoriasis. Alefacept interferes with lymphocyte activation by specifically binding to the lymphocyte antigen, CD2, and inhibiting the interaction between CD2 and its ligand, LFA-3. Alefacept also causes a reduction in subsets of CD2+ T lymphocytes and circulating total CD4+ and CD8+ T-lymphocyte counts (Amevive®, US prescribing information 2011).

Alefacept was evaluated in two randomized, double-blind, placebo-controlled Phase III studies in adults with chronic (≥1 year) plaque psoriasis and a minimum body surface area involvement of 10 % who were candidates for or had previously received systemic therapy or phototherapy. Two 12-week courses of once-weekly intravenous alefacept 7.5 mg or placebo were given in the first randomized double-blind study (Krueger 2003). Patients were followed for 12 weeks after each course. During treatment and follow-up of course 1, a PASI 75 response was achieved by 28 % of alefacept-treated and 8 % of placebo-treated patients ($p<0.001$). Patients who received a single course of alefacept and achieved a PASI 75 response during or after treatment, without the use of phototherapy or systemic therapies, maintained a PASI 50 response for a median duration of more than 7 months. Among patients who received two courses of alefacept, 40 and 71 % of patients achieved a PASI 75 response and PASI 50 response, respectively, during the study period. Alefacept was well tolerated over both courses. In course 1, the incidence of transient chills was higher in the alefacept group compared to the placebo group; more than 90 % of cases occurred within 24 h after the first few doses. The second Phase III study, an international, randomized, double-blind, placebo-controlled, parallel-group trial, provided a basis for comparison of patients treated with either 10 or 15 mg alefacept administered intramuscularly (Lebwohl et al. 2003). Alefacept treatment was associated with dose-related significant improvements in PASI from baseline. Throughout the study, a greater percentage of patients in the 15-mg group (33 %) than in the placebo group (13 %) achieved a PASI 75 response. Of patients in the 15-mg group who achieved a PASI 75 response 2 weeks after the last dose, 71 % maintained at least a PASI 50 response throughout the 12-week follow-up. The only adverse event that occurred at a 5 % or higher incidence among alefacept-treated patients compared to placebo-treated patients was chills (1 % placebo vs. 6 % alefacept), which occurred predominantly with intravenous administration. The adverse reactions which commonly resulted in clinical intervention were cardiovascular events including coronary artery disorder in <1 % of subjects and myocardial infarct in <1 % of subjects. The most common events resulting in discontinuation of treatment with alefacept were CD4+ T-lymphocyte levels below 250 cells/μL.

Alefacept is indicated for the treatment of adult patients with moderate to severe chronic plaque psoriasis who are candidates for systemic therapy or phototherapy.

■ Anti-TNFα Antagonists (Adalimumab, Etanercept, Infliximab)

TNFα antagonists (adalimumab, etanercept, and infliximab) block the binding of TNFα to its receptor, interrupting the subsequent signaling and inflammatory pathways driven by TNFα. This activity suppresses inflammation and the increased activation of T cells that are characteristics of psoriasis (Humira®, US prescribing information 2011; Enbrel®, US prescribing information 2011; Remicade®, US prescribing information 2011).

An evidence-based comparison from clinical trials of three TNFα antagonists (adalimumab, etanercept, and infliximab) has indicated better efficacy of infliximab and adalimumab than etanercept in treating psoriasis (Langley 2012).

A meta-analysis comparing three TNFα antagonists and traditional systemic therapy (e.g., cyclosporine) used in the treatment of moderate to severe psoriasis demonstrated high efficacy and tolerability of TNFα antagonists (Langley 2012). Due to the mode of action, there is a concern that patients receiving TNFα antagonists may become more susceptible to infection; however, a meta-analysis of trial data for these TNFα antagonists showed that serious infection rates were not much higher than those in placebo-treated patients.

Another recent meta-analysis was completed from twenty trials employing anti-psoriatic biologics. Based on an indirect comparison and a placebo PASI 50 response of 13 %, infliximab had the highest predicted mean probability of response of 93, 80, and 54 % for PASI 50, PASI 75, and PASI 90, respectively, followed by ustekinumab 90 mg at 90, 74, and 46 %, respectively, and then ustekinumab 45 mg, adalimumab, etanercept, and efalizumab (Reich et al. 2012).

A head-to-head controlled trial was conducted comparing the efficacy and safety of a TNFα antagonist, etanercept, and ustekinumab in patients with moderate to severe psoriasis (Griffiths et al. 2010). Ustekinumab (see below) was administered subcutaneously at either 45 or 90 mg at weeks 0 and 4, and high-dose etanercept was administered 50 mg twice weekly for 12 weeks. There was at least 75 % improvement in the PASI at week 12 in 67.5 % of patients who received 45 mg of ustekinumab and 73.8 % of patients who received 90 mg of ustekinumab, as compared to 56.8 % of those who received etanercept ($p = 0.01$ and $p < 0.001$, respectively). The efficacy of ustekinumab at 45 or 90 mg was superior to that of high-dose etanercept over a 12-week period while the safety profiles were similar.

■ Ustekinumab

Ustekinumab (Stelara®) is a human IgG1κ monoclonal antibody that binds with high affinity and specificity to the p40 protein subunit used by both the interleukin (IL)-12 and IL-23 cytokines. IL-12 and IL-23 are naturally occurring cytokines that are involved in inflammatory and immune responses, such as natural killer cell activation and CD4+ T-cell differentiation and activation. In in vitro models, ustekinumab was shown to disrupt IL-12- and IL-23-mediated signaling and cytokine cascades by disrupting the interaction of these cytokines with a shared cell surface receptor chain, IL-12β1 (Stelara®, US prescribing information 2011).

The safety and efficacy of ustekinumab were assessed in two Phase III trials (PHOENIX 1 and PHOENIX 2) in patients with moderate to severe psoriasis. The results from the Phase III trials demonstrated that ustekinumab was effective in ameliorating psoriatic plaques, pruritus, and nail psoriasis (Leonardi et al. 2008; Papp et al. 2008). Within 12 weeks of initiating ustekinumab treatment (45 or 90 mg/kg at weeks 0 and 4), more than two-thirds of patients exhibited ≥75 % reduction in PASI (PASI 75). Maximum efficacy was achieved at approximately 24 weeks after initiation of therapy, with approximately 75 % of ustekinumab-treated patients achieving a PASI 75 response. Clinical response to ustekinumab was associated with serum ustekinumab concentrations that were somewhat correlated with patient body weight. While efficacy of the 45 and 90 mg doses of ustekinumab was similar in patients weighing ≤100 kg, the 90 mg dose was more effective than the 45 mg dose in patients weighing >100 kg, who represented approximately one-third of the combined PHOENIX 1 and 2 population. Thus, to optimize efficacy in all patients while minimizing unnecessary drug exposure, fixed dose administration of ustekinumab based on body weight is indicated, i.e., for patients weighing >100 kg, the recommended dose is 90 mg initially and 4 weeks later, followed by 90 mg every 12 weeks; for patients weighing ≤100 kg, the recommended dose is 45 mg initially and 4 weeks later, followed by 45 mg every 12 weeks.

Results of the Phase III psoriasis clinical trials indicated that ustekinumab was generally well tolerated. Rates and types of adverse events (AEs), serious AEs, AEs leading to treatment discontinuation, and laboratory abnormalities were generally comparable among patients receiving placebo, ustekinumab 45 or 90 mg during the 12-week placebo-controlled Phases of PHOENIX 1 and 2. Dose–response relationships for safety events were not apparent. Immunogenicity rates were low, with approximately 5 % of patients developing anti-ustekinumab antibodies, and drug administration was well tolerated, with approximately 1 % of injections having an associated injection site reaction (Leonardi et al. 2008; Papp et al. 2008). Blocking IL-12 and/or IL-23 carries theoretical risks of infection and suppression of tumor immune surveillance (Airoldi et al. 2005). For instance, mycobacterial and salmonella infections were reported in individuals congenitally deficient in IL-12p40 or IL-12R. Rates of serious infections and malignancies in PHOENIX 1 and 2 were low and comparable across treatment groups during the placebo-controlled phases, no apparent increase in the frequency of these AEs was observed through 18 months of treatment, and no mycobacterial or salmonella infections were reported (Leonardi et al. 2008; Papp et al. 2008).

INFLAMMATORY BOWEL DISEASE

Inflammatory bowel disease (IBD) refers to a group of chronic inflammatory diseases of the gastrointestinal tract that mainly comprise two well-defined clinical entities, Crohn's disease (CD) and ulcerative colitis (UC). The incidence of IBD has continued to increase in many regions of the world with approximately 1.4 million Americans affected (Centers for Disease Control and Prevention 2011). CD and UC are two idiopathic IBD. They have some similarities and also unique differences (Baumgart and Sandborn 2007).

CD is a relapsing, transmural inflammatory disease of the gastrointestinal mucosa that can affect the entire gastrointestinal tract from the mouth to the anus, while UC is a relapsing, nontransmural inflammatory disease that is restricted to the colon. CD may involve all layers of the intestine, and there can be normal healthy bowel between patches of diseased bowel, while UC does not affect all layers of the bowel but only affects the top layers of the colon in an even and continuous distribution. In CD, pain is commonly experienced in the lower right abdomen, while in UC, in the lower left part of the abdomen. In CD, colon wall may be thickened and may have a rocky appearance, while in UC the colon wall is thinner and shows continuous inflammation. In clinical practice, disease activity of CD is typically described as mild to moderate (ambulatory patients able to tolerate oral alimentation without manifestations of dehydration, toxicity, abdominal tenderness, painful mass, obstruction, or >10 % weight loss), moderate to severe disease (failure to respond to treatment for mild disease, more prominent symptoms of fever, weight loss, abdominal pain or tenderness, intermittent nausea and vomiting without obstruction, or significant anemia), and severe to fulminant disease (persisting symptoms on corticosteroids, high fevers, persistent vomiting, evidence of intestinal obstruction, rebound tenderness, cachexia, or evidence of an abscess). While disease activity of UC is typically described as mild (up to four bloody stools daily and no systemic toxicity), moderate (four to six blood stools daily and minimal toxicity), or severe (more than six stools daily and signs of toxicity, such as fever, tachycardia, anemia, and raised erythrocyte sedimentation rate), patients with fulminant UC usually have more than ten bloody stools daily, continuous bleeding, anemia requiring blood transfusion, abdominal tenderness, and colonic dilation on plain abdominal radiographs (Baumgart and Sandborn 2007).

There are four therapeutic proteins that are approved for the treatment of CD and/or UC; those are anti-TNFα agents (infliximab, adalimumab, certolizumab pegol) and one is an anti-integrin MAB (natalizumab) as of March 2012 (Table 20.1). Not all anti-TNFα agents have been shown to be effective for IBD. For example, infliximab, the first-in-class anti-TNFα biologic agent approved for treating CD, has been shown to be highly effective in the treatment of CD, but etanercept was shown to be ineffective for this disease. A mechanism postulated to explain the differential effects of infliximab and etanercept for CD was that infliximab could bind membrane-associated TNFα and induce apoptosis of activated T cells and macrophages, but etanercept only binds to soluble TNFα (Van den Brande et al. 2003). However, this theory is questioned by later data showing induction of apoptosis by etanercept and clinical efficacy of certolizumab pegol, a non-apoptotic anti-TNFα agent (Chaudhary et al. 2006; Sandborn et al. 2006). Compared to the anti-TNFα agents, natalizumab has a more restricted use for the treatment of CD with a requirement for patient registration due to its potential risk of progressive multifocal leukoencephalopathy (PML). Natalizumab is approved as a second-line biologic agent for patients with refractory disease not receiving other immunomodulators, including anti-TNFα agents. As of March 2012, infliximab is the only antibody-based therapeutic protein approved for pediatric patients with CD or UC.

Conventional pharmacologic treatments for inflammatory bowel disease include aminosalicylates, corticosteroids, immunomodulators (azathioprine, 6-mecaptopurine, methotrexate, cyclosporine), and antibiotics (metronidazole, ciprofloxacin, clarithromycin). The aim of traditional therapy is to induce and maintain remission in patients. Treatment guidelines generally recommend initiating treatment with first-line agents such as sulfasalazine and systemic corticosteroids, followed by immunomodulators. The aforementioned pharmacologic therapies are often effective in patients with IBD, particularly in those with mildly to moderately active disease; however, a significant proportion of patients have severely active disease that is often refractory to these conventional therapies. Furthermore, these small molecule drugs have limitations in the treatment of IBD. Corticosteroids have many side effects and are not suitable for long-term maintenance therapy. Corticosteroids are also ineffective for healing bowel ulcerations (Modigliani et al. 1990). Immunomodulators promote mucosal healing, but the onset of action is slow. The use of anti-TNFα agents can overcome the shortcomings of the conventional treatment options and provide greater improvement for severe or refractory IBD. Anti-TNFα therapy can rapidly improve signs and symptoms (i.e., induce and maintain clinical response and clinical remission), promote mucosal healing, eliminate corticosteroid use, and has the potential to alter the natural history of IBD.

Historically, therapeutic proteins have been used as rescue therapy for patients with IBD refractory to conventional therapies. Recently, evidence has emerged that early use of anti-TNFα therapy in patients at high risk may induce a greater response and prevent irreversible damage to the intestine (D'Haens et al. 2008). There are also concerns with respect to increased risks of infections and malignancy associated with the use of anti-TNFα agents in patients with IBD (Hoentjen and van Bodegraven 2009). The timing of initiating therapy with therapeutic proteins and the identification of the subset of patients who can achieve maximal benefit from treatment using therapeutic proteins remain active areas of debate, and further clinical research is required to provide evidence-based guidelines.

■ Adalimumab

An overview of adalimumab has been provided earlier in this chapter (Humira®, US prescribing information 2011). In addition to the use of this therapeutic protein for the treatment of arthritides, the efficacy and safety of adalimumab have been assessed in adult patients with moderately to severely active CD (Crohn's Disease Activity Index (CDAI) ≥220 and ≤450). In a randomized, double-blind, controlled Phase III study in patients with active CD, treatment with subcutaneous adalimumab at the recommended induction dosing regimen (160 and 80 mg at weeks 0 and 2, respectively) resulted in 58 and 36 % of patients achieving clinical response (defined as a decrease in CDAI ≥70) and clinical remission (defined as CDAI <150), respectively, at week 4, while the control group (placebo) had 34 and 12 % of patients with clinical response and clinical remission, respectively. Among clinical responders at week 4, further maintenance treatment with 40 mg adalimumab every other week demonstrated greater efficacy than the placebo maintenance group. The adalimumab maintenance group had 43 and 36 % of patients who were in clinical response and clinical remission, respectively, at week 56, while the placebo maintenance group had 18 and 12 % of patients in clinical response and clinical remission, respectively. Most recently, adalimumab was approved in European Union for the treatment of moderately and severely active UC in adult patients who had had an inadequate response to conventional therapy including corticosteroids and 6-mercaptopurine (6-MP) or azathioprine (AZA) or who are intolerant to or have medical contraindications for such therapies (Humira®, Summary of Product Characteristics 2012).

■ Certolizumab Pegol

An overview of certolizumab pegol has been provided earlier in this chapter (Cimzia®, US prescribing information 2011). In addition to its use for the treatment of RA, the efficacy and safety of certolizumab pegol have been assessed in adult patients with moderately to severely active CD (CDAI ≥220 and ≤450). In a randomized, double-blind, controlled Phase III study in patients with active CD, treatment with subcutaneous certolizumab pegol at the recommended induction dosing regimen (400 mg each at weeks 0, 2, and 4, followed by 400 mg every 4 weeks thereafter) resulted in 35 and 22 % of patients achieving clinical response (defined as a decrease in CDAI ≥100) and clinical remission (defined as CDAI ≤150), respectively, at week 6, while the control group (placebo) had 27 and 17 % of patients with clinical response and clinical remission, respectively. Among clinical responders at week 6, further maintenance treatment with 400 mg certolizumab pegol every 4 weeks demonstrated greater efficacy than the placebo maintenance group. The certolizumab pegol maintenance group had 63 and 48 % of patients who were in clinical response and clinical remission, respectively, at week 26, while the placebo maintenance group had 36 and 29 % of patients in clinical response and clinical remission, respectively.

■ Infliximab

An overview of infliximab has been provided earlier in this chapter (Remicade®, US prescribing information 2011). The pharmacokinetics of infliximab in IBD is generally comparable to those in other indications including arthritides and psoriasis (Table 20.1). In addition to its use for the treatment of arthritides, the efficacy and safety of infliximab have been assessed in both adult and pediatric patients with moderately to severely active CD and UC.

In a randomized, double-blind, controlled Phase III study in patients with moderately to severely active Crohn's disease (CDAI ≥220 and ≤400), treatment with an initial dose of 5 mg/kg infliximab at week 0 resulted in 57 % of patients achieving clinical response (defined as a decrease in CDAI ≥70) at week 2. All of the 545 patients who received 5 mg/kg infliximab at week 0 were then randomized to placebo or infliximab maintenance groups (5 or 10 mg/kg at weeks 2 and 6, followed by every 8 weeks). Maintenance treatment with infliximab demonstrated greater efficacy than placebo maintenance treatment. The 5 and 10 mg/kg infliximab maintenance groups had 25 and 34 % of patients who were in clinical remission and discontinued corticosteroid use, respectively, at week 54, while the placebo maintenance group had 11 % of patients in clinical remission with corticosteroid discontinuation.

In another randomized, double-blind, controlled Phase III study in patients with moderately to severely active UC (Mayo score of 6–12, Endoscopy subscore ≥2), patients were randomized at week 0 to receive either placebo or infliximab at weeks 0, 2, and 6 and every 8 weeks thereafter. At week 8, a greater proportion of patients in the 5 mg/kg infliximab treatment group were in clinical response (defined as a decrease in Mayo score by ≥30 % and ≥3 points;) and clinical remission (defined as a Mayo score ≤2 points with no individual subscore >1) compared to the placebo treatment group (69 % vs. 37 % for clinical response; 39 % vs. 15 % for clinical remission). The clinical efficacy was maintained over time. At week 54, the 5 mg/kg infliximab maintenance group had 45 and 35 % of patients who were in clinical response and clinical remission, respectively, while the placebo maintenance group had 20 and 17 % of patients in clinical response and clinical remission, respectively.

For both CD and UC, maintenance therapy with infliximab every 8 weeks significantly reduced disease-related hospitalizations and surgeries. A reduction in corticosteroid use and improvements in quality of life were observed. In addition, the safety and efficacy of infliximab for pediatric patients 6 years of age or older with CD or UC have also been established.

■ Natalizumab

Natalizumab (Tysabri®) is a recombinant humanized IgG4κ monoclonal antibody produced in murine myeloma cells. Natalizumab binds to the α4-subunit of α4β1 and α4β7 integrins expressed on the surface of all leukocytes, and inhibits the α4-mediated adhesion of leukocytes to their counter-receptor(s). Disruption of these molecular interactions prevents transmigration of leukocytes across the endothelium into inflamed parenchymal tissue (Tysabri®, US prescribing information 2012).

The efficacy and safety of natalizumab have been assessed in adult patients with moderately to severely active CD (CDAI ≥220 and ≤450). In a randomized, double-blind, controlled Phase III study in patients with active CD, treatment with intravenous natalizumab at the recommended induction dosing regimen (300 mg every 4 weeks) resulted in 56 % of patients achieving clinical response (defined as a decrease in CDAI ≥70) at week 10, while the control group (placebo) had 49 % of patients achieving clinical response ($p=0.067$). Among clinical responders at both weeks 10 and 12, maintenance treatment with 300 mg natalizumab every 4 weeks demonstrated greater efficacy than that observed in the placebo maintenance group. The natalizumab maintenance group had 54 and 40 % of patients who were in clinical response and clinical remission at month 15, respectively, while the placebo maintenance group had 20 and 15 % of patients in clinical response and clinical remission, respectively.

Natalizumab is indicated for the induction and maintenance of clinical response and remission in adult patients with moderately to severely active CD who have evidence of inflammation and an inadequate response to, or an inability to tolerate, conventional CD therapies and TNF-α inhibitors.

ALLERGIC ASTHMA

Asthma is a complex chronic inflammatory syndrome of the airways and is characterized by variable symptoms of cough, breathlessness, and wheezing. These episodes may be punctuated by periods of more severe and sustained deterioration in control of symptoms, termed exacerbations, which can result in potentially life-threatening bronchospasm. Asthma affects almost 20 million individuals in the USA, six million of which are children. Pharmacotherapeutic management of the disease has progressed but is suboptimal for a subset of moderately to severely affected patients. Treatment with inhaled corticosteroids and short- and long-acting β-adrenoceptor agonists (SABAs and LABAs) is considered the standard of care and is generally effective at attenuating symptoms, particularly in mild to moderate asthma; however, these therapeutic modalities do not necessarily address the underlying pathologies of the disease. A subset of patients with moderate to severe asthma remain symptomatic despite treatment with corticosteroids, suggesting persistent inflammation of the airways. The limitations of existing asthma therapies justify continued research into novel interventions, particularly those that modify disease processes. To that aim, a number of MABs targeting cytokines linked to the underlying pathology of the disease (e.g., IL-13, tralokinumab; IL-5, mepolizumab) are currently being evaluated in the clinic (www.clinicaltrials.gov: NCT01402986 and NCT01366521). One of these agents, omalizumab, is currently approved for patients with moderate to severe asthma who remain symptomatic despite treatment with systemic or inhaled corticosteroids.

■ Omalizumab

IgE plays a seminal role in increasing allergen uptake by dendritic cells, activated mast cells, and basophils. Omalizumab (Xolair®) is a recombinant human IgG1κ MAB that selectively binds IgE and inhibits the binding of IgE to the high-affinity IgE receptor (FcεRI) on the surface of mast cells and basophils. Reduction in surface-bound IgE on FcεRI-bearing cells limits the degree of the allergic response. Omalizumab also reduces the number of FcεRI receptors on basophils in atopic patients and is a first-in-class selective IgE inhibitor approved for use by patients with allergic asthma inadequately controlled by inhaled corticosteroids. It is also the first biologic therapy approved for asthma. Omalizumab is indicated for adults and adolescents (12 years of age and older) with moderately to severely persistent asthma who have a positive skin test or in vitro reactivity to a perennial aeroallergen and whose symptoms are inadequately controlled with inhaled corticosteroids (Xolair®, US prescribing information 2010).

Omalizumab is administered subcutaneously (SC) at doses of 150–375 mg every 2–4 weeks using a table that adjusts target exposure based upon body weight and baseline levels of total IgE, aiming to achieve at least 0.016 mg/kg/IU in order to suppress free IgE levels below at least 50 ng/mL (Hochhaus et al. 2003). Using the recommended dosing table, serum free IgE levels decreased greater than 96 %, with 98 % of individuals having free IgE levels at or below 50 ng/mL by 3 months of multiple dosing. Because omalizumab forms complexes with IgE and these complexes have a slower elimination rate than free (unbound, monomeric)

IgE, treatment with omalizumab produces an apparent increase in total IgE using standard laboratory tests that cannot distinguish bound from free IgE. Following discontinuation of omalizumab there is a slow (up to 12 months) washout of drug-IgE complexes without an apparent rebound in free IgE levels. Pivotal safety and efficacy trials evaluated patients with moderately to severely persistent asthma, according to National Heart, Lung, and Blood Institute criteria, who were defined as having allergic-type asthma based upon a positive test to a perennial allergen. For dose determinations using the recommended table, total baseline IgE had to be between 30 and 700 IU/mL with a body weight of not more than 150 kg. Although some therapeutic effects were observed on allergic and asthma symptoms, pulmonary function, and the need for concomitant medications, omalizumab efficacy was based primarily on the reduction of asthma exacerbations, which were defined as a worsening of asthma that required treatment with systemic corticosteroids or a doubling of baseline-inhaled corticosteroid dose. Omalizumab was very well tolerated in the clinical trials, with only a slight increase in injection site reactions compared to controls. The incidence of antidrug antibodies was reported as 1/1723 (<0.1 %), which is an extremely low immunogenicity rate compared to nearly every other biologic therapy. Malignancy and anaphylaxis warnings were placed on the label based on a numerically higher incidence of these events in the active vs. control groups. Malignant neoplasms were seen in 20 of 4,127 (0.5 %) of omalizumab-treated patients vs. 5 of 2,236 (0.2 %) of controls. No specific tumor patterns were noted. In premarketing clinical trials, the frequency of anaphylaxis attributed to omalizumab use was estimated to be 0.1 %. In post-marketing spontaneous reports, the frequency of anaphylaxis attributed to omalizumab use was estimated to be at least 0.2 % based on an estimated exposure of about 57,300 patients from June 2003 through December 2006. Anaphylaxis has occurred as early as after the first dose but has also occurred beyond 1 year of regularly scheduled treatment. Therefore, omalizumab should be administered under appropriate medical supervision with medications used to treat severe hypersensitivity reactions, including anaphylaxis, readily available.

MULTIPLE SCLEROSIS

Multiple sclerosis (MS) is a chronic demyelinating disease of the CNS. The main pathological findings in MS are inflammation, demyelination, and axonal degeneration. Inflammation and demyelination are responsible for the acute symptoms of the disease, and axonal degeneration is the underlining cause of the progressive disability associated with MS. Although the etiology of MS remains undetermined, it is considered to be an autoimmune disorder. Blood autoreactive T and B lymphocytes, once activated against myelin constituents, migrate across the blood–brain barrier and initiate inflammatory and demyelinating processes within the CNS leading to MS lesions.

■ Natalizumab

Natalizumab (Tysabri®) was the first monoclonal antibody developed for the treatment of MS. The mechanism of action has been described earlier as an antibody targeting integrins; however, the specific mechanism(s) by which natalizumab exerts its effects in MS has not been fully defined (Tysabri®, US prescribing information 2012).

Natalizumab was first approved in November 2004 and was suspended soon after (February 2005) because of the occurrence of three cases of progressive multifocal leukoencephalopathy (PML). Two of the three PML cases were reported in MS patients and one in a patient with CD. In June 2006, the US FDA and European Medicine Agency (EMA) granted approval for the reintroduction of natalizumab under a specific risk management plan designed to redefine the safety profile of natalizumab.

The safety and efficacy of natalizumab were evaluated in two large randomized, double-blind, placebo-controlled, multicenter Phase III studies: AFFIRM (natalizumab alone) (Polman et al. 2006) and SENTINEL (natalizumab in combination with intramuscular IFN-β-1a) (Rudick et al. 2006). In the AFFIRM study, patients with RRMS were randomized to natalizumab 300 mg or placebo intravenous infusions every 4 weeks for 2 years. The annualized relapse rate (ARR) was reduced by 68 % in the first year ($p < 0.001$) in the natalizumab group compared to the placebo group. This effect was still evident in the second year. The cumulative probability of disability progression over the 2 years of follow-up was reduced by 42 % ($p < 0.001$) in the natalizumab group compared to patients receiving placebo. In addition, based on MRI assessments, natalizumab also suppressed the formation of new Gd-enhancing lesions and reduced the mean number of active lesions. In the SENTINEL study, natalizumab 300 mg or placebo was evaluated in combination with IFN-β-1a. The clinical and MRI-associated efficacies were similar to the results from the AFFIRM study; however, this study ended a month earlier than planned because of the occurrence of PML in two patients receiving natalizumab plus IFN-β-1a.

PML is a demyelinating infectious disease of the CNS caused by reactivation of the John Cunningham virus (JCV). PML may be fatal or result in severe disability. The pathogenesis of natalizumab-associated PML is not fully understood, yet it could be a direct consequence of the natalizumab mechanisms of action. As of July 2001, 145 cases of PML have been reported among 88,100 patients treated with natalizumab worldwide in the post-marketing setting (Laffaldano et al. 2011).

Natalizumab is used as monotherapy for the treatment of patients with relapsing forms of multiple sclerosis to delay the accumulation of physical disability and reduce the frequency of clinical exacerbation. It is generally recommended for patients who have had an inadequate response to, or are unable to tolerate, an alternate MS therapy. In the United States, administration of natalizumab is closely regulated by governmental regulatory agencies and the TOUCH program (Gold et al. 2007). Natalizumab must be given as monotherapy, and any prior immunomodulator or immunosuppressive therapy must be discontinued prior to use.

CRYOPYRIN-ASSOCIATED PERIODIC SYNDROMES

Cryopyrin-associated periodic syndrome (CAPS) comprises three genetic autoinflammatory disorders including familial cold autoinflammatory syndrome, Muckle-Wells syndrome, and neonatal-onset, multisystem, inflammatory disorder (also known as chronic infantile neurologic, cutaneous, and articular syndrome) (Kubota and Koike 2010). Since the identification of mutations in the NLRP3 gene that encodes the cryopyrin protein, these three phenotypically distinct disorders have been recognized as a clinical continuum of the same disease in an increasing order of severity. NLRP3 mutations result in overactivation of caspase-1, the enzyme that cleaves precursors of IL-1β, IL-18, and IL-33 into active forms. Excessive production of IL-1β is the central pathophysiology of CAPS. This disease is rare with a prevalence of one in about one million people, although many cases of this disease are believed to be undiagnosed.

Treatment of CAPS includes nonsteroidal anti-inflammatory drugs, colchicine, immunosuppressants, corticosteroids, anti-TNF agents, and the recent addition of anti-IL-1β therapy (Kubota and Koike 2010). Anti-IL-1β therapy is very effective since it exerts pharmacologic action against the underlying cause of the disease, while other mediations were only partially effective. Currently, there are two long-acting anti-IL-1β therapeutic proteins (rilonacept and canakinumab) that are approved for the treatment of CAPS, although a short-acting IL-1β receptor antagonist, anakinra, is also effective for this disease (Hawkins and Lachmann 2003). Clinical data have demonstrated that use of an anti-IL-1β agent can achieve rapid and complete control of both clinical manifestations and laboratory parameters. Both rilonacept and canakinumab were generally well tolerated in the Phase III trials with infections being a commonly reported adverse event due to the immunosuppressant effect of anti-IL-1β therapy.

■ Canakinumab

Canakinumab (Ilaris®) is a recombinant human IgG1κ antihuman-IL-1β monoclonal antibody. Excessive production of IL-1β is the central pathophysiology of CAPS. Canakinumab binds to human IL-1β and neutralizes its activity by blocking its interaction with IL-1 receptors.

The efficacy and safety of canakinumab for the treatment of CAPS were demonstrated in a double-blind, placebo-controlled, randomized withdrawal trial. This study consisted of three parts. Part 1 was an 8-week open-label, single-dose period where all patients received canakinumab. Patients who achieved a complete clinical response and did not relapse by week 8 were randomized to receive either placebo or canakinumab every 8 weeks for 24 weeks in Part 2 of the study. Patients who completed Part 2 or experienced a disease flare entered Part 3, a 16-week open-label active treatment phase. In Part 1, a complete clinical response was observed in 71 % of patients 1 week following initiation of treatment and in 97 % of patients by week 8. In the randomized withdrawal period, a total of 81 % of the patients randomized to placebo flared compared to none (0 %) of the patients randomized to canakinumab treatment.

Canakinumab is indicated for the treatment of CAPS, including familial cold autoinflammatory syndrome and Muckle-Wells syndrome, in adults and children 4 years of age and older (Ilaris®, US prescribing information 2012).

■ Rilonacept

Rilonacept (Arcalyst®) is a dimeric fusion protein consisting of the ligand-binding domains of the extracellular portions of the human interleukin-1 receptor component (IL-1RI) and IL-1 receptor accessory protein (IL-1RAcP) linked in-line to the Fc portion of human IgG1. Rilonacept blocks IL-1β signaling by acting as a soluble decoy receptor that binds IL-1β and prevents its interaction with cell surface receptors. Rilonacept also binds IL-1α and IL-1 receptor antagonist (IL-1ra) with reduced affinity.

The safety and efficacy of rilonacept for the treatment of CAPS were demonstrated in a randomized, double-blind, placebo-controlled study. After 6 weeks of treatment at the recommended dosing regimen, a higher proportion of patients in the rilonacept group experienced improvement from baseline in the composite disease score by at least 30 % (96 % vs. 29 % of patients), by at least 50 % (87 % vs. 8 %), and by at least 75 % (70 % vs. 0 %) compared to the placebo group.

Rilonacept is indicated for the treatment of CAPS, including familial cold autoinflammatory syndrome and Muckle-Wells syndrome, in adults and children 12 and older (Arcalyst®, US prescribing information 2009).

CONCLUSION

The introduction of more than a dozen therapeutic MABs and antibody-based biotherapeutics in the last decades has fundamentally changed the pharmacotherapy paradigm in immune-mediated inflammatory diseases such as RA, IBD, and psoriasis. Though these "targeted biotherapies" are expensive compared to traditional "small molecular" therapies such as methotrexate, they have provided effective treatment alternatives with novel mechanisms of action. Some of these biotherapeutics can not only give targeted relief of symptoms similar to traditional medications but also offer an opportunity to modify or even reverse the course of these diseases, as has been demonstrated in RA. With the further advance in protein engineering technology and better understanding of the etiology and disease progression of immune-mediated inflammatory diseases, equipped with more predictive and diagnostic biomarkers, it is reasonable to anticipate that more MABs or antibody-based "targeted biotherapeutics" will be added to the therapeutic armory to successfully treat this class of disorders.

SELF-ASSESSMENT QUESTIONS

■ Questions

1. Are targeted biologic therapies for autoimmune diseases only to be used once drugs like corticosteroids and methotrexate have had an adequate trial of use and have failed to control the patient's symptoms?
2. What is the primary clinical concern with the immunogenicity of biologic therapies?
3. What is the most likely explanation for why a patient who receives a dose of omalizumab might have an increase in their total serum IgE level for many weeks after the first dose?
4. Why do some cell subsets in the peripheral blood increase after dosing with natalizumab?
5. If a trial reports an ACR70 of 20 % on active drug, what does that mean?
6. What does PASI 75 mean?
7. Given that there are currently five anti-TNFα biotherapeutic agents on the market, how would you compare and contrast them?
8. What are the key differences in the indication for use of rituximab vs. abatacept in RA?
9. What is the mechanism of action for ustekinumab in treating plaque psoriasis?

■ Answers

1. Though the standard of care in diseases like RA is still to start with older DMARDs like methotrexate, the decision of when to start or switch therapies is complex and impacted by individual issues linked to clinical response like tolerance/adherence to a particular therapeutic regimen, severity and course of disease and its progression, and concomitant medications and medical issues. It is likely that the standard of care will continue to change and incorporate earlier use of biologic therapies that can modify the disease course with fewer generalized side effects.
2. If a biologic therapy is highly immunogenic, there is a concern that an increasing number of patients exposed to the drug, particularly upon repeated exposure after a hiatus, because their antidrug antibodies could sometimes neutralize the majority of the drug and they would not likely get the full dose or effect. Though less likely there are also rare examples of antidrug antibodies resulting in an autoimmune or allergic-type reaction.
3. Therapeutic monoclonal antibodies that target soluble molecules like IgE form complexes. Though immune complexes are typically cleared from the blood more quickly than monomeric IgG, soluble target molecules typically have a shorter serum half-life than IgG. So an assay detecting the soluble target (in this case IgE) that can detect target even when it is bound to the drug (which is typically an longer-lived IgG) will show more target present in the serum post-dosing as compared to baseline. This is called a carrier effect. Assuming the drug neutralized the bound target, the test detecting the target can be misleading, because the target, though present, is effectively inactive.
4. Natalizumab is a monoclonal antibody that blocks lymphocyte movement between the blood and tissues ("trafficking"); when this movement is effectively blocked in one direction (from the blood into the tissues), an apparent increase in the peripheral lymphocyte population will be evident on assessment by flow cytometry (or perhaps even on a CBC with differential) post-dosing.
5. An ACR70 of 20 % means the 20 % of the patients had a 70 % improvement in their RA disease.
6. The PASI score stands for Psoriasis Area and Severity Index. This tool allows researchers and dermatologists to put an objective number on what would otherwise be a very subjective idea: how bad is a person's psoriasis. The PASI evaluates the degree of erythema, thickness, and scaling of psoriatic plaques and estimates the extent of involvement of each of these components in four separate body areas (head, trunk, upper, and lower extremities).

If in a clinical study a certain proportion of patients experienced a 75 % reduction in their PASI scores, it is reported as a percentage of people achieving "PASI 75."

7. Although the five anti-TNFα biologics have broadly similar efficacy and safety profiles in RA, there are significant differences in the five anti-TNFα agents particularly with respect to dosing characteristics and also in the details of the approved indications for use.

   Infliximab is the only anti-TNFα agent given intravenously, has the longest dosing interval, and is the first FDA-approved anti-TNFα agent for IBD indication. All the other anti-TNFα agents are administered by subcutaneous administration. It is a chimeric monoclonal antibody that neutralizes TNF-α and has approvals in the most indications (rheumatoid arthritis, psoriatic arthritis, ankylosing spondylitis, adult and pediatric ulcerative colitis, adult and pediatric Crohn's disease, and psoriasis).

   Etanercept is a dimeric soluble fusion protein and has approvals for use in several indications (rheumatoid arthritis, polyarticular juvenile rheumatoid arthritis, psoriatic arthritis, ankylosing spondylitis, and psoriasis). It is used as weekly injection.

   Adalimumab is a human monoclonal antibody that neutralizes TNF-α and has FDA approvals for use in patients with rheumatoid arthritis, psoriatic arthritis, ankylosing spondylitis, psoriasis, and Crohn's disease. It is used at a frequency of every week or every other week.

   Golimumab is a human monoclonal antibody that neutralizes TNF-α and has FDA approvals for use in patients with rheumatoid arthritis, psoriatic arthritis, and ankylosing spondylitis. It is used at a frequency of every month.

   Certolizumab pegol is a recombinant, humanized antibody Fab fragment, with specificity for human tumor necrosis factor alpha (TNFα), conjugated to an approximately 40-kDa polyethylene glycol (PEG2MAL40K). It has been approved by FDA for the treatment of rheumatoid arthritis and Crohn's disease.

8. Rituximab in combination with methotrexate is indicated for the treatment of adult patients with moderate to severe rheumatoid arthritis who have had an inadequate response to one or more TNF antagonist therapies. Abatacept is indicated for use as monotherapy or in combination with DMARDS in patients with moderate to severe active rheumatoid arthritis who have had an inadequate response to DMARDs or TNF antagonists.

9. IL-12 and IL-23 are naturally occurring cytokines that are involved in inflammatory and immune responses, such as natural killer cell (NK) activation and CD4+ T-cell differentiation and activation. Ustekinumab, a human IgG1κ monoclonal antibody that binds with high affinity and specificity to the p40 protein subunit used by both the IL-12 and IL-23, can prevent human IL-12 and IL-23 from binding to the IL-12Rβ1 receptor chain of IL-12 (IL-12Rβ1/β2) and IL-23 (IL-12Rβ1/23R) receptor complexes on the surface of NK and T cells.

Acknowledgement The authors would like to express their gratitude to the critical review of Dr. Karen Weiss and editorial support of Dr. Kenneth Graham of Janssen Research and Development, LLC of Johnson & Johnson.

REFERENCES

Actemra (tocilizumab) (2011) US prescribing information. Genentech Inc., South San Francisco

Airoldi I, Di Carlo E, Cocco C et al (2005) Lack of Il12rb2 signaling predisposes to spontaneous autoimmunity and malignancy. Blood 106:3846–3853

Amevive (alefacept) (2011) US prescribing information. Astellas Pharma US, Inc., Deerfield

Arcalyst (rilonacept) (2009) US prescribing information. Regeneron Pharmaceuticals Inc., Tarrytown

Baumgart DC, Sandborn WJ (2007) Inflammatory bowel disease: clinical aspects and established and evolving therapies. Lancet 369:1641–1657

Benlysta (Belimumab) (2012) US prescribing information. Human Genome Sciences, Inc., Rockville

Bongartz T, Sutton AJ, Sweeting MJ et al (2006) Anti-TNF antibody therapy in rheumatoid arthritis and the risk of serious infections and malignancies: systematic review and meta-analysis of rare harmful effects in randomized controlled trials. JAMA 295:2275–2285

Brown SL, Greene MH, Gershon SK et al (2002) Tumor necrosis factor antagonist therapy and lymphoma development: twenty-six cases reported to the Food and Drug Administration. Arthritis Rheum 46:3151–3158

Centers for Disease Control and Prevention (2011) Inflammatory bowel disease. www.cdc.gov/ibd/

Chaudhary R, Butler M, Playford RJ, Ghosh S (2006) Anti-TNF antibody induced stimulated T lymphocyte apoptosis depends on the concentration of the antibody and etanercept induces apoptosis at rates equivalent to infliximab and adalimumab at 10 micrograms per ml concentration. Gastroenterology 130:A696

Christophers E (2001) Psoriasis—epidemiology and clinical spectrum. Clin Exp Dermatol 26:314–320

Cimzia (certolizumab pegol) (2011) US prescribing information. UCB Inc, Smyrna

D'Haens G, Baert F, van Assche G et al (2008) Early combined immunosuppression or conventional management in patients with newly diagnosed Crohn's disease: an open randomised trial. Lancet 371:660–667

Davis JC, Mease PJ (2008) Insights into the pathology and treatment of spondyloarthritis: from the bench to the clinic. Semin Arthritis Rheum 38:83–100

Ehrlich P (1891) Experimentelle untersuchungen über immunität. I. Ueber Ricin. Dtsch Med Wochenschr 17:976–979
Ellerin T, Rubin RH, Weinblatt ME (2003) Infections and anti-tumor necrosis factor alpha therapy. Arthritis Rheum 48:3013–3022
Enbrel (etanercept) (2011) US prescribing information. Immunex Corporation, Thousand Oaks
Feldman SR, Krueger GG (2005) Psoriasis assessment tools in clinical trials. Ann Rheum Dis 64(Suppl II):ii65–ii68
Gold R, Jawad A, Miller DH et al (2007) Expert opinion: guidelines for the use of natalizumab in multiple sclerosis patients previously treated with immunomodulating therapies. J Neuroimmunol 187:156–158
Griffiths CE, Barker JN (2007) Pathogenesis and clinical features of psoriasis. Lancet 370:263–271
Griffiths CE, Strober BE, van de Kerkhof P et al (2010) Comparison of ustekinumab and etanercept for moderate-to-severe psoriasis. N Engl J Med 362:118–128
Gura T (2002) Therapeutic antibodies: magic bullets hit the target. Nature 417:584–586
Haldar P, Brightling CE, Hargadon B et al (2009) Mepolizumab and exacerbations of refractory eosinophilic asthma. N Engl J Med 360:379–384
Hauser SL et al (2008) B-cell depletion with rituximab in relapsing-remitting multiple sclerosis. N Engl J Med 358:676–688
Hawkins PN, Lachmann HJ (2003) Interleukin-1-receptor antagonist in the Muckle–Wells syndrome. N Engl J Med 348:2583–2584
Hochhaus G, Brookman L, Fox H et al (2003) Pharmacodynamics of omalizumab: implications for optimised dosing strategies and clinical efficacy in the treatment of allergic asthma. Curr Med Res Opin 19:491–498
Hoentjen F, van Bodegraven AA (2009) Safety of anti-tumor necrosis factor therapy in inflammatory bowel disease. World J Gastroenterol 15:2067–2073
Humira (adalimumab) (2011) US prescribing information. Abbott Laboratories, North Chicago
Humira (adalimumab) (2012) Summary of product characteristics (SPC). Abbott Laboratories, North Chicago (updated on the eMC: April 11, 2012)
Ilaris (canakinumab) (2012) US prescribing information. Novartis Pharmaceuticals Corporation, East Hanover
Krueger GG (2003) Clinical response to alefacept: results of a phase 3 study of intravenous administration of alefacept in patients with chronic plaque psoriasis. J Eur Acad Dermatol Venereol 17 (Suppl 2):17–24
Kubota T, Koike R (2010) Cryopyrin-associated periodic syndromes: background and therapeutics. Mod Rheumatol 20:213–221
Laffaldano P, Lucchese G, Trojano M (2011) Treating multiple sclerosis with natalizumab. Expert Rev Neurother 11: 1683–1692
Langley RG (2012) Effective and sustainable biologic treatment of psoriasis: what can we learn from new clinical data? J Eur Acad Dermatol Venereol 26(Suppl 2): 21–29
Lebwohl M, Christophers E, Langley R et al (2003) An international, randomized, double-blind, placebo-controlled phase 3 trial of intramuscular alefacept in patients with chronic plaque psoriasis. Arch Dermatol 139:719–727
Leonardi CL, Kimball AB, Papp KA et al (2008) Efficacy and safety of ustekinumab, a human interleukin-12/23 monoclonal antibody, in patients with psoriasis: 76-week results from a randomised, double-blind, placebo-controlled trial (PHOENIX 1). Lancet 371: 1665–1674
McInnes IB, Schett G (2011) The pathogenesis of rheumatoid arthritis. N Engl J Med 365:2205–2219
Modigliani R, Mary JY, Simon JF et al (1990) Clinical, biological, and endoscopic picture of attacks of Crohn's disease: evolution on prednisolone. Gastroenterology 98:811–818
Nestle FO, Kaplan DH, Barker J (2009) Psoriasis. N Engl J Med 361:496–509
O'Dell JR (2004) Therapeutic strategies for rheumatoid arthritis. N Engl J Med 350(25):2591–2602
Orencia (abatacept) (2011) US prescribing information. Bristol-Myers Squibb, Princeton
Papp KA, Langley RG, Lebwohl M et al (2008) Efficacy and safety of ustekinumab, a human interleukin-12/23 monoclonal antibody, in patients with psoriasis: 52-week results from a randomised, double-blind, placebo-controlled trial (PHOENIX 2). Lancet 371:1675–1684
Papp K, Gulliver W, Lynde C et al (2011) Canadian guidelines for the management of plaque psoriasis. J Cutan Med Surg 15(4):210–219
Polman CH, O'Connor PW, Havrdova E et al (2006) A randomized, placebo-controlled trial of natalizumab for relapsing multiple sclerosis. N Eng J Med 354:899–910
Reich K, Burden AD, Eaton JN, Hawkins NS (2012) Efficacy of biologics in the treatment of moderate to severe psoriasis: a network meta-analysis of randomized controlled trials. Br J Dermatol 166:179–188
Remicade (infliximab) (2011) US prescribing information. Janssen Biotech Inc., Horsham
Rituxan (rituximab) (2012) US prescribing information. Genentech Inc., South San Francisco
Rudick RA, Stuart WH, Calabresi PA et al (2006) Natalizumab plus interferon-β-1a for relapsing multiple sclerosis. N Engl J Med 354:911–923
Salliot C, Finckh A, Katchamart W et al (2011) Indirect comparisons of the efficacy of biological antirheumatic agents in rheumatoid arthritis in patients with an inadequate response to conventional disease-modifying antirheumatic drugs or to an anti-tumour necrosis factor agent: a meta-analysis. Ann Rheum Dis 70:266–271
Sandborn WJ, Feagan BG, Stoinov S et al (2006) Certolizumab pegol administered subcutaneously is effective and well tolerated in patients with active Crohn's disease: results from a 26-week, placebo-controlled phase III study (PRECiSE 1). Gastroenterology 130:A-107
Simponi (golimumab) (2011) US prescribing information. Janssen Biotech Inc., Horsham
Smedby KE, Askling J, Mariette X et al (2008) Autoimmune and inflammatory disorders and risk of malignant lymphomas–an update. J Intern Med 264:514–527
Stelara (ustekinumab) (2011) US prescribing information. Janssen Biotech Inc., Horsham

Taylor PC, Steuer A, Gruber J et al (2004) Comparison of ultrasonographic assessment of synovitis and joint vascularity radiographic evaluation in a randomized, placebo-controlled study of infliximab therapy in early rheumatoid arthritis. Arthritis Rheum 50:1107–1116

Tracey D, Klareskog L, Sasso EH et al (2008) Tumor necrosis factor antagonist mechanisms of action: a comprehensive review. Pharmacol Ther 117:244–279

Tysabri® (natalizumab) (2012) US prescribing information. Biogen Idec Inc, Cambridge

Van den Brande JM, Braat H, van den Brink GR et al (2003) Infliximab but not etanercept induces apoptosis in lamina propria T-lymphocytes from patients with Crohn's disease. Gastroenterology 124:1774–1785

Xolair (omalizumab) (2010) US prescribing information. Genentech Inc., South San Francisco

21 Interferons and Interleukins

Jean-Charles Ryff and Sidney Pestka

INTRODUCTION

In 1957 a substance was described (Isaacs and Lindenmann 1957) which was produced by virus-infected cell cultures and "interfered" with infection by other viruses; it was called interferon. Over the following decades it was realized that "interferon" comprises a family of related proteins with several additional properties. Starting in the 1960s various "factors" produced primarily by white blood cell (WBC) as well as other cell supernatants were described which acted in various ways on other WBCs or somatic cells. They were usually given a descriptive name either associated with their cell of origin or their activity on other cells resulting in a myriad of names. The application of molecular technology allowed us to determine that some cytokines had multiple activities and that different cytokines had similar overlapping activities. A systematic classification based on genetic structure and protein characterization has been effective. The interactive networks and cascades of cytokines, interferons (IFN), interleukins (IL), growth factors (GF), chemokines (CK), their receptors (r or R), and signaling pathways are highly complex and will be further explored in this chapter.

Cytokine is a term coined in 1974 by Stanley Cohen in an attempt at a more systematic approach to the numerous regulatory proteins secreted by hematopoietic and non-hematopoietic cells. Cytokines play a critical role in modulating the innate and adaptive immune systems. They are multifunctional peptides that are now known to be produced by normal and neoplastic cells, as well as from cells of the immune system. These local messengers and signaling molecules are involved in the development of the immune system, cell growth and differentiation, repair mechanisms, and the inflammatory cascade. Traditionally, interleukins can be classified as T-helper cells type 1 (Th1; pro-inflammatory), e.g., IL-2, IL-12, IL-18, and IFN-γ, or type 2 (Th2; anti-inflammatory) stimulating, e.g., IL-4, IL-10, IL-13, and TGF-β. More recently a third category T-helper cells 17 (Th17) have been described which are associated with autoimmunity (Hu et al. 2011). A review of the Th1/Th2 and Th17 concept is provided by Steinmann (2007).

(a) *Interferons*: proteins produced by eukaryotic cells in response to viral infections, tumors, and other biological inducers. They promote an antiviral state in other neighboring cells and also help to regulate the immune response. They exhibit a variety of activities and represent a wide family of proteins.

(b) *Interleukins*: a group of cytokines mainly secreted by leukocytes and primarily affecting growth and differentiation of hematopoietic and immune cells. They are also produced by other normal and malignant cells and are of central importance in the regulation of hematopoiesis, immunity, inflammation, tissue remodeling, and embryonic development.

Thus, all interleukins are cytokines; however, not all cytokines are interleukins.

(c) *Growth factors: proteins that activate cellular proli*feration and/or differentiation. Many growth factors stimulate cellular division in numerous different cell types. Others are specific to a particular cell type. They also promote proliferation of connective tissue and glial and smooth muscle cells, enhance normal wound healing, and promote proliferation and differentiation of erythrocytes (erythropoietin). Hematopoietic growth factors are reviewed in Chap. 18. Some ILs have a function overlap with growth factor, e.g., IL-2, IL-3, and IL-11 (see Table 21.2).

(d) Chemokines: (chemotactic cytokines) a large family of structurally related low molecular weight proteins with potent leukocyte activation and/or chemotactic activity. "CXC" (or α) and "C-C" (or β)

J.-C. Ryff, M.D. (✉)
Biotechnology Research and Innovation Network, Rittergasse 14, Basel CH-4051, Switzerland
e-mail: jean-charles.ryff@bluewin.ch

S. Pestka, M.D.
PBL Interferon Source, Piscataway, NJ, USA
e-mail: www.interferonsource.com

D.J.A. Crommelin, R.D. Sindelar, and B. Meibohm (eds.), *Pharmaceutical Biotechnology*,
DOI 10.1007/978-1-4614-6486-0_21,

chemokine subsets are based on presence or absence of an amino acid between the first two of four conserved cysteins. A third subset, "C," has only two cysteins and to date only one member, IL-16, has been identified. The fourth subgroup, the C-X3-C chemokine, has three amino acid residues between the first two cysteins.

(e) Others, such as tumor necrosis factors (TNF)-α and (TNF)-β and transforming growth factor (TGF)-α, (TGF)-β, and (TGF)-γ.

All cytokines including interferons and interleukins act by binding to specific transmembrane receptors. In general these receptors have two main components: a low affinity ligand-binding domain that ensures ligand specificity and a high-affinity effector domain activating target gene promoters via an intracellular signaling pathway. Because cytokines can bind to their receptors only, where these are expressed on the cell membrane, a functional tissue or cell specificity is ensured.

Cytokine signaling is tightly controlled within the cell through the action of multiple different negative regulators. Members of the suppressors of cytokine signaling (SOCS) family specifically interfere with cytokine signaling by several different mechanisms including direct binding and inhibition of JAK proteins, competition with STAT for binding sites on the cytokine receptor, and activation of proteasomal degradation of signaling components.

Their action is described as:

- *Autocrine*, if the cytokine acts on the cell that secretes it
- *Paracrine*, if the action is restricted to the immediate vicinity of a cytokine's secretion
- *Endocrine*, if the cytokine diffuses to distant regions of the body to affect different tissues

They can act on many targets, can act in concert, or can antagonize one another:

- *Synergy* - action together to induce a different response than either can induce alone
- *Antagonism* - cytokines can counteract one another
- *Pleiotropy* - action in a similar way on more than one "target" cell
- *Redundancy* - more than one cytokine triggers identical or similar responses in a given "target" cell
- *Pathway activation* - triggered sequential induction or "cascade"

INTERFERONS: NOMENCLATURE AND FUNCTIONS

Interferons are a family of naturally occurring proteins and glycoproteins with molecular weights of 16,776–22,093 Da produced and secreted by cells in response to viral infections and to synthetic or biological inducers. By interacting with their specific heterodimeric receptors on the surface of cells, the interferons initiate a broad and varied array of signals that induce cellular antiviral states, modulate inflammatory responses, inhibit or stimulate cell growth, produce or inhibit apoptosis, and modulate many components of the immune system. Structurally, they are part of the helical cytokine family (Fig. 21.1). During the past 25 years, major research efforts have been undertaken to understand the signaling mechanisms through which these cytokines induce their effects. Figure 21.2 as a generic example illustrates the JAK-STAT (Janus-activated kinase-signal transducer and activator of transcription), the best characterized IFN signaling pathway. However, coordination and cooperation of multiple distinct signaling cascades, including the mitogen-activated protein kinase p38

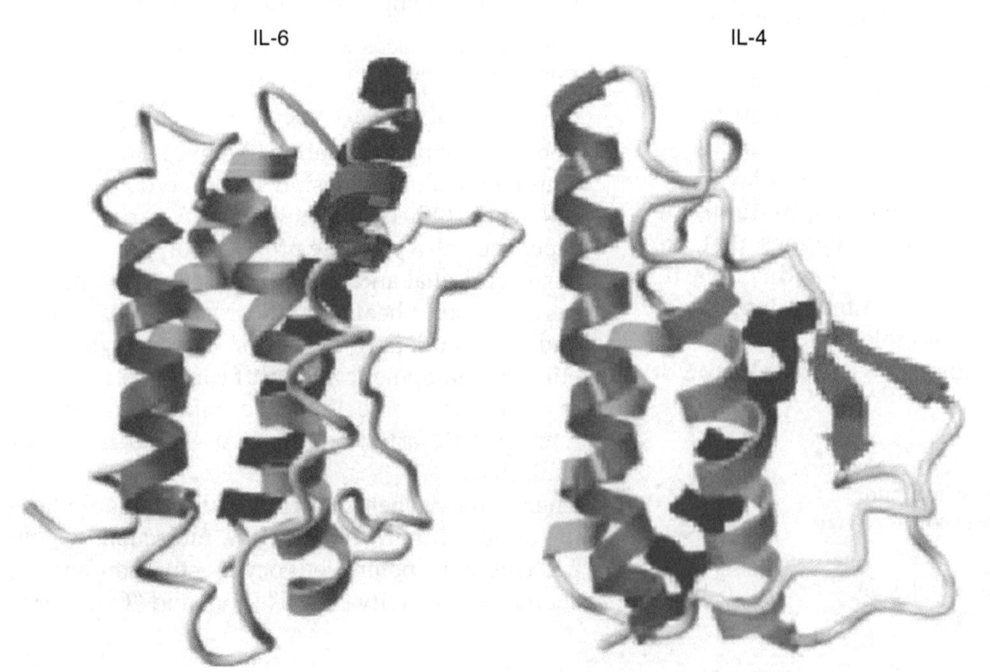

Figure 21.1 ■ Class I helical cytokines. Class I helical cytokines fold into a bundle of four tightly packed α-helices. On the basis of their helix length, class I helical cytokines are characterized as (**a**) long chain, such as IL-6, or (**b**) short chain, such as IL-4 (From: Huising et al. (2006), with permission).

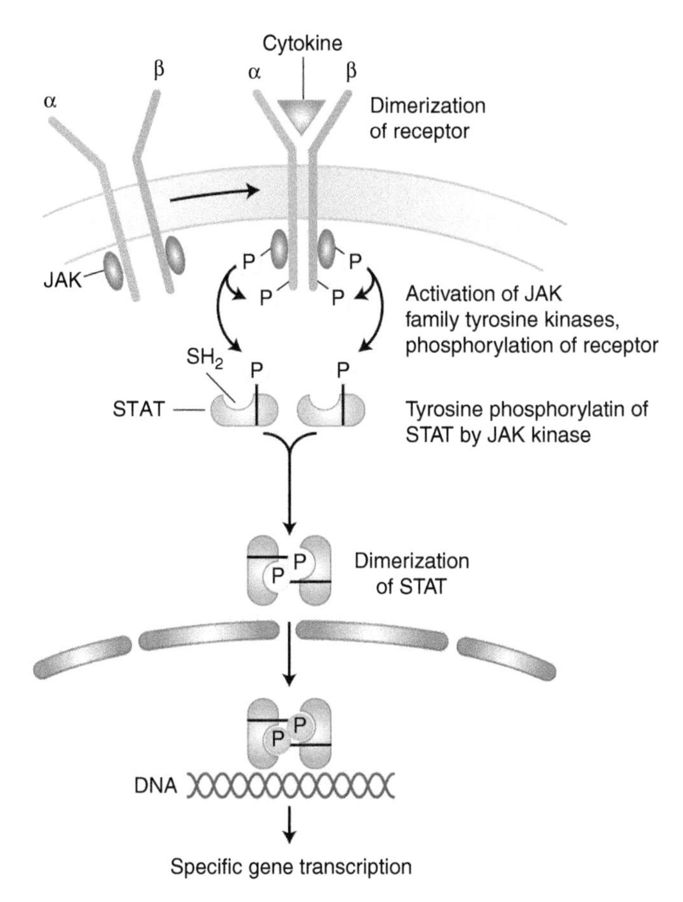

Figure 21.2 ■ Generic JAK-STAT signaling pathway mediated by most cytokine receptors.

cascade and the phosphatidylinositol 3-kinase cascade, are required for the generation of responses to interferons (Platanias 2005). For a review of the IFN signaling pathways, see *Journal of Interferon and Cytokine Research* 2005. Many of the symptoms of acute viral infections are the consequence of the high systemic IFN-α response induced by the infecting viruses particularly during the viremic phase.

Human type I interferons comprise 13 different IFN-α isoforms or subtypes with varying specificities, e.g., affinities to different cell types, and downstream activities. Although there are 13 human IFN-α proteins, two of them (IFN-α1 and IFN-α13) are identical proteins so that the total number of type I IFNs are often listed as 12 (Pestka 1981a, b, 1986). There is also one subtype each for IFN-β (beta), IFN-ε (epsilon), IFN-κ (kappa), and IFN-ω (omega). Their ability to establish an "antiviral state" is the distinctive fundamental property of type I IFNs. They are produced by most cells; however, certain types seem to be more selectively expressed, e.g., IFN-κ by keratinocytes (LaFleur et al. 2001).

Type II IFN consists of a single representative: IFN-γ (gamma) (Pestka 1981a, b). IFN-γ or immune interferon plays an essential role in cell-mediated immune responses. It is produced by NK-cells, dendritic cells, cytotoxic T-cells, progenitor Th0 cells, and Th1 cells.

IFN-α2, IFN-β, and IFN-γ are the most extensively studied to date. All IFNs and IFN-like cytokines have been reviewed in Pestka et al. (2004a).

The names for the human IFNs presently approved by the Human Genome Nomenclature Committee (HGNC) are listed in Table 21.1. For an exhaustive review see Meager (2006).

INTERLEUKINS: NOMENCLATURE AND FUNCTIONS

Interleukins are primarily a collection of immune cell growth, differentiation, and maturation factors. Collectively they orchestrate a precise and efficient immune response to toxins and pathogens, including cancer cells, recognized as foreign. As is the case for IFNs, ILs bind to related specific cell surface receptors which activate similar intracellular signaling cascades (Huising et al. 2006; Lutfalla et al. 2003; Pestka et al. 2004b). Many interleukins, primarily those with pro-inflammatory function, are intrinsically toxic either directly or indirectly, i.e., through induction of toxic gene products. Therefore, the human body has an elaborate system of checks and balances that, under (patho) physiological conditions, regulates the magnitude and duration of an immune response. Under biological conditions, ILs usually have a short circulation time, and their production is regulated by positive and negative feedback loops. Furthermore, their effect is mostly localized, and in some cases soluble receptors or neutralizing antibodies limit their dissemination. Specific receptor antagonists can also control their activity.

Table 21.2 lists the ILs for which the protein and gene structure have been characterized. Their names and symbols have been approved by the HGNC.

Under physiological conditions, the relative concentrations of agonistic and antagonistic interleukins establish a delicate balance in driving pro- and anti-inflammatory processes. This balance can be disturbed by various pathogenic agents or mechanisms:

- Infectious agents or toxins
- Allergens
- Malignant tumors
- Genetic variants

These pathogenic agents or mechanisms result in a self-limited or protracted disequilibrium. Symptoms of disease can be the consequence of an adequate immune response at the end of which the steady state is reestablished. A brisk inflammatory response is the sign of a healthy immune reaction. In some instances an inadequate response can manifest itself as relapsing-remitting progressive disease, e.g., rheumatoid arthritis, asthma, psoriasis, chronic inflammatory bowel disease, multiple sclerosis, chronic hepatitis, or chronic insulitis leading to diabetes mellitus. All have in common that they need a genetic predisposition and an environmental trigger factor to become active and are

Table 21.1 ■ HGNC-approved human interferon names.

Symbol	Name	Symbol	Name
IFN-α1	Interferon, alpha 1	IFN-β1	Interferon, beta 1, fibroblast
IFN-α2	Interferon, alpha 2	IFN-ε1	Interferon, epsilon 1
IFN-α4	Interferon, alpha 4	IFN-κ	Interferon, kappa
IFN-α5	Interferon, alpha 5	IFN-ω1	Interferon, omega 1
IFN-α6	Interferon, alpha 6	IFN-γ	Interferon, gamma
IFN-α7	Interferon, alpha 7		
IFN-α8	Interferon, alpha 8		
IFN-α10	Interferon, alpha 10		
IFN-α13	Interferon, alpha 13[a]		
IFN-α14	Interferon, alpha 14		
IFN-α16	Interferon, alpha 16		
IFN-α17	Interferon, alpha 17		
IFN-α21	Interferon, alpha 21		

Adapted from ExPASy and HGNC data base

P pseudogene

[a]IFN-α13 sequence identical to IFN-α1

There are in addition a number of interferon pseudogenes (nonfunctional and related to interferon genes) mentioned for completion's sake: IFN-α22, IFN-ν (nu) 1, IFNPs 11, 12, 20, 23, and 24, IFN-ωP2, 4, 5, 9, 15, 18, and 19

Symbol	Approved name	Previous symbol	Aliases
IL-1A	Interleukin-1, alpha	IL-1	IL-1 alpha, hematopoietin-1, interleukin-1 family member (IL-1 F) 1
IL-1B	Interleukin-1, beta		IL-1 beta, IL-1 F2, catabolin
IL-1 F3	Interleukin-1 family, member 3		IL-1 delta, IL-1 receptor antagonist homolog 1, IL-1-related protein 3
IL-RN	Interleukin-1 receptor antagonist	IL1F3	IL1RA, ICIL-1RA, IRAP, MGC10430
IL-2	Interleukin-2		T-cell growth factor (TCGF) aldesleukin
IL-3	Interleukin-3		Multi-CSF
IL-4	Interleukin-4		BSF1
IL-5	Interleukin-5		TRF, EDF, BCDF 1
IL-6	Interleukin-6	IFNB2	BCSF2, HSF, HGF, CTL differentiation factor, MGI-2
IL-7	Interleukin-7		
IL-8	Interleukin-8		CXCL8 (chemokine), MDNCF, TCCF, NAP1, GCP1, MONAP, emoctakin
IL-9	Interleukin-9		TCGF P40, P40 cytokine
IL-10	Interleukin-10		CSIF, TGIF, IL-10A
IL-11	Interleukin-11		AGIF, oprelvekin (see Chap. 18)
IL-12A	Interleukin-12A	NKSF1	CLMF p35, CLMF1, IL-12 p35

Table 21.2 ■ HGNC-approved interleukin names.

Symbol	Approved name	Previous symbol	Aliases
IL-12B	Interleukin-12B	NKSF2	CLMF p40, CLMF2, IL-12 p40
IL-13	Interleukin-13		
IL-15	Interleukin-15		
IL-16	Interleukin-16		LCF, pro-IL-16
IL-17A	Interleukin-17A	IL-17	CTLA-8
IL-17B	Interleukin-17B		Cytokine Zcyto7, neuronal interleukin-17-related factor, interleukin-20
IL-17C	Interleukin-17C		Cytokine CX2
IL-17D	Interleukin-17D		Interleukin-27
IL-17 F	Interleukin-17 F		Interleukin-24, cytokine ML-1
IL-18	Interleukin-18	IL-1 F4	IFN-gamma-inducing factor, IL-1 gamma, iboctadekin
IL-19	Interleukin-19		Melanoma differentiation-associated protein-like protein, IL-10C
IL-20	Interleukin-20		Zcyto10
IL-21	Interleukin-21		Za11
IL-22	Interleukin-22		Zcyto18, IL-TIF
IL-23A	Interleukin-23A		IL-23 subunit p19, SGRF
IL-24	Interleukin-24		MDA-7, suppression of tumorigenicity 16 protein
IL-25	Interleukin-25	IL-17E	Interleukin-17E
IL-26	Interleukin-26		AK155 protein
IL-27	Interleukin-27	IL-30	IL-27A, p28
IL-28A	Interleukin-28A		IFN lambda-2, Zcyto20
IL-28B	Interleukin-28B		IFN lambda-3, IFN lambda-4 Zcyto22
IL-29	Interleukin-29		IFN lambda-1, Zcyto21
IL-31	Interleukin-31		
IL-32	Interleukin-32		NK cell protein 4, TAIF
IL-33	Interleukin-33	IL-1 F11	NF-HEV
IL-34	Interleukin-34		
IL-35	Interleukin-35		
IL-36A	Interleukin-36 alpha	IL-1 F6	FIL-1 epsilon
IL-36B	Interleukin-36 beta	IL-1 F8	Interleukin-1 eta, interleukin-1 homolog 2
IL-36G	Interleukin-36 gamma	IL-1 F9	Interleukin-1 epsilon, interleukin-1 homolog 1, IL-1-related protein 2
IL-36RN	Interleukin-36 receptor antagonist	IL-1 F5	FIL1 delta, FIL1D, IL1HY1, IL1RP3, IL1L1, IL-1 F5, IL36Ra, MGC29840
IL-37	Interleukin-37	IL-1 F7	FIL-1 zeta, IL-1 zeta, IL-1 homolog 4, IL-1-related protein 1
IL-38	Interleukin-38[a]	IL-1 F10	Interleukin-1 receptor antagonist-like, FIL-1 theta, IL-1 theta, IL-1 HY2

Adapted from ExPASy and HGNC

Note: The symbols IL-14 and IL-30 are no longer used as approved nomenclature

[a]At the time of completion of this chapter, IL-38 was still only approved by the HGNC as IL-1 F10

Table 21.2 ■ (continued)

at best only partially understood. In many cases these diseases are caused by either insufficient production or overproduction of key interleukins. Thus, in principle, once the diagnosis is made, these interleukins can be therapeutically supplemented or suppressed to restore proper balance (Ryff 1996).

Our current knowledge of the interleukins listed in Table 21.2 is briefly summarized below and each reference selected expands on the subject. Readers interested in the current knowledge about the protein, DNA, RNA, gene, chromosome location, etc. for individual interferons or interleukins are referred to the following databases:

1. www.genatlas.org (with a links to other databases)
2. http://au.expasy.org/sprot/
3. www.rcsb.org/pdb/for the 3D models of individual IFNs or ILs

Cytokines and in particular also interleukins can be classified into various "families" according to gene clustering on chromosomes, gene sequence homologies, secondary and tertiary structure, use of related receptors, and also to their function.

■ Interleukin-1 Family

The IL-1 (Dinarello 2011) family comprises 11 different members: IL-1α, IL-1β, IL-1RN, IL-18, IL-33, IL-36A, B, C, IL-36RN, IL-37, and IL-38. All are thought to have arisen from a common ancestral gene that underwent multiple duplications.

■ Interleukin-1

IL-1 (Towne et al. 2004) is generally used to describe IL-1α and IL-1β, both of which have the same biological effects and play a primordial role in the innate and adaptive immune response. Although IL-1 is the prototypical pro-inflammatory cytokine, it also plays a key role in hematopoiesis, appetite control, and bone metabolism. IL-1 is released as part of the acute-phase reaction of hepatocytes. The primary producers of IL-1 are macrophages, B-cells, and neutrophils. IL-1α and IL-1β are synthesized as pro-peptides of approximately 30 kDa and are then cleaved to produce products of 159 and 153 amino acids. Differences in glycosylation are responsible for the wide variation of reported molecular weights.

■ Interleukin-1Ra

IL-1Ra (Towne et al. 2004) is a naturally occurring IL-1 receptor antagonist (IL-1Ra), an inhibitor of IL-1. It has limited sequence similarity to either IL-1α or IL-1β but does have the ability to bind to the IL-1 receptors. Lacking IL-1 activity, it acts as a useful blocker of the receptor. A recombinant IL-1Ra has been investigated for its potential use in sepsis; the clinical trials were inconclusive. Recombinant IL-1Ra has however been used successfully for the treatment of rheumatoid arthritis and is marketed under the name of Kineret® (see section "Therapeutic Use of Recombinant Interleukins" below).

■ Interleukin-18

IL-18 (Liu et al. 2000) shares unique structural features with the IL-1 family, but it does not have the usual four-helix structure rather an all β-pleated sheet structure. It is produced by activated macrophages such as Kupffer cells of the liver and other resident macrophages from which, after cleavage of its precursor pro-IL-18, the mature protein is released. IL-18 is an early inducer of the Th1 response, co-stimulating, with IL-12, the production of IFN-γ, TNF-α, GM-CSF, and IL-2. IL-18 is associated with the metabolic syndrome and coronary vascular disease (Trøseid et al. 2010).

■ Interleukin-33

IL-33 (Schmitz et al. 2005), unlike other members of the IL-1 family which are all pro-inflammatory, has a major role in the development of a Th2-type immune response by inducing IL-5 and IL-13. Human smooth muscle cells as well as epithelial cells forming the bronchus and small airways show constitutive expression of IL-33 mRNA. In lung or dermal fibroblasts and keratinocytes, IL-33 mRNA is induced after activation with TNF-α and IL-1β. Activated dendritic cells and macrophages are the only hematopoietic cells showing low quantities of IL-33 mRNA. In addition, IL-33 and IL-18 are the only known IL-1 family member genes not located on chromosome 2. IL-33 is thought to play a key role in mediating anaphylactic shock; this effect can be completely neutralized by anti-IL-33 antibodies in an experimental model. Thus, IL-33 may be a potential target for the treatment of anaphylactic shock (Pushparay et al. 2009) and prevention or treatment of atherosclerosis (McLaren et al. 2010).

■ Interleukin-36A, B, and G

IL-36 A, B, and G (or α, β, and γ) (Towne et al. 2011) were previously classified as interleukin-1 family member (IL-1 F) 6, IL-1 F8, and IL-1 F9, respectively.

IL-36A (IL-36 alpha) is a member of the IL-1 family of proteins. Cells reported to express IL-36 alpha include monocytes, B-cells, and T-cells. Notably, IL-36 alpha is the only novel IL-1 family member expressed on T-cells. It is expressed in immune system and fetal brain, but not in other tissues tested or in multiple hematopoietic cell lines.

IL-36B (IL-36 beta) is expressed at low levels in the tonsil, bone marrow, heart, placenta, lung, testis, and colon, but not in any of the hematopoietic cell lines or in adipose tissue. It is expressed at higher levels in psoriatic plaques than in symptomless psoriatic skin or healthy control skin. Increased levels are not detected in inflamed joint tissue. It is induced by pro-inflammatory cytokines IL-1α, IL-1β, and TNF in synovial fibroblasts and by IL-1α and TNF in keratinocytes. In articular chondrocytes it is constitutively expressed. IL-36B stimulates the production of interleukin-6 and interleukin-8 in synovial fibroblasts, articular chondrocytes, and mature adipocytes.

IL-36G (IL-36 gamma) is highly expressed in tissues containing epithelial cells: skin, lung, stomach, and esophagus. In the skin it is only expressed in keratinocytes, but not in fibroblasts, endothelial cells, or melanocytes. Upregulated in lesional psoriasis skin, it is induced by TNF and IFN-γ in keratinocytes.

■ Interleukin-36RN

IL-36 receptor antagonist (RN) (Towne et al. 2011) acts as an IL-36R antagonist controlling the activity of IL-36. Cells expressing IL-36RN include monocytes, B-cells, dendritic cells/Langerhans cells, keratinocytes, and gastric fundus parietal and chief cells. IL-36RN is essential for normal skin maintenance. A variant interleukin-36RN structure leads to its impaired IL-36R affinity and failure to adequately regulate the secretion of inflammatory cytokines leading to generalized pustular psoriasis (Marrakchi et al. 2011). IL-36RN has also been documented to suppress inflammation of the brain by enhancing IL-4 response (Collison et al. 2008).

■ Interleukin-37

IL-37 (Nold et al. 2010) expressed in human monocytes and epithelial cells is a fundamental inhibitor of innate immunity. The overexpression of IL-37 in cells of monocytic or epithelial origin almost completely abolishes the production of pro-inflammatory cytokines as IL-1α/β, TNF-α, IL-6, and IL-8 in response to toll-like receptor (TLR) ligands or IL-1β.

■ Interleukin-38

IL-38 (Lin et al. 2008) was formerly known and HGNC approved as IL-1 F10 (Dinarello 2011) or IL-1HY2 and has been shown to be expressed in basal epithelia in fetal skin, in the spleen, and in proliferating B-cells of the tonsil. This tissue-specific expression pattern and the membership of the IL-1 family suggest a role in the normal immune response and inflammatory pathophysiology.

■ Interleukin-2 Family

Interleukin-2 belongs to a family of cytokines, which also includes IL-4, IL-7, IL-9, IL-15, and IL-21 (Liao et al. 2011). These interleukins all share a common receptor γ chain (γc) and are also known as γc-family cytokines.

■ Interleukin-2

IL-2 (Malek and Castro 2010) originally described as T-cell growth factor (TCGF) is synthesized and secreted primarily by T-cells. IL-2 stimulates the growth, differentiation, and activation of T-cells, B-cells, and NK-cells. The major physiological effect is to promote self-tolerance by suppressing T-cell response in vivo. IL-2 signals through a receptor complex consisting of IL-2 specific IL-2 receptor alpha, IL-2 receptor beta, and a common gamma chain, which is shared by all members of this cytokine family. A soluble form of the IL-2R capable of binding IL-2, a truncated version of the α chain without cytoplasmic tail, has been found in human serum (soluble receptor or sR). High levels of IL-2sR have been found in patients with a wide variety of disorders, including chronic hepatitis C, HIV infection, cancer, solid organ transplant rejection, and arthritis. Soluble IL-2R can bind released IL-2 prior to its binding to cells to prevent overflow or overstimulation. Several other cytokine and adhesion molecule receptors also have circulating forms. This is one manner in which the immunological cascade maintains its checks and balances.

■ Interleukin-4

IL-4 (Gilmour and Lavender 2008) is produced by Th2 cells and by mast cells, basophils, and eosinophils and acts as an antagonist to IFN-γ. It stimulates B-cell proliferation and activation and induces class switch to IgE and IgG4 expression from B-cells, as well as class II major histocompatibility complex (MHC) expression. In addition it induces the differentiation of eosinophils and activity of cytotoxic T-cells. IL-4 regulates the differentiation of helper T-cells to the Th2 type. These T-cells produce the cytokines IL-4, IL-5, IL-9, and IL-13, which can all participate in the allergic response. IL-4 regulates the production of IgE by B-lymphocytes. It also has the ability to stimulate chemokine production and mucus hypersecretion by epithelial cells. Overproduction of IL-4 is associated with allergy and asthma.

■ Interleukin-7

IL-7 (Fry and Mackall 2002) is an essentially tissue-derived cytokine. Its primary sources are stromal and epithelial cells in various locations including the intestinal epithelium, liver, and to a lesser degree dendritic cells. IL-7 acts primarily on pre-B-cells to stimulate their differentiation. It can also stimulate the development of human T-cells. IL-7 is classified as a type I short-chain cytokine of the hematopoietin family which also includes IL-2, IL-3, IL-4, IL-5, GM-CSF, IL-9, IL-13, IL-15, M-CSF, and stem cell factor (SCF).

■ Interleukin-9

IL-9 (Noelle and Nowak 2010) is a Th2 cytokine originally characterized as a factor produced by activated T-cells and able to support the long-term growth of some T-helper clones. IL-9 activities extend to various cell types including mast cells, B-lymphocytes, hematopoietic progenitors, eosinophils, lung epithelial cells, neuronal precursors, and T-lymphocytes. Increased

IL-9 production has been implicated in major pathologies such as asthma supported by its effects on IgE production, mucus production, mast cell differentiation, eosinophil activation, and bronchial hyperresponsiveness. IL-9 stimulates the growth of murine thymic lymphomas and an autocrine loop has been suggested in Hodgkin lymphoma. Finally, IL-9 is required for an efficient immune response against intestinal parasites. IL-9 exerts its effects through a receptor that belongs to the hematopoietic receptor superfamily and consists of two chains, also involved in IL-2, IL-4, IL-7, IL-15, and IL-21 signaling.

■ Interleukin-15

IL-15 (Waldmann 2006) shares the IL-2βγ receptor complex components IL-2Rβ and IL-2Rγ. However, specificity is conferred by a unique α chain (IL-15Rα) completing the IL-15Rαβγ heterotrimeric high-affinity receptor complex. While the role of interleukin-2 is in the elimination of self-reactive T-cells to prevent autoimmunity, interleukin-15 is dedicated to the prolonged maintenance of memory T-cell responses to invading pathogens. In addition, IL-15 and its receptor have a much wider tissue distribution than IL-2 and its receptor.

■ Interleukin-21

IL-21 (Yi et al. 2010) is the most recently discovered member of the IL-2 family of cytokines that utilize the common γ-chain receptor subunit for signal transduction. Structurally, it shows homology to the other interleukins of the IL-2 family. The heterodimeric IL-21R has an IL-21-specific subunit besides the γ-chain. IL-21 expression is restricted primarily to activated $CD4^+$ T-cells. IL-21 expression seems transient and stage specific during T-cell differentiation. It is required for normal humoral immunity and regulates antibody production in cooperation with IL-4. IL-21 also regulates cell-mediated immunity by inducing IFN-γ, TNF-α, and synthesis of perforin and granzyme B leading to cytolytic activity. It can cooperate with other cytokines to generate potent killer T-cells and thus has antitumor activity. Lastly, it also has inhibitory activity by inducing IL-10. Thus, altogether, it is responsible for the coordination of the initiation and cessation of an efficient immune response.

■ Interleukin-10 Family

The IL-10 family (Pestka et al. 2004b) besides IL-10 includes IL-19, IL-20, IL-22, IL-24, IL-26, IL-28A, IL-28B, and IL-29. They share a classical four-helix bundle, a signature element of all helical cytokines (Fig. 21.1), and all share the IL-10R2 or α chain of their dimeric receptor, while each has its own R1 or α chain (Fig. 21.1).

■ Interleukin-10

IL-10 (Pestka et al. 2004b) is a major endogenous anti-inflammatory mediator which acts by profoundly inhibiting the synthesis of pro-inflammatory molecules such as IFN-γ, IL-2, IL-12, and TNF-α. Macrophages are the major source of IL-10, a homodimer. Th2 cell subsets, monocytes, and several other cells can also synthesize. A number of molecules produced under stress conditions including reactive oxygen species stimulate IL-10 synthesis. Recombinant human IL-10 has been tested in clinical trials in rheumatoid arthritis, inflammatory bowel disease, psoriasis, organ transplantation, and chronic hepatitis C. To date the results are mixed or disappointing; however, they give new insight into the immunobiology of IL-10.

■ Interleukin-19

IL-19 (Azuma et al. 2010) is a member of the IL-10 family that includes IL-20, IL-22, IL-24, IL-26, IL-28A, IL-28B, and IL-29. The induction of IL-19 in human monocytes is downregulated by IFN-γ and upregulated by IL-4. IL-19 influences the balance of Th1/Th2 cells in favor of Th2 cells by upregulating IL-4 and downregulating IFN-γ. IL-19 is essential for the induction and maintenance of endotoxin tolerance and appears to play a key role in innate immunity. IL-19 together with IL-20 with whom it shares the same receptor complex has been associated with psoriasis and is thought to be involved in regulating inflammatory response in various tissues and be of particular importance for proper skin development and function.

■ Interleukin-20

IL-20 (Xu 2004) was originally identified from a keratinocyte library, mRNA isolated from skin and trachea. It is classified as a helical cytokine member of the IL-10 family. Keratinocytes and activated monocytes synthetize IL-20. IL-1β, TGF-α, and epidermal growth factor (EGF), factors known to be involved with proliferative and pro-inflammatory signals in the skin, enhance the response to IL-20. It binds to two cell surface receptors: IL-20Rα and IL-20Rβ on keratinocytes and other epithelial cells. IL-20 mediates the hyperproliferation of keratinocytes associated with cutaneous inflammation and has a central role in inflammatory skin diseases such as psoriasis and eczema. It also promotes the expansion of pluripotential hematopoietic progenitor cells indicating a role beyond the response of epithelial cells to inflammation.

■ Interleukin-22

IL-22 (Kotenko et al. 2001a) also belongs to the family of cytokines structurally related to IL-10. In contrast to IL-10, it has pro-inflammatory activities: it upregulates the production of acute-phase proteins. The

IL-22 receptor is composed of an IL-22-binding chain, IL-22R1, and the IL-10R2 subunit, which is shared with the IL-10R. IL-22 is produced by activated human T-helper cells and mast cells. A soluble IL-22-binding protein, IL-22RBP, encoded by a distinct gene, has been identified. This soluble receptor, which has 34 % amino acid identity to the extracellular domain of the IL-22R1, binds IL-22 and antagonizes its functional activities (Kotenko et al. 2001b). The skin is also a target for IL-22; high IL-22 expression has been detected in the skin of patients with T-cell-mediated dermatoses. Normal human epidermal keratinocytes express a functional receptor for IL-22, but not for IL-10. IL-22 plays a role in skin inflammatory processes and wound healing.

▪ Interleukin-24

IL-24 (Wang and Liang 2005) is a novel member of the IL-10 family secreted by activated peripheral blood mononuclear cells and the ligand for two heterodimeric receptors, IL-22R1/IL-20R2 and IL-20R1/IL-20R2. The latter is also a receptor chain for IL-20. Under physiological conditions, the major sources of IL-24 are activated monocytes and Th2 cells, whereas the major IL-24 target tissues, based on the receptor expression pattern, are non-hematopoietic in origin and include the skin, lung, and reproductive tissues. Structurally and functionally, IL-24 is highly conserved across species. It has shown antiangiogenic activity and its gene is a tumor suppressor gene (Dent et al. 2010).

▪ Interleukin-26

IL-26 (Donnelly et al. 2010) is part of the IL-10 family and produced Th17 cells, to some extent in NK-cells. It binds to a heterodimeric receptor composed of the IL-20R1 and IL-10R2 chains and is frequently co-expressed with IL-17 and IL-22. Targeting epithelial cells which express IL-20R1, IL-26 is likely to play a role in local mechanisms of mucosal and cutaneous immunity. Furthermore, IL-26 appears to play a central role in autoimmune disease.

▪ Interleukin-28A and B and Interleukin-29

Recently, the human genomic sequence for a family of three cytokines, designated IL-28A, IL-28B, and IL-29 (Donnelly and Kotenko 2010), that are distantly related to type I IFNs (IFN-λ1–3) (Pestka et al. 2004a, b) and the IL-10 family has been described. Like type I IFNs, IL-28 and IL-29 are induced by viral infection and have antiviral activity. However, IL-28 and IL-29 interact with a heterodimeric class II cytokine receptor that consists of the IL-10 receptor 2 (IL-10R2) and an orphan class II receptor chain, designated IL-28R1. This newly described cytokine family may serve as an alternative to type I IFNs in providing resistance to viral infection and antitumor activity.

▪ Interleukin-12 Family

The IL-12 family (Collison et al. 2008) which includes IL-12, IL-23, IL-27, and IL-35 is a mediator of inflammation. Each member is a heterodimeric complex composed of two subunits whose expression is regulated independently (see also IL-23, IL-27, and IL-35 below).

▪ Interleukin-12

IL-12 (Trinchieri 2003) is a 70 kDa heterodimeric proinflammatory cytokine composed of two covalently linked glycosylated chains: p35 and p40. It is mainly produced by activated monocytes, macrophages, and dendritic cells (DCs), enhances proliferation and cytolytic activity of NK- and T-cells, and stimulates their IFN-γ production towards a Th1 response while it inhibits Th2 cells. Dysregulation of IL-12 production can have a major impact on the modulation of immune and allergic responses. Recombinant IL-12 has several potential therapeutic uses in infectious diseases, allergy, and cancer.

▪ Interleukin-23

IL-23 (Aggarwal et al. 2003) is a heterodimeric cytokine comprising the IL-12 p40 subunit of IL-12 and an IL-23-specific p19 subunit. It is produced by activated dendritic cells and acts on memory CD4+ T-cells. IL-23 induces IL-17 and thus plays an early role in the defense against gram-negative infection. It is also pivotal for establishing and maintaining organ-specific inflammatory autoimmune disease. IL-23 and IL-27 both have potent antitumor activity even against poorly immunogenic tumors using different effector mechanisms.

▪ Interleukin-27

IL-27 (Larousserie et al. 2006) is a novel heterodimeric cytokine of the IL-12 family that consists of EBI3, an IL-12p40 homologous protein (IL-27B), and p28, a newly discovered IL-12p35 analogous polypeptide (IL-27A). It is produced by antigen-presenting cells and specifically acts on naive T-cells. IL-27 synergizes with IL-12 to produce IFN-γ and does not support Th2 cytokine production by activated T-cells. Recent evidence, however, suggests that this receptor/ligand pair is also required to suppress a variety of immune cell effector processes, including proliferation and cytokine production. IL-27 is also an inhibitor of Th17 cell development and presents itself as a potential target for treating inflammatory diseases mediated by these cells (Stumhofer et al. 2006).

▪ Interleukin-35

IL-35 (Collison et al. 2008) is a member of the IL-12 cytokine family which is linked to the IL-6 cytokine superfamily. The IL-12 family comprises IL-12, IL-23, IL-27, and IL-35. Unlike the other three family members, IL-35 is an anti-inflammatory cytokine produced by regulatory T-cells (T-reg), which are a critical subpopulation of CD4$^+$ T-cells essential for maintaining self-tolerance and preventing autoimmunity. IL-35 is a heterodimeric protein composed of the IL-12α and IL-27β chains.

▪ Interleukin-17 Family

IL-17 (Hu et al. 2011) a homodimeric glycoprotein more recently renamed IL-17A can also form a heterodimer with IL-17 F to which it is the most closely related family member. Four additional members, IL-17B to IL-17E, have been discovered, whereby IL-17E has been renamed IL-25 (see below). IL-17A and IL-17 F are predominantly produced upon stimulation of Th17 cells (CD$^+$ T-helper cells type 17) by IL-23 after differentiation of naive T-cells into Th17 has been induced by IL-6 and TGF-β. IL-17C has a very restricted expression pattern but has been detected in adult prostate and fetal kidney libraries. The importance of this family of interleukins and their receptors expressed in disparate tissues goes beyond the modulation of T-cell-mediated inflammatory response and importance in effective host defense against pathogen infection. IL-17s also have a role in the homeostasis of tissues, and the IL-17A/F pathway is implicated in the progression of autoimmune diseases such as rheumatoid arthritis, multiple sclerosis, inflammatory bowel disease, and psoriasis. The IL-17 pathway therefore has become an interesting target for blocking strategies by either monoclonal antibodies against IL-17A or its receptor IL-17RA.

▪ Interleukin-25

IL-25 (Fort et al. 2001) is a cytokine that shares sequence similarity with IL-17 and was previously called IL-17E. It is produced by Th2 cells, and its biological effects differ markedly from those of the other described IL-17 family members and have been implicated in the promotion of Th2 immunity. IL-25 induces IL-4, IL-5, and IL-13 and causes histological changes in the lungs and GI tract, including eosinophilic and mononuclear infiltrates, increased mucus production, and epithelial cell hyperplasia and hypertrophy. IL-25 appears to be a key cytokine for the development of Th2-associated pathologies such as asthma and other allergic reactions, as well as antiparasitic response.

▪ Hematopoietin Family

Because many cytokines are multifunctional and have overlapping activities, several members of the hematopoietin family overlap with other classifications. The hematopoietins (Metcalf 2008) constitute a family of structurally related proteins that includes various interleukins IL-3, IL-5, IL-6, IL-11, and IL-13; growth factors including G-CSF, GM-CSF, M-CSF and thrombopoietin, erythropoietin, SCF (stem cell factor), and SCPF (stem cell proliferation factor); and other proteins identified initially by some biological activities not related to hematopoiesis IL-2, IL-4, IL-9, and IL-12 (see also Chap. 18).

▪ Interleukin-3

IL-3 (Martinez-Moczygemba 2003) is produced by activated T-cells, monocytes/macrophages, and stromal cells. It is a multicolony-stimulating hematopoietic growth factor which stimulates the generation of hematopoietic progenitors of every lineage. Administration of IL-3 produces an increase in erythrocytes, neutrophils, eosinophils, monocytes, and platelets. IL-3, however, is not involved in constitutive hematopoiesis but rather in inductive hematopoiesis upon exposure to immunological stress. IL-3 can act synergistically or additively with other hematopoietic growth factors such as GM-CSF, IL-5, and EPO.

▪ Interleukin-5

IL-5 (Greenfeder et al. 2001) acts as a homodimer originally known as T-cell replacement factor (TRF), eosinophil differentiation factor (EDF), and B-cell growth factor (BCGF) II. It is produced by Th2 helper and mast cells. It acts on the eosinophilic lineage, stimulating eosinophil expansion and chemotaxis, and also has activity on basophils. In humans IL-5 is a very selective cytokine as only eosinophils and basophils express IL-5 receptors. Interleukin-5 has been associated with the cause of several allergic diseases including allergic rhinitis and asthma and is therefore a target for the treatment of severe asthma.

▪ Interleukin-6

IL-6 (Kamimura et al. 2003) is a pro-inflammatory cytokine that not only affects the immune system but also acts in many physiological events in various organs. It is produced by lymphoid and nonlymphoid cells and was formerly known as interferon-β_2 for its weak antiviral activity. By stimulating hepatocytes to produce "acute-phase proteins" it plays a central role in the "acute-phase reaction." It is also responsible for the reactive thrombocytosis seen in acute inflammatory processes by stimulating thrombopoietin (Kaushansky 2005). Furthermore, IL-6 is associated with insulin resistance in type 2 diabetes mellitus (Kristiansen and Mandrup-Poulsen 2005). Together with IL-11 (below) and IL-27, IL-6 is also a member of the gp130 receptor cytokine family (White and Stephens 2011) which also includes other cytokines not classified as interleukins.

▪ Interleukin-11

IL-11 (Du and Williams 1997) initially described as hematopoietic factor with thrombopoietic activity has subsequently been shown to be expressed and active in many other tissues including the brain, spinal cord neurons, gut, and testes. IL-11 acts synergistically with other cytokines such as IL-3, IL-4, IL-7, IL-12, IL-13, SCF, and GM-CSF to stimulate various stages and lineages of hematopoiesis and in particular with IL-3 and thrombopoietin (TPO), also termed megakaryocyte growth and development factor (MGDF), to stimulate various stages of megakaryocytopoiesis and thrombopoiesis. Treatment with IL-11 results in production, differentiation, and maturation of megakaryocytes. IL-11 also has a direct effect on erythroid progenitors and also modulates the differentiation and maturation of myeloid progenitor cells. Alveolar and bronchial epithelial cells produce IL-11, which is upregulated by inflammatory cytokines and respiratory syncytial virus (RSV) suggesting that it plays a role in pulmonary inflammation. IL-11 also is an important regulator of bone metabolism. Evidence indicates that IL-11 together with transforming growth factor (TGF)-β, IL-1, and IL-15 are crucial for successful human implantation and placentation (Guzeloglu-Kayisli et al. 2009).

▪ Interleukin-13

IL-13 (Wills-Karp 2004) is a glycoprotein cloned from activated T-cells. IL-13 was first recognized for its effects on B-cells and monocytes, where it upregulated MHC class II expression, promoted IgE class switching, and inhibited inflammatory cytokine production. The functions of IL-13 overlap considerably with those of IL-4, especially with regard to changes induced on hematopoietic cells. IL-13 also has several unique effector functions that distinguish it from IL-4. Resistance to most gastrointestinal nematodes is mediated by type 2 cytokine responses, in which IL-13 plays a dominant role. By regulating cell-mediated immunity, IL-13 modulates resistance to intracellular organisms. In the lung, IL-13 is the central mediator of allergic asthma, where it regulates eosinophilic inflammation, mucus secretion, and airway hyperresponsiveness. IL-13 can also inhibit tumor immune surveillance. Thus, inhibitors of IL-13 might be effective as cancer immunotherapeutics by boosting type-1-associated antitumor defenses. Investigations into the mechanisms that regulate IL-13 production and/or function have shown that IL-4, IL-9, IL-10, IL-12, IL-18, IL-25, IFN-γ, TGF-β, TNF-α, and the IL-4/IL-13 receptor complex are essential for these processes.

OTHERS NOT (YET) ASSIGNED TO A FAMILY

▪ Interleukin-8

IL-8 (Remick 2005) is a 6–8 kDa CXC chemokine, a potent chemoattractant for neutrophils. It affects the pro-inflammatory effector side, including the stimulation of neutrophil degranulation and the enhancement of neutrophil adherence to endothelial cells. It is produced by monocytes, macrophages, fibroblasts, keratinocytes, and endothelial cells. Elevated levels of IL-8 have been found in psoriatic arthritis, synovial fluid, and synovium. IL-8 has been shown to contribute to human cancer progression through its potential functions as a mitogenic, angiogenic, and motogenic factor, and elevated IL-8 serum concentrations could be useful as a predictor for lung cancer.

▪ Interleukin-16

IL-16 (Cruikshank and Little 2008) is a pro-inflammatory cytokine produced by a variety of immune (T-cells, eosinophils, dendritic cells [DCs]) and nonimmune (fibroblasts, epithelial, and neuronal) cells and induces chemotaxis of not only CD4+ T-cells but also monocyte/macrophages and eosinophils. It is synthesized as a precursor molecule (pro-IL-16), cleaved in the cell cytoplasm and secreted as mature IL-16. It regulates T-cell growth and primes $CD4^+$ T-cells for IL-2 and IL-15. IL-16 has been shown to play a role in asthma (El Bassam et al. 2005), Crohn's disease (CD), and systemic lupus erythematosus (SLE) (Lee et al. 1998). IL-16 also inhibits human (HIV) and simian (SIV) immunodeficiency virus. A neuronal form of IL-16 detected in neurons of the cerebellum and hippocampus has been described (Kurschner and Yuzaki 1999).

▪ Interleukin-31

IL-31 (Bilsborough et al. 2006) is a 4-helix bundle cytokine preferentially expressed by activated T-cells with a Th2 bias. Together with IL-4 and IL-13, IL-31 has been implicated in the pathogenesis of atopic dermatitis because they are produced by a subset of T-cells that home to the skin. IL-31 signals through a heterodimeric receptor constitutively expressed by epithelial cells including keratinocytes. IL-31-stimulated keratinocytes induce a whole array of inflammatory chemokines which also facilitate the recruitment of lymphocytes, monocytes, and polymorphonuclear cells to the epidermis.

▪ Interleukin-32

IL-32 (Kim et al. 2005) is a polypeptide which was described several years ago as natural killer cell transcript 4 (NK4) of activated T-cells and NK-cells and belongs to the pro-inflammatory cytokines. It induces TNF-α and MIP-2, a chemokine, in different cells via the signal pathway of pro-inflammatory cytokines. To date it has been detected in higher concentration in some of the patients with sepsis compared to healthy individuals.

▪ Interleukin-34

IL-34 (Lin et al. 2008) forms homodimers and promotes survival and differentiation of monocytes and macrophages. It elicits its activity by binding to the shared (macrophage) colony-stimulating factor 1 receptor (CSF-1R). Messenger RNA (mRNA) expression of human IL-34 is found mostly in the spleen but occurs in several other tissues as well: the thymus, liver, small intestine, colon, prostate gland, lung, heart, brain, kidney, testes, and ovary. IL-34 also plays an important role in the regulation of osteoclast proliferation and differentiation and in the regulation of bone resorption (Baud'huin et al. 2010).

THERAPEUTIC USE OF RECOMBINANT INTERFERONS

■ IFN-α Therapeutics

Together with recombinant human insulin and growth hormone, recombinant IFN-α was one of the first rDNA-derived pharmaceuticals. The drive to produce recombinant interferon and other rDNA-derived pharmaceuticals developed from the need to obtain large amounts of a well-defined, purified protein for large-scale therapeutic use. Availability of the necessary basic technologies (see Chaps. 1, 2, and 3) made this possible. Starting in the early 1980s, a number of cytokines produced by recombinant gene technology were developed to become innovative therapeutic modalities called biologicals or biopharmaceuticals. Table 21.3 summarizes the recombinant IFNs approved for therapeutic use.

Interferon alfa-2 (a modified generic name for IFN-α2) was developed independently by Hoffmann-LaRoche Ltd. (interferon alfa-2a; Roferon®A) and Schering-Plough Corporation (interferon alfa-2b; Intron®A). Both were obtained by recombinant DNA technology in *E. coli*, consist of 165 amino acids with an approximate molecular weight of 19 kDa, and differ by one amino acid in position 23: Lys for interferon alfa-2a and Arg for interferon alfa-2b (Pestka 1986). For all practical purposes there is no difference between these two products in terms of pharmacological properties or clinical application.

The metabolism of interferon alfa-2a is consistent with that of alfa interferons in general and is therefore used as example. Alfa interferons are totally filtered through the glomeruli and undergo rapid proteolytic degradation during tubular reabsorption (see Chap. 5). Liver metabolism and subsequent biliary excretion are considered minor pathways of elimination for alfa interferons. After intramuscular (IM) and subcutaneous (SC) administrations of 36 MIU, peak serum concentrations range from 1,500 to 2,580 pg/mL (mean 2,020 pg/mL) at a mean time to peak of 3.8 h and from 1,250 to 2,320 pg/mL (mean 1,730 pg/mL) at a mean time to peak of 7.3 h, respectively. The apparent fraction of the dose absorbed after intramuscular injection is >80 %. The pharmacokinetics of interferon alfa-2a after single intramuscular doses to patients with disseminated cancer are similar to those found in healthy volunteers. Dose proportional increases in

Recombinant interferons	Company	First Indication	First approval
Interferon-α			
IFN-α2a produced in *E. coli*; Roferon A®	Hoffmann–La Roche (Basel, Switzerland)	Hairy cell leukemia	1986 (EU and US)
IFN-α2b produced in *E. coli*; Intron® A; Viraferon®; Alfatronol®	Schering-Plough (Kenilworth NJ, USA)	Hairy cell leukemia	1986 (US and EU)
IFN-αcon1, synthetic type I IFN produced in *E. coli*; Infergen®	Amgen (Thousand Oaks, US), Yamanouchi Europe (Leiderdorp, The Netherlands, EU)	Chronic hepatitis C	2001 (US)
Interferon-β			
IFN-β1a produced in CHO cells; Rebif®	Serono } (Geneva, Switzerland)	Relapsing/remitting multiple sclerosis	1998 (EU), 2002 (US)
IFN-β1a produced in CHO cells; Avonex®	Biogen (Cambridge, MA, USA)	Relapsing/remitting multiple sclerosis	1997 (EU), 1996 (US)
IFN-β1b, Cys17 Ser substitution; produced in *E. coli*; Betaferon®	Schering AG (Berlin, Germany)	Relapsing/remitting multiple sclerosis	1995 (EU)
IFN-β1b, Cys17 Ser substitution; produced in *E. coli*; Betaseron®	Berlex Labs/Chiron (Richmond/ Emeryville, CA, USA)	Relapsing/remitting multiple sclerosis	1993 (US)
Interferon-γ			
Actimmune® (IFN-γ1b; produced in *E. coli*)	Genentech (San Francisco CA, USA), InterMune (Palo Alto, CA, USA)	Chronic granulomatous disease	1990 (US)

Adapted from Walsh G. Biopharmaceutical Benchmarks 2008. *Nature Biotechnology* (2006) 24:769–776

Table 21.3 ■ Interferons approved as biopharmaceuticals in the United States and Europe.

serum concentrations are observed after single doses up to 198 MIU. There are no changes in the distribution or elimination of interferon alfa-2a during twice daily (0.5–36 MIU), once daily (1–54 MIU), or three times weekly (1–136 MIU) dosing regimens up to 28 days of dosing. Multiple IM doses of interferon alfa-2a result in an accumulation of two to four times the serum concentrations seen after a single dose.

Roferon®A and Intron®A are approved for the following indications: chronic hepatitis B and C, Kaposi's sarcoma, renal cell carcinoma, malignant melanoma, carcinoid tumor, multiple myeloma, non-Hodgkin lymphoma, hairy cell leukemia, chronic myelogenous leukemia, thrombocytosis associated with chronic myelogenous leukemia, and other myeloproliferative disorders. The approved indications vary depending on company and regulatory policies; for detailed information as well as for the recommended dosing, the reader is referred to the respective product information current in their countries.

The adverse event profile for the three IFN-α is the same; it is generally more or less well tolerated depending on the dose regimen used and subjectively consists primarily of the "influenza-like symptoms" named as such because they mimic the symptoms of early influenza. This, of course, should come as no surprise as these symptoms are caused by peaks of endogenous interferon stimulated by the influenza virus infection. For a detailed reporting of all adverse events, the reader is referred to the product information for each product.

Given the principle that the toxicity of a given medication is defined by its peak, i.e., by the time it is above a toxic threshold concentration and the efficacy by the trough concentration, i.e., the time the substance is below the therapeutic level, it would be desirable to obtain a therapeutic regimen which minimizes fluctuations. A constant therapeutic drug concentration would be an ideal goal. The first step towards that goal, as a proof of concept, was to model a long-acting interferon using an insulin pump to inject patients with chronic hepatitis C with interferon α-2a at predetermined rates per hour for 28 days. A similar study was performed in patients with renal cell carcinoma. These studies indicated that interferon α-2a at a constant dose was indeed better tolerated while showing activity when administered by continuous SC infusion (Carreño et al. 1992; Ludwig et al. 1990). The next step therefore was to develop a new longer-acting molecule by attaching several polyethylene glycol (PEG) chains to the native interferon molecule (see section "Pegylated Interferons and Interleukins: The Next Generation" below).

Roferon® A is supplied as prefilled syringes containing 3 MIU, 4.5 MIU, 6 MIU, or 9 MIU in 0.5 mL; or as cartridges containing 18 MIU per mL for SC injection only; or as vials each containing 3 MIU, 6 MIU, 9 MIU, or 36 MIU in 1 mL; or multidose injectable solution containing 9 MIU (each 0.3 mL contains 3 MIU) or 18 MIU of interferon α-2a (each mL contains 6 MIU) for SC or IM injection. All presentations are human HSA (human serum albumin)-free liquid formulations with 7.21 mg sodium chloride, 0.2 mg polysorbate 80, 10 mg benzyl alcohol (as a preservative), 0.77 mg ammonium acetate, and sterile water for injections.

Intron A® is supplied as vials containing 10 MIU, 15 MIU, or 50 MIU as lyophilisate and a vial with 1 mL of diluent for reconstitution containing 20 mg glycine, 2.3 mg sodium phosphate dibasic, 0.55 mg sodium phosphate monobasic, and 1.0 mg HSA; or as solution vials containing 10 MIU as single dose and 18 MIU or 25 MIU as multidose with 7. 5 mg sodium chloride, 1.8 mg sodium phosphate dibasic, 3 mg sodium phosphate monobasic, 0.1 mg edetate disodium, 0.1 mg polysorbate 80, and 1.5 mg m-cresol as a preservative per mL for SC, IM, or intralesional injection; or solution in multidose pens containing 6 doses of 3 MIU, 5 MIU, or 10 MIU interferon α-2b per 0.2 mL and adjuvants as above for SC injection.

Infergen® (interferon alfacon-1) is a synthetic "consensus" interferon consisting of 166 amino acids and not occurring in nature. It was genetically engineered in *E. coli* by Amgen. The amino acid sequence of the product is derived by comparison of the sequences of several natural interferon-α subtypes and assigning the most frequently observed amino acid in each corresponding position. Infergen® is supplied as single-dose, preservative-free vials containing either 9 μg (0.3 mL) or 15 μg (5 mL) of interferon alfacon-1 for SC injection.

▪ IFN-β Therapeutics

Three IFN-β products (Table 21.3) are marketed worldwide for the treatment of multiple sclerosis: the first was Berlex's Betaseron®, marketed by Schering AG as Betaferon® in Europe. It is interferon-β1b with 165 amino acids and an approximate molecular weight of 18,500 Da, with a cystein-17-serine substitution. It is produced in *E. coli*, which was then the standard method. It is non-glycosylated, as without further engineering glycosylation is not possible in the *E. coli* system (Wacker et al. 2002) (see Chap. 3). Independently, Biogen and Serono developed a glycosylated IFN-β1a produced in Chinese hamster ovary cells. Thus, not only is the amino acid sequence of these IFN-βs identical to that of natural fibroblast-derived human interferon beta, but they are also glycosylated, each containing a single N-linked complex carbohydrate moiety. The two products are marketed as Avonex® and Rebif®, respectively. All three products are indicated for the treatment of multiple sclerosis.

Glycosylating proteins fundamentally alter their pharmacokinetic and pharmacodynamic properties. The non-glycosylated interferon-β1b ($IFN\text{-}\beta_{ser17}$) has the expected short circulation time: time to peak concentration (C_{max}) between 1–8 h with a mean peak serum interferon concentration of 40 IU/mL after a single SC injection of 0.5 mg (16 MIU). Bioavailability is about 50 %. Patients receiving single intravenous (IV) doses up to 2.0 mg (64 MIU) show an increase in serum concentrations which is dose proportional. Mean terminal elimination half-life values ranged from 8.0 min to 4.3 h. Thrice weekly IV dosing for 2 weeks resulted in no accumulation of IFN-β1b in sera of patients. Pharmacokinetic parameters after single and multiple IV doses were comparable. Following every other day SC administration of 0.25 mg (8 MIU) IFN-β1b in healthy volunteers, biologic response marker levels (neopterin, β2-microglobulin, MxA protein, and IL-10) increased significantly above baseline 6–12 h after the first dose. Biologic response marker levels peaked between 40 and 124 h and remained elevated above baseline throughout the 7-day (168-h) study.

Glycosylated IFN-β1a such as Rebif®, on the other hand, is slower to reach C_{max}, with a median of 16 h and the serum elimination half-life is 69±37 h (mean±SD). In healthy volunteers a single SC injection of 60 μg (~18MIU) of interferon-β1a resulted in a C_{max} of 5.1±1.7 IU/mL. Following every other day SC injections in healthy volunteers, an increase in AUC of approximately 240 % was observed, suggesting that accumulation of IFN-β1a occurs after repeated administration. Biological response markers (e.g., 2′,5′-oligoadenylate synthetase (OAS), neopterin, and β_2-microglobulin) are induced by IFN-β1a following a single SC administration of 60 μg. Intracellular 2′,5′-OAS peaked between 12 and 24 h and β_2-microglobulin and neopterin serum concentrations showed a maximum at approximately 24–48 h. All three markers remained elevated for up to 4 days. Administration of 22 μg (6MIU) IFN-β1a three times per week inhibited mitogen-induced release of pro-inflammatory cytokines (IFN-γ, IL-1, IL-6, TNF-α, and TNF-β) by peripheral blood mononuclear cells that, on average, was near double that observed with IFN-β1a administered once per week at either 22 (6 MIU) or 66 μg (12 MIU).

Betaseron®/Betaferon® is formulated as a sterile powder with a 0.54 % sodium chloride solution as diluent. Reconstituted it presents as 0.25 mg (8 MIU of antiviral activity) per mL. The recommended dose is 0.25 mg injected SC every other day.

Avonex® is formulated as a lyophilized powder for IM injection. After reconstitution with the supplied diluent (sterile water for injection), each vial contains 30 μg of IFN-β1a, 15 mg HSA, 5.8 mg sodium chloride, 5.7 mg dibasic sodium phosphate, and 1.2 mg monobasic sodium phosphate in 1.0 mL at a pH of approximately 7.3 or as a prefilled syringe with a sterile solution for IM injection containing 0.5 mL with 30 μg of interferon-β1a, 0.79 mg sodium acetate trihydrate, 0.25 mg glacial acetic acid, 15.8 mg arginine hydrochloride, and 0.025 mg polysorbate 20 in water for injection at a pH of approximately 4.8. The recommended dosage is 30 μg injected IM once a week.

Rebif® is supplied in prefilled 0.5 mL syringes: each 0.5 mL contains either 22 μg (6 MIU) or 44 μg (12 MIU) of IFN-β1a, 2 or 4 mg HSA, 27.3 mg mannitol, 0.4 mg sodium acetate, and water for injection. The recommended dosage is 22 μg (6 MIU) given three times per week by SC injection. This dose is effective in the majority of patients to delay progression of the disease. Patients with a higher degree of disability (EDSS (Kurtzke Expanded Disability Status Scale) (Kurtzke 1983) of 4 or higher) may require a dose of 44 μg (12 MIU) three times per week.

The adverse event profile for the three IFN-βs is similar to IFN-α. It is generally reasonably well tolerated and subjectively again consists primarily of the "influenza-like symptoms." For a detailed reporting of all adverse events, the reader is referred to the product information for each biopharmaceutical.

■ IFN-γ Therapeutics

Actimmune® (recombinant interferon-γ1b; immune IFN) is a single-chain polypeptide containing 140 amino acids. It is produced by genetically engineered *E. coli* containing the DNA which encodes for the human protein. It is a highly purified sterile solution consisting of non-covalent dimers of two identical 16,465 Da monomers. Actimmune® is slowly absorbed after IM injection of 100 μg/m². A C_{max} of 1.5 ng/mL is reached in approximately 4 h and after SC injection a C_{max} of 0.6 ng/mL is reached in 7 h. The apparent fraction of dose absorbed is >89 %. The mean half-life after IV administration was 38 min and after IM and SC dosing with 100 μg/m² were 2.9 and 5.9 h, respectively. Multiple-dose SC pharmacokinetics showed no accumulation of Actimmune® after 12 consecutive daily injections of 100 μg/m².

Actimmune® is a solution filled in a single-dose vial for SC injection. Each 0.5 mL contains: 100 μg (two million IU) of IFN-γ1b, formulated in 20 mg mannitol, 0.36 mg sodium succinate, 0.05 mg polysorbate 20, and sterile water for injection. The dosage for the treatment of patients with chronic granulomatous disease or severe malignant osteopetrosis is 50 μg/m² (1 million IU/m²) for patients whose body surface area is greater than 0.5 m² and 1.5 mcg/kg/dose for patients whose body surface area is equal to or less than 0.5 m².

The adverse event profile of IFN-γ is similar to IFN-α. It is generally well tolerated and subjectively consists primarily of the "influenza-like symptoms." For a detailed reporting of all adverse events, the reader is referred to the Actimmune® product information.

THERAPEUTIC USE OF RECOMBINANT INTERLEUKINS

In general, the development of interleukins as a therapeutic modality is even more complex than for IFNs. Most interleukins are embedded in a regulatory network and so far the pharmacological use of interleukins has been somewhat disappointing. This was largely

Recombinant interleukins	Company	First Indication	First approval
Proleukin® (aldesleukin; IL-2, lacking N-terminal alanine, C125 S substitution, produced in *E. coli*)	Chiron Therapeutics (Emeryville, CA)	RCC (renal cell carcinoma)	1992 (EU and US)
Neumega® (oprelvekin; IL-11, lacking N-terminal proline produced in *E. coli*)	Genetics Institute (Cambridge, MA) now Pfizer Inc	Prevention of chemotherapy-induced thrombocytopenia	1997 (US)
Kineret® (anakinra; IL-1 receptor antagonist (produced in *E. coli*))	Amgen (Thousand Oaks, CA)	RA (rheumatoid arthritis)	2001 (US)

Adapted from Walsh G. Biopharmaceutical Benchmarks 2008. *Nature Biotechnology* (2006) 24:769–776

Table 21.4 ■ Interleukins approved as biopharmaceuticals worldwide.

due to our lack of understanding of the role of these molecules and of the best way to use them; they are less well studied than IFNs. IL-2, for example, was initially developed by oncologists in the days when "go in fast, hit them hard and get out" was the prevalent strategy. Terms like maximal tolerated dose (which we called minimal poisonous dose) actually defined the dose at which a given drug was in most cases no longer tolerated. Thus, IL-2 was given an undeserved bad reputation. Similar thinking nearly killed the development of IFN-α for the treatment of chronic viral hepatitis and was ultimately the main reason for discontinuing the development of IL-2 in chronic hepatitis B (Pardo et al. 1997; Artillo et al. 1998) and IL-12 in chronic hepatitis B and C (Zeuzem et al. 1999; Carreño et al. 2000; Pockros et al. 2003). In spite of this, progress has been made and our understanding of the complexities of such substances and their antagonists is growing. Table 21.4 lists the interleukins that are currently approved.

■ Aldesleukin

Proleukin® (aldesleukin), a non-glycosylated human recombinant interleukin-2 product, is a highly purified protein with a molecular weight of approximately 15 kDa. The chemical name is des-alanyl-1, serine-125 human interleukin-2. It is produced by recombinant DNA technology using a genetically engineered *E. coli* containing an analog of the human interleukin-2 gene. The modified human IL-2 gene encodes a modified human IL-2 differing from the native form: the molecule has no N-terminal alanine—the codon for this amino acid was deleted during the genetic engineering procedure. And, serine was substituted for cysteine at amino acid position 125. Aldesleukin exists as biologically active, non-covalently bound microaggregates with an average size of 27 recombinant interleukin-2 molecules. The pharmacokinetic profile of aldesleukin is characterized by high plasma concentrations following a short IV infusion, rapid distribution into the extravascular space, and elimination from the body by metabolism in the kidneys with little or no bioactive protein excreted in the urine. Studies of IV aldesleukin indicate that upon completion of infusion, approximately 30 % of the administered dose is detectable in plasma. Observed serum levels are dose proportional. The distribution and elimination half-life after a 5-min IV infusion are 13 and 85 min, respectively. In humans and animals, aldesleukin is cleared from the circulation by both glomerular filtration and peritubular extraction in the kidney. The rapid clearance of aldesleukin has led to dosage schedules characterized by frequent, short infusions. The adverse event profile of IL-2 is similar to that seen for IFNs and many ILs. It is generally reasonably well tolerated and subjectively consists primarily of the "influenza-like symptoms." For a detailed reporting of all adverse events, rarely severe, the reader is referred to the product information for Proleukin®.

Proleukin® is supplied as a sterile, lyophilized cake in single-use vials intended for IV injection. After reconstitution with 1.2 mL sterile water for injection, each mL contains 18 million IU (1.1 mg) aldesleukin, 50 mg mannitol, and 0.18 mg sodium dodecyl sulfate, without preservatives, buffered with approximately 0.17 mg monobasic and 0.89 mg dibasic sodium phosphate to a pH of 7.5. It is indicated for the treatment of adults with metastatic renal cell carcinoma or metastatic melanoma. Each treatment course consists of two 5-day treatment cycles: 600,000 IU/kg (0.037 mg/kg) is administered every 8 h by a 15-min IV infusion for a maximum of 14 doses. Following 9 days of rest, the schedule is repeated for another 14 doses or a maximum of 28 doses per course, as tolerated.

■ Oprelvekin

Neumega® (oprelvekin) a non-glycosylated IL-11 is produced in *E. coli* by recombinant DNA technology and has 177 amino acids in length and a molecular mass of approximately 19 kDa. It differs from the 178 amino acid length of native IL-11 in lacking the amino-terminal proline residue. It is used as a thrombopoietic growth factor that directly stimulates the proliferation of hematopoietic stem cells and megakaryocyte progenitor cells and induces megakaryocyte maturation resulting in increased platelet production. Pharmacokinetics show a rapid clearance from the serum and distribution to highly perfused organs. The kidneys are the primary route of elimination and little intact product can be found in the urine (see Chap. 5). After subcutaneous injection of 50 μg/kg, the C_{max} of 17.4 ± 5.4 ng/mL is reached after 3.2 ± 2.4 h (T_{max}) with a half-life of 6.9 ± 1.7 h. The absolute bioavailability is >80 %. There is no accumulation after multiple doses. Patients with severally impaired renal function show a marked decrease in clearance to 40 % of that seen in subjects with normal renal function.

Neumega® is supplied as single-use vials containing 5 mg of oprelvekin (specific activity approximately 8×10^6 U/mg) as a sterile lyophilized powder with 23 mg of glycine, 1.6 mg of dibasic sodium phosphate heptahydrate, and 0.55 mg monobasic sodium phosphate monohydrate. When reconstituted with 1 mL of sterile water for injection, the solution has a pH of 7.0. It is indicated for the prevention of severe thrombocytopenia following myelosuppressive chemotherapy. The recommended dose is 50 μg/kg given once daily by SC injection after a chemotherapy cycle in courses of 10–21 days. Platelet counts should be monitored to assess the optimal course of therapy. Treatment beyond 21 days is not recommended. Oprelvekin is generally well tolerated. Reported adverse events, mainly as a consequence of fluid retention, include edema, tachycardia/palpitations, dyspnea, and oral moniliasis. For a detailed reporting of all adverse events, rarely severe, the reader is referred to the product information for Neumega®.

■ Anakinra

Kineret® (anakinra) is a recombinant, non-glycosylated form of the human interleukin-1 receptor antagonist (IL-1Ra) produced using an *E. coli* bacterial expression system. It consists of 153 amino acids and has a molecular weight of 17.3 kDa differs from native human IL-1Ra in that it has the addition of a single methionine residue at its amino terminus. The absolute bioavailability of Kineret® after a 70 mg SC bolus injection is 95 %. C_{max} occurs 3–7 h after SC administration at clinically relevant doses (1–2 mg/kg) with half-life ranging from 4 to 6 h. There is no accumulation of Kineret® after daily SC doses for up to 24 weeks. The mean plasma clearance with mild and moderate (creatinine clearance 50–80 mL/min and 30–49 mL/min) renal insufficiency was reduced by 16 % and 50 %, respectively. In severe renal insufficiency and end-stage renal disease (creatinine clearance <30 mL/min), mean plasma clearance declined by 70 and 75 %, respectively. Less than 2.5 % of the administered dose is removed by hemodialysis or continuous peritoneal dialysis. A dose schedule change should be considered for subjects with severe renal insufficiency or end-stage renal disease.

Kineret® is supplied in single-use prefilled glass syringes with 27 gauge needles as a sterile, clear, preservative-free solution for daily SC administration. Each prefilled glass syringe contains 0.67 mL (100 mg) of anakinra in a solution (pH 6.5) containing 1.29 mg sodium citrate, 5.48 mg sodium chloride, 0.12 mg disodium EDTA, and 0.70 mg polysorbate 80 in water for injection. It is indicated for the reduction in signs and symptoms and slowing the progression of structural damage in moderately to severely active rheumatoid arthritis and can be used alone or in combination with disease-modifying antirheumatic drugs (DMARD) other than TNF-blocking agents (see also Chap. 20). The recommended dose for the treatment of patients with rheumatoid arthritis is 100 mg/day. Patients with severe renal insufficiency or end-stage renal disease should receive 100 mg every other day. Anakinra is generally well tolerated. The most common adverse reaction is injection-site reactions; the most serious adverse reaction is neutropenia, particularly when used in combination with TNF-blocking agents, and serious infections. For a detailed reporting of all adverse events, rarely severe, the reader is referred to the product information for Kineret®.

PEGYLATED INTERFERONS AND INTERLEUKINS: THE NEXT GENERATION

Since 1977 it has been known that polyethylene glycol (PEG) conjugated proteins are frequently more effective than their native parent molecule. Our understanding of PEG chemistry and how it affects the behavior of a biopharmaceutical has increased with the number of PEGylated proteins developed as therapeutic agents (Table 21.5 gives some examples). PEG is

PEGylated recombinant interferons			
Pegasys® (PEGylated IFN-α2a produced in *E. coli*)	Hoffman–La Roche (Basel, Switzerland)	Chronic hepatitis B and C	2002 (EU and US)
ViraferonPeg® (PEGylated IFN-α2b produced in *E. coli*)	Schering-Plough (Kenilworth NJ, USA)	Chronic hepatitis C	2000 (EU)
PegIntron® (PEGylated IFN-α2b produced in *E. coli*)	Schering-Plough (Kenilworth NJ, USA)	Chronic hepatitis C	2000 (EU), 2001 (US)

Adapted from *Nature Biotechnology* (2006) 24:769–776

Table 21.5 ■ PEGylated interferons approved in the United States and Europe.

hydrophilic, inert, nontoxic, non-immunogenic, and in its most common form either linear or branched, terminated with hydroxyl groups that can be activated to couple to the desired target protein. It has been approved for human administration by mouth, injection, and topical application. Its general structure is

$HO\text{-}(CH_2CH_2O)_n\text{-}CH_2CH_2\text{-}OH$ Bifunctional linear PEG (diol)

For polypeptide modification one hydroxyl group is usually inactivated by conversion to monomethoxy or mPEG, and it becomes monofunctional, i.e., only one hydroxyl group is activated during the PEGylation process, thus avoiding the formation of interprotein (oligomerization) or intraprotein bridges:

$CH_3O\text{-}(CH_2CH_2O)_n\text{-}CH_2CH_2\text{-}OH$ Monofunctional linear mPEG

To couple PEG to a molecule such as polypeptides, polysaccharides, polynucleotides, or small organic molecules, it is necessary to chemically activate it. This is done by preparing a PEG derivative with a functional group chosen according to the desired profile for the final product. In addition to the linear PEGs, branched structures have proven useful for peptide and protein modifications:

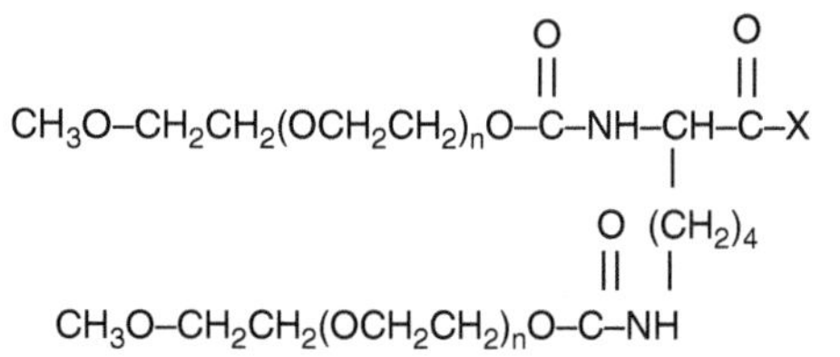

```
                                 O       O
                                 ||      ||
CH3O–CH2CH2(OCH2CH2)nO–C–NH–CH–C–X
                                         |
                               O (CH2)4          branched PEG
                               ||  |
  CH3O–CH2CH2(OCH2CH2)nO–C–NH
```

Branched PEG or PEG2 have a number of advantages over linear structures:

- Attached to proteins they "act" much larger than a linear mPEG of the same MW.
- Two PEG chains are added per attachment site, reducing the chance of protein inactivation.
- They are more effective in protecting proteins from proteolysis, reducing antigenicity, and immunogenicity.

Depending on the desired use for the PEG-modified molecule, different PEGylation strategies can be chosen, for example:

- Multiple shorter-chain PEGylation if the biological activity should be preserved
- A weak PEG-protein bond if a slow release effect is desired
- A branched chain with high MW and a strong bond if prolonged circulation and receptor saturation is the goal

Table 21.5 lists some of the PEGylated protein pharmaceuticals on the market or in various phases of development with appropriate references. For a more in-depth review of PEG chemistries and characteristics, the interested reader is referred to Roberts et al. (2002); Bailon et al. (2001).

The development of rhIFN-α from the native, unmodified molecule to the PEGylated form with the desired pharmacological profile, summarized in Fig. 21.3, is an example of how the understanding of PEG chemistry progressed with experience (Zeuzem et al. 2003). Increasing the length of the PEG chain resulted in progressively longer circulating half-lives due to protracted resorption and lower clearance, ultimately resulting in a near constant serum concentration over an entire week.

The first PEGylated interferon, IFN alfa-2a, used a linear, 5 kDa mPEG with a weak urethane PEG-IFN alfa-2a link. Clinical trials conducted with this compound were unsuccessful because the blood circulation half-life for the conjugate (Fig. 21.3b) was only slightly improved relative to that of the native protein (Fig. 21.3a) (Wills 1990). Development of the product was therefore halted at Phase II clinical trials (Zeuzem et al. 2003). The second compound was developed by Schering-Plough, Kenilworth, NJ, in collaboration with Enzon Pharmaceutical Inc, Bridgewater, NJ. It made use of a longer (12 kDa), linear PEG with a urethane linkage to IFN alfa-2b. The chosen strategy was to combine the advantages of high specific activity with lower serum clearance resulting in PegIntron® (Wang et al. 2002) with markedly improved pharmacological properties allowing once a week administration (Fig. 21.3c) (Glue et al. 2000). PegIntron®, also marketed as Viraferon® in some countries, is approved worldwide for the treatment of chronic hepatitis C.

The development of the third PEGylated interferon, IFN alfa-2a, took a different approach. The strategic goal was to achieve lasting and constant serum concentrations over an entire week. In a collaboration of Roche with Shearwater Polymers in Huntsville, AL, now Nektar, San Carlos, CA, IFN-α-2a was linked by a stable amide bond to four different PEG chains of various sizes, structures, and site-attachment numbers. The resulting products were tested for antiviral activity and a variety of pharmacokinetic parameters including half-life, absorption rate, and mean residence time:

- 20-kDa linear mono-PEG-IFN alfa-2a
- 40-kDa linear di-PEG-IFN alfa-2a
- 20-kDa branched mono-PEG-IFN alfa-2a
- 40-kDa branched mono-PEG-IFN alfa-2a

The 40-kDa, branched PEGylated molecule (later named Pegasys®) exhibited sustained absorption, decreased systemic clearance, and an approximate tenfold increase in serum half-life over regular interferon. The biological activity was similarly prolonged resulting in an optimal pharmacological profile (Fig. 21.3d)

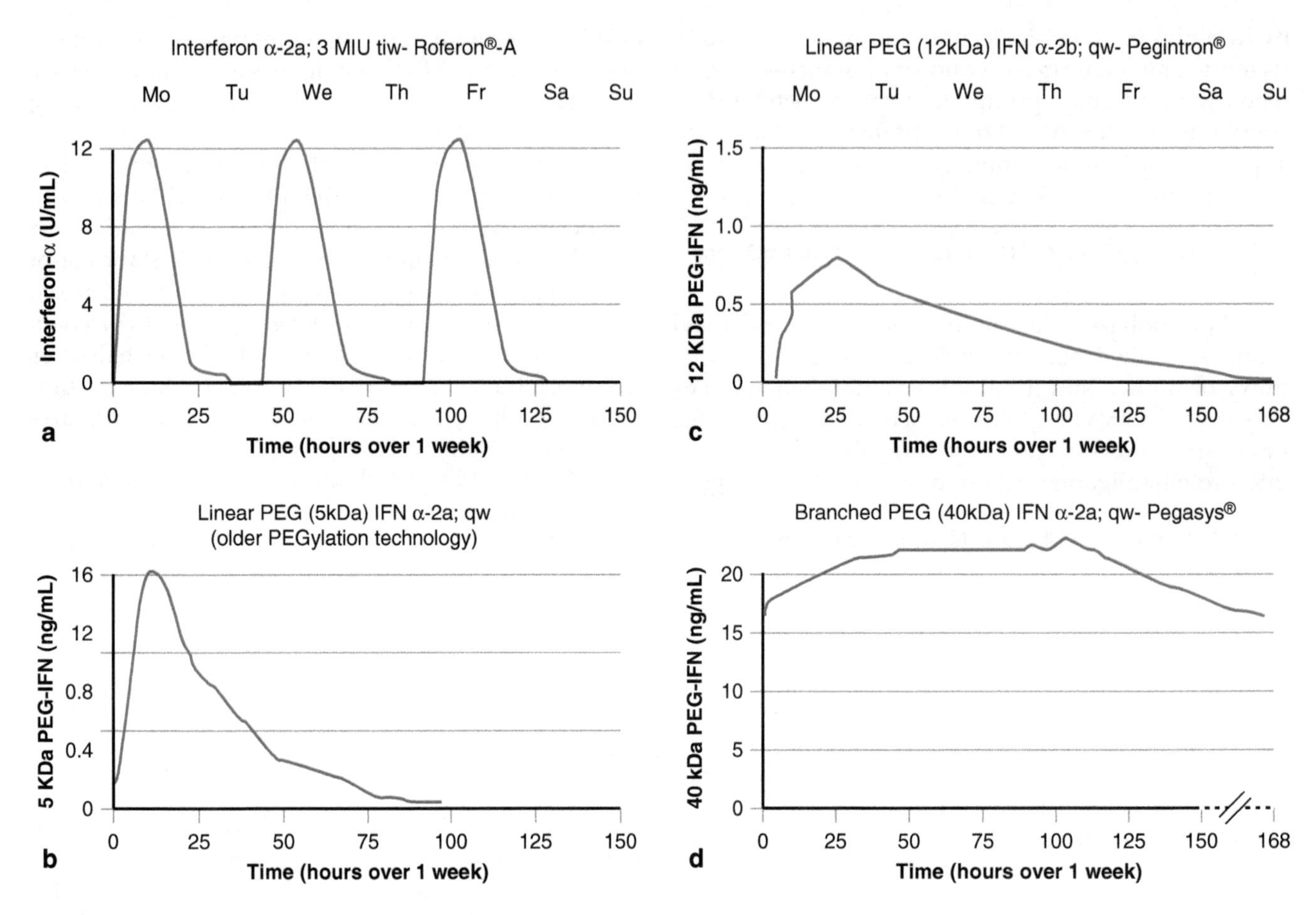

Figure 21.3 ■ (a–d) Pharmacokinetic profiles for IFN and PEG-IFN (Repeated dosing).

(Algranati and Moeli 1999). It was therefore chosen for further clinical development (Reddy et al. 2002) leading to its approval worldwide for the treatment of chronic hepatitis B and C. Pegasys® is being tested for the treatment of renal cell carcinoma (Motzer et al. 2002), malignant melanoma, chronic myeloid leukemia (Talpaz et al. 2005), and non-Hodgkin lymphoma (NHL).

The rapidly growing understanding of the potential of advanced PEGylation chemistry to improve the stability and pharmacological properties of biopharmaceuticals has fostered the development of an increasing number of PEG-biopharmaceuticals. Several of those have proven to offer significant advantages over their native counterparts and found their place in our therapeutic armamentarium. PEG is also used for a variety of other (non-bio)pharmaceutical applications. Table 21.6 lists several examples of different marketed products and others still in development (cf. Chaps. 12, 14 and 18).

OUTLOOK AND CONCLUSIONS

There is a very precise and organized order in the intricate function of the immune system to make it work effectively and we are well on our way to map it. For us to fully understand what still appears as complex to most, a lot of hard work still lies ahead. The fundamental approach to cytokine or cytokine antagonist therapy with biopharmaceuticals is to identify diseases caused by insufficient or excessive cytokine production. In the first case, e.g., certain chronic viral diseases or cancers, appropriate cytokines are used pharmacologically to boost the immune response. Examples include IFN-α with antiviral as well as immunomodulatory properties in chronic viral hepatitis or IL-2 or IL-12 in renal cell cancer and malignant melanoma. For chronic inflammatory or atopic diseases caused by unchecked overproduction of interleukins, two options are available: either interleukin or receptor antagonists, e.g., humanized monoclonal antibodies (Remicade® or Enbrel®; see

Protein name	PEGylation	Product name	Reference
IFN-α2a	Branched, 40 kDa	Pegasys®	Reddy et al. (2002)
IFN-α2b	Linear, 12 kDa	PegIntron®	Wang et al. 2002
Interferon-β	Linear, 20 kDa	mPEGIFNβ1b$_{20,000}$	Baker et al. (2006)
Interleukin-2	Linear, 5 kDa	multiPEGIL-2	Pettit et al. (1997)
Interleukin-6	Linear, 5 kDa	multiPEGIL-6	Tsunoda et al. (2001)
Interleukin-15	Linear, 5 kDa	multiPEGIL-15	Pettit et al. (1997)
TNF-αLys(–)	Linear, 5 kDa	monoPEG TNFα$_{5,000}$	Yamamoto et al. (2003)
Erythropoietin[1]	60 kDa	Mircera® (Ro 50–3821; mPEG epoetinβ$_{60,000}$)	Schellekens (2006)
G-CSF[1]	Linear, 20 kDa	Neulasta® (pegfilgrastim)	Lyman (2005)
GM-CSF[1]	Linear, 5 kDa	Peg-sargramostim	Doherty et al. (2005)
MGDF[1]	Linear, 20 kDa	PEG rHuMGDF	Kuter and Begley (2002)
Adenosine deaminase	Linear, 5 kDa	Adagen® (pegademase)	FDA Drug Label
Arginine deiminase	Linear, 20 kDa	ADI-SS PEG$_{20,000}$	Holtsberg et al. (2002)
Asparaginase	Linear, 5 kDa	Oncaspar®(pegaspargase)	Cao et al. (1990)
Leptin	Branched, 42 kDa	PEG-OB	Hukshorn et al. (2000)
rhGH analog[2]	Linear, 5 kDa	Somavert® (pegvisomant)	Ross et al. (2001)
RNA aptamer	40 kDa	Macugen® (pegaptanib)	Ng et al. (2006)
Doxorubicin	Linear, 2 kDa	Doxil® (doxorubicin HCl liposome injection)	Product information
PEG hydrogel	<20 kDa	SprayGel™ Adhesion Barrier	Ferland (2001)
PEG hydrogel	Unknown	FocalSeal™ Tissue Sealant	Henney (2000)

[1]See Chap. 18
[2]See Chap. 14

Table 21.6 ■ Examples of PEGylated biopharmaceuticals.

Chap. 20) or IL-1R antagonist (Kineret®, see above) in rheumatoid arthritis, or downregulation of the excessively produced interleukin using its antagonistic cytokine, e.g., PEGylated IL-12 in asthma (Leonard and Sur 2003) or IL-10 in psoriasis (Asadullah et al. 2004). To date it appears that the first option is more successful than downregulation by antagonistic cytokines which has so far not resulted in any approved product.

Although, in relation to the magnitude of the potential of cytokines and anti-cytokines, the undeniable success stories to date may appear modest, they do set the scene. In parallel with the exponential boost of basic knowledge initiated by mastering the tools of biotechnology, our understanding of the complex systems we are dealing with has progressed. Diagnostic and pharmacological applications are following closely behind, as well as the capability to monitor the effect of our interventions accurately. As a consequence, an interesting paradigm shift in our approach to many diseases has taken place. Atherosclerosis (von der Thüsen et al. 2003), psoriasis (Barry and Kirby 2004), insulitis (In'tVeld 2011), insulin resistance (Fève and Bastard 2009), and asthma (Wisniewski and Borish 2011) as examples for chronic inflammatory diseases in which interleukins and other cytokines play central roles are becoming therapeutic targets for treatment with biopharmaceuticals. A huge amount of knowledge and experience is available; what is sometimes missing is an integrative view of the many islands of knowledge: "Join the dots to see the greater picture." We must look at the network of immune and other cells or tissues, cytokines, their receptors, and the cascade of events their interaction triggers, and, in addition, how this constellation changes over time. For example, depending on context and time when IL-12 is given, it may induce Th1 cytokines or boost a Th2 response (Biedermann et al. 2006). We need to understand the dynamic processes of acute self-limited and relapsing-remitting progressive diseases consequent to unbalanced cytokine response in order to optimally intervene and reestablish a state of health. The tools are there: polymerase chain reaction, genomics, sequencing, proteomics, and microarrays (see Chaps. 1 and 8). Time is an essential factor as we need to learn to recognize potential or established disease early when intervention is often more effective. Time is lastly also a consideration when treating patients, as a beneficial response can require weeks, months, years, or a lifetime of therapy.

There are still issues in need of solutions: how to manage toxicities of some, mainly the pro-inflammatory cytokines, particularly for their therapeutic use in cancer. A better understanding of the interaction with their receptors, where those receptors are expressed, the dynamics of that expression, and the actions of the cascade their interaction induces is needed. Can we develop a computer model to visualize and help us understand the intricacies of the immune system better? Can biopharmaceuticals be targeted for better efficacy and less toxicity? How much can cell and animal models tell us? Can we predict individuals at risk for certain chronic inflammatory diseases or cancer due to allelic variants of interferon, interleukin, or their receptor genes? Will gene therapy ultimately displace pharmacological replacement or inhibition of cytokines?

In conclusion, interferons as well as interleukins and their antagonists have shown their usefulness in clinical medicine establishing a proof of concept. Ongoing research is looking for innovative approaches to the treatment of various diseases by developing the full potential of this promising biopharmaceutical approach. In addition to their use for therapeutic interventions, future research will also focus the application of interferons and interleukins towards prevention of diseases. Success in terms of marketable products, however, will require hard work, creativity, and persistence.

SELF-ASSESSMENT QUESTIONS

Decide whether each of the statements below is true or false. If you believe a statement is false, explain why.

■ Questions

1. Interferons are defined:
   (a) By the cell type which produces them
   (b) By their anti-inflammatory properties
   (c) By their antiviral activity
   (d) By their protein structure
   (e) By their genetic structure
2. Human interferon alpha:
   (a) Is produced selectively by leukocytes
   (b) Is a virucidal substance
   (c) Triggers antiviral effects in cells expressing appropriate receptors
   (d) Acts on the immune system to booster specific antiviral response
   (e) Comprises twelve subtypes
3. Interleukins are characterized by:
   (a) Their action on target cells
   (b) Their protein structure
   (c) Their genetic structure
   (d) Pro- or anti-inflammatory effect
   (e) Their cell of origin
4. The following interleukins are generally considered to be "pro-inflammatory," i.e., induce and/or be part of a Th1 response:
   (a) The IL-1 family, IL-2, IL-8, and IL-12
   (b) IL-3
   (c) IL-4, IL-5, and IL-9
   (d) IL-10, IL-19, IL-20, IL-22, IL-24, IL-26, IL-28A, IL-28B, and IL-29
   (e) IL-15, IL-16, IL-17, IL-18, IL-22, IL-23, and IL-32
5. Interleukins are:
   (a) Secreted specifically by leukocytes to act on other leukocytes
   (b) Bound to a specific receptor complex to exert their effect
   (c) A family of proteins which regulate the immune response
   (d) Nontoxic products of the body in response to pathogens and other potentially harmful agents
   (e) Long-acting immunomodulators
6. Interferons and interleukins can be toxic; several (patho)physiological containment mechanisms exist to counteract excessive production:
   (a) Soluble receptors
   (b) Binding to cell surface receptors
   (c) Neutralizing antibodies
   (d) Negative feedback mechanisms
   (e) Naturally occurring IL receptor antagonists
7. The following interferons are used as approved therapy:
   (a) IFN-α2
   (b) IFN-β
   (c) IFN-γ
   (d) IFN-ω
   (e) IFN-α8

   Where appropriate, specify some of the indications they are used for.
8. The following interleukins are approved for therapeutic use:
   (a) IL-1
   (b) IL-2

(c) IL-10
(d) IL-11
(e) IL-12
Where appropriate, specify some of the indications they are used for.

9. Protein PEGylation:
(a) Prolongs circulation half-life of the PEGylated protein
(b) Decreases antigenicity of the PEGylated protein
(c) Protects the protein from proteolysis
(d) Is difficult due to the toxicity of polyethylene glycol
(e) Improves the therapeutic efficacy of the PEGylated protein

10. The following PEGylated IFNs and ILs have been approved for therapeutic use:
(a) Interferon-α2
(b) Interferon-β
(c) Interleukin-1
(d) Interleukin-2
(e) Interleukin-12

■ Answers

1. Interferons are defined:
(a) False. Although IFN-α used to be called "leukocyte interferon" and IFN-β "fibroblast interferon" because they were initially produced from buffy coats (leukocytes) infected with Sendai virus and human diploid fibroblasts stimulated with poly(I)-poly(C) or Newcastle disease virus (NDV), respectively, interferons and their Units (IU) are *defined* by their antiviral activity.
(b) False. While they can act as immunomodulators and on occasion have anti-inflammatory properties (e.g., IFN-β for the treatment of multiple sclerosis), they will more often induce a Th1 or pro-inflammatory response. IFN-γ is one of the classical pro-inflammatory markers.
(c) True.
(d and e) False. The full protein and genetic sequences of the different interferons and their subtypes were only defined long after the initial crude IFN mixtures had been tested in the clinic initially against viral diseases and subsequently against cancers.

Today, however, the protein and genetic sequences are necessary to specify an interferon and its purity during the production by biotechnologies. Also new interferons or interleukins will be accepted as such by the Human Genome Nomenclature Committee (HGNC) based on their function and a previously unknown genetic sequence.

2. Human interferon alpha:
(a) False. Interferon alpha is produced by many cell types, including T-cells and B-cells, macrophages, fibroblasts, endothelial cells, and osteoblasts among others.
(b) False. By interacting with their specific heterodimeric receptors on the surface of cells, the interferons initiate a broad and varied array of signals that induce antiviral state.
(c) True.
(d) True.
(e) True. See Table 21.1. Each IFN-α subtype has a distinct antiviral, antiproliferative, and stimulation of cytotoxic activities of NK- and T-cells. To date only one recombinant subtype, IFN-α2, has been predominantly used therapeutically.

3. Interleukins are classified by:
(a and e) False. ILs are characterized by their protein and gene structures registered in the HCGN database (and similar centralized databases). Their names and symbols must be approved by the HGNC.
(b) True.
(c) True.
(d) False. While some ILs can be classified as pro- or anti-inflammatory, this is not what basically defines them.

4. The following interleukins are generally considered to be "anti-inflammatory," i.e., induce and/or be part of a Th2 response:
(a) True.
(b) False. IL-3 is a multicolony-stimulating hematopoietic growth factor which stimulates the generation of hematopoietic progenitors of every lineage.
(c) False. These three interleukins all play a role in the differentiation and activation of basophils and eosinophils leading to a Th2 response.
(d) False. These interleukins are all part of the IL-10 family. However, IL-10, IL-19, and IL-20 are "anti-inflammatory," and IL-22, IL-24, IL-26, IL-28A, IL-28B, and IL-29 are considered "pro-inflammatory."
(e) True.

5. Interleukins are:
(a) False. Interleukins are mainly secreted by leukocytes and primarily affecting growth and differentiation of hematopoietic and immune cells. They are also produced by other normal and malignant cells and are of central importance in the regulation of hematopoiesis, immunity, inflammation, tissue remodeling, and embryonic development.
(b) True.
(c) True.

(d) False. Many interleukins, primarily those with pro-inflammatory function, are intrinsically toxic either directly or indirectly, i.e., through induction of toxic gene products.
(e) False. Interleukins usually have a short circulation time, and their production is regulated by positive and negative feedback loops.

6. Interferons and interleukins can be toxic; several (patho)physiological containment mechanisms exist to counteract excessive production.
   (a) True.
   (b) False. Binding to cell surface receptor is a physiological process and has negligible effect on "circulating" interferons or interleukins.
   (c–e) True.
7. The following interferons are used as approved therapy.
   (a) True. IFN-α (Roferon® A, IntronA®, Infergen®) is indicated for the treatment of chronic hepatitis B and C, Kaposi's sarcoma, renal cell carcinoma, malignant melanoma, carcinoid tumor, multiple myeloma, non-Hodgkin lymphoma, hairy cell leukemia, chronic myelogenous leukemia, thrombocytosis associated with chronic myelogenous leukemia, and other myeloproliferative disorders.
   (b) True. IFN-β (Betaseron®, Betaferon®, Avonex®, Rebif®) is indicated for the treatment of multiple sclerosis.
   (c) True. IFN-γ (Actimmune®) is indicated for the treatment of chronic granulomatous disease and severe malignant osteopetrosis.
   (d) False. IFN-ω has only been studied in vitro and in the nude mouse model where it has shown anticancer activity against several tumor cell lines and transplants.
   (e) False. IFN-α8 has only been studied in various cell lines where it has however consistently shown the most powerful antiviral effect of the subtypes tested.
8. The following interleukins are approved for therapeutic use:
   (a) True. An IL-1 analog/antagonist (Kineret®) is indicated for the treatment of rheumatoid arthritis.
   (b) True. IL-2 (Proleukin®) is indicated for the treatment of adults with metastatic renal cell carcinoma or metastatic melanoma.
   (c) False. Clinical development IL-10 (Tenovil™) as an anti-inflammatory drug for several indications such as psoriasis, Crohn's disease, and rheumatoid arthritis was discontinued in phase III due to insufficient efficacy to warrant further development.
   (d) True. IL-11 (Neumega®) is indicated for the prevention of severe thrombocytopenia following myelosuppressive chemotherapy.
   (e) False. Early clinical trials have been performed in patients with chronic hepatitis C. The program was however discontinued in early phase II due to toxicity.
9. Protein PEGylation:
   (a) True.
   (b) True.
   (c) True.
   (d) False. PEG is inert, nontoxic, non-immunogenic, and in its most common form either linear or branched terminated with hydroxyl groups that can be activated to couple to the desired target protein.
   (e) True.
10. The following PEGylated IFNs and ILs have been approved for therapeutic use:
   (a) True, for chronic hepatitis C and B. Limited clinical trials have also been conducted in renal cell carcinoma, malignant melanoma, and non-Hodgkin lymphoma.
   (b, c, d and e) False although early clinical trials have been conducted with PEGylated IL-2 in RCC and malignant melanoma and pharmacokinetic studies with PEGylated IFN-β in animal models.

REFERENCES

Aggarwal S, Ghilardi N, Xie M-H et al (2003) Interleukin-23 promotes a distinct CD4 T cell activation state characterised by the production of IL-17. J Biol Chem 278:1910–1914

Algranati NE, Sy S, Modi M (1999) A branched methoxy 40-kDa polyethylene glycol (PEG) moiety optimizes the pharmacokinetics of peginterferon alpha-2a (PEG-IFN) and may explain its enhanced efficacy in chronic hepatitis C. Hepatology 40 (Suppl): 190A

Artillo S, Pastore G, Alberti A et al (1998) Double-blind, randomized controlled trial of interleukin-2 for the treatment of chronic hepatitis B. J Med Virol 54:167–172

Asadullah K, Sabat R, Friederich M et al (2004) Interleukin-10: an important immunoregulatory cytokine with major impact on psoriasis. Curr Drug Targets Inflamm Allergy 3:185–192

Azuma YT, Matsuo Y, Kuwamura M, Yancopoulos GD, Valenzuela DM, Murphy AJ, Nakajima H, Karow M, Takeuchi T (2010) Interleukin-19 protects mice from innate-mediated colonic inflammation. Inflamm Bowel Dis 16(6):1017–1028

Baker DP, Lin EY, Lin K et al (2006) N-terminally PEGylated human interferon-beta-1a with improved pharmacokinetic properties and in vivo efficacy in a melanoma angiogenesis model. Bioconjug Chem 17:179–188

Barry J, Kirby B (2004) Novel biologic therapies for psoriasis. Expert Opin Biol Ther 4:975–987

Baud'huin M, Renault R, Charrier C et al (2010) Interleukin-34 is expressed by giant cell tumours of bone and plays a key role in RANKL-induced osteoclastogenesis. J Pathol 221:77–86

Biedermann T, Lametschwandtner G, Tangemann K et al (2006) IL-12 instructs skin homing of human Th2 cells. J Immunol 177:3763–3770

Bilsborough J, Leung DYM, Maurer M et al (2006) IL-31 is associated with cutaneous lymphocyte antigen-positive skin homing T cells in patients with atopic dermatitis. J Allergy Clin Immunol 117:418–425

Cao SG, Zhao QY, Ding ZT et al (1990) Chemical modification of enzyme molecules to improve their characteristics. Ann N Y Acad Sci 613:460–467

Carreño V, Tapia L, Ryff JC et al (1992) Treatment of chronic hepatitis C by continuous subcutaneous infusion of interferon-alpha. J Med Virol 37:215–219

Carreño V, Zeuzem S, Hopf U et al (2000) A phase I/II study of recombinant human interleukin-12 in patients with chronic hepatitis B. J Hepatol 32:317–324

Collison LW, Workman CJ, Kuo TT, Boyd K, Wang Y, Vignali KM, Cross R, Sehy D, Costelloe C, Watson M, Murphy A, McQuillan K, Loscher C, Armstrong ME, Garlanda C, Mantovani A, O'Neill LA, Mills KH, Lynch MA (2008) IL-1F5 mediates anti-inflammatory activity in the brain through induction of IL-4 following interaction with SIGIRR/TIR8. J Neurochem 105(5):1960–1969

Cruikshank WW, Little F (2008) Interleukin-16: the ins and outs of regulating T-cell activation. Crit Rev Immunol 28(6):467–483

Dent P, Yacoub A, Hamed HA et al (2010) The development of MDA-7/IL-24 as a cancer therapeutic. Pharmacol Ther 128(2):375–384

Dinarello CA (2011) Interleukin-1 in the pathogenesis and treatment of inflammatory disease. Blood 117(14):3720–3732

Doherty DH, Rosendahl MS, Smith DJ et al (2005) Site-specific PEGylation of engineered cysteine analogs of recombinant human granulocyte-macrophage colony-stimulating factor. Bioconjug Chem 16:1291–1298

Donnelly RP, Kotenko SV (2010) Interferon-lambda: a new addition to an old family. J Interferon Cytokine Res 30(8):555–564

Donnelly RP, Sheikh F, Dickensheets H, Savan R, Young HA, Walter MR (2010) Interleukin-26: an IL-10-related cytokine produced by th17 cells. Cytokine Growth Factor Rev 21(5):393–401

Du X, Williams DA (1997) Interleukin-11: review of molecular, cell biology, and clinical use. Blood 89:3897–3908

El Bassam S, Pinsonneault S, Kornfeld H et al (2005) Interleukin-16 inhibits interleukin-13 production by allergen-stimulated blood mononuclear cells. Immunology 117:89–96

ExPASy (2012) http://au.expasy.org/uniprot/Q9UBH0 - (Expert Protein Analysis System) proteomics server of the Swiss Institute of Bioinformatics (SIB)

Ferland R, Mulani D, Campbell PK (2001) Evaluation of a sprayable polyethylene glycol adhesion barrier in a porcine efficacy model. Human Reproduction 16(12): 2718–2723

Fève B, Bastard JP (2009) The role of interleukins in insulin resistance and type 2 diabetes mellitus. Nat Rev Endocrinol 5(6):305–311

Fort MM, Cheung J, Yen D et al (2001) IL-25 induces IL-4, IL-5, and IL-13 and Th2-associated pathologies in vivo. Immunity 15:985–995

Fry TJ, Mackall CL (2002) Interleukin-7: from bench to clinic. Blood 99:3892–3904

Gilmour J, Lavender P (2008) Control of IL-4 expression in T-helper 1 and 2 cells. Immunology 124:437–444

Glue P, Fang JWS, Rouzier-Panis R et al (2000) Pegylated interferon-α2b: pharmacokinetics, pharmacodynamics, safety, and preliminary efficacy data. Clin Pharmacol Ther 68:556–567

Greenfeder S, Umland SP, Cuss FM et al (2001) Th2 cytokines and asthma: the role of interleukin-5 in allergic eosinophilic disease. Respir Res 2:71–79

Guzeloglu-Kayisli O, Kayisli UA, Taylor HS (2009) The role of growth factors and cytokines during implantation: endocrine and paracrine interactions. Semin Reprod Med 27(1):62–79

Henney JM (2000) Surgical sealant for lung cancer. JAMA 284:685

HGNC (2012) - www.gene.ucl.ac.uk/nomenclature/ - Gene Families and Grouping – Interferons (IFN) – Interleukins and interleukin receptor genes (IL)

Holtsberg FW, Ensor CM, Steiner MR et al (2002) Poly(ethylene glycol) (PEG) conjugated arginine deiminase: effects of PEG formulations on its pharmacological properties. J Control Release 80:259–271

Hu Y, Shen F, Crellin NK, Ouyang W (2011) The IL-17 pathway as a major therapeutic target in autoimmune disease. Ann N Y Acad Sci 2010(1217):60–76

Huising MO, Cruiswijck CP, Flik G (2006) Phylogeny and evolution of class-I helical cytokines. J Endocrinol 189:1–25

Hukshorn CJ, Saris WHM, Westerterp-Plantenga MS et al (2000) Weekly subcutaneous pegylated recombinant native human leptin (PEG-OB) administration in obese men. J Clin Endocrinol Metab 85:4003–4009

In't VP (2011) Insulitis in human type 1 diabetes. Islets 3(4):131–138

Isaacs A, Lindenmann J (1957) Virus interference I: the interferon. Proc R Soc Ser B 147:258–267

Kamimura D, Ishihara K, Hirano T (2003) IL-6 signal transduction and its physiological roles: the signal orchestration model. Rev Physiol Biochem Pharmacol 149:1–38

Kaushansky K (2005) The molecular mechanisms that control thrombopoiesis. J Clin Invest 115:3339–3345

Kim S-H, Han S-Y, Azam T et al (2005) Interleukin 32: a cytokine and inducer of TNFα. Immunity 22: 131–142

Kotenko SV, Izotova LS, Mirochnitchenko OV et al (2001a) Identification of the functional IL-TIF (IL-22) receptor complex: the IL-10R2 chain (IL-10Rβ) is a common chain of both IL-10 and IL-TIF (IL-22) receptor complexes. J Biol Chem 276:2725–2732

Kotenko SV, Izotova LS, Mirochnitchenko OV et al (2001b) Identification, cloning and characterization of a novel soluble receptor which binds IL-22, and neutralizes its activity. J Immunol 166:7096–7103

Kristiansen OF, Mandrup-Poulsen T (2005) Interleukin-6 and diabetes: the good, the bad or the indifferent. Diabetes 54(Suppl 2):S114–S124

Kurschner C, Yuzaki M (1999) Neuronal interleukin-16 (NIL-16): a dual function PDZ domain protein. J Neurosci 19:7770–7780

Kurtzke JF (1983) Rating neurologic impairment in multiple sclerosis: an expanded disability status scale (EDSS). Neurology 33:1444–1452

Kuter DJ, Begley CG (2002) Recombinant human thrombopoietin: basic biology and evaluation of clinical studies. Blood 100:3457–3469

LaFleur DW, Nardelli B, Tsareva T et al (2001) Interferon-κ, a novel type I interferon expressed in human keratinocytes. J Biol Chem 276:39765–39771

Larousserie F, Charlot P, Bardel E et al (2006) Differential effects of IL-27 on human B cell subsets. J Immunol 176(10):5890–5897

Lee S, Kaneko H, Sekigawa I et al (1998) Circulating interleukin-16 in systemic lupus erythematosus. Br J Rheumatol 37:1334–1337

Leonard P, Sur S (2003) Interleukin-12: potential role in asthma therapy. BioDrugs 17:1–7

Liao W, Lin JX, Leonard WJ (2011) IL-2 family cytokines: new insights into the complex roles of IL-2 as a broad regulator of T helper cell differentiation. Curr Opin Immunol 23(5):598–604

Lin H, Lee E, Hestir K et al (2008) Discovery of a cytokine and its receptor by functional screening of the extracellular proteome. Science 320(5877):807–811

Liu B, Nivick D, Kim SH, Rubinstein M (2000) Production of a biologically active human interleukin 18 requires its prior synthesis as PRO-IL-18. Cytokine 12(10): 1519–1525

Ludwig CU, Ludwig-Habemann R, Obrist R et al (1990) Improved tolerance of interferon alpha-2a by continuous subcutaneous infusion. Onkologie 13:117–122

Lutfalla G, Crollius HR, Stange-Thomman N et al (2003) Comparative genomic analysis reveals independent expansion of a lineage-specific gene family in vertebrates: the class II cytokine receptors and their ligands in mammals and fish. BMC Genomics 4:29–44

Lyman GH (2005) Pegfilgrastim: a granulocyte colony-stimulating factor with sustained duration of action. Expert Opin Biol Ther 5:1635–1646

Malek TR, Castro I (2010) Interleukin-2 receptor signaling: at the interface between tolerance and immunity. Immunity 33(2):153–165

Marrakchi S, Guigue P, Renshaw BL et al (2011) Interleukin-36–receptor antagonist deficiency and generalized pustular psoriasis. N Engl J Med 365:620–628

Martinez-Moczygemba M, Huston DP (2003) Biology of common beta receptor-signaling cytokines: IL-3, IL-5, and GM-CSF. J Allergy Clin Immunology 112(4):653–665

McLaren JE, Michael DR, Salter RC, Ashlin TJ, Calter CJ, Miller AM, Liew FY, Ramji DP (2010) IL-33 reduces macrophage foam cell formation. J Immunol 185(2):1222–1229

Metcalf D (2008) Hematopoietic cytokines. Blood 111(2): 485–491

Motzer RJ, Ashok R, Thompson J et al (2002) Phase II trial of branched peginterferon-α 2a (40 kDa) for patients with advanced renal cell carcinoma. Ann Oncol 13: 1799–1805

Ng EWM, Shima DT, Calias P et al (2006) Pegaptanib, a targeted anti-VEGF aptamer for ocular vascular disease. Nat Rev Drug Discov 5:123–132

Noelle RJ, Nowak EC (2010) Cellular source and immune functions of interleukin-9. Nat Rev Immunol 10:683–687

Nold MF, Nold-Petry CA, Zepp JA, Palmer BE, Bufler P, Dinarello CA (2010) IL-37 is a fundamental inhibitor of innate immunity. Nat Immunol 11:1014–1022

Pardo M, Castillo I, Oliva H et al (1997) A pilot study of recombinant interleukin-2 for treatment of chronic hepatitis C. Hepatology 26(5):1318–1321

Pettit DK, Bonnert TP, Eisenman J et al (1997) Structure-function studies of interleukin 15 using site-specific mutagenesis, polyethylene glycol conjugation, and homology modeling. J Biol Chem 272:2312–2318

Platanias L (2005) Mechanism of type I- and type II-interferon mediated signaling. Nat Rev Immunol 5:375–386

Pockros P, Patel K, O'Brien CB (2003) A multicenter study of recombinant human interleukin-12 for the treatment of chronic hepatitis C infection in patients with non-responsiveness to previous therapy. Hepatology 37:1368–1374

Pushparay PN, Tay HK, H'ng SC et al (2009) The cytokine interleukin-33 mediates anaphylactic shock. Proc Natl Acad Sci U S A 106(24):9773–9778

Reddy KR, Modi WM, Pedder S (2002) Use of peginterferon alfa-2a (40 KD) (Pegasys®) for the treatment of hepatitis C. Adv Drug Deliv Rev 54:571–586

Remick DG (2005) Interleukin-8. Crit Care Med 33(Suppl): S466–S467

Ross RJM, Leung KC, Maamra M et al (2001) Binding and functional studies with the growth hormone receptor antagonist, B2036-PEG (Pegvisomant), reveal effects of pegylation and evidence that it binds to a receptor dimer. J Clin Endocrinol Metab 86:1716–1723

Ryff JC (1996) Both cytokines and their antagonists have a place in clinical medicine. Eur Cytokine Netw 7:437 (Abstract 40)

Schellekens H (2006) Erythropoiesis-stimulating Agents — Present and Future. Business Briefing: European Endocrine Review 2006. Touch Briefings Publishers, London

Schmitz J, Owyang A, Oldham E et al (2005) IL-33, an interleukin-1-like cytokine that signals via the IL-1 receptor-related protein ST2 and induces T helper type 2-associated cytokines. Immunity 23:479–490

Steinman L (2007) A brief history of TH17, the first major revision in the TH1/TH2 hypothesis of T cell-mediated tissue damage. Nat Med 13(2):139–145

Stumhofer JS, Laurence A, Wilson HW et al (2006) Interleukin 27 negatively regulates the development of interleukin 17–producing T helper cells during chronic inflammation of the central nervous system. Nat Immunol 7:937–945

Talpaz M, Rakhit A, Rittweger K et al (2005) Phase I evaluation of a 40-kDa branched-chain long-acting pegylated IFN-a-2a with and without cytarabine in patients with chronic myelogenous leukemia. Clin Cancer Res 11:6247–6455

Towne JE, Garka KE, Renshaw BR, Virca GD, Sims JE (2004) Interleukin (IL)-F6m IL-1F8, and IL-1F9 signal through IL-Rrp2 and IL-1RacP to activate the pathway leading to NF-κB and MAPKs. J Biol Chem 279:13677–13688

Towne JE, Blair R, Renshaw BR, Douangpanya J, Lipsky BP, Shen M, Gabel CA, John E, Sims JE (2011) Interleukin-36 (IL-36) ligands require processing for full agonist (IL-36α, IL-36β and IL-36γ) or antagonist (IL-36Ra) activity. J Biol Chem 286:42594–42602

Trinchieri G (2003) Interleukin-12 and the regulation of innate resistance and adaptive immunity – review. Nat Rev Immunol 3:133–146

Trøseid M, Seljeflot I, Amesen H (2010) The role of interleukin-18 in the metabolic syndrome. Cardiovasc Diabetol 9:11–19

Tsunoda S, Ishikawa T, Watanabe M et al (2001) Selective enhancement of thrombopoietic activity of PEGylated interleukin 6 by a simple procedure using a reversible amino-protective reagent. Br J Haematol 112:181–188

von der Thüsen J, Kuiper J, van Berkel TJC, Biessen EAL (2003) Interleukins in atherosclerosis: molecular pathways and therapeutic potential. Pharmacol Rev 55: 133–166

Wacker A, Linton D, Hitchen P et al (2002) N-linked glycosylation in campylobacter jejuni and its functional transfer into E. coli. Science 298:1790–1793

Waldmann TA (2006) The biology of interleukin-2 and interleukin-15: implications for cancer therapy and vaccine design. Nat Rev Immunol 6:595–601

Wang M, Liang P (2005) Interleukin-24 and its receptors. Immunology 114:166–170

White UA, Stephens JM (2011) The gp130 receptor cytokine family: regulators of adipocyte development and function. Curr Pharm Des 17(4):340–346

Wills RJ (1990) Clinical pharmacokinetics of interferons. Clin Pharmacokinet 19:390–399

Wills-Karp M (2004) Interleukin-13 in asthma pathogenesis. Immunol Rev 202:175–190

Wisniewski JA, Borish L (2011) Novel cytokines and cytokine-producing T cells in allergic disorders. Allergy Asthma Proc 32(2):83–94

Xu W (2004) Interleukin-20. Int Immunopharmacol 4:527–633

Yamamoto Y, Tsutsumi Y, Yoshioka Y et al (2003) Site-specific PEGylation of a lysine-deficient TNF-α with full bioactivity. Nat Biotechnol 21:546–552

Yi JS, Cox MA, Zajac AJ (2010) Interleukin-21: a multifunctional regulator of immunity to infections. Microbes Infect 12(14–15):1111–1119

Zeuzem S, Hopf U, Carreno V et al (1999) A phase I/II study of recombinant human interleukin-12 in patients with chronic hepatitis C. Hepatology 29:1280–1286

Zeuzem S, Welsch C, Herrmann E (2003) Pharmacokinetics of peginterferons. Semin Liver Dis 23(Suppl 1):23–28

FURTHER READING

Reviews which summarize the referenced subject in more detail:

INTERFERONS

Meager A (ed) (2006) The interferons: characterization and application. WILEY-VCH, Weinheim. ISBN 9783527311804 Online ISBN: 3-527-31180-7

Pestka S, Krause CD, Walter M (2004a) Interferons, interferon-like cytokines, and their receptors. Immunol Rev 202:8–32

Pestka S (1981a) Interferons. In: Pestka S (ed) Methods in enzymology, vol 78. Academic Press, New York, p 632

Pestka S (1981b) Interferons. In: Pestka S (ed) Methods in enzymology, vol 79. Academic Press, New York, p 677

Pestka S (1986) Interferons. In: Pestka S (ed) Methods in enzymology, vol 119. Academic Press, New York, p 845

Special issue: the neoclassical pathways of interferon signaling (2005) J Interferon Cytokine Res. 25:731–811

INTERLEUKINS

Pestka S, Krause CD, Sarkar C et al (2004b) IL-10 and related cytokines and receptors. Annu Rev Immunol 22:929–979

Sigal LH (2004a) Interleukins of current clinical relevance (part I). J Clin Rheumatol 10:353–359

Sigal LH (2004b) Interleukins of current clinical relevance (part II). J Clin Rheumatol 11:34–39

PEGYLATION

Bailon P, Palleroni A, Schaffer CA et al (2001) Rational design of a potent, long-lasting form of interferon: a 40 kDa branched polyethylene glycol. Conjugated interferon α-2a for the treatment of hepatitis C. Bioconjug Chem 12:195–202

Roberts MJ, Bentley MD, Harris JM (2002) Chemistry for peptide and protein PEGylation. Adv Drug Deliv Rev 54:459–476

Wang Y-S, Youngster S, Grace M et al (2002) Structural and biological characterization of pegylated recombinant interferon alpha-2b and its therapeutic implications. Adv Drug Deliv Rev 54:547–570

22 Vaccines

Wim Jiskoot, Gideon F.A. Kersten, and Enrico Mastrobattista

INTRODUCTION

Since vaccination was documented by Edward Jenner in 1798, it has become the most successful means of preventing infectious diseases, saving millions of lives every year. However, application of vaccines is currently not limited to the prevention of infectious diseases. Vaccines in the pipeline include anti-drug abuse vaccines (nicotine, cocaine) and vaccines against allergies, cancer, and Alzheimer's disease.

Modern biotechnology has an enormous impact on current vaccine development. The elucidation of the molecular structures of pathogens and the tremendous progress made in immunology as well as developments in proteomics and bioinformatics have led to the identification of protective antigens and ways to deliver them. Together with technological advances, this has caused a move from empirical vaccine development to more rational approaches. A major goal of modern vaccine technology is to fulfill all requirements of the ideal vaccine as summarized in Fig. 22.1, by expressing antigen epitopes (= the smallest molecular structures recognized by the immune system) and/or isolating those antigens that confer an effective immune response and eliminating structures that cause deleterious effects. Thus, better-defined products can be obtained, resulting in improved safety. In addition, modern methodologies may provide simpler production processes for selected vaccine components.

In the following section, immunological principles that are important for vaccine design are summarized. Subsequently, classical vaccines which are not a result of modern genetic or chemical engineering technologies will be addressed. Classical and modern vaccines are listed in Table 22.1. Current strategies used in the development and manufacture of new vaccines are discussed in the section "Modern Vaccine Technologies." It is not our intent to provide a comprehensive review. Rather, we will explain modern approaches to vaccine development and illustrate these approaches with representative examples. In the last section, pharmaceutical aspects of vaccines are dealt with.

The ideal vaccine

- Is 100 % efficient in all individuals of any age
- Provides lifelong protection after single administration
- Does not evoke an adverse reaction
- Is stable under various conditions (temperature, light, transportation)
- Is easy to administer, preferably orally
- Is available in unlimited quantities
- Is cheap

Figure 22.1 ■ Characteristics of the (hypothetical) ideal vaccine.

W. Jiskoot, Ph.D. (✉)
Division of Drug Delivery Technology,
Leiden/Amsterdam Center for Drug Research,
Leiden University, Einsteinweg 55,
Leiden 2333 CC, The Netherlands
e-mail: w.jiskoot@lacdr.leidenuniv.nl

G.F.A. Kersten, Ph.D.
Institue for Translational Vaccinology, Bilthoven,
The Netherlands

E. Mastrobattista, Ph.D.
Department of Pharmaceutics,
Utrecht Institute for Pharmaceutical Sciences,
Utrecht University, Utrecht, The Netherlands

D.J.A. Crommelin, R.D. Sindelar, and B. Meibohm (eds.), *Pharmaceutical Biotechnology*,
DOI 10.1007/978-1-4614-6486-0_22,

Category	Technology	Live/nonliving	Characteristics
Attenuated vaccines	Classical	Live	Bacteria or viruses attenuated in culture; empirically developed
Inactivated vaccines	Classical	Nonliving	Heat-inactivated or chemically inactivated bacteria or viruses; empirically developed
Subunit vaccines	Classical	Nonliving	Extracts of pathogens; combination of purified proteins with killed suspension; purified single components (proteins, polysaccharides); combination of purified components with adjuvant; purified components in a suitable presentation form; polysaccharide-protein conjugates
Genetically improved live vaccines	Modern	Live	Genetically attenuated microorganisms; live viral or bacterial vectors
Genetically improved subunit vaccines	Modern	Nonliving	Proteins expressed in host cells; recombinant protein/peptide vaccines
Recombinant subunit vaccines identified by reverse vaccinology	Modern	Nonliving	Recombinant antigenic proteins obtained from the genomic sequence of the pathogen
Synthetic peptide-based vaccines	Modern	Nonliving	Linear or cyclic peptides; multiple antigen peptides; peptide-protein conjugates
Nucleic acid-based vaccines	Modern	Nonliving	DNA or mRNA coding for antigen

Table 22.1 ■ Categories of classical vaccines and vaccines obtained by modern technologies.

IMMUNOLOGICAL PRINCIPLES

■ Introduction

After a natural infection, the human immune system in most cases launches an immunological response to the particular pathogen. After recovery from the disease, the immunological response indeed protects the affected individual from that disease, in the ideal case forever. This phenomenon is called specific immunity and is caused by the presence of circulating antibodies, cytotoxic cells, and memory cells. Memory cells become active when the same type of antigenic material enters the body on a later occasion. Unlike the primary response after the first infection, the response after repeated infection is very fast and usually sufficiently strong to prevent reoccurrence of the disease.

The principle of vaccination is mimicking an infection in such a way that the natural specific defense mechanism of the host against the pathogen will be activated, but the host will remain free of the disease that normally results from a natural infection. This is effectuated by administration of antigenic components that consist of, are derived from, or are related to the pathogen. The success of vaccination relies on the induction of a protective immune response and a long-lasting immunological memory. Vaccination is also referred to as active immunization, because the host's immune system is activated to respond to the "infection" through humoral and cellular immune responses, resulting in adaptive immunity against the particular pathogen. The immune response is generally highly specific: it discriminates not only between pathogen species but often also between different strains within one species (e.g., strains of meningococci, poliovirus, influenza virus). Albeit sometimes a hurdle for vaccine developers, this high specificity of the immune system allows an almost perfect balance between response to foreign antigens and tolerance with respect to self-antigens. Apart from active immunization, administration of specific antibodies can be utilized for short-lived immunological protection of the host. This is termed passive immunization (Fig. 22.2).

Traditionally, active immunization has mainly served to prevent infectious diseases, whereas passive immunization has been applied for both prevention and therapy of infectious diseases. Through recent developments new potential applications of vaccines for active immunization have emerged, such as the prevention of other diseases than infectious diseases (e.g., cancer) and for the treatment of substance abuse (e.g., nicotine addiction). Such vaccines are referred to as therapeutic vaccines. The difference between passive and active immunization is outlined in Fig. 22.2. Since antibody preparations for passive immunization do not fall under the strict definition of a vaccine, they are not further discussed here.

■ Active Immunization: Generation of an Immune Response

The generation of an immune reaction against a pathogen by vaccination follows several distinct steps that should ultimately lead to long-lasting protection against the pathogen through memory cells. These steps are

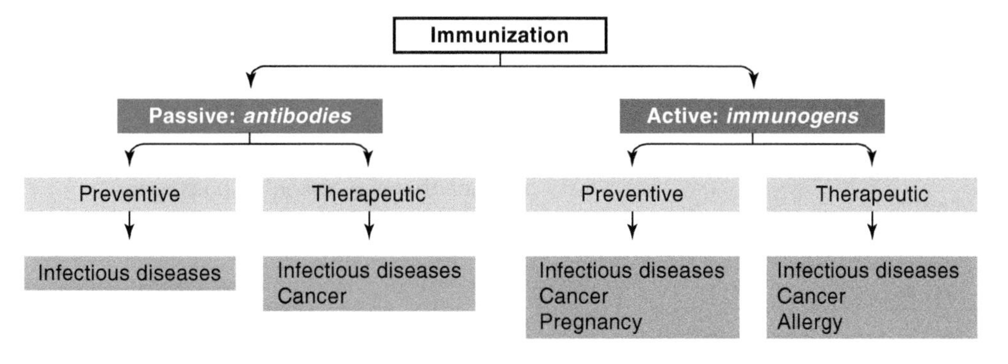

Figure 22.2 ■ Scheme of active immunization (= vaccination) and passive immunization and examples of their fields of application.

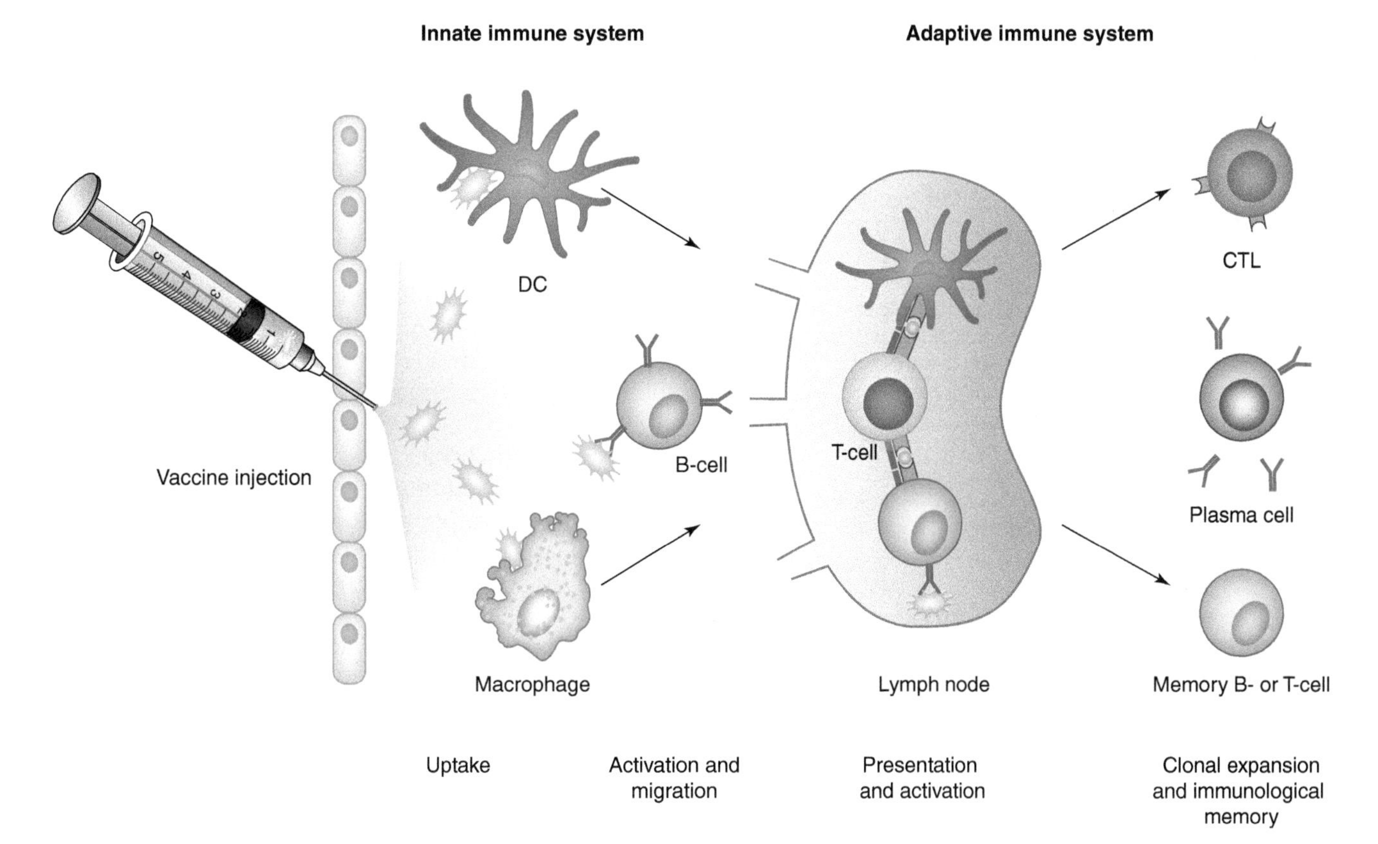

Figure 22.3 ■ Overview of the steps leading to immunization after administration of a vaccine. Upon subcutaneous or intramuscular administration, the vaccine components are taken up by phagocytic cells such as macrophages and dendritic cells (*DCs*) that reside in the peripheral tissue and express pattern recognition receptors (PRRs) that recognize pathogen-associated molecular patterns (PAMPs). Professional antigen-presenting cells (APCs) that have taken up antigens become activated and start migrating towards nearby lymph nodes. Inside the lymph nodes, the antigen processed by the APCs is presented to lymphocytes, which, when recognizing the antigen and receiving the appropriate co-stimulatory signals, become activated. These antigen-specific B and T lymphocytes clonally expand to produce multiple progenitors recognizing the same antigen. In addition, memory B and T cells are formed that provide long-term (sometimes lifelong) protection against infection with the pathogen.

uptake of the vaccine (consisting of either the entire pathogen or antigenic components thereof) by phagocytic cells, activation and migration of professional antigen-presenting cells (APCs) from infected tissue to peripheral lymphoid organs, antigen presentation to T lymphocytes, and finally activation (or inhibition) of T and B lymphocytes. The entire process is illustrated in Fig. 22.3. Below we describe the successive steps leading to an immune response to a pathogen, which are important for the design of vaccines against the pathogen.

■ Innate Response

Every immune reaction against a pathogen starts with activation of the innate immune system. This is a nonspecific fast response against antigens. Important constituents of the innate system are antigen sampling

PRR	PAMP
TLR-1	Triacyl lipoproteins
TLR-2	Peptidoglycans Lipoproteins Lipoarabinomannan Zymosan
TLR-3	Viral dsRNA
TLR-4	Lipopolysaccharide, lipid A, Taxol
TLR-5	Flagellin
TLR-6	Diacyl lipoproteins
TLR-7	Small, synthetic compounds, ssRNA
TLR-8	Small, synthetic compounds
TLR-9	Unmethylated CpG DNA
TLR-10	Unknown
TLR-11	Components from uropathogenic bacteria
Scavenger receptors	Polyanionic ligands
C-type lectin receptors	Sulfated sugars and mannose-, fucose-, and galactose-modified polysaccharides and proteins
NOD-1, NOD-2	Peptidoglycans
Type 3 complement receptors	Zymosan particles, β-glucan

Adapted from Pashine et al. (2005)

Table 22.2 ■ Examples of pattern recognition receptors (PRRs) and their ligands (PAMPs).

cells like macrophages and dendritic cells (DCs). The innate response does not lead to immunological memory. Phagocytic cells sense conserved microbial structures called pathogen-associated molecular patterns (PAMPs). They do this via pattern recognition receptors (PRRs) on the cell surface (for bacterial PAMPs) or in the cytoplasm (for viral PAMPs). Examples of PRRs are toll-like receptors (TLRs), scavenger receptors, and C-type lectins (Table 22.2). TLRs consist of a family of receptors, with each member recognizing different patterns of pathogens (Kawai and Akira 2010). TLRs can be found on many cells including macrophages and DCs. DCs can also engulf materials from their extracellular environment by a receptor-independent process called pinocytosis.

■ Activation and Migration

Besides mediating uptake of antigenic material from the surrounding tissue, PRRs also play an important role in triggering the cytokine network that will eventually influence the type of adaptive immune response that will be evoked against the pathogen. The phagocytic cells that have taken up pathogens from the infected tissue become activated and start to produce proinflammatory cytokines such as interleukin-1β, interleukin-6, and tumor necrosis factor-α as well as chemokines. The chemokines recruit more phagocytic cells such as neutrophils and monocytes to the infection site, whereas the proinflammatory cytokines induce fever and the production of acute-phase response proteins that can opsonize pathogens.

Most phagocytic cells, including DCs and macrophages, and also B cells can serve as APCs to present processed antigenic determinants to lymphocytes in the peripheral lymphoid organs. For instance, DCs that have taken up antigens from infected tissue become activated and migrate via the afferent lymphatic vessels towards nearby lymph nodes where the encounter with pathogen-specific lymphocytes can take place.

■ Antigen Presentation and Lymphocyte Activation

The peripheral lymphoid organs are the primary meeting place between cells of the innate immune system (APCs) and cells of the adaptive immune system (T cells and B cells). Upon interaction with APCs, pathogen-specific T cells and B cells will be activated, provided that they acquire the appropriate signals from the APCs. Besides antigen-specific binding via their antigen receptors, lymphocytes require co-stimulatory signals via interaction of accessory and co-stimulatory molecules between lymphocytes and APCs. This cell-cell interaction is essential for proper stimulation of lymphocytes, and without those accessory signals, antigen-specific T cells may become anergic. Lymphocytes receiving the appropriate signals for activation will clonally expand and generate multiple progenitors all recognizing the same antigen. Clonal expansion is a typical feature of the adaptive immune system, which will be discussed in more detail below.

■ The Adaptive Immune System

The adaptive immune system is involved in elimination of pathogens in the late phase of infection and in the generation of immunological memory. It comprises B and T lymphocytes both bearing antigen-specific receptors. The adaptive immune system can be divided into humoral immunity and cell-mediated immunity (CMI) (see Fig. 22.4 and Table 22.3). The humoral response results in antibody formation (but contains cell-mediated events, see Fig. 22.4, panels a, b); CMI results in the generation of cytotoxic cells (see Fig. 22.4, panels a, c). The action of antibodies and T cells is dependent on accessory factors, some of which are mentioned in Table 22.3. In general, after infection with a pathogen or a protective vaccine, both humoral and cellular responses are generated. This indicates that both are needed for efficient protection. The balance between humoral and cellular responses, however, can differ widely between pathogens and is dependent on how the pathogen is presented to the adaptive immune

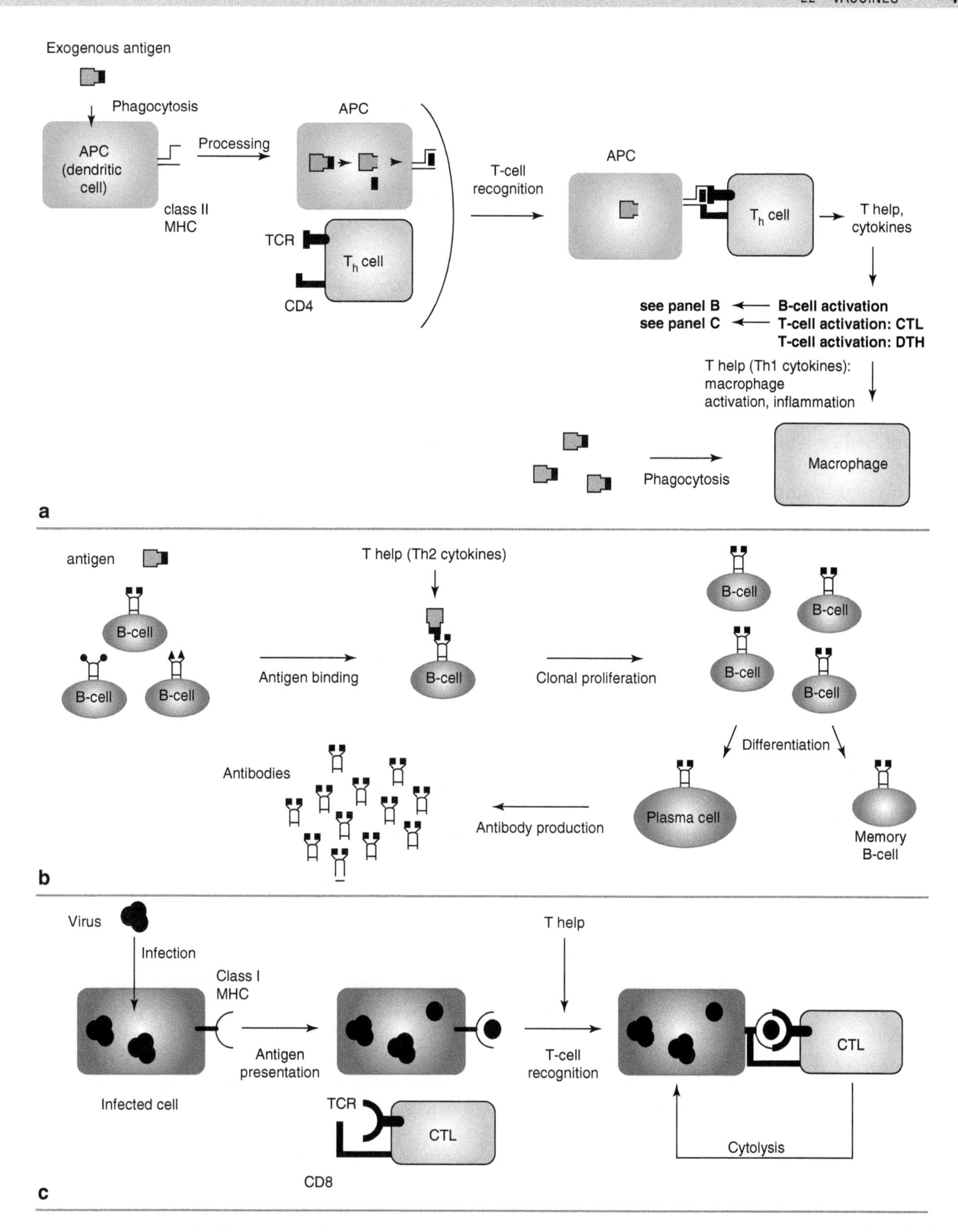

Figure 22.4 ■ Schematic representation of antigen-dependent immune responses. (**a**) Activation of T-helper cells (Th-cells). An antigen-presenting cell (*APC*), e.g., a dendritic cell, phagocytozes exogenous antigens (bacteria or soluble antigens) and degrades them partially. Antigen fragments are presented by MHC class II molecules to a CD4-positive Th-cell; the MHC-antigen complex on the APC is recognized by the T-cell receptor (*TCR*) and CD4 molecules on the Th-cell. The APC-Th-cell interaction leads to activation of the Th-cell. The activated Th-cell produces cytokines, resulting in the activation of macrophages (Th1 help), B cells (Th2 help; panel *b*), or cytotoxic T cells (panel *c*). (**b**) Antibody production. The presence of antigen and Th2-type cytokines causes proliferation and differentiation of B cells. Only B cells specific for the antigen become activated. The B cells, now called plasma cells, produce and secrete large amounts of antibody. Some B cells differentiate into memory cells. (**c**) Activation of cytotoxic T lymphocytes (*CTLs*). CTLs recognize nonself antigens expressed by MHC class I molecules on the surface of virally infected cells or tumor cells. Cytolytic proteins are produced by the CTL upon interaction with the target cell.

Immune response	Immune product	Accessory factors	Infectious agents
Humoral	IgG	Complement, neutrophils	Bacteria and viruses
	IgA	Alternative complement pathway	Microorganisms causing respiratory and enteric infections
	IgM	Complement, macrophages	(Encapsulated) bacteria
	IgE	Mast cells	Parasites
Cell mediated	CTL	Cytolytic proteins	Viruses and mycobacteria
	T_{DTH}	Macrophages	Viruses, mycobacteria, treponema (syphilis), fungi

Table 22.3 ■ Important immune products protecting against infectious diseases.

system by APCs. This may have consequences for the design of a particular vaccine (see "Vaccine Design in Relation with the Immune Response").

Antibodies are the typical representatives of humoral immunity. An antibody belongs to one of four different immunoglobulin classes (IgM, IgG, IgA, or IgE) (cf. Chap. 7). Upon immunization, B cells expressing specific antibodies on their cell surface (representing a fifth immunoglobulin class, IgD) bind intact antigen and are activated. The surface-bound antibodies bind specific epitopes of the pathogen, and in close cooperation with T-helper cells (T_h-cells), the B cell becomes activated eventually resulting in massive clonal proliferation. The proliferated B cells are called plasma cells and excrete large amounts of soluble antibodies (Fig. 22.4 panel b). Antibodies are able to prevent infection or disease by several mechanisms:

1. Binding of antibody covers the antigen with Fc (constant fragment), the "rear end" of immunoglobulins. Phagocytic cells, like macrophages, express surface receptors for Fc. This allows targeting of the opsonized (antibody-coated) antigen to these cells, followed by enhanced phagocytosis.
2. Immune complexes (i.e., antibodies bound to target antigens) can activate complement, a system of proteins which then becomes cytolytic to bacteria, enveloped viruses, or infected cells.
3. Phagocytic cells may express receptors for complement factors associated with immune complexes. Binding of these activated complement factors enhances phagocytosis.
4. Viruses can be neutralized by antibodies through binding at or near receptor binding sites on the virus surface. This may prevent binding to and entry into the host cell.

Antibodies are effective against certain but not all infectious microorganisms. They may have limited value when CMI is the major protective mechanism. Of the cell types that are known to exhibit cytotoxicity, two are antigen sensitized. Because of their specificity, they are of special importance with respect to vaccine design:

1. Cytotoxic T lymphocytes (CTLs) react with target cells and kill them by release of cytolytic proteins like perforin. Target cells express nonself antigens like viral proteins or tumor antigens, by which they are identified. CTL responses, as antibody responses, are highly specific.
2. T cells involved in delayed-type hypersensitivity (T_{DTH}) are able to kill target cells as CTLs do but also have helper (T_{h1}-type, see below) functions that enable them to activate macrophages.

Other less specific cells involved in cytotoxic immune responses are natural killer cells (NK cells). They play a role in antibody-dependent cellular cytotoxicity (ADCC). NK cells recognize opsonized (antibody-coated) cells with their Fc receptors.

Besides plasma cells and cytotoxic cells, in many cases, memory B and T cells develop. Memory B cells do not produce soluble antibody, but on repeated antigen contact, their response time to develop into antibody-excreting plasma cells is shorter compared to naïve B cells.

The occurrence of different types of immune response to vaccines is the result of differences in antigen processing of the vaccine by APCs and, as a result, in the activation of T_h-cells (Figs. 22.3 and 22.4). Major histocompatibility complex (MHC) molecules play an important role in the presentation of processed antigens to T cells. Most cells expose MHC class I molecules and some also MHC class II molecules on their surface.

APCs carrying class II molecules process soluble, exogenous (extracellular) proteins or more complicated structures such as microorganisms (see Fig. 22.4, panel a). After their endocytosis, the proteins are subject to limited proteolysis before they return as peptides to the surface of the APC in combination with the class II molecules for presentation to a T-cell receptor of CD4-positive T_h-cells. The T_h-cells provide type 2 help necessary for the effector function of B cells. This type 2 help is characterized by the lymphokine pattern produced: interleukin 4 (IL-4), IL-5, IL-6, IL-10, and IL-13. These lymphokines trigger B cells, which eventually results in the production of IgM and IgG antibodies.

Cells carrying MHC class I molecules process endogenous (intracellularly produced) antigens like viral and tumor antigens and present them in combination with class I molecules on the cell surface (see Fig. 22.4, panel c). The class I-antigen combination on the APC is recognized by the T-cell receptor of CD8-positive CTLs. Th-cells provide help for the CTLs. For the induction of CMI (Fig. 22.4, panels a, c), type 1 help is needed (production of IL-2 and IL-12, interferon-γ, and tumor necrosis factor). Th-cells are CD4 positive, regardless whether they have Th1 or Th2 functions. There is increasing evidence that the Th1/Th2 balance is an important immunological parameter since some diseases coincide with Th1 (autoimmunity)- or Th2 (allergy)-type responses. Other T-helper cell phenotypes have been identified. Some play a role in autoimmune disease (e.g., T_{h17}-cells) or suppression of the immune response (e.g., T_{reg}-cells).

■ Vaccine Design in Relation with the Immune Response

For the rational design of a new vaccine, understanding of the mechanisms of the protective immunity to the pathogen against which the vaccine is developed is crucial. For instance, to prevent tetanus a high blood titer of antibody against tetanus toxin is required; in mycobacterial diseases such as tuberculosis, a macrophage-activating CMI is most effective; in case of an influenza virus infection, CTLs probably play a significant role besides antibodies. Importantly, the immune effector mechanisms triggered by a vaccine and, hence, the success of immunization depend not only on the nature of the protective components but also on their presentation form, the presence of adjuvants, and the route of administration.

The presentation form of the vaccine is one of the determinants that influence the extent and type of immune response that will be evoked (Pashine et al. 2005; Pulendran and Ahmed 2006). DCs and other APCs play a pivotal role in how the antigenic determinants of a vaccine will be processed and presented to T cells in the peripheral lymphoid organs. Through various PRRs, DCs are more or less able to "sense" the type of pathogen that is encountered. This determines the set of co-stimulatory signals and proinflammatory cytokines that will be generated by APCs when presenting the antigen to Th-cells in the peripheral lymphoid organs. For instance, pathogens or vaccines containing lipoproteins or peptidoglycans will trigger DCs via TLR-2, which predominantly generates a T_{h2} response, whereas stimulation of DCs through TLR-3, TLR-4, TLR-5, or TLR-8 is known to yield robust T_{h1} responses. Therefore, vaccines should be formulated in such a way that the appropriate T_h response will be triggered. This can be done by presenting the antigen in its native format, as is the case for the classical vaccines, or by adding adjuvants that stimulate the desired response (see below).

The response by B cells is dependent upon the nature of the antigen and two types of antigens can be distinguished:

1. Thymus-independent antigens include certain linear antigens that are not readily degraded in the body and have a repeating determinant, such as bacterial polysaccharides. They are able to stimulate B cells without the T_h-cell involvement. Thymus-independent antigens do not induce immunological memory.
2. Thymus-dependent antigens provoke little or no antibody response in T-cell-depleted animals. Proteins are the typical representatives of thymus-dependent antigens. A prerequisite for thymus dependency is that a physical linkage exists between the sites recognized by B cells and those by T_h-cells. When a thymus-independent antigen is coupled to a carrier protein containing T_h-epitopes, it becomes thymus dependent. As a result, these conjugates are able to induce memory.

When the antigen is a protein, the epitopes can be continuous or discontinuous. Continuous epitopes involve linear peptide sequences (usually consisting of up to ten amino acid residues) of the protein (see Fig. 22.5, panel a). Discontinuous epitopes comprise amino acid residues sometimes far apart in the primary sequence, which are brought together through the unique folding of the protein (see Fig. 22.5, panel b). Antibody recognition of B-cell epitopes, whether continuous or discontinuous, is usually dependent on the conformation (= three-dimensional structure). T-cell epitopes, on the other hand, are continuous peptide sequences, the conformation of which does not seem to play a role in T-cell recognition.

■ Route of Administration

The immunological response to a vaccine is dependent on the route of administration. Most current vaccines are administered intramuscularly or subcutaneously. Parenteral immunization usually induces systemic immunity. However, mucosal (e.g., oral, intranasal, or intravaginal) immunization may be preferred, because it may induce both mucosal and systemic immunity. Mucosal surfaces are the common entrance of many pathogens, and the induction of a mucosal secretory IgA response may prevent the attachment and entry of pathogens into the host. For example, antibodies against cholera need to be in the gut lumen to inhibit adherence to and colonization of the intestinal wall. Also, orally administered *Salmonella typhi* not only invades the mucosal lining of the gut but also infects cells of the phagocytic system throughout the body,

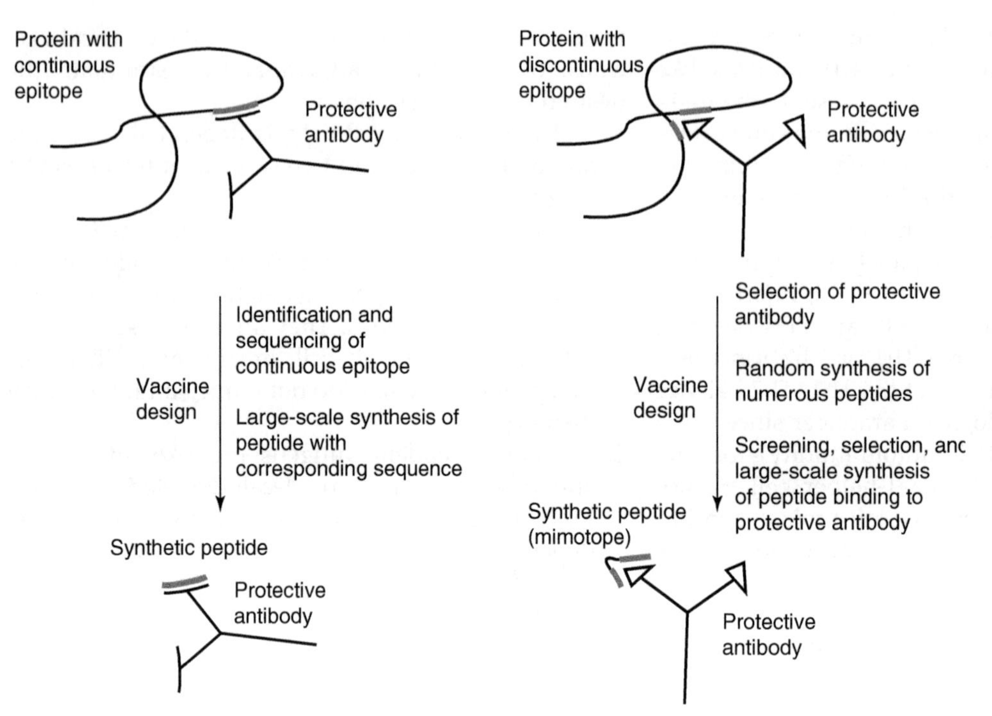

Figure 22.5 ■ Two approaches for the design of synthetic peptide vaccines. Panel *a*: identification and sequencing of a continuous epitope on an immunogenic protein is followed by the synthesis of peptides with the amino acid sequence corresponding to that of the epitope. Panel *b*: synthesis of peptides mimicking discontinuous epitopes that are determined by the three-dimensional structure of the immunogenic protein; a peptide that strongly binds to a protective antibody recognizing the discontinuous epitope is selected. The peptide (mimotope) does not necessarily contain the exact amino acid sequence of the constituent fragments that form the epitope.

thereby stimulating the production of both secretory and systemic antibodies, as well as CMI. Additional advantages of mucosal immunization are the ease of administration and the avoidance of systemic side effects (Holmgren and Czerkinsky 2005; Czerkinsky and Holmgren 2012). Up to now, however, successful mucosal immunization has only been achieved with a limited number of oral vaccines (e.g., oral polio, cholera, typhoid fever, and rotavirus vaccines) and a nasal influenza vaccine (FluMist). Most of these vaccines are based on attenuated (see later) versions of the pathogens for which the route of administration is the same as the natural route of infection.

Apart from mucosal routes, research groups are working on needle-free jet injection of powders and fluids and dermal delivery with microneedles (Kersten and Hirschberg 2004; Bal et al. 2010) (see Chap. 4). A prerequisite of these approaches is that they must be painless. In that case several immunizations can be given with monovalent vaccines, replacing one multivalent vaccine. Up to now, these products have not yet been registered.

CLASSICAL VACCINES

■ Classification

Classical vaccines originate from viruses or bacteria and can be divided in live attenuated vaccines and nonliving vaccines. In addition, three vaccine generations can be distinguished for nonliving vaccines. First-generation vaccines consist of an inactivated suspension of the pathogenic microorganism. Little or no purification is applied. For second-generation vaccines, purification steps are applied, varying from the purification of a pathogenic microorganism (e.g., improved nonliving polio vaccine) to the complete purification of the protective component (e.g., polysaccharide vaccines). Third-generation vaccines are either a well-defined combination of protective components (e.g., acellular pertussis vaccine) or the protective component with the desired immunological properties (e.g., polysaccharides conjugated with a carrier protein). An overview of classical vaccines and their generations is given in Table 22.4.

■ Live Attenuated Vaccines

Before the introduction of recombinant DNA (rDNA) technology, a first step to improved live vaccines was the attenuation of virulent microorganisms by serial passage and selection of mutant strains with reduced virulence or toxicity. Examples are vaccine strains for oral polio vaccine, measles-mumps-rubella (MMR) combination vaccine, and tuberculosis vaccine consisting of bacille Calmette-Guérin (BCG). An alternative approach is chemical mutagenesis. For instance, by treating *Salmonella typhi* with nitrosoguanidine, a mutant strain lacking some enzymes that are responsible for the virulence was isolated (Germanier and Fuer 1975).

Live attenuated organisms have a number of advantages as vaccines over nonliving vaccines. After administration, live vaccines may replicate in the host

Type	Example	Marketed	Characteristics[a]
Live			
Viral	Adenovirus	Yes	Oral vaccine, US military services only
	Poliovirus (Sabin)	Yes	Oral vaccine
	Hepatitis A virus	No	
	Measles virus	Yes	
	Mumps virus	Yes	
	Rubella virus	Yes	
	Varicella zoster virus	Yes	
	Vaccinia virus	Yes	
	Yellow fever virus	Yes	
	Rotavirus	No	
	Influenza virus	No	
Bacterial	Bacille Calmette-Guérin	Yes	Whole organism
	Salmonella typhi	Yes	Whole organism, oral vaccine
Nonliving (first-generation products)			
Viral	Poliovirus (Salk)	Yes	Inactivated whole organisms
	Influenza virus	Yes	
	Japanese B encephalitis virus	Yes	
Bacterial	Bordetella pertussis	Yes	Purified, inactivated whole organisms
	Vibrio cholerae	Yes	
	Salmonella typhi	Yes	
Nonliving (second-generation products)			
Viral	Poliovirus	Yes	
	Rabies virus	Yes	
	Hepatitis A virus	Yes	
	Influenza virus	Yes	Subunit vaccine
	Hepatitis B virus	Yes	Plasma-derived hepatitis B surface antigen
Bacterial	Bordetella pertussis	Yes	Bacterial protein extract
	Haemophilus influenzae type b	Yes	Capsular polysaccharides
	Neisseria meningitidis	Yes	Capsular polysaccharides
	Streptococcus pneumoniae	Yes	Capsular polysaccharides
	Vibrio cholerae	Yes	Bacterial suspension + B subunit of cholera toxin
	Corynebacterium diphtheriae	Yes	Diphtheria toxoid
	Clostridium tetani	Yes	Tetanus toxoid
Nonliving (third-generation products)			
Viral	Measles virus	No	Subunit vaccine, ISCOM formulation
Bacterial	Bordetella pertussis	Yes	Mixture of purified protein antigens
	Haemophilus influenzae type b	Yes	Polysaccharide-protein conjugates
	Neisseria meningitidis	No	Polysaccharide-protein conjugates
	Streptococcus pneumoniae	No	Polysaccharide-protein conjugates

Source: Plotkin et al. (2008)

[a]Unless mentioned otherwise, the vaccine is administered via the needle

Table 22.4 ■ Classical vaccines.

similar to their pathogenic counterparts. This confronts the host with a larger and more sustained dose of antigen, which means that few and low doses are required. In general, the vaccines give long-lasting humoral and cell-mediated immunity.

Live vaccines also have drawbacks. Live viral vaccines bear the risk that the nucleic acid sequence is incorporated into the host's genome. Moreover, reversion to a virulent form may occur, although this is unlikely when the attenuated seed strain contains several mutations. Nevertheless, for diseases such as viral hepatitis, AIDS, and cancer, this drawback makes the use of classical live vaccines virtually unthinkable. Furthermore, it is important to recognize that immunization of immune-deficient children with live organisms can lead to serious complications. For instance, a child with T-cell deficiency may become overwhelmed with BCG and die.

■ Nonliving Vaccines: Whole Organisms

An early approach for preparing vaccines is the inactivation of whole bacteria or viruses. A number of reagents (e.g., formaldehyde, glutaraldehyde) and heat are commonly used for inactivation. Examples of this first-generation approach are pertussis, cholera, typhoid fever, and inactivated polio vaccines. These nonliving vaccines have the disadvantage that little or no CMI is induced. Moreover, they more frequently cause adverse effects as compared to live attenuated vaccines and second- and third-generation nonliving vaccines.

■ Nonliving Vaccines: Subunit Vaccines

Diphtheria and Tetanus Toxoids

- Some bacteria such as *Corynebacterium diphtheriae* and *Clostridium tetani* form toxins. Antibody-mediated immunity to the toxins is the main protection mechanism against infections with these bacteria. Both toxins are proteins and are inactivated with formaldehyde for inclusion in vaccines. The immunogenicity of such toxoids is relatively low and was improved by adsorption of the toxoids to a suspension of aluminum salts. This combination of an antigen and an adjuvant is still used in combination vaccines.

Acellular Pertussis Vaccines

The relatively frequent occurrence of side effects of whole-cell pertussis vaccine was the main reason to develop subunit vaccines. The development of third-generation acellular pertussis vaccines in the 1980s exemplifies how a better insight into factors that are important for pathogenesis and immunogenicity can lead to an improved vaccine. It was conceived that a subunit vaccine consisting of a limited number of purified immunogenic components and devoid of (toxic) lipopolysaccharide would significantly reduce undesired effects. Four protein antigens important for protection have been identified. However, as yet there exists no consensus about the optimal composition of an acellular pertussis vaccine. Current vaccines contain different amounts of two to four of these proteins.

Polysaccharide Vaccines

Bacterial capsular polysaccharides consist of pathogen-specific multiple repeating carbohydrate epitopes, which are isolated from cultures of the pathogenic species. Plain capsular polysaccharides (second-generation vaccines) are thymus-independent antigens that are poorly immunogenic in infants and show poor immunological memory when applied in older children and adults. The immunogenicity of polysaccharides is highly increased when they are chemically coupled to carrier proteins containing T_h-epitopes. This coupling makes them T cell dependent, which is due to the participation of T_h-cells that are activated during the response to the carrier. Examples of such third-generation polysaccharide conjugate vaccines include meningococcal type C, pneumococcal, and *Haemophilus influenzae* type b (Hib) polysaccharide vaccines that are included in many national immunization programs.

MODERN VACCINE TECHNOLOGIES

■ Modern Live Vaccines

Genetically Attenuated Microorganisms

Emerging insights in molecular pathogenesis of many infectious diseases make it possible to attenuate microorganisms very efficiently nowadays. By making multiple deletions, the risk of reversion to a virulent state during production or after administration can be virtually eliminated. A prerequisite for attenuation by genetic engineering is that the factors responsible for virulence and the life cycle of the pathogen are known in detail. It is also obvious that the protective antigens must be known: attenuation must not result in reduced immunogenicity.

An example of an improved live vaccine obtained by homologous genetic engineering is an experimental, oral cholera vaccine. An effective cholera vaccine should induce a local, humoral response in order to prevent colonization of the small intestine. Initial trials with *Vibrio cholerae* cholera toxin (CT) mutants caused mild diarrhea, which was thought to be caused by the expression of accessory toxins. A natural mutant was isolated that was negative for these toxins. Next, CT was detoxified by rDNA technology. The resulting vaccine strain, called CVD 103, is well tolerated by volunteers (Suharyono et al. 1992; Tacket et al. 1999) and challenge experiments with adult volunteers showed protection (Garcia et al. 2005).

Genetically attenuated live vaccines have the general drawbacks mentioned in the section about classically attenuated live vaccines. For these reasons, it is not surprising that homologous engineering is mainly restricted to pathogens that are used as starting materials for the production of subunit vaccines (see the section "Subunit Vaccines," below).

Live Vectored Vaccines

A way to improve the safety or efficacy of vaccines is to use live, avirulent, or attenuated organisms as

Vector	Antigens from	Advantages of vector	Disadvantages of vector
Viral			
Vaccinia	RSV, HIV, VSV, rabies virus, HSV, influenza virus, EBV, Plasmodium spp. (malaria)	Widely used in man (safe) Large insertions possible (up to 41 kB)	Sometimes causing side effects Very immunogenic: repeated use difficult
Avipoxviruses (canarypox, fowlpox)	Rabies virus, measles virus	Abortive replication in man Low immunogenicity	
Poliovirus	Vibrio cholerae, influenza virus, HIV, chlamydia	Widely used in man (safe) Live/oral and inactivated/parenteral forms possible	Small genome
Adenoviruses	RSV, HBV, EBV, HIV, CMV	Oral route applicable	Small genome
Herpes viruses (HSV, CMV, varicella virus)	EBV, HBV	Large genome	
Bacterial			
Salmonella spp.	B. pertussis, HBV, Plasmodium spp., E. coli, influenza virus, streptococci, Vibrio cholerae, Shigella spp.	Strong mucosal responses	
Mycobacteria (BCG)	Borrelia burgdorferi (lyme disease)	Widely used in man (safe) Large insertions possible	
E. coli	Bordetella pertussis Shigella flexneri		

BCG bacille Calmette-Guérin, *CMV* cytomegalovirus, *EBV* Epstein-Barr virus, *HBV* hepatitis B virus, *HIV* human immunodeficiency virus, *HSV* herpes simplex virus, *RSV* respiratory syncytial virus, *VSV* vesicular stomatitis virus

Table 22.5 ■ Examples of recombinant live vaccines.

a carrier to express protective antigens from a pathogen. Both bacteria and viruses can be used for this purpose; some of them are listed in Table 22.5. Live vectored vaccines are created by recombinant technology, wherein one or more genes of the vector organism are replaced by one or more protective genes from the pathogen. Administration of such live vectored vaccines results in efficient and prolonged expression of the antigenic genes either by the vaccinated individual's own cells or by the vector organism itself (e.g., in case of bacteria as carriers).

Most experience has been acquired with vaccinia virus by using the principle that is schematically shown in Fig. 22.6. Advantages of vaccinia virus as vector include (i) its proven safety in humans as a smallpox vaccine, (ii) the possibility for multiple immunogen expression, (iii) the ease of production, (iv) its relative heat resistance, and (v) its various possible administration routes. A multitude of live recombinant vaccinia vaccines with viral and tumor antigens have been constructed, several of which have been tested in the clinic (Jaoko et al. 2008; Jacobs et al. 2009). It has been demonstrated that the products of genes coding for viral envelope proteins can be correctly processed and inserted into the plasma membrane of infected cells. Problems related with the side effects or immunogenicity of vaccinia virus may be circumvented by the use of attenuated strains or poxviruses with a nonhuman natural host.

Adenoviruses can also be used as vaccine vectors (see also Chap. 24). Adenoviruses have several characteristics that make them suitable as vaccine vectors: (i) they can infect a broad range of both dividing and nondividing mammalian cells; (ii) transgene expression is generally high and can be further increased by using heterologous promoter sequences; (iii) adenovirus vectors are mostly replication deficient and do not integrate their genomes into the chromosomes of host cells, making these vectors very safe to use; and (iv) upon parenteral administration, adenovirus vectors induce strong immunity and evoke both humoral and cellular responses against the expressed antigen. A number of clinical trials with human adenovirus vectors (HAd5) expressing antigens of Ebola virus, human immunodeficiency virus (HIV), and severe acute respiratory syndrome (SARS) as vaccines against these diseases are currently in progress or have been terminated (Nayak and Herzog 2010). In a double-blind, phase II clinical trial to study the effectiveness of a Had5-based vaccine against HIV-1 infection, 3,000 HIV-1 seronegative volunteers were either given the Ad5 vaccine or a placebo. Strikingly, there seemed to be an increased HIV-1 infection rate in the group that had received the Ad5 vaccine (Buchbinder et al. 2008).

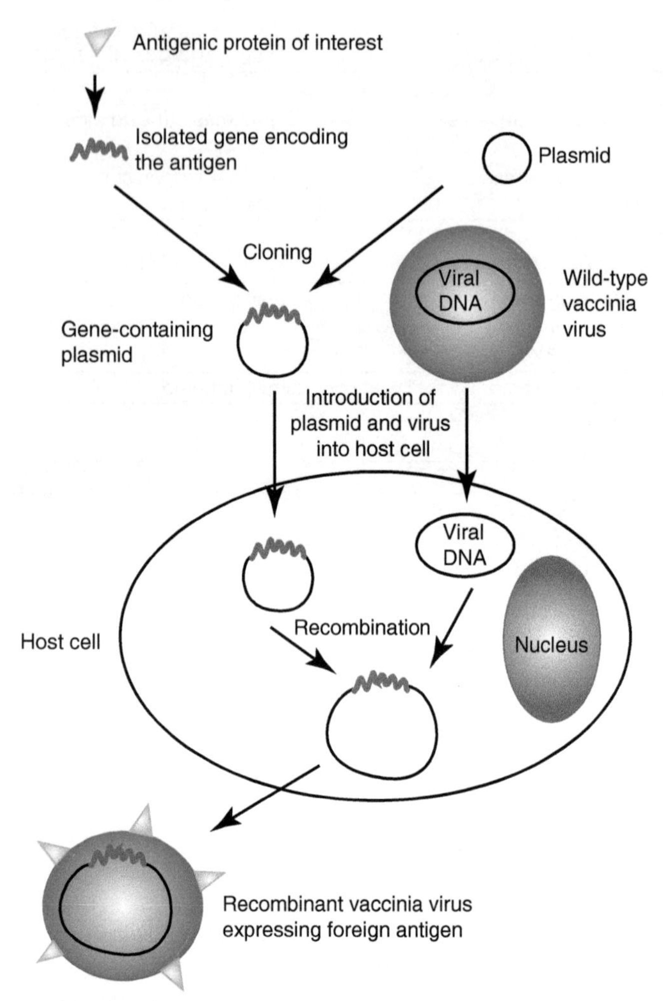

Figure 22.6 ■ Construction of recombinant vaccinia virus as a vector of foreign protein antigens. The gene of interest encoding an immunogenic protein is inserted into a plasmid. The plasmid containing the protein gene and wild-type vaccinia virus are then simultaneously introduced into a host cell line to undergo recombination of viral and plasmid DNA, after which the foreign protein is expressed by the recombinant virus.

A major limitation of the use of live vectored vaccines is the prevalence of preexisting immunity against the vector itself, which could neutralize the vaccine before the immune system can be primed. Such preexisting immunity has been described for adenoviral vectors, for which the prevalence of neutralizing antibodies can be as high as 90 % of the total population. The use of strains with no or low prevalence of preexisting immunity as live vectors is therefore recommended (Nayak and Herzog 2010; Ahi et al. 2011).

■ Modern Subunit Vaccines

Recombinant Protein Vaccines

To improve the yield, facilitate the production, and/or improve the safety of protein-based vaccines, protein antigens are nowadays often produced recombinantly, i.e., expressed by host cells that are safe to handle and/or allow high expression levels.

Heterologous hosts used for the expression of immunogenic proteins include yeasts, bacteria, insect cells, plant cells, and mammalian cell lines. Hepatitis B surface antigen (HBsAg), which previously was obtained from plasma of infected individuals, has been expressed in bakers' yeast, *Saccharomyces cerevisiae* (Valenzuela et al. 1982; Vanlandschoot et al. 2002), and in mammalian cells, Chinese hamster ovary cells (Burnette et al. 1985; Raz et al. 2001), by transforming the host cell with a plasmid containing the HBsAg-encoding gene. Both expression systems yield 22-nm HBsAg particles (also called virus-like particles or VLPs) that are structurally identical to the native virus. Advantages are safety, consistent quality, and high yields. The yeast-derived vaccine has become available worldwide and appears to be as safe and efficacious as the classical plasma-derived vaccine.

The two human papillomavirus (HPV) vaccines currently on the market are produced as recombinant proteins which, like HBsAg, assemble spontaneously into virus-like particles. Antigens for Gardasil, a quadrivalent HPV vaccine, are produced in yeast, whereas antigens for the bivalent vaccine Cervarix are produced in insect cells.

Recombinant Peptide Vaccines

After identification of a protective epitope, it is possible to incorporate the corresponding peptide sequence through genetic fusion into a carrier protein, such as HBsAg, hepatitis B core antigen, and β-galactosidase (Francis and Larche 2005). The peptide-encoding DNA sequence is synthesized and inserted into the carrier protein gene. An example of the recombinant peptide approach is a malaria vaccine based on a 16-fold repeat of the Asn-Ala-Asn-Pro sequence of a *Plasmodium falciparum* surface antigen. The gene encoding this peptide was fused with the HBsAg *gene*, and the fusion product was expressed by yeast cells (Vreden et al. 1991). Genetic fusion of peptides with proteins offers the possibility to produce protective epitopes of toxic antigens derived from pathogenic species as part of nontoxic proteins expressed by harmless species. Furthermore, a uniform product is obtained in comparison with the variability of chemical conjugates (see the section "Synthetic Peptide-Based Vaccines," below).

Synthetic Peptide-Based Vaccines

In principle, a vaccine could consist of only the relevant epitopes instead of intact pathogens or proteins. Epitopes are small enough to be produced synthetically as peptides and a peptide-based vaccine would be much better defined than classical vaccines, making the concept of peptide vaccines attractive. However, it turned out to be difficult to develop these vaccines, and today there are no licensed peptide-based vaccines available yet. Nevertheless, important progress has been made, and some synthetic peptide vaccines have now entered the clinic, e.g., for immunotherapy of cancer (Melief and van der Burg 2008). To understand the complexity of peptide vaccines, one has to distinguish the different types of epitopes.

Epitopes recognized by antibodies or B cells are very often conformation dependent (Van Regenmortel 2009). For this reason, it is difficult to identify them accurately. Manipulation of the antigen, such as digestion or the cloning of parts of the gene, will often affect B-cell epitope integrity. An accurate way of identifying epitopes is to elucidate the crystal structure of antigen-antibody complexes. This is difficult and time consuming, and although crystallography can reveal molecular interactions with unsurpassed detail, the molecular complex likely is much more dynamic in solution. Once the epitope is identified, synthesizing it as a functional peptide has proven to be difficult. The peptides need to be conformationally restrained. This can be achieved by cyclization of the peptide (Oomen et al. 2005) or by the use of scaffolds to synthesize complex peptide structures (Timmerman et al. 2009).

Regarding conformation, T-cell epitopes are less demanding because they are presented naturally as processed peptides by APCs to T cells. As a result, T-cell epitopes are linear. Here, we discern CD8 epitopes (8–10 amino acid residues; MHC class I restricted) and CD4 epitopes (>12 amino acid residues; MHC class II restricted). The main requirement is that they fit into binding grooves of MHC molecules with high enough affinity. Studies with peptide-based cancer vaccines have shown that these should contain both CD8 and CD4 epitopes in order to elicit a protective immune response. Furthermore, minimal peptides that can be externally loaded on MHC molecules of cells have been shown to induce less robust responses than longer peptides that require intracellular processing after uptake by DCs. Another point to consider is the variable repertoire of MHC molecules in a patient population, implying that a T-cell epitope-based peptide vaccine should contain several T-cell epitopes. Following these concepts, clinical trials with overlapping long peptide vaccines have shown promising results in the immunotherapy of patients with HPV-induced malignancies (Melief and van der Burg 2008).

Nucleic Acid Vaccines

Immunization with nucleic acid vaccines involves the administration of genetic material, plasmid DNA or messenger RNA, encoding the desired antigen. The encoded antigen is then expressed by the host cells and after which an immune response against the expressed antigen is raised (Donnelly et al. 2005).

Plasmid DNA is produced by replication in *E. coli* or other bacterial cells and purification by established methods (e.g., density gradient centrifugation, ion-exchange chromatography). Up until now, plasmid DNA has been administered to animals and humans mostly via intramuscular injection. The favorable properties of muscle cells for DNA expression are probably due to their relatively low turnover rate, which prevents that plasmid DNA is rapidly dispersed in dividing cells. After intracellular uptake of the DNA, the encoded protein is expressed on the surface of host cells. After a single injection, the expression can last for more than 1 year. Besides intramuscular injection, subcutaneous, intradermal, and intranasal administrations also seem to be effective. Needleless injection into the skin of DNA-coated gold nanoparticles via a gene gun has been reported to require up to 1,000-fold less DNA when compared to intramuscular administration.

Nucleic acid vaccines offer the safety of subunit vaccines and the advantages of live recombinant vaccines. They can induce strong CTL responses against the encoded antigen. In addition, bacterial plasmids are also ideal for activating innate immunity as TLR-9 expressed on many phagocytic cells can recognize unmethylated bacterial DNA. The main disadvantage of nucleic acid immunization is the poor immunogenicity in man. Therefore, they often require, like subunit vaccines, adjuvants or delivery systems to boost the immune response against the DNA-encoded antigen(s). Nevertheless, DNA has proven to be very effective when used in combination with protein antigens in heterologous DNA-prime/protein-boost strategies. The long-term safety of nucleic acid vaccines remains to be established. The main pros and cons of nucleic acid vaccines are listed in Table 22.6. An advantage of RNA over DNA is that it is not able to incorporate into host DNA. A drawback of RNA, however, is that it is less stable than DNA. Nucleic acids coding for a variety of antigens have shown to induce protective, long-lived humoral and cellular immune responses in various species including man (Liu 2011). Examples of DNA vaccines that have been tested in clinical trials comprise plasmids encoding HIV-1 antigens and malaria antigens.

■ Reverse Vaccinology

Nowadays vaccines can be designed based on the information encoded by the genome of a particular pathogen (Masignani et al. 2002; Rappuoli and Covacci 2003). From many pathogens, the entire genomes have been sequenced and this number is growing (http://cmr.tigr.org). The genome sequence of a pathogen provides a complete picture of all proteins that can be

Advantages	Disadvantages
Low intrinsic immunogenicity of nucleic acids	Effects of long-term expression unknown
Induction of long-term immune responses	Formation of antinucleic acid antibodies possible
Induction of both humoral and cellular immune responses	Possible integration of the vaccine DNA into the host genome
Possibility of constructing multiple epitope plasmids	Concept restricted to peptide and protein antigens
Heat stability	Poor delivery
Ease of large-scale production	Poorly immunogenic in man

Table 22.6 ■ Advantages and disadvantages of nucleic acid vaccines.

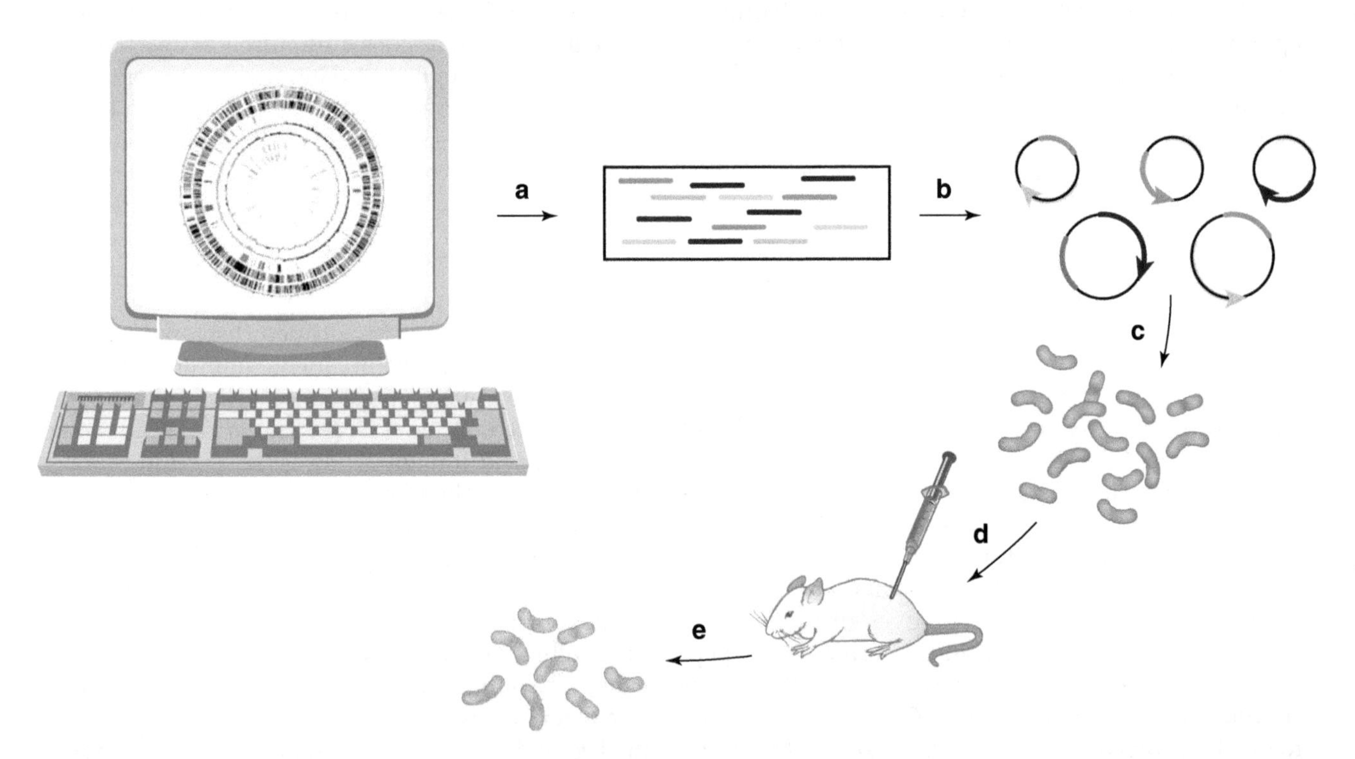

Figure 22.7 ■ Reverse vaccinology involves the analysis of genome sequences of pathogens in silico with the aim to identify potential antigens (**a**). These potential antigens can then be cloned (**b**), produced recombinantly (**c**), and subsequently used for immunological screening (**d**). The entire process leads to a quick identification of a limited number of vaccine candidates that can give protection against infection with the pathogen without the need to test all proteins produced by this pathogen (**e**) (Adapted from Scarselli et al. (2005)).

produced by the pathogens at any given time. Using computer algorithms, proteins that are either excreted or expressed on the surface of the pathogen, and thus most likely available for recognition by the host's immune system, can be identified. After recombinant production and purification, these vaccine candidates can be screened for immunogenicity in mice. From these, the best candidates can be selected and used as subunit vaccines (Fig. 22.7).

A big advantage of reverse vaccinology is the ease at which novel candidate antigens can be selected without the need to cultivate the pathogen. Furthermore, by comparing genomes of different strains of a pathogen, conserved antigens can be identified that can serve as a "broad spectrum" vaccine, giving protection against all strains or serotypes of a given pathogen. One drawback of this approach is that it is limited to the identification of protein-based antigens.

Reverse vaccinology has been successfully used to identify novel antigens for a variety of pathogens, including *Neisseria meningitidis*, *Bacillus anthracis*, *Streptococcus pneumoniae*, *Staphylococcus aureus*, *Chlamydia pneumoniae*, and *Mycobacterium tuberculosis* (Sette and Rappuoli 2010).

■ Therapeutic Vaccines

Most classical vaccine applications are prophylactic: they prevent an infectious disease from developing. Besides prophylactic applications, vaccines may be used to treat already established diseases, such as infectious diseases, cancer, or drug addiction. Although the development of therapeutic vaccines is still in its infancy, some examples will be highlighted here.

Cancer Vaccines

Immunotherapy of cancer requires the activation of tumor-specific T cells, although humoral responses may in some cases (e.g., non-Hodgkin lymphoma) also be effective. Besides the synthetic long peptide-based approach already described above, there are several strategies to boost such a tumor-specific CTL response. Vaccines can be prepared from the patient's tumor itself by mixing irradiated tumor cells or cell extracts with bacterial adjuvants such as BCG to enhance their immunogenicity. Alternatively, heat shock proteins isolated from a patient's tumor that contain associated tumor antigens can be used as a tumor-specific vaccine. In combination with adjuvants, these heat shock proteins can be very potent in stimulating CTL responses against tumor cells as has been demonstrated in several clinical trials (Liu et al. 2002). Another approach is to genetically alter tumor cells in order to make them more immunogenic. Transfection of tumor cells with the gene encoding the co-stimulatory molecule B7 has resulted in direct activation of tumor-specific CTLs by the transformed tumor cells (Garnett et al. 2006). Similar results can be achieved by transforming tumor cells with gene encoding cytokines (e.g., GM-CSF, IL-2, IL-4, and IL-12).

Alternatively, the first step of the immune response can be optimized with an in vitro procedure. Dendritic cells are isolated from the patient and cultured in vitro in the presence of tumor antigen. The antigen loaded cells are subsequently given back to the patient. A licensed therapeutic prostate cancer vaccine is based on this principle (Cheever and Higano 2011). The potency of the vaccine is limited (a few months extension of life expectancy) and the costs are high (almost $100,000 per patient), so there is substantial room for improvement.

Vaccines Against Drug Abuse

Therapeutic vaccines are also being developed for the treatment of drug abuse, such as addiction to nicotine, cocaine, or methamphetamine (Moreno and Janda 2009). The idea is to evoke a humoral immune reaction against the drug molecules. As most of these drugs have their addictive action within the central nervous system, antibodies raised against the drug molecules can prevent the passage of these molecules over the blood–brain barrier and thus prevent the addictive effects. Many abused drugs are small nonprotein substances, which generally do not elicit an immune response as such. In order to activate the host immune system against these substances, they need to be conjugated to proteins, such as ovalbumin or diphtheria toxin. This approach has been effective in animal models. Late and mid-stage clinical studies, however, have shown disappointing results. A vaccine consisting of nicotine conjugated to virus-like particles failed in a phase 2 study, and a nicotine-protein conjugate vaccine failed in phase 3. This demonstrates the current lack of tools to predict or at least minimize the risk for late stage failures.

■ Systems Biology and Vaccines

One of the biggest problems in vaccine development is the inability to predict the efficacy and safety of new vaccines with other methods than phase 3 studies. Animal models are poorly predictive, and even immunological parameters in humans, like the induction of pathogen-neutralizing antibody responses, are not fully predictive. As a result, vaccine efficacy has to be measured in terms of reduction of disease. This is sometimes very difficult because symptoms of a disease can be caused by more than one pathogen (e.g., influenza-like illness). To measure protection, one has to detect the influenza virus in the group with influenza-like illness. In addition, to measure reduction of disease, the groups in the trial need to be very large because it is unknown who will get the disease. Sometimes tens of thousands of people are included in phase 3 trials.

Systems biology approaches that may limit these problems in the future are under development. The idea is to identify gene signatures that correlate with a protective immune response. This is done by a combination of gene expression analysis for, e.g., lymphocytes in the blood and functional assays like measurement of antibodies, cytokines, and cellular responses. Bioinformaticians try to unravel pathways and networks of genes involved in immune responses. Eventually, it may be possible to assign the activity of a limited number of genes to protective immune responses (Nakaya et al. 2011). Perhaps that such gene signatures can be used in future clinical trials (Pulendran et al. 2010). This approach would allow for reduction of the size of clinical studies, reduce the risk of late stage failure, and assess the significance of animal models in the preclinical phase of vaccine development.

PHARMACEUTICAL ASPECTS

■ Production

Except for synthetic peptides, the antigenic components of vaccines are derived from microorganisms or animal cells. For optimal expression of the required vaccine component(s), these microorganisms or animal cells can be genetically modified. Animal cells are used for the cultivation of viruses and for the production of some subunit vaccine components and have the advantage that the vaccine components are released into the culture medium.

Three stages can be discerned in the manufacture of cell-derived vaccines: (1) cultivation or upstream processing, (2) purification or downstream processing, and (3) formulation. For the first two stages, the reader is referred to Chap. 3, whereas the formulation is addressed in the following section.

Formulation

Adjuvants, Immune Potentiators, and Delivery Systems

The success of immunization is not only dependent on the nature of the immunogenic components but also on their presentation form. Therefore, the search for effective and acceptable adjuvants is an important issue in modern vaccine development (Guy 2007). Adjuvants are defined as any material that can increase the humoral and cellular immune response against an antigen. Adjuvants can stimulate the immune system by several, not mutually exclusive mechanisms (Guy 2007): (i) a depot effect leading to slow antigen release and prolonged antigen presentation, (ii) attraction and stimulation of APCs by some local tissue damage and binding to PRRs present on APCs, and (iii) delivery of the antigen to regional lymph nodes by improved antigen uptake, transport, and presentation by APCs.

Colloidal aluminum salts (hydroxide, phosphate) are widely used in many classical vaccine formulations. A few other adjuvants, e.g., monophosphoryl lipid A in HPV vaccine and oil-in-water emulsions in influenza vaccines, have been introduced recently but most are in experimental testing or are used in veterinary vaccines. Table 22.7 shows a list of some well-known adjuvants.

Adjuvant	Characteristics
Aluminum salts	Antigen adsorption is crucial
Lipid A and derivatives	Fragment of lipopolysaccharide, a bacterial endotoxin
MF59	Squalene-based oil-in-water emulsion
Muramyl peptides	Active fragments of bacterial cell walls
Saponins	Plant triterpene glycosides
NBP	Synthetic amphiphiles
DDA	Synthetic amphiphile
CpG	Non-methylated DNA sequences containing CpG-oligodinucleotides
Cytokines	Interleukins (1, 2, 3, 6, 12), interferon-γ, tumor necrosis factor
Cholera toxin, B subunit	Mucosal adjuvant
Emulsions	Both water-in-oil and oil-in-water emulsions are used; often contain amphiphilic adjuvants
Liposomes	Phospholipid membrane vesicles; aqueous interior as well as lipid bilayer may contain antigens and/or adjuvants
ISCOMs	Micellar lipid-saponin complex; not suitable for soluble antigens
Microspheres	Biodegradable polymeric spheres, often poly(lactide-co-glycolide)

DDA dioctadecyldimethylammonium bromide, *ISCOM* immune stimulating complex, *NBP* nonionic block copolymers

Table 22.7 ■ Examples of adjuvants.

Combination Vaccines

Since oral immunization is not possible for most available vaccines (see the section "Route of Administration" above), the strategy to mix individual vaccines in order to limit the number of injections has been common practice since many decades. Currently, vaccines are available containing up to six nonrelated antigens: diphtheria-tetanus-pertussis-hepatitis B-polio-Haemophilus *influenzae* type b vaccine. Another example is measles-mumps-rubella (MMR) vaccine, alone or in combination with varicella vaccine. Sometimes a vaccine contains antigens from several subtypes of a particular pathogen. Pneumococcal conjugate vaccine 13 (PVC13) is an example. This vaccine contains polysaccharides from thirteen pneumococcal strains, conjugated to a carrier protein to improve immunogenicity.

Combining vaccine components sometimes results in pharmaceutical as well as immunological problems. For instance, formaldehyde-containing components may chemically react with other components; an unstable antigen may need freeze drying, whereas other antigens should not be frozen. Components that are not compatible can be mixed prior to injection, if there is no short-term incompatibility. To this end, dual-chamber syringes have been developed.

From an immunological point of view, the immunization schedules of the individual components of combination vaccines should match. Even when this condition is met and the components are pharmaceutically compatible, the success of a combination vaccine is not warranted. Vaccine components in combination vaccines may exhibit a different behavior in vivo compared to separate administration of the components. For instance, enhancement (Paradiso et al. 1993) as well as suppression (Mallet et al. 2004) of humoral immune responses has been reported.

■ Characterization

Second- and third-generation classical vaccines and modern vaccines are better-defined products in terms of immunogenicity, structure, and purity. This means that the products can be characterized with a combination of appropriate biochemical, physicochemical, and immunochemical techniques (see Chap. 2). Vaccines have to meet similar standards as other biotechnological pharmaceuticals. The use of modern analytical techniques for the design and release of new vaccines is gaining importance. Currently, animal experiments are needed for quality control of many vaccines but in vitro analytical techniques may eventually (partly) substitute preclinical tests in vivo. During the development of the production process of a vaccine component, a combination of suitable assays can be defined. These assays can subsequently be applied during its routine production.

Column chromatographic (HPLC) and electrophoretic techniques like gel electrophoresis and capillary electrophoresis provide information about the purity, molecular weight, and electric charge of the vaccine component. Physicochemical assays comprise mass spectrometry and spectroscopy, including circular dichroism and fluorescence spectroscopy. Information is obtained mainly about the molecular weight and the conformation of the vaccine component. Immunochemical assays, such as enzyme-linked immunoassays and radioimmunoassays, are powerful methods for the quantification of the vaccine component. By using monoclonal antibodies (preferably with the same specificity as those of protective human antibodies) information can be obtained about the conformation and accessibility of the epitope to which the antibodies are directed. Moreover, the use of biosensors makes it possible to measure antigen-antibody interactions momentarily, allowing accurate determination of binding kinetics and affinity constants.

■ Storage

Depending on their specific characteristics, vaccines are stored as solution or as a freeze-dried formulation, usually at 2–8 °C. Their shelf life depends on the composition and physicochemical characteristics of the vaccine formulation and on the storage conditions and typically is in the order of several years. The quality of the container can influence the long-term stability of vaccines, e.g., through adsorption or pH changes resulting from contact with the vial wall. The use of pH indicators or temperature- or time-sensitive labels ("vial vaccine monitors," which change color when exposed to extreme temperatures or after the expiration date) can avoid unintentional administration of inappropriately stored or expired vaccine.

CONCLUDING REMARKS

Despite the tremendous success of the classical vaccines, there are still many infectious diseases and other diseases (e.g., cancer) against which no effective vaccine exists. Although modern vaccines – like other biopharmaceuticals – are expensive, calculations may indicate cost-effectiveness for vaccination against many of these diseases. In addition, the growing resistance to the existing arsenal of antibiotics increases the need to develop vaccines against common bacterial infections. It is expected that novel vaccines against several of these diseases will become available, and in these cases, the preferred type of vaccine will be chosen from one of the different options described in this chapter.

SELF-ASSESSMENT QUESTIONS

■ Questions

1. What are the characteristics of the ideal vaccine? Which aspects should be addressed in the design of a vaccine in order to approach these characteristics?
2. How do antibodies neutralize antigens?
3. How do T cells discriminate between exogenous (extracellular) and endogenous (intracellular) antigens? What is the eventual result of these differences in responsiveness?
4. Which categories of classical vaccines exist and what are their characteristics?
5. Mention two main problems related with the immunogenicity of peptide-based vaccines. How are these problems dealt with?
6. Mention at least three advantages and three disadvantages of nucleic acid vaccines. Give one advantage and one disadvantage of RNA vaccines over DNA vaccines.
7. Which stages are discerned in the manufacture of cell-derived vaccines?
8. Mention two or more examples of currently available combination vaccines. Which pharmaceutical and immunological conditions have to be fulfilled when formulating combination vaccines?

■ Answers

1. The characteristics of the ideal vaccine are listed in Fig. 22.1.
2. Antibodies are able to neutralize antigens by at least four mechanisms:
   (a) Fc-mediated phagocytosis
   (b) Complement activation resulting in cytolytic activity
   (c) Complement-mediated phagocytosis
   (d) Competitive binding on sites that are crucial for the biological activity of the antigen
3. T cells are able to distinguish exogenous from endogenous antigens by the type of self-antigen

(MHC antigen) that is associated with processed antigen on the surface of the antigen-presenting cell. Processed antigen binds to MHC molecules, resulting in a cell surface located antigen/MHC complex. The complex is recognized by the T-cell receptor/CD4 or CD8 complex. A cell infected with a virus presents partially degraded viral antigen (i.e., endogenous antigen) complexed with class I MHC. The complex is recognized by CD8-positive T cells, resulting in the induction of cytotoxic T cells. Professional antigen-presenting cells like macrophages phagocytose exogenous antigen and present it in conjunction with class II MHC. CD4-positive T cells bind to the MHC-antigen complex. Subsequent B-cell or macrophage activation leads to antibody or inflammatory responses, respectively.

4. Classical vaccines consist of either live attenuated vaccines or nonliving vaccines. For nonliving vaccines, we discern three generations. The first generation comprises suspensions of inactivated, pathogenic organisms. Second-generation vaccines contain purified components, varying from whole organisms or extracts of organisms to purified single components. Third-generation vaccines are either well-defined mixtures of purified components or protective components formulated in an immunogenic presentation form. Examples of these categories are given in Table 22.4.
5. The first problem concerns the low immunogenicity of plain peptide vaccines. The immunogenicity can be improved by constructing multiple antigen peptides or by chemical coupling of peptides to carrier proteins. Alternatively, peptide epitopes can be incorporated into carrier proteins through genetic fusion of the peptide DNA with that of the carrier protein. The second problem of peptide antigens is that their conformation does not necessarily correspond to that of the epitope in the native protein, which in case of B-cell epitopes may lead to poor immune responses or responses to irrelevant peptide conformations. Solutions to this problem are sought in constraining the conformation of the synthetic peptide by chemical cyclization methods.
6. The advantages and disadvantages of nucleic acid vaccines are listed in Table 22.6. An advantage of RNA is that there is no risk of incorporation into host DNA. On the other hand, RNA is less stable than DNA.
7. The three production stages are (a) cultivation of cells and/or virus, (b) purification of the desired components, and (c) formulation of the vaccine.
8. Examples of combination vaccines include diphtheria-tetanus-pertussis(−polio) vaccines and measles-mumps-rubella(−varicella) vaccines. Prerequisites for combining vaccine components are:
   (a) Pharmaceutical compatibility of vaccine components and additives
   (b) Compatibility of immunization schedules
   (c) No interference between immune responses to individual components

REFERENCES

Ahi YS, Bangari DS, Mittal DS (2011) Adenoviral vector immunity: its implications and circumvention strategies. Curr Gene Ther 11:307–320

Bal SM, Ding Z, van Riet E, Jiskoot W, Bouwstra JA (2010) Advances in transcutaneous vaccine delivery: do all ways lead to Rome? J Control Release 148:266–282

Buchbinder SP, Mehrotra DV, Duerr A, Fitzgerald DW, Mogg R, Li D, Gilbert PB et al (2008) Efficacy assessment of a cell-mediated immunity HIV-1 vaccine (the step study): a double-blind, randomised, placebo-controlled, test-of-concept trial. Lancet 372:1881–1893

Burnette WN, Samal B, Browne J, Ritter GA (1985) Properties and relative immunogenicity of various preparations of recombinant DNA-derived hepatitis B surface antigen. Dev Biol Stand 59:113–120

Cheever MA, Higano CS (2011) PROVENGE (Sipuleucel-T) in prostate cancer: the first FDA-approved therapeutic cancer vaccine. Clin Cancer Res 17:3520–3526

Czerkinsky C, Holmgren J (2012) Mucosal delivery routes for optimal immunization: targeting immunity to the right tissues. Curr Top Microbiol Immunol 354:1–18

Donnelly JJ, Wahren B, Liu MA (2005) DNA vaccines: progress and challenges. J Immunol 175:633–639

Francis JN, Larche M (2005) Peptide-based vaccination: where do we stand? Curr Opin Allergy Clin Immunol 5:537–543

Garcia L, Jidy MD, Garcia H, Rodriguez BL, Fernandez R, Ano G et al (2005) The vaccine candidate Vibrio cholerae 638 is protective against cholera in healthy volunteers. Infect Immun 73:3018–3024

Garnett CT, Greiner JW, Tsang KY, Kudo-Saito C, Grosenbach DW, Chakraborty M et al (2006) TRICOM vector based cancer vaccines. Curr Pharm Des 12:351–361

Germanier R, Fuer E (1975) Isolation and characterization of Gal E mutant Ty 21a of Salmonella typhi: a candidate strain for a live, oral typhoid vaccine. J Infect Dis 131: 553–558

Guy B (2007) The perfect mix: recent progress in adjuvant research. Nat Rev Microbiol 5:501–517

Holmgren J, Czerkinsky C (2005) Mucosal immunity and vaccines. Nat Med 4(Supplement):S45–S53

Jacobs BL, Langland JO, Kibler KV, Denzler KL, White SD, Holechek SA et al (2009) Vaccinia virus vaccines: past, present and future. Antiviral Res 84:1–13

Jaoko W, Nakwagala FN, Anzala O, Manyonyi GO, Birungi J, Nanvubya A et al (2008) Safety and immunogenicity of recombinant low-dosage HIV-1 a vaccine candidates vectored by plasmid pTHr DNA or modified vaccinia virus Ankara (MVA) in humans in East Africa. Vaccine 26:2788–2795

Kawai T, Akira S (2010) The role of pattern-recognition receptors in innate immunity: update on toll-like receptors. Nat Immunol 11:373–384

Kersten G, Hirschberg H (2004) Antigen delivery systems. Expert Rev Vaccines 3:453–462

Liu MA (2011) DNA vaccines: an historical perspective and view to the future. Immunol Rev 239:62–84

Liu B, DeFilippo AM, Li Z (2002) Overcoming immune tolerance to cancer by heat shock protein vaccines. Mol Cancer Ther 1:1147–1151

Mallet E, Belohradsky BH, Lagos R, Gothefors L, Camier P, Carriere JP et al (2004) A liquid hexavalent combined vaccine against diphtheria, tetanus, pertussis, poliomyelitis, Haemophilus influenzae type B and hepatitis B: review of immunogenicity and safety. Vaccine 22:1343–1357

Masignani V, Rappuoli R, Pizza M (2002) Reverse vaccinology: a genome-based approach for vaccine development. Expert Opin Biol Ther 2:895–905

Melief CJM, van der Burg SH (2008) Immunotherapy of established (pre)malignant disease by synthetic long peptide vaccines. Nat Rev Cancer 8:351–360

Moreno AY, Janda KD (2009) Immunopharmacotherapy: vaccination strategies as a treatment for drug abuse and dependence. Pharmacol Biochem Behav 92:199–205

Nakaya HI, Wrammert J, Lee EK, Racioppi L, Marie-Kunze S, Haining WN et al (2011) Systems biology of vaccination for seasonal influenza in humans. Nat Immunol 12:786–795

Nayak S, Herzog RW (2010) Progress and prospects: immune responses to viral vectors. Gene Ther 17:295–304

Oomen CJ, Hoogerhout P, Kuipers B, Vidarsson G, van Alphen L, Gros P (2005) J Mol Biol 351:1070–1080

Paradiso PR, Hogerman DA, Madore DV, Keyserling H, King J, Reisinger KS et al (1993) Safety and immunogenicity of a combined diphtheria, tetanus, pertussis and Haemophilus influenzae type b vaccine in young infants. Pediatrics 92:827–832

Pashine A, Valiante NM, Ulmer JB (2005) Targeting the innate immune response with improved vaccine adjuvants. Nat Med 11:S63–S68

Plotkin SA, Orenstein WA, Offit PA (2008) Vaccines, 5th edn. WB Saunders Company, Philadelphia

Pulendran B, Ahmed R (2006) Translating innate immunity into immunological memory: implications for vaccine development. Cell 124:849–863

Pulendran B, Li SZ, Nakaya HI (2010) Systems vaccinology. Immunity 33:516–529

Rappuoli R, Covacci A (2003) Reverse vaccinology and genomics. Science 302:602

Raz R, Koren R, Bass D (2001) Safety and immunogenicity of a new mammalian cell-derived recombinant hepatitis B vaccine containing Pre-S1 and Pre-S2 antigens in adults. Isr Med Assoc J 3:328–332

Scarselli M, Giuliani MM, Adu-Bobie J, Pizza M, Rappuoli R (2005) The impact of genomics on vaccine design. Trends Biotechnol 23:84–91

Sette A, Rappuoli R (2010) Reverse vaccinology: developing vaccines in the era of genomics. Immunity 33:530–541

Suharyono, Simanjuntak C, Witham N, Punjabi N, Heppner DG, Losonsky G et al (1992) Safety and immunogenicity of single-dose live oral cholera vaccine CVD 103-HgR in 5-9-year-old Indonesian children. Lancet 340:689–694

Tacket CO, Cohen MB, Wasserman SS, Losonsky G, Livio S, Kotloff K et al (1999) Randomized, double-blind, placebo-controlled, multicentered trial of the efficacy of a single dose of live oral cholera vaccine CVD 103-HgR in preventing cholera following challenge with Vibrio cholerae O1 El tor inaba three months after vaccination. Infect Immun 67:6341–6345

Timmerman P, Puijk WC, Boshuizen RS, van Dijken P, Slootstra JW, Beurskens FJ et al (2009) Functional reconstruction of structurally complex epitopes using CLIPS® technology. Open Vaccine J 2:56–67

Valenzuela P, Medina A, Rutter WJ, Ammerer G, Hall BD (1982) Synthesis and assembly of hepatitis B virus surface antigen particles in yeast. Nature 298:347–350

Van Regenmortel MHV (2009) Synthetic peptide vaccines and the search for neutralization B cell epitopes. Open Vaccine J 2:33–44

Vanlandschoot P, Roobrouck A, Van Houtte F, Leroux-Roels G (2002) Recombinant HBsAg, an apoptotic-like lipoprotein, interferes with the LPS-induced activation of ERK-1/2 and JNK-1/2 in monocytes. Biochem Biophys Res Commun 297:486–491

Vreden SG, Verhave JP, Oettinger T, Sauerwein RW, Meuwissen JH (1991) Phase I clinical trial of a recombinant malaria vaccine consisting of the circumsporozoite repeat region of Plasmodium falciparum coupled to hepatitis B surface antigen. Am J Trop Med Hyg 45: 533–538

FURTHER READING

Delves PJ, Martin SJ, Burton DR, Roitt IM (2011) Roitt's essential immunology, 12th edn. Blackwell Scientific Publications, London

Levine MM, Dougan G, Good MF, Liu MA, Nabel GJ, Nataro JP, Rappuoli R (2009) New generation vaccines, 4th edn. Informa Healthcare, London

Plotkin SA, Orenstein WA, Offit PA (2008) Vaccines, 5th edn. WB Saunders Company, Philadelphia

Murphy K (2011) Janeway's immunobiology, 8th edn. Garland Science Publishing, London

23 Oligonucleotides

Raymond M. Schiffelers and Enrico Mastrobattista

INTRODUCTION

Oligonucleotides are (short) chains of (chemically modified) ribo- or deoxyribonucleotides. Their ability to bind to chromosomal DNA, mRNA, or non-coding RNA (ncRNA) through Watson-Crick and Hoogsteen base pairing offers possibilities for highly specific intervention in gene transcription, mRNA translation, gene repair, and recombination for therapeutic applications. In theory, a sequence of 15–17 bases occurs only once in the human genome, which would allow specific manipulation of single genes for oligonucleotides in this size range. In addition, therapeutic effects of oligonucleotides can be obtained through sequence-specific binding of transcription factors and intramolecular folding into structures that can bind to and interfere with the function of various biomolecules. Finally, cells display specific receptors for oligonucleotides. These receptors can activate a variety of immunological responses that can be of therapeutic value.

Due to this multitude of possible effects of nucleic acids, oligonucleotides can be very potent molecules, yet interpretation of the mechanism of therapeutic action of a specific oligonucleotide sequence is not straightforward (Stein et al. 2005). Apart from the desired activity, oligonucleotides are inclined to display (sequence-specific) unintended effects. Partial sequence homology may affect expression of genes other than the targeted gene (known as off-target effects), stimulation of immune responses may occur, and binding to proteins and peptides can alter their activity.

Other characteristics of oligonucleotides also impede clear-cut application as therapeutics. Their physicochemical characteristics induce rapid uptake by macrophages and excretion by the kidneys hindering target tissue accumulation (Geary 2009). In addition, spontaneous passage over cell membranes for these large and charged molecules for intracellular applications is difficult. Finally, oligonucleotides are sensitive to the action of nucleases leading to their degradation. Over the years a number of different modifications have been developed that overcome (part of) these problems (Fig. 23.1) (Kurreck 2003). Many of the clinically studied oligonucleotides contain one or more of these modified nucleotides.

In this chapter, we describe the classes of therapeutic oligonucleotides, categorized according to their mechanism of action. We start with oligonucleotides that are designed to bind to proteins either through a specific sequence and intramolecular folding (aptamers/riboswitches) or immune stimulatory activity by binding to nucleic acid receptors, in particular Toll-like receptors. In the subsequent sections, we discuss oligonucleotides that act by binding to complementary nucleic acids inside the cell. First, we discuss oligonucleotides that correct mutated DNA by changing DNA sequence or structure, or change mRNA by skipping unwanted mRNA fragments. After that, oligonucleotides that regulate the concentrations of mRNA are introduced.

The challenges to apply these oligonucleotides as therapeutics are faced by essentially all classes: rapid clearance, poor stability, and limited cellular uptake. These issues are discussed in the final section of this chapter, together with approaches to overcome these challenges and perspectives for future research.

DIRECT BINDING TO NON-NUCLEIC ACIDS

Sometimes therapeutic oligonucleotides can have applications that are not based on base pairing with endogenous nucleic acids. This can be the result from

R.M. Schiffelers, Ph.D. (✉)
Laboratory Clinical Chemistry & Haematology,
University Medical Center Utrecht,
Room G03.647 UMC Utrecht, Heidelberglaan 100,
Utrecht 3584 GX, The Netherlands
e-mail: r.schiffelers@umcutrecht.nl; r.m.schiffelers@uu.nl

E. Mastrobattista, Ph.D.
Department of Pharmaceutics,
Utrecht Institute for Pharmaceutical Sciences,
Utrecht University, Utrecht, The Netherlands

D.J.A. Crommelin, R.D. Sindelar, and B. Meibohm (eds.), *Pharmaceutical Biotechnology*,
DOI 10.1007/978-1-4614-6486-0_23,

First generation

Second generation

Phosphorothioate DNA (PS)

2'-O-methyl RNA (OMe)

2'-O-methoxy-ethyl RNA (OME)

Third generation

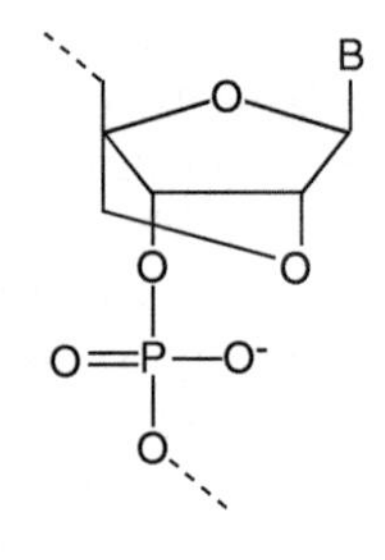

Peptide nucleic acid (PNA)

N3'-P5' phosphoroamidate (NP)

2'-O-fluoro-arabino nucleic acid (FANA)

Locked nucleic acid (LNA)

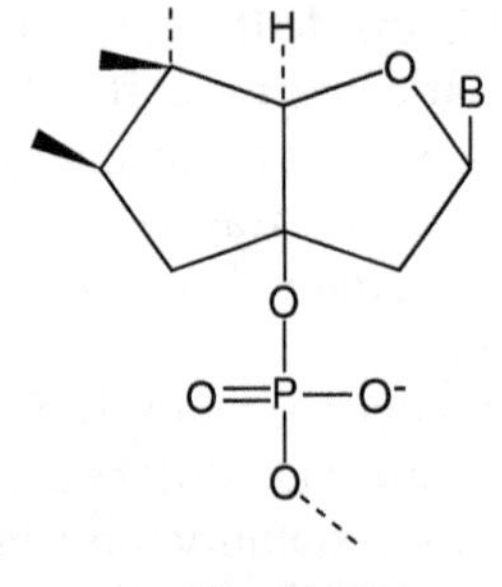

Morpholino phosphoroamidate (MF)

Cyclohexene nucleic acid (CeNA)

Tricyclo DNA (tcDNA)

Figure 23.1 ■ Popular chemical modifications to improve nuclease resistance and distribution profile of oligonucleotides.

binding to specific immune receptors that recognize nucleic acids at unexpected locations (e.g., extracellular DNA) or specific structural qualities (e.g., CpG motifs). In addition, the ability of nucleic acids to fold into complex three-dimensional structures through internal regions of (partial) complementarity allows them to bind to virtually any molecule with nano- to picomolar affinity (Weigand and Suess 2009). This high affinity is supported by data on their extreme specificity. A nucleic acid sequence specifically binding theophylline has a million times higher affinity for theophylline than caffeine, molecules which differ by only one methyl group (Zimmermann et al. 2000).

■ Aptamers/Riboswitches

Aptamers and riboswitches are single-stranded oligonucleotides of either DNA or RNA, generally about 60 nucleotides long, which fold into well-defined three-dimensional structures. They bind to their target molecule by complementary shape interactions accompanied by charge and hydrophobic interactions and hydrogen bridges. The target can be small molecules or macromolecules. Aptamers are isolated artificially, whereas riboswitches occur naturally. Several viruses have been shown to encode small, structured RNAs that bind to viral or cellular proteins with high affinity and specificity. It was demonstrated that these RNAs could modulate the activity of proteins essential for viral replication or inhibit the activity of proteins involved in cellular antiviral responses. Also the genomes of prokaryotes have been shown to contain nutrient responsive riboswitches to regulate gene expression.

Synthetically, such compounds can be identified by subjecting large libraries of nucleic acid molecules to a panning procedure (Fig. 23.2). This selection process has been named SELEX (systematic evolution of ligands by exponential enrichment) (Stoltenburg et al. 2007). The resulting ligands are called aptamers. The SELEX process starts by generating a large library of randomized RNA sequences. This library contains up to 10^{15} different nucleic acid molecules that fold into different structures depending on their sequence. The library is incubated with the structure of interest, and those RNAs present in the library that bind the protein are separated from those that do not. The obtained RNAs are then amplified by reverse transcriptase-PCR and in vitro transcribed to generate a pool of RNAs that have been enriched for those that bind the target of interest. This selection and amplification process is repeated (usually 8–12 rounds) under increasingly stringent binding conditions to promote Darwinian selection until the RNA ligands with the highest affinity for the target protein are isolated. This molecular evolution process can also be performed with DNA, circumventing the need for reverse transcription before PCR and in vitro transcription. Automation has reduced aptamer in vitro selection times from months to days, making aptamers suitable for application in high throughput target validation. Aptamers can bind to proteins for therapeutic use, like antibodies do. However, antibody selection requires a biological system for their production. The selection of aptamers is a chemical process, the oligonucleotides are selected and amplified in PCR-reactions, and therefore it can

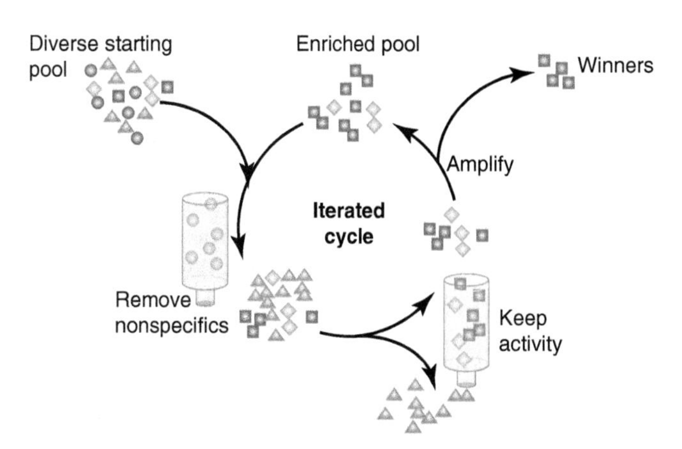

Figure 23.2 ■ General scheme for SELEX (systematic evolution of ligands by exponential enrichment)-based selection of aptamers. The SELEX process starts by generating a large library of randomized RNA sequences. This library contains up to 10^{15} different nucleic acid molecules that fold into different structures depending on their sequence. The library is incubated with the structure of interest and those RNAs present in the library that bind the protein are separated from those that do not. The obtained RNAs are then amplified by reverse transcriptase-PCR and in vitro transcribed to generate a pool of RNAs that have been enriched for those that bind the target of interest. This selection and amplification process is repeated (usually 8–12 rounds) under increasingly stringent binding conditions to promote Darwinian selection until the RNA ligands with the highest affinity for the target protein are isolated (in the Figure: winners). This molecular evolution process can also be performed with DNA, circumventing the need for reverse transcription before PCR and in vitro transcription.

target in principle any protein. In contrast to antibodies, aptamers are prone to bind to functional domains of the target protein (for reasons unknown), such as substrate binding pockets or allosteric sites, thereby modulating the biological function of the molecule. A common problem upon the therapeutic development of aptamers is that they can be so specific for the human version of a target protein that they have poor cross-reactivity with orthologs of the target from other animal species, making their preclinical evaluation difficult. Performing the SELEX procedure by switching between preclinical and clinical target favors selection of cross-reactive aptamers. This may overcome this problem.

One aptamer targeting VEGF, pegaptanib (Macugen®), is marketed for wet age-related macular degeneration (Vinores 2006). The PEGylated aptamer (for PEGylation see also Chap. 21) is injected in the vitreous at a dose of 1.65 mg (0.3 mg of which is aptamer)/eye in 90 μl every 6 weeks. Adverse effects included endophthalmitis and bleeding events in the eye, likely related to the injection procedure. The compound also appeared to exhibit effects in diabetic retinopathy. To increase stability several nucleotides are 2′-O-methyl and 2′-O-fluoro modified and the aptamer is conjugated to polyethylene glycol, which stabilizes siRNA in solution and facilitates clinical delivery.

A number of other aptamers are in clinical development. Anti-C5 factor and antiplatelet-derived growth factor (PDGF) aptamers are developed for the same indication as pegaptanib (Ni et al. 2010). Others target various growth factors and some target proteins involved in coagulation such as von Willebrand factor or tissue factor pathway inhibitor (TFPI).

■ Stimulating Immune Responses

Some differences in chemical structure between genetic information of pathogenic microorganisms and mammals form a recognition signal for immune activation. Specific receptors exist that recognize pathogenic DNA or RNA that subsequently activate a series of genetic programs. This broad proinflammatory activation can have applications in antiviral, immune activating, vaccine adjuvant, and antitumor applications (Underhill 2003).

Prokaryotic DNA contains many CpG dinucleotide sequences, while mammalian DNA has very few, which are usually methylated. Synthetic oligonucleotides containing CpG motifs can mimic prokaryotic DNA and induce immune responses (Murad and Clay 2009). The CpG sequence is a strong recognition signal for mammalian cells through interaction with Toll-like receptor 9 in the endosomes leading to B-cell proliferation and activation of cells of myeloid lineage. CPG 7909 is currently in clinical trials for cancer, while CPG 10101 is developed for hepatitis C virus by Coley Pharmaceutical Group Inc. The difference between the two sequences, which essentially share the same mechanism of action, is that CPG 10101 appears a more potent inducer of the interferon-α pathway, particularly important to fight viral infections. The same firm also develops VaxImmune™ as a support for vaccination protocols, which is based on the same immunostimulatory principle.

In a similar manner, dsRNA can be a predictor of viral infection and Toll-like receptor 3 recognizes dsRNA in the endosomes. In particular, synthetic dsRNA composed of polyinosinic and polycytidylic acids (poly-IC) is a strong activator (Barchet et al. 2008). It is clinically tested as poly-ICLC (Hiltonol™) by Oncovir Inc., a polylysine-complexed formulation of poly-IC, to protect the dsRNA from nuclease-mediated degradation. The system is being developed for the treatment of glioma patients with or without supportive chemotherapy.

GENE REPAIR AND CHROMOSOMAL CHANGE

In eukaryotes, transcription takes place in the nucleus, and translation is located in the cytoplasm. To initiate transcription of a gene, promoters and transcription factors are required, whose action is further complemented by the action of enhancers, binding of specific proteins to regulatory DNA sequences, and methylation of CpG-islands in the promoter region. RNA polymerase produces pre-mRNA molecules that are capped at their 5′-ends and receive a poly (A) tail at their 3′-ends. Nearly all mRNA precursors are spliced. Introns are excised and exons are joined to form the final mRNA sequence (see Chap. 1). Different splicing can form alternative sequences. Subsequently, mRNA binds to the ribosomes. Ribosomes are composed of rRNA and proteins and "read" the mRNA sequence. With the help of tRNA carrying the appropriate amino acids, the protein is formed. The process of transcription and translation is shown in Fig. 23.3.

■ Triplex Helix-Forming Oligonucleotides

Triple helix formation occurs when a polypurine or polypyrimidine DNA or RNA oligonucleotide binds to a polypurine/polypyrimidine region of genomic DNA. Twin helical strands form the DNA backbone. Between the strands grooves exist. These voids are unequally sized as the strands are not directly opposite each other. Triple helix-forming oligonucleotides can bind specifically in the major groove of such stretches of DNA to the polypurine strand, forming (reverse) Hoogsteen hydrogen bonds (Figs. 23.4 and 23.5).

The triplex-forming oligonucleotides have been used for site-directed mutagenesis, in which a mutation is created at a specific site in the chromosomes with or without the use of coupled mutagens, as well as homologous-site-specific recombination using triplex-forming oligonucleotides alone or in combination with a donor fragment to correct genetic disorders (Chin et al. 2007). Although the site specificity is an important benefit for this technique, a complete understanding of the molecular mechanisms involved is still lacking, and rates of mutagenesis or gene correction are still too poor ($\leq$0.1 %) to warrant clinical development.

■ Antisense-Induced Exon Skipping

The exon-skipping technique tries to restore the reading frame by artificially removing one or more exons before or after the deletion or point mutation in the mRNA (Nakamura and Takeda 2009). The most popular disease target to work on is Duchenne's muscular dystrophy which using this technique could be changed into the much milder Becker's dystrophy (Fig. 23.6).

Exons can be skipped from the mRNA with antisense oligoribonucleotides. They attach inside the exon to be removed, or at its borders. The oligonucleotides interfere with the splicing machinery so that the targeted exons are no longer included in the mRNA. In Duchenne's disease the central region of the dystrophin protein is often not essential, and the resulting shorter protein can still perform its stabilizing role of the muscle

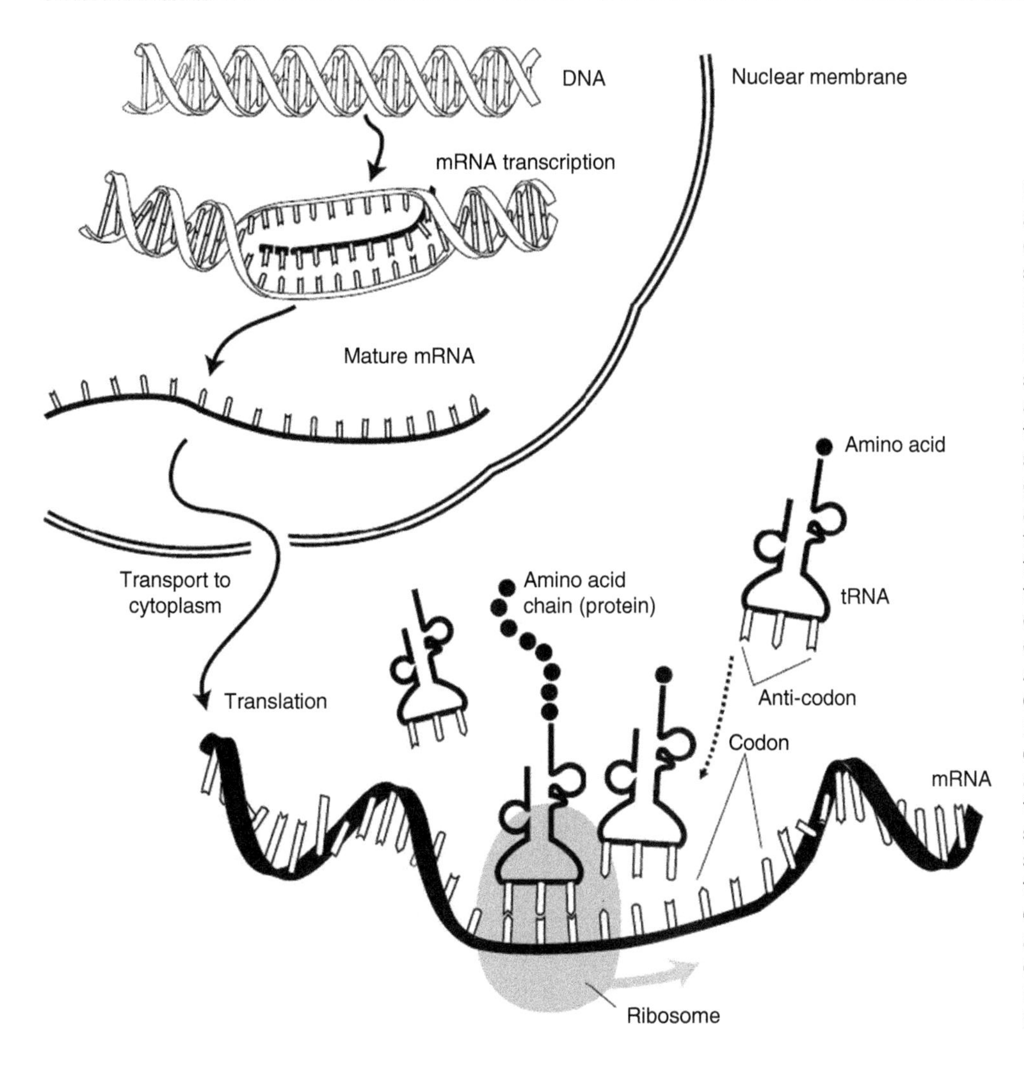

Figure 23.3 ■ Transcription and translation. The synthesis of proteins starts with transcription. DNA is transcribed to RNA to produce mRNA. This happens in the nucleus by RNA polymerase. During transcription the chromosomes are locally decondensed, so that the genes present at this site can be read. Only certain genes are actively transcribed at a certain time depending on the cell's needs. The activity of the genes is determined by transcription factors/enhancers and promoters as well as other internal cellular signals. After splicing of the mRNA, capping and tailing, the mature mRNA is transported to the cytoplasm, where it becomes active in the translation. The translation starts when ribosomes bind mRNA. The ribosomes mediate mRNA codon to tRNA-anticodon binding and couple the amino acids to form a polypeptide (Figure courtesy of National Human Genome Research Institute, NIH, Bethesda, USA).

Figure 23.4 ■ Triple helices are formed through Watson-Crick-base pairing combined with Hoogsteen base pairing, here shown for guanosine = guanosine v cytosine.

cell membrane. The technique is currently in clinical trials for Duchenne's dystrophy, after demonstrating preclinical efficacy in mice and dogs. PRO-044 developed by Prosensa is aiming at exon 44, whereas Prosensa/GlaxoSmithKline's PRO-051/GSK2402968 and AVIPharma AVI-4658 aim at exon 51. Results from a Phase 1-2a study on GSK2402968 show that there were no serious adverse events. The terminal half-life of the compound in the circulation was 1 month. Importantly, GSK2402968 induced detectable, specific exon 51 skipping at doses of 2.0 mg/kg or more resulting in new dystrophin expression leading to a modest improvement in a walk test (Hammond and Wood 2011).

■ Antisense-Induced Ribonucleoprotein Inhibition

Antisense oligonucleotides can be used to inhibit or alter the functions of ribonucleoproteins by specifically binding to the RNA part of the ribonucleoprotein. For example, telomerase, the enzyme involved in preventing the shortening of telomere ends after each cell division, can be inhibited by using oligonucleotides directed against the hTERT domain (i.e., RNA binding domain) (Rankin et al. 2008). Antisense-induced telomerase inhibition resulted in progressive shortening of telomere ends and in some cases induction of apoptosis. As telomerase activity is found in many types of cancer, antisense inhibition of telomerase may

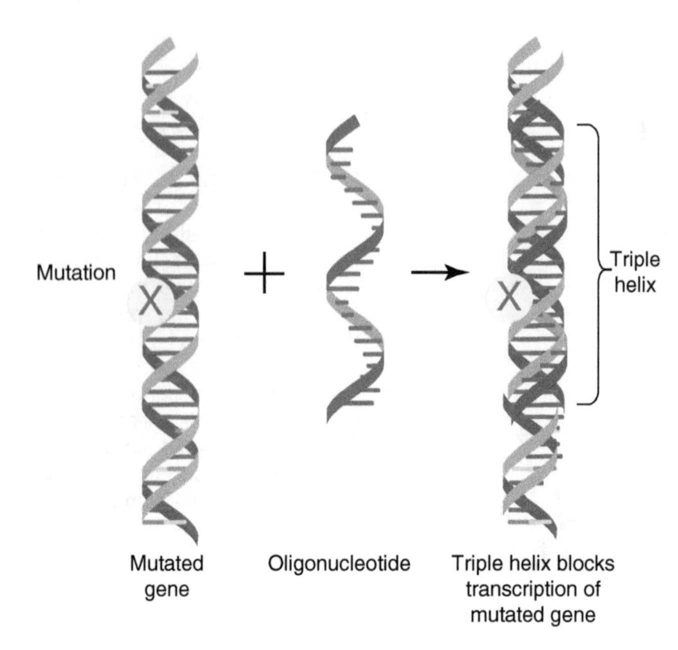

Figure 23.5 ■ Mechanism of action of triple helix-forming oligonucleotides. The oligonucleotide interacts with DNA via Hoogsteen base pairing and thereby prevents mRNA transcription. The 4 bases are indicated by different colors (i.e., *red/blue, yellow/green*). However, the region where the three strands bind should be of one type, i.e., blue-red-blue or red-blue-red (or the same with *yellow-green*) but not mixed. So only two colors for every three-base-bond-combination in the triple helix region.

be an effective approach for cancer treatment. The compound imetelstat (GRN163L) is a lipid-conjugated thioate-modified DNA oligonucleotide developed by Geron corporation and is currently in Phase II clinical trials. Furthermore, antisense oligonucleotides have been used for programmed ribosomal frame shifting to produce different proteins from a single open reading frame (Henderson et al. 2006).

INTERFERING WITH GENE EXPRESSION

Triple helix-forming oligonucleotides, antisense, siRNA, miRNA, transcription factor decoys, ribozymes, DNAzymes, and external guide sequences are all members of the class of oligonucleotides that can knock down gene expression, but they function at different stages of the gene expression process.

■ Triple Helix-Forming Oligonucleotides

Triple helix-forming oligonucleotides, also known as TFO, can also act at the level of transcription of mRNA (Chin et al. 2007). By binding, triplex-forming oligonucleotides can prevent transcription initiation or elongation by binding to promoter, gene or regulatory DNA-regions. The concept has been validated in vivo, but suffers some drawbacks for straightforward application. Because insertion of the third strand in the duplex requires the negatively charged backbones of the nucleic acid strands to come close, it is often difficult to find sufficiently long uninterrupted polypurine sequences in the genome that overcome the electrostatic repulsion and provide stable triplex binding. The use of chemically modified nucleic acids, like peptide nucleic acids (PNA) (Fig. 23.1) that bear no charge in their backbone, strongly facilitates triplex formation and seems especially important for this application, but the approach has not been tested clinically.

■ Transcription Factor Decoys

Transcription factors are nuclear proteins that usually stimulate and occasionally down regulate gene expression by binding to specific DNA sequences, approximately 6–10 base pairs in length, in promoter, or in enhancer regions of the genes that they influence. The corresponding decoys are oligonucleotides that match the attachment site for the transcription factor, known as consensus sequence, thus luring the transcription factor away from its natural target and thereby altering gene expression (Fig. 23.7) (Gambari 2004).

The fact that many transcription factors are involved in regulation of a certain gene and that many genes are controlled by a single transcription factor represent important limitations to the decoy approach, especially when decoy action is only desired in the pathological tissue.

Clinically, this strategy has been evaluated in patients at risk of postoperative neo-intimal hyperplasia after bypass vein grafting. The oligonucleotide, edifoligide, was delivered to grafts intraoperatively by ex vivo pressure-mediated transfection and was designed to target E2F, a transcription factor that regulates a family of genes involved in smooth muscle cell proliferation. While preclinical studies demonstrated beneficial effects, a series of clinical trials yielded mixed results from reduced graft failure to no benefit compared to placebo. The studies did indicate good safety of this local ex vivo treatment strategy. Current clinical studies focus on topical administration of NFκB-decoys for atopic dermatitis (Dajee et al. 2006) and of intratumoral injection of signal transducer and activator of transcription (STAT) 3 hairpin decoys in head and neck cancer (Xi et al. 2005).

■ Antisense/Ribozymes/External Guide Sequences

The function of oligonucleotides to act as antisense molecules was discovered by Zamecnik and Stephenson in 1978, making it the oldest oligonucleotide-based therapeutic approach (Stephenson and Zamecnik 1978). Many of the difficulties associated with the use of oligonucleotides for medical applications have consequently been encountered for antisense molecules

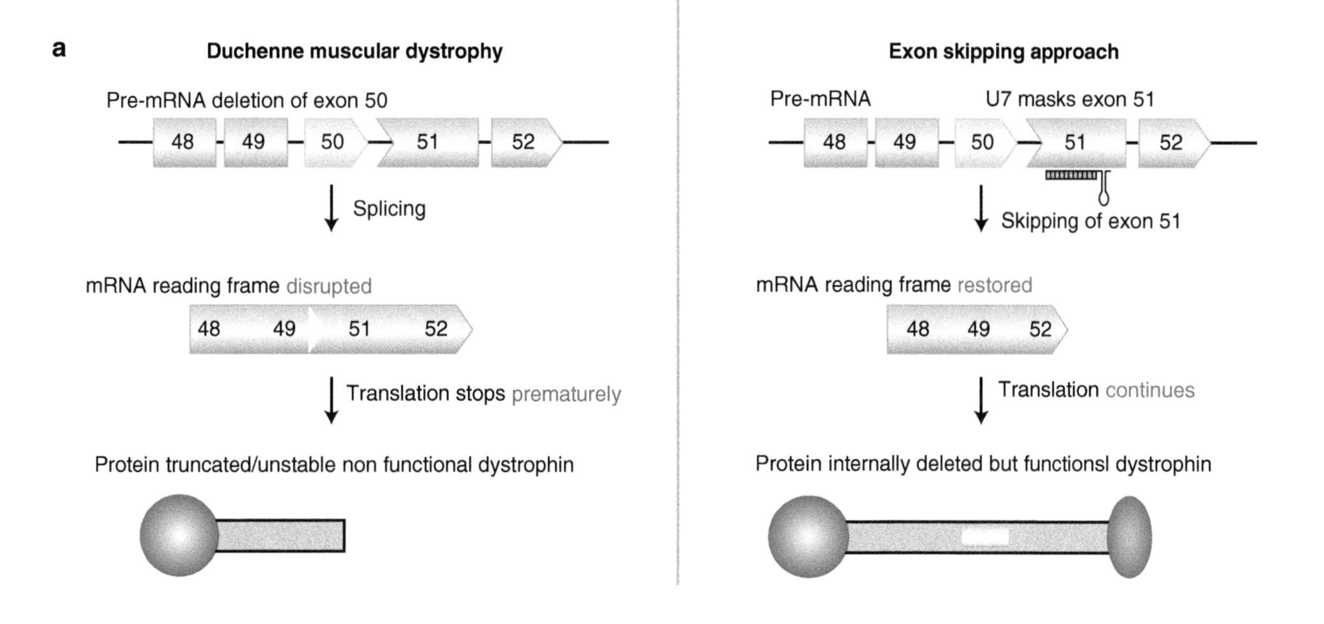

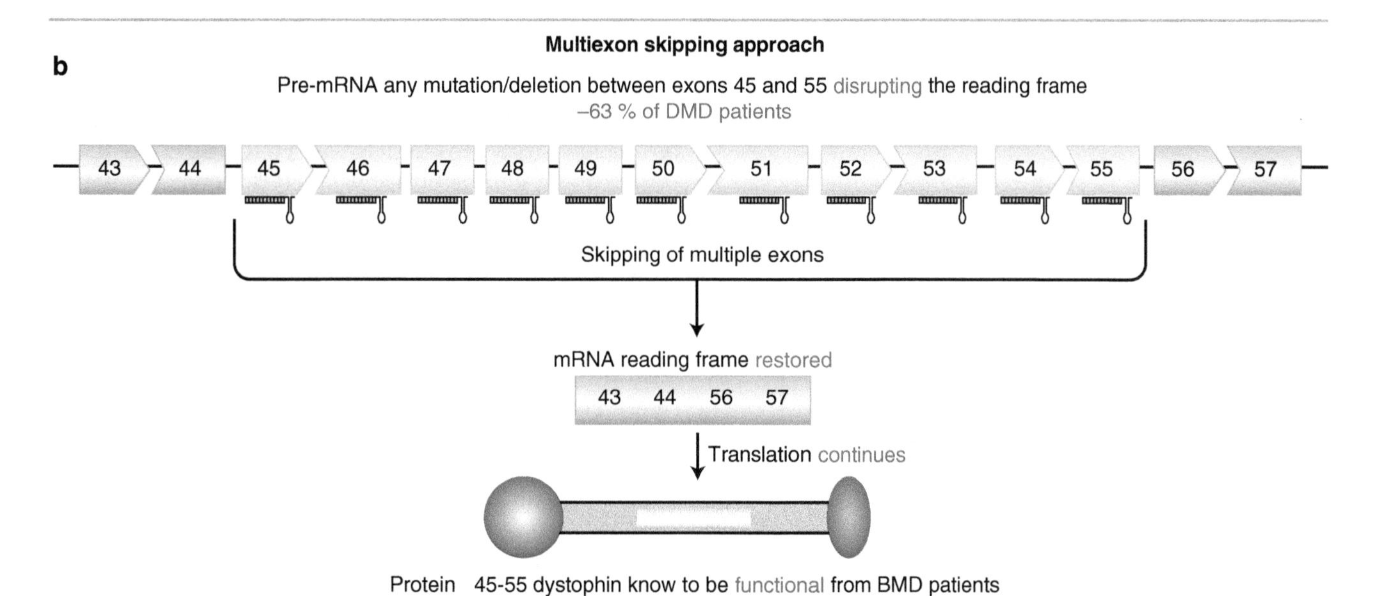

Figure 23.6 ■ Exon-skipping in Duchenne's muscular dystrophy. (**a**) left panel. In Duchenne's muscular dystrophy the mutations in the DMD gene encoding dystrophin cause premature termination of translation. This leads to a non-functional protein that misses the second attachment point to the cytoskeleton. Right panel. By adding an oligonucleotide that preventing splicing factors from interacting with the pre-mRNA, the affected exon is skipped and the mutated region is not incorporated in the mRNA, leading to translation of a functional (albeit shorter) dystrophin molecule that contains both attachment points to the cytoskeleton. (**b**) The same strategy can also be followed to correct mutations in multiple exons, which is the case in the majority of DMD patients.

first, explaining why clinical progress has been difficult. Improvements in synthetic chemistry, knowledge of genome, transcriptome and proteome, and new delivery strategies have revived interest in the technology (Fattal and Barratt 2009). "Classical" antisense oligonucleotides are single-stranded DNA or RNA molecules that generally consist of 13–25 nucleotides. They are complementary to a sense mRNA sequence and can hybridize to it through Watson-Crick base pairing. Three classes of translation inhibiting oligonucleotides can be distinguished based on their mechanism of action:

- mRNA-blocking oligonucleotides, which physically prevent or inhibit the progression of splicing or translation through binding of complementary mRNA sequences (Fig. 23.8)
- mRNA-cleaving oligonucleotides, which induce degradation of mRNA by binding complementary mRNA sequences and recruiting the cytoplasmic nuclease RNase H

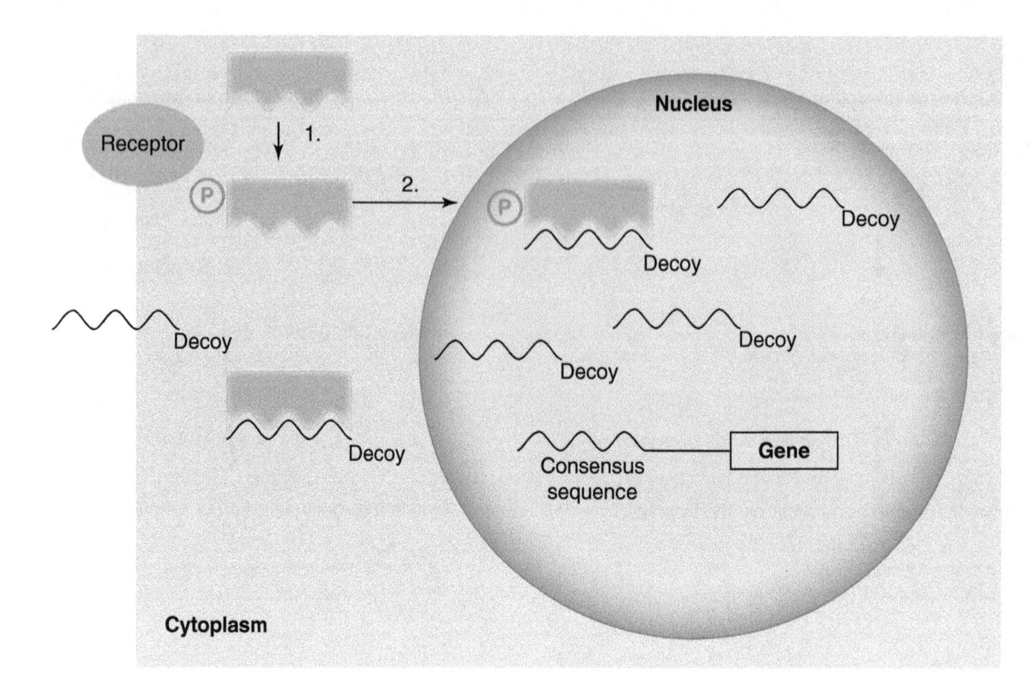

Figure 23.7 ■ Mechanism of action of transcription factor decoys. Transcription factor decoys match the consensus attachment site of the factor and thereby prevent it from binding to the DNA, inhibiting the factor's modulating activity on gene expression level.

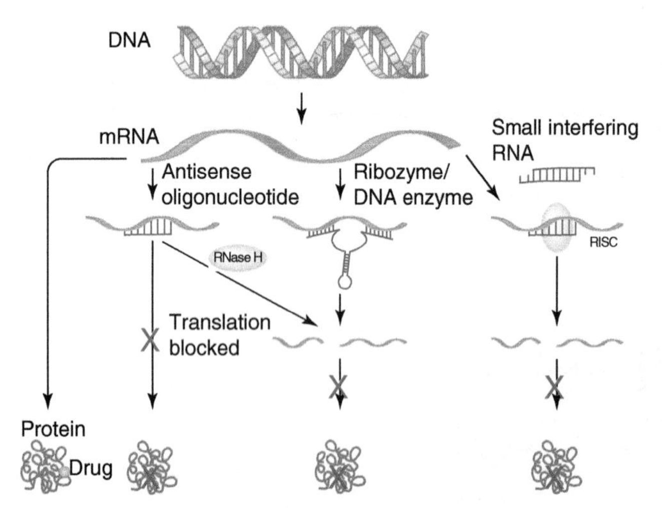

Figure 23.8 ■ Mechanism of action of oligonucleotides that act at the level of mRNA. Antisense oligonucleotides can inhibit mRNA translation by sterically blocking translation or inducing RNaseH-mediated mRNA cleavage. In the ribzoyme/DNAzyme approach, the oligonucleotide possesses mRNA degradative properties or functions as a recruiting factor (known as external guide sequences) for endogenous ribozymes (RNase P). Finally, siRNA- and miRNA-based strategies make use of short double-stranded RNA molecules which unwind and bind complementary mRNA in the RNA-induced silencing complex (RISC), which subsequently cleaves the mRNA.

- mRNA-cleaving oligonucleotides, which induce degradation of mRNA by recruiting nuclear RNase P via external guide sequences or by nuclease activity of the nucleic acid itself (ribozymes/DNAzymes)

The majority of the clinically studied antisense oligonucleotides act through RNase H. RNase H-mediated knockdown generally reaches >80 % down regulation of protein and mRNA expression. In contrast to blocking oligonucleotides, RNase H-recruiting antisense oligonucleotides can inhibit protein expression without a priori restrictions to the region of the mRNA that is targeted. Most blocking oligonucleotides, however, require targeting regions within the 5′-untranslated region or AUG initiation codon region as the ribosome is apparently able to remove bound antisense molecules in the coding region.

One antisense drug has so far been introduced to the market: fomivirsen (Vitravene®), for the treatment of cytomegalovirus-induced retinitis in AIDS patients (Geary et al. 2002). Vitravene® is injected into the vitreous at a dose of 165 μg or 330 μg/eye in 25 μl, once weekly for 3 weeks, followed by 2-week administrations. Reported side effects are related to irritation and inflammation of the eye likely caused by the injection procedure. The oligonucleotide has a thioate backbone which limits nuclease degradation. Local injection at the pathological site improves target cell accumulation.

This successful introduction is overshadowed by failures in Phase III clinical trials of a number of products in recent years. Alicaforsen did not induce clinical remissions compared to placebo in patients with Crohn's disease, and Affinitak™ (targeting protein kinase C-alpha) and oblimersen (Genasense®; Bcl-2 antisense oligonucleotide) failed to substantially prolong survival in cancer patients (Wacheck and

Zangemeister-Wittke 2006). Nevertheless, the experiences from these trials have provided new insights into the use of alicaforsen and Genasense®. Consequently, these compounds are still in clinical trials for other or more specific indications. Interestingly, one antisense molecule is designed to target an endogenous miRNA. These miRNAs (see also the next section) down regulate gene expression by binding to a (partially) complementary mRNA sequence. By inactivating a miRNA through complementary binding to the antisense molecule, gene expression can be restored (Branch and Rice 2010). In this particular case the antisene miravirsen targets miR-122 in hepatitis C virus infection. Table 23.1 shows examples of antisense molecules that are currently in clinical trials.

Although sequence specificity is one of the most attractive features for antisense application, there are reports that show that knockdown of related genes with only limited sequence homology can occur. In addition, the effects of oblimersen appear only partly due to Bcl-2 down regulation but can also be attributed to immune stimulation and mitochondrial apoptosis activation independent of Bcl-2 (Gjertsen et al. 2007).

Ribozymes and DNAzymes are molecules that combine a mRNA binding sequence with a catalytic domain and are capable of cleaving mRNA molecules (see Fig. 23.8). A single molecule is potentially capable to cleave multiple targets capable of multiple turnovers. Several different types of ribozymes are found in nature: the hammerhead, hairpin, hepatitis delta virus (HDV), Varkud satellite RNA, group I and II introns, and the RNA subunit of RNase P. Smaller ribozymes, like hammerhead and hairpin, consist of 40–150 nucleotides. But other ribozymes can be hundreds of nucleotides long and fold into protein-like structures containing single- and double-stranded regions, base triplets, loops, bulges, and junctions (like RNase P).

Ribozymes have been applied clinically (Rossi 1999). In a study in HIV patients, hematopoietic progenitor cells were transduced ex vivo using a retroviral vector carrying an anti-HIV-1 ribozyme. Sustained output of ribozyme expressing mature myeloid and T-lymphoid cells was detected, showing that the concept may work.

DNAzymes structurally and functionally resemble ribozymes, but are made of DNA (Chan and Khachigian 2009). They are artificial molecules and have, so far, not been found in nature. The mechanism of action of DNAzymes is similar to that of ribozymes, but they offer some advantages. Because they are made of DNA, they are easier and less expensive to synthesize, they are much more resistant to degradation, and they possess improved catalytic efficiency. DNAzyme recently produced therapeutic effects in animal models of ischemia, inflammation, and cancer.

RNase P is an endogenous nuclear ribozyme that is substantially larger (several hundred bases) than hammerhead and hairpin ribozymes, which makes it far more difficult to apply exogenously (Ellis and Brown 2009). However, by making use of oligonucleotides that function as so-called small external guide sequences (EGS), they can form a structure together with the mRNA that resembles the endogenous target of RNase P and thereby recruits the enzyme to digest the mRNA-oligonucleotide sequence combination. This concept has not yet been validated in vivo.

■ siRNA/miRNA

MicroRNA (miRNA) and small interfering RNA (siRNA) are double-stranded RNA oligonucleotides of 21–26 base pairs that can cause gene silencing, a process known as RNA interference (Fig. 23.8) (RNAi) (Tiemann and Rossi 2009). In 1998, RNAi was first described in the nematode *Caenorhabditis elegans* (Fire et al. 1998). This silencing phenomenon also occurs in plants, protozoa, fungi, and animals and appears to be conserved in all eukaryotes and may even play a role in prokaryotic cells. It is an important process in endogenous gene expression/translation regulation and defense against pathogens.

miRNAs are produced from transcripts that form stem-loop structures. These are processed in the nucleus into 65–75 nucleotides long pre-miRNA followed by transport to the cytoplasm. Pre-miRNA is further cleaved by an enzyme complex known as Dicer complex to form miRNA, which is loaded into the RNA-induced silencing complex (RISC) that can bind and cleave homologous mRNA. siRNA are produced from endogenous or exogenous long double-stranded RNA (dsRNA) precursors that are also cleaved by Dicer and loaded into RISC. The presence of long dsRNA in mammalian cells can induce an interferon response, which results in nonspecific inhibition of translation and cell death. Therefore, in mammalian systems, the shorter siRNA is used which largely circumvents this response. Next to the direct endonucleolytic cleavage of mRNAs via RISC, miRNA and siRNA appear also to act at other levels. They have been shown to affect methylation of promoters, increase degradation of mRNA (not mediated by RISC), block protein translation, and enhance protein degradation.

Since the discovery of the process, a remarkably rapid progress has been made, and several compounds are currently clinically investigated. This rapid progress is partly due to the strong potency of the RNAi technique which seems to silence gene expression far more efficiently than antisense approaches and partly, also, because much has been learned from previous nucleic acid-based clinical trials. Initial clinical studies focused on macular degeneration as local injection of

Antisense	Target	Indications	Chemistry	Company
Fomivirsen	IE2	CMV infection	Thioate	Isis
Genasense	Bcl-2	Cancer	Thioate	Genta
OGX-011 Custirsen	Clusterin	Cancer	2′O-MOE	OncoGenex
GS101	Insulin receptor substrate	Neovascularization	Thioate	Gene Signal
Monarsen	AcChol esterase	Myasthenia gravis	2′O-ME	Ester Neurosciences
LOR2040	Ribonucleotide reductase	Cancer	Thioate	Lorus Therapeutics
ATL1102	VLA-4/CD49D	Multiple sclerosis	2′O-MOE	Antisense Therapeutics
Archexin	Akt-1	Cancer	Thioate	Rexahn Pharmaceuticals
TPI ASM8	CCR3 receptor	Asthma	Combination	Topigen
	IL5 receptor		Thioate	Pharmaceuticals
LY2181308	Survivin	Cancer	2′O-MOE	Eli Lilly
LY2275796	eIF-4E	Cancer	2′O-MOE	Eli Lilly
AIR 645	Il-4 receptor	Asthma	2′O-MOE	Altair
ISIS-CRP_{RX}	CRP	Cardiovascular	2′O-MOE	Isis
OMJP-$GCGR_{RX}$	Glucagon receptor	Diabetes	2′O-MOE	Ortho-McNeil-Janssen Pharmaceuticals/Isis
ISIS-$SGLT2_{Rx}$	Sodium-dependent glucose transporter 2	Diabetes	2′O-MOE	Isis
Alicaforsen	ICAM-1	Ulcerative colitis	Thioate	Isis
Mipomersen	ApoB100	Hypercholesterolemia	2′O-MOE	Isis
ISIS 113715	PTP-1B	Type 2 diabetes	2′O-MOE	Isis
ISIS 104838	TNF-alpha	Rheumatoid arthritis	2′O-MOE	Isis
AVI-4126	c-Myc	Restenosis	Morpholino	AVI Biopharma
AVI-4065	RdRP	Hepatitis C virus	Morpholino	AVI Biopharma
AVI-4557	CYP450 3A4	Drug metabolism	Morpholino	AVI Biopharma
LErafAON	c-raf	Cancer irradiation	Liposome encapsulated	NeoPharm
AEG35156	XIAP	Cancer	Thioate and 2′O-MOE	Aegera Therapeutics
OGX-427	Hsp27	Bladder cancer	2′O-MOE	OncoGeneX
Trabedersen	TGF-beta-2	Cancer	Thioate	Antisense Pharma
Miravirsen	miR-122	HCV	LNA	Santaris Pharma
SPC2996	Bcl-2	Cancer	LNA	Santaris Pharma
iCO-007	c-raf	Macular degeneration	2′O-MOE	iCO Therapeutics
EZN2968	HIF-1alpha	Cancer	LNA	Enzon Pharmaceuticals
EZN3042	Survivin	Cancer	LNA	Enzon Pharmaceuticals

2′O-MOE 2′-O-methoxy ethyl

Table 23.1 ■ Antisense molecules currently in clinical trials.

oligonucleotides in this immunopriviliged environment allows relative high doses to be administered with minimal immune reactions (as exemplified by the clinically used pegaptanib, see above). An example is PF-4523655 targeting DNA-damage inducible transcript 4 (also known as REDD1) which is currently in clinical trials and developed by Quark Pharmaceuticals. The advantage of this approach appears to be that it targets a novel mediator in the neovascularization route. In general these siRNAs targeting the VEGF-route are well tolerated, but they appear not markedly superior to already existing therapies. At the same time these siRNAs face considerable cost-of-goods. As a result several trials have been halted. The first study demonstrating siRNA-mediated reduction of disease was ALN-RSV01, a siRNA targeting nucleocapsid N gene during a respiratory syncytial virus infection. In the Phase II GEMINI study, a decrease in infection rate in adults experimentally infected with the virus was shown.

At the same time, some intravenous formulations have progressed to clinical stages and have reported positive results.

Calando Pharmaceuticals was the first to show RNAi-mediated knockdown of the M2 subunit of ribonucleotide reductase in melanoma patients after iv administration (Davis et al. 2010). The siRNA was delivered by their proprietary RONDEL technology, consisting of a targeted sterically stabilized cyclodextrin polymer.

Early 2012 positive preliminary results on ALN-PCS, a siRNA targeting PCSK9 for the treatment of severe hypercholesterolemia, were published. ALN-PCS, iv administered in a proprietary cationic lipid formulation, demonstrated silencing of PCSK9 of up to 66 % and reductions of up to over 50 % in levels of low-density lipoprotein cholesterol. Alnylam Pharmaceu-ticals also develops ALN-VSP targeting VEGF and kinesin member 11, also in a lipid formulation for systemic administration in cancer patients. Quark Pharmaceuticals is currently in Phase I clinical trials with a siRNA targeting p53 for treatment of acute kidney injury via the systemic route (Molitoris et al. 2009).

Companies developing miRNA-based therapeutics have not progressed to clinical trials as yet.

PHARMACOKINETICS OF OLIGONUCLEOTIDE-BASED THERAPEUTICS

Pharmacokinetic studies with different types of oligonucleotides (in particular phosphorothioate oligonucleotides) have demonstrated that oligonucleotides are rapidly absorbed from injection sites. Bioavailability of oligonucleotides can be as high as 90 % after intradermal injections. Oral bioavailability, however, is generally very low due to their large molecular weight, multiple charges at physiological pH, and limited stability in the gastrointestinal tract due to nuclease digestion.

Oligonucleotides broadly distribute to peripheral tissues, with highest accumulation in liver, kidney, bone marrow, skeletal muscle, and skin. Passage over the blood–brain barrier has not been reported. Distribution is often fast, with reported distribution half-lives of less than an hour.

Due to the small size of oligonucleotide therapeutics (10–13 kDa), they are normally rapidly cleared from the circulation by renal filtration, with plasma elimination half-lives of <10 min. However, many types of oligonucleotides, especially the phosphorothioate oligonucleotides (see below), bind extensively to plasma proteins. This high plasma protein binding protects oligonucleotides from renal filtration, so that urinary excretion of intact compound is only a minor elimination pathway for highly bound oligonucleotides and plasma elimination half-lives are much longer. Plasma protein binding is also enhanced for lipid-modified oligonucleotides, such as cholesterol-oligonucleotide conjugates. Furthermore, renal filtration can be prevented by modifying the oligonucleotides with large molecules such as polyethylene glycol as long as modification does not hamper its function. PEGylation of aptamers, for instance, results in increased blood residence times of these aptamers, without hampering the ability to bind protein targets. In addition to renal elimination, metabolism by exo- and endonucleases plays an important role in the elimination of oligonucleotides. Nuclease-mediated metabolism is the predominant elimination route for oligonucleotides that have been extensively distributed to peripheral tissues and/or are protected from renal elimination.

IMPROVING OLIGONUCLEOTIDE STABILITY

Nuclease resistance of oligonucleotides can be improved by modifying the backbone of the oligonucleotides. Since the early 1960s of the last century, several chemical modifications have been introduced to prevent such enzymatic degradation. The first generation DNA analogs consisted of the phosphorothioate oligonucleotides, in which one of the non-bridging oxygen atoms in the phosphodiester bond is replaced by sulfur (Fig. 23.1). These phosphorothioate oligonucleotides are more stable in serum, but also show higher binding to proteins than those unmodified oligonucleotides, which can cause toxicity problems. However, protein binding also contributes to the increased circulation half-life seen for PS-oligonucleotide (40–60 h), presumably due to reduced renal excretion. The second generation oligonucleotides are those containing alkyl modifications at the 2′ position of the ribose unit (Fig. 23.1). Although less toxic than the PS-oligonucleotides, they have the disadvantage to be a poor substrate for RNase H and thus can only inhibit translation by forming a steric block. However, oligonucleotides consisting of 2′-O-(2-methoxyethyl) oligonucleotides have greatly improved plasma and tissue half-lives (up to 30 days), presumably due to reduced clearance by nuclease attack. Third generation oligonucleotides consist of a variety of different chemical modifications all aimed to improve stability, pharmacokinetics, and interaction with RNA (Fig. 23.1). Examples are protein nucleic acids (PNAs) that have a polyamide backbone rather than a deoxyribose phosphate backbone, locked nucleic acids (LNAs) that have a methylene bridge between the 2′-oxygen of the ribose and the 4′-carbon, and morpholino nucleic acids containing a nonionic morpholino subunit instead of a ribose interlinked by a phosphoroamidate bond (Fig. 23.1). These third generation oligonucleotides all have in common the superior stability and RNA binding properties, but all lack RNase H activation. Therefore, chimeras of third generation oligonucleotides with DNA have been made that combine good stability and effective RNase H activation.

Modification of ribozymes or aptamers to obtain better stability and resistance against nucleolytic degradation is even more challenging, as such alterations most likely result in loss of enzymatic activity or binding of the complementary strand. The effect of each nucleotide modification on the stability and activity of the ribozyme or aptamer often has to be established empirically. The serum half-life of a DNA ribozyme could be ten-fold increased by protecting the 3′-end with 3′-3′inverted nucleotides (Schubert et al. 2003).

Stability of siRNA in cell culture and serum is generally not a limiting factor due to thermal instability of RNase A. However, for protection against intracellular nucleases, further enhancement of stability might be required. This can be achieved by inserting modified nucleotides on either ends of the siRNA. However, one needs to be aware that the amount and chemical nature of modified nucleotides may influence the gene silencing efficacy of siRNA.

IMPROVING CELLULAR UPTAKE

Besides metabolic elimination by nucleases, the poor cellular uptake of oligonucleotides poses a problem for therapeutic application of oligonucleotides. Compared to conventional drugs, oligonucleotides are relatively large and polyanionic, making passage over cellular membranes virtually impossible (cf. Chap. 4). However, scarce evidence exists that oligonucleotides bound to plasma proteins are taken up by endocytosis. At least the following two distinct uptake mechanisms have been identified: (i) a nonproductive uptake pathway that leads to lysosomal degradation of the plasma protein-bound oligonucleotides and which is the dominant uptake route and (ii) a productive pathway that appears to be clathrin and caveolin independent and which is cell type dependent and accounts for only a minor fraction of the internalized oligonucleotides. This uptake pathway leads to cytosolic release of oligonucleotides via an as yet to identify endosomal release or transport mechanism (Geary 2009). Double-stranded RNA also appears to be transported into (nematode) cells by a transmembrane channel: SID-1(Feinberg and Hunter 2003). Cellular uptake of oligonucleotides can be improved in vitro and in vivo by physical methods, chemically modifying the oligonucleotides, or by making use of specialized delivery systems.

Electroporation of tissue after local injection of oligonucleotides results in improved cellular uptake. Due to high-voltage pulses, transient perforations in the cell membrane occur that allow passage of oligonucleotides into the cytosol. This technique can of course only be applied for delivery of oligonucleotides in vitro and in vivo to tissues readily available for electroporation (e.g., skin, skeletal muscle, or superficial tumor tissue).

Grafting oligonucleotides with cationic groups in order to reduce the ionic repulsion between oligonucleotide and the negatively charged cell membrane represents an alternative strategy to enhance cellular uptake. Synthetic guanidinium-containing oligonucleotides (GONs) showed improved duplex and triplex stability in addition to enhanced cellular uptake. The uptake pattern suggests that these cationic oligonucleotides are internalized by endocytosis, although cytosolic localization could also be observed.

Lipid modification of siRNA has also been proven to be beneficial for cellular uptake and subsequent gene silencing. siRNA against apoB mRNA, which was modified by attaching a cholesterol group to the 3′ terminus of the sense strand, showed increased silencing of the gene encoding for apolipoprotein B compared to the unmodified siRNA after intravenous injection into mice.

Alternatively, cell-penetrating peptides (CPPs) can be conjugated to oligonucleotides with the purpose to enhance membrane translocation. CPPs are small basic peptides derived from protein transduction domains present in a variety of proteins which have strong membrane translocating properties. Conjugation of such a CPP called transportan to PNAs resulted in enhanced cellular uptake in a dose-dependent manner with preservation of antisense activity of the PNA (Chaubey et al. 2005).

Another strategy involves the use of sophisticated delivery systems to enhance cellular uptake and to target oligonucleotides to specific tissues or cells. Most of the delivery systems for oligonucleotides are based on complexation of oligonucleotides with cationic molecules, of which cationic lipids (e.g., Lipofectamine™) are the most common.

The complexation has a dual function: it protects the oligonucleotides from nuclease attack and enhances cellular internalization. By shielding the cationic complexes with large polyethylene glycol polymers displaying targeting ligands, such complexes can be targeted to specific cell types. This has been demonstrated for the delivery of siRNA to angiogenic vascular endothelial cells (Schiffelers et al. 2004). Targeting can also be accomplished by covalently coupling the oligonucleotides to antibodies. In this way, PNAs conjugated to an anti-transferrin receptor monoclonal antibody could be transported over the blood–brain barrier (Pardridge et al. 1995).

The intravenous formulations of siRNA that are in clinical trials and have shown target protein knockdown are based on sterically stabilized cyclodextrin polymers and cationic lipids. The fact that these formulations can specifically inhibit target protein production shows that these challenging nanomedicine formulations can be developed for clinical trials and can perform as designed. It is clear, however, that only a minute fraction of the injected dose arrives at the target site and even more importantly within the target cell cytoplasm.

One of the most intriguing findings reported recently is the observation that miRNA (and mRNA) can be transported between cells through endogenous carrier systems. Cellular export of miRNAs via HDL was demonstrated to be regulated by neutral sphingomyelinase. HDL-mediated delivery of both exogenous and endogenous miRNAs was shown to inhibit mRNA and dependent on cellular uptake via scavenger receptor class B type I (Vickers et al. 2011).

In addition, a couple of extracellular, cell-derived membrane vesicle classes, called exosomes and microvesicles, have been shown to contain miRNA and mRNA in their aqueous interior. Also these RNAs could be functionally delivered to recipient cells. Mimicking these endogenous delivery systems may enable more efficient functional RNA delivery than those based on synthetic approaches. A recent study showed that exosomal delivery of siRNA over the blood–brain barrier could be achieved (Alvarez-Erviti et al. 2011).

DIAGNOSTIC APPLICATIONS

Apart from the direct application of oligonucleotides in medicine as therapeutics, they also fulfill an increasingly important role as diagnostic agents. The highly specific binding of complementary oligonucleotide sequences can be used to detect gene expression profiles and mutations, whereas aptamers can be used to detect the presence of specific compounds. PCR amplification of specific nucleic sequences can provide information on the presence and abundance of this particular sequence. For example, the Amplicor HIV-1 Monitor v1.5 assays manufactured by Roche Diagnostics are currently approved for in vitro diagnostic use to determine viral load in blood samples.

Other applications, like microarrays, do not focus on single genes but provide an overview of the "transcriptome" (Fig. 23.9). A DNA microarray consists of oligonucleotides of approximately 25 bases that are spotted on a chip in an orderly arrangement, representing the genes of interest of an organism. Each oligonucleotide is spotted at a specific location on the array so that the location of each oligonucleotide with corresponding gene is known. Robotic spotters can currently place tens of thousands of oligonucleotides accurately on one slide of a few square centimeters. Each spot contains identical single-stranded oligonucleotides that are attached to the slide surface, allowing cellular DNA or RNA to be labeled and hybridized to the complementary sequence on the array. By quantifying the binding of the labeled DNA or RNA to the specific spots, the abundance of each species can be determined and related to the corresponding gene.

In 2004, the FDA approved the first microarray AmpliChip CYP450 for clinical use. The AmpliChip CYP450 provides complete coverage of the gene variations, including duplications and deletions, of the cytochrome P450 enzymes 2D6 and 2C19. These genes are involved in the metabolism of approximately 25 % of all prescription drugs. It could be regarded as an important step to bring individualization of therapy closer to the patient.

The inherent specificity and selectivity of aptamers makes these oligonucleotides very useful to detect disease-associated molecules, in a similar manner as antibody based immunoassays. However, they are not in clinical use yet.

Padlock probes can also be used for diagnostic applications (Nilsson et al. 1994). They consist of long oligonucleotides, whose ends are complementary to adjacent target sequences. Upon hybridization, the ends of the oligonucleotides are brought together, allowing ligation of the oligonucleotide ends into a closed and intertwined circle that cannot be replaced by the complementary DNA strand. This closed circle can then be amplified by rolling circle amplification, a powerful and robust DNA amplification method based on the mesophilic Phi29 DNA polymerase, allowing amplification of the padlock signal to detectable levels (Fig. 23.10). Padlock probes are more specific than conventional antisense oligonucleotides, as mis-annealing of either one of the ends does not result in proper ligation of the ends and thus prevents circularization of the padlock probe. It has been used for detection of pathogens in biological samples, single nucleotide

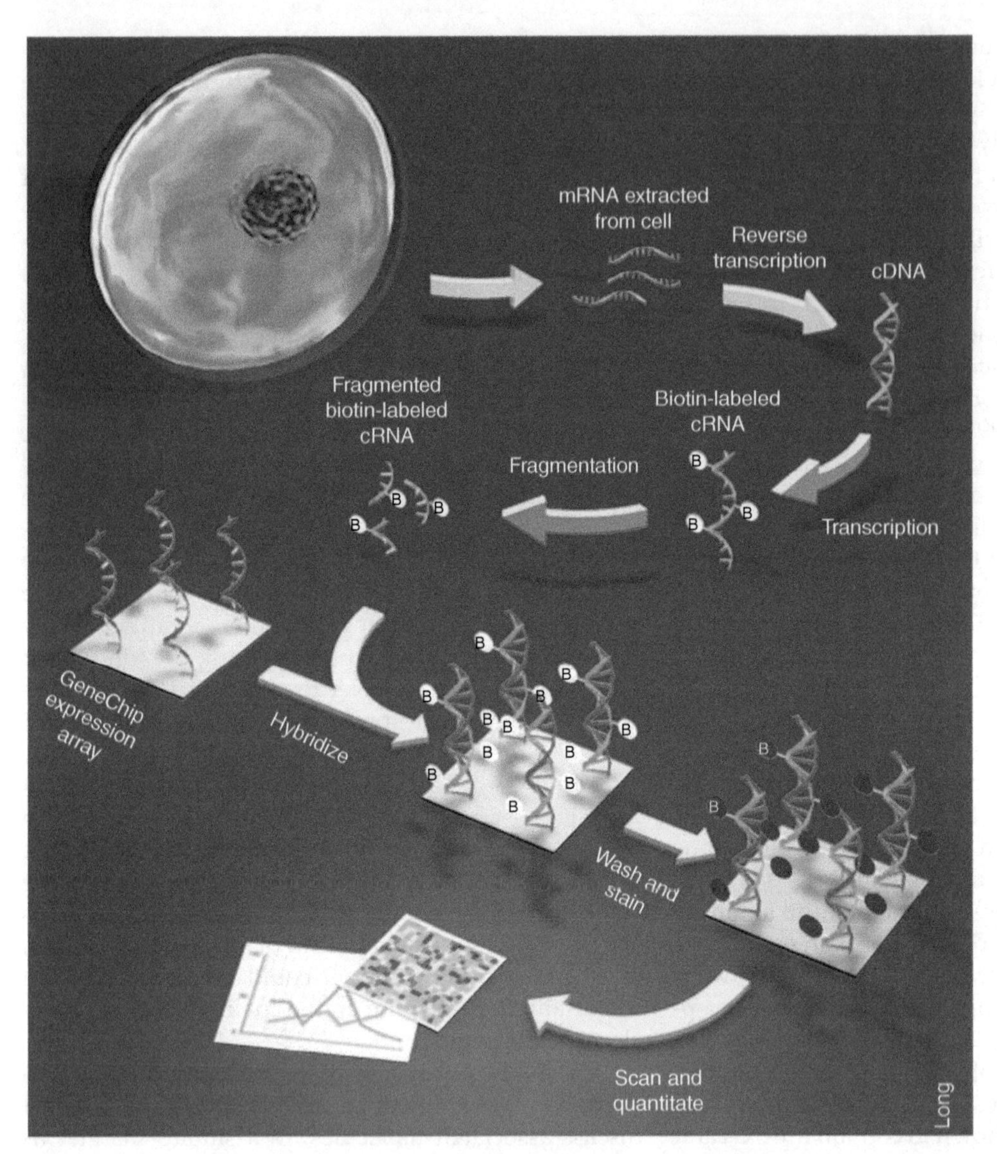

Figure 23.9 ■ Application of microarrays. Isolated mRNA is reverse transcribed into cDNA and subsequently transcribed into labeled cRNA. Fragments of the cRNA can hybridize to complementary oligonucleotides from genes of interest, which are spotted at known locations. Quantification of bound cRNA yields a gene expression profile.

polymorphisms (SNPs), and microRNA, but also for in situ genotyping of individual DNA molecules (Larsson et al. 2004).

PERSPECTIVES

At present, two oligonucleotide-based drugs are marketed, Vitravene® and Macugen®. Both contain chemically modified nucleotides and both are injected in the vitreous, i.e., at the site of the disease process. These choices reflect two of the main difficulties in applying oligonucleotides as therapeutics: (1) oligonucleotides are sensitive towards nucleases, and (2) oligonucleotides have difficulties in reaching the target site. As one of the marketed drugs is an aptamer against an extracellular growth factor, this target site accumulation is apparently even complicated for oligonucleotides that act extracellularly, let alone for oligonucleotide classes that need to interact with intracellular machinery for their action. Current chemical modifications are increasingly able to circumvent these problems by introducing functional groups that offer nuclease resistance and enhanced cellular uptake. Nevertheless, these chemical modifications usually do not enhance specificity of delivery to the target cell population and can distort interaction with the normal cellular machinery, and the chemically modified bases may be incorporated of into normal metabolism.

One of the most promising strategies to circumvent heavy chemical modifications and improve target cell delivery appears the use of nanotechnological vehicles (cf. Chap. 4). Complexing nucleic acids within nanoparticles increases their apparent molecular weight preventing renal excretion, protects against nuclease digestion, and improves target cell recognition and uptake. Targeted delivery seems especially important in view of the plethora of activities nucleic acids can display. It seems hardly possible to find nucleic acid

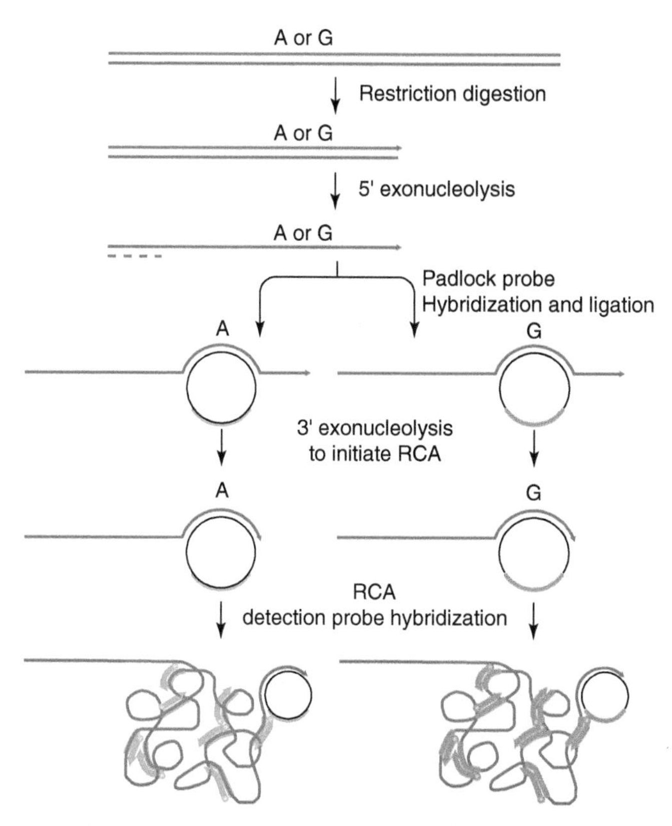

Figure 23.10 In situ genotyping using allele-specific padlock probes and rolling circle amplification (RCA). After fixation of the cells, genomic DNA is cut and made single stranded at the allele position using restriction enzymes and a 5′-specific exonuclease, respectively. Padlock probes are allowed to anneal and ends are closed by ligation to form a circle. Rolling circle amplification is initiated using the target strand as primers, after the protruding 3′ ends of the target strand were removed by the 3′-exonucleolytic activity of the DNA polymerase. In this way multiple copies of the padlock probe are appended onto the 3′ end of the genomic DNA. By using fluorescently labeled oligonucleotides, the 3′ tag can be visualized in cells, allowing in situ detection of single DNA molecules.

sequences that will act through a single pathway and not interact with any of the other pathways. Limiting the number of cell types where the nucleic acids distribute to will likely contribute to reduce side effects.

SELF-ASSESSMENT QUESTIONS

■ Questions

1. Which AS-oligonucleotide modifications are able to recruit RNase H, and which not?
2. Explain the principle of RNAi.
3. What are the major obstacles in therapeutic applications of oligonucleotides?
4. What is the difference between gene correction and gene silencing?
5. Are there any structural and functional differences between DNAzymes and aptamers? If so, name them.
6. What are the different requirements for antisense oligonucleotides that are made for exon skipping and those that are made to inhibit translation of a mutated gene?

■ Answer

1. Only charged antisense oligodeoxyribonucleotide phosphodiesters and phosphorothioates elicit efficient RNase H activity. Non-charged oligonucleotides, including, for example, the peptide nucleic acids, morpholino-oligos, and 2′-O-alkyloligoribonucleotides, do not recruit RNAse H activity and act by physical mRNA-blockade.
2. RNAi is a mechanism for RNA-guided regulation of gene expression in which double-stranded ribonucleic acid inhibits the expression of genes with complementary nucleotide sequences. The RNAi pathway is initiated by the enzyme Dicer, which cleaves double-stranded RNA to short double-stranded fragments of 20–25 base pairs. One of the two strands of each fragment, known as the guide strand, is then incorporated into the RNA-induced silencing complex and base pairs with complementary sequences. The most well-studied outcome of this recognition event is a form of posttranscriptional gene silencing; however, the process also affects methylation of promoters, increases degradation of mRNA (not mediated by RISC), blocks protein translation, and enhances protein degradation.
3. Poor pharmacokinetics, instability, and inability to cross membranes.
4. Gene correction makes use of a homologous recombination process to permanently correct single or multiple point mutations within a region of the gene of interest and therefore acts at the level of DNA. Gene silencing aims at reducing the level of active transcripts of the gene of interest by targeted degradation of the mRNA.
5. Both oligonucleotides form complex three-dimensional structures that can bind their target molecule. However, DNAzymes possess a domain that catalyzes target molecule conversion, whereas aptamers act by physical blockade.
6. Exons can be skipped from the mRNA with antisense oligoribonucleotides. They attach inside the exon to be removed, or at its borders. The oligonucleotides interfere with the splicing machinery so that the targeted exons are no longer included in the mRNA. Classical antisense oligonucleotides are single-stranded DNA or RNA molecules that generally consist of 13–25 nucleotides. They are complementary to a sense mRNA sequence and can hybridize to it through Watson-Crick base pairing. Translation inhibition can be achieved through physical mRNA blockade, or mRNA cleavage by recruitment of RNase H.

REFERENCES

Alvarez-Erviti L, Seow Y et al (2011) Delivery of siRNA to the mouse brain by systemic injection of targeted exosomes. Nat Biotechnol 29(4):341–345

Barchet W, Wimmenauer V et al (2008) Accessing the therapeutic potential of immunostimulatory nucleic acids. Curr Opin Immunol 20(4):389–395

Branch AD, Rice CM (2010) Antisense gets a grip on miR-122 in chimpanzees. Sci Transl Med 2(13):13ps1

Chan CW, Khachigian LM (2009) DNAzymes and their therapeutic possibilities. Intern Med J 39(4):249–251

Chaubey B, Tripathi S et al (2005) A PNA-transportan conjugate targeted to the TAR region of the HIV-1 genome exhibits both antiviral and virucidal properties. Virology 331(2):418–428

Chin JY, Schleifman EB et al (2007) Repair and recombination induced by triple helix DNA. Front Biosci 12:4288–4297

Dajee M, Muchamuel T et al (2006) Blockade of experimental atopic dermatitis via topical NF-kappaB decoy oligonucleotide. J Invest Dermatol 126(8):1792–1803

Davis ME, Zuckerman JE et al (2010) Evidence of RNAi in humans from systemically administered siRNA via targeted nanoparticles. Nature 464(7291):1067–1070

Ellis JC, Brown JW (2009) The RNase P family. RNA Biol 6(4):362–369

Fattal E, Barratt G (2009) Nanotechnologies and controlled release systems for the delivery of antisense oligonucleotides and small interfering RNA. Br J Pharmacol 157(2):179–194

Feinberg EH, Hunter CP (2003) Transport of dsRNA into cells by the transmembrane protein SID-1. Science 301(5639):1545–1547

Fire A, Xu S et al (1998) Potent and specific genetic interference by double-stranded RNA in Caenorhabditis elegans. Nature 391(6669):806–811

Gambari R (2004) New trends in the development of transcription factor decoy (TFD) pharmacotherapy. Curr Drug Targets 5(5):419–430

Geary RS (2009) Antisense oligonucleotide pharmacokinetics and metabolism. Expert Opin Drug Metab Toxicol 5(4):381–391

Geary RS, Henry SP et al (2002) Fomivirsen: clinical pharmacology and potential drug interactions. Clin Pharmacokinet 41(4):255–260

Gjertsen BT, Bredholt T et al (2007) Bcl-2 antisense in the treatment of human malignancies: a delusion in targeted therapy. Curr Pharm Biotechnol 8(6):373–381

Hammond SM, Wood MJ (2011) PRO-051, an antisense oligonucleotide for the potential treatment of Duchenne muscular dystrophy. Curr Opin Mol Ther 12(4):478–486

Henderson CM, Anderson CB et al (2006) Antisense-induced ribosomal frameshifting. Nucleic Acids Res 34(15):4302–4310

Kurreck J (2003) Antisense technologies. Improvement through novel chemical modifications. Eur J Biochem 270(8):1628–1644

Larsson C, Koch J et al (2004) In situ genotyping individual DNA molecules by target-primed rolling-circle amplification of padlock probes. Nat Methods 1(3):227–232

Molitoris BA, Dagher PC et al (2009) SiRNA targeted to p53 attenuates ischemic and cisplatin-induced acute kidney injury. J Am Soc Nephrol 20(8):1754–1764

Murad YM, Clay TM (2009) CpG oligodeoxynucleotides as TLR9 agonists: therapeutic applications in cancer. BioDrugs 23(6):361–375

Nakamura A, Takeda S (2009) Exon-skipping therapy for Duchenne muscular dystrophy. Neuropathology 29(4):494–501

Ni X, Castanares M et al (2010) Nucleic acid aptamers: clinical applications and promising new horizons. Curr Med Chem 18(27):4206–4214

Nilsson M, Malmgren H et al (1994) Padlock probes: circularizing oligonucleotides for localized DNA detection. Science 265(5181):2085–2088

Pardridge WM, Boado RJ et al (1995) Vector-mediated delivery of a polyamide ("peptide") nucleic acid analogue through the blood–brain barrier in vivo. Proc Natl Acad Sci U S A 92(12):5592–5596

Rankin AM, Faller DV et al (2008) Telomerase inhibitors and 'T-oligo' as cancer therapeutics: contrasting molecular mechanisms of cytotoxicity. Anticancer Drugs 19(4):329–338

Rossi JJ (1999) The application of ribozymes to HIV infection. Curr Opin Mol Ther 1(3):316–322

Schiffelers RM, Ansari A et al (2004) Cancer siRNA therapy by tumor selective delivery with ligand-targeted sterically stabilized nanoparticle. Nucleic Acids Res 32(19):e149

Schubert S, Gul DC et al (2003) RNA cleaving '10-23' DNAzymes with enhanced stability and activity. Nucleic Acids Res 31(20):5982–5992

Stein CA, Benimetskaya L et al (2005) Antisense strategies for oncogene inactivation. Semin Oncol 32(6):563–572

Stephenson ML, Zamecnik PC (1978) Inhibition of Rous sarcoma viral RNA translation by a specific oligodeoxyribonucleotide. Proc Natl Acad Sci U S A 75(1):285–288

Stoltenburg R, Reinemann C et al (2007) SELEX–a (r)evolutionary method to generate high-affinity nucleic acid ligands. Biomol Eng 24(4):381–403

Tiemann K, Rossi JJ (2009) RNAi-based therapeutics-current status, challenges and prospects. EMBO Mol Med 1(3):142–151

Underhill DM (2003) Toll-like receptors: networking for success. Eur J Immunol 33(7):1767–1775

Vickers KC, Palmisano BT et al (2011) MicroRNAs are transported in plasma and delivered to recipient cells by high-density lipoproteins. Nat Cell Biol 13(4):423–433

Vinores SA (2006) Pegaptanib in the treatment of wet, age-related macular degeneration. Int J Nanomedicine 1(3):263–268

Wacheck V, Zangemeister-Wittke U (2006) Antisense molecules for targeted cancer therapy. Crit Rev Oncol Hematol 59(1):65–73

Weigand JE, Suess B (2009) Aptamers and riboswitches: perspectives in biotechnology. Appl Microbiol Biotechnol 85(2):229–236

Xi S, Gooding WE et al (2005) In vivo antitumor efficacy of STAT3 blockade using a transcription factor decoy approach: implications for cancer therapy. Oncogene 24(6):970–979

Zimmermann GR, Wick CL et al (2000) Molecular interactions and metal binding in the theophylline-binding core of an RNA aptamer. RNA 6(5):659–667

FURTHER READING

Bennet CF, Swayze EE (2010) RNA targeting therapeutics: molecular mechanisms of antisense oligonucleotides as a therapeutic platform. Ann Rev Pharmacol Toxiol 50:259–293

Cho-Chung YS, Gewirtz AM, Stein CA (2005) Therapeutic oligonucleotides: transcriptional and translational strategies for silencing gene expression. Annals of the New York Academy of Sciences New York Academy of Sciences, New York

Kalota A, Dondeti VR, Gewirtz AM (2006) Progress in the development of nucleic acid therapeutics. Handb Exp Pharmacol 173:173–196

Klussmann S (2006) The aptamer handbook: functional oligonucleotides and their applications. Wiley-VCH Verlag, Weinheim

Phillips MI (2004) Antisense therapeutics. Methods in molecular medicine. Humana Press, Totowa

Raz E (2000) Immunostimulatory DNA sequences. Springer, Berlin

Xie FY, Woodle MC, Lu PY (2006) Harnessing in vivo siRNA delivery for drug discovery and therapeutic development. Drug Discov Today 11(1–2):67–73

24 Gene Therapy

Hao Wu and Ram I. Mahato

INTRODUCTION

The human body is composed of a variety of proteins. Almost all human diseases are the results of improper production or function of proteins. Traditional small molecule drugs usually interact with proteins such as enzymes, hormones, and transcriptional factors to exert their therapeutic potential. However, many severe and deliberating diseases (e.g., diabetes, hemophilia, cystic fibrosis) and several chronic diseases (e.g., hypertension, ischemic heart disease, asthma, Parkinson's disease, motor neuron disease, multiple sclerosis) remain inadequately treated by the conventional pharmaceutical approaches.

Gene therapy is the use of nucleic acids as a pharmaceutical agent to treat disease (Fig. 24.1). It derives its name from the idea that DNA can be used to supplement or alter genes within an individual's cells as a therapy to treat disease. Unlike small molecule drugs or protein drugs which are usually formulated in capsule or tablet forms, therapeutic nucleic acids are packaged within a specialized vector to get inside cells within the body. Gene therapy usually targets one or more defected genes without affecting the normal gene at the site of diseases. A gene therapy target can be an abnormal oncogene whose product has the potential to cause a tumor or a defect gene whose product is critical to maintain normal physiological functions.

The potential use of nucleic acids as therapeutics has attracted great attention to treat severe and debilitating genetic diseases. Compared with small molecule drugs, gene therapy per se will not induce drug resistance even after repeated treatments since the targets of gene medicine are not certain receptors but the genes encoding them. Moreover, this technique could be a permanent treatment to give someone who is born with a genetic disease a chance to live a normal life once and for all. However, there are certain limitations to gene therapy as well. The major disadvantage of gene therapy is that gene medicines are not easy to be formulated into conventional dosage forms and delivered for a routine use. Therefore, the clinical applications of gene therapy can so far only be conducted in hospitals with well-trained specialists. The cost of gene therapy is so far also much higher compared with traditional medicines.

The first approved gene therapy case in the United States took place on September 14, 1990, at the National Institute of Health (NIH). The patient was a 4-year-old girl with severe combined immunodeficiency (SCID) disease caused by a defect adenosine deaminase (ADA) gene. In the therapy procedure, the medical group extracted some of the girl's T lymphocytes, exposed them to a genetically engineered retrovirus that had lost its virulence but carried the normal ADA gene, and transfused them back to the girl's bloodstream. The treatment was successful (Blaese et al. 1995). Ten years after treatment, lymphocytes from the patient continued to express the recombinant transgene, indicating that the effects of gene transfer can be long lasting (Muul et al. 2003).

In this chapter, we will discuss the current state of gene therapy and common approaches to gene transfer. The biology and utility of gene transfer systems, recent advances in cell-based gene delivery systems, the diseases currently subjected to gene therapy, and the regulation of gene products will be discussed and reviewed.

VECTORS FOR GENE TRANSFER

■ Basic Components of Plasmid (cf.Chap. 1)

Gene therapy can be classified into viral gene therapy and nonviral gene therapy, both of which rely on the successful construction of a gene expression plasmid.

H. Wu, Ph.D.
Department of Pharmaceutical Sciences,
University of Tennessee Health Science Center,
19 S. Manassas, RM 224, Memphis, TN 38103-3308, USA

R.I. Mahato, Ph.D. (✉)
Department of Pharmaceutical Sciences,
College of Pharmacy, University of Nebraska Medical Center,
986025 Nebraska Medical Center, Omaha, NE 68198-6025, USA
e-mail: ram.mahato@unmc.edu

D.J.A. Crommelin, R.D. Sindelar, and B. Meibohm (eds.), *Pharmaceutical Biotechnology*,
DOI 10.1007/978-1-4614-6486-0_24,

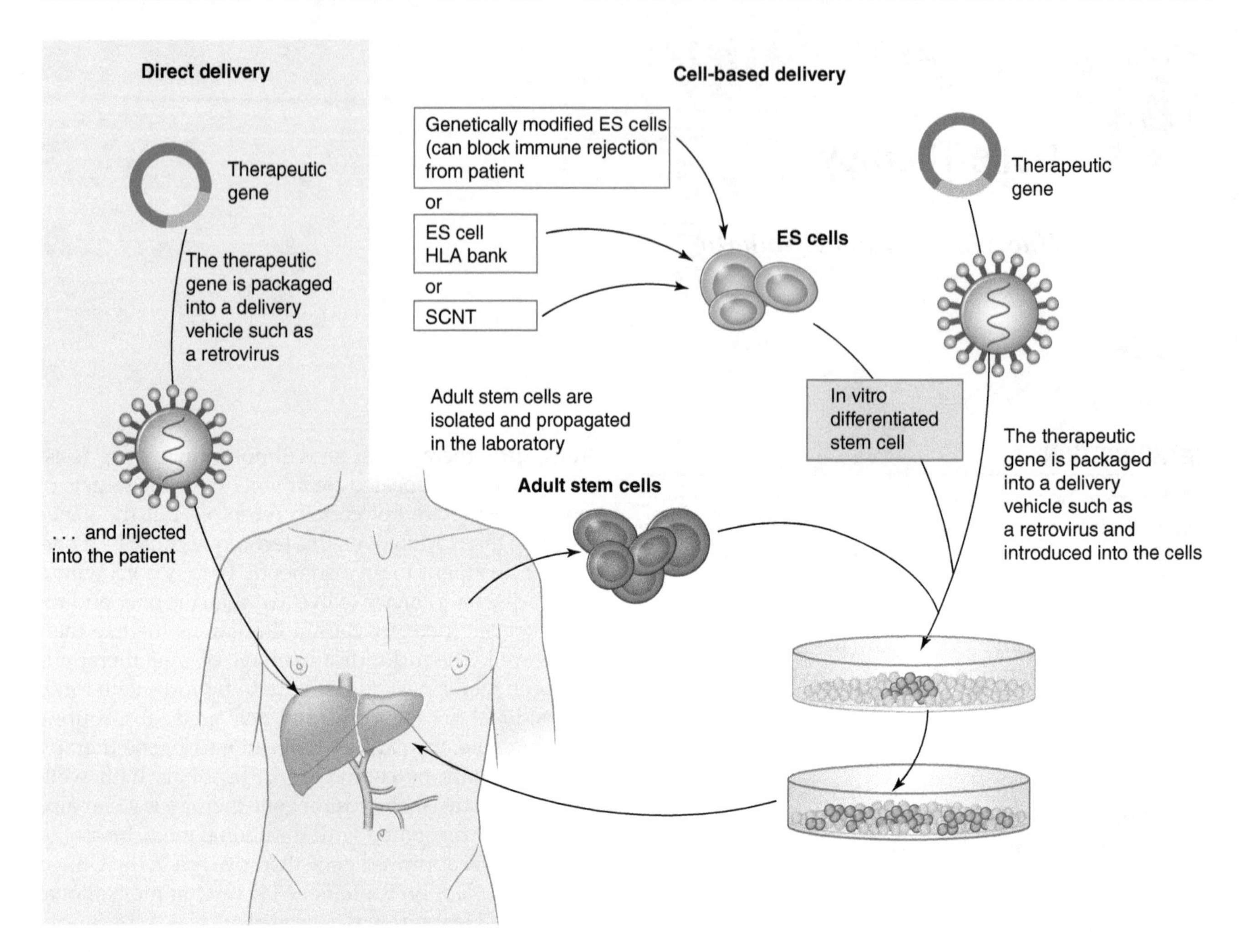

Figure 24.1 ■ Methods of administration of gene therapy vectors. In vivo gene transfer involves direct administration of the vector in the tissue of interest. Ex vivo gene transfer requires collection of cellular targets from the patient. The cells are treated in culture with the vector. Cells expressing the therapeutic transgene are harvested and given back to the patient (From (Zwaka 2006) with permission to reprint (also cf. Chap. 25)).

A plasmid is a circular, double-strand DNA molecule which contains a complementary DNA (cDNA) sequence coding for the therapeutic gene and several other genetic elements including bacterial elements, transcription regulatory elements (TRE), multiple cloning sites (MCS), untranslated regions (UTR), introns, polyadenylation (polyA) sequences, and fusion tags, all of which have great impact on the functioning of the final genetic products. After constructing the plasmid, certain screening methods are needed to validate the construct. For example, DNA sequencing, polymerase chain reaction (PCR), and Southern blot are useful to validate the structure of the construct. Western blot and enzyme-linked immunosorbent assay (ELISA) are useful to confirm the function of the construct (see also Chaps. 1 and 2).

Bacterial Elements

Plasmids have two features that are important for their propagation in bacteria. One is the bacterial origin of replication (Ori), which is a specific DNA sequence that binds to factors that regulate replication of plasmid and in turn control the number of copies of plasmid per bacterium. The second required element is a selectable marker, usually a gene that confers resistance to an antibiotic. The marker helps in the selection of bacteria that have the gene expression plasmid of interest. *Escherichia coli* (E. coli) is a commonly used bacterium for propagating plasmids. It has the property to transfer DNA either by bacterial conjugation, transduction, or transformation. The extensive knowledge about E. coli's physiology and genetics accounts for its preferential use as a host for gene expression. Human insulin was the first product to be produced using recombinant DNA technology from E. coli.

Transcription Regulatory Elements (TRE)

Gene-expressing plasmids contain transcription regulatory elements (TRE) to control transcription. Various TREs (promoters, enhancers, operators, silencers,

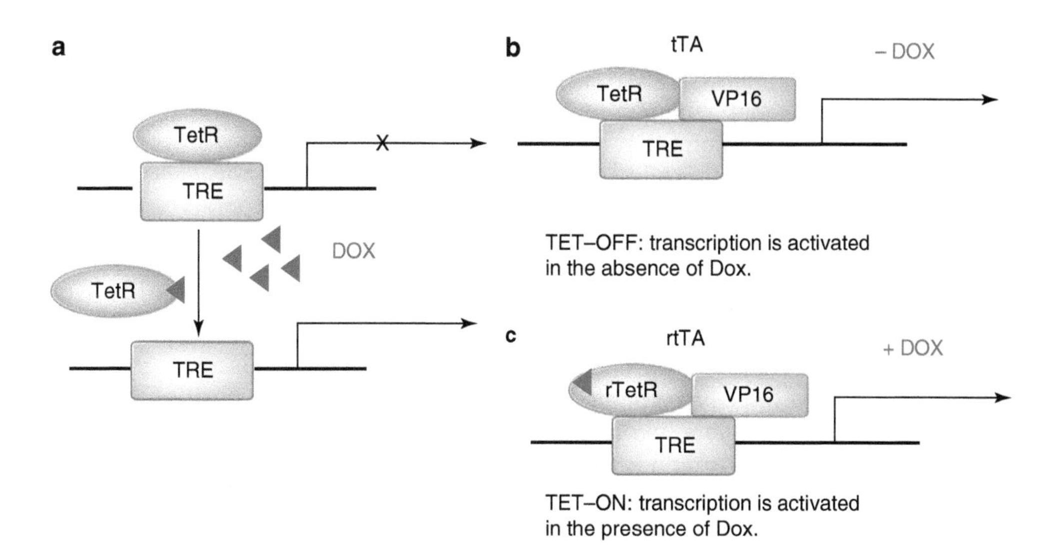

Figure 24.2 ■ Tetracycline (*Tet*)-based reversible inducible gene expression systems. (**a**) Unmodified Tet system. In the absence of doxycycline (*Dox*), tetracycline repressor (*TetR*) binds to Tet response element (*TRE*) and inhibits expression through steric hindrance; after administration of Dox, Dox binds to TetR, removes TetR from TRE, and activates expression. (**b**) Modified Tet-off system. In the Tet-off system, the chimeric tetracycline transactivator protein (*tTA*) consists of the TetR domain fused to the VP16 transactivating domain of herpes simplex virus. tTA binds to TRE to activate gene expression. (**c**) Modified Tet-On system. TetR is mutated to rTetR. Therefore tTA is mutated to rtTA, which must have Dox present in order to bind to TRE and activate gene expression, *r* repressor.

insulators, etc.) interact with the molecular machinery (general transcription factors, activators, co-activators, and repressors) to control the patterns of gene expression.

A promoter is a DNA sequence that enables a gene to be transcribed. The promoter is recognized by RNA polymerase and transcriptional factors. Any mutation in this region will prevent the binding of RNA polymerase and the subsequent transcription and translation. The proper choice of promoter governs the strength and duration of transgene expression. Cytomegalovirus (CMV), Rous sarcoma virus (RSV), and Simian virus 40 (SV40) are some of the strongest known viral promoters. The potency of a promoter can be cell and tissue specific. Overwhelming evidence suggests that the CMV promoter is surprisingly silenced in both embryonic stem cells (ES) and other adult stem cells, making such promoter unsuitable for the new evolving stem-cell-based gene therapy (Qin et al. 2010; McGinley et al. 2011). Other promoters such as EF-1α, chicken β-actin promoter coupled with CMV early enhancer (CAGG), and SV40 are more efficient to drive the transgene expression in stem cells (Qin et al. 2010; McGinley et al. 2011). The distance between the promoter and the transgene cassette also has great impact on gene expression. Several reports have suggested that an insertion between the CMV promoter and transgene cassette surprisingly increases the transgene expression (Li and Mahato 2009).

An enhancer is a short DNA sequence that can bind transcription factors or activators to enhance transcription levels of genes in a gene cluster. While most enhancers are usually close to the promoters and genes, certain enhancers control gene expression from a far distance or even from different chromosomes (Spilianakis et al. 2005). Enhancer-promoter interaction plays a major role to drive gene expression. Enhancers do not directly act on the promoter region, but elicit their effects once they are bound by activators or other transcription factors. These proteins recruit the RNA polymerase and the general transcription factors and stabilize the transcription initiation complex. Different enhancer-promoter combinations have been widely explored to improve the gene transfer efficiency in a variety of tissues and species (Hagstrom et al. 2000).

Other TREs include insulators, operators, and silencers. Insulators are mainly genetic boundary elements to block the enhancer-promoter interaction or more rarely barriers against condensed chromatin proteins. Operators and silencers are usually short DNA sequence close to the promoter with binding affinity to a set of proteins named repressors and inducers. Based on these interactions, an inducible or repressible system can be constructed to either increase or decrease transcription depending on the requirements. For example, the tetracycline-repressor-regulated gene expression system is a popular inducible system in constructing transcription regulatory plasmids (Fig. 24.2a). Based on this system, more advanced Tet-On and Tet-Off systems are constructed for reversible control of the transgene expression (Fig. 24.2b, c).

Multiple Cloning Site (MCS)

A multiple cloning site (MCS), also known as a polylinker, is a short DNA segment which contains many restriction endonuclease recognition sites. Restriction sites within an MCS are typically unique and occur only once within a given plasmid. Within each MCS, there are usually up to 20 restriction sites which can be identified and easily cleaved with commonly used restriction endonucleases. MCS allows the insertion of single cDNA or multiple cDNA depending on the requirement of the therapeutic genes. Generally, the choice of the restriction site for cDNA cloning has no impact on the ultimate transgene expression. However, the choice of the cloning site might occasionally lead to a change in the secondary structure of mRNA and a subsequent translation inhibition. It might call for reengineering MCS for function and convenience (Crook et al. 2011).

Untranslated Regions (UTR)

To express a therapeutic protein, an mRNA must be generated from a cDNA template which is inserted in the MCS and transported into the cytoplasm to be translated. In molecular genetics, untranslated regions (UTR) refer to two sections on each side of a coding sequence on a strand of mRNA. The 5′ UTR is the region of the mRNA transcript that is located between the capsite and the initiation codon. The 5′ UTR contains regulatory elements controlling gene expression. Such elements include the binding sites for proteins to stabilize the mRNA structure, the riboswitches to regulate mRNA's own translation activities, the binding sequence to stabilize or inhibit the translation-initiation complex, and introns to control mRNA splicing and export. The 3′ UTR is the region of the mRNA transcript following the termination codon. 3′ UTR plays an important role in mRNA stability. It contains binding sites for proteins which may affect the mRNA's stability or location in the cell; a polyadenylation tail which is important for the nuclear export, translation, and stability of mRNA; and binding sites for miRNAs which are part of the endogenous gene-silencing machinery.

Introns

The protein coding region in the eukaryotic gene is often interrupted by the stretches of noncoding DNA called introns. In any eukaryotic cells, introns are co-transcribed with protein-encoding exons into a premature mRNA and removed by the mRNA splicing. There is no intron in the cDNA sequence. However, extensive studies have shown that transcripts from the intronless gene are degraded rapidly and that at least one intron should be included in the transcription unit for the optimal transgene expression (Ryu and Mertz 1989). Introns are frequently inserted into the 5′ UTR of the transcript unit (Huang and Gorman 1990).

Polyadenylation (polyA) Sequence

The polyadenylation (polyA) sequence is important for the nuclear export, translation, and stability of mRNA. At the end of transcription, the 3′ segment of the newly made RNA is first cleaved off by a set of proteins. These proteins then synthesize the polyA tail at the RNA's 3′ end. The polyA signal is a recognition site consisting of AAUAAA hexamer positioned 10–30 nucleotides upstream of the 5′ end and a GU- or U-rich element located maximally 30 nucleotides downstream of the 3′end (Mahato et al. 1999). The most important function of polyA sequence is to prevent the mRNA from enzymatic degradation.

Fusion Tag

Fusion tag is a protein or a peptide located either on the C- or N-terminal of the target protein to exert one or several functions such as improving expression, solubility, detection, purification, or localization. For example, fusion of the N-terminus of the target protein to the C-terminus of a highly expressed fusion partner results in high-level expression of the target protein. Maltose binding protein (MBP) is frequently used to increase the solubility of recombinant proteins expressed in E. coli systems (Bedouelle and Duplay 1988). Fluorescent protein tags, such as green fluorescent protein (GFP), provide information about the intracellular location of the transgene expression. Fusion tags like glutathione S-transferase (GST) and MBP made the isolation of recombinant proteins easy using affinity chromatography with specific resins. Other fusion tags such as several peptide sequences of the human c-myc protein were used to increase the nuclear translocation of the target protein to exert physiological functions (Dang and Lee 1988).

VIRAL VECTORS

All viruses hold the inherited advantage to bind to their hosts and introduce their genetic material into the host cell with high efficiency. To construct a viral vector, the genes responsible for the viral replication and pathogenicity are first removed and replaced with a transgene cassette. Then the recombinant viral genome is inserted into a shuttle plasmid and transduced into a packaging cell line which contains the genes responsible for the viral replication to generate the recombinant viral vectors (Fig. 24.3). The vector construct contains the terminal sequences (ITRs or LTRs), the packaging signal (ψ), and the transgene cassette. The packaging signal (ψ) regulates the essential process of

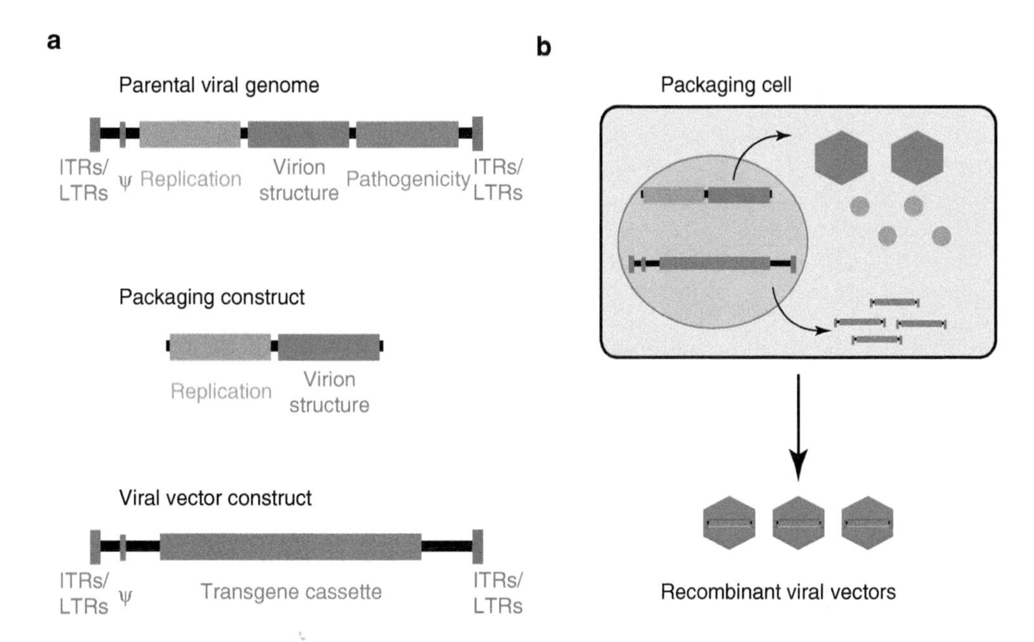

Figure 24.3 ■ General overview of constructing and maintaining a viral vector. (**a**) Schematic overview of a viral genome, the packaging construct and the vector construct. The viral genome containing genes involved in replication, production of the virion structure, and pathogenicity of the virus is flanked by the terminal sequences (ITRs or LTRs) and the packaging signal (ψ). The packaging construct contains only genes that encode replication and structural proteins. The vector construct contains the terminal sequences (ITRs or LTRs), the packaging signal (ψ), and the transgene cassette. (**b**) The packaging and vector constructs are introduced into the packaging cell by transfection, by infection with helper virus, or by generating stable cell lines. Replication-related proteins and viral particles are expressed by the packaging construct. The vector constructs are replicated and encapsidated into virus particles to generate the recombinant viral vector.

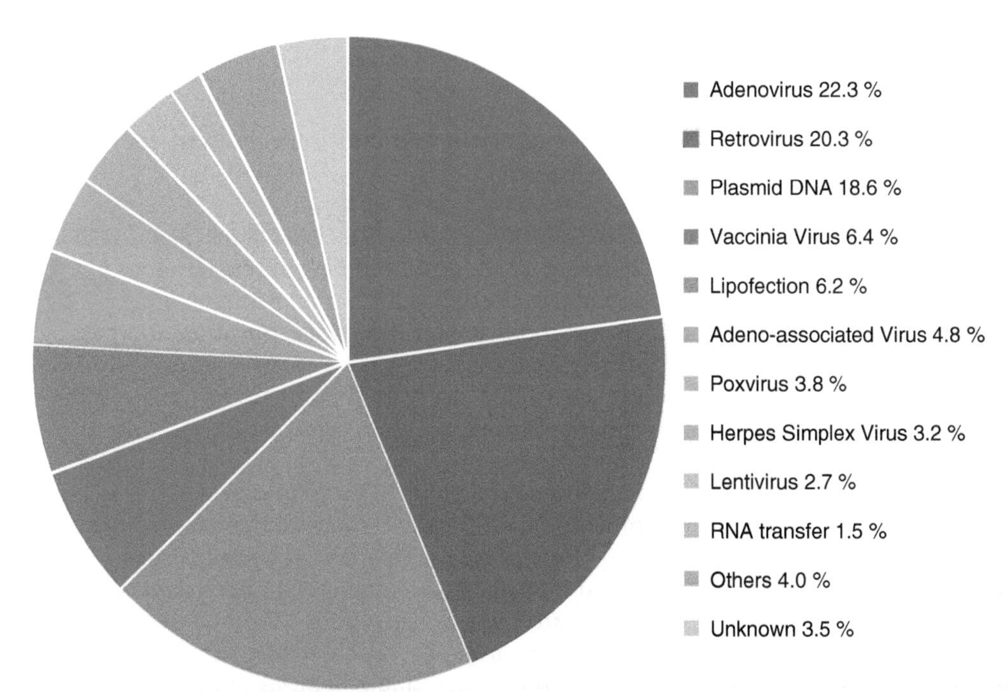

Figure 24.4 ■ Vectors of gene therapy clinical trials (Wiley 2012). Others indicate the clinical trials in which vector used was not reported.

packaging the genetic materials into the viral capsid during replication. Viral vectors typically hold high transduction efficiency and do not need additional carriers for effective gene delivery. To date, approximately 70 % of all gene therapy clinical trials employ viral vectors (Fig. 24.4). Retrovirus, lentivirus, adenovirus, and adeno-associated virus (AAV) are the most extensively studied and used viral vectors for human gene therapy. Their characteristics have been listed and compared in Table 24.1.

	Retrovirus	Lentivirus	Adenovirus	Adeno-associated virus
Genetic material	RNA	RNA	dsDNA	ssDNA
Genome size	7–11 kb	8 kb	26–45 kb	4.7 kb
Cloning capacity	8 kb	8 kb	7[b]–35[c] kb	< 5 kb
Genome forms	Integrated	Integrated	Episomal	Stable/episomal
Diameter	100–145 nm	80–120 nm	80–100 nm	20–22 nm
Tropism	Dividing cells only	Broad, dividing & nondividing cells	Broad, dividing & nondividing cells	Broad, not suitable for hematopoietic cells
Virus Protein Expression	No	Yes/no	Yes b/no c	No
Transgene expression	Slow, constitutive	Slow, constitutive	Rapid, transient	Moderate, constitutive, transient
Delivery method	Ex vivo	Ex vivo	Ex/in vivo	Ex/in vivo
Typical yield (viral particle/ml)	$<10^{8}$	$<10^{7}$	$<10^{14}$	$<10^{13}$
Preexisting immunity	Unlikely	Perhaps, post-entry	Yes	Yes
Immunogenicity	Low	Low	High	Moderate
Potential pathogenicity	Low	High	Low	None
Safety	Insertional mutagenesis	Insertional mutagenesis	Potent inflammatory response	None to date but long term not clear
Physical stability	Poor	Poor	Fair	High

[a]Information compiled from references (Edelstein et al. 2004; Weber and Fussenegger 2006)

[b]First-generation, replication-defective adenovirus

[c]Helper-dependent adenovirus

Table 24.1 ■ Characteristics of viral vectors for gene transfer[a].

■ Retrovirus

Biology

Retroviruses are enveloped RNA viruses containing two copies of a single-stranded RNA genome (Fig. 24.5). Retroviruses are 80–100 nm in diameter and have a genome of about 7–10 kb, composed of group-specific antigen (gag) gene codes for core and structural proteins of the virus; polymerase (pol) gene codes for reverse transcriptase, protease, and integrase; and envelope (env) gene codes for the retroviral coat proteins. The long terminal repeats (LTRs) control the expression of viral genes, hence act as enhancer-promoter. The final element of the genome, the packaging signal (ψ), helps in differentiating the viral RNA from the host RNA (Verma 1990).

After viral binding and introducing the viral RNA into the host cell, reverse transcriptase converts the viral RNA to linear double-stranded DNA that integrates into the host genome with the help of the viral integrase. The integrated construct, the provirus, will later undergo transcription and translation as cellular genes do to produce viral genomic RNA and mRNA encoding viral proteins. Virus particles then assemble in the cytoplasm and bud from the host cell to infect other cells.

Suitability of Retroviruses as Vectors for Gene Transfer

To generate replication-deficient retroviral vectors, the sequences encoding the virion proteins (gag, pol, and env) responsible for the viral replication and pathogenicity are replaced by transgenes. The transgenes can be controlled by the native LTRs or exogenous enhancer-promoter sequences which can be engineered into the genome along with the transgene. The chimeric genome is then introduced into packaging cell lines, mostly HEK293 cells to produce the retroviral vectors.

Retroviral vectors have several features for gene transfer applications (Table 24.1). They can accommodate transgene cassettes of 8 kb. They are capable of integrating into the host genome. Therefore retroviruses can produce stable, long-term transgene expression in dividing cells with low immunogenic potential. Retroviruses can also be used to direct the transdifferentiation of stem cells or reprogram the differentiated somatic cells to have stem-cell-like properties (see later). Such features made retroviruses a valuable tool in a new emerging area named "stem-cell-based gene

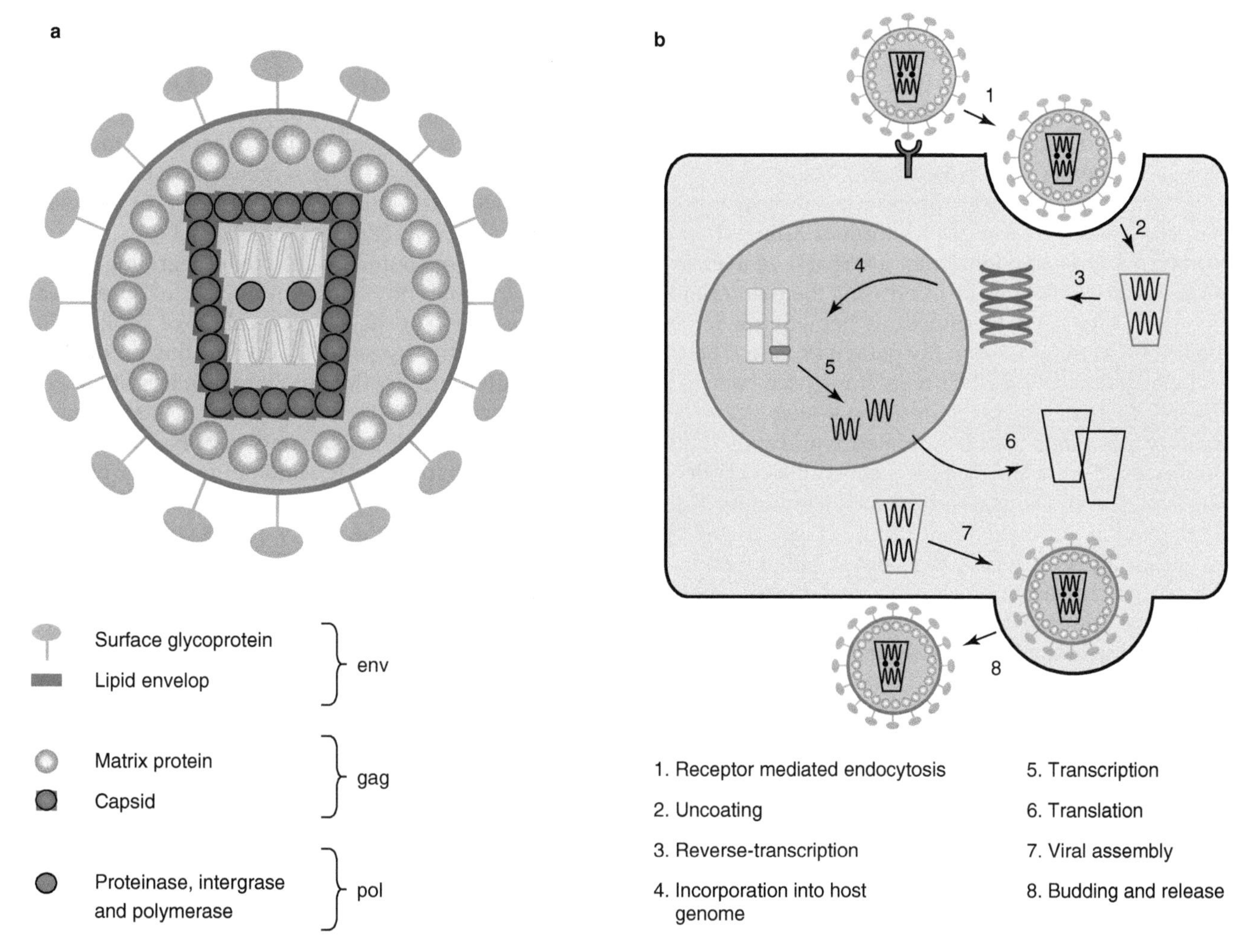

Figure 24.5 ■ The retrovirus. (**a**) Cross section of a retrovirus. (**b**) The retrovirus replication cycle. Retroviruses enter cells by receptor-mediated endocytosis. In the endosome, the lipid envelope and capsid matrix proteins are degraded. The viral RNA is reverse-transcribed into double-stranded DNA which is then shuttled to the nucleus. In the nucleus, the double-stranded viral DNA is inserted into genomic DNA as a provirus. RNA polymerase in the cell copies the viral RNA in the nucleus. These molecules are shuttled out of the nucleus and serve as templates for making additional copies of viral RNA and mRNA that is translated into viral proteins that form the envelope. New virus particles are assembled in the cytoplasm and bud from the cell membrane.

therapy" in the past two decades. However, there are several disadvantages of these vectors. Retroviruses cannot transduce nondividing cells, which are often targets for many gene transfer applications. In addition, current methods of virus production generate preparations in which the virus titer is very low (1×10^5–1×10^7 active virus particles/mL), making its clinical use difficult. Retroviruses are also inactivated by complement systems and rapidly removed from the systemic circulation in response to cellular proteins incorporated in the viral envelope during the budding process.

The major limitation of retrovirus-based gene therapy is that the retrovirus randomly inserts the genetic material into the host genome. If genetic material happens to be inserted in the middle of a gene of the host cell, this gene will be disrupted (insertional mutagenesis). If the gene happens to be one regulating cell division, uncontrolled cell division (i.e., insertional oncogenesis) can occur. Fortunately, this problem has begun to be addressed by custom-designed zinc finger nucleases (ZFNs) [1] or by genetic manipulation of the LTRs of the viral genome (Montini et al. 2009).

Clinical Use of Retrovirus

Approximately 20 % of the currently active clinical trials employ retroviral vectors for gene transfer (Fig. 24.4). The Moloney murine leukemia virus (MoMLV), one of the most thoroughly characterized retroviruses, was the first viral vector to be used in the clinic for treating ADA deficiency caused by SCID, an inherited disease in which the buildup of deoxyadenosine caused by ADA deficiency prohibits the expansion of lymphocytes (Blackburn and

Kellems 2005). MoMLV-expressing recombinant ADA was used to transduce autologous T lymphocytes isolated from the patient ex vivo. Sustained engraftment of cells has been documented 10 years after the last infusion, and no severe adverse effects from this therapy have been reported (Muul et al. 2003).

Another successful clinical trial employing retroviruses was for treating a rare form of X-linked severe combined immunodeficiency (X-SCID) (Cavazzana-Calvo et al. 2000). MoMLV expressing the γc-interleukin receptor was used to transduce hematopoietic stem cells (HSCs) isolated from the patient ex vivo. Then the genetically modified HSCs were transfused back to the patient to reconstitute the immune system. More than twenty patients have been treated worldwide, with a high rate of immune system reconstitution observed. However, a leukemia syndrome was reported in several patients enrolled in the trial (Hacein-Bey-Abina et al. 2003). As a result, the United States Food and Drug Administration (FDA), the Gene Therapy Advisory Committee (GTAC), and Committee of Safety of Medicine in the United Kingdom have declared that this approach should not be first-line therapy for X-SCID, but should be considered in the absence of other therapeutic options.

■ Lentivirus

Biology

Lentiviruses are unique retroviruses being able to replicate in both dividing and nondividing cells. The biology of lentiviruses is quite similar to retroviruses. Apart from the genes gag, pol, and env, lentivirus has six accessory genes such as tat, rev, vpr, vpu, nef, and vif, which regulate the synthesis and processing of viral RNA and other replicative functions.

Human immunodeficiency virus (HIV) is the best known lentivirus. HIV virus has been genetically manipulated to generate viral vectors for efficient gene transfer into human helper T cells and macrophages. Apart from the genes gag, pol, and env, the accessory genes of the lentiviral genome can also be removed to incorporate more genetic materials without affecting the production efficiency of the virus. HEK293 cells were the most frequently used packaging cell lines for lentivirus generation.

Suitability of Lentiviruses as Vectors for Gene Transfer

The significance of lentiviral vectors lies in the fact that they can efficiently transduce nondividing cells or terminally differentiated cells such as neurons, macrophages, hematopoietic stem cells, muscle, and liver cells as well as other cell types for which traditional retrovirus-based gene therapy methods cannot be used. Previous studies have shown that when injected into the rodent brain, liver, muscle, or pancreatic islet cells, lentivirus promoted a sustained gene expression for over 6 months (Miyoshi et al. 1997). Lentiviruses do not elicit significant immune responses and thus can be ideal for in vivo gene expression. Magnetic nanoparticles have been employed for targeted delivery of lentiviral vectors to the endothelial cells even in perfused blood vessels (Hofmann et al. 2009).

Lentiviruses only have limited integrating potential and consequently induce less risk of insertional mutagenesis. However, the generation of replication-competent lentiviruses (RCL) during the production phase or after introduction into target cells is still a primary concern for the clinical use of lentiviruses. Development of self-inactivating vectors that contain deletions within the 3′ LTR, eliminating transcription of the packaging signal to prevent virus assembly, has significantly improved the safety profile of lentiviruses (Zufferey et al. 1998). Another choice is to develop non-integrating lentiviral vectors by point mutations into the catalytic site, chromosome binding site, and viral DNA binding site of the viral integrase (Apolonia et al. 2007).

Clinical Use of Lentiviral Vectors

Because of the perceived risks associated with the use of lentiviruses, clinical trials with these vectors were not initiated until 2001, most of which are for treating HIV infection (MacGregor 2001). In these studies, peripheral blood mononuclear cells (PBMCs) were obtained from the patient. After selective depletion of CD8 T cells, the remaining cells, CD4 T cells were enriched, transduced with the lentiviral vector VRX496, and expanded in culture. The VRX496-transduced cells were then infused back into the patient. The VRX496 vector contains an antisense sequence targeted to the HIV env gene. Expression of the antisense env from an HIV vector transcript would target wild-type HIV env RNA and destroy it and hence, decrease the productive HIV replication from CD4 T cells. The clinical goal for this treatment approach is to decrease HIV viral loads and promote CD4 T cell survival in vivo. Results from this trial showed that although no serious treatment-related adverse events have occurred, no statistically significant anti-HIV effects could be observed in a pilot trial (Manilla et al. 2005).

■ Adenovirus

Biology

Adenoviruses are non-enveloped (without an outer lipid bilayer), icosahedral, lytic DNA viruses composed of a nucleocapsid and a linear double-stranded genome (Fig. 24.6a). Adenoviruses are capable of infecting both dividing and nondividing cells. Fifty-seven serotypes of adenoviruses have been identified

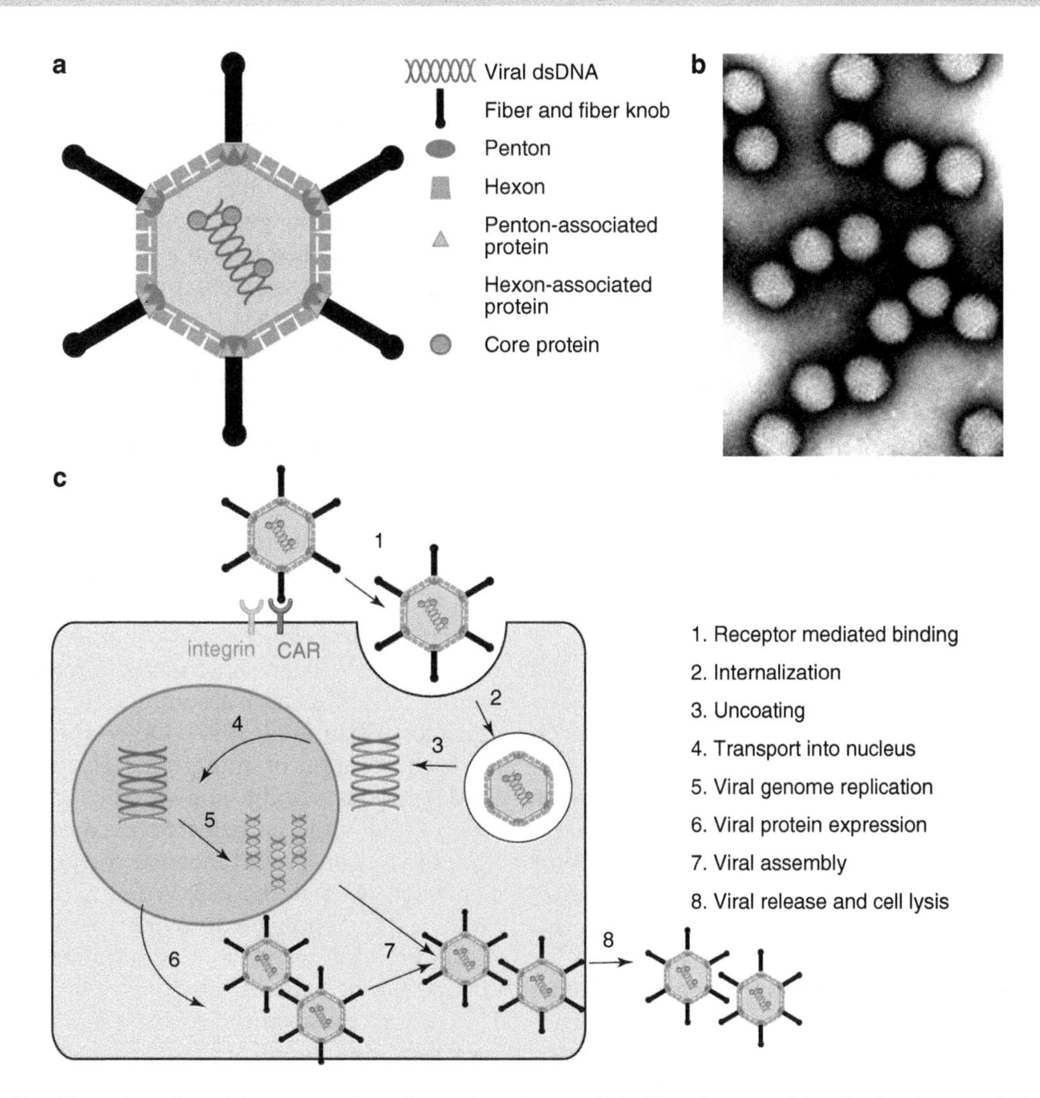

Figure 24.6 ■ The adenovirus. (**a**) Cross section of an adenovirus particle. The virus consists of a double-stranded DNA genome encased in a protein capsid. The capsid is primarily made up of hexon proteins. Penton proteins are positioned at each of the vertices of the icosahedral capsid and serve as the base for each fiber protein. Hexon-associated and penton-associated proteins are the glue that holds these proteins together within and across the facets of the capsid. Core proteins bind to penton proteins and serve as a bridge between the virus core and the capsid. (**b**) Electron micrograph of intact adenovirus serotype 5 particles. (**c**) The adenovirus replication cycle. Adenovirus infection begins with the attachment of fiber proteins to cellular receptors such as coxsackie and adenovirus receptors (*CAR*) and integrins. Through receptor-mediated endocytosis, the virus enters the cytoplasm. In the endosome, capsid proteins are degraded and viral DNA is released into the cytoplasm and transported to the nucleus for replication. After assembly into new viral particles in the cytoplasm, the host cell is lysed and viruses are released. In the case of gene therapy, recombinant replication-defective adenoviruses are used to transduce targeted cells. The inserted transgenes are transcribed to mRNA in the nucleus. Messenger RNA is then transported out of the nucleus and into the cytoplasm where it is translated to therapeutic proteins.

to date. They are grouped into 7 subgroups (A-G) based on genome size, composition, hemagglutinating properties, and oncogenicity. The adenoviruses serotypes 2 and 5 are the most extensively studied and the first to be used as vectors for gene therapy. The adenoviral genome is linear, non-segmented dsDNA, between 26 and 45 kb, composed of six early (E1a, E1b, E2a, E2b, E3, and E4) and five late (L1, L2, L3, L4, L5) genes. The early genes encode proteins necessary for the viral replication, while the late genes encode proteins to assemble into viral particles.

Adenovirus infection begins with binding of the fiber knob on the surface of the viral capsid to the coxsackievirus and adenovirus receptor (CAR) and major histocompatibility complex (MHC) class I (Fig. 24.6c). After initial binding, the penton base interacts with integrin on the cell surface to initiate a series of cell signaling processes allowing internalization via receptor-mediated endocytosis. (Nemerow and Stewart 1999; Medina-Kauwe 2003). Adenovirus particles enter the nucleus as early as 30 min after initial cellular contact. Viral DNA replication and particle assembly in the nucleus starts 8 h after infection and culminates in the release of 10^4–10^5 mature virus particles per cell, 30–40 h post-infection by cell lysis (Majhen and Ambriovic-Ristov 2006).

Suitability of Adenoviruses as Vectors for Gene Transfer

To construct an adenoviral vector for gene therapy, the E1 region and the E3 region of the viral genome were often removed to prevent viral replication and accommodate transgene cassettes. Adenoviruses have a big genome capable to accommodate large transgene cassettes. The adenoviral genome is also easy to be manipulated to generate a vector with multiple deletions and inserts without affecting its transduction efficiency. Recently, adenoviruses with both E1 and E3 inserts to simultaneously express two therapeutic genes have been reported (Panakanti and Mahato 2009). Moreover, adenoviruses with E1, E3, and E4 deletions and even "gutless" adenovirus (adenovirus without viral coding regions) have been constructed to drive transgene expression (Armentano et al. 1995; Chen et al. 1997).

Other favorable characteristics of adenoviruses include that the biology of the virus is well understood, that recombinant virus can be generated with high titer and purity, that transgene expression from adenoviruses is rapid and robust, and that adenoviruses can infect a wide range of dividing and nondividing cells. Adenoviruses do not integrate into the host genome. While this minimizes the risk of insertional mutagenesis, gene expression is transient making adenoviruses unsuitable for long-term correction of genetic defects.

A significant drawback to the use of recombinant adenoviruses is the ability of the virus to elicit a strong immune response including T lymphocyte-mediated "cellular immunity" and antibody-producing "humoral immunity." The cellular response results in the killing of adenovirus transfected cells by T lymphocytes, whereas the humoral response results in the production of antibodies to adenovirus, resulting in the clearance of the adenoviral vectors from the bloodstream. Both actions bring an end to the transgene expression (Dai et al. 1995). Moreover, the preexisting adenovirus serotype 5 immunity in human populations has been shown to significantly reduce the efficacy of these vectors in both preclinical studies and clinical trial (Ertl 2005).

The immunogenicity of recombinant adenovirus also raises intensive safety concern of its clinical applications. The massive immune responses caused by administration of adenovirus could lead to multiple organ failure and brain death. In 1999, a patient died 4 days after an injection with an adenoviral vector, which is the first death of a participant in a clinical trial for gene therapy (Stolberg 1999). Another patient experienced a severe immune response syndrome characterized by multiple organ failure and sepsis and died soon after an adenoviral injection in 2003 (Raper et al. 2003). Preclinical studies also confirmed that the immune response generated by adenoviral vectors must be suppressed before a therapeutic effect can be expected. The transgene expression from adenovirus-transduced cells lasted for about 5–10 days, partially due to the clearance of the transduced cells by the host immune system (Lochmuller et al. 1996). Adenoviruses show extended duration of expression when given to nude mice (mice with an "inhibited" immune system) or when an immunosuppressant is administered (Dai et al. 1995).

A significant effort has been put forth to address the issue of the adenovirus-induced systemic immune response. Adenoviruses with more deletions in the early genes and even "gutless" helper-dependent adenoviruses have been constructed to reduce the inflammatory response and accommodate more transgene cassettes (Chen et al. 1997). Other strategies involved the incorporation of an arginine-glycine-aspartic acid (RGD) sequence and tissue-specific ligands to the surface of the viral particles to decrease the systemic immune responses and increase the gene transfer efficiency (Stewart et al. 1997; Wu et al. 2011). However, none of these strategies provided the full elimination of the immune response. Coadministration of immunosuppressive agents such as cyclophosphamide, FK506, and cyclosporin A extended the duration of transgene expression, but did not prevent development of neutralizing antibodies (Xu et al. 2005; Lochmuller et al. 1996). "Stealth" adenoviruses coated with polyethylene glycol (PEG) or other polymers were also designed to reduce the immunogenicity, increase the blood circulation time, and prolong the transgene expression (Chillon et al. 1998). However, masking adenovirus with PEG or other polymers significantly decreased gene transfer efficiency of adenoviruses.

Generation of replication-competent adenovirus (RCA) is another problem of using adenovirus in the human body. Although the early genes responsible for viral replication and pathogenicity are already removed in the vector construction process, RCA can still be generated by homologous recombination if there is some overlap between sequences in the virus genome and packaging cell. The RCA mixed in the clinically used adenoviral products could be extremely dangerous for the patients. Several groups have observed the production of RCA from HEK293 cells owing to sequence overlap (Louis et al. 1997). Some new packaging cell lines with less overlap have been reported to overcome such problem (Fallaux et al. 1998; 1999).

Clinical Use of Adenoviral Vectors

Today, approximately 23 % of all gene therapy clinical trials involve recombinant adenoviruses, making them the most widely used vector for gene transfer (Fig. 24.4). The safety concern regarding the immunogenicity of adenovirus is the major hurdle for its clinical application. Adenoviral gene therapy faced a major setback in 1999, when a patient died 4 days after injection with an adenoviral vector carrying a corrected gene to test the safety of the procedure (Stolberg 1999). Gendicine, an adenoviral p53-based gene therapy was approved by the Chinese food and drug regulators in 2003 for treatment of head and neck cancer. Advexin, a similar gene therapy approach from Introgen, was turned down by the US Food and Drug Administration (FDA) in 2008. Moreover, despite over 300 clinical trials that have shown it to be well tolerated and efficient in gene transfer, the clinical efficacy of this vector has yet to be proven (Shirakawa 2009).

However, certain breakthroughs in adenovirus-based gene therapy have been made. With the help of the tissue-specific targeted delivery strategies, the new generation of adenoviral vectors is less likely to induce severe systemic immunity. For example, the aerosol administration of a recombinant adenovirus expressing the cystic fibrosis transmembrane conductance regulator (CFTR) to cystic fibrosis patients demonstrated the safety and the proof of concept of adenovirus-based gene therapy (Bellon et al. 1997). In another phase I/II trial, using the gene-directed enzyme-prodrug therapy concept ("suicide gene therapy," see Figs. 24.12 and 24.13 and Disease Targets for Gene Therapy), the intratumoral administration of adenovirus encoding a suicide gene (thymidine kinase, TK) and of intravenous ganciclovir increased the median survival time of patients with malignant glioma from 37.7 to 62.4 weeks without adverse effects (Immonen et al. 2004). Cerepro, a drug composed of thymidine kinase (TK) encoding recombinant adenoviruses, has been granted orphan drug status by the European Committee for Orphan Medicinal Products and by the Office of Orphan Products Development, FDA (see also under: Disease Targets for Gene Therapy).

■ Adeno-Associated Virus (AAV)

Biology

The AAV genome is a 4.7 kb linear, single-stranded DNA molecule composed of two open reading frames (ORF), rep, cap, and two inverted terminal repeats (ITRs) that define the start and end of the viral genome and packaging sequence. The rep genes encode proteins responsible for viral replication, while the cap genes encode structural capsid proteins. ITRs are required for genome replication, packaging, and integration.

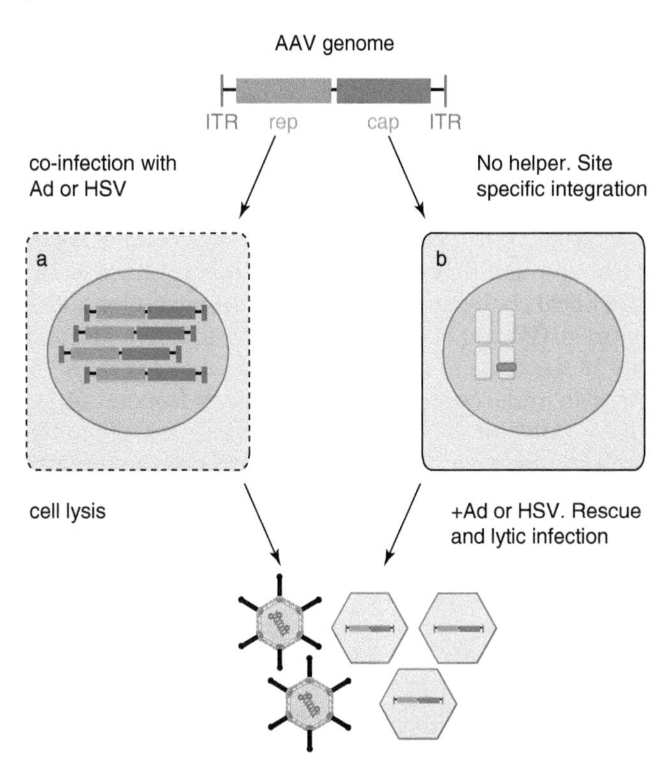

Figure 24.7 ■ Life cycle of adeno-associated virus (*AAV*). AAV can enter cells through receptor-mediated endocytosis. Once in the nucleus, the virus can follow one of two distinct and interchangeable pathways. (**a**) In the presence of helper virus (adenovirus or herpes simplex virus), AAV enters a lytic phase. The AAV genome undergoes DNA replication resulting in amplification of the genome and production of progeny virions. The newly formed AAV viral particles along with helper viruses are released from the cell by helper-induced lysis. (**b**) In the absence of helper virus, it enters a latent phase. During this phase, part of the AAV vectors integrates into host genomic DNA. While the majority of the AAV vector persists in an extrachromosomal latent state without integrating into the host genome. The latent AAV genome cannot undergo replication and production of progeny virions in the absence of helper virus. However, the transgenes carried in the AAV genome are transferred to the host cell and co-express using the host gene expression machinery *ITRs* inverted terminal repeats, *rep* replication.

The icosahedral AAV capsid is 25 nm in diameter. AAV is deficient in replication, and there are no packaging cells which can express all the replication-related proteins of the AAV. Therefore AAV requires coinfection with a helper virus, such as an adenovirus or a herpes simplex virus to replicate (Fig. 24.7). Eleven distinct AAV serotypes have been identified, and over 100 AAV variants have been found in human and nonhuman primate tissues in 2006 (Wu et al. 2006; Mori et al. 2004). The biology of AAV serotype 2 (AAV2) has been the most extensively studied, and this serotype is most often used as a vector for gene transfer.

Suitability of Adeno-Associated Viruses for Gene Transfer

Recombinant AAV vectors have rapidly gained popularity for gene therapy applications within the last decade, due to their lack of pathogenicity and ability to establish long-term gene expression (Table 24.1). The viral genome is simple, making it easy to manipulate. The virus is resistant to physical and chemical challenges during purification and long-term storage (Wright et al. 2003; Croyle et al. 2001). The ability of the virus to integrate in the human chromosome was an initial concern, but eventually it turned out that AAV only integrates into a fixed location of human chromosome and the integration frequency of recombinant AAV is quite low (Surosky et al. 1997).

The AAV vectors are produced by replacing the rep and cap genes with the transgene. Only one out of 100–1,000 viral particles is infectious. Apart from the production of AAV vectors being laborious, these vectors also have the drawback of limited packaging capacity (4.7 kb) for the transgene. Large genes are therefore not suitable for use in a standard AAV vector. To overcome the limited coding capacity, the ITRs of two AAV genomes can anneal to form a head-to-tail structure through trans-splicing between two genomes, almost doubling the capacity of the vector (Yan et al. 2000).

Since recombinant AAV vectors do not contain any viral open reading frames (ORFs), they induce only limited immune responses in humans. Intravenous administration of AAV vectors in mice causes transient production of pro-inflammatory cytokines and limited infiltration of neutrophils, in contrast to an innate response lasting 24 h or longer induced by aggressive viruses (Zaiss et al. 2002). However, despite the limited innate immunity elicited by AAV vectors, the humoral immunity elicited by AAV is still a common event. Up to 80 % of individuals are thought to be positive for AAV2 antibodies in the human population (Erles et al. 1999). The associated neutralizing activity limits the usefulness of the most commonly used serotype AAV2 in certain applications.

Clinical Use of Adeno-Associated Virus Vectors

To date, 86 clinical trials employing recombinant AAV vectors have been initiated worldwide (Fig. 24.4). The first clinical use of recombinant AAV was to transfer the cystic fibrosis transmembrane conductance regulator (CFTR) cDNA to the respiratory epithelium for treating cystic fibrosis (Flotte et al. 1996). This is the first trial to suggest that gene therapy could treat cystic fibrosis in a positive manner. Other phase I and phase II trials have shown that AAV-mediated gene transfer is safe and effective for treating Leber's congenital amaurosis (High and Aubourg 2011; Simonelli et al. 2010), hemophilia (Nathwani et al. 2011), lipoprotein lipase deficiency (Rip et al. 2005), and Parkinson's disease (LeWitt et al. 2011). There are currently four trials using AAV2 vectors in phase III testing for metastatic hormone-resistant prostate cancer (Simons and Sacks 2006).

NONVIRAL VECTORS

The inherent problems with recombinant viruses such as immunogenicity, a.o. reflected in the generation of neutralizing antibodies, and insertional mutagenesis have called for the design of efficient, nonviral vectors for human gene therapy. Nonviral vectors are significantly less immunogenic and are not likely to induce insertional mutagenesis and unwanted homologous recombination after uptake by the cells. They are also relatively easy to be manipulated, produced, and purified in a large scale compared with their viral counterparts. Nonviral gene therapy includes local administration of naked plasmid or using specialized carriers to deliver plasmids to this area. However, their clinical utility is still hampered by the low transfection efficiency, which stems from nonspecific uptake of the vector by epithelial barriers and extracellular matrix and poor delivery into the therapeutic target (Fig. 24.8). The intracellular gene-silencing machinery also prevents the long-term transgene expression. New emerging delivery systems and vector-constructing technologies try to address these issues (cf. Chap. 4)

■ Delivery Methods for Nonviral Gene Transfer

Physical Methods for Gene Transfer

Physical methods involve transfer of naked plasmid by direct disruption of (target) cell membranes. Chemical methods increase the plasmid uptake by the targeted cells using lipid-, peptide-, or polymer-based carriers.

The earliest techniques to deliver recombinant DNA to cellular targets include microinjection, particle bombardment, and electroporation (Table 24.2). Microinjection, direct injection of DNA or RNA into the cytoplasm or nucleus of a single cell, is the simplest and most effective method for physical delivery of genetic material to cells. This transfects 100 % of the treated cells and minimizes waste of plasmid DNA. But, it requires highly specialized equipment and skills. Moreover, microinjection is not suitable for in vivo gene transfer or in vitro gene transfer into tissues or organs composed of a large amount of cells. Particle bombardment, or gene gun treatment, starts with coating tungsten or gold particles with plasmid DNA. The coated particles are loaded into a gene gun barrel, accelerated with gas pressure and shot into targeted cells or tissues in a petri dish. Particle bombardment can be used to introduce a variety of DNA vaccines into desirable cells in vitro. However, particle bombardment has a low penetration capacity, making it unsuitable for in vivo gene delivery apart from easily

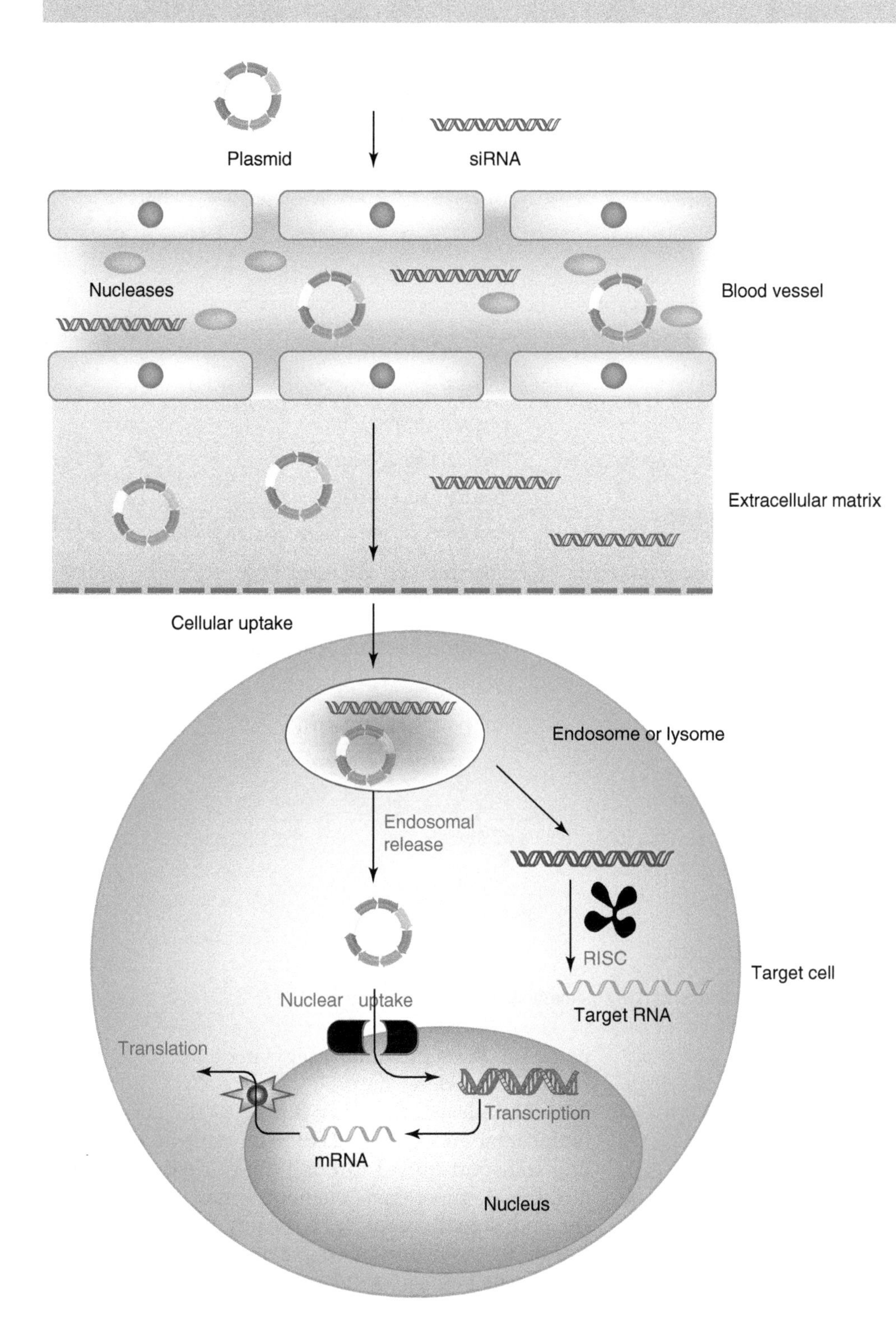

Figure 24.8 ■ Barriers for nonviral gene delivery. Following systemic administration, gene medicine (plasmid or siRNA) comes in contact with blood nucleases. Then they may traverse the blood vessel barrier and the extracellular matrix compartment prior to crossing the plasma membrane barrier. Upon entering the cell via receptor-mediated endocytosis, they are trapped in endosomes and need to be released in the cytosol. Endosomal escape is a major rate-limiting step in gene delivery (From (Singh et al. 2011) with permission to reprint).

accessible tissue, e.g., the skin. Electroporation is used to generate temporary pores in the plasma membrane to transfer plasmid DNA to the cells by an externally applied high-voltage electrical field. Electroporation increases the gene transfer efficiency by 100–1000 folds compared to naked DNA solutions and has met with great success in laboratory practices and clinical trials (Wells 2004). For example, it is frequently used to produce transgenic animals, a powerful tool in preclinical studies (see Chap. 8). Electroporation-mediated gene transfer also demonstrated safety and efficacy in clinical trials to treat melanoma, prostate cancer, and HIV infection (Daud et al. 2008; Vasan et al. 2011). Other physical methods for gene transfer include sonoporation, laser irradiation, magnetofection, and hydroporation. However, because most of the physical methods induce stress and disruption of cellular structure and function, physical methods are less widely studied compared with chemical methods (see below) and are generally restricted to in vitro gene transfer of cultured cells or embryonic stem cells (Table 24.2).

	Advantages	Disadvantages
Naked DNA	No special skills needed Easy to produce	Low transduction efficiency Transient gene expression
Physical methods		
Microinjection	Up to 100 % transduction efficiency (nuclear injection)	Requires highly specialized skills for delivery Limited to ex vivo delivery
Gene gun	Easy to perform Effective immunization with low amount of DNA	Poor tissue penetration
Electroporation	High transduction efficiency	Transient gene expression Toxicity, tissue damage Highly Invasive
Sonoporation	Method well tolerated for other applications	Transient gene expression Toxicity not yet established
Laser irradiation	Can achieve 100 % transduction efficiency	Special skills and expensive equipment necessary
Magnetofection	Safety of method established in the clinic	Poor efficiency with naked DNA
Chemical methods		
Liposomes	Easy to produce	Protein and tissue binding Transient gene expression
Micelles	Easy to produce and manipulate	Unstable Protein and tissue binding
Cationic polymers	High DNA loading Easy to produce and manipulate	Transient gene expression Toxicity
Dendrimers	High DNA loading High transduction efficiency	Extremely toxic
Solid lipid nanoparticles	Low toxicity Controlled release and targeting	NA

Table 24.2 ■ Summary of nonviral methods used for gene transfer.

Cationic Lipids

Since the invention of lipofectamine in 1987, numerous cationic lipids have been synthesized and tested for gene delivery. Most cationic lipids are composed of three parts: (i) a hydrophobic lipid anchor group; (ii) a linker group, such as an ester, amide, or carbamate; and (iii) a positively charged head group, which interacts with the negatively charged plasmid DNA, leading to its condensation and aggregate (nanometer/micrometer range) formation (Fig. 24.9) (Mahato et al. 1997). 2, 3-dioleyloxypropyl-1-trimethyl ammonium bromide (DOTMA) and 3-β[N(NV,NV-dimethylaminoethane)-carbamoyl] cholesterol (DC-Chol) are two commonly used cationic lipids with different structures. Cationic lipids are usually mixed with a neutral co-lipid such as dioleoylphosphatidylethanolamine (DOPE) at a certain molar ratio to reduce the toxicity and enhance gene delivery. PEGylation of polyplexes is frequently used to reduce the plasma binding and increase the circulation time of cationic liposomes. Targeting strategies for these colloidal (pegylated) lipoplexes have been discussed in Chap. 4.

Lipoplexes are taken up by cells through the endosomal route (Fig. 24.8). For endosomal escape and transport to and through the nuclear membrane, additional functional elements may be attached: for endosomal escape (pH sensitive fusiogenic peptides), for transport in the cytoplasm, and nuclear membrane passage (a nuclear translocation peptide).

Peptides

Just like cationic lipids, cationic peptides condense DNA in a similar manner and can be used as gene delivery carriers. Poly (L-lysine) (PLL), a polydisperse, synthetic repeat of the amino acid lysine, was one of the first cationic peptides to deliver genes. However, increase in length of PLL leads to increasing cytotoxicity. Besides, PLL shows limited transfection efficiency and needs the addition of endoosmolytic agents such as fusogenic peptides (see above) to facilitate plasmid

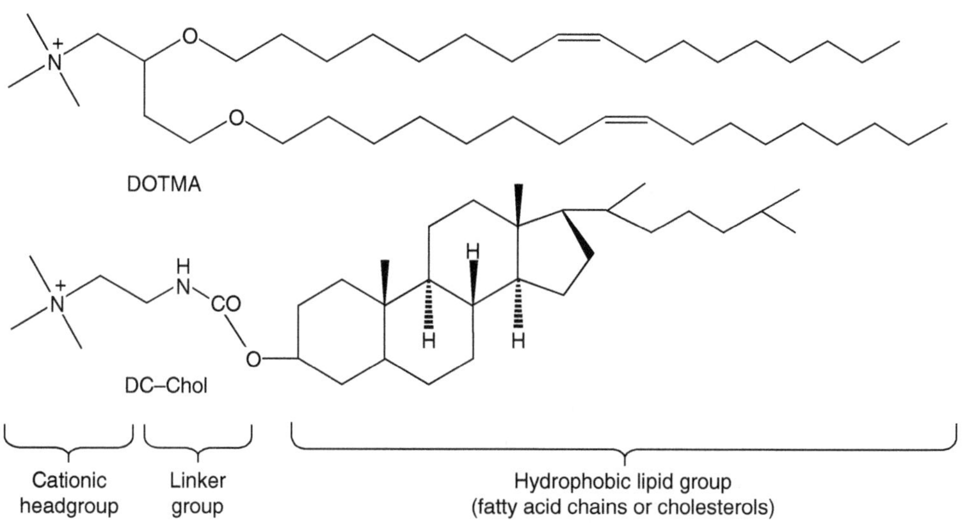

Figure 24.9 ■ Basic components of cationic lipids N-[1-(2,3-dioleyloxy)propyl]-N,N,N-trimethylammonium chloride (DOTMA) and 3-β [N(NV,NV-dimethylaminoethane)-carbamoyl] cholesterol (DC-Chol) (From (Mahato et al. 1997) with permission to reprint).

release into the cytoplasm. Due to these issues, many researchers have turned to the development of PLL-containing "active" peptides and have met with some success (McKenzie et al. 1999). Such peptides offer many advantages over PLL, such as lower toxicity, precise control of synthesis, and homogeneity of peptide length (Martin and Rice 2007).

Peptide-based gene delivery systems have the potential to overcome extracellular and intracellular barriers of gene delivery using a single peptide sequence. However, they suffer from nonspecific plasma protein binding and uptake by the reticuloendothelial system. Here again, PEGylation was demonstrated to be an effective strategy to increase the blood circulation time of the complex (Mannisto et al. 2002). Another unique challenge for peptide-based gene delivery systems is cytosolic proteasomes, which degrade unneeded or damaged proteins by proteolysis. Proteasomes destabilize and degrade the DNA/peptide condensates and prevent effective gene transfer. The involvement of proteasomes in the gene delivery process was first identified in AAV-mediated gene transfer and further confirmed in peptide-based, nonviral gene transfer strategies. In both cases, gene transfer efficiency was significantly higher in the presence of a proteasome inhibitor (Duan et al. 2000; Kim et al. 2005). Coadministration of proteasome inhibitors is the most effective strategy to address this issue.

Polymers

Synthetic and naturally occurring cationic polymers constitute another category of gene carriers. Polyethyleneimine (PEI) has been the most widely used cationic polymer for gene delivery in the last two decades. Boussif et al. first reported that PEI condensed with oligonucleotides and plasmids forms colloid particles (1–1,000 nm) that are a highly efficient delivery system, both in vitro and in vivo (Boussif et al. 1995). However, PEI, especially PEI with a high molecular weight (>25 kD), is extremely cytotoxic. PEI induces the disruption of the cell membrane leading to immediate necrotic cell death and disruption of the mitochondrial membrane after internalization leading to apoptosis. PEI binds to blood components, extracellular matrix, and untargeted cells after intravenous injection. Chemical modification of PEI was proposed to overcome these problems. For example, cholesteryl chloroformate readily formed micelles (10–100 nm) in aqueous solution when conjugated to branched PEI (Wang et al. 2002). This new lipopolymer showed decreased toxicity and optimal gene transfer efficiency.

A simple mixing of plasmid DNA and PEG-b-PLL polymer resulted in the spontaneous formation of polyion complex (PIC) micelles characterized by a small particle size, excellent colloidal stability, and optimal gene transferring ability in serum-containing media (Itaka et al. 2003). However, it should be noted that micelles are in a thermodynamically unstable state. The micelle structure may disintegrate upon the drastic dilution following intravenous injection and lead to inefficient gene transfer. To address this issue, cross-linked micelles were prepared using thiol-modified PEG-b-PLL through the formation of disulfide bonds in the core area (Miyata et al. 2004). These cross-linked micelles are more stable in the blood during circulation and the disulfide bonds are assumed to be cleavable in the cytoplasm (Miyata et al. 2005).

■ Clinical Use of Nonviral Vectors

It is not possible, with current nonviral technologies, to match the high transduction efficiencies and high levels of expression reported with certain viral methods in vivo. Nevertheless, nonviral gene therapies may provide a means for achieving short-term expression of therapeutic gene products in certain tissues with a high degree of safety. Principal approval specifications and recommended assays for assessing the final plasmid DNA preparation purity, safety, and potency for gene therapy and DNA vaccines applications are listed in Table 24.3. There are currently 333 clinical trials using plasmid DNA to treat a number of diseases (Fig. 24.4). Many of these trials are still in phase I testing so far. Collectively, these clinical studies provided "proof-of-principle" for nonviral gene therapy but also highlighted the need for development of formulations with enhanced transfection efficiency and therapeutic efficacy. It should also be mentioned that the majority of these trials were uncontrolled, open label, phase I designs primarily investigating safety and feasibility. The possibility of strong placebo effects cannot be overlooked in these trials. The efficacy results from these studies should be interpreted with caution and can only be assessed by conducting further phase II/III trials.

STEM-CELL-BASED GENE THERAPY

Stem-cell-based gene therapy emerges from recent progress in both cell therapy and gene therapy. Cell therapy describes the process of introducing new cells into a tissue to treat a disease. Recent advances in cell therapy, especially evolving basic insights in stem cell behavior, fostered breakthroughs in regenerative medicine, which is the process of replacing or regenerating human cells, tissues, or organs to restore or establish normal function. Stem-cell-based gene therapy is a multistep process, starting with the isolation of stem cells from the patients. This step is followed by ex vivo expansion of the stem cells and transduction with gene transfer vectors. Finally, the transfected stem cells are infused back into the patient to treat a specific disease.

STEM CELL THERAPY (cf. CHAP. 25)

Stem cells exist in all multicellular organisms and share two characteristic properties. They have prolonged or unlimited self-renewal capacity and the potential to differentiate into a variety of specialized cell types. The earliest stem cells in human life are embryonic stem cells (ESCs), derived from the inner cell mass of the blastocyst and capable to differentiate into all derivatives of the three primary germ layers: ectoderm, endoderm, and mesoderm. Besides the ESCs, which can only be isolated from early embryos, there are other types of stem cells in the mature tissues of all aged mammals, the adult stem cells. Adult stem cells have unlimited self-renewal capacity and a more restricted differentiation potential. They multiply by cell division to replenish dying cells and regenerate damaged tissues. The most famous adult stem cells are bone marrow hematopoietic stem cells (HSCs), which give rise to all the blood cell types and lymphoid lineages. Bone marrow also contains a population of adult stem cells named mesenchymal stem cells (MSCs).

Among all types of stem cells, MSCs have attracted special attentions because of their wide applicability in regenerative medicine. Direct injection of highly pluripotent ESCs into ectopic organs often gives rise to teratoma, a benign tumor containing derivatives of all three germ layers (Nussbaum et al. 2007). MSCs are less potent to induce teratoma or other malignant transformations as they only have restricted differentiation potential (Rubio et al. 2005). Compared with other adult stem cells such as HSCs, mammary stem cells, or neural

Impurity	Recommended assay	Approval specification
Proteins	BCA protein assay	< 3 μg/mg pDNA
RNA	Analytical HPLC	<0.2 μg/mg pDNA
gDNA	Real-time PCR	<0.2 μg/mg pDNA
Endotoxins	LAL assay	< 10 E.U./mg pDNA
Plasmid isoforms (linear, relaxed, denatured)	Analytical HPLC or capillary gel electrophoresis	<3 %
Bacterial and fungal	Method outlined in 21 CFR 610.12	No growth
Biological activity and identity	Restriction endonucleases	Coherent fragments with the plasmid restriction map
	Agarose gel electrophoresis	Expected migration from size and supercoiling
	Transformation efficiency	Comparable with plasmid standards

Information compiled from references (Manthorpe et al. 2005; U.S. Department of Health and Human Services 1998)

Table 24.3 ■ Principal approval specifications and recommended assays for assessing the final plasmid DNA preparation purity, safety, and potency for gene therapy and DNA vaccines applications.

stem cells, MSCs have a well-characterized trophic effect and immunomodulatory property, making them good candidates in treating degenerative diseases. As a trophic mediator, MSC produces bioactive factors which inhibit apoptosis, promote angiogenesis, and stimulate mitosis and differentiation into tissue-specific reparative cells. For example, intravenous transplantation of MSCs was successful in treating systemic diseases such as graft-versus-host disease (GVHD) and osteogenesis imperfecta in humans (Le Blanc et al. 2004). Wakitani et al. also reported several successful clinical cases treating cartilage defects with MSCs (Wakitani et al. 2007). Primary MSCs or genetically modified MSCs have also been employed in regenerating hematocytes, tendon, bone marrow, muscle, and other connective tissues (Phinney and Prockop 2007).

There are two major directions of stem-cell-based gene therapy: (1) stem cells were used as gene delivery vehicles to express therapeutic genes in the target sites and (2) stem cells were reprogrammed or transdifferentiated by genetic modification to replenish the defect cells or tissues (regenerative medicine).

STEM CELLS AS GENE DELIVERY VEHICLES

The advances in gene therapy in the recent two decades have had a great impact on how stem cells could be used to treat certain diseases. Since the first successful gene therapy case in which an ADA gene was inserted into autologous T lymphocytes to treat ADA deficiency-induced SCID, several groups had the ambitious goal to permanently correct ADA deficiency by inserting the ADA gene into hematopoietic progenitor cells from bone marrow and umbilical cord blood (Bordignon et al. 1995; Kohn et al. 1995). Although the overall outcome was disappointing, transgene expression in hematopoietic progenitor cells did provide a selective survival, growth, and differentiation advantage to the lymphocyte descendants.

Stem-cell-based gene therapy also showed promising results in the X-SCID clinical trial in 1999. The protocol consisted of the isolation of CD34-enriched bone marrow progenitor cells that were harvested from the iliac crest of the patients and the ex vivo transduction with retrovirus encoding the common cytokine-receptor γ-chain (γc). In four out of five patients, the infusion of transduced CD34+ cells led to the generation of functional peripheral-transduced T cells similar to those of age-matched controls within 6–12 weeks (Cavazzana-Calvo et al. 2000). Traditional gene therapy (as discussed before) focuses on the introduction of genetic material in mature cells to treat a hereditary, genetic disease, while the stem-cell-based gene therapy may represent a permanent treatment for these genetic diseases.

Recent preclinical studies also demonstrated the promising future of using genetically modified stem cells as therapeutic agents. For example, MSCs were widely reported to be competent trophic mediator for islet transplantation. However, MSCs alone are insufficient to support a rapid and functional revascularization of islet grafts. Genetically modified MSCs not only reversed the incompetence of primary MSCs but also provided MSCs with new functions to target various diseases (Dzau et al. 2005).

STEM CELLS AS REGENERATIVE MEDICINE (cf. CHAP. 25)

Stem cells can be reprogrammed or transdifferentiated by genetic modification to replenish defect cells or tissues. These features make genetically modified stem cells a powerful tool in regenerative medicine. Viral vectors efficiently transduce stem cells and direct their differentiation. Peng and coworkers first demonstrated that muscle-derived stem cells genetically engineered with retrovirus to express human bone morphogenetic protein-4 (BMP4) and VEGF promoted bone formation and bone healing in a mouse model (Peng et al. 2002). This finding was further supported by the finding of Tsuda and coworkers that MSCs genetically engineered with adenovirus to overexpress bone morphogenetic protein-2 (BMP2) enhanced ectopic bone formation in rats (Tsuda et al. 2003). From these studies one can conclude that viable human bone grafts can be engineered under laboratory conditions by using human MSCs and a "biomimetic" scaffold-bioreactor system (Grayson et al. 2010).

Other preclinical studies showed that stem cells can readily transdifferentiate into different types of cells through genetic modification using viral or nonviral vectors. For example, genetically modified induced pluripotent stem cells (iPSCs) and MSCs represent a major source of artificial β-cells and islets for treating diabetes. Karnieli et al. and Li et al. both reported the reversal of hyperglycemia in streptozotocin-induced diabetic mice after transplantation of insulin-producing cells originated from genetically modified Pdx-1 expressing MSCs (Karnieli et al. 2007). In another report, Li et al. demonstrated the in vitro formation of islet-like structures using genetically modified MSCs (Li et al. 2008).

Viral vectors were the most popular tools to direct the differentiation of stem cells in regenerative medicine. However, because of the risks of insertional mutagenesis and generation of replication-competent viruses, nonviral vectors were also studied in stem-cell-based gene therapy. Corsi et al., described a way to transfect MSCs using chitosan-DNA nanoparticles

(Corsi et al. 2003). Kaji et al., introduced a "smart" nonviral transgene expression system to efficiently reprogram somatic cells to iPSCs and shut down the transgene expression once the reprogramming was done (Kaji et al. 2009). Nevertheless, the application of nonviral vectors is still limited because stem cells are typically hard to be transfected by nonviral vectors.

Sources for stem cells are another problem for stem-cell-based gene therapy. iPSCs induced from somatic cells offer potential alternatives to the ESCs and other adult stem cells whose supply is limited nowadays. However, teratoma formation and immunogenicity remain an unsolved issue hampering the wide application of iPSCs (Zhao et al. 2011).

It should be noted that most stem-cell-based gene therapies, especially the ones that use retroviruses or lentiviruses as gene delivery vectors, do not provide a mechanism to shut down the therapeutic gene expression when further expression is unnecessary or to clear the proliferative stem cells when the damaged tissue is fully healed. An inducible system can be added to the vector structures to achieve temporal and spatial control of the transgene expression (Fig. 24.2). Self-inactivating retroviruses were also constructed for safe and efficient in vivo gene delivery (Miyoshi et al. 1998). To clear the excess of stem cells when the damaged tissue is fully healed, Schuldiner et al. reported that human embryonic stem cells genetically engineered to express a "suicide" gene could be eliminated in vivo by administration of the FDA-approved drug ganciclovir (Schuldiner et al. 2003) (cf. Fig. 24.12).

DISEASE TARGETS FOR GENE THERAPY

There are currently 1786 active gene therapy clinical trials worldwide (Wiley 2012). Approximately 65 % of these trials are for cancer treatment. Treatment of monogenetic diseases, cardiovascular diseases, and infectious diseases each take ~10 % of the number of active clinical trials. Treatment of neurological diseases, which has expanded very fast in the last 5 years, is the goal of 2 % of active clinical trials (Fig. 24.10). Currently, gene therapy trials are primarily performed in the United States (64 % of all trials), the United Kingdom (11 %), and Germany (4.5 %). The geographical distribution of gene therapy clinical trials is summarized in Fig. 24.11. General indications for all gene therapy trials in the clinic are summarized in Table 24.4.

CANCER GENE THERAPY

The majority of today's gene therapy clinical trials are devoted to treat cancer. There are two potential benefits of using gene therapy to treat cancer: (a) gene-based treatments can attack existing cancer at the molecular level, eliminating the need for drugs, radiation, or surgery and (b) identifying cancer susceptibility genes in individuals or families may have significant impact in preventing the disease before it occurs.

Strategies to achieve these goals include (a) correction of genetic mutations contributing to the malignant phenotype by replacing missing genes or altered defect genes with healthy genes, (b) enhancement of a patient's immune response to cancer (immunotherapy), (c) insertion of genes into cancer cells to make

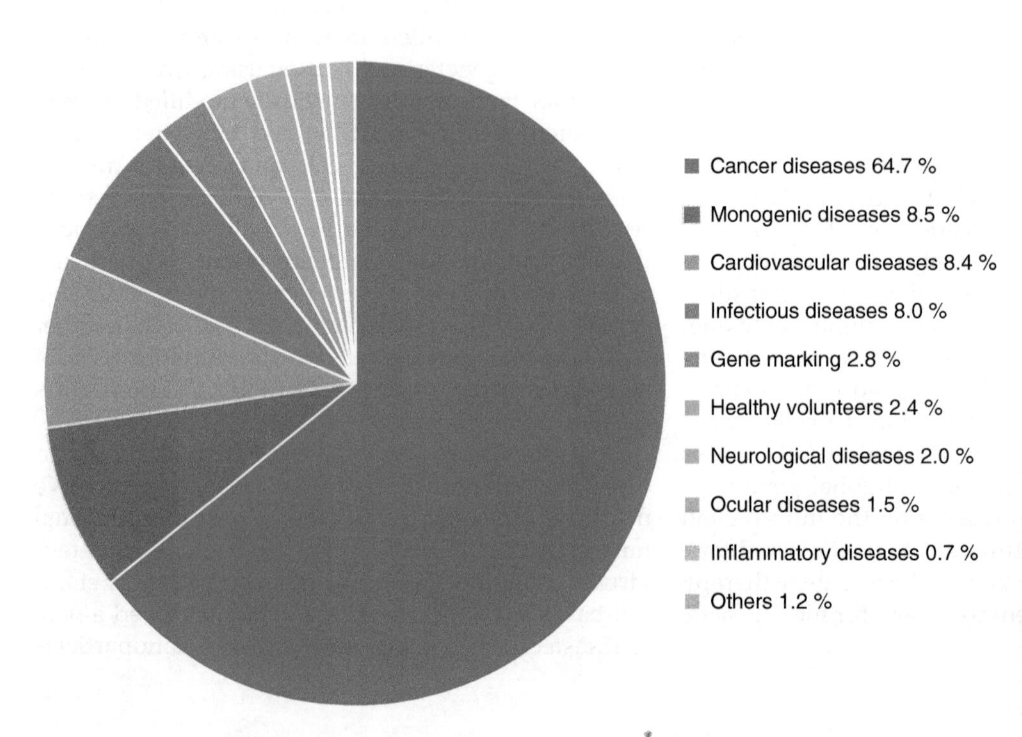

Figure 24.10 ■ Disease targets of gene therapy clinical trials (Wiley 2012). Other diseases include inflammatory bowel disease, rheumatoid arthritis, chronic renal disease, carpal tunnel syndrome, Alzheimer's disease, diabetic neuropathy, Parkinson's disease, erectile dysfunction, retinitis pigmentosa, and glaucoma.

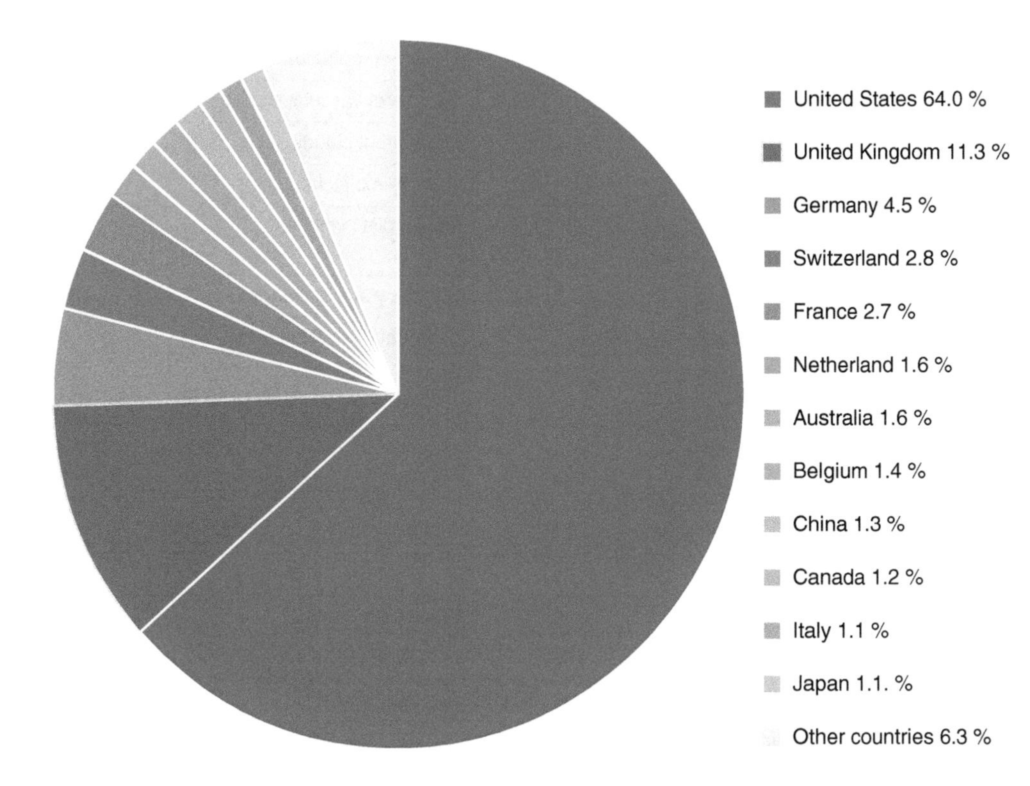

Figure 24.11 ■ International status of gene therapy clinical trials (Wiley 2012). Other countries include Austria, Czech Republic, Denmark, Egypt, Finland, Ireland, Israel, Mexico, New Zealand, Norway, Poland, Romania, Russia, Singapore, South Korea, Spain, Sweden, and Taiwan. Trials held in each of these countries represent less than 1 % of all clinical trials held worldwide.

them more sensitive to conventional chemotherapy and radiotherapy or other treatments, (d) introduction of "suicide genes" into a patient's cancer cells that can enzymatically activate a prodrug in these cells to the destroy them, and (e) direct tumor killing through oncolytic viruses.

■ Correction of Genetic Mutations

In this approach, gene therapy is used to correct genetic mutations contributing to the malignant phenotype by replacing missing genes or removing defect genes. Understanding cancer at the molecular level is the platform for gene correction in cancer therapy. The inactivation or activation of certain genes may contribute to tumor growth. Although the complex process of tumor development and growth limits the utility of this strategy, approximately 12 % of cancer gene therapy clinical trials involve overexpression of tumor-suppressor genes such as p53, MDA-7, and ARF (Majhen and Ambriovic-Ristov 2006). Mutations in the p53 gene are most commonly seen in a wide spectrum of tumors (Roth and Cristiano 1997). Efficient delivery and expression of the wild-type p53 tumor-suppressor gene prevents the growth of human cancer cells in culture, causes regression of established human tumors in nude mice, or sensitizes the existing tumors to the therapeutic effect of conventional chemotherapy and radiotherapy (Roth and Cristiano 1997). The results from clinical trials indicated that the therapeutic effect of gendicine, the first gene therapy product, was promising in patients with head and neck squamous cancers (Peng 2005). However, the results were only validated in China. Efficient delivery of tumor-suppressor genes deep within tumors is difficult, and restriction of gene expression in malignant tissue is challenging. Gene silencing by this approach has also limited success, especially when a prolonged silencing effect is required. Despite these reservations, prostate, lung, and pancreatic tumors have been successfully treated in the clinic with a variety of genes and transfer methods.

■ Immunotherapy

In this approach, gene therapy is used to stimulate the body's natural ability to attack cancer cells. In one study, autologous peripheral blood lymphocytes genetically engineered with retroviral vectors encoding the melanoma specific T cell receptor (TCR) were administered to patients with metastatic melanoma. The genetically engineered T lymphocytes then recognize the antigens on the surface of the tumor cells through TCRs and kill the tumor cells. Cancer regressions were first reported by Morgan et al. and further confirmed by Johnson et al. (Morgan et al. 2006; Johnson et al. 2009). In other clinical studies, expression of either pro-inflammatory cytokines (IL-2, IL-4, and IL-12), co-stimulatory molecules (HLA-B7 LFA-3), or tumor-specific antigens (mucin-1 and CEA) to stimulate antitumor immune responses has also been tested (Majhen and Ambriovic-Ristov 2006).

■ Tumors Sensitization

In this approach, genes are inserted into cancer cells to make them more sensitive to conventional chemotherapy and radiotherapy or other treatments. We

Cancer	Other diseases	Cardiovascular disease
Gynecological	Inflammatory bowel disease	Peripheral vascular disease
Breast, ovary, cervix	Rheumatoid arthritis	Intermittent claudication
Nervous system	Chronic renal disease	Critical limb ischemia
Glioblastoma, leptomeningeal carcinomatosis, glioma, astrocytoma, neuroblastoma	Fractures	Myocardial ischemia
Gastrointestinal	Erectile dysfunction	Coronary artery stenosis
Colon, colorectal, liver metastases, post-hepatitis liver cancer, pancreas	Anemia of end stage renal disease	Stable and unstable angina
Genitourinary	Parotid salivary hypofunction	Venous ulcers
Prostate, renal	Type I diabetes	Vascular complications of diabetes
Skin	Detrusor overactivity	Pulmonary hypertension
Melanoma	Graft-versus-host disease	Heart failure
Head and neck		
Nasopharyngeal carcinoma	**Monogenic disorders**	**Infectious disease**
Lung	Cystic fibrosis	HIV/AIDS
Adenocarcinoma, small cell, non-small cell	Severe combined immunodeficiency (SCID)	Tetanus
Mesothelioma	Alpha-1 antitrypsin deficiency	Epstein-Barr virus
Hematological	Hemophilia A and B	Cytomegalovirus infection
Leukemia, lymphoma, multiple myeloma	Hurler syndrome	Adenovirus infection
Sarcoma	Hunter syndrome	Japanese encephalitis
Germ cell	Huntington's chorea	Hepatitis C
	Duchenne muscular dystrophy	Hepatitis B
Neurological diseases	Becker muscular dystrophy	Influenza
Alzheimer's disease	Canavan disease	
Carpal tunnel syndrome	Chronic granulomatous disease (CGD)	
Cubital tunnel syndrome	Familial hypercholesterolemia	
Diabetic neuropathy	Gaucher disease	
Epilepsy	Fanconi's anemia	
Multiple sclerosis	Purine nucleoside phosphorylase deficiency	
Myasthenia gravis	Ornithine transcarbamylase deficiency	
Parkinson's disease	Leukocyte adherence deficiency	
Peripheral neuropathy	Gyrate atrophy	
	Fabry disease	
Ocular diseases	Familial amyotrophic lateral sclerosis	
Age-related macular degeneration	Junctional epidermolysis bullosa	
Diabetic macular edema	Wiskott-Aldrich syndrome	
Glaucoma	Lipoprotein lipase deficiency	
Retinitis pigmentosa	Late infantile neuronal ceroid lipofuscinosis	
Superficial corneal opacity	RPE65 mutation (retinal disease)	
	Mucopolysaccharidosis	

Information obtained from reference (Wiley 2012)

Table 24.4 ■ Conditions for which human gene transfer trials have been approved.

previously mentioned that transgene expression of p53 sensitized the tumors to the therapeutic effect of conventional chemotherapy and radiotherapy (Lesoon-Wood et al. 1995; Chen et al. 1996). In other studies, the RNAi mechanism (see Chap. 23) was used to overcome multidrug resistance (MDR) in cancer cells. MDR, which typically represents overexpression of P-glycoprotein, a drug efflux transporter on cancer cell membranes, is a frequent impediment to successful chemotherapy. Synthetic siRNA- or vector-mediated MDR1 gene silencing were widely reported to be successful to reduce the chemoresistance of certain types of cancers (Huang et al. 2008).

■ Gene-Directed Enzyme-Prodrug Therapy

In this approach, gene therapy aims to maximize the effect of a toxic drug and minimize its systemic effects by generating the drug in situ within the tumor. In the first step of this procedure, the gene for an exogenous enzyme is delivered and expressed in the tumor cells. Subsequently, a prodrug is administered and converted to the active drug (toxic metabolites) by the foreign enzyme expressed inside or on the surface of tumor cells (Fig. 24.12). The suicide gene is usually of viral or prokaryotic origin with no human homolog. However, this is not an absolute requirement provided the prodrug is not activated to any significant degree by the native cellular enzyme. In preclinical studies, Chen et al. first reported a successful combination of a suicide gene/prodrug system and immunotherapy to treat hepatic metastases in mice (Chen et al. 1996). Uckert et al. further improved the system to a double suicide gene system as a safety mechanism for the elimination of tumor cells in a reliable fashion (Uckert et al. 1998). There are several variants of gene-directed enzyme-prodrug therapy. The herpes simplex virus-thymidine kinase (HSV-tk)/ganciclovir system, the cytosine deaminase/5-fluorocytosine system, the nitroreductase/CB1954 system, and the carboxypeptidase G2/CMDA system are the most "popular" systems (Fig. 24.13). Cerepro, a gene medicine developed by Ark Therapeutics Group PLC, has been granted orphan drug status by the European Medicines Agency and the FDA. Cerepro is an adenovirus containing a herpes simplex type-1 thymidine kinase transgene for treating malignant glioma together with ganciclovir. In a phase II clinical trial with Cerepro, the mean survival time of patients increased to 15 months as compared to 7.4 months in patients treated with retroviral therapy or 8.3 months with a noneffective adenovirus (Sandmair et al. 2000).

■ Oncolytic Viruses (Virotherapy)

In this approach, oncolytic viruses were directly introduced into tumors to induce cell death through viral replication, expression of cytotoxic proteins, and cell lysis. Vaccinia virus, herpes simplex type-I (HSV), reovirus, Newcastle disease virus, poliovirus, and adenovirus are often selected for this application because they naturally target cancers and contain genomes that can be easily manipulated (Aghi and Martuza 2005). Despite some clinical success, significant safety precautions must be taken, making clinical trials with these viruses extremely expensive and cumbersome (Markert et al. 2000; Varghese and Rabkin 2002). The clearance of the virus by cellular immunity and preexisting neutralizing antibodies in the majority of the population also negatively affects the efficacy of virotherapy. In 2006, Oncorine, a drug made of conditionally replicative adenoviruses by Sunway Biotech, Shanghai, China, gained marketing approval in China for treating head and neck cancer. This adenovirus contains a deletion in the E1B 55 K region and only replicates in p53-deficient cancer cells. The company has claimed significant benefits of using Oncorine in clinical trials of lung cancer, liver cancer, pancreatic cancer, and malignant effusion (Xia et al. 2004).

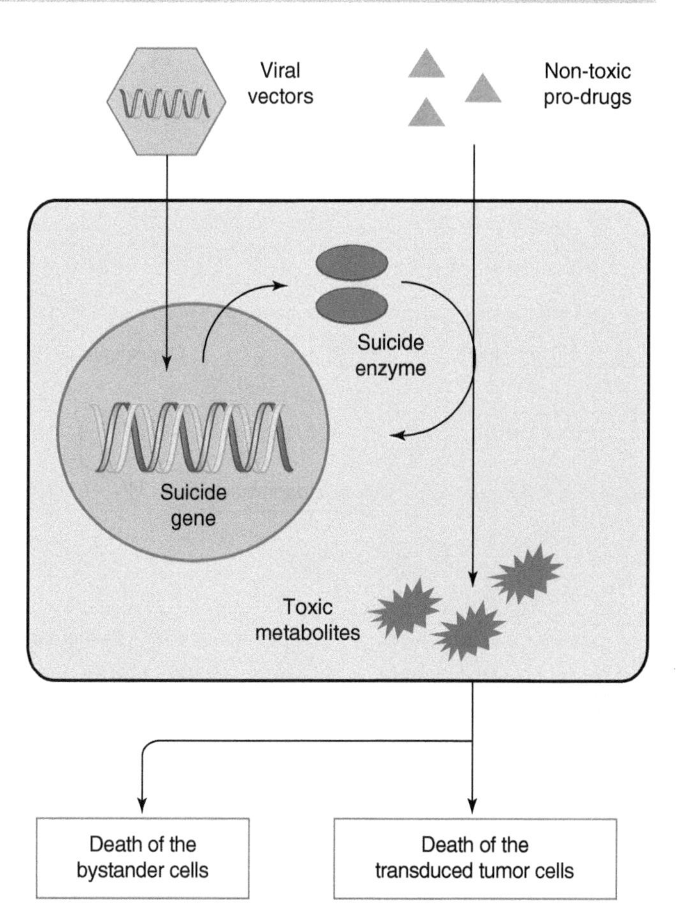

Figure 24.12 ■ A schematic illustration of gene-directed enzyme-prodrug therapy. The suicide gene for an exogenous enzyme is delivered and expressed in the tumor cells. Then a prodrug is administered and converted to the active drug by the foreign enzyme expressed inside or on the surface of tumor cells

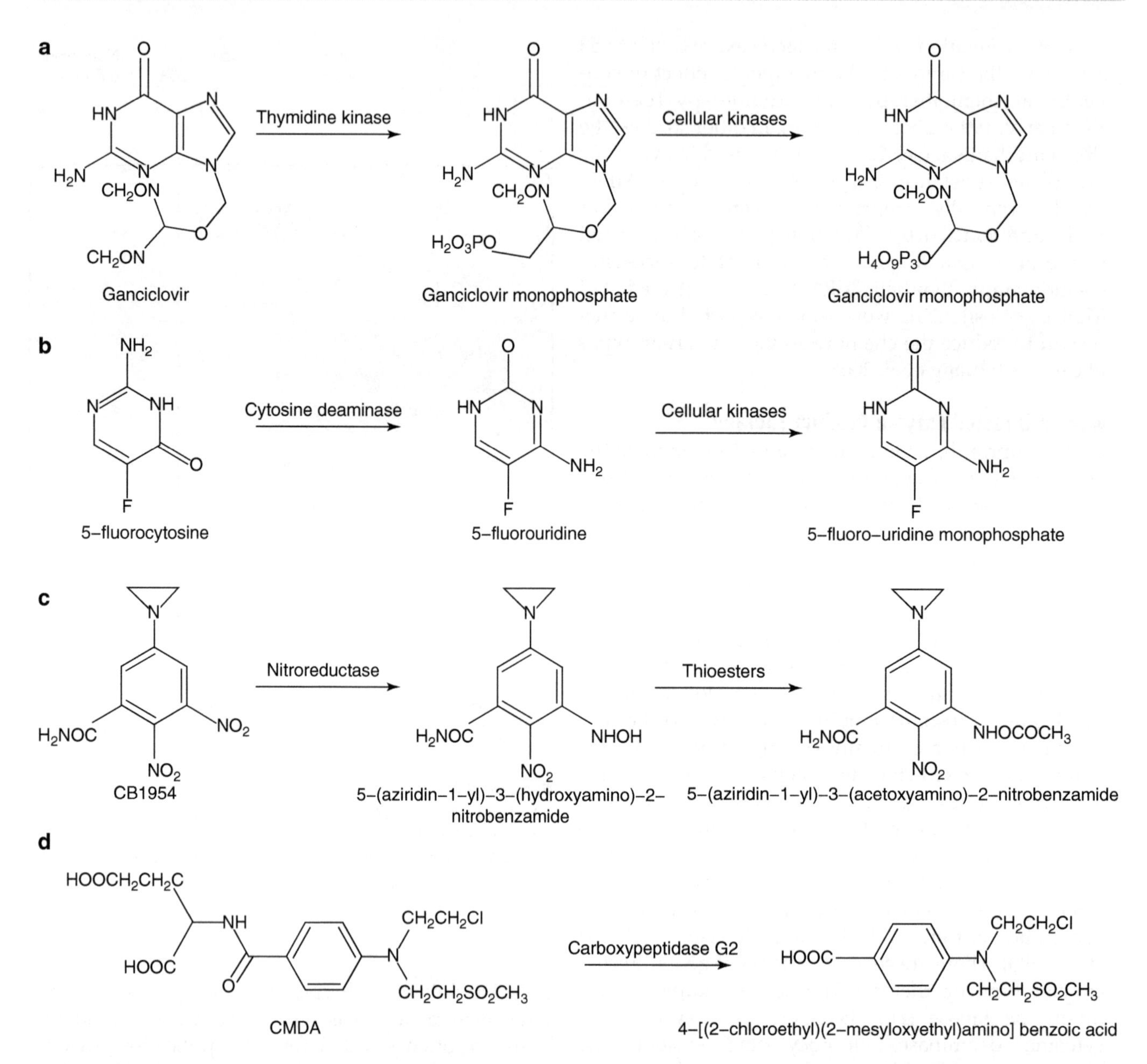

Figure 24.13 ■ Models of gene-directed enzyme-prodrug therapy. (**a**) The herpes simplex virus-thymidine kinase (HSV-tk) system. A vector expressing the gene for HSV-tk enters a cellular target. This enzyme is expressed and phosphorylates the drug ganciclovir (GCV). This is subsequently converted to the di- and triphosphate forms by guanylate kinase and other cellular kinases. The triphosphate is incorporated into cellular DNA during cell division causing single strand breaks. (**b**) The cytosine deaminase/5-fluorocytosine system. A vector expressing cytosine deaminase (CD) enters a cellular target. Overexpression of CD activates 5-fluorocytosine (5-FC) to 5-fluorouridine (5-FU). 5-FU is converted to mono-, di-, and triphosphate forms by cellular kinases. All of these compounds are cytotoxic. (**c**). The nitroreductase/CB1954 system. A vector expressing E. coli nitroreductase (NTR) enters a cellular target. Expression of NTR allows the cell to convert the compound CB1954 to a potent DNA-cross-linking agent. (**d**) The carboxypeptidase G2/CMDA system. The bacterial enzyme carboxypeptidase G2 (CPG2) is able to cleave the glutamic acid moiety from the prodrug releasing the DNA-cross-linking mustard drug 4-[(2-chloroethyl-2-mesyloxyethyl) amino] benzoic acid without further catalytic requirements.

■ Nonviral Gene Therapy

In the last two decades, gene therapy has been widely used in clinical trials for cancer treatment and the results were quite encouraging. Most of the clinical trials employed viral vector-based gene therapeutics, probably because of the high transfection efficiency of this strategy. Despite the fact that nonviral-based gene therapeutics are relatively safer and less tumorigenic, extensive work is still needed to further optimize this strategy (increase transgene expression, reduce plasma protein binding, escape reticuloendothelial system (RES) and endosome, etc.) to make it clinically acceptable. Chapter 4 deals with new insights and strategies to improve targeting efficiency, and the artificial virus approach (for endosomal escape and nuclear membrane passage) describes means to enhance intracellular transport and delivery (Mastrobattista et al. 2004).

MONOGENETIC DISEASES

The greatest successes of gene therapy to date have been achieved in treating monogenetic diseases, which is the second largest disease group treated by gene therapy, comprising 8.5 % of all the active gene therapy clinical trials. The ultimate therapeutic goal of gene therapy for monogenetic disorders is to permanently replace a defect gene with a good copy to restore normal function and permanently reverse disease processes. To date, clinical trials have not met this objective. Of the 151 active clinical gene therapy trials for monogenetic diseases, approximately one third targeted cystic fibrosis (CF), the most common inherited genetic disease in Europe and the United States (Wiley 2012). Up until now, the severe combined immunodeficiency syndromes, comprising 20 % of trials for inherited disorders, is the only group of diseases in which gene therapy has shown a lasting, clinically meaningful therapeutic benefit. Other monogenetic diseases currently in clinical trials are listed in Table 24.4.

Issues which have prevented successful gene transfer for monogenetic diseases to date include (a) lack of suitable gene delivery technologies, (b) unfavorable interactions between the host and gene transfer vector, (c) complex biology and pathology of monogenetic diseases and target organs, and (d) lack of relevant measures to assess the clinical efficacy of gene transfer. The greatest challenges that remain in treating monogenetic disease are to induce gene expression sufficiently to correct the clinical phenotype without precipitation of host immune responses and minimizing the risk of insertional mutagenesis for integrating vectors in dividing cellular targets. Improvements in vector technology and advancements in the understanding of cellular processes will vastly improve methods for correction of genetic disease.

CARDIOVASCULAR DISEASES

Cardiovascular diseases are the third largest group of diseases actively treated by gene therapy clinical trials. The current understanding of molecular mechanisms of cardiovascular diseases has uncovered a large number of genes that could serve as potential targets for molecular therapies. For example, overexpression of genes involved in vasodilation such as endothelial nitric oxide synthase (eNOS) and heme oxygenase-1 (HO-1) or inhibition of molecules involved in vasoconstriction (angiotensin converting enzyme (ACE), angiotensinogen (AGT)) have reduced blood pressure in animal models of hypertension (Melo et al. 2006). Most clinical trials for cardiovascular diseases are designed for treating coronary and peripheral ischemia. Overexpression of pro-angiogenic factors such as vascular endothelial growth factor (VEGF), fibroblast growth factor (FGF), and hepatocyte growth factor (HGF) have been effective in myocardial and peripheral ischemia in preclinical studies (Springer 2006). Despite the lack of significant benefit in several earlier clinical trials, VEGF gene therapy did show an excellent safety profile and improvement of symptoms in patients following adenoviruses or plasmid intramyocardial administration in both pilot studies and long-term follow-ups (Stewart et al. 2006; Reilly et al. 2005). However, limited success was experienced in using gene therapy to treat cardiovascular diseases compared to other areas. Larger, double-blind, randomized, controlled trials are needed to minimize the potential bias for placebo effects suspected to occur in some trials. Stringent criteria for patient selection are needed as many with cardiovascular disease often have underlying conditions that may influence the results. Endpoints for assessing efficacy and measures to assess potential short- and long-term complications must also be standardized among research groups. The efficacy of gene therapy for cardiovascular disease will most likely be enhanced by strategies that incorporate multiple gene targets with cell-based approaches.

INFECTIOUS DISEASES

One hundred and forty-two clinical trials for treating infectious diseases have been initiated, comprising 8 % of the total number of gene therapy clinical trials (Wiley 2012). Gene transfer for acquired immunodeficiency syndrome (AIDS) is the main application in this category. Many gene therapy trials for AIDS involve ex vivo transfer of genetic material to autologous T cells using self-inactivating or conditionally replicating viral vectors to improve the immune system of the patients (Levine et al. 2006; Manilla et al. 2005). Other trials employed overexpression of HIV inhibitors such as RevM10 to increase $CD4^+$ T cell survival in HIV-infected individuals (Morgan et al. 2005; Ranga et al. 1998).

The most important achievement in the gene therapy studies to treat infectious diseases is the development of DNA vaccination, a technique to protect the host from diseases by producing an immunological response through injecting genetically engineered viral DNA (see Chap. 22). Although DNA vaccination was first proposed and studied for HIV, it has achieved little success in the last 10 years (MacGregor et al. 1998). Clinical studies using DNA vaccine for other infectious diseases caused by hepatitis B virus (HBV), influenza virus, and Ebola virus were also reported (Tacket et al. 1999). In 2010, researchers from the USA and France reported the first HIV DNA vaccine which can induce a long-lasting HIV-specific immune response in nonhuman primates, a discovery that could prove significant in the development of HIV vaccines (Arrode-Bruses et al. 2010). Currently, PENNVAX™, a DNA vaccine product for HIV developed by Inovio Pharmaceuticals, is in phase I clinical studies.

NEUROLOGICAL DISEASES

Significant progress has been made in gene therapy for neurological diseases in the last 5 years. The two most common neurological diseases targeted by gene therapy are Alzheimer's disease and Parkinson's disease. In 2005, a phase I trial of ex vivo nerve growth factor (NGF) gene delivery in eight individuals with mild Alzheimer disease was performed at the University of California in San Diego. Briefly, autologous fibroblasts obtained from small-skin biopsies in each individual were genetically modified to produce and secrete human NGF using retroviral vectors and reimplanted into the forebrain. The results indicated improvement in the rate of cognitive decline, significant increases in cortical 18-fluorodeoxyglucose concentrations (PET imaging), and robust nerve growth responses to NGF (Tuszynski et al. 2005). In 2007, the first gene therapy clinical trial was conducted at the New York Presbyterian Hospital. Briefly, serial doses of adeno-associated virus (AAV) encoding glutamic acid decarboxylase (GAD) were infused into the subthalamic nucleus of patients with Parkinson's disease. The results indicated that AAV-GAD gene therapy is safe and well tolerated by patients with advanced Parkinson's disease. Although this open label, non-randomized phase I study did not include a sham group and was not designed to assess the effectiveness of gene therapy, the preliminary data were encouraging, showing substantial improvements in Unified Parkinson's Disease Rating Scale (UPDRS), beginning at 3 months after surgery and continuing until the end of the trial (12 months after surgery) (Kaplitt et al. 2007).

REGULATORY ISSUES OF GENE THERAPY PRODUCTS

Any studies involving humans must be reviewed with great care. Gene therapy presents unique safety and infection control issues, which make it necessary for scientists to take special precautions with gene therapy. In the USA two organizations within the United States Department of Health and Human Services (DHHS), the Office for Human Research Protections (OHRP) and the Food and Drug Administration (FDA), have specific authority described in the Code of Federal Regulations (CFR). The OHRP mandates any gene therapy clinical trial involving human subjects to be reviewed, approved, and monitored by the Institutional Review Board (IRB) at each investigative site. The FDA's Center for Biologics Evaluation and Research (CBER) oversees human gene therapy clinical trials conducted by the manufacturers. Any gene therapy product must be tested extensively to meet the FDA requirements for safety and efficacy before approval for marketing. All gene therapy clinical trial protocols must be conducted under Investigational New Drug (IND) applications. Regulations pertaining to this process appear in Title 21 of the Code of Federal Regulations (CFR), Part 312. Another DHHS organization, the National Institutes of Health (NIH) oversees the gene therapy studies and clinical trials conducted by federally funded investigators. The NIH monitors scientific progress in human genetics research through the Office of Biotechnology Activities (OBA). Inside OBA, the Recombinant DNA Advisory Committee (RAC) was established in 1974 in response to public concerns regarding the safety of manipulating genetic material through the use of recombinant DNA techniques. Any human gene transfer research receiving NIH funding must be registered with OBA and reviewed by the RAC. Another responsibility of RAC is to cooperate with the Institutional Biosafety Committees (IBC) to oversee recombinant DNA research at each investigative site. Figure 24.14 shows the interactions of these different regulatory agencies in the development process of a gene product. Other countries also have a number of committees that must approve gene therapy protocols and address other scientific and ethical concerns associated with clinical trials.

CONCLUDING REMARKS

Within the last 20 years, the field of gene therapy has come a long way from bench to bedside. Many vectors developed for gene transfer have now been tested in the clinic. Three products (Gendicine and Oncorine in China, Cerepro in Europe) have been given marketing approval and several others are in late phases of testing. Although the biology of gene transfer vectors is well understood, several barriers must be overcome for turning genes into therapeutics. The immune responses and the insertional mutagenesis of viral vectors and the lack of transgene efficiency of nonviral vectors are the most significant barriers of gene therapy.

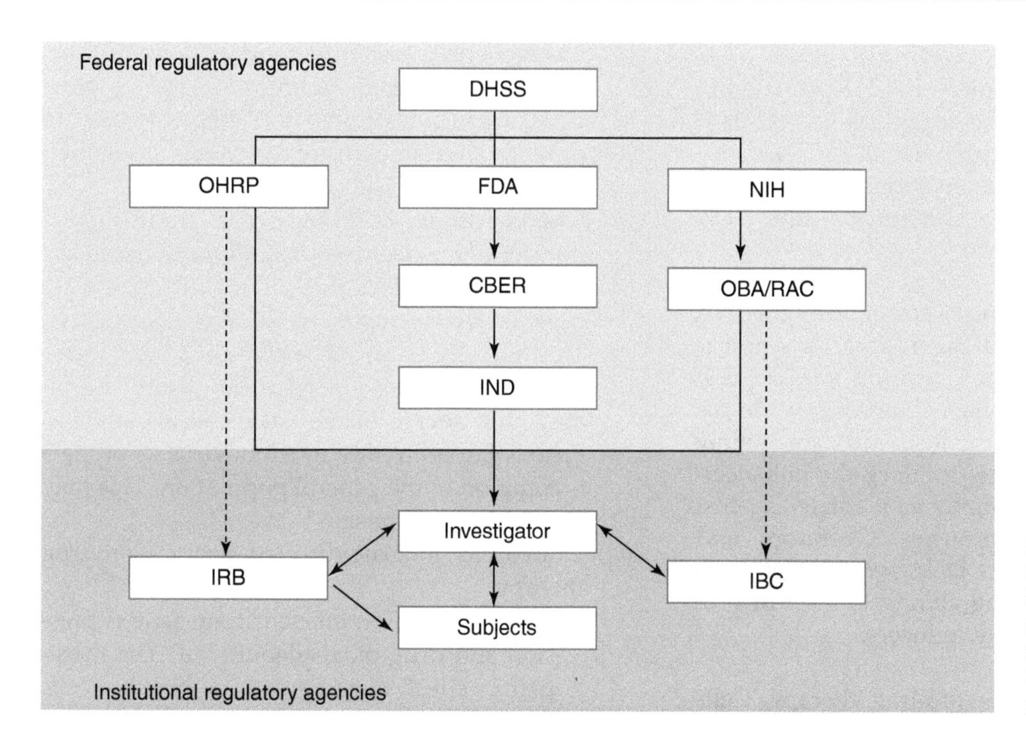

Figure 24.14 ■ The interactions of the regulatory agencies involved in the implementation of a gene therapy protocol (From Mendell and Miller (2004). with permission to reprint).

Targeted delivery of gene expression systems and spatial and temporal control of transgene expression in target tissues based on the severity and the process of the disease are also critical to the success of many gene therapy applications. Although clinical trials have shown short-term safety and efficacy, long-term surveillance over a period of decades is lacking, and the safety and efficacy of genetic medicines have so far only been validated in limited patient populations. Other factors such as the use of concurrent medications and concurrent medical conditions, objective assessment of improvements and endpoints, and the assessment of placebo effects need to be standardized to get reliable and reproducible results among different research groups. In addition, cost-effectiveness analyses have to be considered as the production of gene therapy vectors itself is costly and requires specialized equipment and personnel. In the future, the further development of genetic medicines that can be widely used will heavily rely on collaborations between academic institutions and commercial partners from the pharmaceutical and biotechnology industries.

SELF-ASSESSMENT QUESTIONS AND ANSWERS

■ Questions

1. What was the disease target for the first gene therapy clinical trial? What vector was selected for gene transfer?
2. Identify and describe five transcription regulatory elements (TRE) discussed in the chapter.
3. Several clinical trials involve gene transfer for treating malignant glioma. One approach involves the use of a recombinant retrovirus expressing the HSV-tk transgene. Another involves the use of a recombinant adenovirus expressing the p53 transgene.
   (A) Which of the five current strategies to treat cancer by viral gene therapy does each of these trials employ? Describe the principle behind each strategy.
   (B) List 2 advantages and 2 disadvantages associated with the vector used in each of these trials.
   (C) Outline potential drawbacks to the use of each of these strategies for cancer therapy.
   (D) What other approaches could have been selected to prevent the growth and spread of malignant tissue? Explain the principle behind each.
4. What is the purpose of the packaging cell line during the production of recombinant viral vectors for gene transfer? What is the risk associated with using packaging cell lines for vector production?
5. Provide two examples of how gene therapy is used to modulate the immune system to fight infection.
6. Describe one clinical trial for retrovirus-based gene therapy and adenovirus-based gene therapy, and identify the most significant adverse effects that have been reported for each trial.
7. Identify the three marketed gene therapy products in the world and describe the mechanism of actions of each product.

Answers

1. The first gene therapy clinical trial was initiated in 1990 for treating adenosine deaminase (ADA) deficiency. In this trial, patients with ADA deficiency were given peripheral blood lymphocytes treated with a retroviral vector expressing the ADA transgene.
2. Promoter is a DNA sequence that enables a gene to be transcribed. The promoter is recognized by RNA polymerase and transcriptional factors. Enhancer is a short DNA sequence that can bind transcription factors or activators to enhance transcription levels of a gene from a distance. Insulators are mainly genetic boundary elements to block the enhancer-promoter interaction or rarely as a barrier against condensed chromatin proteins. Operators and silencers are usually short DNA sequence close to the promoter with binding affinity to a set of proteins named repressors and inducers.
3. (A) Retrovirus trial

   Gene-directed enzyme-prodrug therapy. Cells transduced by the virus express an enzyme capable of converting a prodrug (in this case ganciclovir) to a cytotoxic metabolite. This conversion cannot occur in cells that do not express the transgene, limiting the cytotoxic effect to transduced cells and their neighbors through the bystander effect.

   Adenovirus trial

   Correction of genetic mutations that contribute to a malignant phenotype. Cells transduced by the virus express a gene such as p53 that is necessary for controlled cell division and development. This prevents the uncontrolled growth and division associated with malignant disease.

   (B) Retrovirus

   Advantages – (i) Retroviruses can infect dividing cells which are the therapeutic target in this trial. Despite this fact, transduction efficiency of this vector in vivo has been low. (ii) Retroviruses are capable of inducing long-term gene expression which should be sufficient for effective removal of malignant tissue.

   Disadvantages – (i) Retroviruses have the potential for inducing insertional mutagenesis in normal, healthy cells. (ii) Transgene expression is sometimes limited by the host immune response to cellular components acquired by the virus during large scale production.

   Adenovirus

   Advantages – (i) Adenoviruses can infect dividing cells which are the therapeutic target in this trial. (ii) Adenoviruses can induce high levels of transgene expression in short periods of time. (iii) Adenoviruses do not have the risk of insertional mutagenesis. (iv) It is relatively easy to produce large amounts of recombinant adenovirus sufficient for clinical use.

   Disadvantages – (i) Transgene expression is transient, making readministration necessary for continued effect. (ii) Adenoviral vectors are capable of inducing a potent immune response. This not only limits the success of gene transfer after a second dose of virus but also is associated with severe toxicity at certain doses. (iii) Preexisting immunity to adenovirus serotype 5 is common in the general population. This may also limit gene transfer.

   (C) Drawbacks to gene-directed enzyme-prodrug therapy.

   (i) Efficacy relies on efficient transgene expression and drug bioavailability. (ii) The therapeutic effect may spread to healthy cells through the bystander effect.

   Drawbacks to gene correction therapy.

   (i) Gene correction may stop tumor growth but not eliminate it. (ii) Expression is not limited to malignant tissue.

   (D) Other approaches for cancer gene therapy

   (i) Immunotherapy. A vector expressing pro-inflammatory cytokines, co-stimulatory molecules, or tumor-specific antigens is injected directly into the tumor mass. This facilitates the formation of an antitumor immune response that targets and destroys malignant cells.

   (ii) Virotherapy. A replication-competent virus that naturally targets cancers is directly injected in the tumor mass. The virus can induce cell death during replication in malignant tissue by producing cytotoxic proteins and subsequent cell lysis.
4. (i) The primary purpose of the packaging cell line is to provide genetic elements that support virus replication and assembly. These have been eliminated from the vector to prevent it from causing disease in the patient. (ii) The recombinant virus can incorporate elements for replication into its genome through homologous recombination during the production process. The potential for generation of replication-competent virus (RCV) in this manner does exist for each vector but can vary due to specific features of a given packaging cell line.
5. (i) Gene transfer into autologous immunocytes to increase the immune system of a patient. (ii)

Overexpression of protein inhibitors that interfere with virus infection and replication. (iii) Overexpression of known antigenic epitopes of the pathogen by DNA vaccination to stimulate an immune response.

6. (i) One trial employed aerosol administration of a recombinant adenovirus expressing cystic fibrosis transmembrane conductance regulator (CFTR) to treat cystic fibrosis (CF). Another trial employed a recombinant retrovirus expressing recombinant adenosine deaminase (ADA) to transduce autologous T lymphocytes isolated from patients for treating ADA deficiency-induced severe combined immunodeficiency (ADA-SCID). (ii) CF trial. Massive immune response to the recombinant viral vector.

   ADA-SCID trial. Lymphoproliferative leukemia caused by insertional mutagenesis.

7. Gendicine is a recombinant adenoviral vector which expresses p53 tumor suppressor and is used to treat patients with head and neck squamous cancers.

   Oncorine is a recombinant adenoviral vector which contains a deletion in the E1B 55 K region and only replicates in p53 deficient cancer cells. Oncorine kills tumor cells through viral replication, expression of cytotoxic proteins, and cell lysis.

   Cerepro is a recombinant adenoviral vector encoding herpes simplex type-1 thymidine kinase (TK). Cerepro is used for treating malignant glioma together with ganciclovir through gene-directed enzyme-prodrug therapy.

Acknowledgements We would like to thank the National Institutes of Health (NIH) for the financial support (R01DK69968).

REFERENCES

Aghi M, Martuza RL (2005) Oncolytic viral therapies - the clinical experience. Oncogene 24(52):7802–7816

Apolonia L, Waddington SN, Fernandes C, Ward NJ, Bouma G, Blundell MP, Thrasher AJ, Collins MK, Philpott NJ (2007) Stable gene transfer to muscle using non-integrating lentiviral vectors. Mol Ther 15(11): 1947–1954. doi:10.1038/sj.mt.6300281

Armentano D, Sookdeo CC, Hehir KM, Gregory RJ, St George JA, Prince GA, Wadsworth SC, Smith AE (1995) Characterization of an adenovirus gene transfer vector containing an E4 deletion. Hum Gene Ther 6(10):1343–1353. doi:10.1089/hum.1995.6.10-1343

Arrode-Bruses G, Sheffer D, Hegde R, Dhillon S, Liu Z, Villinger F, Narayan O, Chebloune Y (2010) Characterization of T-cell responses in macaques immunized with a single dose of HIV DNA vaccine. J Virol 84(3): 1243–1253. doi:10.1128/JVI.01846-09

Bedouelle H, Duplay P (1988) Production in Escherichia coli and one-step purification of bifunctional hybrid proteins which bind maltose. Export of the klenow polymerase into the periplasmic space. Eur J Biochem 171(3):541–549

Bellon G, Michel-Calemard L, Thouvenot D, Jagneaux V, Poitevin F, Malcus C, Accart N et al (1997) Aerosol administration of a recombinant adenovirus expressing CFTR to cystic fibrosis patients: a phase I clinical trial. Hum Gene Ther 8(1):15–25. doi:10.1089/hum.1997.8.1-15

Blackburn MR, Kellems RE (2005) Adenosine deaminase deficiency: metabolic basis of immune deficiency and pulmonary inflammation. Adv Immunol 86:1–41

Blaese RM, Culver KW, Miller AD, Carter CS, Fleisher T, Clerici M, Shearer G et al (1995) T lymphocyte-directed gene therapy for ADA- SCID: initial trial results after 4 years. Science 270(5235):475–480

Bordignon C, Notarangelo LD, Nobili N, Ferrari G, Casorati G, Panina P, Mazzolari E et al (1995) Gene therapy in peripheral blood lymphocytes and bone marrow for ADA- immunodeficient patients. Science 270(5235):470–475

Boussif O, Lezoualc'h F, Zanta MA, Mergny MD, Scherman D, Demeneix B, Behr P (1995) A versatile vector for gene and oligonucleotide transfer into cells in culture and in vivo: polyethylenimine. Proc Natl Acad Sci U S A 92(16):7297–7301

Cavazzana-Calvo M, Hacein-Bey S, de Saint Basile G, Gross F, Yvon E, Nusbaum P, Selz F et al (2000) Gene therapy of human severe combined immunodeficiency (SCID)-X1 disease. Science 288(5466):669–672

Chen HH, Mack LM, Kelly R, Ontell M, Kochanek S, Clemens PR (1997) Persistence in muscle of an adenoviral vector that lacks all viral genes. Proc Natl Acad Sci U S A 94(5):1645–1650

Chen SH, Kosai K, Xu B, Pham-Nguyen K, Contant C, Finegold MJ, Woo SL (1996) Combination suicide and cytokine gene therapy for hepatic metastases of colon carcinoma: sustained antitumor immunity prolongs animal survival. Cancer Res 56(16):3758–3762

Chillon M, Lee JH, Fasbender A, Welsh MJ (1998) Adenovirus complexed with polyethylene glycol and cationic lipid is shielded from neutralizing antibodies in vitro. Gene Ther 5(7):995–1002. doi:10.1038/sj.gt.3300665

Corsi K, Chellat F, Yahia L, Fernandes JC (2003) Mesenchymal stem cells, MG63 and HEK293 transfection using chitosan-DNA nanoparticles. Biomaterials 24(7):1255–1264

Crook NC, Freeman ES, Alper HS (2011) Re-engineering multicloning sites for function and convenience. Nucleic Acids Res 39(14):e92. doi:10.1093/nar/gkr346

Croyle MA, Cheng X, Wilson JM (2001) Development of formulations that enhance the physical stability of viral vectors for human gene therapy. Gene Ther 8(17):1281–1291

Dai Y, Schwarz EM, Gu D, Zhang WW, Sarvetnick N, Verma IM (1995) Cellular and humoral immune responses to adenoviral vectors containing factor IX gene: tolerization of factor IX and vector antigens allows for long-term expression. Proc Natl Acad Sci U S A 92(5):1401–1405

Dang CV, Lee WM (1988) Identification of the human c-myc protein nuclear translocation signal. Mol Cell Biol 8(10):4048–4054

Daud AI, DeConti RC, Andrews S, Urbas P, Riker AI, Sondak VK, Munster PN et al (2008) Phase I trial of interleukin-12 plasmid electroporation in patients with metastatic melanoma. J Clin Oncol 26(36):5896–5903. doi:10.1200/JCO.2007.15.6794

Duan D, Yue Y, Yan Z, Yang J, Engelhardt JF (2000) Endosomal processing limits gene transfer to polarized airway epithelia by adeno-associated virus. J Clin Invest 105(11):1573–1587. doi:10.1172/JCI8317

Dzau VJ, Gnecchi M, Pachori AS (2005) Enhancing stem cell therapy through genetic modification. J Am Coll Cardiol 46(7):1351–1353. doi:10.1016/j.jacc.2005.07.023

Edelstein ML, Abedi MR, Wixon J, Edelstein RM (2004) Gene therapy clinical trials worldwide 1989-2004-an overview. J Gene Med 6(6):597–602

Erles K, Sebokova P, Schlehofer JR (1999) Update on the prevalence of serum antibodies (IgG and IgM) to adeno-associated virus (AAV). J Med Virol 59(3): 406–411

Ertl HC (2005) Challenges of immune responses in gene replacement therapy. IDrugs 8(9):736–738

Fallaux FJ, et al (1998) New helper cells and matched early region 1-deleted adenovirus vectors prevent generation of replication-competent adenoviruses. Human gene therapy 9:1909–1917

Fallaux FJ, van der Eb AJ, Hoeben, RC (1999) Who's afraid of replication-competent adenoviruses? Gene therapy 6:709–712

Flotte T, Carter B, Conrad C, Guggino W, Reynolds T, Rosenstein B, Taylor G, Walden S, Wetzel R (1996) A phase I study of an adeno-associated virus-CFTR gene vector in adult CF patients with mild lung disease. Hum Gene Ther 7(9):1145–1159

Grayson WL, Frohlich M, Yeager K, Bhumiratana S, Chan ME, Cannizzaro C, Wan LQ, Liu XS, Guo XE, Vunjak-Novakovic G (2010) Engineering anatomically shaped human bone grafts. Proc Natl Acad Sci U S A 107(8): 3299–3304. doi:10.1073/pnas.0905439106

Hacein-Bey-Abina S, von Kalle C, Schmidt M, Le Deist F, Wulffraat N, McIntyre E, Radford I, Villeval JL, Fraser CC, Cavazzana-Calvo M, Fischer A (2003) A serious adverse event after successful gene therapy for X-linked severe combined immunodeficiency. N Engl J Med 348(3):255–256

Hagstrom JN, Couto LB, Scallan C, Burton M, McCleland ML, Fields PA, Arruda VR, Herzog RW, High KA (2000) Improved muscle-derived expression of human coagulation factor IX from a skeletal actin/CMV hybrid enhancer/promoter. Blood 95(8):2536–2542

High KA, Aubourg P (2011) RAAV human trial experience. Methods Mol Biol 807:429–457. doi:10.1007/978-1-61779-370-7_18

Hofmann A, Wenzel D, Becher UM, Freitag DF, Klein AM, Eberbeck D, Schulte M et al (2009) Combined targeting of lentiviral vectors and positioning of transduced cells by magnetic nanoparticles. Proc Natl Acad Sci U S A 106(1):44–49. doi:10.1073/pnas.0803746106

Huang C, Li M, Chen C, Yao Q (2008) Small interfering RNA therapy in cancer: mechanism, potential targets, and clinical applications. Expert Opin Ther Targets 12(5):637–645. doi:10.1517/14728222.12.5.637

Huang MT, Gorman CM (1990) The simian virus 40 small-t intron, present in many common expression vectors, leads to aberrant splicing. Mol Cell Biol 10(4): 1805–1810

Immonen A, Vapalahti M, Tyynela K, Hurskainen H, Sandmair A, Vanninen R, Langford G, Murray N, Yla-Herttuala S (2004) AdvHSV-tk gene therapy with intravenous ganciclovir improves survival in human malignant glioma: a randomised, controlled study. Mol Ther 10(5):967–972. doi:10.1016/j.ymthe.2004.08.002

Itaka K, Yamauchi K, Harada A, Nakamura K, Kawaguchi H, Kataoka K (2003) Polyion complex micelles from plasmid DNA and poly(ethylene glycol)-poly(L-lysine) block copolymer as serum-tolerable polyplex system: physicochemical properties of micelles relevant to gene transfection efficiency. Biomaterials 24(24): 4495–4506

Johnson LA, Morgan RA, Dudley ME, Cassard L, Yang JC, Hughes MS, Kammula US et al (2009) Gene therapy with human and mouse T-cell receptors mediates cancer regression and targets normal tissues expressing cognate antigen. Blood 114(3):535–546. doi:10.1182/blood-2009-03-211714

Kaji K, Norrby K, Paca A, Mileikovsky M, Mohseni P, Woltjen K (2009) Virus-free induction of pluripotency and subsequent excision of reprogramming factors. Nature 458(7239):771–775. doi:10.1038/nature07864

Kaplitt MG, Feigin A, Tang C, Fitzsimons HL, Mattis P, Lawlor PA, Bland RJ et al (2007) Safety and tolerability of gene therapy with an adeno-associated virus (AAV) borne GAD gene for Parkinson's disease: an open label, phase I trial. Lancet 369(9579):2097–2105. doi:10.1016/S0140-6736(07)60982-9

Karnieli O, Izhar-Prato Y, Bulvik S, Efrat S (2007) Generation of insulin-producing cells from human bone marrow mesenchymal stem cells by genetic manipulation. Stem Cells 25(11):2837–2844. doi:10.1634/stemcells.2007-0164, 2007–0164 [pii]

Kim J, Chen CP, Rice KG (2005) The proteasome metabolizes peptide-mediated nonviral gene delivery systems. Gene Ther 12(21):1581–1590. doi:10.1038/sj.gt.3302575

Kohn DB, Weinberg KI, Nolta JA, Heiss LN, Lenarsky C, Crooks GM, Hanley ME et al (1995) Engraftment of gene-modified umbilical cord blood cells in neonates with adenosine deaminase deficiency. Nat Med 1(10):1017–1023

Le Blanc K, Rasmusson I, Sundberg B, Gotherstrom C, Hassan M, Uzunel M, Ringden O (2004) Treatment of severe acute graft-versus-host disease with third party haploidentical mesenchymal stem cells. Lancet 363(9419):1439–1441. doi:10.1016/S0140-6736(04)16104-7, S0140-6736(04)16104-7 [pii]

Lesoon-Wood LA, Kim WH, Kleinman HK, Weintraub BD, Mixson AJ (1995) Systemic gene therapy with p53 reduces growth and metastases of a malignant human breast cancer in nude mice. Hum Gene Ther 6(4): 395–405. doi:10.1089/hum.1995.6.4-395

Levine BL, Humeau LM, Boyer J, Macgregor RR, Rebello T, Lu X, Binder GK, Slepushkin V, Lemiale F, Mascola JR, Bushman FD, Dropulic B, June CH (2006) Gene transfer in humans using a conditionally replicating lentiviral vector. Proc Natl Acad Sci U S A 103(46):17372–17377

LeWitt PA, Rezai AR, Leehey MA, Ojemann SG, Flaherty AW, Eskandar EN, Kostyk SK et al (2011) AAV2-GAD gene therapy for advanced Parkinson's disease: a double-blind, sham-surgery controlled, randomised trial. Lancet Neurol 10(4):309–319. doi:10.1016/S1474-4422(11)70039-4

Li F, Mahato RI (2009) Bipartite vectors for co-expression of a growth factor cDNA and short hairpin RNA against an apoptotic gene. J Gene Med 11(9):764–771. doi:10.1002/jgm.1357

Li L, Li F, Qi H, Feng G, Yuan K, Deng H, Zhou H (2008) Coexpression of Pdx1 and betacellulin in mesenchymal stem cells could promote the differentiation of nestin-positive epithelium-like progenitors and pancreatic islet-like spheroids. Stem Cells Dev 17(4):815–823. doi:10.1089/scd.2008.0060

Lochmuller H, Petrof BJ, Pari G, Larochelle N, Dodelet V, Wang Q, Allen C et al (1996) Transient immunosuppression by FK506 permits a sustained high-level dystrophin expression after adenovirus-mediated dystrophin minigene transfer to skeletal muscles of adult dystrophic (mdx) mice. Gene Ther 3(8):706–716

Louis N, Evelegh C, Graham FL (1997) Cloning and sequencing of the cellular-viral junctions from the human adenovirus type 5 transformed 293 cell line. Virology 233:423–429

MacGregor RR (2001) Clinical protocol. A phase 1 open-label clinical trial of the safety and tolerability of single escalating doses of autologous CD4 T cells transduced with VRX496 in HIV-positive subjects. Hum Gene Ther 12(16):2028–2029

MacGregor RR, Boyer JD, Ugen KE, Lacy KE, Gluckman SJ, Bagarazzi ML, Chattergoon MA et al (1998) First human trial of a DNA-based vaccine for treatment of human immunodeficiency virus type 1 infection: safety and host response. J Infect Dis 178(1):92–100

Mahato RI, Rolland A, Tomlinson E (1997) Cationic lipid-based gene delivery systems: pharmaceutical perspectives. Pharm Res 14(7):853–859

Mahato RI, Smith LC, Rolland A (1999) Pharmaceutical perspectives of nonviral gene therapy. Adv Genet 41:95–156

Majhen D, Ambriovic-Ristov A (2006) Adenoviral vectors–how to use them in cancer gene therapy? Virus Res 119(2):121–133

Manilla P, Rebello T, Afable C, Lu X, Slepushkin V, Humeau LM, Schonely K et al (2005) Regulatory considerations for novel gene therapy products: a review of the process leading to the first clinical lentiviral vector. Hum Gene Ther 16(1):17–25. doi:10.1089/hum.2005.16.17

Mannisto M, Vanderkerken S, Toncheva V, Elomaa M, Ruponen M, Schacht E, Urtti A (2002) Structure-activity relationships of poly(L-lysines): effects of pegylation and molecular shape on physicochemical and biological properties in gene delivery. J Control Release 83(1):169–182

Manthorpe M, Hobart P, Hermanson G, Ferrari M, Geall A, Goff B, Rolland A (2005) Plasmid vaccines and therapeutics: from design to applications. Adv Biochem Eng Biotechnol 99:41–92

Markert JM, Medlock MD, Rabkin SD, Gillespie GY, Todo T, Hunter WD, Palmer CA, Feigenbaum F, Tornatore C, Tufaro F, Martuza RL (2000) Conditionally replicating herpes simplex virus mutant, G207 for the treatment of malignant glioma: results of a phase I trial. Gene Ther 7(10):867–874

Martin ME, Rice KG (2007) Peptide-guided gene delivery. AAPS J 9(1):E18–E29. doi:10.1208/aapsj0901003

McGinley L, McMahon J, Strappe P, Barry F, Murphy M, O'Toole D, O'Brien T (2011) Lentiviral vector mediated modification of mesenchymal stem cells & enhanced survival in an in vitro model of ischaemia. Stem Cell Res Ther 2(2):12. doi:10.1186/scrt53, scrt53 [pii]

McKenzie DL, Collard WT, Rice KG (1999) Comparative gene transfer efficiency of low molecular weight polylysine DNA-condensing peptides. J Pept Res 54(4):311–318

Medina-Kauwe LK (2003) Endocytosis of adenovirus and adenovirus capsid proteins. Adv Drug Deliv Rev 55(11):1485–1496

Melo LG, Pachori AS, Gnecchi M, Dzau VJ (2006) Genetic therapies for cardiovascular diseases. Trends Mol Med 11(5):240–250

Mendell JR, Miller A (2004) Gene transfer for neurologic disease: agencies, policies, and process. Neurology 63(12):2225–2232

Miyata K, Kakizawa Y, Nishiyama N, Harada A, Yamasaki Y, Koyama H, Kataoka K (2004) Block catiomer polyplexes with regulated densities of charge and disulfide cross-linking directed to enhance gene expression. J Am Chem Soc 126(8):2355–2361. doi:10.1021/ja0379666

Miyata K, Kakizawa Y, Nishiyama N, Yamasaki Y, Watanabe T, Kohara M, Kataoka K (2005) Freeze-dried formulations for in vivo gene delivery of PEGylated polyplex micelles with disulfide crosslinked cores to the liver. J Control Release 109(1–3):15–23. doi:10.1016/j.jconrel.2005.09.043

Miyoshi H, Blomer U, Takahashi M, Gage FH, Verma IM (1998) Development of a self-inactivating lentivirus vector. J Virol 72(10):8150–8157

Miyoshi H, Takahashi M, Gage FH, Verma IM (1997) Stable and efficient gene transfer into the retina using an HIV-based lentiviral vector. Proc Natl Acad Sci U S A 94(19):10319–10323

Montini E, Cesana D, Schmidt M, Sanvito F, Bartholomae CC, Ranzani M, Benedicenti F et al (2009) The genotoxic potential of retroviral vectors is strongly modulated by vector design and integration site selection in a mouse model of HSC gene therapy. J Clin Invest 119(4):964–975. doi:10.1172/JCI37630

Morgan RA, Dudley ME, Wunderlich JR, Hughes MS, Yang JC, Sherry RM, Royal RE et al (2006) Cancer regression in patients after transfer of genetically engineered lymphocytes. Science 314(5796):126–129. doi:10.1126/science.1129003

Morgan RA, Walker R, Carter CS, Natarajan V, Tavel JA, Bechtel C, Herpin B, Muul L, Zheng Z, Jagannatha S, Bunnell BA, Fellowes V, Metcalf JA, Stevens R, Baseler M, Leitman SF, Read EJ, Blaese RM, Lane HC (2005) Preferential survival of CD4+ T lymphocytes engineered with anti-human immunodeficiency virus (HIV) genes in HIV-infected individuals. Hum Gene Ther 16(9):1065–1074

Mori S, Wang L, Takeuchi T, Kanda T (2004) Two novel adeno-associated viruses from cynomolgus monkey: pseudotyping characterization of capsid protein. Virology 330(2):375–383. doi:10.1016/j.virol.2004.10.012

Muul LM, Tuschong LM, Soenen SL, Jagadeesh GJ, Ramsey WJ, Long Z, Carter CS, Garabedian EK, Alleyne M, Brown M, Bernstein W, Schurman SH, Fleisher TA, Leitman SF, Dunbar CE, Blaese RM, Candotti F (2003) Persistence and expression of the adenosine deaminase gene for 12 years and immune reaction to gene transfer components: long-term results of the first clinical gene therapy trial. Blood 101(7):2563–2569

Nathwani AC, Tuddenham EG, Rangarajan S, Rosales C, McIntosh J, Linch DC, Chowdary P et al (2011) Adenovirus-associated virus vector-mediated gene transfer in hemophilia B. N Engl J Med 365(25): 2357–2365. doi:10.1056/NEJMoa1108046

Nemerow GR, Stewart PL (1999) Role of a_v integrins in adenovirus cell entry and gene delivery. Microbiol Mol Biol Rev 63(3):725–734

Nussbaum J, Minami E, Laflamme MA, Virag JA, Ware CB, Masino A, Muskheli V, Pabon L, Reinecke H, Murry CE (2007) Transplantation of undifferentiated murine embryonic stem cells in the heart: teratoma formation and immune response. FASEB J 21(7):1345–1357. doi:10.1096/fj.06-6769com, fj.06-6769com [pii]

Panakanti R, Mahato RI (2009) Bipartite adenoviral vector encoding hHGF and hIL-1Ra for improved human islet transplantation. Pharm Res 26(3):587–596. doi:10.1007/s11095-008-9777-y

Peng H, Wright V, Usas A, Gearhart B, Shen HC, Cummins J, Huard J (2002) Synergistic enhancement of bone formation and healing by stem cell-expressed VEGF and bone morphogenetic protein-4. J Clin Invest 110(6):751–759. doi:10.1172/JCI15153

Peng Z (2005) Current status of gendicine in China: recombinant human Ad-p53 agent for treatment of cancers. Hum Gene Ther 16(9):1016–1027. doi:10.1089/hum.2005.16.1016

Phinney DG, Prockop DJ (2007) Concise review: mesenchymal stem/multipotent stromal cells: the state of transdifferentiation and modes of tissue repair–current views. Stem Cells 25(11):2896–2902. doi:10.1634/stemcells.2007-0637

Qin JY, Zhang L, Clift KL, Hulur I, Xiang AP, Ren BZ, Lahn BT (2010) Systematic comparison of constitutive promoters and the doxycycline-inducible promoter. PLoS One 5(5):e10611. doi:10.1371/journal.pone.0010611

Ranga U, Woffendin C, Verma S, Xu L, June CH, Bishop DK, Nabel GJ (1998) Enhanced T cell engraftment after retroviral delivery of an antiviral gene in HIV-infected individuals. Proc Natl Acad Sci U S A 95(3):1201–1206

Raper SE, Chirmule N, Lee FS, Wivel NA, Bagg A, Gao GP, Wilson JM, Batshaw ML (2003) Fatal systemic inflammatory response syndrome in a ornithine transcarbamylase deficient patient following adenoviral gene transfer. Mol Genet Metab 80(1–2):148–158

Reilly JP, Grise MA, Fortuin FD, Vale PR, Schaer GL, Lopez J, VAN Camp JR et al (2005) Long-term (2-year) clinical events following transthoracic intramyocardial gene transfer of VEGF-2 in no-option patients. J Interv Cardiol 18(1):27–31. doi:10.1111/j.1540-8183.2005.04026.x

Rip J, Nierman MC, Sierts JA, Petersen W, Van den Oever K, Van Raalte D, Ross CJ et al (2005) Gene therapy for lipoprotein lipase deficiency: working toward clinical application. Hum Gene Ther 16(11):1276–1286. doi:10.1089/hum.2005.16.1276

Roth JA, Cristiano RJ (1997) Gene therapy for cancer: what have we done and where are we going? J Natl Cancer Inst 89(1):21–39

Rubio D, Garcia-Castro J, Martin MC, de la Fuente R, Cigudosa JC, Lloyd AC, Bernad A (2005) Spontaneous human adult stem cell transformation. Cancer Res 65(8):3035–3039, 65/8/3035 [pii]

Ryu WS, Mertz JE (1989) Simian virus 40 late transcripts lacking excisable intervening sequences are defective in both stability in the nucleus and transport to the cytoplasm. J Virol 63(10):4386–4394

Sandmair AM, Loimas S, Puranen P, Immonen A, Kossila M, Puranen M, Hurskainen H et al (2000) Thymidine kinase gene therapy for human malignant glioma, using replication-deficient retroviruses or adenoviruses. Hum Gene Ther 11(16):2197–2205. doi:10.1089/104303400750035726

Schuldiner M, Itskovitz-Eldor J, Benvenisty N (2003) Selective ablation of human embryonic stem cells expressing a "suicide" gene. Stem Cells 21(3):257–265. doi:10.1634/stemcells.21-3-257

Shirakawa T (2009) Clinical trial design for adenoviral gene therapy products. Drug News Perspect 22(3):140–145. doi:10.1358/dnp.2009.22.3.1354090

Simonelli F, Maguire AM, Testa F, Pierce EA, Mingozzi F, Bennicelli JL, Rossi S et al (2010) Gene therapy for Leber's congenital amaurosis is safe and effective through 1.5 years after vector administration. Mol Ther 18(3):643–650. doi:10.1038/mt.2009.277

Simons JW, Sacks N (2006) Granulocyte-macrophage colony-stimulating factor-transduced allogeneic cancer cellular immunotherapy: the GVAX vaccine for prostate cancer. Urol Oncol 24(5):419–424

Singh S, Narang AS, Mahato RI (2011) Subcellular fate and off-target effects of siRNA, shRNA, and miRNA. Pharm Res 28(12):2996–3015. doi:10.1007/s11095-011-0608-1

Spilianakis CG, Lalioti MD, Town T, Lee GR, Flavell RA (2005) Interchromosomal associations between alternatively expressed loci. Nature 435(7042):637–645. doi:10.1038/nature03574

Springer ML (2006) A balancing act: therapeutic approaches for the modulation of angiogenesis. Curr Opin Investig Drugs 7(3):243–250

Stewart DJ, Hilton JD, Arnold JM, Gregoire J, Rivard A, Archer SL, Charbonneau F et al (2006) Angiogenic gene therapy in patients with nonrevascularizable ischemic heart disease: a phase 2 randomized, controlled trial of AdVEGF(121) (AdVEGF121) versus maximum medical treatment. Gene Ther 13(21):1503–1511. doi:10.1038/sj.gt.3302802

Stewart PL, Chiu CY, Huang S, Muir T, Zhao Y, Chait B, Mathias P, Nemerow GR (1997) Cryo-EM visualization of an exposed RGD epitope on adenovirus that escapes antibody neutralization. EMBO J 16(6):1189–1198. doi:10.1093/emboj/16.6.1189

Stolberg SG (1999) The biotech death of Jesse Gelsinger. N Y Times Mag 136–140:149–150

Surosky RT, Urabe M, Godwin SG, McQuiston SA, Kurtzman GJ, Ozawa K, Natsoulis G (1997) Adeno-associated virus Rep proteins target DNA sequences to a unique locus in the human genome. J Virol 71(10):7951–7959

Tacket CO, Roy MJ, Widera G, Swain WF, Broome S, Edelman R (1999) Phase 1 safety and immune response studies of a DNA vaccine encoding hepatitis B surface antigen delivered by a gene delivery device. Vaccine 17(22):2826–2829

Tsuda H, Wada T, Ito Y, Uchida H, Dehari H, Nakamura K, Sasaki K, Kobune M, Yamashita T, Hamada H (2003) Efficient BMP2 gene transfer and bone formation of mesenchymal stem cells by a fiber-mutant adenoviral vector. Mol Ther 7(3):354–365

Tuszynski MH, Thal L, Pay M, Salmon DP, HS U, Patel P et al (2005) A phase 1 clinical trial of nerve growth factor gene therapy for Alzheimer disease. Nat Med 11(5):551–555. doi:10.1038/nm1239

U.S. Department of Health and Human Services (1998) Guidance for industry: guidance for human somatic cell therapy and gene therapy. Center for Biologics Evaluation and Research, United States Food and Drug Administration, Rockville, MD

Uckert W, Kammertons T, Haack K, Qin Z, Gebert J, Schendel DJ, Blankenstein T (1998) Double suicide gene (cytosine deaminase and herpes simplex virus thymidine kinase) but not single gene transfer allows reliable elimination of tumor cells in vivo. Hum Gene Ther 9(6):855–865. doi:10.1089/hum.1998.9.6-855

Varghese S, Rabkin SD (2002) Oncolytic herpes simplex virus vectors for cancer virotherapy. Cancer Gene Ther 9(12):967–978

Vasan S, Hurley A, Schlesinger SJ, Hannaman D, Gardiner DF, Dugin DP, Boente-Carrera M et al (2011) In vivo electroporation enhances the immunogenicity of an HIV-1 DNA vaccine candidate in healthy volunteers. PLoS One 6(5):e19252. doi:10.1371/journal.pone.0019252

Verma IM (1990) Gene therapy. Sci Am 263(5):68–72, 81–64

Wakitani S, Nawata M, Tensho K, Okabe T, Machida H, Ohgushi H (2007) Repair of articular cartilage defects in the patello-femoral joint with autologous bone marrow mesenchymal cell transplantation: three case reports involving nine defects in five knees. J Tissue Eng Regen Med 1(1):74–79. doi:10.1002/term.8

Wang DA, Narang AS, Kotb M, Gaber AO, Miller DD, Kim SW, Mahato RI (2002) Novel branched poly(ethylenimine)-cholesterol water-soluble lipopolymers for gene delivery. Biomacromolecules 3(6):1197–1207

Weber W, Fussenegger M (2006) Pharmacologic transgene control systems for gene therapy. J Gene Med 8(5):535–556

Wells DJ (2004) Gene therapy progress and prospects: electroporation and other physical methods. Gene Ther 11(18):1363–1369. doi:10.1038/sj.gt.3302337

Wiley (2012) The Journal of Gene Medicine Clinical Trials Worldwide Database http://www.wiley.co.uk/genetherapy/clinical/. Accessed 3 Mar 2012

Wright JF, Qu G, Tang C, Sommer JM (2003) Recombinant adeno-associated virus: formulation challenges and strategies for a gene therapy vector. Curr Opin Drug Discov Devel 6(2):174–178

Wu H, Yoon AR, Li F, Yun CO, Mahato RI (2011) RGD peptide-modified adenovirus expressing HGF and XIAP improves islet transplantation. J Gene Med. doi:10.1002/jgm.1626

Wu Z, Asokan A, Samulski RJ (2006) Adeno-associated virus serotypes: vector toolkit for human gene therapy. Mol Ther 14(3):316–327

Xia ZJ, Chang JH, Zhang L, Jiang WQ, Guan ZZ, Liu JW, Zhang Y et al (2004) Phase III randomized clinical trial of intratumoral injection of E1B gene-deleted adenovirus (H101) combined with cisplatin-based chemotherapy in treating squamous cell cancer of head and neck or esophagus. Ai Zheng 23(12):1666–1670

Xu ZL, Mizuguchi H, Sakurai F, Koizumi N, Hosono T, Kawabata K, Watanabe Y, Yamaguchi T, Hayakawa T (2005) Approaches to improving the kinetics of adenovirus-delivered genes and gene products. Adv Drug Deliv Rev 57(5):781–802

Yan Z, Zhang Y, Duan D, Engelhardt JF (2000) Trans-splicing vectors expand the utility of adeno-associated virus for gene therapy. Proc Natl Acad Sci U S A 97(12):6716–6721

Zaiss AK, Liu Q, Bowen GP, Wong NC, Bartlett JS, Muruve DA (2002) Differential activation of innate immune responses by adenovirus and adeno-associated virus vectors. J Virol 76(9):4580–4590

Zhao T, Zhang ZN, Rong Z, Xu Y (2011) Immunogenicity of induced pluripotent stem cells. Nature 474(7350):212–215. doi:10.1038/nature10135

Zufferey R, Dull T, Mandel RJ, Bukovsky A, Quiroz D, Naldini L, Trono D (1998) Self-inactivating lentivirus vector for safe and efficient in vivo gene delivery. J Virol 72(12):9873–9880

Zwaka T (2006) Use of genetically modified stem cells in experimental gene therapies. In: Regenerative medicine. National Institutes of Health, Bethesda

FURTHER READING

Schleef M et al. (2001) Plasmids for therapy cbrsand vaccination. Wiley-VCH, New York, NY

Schleef M (2005) DNA pharmaceuticals: formulation and delivery in gene therapy, DNA vaccination and immunotherapy. John Wiley & Sons, Hoboken

National Institutes of Health (2006) Regenerative medicine. National Institutes of Health, Bethesda, DC

Narang A, Mahato RI (2010) Targeted delivery of small and macromolecular drugs. CRC press, Boca Raton, FL

25 Stem Cell Technology

Colin W. Pouton

INTRODUCTION

■ Significance of Stem Cell Technology

Advances in stem cell biology over the past decade have given rise to new biotechnologies that will be used in a range of pharmaceutical applications including creation of cellular tools for studying the origins and progression of disease, development of phenotypic screens for use in drug discovery, development of cellular assays of drug toxicology and metabolism, and production of cell-based medicinal products for clinical use in various forms of "cell therapy." A phenotypic screen makes use of an assay that detects a cellular response such as a change in cell morphology, or another downstream event, rather than an initial signal transduction event at the plasma membrane. Stem cells are of particular value in this regard because they can be expanded and then differentiated, providing an unlimited source of mature differentiated human cells (such as neurons or cardiomyocytes). These cells offer the opportunity to develop high-throughput plate-based assays to detect features such as the rate and amplitude of beating of heart cells, or a functional response of a specific subtype of mature neurons, generating potentially more informative assays than current cell-based assays.

■ What Is a Stem Cell?

The fundamental property of a stem cell is the capability to multiply and to give rise to a variety of differentiated cells, but the general term "stem cell" is used in several contexts, each important for different reasons, as shown in Table 25.1. Adult, embryonic, mesenchymal, and induced pluripotent stem cells are currently the subject of intense investigation. Stem cells differ in the breadth of mature cell phenotypes to which they can give rise, and they are characterized by their potency as defined in Table 25.2. Before the practical applications of stem cells can be fully realized, in both the research laboratory and clinic, it will be necessary to understand in detail how to control their differentiation towards mature postmitotic phenotypes.

■ Adult Stem Cells

Adult (or tissue) stem cells are now known to be present in many if not all individual organs in adults and are generally thought to be "multipotent," meaning they can give rise to the cells found in their organ of origin, but not in other organs (Fig. 25.1). The identification of adult stem cells in human tissues has necessitated a repositioning of basic tenets of some biological sciences, most notably in neuroscience, where the prevailing view was that no new neurons were born in humans after birth (Zhao et al. 2008). Adult stem cells are rare and they cannot always be isolated and grown in culture. Even when they can be grown in culture, usually they cannot be grown indefinitely. In tissues, they exist in a defined, organized environment of supporting cells that define the architecture of the "stem cell niche" (Scadden 2006). For example, in the bone marrow there are many hematopoietic stem cell niches, each of which contains stromal cells to support the function of hematopoietic stem cell (HSC). Each HSC is capable of producing the progenitors of all types of blood cells (Taichman 2005). Differentiation of HSCs has been studied extensively and is now well understood (Fig. 25.2), but at present conditions that allow HSCs to be maintained and expanded in vitro have not been established. A hallmark of adult stem cells is their ability to "self-renew" and undergo asymmetric cell division. This means that when they divide they usually give rise to two different cells, one an identical stem cell and the other a partly differentiated progenitor cell (Fig. 25.2), a process that occurs in a polarized

C.W. Pouton, Ph.D.
Drug Delivery, Disposition and Dynamics,
Monash Institute of Pharmaceutical Sciences,
Monash University (Parkville Campus),
381 Royal Parade, Parkville, VIC 3052, Australia
e-mail: colin.pouton@monash.edu

D.J.A. Crommelin, R.D. Sindelar, and B. Meibohm (eds.), *Pharmaceutical Biotechnology*,
DOI 10.1007/978-1-4614-6486-0_25,

Type of stem cell	Origin	Characteristic potential	Practical uses
Adult (tissue) stem cells	Exist in small number in many tissues, often in a well-defined and supportive niche	Multipotent: give rise to cells of the relevant tissue or local environment	Some adult stem cells can be expanded in vitro (e.g., neural stem cells). Characterization of phenotype is challenging
Mesenchymal stem cells (MSCs)	A collective term for stem cells sourced from stromal or connective tissue (bone marrow, adipose tissue, or umbilical cord tissue)	Yet to be fully determined. MSCs can differentiate into cells of connective tissues, e.g., chondrocytes, osteoblasts, and adipocytes, but they have also been reported to give rise to many other unrelated cell types	MSCs from specific sources are in clinical development for several applications (e.g., see www.mesoblast.com)
Cord blood-derived MSCs	A specific source of MSCs. Extracted at birth from umbilical cord blood	Yet to be fully determined. The hope is that they will be a source of many cell types for individual patients	Private cell banks are already established for cryopreservation of cord blood samples
Embryonic stem cells	Result from in vitro culture of the inner cell mass (embryoblast)	Pluripotent: have the potential to produce all cell types of the adult organism	Vital source of differentiated cells for research and possibly cell therapy
Induced pluripotent stem (iPS) cells	Derived by reprogramming of somatic cells (often skin fibroblasts) taken from an adult biopsy	Pluripotent, although methods for full reprogramming are still in development	May allow pluripotent cells from individuals to be obtained without the need for fertilized human eggs

Table 25.1 ■ Origin, characteristics, and uses of stem cells.

Totipotent (or omnipotent) cells	Can differentiate into all embryonic and extraembryonic cell types (i.e., in humans they give rise to the fetus and the placenta)
Pluripotent cells	Can differentiate into all three germ cell types (endoderm, mesoderm, or ectoderm) and subsequently into all embryonic cell types
Multipotent cells	Can differentiate into closely related cells, such as all cells in a particular organ. Typical of adult tissue stem cells
Oligopotent cells	Can differentiate into a restricted closely related group, such as a hematopoietic progenitor cell that can produce a subset of blood cell types
Unipotent cells	Have the property of self-renewal but can only give rise to cells of their own phenotype, such as muscle stem cells

Table 25.2 ■ Definitions of cell potency.

manner controlled by the niche. The common pattern in adult tissues is that the resulting progenitor cells, sometimes referred to as "transit amplifying" cells, are capable of expansion by symmetric division and can subsequently differentiate to form the various cell types needed for repair or replenishment of the relevant tissue. Such mechanisms are well documented in tissues that are regenerated continuously in the adult, such as the epithelia of the skin, intestine and other mucosal tissues, and the bone marrow (Lander et al. 2012). It has now become apparent that similar processes are also found in organs that are not continuously replenished, such as the brain. The realization that adult stem cells are present in many organs offers the possibility that repair and regeneration could be stimulated and controlled in degenerative diseases by drug therapy, but whether this will be possible remains to be seen. In the brain, neural stem cells have been identified in the subventricular zone and in the dentate gyrus (part of the hippocampus) (Alvarez-Buylla et al. 2001), but whether they are present in other regions of the brain remains to be investigated.

Adult stem cells have been used since the 1950s to treat cancers of blood cells, as one of the components of bone marrow transplants (Santos 1983). This procedure involves whole body irradiation to kill malignant cells in multiple myelomas and leukemia. The patient then receives a bone marrow transplant, not in itself a stem cell product, but the transplant contains a few HSCs which subsequently home to the bone marrow stem cell niches and begin to replenish the blood (Fig. 25.3). Rejection and graft-versus-host disease are

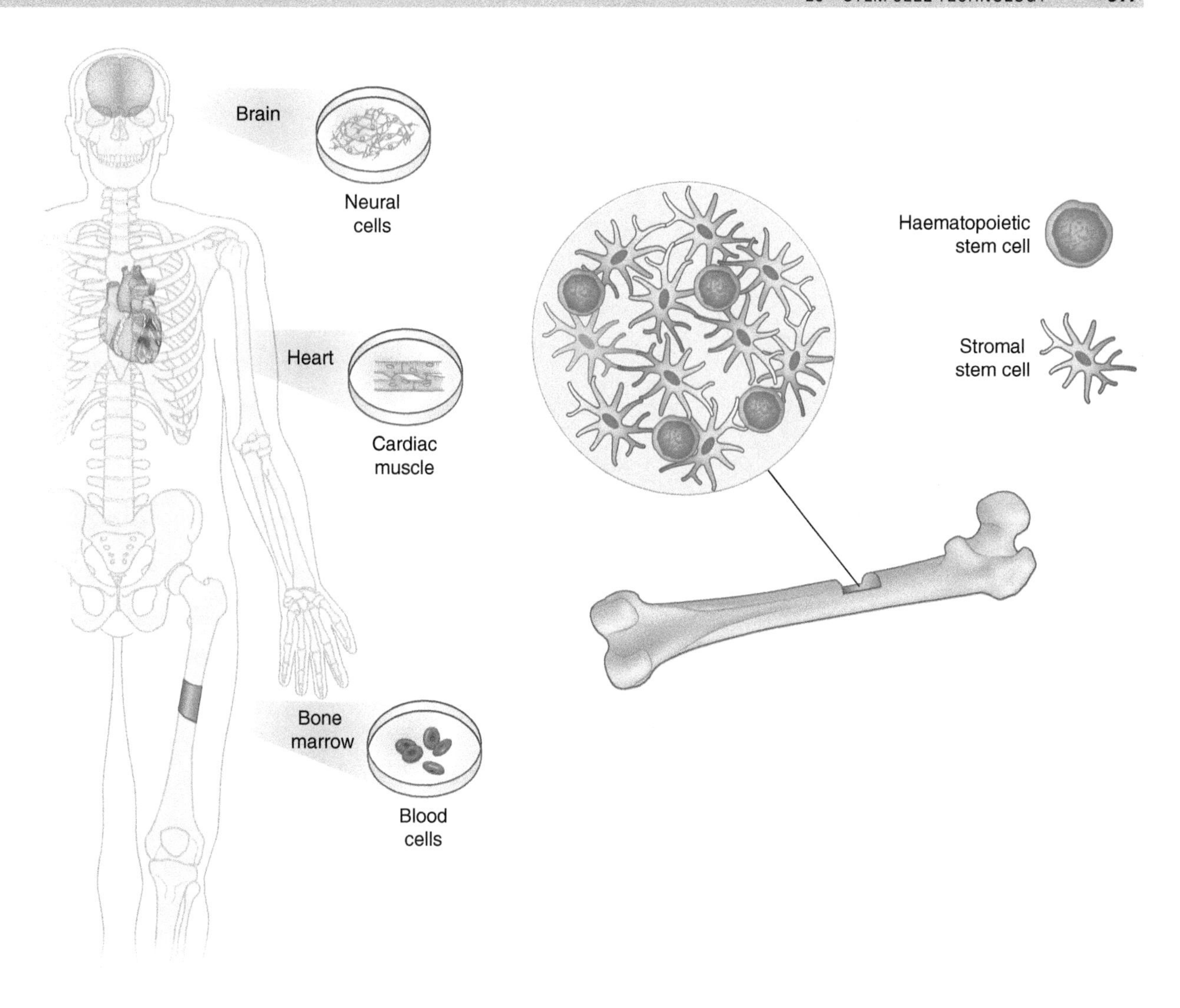

Figure 25.1 ■ Adult stem cells are present in many tissues in specific stem cell niches, giving rise to a specific group of cells found in the relevant tissue. The examples shown have been studied in detail but adult stem cells, yet to be defined, may be present in many other tissues.

still threatening complications of this form of therapy, but its practice can now be considered to be routine.

Mesenchymal stem cells (MSCs), sometimes called multipotent stromal cells, have generated considerable interest in recent years for cell therapy applications (Bianco et al. 2008). MSCs can be isolated from bone marrow, adipose tissue, and umbilical cord tissue (from the particularly rich source of Wharton's jelly and also from umbilical cord blood). Because cord blood can be sampled, frozen, and banked at birth, this source of MSCs has been identified as a potential source of cells for use in a regenerative capacity in later life. There are now several private companies that offer personal cell banking services, and public cord blood banks that supply pooled cord blood samples for clinical use. Whether cord blood banking will prove to be useful remains to be seen. MSCs have been reported to differentiate into various phenotypes (including chondrocytes, osteoblasts, and adipocytes) as well as other phenotypes, suggesting that MSCs have wider potential than one would expect. MSCs have been investigated in preclinical models of many applications and have been reported to home to damaged tissues and tumors from the vasculature. MSCs are in clinical development using direct injection for treatment of bone and joint diseases, heart disease, for repair of muscle and ligament damage, and even for repair of ischemic brain tissue. Which of these applications will prove to be successful is difficult to predict at this stage of clinical development.

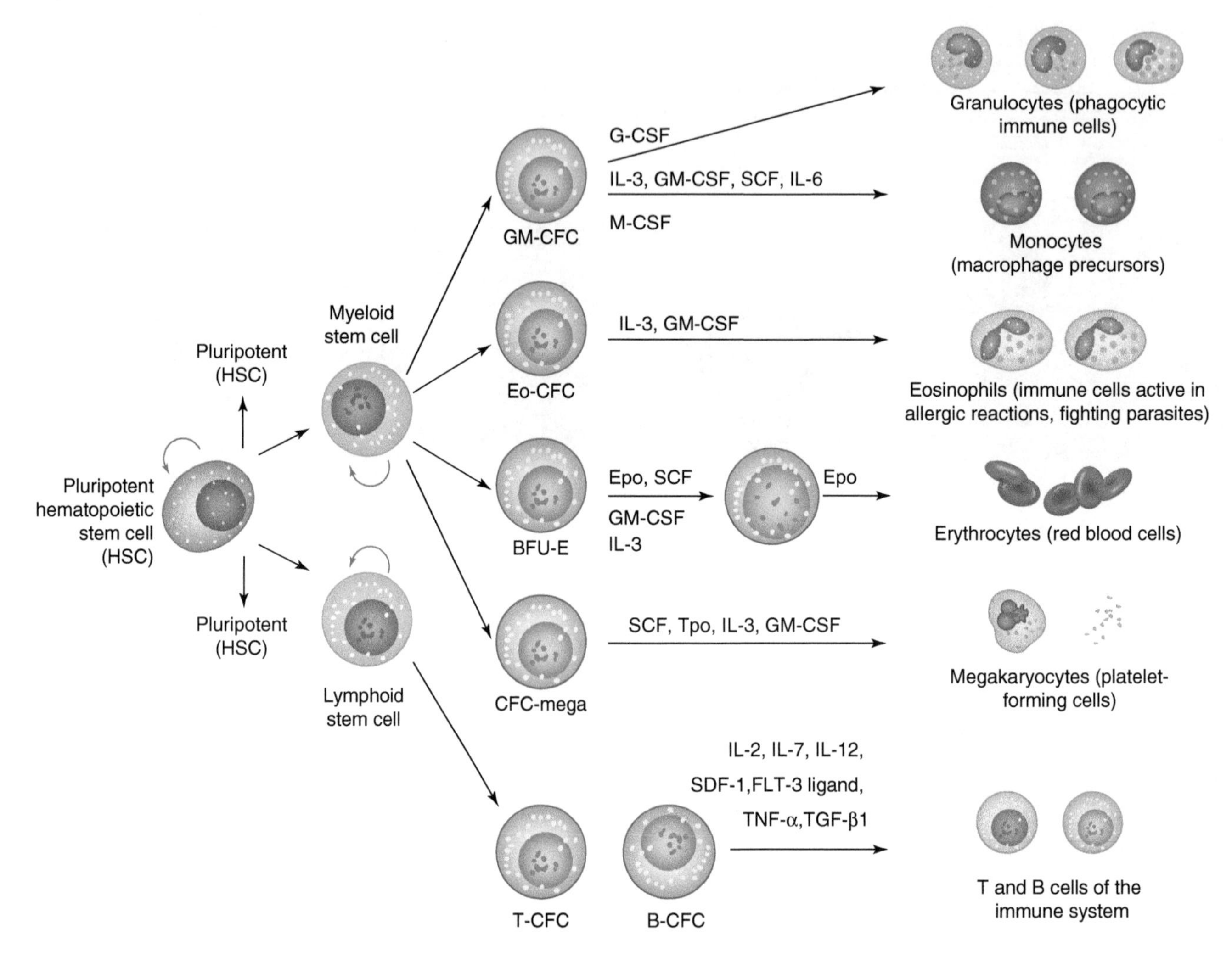

Figure 25.2 ■ Asymmetric division of adult hematopoietic stem cells (HSCs), to produce myeloid or lymphoid stem cells, further differentiation to form mitotic progenitors, and subsequently under the control of specific growth factors and cytokines, to form fully differentiated blood cells. The differentiation pathways of the hematopoietic system are better characterized than those of other tissues, but the pattern of differentiation is typical of other tissues.

■ Embryonic Stem Cells

During the earliest stages of mammalian development, soon after egg and sperm combine, the resulting diploid cells are said to be "totipotent," i.e., they can give rise to both the embryo and placental tissue. At the blastocyst stage of embryogenesis (day 5 in humans), the "inner cell mass" or "embryoblast" is compacted and separated from the surrounding "trophoblast." The latter combines with the maternal endometrium to form the placenta. The inner cell mass can be extracted and grown in vitro as embryonic stem (ES) cells, which can give rise to all three germ cell types (mesoderm, endoderm, and ectoderm), and therefore potentially any cell type found in the adult (Fig. 25.4). Mouse ES cells were first isolated in 1981 (Evans and Kaufman 1981; Martin 1981), but it took until 1998 for a similar procedure to be described allowing human ES cells to be grown in culture (Thomson et al. 1998). ES cells can now be grown for many cell divisions, limited only by genetic damage that occurs by mutation after extensive culture. The pluripotency of ES cells can be demonstrated in mice by injecting cells into a fertilized egg, resulting in the production of chimeric mice (i.e., mice made up of cells derived from both the donor and the injected ES cells). This process has been used routinely over the past 20 years to produce transgenic mice for research purposes. Human ES cells are usually identified by their ability to produce a teratoma (a tumor containing cells from all three germ cell types) when cells are injected into immunodeficient mice, but this not as robust a method for validation of pluripotency.

■ Maintenance and Differentiation of ES Cells in Culture

Mouse ES cells were first grown as compact colonies on a feeder layer of mouse embryonic fibroblasts, in media

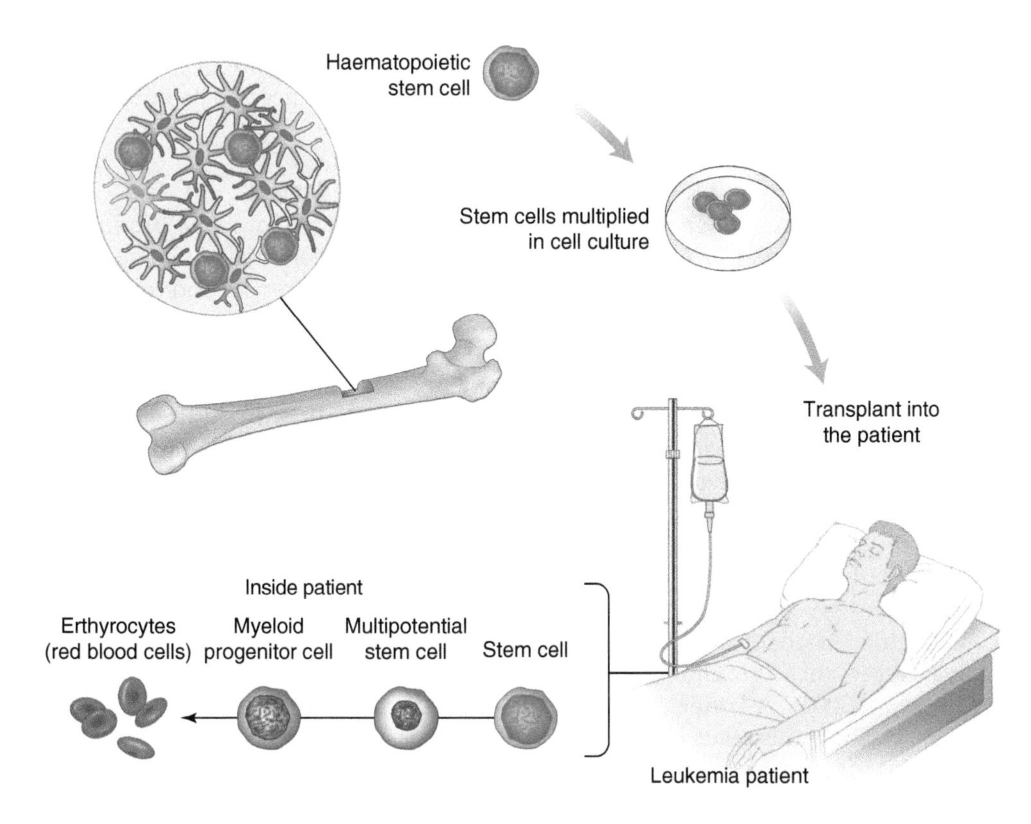

Figure 25.3 ■ Schematic representation of bone marrow transplantation, a form of stem cell therapy that was first used over 50 years ago. The transplant contains hematopoietic stem cells from the donor. These cells repopulate niches in the recipient bone marrow.

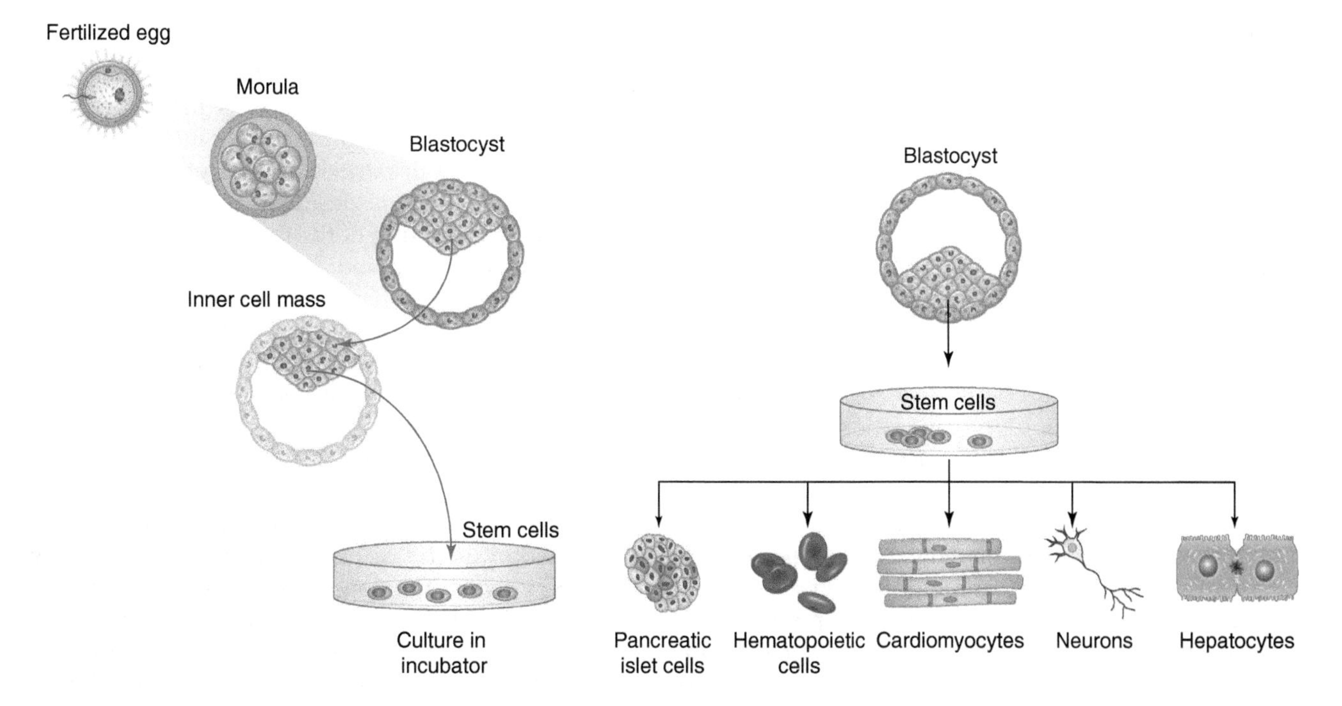

Figure 25.4 ■ Extraction of the inner cell mass of the blastocyst gives rise to embryonic stem cells (ES cells), which have the capacity to differentiate into all 200+ somatic cell types found in the adult human.

containing leukemia inhibitory factor (LIF) and fetal bovine serum. Efforts to simplify culture methods soon established that the feeders could be substituted with gelatin-coated culture plates, though differentiation occurs to some extent in the absence of the feeder layer. The vital component in serum was found to be bone morphogenetic protein (BMP). Thus, mouse ES cells can be grown in chemically defined medium with LIF and BMP4 (Ying et al. 2003). Human ES cells are grown in the presence of high concentrations of basic fibroblast growth factor (FGF2) and are unresponsive to LIF (Levenstein et al. 2006). The difference in responsiveness between mouse and human ES cells has been extensively studied and debated. The two methods of derivation may result in isolation of cells from slightly different stages of development. Human ES cells are thought to resemble cells from the later epiblast stage. More recently, it has been demonstrated that mouse ES cells can be maintained and grown very efficiently in the presence of small molecule inhibitors of mitogen-activated protein kinase (MEK1/2) and glycogen synthase kinase-3β (GSK-3β). This medium changes their phenotype slightly to what may be represent a "ground state" for mouse ES cells (Ying et al. 2008). A better understanding of the ground state and how this relates to human ES cells will be an important step forward and will allow human ES technology to be reproduced more effectively.

The technical challenge, now that human ES cells can be maintained and expanded, is to develop robust methods to control and direct ES cell differentiation, so that human cells of any desired phenotype can be obtained (Keller 2005; Murry and Keller 2008). In the context of cell therapy, it is also important to ensure that no undesired cells are present in a product for clinical use, such as undifferentiated cells, or cells that are capable of dedifferentiation, either of which could cause tumor formation after implantation. This science is immature at present and will remain a priority for investigation for several years. Thus, far attention has focused on the differentiation of human ES cells towards products that could be of obvious use for cell therapy, e.g., midbrain dopaminergic neurons for Parkinson's disease, cardiomyocytes for reinforcement of damaged heart tissue, and pancreatic β-islet cells for implantation in Type I diabetes. From a fundamental view, to improve our understanding of cell phenotype, for screening of small molecules that modulate cell function, and for disease modeling, it will be important to research ways of producing many other cell types. At present, fine tuning of differentiation programs is beyond our control. Differentiation usually results in mixed populations of cells. For example, neural differentiation can be induced quite effectively, but the result of further differentiation is a mixed population of cells that often include both neurons and glia, and the neurons are comprised of a variety of neuronal subtypes. Timing, duration, and concentration of exposure to specific morphogens are of critical importance to the outcome and will need to be optimized in each case.

■ Cell Therapy: The Broader Context

The potential of stem cells as a source of products for cell therapy needs to be understood in the context of alternative approaches to cell therapy and transplantation. A few cell therapies are already in clinical use. There are long-standing clinical practices that involve cell transplantation, including bone marrow transplants, and in vitro fertilization. In addition the FDA and EMA have both approved cell therapy products for niche applications. The chondrocyte product ChondroCelect (TiGenix) is currently the only cell therapy product approved in Europe by the EMA. Provenge (Dendreon) is a patient-specific cell therapy for prostate cancer immunotherapy that has been approved by the FDA. In this case, a sample of the patient's own white blood cells is treated with an engineered fusion polypeptide to produce a vaccine for reinjection. The intention is to deliver a cancer vaccine to professional antigen-presenting cells, i.e., dendritic cells. These applications of cell therapy are outside the scope of this chapter and will not be discussed in detail, but their development has done much to define a framework for development of stem cell-based products and their evaluation by regulatory agencies.

The term "regenerative medicine" is often used to describe the current interest in repair, restoration, or replacement of damaged tissue. The strong interest in use of stem cell-derived products is based on their potential to expand and differentiate in vivo, giving them the potential to participate actively in repair of damaged tissue. This is desirable but will require a complete understanding of the fate of stem cells, early and late progenitors, and differentiated cells after transplantation in each specific clinical application. This science is in its infancy at present. Differentiated cells that are incapable of dividing or further differentiation may also have useful roles in cell therapy and could be used to secrete protective or regenerative proteins (i.e., growth factors or cytokines) to the local environment ("gene therapy by cell therapy," cf. Chap. 24).

Another fundamental consideration in all forms of cell therapy is the distinction between autologous cell therapy (when the donor is also the recipient) and allogeneic cell therapy (when the donor cells are delivered to one or more different recipients). Allogeneic therapies and xenogenic therapies (those derived from animal sources) introduce immunological complications that need to be managed.

The general interest in stem cell therapies and regenerative medicine around the world has allowed unregulated practice of cell therapy to develop in some countries. This is a major concern for stem cell scientists, because treatments are being offered in the absence of any proven efficacy. In addition there is suspicion that the products in use have been manufactured with insufficient attention to quality control. Patients are travelling to private clinics and paying large sums of money for unproven treatments, creating a phenomenon that has been referred to as "stem cell tourism." It is very important that patients are warned of the dangers of falling prey to unethical operations. An up-to-date source of information on private clinics and stem cell tourism is available at the website of the International Society for Stem Cell Research (www.isscr.org).

IMMUNOLOGICAL CONSIDERATIONS IN CELL THERAPY

The potential application of cellular products derived from ES cells in cell therapy is limited by graft-host rejection issues, as with all therapeutic strategies based on transplantation, unless the transplant is derived from "self." Administration of drugs to suppress the immune response is standard practice for patients undergoing transplantation, but with immunosuppression come side effects and uncertainty. The hope is that induced pluripotent stem (iPS) cell technology (see below) may overcome rejection problems but it is too early to be sure at this stage. Another approach is to bank a collection of ES cell lines that allows selection of a matched HLA haplotype or a close match (Lui et al. 2009). It has been estimated that with a bank of 70–100 ES cell lines, a partially matched ES cell line can be chosen that is adequate for each recipient.

An alternative, particularly when a match cannot be found, is to produce ES cells for individual patients, by somatic cell nuclear transfer (SCNT) (Wilmut et al. 2002). This process, also known as "therapeutic cloning," involves implantation of a cell nucleus from the patient (i.e., genomic DNA extracted from a skin biopsy) into a human egg which has undergone removal of its own DNA. The environment in the enucleated egg is able to reprogram the DNA from the patient, removing epigenetic marks and restoring the DNA to an embryonic state. The development of an inner cell mass in the egg, after a period of incubation, allows extraction of ES cells that have the patient's exact genotype. These cells could be used subsequently for production of implants for cell therapy (Fig. 25.5). SCNT is also the first step in the process by which animals are cloned by "reproductive cloning," which involves implantation of the engineered egg into a surrogate mother (Fig. 25.6) (Campbell et al. 1996). Reproductive cloning of humans is illegal but is also likely to be impractical. It is known from experience with animal cloning that SCNT is an inefficient process. Most eggs that have undergone SCNT are unable to completely reprogram the donor DNA, and as a result the surrogate pregnancy is usually unproductive. Even when the pregnancy comes to term, the cloned offspring is known to carry many epigenetic marks that may compromise normal development, and the famous sheep, "Dolly," the first large animal to be cloned by way of SCNT, is known to have had several developmental defects (Wilmut et al. 2009). Second-generation animals, produced by mating a clone with another parent, are usually unaffected by such defects, indicating that SCNT is much less efficient than the natural process of reprogramming of DNA in a fertilized egg. Given that defects are known to occur after SCNT, the subsequent derivation of cells for clinical uses might also be prone to failure due to defects in ES cell differentiation. There is insufficient data available at this stage to judge whether this will be a limitation in practice. There are significant ethical concerns that have limited the practice of SCNT. A human egg donor is required, and unless the process becomes more efficient, women who are prepared to donate eggs would need to provide several eggs to produce a single ES cell line. There is concern that women could be exploited, particularly women from low economic backgrounds, and as a result SCNT is not supported by government funding at present in most countries. A restricted number of ES cell lines have been produced using spare eggs from in vitro fertilization programs, but the status of SCNT remains a controversial topic and is subject to legal constraints that vary from country to country. An alternative source of cells for regenerative medicine in the future may be umbilical cord blood stem cells, which are now being banked at childbirth, at least in private practice. Whether cord blood cells can be harnessed to produce all cell phenotypes is not clear at present. However, many of the ethical issues surrounding SCNT, and uncertainty of cord blood stem cell potency, may become irrelevant if the promise of induced pluripotent stem (iPS) cells can be realized.

IPS CELL TECHNOLOGY

In 2006 stem cell scientists were surprised by a remarkable discovery that has revolutionized the field and its potential practical application. Two Japanese scientists reported that mouse skin fibroblasts could be reprogrammed to produce pluripotent cells by forcing expression of just four genes (*Sox2*, *Oct4*, *Klf4*, and *cMyc*) using lentiviral vectors (Takahashi and Yamanaka 2006). A year later similar methods were

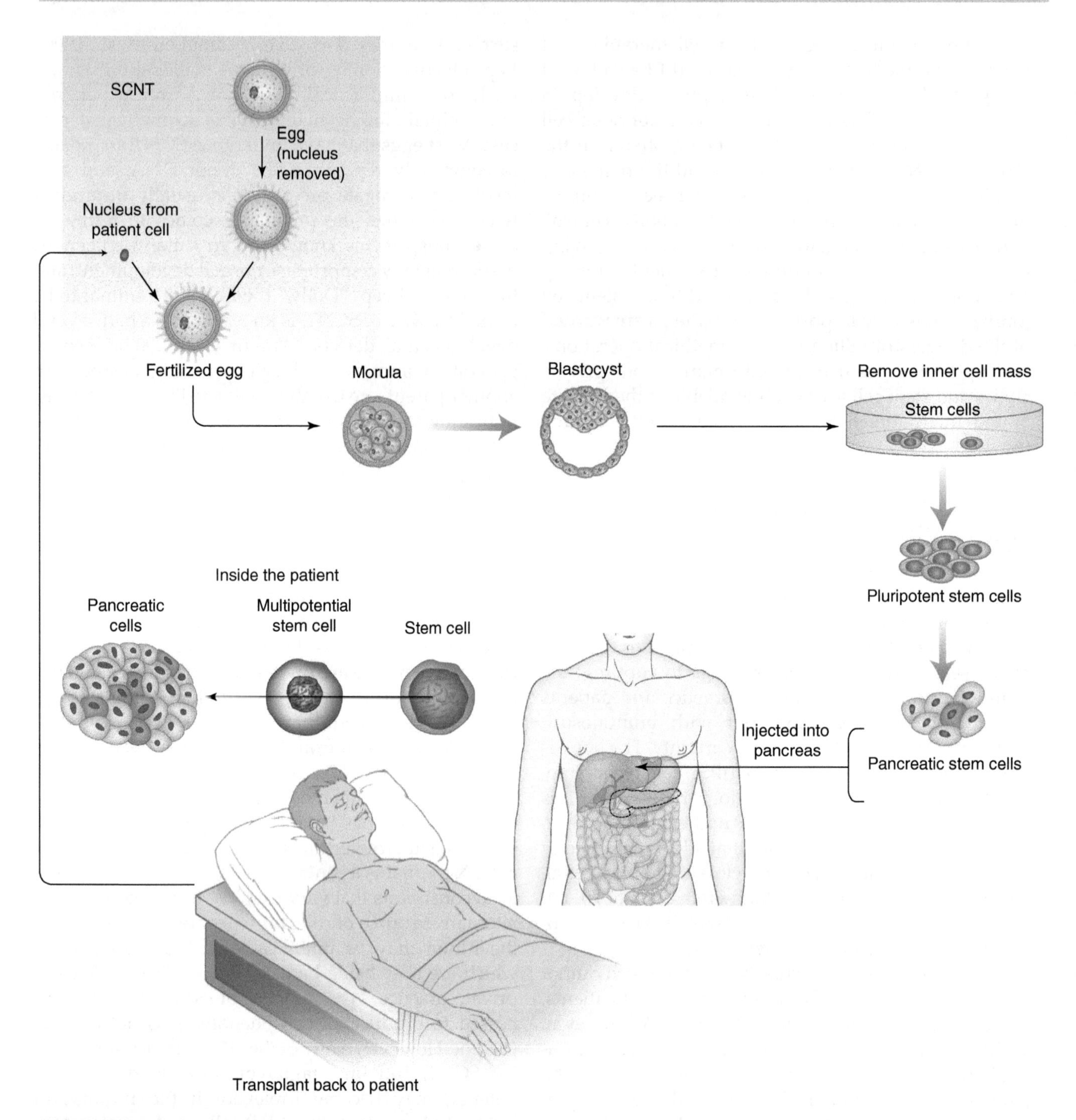

Figure 25.5 ■ Schematic diagram of the production and clinical use of cell therapies derived using somatic cell nuclear transfer (therapeutic cloning). The example given is for possible treatment of Type I insulin-dependent diabetes.

published for production of iPS cells from human fibroblasts (Takahashi et al. 2007; Yu et al. 2007). This indicated that patient-specific pluripotent stem cells could be produced without the need for human eggs, using cells extracted from a simple skin biopsy. The significance of this discovery to regenerative medicine cannot be overestimated. Over the last 5 years, the iPS cell field has exploded with activity, and the technology is now in use in hundreds of stem cell biology laboratories around the world. The four genes initially identified can be partly substituted by alternatives, and several experiments have shown that integrated lentiviral constructs can be avoided to reduce safety concerns, by using nonviral plasmids (Jia et al. 2010), microRNAs (Yang et al. 2011), protein transduction, and even by substituting some of the factors with

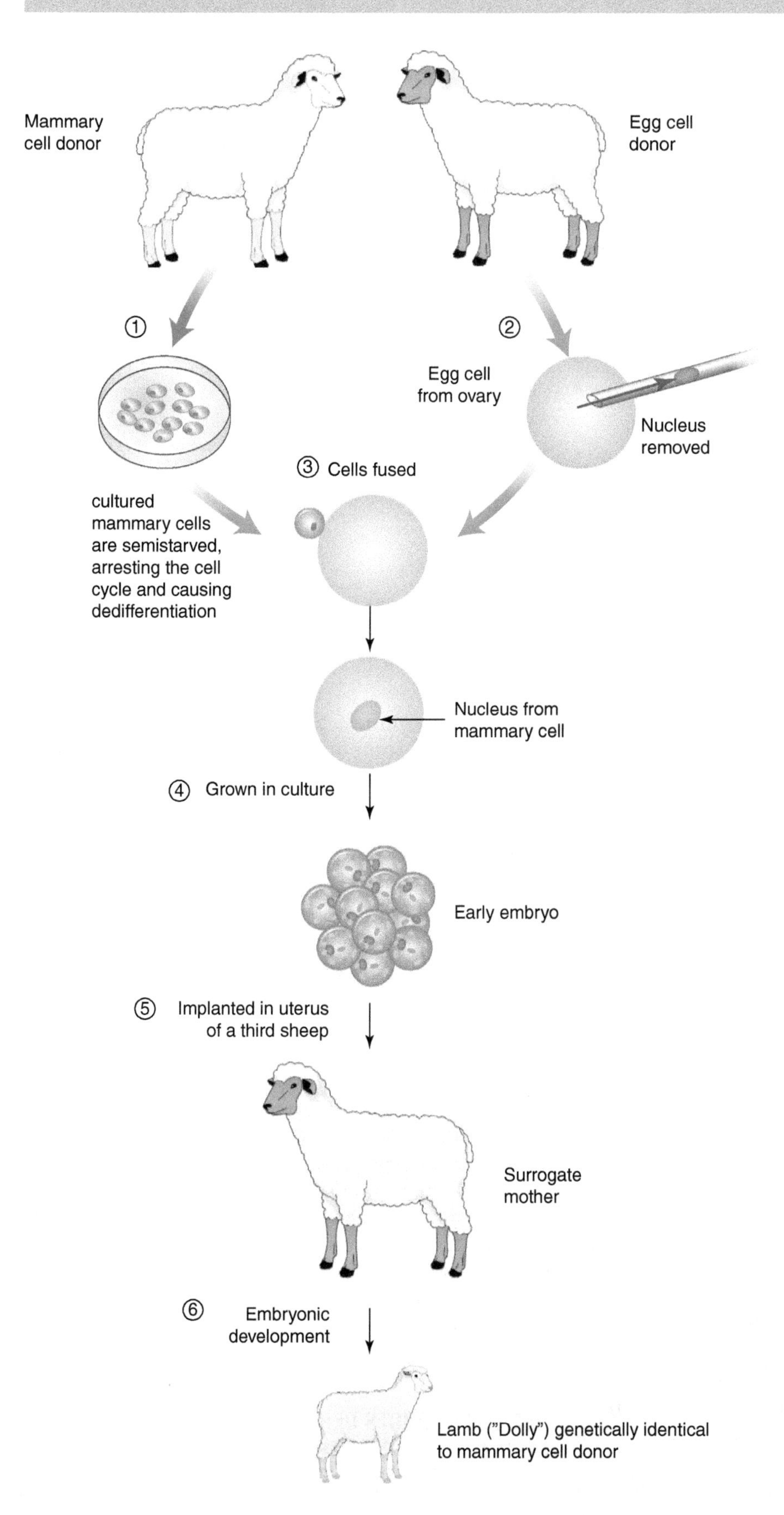

Figure 25.6 ■ Schematic diagram of the concept of reproductive cloning, as used to produce the cloned sheep "Dolly".

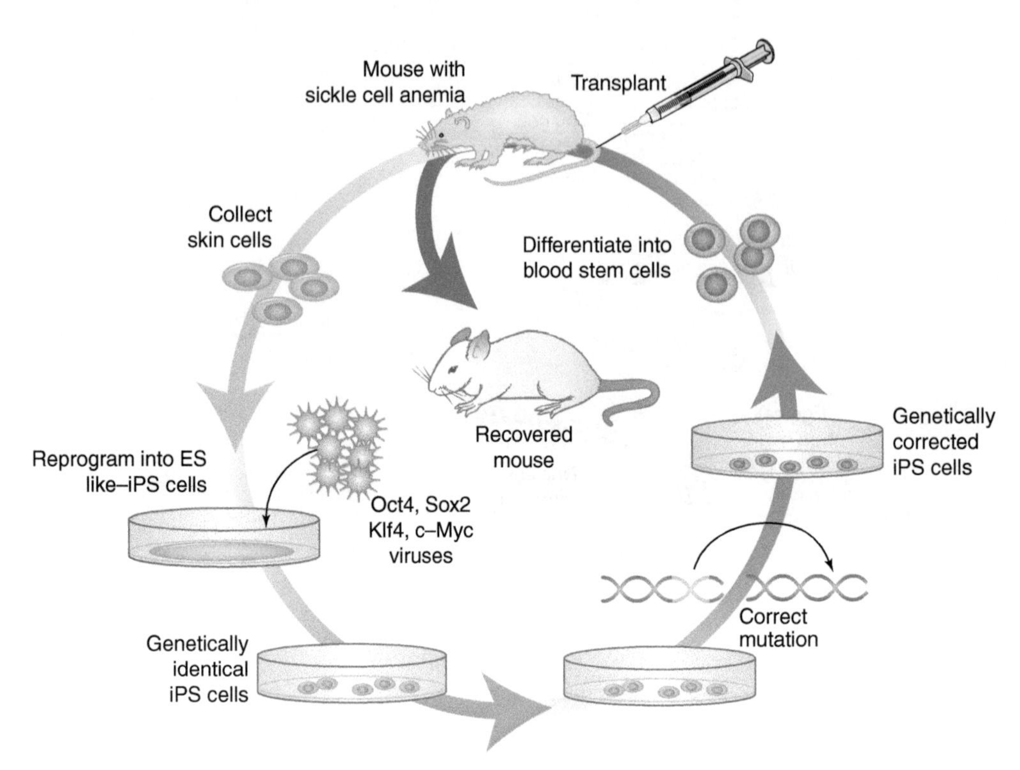

Figure 25.7 ■ Method used to produce iPS cells, correct a genetic defect responsible for sickle cell anemia, and implant the corrected stem cells into mice to cure sickle cell anemia in an animal model.

small molecules (Yuan et al. 2011). Often the safer alternative methods work with reduced efficiency but nevertheless produce the same result. The technology is still in the early years of its development, but if it delivers its potential, iPS technology will have profound effects on the understanding of disease, correction of genetic defects, and cell therapy. Already iPS cells have been used to correct defects in mouse models of Parkinson's disease (Hargus et al. 2010) and to cure a model of sickle cell anemia in mice (Fig. 25.7) (Hanna et al. 2007).

Considerable effort has been directed at investigating how iPS cells differ from ES cells and investigating whether reprogramming is complete enough to produce truly pluripotent cells. True pluripotency is difficult to demonstrate unequivocally in human iPS cells so the development of methods to measure the extent of reprogramming will be important for practical applications. There are indications that iPS cells can have chromosomal defects and are not fully reprogrammed (Chin et al. 2010). Female human iPS cells appear to maintain the inactivated X chromosome that was present in the skin fibroblasts, though this has not been a problem with mouse iPS cells (Tchieu et al. 2010). In a recent report in mice, iPS cells induced an immune response in a genetically identical host from which the cells were derived (Zhao et al. 2011). The mechanisms causing this immunogenicity need to be studied in more detail to investigate whether this is a widespread problem. It is possible that the above unfavorable reports may be the result of inadequate control over reprogramming. Studies of the properties of iPS cell generated from multiple laboratories will address these important issues in the coming years.

DIRECT REPROGRAMMING

During the last 2–3 years, forced expression of genes has been used to convert fibroblasts directly into unrelated differentiated cells, including neurons (Ambasudhan et al. 2011; Wernig et al. 2008) and cardiomyocytes (Burridge et al. 2012). The technique used is analogous to that used to derive iPS cells, except that genes associated with the desired somatic cell are expressed instead of pluripotency genes. The realization that cellular phenotypes can be transformed in this way has been met with astonishment and is certainly breakthrough technology. It raises the possibility that interconversion could be performed in vivo, though it does not allow for expansion of cells in preparation for an implant. However, direct reprogramming of fibroblasts to neural stem cells, as reported in 2012 (Han et al. 2012; Thier et al. 2012), may be a short cut to neurons. This approach may offer some advantages over production of neurons by way of iPS cells.

USE OF PRODUCTS DERIVED FROM STEM CELLS IN CELL THERAPY

There is much work to be done to develop effective cell therapies. From a clinical perspective, it is not always clear what type of cell needs to be implanted, where or how and how many cells are needed for a clinically relevant dose. For example, the major symptoms of Parkinson's disease are caused by loss of A9 dopaminergic neurons that project from the substantia nigra, releasing dopamine at synapses in the striatum. To reduce or replace this loss of A9 neurons by cell therapy could be achieved in at least two ways. Firstly, the cells of the substantia nigra could be supported with an implant of neural stem cells, which appear to protect the remaining neurons in animal models. This effect has been termed the "chaperone" effect (Redmond et al. 2007) and may be the result of secretion of supporting factors such as brain- and/or glial-derived neurotrophic factors (BDNF and GDNF) from the implant. The alternative or perhaps complementary approach would be to replace the dying neurons with new dopaminergic neurons. In this case the neurons would need to be integrated and form useful interactions with the relevant neural pathways in the brain. If the strategy is to provide new neurons, should the cells be neural stem cells, progenitors, dopaminergic precursors or mature dopaminergic neurons, or a mix? Are any dopaminergic neurons effective or must they have the exact phenotype of A9 neurons? Should the cells be implanted into the striatum or the substantia nigra or both? Will the damaged local environment be toxic to the implant? If so, what can be done to prepare the tissue for an implant? These are all questions that will need to be addressed as the technology progresses.

Each opportunity for cell therapy raises specific questions that will need to be addressed before such therapies can be used in a widespread manner. Several diseases and traumas of the CNS, other than Parkinson's disease, are under investigation as targets for cell therapy. These include Huntingdon's disease, Alzheimer's disease, and stroke (Lindvall and Kokaia 2010). In animal models of Alzheimer's disease, neural stem cell implants improved cognitive defects, probably by secretion of BDNF and protection of forebrain cholinergic neurons (Blurton-Jones et al. 2009). This may represent a similar mechanism of action and outcome to the studies carried out in models of Parkinson's disease in primates (Redmond et al. 2007). In stroke and neurotrauma, the most important objectives may be to reduce inflammation in the short term, and a gene therapy approach delivered by cell therapy may be a good option. There is considerable activity in cardiac cell implantation (Freund and Mummery 2009). In cardiac cell therapy for myocardial infarction, the treatment would ideally provide structure, strength, and elasticity, to regenerate heart muscle function. However, it is vital that cardiomyocyte implants are integrated with the heart conduction system and do not result in ectopic beating of heart tissue, which could be life-threatening. The extent of productive integration and reduction in scar tissue in the heart has been limited thus far and will need to be improved. In another busy research field, which aims to devise a cell therapy to treat Type I (and potentially Type II) diabetes, the search is on to find a mechanism to replace glucose sensitive, insulin-producing pancreatic β-islet cells. A simple injection of islet cells into the pancreas is unlikely to succeed because the implanted cells would be placed in the same destructive environment that led to loss of the patient's own islet cells, usually by an autoimmune reaction. In this case new islet cells are being tested in an encapsulated form, effectively in a polymeric delivery system, so that the cells are spared from immediate destruction (Kroon et al. 2008). This approach may also allow the implant to be removed if there are safety concerns after implantation. It can be anticipated that many other therapeutic opportunities are currently under investigation and will be explored in the future. Perhaps the most significant advances have been made in construction of implants for treatment of macular degeneration, which causes blindness due to retinal damage (Idelson et al. 2009; Schwartz et al. 2012). Clinical studies have commenced, using injection of retinal pigment epithelial cells derived from ES cells into the eye, with early signs of success.

DISEASE MODELING AND DRUG DISCOVERY

Whether or not iPS cells have a future in cell therapy, they will undoubtedly have a bright future in disease modeling and in drug discovery (McKernan et al. 2010; Rowntree and McNeish 2010). Since iPS cells can be expanded in a similar manner to ES cells and can be differentiated to obtain mature somatic cells, it is now possible to obtain differentiated cells with the exact genotype of patients who are suffering from specific diseases. This is particularly exciting in relation to neurological and neurodegenerative diseases of the central nervous system (CNS). Generally, CNS diseases can only be studied using postmortem tissue, when the brain tissue is too badly damaged to extract useful information on the origins and progression of the disease. Using iPS technology it is now possible to compare neurons from unaffected and diseased patients before the disease has emerged (Fig. 25.8). The ability to expand pluripotent cells prior to differentiation means that a sufficient number of cells can be generated to consider developing plate-based screening assays. Over the past 2 years, this strategy has been used to study neurons derived from patients with schizophrenia (Brennand et al. 2011), Rett syndrome (Marchetto et al. 2010), and Alzheimer's disease (Israel et al. 2012). Defects were observed in all cases, and markers of disease were upregulated in the

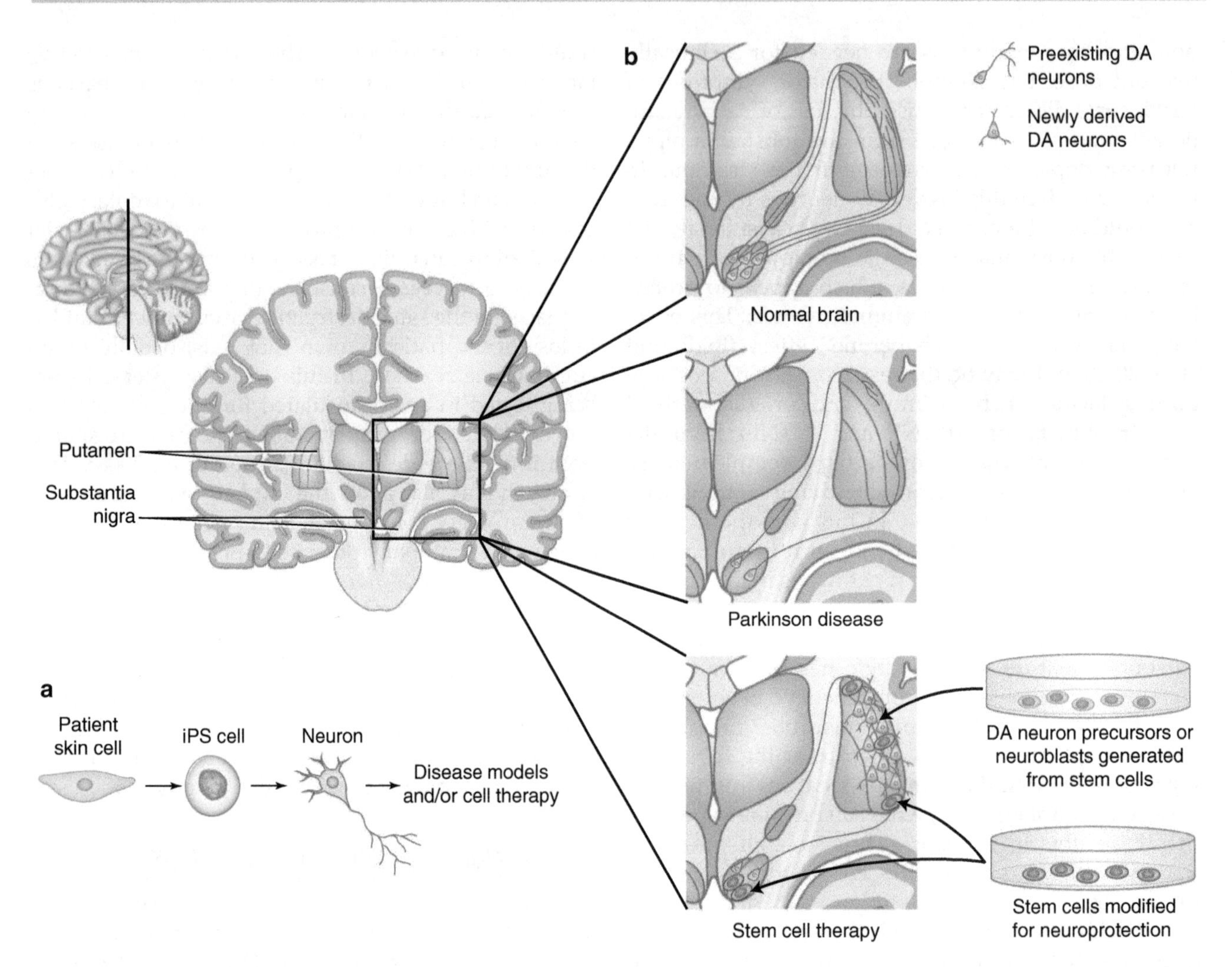

Figure 25.8 ■ (**a**) Derivation of iPS cells from patients suffering from disease, such as Parkinson's disease (PD) has the potential to lead to a new generation of cell culture disease models, allowing a comparison of functional properties and phenotype of patient-derived neurons with control neurons. (**b**) ES cells, iPS cells, or corrected patient-derived iPS cells could be used for cell therapy, shown here for PD. Loss of A9 dopaminergic neurons in the substantia nigra could be prevented by protective implants into the substantia nigra as well as implantation of precursor cells into the striatum.

patient-derived cells. Remarkably, defects in synapse formation in schizophrenia-derived cells were reversed by exposure to antipsychotic drugs, illustrating how such cells will be valuable for drug discovery and also suggesting that antipsychotic drugs may act to modulate synaptic connections between neurons, as well as producing short-term benefits.

Pharmaceutical companies are beginning to explore the use of stem cell-derived assays to study specific functional responses in screening experiments (Pouton and Haynes 2007). Assays of this type require more skill and effort to set up, qualify, and validate than simple cell-based assays, such as stably overexpressed receptor models in immortalized cells. But the rewards could be the discovery of more relevant hits and lead compounds with improved activity in vivo. A phenotypic response in a cell line that has the appropriate signaling systems expressed at appropriate levels is likely to be a smarter approach to drug discovery than use of existing screening assays. The availability of appropriate disease models using iPS technology will add value and may have the power to differentiate between desired activities in diseased cells and side effects in normal cells.

CANCER STEM CELLS

An added consequence of advances in the understanding of stem cell biology, particularly in relation to adult stem cell biology, has led to the realization that many cancers, particularly leukemia (Majeti et al. 2009) and solid tumors of epithelial tissue, may have their origins in mutations and familial polymorphisms in tissue stem cells. Many authors have suggested that the failure of cancer chemotherapy can be explained by its inability to kill slowly dividing cancer stem cells.

This has become a hotly debated topic in cancer biology. Much of the argument concerns the difficulty in establishing the proportion of cells within a tumor that have the capacity to give rise to a new tumor, when transplanted into a healthy animal. The need to use tumor xenografts in immunodeficient mice makes this a difficult assay to interpret when examining human tumors. The transplanted cells may be unable to give rise to a tumor because the local environment of the transplant is not appropriate, being the wrong species and most likely the wrong tissue. Nevertheless, there is mounting evidence that many tumors are initiated in stem or progenitor cells. To accept the cancer stem cell model implies a paradigm shift and implies that the target of drug therapy to eradicate the disease should be the cancer stem cell. Some cancers may arise not from adult stem cells, but from a downstream progenitor cell. This appears to be the case for some prostate cancers (Wang et al. 2009) and also basal cell-like "triple negative" breast cancer (Lim et al. 2009), which has the appearance of a basal tumor but in fact arises from a mammary progenitor cell. Different approaches to drug therapy of cancer are certain to emerge from research on cancer stem cells, but at present it is not clear how these cells can be targeted effectively. Study of gene expression in cancer stem cells, for example, by transcriptomics (cf. Chap. 8) using DNA array technology, may identify specific targets for drug discovery, either to inhibit cancer stem cell activity or to cause their differentiation into "normal" cells. For example, the invasive brain tumor, glioblastoma multiforme, could potentially be treated using a strategy that reduced the cancer stem cell phenotype to that of normal glial cells, as an alternative or additive to traditional chemotherapy.

REGULATORY ISSUES

Within the US Food and Drug Administration (FDA), the responsibility for regulation of cell therapy products lies with the Center for Biologics Evaluation and Research (CBER). Cell therapy products are defined by the FDA as "human cell, tissue, and cellular and tissue-based products (HCT/Ps)." Although there is a considerable amount of activity in research and development, both in the academic and commercial sectors, no stem cell therapy or gene therapy products have been approved by the FDA (June 2012). To monitor activity, review data, and anticipate future needs, the FDA operates the Cellular, Tissue and Gene Therapies Advisory Committee. Transcripts of meetings can be viewed at the FDA website, which provides a useful insight into current thinking on regulatory aspects of regenerative medicine. Details of the product development and approval process for HCT/Ps are posted at the CBER site (www.fda.gov/biologicsbloodvaccines/cellulargenetherapyproducts). At the clinical trials stage, cell therapy products require an Investigational New Drug Application (IND- clinical trials) and for product approval, a Biologics License Application (BLA- marketing) (see the PHS Acts 42 USC 262 and 21 CFR 1271). European Union countries and some other countries in Europe are guided by the European Medicine Agency (EMA – www.ema.europa.eu/ema), which drafts guidelines for cell therapies in a product category called advanced therapies. This includes gene therapy, cell therapy, and tissue-engineered products. The majority of cell therapy products will be classified as Advanced Therapy Medicinal Products (ATMP) in Europe and will therefore be regulated by medicinal product Directive 2001/83/EC and Regulation EC (No 1394/2007). They will require a clinical trial application (CTA), which is the responsibility of the appropriate national competent authority and subsequently will submit a Marketing Authorization Application (MAA) to the EMA. The Committee for Advanced Therapies (CAT) at the EMA is responsible for evaluation of product license applications and makes recommendations to the Committee for Medicinal Products for Human Use (CHMP). A landmark first gene therapy product (Glybera) was approved by the EMA in 2012.

Both the FDA and EMA, and other regulatory agencies around the world, are grappling with new paradigms in terms of the balance between risk and benefit. Risks in cell therapy are difficult to anticipate, so risk-benefit analysis will be a considerable challenge to committees going forward. There are also complex safety issues relating to the quality control of cell therapy products, which will require examination by regulatory authorities (Herberts et al. 2011). Each product will require detailed assessment on a case-by-case basis. Publicly Available Specification documents published by the British Standards Institute (BSI PAS 83:2012; BSI PAS 84:2012, and BSI PAS 93:2011) are freely available and provide valuable guidelines on quality control and development of cell therapy products.

CONCLUDING REMARKS

There is a general view that advances in biomedical science, including stem cell biology, will be important stimuli for change in our approach to many diseases and healthcare in general. Practical application of stem cell technology will require highly trained practitioners, both at the technical level and with regard to advising and counselling patients. Pharmaceutical scientists and pharmacists will be important members of the team of professionals that deliver these changes in healthcare, a challenging and exciting prospect. Much can be learned from the R&D processes used by traditional biotech (e.g., during development of therapeutic monoclonal antibodies and vaccines). Pharmacists can play a key role in development of stem cell therapies, as many applications are conceived by academic groups and small spin-off companies, who do not necessarily know how to translate a concept into a medicinal product. Pharmacy professionals can provide valuable experience in relation to the application of the principles of GLP, GMP, and GCP.

SELF-ASSESSMENT QUESTIONS

■ Questions

1. What is the difference between embryonic and adult stem cells?
2. What are the possible advantages in using embryonic stem cell technology in drug discovery?
3. How is somatic cell nuclear transfer carried out and what are the problems with this technique?
4. What are induced pluripotent stem cells and why are they important?
5. Which diseases could potentially be treated with cell therapy?
6. What problems could arise in use of stem cell-derived products for cell therapy?

■ Answers

1. Embryonic stem cells are grown in vitro after extraction of the inner cell mass from a blastocyst. Adult stem cells are found in vivo in many tissues, usually in the specialized environment of a stem cell niche, that support their asymmetric cell division.
2. Cell culture models used in drug discovery are often immortalized cells that are used to assay for receptor activation using stably transformed cells. The advantage of embryonic stem cells is that they can be expanded in a pluripotent state and then encouraged to differentiate into specialized mature somatic cells. These fully differentiated cells are likely to express the appropriate signaling systems which will allow a sophisticated functional experiment to been designed. This approach will result in more powerful data on the efficacy of drug candidates.
3. Somatic cell nuclear transfer (SCNT) involves the injection of a donor genome into an enucleated egg, such that the embryo develops as a clone of the donor genome. This allows embryonic stem cells to be derived using the donor genome and in principle allows implantation into the uterus of a recipient female leading to pregnancy. There are ethical problems concerned with supply of fertilized human eggs and also technical problems caused by incomplete reprogramming of the donor nucleus.
4. iPS cells are produced by transient expression of pluripotency genes in somatic cells, leading to reprogramming to form pluripotent cells resembling embryonic stem cells. The production of iPS cells allows pluripotent cells to be obtained from a patient without the need for SCNT. iPS cells could be used to derive differentiated cells for implantation therapy or to produce models of disease.
5. A variety of diseases may 1 day be treatable with cell therapy. Examples of current "test beds" for cell therapy are Parkinson's disease, myocardial infarction, and macular degeneration. A number of other neural conditions are under investigation including Huntingdon's disease, Alzheimer's disease, stroke, and spinal injury.
6. One of the concerns with stem cell-based therapy is the possibility that rare pluripotent or multipotent cells in the implant could give rise to tumors. Thus, the quality control of the implant will be of paramount importance. Often, in particular in treatment of CNS diseases, it is not clear whether a progenitor, precursor, or fully mature cell should be implanted. Careful preclinical work will be needed in each clinical indication to establish the most effective approach. Where the strategy is designed to replace a cell that is lost in a particular disease, the environment into which the implant is placed may not be supportive of cell survival and integration. In general, attention will need to be paid to repairing the tissue to provide a protective environment for the implant.

REFERENCES

Alvarez-Buylla A, Garcia-Verdugo JM, Tramontin AD (2001) A unified hypothesis on the lineage of neural stem cells. Nat Rev Neurosci 2:287–293

Ambasudhan R, Talantova M, Coleman R, Yuan X, Zhu S, Lipton SA, Ding S (2011) Direct reprogramming of adult human fibroblasts to functional neurons under defined conditions. Cell Stem Cell 9:113–118

Bianco P, Robey PG, Simmons PJ (2008) Mesenchymal stem cells: revisiting history, concepts, and assays. Cell Stem Cell 2:313–319

Blurton-Jones M, Kitazawa M, Martinez-Coria H, Castello NA, Muller FJ, Loring JF, Yamasaki TR, Poon WW, Green KN, LaFerla FM (2009) Neural stem cells improve cognition via BDNF in a transgenic model of Alzheimer disease. Proc Natl Acad Sci U S A 106: 13594–13599

Brennand KJ, Simone A, Jou J, Gelboin-Burkhart C, Tran N, Sangar S, Li Y, Mu Y, Chen G, Yu D, McCarthy S, Sebat J, Gage FH (2011) Modelling schizophrenia using human induced pluripotent stem cells. Nature 473:221–225

BSI PAS 83:2012 (2012) Developing human cells for clinical applications in the Euopean Union and the United States of America – guide. Publicly Available Specification PAS83:2012, The British Standards Institution, ISBN 978 0 580 71052 0

BSI PAS 84:2012 (2012) Cell therapy and regenerative medicine – glossary. Publicly Available Specification PAS84:2012, The British Standards Institution, ISBN 978 0 580 74904 9

BSI PAS 93:2011 (2011) Characterization of human cells for clinical applications – guide. Publicly Available Specification PAS93:2011, The British Standards Institution, ISBN 978 0 580 69850 7

Burridge PW, Keller G, Gold JD, Wu JC (2012) Production of de novo cardiomyocytes: human pluripotent stem cell differentiation and direct reprogramming. Cell Stem Cell 10:16–28

Campbell KH, McWhir J, Ritchie WA, Wilmut I (1996) Sheep cloned by nuclear transfer from a cultured cell line. Nature 380:64–66

Chin MH, Pellegrini M, Plath K, Lowry WE (2010) Molecular analyses of human induced pluripotent stem cells and embryonic stem cells. Cell Stem Cell 7:263–269

Evans MJ, Kaufman MH (1981) Establishment in culture of pluripotential cells from mouse embryos. Nature 292:154–156

Freund C, Mummery CL (2009) Prospects for pluripotent stem cell-derived cardiomyocytes in cardiac cell therapy and as disease models. J Cell Biochem 107: 592–599

Han DW, Tapia N, Hermann A, Hemmer K, Hoing S, Arauzo-Bravo MJ, Zaehres H, Wu G, Frank S, Moritz S, Greber B, Yang JH, Lee HT, Schwamborn JC, Storch A, Scholer HR (2012) Direct reprogramming of fibroblasts into neural stem cells by defined factors. Cell Stem Cell 10:465–472

Hanna J, Wernig M, Markoulaki S, Sun CW, Meissner A, Cassady JP, Beard C, Brambrink T, Wu LC, Townes TM, Jaenisch R (2007) Treatment of sickle cell anemia mouse model with iPS cells generated from autologous skin. Science 318:1920–1923

Hargus G, Cooper O, Deleidi M, Levy A, Lee K, Marlow E, Yow A, Soldner F, Hockemeyer D, Hallett PJ, Osborn T, Jaenisch R, Isacson O (2010) Differentiated Parkinson patient-derived induced pluripotent stem cells grow in the adult rodent brain and reduce motor asymmetry in Parkinsonian rats. Proc Natl Acad Sci U S A 107: 15921–15926

Herberts CA, Kwa MS, Hermsen HP (2011) Risk factors in the development of stem cell therapy. J Transl Med 9:29

Idelson M, Alper R, Obolensky A, Ben-Shushan E, Hemo I, Yachimovich-Cohen N, Khaner H, Smith Y, Wiser O, Gropp M, Cohen MA, Even-Ram S, Berman-Zaken Y, Matzrafi L, Rechavi G, Banin E, Reubinoff B (2009) Directed differentiation of human embryonic stem cells into functional retinal pigment epithelium cells. Cell Stem Cell 5:396–408

Israel MA, Yuan SH, Bardy C, Reyna SM, Mu Y, Herrera C, Hefferan MP, Van Gorp S, Nazor KL, Boscolo FS, Carson CT, Laurent LC, Marsala M, Gage FH, Remes AM, Koo EH, Goldstein LS (2012) Probing sporadic and familial Alzheimer's disease using induced pluripotent stem cells. Nature 482:216–220

Jia F, Wilson KD, Sun N, Gupta DM, Huang M, Li Z, Panetta NJ, Chen ZY, Robbins RC, Kay MA, Longaker MT, Wu JC (2010) A nonviral minicircle vector for deriving human iPS cells. Nat Methods 7:197–199

Keller G (2005) Embryonic stem cell differentiation: emergence of a new era in biology and medicine. Genes Dev 19:1129–1155

Kroon E, Martinson LA, Kadoya K, Bang AG, Kelly OG, Eliazer S, Young H, Richardson M, Smart NG, Cunningham J, Agulnick AD, D'Amour KA, Carpenter MK, Baetge EE (2008) Pancreatic endoderm derived from human embryonic stem cells generates glucose-responsive insulin-secreting cells in vivo. Nat Biotechnol 26:443–452

Lander AD, Kimble J, Clevers H, Fuchs E, Montarras D, Buckingham M, Calof AL, Trumpp A, Oskarsson T (2012) What does the concept of the stem cell niche really mean today? BMC Biol 10:19

Levenstein ME, Ludwig TE, Xu RH, Llanas RA, VanDenHeuvel-Kramer K, Manning D, Thomson JA (2006) Basic fibroblast growth factor support of human embryonic stem cell self-renewal. Stem Cells 24:568–574

Lim E, Vaillant F, Wu D, Forrest NC, Pal B, Hart AH, Asselin-Labat ML, Gyorki DE, Ward T, Partanen A, Feleppa F, Huschtscha LI, Thorne HJ, Fox SB, Yan M, French JD, Brown MA, Smyth GK, Visvader JE, Lindeman GJ (2009) Aberrant luminal progenitors as the candidate target population for basal tumor development in BRCA1 mutation carriers. Nat Med 15:907–913

Lindvall O, Kokaia Z (2010) Stem cells in human neurodegenerative disorders–time for clinical translation? J Clin Invest 120:29–40

Lui KO, Waldmann H, Fairchild PJ (2009) Embryonic stem cells: overcoming the immunological barriers to cell replacement therapy. Curr Stem Cell Res Ther 4:70–80

Majeti R, Chao MP, Alizadeh AA, Pang WW, Jaiswal S, Gibbs KD Jr, van Rooijen N, Weissman IL (2009) CD47 is an adverse prognostic factor and therapeutic antibody target on human acute myeloid leukemia stem cells. Cell 138:286–299

Marchetto MC, Carromeu C, Acab A, Yu D, Yeo GW, Mu Y, Chen G, Gage FH, Muotri AR (2010) A model for neural development and treatment of Rett syndrome using human induced pluripotent stem cells. Cell 143: 527–539

Martin GR (1981) Isolation of a pluripotent cell line from early mouse embryos cultured in medium conditioned by teratocarcinoma stem cells. Proc Natl Acad Sci U S A 78:7634–7638

McKernan R, McNeish J, Smith D (2010) Pharma's developing interest in stem cells. Cell Stem Cell 6:517–520

Murry CE, Keller G (2008) Differentiation of embryonic stem cells to clinically relevant populations: lessons from embryonic development. Cell 132:661–680

Pouton CW, Haynes JM (2007) Embryonic stem cells as a source of models for drug discovery. Nat Rev Drug Discov 6:605–616

Redmond DE Jr, Bjugstad KB, Teng YD, Ourednik V, Ourednik J, Wakeman DR, Parsons XH, Gonzalez R,

Blanchard BC, Kim SU, Gu Z, Lipton SA, Markakis EA, Roth RH, Elsworth JD, Sladek JR Jr, Sidman RL, Snyder EY (2007) Behavioral improvement in a primate Parkinson's model is associated with multiple homeostatic effects of human neural stem cells. Proc Natl Acad Sci U S A 104:12175–12180

Rowntree RK, McNeish JD (2010) Induced pluripotent stem cells: opportunities as research and development tools in 21st century drug discovery. Regen Med 5:557–568

Santos GW (1983) History of bone marrow transplantation. Clin Haematol 12:611–639

Scadden DT (2006) The stem-cell niche as an entity of action. Nature 441:1075–1079

Schwartz SD, Hubschman JP, Heilwell G, Franco-Cardenas V, Pan CK, Ostrick RM, Mickunas E, Gay R, Klimanskaya I, Lanza R (2012) Embryonic stem cell trials for macular degeneration: a preliminary report. Lancet 379: 713–720

Taichman RS (2005) Blood and bone: two tissues whose fates are intertwined to create the hematopoietic stem-cell niche. Blood 105:2631–2639

Takahashi K, Yamanaka S (2006) Induction of pluripotent stem cells from mouse embryonic and adult fibroblast cultures by defined factors. Cell 126:663–676

Takahashi K, Tanabe K, Ohnuki M, Narita M, Ichisaka T, Tomoda K, Yamanaka S (2007) Induction of pluripotent stem cells from adult human fibroblasts by defined factors. Cell 131:861–872

Tchieu J, Kuoy E, Chin MH, Trinh H, Patterson M, Sherman SP, Aimiuwu O, Lindgren A, Hakimian S, Zack JA, Clark AT, Pyle AD, Lowry WE, Plath K (2010) Female human iPSCs retain an inactive X chromosome. Cell Stem Cell 7:329–342

Thier M, Worsdorfer P, Lakes YB, Gorris R, Herms S, Opitz T, Seiferling D, Quandel T, Hoffmann P, Nothen MM, Brustle O, Edenhofer F (2012) Direct conversion of fibroblasts into stably expandable neural stem cells. Cell Stem Cell 10:473–479

Thomson JA, Itskovitz-Eldor J, Shapiro SS, Waknitz MA, Swiergiel JJ, Marshall VS, Jones JM (1998) Embryonic stem cell lines derived from human blastocysts. Science 282:1145–1147

Wang X, Kruithof-de Julio M, Economides KD, Walker D, Yu H, Halili MV, Hu YP, Price SM, Abate-Shen C, Shen MM (2009) A luminal epithelial stem cell that is a cell of origin for prostate cancer. Nature 461:495–500

Wernig M, Zhao JP, Pruszak J, Hedlund E, Fu D, Soldner F, Broccoli V, Constantine-Paton M, Isacson O, Jaenisch R (2008) Neurons derived from reprogrammed fibroblasts functionally integrate into the fetal brain and improve symptoms of rats with Parkinson's disease. Proc Natl Acad Sci U S A 105:5856–5861

Wilmut I, Beaujean N, de Sousa PA, Dinnyes A, King TJ, Paterson LA, Wells DN, Young LE (2002) Somatic cell nuclear transfer. Nature 419:583–586

Wilmut I, Sullivan G, Taylor J (2009) A decade of progress since the birth of Dolly. Reprod Fertil Dev 21:95–100

Yang CS, Li Z, Rana TM (2011) microRNAs modulate iPS cell-generation. RNA 17:1451–1460

Ying QL, Nichols J, Chambers I, Smith A (2003) BMP induction of Id proteins suppresses differentiation and sustains embryonic stem cell self-renewal in collaboration with STAT3. Cell 115:281–292

Ying QL, Wray J, Nichols J, Batlle-Morera L, Doble B, Woodgett J, Cohen P, Smith A (2008) The ground state of embryonic stem cell self-renewal. Nature 453: 519–523

Yu J, Vodyanik MA, Smuga-Otto K, Antosiewicz-Bourget J, Frane JL, Tian S, Nie J, Jonsdottir GA, Ruotti V, Stewart R, Slukvin II, Thomson JA (2007) Induced pluripotent stem cell lines derived from human somatic cells. Science 318:1917–1920

Yuan X, Wan H, Zhao X, Zhu S, Zhou Q, Ding S (2011) Brief report: combined chemical treatment enables Oct4-induced reprogramming from mouse embryonic fibroblasts. Stem Cells 29:549–553

Zhao C, Deng W, Gage FH (2008) Mechanisms and functional implications of adult neurogenesis. Cell 132:645–660

Zhao T, Zhang ZN, Rong Z, Xu Y (2011) Immunogenicity of induced pluripotent stem cells. Nature 474:212–215

Index

D.J.A. Crommelin, R.D. Sindelar, and B. Meibohm (eds.), *Pharmaceutical Biotechnology*,
DOI 10.1007/978-1-4614-6486-0, © Springer Science+Business Media New York 2013

CPI Antony Rowe
Chippenham, UK
2018-05-16 10:00